Vehicle Dynamics

Reza N. Jazar

Vehicle Dynamics

Theory and Application

Second Edition

 Springer

Reza N. Jazar
School of Aerospace, Mechanical and
  Manufacturing Engineering
RMIT University
Bundoora, VIC
Australia

ISBN 978-1-4614-8543-8   ISBN 978-1-4614-8544-5 (eBook)
DOI 10.1007/978-1-4614-8544-5
Springer New York Heidelberg Dordrecht London

Library of Congress Control Number: 2013951659

Printed on acid-free paper

Springer is part of Springer Science+Business Media (www.springer.com)

Contents

IV Vehicle Vibration   795

Dedicated to
my son, *Kavosh,*
my daughter, *Vazan,*
and my wife, *Mojgan.*

Nature is not perfect, not even optimum.

Preface to the Second Edition

The second edition of this textbook would not be possible without comments and contribution of my students and colleagues, especially those at Columbia University in New York and RMIT University in Melbourne. I am deeply thankful of my friend Mr. Stefan Anthony who shared with me his 50 years of experiences in automotive industries. Special thanks to Andy Fu and Hormoz Marzbani who kindly read the book carefully and caught typos as well as logical mismatches.

New topics introduced in this edition are results of students' feedback which assisted me in clarifying and better presenting some aspects of this book.

The intent of this book is to explain vehicle dynamics in a manner I would have liked it explained to me as a student. This book can now help students and engineers by being a great reference that covers all aspects of vehicle dynamics and providing students with detailed explanations and information.

The first edition of this book was published in 2008 by Springer. Soon after its publication, the book become very popular in the automotive field. It was appreciated by many students and instructors in addition to my own students and colleagues. Their questions, comments, and suggestions have helped me in creating the second edition.

Preface

This text is for engineering students. It introduces the fundamental knowledge used in *vehicle dynamics*. This knowledge can be utilized to develop computer programs for analyzing the ride, handling, and optimization of road vehicles.

Vehicle dynamics has been in the engineering curriculum for more than a hundred years. Books on the subject are available, but most of them are written for specialists and are not suitable for a classroom application. A new student, engineer, or researcher would not know where and how to start learning vehicle dynamics. So, there is a need for a textbook for beginners. This textbook presents the fundamentals with a perspective on future trends.

The study of classical vehicle dynamics has its roots in the work of great scientists of the past four centuries and creative engineers in the past century who established the methodology of dynamic systems. The development of vehicle dynamics has moved toward modeling, analysis, and optimization of multi-body dynamics supported by some compliant members. Therefore, merging dynamics with optimization theory was an expected development. The fast-growing capability of accurate positioning, sensing, and calculations, along with intelligent computer programming are the other important developments in vehicle dynamics. So, a textbook help the reader to make a computer model of vehicles, which this book does.

Level of the Book

This book has evolved from nearly a decade of research in nonlinear dynamic systems and teaching courses in vehicle dynamics. It is addressed primarily to the last year of undergraduate study and the first year graduate student in engineering. Hence, it is an intermediate textbook. It provides both fundamental and advanced topics. The whole book can be covered in two successive courses, however, it is possible to jump over some sections and cover the book in one course. Students are required to know the fundamentals of kinematics and dynamics, as well as a basic knowledge of numerical methods.

The contents of the book have been kept at a fairly theoretical-practical level. Many concepts are deeply explained and their application emphasized, and most of the related theories and formal proofs have been explained. The book places a strong emphasis on the physical meaning and applications of the concepts. Topics that have been selected are of high interest in the field. An attempt has been made to expose students to a broad range of topics and approaches.

There are four special chapters that are indirectly related to vehicle dynamics: *Applied Kinematics*, *Applied Mechanisms*, *Applied Dynamics*, and *Applied Vibrations*. These chapters provide the related background to understand vehicle dynamics and its subsystems.

Organization of the Book

The text is organized so it can be used for teaching or for self-study. Chapter 1 "Fundamentals," contains general preliminaries about tire and rim with a brief review of road vehicle classifications.

Part *I* "One Dimensional Vehicle Dynamics," presents forward vehicle dynamics, tire dynamics, and driveline dynamics. Forward dynamics refers to weight transfer, accelerating, braking, engine performance, and gear ratio design.

Part *II* "Vehicle Kinematics," presents a detailed discussion of vehicle mechanical subsystems such as steering and suspensions.

Part *III* "Vehicle Dynamics," employs Newton and Lagrange methods to develop the maneuvering dynamics of vehicles.

Part *IV* "Vehicle Vibrations," presents a detailed discussion of vehicle vibrations. An attempt is made to review the basic approaches and demonstrate how a vehicle can be modeled as a vibrating multiple degree-of-freedom system. The concepts of the Newton-Euler dynamics and Lagrangian method are used equally for derivation of equations of motion. The RMS optimization technique for suspension design of vehicles is introduced and applied to vehicle suspensions. The outcome of the optimization technique is the optimal stiffness and damping for a car or suspended equipment.

Method of Presentation

This book uses a *"fact-reason-application"* structure. The "fact" is the main subject we introduce in each section. Then the reason is given as a "proof." The application of the fact is examined in some "examples." The "examples" are a very important part of the book because they show how to implement the "facts." They also cover some other facts that are needed to expand the subject.

Prerequisites

Since the book is written for senior undergraduate and first-year graduate-level students of engineering, the assumption is that users are familiar with matrix algebra as well as basic dynamics. Prerequisites are the fundamentals of kinematics, dynamics, vector analysis, and matrix theory. These basics are usually taught in the first three undergraduate years.

Unit System

The system of units adopted in this book is, unless otherwise stated, the international system of units (SI). The units of degree (deg) or radian (rad) are utilized for variables representing angular quantities.

Symbols

- Lowercase bold letters indicate a vector. Vectors may be expressed in an n dimensional Euclidian space. Example:

$$\mathbf{r} \quad, \quad \mathbf{s} \quad, \quad \mathbf{d} \quad, \quad \mathbf{a} \quad, \quad \mathbf{b} \quad, \quad \mathbf{c}$$
$$\mathbf{p} \quad, \quad \mathbf{q} \quad, \quad \mathbf{v} \quad, \quad \mathbf{w} \quad, \quad \mathbf{y} \quad, \quad \mathbf{z}$$
$$\boldsymbol{\omega} \quad, \quad \boldsymbol{\alpha} \quad, \quad \boldsymbol{\epsilon} \quad, \quad \boldsymbol{\theta} \quad, \quad \boldsymbol{\delta} \quad, \quad \boldsymbol{\phi}$$

- Uppercase bold letters indicate a dynamic vector or a dynamic matrix, such as force and moment. Example:

$$\mathbf{F} \qquad \mathbf{M}$$

- Lowercase letters with a hat indicate a unit vector. Unit vectors are not bold. Example:

$$\hat{\imath} \quad, \quad \hat{\jmath} \quad, \quad \hat{k} \quad, \quad \hat{u} \quad, \quad \hat{u} \quad, \quad \hat{n}$$
$$\hat{I} \quad, \quad \hat{J} \quad, \quad \hat{K} \quad, \quad \hat{u}_\theta \quad, \quad \hat{u}_\varphi \quad, \quad \hat{u}_\psi$$

- Lowercase letters with a tilde indicate a 3×3 skew symmetric matrix associated to a vector. Example:

$$\tilde{a} = \begin{bmatrix} 0 & -a_3 & a_2 \\ a_3 & 0 & -a_1 \\ -a_2 & a_1 & 0 \end{bmatrix} \qquad \mathbf{a} = \begin{bmatrix} a_1 \\ a_2 \\ a_3 \end{bmatrix}$$

- An arrow above two uppercase letters indicates the start and end points of a position vector. Example:

$$\overrightarrow{ON} = \text{a position vector from point } O \text{ to point } N$$

- The length of a vector is indicated by a non-bold lowercase letter. Example:

$$r = |\mathbf{r}| \qquad a = |\mathbf{a}| \qquad b = |\mathbf{b}| \qquad s = |\mathbf{s}|$$

- Capital letter B is utilized to denote a body coordinate frame. Example:

$$B(oxyz) \qquad B(Oxyz) \qquad B_1(o_1 x_1 y_1 z_1)$$

- Capital letter G is utilized to denote a global, inertial, or fixed coordinate frame. Example:

$$G \qquad G(XYZ) \qquad G(OXYZ)$$

- Right subscript on a transformation matrix indicates the *departure* frames. Example:

$$R_B = \text{transformation matrix from frame } B(oxyz)$$

- Left superscript on a transformation matrix indicates the *destination* frame. Example:

$$^G R_B = \text{transformation matrix from frame } B(oxyz)$$
$$\text{to frame } G(OXYZ)$$

- Capital letter R indicates rotation or a transformation matrix, if it shows the beginning and destination coordinate frames. Example:

$$^G R_B = \begin{bmatrix} \cos\alpha & -\sin\alpha & 0 \\ \sin\alpha & \cos\alpha & 0 \\ 0 & 0 & 1 \end{bmatrix}$$

- Whenever there is no sub or superscript, the matrices are shown in a bracket. Example:

$$[T] = \begin{bmatrix} \cos\alpha & -\sin\alpha & 0 \\ \sin\alpha & \cos\alpha & 0 \\ 0 & 0 & 1 \end{bmatrix}$$

- Left superscript on a vector denotes the frame in which the vector is expressed. That superscript indicates the frame that the vector belongs to; so the vector is expressed using the unit vectors of that frame. Example:

$$^G \mathbf{r} = \text{position vector expressed in frame } G(OXYZ)$$

- Right subscript on a vector denotes the tip point that the vector is referred to. Example:

$$^G \mathbf{r}_P = \text{position vector of point } P$$
$$\text{expressed in coordinate frame } G(OXYZ)$$

- Right subscript on an angular velocity vector indicates the frame that the angular vector is referred to. Example:

$$\boldsymbol{\omega}_B = \text{angular velocity of the body coordinate frame } B(oxyz)$$

- Left subscript on an angular velocity vector indicates the frame that the angular vector is measured with respect to. Example:

$$_G\omega_B \;=\; \text{angular velocity of the body coordinate frame } B(oxyz)$$
$$\text{with respect to the global coordinate frame } G(OXYZ)$$

- Left superscript on an angular velocity vector denotes the frame in which the angular velocity is expressed. Example:

$$_G^{B_2}\omega_{B_1} \;=\; \text{angular velocity of the body coordinate frame } B_1$$
$$\text{with respect to the global coordinate frame } G,$$
$$\text{and expressed in body coordinate frame } B_2$$

Whenever the subscript and superscript of an angular velocity are the same, we usually drop the left superscript. Example:

$$_G\omega_B \equiv {}_G^G\omega_B$$

Also for position, velocity, and acceleration vectors, we drop the left subscripts if it is the same as the left superscript. Example:

$$_B^B\mathbf{v}_P \equiv {}^B\mathbf{v}_P$$

- Left superscript on derivative operators indicates the frame in which the derivative of a variable is taken. Example:

$$\frac{^Gd}{dt}x \qquad \frac{^Gd}{dt}{}^B\mathbf{r}_P \qquad \frac{^Bd}{dt}{}_B^G\mathbf{r}_P$$

If the variable is a vector function, and also the frame in which the vector is defined is the same frame in which a time derivative is taken, we may use the following short notation,

$$\frac{^Gd}{dt}{}^G\mathbf{r}_P = {}^G\dot{\mathbf{r}}_P \qquad \frac{^Bd}{dt}{}_o^B\mathbf{r}_P = {}_o^B\dot{\mathbf{r}}_P$$

and write equations simpler. Example:

$$^G\mathbf{v} = \frac{^Gd}{dt}{}^G\mathbf{r}(t) = {}^G\dot{\mathbf{r}}$$

- If followed by angles, lowercase c and s denote cos and sin functions in mathematical equations. Example:

$$c\alpha = \cos\alpha \qquad s\varphi = \sin\varphi$$

- Capital bold letter $\mathbf{I}$ indicates a unit matrix, which, depending on the dimension of the matrix equation, could be a 3×3 or a 4×4 unit matrix. $\mathbf{I}_3$ or $\mathbf{I}_4$ are also being used to clarify the dimension of $\mathbf{I}$. Example:

$$\mathbf{I} = \mathbf{I}_3 = \begin{bmatrix} 1 & 0 & 0 \\ 0 & 1 & 0 \\ 0 & 0 & 1 \end{bmatrix}$$

- An asterisk ★ indicates a more advanced subject or example that is not designed for undergraduate teaching and can be dropped in the first reading.

1

Tire and Rim Fundamentals

Tires are the only component of a vehicle to transfer forces between the road and the vehicle. Tire parameters such as dimensions, maximum load-carrying capacity, and maximum speed index are usually indicated on its sidewall. In this chapter, we review some topics about tires, wheels, roads, vehicles, and their interactions.

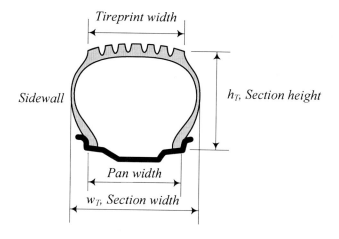

FIGURE 1.1. Cross section of a tire on a rim to show tire height and width.

1.1 Tires and Sidewall Information

Pneumatic tires are the only means to transfer forces between the road and the vehicle. Tires are required to produce the forces necessary to control the vehicle, and hence, they are an important component of a vehicle. Figure 1.1 illustrates a cross section view of a tire on a rim to show the dimension parameters that are used to standard tires.

The *section height, tire height,* or simply *height*, h_T, is a number that must be added to the rim radius to make the wheel radius. The *section width,* or *tire width,* w_T, is the widest dimension of a tire when the tire is not loaded.

Tires are required to have certain information printed on the tire *sidewall.* Figure 1.2 illustrates a side view of a sample tire to show the important information printed on a tire sidewall.

R.N. Jazar, *Vehicle Dynamics: Theory and Application,*
DOI 10.1007/978-1-4614-8544-5_1, © Springer Science+Business Media New York 2014

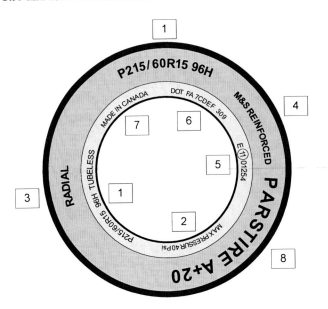

FIGURE 1.2. Side view of a tire and the most important information printed on a tire sidewall.

The codes in Figure 1.2 are:

1 Size number.
2 Maximum allowed inflation pressure.
3 Type of tire construction.
4 M&S denotes a tire for mud and snow.
5 E-Mark is the Europe type approval mark and number.
6 US Department of Transport (DOT) identification numbers.
7 Country of manufacture.
8 Manufacturers, brand name, or commercial name.

The most important information on the sidewall of a tire is the *size number*, indicated by 1 . To see the format of the size number, an example is shown in Figure 1.3 and their definitions are explained as follows.

P *Tire type.* The first letter indicates the proper type of car that the tire is made for. P stands for passenger car. The first letter can also be ST for special trailer, T for temporary, and LT for light truck.

215 *Tire width.* This three-number code is the width of the unloaded tire from sidewall to sidewall measured in [mm].

60 *Aspect ratio.* This two-number code is the ratio of the tire section height h_T to tire width w_T, expressed as a percentage. Aspect ratio is shown

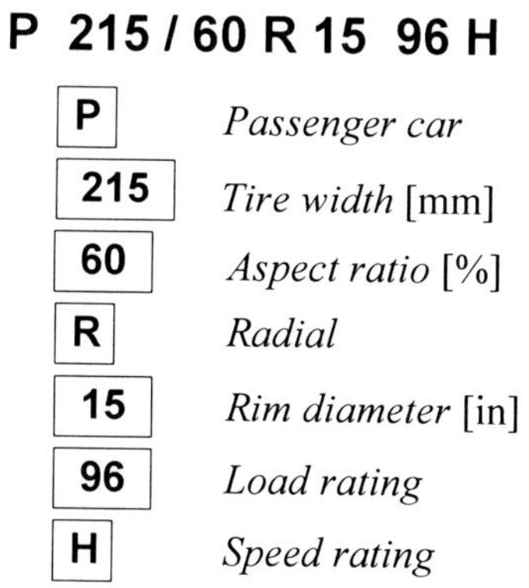

P 215 / 60 R 15 96 H

P	*Passenger car*
215	*Tire width* [mm]
60	*Aspect ratio* [%]
R	*Radial*
15	*Rim diameter* [in]
96	*Load rating*
H	*Speed rating*

FIGURE 1.3. A sample of a tire size number and its meaning.

by s_T.

$$s_T = \frac{h_T}{w_T} \times 100 \tag{1.1}$$

Generally speaking, tire aspect ratios range from 35, for race car tires, to 75 for tires used on utility vehicles.

$\boxed{R}$ *Tire construction type.* The letter $\boxed{R}$ indicates that the tire has a radial construction. It may also be $\boxed{B}$ for bias belt or bias ply, and $\boxed{D}$ for diagonal.

$\boxed{15}$ *Rim diameter.* This is a number in [in] to indicate diameter of the rim that the tire is designed to fit on.

$\boxed{96}$ *Load rate or load index.* Many tires come with a service description at the end of the tire size. The service description is made of a two-digit number (load index) and a letter (speed rating). The load index is a representation of the maximum load each tire is designed to support. Table 1.1 shows some of the most common load indices and their load-carrying capacities. The load index is generally valid for speeds under 210 km/ h ($\approx$ 130 mi/ h).

$\boxed{H}$ *Speed rate.* Speed rate indicates the maximum speed that the tire can sustain for a ten minute endurance without breaking down.

Table 1.2 shows the most common speed rate indices and their meanings. The maximum speed of a tire is related to its typical application as is shown in Table 1.3.

Table 1.2 indicates a confusing point as why we need indices W and Y while we have Z. It is because when the Z speed rate were first introduced, It was assumed to be the highest tire speed rating that would ever be required. The Z indexed tire is capable of speeds in excess of +240 km/ h $\approx$ +149 mi/ h, however it is not clear how far above 240 km/ h is possible. It was expected that vehicles will not be driven much higher than 240 km/ h.

Very soon vehicles with higher speeds capability appeared and therefore, automotive industry add W and Y indices to identify the tires that meet the needs of new vehicles that have very high speed capabilities.

Example 1 *Weight of a car and load index of its tire.*

For a car that weighs $2\,tons = 2000\,\mathrm{kg}$, *we need a tire with a load index higher than* 84. *This is because we have about* $500\,\mathrm{kg}$ *per tire and it is in a load index of* 84 *as indicated in Table* 1.1.

Table 1.1 - Maximum load-carrying capacity tire index.

Index	Maximum load	Index	Maximum load
0	$45\,\mathrm{kg} \approx 99\,\mathrm{lbf}$		
...	...	100	$800\,\mathrm{kg} \approx 1764\,\mathrm{lbf}$
71	$345\,\mathrm{kg} \approx 761\,\mathrm{lbf}$	101	$825\,\mathrm{kg} \approx 1819\,\mathrm{lbf}$
72	$355\,\mathrm{kg} \approx 783\,\mathrm{lbf}$	102	$850\,\mathrm{kg} \approx 1874\,\mathrm{lbf}$
73	$365\,\mathrm{kg} \approx 805\,\mathrm{lbf}$	103	$875\,\mathrm{kg} \approx 1929\,\mathrm{lbf}$
74	$375\,\mathrm{kg} \approx 827\,\mathrm{lbf}$	104	$900\,\mathrm{kg} \approx 1984\,\mathrm{lbf}$
75	$387\,\mathrm{kg} \approx 853\,\mathrm{lbf}$	105	$925\,\mathrm{kg} \approx 2039\,\mathrm{lbf}$
76	$400\,\mathrm{kg} \approx 882\,\mathrm{lbf}$	106	$950\,\mathrm{kg} \approx 2094\,\mathrm{lbf}$
77	$412\,\mathrm{kg} \approx 908\,\mathrm{lbf}$	107	$975\,\mathrm{kg} \approx 2149\,\mathrm{lbf}$
78	$425\,\mathrm{kg} \approx 937\,\mathrm{lbf}$	108	$1000\,\mathrm{kg} \approx 2205\,\mathrm{lbf}$
79	$437\,\mathrm{kg} \approx 963\,\mathrm{lbf}$	109	$1030\,\mathrm{kg} \approx 2271\,\mathrm{lbf}$
80	$450\,\mathrm{kg} \approx 992\,\mathrm{lbf}$	110	$1060\,\mathrm{kg} \approx 2337\,\mathrm{lbf}$
81	$462\,\mathrm{kg} \approx 1019\,\mathrm{lbf}$	111	$1090\,\mathrm{kg} \approx 2403\,\mathrm{lbf}$
82	$475\,\mathrm{kg} \approx 1047\,\mathrm{lbf}$	113	$1120\,\mathrm{kg} \approx 2469\,\mathrm{lbf}$
83	$487\,\mathrm{kg} \approx 1074\,\mathrm{lbf}$	113	$1150\,\mathrm{kg} \approx 2581\,\mathrm{lbf}$
84	$500\,\mathrm{kg} \approx 1102\,\mathrm{lbf}$	114	$1180\,\mathrm{kg} \approx 2601\,\mathrm{lbf}$
85	$515\,\mathrm{kg} \approx 1135\,\mathrm{lbf}$	115	$1215\,\mathrm{kg} \approx 2679\,\mathrm{lbf}$
86	$530\,\mathrm{kg} \approx 1163\,\mathrm{lbf}$	116	$1250\,\mathrm{kg} \approx 2806\,\mathrm{lbf}$
87	$545\,\mathrm{kg} \approx 1201\,\mathrm{lbf}$	117	$1285\,\mathrm{kg} \approx 2833\,\mathrm{lbf}$
88	$560\,\mathrm{kg} \approx 1235\,\mathrm{lbf}$	118	$1320\,\mathrm{kg} \approx 2910\,\mathrm{lbf}$
89	$580\,\mathrm{kg} \approx 1279\,\mathrm{lbf}$	119	$1360\,\mathrm{kg} \approx 3074\,\mathrm{lbf}$
90	$600\,\mathrm{kg} \approx 1323\,\mathrm{lbf}$	120	$1400\,\mathrm{kg} \approx 3086\,\mathrm{lbf}$
91	$615\,\mathrm{kg} \approx 1356\,\mathrm{lbf}$	121	$1450\,\mathrm{kg} \approx 3197\,\mathrm{lbf}$
92	$630\,\mathrm{kg} \approx 1389\,\mathrm{lbf}$	122	$1500\,\mathrm{kg} \approx 3368\,\mathrm{lbf}$
93	$650\,\mathrm{kg} \approx 1433\,\mathrm{lbf}$	123	$1550\,\mathrm{kg} \approx 3417\,\mathrm{lbf}$
94	$670\,\mathrm{kg} \approx 1477\,\mathrm{lbf}$	124	$1600\,\mathrm{kg} \approx 3527\,\mathrm{lbf}$
95	$690\,\mathrm{kg} \approx 1521\,\mathrm{lbf}$	125	$1650\,\mathrm{kg} \approx 3690\,\mathrm{lbf}$
96	$710\,\mathrm{kg} \approx 1565\,\mathrm{lbf}$	126	$1700\,\mathrm{kg} \approx 3748\,\mathrm{lbf}$
97	$730\,\mathrm{kg} \approx 1609\,\mathrm{lbf}$	127	$1750\,\mathrm{kg} \approx 3858\,\mathrm{lbf}$
98	$750\,\mathrm{kg} \approx 1653\,\mathrm{lbf}$	128	$1800\,\mathrm{kg} \approx 3968\,\mathrm{lbf}$
99	$775\,\mathrm{kg} \approx 1709\,\mathrm{lbf}$	...	...
		199	$13600\,\mathrm{kg} \approx 30000\,\mathrm{lbf}$

Example 2 *Height of a tire based on tire numbers.*

A tire has the size number P215/60R15 96H. The aspect ratio 60 means the height of the tire is equal to 60% of the tire width. To calculate the tire height in [mm], *we should multiply the first number* (215) *by the second number* (60) *and divide by* 100.

$$h_T = 215 \times \frac{60}{100} = 129\,\text{mm} \tag{1.2}$$

This is the tire height from rim to tread.

Table 1.2 - Maximum speed tire index.

Index	Maximum speed	Index	Maximum speed
A1	$5\,\text{km}/\text{h} \approx 3\,\text{mi}/\text{h}$	K	$110\,\text{km}/\text{h} \approx 68\,\text{mi}/\text{h}$
A2	$10\,\text{km}/\text{h} \approx 6\,\text{mi}/\text{h}$	L	$120\,\text{km}/\text{h} \approx 75\,\text{mi}/\text{h}$
A3	$15\,\text{km}/\text{h} \approx 9\,\text{mi}/\text{h}$	M	$130\,\text{km}/\text{h} \approx 81\,\text{mi}/\text{h}$
A4	$20\,\text{km}/\text{h} \approx 12\,\text{mi}/\text{h}$	N	$140\,\text{km}/\text{h} \approx 87\,\text{mi}/\text{h}$
A5	$25\,\text{km}/\text{h} \approx 16\,\text{mi}/\text{h}$	P	$150\,\text{km}/\text{h} \approx 93\,\text{mi}/\text{h}$
A6	$30\,\text{km}/\text{h} \approx 19\,\text{mi}/\text{h}$	Q	$160\,\text{km}/\text{h} \approx 100\,\text{mi}/\text{h}$
A7	$35\,\text{km}/\text{h} \approx 22\,\text{mi}/\text{h}$	R	$170\,\text{km}/\text{h} \approx 106\,\text{mi}/\text{h}$
A8	$40\,\text{km}/\text{h} \approx 25\,\text{mi}/\text{h}$	S	$180\,\text{km}/\text{h} \approx 112\,\text{mi}/\text{h}$
B	$50\,\text{km}/\text{h} \approx 31\,\text{mi}/\text{h}$	T	$190\,\text{km}/\text{h} \approx 118\,\text{mi}/\text{h}$
C	$60\,\text{km}/\text{h} \approx 37\,\text{mi}/\text{h}$	U	$200\,\text{km}/\text{h} \approx 124\,\text{mi}/\text{h}$
D	$65\,\text{km}/\text{h} \approx 40\,\text{mi}/\text{h}$	H	$210\,\text{km}/\text{h} \approx 130\,\text{mi}/\text{h}$
E	$70\,\text{km}/\text{h} \approx 43\,\text{mi}/\text{h}$	V	$240\,\text{km}/\text{h} \approx 150\,\text{mi}/\text{h}$
F	$80\,\text{km}/\text{h} \approx 50\,\text{mi}/\text{h}$	W	$270\,\text{km}/\text{h} \approx 168\,\text{mi}/\text{h}$
G	$90\,\text{km}/\text{h} \approx 56\,\text{mi}/\text{h}$	Y	$300\,\text{km}/\text{h} \approx 188\,\text{mi}/\text{h}$
J	$100\,\text{km}/\text{h} \approx 62\,\text{mi}/\text{h}$	Z	$+240\,\text{km}/\text{h} \approx +149\,\text{mi}/\text{h}$

Table 1.3 - Maximum speed index and vehicle classification.

Index	Vehicle classification
L	Off-Road and Light Truck Tires
N	Temporary Spare Tires
Q	Off-Road, Studless and Studdable Winter Tires
R	Passenger cars and Light Truck Tires
S	Family Sedans and Vans
T	Family Sedans and Vans
H	Sport Sedans and Coupes
V	Sport Sedans, Coupes and Sports Cars

Example 3 *Alternative tire size indication.*

If the load index is not indicated on the tire, then a tire with a size number such as 255/50R17 V may also be numbered by 255/50V R17.

Example 4 *Tire and rim widths.*

The dimensions of a tire are dependent on the rim on which it is mounted. For tires with an aspect ratio of 50 and above, the rim width is approximately 70% of the tire's width, rounded to the nearest 0.5 in. As an example, a P255/50R16 tire has a design width of 255 mm = 10.04 in however, 70% of 10.04 in is 7.028 in, which rounded to the nearest 0.5 in, is 7 in. Therefore, a P255/50R16 tire should be mounted on a 7 × 16 rim.

For tires with aspect ratio 45 and below, the rim width is 85% of the tire's section width, rounded to the nearest 0.5 in. For example, a P255/45R17 tire with a section width of 255 mm = 10.04 in, needs an 8.5 in rim because 85% of 10.04 in is 8.534 in ≈ 8.5 in. Therefore, a P255/45R17 tire should be mounted on an $8\frac{1}{2} \times 17$ rim.

Example 5 *Calculating tire diameter and radius.*

To calculate the overall diameter of a tire, we first find the tire height by multiplying the tire width and the aspect ratio. As an example, we use tire number P235/75R15.

$$h_T = 235 \times 75\% = 176.25 \, \text{mm} \approx 6.94 \, \text{in} \tag{1.3}$$

Then, we add twice the tire height h_T to the rim diameter.

$$D = 2 \times 6.94 + 15 = 28.88 \, \text{in} \approx 733.8 \, \text{mm} \tag{1.4}$$
$$R = D/2 = 366.9 \, \text{mm} \tag{1.5}$$

Example 6 *Speed rating code.*

Two similar tires are coded as P235/70HR15 and P235/70R15 100H. Both tires have speed code of $H \equiv 210 \, \text{km/h}$. However, the second tire can sustain the speed only when it is loaded less than the specified load index, so it states $100H \equiv 800 \, \text{kg} \, 210 \, \text{km/h}$.

Speed ratings generally depend on the type of tire. Off road vehicles usually use Q-rated tires, passenger cars usually use R-rated tires for typical street cars or T-rated for performance cars.

Example 7 *Tire weight.*

The average weight of a tire for passenger cars is $10 - 12 \, \text{kg} \approx 22 - 26 \, \text{lbf}$. The weight of a tire for light trucks is $14 - 16 \, \text{kg} \approx 30 - 35 \, \text{lbf}$, and the average weight of commercial truck tires is $135 - 180 \, \text{kg} \approx 300 - 400 \, \text{lbf}$.

Example 8 *Effects of aspect ratio.*

A higher aspect ratio provides a softer ride and an increase in deflection under the load of the vehicle. However, lower aspect ratio tires are normally used for higher performance vehicles. They have a wider road contact area and a faster response. This results in less deflection under load, causing a rougher ride to the vehicle.

Changing to a tire with a different aspect ratio will result in a different contact area, therefore changing the load capacity of the tire.

$$DOT \quad DNZE \quad ABCD \quad 2315$$

FIGURE 1.4. An example of a US DOT tire identification number.

$$DOT \quad B3CD \quad E52X \quad 2112 \quad \blacksquare \textcolor{red}{\ast} \blacksquare$$

FIGURE 1.5. An example of a Canadian DOT tire identification number.

Example 9 ★ *BMW tire size code.*

BMW, a European car, uses the metric system for sizing its tires. As an example, $TD230/55ZR390$ is a metric tire size code. TD indicates the BMW TD model, 230 is the section width in [mm], 55 is the aspect ratio in percent, Z is the speed rating, R means radial, and 390 is the rim diameter in [mm].

Example 10 ★ *"MS," "M + S," "M/S," and "M&S" signs.*

The sign "MS,"and "M + S," and "M/S," and "M&S" indicate that the tire has some mud and snow capability. Most radial tires have one of these signs.

Example 11 ★ *U.S. DOT tire identification number.*

The US tire identification number is in the format "DOT DNZE ABCD 2315." It begins with the letters DOT to indicate that the tire meets US federal standards. DOT stands for Department of Transportation. The next two characters, DN, after DOT is the plant code, which refers to the manufacturer and the factory location at which the tire was made.

The next two characters, ZE, are a letter-number combination that refers to the specific mold used for forming the tire. It is an internal factory code and is not usually a useful code for customers. The list can be easily found on internet.

The last four numbers, 2315, represents the week and year the tire was built. The other numbers, ABCD, are marketing codes used by the manufacturer or at the manufacturer's instruction. An example is shown in Figure 1.4.

DN is the plant code for Goodyear-Dunlop Tire located in Wittlich, Germany. ZE is the tire's mold size, ABCD is the compound structure code, 23 indicates the 23th week of the year, and 15 indicates year 2015. So, the tire is manufactured in the 23th week of 2015 at Goodyear-Dunlop Tire in Wittlich, Germany.

Example 12 ★ *Canadian tires identification number.*

In Canada, all tires should have a DOT identification number on the sidewall. An example is shown in Figure 1.5.

This identification number provides the manufacturer, time, and place that the tire was made. The first two characters following DOT indicate the manufacturer and plant code. In this case, B3 indicates Group Michelin located at Bridgewater, Nova Scotia, Canada. The third and fourth characters, CD, are the tire's mold size code. The fifth, sixth, seventh, and eighth characters, E52X, are optional and are used by the manufacturer. The final four numbers, 2112, indicates the manufacturing date. For example, 2112 indicate the twenty first week of year 2012. Finally, the maple leaf sign or the flag sign following the identification number certifies that the tire meets Transport Canada requirements.

Table 1.4 - European county codes for tire manufacturing.

Code	Country	Code	Country
E1	Germany	E14	Switzerland
E2	France	E15	Norway
E3	Italy	E16	Finland
E4	Netherlands	E17	Denmark
E5	Sweden	E18	Romania
E6	Belgium	E19	Poland
E7	Hungary	E20	Portugal
E8	Czech Republic	E21	Russia
E9	Spain	E22	Greece
E10	Yugoslavia	E23	Ireland
E11	United Kingdom	E24	Croatia
E12	Austria	E25	Slovenia
E13	Luxembourg	E26	Slovakia

Example 13 ★ *E-Mark and international codes.*

All tires sold in Europe after July 1997 must carry an E-mark. An example is shown by $\boxed{5}$ *in Figure 1.2. The mark itself is either an upper or lower case "E" followed by a number in a circle or rectangle, followed by a further number. An "E" indicates that the tire is certified to comply with the dimensional, performance and marking requirements of ECE regulation. ECE or UNECE stands for the united nations economic commission for Europe. The number in the circle or rectangle is the country code. Example: 11 is the UK. The first two digits outside the circle or rectangle indicate the regulation series under which the tire was approved. Example: "02" is for ECE regulation 30 governing passenger tires, and "00" is for ECE regulation 54 governing commercial vehicle tires. The remaining numbers represent the ECE mark type approval numbers. Tires may have also been tested and met the required noise limits. These tires may have a second ECE branding followed by an "−s" for sound.*

Table 1.4 indicates the European country codes for tire manufacturing.

Besides the DOT and ECE codes for US and Europe, we may also see the other country codes such as: ISO − 9001 for international standards organization, C.C.C for China compulsory product certification, JIS D 4230 for Japanese industrial standard and so on.

Example 14 ★ *Light truck tires.*
The tire sizes for a light truck may be shown in two formats:

$$LT245/70R16$$

or

$$32 \times 11.50R16LT$$

In the first format, LT ≡light truck, 245 ≡tire width in millimeters, 70 ≡aspect ratio in percent, R ≡radial structure, and 16 ≡rim diameter in inches.

In the second format, 32 ≡tire diameter in inches, 11.50 ≡tire width in inches, R ≡radial structure, 16 ≡rim diameter in inches, and LT ≡light truck.

Example 15 ★ *UTQG ratings.*
Tire manufacturers may put some other symbols, numbers, and letters on their tires supposedly rating their products for wear, wet traction, and heat resistance. These characters are referred to as UTQG (Uniform Tire Quality Grading), although there is no uniformity and standard in how they appear. There is an index for wear to show the average wearing life time in mileage. The higher the wear number, the longer the tire lifetime. An index of 100 is equivalent to approximately 20000 miles or 30000 km. Other numbers are indicated in Table 1.5.

Table 1.5 - Tread wear rating index.

Index	Life (Approximate)	
100	32000 km	20000 mi
150	48000 km	30000 mi
200	64000 km	40000 mi
250	80000 km	50000 mi
300	96000 km	60000 mi
400	129000 km	80000 mi
500	161000 km	100000 mi

The UTQG also rates tires for wet traction and heat resistance. These are rated in letters between "A" to "C," where "A" is the best, "B" is intermediate and "C" is acceptable. An "A" wet traction rating is typically an indication that the tire has a deep open tread pattern with lots of sipping, which are the fine lines in the tread blocks.

An "A" heat resistance rating indicates two things: first, low rolling resistance due to stiffer tread belts, stiffer sidewalls, or harder compounds; second, thinner sidewalls, more stable blocks in the tread pattern. Temperature rating is also indicated by a letter between "A" to "CM" where "A" is the best, "B" is intermediate, and "C" is acceptable.

There might also be a traction rating to indicate how well a tire grips the road surface. This is an overall rating for both dry and wet conditions. Tires are rated as: "AA" for the best, "A" for better, "B" for good, and "C" for acceptable.

Example 16 ★ *Tire sidewall additional marks.*

$TL \equiv$ *Tubeless*

$TT \equiv$ *Tube type, tire with an inner-tube*

Made in Country $\equiv$ *Name of the manufacturing country*

$C \equiv$ *Commercial tires made for commercial trucks; Example:* $185R14C$

$B \equiv$ *Bias ply*

$SFI \equiv$ *Side facing inwards*

$SFO \equiv$ *Side facing outwards*

$TWI \equiv$ *Tire wear index*

It is an indicator in the main tire profile, which shows when the tire is worn down and needs to be replaced.

$SL \equiv$ *Standard load; Tire for normal usage and loads*

$XL \equiv$ *Extra load; Tire for heavy loads*

RF *or* $rf \equiv$ *Reinforced tires*

Arrow $\equiv$ *Direction of rotation*

Some tread patterns are designed to perform better when driven in a specific direction. Such tires will have an arrow showing which way the tire should rotate when the vehicle is moving forwards.

Example 17 ★ *Plus one (+1) concept.*

The plus one (+1) concept describes the sizing up of a rim and matching it to a proper tire. Generally speaking, each time we add 1 in to the rim diameter, we should add 20 mm to the tire width and subtract 10% from the aspect ratio. This compensates the increases in rim width and diameter, and provides the same overall tire radius. Figure 1.6 illustrates the idea.

By using a tire with a shorter sidewall, we get a quicker steering response and better lateral stability. However, we will have a stiffer ride.

Example 18 ★ *Under- and over-inflated tire.*

Overheat caused by improper inflation of tires is a common tire failure. An under-inflated tire will support less of the vehicle weight with the air pressure in the tire; therefore, more weight will be supported by the tire wall. This tire load increase causes the tire to have a larger tireprint that creates more friction and more heat.

In an over-inflated tire, too much of the vehicle weight is supported by the tire air pressure. The vehicle will be bouncy and hard to steer because the tireprint is small and only the center portion of the tireprint is contacting the road surface.

In a properly-inflated tire, approximately 95% of the vehicle weight is supported by the air pressure in the tire and 5% is supported by the tire wall.

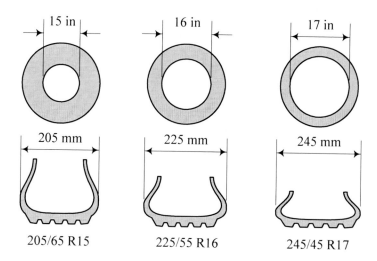

205/65 R15   225/55 R16   245/45 R17

FIGURE 1.6. The plus one (+1) concept is a rule to find the tire to a rim with a 1 inch increase in diameter.

1.2 Tire Components

A tire is an advanced engineering product made of rubber and a series of synthetic materials cooked together. Fiber, textile, and steel cords are some of the components that go into the tire's inner liner, body plies, bead bundle, belts, sidewalls, and tread. Figure 1.7 illustrates a sample of tire interior components and their arrangement.

The main components of a tire are:

Bead or *bead bundle* is a loop of high strength steel cable coated with rubber. It gives the tire the strength it needs to stay seated on the wheel rim and to transfer the tire forces to the rim.

Inner layers are made up of different fabrics, called plies. The most common ply fabric is polyester cord. The top layers are also called cap plies. *Cap plies* are polyesteric fabric that help hold everything in place. Cap plies are not found on all tires; they are mostly used on tires with higher speed ratings to help all the components stay in place at high speeds.

An *inner liner* is a specially compounded rubber that forms the inside of a tubeless tire. It inhibits loss of air pressure.

Belts or *belt buffers* are one or more rubber-coated layers of steel, polyester, nylon, Kevlar (para-aramid synthetic fiber) or other materials running circumferentially around the tire under the tread. They are designed to reinforce body plies to hold the tread flat on the road and make the best contact with the road. Belts reduce squirm to improve tread wear and resist damage from impacts and penetration.

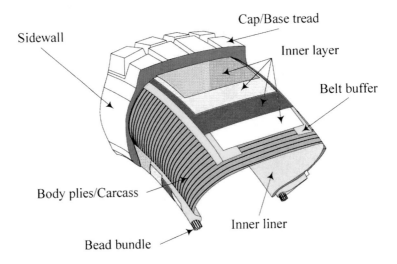

FIGURE 1.7. Illustration of a sample radial tire interior components and arrangement.

The *carcass* or *body plies* are the main part in supporting the tension forces generated by tire air pressure. The carcass is made of rubber-coated steel or other high strength cords tied to bead bundles. The cords in a radial tire, as shown in Figure 1.7, run perpendicular to the tread. The plies are coated with rubber to help them bond with the other components and to seal in the air.

A tire's strength is often described by the number of carcass plies. Most car tires have two carcass plies. By comparison, large commercial jetliners often have tires with 30 or more carcass plies.

The *sidewall* provides lateral stability for the tire, protects the body plies, and helps to keep the air in the tire. It may contain additional components to help increase the lateral stability.

The *tread* is the portion of the tire that comes in contact with the road. Tread designs vary widely depending on the specific purpose of the tire. The tread is made from a mixture of different kinds of natural and synthetic rubbers. The outer perimeter of a tire is called the *crown*.

The *tread groove* is the space or area between two tread rows or blocks. The tread groove gives the tire traction and is especially useful during rain or snow.

Example 19 *Tire rubber main material.*

There are two major ingredients in a rubber compound: the rubber and the filler. They are combined in such a way to achieve different objectives. The objective may be performance optimization, traction maximization, or less rolling resistance. The most common fillers are different types of carbon black and silica. The other tire ingredients are antioxidants, antiozonant, and anti-aging agents.

Tires are combined with several components and cooked with a heat treatment. The components must be formed, combined, assembled, and cured together. Tire quality depends on the ability to blend all of the separate components into a cohesive product that satisfies the driving needs. A modern tire is basically a mixture of steel, fabric, and rubber. Generally speaking, the weight percentage of the components of a tire are:

1. *Reinforcements: steel, rayon, nylon, 16%*
2. *Rubber: natural/synthetic, 38%*
3. *Compounds: carbon, silica, chalk, 30%*
4. *Softener: oil, resin, 10%*
5. *Vulcanization: sulfur, zinc oxide, 4%*
6. *Miscellaneous, 2%*

Example 20 *Tire cords.*

Because tires have to carry heavy loads, steel and fabric cords are used in their construction to reinforce the rubber compound and provide strength. The most common materials suitable for the tire application are cotton, rayon, polyester, steel, fiberglass, and aramid.

Example 21 *Bead components and preparation.*

The bead component of tires is a non-extensible composite loop that anchors the carcass and locks the tire into the rim. The tire bead components include the steel wire loop and apex or bead filler. The bead wire loop is made from a steel wire covered by rubber and wound around the tire with several continuous loops. The bead filler is made from a very hard rubber compound, which is extruded to form a wedge.

Example 22 *Tire ply construction.*

The number of plies and cords indicates the number of layers of rubber-coated fabric or steel cords in the tire. In general, the greater the number of plies, the more weight a tire can support. Tire manufacturers also indicate the number and type of cords used in the tire.

Example 23 ★ *Tire tread extrusion.*

Tire tread, or the portion of the tire that comes in contact with the road, consists of the tread, tread shoulder, and tread base. There are at least three different rubber compounds used in forming the tread profile. The three rubber compounds are extruded simultaneously into a shared extruder head.

Example 24 ★ *Different rubber types used in tires.*

There are five major rubbers used in tire production: natural rubber, styrene-butadiene rubber (SBR), polybutadiene rubber (BR), butyl rubber, and halogenated butyl rubber. The first three are primarily used for tread and sidewall compounds, while butyl rubber and halogenated butyl rubber

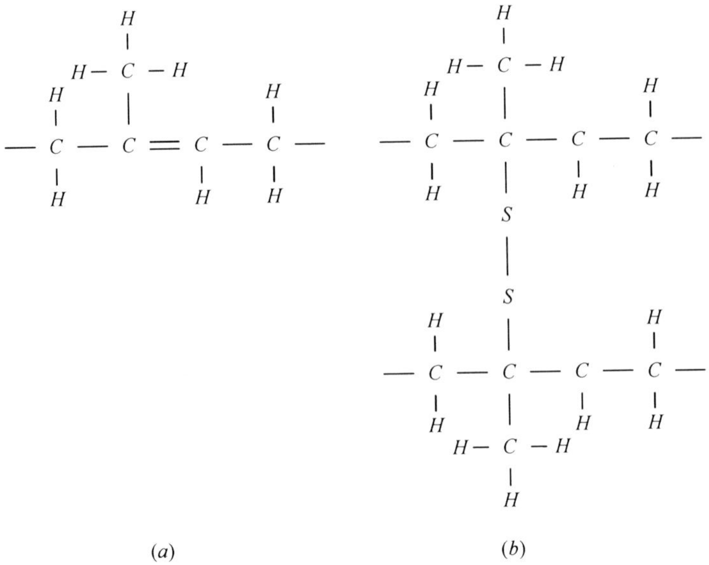

FIGURE 1.8. Natural rubber chemical compond: (a) the monomer unit, (b) the vulcanized rubber.

are primarily used for the inner liner and the inside portion that holds the compressed air inside the tire.

The natural rubber latex comes from the Pará rubber tree. The para rubber tree initially grew in South America, but today Malaysia is the main producer of this material. When rubber first arrived in England around 1770, it was observed that a piece of the material was good for rubbing off pencil marks on paper, and hence the name rubber appeared.

Styrene-butadiene rubber (SBR) provides aging stability, polybutadiene rubber (BR) provides wear resistance.

Example 25 ★ History of rubber.

About 2500 years ago, people living in Central and South America used the sap and latex of a local tree to waterproof their shoes and clothes. This material was introduced to the first pilgrim travelers in the 17th century. The first application of this new material was discovered by the English as an eraser. This application supports the name **rubber**, because it was used for rubbing out pencil marks. The rubber pneumatic tires were invented in 1845 and its production began in 1888.

The natural rubber is a mixture of polymers and isomers. The main rubber isomer is shown in Figure 1.8 (a) and is called **isoprene**. The natural rubber may be vulcanized to make longer and stronger polyisopren, suitable for tire production. Vulcanization is usually done by sulfur as cross-links. Figure 1.8 (b) illustrates a vulcanized rubber polymer.

Example 26 ★ *A world without rubber.*

Rubber is the main material used to make a tire compliant. A compliant tire can stick to the road surface while it goes out of shape and provides distortion to move in another direction. The elastic characteristic of a tire allows the tire to be pointed in a direction different than the direction the car is pointed. There is no way for a vehicle to turn without rubber tires, unless it moves at a very low speed. If vehicles were equipped with only noncompliant wheels then trains moving on railroads would be the main travelling vehicles. People could not live too far from the railways and there would not be much use for bicycles and motorcycles.

Example 27 *Tire information tips.*

A new front tire with a worn rear tire can cause instability.

Tires stored in direct sunlight for long periods of time will harden and age more quickly than those kept in a dark area.

Prolonged contact with oil or gasoline causes contamination of the rubber compound, making the tire life short.

1.3 Radial and Non-Radial Tires

Tires are divided into two classes: *radial* and *non-radial*, depending on the angle between carcass metallic cords and the tire-plane. Each type of tire construction has its own set of characteristics that are the key to its performance.

The radial tire is constructed with reinforced steel cable belts that are assembled in parallel and run side to side, from one bead to another bead at an angle of 90 deg to the circumferential centerline of the tire. This makes the tire more flexible radially, which reduces rolling resistance and improves cornering capability. Figure 1.7 shows the interior structure and the carcass arrangement of a radial tire.

The non-radial tires are also called *bias-ply* and *cross-ply* tires. The plies are layered diagonal from one bead to the other bead at about a 30 deg angle, although any other angles may also be applied. One ply is set on a bias in one direction as succeeding plies are set alternately in opposing directions as they cross each other. The ends of the plies are wrapped around the bead wires, anchoring them to the rim of the wheel. Figure 1.9 shows the interior structure and the carcass arrangement of a non-radial tire.

The most important difference in the dynamics of radial and non-radial tires is their different ground sticking behavior when a lateral force is applied on the wheel. This behavior is shown in Figure 1.10. The radial tire, shown in Figure 1.10(a), flexes mostly in the sidewall and keeps the tread flat on the road. The bias-ply tire, shown in Figure 1.10(b) has less contact with the road as both tread and sidewalls distort under a lateral load.

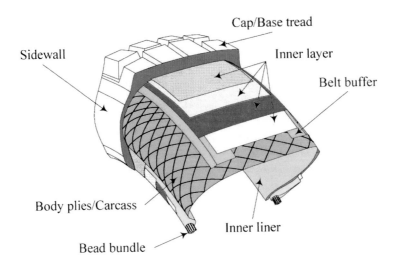

FIGURE 1.9. Illustration of a non-radial tire's interior components and arrangement.

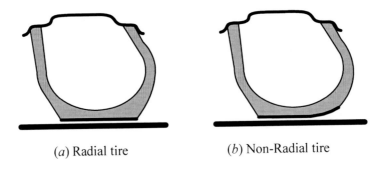

(a) Radial tire   (b) Non-Radial tire

FIGURE 1.10. Ground-sticking behavior of radial and non-radial tires in the presence of a lateral force.

The radial arrangement of carcass in a radial tire allows the tread and sidewall act independently. The sidewall flexes more easily under the weight of the vehicle. So, more vertical deflection is achieved with radial tires. As the sidewall flexes under the load, the belts hold the tread firmly and evenly on the ground which reduces tread scrub. In a cornering maneuver, the independent action of the tread and sidewalls keeps the tread flat on the road. This allows the tire to hold its path. Radial tires are the preferred tire in most applications today.

The cross arrangement of carcass in bias-ply tires allows it act as a unit. When the sidewalls deflect or bend under load, the tread squeezes in and distorts. This distortion affects the tireprint and decrease traction. Because of the bias-ply inherent construction, sidewall strength is less than that of a radial tire's construction and cornering is less effective.

Example 28 *Increasing the strength of tires.*

The strength of bias-ply tires increases by increasing the number of plies and bead wires. However, more plies means more mass, which increases heat and reduces tire life. To increase a radial tire's strength, larger diameter steel cables are used in the tire's carcass.

Example 29 *Tubeless and tube-type tire construction.*

A tubeless tire is similar in construction to a tube-type tire, except that a thin layer of air and moisture-resistant rubber is used on the inside of the tubeless tire from bead to bead to obtain an internal seal of the casing. This eliminates the need for a tube and flap. Both tires, in equivalent sizes, can carry the same load at the same inflation pressure.

Example 30 ★ *New shallow tires.*

Low aspect ratio tires are radial tubeless tires that have a section width wider than their section height. The aspect ratio of these tires is between 50% to 30%. These shallow tires have shorter sidewall heights and wider tread widths. This feature improves stability and handling from a higher lateral stiffness rates.

Example 31 ★ *Tire function.*

A tire is a pneumatic system to support a vehicle's load. Tires support a vehicle's load by using compressed air to create tension in the carcass plies. Tire carcass are a series of cords that have a high tension strength, and almost no compression strength. The air pressure creates tension in the carcass and carries the load. In an inflated and unloaded tire, the cords pull equally on the bead wire all around the tire. When the tire is loaded, the tension in the cords between the rim and the ground is relieved while the tension in other cords is unchanged. Therefore, the cords opposite the ground pull the bead upwards. This is how pressure is transmitted from the ground to the rim.

Besides vertical load carrying, a tire must transmit acceleration, braking,

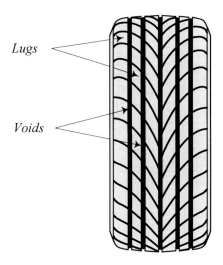

FIGURE 1.11. A sample of tire tread to show lugs and voids.

and cornering forces to the road. These forces are transmitted to the rim in a similar manner. Acceleration and braking forces also depend on the friction between the rim and the bead. A tire also acts as a spring between the rim and the road.

1.4   Tread

The tread pattern is made up of tread *lugs* and tread *voids*. The lugs are the sections of rubber that make contact with the road and voids are the spaces that are located between the lugs. Lugs are also called *slots* or *blocks*, and voids are also called *grooves*. The tread pattern of block-groove configurations affects the tire's traction and noise level. Wide and straight grooves running circumferentially have a lower noise level and high lateral friction. More lateral grooves running from side to side increase traction and noise levels. A sample of a tire tread is shown in Figure 1.11.

Tires need both circumferential and lateral grooves. The water on the road is compressed into the grooves by the vehicle's weight and is evacuated from the tireprint region, providing better traction at the tireprint contact. Without such grooves, the water would not be able to escape out to the sides of the wheel. This would cause a thin layer of water to remain between the road and the tire, which causes a loss of friction with the road surface. Therefore, the grooves in the tread provide an escape path for water.

On a dry road, the tire treads reduce grip because they reduce the contact area between the rubber and the road. This is the reason for using treadless or slick tires at smooth and dry race tracks.

The mud-terrain tire pattern is characterized by large lugs and large voids. The large lugs provide large bites in poor traction conditions and the large voids allow the tire to clean itself by releasing and expelling the mud and dirt. The all-terrain tire pattern is characterized by smaller voids and lugs when compared to the mud terrain tire. A denser pattern of lugs and smaller voids make all-terrain tires quieter on the street. However, smaller voids cannot clean themselves easily and if the voids fill up with mud, the tire loses some of it's traction. The all-terrain tire is good for highway driving.

Example 32 *Asymmetrical and directional tread design.*

The design of the tread pattern may be asymmetric and change from one side to the other. Asymmetric patterns are designed to have two or more different functions and provide a better overall performance.

A directional tire is designed to rotate in only one direction for maximum performance. Directional tread pattern is especially designed for driving on wet, snowy, or muddy roads. A non-directional tread pattern is designed to rotate in either direction without sacrificing in performance.

Example 33 *Self-cleaning.*

Self-cleaning is the ability of a tire's tread pattern to release mud or material from the voids of tread. This ability provides good bite on every rotation of the tire. A better mud tire releases the mud or material easily from the tread voids.

Example 34 ★ *Hydroplaning*

Hydroplaning *is sliding of a tire on a film of water. Hydroplaning can occur when a car drives through standing water and the water cannot totally escape out from under the tire. This causes the tire to lift off the ground and slide on the water. The hydroplaning tire will have little traction and therefore, the car will not obey the driver's command.*

Deep grooves running from the center front edge of the tireprint to the corners of the back edges, along with a wide central channel help water to escape from under the tire. Figure 1.12 illustrates the hydroplaning phenomena when the tire is riding over a water layer.

There are three types of hydroplaning: dynamic, viscous, and rubber hydroplaning. Dynamic hydroplaning occurs when standing water on a wet road is not displaced from under the tires fast enough to allow the tire to make pavement contact over the total tireprint. The tire rides on a wedge of water and loses its contact with the road. The speed at which hydroplaning happens is called hydroplaning speed.

Viscous hydroplaning occurs when the wet road is covered with a layer of oil, grease, or dust. Viscous hydroplaning happens with less water depth and at a lower speed than dynamic hydroplaning.

Rubber hydroplaning is generated by superheated steam at high pressure in the tireprint, which is caused by the friction-generated heat in a hard

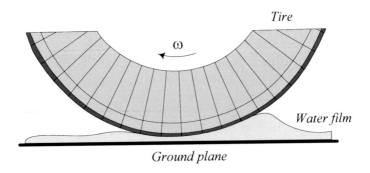

FIGURE 1.12. Illustration of hydroplaning phnomena.

braking.

Example 35 ★ *Aeronautic hydroplaning speed.*
   In aerospace engineering the hydroplaning speed is estimated in [*knots*]
by

$$v_H = 9\sqrt{p} \tag{1.6}$$

where, p is tire inflation pressure in [*psi*].
   For main wheels of a B757 aircraft, the hydroplaning speed would be

$$v_H = 9\sqrt{144} = 108\,knots \approx 55.5\,\mathrm{m/\,s} \tag{1.7}$$

Equation (1.6) for a metric system would be

$$v_x = 5.5753 \times 10^{-2}\sqrt{p} \tag{1.8}$$

where v_x is in [m/ s] *and p is in* [Pa]. *As an example, the hydroplaning
speed of a car using tires with pressure 28psi $\approx$ 193053 Pa is*

$$\begin{aligned} v_x &= 5.5753 \times 10^{-2}\sqrt{193053} \approx 24.5\,\mathrm{m/\,s} \\ &\approx 47.6\,knots \approx 88.2\,\mathrm{km/\,h} \approx 54.8\,\mathrm{mi/\,h} \end{aligned} \tag{1.9}$$

1.5   Tireprint

The contact area between a tire and the road is called the *tireprint* and is
shown by A_P. At any point of a tireprint, the normal and friction forces are
transmitted between the road and tire. The effect of the contact forces can
be described by a resulting force system including force and torque vectors
applied at the center of the tireprint.
   The tireprint is also called *contact patch, contact region,* or *tire footprint.*
A model of tireprint is shown in Figure 1.13.

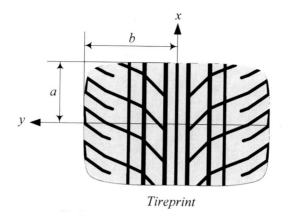

Tireprint

FIGURE 1.13. A tireprint.

The area of the tireprint is inversely proportional to the tire pressure. Lowering the tire pressure is a technique used for off-road vehicles in sandy, muddy, or snowy areas, and for drag racing. Decreasing the tire pressure causes the tire to slump so more of the tire is in contact with the surface, giving better traction in low friction conditions. It also helps the tire grip small obstacles as the tire conforms more to the shape of the obstacle, and makes contact with the object in more places.

Example 36 *Uneven wear in front and rear tires.*

In most vehicles, the front and rear tires will wear at different rates. So, it is advised to swap the front and rear tires as they wear down to even out the wear patterns. This is called rotating the tires.

Front tires, especially on front-wheel drive vehicles, wear out more quickly than rear tires.

1.6   Wheel and Rim

When a tire is installed on a rim and is inflated, it is called a wheel. A *wheel* is a combined tire and *rim*. The rim is the cylindrical part on which the tire is installed. Most passenger cars are equipped with steel or metallic rims. The steel rim is made by welding a disk to a shell. However, light alloy rims made with light metals such as aluminium and magnesium are also popular. Figure 1.14 illustrates a wheel and the most important dimensional names.

A rim has two main parts: *flange* and *spider*. The flange or *hub* is the ring or shell on which the tire is mounted. The spider or *center section* is the disc section that is attached to the hub. The rim width is also called pan width and measured from inside to inside of the bead seats of the flange. Flange provides lateral support to the tire. A flange has two *bead seats*

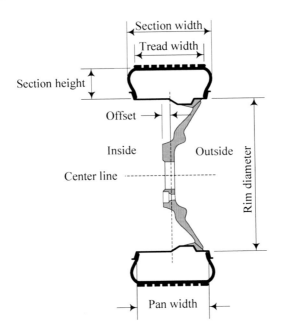

FIGURE 1.14. Illustration of a wheel and its dimensions.

providing radial support to the tire. The *well* is the middle part between the bead seats with sufficient depth and width to enable the tire beads to be mounted and demounted on the rim. The *rim hole* or *valve aperture* is the hole or slot in the rim that accommodates the valve for tire inflation.

There are two main rim shapes:

1. drop center rim (DC) and,
2. wide drop center rim (WDC).

The WDC may also come with a *hump*. The humped WDC may be called $WDCH$. Their cross sections are illustrated in Figure 1.15.

Drop center (DC) rims usually are symmetric with a well between the bead seats. The well is built to make mounting and demounting the tire easier. The bead seats are around $5\,\deg$ tapered. Wide drop center rims (WDC) are wider than DC rims and are built for low aspect ratio tires. The well of WDC rims are shallower and wider. Today, most passenger cars are equipped with WDC rims. The WDC rims may be manufactured with a hump behind the bead seat area to prevent the bead from slipping down.

A sample of rim numbering and its meaning is shown in Figure 1.16. Rim width, rim diameter, and offset are shown in Figure 1.14. Offset is the distance between the inner plane and the center plane of the rim. A rim may be designed with a negative, zero, or positive offset. A rim has a positive offset if the spider is outward from the center plane.

Rim width
Rim diameter
DC rim
Rim width
Rim diameter
WDC rim
Rim width
Rim diameter
WDCH rim
Hump

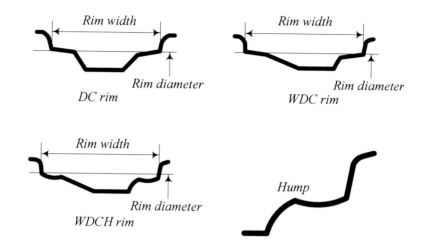

FIGURE 1.15. Illustration of DC, WDC, and $WDCH$ rims and their geometry.

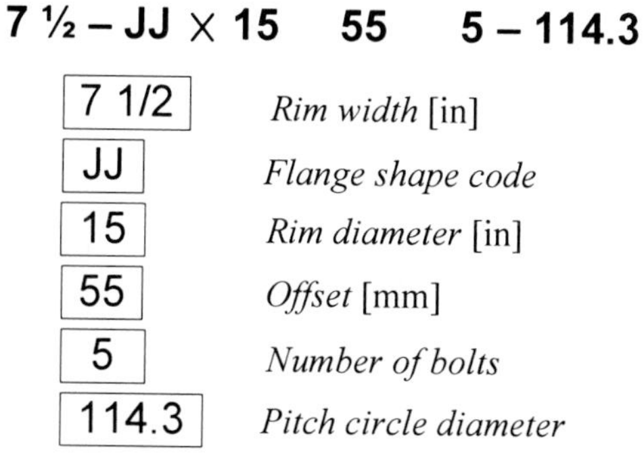

7 ½ – JJ × 15 55 5 – 114.3
7 1/2
Rim width [in]
JJ
Flange shape code
15
Rim diameter [in]
55
Offset [mm]
5
Number of bolts
114.3
Pitch circle diameter

FIGURE 1.16. A sample rim number.

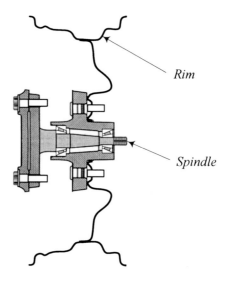

FIGURE 1.17. Illustration of a wheel attched to the spindle axle.

The flange shape code signifies the tire-side profile of the rim and can be B, C, D, E, F, G, J, JJ, JK, and K. Usually the profile code follows the nominal rim width but different arrangements are also used. Figure 1.17 illustrates how a wheel is attached to the spindle axle.

Example 37 *Wire spoke wheel.*

A rim that uses wires to connect the center part to the exterior flange is called a wire spoke wheel, or simply a wire wheel. The wires are called spokes. This type of wheel is usually used on classic vehicles. The high-power cars do not use wire wheels because of safety. Figure 1.18 depicts two examples of wire spoke wheels.

Example 38 *Light alloy rim material.*

Metal is the main material for manufacturing rims, however, new composite materials are also used for rims. Composite material rims are usually thermoplastic resin with glass fiber reinforcement, developed mainly for low weight. Their strength and heat resistance still need improvement before being a proper substitute for metallic rims.

Other than steel and composite materials, light alloys such as aluminum, magnesium, and titanium are used for manufacturing rims.

Aluminum is very good for its weight, thermal conductivity, corrosion resistance, easy casting, low temperature, easy machine processing, and recycling. Magnesium is about 30% lighter than aluminum, and is excellent for size stability and impact resistance. However, magnesium is more expensive and it is used mainly for luxury or racing cars. The corrosion resistance of magnesium is not as good as aluminum. Titanium is much stronger than

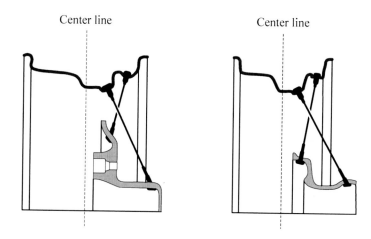

FIGURE 1.18. Two samples of wire spoke wheel.

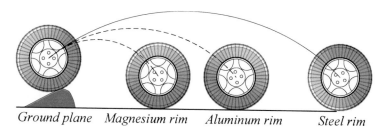

Ground plane   Magnesium rim   Aluminum rim   Steel rim

FIGURE 1.19. The difference between aluminum, magnesium, and steel rims in regaining road contact after a jump.

aluminum and magnesium with excellent corrosion resistance. However, titanium is expensive and hard to be machine processed.

The difference between aluminum, magnesium, and steel rims is illustrated in Figure 1.19. Light weight wheels regain contact with the ground quicker than heavier wheels.

Example 39 *Spare tire.*

Road vehicles typically carry a spare tire, which is already mounted on a rim ready to use in the event of flat tire. Since 1980, some cars have been equipped with spare tires that are smaller than normal size. These spare tires are called doughnuts or space-saver spare tires. Although the doughnut spare tire is not very useful or popular, it can help to save a little space, weight, cost, and gas mileage. Doughnut spare tires can not be driven far or fast.

Example 40 *Wheel history.*

Stone and wooden wheels were invented and used somewhere in the Mid-

dle East about 5000 years ago. Hard wheels have some inefficient character-istics namely poor traction, low friction, harsh ride, and poor load carrying capacity.

Solid rubber tires and air tube tires began to be used in the late nineteen and early twentieth century.

1.7   Vehicle Classifications

Road vehicles are usually classified based on their size and number of axles. Although there is no standard or universally accepted classification method, there are a few important and applied vehicle classifications.

1.7.1   ISO and FHWA Classification

ISO3833 classifies ground vehicles in 7 groups:

1. Motorcycles
2. Passenger cars
3. Busses
4. Trucks
5. Agricultural tractors
6. Passenger cars with trailer
7. Truck trailer/semi trailer road trains

The Federal Highway Administration ($FHWA$) classifies road vehicles based on size and application. All road vehicles are classified in 13 classes as described below:

1. Motorcycles
2. Passenger cars, including cars with a one-axle or two-axle trailer
3. Other two-axle vehicles, including: pickups, and vans, with a one-axle or two-axle trailer
4. Buses
5. Two axle, six-tire single units
6. Three-axle single units
7. Four or more axle single units
8. Four or fewer axle single trailers
9. Five-axle single trailers
10. Six or more axle single trailers
11. Five or less axle multi-trailers
12. Six-axle multi-trailers
13. Seven or more axle multi-trailers

Figure 1.20 illustrates the $FHWA$ classification. The definition of $FHWA$ classes follow.

FIGURE 1.20. The *FHWA* vehicle classification.

FIGURE 1.21. A three-wheel motorcycle.

Motorcycles: Any motorvehicle having a seat or saddle and no more than three wheels that touch the ground is a motorcycle. Motorcycles, motor scooters, mopeds, motor-powered or motor-assisted bicycles, and three-wheel motorcycles are in this class. Motorcycles are usually, but not necessarily, steered by handlebars. Figure 1.21 depicts a three-wheel motorcycle.

Passenger Cars: Street cars, including sedans, coupes, and station wagons manufactured primarily for carrying passengers, are in this class. Passenger cars are also called street cars, *automobiles*, or *autos*.

Other Two-Axle, Four-Tire Single-Unit Vehicles: All two-axle, four-tire vehicles other than passenger cars make up this class. This class includes pickups, panels, vans, campers, motor homes, ambulances, hearses, carryalls, and minibuses. Other two-axle, four-tire single-unit vehicles pulling recreational or light trailers are also included in this class. Distinguishing class 3 from class 2 is not clear, so these two classes may sometimes be combined into class 2.

Buses: A motor vehicle able to carry more than ten persons is a bus. Buses are manufactured as traditional passenger-carrying vehicles with two axles and six tires. However, buses with three or more axles are also manufactured.

Two-Axle, Six-Tire, Single-Unit Trucks: Vehicles on a single frame including trucks, camping and recreational vehicles, motor homes with two axles, and dual rear wheels.

Three-Axle Single-Unit Trucks: Vehicles having a single frame including trucks, camping, recreational vehicles, and motor homes with three axles.

Four-or-More-Axle-Single-Unit Trucks: All trucks on a single frame with four or more axles make up this class.

Four-or-Fewer-Axle Single-Trailer Trucks: Vehicles with four or fewer axles consisting of two units, one of which is a tractor or straight truck power unit.

Five-Axle Single-Trailer Trucks: Five-axle vehicles consisting of two units, one of which is a tractor or straight truck power unit.

Six-or-More-Axle Single-Trailer Trucks: Vehicles with six or more axles consisting of two units, one of which is a tractor or straight truck power unit.

Five-or-Fewer-Axle Multi-Trailer Trucks: Vehicles with five or fewer axles consisting of three or more units, one of which is a tractor or straight truck power unit.

Six-Axle Multi-Trailer Trucks: Six-axle vehicles consisting of three or more units, one of which is a tractor or straight truck power unit.

Seven or More Axle Multi-Trailer Trucks: Vehicles with seven or more axles consisting of three or more units, one of which is a tractor or straight truck power unit.

The classes 6 to 13 are also called truck. A *truck* is a motor vehicle designed primarily for carrying load or property.

1.7.2 *Passenger Car Classifications*

A passenger car or automobile is a motorvehicle designed for carrying ten or fewer persons. Automobiles may be classified based on their size and weight. Size classification is based on wheelbase, the distance between front and rear axles. Weight classification is based on *curb weight*, the weight of an automobile with standard equipment, and a full complement of fuel and other fluids, but with no load, persons, or property. The wheelbase is rounded to the nearest inch and the curb weight to the nearest 100 lb ≈ 50 kg before classification.

For a size classification, passenger car may be classified as a *small, midsize*, and *large* car. *Small cars* have a wheelbase of less than 99 in ≈ 2.5 m, *midsize cars* have a wheelbase of less than 109 in ≈ 2.8 m and greater than 100 in ≈ 2.5 m, and *large cars* have a wheelbase of more than 110 in ≈ 2.8 m. Each class may also be divided further.

For a weight classification, passenger car may be classified as *light, midweight*, and *heavy*. *Light weight cars* have a curb weight of less than 2400 lb ≈ 1100 kg, *midweight cars* have a curb weight of less than 3400 lb ≈ 1550 kg and more than 2500 lb ≈ 1150 kg, and *heavy cars* have a curb weight of more than 3500 lb ≈ 1600 kg. Each class may also be divided in some subdivisions.

Dynamically, passenger cars may be classified by their type of suspension, engine, driveline arrangement, weight distribution, or any other parameters that affect the dynamics of a car. However, in the market, passenger cars are usually divided into the following classes according to the number of passengers and load capacity.

1. Economy

2. Compact
3. Intermediate
4. Standard Size
5. Full Size
6. Premium Luxury
7. Convertible Premium
8. Convertible
9. Minivan
10. Midsize
11. SUV

In another classification, cars are divided according to size and shape. However, using size and shape to classify passenger cars is not clear-cut; many vehicles fall in between classes. Also, not all are sold in all countries, and sometimes their names differ between countries. Common entries in the shape classification are the *sedan, coupe, convertible, minivan/van, wagon,* and *SUV.*

A *sedan* is a car with a four-door body configuration and a conventional trunk or a sloping back with a hinged rear cargo hatch that opens upward.

A *coupe* is a two-door passenger car.

A *convertible* is a car with a removable or retractable top.

A *minivan/van* is a vehicle usually with a box-shaped body enclosing a large cargo or passenger area. The identified gross weight of a van is less than 10000 lb ≈ 4500 kg. Vans can be identifiable by their enclosed cargo or passenger area, short hood, and box shape. Vans can be divided into *mini van, small van, midsize van, full-size van,* and *large van.* The van subdivision has the same specifications as SUV subdivisions.

A *wagon* is a car with an extended body and a roofline that extends past the rear doors.

An *SUV* (Sport Utility Vehicle) is a vehicle with off-road capability. SUV is designed for carrying ten or fewer persons, and generally considered a multi-purpose vehicle. Most SUVs are four-wheel-drive with and increased ground clearance. The SUV is also known as 4 *by* 4, $4WD$, 4 × 4 or 4*x*4. SUVs can be divided into *mini, small, midsize, full-size,* and *large SUV.*

Mini SUVs are those with a wheelbase of less than or equal to 88 in ≈ 224 cm. A mini SUV is typically a microcar with a high clearance, and off-road capability. Small SUVs have a wheelbase of greater than 88 in ≈ 224 cm with an overall width of less than 66 in ≈ 168 cm. Small SUVs are short and narrow 4 × 4 multi-purpose vehicles. Midsize SUVs have a wheelbase of greater than 88 in ≈ 224 cm with an overall width greater than 66 in ≈ 168 cm, but less than 75 in ≈ 190 cm. Midsize SUVs are 4 × 4 multi-purpose vehicles designed around a shortened pickup truck chassis. Full-size SUVs are made with a wheelbase greater than 88 in ≈ 224 cm and a width between 75 in ≈ 190 cm and 80 in ≈ 203 cm. Full-size SUVs are 4×4 multi-purpose vehicles designed around an enlarged pickup truck chassis. Large SUVs are made with a wheelbase of greater than $88\,\text{in} \approx 224\,\text{cm}$ and a width more than $80\,\text{in} \approx 203\,\text{cm}$.

Because of better performance, the vehicle manufacturing companies are going to make more cars four-wheel-drive. So, four-wheel-drive does not refer to SUV a specific class of cars anymore.

A *truck* is a vehicle with two or four doors and an exposed cargo box. A light truck has a gross weight of less than $10\,000\,\text{lb} \approx 4\,500\,\text{kg}$. A medium truck has a gross weight from $10\,000\,\text{lb} \approx 4\,500\,\text{kg}$ to $26\,000\,\text{lb} \approx 12\,000\,\text{kg}$. A heavy truck is a truck with a gross weight of more than $26\,000\,\text{lb} \approx 12\,000\,\text{kg}$.

Gross weight is the maximum weight of a vehicle as specified by the manufacturer. Gross weight includes chassis, body, engine, fluids, fuel, accessories, driver, passengers and cargo.

1.7.3 Passenger Car Body Styles

Passenger cars are manufactured in so many different styles and shapes. Not all of those classes are made today, and some have new shapes and still carry the same old names. Some of them are as follows:

Convertible or *cabriolet* cars are automobiles with removable or retractable roofs. There are also the subdivisions *cabrio coach* or *semi-convertible* with partially retractable roofs.

Coupé or *coupe* are two-door automobiles with two or four seats and a fixed roof. In cases where the rear seats are smaller than regular size, it is called a two-plus-two or $2 + 2$. Coupé cars may also be convertible.

Crossover SUV or *XUV* cars are smaller sport utility vehicles based on a car platform rather than truck chassis. Crossover cars are a mix of SUV, minivan, and wagon to encompass some of the advantages of each.

Estate car or just *estate* is the British/English term for what North Americans call a station wagon.

Hardtop cars are those having a removable solid roof on a convertible car. However, today a fixed-roof car whose doors have no fixed window frame are also called a hardtops.

Hatchback cars are identified by a rear door, including the back window that opens to access a storage area that is not separated from the rest of the passenger compartment. A hatchback car may have two or four doors and two or four seats. They are also called three-door, or five-door cars. A hatchback car is called a *liftback* when the opening area is very sloped and is lifted up to open.

A *limousine* is a chauffeur-driven car with a glass-window dividing the front seats from the rear. Limousines are usually an extended version of a luxury car.

P 215 / 60 R 15 96 H

FIGURE 1.22. A sample of tire size.

Minivans are boxy wagon cars usually containing three rows of seats, with a capacity of six or more passengers and extra luggage space.

An *MPV* (*Multi-Purpose Vehicle*) is designed as large cars or small buses having off-road capability and easy loading of goods. However, the idea for a car with a multi-purpose application can be seen in other classes, especially SUVs.

Notchback cars are something between the hatchback and sedan. Notchback is a sedan with a separate trunk compartment.

A *pickup truck* (or simply *pickup*) is a small or medium-sized truck with a separate cabin and rear cargo area. Pickups are made to act as a personal truck, however they might also be used as light commercial vehicles.

Sedan is the most common body style that are cars with four or more seats and a fixed roof that is full-height up to the rear window. Sedans can have two or four doors.

Station wagon or *wagon* is a car with a full-height body all the way to the rear; the load-carrying space created is accessed via a rear door or doors.

1.8   Summary

A sample of tire size and performance code is shown in Figure 1.22 and their definitions are explained as follows:
$\boxed{P}$ stands for passenger car. $\boxed{215}$ is the unloaded tire width, in [mm]. $\boxed{60}$ is the aspect ratio of the tire, $s_T = \frac{h_T}{w_T} \times 100$, which is the section height to tire width, expressed as a percentage. $\boxed{R}$ stands for radial. $\boxed{15}$ is the rim diameter that the tire is designed to fit in [in]. $\boxed{96}$ is the load index, and $\boxed{H}$ is the speed rate index.

Road vehicles are usually classified based on their size and number of axles. There is no universally accepted standard classification, however, ISO and FHWA present two important classifications in North America.

ISO3833 classifies ground vehicles into seven groups:

        1.Motorcycles
        2.Passenger cars
        3.Busses
        4.Trucks
        5.Agricultural tractors
        6.Passenger cars with trailer
        7.Truck trailer/semitrailer road trains

FHWA classifies all road vehicles into 13 classes:

        1.Motorcycles
        2.Passenger cars with one or two axles trailer
        3.Other two-axle four-wheel single units
        4.Buses
        5.Two-axle six-wheel single units
        6.Three-axle single units
        7.Four-or-more-axle single units
        8.Four-or-less-axle single trailers
        9.Five-axle single trailers
        10.Six-or-more-axle single trailers
        11.Five-or-less-axle multi-trailers
        12.Six-axle-multi-trailers
        13.Seven-or-more-axle multi-trailers

1.9   Key Symbols

A_P	tireprint area
B	bias ply tire
D	tire diameter
D	diagonal
DC	drop center rim
DOT	Department of Transportation
$FHWA$	Federal Highway Administration
h_T	section height
H	speed rate
$WDCH$	humped wide drop center rim
LT	light truck
$M\&S$	mud and snow
p	tire inflation pressure
P	passenger car
R	radial tire
s_T	aspect ratio
ST	special trailer
SUV	sports utility vehicle
T	temporary tire
v_H	hydroplaning speed
v, v_x	forward velocity of vehicle
w_T	tire width
WDC	wide drop center rim

Exercises

1. Tire size codes.

   Explain the meaning of the following tire size codes:

   a. $10.00R20\ 14(G)$

   b. $18.4R46$

   c. $480/80R46155A8$

   d. $18.4 - 38(10)$

   e. $76 \times 50.00B32 = 1250/45B32$

   f. $LT255/85B16$

   g. $33x12.50R15LT$

2. Tire height and diameter.

   Find the tire height h_T and diameter D for the following tires.

   a. $480/80R46\ 155A8$

   b. $P215/65R15\ 96H$

3. ★ Plus one.

   Increase 1 in to the diameter of the rim of the following tires and find a proper tire for the new rim.

   a. $P215/65R15\ 96H$

   b. $P215/60R15\ 96H$

4. ★ Problem of tire beads.

   Explain what would be the possible problem for a tire that has tight or loose beads.

5. Tire of Porsche 911 turbo^{TM}.

   A model of Porsche 911 turbo^{TM} uses the following tires.

   $$front:\quad 235/35ZR19 \qquad rear:\quad 305/30ZR19$$

   Determine and compare h_T, and D for the front and rear tires.

6. Tire of Porsche Cayenne turbo^{TM}.

   A model of Porsche Cayenne turbo^{TM} is an all-wheel-drive that uses the following tire.

   $$255/55R18$$

   What is the angular velocity of its tires when it is moving at the top speed $v = 171\,\text{mi}/\,\text{h} \approx 275\,\text{km}/\,\text{h}$?

7. Tire of Ferrari P 4/5 by Pininfarina^{TM}.

   A model of Ferrari P 4/5 by Pininfarina^{TM} is a rear-wheel-drive sport car that uses the following tires.

   $$front:\quad 255/35ZR20 \qquad rear:\quad 335/30ZR20$$

   What is the angular velocity of its tires when it is moving at the top speed $v = 225\,\text{mi/h} \approx 362\,\text{km/h}$?

8. Tire of Mercedes-Benz SLR 722 Edition^{TM}.

   A model of Mercedes-Benz SLR 722 Edition^{TM} uses the following tires.

   $$front\quad 255/35R19 \qquad rear\quad 295/30R19$$

   What is the speed of this car if its rear tires are turning at

   $$\omega = 2000\ rpm$$

   At that speed, what would be the angular velocity of the front tires?

9. Tire of Chevrolet Corvette Z06^{TM}.

   A model of Chevrolet Corvette Z06^{TM} uses the following tires.

   $$front\quad 275/35ZR18 \qquad rear\quad 325/30ZR19$$

   What is the speed of this car if its rear tires are turning at

   $$\omega = 2000\ rpm$$

   At that speed, what would be the angular velocity of the front tires?

10. Tire of Koenigsegg CCX^{TM}.

    Koenigsegg CCX^{TM} is a sport car, equipped with the following tires.

    $$front\quad 255/35R19 \qquad rear\quad 335/30R20$$

    What is the angular speed ratio of the rear tire to the front tire?

11. ★ Vehicle speed measurement.

    Let us assume that the speed indicator of a vehicle is showing the speed proportional to the angular speed of its wheel. The speed meter is correct when the tire is new and it is as indicated below:

    $$front\quad 255/35R19$$

    (a) What is the angular speed of the tire when the car is moving at $100\,\text{km/h}$?

    (b) What would be the angular speed of the tire if it loses its tread such that the radius reduces by 5%?

    (c) Let us assume to change the tires to $335/30R20$ while using the same speed meter with the same calibration. What the speed indicator shows when the vehicle is moving at $100\,\text{km/h}$?

Part I

Vehicle Motion

2

Forward Vehicle Dynamics

Straight motion of an ideal rigid vehicle is the subject of this chapter.
We ignore air friction and examine the load variation under the tires to
determine the vehicle's limits of acceleration, road grade, and kinematic
capabilities.

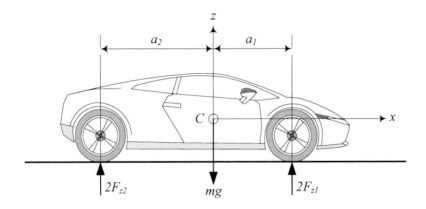

FIGURE 2.1. A parked car on level pavement.

2.1 Parked Car on a Level Road

When a car is parked on level pavement, the normal force, F_z, under each
of the front and rear wheels, F_{z_1}, F_{z_2}, are

$$F_{z_1} = \frac{1}{2}mg\frac{a_2}{l} \tag{2.1}$$

$$F_{z_2} = \frac{1}{2}mg\frac{a_1}{l} \tag{2.2}$$

where, a_1 is the distance of the car's mass center, C, from the front axle,
a_2 is the distance of C from the rear axle, and l is the wheel base.

$$l = a_1 + a_2 \tag{2.3}$$

Proof. Consider a longitudinally symmetrical car as shown in Figure 2.1.
It can be modeled as a two-axel vehicle which is equivalent to a rigid beam

R.N. Jazar, *Vehicle Dynamics: Theory and Application*,
DOI 10.1007/978-1-4614-8544-5_2, © Springer Science+Business Media New York 2014

having two supports. The vertical force under the front and rear wheels can be determined using planar static equilibrium equations.

$$\sum F_z = 0 \tag{2.4}$$

$$\sum M_y = 0 \tag{2.5}$$

Applying the equilibrium equations

$$2F_{z_1} + 2F_{z_2} - mg = 0 \tag{2.6}$$
$$-2F_{z_1}a_1 + 2F_{z_2}a_2 = 0 \tag{2.7}$$

provides the reaction forces under the front and rear tires.

$$F_{z_1} = \frac{1}{2}mg\frac{a_2}{a_1 + a_2} = \frac{1}{2}mg\frac{a_2}{l} \tag{2.8}$$

$$F_{z_2} = \frac{1}{2}mg\frac{a_1}{a_1 + a_2} = \frac{1}{2}mg\frac{a_1}{l} \tag{2.9}$$

∎

Example 41 *Reaction forces under wheels.*

A model of Lamborghini Diablo^{TM} has 1576 kg mass. Its mass center, C, is 108.65 cm behind the front wheel axis, and it has a 265 cm wheel base.

$$a_1 = 1.086 \text{ m} \qquad l = 2.65 \text{ m} \qquad m = 1576 \text{ kg} \tag{2.10}$$

The force under each front wheel is

$$F_{z_1} = \frac{1}{2}mg\frac{a_2}{l} = \frac{1}{2} \times 1576 \times 9.81 \times \frac{2.65 - 1.086}{2.65} = 4560.9 \text{ N} \tag{2.11}$$

and the force under each rear wheel is

$$F_{z_2} = \frac{1}{2}mg\frac{a_1}{l} = \frac{1}{2} \times 1576 \times 9.81 \times \frac{1.086}{2.65} = 3169.4 \text{ N} \tag{2.12}$$

The front/rear load distribution is usually a reported number to be used to determine the position of the mass center. The front/rear load distribution of the Lamborghini Diablo^{TM} is 41/59%.

Example 42 *Mass center position.*

Equations (2.1) and (2.2) can be rearranged to calculate the position of mass center.

$$a_1 = \frac{2l}{mg}F_{z_2} \qquad a_2 = \frac{2l}{mg}F_{z_1} \tag{2.13}$$

Reaction forces under the front and rear wheels of a horizontally parked car, with a wheel base $l = 2.65$ m, are:

$$F_{z_1} = 4560.9 \text{ N} \qquad F_{z_2} = 3169.4 \text{ N} \tag{2.14}$$

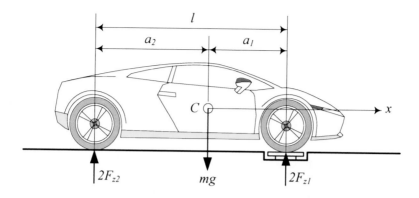

FIGURE 2.2. Measuring the force under the front wheels.

Therefore, the longitudinal position of the car's mass center is at

$$a_1 = \frac{2l}{mg} F_{z_2} = 2\frac{2.65}{2(4560.9 + 3169.4)} \times 3169.4 = 1.086\,\text{m} \quad (2.15)$$

$$a_2 = \frac{2l}{mg} F_{z_1} = 2\frac{2.65}{2(4560.9 + 3169.4)} \times 4560.9 = 1.563\,\text{m} \quad (2.16)$$

Example 43 *Longitudinal mass center determination.*

The position of mass center C can be determined experimentally. To determine the longitudinal position of C, we should measure the total weight of the car as well as the force under the front or the rear wheels. Figure 2.2 illustrates a situation in which we measure the force under the front wheels.

Assuming the force under the front wheels is $2F_{z_1}$, the position of the mass center is calculated by static equilibrium conditions

$$\sum F_z = 0 \qquad \sum M_y = 0 \quad (2.17)$$

Applying the equilibrium equations

$$2F_{z_1} + 2F_{z_2} - mg = 0 \qquad -2F_{z_1}a_1 + 2F_{z_2}a_2 = 0 \quad (2.18)$$

provides the longitudinal position of C and the reaction forces under the rear wheels.

$$a_1 = \frac{2l}{mg} F_{z_2} = \frac{l}{mg}(mg - 2F_{z_1}) = l\left(1 - \frac{2F_{z_1}}{mg}\right) \quad (2.19)$$

$$F_{z_2} = \frac{1}{2}(mg - 2F_{z_1}) \quad (2.20)$$

Example 44 *Lateral mass center determination.*

Most cars are approximately symmetrical about the longitudinal center plane and therefore, the lateral position of the mass center C is close to

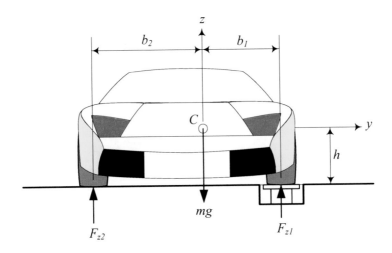

FIGURE 2.3. A laterally asymmetric car has an off center mass center.

the center plane. However, the lateral position of C may be calculated by weighing one side of the car. The problem is that although the left and right sides of cars are almost symmetric, the front and rear of cars are not symmetric and there is an unequal weight distribution among the front and rear. Therefore, when we measure the weight of the left and right sides there would be a significant difference between the front and rear, and they must be analyzed separately.

As an example let us consider the Lamborghini Diablo^{TM} with total mass of 1576 kg, and front/rear load distribution of 41/59%. Let us assume there is a 4% difference between the left and right sides in the front of the car. The load distribution indicates that the front of the car weighs

$$2F_{z_f} = 0.59 \times 1576g = 9121.7\,\text{N} \tag{2.21}$$

As depicted in Figure 2.3, let us name the vertical force under the tire at left and right sides by F_{z_1} and F_{z_2}, respectively. If the left side of the car is 4% heavier than the right, then

$$
\begin{aligned}
F_{z_1} &= 0.52 \times 2F_{z_f} = 0.52 \times 9121.7 = 4743.3\,\text{N} \tag{2.22}\\
F_{z_2} &= 0.48 \times 2F_{z_f} = 0.48 \times 9121.7 = 4378.4\,\text{N} \tag{2.23}
\end{aligned}
$$

The car's front and rear tracks are reported as

$$w_f = 1540\,\text{mm} \approx 60.52\,\text{in} \qquad w_r = 1640\,\text{mm} \approx 64.45\,\text{in} \tag{2.24}$$

therefore

$$
\begin{aligned}
b_1 &= 0.48 \times w_f = 0.48 \times 1540 = 739.2\,\text{mm} \tag{2.25}\\
b_2 &= 0.52 \times w_f = 0.52 \times 1540 = 800.8\,\text{mm} \tag{2.26}
\end{aligned}
$$

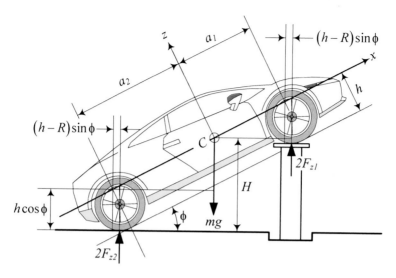

FIGURE 2.4. Measuring the force under the wheels to find the height of the mass center.

To determine the exact position of the mass center, we should measure the vertical force under all tires simultaneously.

Example 45 *Height mass center determination.*

To determine the height of mass center C, we should measure the force under the front or rear wheels while the car is on an inclined surface. Experimentally, we use a device such as is shown in Figure 2.4. The car is parked on a level surface such that the front wheels are on a scale jack. The front wheels will be locked and anchored to the jack, while the rear wheels will be left free to turn. The jack lifts the front wheels and the required vertical force applied by the jacks is measured by a load cell.

Assume that we have the longitudinal position of C and the jack is lifted such that the car makes an angle ϕ with the horizontal plane. The slope angle ϕ is measurable using level meters. Assuming the force under the front wheels is measured as $2F_{z_1}$, the height of the mass center can be calculated by static equilibrium conditions

$$\sum F_Z = 0 \qquad \sum M_y = 0 \tag{2.27}$$

Applying the equilibrium equations

$$2F_{z_1} + 2F_{z_2} - mg = 0 \tag{2.28}$$

$$\begin{aligned} -2F_{z_1}\left(a_1 \cos \phi + (h-R) \sin \phi\right) \\ +2F_{z_2}\left(a_2 \cos \phi - (h-R) \sin \phi\right) &= 0 \end{aligned} \tag{2.29}$$

provides the vertical position of C and the reaction forces under the rear *wheels.*

$$F_{z_2} = \frac{1}{2}mg - F_{z_1} \tag{2.30}$$

$$h = \frac{F_{z_1}\left(R\sin\phi - a_1\cos\phi\right) + F_{z_2}\left(R\sin\phi + a_2\cos\phi\right)}{0.5mg\sin\phi}$$

$$= R - \frac{a_1 F_{z_1} - a_2 F_{z_2}}{0.5mg}\cot\phi = R + \left(a_2 - \frac{2F_{z_1}}{mg}l\right)\cot\phi \tag{2.31}$$

There are three main assumptions in this calculation:
1. the tires are assumed to be rigid disks with radius R,
2. fluid shift, such as fuel, coolant, and oil, are ignored,
3. suspension deflections are assumed to be zero.

Suspension deflection generates the maximum effect on height determination error. To eliminate the suspension deflection, we should lock the suspension, usually by replacing the shock absorbers with rigid rods to keep the vehicle at its ride height.

Let us consider a car with the following measured specifications

$$
\begin{array}{lll}
m & = & 1576\,\text{kg} \quad 2F_{z_1} = 6300\,\text{N} \\
\phi & = & 30\,\text{deg} \approx 0.5236\,\text{rad} \\
a_1 & = & 108.6\,\text{cm} \quad l = 265\,\text{cm} \quad Tire:245/40ZR17
\end{array}
\tag{2.32}
$$

If the tire has a radius of

$$R \approx 30\,\text{cm} \tag{2.33}$$

then the car has a C at height h.

$$h = 65.2\,\text{cm} \tag{2.34}$$

Example 46 *Different front and rear tires.*

Depending on the application, it is sometimes necessary to use different type of tires and wheels for front and rear axles. When the longitudinal position of C for a symmetric vehicle is determined, we can find the height of C by measuring the load on only one axle. As an example, consider the motorcycle in Figure 2.5. It has different front and rear tires.

Assume the load under the rear wheel of the motorcycle F_z is known. The height h of C can be found by taking a moment of the forces about the tireprint of the front tire.

$$h = R_r + \left(a_2 - \frac{F_{z_1}}{mg}l\right)\cot\phi - \frac{F_{z_1}}{mg}\left(R_r - R_f\right) \tag{2.35}$$

In case of a four wheel vehicle, the equation will be

$$h = R_r + \left(a_2 - \frac{2F_{z_1}}{mg}l\right)\cot\phi - \frac{2F_{z_1}}{mg}\left(R_r - R_f\right) \tag{2.36}$$

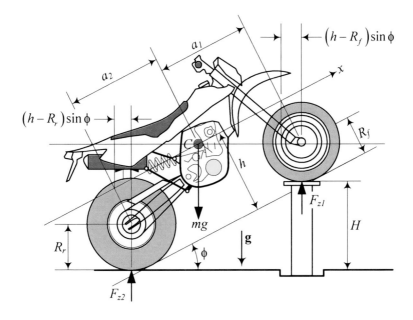

FIGURE 2.5. A motorcycle with different front and rear tires.

Example 47 *Statically indeterminate.*

A vehicle with more than three wheels is statically indeterminate. To determine the vertical force under each tire, we need to know the mechanical properties and conditions of the tires, such as the value of deflection at the center of the tire, and its vertical stiffness.

2.2   Parked Car on an Inclined Road

Passenger cars are usually equipped with rear wheels hand brakes or park brakes. When a car is parked on an inclined pavement as shown in Figure 2.6, the normal force, F_z, under each of the front and rear wheels, F_{z_1}, F_{z_2}, is

$$F_{z_1} = \frac{1}{2}mg\frac{a_2}{l}\cos\phi - \frac{1}{2}mg\frac{h}{l}\sin\phi \qquad (2.37)$$

$$F_{z_2} = \frac{1}{2}mg\frac{a_1}{l}\cos\phi + \frac{1}{2}mg\frac{h}{l}\sin\phi \qquad (2.38)$$

$$l = a_1 + a_2$$

and the brake force, F_{x_2}, is

$$F_{x_2} = \frac{1}{2}mg\sin\phi \qquad (2.39)$$

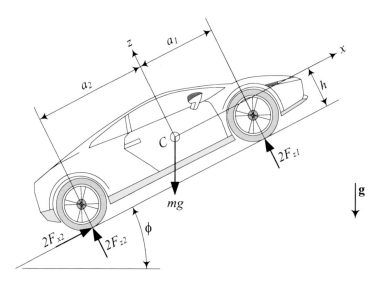

FIGURE 2.6. A parked car on inclined pavement.

where, ϕ is the angle of the road with the horizon. The horizon is perpendicular to the gravitational acceleration $\mathbf{g}$.

Proof. Consider the car shown in Figure 2.6. Let us assume the parking brake forces are applied on only the rear tires. It means the front tires are free to spin. Applying the planar static equilibrium equations

$$\sum F_x = 0 \tag{2.40}$$

$$\sum F_z = 0 \tag{2.41}$$

$$\sum M_y = 0 \tag{2.42}$$

shows that

$$2F_{x_2} - mg\sin\phi = 0 \tag{2.43}$$

$$2F_{z_1} + 2F_{z_2} - mg\cos\phi = 0 \tag{2.44}$$

$$-2F_{z_1}a_1 + 2F_{z_2}a_2 - 2F_{x_2}h = 0 \tag{2.45}$$

These equations provide the brake force and reaction forces under the front and rear tires.

$$F_{z_1} = \frac{1}{2}mg\frac{a_2}{l}\cos\phi - \frac{1}{2}mg\frac{h}{l}\sin\phi \tag{2.46}$$

$$F_{z_2} = \frac{1}{2}mg\frac{a_1}{l}\cos\phi + \frac{1}{2}mg\frac{h}{l}\sin\phi \tag{2.47}$$

$$F_{x_2} = \frac{1}{2}mg\sin\phi \tag{2.48}$$

■

Example 48 *Tilting angle.*

*When $\phi = 0$, Equations (2.37) and (2.38) reduce to (2.1) and (2.2). By increasing the inclination angle, the normal force under the front tires of a parked car decreases and the normal force and braking force under the rear tires increase. The limit for increasing ϕ is where the weight vector mg goes through the contact point of the rear tire with the ground. Such an angle is called a **tilting angle**, and shown by ϕ_T.*

At $\phi = \phi_T$, we have $F_{z_1} = 0$, and therefore,

$$\tan \phi_T = \frac{a_2}{h} \tag{2.49}$$

Example 49 *Ultimate angle.*

*The required braking force F_{x_2} increases by the inclination angle. Because F_{x_2} is equal to the friction force between the tire and pavement, its maximum depends on the tire and pavement conditions. There is an **ultimate angle** ϕ_M at which the braking force F_{x_2} will saturate and cannot increase any more. At this maximum angle, the braking force is proportional to the normal force F_{z_2}*

$$F_{x_2} = \mu_{x_2} F_{z_2} \tag{2.50}$$

where, the coefficient μ_{x_2} is the x-direction friction coefficient for the rear wheels. At $\phi = \phi_M$, the equilibrium equations (2.43)-(2.45) will reduce to

$$2\mu_{x_2} F_{z_2} - mg \sin \phi = 0 \tag{2.51}$$
$$2F_{z_1} + 2F_{z_2} - mg \cos \phi = 0 \tag{2.52}$$
$$-2F_{z_1} a_1 + 2F_{z_2} a_2 - 2\mu_{x_2} F_{z_2} h = 0 \tag{2.53}$$

These equations provide

$$F_{z_1} = \frac{1}{2} mg \frac{a_2}{l} \cos \phi_M - \frac{1}{2} mg \frac{h}{l} \sin \phi_M \tag{2.54}$$
$$F_{z_2} = \frac{1}{2} mg \frac{a_1}{l} \cos \phi_M + \frac{1}{2} mg \frac{h}{l} \sin \phi_M \tag{2.55}$$
$$\tan \phi_M = \frac{a_1 \mu_{x_2}}{l - \mu_{x_2} h} \tag{2.56}$$

showing that there is a relation between the friction coefficient μ_{x_2}, maximum inclination or ultimate angle ϕ_M, and the geometrical position of the mass center C. The angle ϕ_M increases by decreasing h.

For a car having the specifications

$$\mu_{x_2} = 1 \qquad a_1 = 110\,\text{cm} \qquad l = 230\,\text{cm} \qquad h = 35\,\text{cm} \tag{2.57}$$

the ultimate angle is

$$\phi_M \approx 0.514\,\text{rad} \approx 29.43\,\text{deg} \tag{2.58}$$

Example 50 *Front wheel braking.*
   When the front wheels are the park braking wheels $F_{x_2} = 0$ and $F_{x_1} \neq 0$. In this case, the equilibrium equations (2.40)-(2.42) will be

$$2F_{x_1} - mg\sin\phi = 0 \tag{2.59}$$

$$2F_{z_1} + 2F_{z_2} - mg\cos\phi = 0 \tag{2.60}$$

$$-2F_{z_1}a_1 + 2F_{z_2}a_2 - 2F_{x_1}h = 0 \tag{2.61}$$

These equations provide the brake force F_{x_1} and reaction forces under the front and rear tires.

$$F_{z_1} = \frac{1}{2}mg\frac{a_2}{l}\cos\phi - \frac{1}{2}mg\frac{h}{l}\sin\phi \tag{2.62}$$

$$F_{z_2} = \frac{1}{2}mg\frac{a_1}{l}\cos\phi + \frac{1}{2}mg\frac{h}{l}\sin\phi \tag{2.63}$$

$$F_{x_1} = \frac{1}{2}mg\sin\phi \tag{2.64}$$

At the ultimate angle $\phi = \phi_M$, we have

$$F_{x_1} = \mu_{x_1}F_{z_1} \tag{2.65}$$

and therefore,

$$2\mu_{x_1}F_{z_1} - mg\sin\phi_M = 0 \tag{2.66}$$

$$2F_{z_1} + 2F_{z_2} - mg\cos\phi_M = 0 \tag{2.67}$$

$$2F_{z_1}a_1 - 2F_{z_2}a_2 + 2\mu_{x_1}F_{z_1}h = 0 \tag{2.68}$$

These equations provide

$$F_{z_1} = \frac{1}{2}mg\frac{a_2}{l}\cos\phi_M - \frac{1}{2}mg\frac{h}{l}\sin\phi_M \tag{2.69}$$

$$F_{z_2} = \frac{1}{2}mg\frac{a_1}{l}\cos\phi_M + \frac{1}{2}mg\frac{h}{l}\sin\phi_M \tag{2.70}$$

$$\tan\phi_M = \frac{a_2\mu_{x_1}}{l - \mu_{x_1}h} \tag{2.71}$$

Let us name the ultimate angle for the front wheel brake in Equation (2.71) as ϕ_{M_f}, and the ultimate angle for the rear wheel brake in Equation (2.56) as ϕ_{M_r}. Comparing ϕ_{M_f} and ϕ_{M_r} shows that

$$\frac{\phi_{M_f}}{\phi_{M_r}} = \frac{a_2\mu_{x_1}\left(l - \mu_{x_2}h\right)}{a_1\mu_{x_2}\left(l - \mu_{x_1}h\right)} \tag{2.72}$$

We may assume the front and rear tires are the same,

$$\mu_{x_1} = \mu_{x_2} \tag{2.73}$$

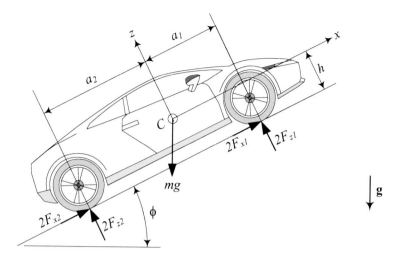

FIGURE 2.7. A four wheel brake car, parked uphill.

and therefore,

$$\frac{\phi_{M_f}}{\phi_{M_r}} = \frac{a_2}{a_1} \tag{2.74}$$

Hence, if $a_2 < a_1$ then $\phi_{M_f} < \phi_{M_r}$ and therefore, a rear brake is more effective than a front brake on uphill parking as long as ϕ_{M_r} is less than the tilting angle, $\phi_{T_r} < \arctan \frac{a_1}{h}$. At the tilting angle, the weight vector passes through the contact point of the rear wheel with the ground.

Similarly we may conclude that when parked on a downhill road, the front brake is more effective than the rear brake.

Example 51 *Four-wheel braking.*

Consider a four-wheel brake car, parked uphill as shown in Figure 2.7. In these conditions, there will be two brake forces F_{x_1} on the front wheels and two brake forces F_{x_2} on the rear wheels.

The equilibrium equations for this car are

$$2F_{x_1} + 2F_{x_2} - mg \sin \phi = 0 \tag{2.75}$$

$$2F_{z_1} + 2F_{z_2} - mg \cos \phi = 0 \tag{2.76}$$

$$-2F_{z_1}a_1 + 2F_{z_2}a_2 - (2F_{x_1} + 2F_{x_2})h = 0 \tag{2.77}$$

These equations provide the brake force and reaction forces under the front and rear tires.

$$F_{z_1} = \frac{1}{2}mg\frac{a_2}{l}\cos\phi - \frac{1}{2}mg\frac{h}{l}\sin\phi \tag{2.78}$$

$$F_{z_2} = \frac{1}{2}mg\frac{a_1}{l}\cos\phi + \frac{1}{2}mg\frac{h}{l}\sin\phi \tag{2.79}$$

$$F_{x_1} + F_{x_2} = \frac{1}{2}mg\sin\phi \tag{2.80}$$

At the ultimate angle $\phi = \phi_M$, *all wheels will begin to slide simultaneously and therefore,*

$$F_{x_1} = \mu_{x_1}F_{z_1} \qquad F_{x_2} = \mu_{x_2}F_{z_2} \tag{2.81}$$

The equilibrium equations show that

$$2\mu_{x_1}F_{z_1} + 2\mu_{x_2}F_{z_2} - mg\sin\phi_M = 0 \tag{2.82}$$

$$2F_{z_1} + 2F_{z_2} - mg\cos\phi_M = 0 \tag{2.83}$$

$$-2F_{z_1}a_1 + 2F_{z_2}a_2 - \left(2\mu_{x_1}F_{z_1} + 2\mu_{x_2}F_{z_2}\right)h = 0 \tag{2.84}$$

Assuming

$$\mu_{x_1} = \mu_{x_2} = \mu_x \tag{2.85}$$

will provide

$$F_{z_1} = \frac{1}{2}mg\frac{a_2}{l}\cos\phi_M - \frac{1}{2}mg\frac{h}{l}\sin\phi_M \tag{2.86}$$

$$F_{z_2} = \frac{1}{2}mg\frac{a_1}{l}\cos\phi_M + \frac{1}{2}mg\frac{h}{l}\sin\phi_M \tag{2.87}$$

$$\tan\phi_M = \mu_x \tag{2.88}$$

2.3   Accelerating Car on a Level Road

When a car is speeding with acceleration a on a level road as shown in Figure 2.8, the vertical forces under the front and rear wheels are

$$F_{z_1} = \frac{1}{2}mg\frac{a_2}{l} - \frac{1}{2}ma\frac{h}{l} \tag{2.89}$$

$$F_{z_2} = \frac{1}{2}mg\frac{a_1}{l} + \frac{1}{2}ma\frac{h}{l} \tag{2.90}$$

The first terms, $\frac{1}{2}mg\frac{a_2}{l}$ and $\frac{1}{2}mg\frac{a_1}{l}$, are called *static parts*, and the second terms, $\pm\frac{1}{2}ma\frac{h}{l}$, are called *dynamic parts* of the normal forces.

Proof. The vehicle is considered as a rigid body that moves along a horizontal road. The force at the tireprint of each tire may be decomposed

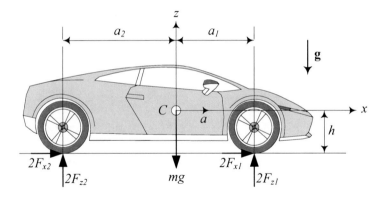

FIGURE 2.8. An accelerating car on a level pavement.

to a normal and a longitudinal force. The equations of motion for the accelerating car come from Newton's equation in x-direction and two static equilibrium equations in y and z-direction.

$$\sum F_x = ma \tag{2.91}$$

$$\sum F_z = 0 \tag{2.92}$$

$$\sum M_y = 0 \tag{2.93}$$

Expanding the equations produces three equations for four unknowns F_{x_1}, F_{x_2}, F_{z_1}, F_{z_2}.

$$2F_{x_1} + 2F_{x_2} = ma \tag{2.94}$$

$$2F_{z_1} + 2F_{z_2} - mg = 0 \tag{2.95}$$

$$-2F_{z_1}a_1 + 2F_{z_2}a_2 - 2\left(F_{x_1} + F_{x_2}\right)h = 0 \tag{2.96}$$

However, it is possible to eliminate $(F_{x_1} + F_{x_2})$ between the first and third equations, and solve for the normal forces F_{z_1}, F_{z_2}.

$$F_{z_1} = \left(F_{z_1}\right)_{st} + \left(F_{z_1}\right)_{dyn} = \frac{1}{2}mg\frac{a_2}{l} - \frac{1}{2}ma\frac{h}{l} \tag{2.97}$$

$$F_{z_2} = \left(F_{z_2}\right)_{st} + \left(F_{z_2}\right)_{dyn} = \frac{1}{2}mg\frac{a_1}{l} + \frac{1}{2}ma\frac{h}{l} \tag{2.98}$$

The static parts

$$\left(F_{z_1}\right)_{st} = \frac{1}{2}mg\frac{a_2}{l} \tag{2.99}$$

$$\left(F_{z_2}\right)_{st} = \frac{1}{2}mg\frac{a_1}{l} \tag{2.100}$$

are weight distribution for a stationary car and depend on the horizontal position of the mass center. However, the dynamic parts

$$(F_{z_1})_{dyn} = -\frac{1}{2}ma\frac{h}{l} \tag{2.101}$$

$$(F_{z_2})_{dyn} = \frac{1}{2}ma\frac{h}{l} \tag{2.102}$$

indicate the weight distribution because of horizontal acceleration, and depend on the vertical position of the mass center.

When accelerating $a > 0$, the normal forces under the front tires are less than the static load, and under the rear tires are more than the static load.
■

Example 52 *Front-wheel-drive accelerating on a level road.*
When the car is front-wheel-drive, $F_{x_2} = 0$. Equations (2.94) to (2.96) will provide the same vertical tireprint forces as (2.89) and (2.90). However, the required horizontal force, $2F_{x_1}$, to achieve the same acceleration, a, must be provided by solely the front wheels.

$$2F_{x_1} = ma \tag{2.103}$$

Example 53 *Rear-wheel drive accelerating on a level road.*
If a car is rear-wheel drive then, $F_{x_1} = 0$ and the required force to achieve the acceleration, a, must be provided only by the rear wheels.

$$2F_{x_2} = ma \tag{2.104}$$

The vertical force under the wheels will still be the same as (2.89) and (2.90).

Example 54 *Maximum acceleration on a level road.*
The maximum acceleration of a car is proportional to the friction under its tires. We assume the friction coefficients at the front and rear tires are equal and all tires reach their maximum tractions at the same time.

$$F_{x_1} = \pm\mu_x F_{z_1} \qquad F_{x_2} = \pm\mu_x F_{z_2} \tag{2.105}$$

Newton's equation (2.94) can now be written as

$$ma = \pm 2\mu_x (F_{z_1} + F_{z_2}) \tag{2.106}$$

Substituting F_{z_1} and F_{z_2} from (2.97) and (2.98) results in

$$a = \pm\mu_x g \tag{2.107}$$

Therefore, the maximum acceleration and deceleration depend directly on the friction coefficient.

Example 55 *Maximum acceleration for a single-axle drive car.*

The maximum acceleration a_{rwd} for a rear-wheel-drive car is achieved when F_{x_2} reaches its maximum frictional force. Substituting $F_{x_1} = 0$, and $F_{x_2} = \mu_x F_{z_2}$ in Equation (2.94) and using Equation (2.90)

$$\mu_x mg \left(\frac{a_1}{l} + \frac{h}{l} \frac{a_{rwd}}{g} \right) = m a_{rwd} \qquad (2.108)$$

yields

$$\frac{a_{rwd}}{g} = \frac{a_1 \mu_x}{l - h\mu_x} = \frac{\mu_x}{1 - \mu_x \dfrac{h}{l}} \frac{a_1}{l} \qquad (2.109)$$

If friction is strong enough, the front wheels can leave the ground at $F_{x_2} < \mu_x F_{z_2}$ and we have $F_{z_1} = 0$. Substituting $F_{z_1} = 0$ in Equation (2.89) provides the maximum acceleration at which the front wheels are still on the road.

$$\frac{a_{rwd}}{g} = \frac{a_2}{h} \qquad (2.110)$$

Therefore, the maximum attainable acceleration of a rear-wheel-drive car would be the less value of Equation (2.109) or (2.110).

Similarly, the maximum acceleration a_{fwd} for a front-wheel drive car is achieved when we substitute $F_{x_2} = 0$, $F_{x_1} = \mu_x F_{z_1}$ in Equation (2.94) and use Equation (2.89).

$$\frac{a_{fwd}}{g} = \frac{a_2 \mu_x}{l + h\mu_x} = \frac{\mu_x}{1 + \mu_x \dfrac{h}{l}} \left(1 - \frac{a_1}{l} \right) \qquad (2.111)$$

To visualize the effect of changing the position of mass center on the maximum achievable acceleration, we plot Figure 2.9 for a sample car with

$$\mu_x = 1 \qquad h = 0.56 \, \text{m} \qquad l = 2.6 \, \text{m} \qquad (2.112)$$

Passenger cars are usually in the range $0.4 < (a_1/l) < 0.6$. In this range, $a_{rwd} > a_{fwd}$ and rear-wheel-drive cars can reach higher forward acceleration than front-wheel-drive cars. It is an important applied fact, especially for race cars.

For the race cars, the maximum acceleration may also be limited by the lift off condition (2.110).

Example 56 *Minimum time for $0 - 100 \, \text{km/h}$ on a level road.*

Consider a car with the following characteristics:

$$
\begin{array}{ll}
length = 4245 \, \text{mm} & wheel \ base = 2272 \, \text{mm} \\
width = 1795 \, \text{mm} & front \ track = 1411 \, \text{mm} \\
height = 1285 \, \text{mm} & rear \ track = 1504 \, \text{mm} \\
h = 220 \, \text{mm} & net \ weight = 1500 \, \text{kg} \\
\mu_x = 1 & a_1 = a_2
\end{array}
\qquad (2.113)
$$

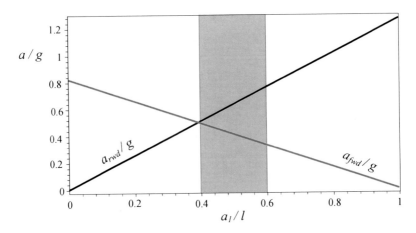

FIGURE 2.9. Effect of mass center position on the maximum achievable acceleration of a front- and a rear-wheel drive car.

Assume the car is rear-wheel-drive and its engine can provide the maximum traction supported by friction. Equation (2.90) determines the load on the rear wheels and therefore, the forward equation of motion (2.94) becomes

$$2F_{x_2} = 2\mu_x F_{z_2} = \mu_x m g \frac{a_1}{l} + \mu_x m a \frac{h}{l}$$

$$= ma \tag{2.114}$$

Rearrangement provides the following differential equation to calculate velocity and displacement:

$$a = \ddot{x} = \frac{\mu_x g \frac{a_1}{l}}{1 - \mu_x \frac{h}{l}} = \frac{a_1 \mu_x}{l - h \mu_x} g \tag{2.115}$$

Taking an integral between $v = 0$ and $v = 100\,\mathrm{km/h} \approx 27.78\,\mathrm{m/s}$

$$\int_0^{27.78} dv = \int_0^t a\, dt \tag{2.116}$$

shows that the minimum time for $0 - 100\,\mathrm{km/h}$ on a level road for the rear-wheel-drive is

$$t = \frac{27.78}{g\mu_x \dfrac{a_1}{l - h\mu_x}} \approx 5.11\,\mathrm{s} \tag{2.117}$$

If the same car was front-wheel-drive, then the traction force would be

$$2F_{x_1} = 2\mu_x F_{z_1} = \mu_x m g \frac{a_2}{l} - \mu_x m a \frac{h}{l}$$

$$= ma \tag{2.118}$$

and the equation of motion would reduce to

$$a = \ddot{x} = \frac{\mu_x g \dfrac{a_2}{l}}{1 + \mu_x g \dfrac{h}{l}\dfrac{1}{g}} = g\mu_x \frac{a_2}{l + h\mu_x} \tag{2.119}$$

The minimum time for $0 - 100 \, \text{km/h}$ *on a level road for the front-wheel-drive car would be*

$$t = \frac{27.78}{g\mu_x \dfrac{a_2}{l + h\mu_x}} \approx 6.21 \, \text{s} \tag{2.120}$$

Now consider the same car to be four-wheel-drive with the assumption that all wheels reach their maximum possible traction at the same time. Then, the traction force is

$$\begin{aligned} 2F_{x_1} + 2F_{x_2} &= 2\mu_x \left(F_{z_1} + F_{z_2} \right) = \frac{g}{l} m \left(a_1 + a_2 \right) \\ &= m\,a \end{aligned} \tag{2.121}$$

and the minimum time for $0 - 100 \, \text{km/h}$ *on a level road for this four-wheel-drive car can theoretically be reduced to*

$$t = \frac{27.78}{g} \approx 2.83 \, \text{s} \tag{2.122}$$

2.4   Accelerating Car on an Inclined Road

When a car is accelerating on an inclined pavement with angle ϕ as shown in Figure 2.10, the normal force under each of the front and rear wheels, F_{z_1}, F_{z_2}, would be:

$$F_{z_1} = \frac{1}{2} mg \left(\frac{a_2}{l} \cos\phi - \frac{h}{l} \sin\phi \right) - \frac{1}{2} ma\frac{h}{l} \tag{2.123}$$

$$F_{z_2} = \frac{1}{2} mg \left(\frac{a_1}{l} \cos\phi + \frac{h}{l} \sin\phi \right) + \frac{1}{2} ma\frac{h}{l} \tag{2.124}$$

$$l = a_1 + a_2$$

The dynamic parts, $\pm\frac{1}{2}ma\frac{h}{l}$, depend on acceleration a and height h of mass center C and not on the slope ϕ, while the static parts are influenced by the slope angle ϕ as well as the longitudinal and vertical positions of the mass center.

Proof. The Newton's equation in x-direction and two static equilibrium equations must be examined to find the equation of motion and ground

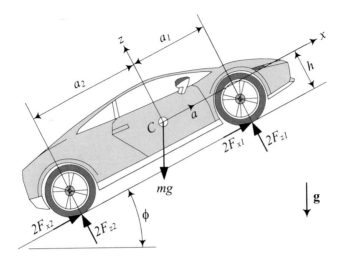

FIGURE 2.10. An accelerating car on inclined pavement.

reaction forces.

$$\sum F_x = ma \tag{2.125}$$

$$\sum F_z = 0 \tag{2.126}$$

$$\sum M_y = 0 \tag{2.127}$$

Expanding these equations produces three equations for four unknowns F_{x_1}, F_{x_2}, F_{z_1}, F_{z_2}.

$$2F_{x_1} + 2F_{x_2} - mg\sin\phi = ma \tag{2.128}$$

$$2F_{z_1} + 2F_{z_2} - mg\cos\phi = 0 \tag{2.129}$$

$$-2F_{z_1}a_1 + 2F_{z_2}a_2 - 2\left(F_{x_1} + F_{x_2}\right)h = 0 \tag{2.130}$$

It is possible to eliminate $(F_{x_1} + F_{x_2})$ between the first and third equations, and solve for the normal forces F_{z_1}, F_{z_2}.

$$
\begin{aligned}
F_{z_1} &= \left(F_{z_1}\right)_{st} + \left(F_{z_1}\right)_{dyn} \\
&= \frac{1}{2}mg\left(\frac{a_2}{l}\cos\phi - \frac{h}{l}\sin\phi\right) - \frac{1}{2}ma\frac{h}{l} \tag{2.131}
\end{aligned}
$$

$$
\begin{aligned}
F_{z_2} &= \left(F_{z_2}\right)_{st} + \left(F_{z_2}\right)_{dyn} \\
&= \frac{1}{2}mg\left(\frac{a_1}{l}\cos\phi + \frac{h}{l}\sin\phi\right) + \frac{1}{2}ma\frac{h}{l} \tag{2.132}
\end{aligned}
$$

■

Example 57 *Front-wheel-drive car, accelerating on inclined road.*

For a front-wheel-drive car, we should substitute $F_{x_1} = 0$ in Equations (2.128) and (2.130) to have the governing equations. However, it does not affect the expression of the ground reaction forces under the tires (2.131) and (2.132) as long as the car is driven under its frictional limit conditions.

Example 58 *Rear-wheel-drive car, accelerating on inclined road.*

Substituting $F_{x_2} = 0$ in Equations (2.128) and (2.130) and solving for the normal reaction forces under each tire provides the same results as (2.131) and (2.132). Hence, the expression of the normal forces applied on the tires do not make any difference if the car is front-, rear-, or all-wheel drive. As long as we drive in a straight inclined path at low acceleration, the drive wheels can be the front or the rear ones. However, the advantages and disadvantages of front-, rear-, or all-wheel drive cars appear in maneuvering, slippery roads, or when the maximum acceleration is required.

Example 59 *Maximum acceleration on an inclined road.*

The maximum acceleration depends on the friction under the tires. Let us assume the friction coefficients at the front and rear tires are equal. Then, the front and rear traction forces are

$$F_{x_1} \leq \mu_x F_{z_1} \qquad F_{x_2} \leq \mu_x F_{z_2} \tag{2.133}$$

If we assume the front and rear wheels reach their traction limits at the same time, then

$$F_{x_1} = \pm \mu_x F_{z_1} \qquad F_{x_2} = \pm \mu_x F_{z_2} \tag{2.134}$$

and we may rewrite Newton's equation (2.125) as

$$m a_M = \pm 2 \mu_x \left(F_{z_1} + F_{z_2} \right) - mg \sin \phi \tag{2.135}$$

where, a_M is the maximum achievable acceleration.

Now substituting F_{z_1} and F_{z_2} from (2.131) and (2.132) results in

$$\frac{a_M}{g} = \pm \mu_x \cos \phi - \sin \phi \tag{2.136}$$

Accelerating on an uphill road and braking on a downhill road are the extreme cases in which the car can stall, $a = 0$. In these cases, the car can move as long as

$$\mu_x \geq |\tan \phi| \tag{2.137}$$

Following the directions of the body coordinate frame of the vehicle, uphill road should be assigned by $\phi < 0$ and downhill by $\phi > 0$, as the slope angle ϕ should be measured about the y-axis. However, in this section we have used the absolute value of the slope angle for calculations.

Example 60 ★ *Limits of acceleration and inclination angle.*
  Considering $F_{z_1} > 0$ and $F_{z_2} > 0$, we can write Equations (2.123) and (2.124) as

$$\frac{a}{g} \leq \frac{a_2}{h}\cos\phi - \sin\phi \tag{2.138}$$

$$\frac{a}{g} \geq -\frac{a_1}{h}\cos\phi - \sin\phi \tag{2.139}$$

Hence, if there is limit on friction, the maximum achievable acceleration $(a > 0)$ is limited by a_2, h, ϕ; while the maximum deceleration $(a < 0)$ is limited by a_1, h, ϕ. These two equations can be combined to result in

$$-\frac{a_1}{h}\cos\phi \leq \frac{a}{g} + \sin\phi \leq \frac{a_2}{h}\cos\phi \tag{2.140}$$

If $a \to 0$, then the limits of the inclination angle would be

$$-\frac{a_1}{h} \leq \tan\phi \leq \frac{a_2}{h} \tag{2.141}$$

This is the maximum and minimum road inclination angles that the car can stay on without tilting or falling.
  Figure 2.11 illustrates the boundaries of maximum acceleration and deceleration at each uphill or downhill angles for a sample car with

$$a_1 = 1.5\,\mathrm{m} \qquad a_1 = 1.7\,\mathrm{m} \qquad h = 0.7\,\mathrm{m} \tag{2.142}$$

On a flat road, $\phi = 0$ and the possible acceleration and deceleration are limited to

$$-\frac{a_1}{h} \leq \frac{a}{g} \leq \frac{a_2}{h} \tag{2.143}$$

$$-2.14 \leq \frac{a}{g} \leq 2.43 \tag{2.144}$$

By increasing the slope angle of the uphill road, the maximum possible acceleration drops according to the upper line in Figure 2.11. The friction cannot accelerate the car anymore at $\phi = \arctan\frac{a_2}{h} = 67.62\,\mathrm{deg}$. The car will fall back if the slope is higher. Decreasing the slope of the downhill road, the maximum possible deceleration drops according to the lower line in the figure. The car cannot decelerate anymore at downhill slope of $\phi = \arctan\frac{a_1}{h} = 64.98\,\mathrm{deg}$ and the car will fall down if ϕ is higher. The limit of Equation (2.141) is illustrated in the magnified figure.

Example 61 *Maximum deceleration for a single-axle-brake car.*
  We can find the maximum braking deceleration a_{fwb} of a front-wheel-brake car on a horizontal road by substituting $\phi = 0$, $F_{x_2} = 0$, $F_{x_1} = -\mu_x F_{z_1}$ in Equation (2.128) and using Equation (2.123)

$$-\mu_x g\left(\frac{a_2}{l} - \frac{h}{l}\frac{a_{fwb}}{g}\right) = a_{fwb} \tag{2.145}$$

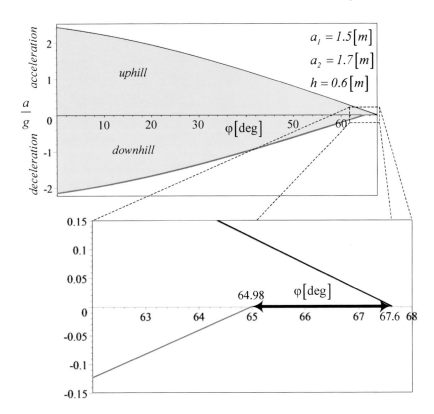

FIGURE 2.11. Limits of acceleration and deceleration versus inclination angle.

which provides us with

$$\frac{a_{fwb}}{g} = -\frac{\mu_x}{1 - \mu_x \dfrac{h}{l}}\left(1 - \frac{a_1}{l}\right) \tag{2.146}$$

Similarly, the maximum braking deceleration a_{rwb} of a rear-wheel-brake car can be achieved when we substitute $\phi = 0$, $F_{x_1} = 0$, $F_{x_2} = -\mu_x F_{z_2}$ and use Equation (2.124).

$$\frac{a_{rwb}}{g} = -\frac{\mu_x}{1 + \mu_x \dfrac{h}{l}}\frac{a_1}{l} \tag{2.147}$$

The effect of changing the position of the mass center on the maximum achievable braking deceleration is shown in Figure 2.12 for a sample car with

$$\mu_x = 1 \qquad h = 0.56\,\text{m} \qquad l = 2.6\,\text{m} \tag{2.148}$$

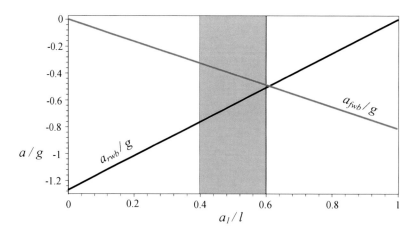

FIGURE 2.12. Effect of mass center position on the maximum achievable deccel-eration of a front-wheel and a rear-wheel-drive car.

Passenger cars are usually in the range $0.4 < (a_1/l) < 0.6$. In this range, $(a_{fwb}/g) < (a_{rwb}/g)$ and therefore, front-wheel-brake cars can reach better forward deceleration than rear-wheel-brake cars. Hence, front brakes are much more important than the rear brakes.

Example 62 ★ *A car with a trailer.*

Figure 2.13 depicts a car moving on an inclined road while pulling a trailer. To analyze the car-trailer motion, we need to separate the car and trailer to see the forces at the hinge, as shown in Figure 2.14. We assume the mass center of the trailer C_t is at distance b_3 in front of the only axle of the trailer. If C_t is behind the trailer axle, then b_3 should be negative in the following equations.

For an ideal hinge between a car and a trailer moving in a straight path, there must be a horizontal force F_{x_t} and a vertical force F_{z_t}, but there would be no moment. Writing the Newton's equation in x-direction and two static equilibrium equations for both the trailer and the vehicle

$$\sum F_x = m_t a \qquad \sum F_z = 0 \qquad \sum M_y = 0 \qquad (2.149)$$

we find the following set of equations:

$$2F_{x_1} + 2F_{x_2} - F_{x_t} - mg\sin\phi = ma \qquad (2.150)$$

$$2F_{z_1} + 2F_{z_2} - F_{z_t} - mg\cos\phi = 0 \qquad (2.151)$$

$$-2F_{z_1}a_1 + 2F_{z_2}a_2 - 2\left(F_{x_1} + F_{x_2}\right)h$$
$$+F_{x_t}\left(h - h_1\right) - F_{z_t}\left(b_1 + a_2\right) = 0 \qquad (2.152)$$

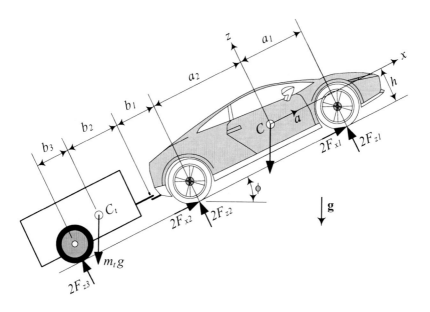

FIGURE 2.13. A car moving on an inclined road and pulling a trailer.

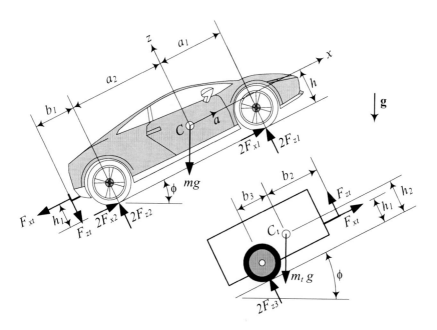

FIGURE 2.14. Free-body-diagram of a car and the trailer when moving on an uphill road.

$$F_{x_t} - m_t\, g \sin\phi = m_t\, a \tag{2.153}$$

$$2F_{z_3} + F_{z_t} - m_t\, g \cos\phi = 0 \tag{2.154}$$

$$2F_{z_3} b_3 - F_{z_t} b_2 - F_{x_t}\left(h_2 - h_1\right) = 0 \tag{2.155}$$

If the value of traction forces F_{x_1} and F_{x_2} are given, then these are six equations for six unknowns: a, F_{x_t}, F_{z_t}, F_{z_1}, F_{z_2}, F_{z_3}. Solving these equations provides the following solutions:

$$a = \frac{2}{m + m_t}\left(F_{x_1} + F_{x_2}\right) - g \sin\phi \tag{2.156}$$

$$F_{x_t} = \frac{2m_t}{m + m_t}\left(F_{x_1} + F_{x_2}\right) \tag{2.157}$$

$$F_{z_t} = \frac{h_1 - h_2}{b_2 - b_3}\frac{2m_t}{m + m_t}\left(F_{x_1} + F_{x_2}\right) + \frac{b_3}{b_2 - b_3}m_t\, g \cos\phi \tag{2.158}$$

$$F_{z_1} = \frac{b_3}{2l}\left(\frac{2a_2 - b_1}{b_2 - b_3}m_t + \frac{a_2}{b_3}m\right) g \cos\phi$$
$$+ \left[\frac{2a_2 - b_1}{b_2 - b_3}\left(h_1 - h_2\right)m_t - h_1 m_t - hm\right]\frac{F_{x_1} + F_{x_2}}{l\left(m + m_t\right)} \tag{2.159}$$

$$F_{z_2} = \frac{b_3}{2l}\left(\frac{a_1 - a_2 + b_1}{b_2 - b_3}m_t + \frac{a_1}{b_3}m\right) g \cos\phi$$
$$+ \left[\frac{a_1 - a_2 + b_1}{b_2 - b_3}\left(h_1 - h_2\right)m_t + h_1 m_t + hm\right]\frac{F_{x_1} + F_{x_2}}{l\left(m + m_t\right)} \tag{2.160}$$

$$F_{z_3} = \frac{1}{2}\frac{b_2}{b_2 - b_3}m_t\, g \cos\phi + \frac{h_1 - h_2}{b_2 - b_3}\frac{m_t}{m + m_t}\left(F_{x_1} + F_{x_2}\right) \tag{2.161}$$

$$l = a_1 + a_2 \tag{2.162}$$

However, if the value of acceleration a is known, then unknowns are: $F_{x_1} + F_{x_2}$, F_{x_t}, F_{z_t}, F_{z_1}, F_{z_2}, F_{z_3}.

$$F_{x_1} + F_{x_2} = \frac{1}{2}\left(m + m_t\right)\left(a + g \sin\phi\right) \tag{2.163}$$

$$F_{x_t} = m_t\left(a + g \sin\phi\right) \tag{2.164}$$

$$F_{z_t} = \frac{h_1 - h_2}{b_2 - b_3}m_t\left(a + g \sin\phi\right) + \frac{b_3}{b_2 - b_3}m_t\, g \cos\phi \tag{2.165}$$

$$F_{z_1} = \frac{b_3}{2l}\left(\frac{2a_2 - b_1}{b_2 - b_3}m_t + \frac{a_2}{b_3}m\right) g \cos\phi \tag{2.166}$$
$$+ \frac{1}{2l}\left[\frac{2a_2 - b_1}{b_2 - b_3}\left(h_1 - h_2\right)m_t - h_1 m_t - hm\right]\left(a + g \sin\phi\right)$$

$$F_{z_2} = \frac{b_3}{2l}\left(\frac{a_1 - a_2 + b_1}{b_2 - b_3}m_t + \frac{a_1}{b_3}m\right)g\cos\phi \qquad (2.167)$$

$$+\frac{1}{2l}\left[\frac{a_1 - a_2 + b_1}{b_2 - b_3}(h_1 - h_2)m_t + h_1 m_t + hm\right](a + g\sin\phi)$$

$$F_{z_3} = \frac{1}{2}\frac{m_t}{b_2 - b_3}(b_2 g\cos\phi + (h_1 - h_2)(a + g\sin\phi)) \qquad (2.168)$$

$$l = a_1 + a_2$$

Example 63 ★ *Maximum inclination angle for a car with a trailer.*
For a car and trailer as shown in Figure 2.13, the maximum inclination angle ϕ_M is the angle at which the car cannot accelerate the vehicle. Substituting $a = 0$ and $\phi = \phi_M$ in Equation (2.156) shows that

$$\sin\phi_M = \frac{2}{(m + m_t)g}(F_{x_1} + F_{x_2}) \qquad (2.169)$$

The value of maximum inclination angle ϕ_M increases by decreasing the total weight of the vehicle and trailer $(m + m_t)g$ or increasing the traction force $F_{x_1} + F_{x_2}$.
The traction force is limited by the maximum torque on the drive wheels and the friction under the drive tires. Let us assume the vehicle is four-wheel-drive and friction coefficients at the front and rear tires are equal. Then, the front and rear traction forces are

$$F_{x_1} \le \mu_x F_{z_1} \qquad F_{x_2} \le \mu_x F_{z_2} \qquad (2.170)$$

If we assume the front and rear wheels reach their traction limits at the same time, then

$$F_{x_1} = \mu_x F_{z_1} \qquad F_{x_2} = \mu_x F_{z_2} \qquad (2.171)$$

and we may rewrite the Equation (2.169) as

$$\sin\phi_M = \frac{2\mu_x}{(m + m_t)g}(F_{z_1} + F_{z_2}) \qquad (2.172)$$

Now substituting F_{z_1} and F_{z_2} from (2.159) and (2.160) yields

$$(mb_3 - mb_2 - m_t b_3)\mu_x\cos\phi_M + (b_2 - b_3)(m + m_t)\sin\phi_M$$
$$= 2\mu_x\frac{m_t(h_1 - h_2)}{m + m_t}(F_{x_1} + F_{x_2}) \qquad (2.173)$$

If we arrange Equation (2.173) as

$$A\cos\phi_M + B\sin\phi_M = C \qquad (2.174)$$

then

$$\phi_M = \text{atan2}(\frac{C}{\sqrt{A^2 + B^2}}, \pm\sqrt{1 - \frac{C^2}{A^2 + B^2}}) - \text{atan2}(A, B) \qquad (2.175)$$

and

$$\phi_M = \text{atan2}(\frac{C}{\sqrt{A^2 + B^2}}, \pm\sqrt{A^2 + B^2 - C^2}) - \text{atan2}(A, B) \quad (2.176)$$

where

$$A = (mb_3 - mb_2 - m_t b_3)\,\mu_x \quad (2.177)$$
$$B = (b_2 - b_3)(m + m_t) \quad (2.178)$$
$$C = 2\mu_x \frac{m_t(h_1 - h_2)}{m + m_t}(F_{x_1} + F_{x_2}) \quad (2.179)$$

For a rear-wheel-drive car pulling a trailer with the following character-istics:

$$\begin{array}{llll}
l = 2272\,\text{mm} & w = 1457\,\text{mm} & h = 230\,\text{mm} & \\
a_1 = a_2 & h_1 = 310\,\text{mm} & h_2 = 560\,\text{mm} & \\
b_1 = 680\,\text{mm} & b_2 = 610\,\text{mm} & b_3 = 120\,\text{mm} & (2.180) \\
m = 1500\,\text{kg} & m_t = 150\,\text{kg} & a = 1\,\text{m/s}^2 & \\
\mu_x = 1 & \phi = 10\,\text{deg} & &
\end{array}$$

we find

$$\begin{array}{llll}
F_{z_1} &=& 3441.78\,\text{N} & F_{z_2} = 3877.93\,\text{N} \\
F_{z_3} &=& 798.57\,\text{N} & F_{z_t} = 147.99\,\text{N} \quad (2.181) \\
F_{x_t} &=& 405.52\,\text{N} & F_{x_2} = 2230.37\,\text{N}
\end{array}$$

To check if the required traction force F_{x_2} is applicable, we should compare it to the maximum available friction force μF_{z_2} and it must be

$$F_{x_2} \le \mu F_{z_2} \quad (2.182)$$

Example 64 ★ *Solution of equation $a\cos\theta + b\sin\theta = c$.*
   The first type of trigonometric equation is

$$a\cos\theta + b\sin\theta = c \quad (2.183)$$

It can be solved by introducing two new variables r and η such that

$$a = r\sin\eta \qquad b = r\cos\eta \quad (2.184)$$

and therefore,

$$r = \sqrt{a^2 + b^2} \qquad \eta = \text{atan2}(a, b) \quad (2.185)$$

Substituting the new variables shows that

$$\sin(\eta + \theta) = \frac{c}{r} \qquad \cos(\eta + \theta) = \pm\sqrt{1 - \frac{c^2}{r^2}} \quad (2.186)$$

Hence, the solutions of the problem are

$$\theta = \text{atan2}(\frac{c}{r}, \pm\sqrt{1 - \frac{c^2}{r^2}}) - \text{atan2}(a, b) \qquad (2.187)$$

and

$$\theta = \text{atan2}(\frac{c}{r}, \pm\sqrt{r^2 - c^2}) - \text{atan2}(a, b) \qquad (2.188)$$

Therefore, the equation $a\cos\theta + b\sin\theta = c$ has two solutions if $r^2 = a^2 + b^2 > c^2$, one solution if $r^2 = c^2$, and no solution if $r^2 < c^2$.

Example 65 ★ *The function $\arctan_2 \frac{y}{x} = \text{atan2}(y, x)$.*

There are many situations in kinematics in which we need to find an angle based on the sin and cos functions of the angle. However, regular arctan cannot show the effect of the individual sign for the numerator and denominator. It always represents an angle in the first or fourth quadrant, $-\frac{\pi}{2} \leq \theta \leq \frac{\pi}{2}$. To overcome this problem and determine the angle in the correct quadrant, the atan2 function is

$$\text{atan2}(y, x) = \begin{cases} \text{sgn}\, y \, \tan^{-1}\left|\frac{y}{x}\right| & x > 0, y \neq 0 \\ \frac{\pi}{2}\,\text{sgn}\, y & x = 0, y \neq 0 \\ \text{sgn}\, y \left(\pi - \tan^{-1}\left|\frac{y}{x}\right|\right) & x < 0, y \neq 0 \\ \pi - \pi\,\text{sgn}\, x & x \neq 0, y = 0 \end{cases} \qquad (2.189)$$

where sgn represents the signum function:

$$\text{sgn}\, absolute x value = of \begin{cases} 1 & x > 0 \\ 0 & x = 0 \quad the \\ -1 & x < 0 \end{cases} \qquad (2.190)$$

In this text, wherever $\tan^{-1}\frac{y}{x}$ is used, it should be calculated based on atan2(y, x).

Example 66 *Zero vertical force at the hinge.*

We can make the vertical force at the hinge equal to zero by examining Equation (2.158) for the hinge vertical force F_{z_t}.

$$F_{z_t} = \frac{h_1 - h_2}{b_2 - b_3}\frac{2m_t}{m + m_t}(F_{x_1} + F_{x_2}) + \frac{b_3}{b_2 - b_3}m_t g \cos\phi \qquad (2.191)$$

To make $F_{z_t} = 0$, it is enough to adjust the position of trailer mass center C_t exactly on top of the trailer axle and at the same height as the hinge. In these conditions we have

$$h_1 = h_2 \qquad b_3 = 0 \qquad (2.192)$$

that makes

$$F_{z_t} = 0 \qquad (2.193)$$

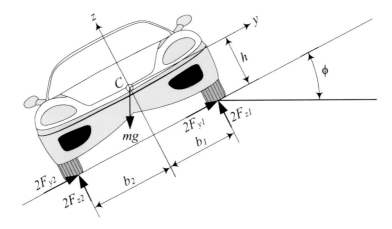

FIGURE 2.15. Normal force under the uphill and downhill tires of a vehicle, parked on banked road.

However, to increase safety, the load should also be distributed evenly throughout the trailer. Heavy items should be loaded as low as possible, mainly over the axle. Bulkier and lighter items should be distributed to give a little positive b_3. Such a trailer is called nose weight at the towing coupling.

2.5 Parked Car on a Banked Road

Figure 2.15 depicts the effect of a bank angle ϕ on the load distribution of a vehicle. A bank causes the load on the lower tires to increase and the load on the upper tires to decrease. The tire reaction forces are:

$$F_{z_1} = \frac{1}{2}\frac{mg}{w}\left(b_2\cos\phi - h\sin\phi\right) \tag{2.194}$$

$$F_{z_2} = \frac{1}{2}\frac{mg}{w}\left(b_1\cos\phi + h\sin\phi\right) \tag{2.195}$$

$$w = b_1 + b_2 \tag{2.196}$$

The maximum bank angle is

$$\tan\phi_M = \mu_y \tag{2.197}$$

at which the car will slides down laterally.

Proof. Starting with equilibrium equations in the body coordinate frame

$(Cxyz)$

$$\sum F_y = 0 \tag{2.198}$$

$$\sum F_z = 0 \tag{2.199}$$

$$\sum M_x = 0 \tag{2.200}$$

we can write

$$2F_{y_1} + 2F_{y_2} - mg \sin \phi = 0 \tag{2.201}$$
$$2F_{z_1} + 2F_{z_2} - mg \cos \phi = 0 \tag{2.202}$$
$$2F_{z_1} b_1 - 2F_{z_2} b_2 + 2\left(F_{y_1} + F_{y_2}\right) h = 0 \tag{2.203}$$

We assumed the force under the lower tires, front and rear, are equal, and also the forces under the upper tires, front and rear are equal. To calculate the reaction forces under each tire, we may assume the overall lateral force $F_{y_1} + F_{y_2}$ as an unknown. The solution of these equations provide the lateral and reaction forces under the upper and lower tires.

$$F_{z_1} = \frac{1}{2} mg \frac{b_2}{w} \cos \phi - \frac{1}{2} mg \frac{h}{w} \sin \phi \tag{2.204}$$

$$F_{z_2} = \frac{1}{2} mg \frac{b_1}{w} \cos \phi + \frac{1}{2} mg \frac{h}{w} \sin \phi \tag{2.205}$$

$$F_{y_1} + F_{y_2} = \frac{1}{2} mg \sin \phi \tag{2.206}$$

At the ultimate angle $\phi = \phi_M$, all wheels will begin to slide simultaneously and therefore,

$$F_{y_1} = \mu_{y_1} F_{z_1} \tag{2.207}$$
$$F_{y_2} = \mu_{y_2} F_{z_2} \tag{2.208}$$

The equilibrium equations show that

$$2\mu_{y_1} F_{z_1} + 2\mu_{y_2} F_{z_2} - mg \sin \phi = 0 \tag{2.209}$$
$$2F_{z_1} + 2F_{z_2} - mg \cos \phi = 0 \tag{2.210}$$
$$2F_{z_1} b_1 - 2F_{z_2} b_2 + 2\left(\mu_{y_1} F_{z_1} + \mu_{y_2} F_{z_2}\right) h = 0 \tag{2.211}$$

Assuming

$$\mu_{y_1} = \mu_{y_2} = \mu_y \tag{2.212}$$

provides us with

$$F_{z_1} = \frac{1}{2} mg \frac{b_2}{w} \cos \phi_M - \frac{1}{2} mg \frac{h}{w} \sin \phi_M \tag{2.213}$$

$$F_{z_2} = \frac{1}{2} mg \frac{b_1}{w} \cos \phi_M + \frac{1}{2} mg \frac{h}{w} \sin \phi_M \tag{2.214}$$

$$\tan \phi_M = \mu_y \tag{2.215}$$

These calculations are correct as long as

$$\tan \phi_M \leq \frac{b_2}{h} \tag{2.216}$$

$$\mu_y \leq \frac{b_2}{h} \tag{2.217}$$

If the lateral friction μ_y is higher than b_2/h then the car will roll downhill. To increase the capability of a car moving on a banked road, the car should be as wide as possible with a mass center as low as possible. ■

Example 67 *Tire forces of a parked car in a banked road.*
    A car having

$$m = 980\,\mathrm{kg} \qquad h = 0.6\,\mathrm{m} \qquad w = 1.52\,\mathrm{m} \qquad b_1 = b_2 \tag{2.218}$$

is parked on a banked road with $\phi = 4\,\mathrm{deg}$. *The forces under the lower and upper tires of the car are:*

$$F_{z_1} = 2265.2\,\mathrm{N} \qquad F_{z_2} = 2529.9\,\mathrm{N} \qquad F_{y_1} + F_{y_2} = 335.3\,\mathrm{N} \tag{2.219}$$

    The ratio of the uphill force F_{z_1} *to downhill force* F_{z_2} *depends on only the mass center position.*

$$\frac{F_{z_1}}{F_{z_2}} = \frac{b_2 \cos \phi - h \sin \phi}{b_1 \cos \phi + h \sin \phi} \tag{2.220}$$

Assuming a symmetric car with $b_1 = b_2 = w/2$ *simplifies the equation to*

$$\frac{F_{z_1}}{F_{z_2}} = \frac{w \cos \phi - 2h \sin \phi}{w \cos \phi + 2h \sin \phi} \tag{2.221}$$

Figure 2.16 illustrates the behavior of force ratio F_{z_1}/F_{z_2} *as a function of* ϕ *for* $h = 0.6\,\mathrm{m}$ *and* $w = 1.52\,\mathrm{m}$. *The rolling down angle* $\phi_M = \tan^{-1}(b_2/h) = 51.71\,\mathrm{deg}$ *indicates the bank angle at which the force under the uphill wheels become zero and the car rolls down. The negative part of the curve indicates the required force to keep the car on the road, which is not applicable in real situations.*

Example 68 *Vehicle on a banked round road.*
    When a vehicle of mass m is moving with speed v on a flat round path of radius R, the direction of the wheels lateral force are inward and provide the required centripetal acceleration.

$$2F_{y_1} + 2F_{y_2} = m\frac{v^2}{R} \tag{2.222}$$

Knowing that the lateral wheels lateral force are limited by the maximum friction force between tire and road, we conclude that there is a maximum

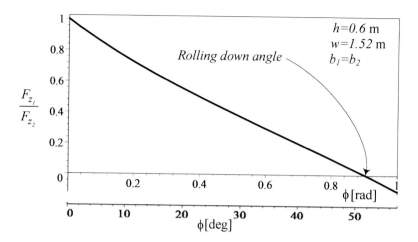

FIGURE 2.16. Illustration of the force ratio F_{z_1}/F_{z_2} as a function of road bank angle ϕ.

speed v_M at which the required lateral force will not be produced by tires and vehicle slides out of the road.

To have a safe road, we have to design round roads such that vehicle do not need any wheel lateral force to provide the required centripetal force. Designing roads with a bank angle is a good approximate solution; so a component of weight force provides the required centripetal force. Figure 2.17 illustrates a moving car on a round banked road. Assuming equal vertical force under each tire, the balance of the applied forces on the vehicle in the body coordinate frame provides us with

$$mg\cos\phi - 4F_{z_1} - m\frac{v^2}{R}\sin\phi = 0 \qquad (2.223)$$

$$mg\sin\phi - m\frac{v^2}{R}\cos\phi = 0 \qquad (2.224)$$

The second equation indicates the required bank angle as a function of speed

$$\tan\phi = \frac{v^2}{Rg} \qquad (2.225)$$

The bank angle is independent of the vehicle mass m and is a function of the road radius of turn R and velocity of the vehicle v. Assuming that the road radius of turn R is not variable, the bank angle must vary with the speed of the vehicle. To design the road, we have to decide about the proper velocity of vehicles on the road and calculate the bank angle based on (2.225). Because the angle is only a function of the vehicle velocity, the proper banked road works well for all types of vehicles as long as they keep their velocity as recommended. Any lower or higher speed would respectively

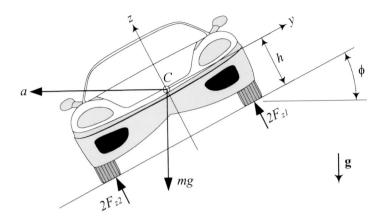

FIGURE 2.17. A moving car on a round banked road.

need some positive or negative lateral force to be generated by tires. The lack or excessive tire lateral force will be provided by steering and sideslip angles of vehicles, or by roll and camber angles of bicycles and motorcycles.

As of today, we do not make roads with variable ϕ. However, imagine that roads on a circular path were smart to adjust their bank angle according to the speed of vehicles. Then, Equation (2.225) is the function the road must follow. Figure 2.18 depicts the bank angle as a function of vehicle speed on a circular path with radius $R = 100\,\mathrm{m}$. The angle is an increasing function of speed with an asymptotic value of $\phi = 90\,\mathrm{deg}$ regardless of R provided that $R \neq 0$.

$$\lim_{v \to \infty} \phi = 90\,\mathrm{deg} \qquad (2.226)$$

Figure 2.19 illustrates the sensitivity of the bank angle to the radius of turn at different speed.

2.6 ★ Optimal Drive and Brake Force Distribution

A certain acceleration a can be achieved by adjusting and controlling the longitudinal forces F_{x_1} and F_{x_2}. The optimal longitudinal forces under the front and rear tires of a four wheel drive to achieve the maximum acceleration are

$$
\begin{aligned}
\frac{F_{x_1}}{mg} &= -\frac{1}{2}\frac{h}{l}\left(\frac{a}{g}\right)^2 + \frac{1}{2}\frac{a_2}{l}\frac{a}{g} \\
&= -\frac{1}{2}\mu_x^2\frac{h}{l} + \frac{1}{2}\mu_x\frac{a_2}{l}
\end{aligned}
\qquad (2.227)
$$

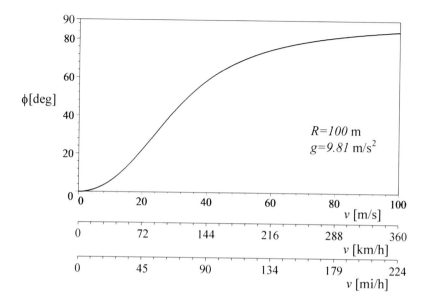

FIGURE 2.18. The required bank angle ϕ as a function of vehicle speed on a circular path with radius $R = 100$ m.

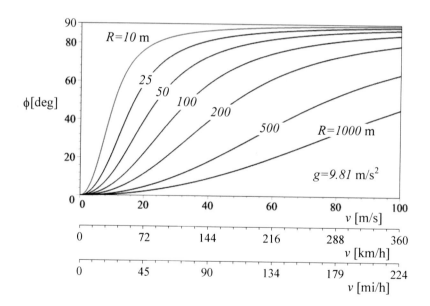

FIGURE 2.19. The required road bank angle ϕ for different radius of turn and different speed.

$$\frac{F_{x_2}}{mg} = \frac{1}{2}\frac{h}{l}\left(\frac{a}{g}\right)^2 + \frac{1}{2}\frac{a_1}{l}\frac{a}{g}$$

$$= \frac{1}{2}\mu_x^2\frac{h}{l} + \frac{1}{2}\mu_x\frac{a_1}{l} \tag{2.228}$$

Proof. The longitudinal equation of motion for a car on a horizontal road is

$$2F_{x_1} + 2F_{x_2} = ma \tag{2.229}$$

and the maximum traction forces under each tire is a function of normal force and the friction coefficient.

$$F_{x_1} \leq \pm\mu_x F_{z_1} \tag{2.230}$$

$$F_{x_2} \leq \pm\mu_x F_{z_2} \tag{2.231}$$

However, the normal forces are a function of the car's acceleration and geometry.

$$F_{z_1} = \frac{1}{2}mg\frac{a_2}{l} - \frac{1}{2}mg\frac{h}{l}\frac{a}{g} \tag{2.232}$$

$$F_{z_2} = \frac{1}{2}mg\frac{a_1}{l} + \frac{1}{2}mg\frac{h}{l}\frac{a}{g} \tag{2.233}$$

We may generalize the equations by making them dimensionless. Under the best conditions, we should adjust the traction forces to their maximum

$$\frac{F_{x_1}}{mg} = \frac{1}{2}\mu_x\left(\frac{a_2}{l} - \frac{h}{l}\frac{a}{g}\right) \tag{2.234}$$

$$\frac{F_{x_2}}{mg} = \frac{1}{2}\mu_x\left(\frac{a_1}{l} + \frac{h}{l}\frac{a}{g}\right) \tag{2.235}$$

and therefore, the longitudinal equation of motion (2.229) becomes

$$\frac{a}{g} = \mu_x \tag{2.236}$$

Substituting this result back into Equations (2.234) and (2.235) shows that

$$\frac{F_{x_1}}{mg} = -\frac{1}{2}\frac{h}{l}\left(\frac{a}{g}\right)^2 + \frac{1}{2}\frac{a_2}{l}\frac{a}{g} \tag{2.237}$$

$$\frac{F_{x_2}}{mg} = \frac{1}{2}\frac{h}{l}\left(\frac{a}{g}\right)^2 + \frac{1}{2}\frac{a_1}{l}\frac{a}{g} \tag{2.238}$$

Depending on the geometric parameters of the car (h, a_1, a_2), and the acceleration $a > 0$, these two equations determine how much the front and rear driving forces must be. The same equations are applied for deceleration

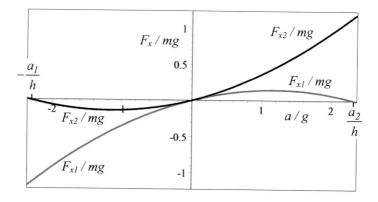

FIGURE 2.20. Optimal driving and braking forces for a sample car.

$a < 0$, to determine the value of optimal front and rear braking forces. Figure 2.20 depicts a graphical illustration of the optimal driving and braking forces for a sample car using the following data:

$$\mu_x = 1 \qquad \frac{h}{l} = \frac{0.56}{2.6} = 0.2154 \qquad \frac{a_1}{l} = \frac{a_2}{l} = \frac{1}{2} \qquad (2.239)$$

When accelerating $a > 0$, the optimal driving force on the rear tire grows rapidly while the optimal driving force on the front tire drops after a maximum at

$$\frac{a}{g} = \frac{1}{2}\frac{a_2}{h} \qquad (2.240)$$

$$\frac{F_{x_1}}{mg} = \frac{1}{8}\frac{a_2}{h}\frac{a_2}{l} \qquad (2.241)$$

The value of $a/g = a_2/h$ is the maximum possible acceleration at which the front tires lose their contact with the ground. The acceleration at which front (or rear) tires lose their ground contact is called *tilting acceleration*. The required force at the rear wheels to achieve $a/g = a_2/h$ would be $F_{x_2}/(mg) = a_2/(2h)$.

The opposite phenomenon happens when decelerating. For $a < 0$, the absolute value of the optimal front brake force increases rapidly and the rear brake force goes to zero after a maximum at

$$\frac{a}{g} = -\frac{1}{2}\frac{a_2}{h} \qquad (2.242)$$

$$\frac{F_{x_1}}{mg} = -\frac{1}{8}\frac{a_1}{h}\frac{a_1}{l} \qquad (2.243)$$

The deceleration $a/g = -a_1/h$ is the maximum possible deceleration at which the rear tires lose their ground contact.

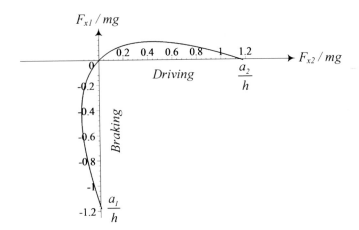

FIGURE 2.21. Optimal traction and braking force distribution between the front and rear wheels.

The graphical representation of the optimal driving and braking forces can be shown better by plotting $F_{x_1}/(mg)$ versus $F_{x_2}/(mg)$ using a/g as a parameter.

$$F_{x_1} = \frac{a_2 - \dfrac{a}{g}h}{a_1 + \dfrac{a}{g}h} F_{x_2} \tag{2.244}$$

$$\frac{F_{x_1}}{F_{x_2}} = \frac{a_2 - \mu_x h}{a_1 + \mu_x h} \tag{2.245}$$

Such a plot is shown in Figure 2.21. This is a design curve describing the relationship between forces under the front and rear wheels to achieve the maximum acceleration or deceleration.

Adjusting the optimal force distribution is not an automatic procedure and needs a force distributor control system to measure and adjust the forces. ∎

Example 69 ★ *Slope at zero.*

The initial optimal traction force distribution is the slope of the optimal curve $(F_{x_1}/(mg), F_{x_2}/(mg))$ *at zero.*

$$\frac{d\dfrac{F_{x_1}}{mg}}{d\dfrac{F_{x_2}}{mg}} = \lim_{a \to 0} \frac{-\dfrac{1}{2}\dfrac{h}{l}\left(\dfrac{a}{g}\right)^2 + \dfrac{1}{2}\dfrac{a_2}{l}\dfrac{a}{g}}{\dfrac{1}{2}\dfrac{h}{l}\left(\dfrac{a}{g}\right)^2 + \dfrac{1}{2}\dfrac{a_1}{l}\dfrac{a}{g}} = \frac{a_2}{a_1} \tag{2.246}$$

Therefore, the initial traction force distribution depends on only the position of mass center C.

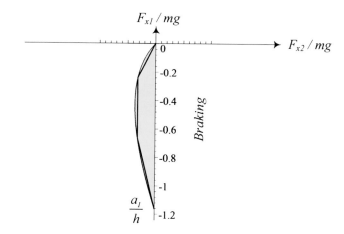

FIGURE 2.22. Optimal braking force distribution between the front and rear wheels, along with a three-line under estimation.

Example 70 ★ *Brake balance.*

When braking, a car is stable if the rear wheels do not lock. Thus, the rear brake forces must be less than the maximum possible braking force at all time. This means the brake force distribution should always be in the shaded area of Figure 2.22, and below the optimal curve. This restricts the achievable deceleration, especially at low friction values, but increases the stability of the car.

Whenever it is easier for a force distributor to follow a line, the optimal brake curve is underestimated using two or three lines, and a control system adjusts the force ratio F_{x_1}/F_{x_2}. A sample of three-line approximation is shown in Figure 2.22.

Distribution of the brake force between the front and rear wheels is called **brake balance**. Brake balance varies with deceleration. The higher the deceleration, the more load will transfer to the front wheels and the more braking effort they can support. Meanwhile the rear wheels are unloaded and they must have less braking force.

Example 71 ★ *Best race car.*

Racecars are always supposed to work at their maximum achievable acceleration to finish their race in minimum time. They are usually equipped with rear-wheel-drive and all-wheel-brake. However, if an all-wheel-drive race car is reasonable to build, then a force distributor, to follow the curve shown in Figure 2.23, is what it needs to race better.

Example 72 ★ *Effect of C location on braking.*

Load is transferred from the rear wheels to the front when the brakes are applied. The higher the C, the more load transfer. So, to improve braking,

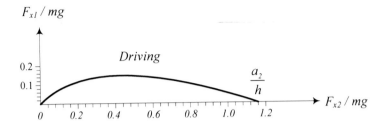

FIGURE 2.23. Optimal traction force distribution between the front and rear wheels.

the mass center C should be as low as possible and as back as possible. This is not feasible for every vehicle, especially for forward-wheel drive street cars. However, this fact should be considered when a car is being designed for better braking performance.

Example 73 ★ *Front and rear wheel locking.*

The optimal brake force distribution is according to Equation (2.245) for an ideal F_{x_1}/F_{x_2} ratio. However, if the brake force distribution is not ideal, then either the front or the rear wheels will lock up first. Locking the rear wheels makes the vehicle unstable, and it loses directional stability. When the rear wheels lock, they slide on the road and they lose their capacity to support lateral force. The resultant shear force at the tireprint of the rear wheels reduces to a dynamic friction force in the opposite direction of the sliding.

A slight lateral motion of the rear wheels, by any disturbance, develops a yaw motion because of unbalanced lateral forces on the front and rear wheels. The yaw moment turns the vehicle about the z-axis until the rear end leads the front end and the vehicle turns 180 deg. Figure 2.24 illustrates a 180 deg sliding rotation of a rear-wheel-locked car.

The lock-up of the front tires does not cause a directional instability, although the car would not be steerable and the driver would lose control.

2.7 ★ Vehicles With More Than Two Axles

If a vehicle has more than two axles, such as the three-axle car shown in Figure 2.25, then the vehicle will be statically indeterminate and the normal forces under the tires cannot be determined by static equilibrium equations. We need to consider the suspensions' deflection to determine their applied forces.

The n normal forces F_{z_i} under the tires can be calculated using the

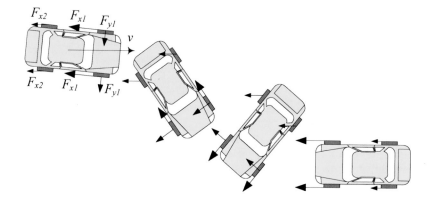

FIGURE 2.24. 180 deg sliding rotation of a rear-wheel-locked car.

following n equations.

$$2\sum_{i=1}^{n} F_{z_i} - mg\cos\phi = 0 \tag{2.247}$$

$$2\sum_{i=1}^{n} F_{z_i} x_i + h\left(a + mg\sin\phi\right) = 0 \tag{2.248}$$

$$\frac{F_{z_i}}{k_i} - \frac{x_i - x_1}{x_n - x_1}\left(\frac{F_{z_n}}{k_n} - \frac{F_{z_1}}{k_1}\right) - \frac{F_{z_1}}{k_1} = 0 \qquad i = 2,3,\cdots,n-1 \tag{2.249}$$

where F_{x_i} and F_{z_i} are the longitudinal and normal forces under the tires attached to the axle number i, and x_i is the distance of mass center C from the axle number i. The distance x_i is positive for axles in front of C, and is negative for the axles in back of C. The parameter k_i is the vertical stiffness of the suspension at axle i.

Proof. For a multiple-axle vehicle, the following equations

$$\sum F_x = ma \tag{2.250}$$

$$\sum F_z = 0 \tag{2.251}$$

$$\sum M_y = 0 \tag{2.252}$$

provide the same sort of equations as (2.128)-(2.130). However, if the total number of axles are n, then the individual forces can be substituted by a summation.

$$2\sum_{i=1}^{n} F_{x_i} - mg\sin\phi = ma \tag{2.253}$$

$$2\sum_{i=1}^{n} F_{z_i} - mg\cos\phi = 0 \tag{2.254}$$

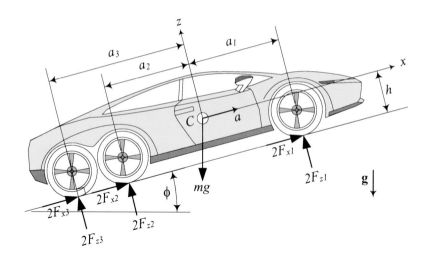

FIGURE 2.25. A three-axle car moving on an inclined road.

$$2\sum_{i=1}^{n} F_{z_i} x_i + 2h \sum_{i=1}^{n} F_{x_i} = 0 \tag{2.255}$$

The overall forward force $F_x = 2\sum_{i=1}^{n} F_{x_i}$ can be eliminated between Equations (2.253) and (2.255) to make Equation (2.248). Then, there remain two equations (2.247) and (2.248) for n unknowns F_{z_i}, $i = 1, 2, \cdots, n$. Hence, we need $n - 2$ extra equations to be able to find the wheel loads. The extra equations come from the compatibility among the suspensions' deflection.

We ignore the tires' compliance, and use z to indicate the vehicle's vertical direction as is shown in Figure 2.25. Then, if z_i is the suspension deflection at the center of axle i, and k_i is the vertical stiffness of the suspension at axle i, the deflections are

$$z_i = \frac{F_{z_i}}{k_i} \tag{2.256}$$

For a flat road, and a rigid vehicle, we must have

$$\frac{z_i - z_1}{x_i - x_1} = \frac{z_n - z_1}{x_n - x_1} \qquad i = 2, 3, \cdots, n - 1 \tag{2.257}$$

which, after substituting with (2.256), reduces to Equation (2.249). The $n - 2$ equations (2.257) along with the two equations (2.247) and (2.248) are enough to calculate the normal load under each tire.

The resultant set of equations is linear and may be arranged in a matrix form

$$[A][X] = [B] \tag{2.258}$$

where

$$[X] = \begin{bmatrix} F_{z_1} & F_{z_2} & F_{z_3} & \cdots & F_{z_n} \end{bmatrix}^T \tag{2.259}$$

$$[A] = \begin{bmatrix} 2 & 2 & \cdots & \cdots & \cdots & \cdots & 2 \\ 2x_1 & 2x_2 & \cdots & \cdots & \cdots & \cdots & 2x_n \\ \dfrac{x_n - x_2}{k_1 l} & \dfrac{1}{k_2} & \cdots & \cdots & \cdots & \cdots & \dfrac{x_2 - x_1}{k_n l} \\ \cdots & \cdots & \cdots & \cdots & \cdots & \cdots & \cdots \\ \dfrac{x_n - x_i}{k_1 l} & \cdots & \cdots & \dfrac{1}{k_i} & \cdots & \cdots & \dfrac{x_i - x_1}{k_n l} \\ \cdots & \cdots & \cdots & \cdots & \cdots & \cdots & \cdots \\ \dfrac{x_n - x_{n-1}}{k_1 l} & \cdots & \cdots & \cdots & \cdots & \dfrac{1}{k_{n-1}} & \dfrac{x_{n-1} - x_1}{k_n l} \end{bmatrix} \tag{2.260}$$

$$l = x_1 - x_n \tag{2.261}$$

$$[B] = \begin{bmatrix} mg\cos\phi & -h\,(a + mg\sin\phi) & 0 & \cdots & 0 \end{bmatrix}^T \tag{2.262}$$

∎

Example 74 ★ *Wheel reactions for a three-axle car.*

Figure 2.25 illustrates a three-axle car moving on an inclined road. We start counting the axles of a multiple-axle vehicle from the front axle as axle-1, and move sequentially to the back as shown in the figure.

The set of equations for the three-axle car is

$$2F_{x_1} + 2F_{x_2} + 2F_{x_3} - mg\sin\phi = ma \tag{2.263}$$

$$2F_{z_1} + 2F_{z_2} + 2F_{z_3} - mg\cos\phi = 0 \tag{2.264}$$

$$2F_{z_1}x_1 + 2F_{z_2}x_2 + 2F_{z_3}x_3 + 2h\,(F_{x_1} + F_{x_2} + F_{x_3}) = 0 \tag{2.265}$$

$$\frac{1}{x_2 - x_1}\left(\frac{F_{z_2}}{k_2} - \frac{F_{z_1}}{k_1}\right) - \frac{1}{x_3 - x_1}\left(\frac{F_{z_3}}{k_3} - \frac{F_{z_1}}{k_1}\right) = 0 \tag{2.266}$$

which can be simplified to

$$2F_{z_1} + 2F_{z_2} + 2F_{z_3} - mg\cos\phi = 0 \tag{2.267}$$

$$2F_{z_1}x_1 + 2F_{z_2}x_2 + 2F_{z_3}x_3 + hm\,(a + g\sin\phi) = 0 \tag{2.268}$$

$$(x_2 k_2 k_3 - x_3 k_2 k_3)\,F_{z_1} + (x_1 k_1 k_2 - x_2 k_1 k_2)\,F_{z_3}$$
$$- (x_1 k_1 k_3 - x_3 k_1 k_3)\,F_{z_2} = 0 \tag{2.269}$$

The set of equations for wheel loads is linear and may be rearranged in a matrix form

$$[A]\,[X] = [B] \tag{2.270}$$

where

$$[A] = \begin{bmatrix} 2 & 2 & 2 \\ 2x_1 & 2x_2 & 2x_3 \\ k_2 k_3 (x_2 - x_3) & k_1 k_3 (x_3 - x_1) & k_1 k_2 (x_1 - x_2) \end{bmatrix} \qquad (2.271)$$

$$[X] = \begin{bmatrix} F_{z_1} \\ F_{z_2} \\ F_{z_3} \end{bmatrix} \qquad (2.272)$$

$$[B] = \begin{bmatrix} mg \cos \phi \\ -hm (a + g \sin \phi) \\ 0 \end{bmatrix} \qquad (2.273)$$

The unknown vector may be found using matrix inversion

$$[X] = [A]^{-1} [B] \qquad (2.274)$$

The solution of the equations are

$$\frac{1}{k_1 m} F_{z_1} = \frac{Z_1}{Z_0} \qquad \frac{1}{k_2 m} F_{z_2} = \frac{Z_2}{Z_0} \qquad \frac{1}{k_2 m} F_{z_3} = \frac{Z_3}{Z_0} \qquad (2.275)$$

where,

$$Z_0 = -4 k_1 k_2 (x_1 - x_2)^2 - 4 k_2 k_3 (x_2 - x_3)^2 - 4 k_1 k_3 (x_3 - x_1)^2 \qquad (2.276)$$

$$\begin{aligned} Z_1 = \; & g (x_2 k_2 - x_1 k_3 - x_1 k_2 + x_3 k_3) h \sin \phi \\ & + a (x_2 k_2 - x_1 k_3 - x_1 k_2 + x_3 k_3) h \\ & + g (k_2 x_2^2 - x_1 k_2 x_2 + k_3 x_3^2 - x_1 k_3 x_3) \cos \phi \end{aligned} \qquad (2.277)$$

$$\begin{aligned} Z_2 = \; & g (x_1 k_1 - x_2 k_1 - x_2 k_3 + x_3 k_3) h \sin \phi \\ & + a (x_1 k_1 - x_2 k_1 - x_2 k_3 + x_3 k_3) h \\ & + g (k_1 x_1^2 - x_2 k_1 x_1 + k_3 x_3^2 - x_2 k_3 x_3) \cos \phi \end{aligned} \qquad (2.278)$$

$$\begin{aligned} Z_3 = \; & g (x_1 k_1 + x_2 k_2 - x_3 k_1 - x_3 k_2) h \sin \phi \\ & + a (x_1 k_1 + x_2 k_2 - x_3 k_1 - x_3 k_2) h \\ & + g (k_1 x_1^2 - x_3 k_1 x_1 + k_2 x_2^2 - x_3 k_2 x_2) \cos \phi \end{aligned} \qquad (2.279)$$

$$x_1 = a_1 \qquad x_2 = -a_2 \qquad x_3 = -a_3 \qquad (2.280)$$

2.8 ★ Vehicles on a Crest and Dip

When a road has an outward or inward curvature, we call the road is a crest or a dip. The road curvature can decrease or increase the normal forces under the wheels. In this section we assume the radius of curvature of the road R_H is much bigger than the height of the vehicle mass center, $R_H \gg h$.

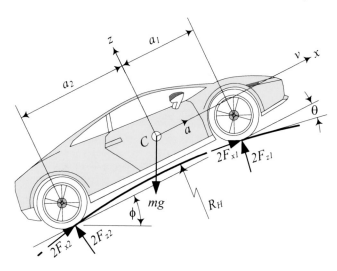

FIGURE 2.26. A cresting vehicle at a point where the hill has a radius of curvature R_h.

2.8.1 ★ Vehicles on a Crest

Moving on the convex curve of a hill is called *cresting*. The normal force under the wheels of a cresting vehicle is less than the force on a flat inclined road with the same slope, because of the developed centrifugal force mv^2/R_H in the $-z$-direction.

Figure 2.26 illustrates a cresting vehicle at the point on the hill with a radius of curvature R_H and average slope ϕ with respect to horizon. The traction and normal forces under its tires are approximately equal to

$$F_{x_1} + F_{x_2} \approx \frac{1}{2}m\left(a + g\sin\phi\right) \qquad (2.281)$$

$$F_{z_1} \approx \frac{1}{2}mg\left[\left(\frac{a_2}{l}\cos\phi + \frac{h}{l}\sin\phi\right)\right] - \frac{1}{2}ma\frac{h}{l} - \frac{1}{2}m\frac{v^2}{R_H}\frac{a_2}{l} \qquad (2.282)$$

$$F_{z_2} \approx \frac{1}{2}mg\left[\left(\frac{a_1}{l}\cos\phi - \frac{h}{l}\sin\phi\right)\right] + \frac{1}{2}ma\frac{h}{l} - \frac{1}{2}m\frac{v^2}{R_H}\frac{a_1}{l} \qquad (2.283)$$

$$l = a_1 + a_2 \qquad (2.284)$$

Proof. For the cresting car shown in Figure 2.26, the normal and tangential directions are equivalent to the $-z$ and x directions respectively. Hence, the governing equation of motion for the car is

$$\sum F_x = ma \tag{2.285}$$

$$-\sum F_z = m\frac{v^2}{R_H} \tag{2.286}$$

$$\sum M_y = 0 \tag{2.287}$$

Expanding these equations yields

$$2F_{x_1}\cos\theta + 2F_{x_2}\cos\theta - mg\sin\phi = ma \tag{2.288}$$

$$-2F_{z_1}\cos\theta - 2F_{z_2}\cos\theta + mg\cos\phi = m\frac{v^2}{R_H} \tag{2.289}$$

$$2F_{z_1}a_1\cos\theta - 2F_{z_2}a_2\cos\theta + 2\left(F_{x_1} + F_{x_2}\right)h\cos\theta$$
$$+2F_{z_1}a_1\sin\theta - 2F_{z_2}a_2\sin\theta - 2\left(F_{x_1} + F_{x_2}\right)h\sin\theta = 0 \tag{2.290}$$

We may eliminate $\left(F_{x_1} + F_{x_2}\right)$ between the first and third equations, and solve for the total traction force $F_{x_1} + F_{x_2}$ and wheel normal forces F_{z_1}, F_{z_2}.

$$F_{x_1} + F_{x_2} = \frac{ma + mg\sin\phi}{2\cos\theta} \tag{2.291}$$

$$F_{z_1} = \frac{1}{2}mg\left[\left(\frac{a_2}{l\cos\theta}\cos\phi + \frac{h\left(1 - \sin 2\theta\right)}{l\cos\theta\cos 2\theta}\sin\phi\right)\right]$$
$$-\frac{1}{2}ma\frac{h\left(1 - \sin 2\theta\right)}{l\cos\theta\cos 2\theta} - \frac{1}{2}m\frac{v^2}{R_H}\frac{a_2}{l\cos\theta} \tag{2.292}$$

$$F_{z_2} = \frac{1}{2}mg\left[\left(\frac{a_1}{l\cos\theta}\cos\phi - \frac{h\left(1 - \sin 2\theta\right)}{l\cos\theta\cos 2\theta}\sin\phi\right)\right]$$
$$+\frac{1}{2}ma\frac{h\left(1 - \sin 2\theta\right)}{l\cos\theta\cos 2\theta} - \frac{1}{2}m\frac{v^2}{R_H}\frac{a_1}{l\cos\theta\cos\theta} \tag{2.293}$$

If the car's wheel base is much smaller than the radius of curvature, $l \ll R_H$, then the slope angle θ is too small, and we may use trigonometric approximations

$$\cos\theta \approx \cos 2\theta \approx 1 \tag{2.294}$$

$$\sin\theta \approx \sin 2\theta \approx 0 \tag{2.295}$$

Substituting these approximations in Equations (2.291)-(2.293) produces the following approximate results:

$$F_{x_1} + F_{x_2} \approx \frac{1}{2}m\left(a + g\sin\phi\right) \tag{2.296}$$

$$2\frac{F_{z_1}}{mg} \approx \left(\frac{a_2}{l}\cos\phi + \frac{h}{l}\sin\phi\right) - \frac{a\,h}{g\,l} - \frac{v^2/R_H}{g}\frac{a_2}{l} \tag{2.297}$$

$$2\frac{F_{z_2}}{mg} \approx \left(\frac{a_1}{l}\cos\phi - \frac{h}{l}\sin\phi\right) + \frac{a\,h}{g\,l} - \frac{v^2/R_H}{g}\frac{a_1}{l} \tag{2.298}$$

■

Example 75 ★ *Wheel loads of a cresting car.*
  Consider a car with these specifications:

$$l = 2272\,\text{mm} \quad w = 1457\,\text{mm} \quad h = 230\,\text{mm} \quad a_1 = a_2$$
$$m = 1500\,\text{kg} \quad v = 15\,\text{m/s} \quad a = 1\,\text{m/s}^2 \tag{2.299}$$

which is cresting a hill at a point where the road has

$$R_H = 40\,\text{m} \qquad \phi = 30\,\text{deg} \qquad \theta = 2.5\,\text{deg} \tag{2.300}$$

The force information on the car is:

$$F_{x_1} + F_{x_2} = 4432.97\,\text{N} \quad F_{z_1} = 666.33\,\text{N} \quad F_{z_2} = 1488.75\,\text{N}$$
$$F_{z_1} + F_{z_2} = 2155.08\,\text{N} \quad mg = 14715\,\text{N} \quad m\frac{v^2}{R_H} = 8437.5\,\text{N} \tag{2.301}$$

  If we simplifying the results by assuming small θ, the approximate values of the forces would be

$$F_{x_1} + F_{x_2} = 4428.75\,\text{N} \quad F_{z_1} \approx 628.18\,\text{N} \quad F_{z_2} \approx 1524.85\,\text{N}$$
$$F_{z_1} + F_{z_2} \approx 2153.03\,\text{N} \quad mg = 14715\,\text{N} \quad m\frac{v^2}{R_H} = 8437.5\,\text{N} \tag{2.302}$$

Example 76 ★ *Losing the road contact on a crest.*
  *When a car moves too fast, it can lose its road contact on a crest. Such a car is called a **flying car**. The condition to have a flying car is $F_{z_1} = 0$ and $F_{z_2} = 0$.*
  Assuming a symmetric flying car $a_1 = a_2 = l/2$ with no acceleration, and using the approximate Equations (2.282) and (2.283), we have

$$\frac{1}{2}mg\left[\left(\frac{a_2}{l}\cos\phi + \frac{h}{l}\sin\phi\right)\right] - \frac{1}{2}m\frac{v^2}{R_H}\frac{a_2}{l} = 0 \tag{2.303}$$

$$\frac{1}{2}mg\left[\left(\frac{a_1}{l}\cos\phi - \frac{h}{l}\sin\phi\right)\right] - \frac{1}{2}m\frac{v^2}{R_H}\frac{a_1}{l} = 0 \tag{2.304}$$

and we can find the critical minimum speed v_c to start flying. There are two critical speeds v_{c_1} and v_{c_2} for losing the contact of the front and rear wheels, respectively.

$$v_{c_1} = \sqrt{2gR_H\left(\frac{h}{l}\sin\phi + \frac{1}{2}\cos\phi\right)} \tag{2.305}$$

$$v_{c_2} = \sqrt{-2gR_H\left(\frac{h}{l}\sin\phi - \frac{1}{2}\cos\phi\right)} \tag{2.306}$$

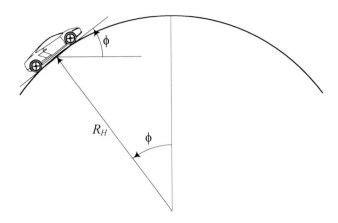

FIGURE 2.27. A cresting car over a circular hill.

For any car, the critical speeds v_{c_1} and v_{c_2} are functions of the hill's radius of curvature R_H and the angular position on the hill, indicated by ϕ. The angle ϕ cannot be out of the tilting angles given by Equation (2.141).

$$-\frac{a_1}{h} \leq \tan \phi \leq \frac{a_2}{h} \tag{2.307}$$

Figure 2.27 illustrates a cresting car over a circular hill, and Figure 2.28 depicts the critical speeds v_{c_1} and v_{c_2} at a different angle ϕ for -1.371 rad $\leq \phi \leq 1.371$ rad. The specifications of the car and the hill are:

$$\begin{array}{lll} l = 2272\,\text{mm} & h = 230\,\text{mm} & a_1 = a_2 \\ a = 0\,\text{m/s}^2 & R_H = 100\,\text{m} \end{array} \tag{2.308}$$

At the maximum uphill slope $\phi = 1.371$ rad ≈ 78.5 deg, the front wheels can leave the ground at zero speed while the rear wheels are on the ground. When the car moves over the hill and reaches the maximum downhill slope $\phi = -1.371$ rad ≈ -78.5 deg the rear wheels can leave the ground at zero speed while the front wheels are on the ground. As long as the car is moving uphill, the front wheels can leave the ground at a lower speed while going downhill the rear wheels leave the ground at a lower speed. Hence, at each slope angle ϕ the lower curve of Figure 2.28 determines the critical speed v_c.

To have a general image of the critical speed, we may plot the lower values of v_c as a function of ϕ using R_H or h/l as a parameter. Figure 2.29 shows the effect of hill radius of curvature R_H on the critical speed v_c for a car with $h/l = 0.10123$ mm/mm and Figure 2.30 shows the effect of a car's height factor h/l on the critical speed v_c for a circular hill with $R_H = 100$ m.

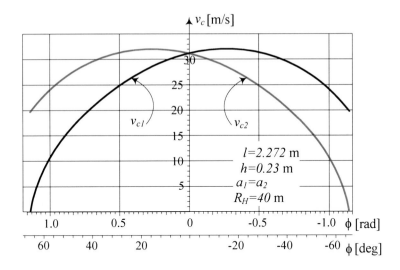

FIGURE 2.28. Critical speeds v_{c_1} and v_{c_2} at different angle ϕ for a specific car and hill.

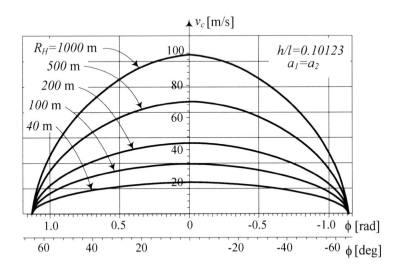

FIGURE 2.29. Effect of hill radius of curvature R_h on the critical speed v_c for a car.

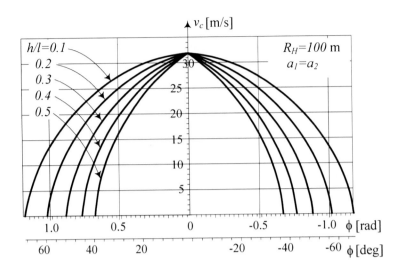

FIGURE 2.30. Effect of a car's height factor h/l on the critical speed v_c for a circular hill.

2.8.2 ★ Vehicles on a Dip

Moving on the concave curve of a hill is called *dipping*. The normal force under the wheels of a dipping vehicle is more than the force on a flat inclined road with the same slope, because of the developed centrifugal force mv^2/R_H in the z-direction.

Figure 2.31 illustrates a dipping vehicle at a point where the hill has a radius of curvature R_H and equivalent slop ϕ. The traction and normal forces under the tires of the vehicle are approximately equal to

$$F_{x_1} + F_{x_2} \approx \frac{1}{2} m \left(a + g \sin \phi \right) \qquad (2.309)$$

$$F_{z_1} \approx \frac{1}{2} mg \left[\left(\frac{a_2}{l} \cos \phi + \frac{h}{l} \sin \phi \right) \right] - \frac{1}{2} ma \frac{h}{l} + \frac{1}{2} m \frac{v^2}{R_H} \frac{a_2}{l} \qquad (2.310)$$

$$F_{z_2} \approx \frac{1}{2} mg \left[\left(\frac{a_1}{l} \cos \phi - \frac{h}{l} \sin \phi \right) \right] + \frac{1}{2} ma \frac{h}{l} + \frac{1}{2} m \frac{v^2}{R_H} \frac{a_1}{l} \qquad (2.311)$$

$$l = a_1 + a_2 \qquad (2.312)$$

Proof. To develop the equations for the traction and normal forces under the tires of a dipping car, we follow the same procedure as a cresting car. The normal and tangential directions of a dipping car, shown in Figure 2.31, are equivalent to the z and x directions respectively. Hence, the governing

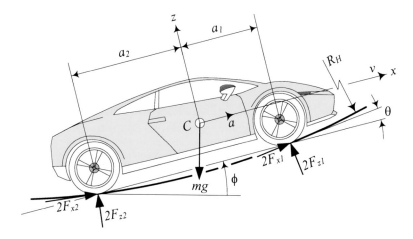

FIGURE 2.31. A dipping vehicle at a point where the hill has a radius of curvature R_h.

equations of motion for the car are

$$\sum F_x = ma \tag{2.313}$$

$$\sum F_z = m\frac{v^2}{R_H} \tag{2.314}$$

$$\sum M_y = 0 \tag{2.315}$$

Expanding these equations yields

$$2F_{x_1}\cos\theta + 2F_{x_2}\cos\theta - mg\sin\phi = ma \tag{2.316}$$

$$-2F_{z_1}\cos\theta - 2F_{z_2}\cos\theta + mg\cos\phi = m\frac{v^2}{R_H} \tag{2.317}$$

$$2F_{z_1}a_1\cos\theta - 2F_{z_2}a_2\cos\theta + 2\left(F_{x_1}+F_{x_2}\right)h\cos\theta$$
$$+2F_{z_1}a_1\sin\theta - 2F_{z_2}a_2\sin\theta - 2\left(F_{x_1}+F_{x_2}\right)h\sin\theta = 0 \tag{2.318}$$

The total traction force $(F_{x_1} + F_{x_2})$ may be eliminated between the first and third equations. Then, the resultant equations provide us with the total traction force $F_{x_1} + F_{x_2}$ and wheel normal forces F_{z_1}, F_{z_2}:

$$F_{x_1} + F_{x_2} = \frac{ma + mg\sin\phi}{2\cos\theta} \tag{2.319}$$

$$F_{z_1} = \frac{1}{2}mg\left[\left(\frac{a_2}{l\cos\theta}\cos\phi + \frac{h(1-\sin 2\theta)}{l\cos\theta\cos 2\theta}\sin\phi\right)\right]$$
$$-\frac{1}{2}ma\frac{h(1-\sin 2\theta)}{l\cos\theta\cos 2\theta} + \frac{1}{2}m\frac{v^2}{R_H}\frac{a_2}{l\cos\theta} \qquad (2.320)$$

$$F_{z_2} = \frac{1}{2}mg\left[\left(\frac{a_1}{l\cos\theta}\cos\phi - \frac{h(1-\sin 2\theta)}{l\cos\theta\cos 2\theta}\sin\phi\right)\right]$$
$$+\frac{1}{2}ma\frac{h(1-\sin 2\theta)}{l\cos\theta\cos 2\theta} + \frac{1}{2}m\frac{v^2}{R_H}\frac{a_1}{l\cos\theta\cos\theta} \qquad (2.321)$$

Assuming very small θ, these forces can be approximated to

$$F_{x_1} + F_{x_2} \approx \frac{1}{2}m(a + g\sin\phi) \qquad (2.322)$$

$$2\frac{F_{z_1}}{mg} \approx \left(\frac{a_2}{l}\cos\phi + \frac{h}{l}\sin\phi\right) - \frac{a}{g}\frac{h}{l} + \frac{v^2/R_H}{g}\frac{a_2}{l} \qquad (2.323)$$

$$2\frac{F_{z_2}}{mg} \approx \left(\frac{a_1}{l}\cos\phi - \frac{h}{l}\sin\phi\right) + \frac{a}{g}\frac{h}{l} + \frac{v^2/R_H}{g}\frac{a_1}{l} \qquad (2.324)$$

■

Example 77 ★ *Wheel loads of a dipping car.*
Consider a car with the following specifications:

$$l = 2272\,\text{mm} \quad w = 1457\,\text{mm} \quad h = 230\,\text{mm} \quad a_1 = a_2$$
$$m = 1500\,\text{kg} \quad v = 15\,\text{m/s} \quad a = 1\,\text{m/s}^2 \qquad (2.325)$$

that is dipping on a hill at a point where the road has

$$R_H = 40\,\text{m} \qquad \phi = 30\,\text{deg} \qquad \theta = 2.5\,\text{deg} \qquad (2.326)$$

The force information on the car is:

$$F_{x_1} + F_{x_2} = 4432.97\,\text{N} \quad F_{z_1} = 4889.1\,\text{N} \quad F_{z_2} = 5711.52\,\text{N}$$
$$F_{z_1} + F_{z_2} = 10600.62\,\text{N} \quad mg = 14715\,\text{N} \quad m\frac{v^2}{R_H} = 8437.5\,\text{N} \qquad (2.327)$$

If we ignore the effect of θ by assuming a very value, then the approximate value of the forces are

$$F_{x_1} + F_{x_2} = 4428.75\,\text{N} \quad F_{z_1} \approx 4846.93\,\text{N} \quad F_{z_2} \approx 5743.6\,\text{N}$$
$$F_{z_1} + F_{z_2} \approx 10590.53\,\text{N} \quad mg = 14715\,\text{N} \quad m\frac{v^2}{R_H} = 8437.5\,\text{N} \qquad (2.328)$$

2.9 Summary

For straight motion of a symmetric rigid vehicle, we may assume the forces on the left wheels are equal to the forces on the right wheels, and simplify the tire force calculation.

When a car is accelerating on an inclined road with slop angle ϕ, the normal forces under the front and rear wheels, F_{z_1}, F_{z_2}, are

$$F_{z_1} = \frac{1}{2}mg\left(\frac{a_2}{l}\cos\phi - \frac{h}{l}\sin\phi\right) - \frac{1}{2}ma\frac{h}{l} \qquad (2.329)$$

$$F_{z_2} = \frac{1}{2}mg\left(\frac{a_1}{l}\cos\phi + \frac{h}{l}\sin\phi\right) + \frac{1}{2}ma\frac{h}{l} \qquad (2.330)$$

$$l = a_1 + a_2 \qquad (2.331)$$

The first parenthesis, $\frac{1}{2}mg\left(\frac{a_1}{l}\cos\phi \pm \frac{h}{l}\sin\phi\right)$ is the static force which depends on slope and mass center position. The term $\pm\frac{1}{2}mg\frac{h}{l}\frac{a}{g}$ is the dynamic force, because it depends on the acceleration a.

2.10   Key Symbols

$a \equiv \ddot{x}$	acceleration
a_{fwd}	front wheel drive acceleration
a_{rwd}	rear wheel drive acceleration
a_1	distance of first axle from mass center
a_2	distance of second axle from mass center
a_i	distance of axle number i from mass center
a_M	maximum acceleration
a, b	arguments for $atan2(a, b)$
A, B, C	constant parameters
b_1	distance of left wheels from mass center
b_1	distance of hinge point from rear axle
b_2	distance of right wheels from mass center
b_2	distance of hinge point from trailer mass center
b_3	distance of trailer axle from trailer mass center
C	mass center of vehicle
C_t	mass center of trailer
d	height
F	force
F_x	traction or brake force under a wheel
F_{x_1}	traction or brake force under front wheels
F_{x_2}	traction or brake force under rear wheels
F_{x_t}	horizontal force at hinge
F_z	normal force under a wheel
F_{z_1}	normal force under front wheels
F_{z_2}	normal force under rear wheels
F_{z_3}	normal force under trailer wheels
F_{z_t}	normal force at hinge
g, $\mathbf{g}$	gravitational acceleration
h	height of C
H	height
I	mass moment of inertia
k_i	vertical stiffness of suspension at axle number i
l	wheel base
m	car mass
m_t	trailer mass
M	moment
R	tire radius
R_f	front tire radius
R_r	rear tire radius
R_H	radius of curvature
t	time
$v \equiv \dot{x}$, $\mathbf{v}$	velocity
v_c	critical velocity

w	track
z_i	deflection of axil number i
x, y, z	vehicle coordinate axes
X, Y, Z	global coordinate axes
θ	road slope
ϕ	road angle with horizon
ϕ_M	ultimate angle, maximum slope angle
ϕ_T	tilting angle
μ	friction coefficient

Subscriptions

dyn	dynamic
f	front
fwd	front-wheel-drive
M	maximum
r	rear
rwd	rear-wheel-drive
st	statics

Exercises

1. Axle load.

   Consider a car is parked on a level road. Find the load on the front and rear axles if

   (a) $m = 1765\,\text{kg}$ $a_1 = 1.22\,\text{m}$ $a_2 = 1.62\,\text{m}$
   (b) $m = 1665\,\text{kg}$ $a_1 = 1.42\,\text{m}$ $a_2 = 1.42\,\text{m}$
   (c) $m = 1245\,\text{kg}$ $a_1 = 1.62\,\text{m}$ $a_2 = 1.22\,\text{m}$

2. Mass center distance ratio.

   Peugeot 907 Concept^{TM} approximately has

   $$m = 1400\,\text{kg} \qquad l = 97.5\,\text{in}$$

   Assume $a_1/a_2 \approx 1.131$ and determine the axles load.

3. Parked car on a level road.

   Consider a parked car on a level road and show that

   $$\frac{F_{x_1}}{F_{x_2}} = \frac{a_2}{a_1}$$

4. Axle load ratio.

   (a) Jeep Commander XK^{TM} approximately has

   $$mg = 5091\,\text{lb} \qquad l = 109.5\,\text{in}$$

   Assume $F_{z_1}/F_{z_2} \approx 1.22$ and determine the axles load.

   (b) BMW $X5^{TM}$ approximately has

   $$m = 2900\,\text{kg} \qquad l = 2933\,\text{mm}$$

   if the front/rear load ratio is $1400/1500$, determine a_1 and a_2.

5. Axle load and mass center distance ratio.

   The wheelbase of the 1981 DeLorean Sportscar^{TM} is

   $$l = 94.89\,\text{in}$$

   Find the axles load if we assume:

   $$a_1/a_2 \approx 0.831 \qquad mg = 3000\,\text{lb}$$

6. Mass center lateral position.

   In Example 44, determine b_1 and b_2 for the rear of the car. Assume that the left side of the car is 4% heavier than the right side.

7. ★ Different front and rear tires.

   Prove Equations (2.35) and (2.36).

8. Mass center height.

   McLaren SLR 722 Sportscar^{TM} has the following specifications.

$$m \;=\; 1649\,\text{kg} \qquad\qquad l = 2700\,\text{mm}$$
$$\text{front tire} \;:\; 255/35ZR19 \qquad \text{rear tire} : 295/30ZR19$$

   When the front axle is lifted $H = 540\,\text{mm}$, assume that

$$a_1 = a_2 \qquad F_{z_2} = 0.68 mg$$

   What is the height h of the mass center?

9. ★ Mass center height limit.

   In the equation of the height of the mass center, h

$$h = R + \left(a_2 - \frac{2F_{z_1}}{mg} l \right) \cot \phi$$

   when $\phi \to 0$ then, $\cot \phi \to \infty$. Explain why it is wrong to say that $h \to \infty$.

10. ★ Mass center height in terms of H.

    It may be more practical if in the height calculation experiment of Example 45, we measure d, the height of the cylinder, instead of measuring the angle ϕ. Use Figure 2.32 and re-calculate h in terms of d instead of ϕ.

11. A parked car on an uphill road.

    Specifications of Lamborghini Gallardo^{TM} are

$$m = 1430\,\text{kg} \qquad l = 2560\,\text{mm}$$

    (a) Assume

$$a_1 = a_2 \qquad h = 520\,\text{mm}$$

    and determine the forces F_{z_1}, F_{z_2}, and F_{x_2} if the car is parked on an uphill with $\phi = 30\,\text{deg}$ and the hand brake is connected to the rear wheels.

    (b) What would be the maximum road grade ϕ_M, that the car can be parked, if $\mu_{x_2} = 1$.

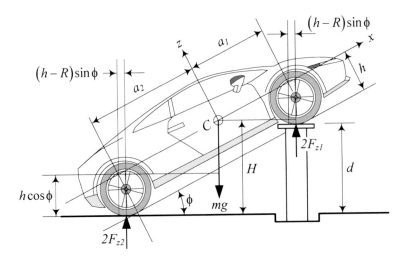

FIGURE 2.32. Determination of the height of mass center using the height of cylinder d instead of the angle ϕ.

12. Parked on an uphill road.

Rolls-Royce Phantom^{TM} has

$$m = 2495 \, \text{kg} \qquad l = 3570 \, \text{mm}$$

Assume the car is parked on an uphill road and

$$a_1 = a_2 \qquad h = 670 \, \text{mm} \qquad \phi = 30 \, \text{deg}$$

Determine the forces under the wheels if the car is

(a) front wheel braking
(b) rear wheel braking
(c) four wheel braking.

13. A parked car on a downhill road.

Solve Exercise 11 if the car is parked on a downhill road.

14. Forces at tilting angle.

Determine the brake force $2F_{x_2}$ and normal force F_{z_2} when a rear park brake car is parked uphill.

15. ★ Angle of an uphill road.

Consider a parked car on an uphill road and show that if the forces under the front and rear wheels, F_{z_1} and F_{z_2}, are given then the inclined angle of the road is:

$$\phi = \arctan\left(\frac{a_2 - a_1\left(F_{z_1}/F_{z_2}\right)}{h + h\left(F_{z_1}/F_{z_2}\right)}\right)$$

16. Maximum acceleration.

    Honda CR-V^{TM} is a midsize SUV car with

    $$m = 1550\,\text{kg} \qquad l = 2620\,\text{mm}$$

    Assume

    $$a_1 = a_2 \qquad h = 720\,\text{mm} \qquad \mu_x = 0.8$$

    and determine the maximum acceleration of the car if

    (a) the car is rear-wheel drive

    (b) the car is front-wheel drive

    (c) the car is four-wheel drive.

17. ★ Acceleration on a level road.

    Consider an accelerating car on a level road. Show that if the forces under the front and rear wheels, F_{z_1} and F_{z_2}, are given then the acceleration of the car is

    $$a = \frac{a_2 - a_1\left(F_{z_1}/F_{z_2}\right)}{h + h\left(F_{z_1}/F_{z_2}\right)}\,g$$

18. ★ Slope of a/g.

    Compare the slopes of a_{fwd}/g, a_{rwd}/g, a_{fwb}/g, and a_{rwb}/g to determine if there is any equation among them.

19. ★ Acceleration on an inclined road.

    Consider an accelerating car on an inclined road. Assume that the forces under the front and rear wheels, F_{z_1} and F_{z_2}, are given. Determine the acceleration of the car a in terms of F_{z_1}/F_{z_2}.

20. Minimum time for $0 - 100\,\text{km/h}$.

    RoadRazer^{TM} is a light weight rear-wheel drive sportscar with

    $$m = 300\,\text{kg} \qquad l = 2286\,\text{mm} \qquad h = 260\,\text{mm}$$

    Assume $a_1 = a_2$. If the car can reach the speed $0 - 100\,\text{km/h}$ in $t = 3.2\,\text{s}$, what would be the minimum friction coefficient?

21. Axle load of an all-wheel drive car.

   Acura Courage^{TM} is an all-wheel drive car with

$$m = 2058.9 \, \text{kg} \qquad l = 2750.8 \, \text{mm}$$

   Assume $a_1 = a_2$ and $h = 760 \, \text{mm}$. Determine the axles load if the car is accelerating at $a = 1.7 \, \text{m}/\text{s}^2$.

22. ★ A car with a trailer.

   Volkswagen Touareg^{TM} is an all-wheel drive car with

$$m = 2268 \, \text{kg} \qquad l = 2855 \, \text{mm}$$

   Assume $a_1 = a_2$ and the car is pulling a trailer with

$$
\begin{aligned}
m_t &= 600 \, \text{kg} & h_1 &= h_2 \\
b_1 &= 855 \, \text{mm} & b_2 &= 1350 \, \text{mm} & b_3 &= 150 \, \text{mm}
\end{aligned}
$$

   If the car is accelerating on a level road with acceleration $a = 2 \, \text{m}/\text{s}^2$, what would be the forces at the hinge.

23. A parked car on a banked road.

   Cadillac Escalade^{TM} is a SUV car with

$$
\begin{aligned}
m &= 2569.6 \, \text{kg} & l &= 2946.4 \, \text{mm} \\
w_f &= 1732.3 \, \text{mm} & w_r &= 1701.8 \, \text{mm}
\end{aligned}
$$

   Assume $b_1 = b_2$, $h = 940 \, \text{mm}$, and use an average track to determine the wheels load when the car is parked on a banked road with $\phi = 12 \, \text{deg}$.

24. ★ A parked car on a banked road with $w_f \neq w_r$.

   Determine the wheels load of a parked car on a banked road, if the front and rear tracks of the car are different.

25. ★ Angle of a banked road.

   Consider a parked car on a banked road and show that if the forces under the left and right wheels, F_{z_1} and F_{z_2}, are given then the bank angle of the road is:

$$\phi = \arctan \left(\frac{b_2 - b_1 \left(F_{z_1}/F_{z_2} \right)}{h + h \left(F_{z_1}/F_{z_2} \right)} \right)$$

26. Optimal traction force.

   Mitsubishi Outlander^{TM} is an all-wheel drive SUV car with

$$m = 1599.8 \, \text{kg} \qquad l = 2669.6 \, \text{mm} \qquad w = 1539.3 \, \text{mm}$$

Assume
$$a_1 = a_2 \qquad h = 760 \, \text{mm} \qquad \mu_x = 0.75$$
and find the optimal traction force ratio F_{x_1}/F_{x_2} to reach the maximum acceleration.

27. ★ A three-axle car.

Citroën Cruise Crosser^{TM} is a three-axle off-road pick-up car. Assume

$$
\begin{aligned}
m &= 1800 \, \text{kg} \\
a_1 &= 1100 \, \text{mm} & a_2 &= 1240 \, \text{mm} & a_3 &= 1500 \, \text{mm} \\
k_1 &= 12800 \, \text{N/m} & k_2 &= 14000 \, \text{N/m} & k_3 &= 14000 \, \text{N/m}
\end{aligned}
$$

and find the axles load on a level road when the car is moving with no acceleration.

28. ★ Car on top of a crest.

Consider a car on top of a crest and show that

$$\frac{F_{z_1}}{F_{z_2}} \approx \frac{a_2 \left(v^2/R_H - g \right) + ah}{a_1 \left(v^2/R_H - g \right) - ah}$$

29. ★ Car at the bottom of a dip.

Consider a car at the bottom of a dip and show that

$$\frac{F_{z_1}}{F_{z_2}} \approx \frac{a_2 \left(v^2/R_H + g \right) - ah}{a_1 \left(v^2/R_H + g \right) + ah}$$

3

Tire Dynamics

The tire is the main component of a vehicle interacting with the road. The performance of a vehicle is mainly influenced by the characteristics of its tires. Tires affect a vehicle's handling, traction, ride comfort, and fuel consumption. To understand its importance, it is enough to remember that a vehicle can maneuver only by longitudinal, vertical, and lateral force systems generated under the tires.

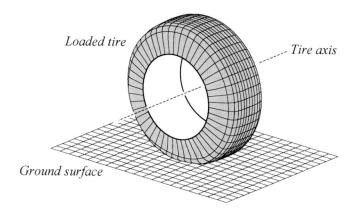

FIGURE 3.1. A vertically loaded stationary tire.

Figure 3.1 illustrates a vertically loaded stationary tire. To model the tire-road interactions, we determine the tireprint and describe the forces distributed on the tireprint.

3.1 Tire Coordinate Frame and Tire Force System

To describe the tire-road interaction and its force system, we assume a flat ground and attach a Cartesian coordinate frame at the center of the tireprint as shown in Figure 3.2. The x-axis is along the intersection line of the tire-plane and the ground. The *tire plane* is the plane made by narrowing the tire to a flat disk. The z-axis is perpendicular to the ground and upward, and the y-axis makes the coordinate system a right-hand triad.

To show the tire orientation, we use two angles: camber angle γ and sideslip angle α. The *camber angle* is the angle between the tire-plane and

R.N. Jazar, *Vehicle Dynamics: Theory and Application*,
DOI 10.1007/978-1-4614-8544-5_3, © Springer Science+Business Media New York 2014

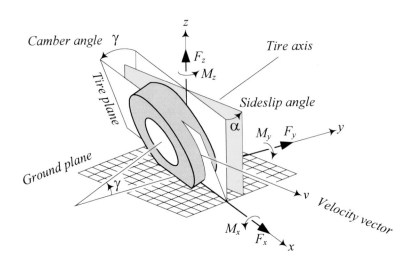

FIGURE 3.2. Tire coordinate system.

the vertical plane measured about the x-axis. The camber angle can be recognized better in a front view as shown in Figure 3.3(a). The *sideslip angle* α, or simply *sideslip*, is the angle between the velocity vector **v** and the x-axis measured about the z-axis. The sideslip can be recognized better in a top view, as shown in Figure 3.3(b).

The resultant force system that a tire receives from the ground is assumed to be located at the center of the tireprint and can be decomposed along x, y, and z axes. Therefore, the interaction of a tire with the road generates a three dimensional (3D) force system including three forces and three moments, as shown in Figure 3.2.

1. *Longitudinal force F_x.* It is the force acting along the x-axis. The resultant longitudinal force $F_x > 0$ if the car is accelerating, and $F_x < 0$ if the car is braking. Longitudinal force is also called *forward force*.

2. *Normal force F_z.* It is the vertical force, normal to the ground plane. The resultant normal force $F_z > 0$ if it is upward. The traditional tires and pavements are unable to provide $F_z < 0$. Normal force is also called *vertical force* or *wheel load*.

3. *Lateral force F_y.* It is the force, tangent to the ground and orthogonal to both F_x and F_z. The resultant lateral force $F_y > 0$ when it is in the y-direction.

4. *Roll moment M_x.* It is the longitudinal moment about the x-axis. The resultant roll moment $M_x > 0$ when it tends to turn the tire

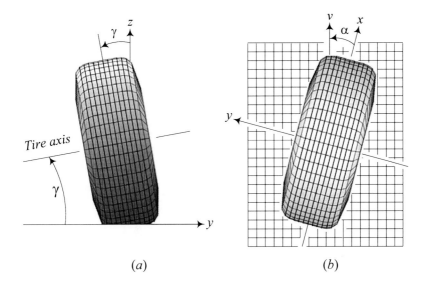

(a)   (b)

FIGURE 3.3. Camber and sideslip angles illustration. (a) Front view of a tire and measurment of the camber angle. (b) Top view of a tire and measurment of the sideslip angle.

about the x-axis. The roll moment is also called the *bank moment*, *tilting torque*, or *overturning moment*.

5. *Pitch moment M_y.* It is the lateral moment about the y-axis. The resultant pitch moment $M_y > 0$ when it tends to turn the tire about the y-axis and move it forward. The pitch moment is also called *rolling resistance torque*.

6. *Yaw moment M_z.* It is the upward moment about the z-axis. The resultant yaw moment $M_z > 0$ when it tends to turn the tire about the z-axis. The yaw moment is also called the *aligning moment, self aligning moment*, or *bore torque*.

These are the force system that are applied on the tire from the ground. All other possible forces on a wheel are at the wheel axle. The driving or braking moment applied to the tire from the vehicle about the tire axis is called *wheel torque T.*

Example 78 *Origin of tire coordinate frame.*

For a cambered tire, it is not always possible to find or define a center point for the tireprint to be used as the origin of the tire coordinate frame. It is more practical to set the origin of the tire coordinate frame at the center of the intersection line between the tire-plane and the ground at zero camber and sideslip angles. So, the origin of the tire coordinate frame is at

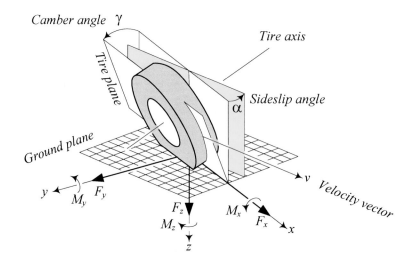

FIGURE 3.4. SAE tire coordinate system.

the center of the tireprint when the tire is standing upright and stationary on a flat road.

Example 79 ★ *SAE tire coordinate system.*

The tire coordinate system adopted by the Society of Automotive Engineers (SAE) is shown in Figure 3.4. The origin of the coordinate system is at the center of the tireprint when the tire is standing stationary. The x-axis is at the intersection of the tire-plane and the ground plane. The z-axis is downward and perpendicular to the tireprint. The y-axis is on the ground plane and goes to the right to make the coordinate frame a right-hand frame.

Because of this coordinate arrangement, the positive direction of some angles and forces are in negative direction of the coordinate axes. Besides some miss-match and confusion, sometimes authors consider forces positive when they are in the negative direction of the coordinate axes. However, the signs and positive directions of forces and angles are not commonly accepted among authors who are using SAE coordinate and they usually refer to right and left direction instead of coordinate axes. For example, the sideslip angle α is considered positive if the tire is slipping to the right. Furthermore, in SAE convention, the camber angle for the left and right tires of a vehicle have opposite signs. So, the camber angle of the left tire is positive when the tire leans to the right and the camber angle of the right tire is positive when the tire leans to the left.

The SAE coordinate system is as good as the coordinate system in Figure 3.2 and may be used alternatively. However, having the z-axis directed downward is sometimes inefficient and confusing.

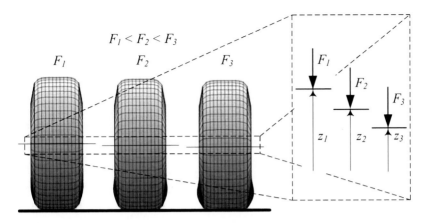

FIGURE 3.5. Vertically loaded tire at zero camber.

3.2   Tire Stiffness

As an applied approximation, the vertical tire force F_z can be calculated as a linear function of the normal tire deflection $\triangle z$ measured at the tire center.

$$F_z = k_z \triangle z \tag{3.1}$$

The coefficient k_z is called *tire stiffness* in the z-direction. Similarly, the reaction of a tire to a lateral and a longitudinal force can be approximated as

$$F_x = k_x \triangle x \tag{3.2}$$
$$F_y = k_y \triangle y \tag{3.3}$$

where the coefficient k_x and k_y are called *tire stiffness* in the x and y directions.

Proof. The deformation behavior of tires to the applied forces in any three directions x, y, and z are the first important tire characteristics in tire dynamics. Calculating the tire stiffness is generally based on experiment and therefore, they are dependent on the tire's mechanical properties, as well as environmental characteristics.

Consider a vertically loaded tire on a stiff and flat ground as shown in Figure 3.5. The tire will deflect under the load and generate a pressurized contact area to balance the vertical load.

Figure 3.6 depicts a sample of experimental stiffness curve in the $(F_z, \triangle z)$ plane. The curve may be expressed by a curve fitted mathematical function

$$F_z = f(\triangle z) \tag{3.4}$$

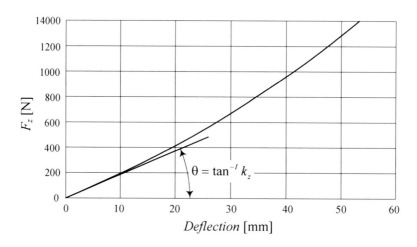

FIGURE 3.6. A sample tire vertical stiffness curve.

however, we usually use a linear approximation for the range of the usual application.

$$F_z = \frac{\partial f}{\partial (\triangle z)} \triangle z \qquad (3.5)$$

The coefficient $\frac{\partial f}{\partial(\triangle z)}$ is the slope of the experimental stiffness curve at zero and is shown by the stiffness coefficient k_z

$$k_z = \tan \theta = \lim_{\triangle z \to 0} \frac{\partial f}{\partial (\triangle z)} \qquad (3.6)$$

Therefore, the normal tire deflection $\triangle z$ remains proportional to the vertical tire force F_z by definition.

$$F_z = k_z \triangle z \qquad (3.7)$$

The stiffness curve can be influenced by many parameters. The most effective one is the tire inflation pressure. The tire can apply only pressure forces to the road, so normal force is restricted to $F_z > 0$.

Lateral and longitudinal force/deflection behavior is also determined experimentally by applying a force in the appropriate direction. The lateral and longitudinal forces are limited by the sliding force when the tire is vertically loaded. Figure 3.7 depicts a sample of longitudinal and lateral stiffness curves compared to a vertical stiffness curve.

The practical range of tires' longitudinal and lateral stiffness curves is the linear part and may be estimated by linear equations.

$$F_x = k_x \triangle x \qquad (3.8)$$
$$F_y = k_y \triangle y \qquad (3.9)$$

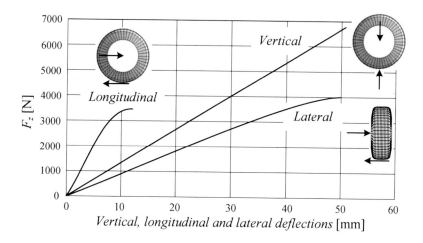

FIGURE 3.7. Vertical, longitudinal, and lateral stiffness curves.

The coefficients k_x and k_y are called the *tire stiffness* in the x and y directions. They are measured by the slope of the experimental stiffness curves in the $(F_x, \triangle x)$ and $(F_y, \triangle y)$ planes.

$$k_x = \lim_{\triangle x \to 0} \frac{\partial f}{\partial (\triangle x)} \tag{3.10}$$

$$k_y = \lim_{\triangle y \to 0} \frac{\partial f}{\partial (\triangle y)} \tag{3.11}$$

When the longitudinal and lateral forces increase, parts of the tireprint creep and slide on the ground until the whole tireprint starts sliding. At this point, the applied force saturates and reaches its maximum supportable value.

When we measure displacements from the equilibrium position at zero deflection, we may respectively show $\triangle x$, $\triangle y$, $\triangle z$ by x, y, z. Generally, a tire is most stiff in the longitudinal direction, x, and least stiff in the lateral direction, y.

$$k_x > k_z > k_y \tag{3.12}$$

Figure 3.8 illustrates tire deformation under a lateral and a longitudinal force. ■

Example 80 ★ *Nonlinear tire stiffness.*
    In a better modeling, the vertical tire force F_z is a function of the normal tire deflection z, and deflection rate $\dot{z}$.

$$F_z = F_z\,(z, \dot{z}) = F_{z_s} + F_{z_d} \tag{3.13}$$

In a first approximation we may assume F_z is a linear combination of a static and a dynamic part. The static part is a nonlinear, usually a polynomial

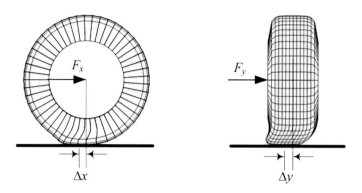

FIGURE 3.8. Illustration of laterally and longitudinally tire deformation.

function of the vertical tire deflection and the dynamic part is proportional to the vertical speed of the tire.

$$F_{z_s} = k_1 z + k_2 z^2 + k_3 z^3 \qquad (3.14)$$

$$F_{z_d} = k_4 \dot{z} \qquad (3.15)$$

The constants k_1, k_2, and k_3 are calculated from the first, second, and third slopes of the experimental stiffness curve in the (F_z, z) plane. The constant k_4 is the first slope of the curve in the $(F_z, \dot{z})$ plane, which indicates the tire damping.

$$k_1 = \frac{\partial F_z(0)}{\partial z} \qquad k_2 = \frac{1}{2!} \frac{\partial^2 F_z(0)}{\partial z^2} \qquad k_2 = \frac{1}{3!} \frac{\partial^2 F_z(0)}{\partial z^3} \qquad (3.16)$$

$$k_4 = \frac{\partial F_z(0)}{\partial \dot{z}} \qquad (3.17)$$

The value of $k_1 = 200000 \, \mathrm{N/m}$ is a good approximation for a $205/50R15$ passenger car tire, and $k_1 = 1200000 \, \mathrm{N/m}$ is a good approximation for a $X31580R22.5$ truck tire.

Tires with a larger number of plies have higher damping, because the damping is a result of the plies' internal friction. Tire damping decreases by increasing speeds.

Example 81 ★ *Hysteresis effect.*

Rubber is a viscoelastic material, and therefore, their loading and unloading stiffness curves are not exactly the same. Figure 3.9 illustrates a sample of loading and unloading of a rubbery tire. The curves make a loop with the unloading curve sitting below the loading. The area within the loop is the amount of dissipated energy during loading and unloading. As a rolling tire rotates under the weight of a vehicle, it experiences repeated cycles of deformation and recovery, and it dissipates energy loss as heat. Such a behavior is a common property of hysteretic material and is called hysteresis.

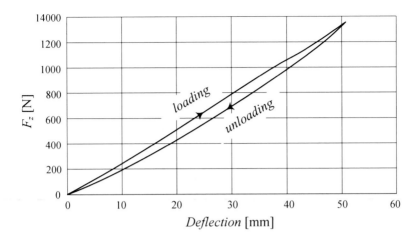

FIGURE 3.9. Histeresis loop in a vertically loading and unloading tire.

*So, **hysteresis** is a characteristic of a deformable material such as rubber, that the energy of deformation is greater than the energy of recovery. The amount of dissipated energy depends on the mechanical characteristics of the tire. Hysteretic energy loss in rubber decreases as temperature increases.*

The hysteresis effect causes a loaded rubber not to rebound fully after load removal. Consider a high hysteresis race car tire turning over road irregularities. The deformed tire recovers slowly, and therefore, it cannot push the tireprint tail on the road as hard as the tireprint head. The difference in head and tail pressures causes a resistance force, which is reason for the rolling resistance.

Race cars have high hysteresis tires to increase friction and traction. Street cars have low hysteresis tires to reduce the rolling resistance and low operating temperature. Hysteresis level of tires inversely affect the stopping distance. A high hysteresis tire makes the stopping shorter, however, it wears rapidly and has a shorter life time.

3.3 Effective Radius

Consider a vertically loaded wheel that is turning on a flat surface as shown in Figure 3.10. The *effective radius* of the wheel R_w, which is also called a *rolling radius*, is defined by

$$R_w = \frac{v_x}{\omega} \tag{3.18}$$

where, v_x is the forward velocity and $\omega = \omega_w$ is the angular velocity of the wheel. The effective radius R_w is approximately equal to

$$R_w \approx R_g - \frac{R_g - R_h}{3} \tag{3.19}$$

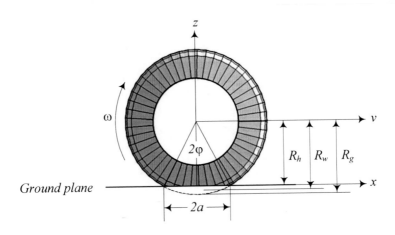

FIGURE 3.10. Effective radius R_w compared to tire radius R_g and loaded height R_h.

and is a number between the unloaded or *geometric radius* R_g and the *loaded height* R_h.

$$R_h < R_w < R_g \tag{3.20}$$

Proof. An effective radius $R_w = v_x/\omega_w$ is defined by measuring the tire's angular velocity $\omega = \omega_w$ and forward velocity v_x. When the tire rolls, each part of the circumference is flattened as it passes through the contact area. A practical estimate of the effective radius can be made by substituting the arc with the straight length of tireprint. The tire vertical deflection is

$$R_g - R_h = R_g \left(1 - \cos\varphi\right) \tag{3.21}$$

and

$$R_h = R_g \cos\varphi \tag{3.22}$$
$$a = R_g \sin\varphi \tag{3.23}$$

If the motion of the tire is compared to the rolling of a rigid disk with radius R_w, then the tire must move a distance $a \approx R_w\varphi$ for an angular rotation φ.

$$a = R_g \sin\varphi \approx R_w\varphi \tag{3.24}$$

Hence,

$$R_w = \frac{R_g \sin\varphi}{\varphi} \tag{3.25}$$

Expanding $\frac{\sin\varphi}{\varphi}$ in a Taylor series shows that

$$R_w = R_g \left(1 - \frac{1}{6}\varphi^2 + O\left(\varphi^4\right)\right) \tag{3.26}$$

Using Equation (3.21) we may approximate

$$\cos\varphi \approx 1 - \frac{1}{2}\varphi^2 \tag{3.27}$$

$$\varphi^2 \approx 2\left(1 - \cos\varphi\right) \approx 2\left(1 - \frac{R_h}{R_g}\right) \tag{3.28}$$

and therefore,

$$R_w \approx R_g\left(1 - \frac{1}{3}\left(1 - \frac{R_h}{R_g}\right)\right) = \frac{2}{3}R_g + \frac{1}{3}R_h \tag{3.29}$$

Because R_h is a function of tire load F_z,

$$R_h = R_h\left(F_z\right) = R_g - \frac{F_z}{k_z} \tag{3.30}$$

the effective radius R_w is also a function of the tire load. The angle φ is called *tireprint angle* or *tire contact angle*.

The vertical stiffness of radial tires is less than non-radial tires under the same conditions. So, the loaded height of radial tires, R_h, is less than the non-radials'. However, the effective radius of radial tires R_w, is closer to their unloaded radius R_g. As a good estimate, for a non-radial tire, $R_w \approx 0.96R_g$, and $R_h \approx 0.94R_g$, while for a radial tire, $R_w \approx 0.98R_g$, and $R_h \approx 0.92R_g$.

Generally speaking, the effective radius R_w depends on the type of tire, stiffness, load conditions, inflation pressure, and tire's forward velocity. ■

Example 82 *Tire rotation.*

The geometric radius of a tire P235/75R15 is $R_g = 366.9$ mm, because

$$h_T = 235 \times 75\% = 176.25\,\text{mm} \approx 6.94\,\text{in} \tag{3.31}$$

and therefore,

$$R_g = \frac{2h_T + 15}{2} = \frac{2 \times 6.94 + 15}{2} = 14.44\,\text{in} \approx 366.9\,\text{mm} \tag{3.32}$$

Consider a vehicle with such a tire is traveling at a high speed such as $v = 50$ m/s $= 180$ km/h ≈ 111.8 mi/h. The tire is radial, and therefore the effective tire radius R_w is approximately equal to

$$R_w \approx 0.98R_g \approx 359.6\,\text{mm} \tag{3.33}$$

After moving a distance $d = 100$ km, this tire must have been turned $n_1 = 44259$ times because

$$n_1 = \frac{d}{\pi D} = \frac{100 \times 10^3}{2\pi \times 359.6 \times 10^{-3}} = 44259 \tag{3.34}$$

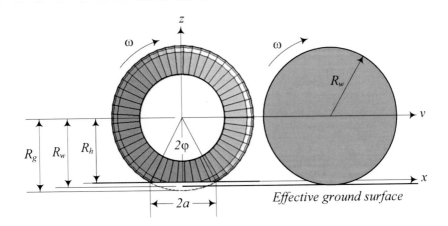

FIGURE 3.11. Equivalent tire is rolling under on the equivalent ground surface below the ground.

Now assume the vehicle travels the same distance $d = 100\,\text{km}$ at a low inflation pressure, such that the effective radius of the tire remained close to the loaded radius

$$R_w \approx R_h \approx 0.92 R_g = 330.8\,\text{mm} \tag{3.35}$$

This tire must turn $n_2 = 48112$ times to travel $d = 100\,\text{km}$, because,

$$n_2 = \frac{d}{\pi D} = \frac{100 \times 10^3}{2\pi \times 330.8 \times 10^{-3}} = 48112 \tag{3.36}$$

Example 83 *Tire is rolling under the ground.*
The equivalent radius of a rolling loaded tire, R_w, is, by definition, the radius of a solid disc when roles with the same angular velocity of the tire, ω, moves with the same velocity, v_x, as the tire.

$$v_x = R_w \omega \tag{3.37}$$

However, because $R_w > R_h$ the equivalent disc must be rolling on a solid surface below the ground by the distance $R_w - R_h$ as is shown in Figure 3.11.

Example 84 *Compression and expansion of tires in the tireprint zone.*
Because of longitudinal deformation, the peripheral velocity of any point of the tread varies periodically. When it gets close to the starting point of the tireprint, it slows down and a circumferential compression occurs. The tire treads are compressed in the first half of the tireprint and gradually expanded in the second half. The treads in the tireprint contact zone almost stick to the ground. Assuming the relative velocity of the tire with respect

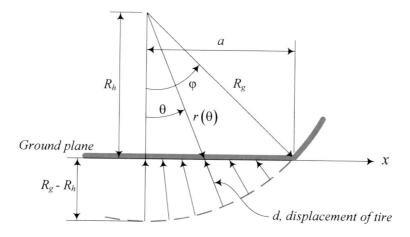

FIGURE 3.12. Radial motion of the tire peripheral points in the contact area.

to the ground is zero at the centerpoint of the tireprint, the relative speed of the circumferential elements of the tire varies between zero to almost double of the forward velocity of the tire center v_x at the top.

Example 85 ★ *Radial motion of tire's peripheral points in tireprint.*

The radial displacement of a tire's peripheral elements during road contact may be modeled by a function

$$d = d(\theta) \tag{3.38}$$

We assume that a peripheral element of the tire moves along only the radial direction during contact with the ground, as shown in Figure 3.12.

Let us show a radius at an angle θ, by $r = r(\theta)$. Knowing that

$$\cos\theta = \frac{R_h}{r} \qquad \cos\varphi = \frac{R_h}{R_g} \tag{3.39}$$

where φ is the half of the contact angle, we find

$$r = R_g \frac{\cos\varphi}{\cos\theta} \tag{3.40}$$

Thus, the displacement function is

$$d = R_g - r(\theta) = R_g \left(1 - \frac{\cos\varphi}{\cos\theta}\right) \qquad -\varphi < \theta < \varphi \tag{3.41}$$

Example 86 ★ *Tread travel.*

Let us follow an element of tire tread with respect to the center of the tire when it travels around the spin axis and the vehicle moves at a constant speed. Although the wheel is turning at constant angular velocity ω, the tread does not travel at constant speed. At the top of the tire, the radius is equal to

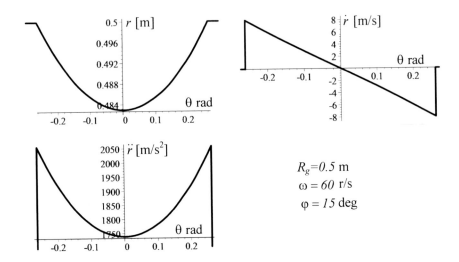

FIGURE 3.13. Radial displacement, velocity, and acceleration of tire treads in the tireprint.

the unloaded geometric radius R_g and the speed of the tread is $R_g\omega$ relative to the wheel center. As the tire turns, the tread approaches the leading edge of tireprint, and slows down. The tread is compacted radially, and gets squeezed in the heading part of the tireprint area. Then, it is stretched out and unpacked in the tail part of the tireprint as it moves to the tail edge. At the center of the tireprint, the tread velocity is $R_h\omega$.

The variable radius of a tire during the motion through the tireprint is

$$r = R_g \frac{\cos\varphi}{\cos\theta} \qquad -\varphi < \theta < \varphi \qquad (3.42)$$

where φ is the half of the contact angle, and θ is the angular rotation of the tire, as shown in Figure 3.12. The angular velocity of the tire is $\omega = -\dot\theta$ and is assumed to be constant. Then, the radial velocity $\dot r$ and acceleration $\ddot r$ of the tread with respect to the wheel center are

$$\dot r = -R_g\omega\cos\varphi\frac{\sin\theta}{\cos^2\theta} \qquad (3.43)$$

$$\ddot r = R_g\omega^2\frac{\cos\varphi}{\cos^3\theta}\left(2 - \cos^2\theta\right) \qquad (3.44)$$

Figure 3.13 depicts r, $\dot r$, and $\ddot r$ for a sample tire with

$$R_g = 0.5\,\mathrm{m} \qquad \varphi = 15\,\mathrm{deg} \qquad \omega = 60\,\mathrm{rad/s} \qquad (3.45)$$

The minimum r happens at $\theta = 0$ which is $1.7\,\mathrm{cm}$ shorter than R_g.

$$R_h = r_{\min} = 0.5\frac{\cos\dfrac{\pi}{12}}{\cos 0} = 0.483\,\mathrm{m} \qquad (3.46)$$

Considering the time $t = \theta/\omega = \frac{\pi}{12}/60 = 4.363 \times 10^{-3}$ s that takes for a tire element to pass the tireprint, the maximum radial speed and acceleration of the tire tread elements would respectively be $8\,\mathrm{m/s}$ and $2050\,\mathrm{m/s^2}$.

Example 87 ★ *Impossible kinematics for sticking tire.*

Consider a tread element of a rolling tire with a constant angular velocity. Ideally the tread element sticks to the ground in the tireprint and keeps its constant velocity with respect to the center of the tire. These conflicting conditions are expressed by

$$\omega = -\dot{\theta} = cte \qquad \dot{x} = v = R_h\omega = cte \qquad (3.47)$$

Taking a derivative of the kinematic constraint equations

$$x = r\sin\theta \qquad R_h = r\cos\theta \qquad -\varphi < \theta < \varphi \qquad (3.48)$$

provides us with

$$v = \dot{r}\sin\theta - r\omega\cos\theta \qquad 0 = \dot{r}\cos\theta + r\omega\sin\theta \qquad (3.49)$$

Eliminating $\dot{r}$ between them yields

$$r = -\frac{v\cos\theta}{\omega} = -R_h\cos\theta = -R_g\cos\varphi\cos\theta \qquad (3.50)$$

Equation (3.50) is not compatible with (3.42) and (3.47), indicating that the conditions (3.47) cannot both be satisfied. So, either $\omega = -\dot{\theta} \neq cte$ or $\dot{x} = v \neq cte$. As, it is possible to keep the angular velocity of the tire constant, $\omega = cte$, the tire tread cannot stick to the ground and will slide on the ground. Therefore,

$$\dot{x} = v \neq cte \qquad (3.51)$$

Example 88 ★ *Velocity analysis in tireprint.*

Consider an element of a rolling tire tread. The element's velocity relative to the center of the tire is a tangential vector with speed $v = R_g\omega$ as long as it is not in the tireprint zone. As soon as the element enters the tireprint, its radius changes with the rate of $\dot{r} = -R_g\omega\cos\varphi\frac{\sin\theta}{\cos^2\theta}$. The element's velocity has then two components: a radial speed of $\dot{r}$ and a tangential speed of $v = r\omega$. As the element must lay on the ground surface, the resultant velocity vector of the element must be along the surface. Figure 3.14 illustrates the kinematics of the system.

The speed of the tread element in tireprint is therefore equal to

$$\begin{aligned} v &= \sqrt{\dot{r}^2 + r^2\omega^2} = \sqrt{\left(-R_g\omega\cos\varphi\frac{\sin\theta}{\cos^2\theta}\right)^2 + \left(R_g\frac{\cos\varphi}{\cos\theta}\omega\right)^2} \\ &= R_g\omega\frac{\cos\varphi}{\cos^2\theta} \qquad -\varphi < \theta < \varphi \end{aligned} \qquad (3.52)$$

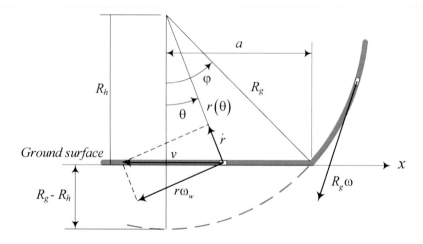

FIGURE 3.14. Velocity analysis of tire tread in tireprint.

There is a constraint equation between the velocity of the ground, v_G, and the angular velocity ω, which allows us to calculate the ground velocity.

$$\frac{\omega}{\varphi} = \frac{v_G}{a} \qquad a = R_g \sin \varphi \tag{3.53}$$

Showing the constant speed of the ground by $v_G = a\omega/\varphi$, we may determine the relative velocity v_{rel} of the tire tread and the ground as

$$v_{rel} = v - v_G = R_g \omega \frac{\cos \varphi}{\cos^2 \theta} - R_g \omega \frac{\sin \varphi}{\varphi} = R_g \omega \left(\frac{\cos \varphi}{\cos^2 \theta} - \frac{\sin \varphi}{\varphi} \right) \tag{3.54}$$

Figure 3.15 depicts velocity v for a sample tire at a high speed with

$$R_g = 0.5 \, \text{m} \qquad \varphi = 15 \, \text{deg} \qquad \omega = 60 \, \text{rad/s} \tag{3.55}$$

and Figure 3.16 illustrates the relative velocity v_{rel} for the same tire with different tireprint angle φ. Theoretically, as long as the tread is not in tireprint zone the velocity of tire tread is $v = R_g \omega$. As soon as the element enters the tireprint, its speed jumps to the higher value $v = R_g \omega / \cos \varphi$. Then the speed gradually decreases to the lower value $v = R_g \omega \cos \varphi$ at the center of tireprint. The speed will then increase to $v = R_g \omega / \cos \varphi$ at the moment it exits the tireprint and jumps back to $v = R_g \omega$ after that. Practically, the sharp edges of the velocity profile would be substituted by small curves.

Example 89 ★ *Displacement analysis in tireprint.*
  We may express Equation (3.52) as a differential equation

$$dx = R_g \omega \frac{\cos \varphi}{\cos^2 \theta} dt = R_g \frac{\cos \varphi}{\cos^2 \theta} d\theta \tag{3.56}$$

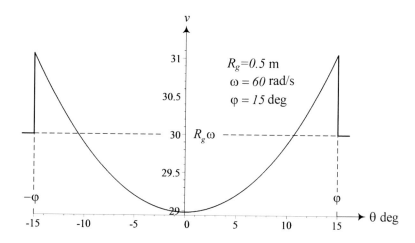

FIGURE 3.15. Velocity profile of a tire tread element in tireprint zone.

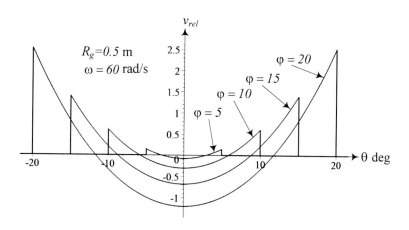

FIGURE 3.16. Relative velocity of a tire tread element with respect to the ground in tireprint zone.

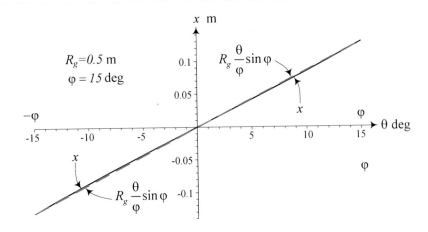

FIGURE 3.17. Horizontal displacement of tire tread, x, and the contact-free displacement, $R_g\theta$.

and determine the horizontal displacement of the tire tread by integration.

$$x = R_g \cos\varphi \tan\theta \qquad -\varphi < \theta < \varphi \qquad (3.57)$$

Comparing the tread displacement x with respect to the ground displacement x_G

$$x_G = R_g \frac{\theta}{\varphi} \sin\varphi \qquad (3.58)$$

provides us with the relative displacement of tire and ground, x_{rel}.

$$
\begin{aligned}
x_{rel} &= x - x_G = R_g \cos\varphi \tan\theta - R_g \frac{\theta}{\varphi}\sin\varphi \\
&= R_g \cos\varphi \left(\tan\theta - \frac{\theta}{\varphi}\tan\varphi \right) \qquad (3.59)
\end{aligned}
$$

Interestingly, the tire tread displacement in tireprint is independent of the angular velocity. Therefore, the horizontal displacement of tire tread remains the same at different time rate for different angular velocity. Figure 3.17 illustrates the tread displacement, x, for a sample tire, and Figure 3.18 depicts the relative displacement of tire tread and the ground, x_{rel}.

When x_{rel} is positive, the tread of the tire is traveling ahead of the road; and when x_{rel} is negative, the tread is behind the corresponding point on the road. Figure 3.19 illustrates how the treads of the tire stretch relative to the ground. The tread stretch of free rolling tire would be symmetric with respect to the center of the tireprint.

Example 90 ★ *Velocity centers of tireprint.*
    The relative velocity of tire tread with respect to the ground has two zeros. The zero relative points indicate the velocity centers of the tireprint at which

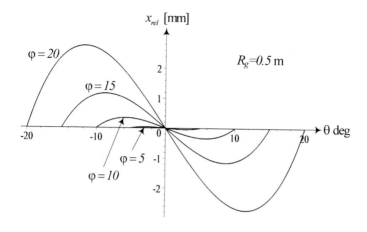

FIGURE 3.18. Horizontal displacement of tire tread with respect to the ground.

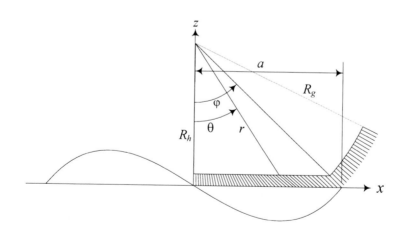

FIGURE 3.19. Tread displacement illustration of a rolling tire.

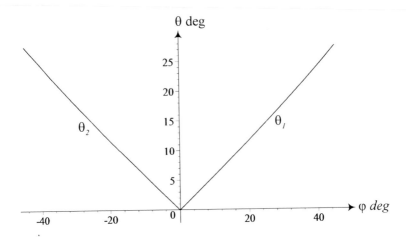

FIGURE 3.20. The velocity centers θ_1 and θ_2 as functions of φ.

tire tread is going to start going faster or slower than the road. Solving the relative velocity Equation (3.54) for the angular position of the velocity centers yields

$$\theta_1 = \cos^{-1} \sqrt{\frac{\varphi}{\tan\varphi}} \qquad \varphi > 0 \qquad (3.60)$$

$$\theta_2 = \pi - \cos^{-1} \sqrt{\frac{\varphi}{\tan\varphi}} \qquad \varphi < 0 \qquad (3.61)$$

Figure depicts θ_1 and θ_2 as functions of φ. Interestingly, the position of velocity centers are independent from the size and kinematics of the tire as they are only functions of the tireprint angle φ.

The velocity centers θ_1 and θ_2 also determine the angular positions of the maximum and minimum point of the relative displacement of Figure 3.18.

Example 91 ★ *Longitudinal strain and stress.*

Having the horizontal stretch of treads of a rolling tire as Equation (3.59), we are able to determine the longitudinal strain ε in tire treads.

$$\varepsilon = \frac{\Delta x}{x} = \frac{R_g \cos\varphi \left(\tan\theta - \dfrac{\theta}{\varphi} \tan\varphi \right)}{R_g \cos\varphi \tan\theta} = 1 - \frac{\theta \tan\varphi}{\varphi \tan\theta} \qquad (3.62)$$

Assuming that the strain is proportional to the tangential stress, we may estimate the longitudinal shear stress τ in tireprint as

$$\tau \approx G\varepsilon = G \left(1 - \frac{\theta \tan\varphi}{\varphi \tan\theta} \right) \qquad (3.63)$$

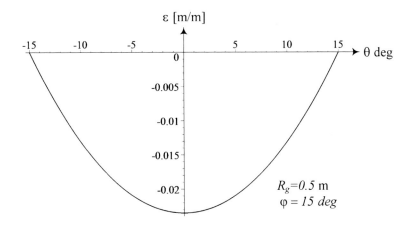

FIGURE 3.21. Strain distribution in the tireprint.

However, rubbery materials do not behave as linear Hookean material and their stress-strain relationship does not follow the linear Hooke's law. The tangential stress is almost symmetric for a free rolling tire. Applying a driving moment will shift the symmetric point of the tread displacement of Figures 3.18 and 3.19 backward. The imbalance tread displacement yields the resultant forward shear stresses become higher than the resultant backward shear stresses, and providing a forward traction force. The reverse situation happens in braking. Figure 3.21 illustrates the strain distribution in the tireprint for a sample tire.

Example 92 ★ *Acceleration analysis in tireprint.*

Consider an element of a rolling tire tread. The element's acceleration relative to the center of the tire is a combination of tangential and radial components.

$$
\begin{aligned}
\mathbf{a} &= a_r\hat{u}_r + a_\theta\hat{u}_\theta = \left(\ddot{r} - r\dot{\theta}^2\right)\hat{u}_r + \left(2\dot{r}\dot{\theta} + r\ddot{\theta}\right)\hat{u}_\theta \\
&= \left(\ddot{r} - r\omega^2\right)\hat{u}_r - 2\dot{r}\omega\hat{u}_\theta
\end{aligned}
\tag{3.64}
$$

Substituting the values of variables from (3.42)-(3.44) provides us with

$$
a_r = 2R_g\omega^2\cos\varphi\,\frac{\sin^2\theta}{\cos^3\theta}
\tag{3.65}
$$

$$
a_\theta = 2R_g\omega^2\cos\varphi\,\frac{\sin\theta}{\cos^2\theta}
\tag{3.66}
$$

$$
a = \sqrt{a_r^2 + a_\theta^2} = 2R_g\omega^2\cos\varphi\,\frac{\sin\theta}{\cos^3\theta}
\tag{3.67}
$$

The absolute acceleration, a, of the element must be on x-axis, as the element only moves along the ground surface. Figure 3.22 depicts a_r, a_θ, and

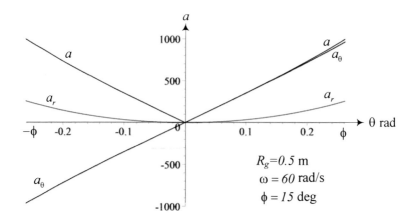

FIGURE 3.22. Radial and tangential components of the acceleration of a tire tread element in tireprint zone.

a for a sample tire with

$$R_g = 0.5\,\mathrm{m} \qquad \varphi = 15\,\mathrm{deg} \qquad \omega = 60\,\mathrm{rad/s} \tag{3.68}$$

Figure 3.23 illustrates the acceleration components of a tread element in the tireprint.

★ *Jerk analysis in tireprint.*

Having the absolute acceleration of the treads in tireprint as given in Equation (3.67) is enough to calculate the jerk j of the treads.

$$j = \frac{da}{dt} = \omega \frac{da}{d\theta} = -2\omega^3 \frac{R_g}{\cos^4 \theta} (\cos 2\theta - 2)\cos\varphi \tag{3.69}$$

Figure 3.24 illustrates $j/(R_g\omega^3)$ as a function of θ for a tire with $\varphi = 15$ deg.

The jerk j of the treads is proportional to $R_g\omega^3$. If we accept that the noise of a tire is proportional to the treads' jerk, then to reduce the noise of a given tire, we should reduce the size of the tire or the angular velocity of the tire, ω. However, these reduction must be along with keeping $R_g\omega$ constant, as $R_g\omega$ is almost equal to the speed of the car. Therefore, to reduce the jerk of a tire for a given speed v, we should use a bigger tire to have a smaller ω.

$$\omega \approx \frac{v}{R_g} \tag{3.70}$$

$$j = -2\omega^2 \frac{v}{\cos^4 \theta} (\cos 2\theta - 2)\cos\varphi \tag{3.71}$$

By increasing R_g while keeping v constant, we reduce jerk j significantly. First because j drops by the order of ω^2, and second, because φ drops by

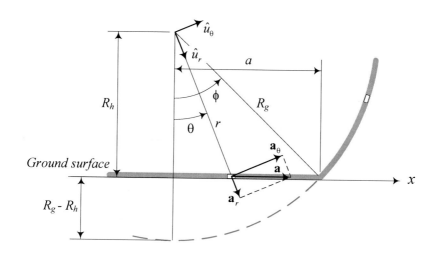

FIGURE 3.23. The acceleration components of a tread element in the tireprint.

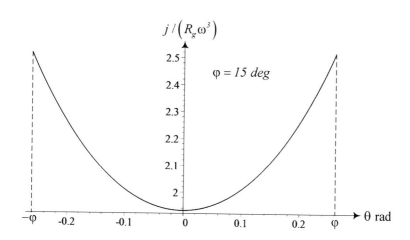

FIGURE 3.24. Nondimensional tire tread jerk $j/\left(R_g\omega^3\right)$ as a function of θ for a tire with $\varphi = 15\,\mathrm{deg}$.

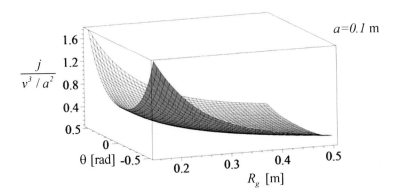

FIGURE 3.25. A plot of the tire tread jerk $j/\left(v^3/a^2\right)$ as a function of R_g and θ for $a = 0.1\,\mathrm{m}$.

the order of $\sin^{-1}\frac{1}{R_g}$ *as*

$$\varphi = \sin^{-1}\frac{a}{R_g} \tag{3.72}$$

The half length of the tireprint, a is a function of the total weight on a tire and is almost constant for a given car using every tire with the same width w. The jerk as a function of the tire radius R_g is

$$j = -2\frac{v^3}{R_g^2}\sqrt{1 - \frac{a^2}{R_g^2}\frac{\cos 2\theta - 2}{\cos^4\theta}} \tag{3.73}$$

A plot of $j/\left(v^3/a^2\right)$ versus R_g and θ is given in Figure 3.25 for $a = 0.1\,\mathrm{m}$ to show the effect of increasing R_g on the level of tire tread jerk.

3.4   ★ Tireprint Forces of a Static Tire

The force per unit area applied on a tire in the tireprint can be decomposed into a component normal to the ground and a tangential component on the ground. The normal component is the contact pressure σ_z, while the tangential component can be further decomposed in the x and y directions to make the longitudinal and lateral shear stresses τ_x and τ_y.

For a stationary tire under normal load, the tireprint is symmetrical. Due to equilibrium conditions, the overall integral of the normal stress over the tireprint area A_P must be equal to the normal load F_z, and the integral of

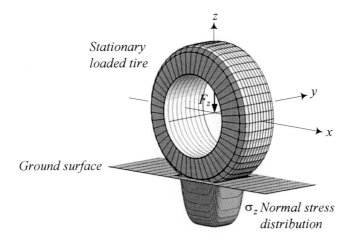

FIGURE 3.26. Normal stress σ_z applied on the round because of a stationary tire under a normal load F_z.

shear stresses must be equal to zero.

$$\int_{A_P} \sigma_z(x,y)\,dA = F_z \tag{3.74}$$

$$\int_{A_P} \tau_x(x,y)\,dA = 0 \tag{3.75}$$

$$\int_{A_P} \tau_y(x,y)\,dA = 0 \tag{3.76}$$

3.4.1 ★ Static Tire, Normal Stress

Figure 3.26 illustrates a stationary tire under a normal load F_z along with the generated normal stress σ_z applied on the ground. The applied loads on the tire are illustrated in the side view shown in Figure 3.27. For a stationary tire, the shape of normal stress $\sigma_z(x,y)$ over the tireprint area depends on the tire and load conditions, however its distribution over the tireprint is generally in the shape shown in Figure 3.28.

The normal stress $\sigma_z(x,y)$ may be approximated by the function

$$\sigma_z(x,y) = \sigma_{zM}\left(1 - \frac{x^6}{a^6} - \frac{y^6}{b^6}\right) \tag{3.77}$$

where a and b indicate the dimensions of the tireprint, as shown in Figure 3.29. The tireprints may approximately be modeled by a mathematical function

$$\frac{x^{2n}}{a^{2n}} + \frac{y^{2n}}{b^{2n}} = 1 \qquad n \in N \tag{3.78}$$

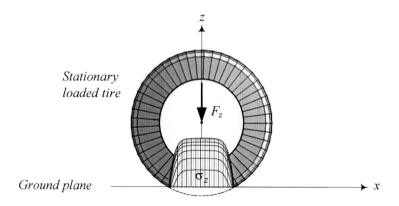

FIGURE 3.27. Side view of a normal force F_z and stress σ_z applied on a stationary tire.

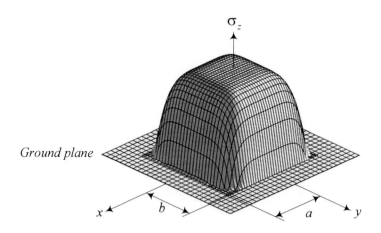

FIGURE 3.28. A model of normal stress $\sigma_z(x, y)$ in the tireprint area for a stationary tire.

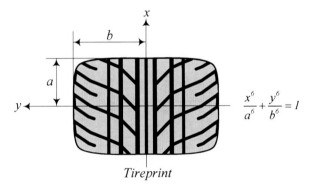

$$\frac{x^6}{a^6} + \frac{y^6}{b^6} = 1$$

Tireprint

FIGURE 3.29. A mode for tireprint of stationary radial tires under normal load.

For radial tires, $n = 3$ or $n = 2$ may be used,

$$\frac{x^6}{a^6} + \frac{y^6}{b^6} = 1 \tag{3.79}$$

while for non-radial tires $n = 1$ is a better approximation.

$$\frac{x^2}{a^2} = \frac{y^2}{b^2} = 1 \tag{3.80}$$

Example 93 *Normal stress in tireprint.*
   A car weighs 800 kg. *If* $a_1 = a_2$ *and the tireprint of each radial tire is* $A_P = 4 \times a \times b = 4 \times 5\,\text{cm} \times 12\,\text{cm}$, *then the normal stress distribution under each radial tire,* σ_z, *must satisfy the equilibrium equation.*

$$\begin{aligned} F_z &= \frac{1}{4} 800 \times 9.81 = \int_{A_P} \sigma_z(x, y)\, dA \\ &= \int_{-0.05}^{0.05} \int_{-0.12}^{0.12} \sigma_{zM} \left(1 - \frac{x^6}{0.05^6} - \frac{y^6}{0.12^6} \right) dy\, dx \\ &= 1.7143 \times 10^{-2} \sigma_{zM} \end{aligned} \tag{3.81}$$

Therefore, the maximum normal stress is

$$\sigma_{zM} = \frac{F_z}{1.7143 \times 10^{-2}} = 1.1445 \times 10^5\,\text{Pa} \tag{3.82}$$

and the stress distribution over the tireprint is

$$\sigma_z(x, y) = 1.1445 \times 10^5 \left(1 - \frac{x^6}{0.05^6} - \frac{y^6}{0.12^6} \right) \text{Pa} \tag{3.83}$$

Example 94 *Normal stress in tireprint for n = 2.*

The maximum normal stress σ_{zM} for an 800 kg car having $a_1 = a_2$ and $A_P = 4 \times a \times b = 4 \times 5\,\text{cm} \times 12\,\text{cm}$, can be found for non-radial tires with $n = 2$ as

$$F_z = \frac{1}{4}800 \times 9.81 = \int_{A_P} \sigma_z(x, y)\, dA$$

$$= \int_{-0.05}^{0.05} \int_{-0.12}^{0.12} \sigma_{zM} \left(1 - \frac{x^4}{0.05^4} - \frac{y^4}{0.12^4}\right) dy\, dx$$

$$= 1.44 \times 10^{-2}\sigma_{zM} \tag{3.84}$$

$$\sigma_{zM} = \frac{F_z}{1.44 \times 10^{-2}} = 1.3625 \times 10^5\,\text{Pa} \tag{3.85}$$

Comparing $\sigma_{zM} = 1.3625 \times 10^5$ Pa for $n = 2$ to $\sigma_{zM} = 1.1445 \times 10^5$ Pa for $n = 3$ shows that maximum stress for non-radial tires, $n = 2$, is $\left(1 - \frac{1.1445}{1.3625}\right) \times 100 = 16\%$ more than the radial tires, $n = 3$.

3.4.2 ★ Static Tire, Tangential Stresses

Because of tire deformation in contact with the ground, a three-dimensional stress distribution will appear in the tireprint even for a stationary tire. The *tangential stress* τ on the tireprint can be decomposed into x and y directions. The tangential stress is also called *shear stress* or *friction stress*.

For a static tire under a normal load, the tangential stresses are inward in x direction and outward in y direction. Hence, the tire tries to stretch the ground in the x-axis and compact the ground on the y-axis. Figure 3.30 depicts the shear stresses on a vertically loaded stationary tire. The force distribution on the tireprint is not constant and is influenced by tire structure, load, inflation pressure, and environmental conditions.

The tangential stress τ_x in the x-direction may be modeled and expressed by

$$\tau_x(x, y) = -\tau_{xM} \left(\frac{x^{2n+1}}{a^{2n+1}}\right) \sin^2\left(\frac{x}{a}\pi\right) \cos\left(\frac{y}{2b}\pi\right) \qquad n \in N \tag{3.86}$$

τ_x is negative for $x > 0$ and is positive for $x < 0$, showing an inward longitudinal stress. Figure 3.31 illustrates the absolute value of a τ_x distribution for $n = 1$.

The y-direction tangential stress τ_y may be modeled by the equation

$$\tau_y(x, y) = -\tau_{yM} \left(\frac{x^{2n}}{a^{2n}} - 1\right) \sin\left(\frac{y}{b}\pi\right) \qquad n \in N \tag{3.87}$$

where τ_y is positive for $y > 0$ and negative for $y < 0$, showing an outward lateral stress. Figure 3.32 illustrates the absolute value of a τ_y distribution for $n = 1$.

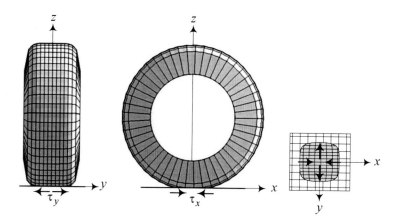

FIGURE 3.30. Direction of tangential stresses on the tireprint of a stationary vertically loaded tire.

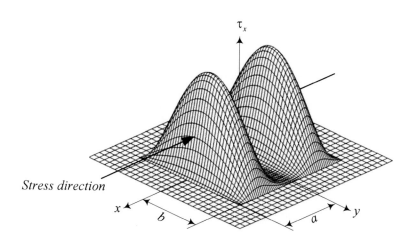

FIGURE 3.31. Absolute value of a τ_x distribution model for $n = 1$.

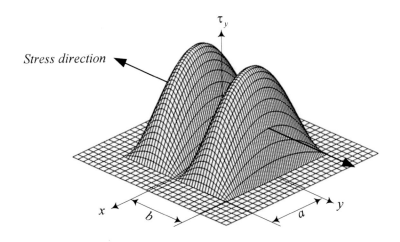

FIGURE 3.32. Absolute value of a τ_y distribution model for $n = 1$.

3.5 Rolling Resistance

A turning tire on the ground generates a longitudinal force $\mathbf{F}_r$ called *rolling resistance*. The force is opposite to the direction of motion and is proportional to the normal force on the tireprint.

$$\mathbf{F}_r = -F_r \hat{\imath} \qquad (3.88)$$
$$F_r = \mu_r F_z \qquad (3.89)$$

The parameter μ_r is called the *rolling friction coefficient*. The value of μ_r depends on tire speed, inflation pressure, sideslip and camber angles. It also depends on mechanical properties, speed, wear, temperature, load, size, driving and braking forces, and road condition.

Proof. When a tire is turning on the road, that portion of the tire's circumference that passes over the pavement undergoes a deflection. Part of the energy that is spent in deformation will not be restored in the following relaxation. Hence, a change in the distribution of the contact pressure makes normal stress σ_z in the heading part of the tireprint higher than the tailing part. The dissipated energy and stress distortion cause the rolling resistance.

Figures 3.33 and 3.34 illustrate a model of normal stress distribution across the tireprint and their resultant force F_z for a turning tire.

Because of higher normal stress in the front part of the tireprint, the resultant normal force moves forward. This forward shift of the normal force makes a resistance moment in the $-y$ direction, opposing the forward

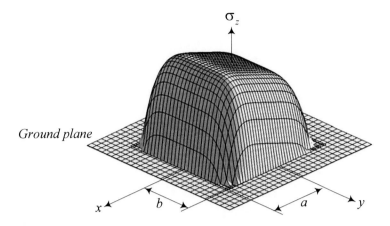

FIGURE 3.33. A model of normal stress $\sigma_z(x, y)$ in the tireprint area for a rolling tire.

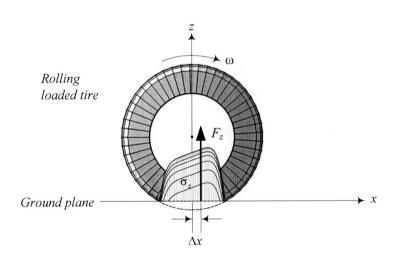

FIGURE 3.34. Side view of a normal stress σ_z distribution and its resultant force F_z on a rolling tire.

rotation.

$$\mathbf{M}_r = -M_r \hat{\jmath} \tag{3.90}$$

$$M_r = F_z \Delta x \tag{3.91}$$

The rolling resistance moment $\mathbf{M}_r$ may be assumed to be the result of a rolling resistance force $\mathbf{F}_r$ parallel to the x-axis.

$$\mathbf{F}_r = -F_r \hat{\imath} \tag{3.92}$$

$$F_r = \frac{1}{R_h} M_r = \frac{\Delta x}{R_h} F_z \tag{3.93}$$

As the rolling resistance force F_r is directly proportional to the normal load, we may define the proportionality by using a rolling friction coefficient μ_r.

$$F_r = \mu_r F_z \tag{3.94}$$

■

Example 95 *A model for normal stress of a turning tire.*
We may assume that the normal stress of a turning tire is expressed by

$$\sigma_z = \sigma_{zm} \left(1 - \frac{x^{2n}}{a^{2n}} - \frac{y^{2n}}{b^{2n}} + \frac{x}{4a} \right) \tag{3.95}$$

where $n = 3$ or $n = 2$ for radial tires and $n = 1$ for non-radial tires. We may determine the stress mean value σ_{zm} by knowing the total load on the tire. As an example, using $n = 3$ for an 800 kg car with a tireprint $A_P = 4 \times a \times b = 4 \times 5\,cm \times 12\,cm$, we have

$$
\begin{aligned}
F_z &= \frac{1}{4} 800 \times 9.81 = \int_{A_P} \sigma_z(x,y)\, dA \\
&= \int_{-0.05}^{0.05} \int_{-0.12}^{0.12} \sigma_{zm} \left(1 - \frac{x^6}{0.05^6} - \frac{y^6}{0.12^6} + \frac{x}{4 \times 0.05} \right) dy\, dx \\
&= 1.7143 \times 10^{-2} \sigma_{zm}
\end{aligned}
\tag{3.96}
$$

and therefore,

$$\sigma_{zm} = \frac{F_z}{1.7143 \times 10^{-2}} = 1.1445 \times 10^5 \, \text{Pa} \tag{3.97}$$

Example 96 *Deformation and rolling resistance.*
The distortion of stress distribution is proportional to the tire-road deformation that is the reason for shifting the resultant force forward. Hence, the rolling resistance increases with increasing deformation. A high pressure tire on concrete has lower rolling resistance than a low pressure tire on soil.

To model the mechanism of dissipation energy for a turning tire, we assume there are many small dampers and springs in the tire structure. Pairs of parallel dampers and springs are installed radially and circumstantially. Figures 3.35 illustrates the spring and damping structures of a tire.

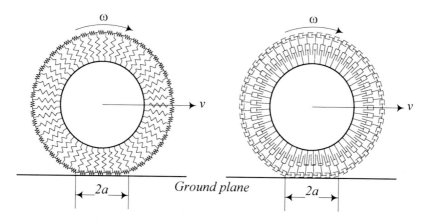

FIGURE 3.35. Spring and damping structures of a tire.

3.5.1   Effect of Speed on the Rolling Friction Coefficient

The rolling friction coefficient μ_r increases with a second degree of speed. Experiment shows that we may express $\mu_r = \mu_r(v_x)$ by the function

$$\mu_r = \mu_0 + \mu_1 v_x^2 \tag{3.98}$$

Proof. The rolling friction coefficient increases by increasing speed experimentally. We can use a polynomial function

$$\mu_r = \sum_{i=0}^{n} \mu_i v_x^i \tag{3.99}$$

to fit the experimental data. Practically, two or three terms of the polynomial would be enough to express $\mu_r = \mu_r(v_x)$. The function

$$\mu_r = \mu_0 + \mu_1 v_x^2 \tag{3.100}$$

is simple and good enough for representing experimental data and analytic calculation. The values of

$$\mu_0 = 0.015 \tag{3.101}$$
$$\mu_1 = 7 \times 10^{-6}\,\mathrm{s^2/m^2} \tag{3.102}$$

are applicable for most passenger car tires. However, μ_0 and μ_1 should be determined experimentally for any individual tire. Figure 3.36 depicts a comparison between Equation (3.98) and experimental data for a radial tire.

Generally speaking, the rolling friction coefficient of radial tires show to be less than non-radials. Figure 3.37 illustrates a sample comparison.

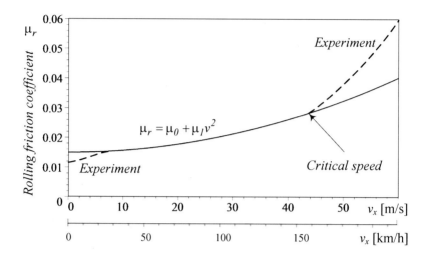

FIGURE 3.36. Comparison between the analytic equation and experimental data for the rolling friction coefficient of a radial tire.

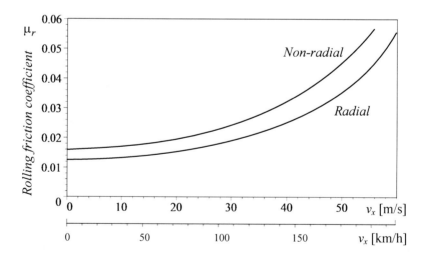

FIGURE 3.37. Comparison of the rolling friction coefficient between radial and non-radial tires.

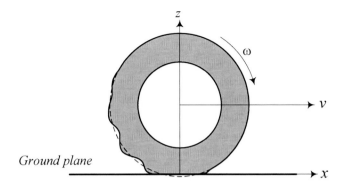

FIGURE 3.38. Illustration of circumferential waves in a rolling tire at its critical speed.

Equation (3.98) is applied when the speed is below the tire's critical speed. The *critical speed* is the speed at which standing circumferential waves appear and the rolling friction increases rapidly. The wavelength of the standing waves are close to the length of the tireprint. Above the critical speed, overheating happens and tire fails very soon. Figure 3.38 illustrates the circumferential waves in a rolling tire at its critical speed. ∎

Example 97 *Rolling resistance force and vehicle velocity.*
  For computer simulation purposes, a fourth degree equation is presented to evaluate the rolling resistance force F_r

$$F_r = C_0 + C_1\, v_x + C_2\, v_x^4 \qquad (3.103)$$

The coefficients C_i are dependent on the tire characteristics, however, the following values can be used for a typical radial passenger car tire:

$$\begin{aligned}
C_0 &= 9.91 \times 10^{-3} \\
C_1 &= 1.95 \times 10^{-5} \\
C_2 &= 1.76 \times 10^{-9}
\end{aligned} \qquad (3.104)$$

Example 98 ★ *Wave occurrence justification.*
  The normal stress will move forward when the tire is turning on a road. By increasing the speed, the normal stress will shift more and concentrate in the first half of the tireprint, causing low stress in the second half of the tireprint. High stress in the first half along with no stress in the second half is similar to hammering the tire repeatedly. When the tire is moving faster than its critical speed, the hammered tire will not be fully recovered before the next impact with the grounds.

Example 99 ★ *Race car tires.*
   Racecars have very smooth tires, known as **slicks***. Smooth tires reduce the rolling friction and maximize straight line speed. The slick racing tires are also pumped up to high pressure. High pressure reduces the tireprint area. Hence, the normal stress shift reduces and the rolling resistance decreases.*

Example 100 *Road pavement and rolling resistance.*
   The effect of the pavement and road condition is introduced by assigning a value for μ_0 in equation $\mu_r = \mu_0 + \mu_1 v_x^2$. Table 3.1 is a good reference.

Table 3.1 - The value of μ_0 on different pavements.

Road and pavement condition	μ_0
Very good concrete	$0.008 - 0.01$
Very good tarmac	$0.01 - 0.0125$
Average concrete	$0.01 - 0.015$
Very good pavement	0.015
Very good macadam	$0.013 - 0.016$
Average tarmac	0.018
Concrete in poor condition	0.02
Good block paving	0.02
Average macadam	$0.018 - 0.023$
Tarmac in poor condition	0.023
Dusty macadam	$0.023 - 0.028$
Good stone paving	$0.033 - 0.055$
Good natural paving	0.045
Stone pavement in poor condition	0.085
Snow shallow (5 cm)	0.025
Snow thick (10 cm)	0.037
Unmaintained natural road	$0.08 - 0.16$
Sand	$0.15 - 0.3$

Example 101 ★ *Effect of tire structure, size, wear, and temperature on μ_r.*
   The tire material and the arrangement of tire plies affect the rolling friction coefficient and the critical speed. Radial tires have around 20% lower μ_r, and 20% higher critical speed than a similar non-radial tires.
   Tire radius R_g and aspect ratio h_T/w_T are the two size parameters that affect the rolling resistance coefficient. A tire with larger R_g and smaller h_T/w_T has lower rolling resistance and higher critical speed.
   Generally speaking, the rolling friction coefficient decreases with wear in both radial and non-radial tires, and increases by increasing temperature.

3.5.2 Effect of Inflation Pressure and Load on the Rolling Friction Coefficient

The rolling friction coefficient μ_r decreases by increasing the inflation pressure p. The effect of increasing pressure is equivalent to decreasing normal load F_z.

The following empirical equation has been suggested to show the effects of both pressure p and load F_z on the rolling friction coefficient.

$$\mu_r = \frac{K}{1000}\left(5.1 + \frac{5.5 \times 10^5 + 90F_z}{p} + \frac{1100 + 0.0388F_z}{p}v_x^2\right) \qquad (3.105)$$

The parameter K is equal to 0.8 for radial tires, and is equal to 1.0 for non-radial tires. The value of F_z, p, and v_x must be in $[\mathrm{N}]$, $[\mathrm{Pa}]$, and $[\mathrm{m/s}]$ respectively. Low tire pressure increases roll resistance, fuel consumption, tire wear, and tire temperature.

Example 102 *Motorcycle rolling friction coefficient.*
The following equations are suggested for calculating rolling friction coefficient μ_r of motorcycles. They can be only used as a rough estimate for passenger cars. The equations consider the inflation pressure and forward velocity of the motorcycle, but not considering the load on tire F_z.

$$\mu_r = \begin{cases} 0.0085 + \dfrac{1800}{p} + \dfrac{2.0606}{p}v_x^2 & v_x \le 46\,\mathrm{m/s} \\[2mm] \dfrac{1800}{p} + \dfrac{3.7714}{p}v_x^2 & v_x > 46\,\mathrm{m/s} \end{cases} \qquad (3.106)$$

$$46\,\mathrm{m/s} \approx 165\,\mathrm{km/h}$$

The speed v_x must be expressed in $[\mathrm{m/s}]$ and the pressure p must be in $[\mathrm{Pa}]$. Figure 3.39 illustrates this equation for $v_x \le 46\,\mathrm{m/s}\,(\approx 165\,\mathrm{km/h})$. Increasing the inflation pressure p decreases the rolling friction coefficient μ_r.

Example 103 *Dissipated power because of rolling friction.*
Rolling friction reduces the vehicle's power. The dissipated power because of rolling friction is equal to the rolling friction force F_r times the forward velocity v_x. Using Equation (3.105), the rolling resistance power is

$$\begin{aligned} P &= F_r v_x = -\mu_r v_x F_z \\ &= \frac{-K v_x}{1000}\left(5.1 + \frac{5.5 \times 10^5 + 90F_z}{p} + \frac{1100 + 0.0388F_z}{p}v_x^2\right)F_z \end{aligned} \qquad (3.107)$$

The resultant power P is in $[\mathrm{W}]$ when the normal force F_z is expressed in $[\mathrm{N}]$, velocity v_x in $[\mathrm{m/s}]$, and pressure p in $[\mathrm{Pa}]$.

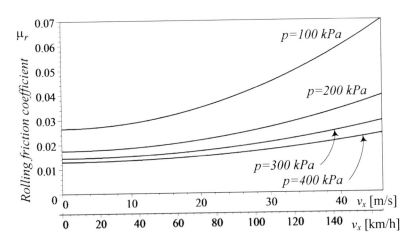

FIGURE 3.39. Motorcycle rolling friction coefficient.

The rolling resistance dissipated power for motorcycles can be found based on Equation (3.106).

$$P = \begin{cases} \left(0.0085 + \dfrac{1800}{p} + \dfrac{2.0606}{p} v_x^2\right) v_x F_z & v_x \leq 46\,\mathrm{m/s} \\[3mm] \left(\dfrac{1800}{p} + \dfrac{3.7714}{p} v_x^2\right) v_x F_z & v_x > 46\,\mathrm{m/s} \end{cases} \tag{3.108}$$

$46\,\mathrm{m/s} \approx 165\,\mathrm{km/h}$

Example 104 *Rolling resistance dissipated power.*

If a vehicle is moving at $100\,\mathrm{km/h} \approx 27.78\,\mathrm{m/s} \approx 62\,\mathrm{mi/h}$ and each radial tire of the vehicle is pressurized up to $220\,\mathrm{kPa} \approx 32\,\mathrm{psi}$ and loaded by $220\,\mathrm{kg}$, then the dissipated power, because of rolling resistance, is

$$\begin{aligned} P &= 4 \times \frac{K\, v_x}{1000} \left(5.1 + \frac{5.5 \times 10^5 + 90 F_z}{p} + \frac{1100 + 0.0388 F_z}{p} v_x^2\right) F_z \\ &= 2424.1\,\mathrm{W} \approx 2.4\,\mathrm{kW} \end{aligned} \tag{3.109}$$

To compare the given equations, assume the vehicle has motorcycle tires with power loss given by Equation (3.108).

$$\begin{aligned} P &= 4 \times \left(0.0085 + \frac{1800}{p} + \frac{2.0606}{p} v_x^2\right) v_x F_z \\ &= 5734.1\,\mathrm{W} \approx 5.7\,\mathrm{kW} \end{aligned} \tag{3.110}$$

It shows that if the vehicle uses motorcycle tires, it dissipates more power.

Proper Inflation   *Over Inflation*   *Under Inflation*

FIGURE 3.40. Tire-road contact of an over- and under-inflated tire compared to a properly inflated tire.

Example 105 *Effects of improper inflation pressure.*

High inflation pressure increases stiffness of tires, which reduces ride comfort and generates vibration. Tireprint area and traction force are reduced when tires are over inflated. Over-inflation causes the tire to transmit shock loads to the suspension, and reduces the tire's ability to support the required load for cornerability, braking, and acceleration.

Under-inflation results in higher internal shear stress, cracking and tire component separation. It also increases sidewall flexing and rolling resistance that cause heat and mechanical failure. A tire's load capacity is largely determined by its inflation pressure. Therefore, under-inflation results in an overloaded tire that operates at high deflection with low fuel economy and low handling.

Figure 3.40 illustrates the effect of over and under inflation on tire-road contact compared to a proper inflated tire. Proper inflation pressure is necessary for optimum tire performance, safety, and fuel economy. Correct inflation is especially significant to the endurance and performance of radial tires because it may not be possible to find a 5 psi $\approx$ 35 kPa under-inflation in a radial tire just by visual observation. However, under-inflation of 5 psi $\approx$ 35 kPa can reduce up to 25% of the tire performance and life.

A tire may lose 1 to 2 psi ($\approx$ 7 to 14 kPa) every month. The inflation pressure can also change by 1 psi $\approx$ 7 kPa for every 10 °F $\approx$ 5 °C of temperature change. As an example, if a tire is inflated to 35 psi $\approx$ 240 kPa on an 80 °F $\approx$ 26 °C summer day, it could have an inflation pressure of 23 psi $\approx$ 160 kPa on a 20 °F $\approx$ −6 °C day in winter. This represents a normal loss of 6 psi $\approx$ 40 kPa over the six months and an additional loss of 6 psi $\approx$ 40 kPa due to the 60 °F $\approx$ 30 °C change. At 23 psi $\approx$ 160 kPa, this tire is functioning under-inflated.

Example 106 *Small / large and soft / hard tires.*

If the driving tires are small, the vehicle becomes twitchy with low traction and low top speed. On the other hand, when the driving tires are big, then the vehicle has slow steering response and high tire distortion in turns, decreasing the stability.

Softer front tires show more steerability, less stability, and more wear while hard front tires show the opposite. Soft rear tires have more rear traction, but they make the vehicle less steerable, more bouncy, and less stable. Hard rear tires have less rear traction, but they make the vehicle more steerable, less bouncy, and more stable.

3.5.3 ★ Effect of Sideslip Angle on Rolling Resistance

When a tire is rolling on the road with a sideslip angle α, a significant increase in rolling resistance occurs. The rolling resistance force F_r would then be

$$F_r = F_x \cos \alpha + F_y \sin \alpha \approx F_x - C_\alpha \alpha^2 \qquad (3.111)$$

where, F_x is the longitudinal force opposing the motion, and F_y is the lateral force, both in tire coordinate frame.

Proof. Figure 3.41 illustrates the top view of a rolling tire on the ground under a sideslip angle α. The rolling resistance force is defined as the force opposite to the velocity vector of the tire, which has angle α with the x-axis. Assume a longitudinal force F_x in $-x$-direction is applied on the tire. Sideslip α increases F_x and generates a lateral force F_y in $-y$-direction. The sum of the components of the longitudinal force F_x and the lateral force F_y makes the rolling resistance force F_r.

$$F_r = F_x \cos \alpha + F_y \sin \alpha \qquad (3.112)$$

For small values of the sideslip angle α, the lateral force is proportional to $-\alpha$ and therefore,

$$F_r \approx F_x - C_\alpha \alpha^2 \qquad (3.113)$$

∎

3.5.4 ★ Effect of Camber Angle on Rolling Resistance

When a tire travels with a camber angle γ, the component of rolling moment $\mathbf{M}_r$ on rolling resistance $\mathbf{F}_r$ will be reduced, however, a component of aligning moment M_z on rolling resistance will appear.

$$\mathbf{F}_r = -F_r \,\hat{\imath} \qquad (3.114)$$

$$F_r = \frac{1}{R_h} M_r \cos \gamma + \frac{1}{R_h} M_z \sin \gamma \qquad (3.115)$$

Proof. Rolling moment M_r appears when a tire is rolling and the normal force F_z shifts forward. However, only the component $M_r \cos \gamma$ is perpendicular to the tire-plane and prevents the tire's spin. Furthermore, when a moment in z-direction is applied on the tire, only the component $M_z \sin \gamma$ will prevent the tire's spin. Therefore, the camber angle γ will affect the rolling resistance according to

$$\mathbf{F}_r = -F_r \,\hat{\imath}$$

$$F_r = \frac{1}{R_h} M_r \cos \gamma + \frac{1}{R_h} M_z \sin \gamma \qquad (3.116)$$

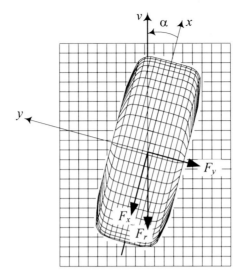

FIGURE 3.41. Effect of sideslip angle α on rolling resistance force F_r.

where R_h is the height of tire center from the road surface as is shown in Figure 3.10. The rolling moment M_r may be substituted by Equation (3.90) to show the effect of normal force F_z.

$$F_r = \frac{\Delta x}{R_h} F_z \cos\gamma + \frac{1}{R_h} M_z \sin\gamma \tag{3.117}$$

■

3.6   Longitudinal Force

The *longitudinal slip ratio* of a tire is defined by

$$s = \frac{R_g \omega_w}{v_x} - 1 \tag{3.118}$$

where, R_g is the tire's geometric and unloaded radius, ω_w is the tire's angular velocity, and v_x is the tire's forward velocity. Slip ratio is positive for driving and is negative for braking.

To accelerate or brake a vehicle, longitudinal forces must develop between the tire and the ground. When a moment is applied to the spin axis of the tire, slip ratio occurs and a longitudinal force F_x is generated at the tireprint. The force F_x is proportional to the normal force,

$$\mathbf{F}_x = F_x \hat{\imath} \tag{3.119}$$

$$F_x = \mu_x(s) F_z \tag{3.120}$$

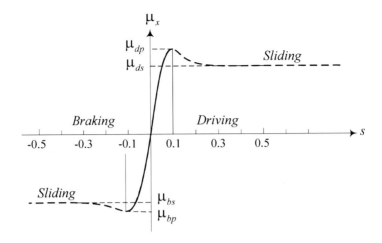

FIGURE 3.42. Longitudinal friction coefficient as a function of slip ratio s, in driving and braking.

where the coefficient $\mu_x(s)$ is called the *longitudinal friction coefficient* and is a function of slip ratio s as shown in Figure 3.42. The friction coefficient reaches a driving peak value μ_{dp} at $s \approx 0.1$, before dropping to an almost steady-state driving slide value μ_{ds}. The friction coefficient $\mu_x(s)$ may be assumed proportional to s when s is very small

$$\mu_x(s) = C_s\, s \qquad s \ll 1 \tag{3.121}$$

where C_s is called the *longitudinal slip coefficient*.

The tire will spin when $s \gtrsim 0.1$ and the friction coefficient remains almost constant. The same phenomena happen in braking at the values μ_{bp} and μ_{bs}.

Proof. Slip ratio, or simply *slip*, is defined as the difference between the actual speed of the tire v_x and the equivalent tire speed $R_w \omega_w$. Figure 3.43 illustrates a rolling tire on the ground. The ideal distance that the tire would freely travel with no slip is denoted by d_F, while the actual distance the tire travels is denoted by d_A. Thus, for a spinning tire, $d_F > d_A$, and for a slipping tire, $d_F < d_A$.

The difference $d_F - d_A$ is the tire slip and therefore, the slip ratio of the tire is

$$s = \frac{d_F - d_A}{d_A} \tag{3.122}$$

To have the instant value of s, we must measure the travel distances in an infinitesimal time period, and therefore,

$$s \equiv \frac{\dot{d}_F - \dot{d}_A}{\dot{d}_A} \tag{3.123}$$

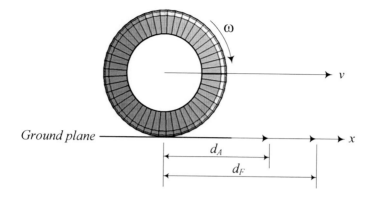

FIGURE 3.43. A turning tire on the ground to show the no slip travel distance d_F, and the actual travel distance d_A.

If the angular velocity of the tire is ω_w then, $\dot{d}_F = R_g\omega_w$ and $\dot{d}_A = R_w\omega_w$ where, R_g is the geometric tire radius and R_w is the effective radius. Therefore, the slip ratio s can be defined based on the actual speed $v_x = R_w\omega_w$, and the freely rolling speed $R_g\omega_w$

$$s = \frac{R_g\omega_w - R_w\omega_w}{R_w\omega_w} = \frac{R_g\omega_w}{v_x} - 1 \qquad (3.124)$$

A tire can exert longitudinal force only if a longitudinal slip is present. During acceleration, the actual velocity v_x is less than the free velocity $R_g\omega_w$, and therefore, $s > 0$. However, during braking, the actual velocity v_x is higher than the free velocity $R_g\omega_w$ and therefore, $s < 0$.

The frictional force F_x between a tire and the road surface is a function of normal load F_z, vehicle speed v_x, and wheel angular speed ω_w. In addition to these variables there are a number of parameters that affect F_x, such as tire pressure, tread design, wear, and road surface conditions. It has been determined empirically that a contact friction force of the form $F_x = \mu_x(\omega_w, v_x)F_z$ can model experimental measurements obtained with constant v_x, ω_w.

Longitudinal slip is also called *circumferential* or *tangential slip*. ∎

Example 107 *Slip ratio is $0 < s < \infty$ in driving.*

When we drive, a driving moment is applied to the tire axis. The tread of the tire will be compressed circumstantially in the tireprint zone. Hence, the tire is moving slower than a free tire

$$R_w\omega_w < R_g\omega_w \qquad (3.125)$$

and therefore $s > 0$. The equivalent radius for a driving tire is less than the geometric radius

$$R_w < R_g \qquad (3.126)$$

Equivalently, we may express the condition using the equivalent angular velocity ω_{eq} and deduce that a driving tire turns faster than a free tire

$$R_g \omega_{eq} < R_g \omega_w \qquad (3.127)$$

The driving moment can be high enough to overcome the friction and turn the tire on pavement while the car is not moving. In this case $v_x = 0$ and therefore, $s = \infty$. It shows that the longitudinal slip would be between $0 < s < \infty$ when accelerating.

$$0 < s < \infty \quad for \quad a > 0 \qquad (3.128)$$

The tire speed $R_w \omega_w$ equals vehicle speed v_x only if acceleration is zero. In this case, the normal force acting on the tire and the size of the tireprint are constant in time.

Applying a driving moment generates a positive $s > 0$ and shifts the symmetric point of the tread displacement of Figures 3.18 and 3.19 backward. This results more forward shear stresses which provides a forward traction force.

Example 108 *Samples for longitudinal friction coefficients μ_{dp} and μ_{ds}.*

Table 3.2 shows the average values of longitudinal friction coefficients μ_{dp} and μ_{ds} for a passenger car tire 215/65R15. It is practical to assume $\mu_{dp} = \mu_{bp}$, and $\mu_{ds} = \mu_{bs}$.

Table 3.2 - Average of longitudinal friction coefficients.

Road surface	Peak value, μ_{dp}	Sliding value, μ_{ds}
Asphalt, dry	$0.8 - 0.9$	0.75
Concrete, dry	$0.8 - 0.9$	0.76
Asphalt, wet	$0.5 - 0.7$	$0.45 - 0.6$
Concrete, wet	0.8	0.7
Gravel	0.6	0.55
Snow, packed	0.2	0.15
Ice	0.1	0.07

Example 109 *Slip ratio is $-1 < s < 0$ in braking.*

When we brake, a braking moment is applied to the wheel axis. The tread of the tire will be stretched circumstantially in the tireprint zone. Hence, the tire is moving faster than a free tire

$$R_w \omega_w > R_g \omega_w \qquad (3.129)$$

and therefore, $s < 0$. The equivalent radius for a braking tire is more than the free radius

$$R_w > R_g \qquad (3.130)$$

Equivalently, we may express the condition using the equivalent angular velocity ω_e and deduce that a braking tire turns slower than a free tire

$$R_g \omega_{eq} > R_g \omega_w \tag{3.131}$$

The brake moment can be high enough to lock the tire. In this case $\omega_w = 0$ and therefore, $s = -1$. It shows that the longitudinal slip would be between $-1 < s < 0$ when braking.

$$-1 < s < 0 \quad for \quad a < 0 \tag{3.132}$$

Example 110 *Slip ratio based on equivalent angular velocity ω_{eq}.*

It is possible to define an effective angular velocity ω_{eq} as an equivalent angular velocity for a tire with radius R_g to proceed with the actual speed $v_x = R_g \omega_{eq}$. Using ω_{eq} we have

$$v_x = R_g \omega_{eq} = R_w \omega_w \tag{3.133}$$

and therefore,

$$s = \frac{R_g \omega_w - R_g \omega_{eq}}{R_g \omega_{eq}} = \frac{\omega_w}{\omega_{eq}} - 1 \tag{3.134}$$

Example 111 *Power and maximum velocity.*

Consider a moving car with power $P = 100\,\text{kW} \approx 134\,\text{hp}$ that can attain $279\,\text{km/h} \approx 77.5\,\text{m/s} \approx 173.3\,\text{mi/h}$. The total driving force must be

$$F_x = \frac{P}{v_x} = \frac{100 \times 10^3}{77.5} = 1290.3\,\text{N} \tag{3.135}$$

If we assume that the car is rear-wheel-drive and the rear wheels are driving at the maximum traction under the load $1600\,\text{N}$, then the longitudinal friction coefficient μ_x is

$$\mu_x = \frac{F_x}{F_z} = \frac{1290.3}{1600} \approx 0.806 \tag{3.136}$$

Example 112 *Slip of hard tire on hard road.*

A tire with no slip cannot create any tangential force. Assume a toy car equipped with steel tires is moving on a glass table. Such a car cannot accelerate or steer easily. If the car can accelerate at very low rate, it is because there is sufficient microscopic slip to generate forces to drive. The glass table and the small contact area of the small metallic tires deform and stretch each other, although such a deformation is in microscopic scale. If there is any friction between the tire and the surface, there must be slip to maneuver.

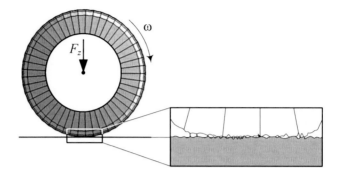

FIGURE 3.44. The molecular binding between the tire and road surfaces.

Example 113 ★ *Friction mechanisms.*

Rubber tires generate friction in three mechanisms: 1. *adhesion,* 2. *deformation, and* 3. *wear.*

$$F_x = F_{ad} + F_{de} + F_{we} \tag{3.137}$$

Adhesion *friction is equivalent to sticking. The rubber resists sliding on the road because adhesion causes it stick to the road surface. Adhesion occurs as a result of molecular binding between the rubber and surfaces. Because the real contact area is much less than the observed contact area, high local pressure make molecular binding, as shown in Figure 3.44. Bound occurs at the points of contact and welds the surfaces together. The adhesion friction is equal to the required force to break these molecular bounds and separate the surfaces. The adhesion is also called* **cold welding** *and is attributed to pressure rather than heat. Higher load increases the contact area, makes more bounds, and increases the friction force. So the adhesion friction confirms the traditional friction equation*

$$F_x = \mu_x(s) \, F_z \tag{3.138}$$

The main contribution to tire traction force on a dry road is the adhesion friction. The adhesion friction decreases considerably on a road covered by water, ice, dust, or lubricant. Water on a wet road prevents direct contact between the tire and road and reduces the formation of adhesion friction. The main contribution to tire friction when it slides on a road surface is the viscoelastic energy dissipation in the tireprint area. This dissipative energy is velocity and is time-history dependent.

Deformation friction *is the result of deforming rubber and filling the microscopic irregularities on the road. The surface of the road has many peaks and valleys called* **asperities**. *Movement of a tire on a rough surface results in the deformation of the rubber by peaks and high points on the surface. A load on the tire causes the peaks of irregularities to penetrate the tire and the tire drapes over the peaks. The deformation friction force,* needed to move the irregularities in the rubber, comes from the local high pressure across the irregularities. Higher load increases the penetration of the irregularities in the tire and therefore increases the friction force. So the deformation friction also confirms the friction equation (3.138).

The main contribution to the tire traction force on a wet road is the deformation friction. The adhesion friction decreases considerably on a road covered by water, ice, dust, or lubricant.

Deformation friction exists in relative movement between any contacted surfaces. No matter how much care is taken to form a smooth surface, the surfaces are irregular with microscopic peaks and valleys. Opposite peaks interact with each other and cause damage to both surfaces.

Wear friction is the result of excessive local stress over the tensile strength of the rubber. High local stresses deform the structure of the tire surface past the elastic point. The polymer bonds break, and the tire surface tears in microscopic scale. This tearing makes the wear friction mechanism. Wear results in separation of material. Higher load eases the tire wear and therefore increases the wear friction force. So the wear friction also confirms the friction equation (3.138).

Example 114 tanh *model of slip friction coefficient* μ_x.

We introduce the tanh describing function to provide a mathematical equation for analytical calculations.

$$\mu_x = \mu_s \tanh\left(\frac{s}{s_c}\right) \tag{3.139}$$

The equation needs two numbers μ_s and s_c which should be measured experimentally. Assuming a symmetric μ_x in driving and braking, μ_s is the sliding value of μ_x, and s_c is the critical value of s at which $\mu_x = 0.76\mu_s$. Figure 3.45 illustrates the mathematical model of slip friction coefficient μ_x.

In case the μ_s and s_c are different in driving and braking, the function

$$\mu_x = \mu_{ds} H(s) \tanh\left(\frac{s}{s_{dc}}\right) + \mu_{bs} H(-s) \tanh\left(\frac{s}{s_{bc}}\right) \tag{3.140}$$

models the friction coefficient μ_x, where $H(s)$ is the Heaviside function, s_{dc} and s_{bc} are respectively the critical values of s in driving and braking.

Example 115 ★ *Empirical slip models.*

Based on experimental data and curve fitting methods, some mathematical equations were presented to simulate the longitudinal tire force as a function of longitudinal slip s. Most of these models are too complicated to be useful in vehicle dynamics. However, a few of them are simple and accurate enough to be applied.

The Pacejka model, which was presented in 1991, has the form

$$F_x(s) = c_1 \sin\left(c_2 \tan^{-1}\left(c_3 s - c_4\left(c_3 s - \tan^{-1}(c_3 s)\right)\right)\right) \tag{3.141}$$

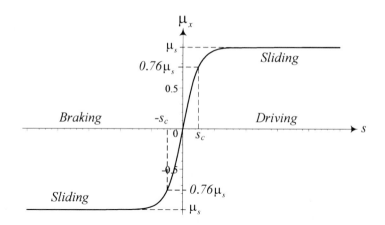

FIGURE 3.45. The mathematical model of slip friction coefficient μ_x.

where c_1, c_2, c_3, and c_4 are four constants based on the tire experimental data.

The 1987 Burckhardt model is a simpler equation that also needs three numbers.

$$F_x\left(s\right) = c_1\left(1 - e^{-c_2 s}\right) - c_3 s \tag{3.142}$$

There is another Burckhardt model that includes the velocity dependency.

$$F_x\left(s\right) = \left(c_1\left(1 - e^{-c_2 s}\right) - c_3 s\right) e^{-c_4 v} \tag{3.143}$$

This model needs four numbers to be measured from experiment.

By expanding and approximating the 1987 Burckhardt model, the simpler model by Kiencke and Daviss was suggested in 1994. This model is

$$F_x\left(s\right) = k_s \frac{s}{1 + c_1 s + c_2 s^2} \tag{3.144}$$

where k_s is the slope of $F_x\left(s\right)$ versus s at $s = 0$

$$k_s = \lim_{s \to 0} \frac{\triangle F_s}{\triangle s} \tag{3.145}$$

and c_1, c_2 are two experimental numbers.

Another simple model is the 2002 De-Wit model

$$F_x\left(s\right) = c_1\sqrt{s} - c_2 s \tag{3.146}$$

that is based on two numbers c_1, c_2.

In either case, we need at least one experimental curve such as shown in Figure 3.42 to find the constant numbers c_i. The constants c_i are the numbers that best fit the associated equation with the experimental curve. The 1997 Burckhardt model (3.143) needs at least two similar tests at two different speeds.

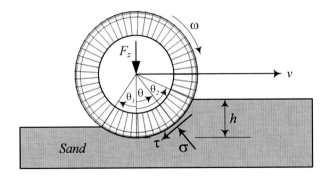

FIGURE 3.46. A tire turning on sand.

Example 116 ★ *Alternative slip ratio definitions.*
An alternative method to define the slip ratio is

$$
s = \begin{cases}
1 - \dfrac{v_x}{R_g \omega_w} & R_g \omega_w > v_x \quad driving \\[3mm]
\dfrac{R_g \omega_w}{v_x} - 1 & R_g \omega_w < v_x \quad braking
\end{cases}
\tag{3.147}
$$

where v_x is the speed of the wheel center, ω_w is the angular velocity of the wheel, and R_g is the tire radius.
In another alternative definition, the following equation is used for longitudinal slip:

$$
s = 1 - \left(\frac{R_g \omega_w}{v_x} \right)^n \quad where \quad n = \begin{cases} +1 & R_g \omega_w \le v_x \\ -1 & R_g \omega_w > v_x \end{cases} \tag{3.148}
$$

$$
s \in [0, 1]
$$

In this definition s is always between zero and one. When $s = 1$, then the tire is either locked while the car is sliding, or the tire is spinning while the car is not moving.

Example 117 ★ *Tire on soft sand.*
Figure 3.46 illustrates a tire rolling on sand. The sand will be packed when the tire passes. The applied stresses from the sand on the tire are developed during the angle $\theta_1 < \theta < \theta_2$ measured counterclockwise from vertical direction.
It is possible to define a relationship between the normal stress σ and tangential stress τ under the tire

$$
\tau = (c + \sigma \tan \theta) \left(1 - e^{\frac{R_g}{k}[\theta_1 - \theta + (1-s)(\sin \theta) - \sin \theta_1]} \right) \tag{3.149}
$$

where s is the slip ratio defined in Equation (3.147), and

$$
\tau_M = c + \sigma \tan \theta \tag{3.150}
$$

is the maximum shear stress in the sand applied on the tire. In this equation, c is the cohesion stress of the sand, and k is a constant.

Example 118 ★ *Lateral slip ratio.*

Analytical expressions can be established for the force contributions in x and y directions using adhesive and sliding concept by defining longitudinal and lateral slip ratios s_x and s_y

$$s_x = \frac{R_g \omega_w}{v_x} - 1 \qquad s_y = \frac{R_g \omega_w}{v_y} \qquad (3.151)$$

where v_x is the longitudinal speed of the wheel and v_y is the lateral speed of the wheel. The unloaded geometric radius of the tire is denoted by R_g and ω_w is the angular velocity of the wheel.

At very low slips, the resulting tire forces are proportional to the slip

$$F_x = C_{s_x} s_x \qquad F_y = C_{s_y} s_y \qquad (3.152)$$

where C_{s_x} is the longitudinal slip coefficient and C_{s_y} is the lateral slip coefficient.

3.7 Lateral Force

When a rolling tire is under a vertical force F_z and a lateral force F_y at the tire axis, its path of motion on the road makes an angle α with respect to the tire-plane. The angle is called *sideslip* angle and is proportional to the *lateral force*

$$\mathbf{F}_y = F_y \hat{j} \qquad (3.153)$$
$$F_y = -C_\alpha \alpha \qquad (3.154)$$

where C_α is called the *cornering stiffness* of the tire.

$$C_\alpha = \lim_{\alpha \to 0} \frac{\partial(-F_y)}{\partial \alpha} = \left| \lim_{\alpha \to 0} \frac{\partial F_y}{\partial \alpha} \right| \qquad (3.155)$$

The lateral force $\mathbf{F}_y$ at tireprint is at a distance a_{x_α} behind the centerline of the tireprint and makes a moment $\mathbf{M}_z$ called *aligning moment*.

$$\mathbf{M}_z = M_z \hat{k} \qquad (3.156)$$
$$M_z = F_y a_{x_\alpha} \qquad (3.157)$$

For small α, the aligning moment $\mathbf{M}_z$ tends to turn the tire about the z-axis and make the x-axis align with the velocity vector $\mathbf{v}$. The aligning moment always tends to reduce α.

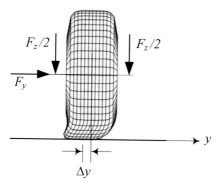

FIGURE 3.47. Front view of a laterally deflected tire.

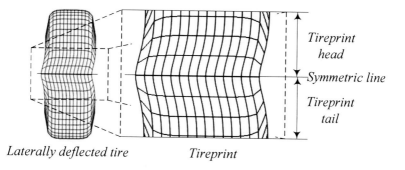

FIGURE 3.48. Bottom view of a laterally deflected tire.

Proof. When a wheel is under a constant load F_z and then a lateral force is applied on the rim, the tire will deflect laterally as is shown in Figure 3.47. The tire acts as a linear spring under small lateral forces F_y with a lateral stiffness k_y.

$$F_y = k_y \, \Delta y \tag{3.158}$$

The wheel will start sliding laterally when the lateral force reaches a maximum value F_{y_M}. At this point, the lateral force approximately remains constant and is proportional to the vertical load

$$F_{y_M} = \mu_y \, F_z \tag{3.159}$$

where, μ_y is the tire friction coefficient in the y-direction. A bottom view of the tireprint of a laterally deflected tire is shown in Figure 3.48.

If the laterally deflected tire is rolling forward on the road, the tireprint will also flex longitudinally. A bottom view of the tireprint for such a laterally deflected and rolling tire is shown in Figure 3.49. Although the tire-plane of such a tire remains perpendicular to the road, the path of the

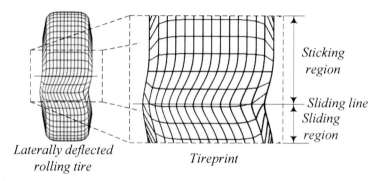

FIGURE 3.49. Bottom view of a laterally deflected and turning tire.

wheel makes an angle α with tire-plane. As the wheel rolls forward, undeflected treads enter the tireprint region and deflect laterally as well as longitudinally. When a tread moves toward the end of the tireprint, its lateral deflection increases until it approaches the tailing edge of the tireprint. The normal load decreases at the tail of the tireprint, so the friction force is lessened and the tread can slide back to its original position when leaving the tireprint region. The point where the laterally deflected tread slides back is called *sliding line*.

A rolling tire under lateral force and the associated *sideslip angle* α are shown in Figure 3.50. Lateral distortion of the tire treads is a result of a tangential stress distribution τ_y over the tireprint. Assuming that the tangential stress τ_y is proportional to the distortion, the resultant lateral force F_y

$$F_y = \int_{A_P} \tau_y \, dA_p \tag{3.160}$$

is at a distance a_{x_α} behind the center line.

$$a_{x_\alpha} = \frac{1}{F_y} \int_{A_P} x \, \tau_y \, dA_p \tag{3.161}$$

The distance a_{x_α} is called the *pneumatic trail*, and the resultant moment $\mathbf{M}_z$ is called the aligning moment.

$$\mathbf{M}_z = M_z \, \hat{k} \tag{3.162}$$

$$M_z = F_y \, a_{x_\alpha} \tag{3.163}$$

The aligning moment tends to turn the tire about the z-axis and make it align with the direction of tire velocity vector $\mathbf{v}$. A stress distribution τ_y, the resultant lateral force F_y, and the pneumatic trail a_{x_α} are illustrated in Figure 3.50.

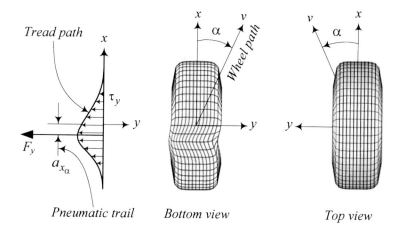

FIGURE 3.50. The stress distribution τ_y, the resultant lateral force F_y, and the pneumatic trail a_y for a turning tire going on a positive slip angle α.

There is also a lateral shift in the tire vertical force F_z because of slip angle α, which generates a *slip moment* M_x about the forward x-axis.

$$\mathbf{M}_x = -M_x\,\hat{\imath} \tag{3.164}$$

$$M_x = F_z\,a_{y_\alpha} \tag{3.165}$$

The slip angle α always increases by increasing the lateral force F_y. However, the sliding line moves toward the tail at first and then moves forward by increasing the lateral force F_y. Slip angle α and lateral force F_y work as action and reaction. A lateral force generates a slip angle, and a slip angle generates a lateral force. Hence, we can steer the tires of a car to make a slip angle and produce a lateral force to turn the car. Steering causes a slip angle in the tires and creates a lateral force. The slip angle $\alpha > 0$ if the tire should be turned about the z-axis to be aligned with the velocity vector $\mathbf{v}$. A positive slip angle α generates a negative lateral force F_y. Hence, steering to the right about the $-z$-axis makes a positive slip angle and produces a negative lateral force to move the tire to the right. Using the velocity vector of the tire, and its components, $\mathbf{v} = v_x\hat{\imath} + v_y\hat{\jmath}$, we may also define the sideslip angle as

$$\alpha = \arctan\frac{v_y}{v_x} \tag{3.166}$$

A sample of measured lateral force F_y as a function of slip angle α for a constant vertical load is plotted in Figure 3.51. The lateral force F_y is linear for small slip angles, however the rate of increasing F_y decreases for higher α. The lateral force remains constant or drops slightly when α

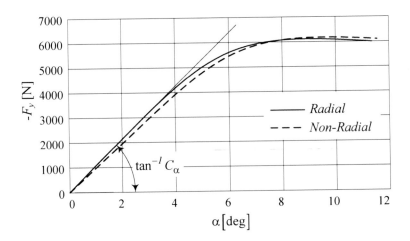

FIGURE 3.51. Lateral force F_y as a function of slip angle α for a constant vertical load.

reaches a critical value at which the tire slides on the road. Therefore, we may assume the lateral force F_y is proportional to the slip angle α for low values of α.

$$F_y \quad = \quad -C_\alpha\,\alpha \qquad\qquad (3.167)$$

$$C_\alpha \quad = \quad -\lim_{\alpha\to 0}\frac{\partial F_y}{\partial \alpha} \qquad\qquad (3.168)$$

The cornering stiffness C_α of radial tires are higher than C_α for non-radial tires. This is because radial tires need a smaller slip angle α to produce the same amount of lateral force F_y.

Examples of aligning moments for radial and non-radial tires are illustrated in Figure 3.52. The pneumatic trail a_{x_α} increases for small slip angles up to a maximum value, and decreases to zero and even negative values for high slip angles. Therefore, the behavior of aligning moment $\mathbf{M}_z$ is similar to what is shown in Figure 3.52.

The lateral force $F_y = -C_\alpha\,\alpha$ can be decomposed to $F_y\cos\alpha$, parallel to the path of motion v, and $F_y\sin\alpha$, perpendicular to v as shown in Figure 3.53. The component $F_y\cos\alpha$, normal to the path of motion, is called *cornering force*, and the component $F_y\sin\alpha$, along the path of motion, is called *drag force*.

The lateral force F_y is also called *side force* or *grip*. We may combine the lateral forces of all tires of a vehicle and have them acting at the car's mass center C. ∎

Example 119 *Effect of tire load on lateral force curve.*
When the wheel load F_z increases, the tire treads can stick to the road better. Hence, the lateral force increases at a constant slip angle α, and the

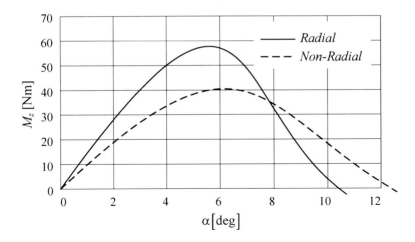

FIGURE 3.52. Aligning moment M_z as a function of slip angle α for a constant vertical load.

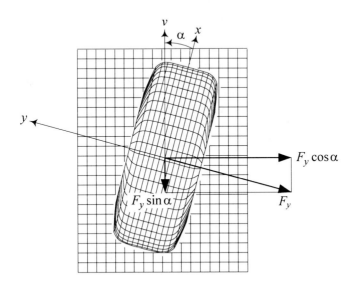

FIGURE 3.53. The cornering and drag components of a lateral force F_y.

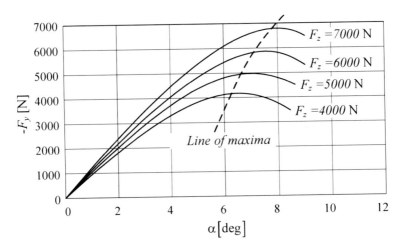

FIGURE 3.54. Lateral force behavior of a sample tire for different normal loads as a function of slip angle α.

slippage occurs at the higher slip angles. Figure 3.54 illustrates the lateral force behavior of a sample tire for different normal loads.

Increasing the load not only increases the maximum attainable lateral force, it also pushes the maximum of the lateral force to higher slip angles.

Sometimes the effect of load on lateral force is presented in a dimensionless variable to make it more practical. Figure 3.55 depicts a sample.

Example 120 ★ *Gough diagram.*

The slip angle α is the main affective parameter on the lateral force F_y and aligning moment $M_z = F_y a_{x_\alpha}$. However, F_z and M_z depend on many other parameters such as speed v, pressure p, temperature, humidity, and road conditions. A better method to show F_z and M_z is to plot them versus each other for a set of parameters. Such a graph is called a **Gough diagram**. Figure 3.56 depicts a sample Gough diagram for a radial passenger car tire. Every tire has its own Gough diagram, although we may use an average diagram for radial or non-radial tires.

Example 121 *Effect of velocity.*

The curve of lateral force as a function of the slip angle $F_y(\alpha)$ decreases as velocity increases. Hence, we need to increase the sideslip angle at higher velocities to generate the same lateral force. Sideslip angle increases by increasing the steer angle. Figure 3.57 illustrates the effect of velocity on F_y for a radial passenger tire. Because of this behavior, the curvature of trajectory of a one-wheel-car at a fixed steer angle increases by increasing the driving speed.

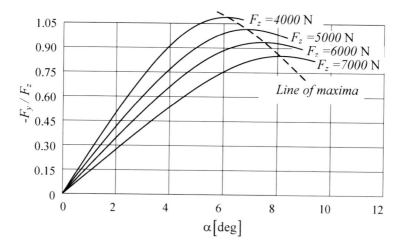

FIGURE 3.55. Effect of load on lateral force as a function of slip angle α presented in a dimensionless fashion.

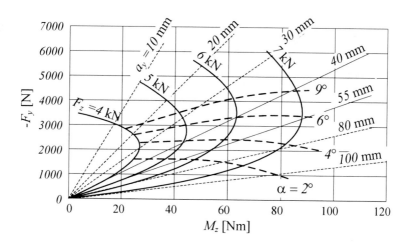

FIGURE 3.56. Gough diagram for a radial passenger car tire.

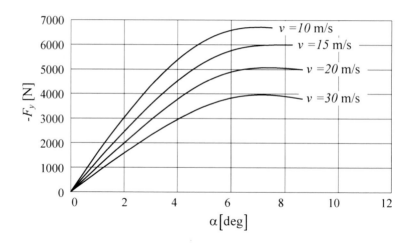

FIGURE 3.57. Effect of velocity on F_y and M_z for a radial tire.

Example 122 ★ *Cubic model for lateral force.*

When the sideslip angle is not small, the linear approximation (3.154) cannot show the tire behavior. Based on a parabolic normal stress distribution on the tireprint, the following third-degree function was presented in the 1950s to calculate the lateral force at high sideslips

$$F_y = -C_\alpha\, \alpha \left(1 - \frac{1}{3}\left|\frac{C_\alpha\, \alpha}{F_{yM}}\right| + \frac{1}{27}\left(\frac{C_\alpha\, \alpha}{F_{yM}}\right)^2\right) \tag{3.169}$$

where F_{yM} is the maximum lateral force that the tire can support. F_{yM} is set by the tire load and the lateral friction coefficient μ_y. Let us show the sideslip angle at which the lateral force F_y reaches its maximum value F_{yM} by α_M. Equation (3.169) shows that

$$\alpha_M = \frac{3F_{yM}}{C_\alpha} \tag{3.170}$$

and therefore,

$$F_y = -C_\alpha\, \alpha\left(1 - \frac{\alpha}{\alpha_M} + \frac{1}{3}\left(\frac{\alpha}{\alpha_M}\right)^2\right) \tag{3.171}$$

$$\frac{F_y}{F_{yM}} = \frac{3\alpha}{\alpha_M}\left(1 - \frac{\alpha}{\alpha_M} + \frac{1}{3}\left(\frac{\alpha}{\alpha_M}\right)^2\right) \tag{3.172}$$

Figure 3.58 shows the cubic curve model for lateral force as a function of sideslip angle. The Equation is applicable only for $0 \le \alpha \le \alpha_M$.

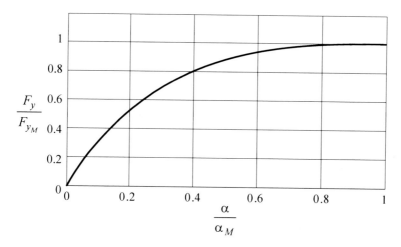

FIGURE 3.58. A cubic curve model for lateral force as a function of the sideslip angle.

Example 123 ★ *A model for lateral stress.*

Consider a tire rolling on a dry road at a low sideslip angle α. Assume the developed lateral stress on tireprint can be expressed by the equation

$$\tau_y(x,y) = c\tau_{y_M}\left(1 - \frac{x}{a}\right)\left(1 - \frac{x^3}{a^3}\right)\cos^2\left(\frac{y}{2b}\pi\right) \tag{3.173}$$

The coefficient c is proportional to the tire load F_z sideslip α, and longitudinal slip s. If the tireprint area is $A_P = 4 \times a \times b = 4 \times 5\,\text{cm} \times 12\,\text{cm}$, then the lateral force under the tire, F_y, for c = 1 is

$$F_y = \int_{A_P} \tau_y(x,y)\,dA$$
$$= \int_{-0.05}^{0.05}\int_{-0.12}^{0.12} \tau_{y_M}\left(1 - \frac{x}{0.05}\right)\left(1 - \frac{x^3}{0.05^3}\right)\cos^2\left(\frac{y\pi}{0.24}\right)dy\,dx$$
$$= 0.0144\tau_{y_M} \tag{3.174}$$

If we calculate the lateral force $F_y = 1000\,\text{N}$ by measuring the lateral acceleration, then the maximum lateral stress is

$$\tau_{y_M} = \frac{F_z}{0.0144} = 69444\,\text{Pa} \tag{3.175}$$

and the lateral stress distribution over the tireprint is

$$\tau_y(x,y) = 69444\left(1 - \frac{x}{0.05}\right)\left(1 - \frac{x^3}{0.05^3}\right)\cos^2\left(\frac{y\pi}{0.24}\right)\text{Pa} \tag{3.176}$$

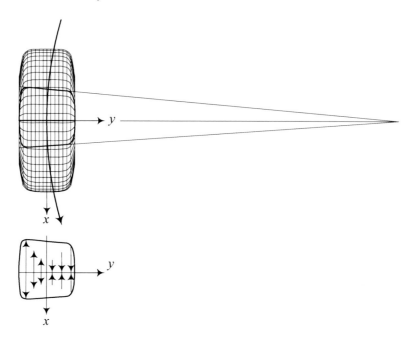

FIGURE 3.59. The unbalanced longitudinal stress distribution on a free rolling tire in a turn.

Example 124 *Tire on a circle.*

Ideally a tire must be a disc to be able to turn slip free on a circle. A free rolling wide tire on a turn will have an unbalanced longitudinal stress distribution as is shown in Figure 3.59. The inner part of the tireprint contracts, while the outer part of the tireprint extracts. The amount of contraction and extraction increases by increasing the distance from the tire plane indicated by the x-axis. Furthermore, the leading part of the tireprint is shorter than the tail part; therefore, the y-axis of the tireprint coordinate frame is not a symmetric line. As a result, there is an asymmetric stress distribution in the tireprint which generates an aligning moment, an anti-spin moment, and a rolling resistance force.

3.8 Camber Force

Camber angle γ is the tilting angle of tire about the longitudinal x-axis. Camber angle generates a lateral force F_y called *camber thrust* or *camber force*. Figure 3.60 illustrates a front view of a cambered tire and the generated camber force F_y. Camber angle is considered positive $\gamma > 0$, if it is in the positive direction of the x-axis, measured from the z-axis to the tire-plane. A positive camber angle generates a camber force along the $-y$-axis.

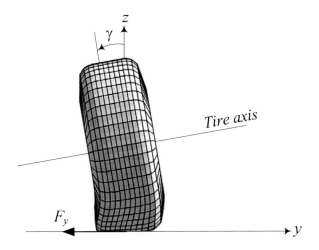

FIGURE 3.60. A front view of a cambered tire and the generated camber force.

The camber force is proportional to γ at low camber angles, and depends directly on the wheel load F_z. Therefore,

$$\mathbf{F}_y = F_y\,\hat{\jmath} \tag{3.177}$$
$$F_y = -C_\gamma\,\gamma \tag{3.178}$$

where C_γ is called the *camber stiffness* of tire.

$$C_\gamma = \lim_{\gamma \to 0} \frac{\partial\left(-F_y\right)}{\partial\gamma} \tag{3.179}$$

In presence of both, camber γ and sideslip α, the overall lateral force F_y on a tire is a superposition of the corner force and camber thrust.

$$F_y = -C_\gamma\,\gamma - C_\alpha\,\alpha \tag{3.180}$$

Proof. When a wheel is under a constant load and then a camber angle is applied on the rim, the tire will deflect laterally such that the tireprint area is longer in the cambered side and shorter in the other side. Figure 3.61 compares the tireprint of a straight and a cambered tire, turning slowly on a flat road. As the wheel rolls forward, undeflected treads enter the tireprint region and deflect laterally as well as longitudinally. However, because of the shape of the tireprint, the treads entering the tireprint closer to the cambered side have more time to be stretched laterally. Because the developed lateral stress is proportional to the lateral stretch, the nonuniform tread stretching generates an asymmetric stress distribution and more lateral stress will be developed on the cambered side. The result of the nonuniform lateral stress distribution over the tireprint of a cambered tire produces the camber thrust F_y in the cambered direction.

$$\mathbf{F}_y = F_y \hat{\jmath} \tag{3.181}$$

$$F_y = \int_{A_P} \tau_y \, dA \tag{3.182}$$

The camber thrust is proportional to the camber angle for small angles.

$$F_y = -C_\gamma \gamma \tag{3.183}$$

When the tire is rolling, the resultant camber thrust F_y shifts forward by a distance a_{x_γ}. The resultant moment in the z-direction is called *camber torque*, and the distance a_{x_γ} is called *camber trail*.

$$\mathbf{M}_z = M_z \hat{k} \tag{3.184}$$

$$M_z = F_y a_{x_\gamma} \tag{3.185}$$

Camber trail is usually very small and hence, the camber torque can be ignored in linear analysis of vehicle dynamics.

Because the tireprint of a cambered tire deforms to be longer in the cambered side, the resultant vertical force F_z that supports the wheel load,

$$F_z = \int_{A_P} \sigma_z \, dA \tag{3.186}$$

shifts laterally by a distance a_{y_γ} from the center of the tireprint.

$$a_{y_\gamma} = \frac{1}{F_z} \int_{A_P} y \sigma_z \, dA_p \tag{3.187}$$

The distance a_{y_γ} is called the *camber arm*, and the resultant moment $\mathbf{M}_x$ is called the *camber moment*.

$$\mathbf{M}_x = M_x \hat{k} \tag{3.188}$$

$$M_x = -F_z a_{y_\gamma} \tag{3.189}$$

The camber moment tends to turn the tire about the x-axis and make the tire-plane align with the z-axis. The camber arm a_{y_γ} is proportional to the camber angle γ for small angles.

$$a_{y_\gamma} = C_{y_\gamma} \gamma \tag{3.190}$$

Figure 3.62 shows the camber force F_y for different camber angle γ at a constant tire load $F_z = 4500\,\text{N}$. Radial tires generate lower camber force due to their higher flexibility.

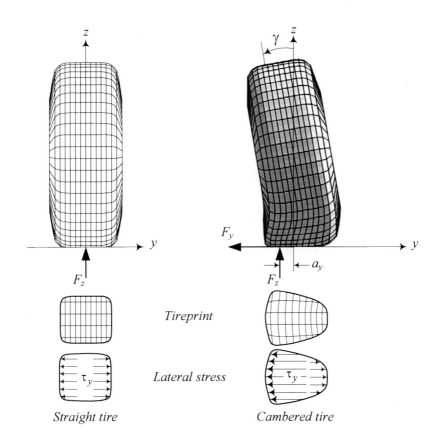

FIGURE 3.61. The tireprint of a straight and a cambered tire, turning slowly on a flat road.

It is better to illustrate the effect of F_z graphically to visualize the camber force. Figure 3.63 depicts the variation of camber force F_y as a function of normal load F_z at different camber angles for a sample radial tire.

If we apply a slip angle α to a rolling cambered tire, the tireprint will distort similar to the shape in Figure 3.64 and the path of treads become more complicated. The resultant lateral force would be at a distance a_{x_γ} and a_{y_γ} from the center of the tireprint. Both distances a_{x_γ} and a_{y_γ} are functions of angles α and γ. Camber force due to γ, along with the corner force due to α, give the total lateral force applied on a tire. Therefore, the lateral force can be calculated as

$$F_y = -C_\alpha\,\alpha - C_\gamma\,\gamma \qquad (3.191)$$

that is acceptable for $\gamma \lesssim 10\,\text{deg}$ and $\alpha \lesssim 5\,\text{deg}$. Presence of both camber angle γ and slip angle α makes the situation interesting because the total lateral force can be positive or negative according to the directions of γ

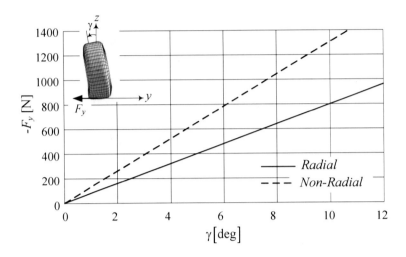

FIGURE 3.62. The camber force F_y for different camber angle γ at a constant tire load.

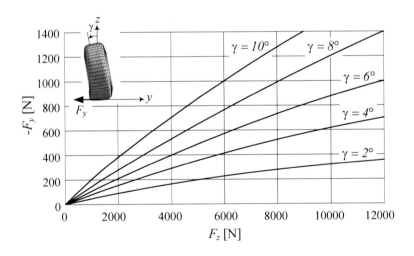

FIGURE 3.63. The variation of camber force F_y as a function of normal load F_z at different camber angles for a sample radial tire.

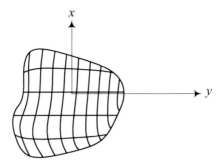

FIGURE 3.64. Tireprint of a cambered tire under a sideslip.

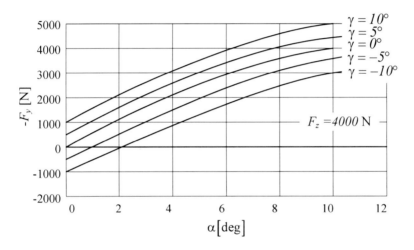

FIGURE 3.65. An example for lateral force as a function of γ and α at a constant load $F_z = 4000\,\mathrm{N}$.

and α. Figure 3.65 illustrates an example of lateral force as a function of γ and α at a constant load $F_z = 4000\,\mathrm{N}$. Similar to lateral force, the aligning moment M_z can be approximated as a combination of the slip and camber angle effects

$$M_z = C_{M_\alpha}\,\alpha + C_{M_\gamma}\,\gamma \qquad (3.192)$$

For a radial tire, $C_{M_\alpha} \approx 0.013\,\mathrm{N\,m/\,deg}$ and $C_{M_\gamma} \approx 0.0003\,\mathrm{N\,m/\,deg}$, while for a non-radial tire, $C_{M_\alpha} \approx 0.01\,\mathrm{N\,m/\,deg}$ and $C_{M_\gamma} \approx 0.001\,\mathrm{N\,m/\,deg}$. ∎

Example 125 ★ *Banked road.*

  Consider a vehicle moving on a road with a transversal slope β, while its tires remain vertical. There is a downhill component of weight, $F_1 = mg\sin\beta$, that pulls the vehicle down. There is also an uphill camber force due to camber $\gamma \approx \beta$ of tires with respect to the road $F_2 = C_\gamma\,\gamma$. The *resultant lateral force* $F_y = C_\gamma \gamma - mg \sin \beta$ *depends on camber stiffness* C_γ *and determines if the vehicle goes uphill or downhill. Since the camber stiffness* C_γ *is higher for non-radial tires, it is more possible for a non-radial than a radial tire to go uphill.*

The effects of cambering are particularly important for motorcycles that produce a large part of the lateral force by cambering. For cars and trucks, the camber angles are much smaller and in many applications their effect can be negligible. However, some suspensions are designed to make the wheels cambered when the axle load varies, or when they steered.

Example 126 *Camber importance and tireprint model.*

Cambering of a tire creates a lateral force, even though there is no sideslip. The effects of cambering are particularly important for motorcycles that produce a large part of the lateral force by camber. The following equations are presented to model the lateral deviation of a cambered tireprint from the straight tireprint, and express the lateral stress τ_y *due to camber*

$$y = -\sin\gamma \left(\sqrt{R_g^2 - x^2} - \sqrt{R_g^2 - a^2} \right) \tag{3.193}$$

$$\tau_y = -\gamma k \left(a^2 - x^2 \right) \tag{3.194}$$

where k is chosen such that the average camber defection is correct in the tireprint

$$\int_{-a}^{a} \tau_y \, dx = \int_{-a}^{a} y \, dx \tag{3.195}$$

Therefore,

$$k = \frac{3\sin\gamma}{4a^3\gamma} \left(-a\sqrt{R_g^2 - a^2} + R_g^2 \sin^{-1} \frac{a}{R_g} \right) \tag{3.196}$$

$$\approx \frac{3}{4} \frac{R_g \sqrt{R_g^2 - a^2}}{a^2} \tag{3.197}$$

and

$$\tau_y = -\frac{3}{4}\gamma \frac{R_g \sqrt{R_g^2 - a^2}}{a^2} \left(a^2 - x^2 \right) \tag{3.198}$$

3.9 Tire Force

Tires may be considered as a force generator with two major outputs: forward force F_x, lateral force F_y, and three minor outputs: aligning moment M_z, roll moment M_x, and pitch moment M_y. The input of the force generator is the tire load F_z, sideslip α, longitudinal slip s, and the camber angle γ.

$$
\begin{aligned}
F_x &= F_x\left(F_z, \alpha, s, \gamma\right) & (3.199) \\
F_y &= F_y\left(F_z, \alpha, s, \gamma\right) & (3.200) \\
M_x &= M_x\left(F_z, \alpha, s, \gamma\right) & (3.201) \\
M_y &= M_y\left(F_z, \alpha, s, \gamma\right) & (3.202) \\
M_z &= M_z\left(F_z, \alpha, s, \gamma\right) & (3.203)
\end{aligned}
$$

Ignoring the rolling resistance and aerodynamic force, the major output forces can be approximated by a set of linear equations for a given load F_z

$$
\begin{aligned}
F_x &= \mu_x\left(s\right) F_z \qquad \mu_x\left(s\right) = C_s\, s \\
F_y &= -C_\alpha\, \alpha - C_\gamma\, \gamma & (3.204)
\end{aligned}
$$

where, C_s is the *longitudinal slip coefficient*, C_α is the *lateral stiffness*, and C_γ is the *camber stiffness*.

When the tire has a combination of tire inputs, α, s, γ, the tire forces are called *tire combined force*. The most important tire combined force is the shear force because of longitudinal slip and sideslips. However, as long as the angles and slips are within the linear range of tire behavior, a superposition can be utilized to estimate the output forces.

Driving and braking forces change the lateral force F_y generated at any sideslip angle α. This is because the longitudinal force pulls the tireprint in the direction of the driving or braking force and hence, the length of lateral displacement of the tireprint will also change.

Figure 3.66 illustrates how a sideslip α affects the longitudinal force ratio F_x/F_z as a function of slip ratio s. Figure 3.67 illustrates the effect of sideslip α on the lateral force ratio F_y/F_z as a function of slip ratio s. Figure 3.68 and 3.69 illustrate the same force ratios as Figures 3.66 and 3.67 when the slip ratio s is a parameter.

Proof. Consider a turning tire under a sideslip angle α. The tire develops a lateral force $F_y = -C_\alpha\, \alpha$. Applying a driving or braking force on this tire will reduce the lateral force while developing a longitudinal force $F_x = \mu_x\left(s\right) F_z$. Experimental data shows that the reduction in lateral force in presence of a slip ratio s is similar to Figure 3.67. Now assume the sideslip α is reduced to zero. Reduction α will increase the longitudinal force while decreasing the lateral force. Increasing the longitudinal force is experimentally similar to Figure 3.68.

A turning tire under a slip ratio s develops a longitudinal force $F_x = \mu_x\left(s\right) F_z$. Applying a sideslip angle α will reduce the longitudinal force while developing a lateral force. Experimental data shows that the reduction in longitudinal force in presence of a sideslip α is similar to Figure 3.66. Now assume the slip ratio s and hence, the driving or breaking force is reduced to zero. Reduction in s will increase the lateral force while decreasing

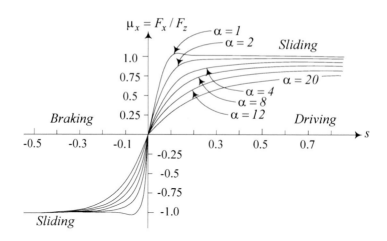

FIGURE 3.66. Longitudinal force ratio F_x/F_z as a function of slip ratio s for different sideslip α.

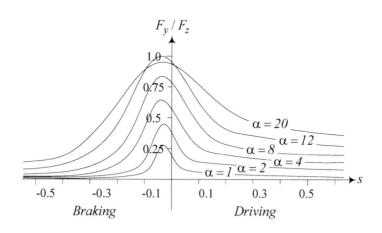

FIGURE 3.67. Lateral force ratio F_y/F_z as a function of slip ratio s for different sideslip α.

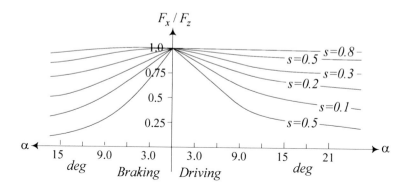

FIGURE 3.68. Longitudinal force ratio F_x/F_z as a function of sideslip α for different slip ratio s.

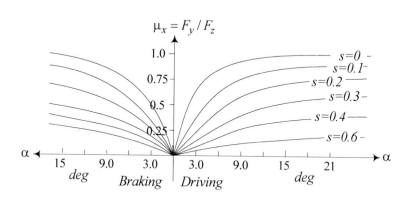

FIGURE 3.69. Lateral force ratio F_y/F_z as a function of sideslip α for different slip ratio s.

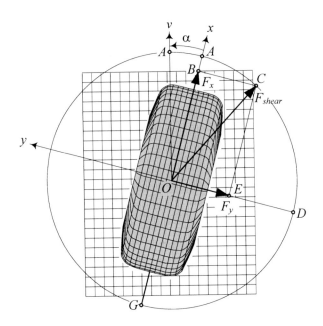

FIGURE 3.70. Friction ellipse.

the longitudinal force. Increasing the lateral force is similar to Figure 3.67.

■

Example 127 *Friction ellipse.*
When the tire is under both longitudinal slip and sideslips, the tire is under combined slip. The shear force on the tireprint of a tire under a combined slip can approximately be found using a friction ellipse model.

$$\left(\frac{F_y}{F_{yM}}\right)^2 + \left(\frac{F_x}{F_{xM}}\right)^2 = 1 \tag{3.205}$$

A friction ellipse is shown in Figure 3.70.

Proof. *The shear force $\mathbf{F}_{shear}$, applied on the tire at tireprint, parallel to the ground surface, has two components: the longitudinal force F_x and the lateral force F_y.*

$$\mathbf{F}_{shear} = F_x\,\hat{\imath} + F_y\,\hat{\jmath} \tag{3.206}$$

$$F_x = C_s\, s\, F_z \tag{3.207}$$

$$F_y = -C_\alpha\, \alpha \tag{3.208}$$

These forces cannot exceed their maximum values F_{yM} and F_{xM}.

$$F_{yM} = \mu_y\, F_z \tag{3.209}$$

$$F_{xM} = \mu_x\, F_z \tag{3.210}$$

The tire shown in Figure 3.70 is moving along the velocity vector v at a sideslip angle α. The x-axis indicates the tire-plane. When there is no sideslip, the maximum longitudinal force is $F_{x_M} = \mu_x F_z = \overrightarrow{OA}$. Now, if a sideslip angle α is applied, a lateral force $F_y = \overrightarrow{OE}$ is generated and the longitudinal force reduces to $F_x = \overrightarrow{OB}$. The maximum lateral force would be $F_{y_M} = \mu_y F_z = \overrightarrow{OD}$ where there is no longitudinal slip.

In presence of the longitudinal and lateral forces, we may assume that the tip point of the maximum shear force vector is on a friction ellipse:

$$\left(\frac{F_y}{F_{y_M}}\right)^2 + \left(\frac{F_x}{F_{x_M}}\right)^2 = 1 \tag{3.211}$$

When $\mu_x = \mu_y = \mu$, the friction ellipse would be a circle and

$$F_{shear} = \mu F_z \tag{3.212}$$

■

Example 128 *Wide tires.*

A wide tire has a shorter tireprint than a narrow tire. Assuming the same vehicle and same tire pressure, the area of tireprint would be equal in both tires. The shorter tireprint at the same sideslip has more of its length stuck to the road than longer tireprint. So, a wider tireprint generates more lateral force than a narrower tireprint for the same tire load and sideslip.

Generally speaking, tire performance and maximum force capability decrease with increasing speed in both wide and narrow tires.

Example 129 *Sinus tire forces model.*

A few decades ago, a series of applied sine functions were developed based on experimental data to model tire forces. The sine functions may effectively be used to model tire forces, especially for computer purposes.

The sine function model for the lateral force of a tire is

$$F_y = A \sin\left(B \tan^{-1}(C\Phi)\right) \tag{3.213}$$
$$\Phi = (1 - E)(\alpha + \delta)\mu F_z \tag{3.214}$$
$$C = \frac{C_\alpha}{AB} \quad C_\alpha = C_1 \sin\left(2\tan^{-1}\frac{F_z}{C_2}\right) \tag{3.215}$$
$$A, B = \text{Shape factors} \tag{3.216}$$
$$C_1 = \text{Maximum cornering stiffness} \tag{3.217}$$
$$C_2 = \text{Tire load at maximum cornering stiffness} \tag{3.218}$$

Example 130 ★ *Empirical tire force model.*

Based on experimental data and curve fitting, the Pacejka model was presented in 1991, to model nonlinear tire forces.

$$F(u) = c_1 \sin\left(c_2 \tan^{-1}\left(c_3 u (1 - c_4) + c_4 \tan^{-1}(c_3 s)\right)\right) \tag{3.219}$$

where c_1, c_2, c_3, and c_4 are three constants based on the tire experimental data. When the variable u is the longitudinal slip s, then $F(s) = F_x$ represents the longitudinal traction force. When the variable u is the sideslip angle α, then $F(\alpha) = F_y$ represents the lateral force. The coefficients c_1, c_2, c_3, and c_4 are referred to as the stiffness, shape, peak, and curvature coefficients respectively.

When the tire is under both longitudinal and lateral forces, then F_x and F_y get coefficients to satisfy the friction ellipse (3.211).

3.10   Summary

We attach a coordinate frame $(oxyz)$ to the tire at the center of the tireprint, called the tire frame. The x-axis is along the intersection line of the tire-plane and the ground. The z-axis is perpendicular to the ground, and the y-axis makes the coordinate system right-hand. We show the tire orientation using two angles: camber angle γ and sideslip angle α. The camber angle γ is the angle between the tire-plane and the vertical plane measured about the x-axis, and the sideslip angle α is the angle between the velocity vector $\mathbf{v}$ and the x-axis measured about the z-axis.

A vertically loaded wheel rolling on a flat surface has an effective radius R_w, called rolling radius

$$R_w = \frac{v_x}{\omega_w} \tag{3.220}$$

where v_x is the forward velocity, and ω_w is the angular velocity of the wheel. The effective radius R_w is approximately equal to

$$R_w \approx R_g - \frac{R_g - R_h}{3} \tag{3.221}$$

and is a number between the unloaded or geometric radius R_g and the loaded height R_h.

$$R_h < R_w < R_g \tag{3.222}$$

The tire force in the x-direction is a combination of the longitudinal force F_x and the rolling resistance F_r. The longitudinal force is

$$F_x = \mu_x(s) F_z \tag{3.223}$$

where s is the longitudinal slip ratio of the tire

$$s = \frac{R_g \omega_w}{v_x} - 1 \tag{3.224}$$

$$\mu_x(s) = C_s s \qquad s \ll 1 \tag{3.225}$$

The wheel force in the tire y-direction, F_y, is a combination of the cornering and camber thrust forces.

$$F_y = -C_\gamma \gamma - C_\alpha \alpha \tag{3.226}$$

The first term, $-C_\gamma \gamma$, is the camber thrust and is proportional to the camber angle γ of the tire. The second term, $C_\alpha \alpha$, is called the cornering force and is proportional to the sideslip angle α of the tire. The sideslip coefficient C_α and the camber coefficient C_γ are proportional to the normal load on the tire and are measured experimentally.

A rolling tire on the ground also generates a longitudinal force called rolling resistance. The force is opposite to the direction of motion and is proportional to the normal force on the tireprint.

$$F_r = \mu_r \, F_z \qquad (3.227)$$

The parameter μ_r is called the rolling friction coefficient and is a function of tire mechanical properties, speed, wear, temperature, load, size, driving and braking forces, and road condition. Because F_r is in $-v$-direction, it affects both F_x and F_y when $\alpha \neq 0$.

3.11   Key Symbols

$a \equiv \ddot{x}$	acceleration
a, b	semiaxes of A_P
a_{x_α}	aligning arm
a_{x_γ}	camber trail
a_{y_γ}	camber arm
A_P	tireprint area
c_1, c_2, c_3, c_4	coefficients of the function $F_x = F_x(s)$
C_0, C_1, C_2	coefficients of the polynomial function $F_r = F_r(v_x)$
C_s	longitudinal slip coefficient
C_{s_x}, C_{s_y}	longitudinal and lateral slip coefficients
C_α	sideslip coefficient, sideslip stiffness
C_γ	camber coefficient, camber stiffness
d	radial displacement of tire elements
d_F	no slip tire travel
d_A	actual tire travel
D	tire diameter
E	Young modulus
f	function
f_k	spring force
$F_r \; \mathbf{F}_r$	rolling resistance force
F_x	longitudinal force, forward force
F_y	lateral force
F_{yM}	maximum lateral force
F_z	normal force, vertical force, wheel load
$g \; \mathbf{g}$	gravitational acceleration
j	jerk of tire tread in tireprint
k	stiffness
k_1, k_2, k_3, k_4	nonlinear tire stiffness coefficients
k_{eq}	equivalent stiffness
k_s	slope of $F_x(s)$ versus s at $s = 0$
k_x	tire stiffness in the x-direction
k_y	tire stiffness in the y-direction
k_z	tire stiffness in the z-direction
K	radial and non-radial tires parameter in $\mu_r = \mu_r(p, v_x)$
m	mass
$M_r \; \mathbf{M}_r$	rolling resistance moment
$M_x, \; \mathbf{M}_x$	roll moment, bank moment, tilting torque,
M_y	pitch moment, rolling resistance torque
M_z	yaw moment, aligning moment, self aligning moment bore torque
n	exponent for shape and stress distribution of A_P
n_1	number of tire rotations
p	tire inflation pressure

P	rolling resistance power
r	radial position of tire periphery
R_g	geometric radius
R_h	loaded height
R_w	rolling radius, effective radius
s	longitudinal slip
s_c	critical longitudinal slip
s_y	lateral slip
T	wheel torque
$v \equiv \dot{x}$, $\mathbf{v}$	velocity, tread velocity in tireprint
v_g	velocity of the ground
v_{rel}	relative velocity of tire tread with respect to ground
x, y, z, $\mathbf{x}$	displacement
x, y, z	coordinate axes
x_g	displacement of the ground
x_{rel}	relative displacement of tread with respect to ground
$\triangle x$	tire deflection in the x-direction, rolling resistance arm
$\triangle y$	tire deflection in the y-direction
$\triangle z$	tire deflection in the z-direction
$\dot{z}$	tire deflection rate in the z-direction
α	sideslip angle
α_M	maximum sideslip angle
β	transversal slope
γ	camber angle
δ	deflection
ε	strain
$\triangle x$	tire deflection in the x-direction, rolling resistance arm
$\triangle y$	tire deflection in the y-direction
$\triangle z$	tire deflection in the z-direction
θ	tire angular rotation
μ_0, μ_1	nonlinear rolling friction coefficient
μ_r	rolling friction coefficient
$\mu_x(s)$	longitudinal friction coefficient
μ_{dp}	friction coefficient driving peak value
μ_{ds}	friction coefficient steady-state value
σ_{zM}	maximum normal stress
$\sigma_z(x, y)$	normal stress over the tireprint
σ_{zm}	normal stress mean value
τ	shear stress
τ_x, τ_y	shear stresses over the tireprint
τ_{xM}, τ_{yM}	maximum shear stresses
φ	contact angle, angular length of A_P
ω_{eq}	equivalent tire angular velocity
ω, ω_w	angular velocity of a wheel, actual tire angular velocity

Exercises

1. Tireprint size and average normal stress.

    The curb weight of a model of Land Rover $LR3^{TM}$ is

    $$m = 2461\,\text{kg} \approx 5426\,\text{lb}$$

    while the gross vehicle weight might be

    $$m = 3230\,\text{kg} \approx 7121\,\text{lb}$$

    Assume a front to rear load ratio

    $$\frac{F_{z_f}}{F_{z_r}} = \frac{1450\,\text{kg}}{1875\,\text{kg}}$$

    and use the following data

    $$l = 2885\,\text{mm} \approx 113.6\,\text{in} \qquad Tires = 255/55R19$$

    to determine the the size parameters of the tireprints a and b, for the front and rear tires as a function of inflation pressure p. Assume a uniform normal stress on tireprints.

2. Tireprint size, radial tire.

    Holden TK Barina^{TM} is a hatchback car with the following characteristics.

    $$m = 860\,\text{kg} \qquad l = 2480\,\text{mm} \qquad Tires = 185/55R15\ 82V$$

    Assume

    $$\frac{a_1}{a_2} = 1.1$$

    and determine the size of its tireprints for $n = 3$ and assuming a maximum pressure in the tireprint equal to σ_{z_M} [Pa].

3. ★ Equivalent viscous damping.

    The loading and unloading of rubbery materials make a loop in the force-displacement, (F, x), plane.

    (a) Show that the area of the loop determines the wasted energy in one cycle of loading and unloading.

    (b) Explain why the loading curve must be higher than the unloading curve.

    (c) Assuming the material is elastic, determine the equivalent viscous damping c_{eq} such that the amount of the wasted energy in a cycle remains the same.

3. Tire Dynamics   175

4. Tireprint size and load on tire.

   Assuming a uniform upward pressure p under the tire that is carrying a load F_z,

   $$pA_P = p(2a \times 2b) = F_z$$

   (a) determine the half length of the tireprint, a, as a function of pressure p.

   (b) determine the loaded height R_h as a function of a, and eventually as a function of F_z.

   (c) determine the vertical stiffness of a tire based on the curve of F_z versus R_h.

   (d) determine the effective rolling radius R_w as a function of tire load F_z.

5. Average strain in tireprint.

   Consider a loaded tire with a tireprint area of $A_P = 2a \times 2b$. The arc $R_g\varphi$ of the tire becomes $2a$ under the load. Determine the contact strain $\varepsilon = \Delta l/l$ of the tire in the tireprint.

6. Radial jerk.

   Having the function $r = r(\theta)$, determine and plot the radial jerk $\dddot{r}$ in tireprint.

7. ★ Tread velocity in tireprint.

   The speed of tread of a rolling tire in tireprint zone is

   $$v = R_g\omega \frac{\cos\varphi}{\cos^2\theta} \qquad -\varphi < \theta < \varphi$$

   (a) Plot $v/R_g\omega$ as a function of θ for $\varphi = 15\deg$.

   (b) Show that v is equal to $R_g\omega$ at two symmetric points and determine their associated θ.

8. Maximum v_{rel} in tireprint.

   The extremes of v_{rel} in tireprint occurs at the two ends of the tireprint and at the middle point $\theta = 0$. Determine and plot v_{rel} at $\theta = 0$ as a function of φ.

9. Rolling resistance coefficient.

   Alfa Romeo Spider^{TM} has the following characteristics.

   $$\begin{aligned} m &= 1690\,\text{kg} \approx 3725.8\,\text{lb} \qquad l = 2530\,\text{mm} \approx 99.6\,\text{in} \\ Tires &= P225/50R17 \end{aligned}$$

Determine the rolling resistance coefficient μ_r for the front and rear tires of the car at the top speed v_M.

$$v_M = 235.0 \, \text{km/h} \approx 146.0 \, \text{mi/h}$$

Assume $a_1/a_2 = 1.2$ and use $p = 27 \, psi$.

10. Rolling resistance power.

A model of Mitsubishi Galant^{TM} has the following specifications.

$$
\begin{array}{rll}
m & = & 1700 \, \text{kg} \qquad l = 2750 \, \text{mm} \qquad v_M \approx 190 \, \text{km/h} \\
Tires & = & P235/45R18
\end{array}
$$

Assume $a_1/a_2 = 1.2$ and $p = 27 \, psi$ to find the rolling resistance power at the maximum speed.

11. Rolling resistance force.

Is the direction of the rolling resistance force $\mathbf{F}_r$, in $-\mathbf{v}$ or $-\hat{\imath}$? Discuss the conditions and assumptions.

12. Longitudinal slip.

(a) Determine the longitudinal slip s for the tire $P225/50R17$ if $R_w = 0.98 R_g$.

(b) If the speed of the wheel is $v_x = 100 \, \text{km/h}$, what would be the wheel angular velocity ω_w and equivalent angular velocity ω_{eq} of the tire.

13. Cornering and drag force on a tire.

Consider the tire for which we have estimated the lateral force behavior shown in Figure 3.54. If the sideslip angle α is 4 deg and $F_z = 5000 \, \text{N}$, calculate the cornering and drag force on the tire.

14. Required camber angle.

Consider the tire for which we have estimated the behavior shown in Figure 3.65. Assume $F_z = 4000 \, \text{N}$ and we need a lateral force $F_y = -3000 \, \text{N}$. If $\alpha = 4 \, \text{deg}$, what would be the required camber angle γ? Estimate the coefficients C_α and C_γ.

15. High camber angle.

Consider a tire with $C_\gamma = 300 \, \text{N/deg}$ and $C_\alpha = 700 \, \text{N/deg}$. If the camber angle is $\gamma = 18 \, \text{deg}$ how much lateral force will develop for a zero sideslip angle? How much sideslip angle is needed to reduce the value of the lateral force to $F_y = -3000 \, \text{N}$?

16. Sideslip and longitudinal slip.

    Consider the tire for which we have estimated the behavior shown in
    Figure 3.67. Assume a vehicle with that tire is turning with a constant
    speed on a circle such that $\alpha = 4\,\text{deg}$. What should be the sideslip
    angle α if we accelerate the vehicle such that $s = 0.05$, or decelerate
    the vehicle such that $s = -0.05$, to provide the same lateral force?

17. ★ Motion of the air in tire.

    What do you think about the motion of the pressurized air within
    the tires, when the vehicle moves with constant velocity or constant
    acceleration?

4

Driveline Dynamics

The maximum achievable acceleration of a vehicle is limited by two factors: maximum torque at driving wheels, and maximum traction force at
tireprints. The first one depends on engine and transmission performance,
and the second one depends on tire-road friction. In this chapter, we examine engine and transmission performance.

4.1 Engine Dynamics

The maximum attainable power P_e of an internal combustion engine is
a function of the engine angular velocity w_e. This function must be de-
termined experimentally, however, the *power performance function, $P_e = P_e(w_e)$*, can be estimated by a third-order polynomial.

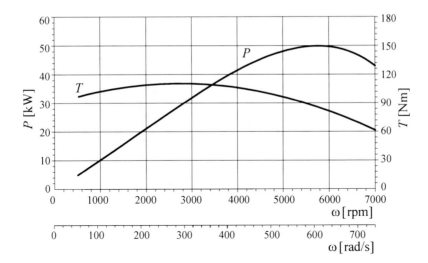

FIGURE 4.1. A sample of power and torque performances for a spark ignition
engine.

$$P_e = \sum_{i=1}^{3} P_i \, \omega_e^i = P_1 \, \omega_e + P_2 \, \omega_e^2 + P_3 \, \omega_e^3 \qquad (4.1)$$

If we indicate the maximum power of an engine by P_M, measured in

R.N. Jazar, *Vehicle Dynamics: Theory and Application*,
DOI 10.1007/978-1-4614-8544-5_4, © Springer Science+Business Media New York 2014

$[W = N m/s]$, which occurs at the angular velocity ω_M, measured in $[rad/s]$ then for spark ignition engines we use

$$P_1 = \frac{P_M}{\omega_M} \qquad P_2 = \frac{P_M}{\omega_M^2} \qquad P_3 = -\frac{P_M}{\omega_M^3} \qquad (4.2)$$

Figure 4.1 illustrates a sample for power performance of a spark ignition engine that provides $P_M = 50\,kW$ at $\omega_M = 586\,rad/s \approx 5600\,rpm$. The curve begins at an angular velocity at which the engine idles smoothly.

For indirect injection Diesel engines we use

$$P_1 = 0.6\frac{P_M}{\omega_M} \qquad P_2 = 1.4\frac{P_M}{\omega_M^2} \qquad P_3 = -\frac{P_M}{\omega_M^3} \qquad (4.3)$$

and for direct injection Diesel engines we use

$$P_1 = 0.87\frac{P_M}{\omega_M} \qquad P_2 = 1.13\frac{P_M}{\omega_M^2} \qquad P_3 = -\frac{P_M}{\omega_M^3} \qquad (4.4)$$

The driving torque T_e of the engine is the torque that provides P_e

$$T_e = \frac{P_e}{\omega_e} = P_1 + P_2\,\omega_e + P_3\,\omega_e^2 \qquad (4.5)$$

Example 131 *Porsche 911^{TM} and Corvette Z06^{TM} engines.*

A model of Porsche 911 turbo has a flat-6 cylinder twin-turbo engine with $3596\,cm^3 \approx 220\,in^3$ total displacement. The engine provides a maximum power $P_M = 353\,kW \approx 480\,hp$ at $\omega_M = 6000\,rpm \approx 628\,rad/s$, and a maximum torque $T_M = 620\,N m \approx 457\,lb\,ft$ at $\omega_e = 5000\,rpm \approx 523\,rad/s$. The car weighs around $1585\,kg \approx 3494\,lb$ and can move from 0 to $96\,km/h \approx 60\,mi/h$ in 3.7 s. Porsche 911 has a top speed of $310\,km/h \approx 193\,mi/h$.

The power performance equation for the Porsche 911 engine has the coefficients

$$P_1 = \frac{P_M}{\omega_M} = \frac{353000}{628} = 562.1\,W\,s \qquad (4.6)$$

$$P_2 = \frac{P_M}{\omega_M^2} = \frac{353000}{628^2} = 0.89507\,W\,s^2 \qquad (4.7)$$

$$P_3 = -\frac{P_M}{\omega_M^3} = -\frac{353000}{628^3} = -1.4253 \times 10^{-3}\,W\,s^3 \qquad (4.8)$$

and, its power performance function is

$$P_e = 562.1\,\omega_e + 0.89507\,\omega_e^2 - 1.4253 \times 10^{-3}\,\omega_e^3 \qquad (4.9)$$

A model of Corvette Z06 uses a V8 engine with $6997\,cm^3 \approx 427\,in^3$ total displacement. The engine provides a maximum power $P_M = 377\,kW \approx 512\,hp$ at $\omega_M = 6300\,rpm \approx 660\,rad/s$ and a maximum torque $T_M =$

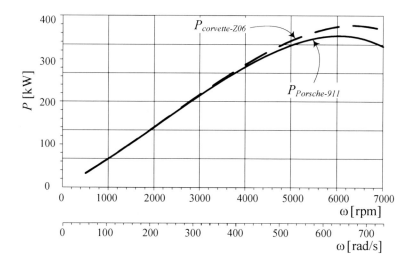

FIGURE 4.2. Power performance curves for the Porsche 911 and Corvette Z06.

$637\,\mathrm{N\,m} \approx 470\,\mathrm{lb\,ft}$ *at* $\omega_e = 4800\,\mathrm{rpm} \approx 502\,\mathrm{rad/s}$. *The Corvette weighs around* $1418\,\mathrm{kg} \approx 3126\,\mathrm{lb}$ *and can move from* 0 *to* $100\,\mathrm{km/h} \approx 62\,\mathrm{mi/h}$ *in* $3.9\,\mathrm{s}$ *in first gear. Its top speed is* $320\,\mathrm{km/h} \approx 198\,\mathrm{mi/h}$.

The power performance equation for the engine of Corvette Z06 has the coefficients

$$P_1 = \frac{P_M}{\omega_M} = \frac{377000}{660} = 571.2\,\mathrm{W\,s} \tag{4.10}$$

$$P_2 = \frac{P_M}{\omega_M^2} = \frac{377000}{660^2} = 0.86547\,\mathrm{W\,s^2} \tag{4.11}$$

$$P_3 = -\frac{P_M}{\omega_M^3} = -\frac{377000}{660^3} = -1.3113 \times 10^{-3}\,\mathrm{W\,s^3} \tag{4.12}$$

and therefore, its power performance function is

$$P_e = 571.2\,\omega_e + 0.86547\,\omega_e^2 - 1.3113 \times 10^{-3}\,\omega_e^3 \tag{4.13}$$

The power performance curves for the Porsche 911 *and Corvette Z06 are plotted in Figure 4.2 for comparison.*

Although there is almost no limit for developing a powerful engine, any engine with power around 100 hp *would have enough for street cars with usual urban applications. It seems that engines with* 1000 hp *pass the limits of application for street cars. However, race cars may have higher power depending on the race regulations. As an example, formula-1 regulations dictate the type of engine permitted. It must be a four-stroke engine, less than* 3000 cm³ *swept volume, no more than ten cylinders, and no more than five valves per cylinder, but there is no limit on power.*

Example 132 *Below the curves* $P_e = P_e(\omega_e)$ *and* $T_e = T_e(\omega_e)$.

An engine can theoretically work at any point under the performance curve $P_e = P_e(\omega_e)$. *Therefore, the power performance curve* $P_e = P_e(\omega_e)$ *indicates the limit of power capacity of an engine. Assume the angular velocity of an engine is kept constant by applying a braking force. Then, by opening the throttle, we produce more power until the throttle is widely open, and the maximum power at that angular velocity is gained.*

Power performance increases by increasing ω_e *and continues to climb up to the maximum power* P_M *at* ω_M. *The power performance then starts decreasing. The torque* $T_e = P_e/\omega_e$ *also increases with* ω_e *but reaches a maximum at a lower angular velocity before the maximum power. Hence, the torque starts decreasing sooner than the power when* ω_e *is increasing. When the power starts decreasing, the torque is very far from its peak value.*

Drivers usually do not feel the engine power, however they may feel the engine torque when accelerating.

Example 133 *Engine efficiency curves.*

Engines are supposed to convert the chemical energy, embedded in the fuel, into mechanical energy at the engine output shaft. Depending on the working conditions, this conversion occurs at a specific efficiency. The constant efficiency contours can be added to the performance map of the engine to show the efficiency at an operating condition. Hence, every point under the curve $P_e = P_e(\omega_e)$ *can be an operating condition at a specific efficiency. The maximum efficiency usually happens around the angular velocity corresponding to the maximum torque when the throttle is almost wide open. A sample of power performance of a spark ignition engine with constant efficiency contours is shown in Figure 4.3.*

Example 134 *Power units.*

There are some different units for expressing power. The metric unit for power is Watt [W].

$$1\,\mathrm{W} = \frac{1\,\mathrm{J}}{1\,\mathrm{s}} = \frac{1\,\mathrm{N\,m}}{1\,\mathrm{s}} \tag{4.14}$$

Horsepower [hp] *is also commonly used in vehicle dynamics and vehicle industries,*

$$1\,\mathrm{W} = 0.001341\,\mathrm{hp} \qquad 1\,\mathrm{hp} = 745.699872\,\mathrm{W} \tag{4.15}$$

However, there are four definitions for horsepower: international, metric, water, and electric. They slightly differ.

$$
\begin{aligned}
1\,\mathrm{hp}(international) &= 745.699872\,\mathrm{W} & (4.16)\\
1\,\mathrm{hp}(electrical) &= 746\,\mathrm{W} & (4.17)\\
1\,\mathrm{hp}(water) &= 746.043\,\mathrm{W} & (4.18)\\
1\,\mathrm{hp}(metric) &= 735.4988\,\mathrm{W} & (4.19)
\end{aligned}
$$

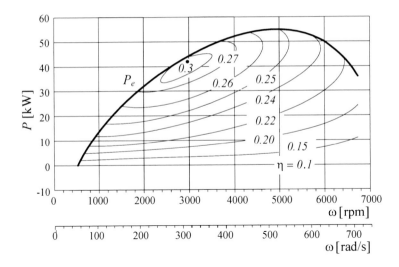

FIGURE 4.3. An example of power performance in a spark ignition engine with constant efficiency contours.

Depending on the application, other units may also be used.

$$1\,\mathrm{W} = 0.239006\,\mathrm{cal/s} \tag{4.20}$$

$$1\,\mathrm{W} = 0.000948\,\mathrm{Btu/s} \tag{4.21}$$

$$1\,\mathrm{W} = 0.737561\,\mathrm{ft\,lb/s} \tag{4.22}$$

James Watt (1736 − 1819) experimented and concluded that a horse can lift a weight of 550 lb for one foot in one second. It means the horse performs work at the rate of 550 ft lb/s ≈ 745.701 W, or 33000 ft lb/ min. Watt then stated that 33000 ft lb/ min of work was equivalent to the power of one horse, or, one horsepower. The following formulas apply for calculating horsepower from a torque measurement in the English unit system:

$$P[\mathrm{hp}] = \frac{T[\mathrm{ft\,lb}]\,\omega[\mathrm{rpm}]}{5252} \qquad P[\mathrm{hp}] = \frac{F[\mathrm{lb}]\,v_x[\mathrm{mi/h}]}{374} \tag{4.23}$$

Example 135 *Fuel consumption at constant speed.*

Consider a vehicle moving straight at a constant speed v_x. The energy required to travel can be calculated by multiplying the power at the drive wheels by time

$$E = Pt = P\frac{d}{v_x} \tag{4.24}$$

where d is the distance traveled and E is the energy needed to turn the wheels. To find the actual energy needed to run the whole vehicle, we should include the coefficients of efficiencies. We use η_e to indicate engine efficiency, H to indicate thermal value of fuel, and ρ_f to indicate density of *the fuel. When the vehicle moves at constant speed, the traction force F_x is equal to the resistance forces. Therefore, the fuel consumption per unit distance, q, is*

$$q = \frac{F_x}{\eta_e \, \eta_t \, \rho_f \, H} \qquad (4.25)$$

The dimension of q in SI is $\left[\,\mathrm{m^3/\,m}\,\right]$, however, liter per $100\,\mathrm{km}$ is more common. In the United States, the fuel consumption of vehicles is called $\left[\,\mathrm{mi/\,gal}\,\right]$.

Example 136 ★ *Changing the curve $P_e = P_e\,(\omega_e)$.*

The whole power performance curve moves up when the engine's compression ratio increases. The angular velocity associated to the engine's peak torque can be moved by changing the cam, header lengths, and intake manifold runner lengths.

The wheel power curve, or the power delivered to the ground, may have a different shape and a different peak ω_e, because of transmission losses. The best result is obtained from a power curve measured by a chassis dynamometer.

Example 137 ★ *Power peak versus torque peak.*

When the engine is operating at its torque peak (say $P_e = 173.4\,\mathrm{kW} \approx 232.5\,\mathrm{hp}$ at $\omega_e = 3600\,\mathrm{rpm}$) in a gear, it is generating some level of torque (say $T_M = 460\,\mathrm{N\,m} \approx 340\,\mathrm{ft\,lb}$ times the overall gearing ratio) at the drive wheels. This is the best performance in that gear. By changing the gear and making the engine to operate at the power peak (say $P_e = 209\,\mathrm{kW} \approx 280\,\mathrm{hp}$ at $\omega_e = 5000\,\mathrm{rpm}$), it delivers less torque $T_e = 400\,\mathrm{N\,m} \approx 295\,\mathrm{ft\,lbf}$. However, it will deliver more torque to the drive wheels, at the same car speed. This is because we gear it up by nearly $39\%(\approx [5000 - 3600]\,/3600)$, while the engine torque is dropped by $13\%(\approx [460 - 400]\,/460)$. Hence, we gain 26% in drive wheel torque at the power peak versus the torque peak, at a given car speed.

As long as the performance curves of engines are similar to those in Figure 4.1, any engine speed, other than the power peak speed ω_M, at a given car speed will provide a lower torque value at the drive wheels. Therefore, theoretically the best top speed will always occur when the vehicle is operating at its power peak.

A car running at its power peak can accelerate no faster at the same vehicle speed. There is no better gear to choose, even if another gear would place the engine closer to its torque peak. A car running at peak power at a given vehicle speed is delivering the maximum possible torque to the tires, although the engine will not be running at its torque peak. The transmission amplifies the torque coming from the engine by a factor equal to the gear ratio.

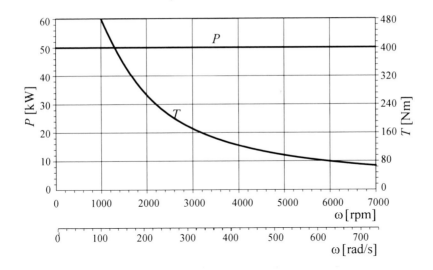

FIGURE 4.4. Power and torque performance curves for an ideal engine.

Example 138 ★ *Ideal engine performance.*

It is said that an ideal engine is one that produces a constant power regardless of speed. For such an ideal engine we have

$$P_e = P_0 \qquad T_e = \frac{P_0}{\omega_e} \qquad (4.26)$$

Figure 4.4 depicts a sample of the power and torque performance curves for an ideal engine having $P_0 = 50\,\text{kW}$.

In vehicle dynamics, we introduce a gearbox to keep the engine running at the maximum power or in a working range around the maximum power. So, practically we keep the power of the engine, and therefore, the power at wheels constant at the maximum value. Hence, the torque at the wheels should be similar to the torque of an ideal engine.

A constant power performance is an applied approximation for electrical motors.

Another idea of an ideal engine is the one that provides a linear torque-speed relationship. For such an ideal engine we have

$$T_e = C_e\,\omega_e \qquad P_e = C_e\,\omega_e^2 \qquad (4.27)$$

However, internal combustion engines do not work like this engine. Figure 4.5 illustrates such an ideal performance for $C_e = 0.14539$.

Example 139 ★ *Maximum power and torque at the same ω_M.*

Ideal performance for an engine would be having maximum power and maximum torque at the same angular velocity ω_M. However, it is impossible to have such an engine because the maximum torque T_M of a spark ignition

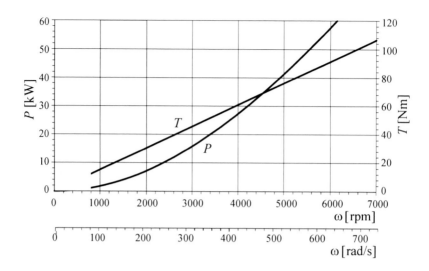

FIGURE 4.5. Performance curves of an ideal engine having a linear torque-speed relationship $T_e = 0.14539\,\omega_e$.

engine occurs at

$$\frac{dT_e}{d\omega_e} = P_2 + 2P_3\,\omega_e = 0 \tag{4.28}$$

$$\omega_e = \frac{-P_2}{2P_3} = \frac{P_M/\omega_M^2}{2P_M/\omega_M^3} = \frac{1}{2}\omega_M \tag{4.29}$$

that is half of the speed at which the power is maximum.
  When the torque is maximum, the power is at

$$P_e = P_1\frac{\omega_M}{2} + P_2\left(\frac{\omega_M}{2}\right)^2 + P_3\left(\frac{\omega_M}{2}\right)^3 = \frac{5}{8}P_M \tag{4.30}$$

However, when the power is maximum at $\omega_e = \omega_M$, the torque is

$$T_e = \frac{1}{\omega_M}P_M \tag{4.31}$$

4.2 Driveline and Efficiency

We use the word *driveline*, equivalent to transmission, to call the systems and devices that transfer torque and power from the engine to the drive wheels of a vehicle. Most vehicles use one of two common transmission types: manual gear transmission, and automatic transmission with torque convertor. A driveline includes the engine, clutch, gearbox, propeller shaft,

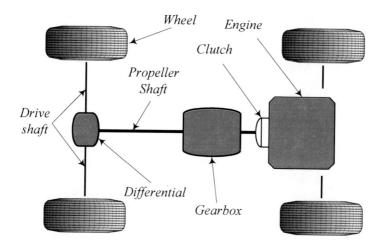

FIGURE 4.6. Driveline components of a rear wheel drive vehicle.

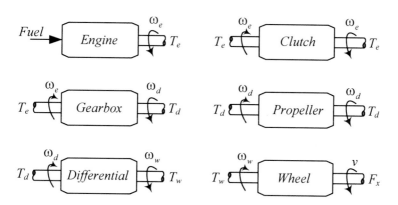

FIGURE 4.7. The input and output torque and angular velocity of each driveline component.

differential, drive shafts, and drive wheels. Figure 4.6 illustrates how the driveline for a rear-wheel-drive vehicle is assembled.

The *engine* is the power source in the driveline. The output from the engine is an engine torque, T_e, at an associated engine speed ω_e.

The *clutch* connects and disconnects the engine to the rest of the driveline when the vehicle is equipped with a manual gearbox.

The *gearbox* is used to change the transmission ratio between the engine and the drive wheels.

The *propeller shaft* connects the gearbox to the differential. The propeller shaft does not exist in front-engined front-wheel-drive and rear-engined rear-wheel-drive vehicles. In those vehicles, the differential is integrated with the gearbox in a unit that is called the *transaxle*.

The *differential* is a constant transmission ratio gearbox that allows the drive wheels to have different speeds. So, they can handle the car in a curve.

The *drive shafts* connect the differential to the drive wheels.

The *drive wheels* transform the engine torque to a traction force on the road.

The input and output torque and angular velocity for each device in a driveline are indicated in Figure 4.7.

The available power P_w at the drive wheels is

$$P_w = \eta P_e \tag{4.32}$$

where $\eta < 1$ indicates the *overall efficiency* between the engine and the drive wheels

$$\eta = \eta_c \, \eta_t \tag{4.33}$$

$\eta_c < 1$ is the *convertor efficiency* and $\eta_t < 1$ is the *transmission efficiency*.

The relationship between the angular velocity of the engine and the velocity of the vehicle is

$$v_x = \frac{R_w \, \omega_e}{n_g \, n_d} \tag{4.34}$$

where n_g is the transmission ratio of the gearbox, n_d is the transmission ratio of the differential, ω_e is the engine angular velocity, and R_w is the effective tire radius.

Transmission ratio or *gear reduction ratio* of a gearing device, n, is the ratio of the input velocity to the output velocity

$$n = \frac{\omega_{in}}{\omega_{out}} \tag{4.35}$$

while the *speed ratio* ω_r is the ratio of the output velocity to the input velocity.

$$\omega_r = \frac{\omega_{out}}{\omega_{in}} \tag{4.36}$$

Proof. The engine is connected to the drive wheels through a *driveline*. Because of friction in the driveline, especially in the gearbox and torque convertor, the power at the drive wheels is always less than the power at the engine output shaft. The ratio of output power to input power is a number called *efficiency*

$$\eta = \frac{P_{out}}{P_{in}} \tag{4.37}$$

If we show the efficiency of transmission by η_t and the efficiency of torque convertor by η_c, then the overall efficiency of the driveline is $\eta = \eta_c\,\eta_t$. The power at the wheel is the output power of driveline $P_{out} = P_w$ and the engine power is the input power to the driveline $P_{in} = P_e$. Therefore,

$$P_w = \eta P_e \tag{4.38}$$

Figure 4.8 illustrates a driving wheel with radius R_w that is rolling with angular velocity ω_w on the ground and moving with velocity v_x.

$$v_x = R_w\,\omega_w \tag{4.39}$$

There are two gearing devices between the engine and the drive wheel: gearbox and differential. Assigning n_g for the transmission ratio of the gearbox and n_d for the transmission ratio of the differential, the overall transmission ratio of the driveline is

$$n = n_g\,n_d \tag{4.40}$$

Therefore, the angular velocity of the engine ω_e is n times of the angular velocity of the drive wheel ω_w.

$$\omega_e = n\,\omega_w = n_g\,n_d\,\omega_w \tag{4.41}$$

These yields

$$v_x = \frac{R_w\,\omega_e}{n_g\,n_d} \tag{4.42}$$

∎

Example 140 *Front and rear-engined, front and rear drive.*

The engine may be installed in the front or back of a car. They are called front-engined and rear-engined vehicle respectively. The driving wheels may also be the front, the rear, or all wheels. Therefore, there are six possible combinations. Out of those six combinations, the front-engined front-wheel-drive FWD, front-engined rear-wheel-drive RWD, and front-engined all-wheel-drive AWD vehicles are the most common. There are only a few manufacturers that make cars with rear-engined rear-wheel-drive. However, there is no rear-engined front-wheel-drive vehicle.

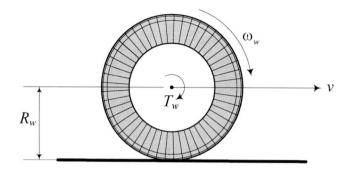

FIGURE 4.8. A tire with radius R_w rolling on the ground and moving with velocity v and angular velocity ω_w.

Example 141 *Torque at the wheel.*

The power at the wheel is $P_w = \eta P_e$, and the angular velocity at the wheel is $\omega_w = \omega_e / (n_g\, n_d)$. Knowing $P = T\omega$, we find out that the available torque at the wheel, T_w, is

$$T_w = \frac{P_w}{\omega_w} = \eta\, n_g\, n_d\, \frac{P_e}{\omega_e} = \eta\, n_g\, n_d\, T_e \qquad (4.43)$$

Example 142 *Power law.*

For any mechanical device in the driveline of a car, there is a simple law to remember.

$$\begin{aligned} Power\ out\ &=\ Power\ in\ minus\ losses \\ P_{out}\ &=\ P_{in} - P_{loss} \end{aligned} \qquad (4.44)$$

Also, because of

$$\begin{aligned} Power\ &=\ Torque \times angular\ velocity \\ P\ &=\ T\omega \end{aligned} \qquad (4.45)$$

any gearing device in the driveline of a car can reduce or increase the input torque by increasing or decreasing the angular velocity.

Example 143 ★ *Volumetric, thermal, and mechanical efficiencies.*

There is an efficiency between the attainable power in fuel and the power available at the engine's output shaft.

$$\eta' = \eta_V\, \eta_T\, \eta_M \qquad (4.46)$$

η_V is the engine **volumetric efficiency**, η_T is the **thermal efficiency**, and η_M is the **mechanical efficiency**.

Volumetric efficiency η_V identifies how much fueled air gets into the cylinder.

The fueled air mixture that fills the cylinder volume in the intake stroke is what will be used to create the power. Volumetric efficiency η_V indicates the amount of fueled air in the cylinder relative to atmospheric air. If the cylinder is filled with fueled air at atmospheric pressure, then the engine has 100% volumetric efficiency. Super and turbo chargers increase the pressure entering the cylinder, giving the engine a volumetric efficiency greater than 100%. However, if the cylinder is filled with less than the atmospheric pressure, then the engine has less than 100% volumetric efficiency. Engines typically run between 80% and 100% of η_V.

Volumetric efficiency η_V can be changed by any occurrence that affects the fueled air flow into the cylinder. The power of an engine is proportionally dependent on the mass ratio of fuel/air that gets into the cylinders of the engine.

Thermal efficiency η_T identifies how much of the fuel is converted to usable power.

Although having more fueled air into the cylinder means more fuel energy is available to make power, not all of the available energy converts to mechanical energy. The best engines can convert only about 1/3 of the chemical energy to mechanical energy.

Thermal efficiency is changed by the compression ratio, ignition timing, plug location, and chamber design. Low compression engines may have $\eta_T \approx 0.26$. A high compression racing engine may have $\eta_T \approx 0.34$. Therefore, racing engines may produce about 30% more power because of their higher η_T.

Any improvement in the thermal efficiency η_T significantly improves the final power that the engine produces. Therefore, a huge investment is expended in research to improve η_T.

Mechanical efficiency η_M identifies how much power is consumed by the engine to run itself.

Some of the produced power is consumed by the engine's moving parts. It takes power to overcome the friction between parts and to run engine accessories. So, depending on how much fuel goes into the cylinder and how much converts to power, some of this power is used by the engine to run itself. The leftover power is what we can measure on an engine dynamometer. The difference between the engine output power and the generated power in the cylinders provides the mechanical efficiency η_M.

Mechanical efficiency is affected by mechanical components of the engine or the devices attached to the engine. It also depends on the engine speed. The greater the speed, the more power it takes to turn the engine. This means that η_M drops with speed. The mechanical efficiency η_M is also called **friction power** because it indicates how much power is needed to overcome the engine friction.

The engine power performance curve supplied by a car manufacturer is usually the gross engine performance and does not include the mechanical efficiency. Therefore, the effective engine power available at the transmis-

sion input shaft is reduced by the power needed for accessories such as the fan, electric alternator, power steering pump, water pump, braking system, and air conditioning compressor.

4.3 Gearbox and Clutch Dynamics

The internal combustion engine cannot operate below a minimum engine speed ω_{min}. Consequently, the vehicle cannot move slower than a minimum speed v_{min} while the engine is engaged to the drive wheels.

$$v_{min} = \frac{R_w\,\omega_{min}}{n_g\,n_d} \tag{4.47}$$

At starting and stopping stages of motion, the vehicle needs to have speeds less than v_{min}. A clutch or a torque converter must be used for starting, stopping, and gear shifting.

Consider a vehicle with only one drive wheel. The forward velocity v_x of the vehicle is proportional to the angular velocity of the engine ω_e, and the tire *traction force* F_x is proportional to the engine torque T_e

$$\omega_e = \frac{n_i\,n_d}{R_w}\,v_x \tag{4.48}$$

$$T_e = \frac{1}{\eta}\frac{R_w}{n_i n_d}F_x \tag{4.49}$$

where R_w is the effective tire radius, n_d is the differential transmission ratio, n_i is the gearbox transmission ratio in gear number i, and η is the overall driveline efficiency. Equation (4.48) is called the *speed equation*, and Equation (4.49) is called the *traction equation*. The speed equation will be used to design the gear ratios of the gearbox of the vehicle.

Proof. The froward velocity v_x of a driving wheel with radius R_w is

$$v_x = R_w\,\omega_w \tag{4.50}$$

and the traction force F_x on the driving wheel is

$$F_x = \frac{T_w}{R_w} \tag{4.51}$$

T_w is the applied spin torque on the wheel, and ω_w is the wheel angular velocity.

The wheel inputs T_w and ω_w are the output torque and angular velocity of differential. The differential input torque T_d and angular velocity ω_d are

$$T_d = \frac{1}{\eta_d\,n_d}T_w \tag{4.52}$$

$$\omega_d = n_d\,\omega_w \tag{4.53}$$

where n_d is the differential transmission ratio and η_d is the differential efficiency.

The differential inputs T_d and ω_d are the output torque and angular velocity to the vehicle's gearbox. The engine's torque T_e and angular velocity ω_e are the inputs of the gearbox. The input-output relationships for a gearbox depend on the engaged gear ratio n_i.

$$T_e = \frac{1}{\eta_g\, n_i} T_d \tag{4.54}$$

$$\omega_e = n_i \omega_d \tag{4.55}$$

η_g is the gearbox efficiency, and n_i is the gear reduction ratio in the gear number i. Therefore, the forward velocity of a driving wheel v_x, is proportional to the engine angular velocity ω_e, and the tire traction force F_x is proportional to the engine torque T_e, when the driveline is engaged to the engine.

$$\omega_e = \frac{n_i\, n_d}{R_w} v_x \tag{4.56}$$

$$T_e = \frac{1}{\eta_g\eta_d}\frac{1}{n_i n_d} T_w = \frac{1}{\eta_g\eta_d}\frac{R_w}{n_i n_d} F_x = \frac{1}{\eta}\frac{R_w}{n_i n_d} F_x \tag{4.57}$$

Having the torque performance function $T_e = T_e\left(\omega_e\right)$ enables us to determine the wheel torque T_w as a function of vehicle speed v_x at each gear ratio n_i.

$$T_w = \eta\, n_i n_d\, T_e\left(\omega_e\right) \tag{4.58}$$

Using the approximate equation (4.5) for T_e provides

$$
\begin{aligned}
T_w &= \eta\, n_i n_d \left(P_1 + P_2 \left(\frac{n_i\, n_d}{R_w} v_x\right) + P_3 \left(\frac{n_i\, n_d}{R_w} v_x\right)^2 \right)\\
&= \eta P_1 n_d n_i + \eta\frac{P_2}{R_w} n_d^2 n_i^2 v_x + \eta\frac{P_3}{R_w^2} n_d^3 n_i^3 v_x^2 \tag{4.59}
\end{aligned}
$$

■

Example 144 *A six-gear gearbox.*
Consider an inefficient passenger car with the following specifications:

$$
\begin{aligned}
m &= 1550\,\text{kg} \qquad R_w = 0.326\,\text{m} \qquad \eta = 0.24\\
torque &= 392\,\text{N m} \;\, at\; 4400\,\text{rpm} \approx 460.7\,\text{rad/s} \tag{4.60}\\
power &= 206000\,\text{W} \;\, at\; 6800\,\text{rpm} \approx 712.1\,\text{rad/s}
\end{aligned}
$$

1st gear ratio $= n_1 = 3.827$ *2nd gear ratio* $= n_2 = 2.36$
3rd gear ratio $= n_3 = 1.685$ *4th gear ratio* $= n_4 = 1.312$
5th gear ratio $= n_5 = 1$ *6th gear ratio* $= n_6 = 0.793$
reverse gear ratio $= n_r = 3.28$ *final drive ratio* $= n_d = 3.5451$

$$\tag{4.61}$$

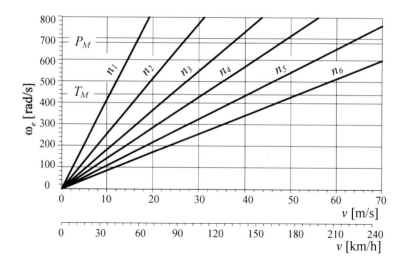

FIGURE 4.9. A sample of a gear-speed plot for a gearbox.

Based on the speed equation (4.48),

$$\omega_e = \frac{n_i\, n_d}{R_w} v_x = \frac{3.5451 n_i}{0.326} v_x = 10.875 n_i\, v_x \qquad (4.62)$$

we find the gear-speed relationships at different gears and plot them as is shown in Figure 4.9.

The angular velocities associated to maximum power and maximum torque are indicated by thin lines. The power and torque performance equations for the engine can be approximated by

$$P_e = 289.29\, \omega_e + 0.40624\, \omega_e^2 - 5.7049 \times 10^{-4}\, \omega_e^3 \qquad (4.63)$$
$$T_e = 289.29 + 0.406\,24 \omega_e - 5.704\,9 \times 10^{-4} \omega_e^2 \qquad (4.64)$$

because

$$P_1 = \frac{P_M}{\omega_M} = \frac{206000}{712.1} = 289.29\ \mathrm{W/s} \qquad (4.65)$$

$$P_2 = \frac{P_M}{\omega_M^2} = \frac{206000}{712.1^2} = 0.40624\ \mathrm{W/s^2} \qquad (4.66)$$

$$P_3 = -\frac{P_M}{\omega_M^3} = -\frac{206000}{712.1^3} = -5.7049 \times 10^{-4}\ \mathrm{W/s^3} \qquad (4.67)$$

Using the torque equation (4.64) and the traction equation (4.58), we plot the wheel torque as a function of vehicle speed at different gears.

$$
\begin{aligned}
T_w &= \eta\, n_i n_d\, T_e \\
&= \eta\, n_i n_d \left(289.29 + 0.406\,24 \omega_e - 5.704\,9 \times 10^{-4} \omega_e^2 \right) \\
&= -5.7405 \times 10^{-2} n_i^3 v_x^2 + 3.7588 n_i^2 v_x + 246.13 n_i \qquad (4.68)
\end{aligned}
$$

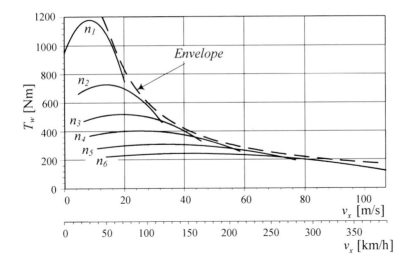

FIGURE 4.10. Wheel torque-speed Equation (4.68) at each gear n_i of a gearbox, and the envelope curve simulating an ideal engine behavior.

Figure 4.10 shows the wheel torque-speed Equation (4.68) at each gear n_i. The envelope curve for the series of torque-speed equations is similar to the torque curve of a constant power ideal engine.

Example 145 ★ *Envelope curve for torque-speed family.*
The torque-speed equation of a car is similar to Equation (4.68) that is a second degree of speed having the gear ratio $n = n_i$ as a parameter.

$$T = an^3v^2 + bn^2v + cn \tag{4.69}$$

A variation of the parameter generates a series of curves called family. An envelope is a curve tangent to all members of the family. To find the envelope of a family, we should eliminate the parameter between the equation of the family and its derivative with respect to the parameter. The derivative of the family (4.69) with respect to the parameter n

$$\frac{\partial T}{\partial n} = 3an^2v^2 + 2bnv + c = 0 \tag{4.70}$$

yields

$$n = \frac{-b \pm \sqrt{b^2 - 3ac}}{3av} \tag{4.71}$$

Substituting the positive parameter back into the equation of the family provides the equation of the envelop.

$$T = \frac{\left(b + \sqrt{b^2 - 3ac}\right)\left(b^2 + b\sqrt{b^2 - 3ac} - 6ac\right)}{27a^2v} \tag{4.72}$$

Therefore, the equation of envelope for the wheel torque-speed family at different gears is equivalent to

$$T \equiv \frac{C}{v} \tag{4.73}$$

where C is a constant.

$$C = \frac{2b^3 - 9abc + 2\left(b^2 - 3ac\right)^{3/2}}{27a^2} \tag{4.74}$$

Such a torque equation belongs to an ideal constant power device introduced in Example 138.

Example 146 *Mechanical and hydraulic clutches.*

Mechanical clutches are widely used in passenger cars and are normally in the form of a dry single-disk clutch. The adhesion between input and output shafts is produced by circular disks that rub against each other.

Engagement begins with the engine running at $\omega_e = \omega_{min}$. The clutch then being released gradually from time $t = 0$ to $t = t_1$. Therefore, the transmitted torque T_c from the engine to the gearbox increases almost linearly in time from $T_c = 0$ to the maximum value of $T_c = T_{c_1}$ that can be handled in slipping mode. The transmitted torque remains constant until the input and output disks stick together and a speed equality is achieved. At this time, the clutch is fully engaged and $T_c = T_e$.

The transmitted torque T_c should overcome the resistance force and the vehicle should accelerate sometime in $0 < t \leq t_1$. The magnitude of the transferable torque depends on the applied force between the disks, the frictional coefficient between clutch disks, the effective frictional area, and the number of frictional pairs. The axial force is generally produced by a preloaded spring. The driver can control the spring force by using the clutch pedal, and adjust the transferred torque.

The hydraulic clutch consists of a pump wheel connected to the engine and a clutch-ended turbine that is equipped with radial vanes. A torque is transferred between the pump wheel and the turbine over fluid, which is accelerated by the pump and decelerated in the turbine. The hydraulic clutch is also called Foettinger clutch.

The transferred torque can be calculated according to the Foettinger's law

$$T_c = C_c \rho \omega_p^2 D^2 \tag{4.75}$$

where C_c is the slip factor, ρ is the oil density, ω_p is the pump angular velocity, and D is the clutch diameter.

Example 147 *Acceleration capacity at different speed.*

Assume an engine is working at speed ω_M associated to the maximum power P_M.

$$P_M = T_e\, \omega_M = \frac{1}{\eta} F_x\, v_x \tag{4.76}$$

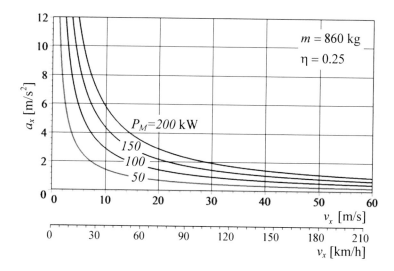

FIGURE 4.11. An example for the acceleration capacity a_x as a fucntion of forward speed v_x.

Substituting

$$F_x = ma_x \tag{4.77}$$

indicates that

$$P_M = \frac{m}{\eta} a_x\, v_x \tag{4.78}$$

and therefore,

$$a_x = P_M \frac{\eta}{m} \frac{1}{v_x} \tag{4.79}$$

*Equation (4.79) is called **acceleration capacity** and expresses the achievable acceleration of a vehicle at speed v_x. The acceleration capacity decreases by increasing velocity and increases by the maximum power. As an example, Figure 4.11 depicts the acceleration capacity a_x as a function of the forward speed v_x for a vehicle with mass $m = 860\,kg$, and a low efficiency $\eta = 0.25$ for the maximum powers of $P_M = 50\,kW \approx 67\,hp$, $P_M = 100\,kW \approx 134\,hp$, $P_M = 150\,kW \approx 201\,hp$, and $P_M = 200\,kW \approx 268\,hp$.*

Example 148 *Power-limited and traction-limited accelerations.*

   Acceleration capacity is power-limited and is based on the assumption that the driving force does not reach the tire traction limit. Therefore, the vehicle reaches its peak acceleration because the engine cannot deliver any more power.

   The traction-limited acceleration happens when the engine delivers more power, but vehicle acceleration is limited because the tires cannot transmit any more driving force to the ground. Equation $F_x = \mu_x F_z$ gives the maximum transmittable force. If more driving torque is applied to the wheel, *the tire slips and enters the dynamic friction regime where the coefficient of friction, and hence the traction force, are less.*

Example 149 ★ *Gearbox stability condition.*

Consider a vehicle moving at speed v_x when the gearbox is engaged in gear number i with transmission ratio n_i. To be safe, we have to select the transmission ratios such that when the engine reaches the maximum torque it can shift to a lower gear n_{i-1} without reaching the maximum permissible engine speed. The maximum permissible engine speed is usually indicated by a red line or red region on engine rpm indicator.

Let us show the engine speed for the maximum torque T_M by $\omega_e = \omega_T$. The speed of the vehicle at $\omega_e = \omega_T$ is

$$v_x = \frac{R_w}{n_i \, n_d} \omega_T \tag{4.80}$$

When we shift the gear to n_{i-1} the engine speed ω_e jumps to a higher speed $\omega_e = \omega_{i-1} > \omega_T$ at the same vehicle speed

$$\omega_{i-1} = \frac{n_{i-1} \, n_d}{R_w} v_x \tag{4.81}$$

The stability condition requires that ω_{i-1} be less than the maximum permissible engine speed ω_{Max}

$$\omega_{i-1} \leq \omega_{Max} \tag{4.82}$$

Using Equations (4.80) and (4.81), we may define the following condition between transmission ratios at two successive gears and the engine speed:

$$\frac{\omega_{i-1}}{\omega_i} = \frac{\omega_{Max}}{\omega_T} = \frac{n_{i-1}}{n_i} \tag{4.83}$$

A constant relative gear ratio, at a constant vehicle speed, is a simple rule for a stable gearbox design

$$\frac{n_{i-1}}{n_i} = c_g \tag{4.84}$$

Example 150 *Transmission ratios and stability condition.*

Consider a passenger car with the following gearbox transmission ratios:

1st gear ratio $= n_1 = 3.827$ *2nd gear ratio* $= n_2 = 2.36$
3rd gear ratio $= n_3 = 1.685$ *4th gear ratio* $= n_4 = 1.312$
5th gear ratio $= n_5 = 1$ *6th gear ratio* $= n_6 = 0.793$
final drive ratio $= n_d = 3.5451$

$$\tag{4.85}$$

The stability condition requires that $n_{i-1}/n_i = cte$. Examination of the gear ratios indicates that the relative gear ratios are not constant.

$$\frac{n_5}{n_6} = \frac{1}{0.793} = 1.261 \qquad \frac{n_4}{n_5} = \frac{1.312}{1} = 1.312$$

$$\frac{n_3}{n_4} = \frac{1.685}{1.312} = 1.2843 \qquad \frac{n_2}{n_3} = \frac{2.36}{1.685} = 1.4 \qquad (4.86)$$

$$\frac{n_1}{n_2} = \frac{3.827}{2.36} = 1.6216$$

We may change the gear ratios to have $n_{i-1}/n_i = cte$. Let us start from the highest gears and find the lower gears using $c_g = n_6/n_5 = 1.261$.

$$n_5 = 1$$
$$n_6 = 0.793$$
$$n_4 = c_g n_5 = 1.261$$
$$n_3 = c_g n_4 = 1.261 \times 1.261 = 1.59$$
$$n_2 = c_g n_3 = 1.261 \times 1.59 = 2$$
$$n_1 = c_g n_2 = 1.261 \times 2 = 2.522 \qquad (4.87)$$

We may also start from the first two gears and find the higher gears using $c_g = n_1/n_2 = 3.827/2.36 = 1.6216$.

$$n_1 = 3.827$$
$$n_2 = 2.36$$
$$n_3 = \frac{n_2}{c_g} = \frac{2.36}{1.6216} = 1.455$$
$$n_4 = \frac{n_3}{c_g} = \frac{1.455}{1.6216} = 0.897$$
$$n_5 = \frac{n_4}{c_g} = \frac{0.897}{1.6216} = 0.553$$
$$n_6 = \frac{n_5}{c_g} = \frac{0.553}{1.6216} = 0.341 \qquad (4.88)$$

None of these two sets shows a practical design. The best way to apply a constant relative ratio is to use the first and final gears and fit four intermittent gears such that $n_{i-1}/n_i = cte$. Using n_1 and n_6 we have,

$$\frac{n_1}{n_6} = \frac{3.827}{0.793} = \frac{n_1}{n_2}\frac{n_2}{n_3}\frac{n_3}{n_4}\frac{n_4}{n_5}\frac{n_5}{n_6} = c_g^5 \qquad (4.89)$$

and therefore,

$$c_g = 1.37 \qquad (4.90)$$

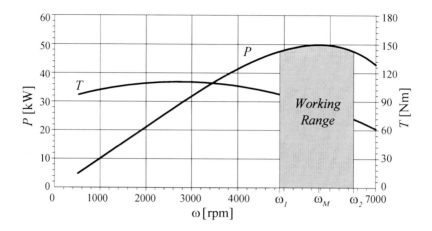

FIGURE 4.12. The angular velocity range (ω_1, ω_2) around ω_M, and engine's working range.

Now we are able to find the gear ratios to meet the first and sixth gear ratios requirements.

$$
\begin{aligned}
n_1 &= 3.827 \\
n_2 &= \frac{n_1}{c_g} = \frac{3.827}{1.37} = 2.793 \\
n_3 &= \frac{n_2}{c_g} = \frac{2.793}{1.37} = 2.039 \\
n_4 &= \frac{n_3}{c_g} = \frac{2.039}{1.37} = 1.488 \\
n_5 &= \frac{n_4}{c_g} = \frac{1.488}{1.37} = 1.086 \\
n_6 &= 0.793
\end{aligned}
\tag{4.91}
$$

4.4   Gearbox Design

The speed and traction equations (4.48) and (4.49) can be used to calculate the gear ratios of a gearbox and the vehicle performance. Theoretically the engine should work at its maximum power to have the best performance. However, to control the speed of the vehicle, we need to vary the engine's angular velocity. Hence, we pick an angular velocity range (ω_1, ω_2) around ω_M, which is associated to the maximum power P_M, and sweep the range (ω_1, ω_2) repeatedly at different gears. The range (ω_1, ω_2) is called the engine's *working range* as is shown in Figure 4.12.

As a general guideline, we may use the following recommendations to design the transmission ratios of a vehicle gearbox:

1. We may design the differential transmission ratio n_d and the final gear n_n such that the final gear n_n is a direct gear, $n_n = 1$, when the vehicle is moving at the moderate highway speed. Using $n_n = 1$ implies that the input and output of the gearbox are directly connected with each other. Direct engagement maximizes the mechanical efficiency of the gearbox.

2. We may design the differential transmission ratio n_d and the final gear n_n such that the final gear n_n is a direct gear, $n_n = 1$, when the vehicle is moving at the maximum attainable speed.

3. The first gear n_1 may be designed by the maximum desired torque at driving wheels. Maximum torque is determined by the slope of a desired climbing road.

4. We can find the intermediate gears using the gear stability condition. Stability condition provides that the engine speed must not exceed the maximum permissible speed if we gear down from n_i to n_{i-1}, when the engine is working at the maximum torque in n_i.

5. The value of c_g for relative gear ratios

$$\frac{n_{i-1}}{n_i} = c_g \tag{4.92}$$

can be chosen in the range.

$$1 \leq c_g \leq 2 \tag{4.93}$$

To determine the middle gear ratios, there are two recommended methods:

1. Geometric ratios
2. Progressive ratios

4.4.1 Geometric Ratio Gearbox Design

When the jump of engine speed in any two successive gears is constant at a vehicle speed, we call the gearbox *geometric*. The design condition for a geometric gearbox is

$$n_i = \frac{n_{i-1}}{c_g} \tag{4.94}$$

where c_g is the constant relative gear ratio and is called *step jump*.

Proof. A geometric gearbox has constant engine speed jump in any gear shift. So, a geometric gearbox must have a gear-speed plot such as that shown in Figure 4.13.

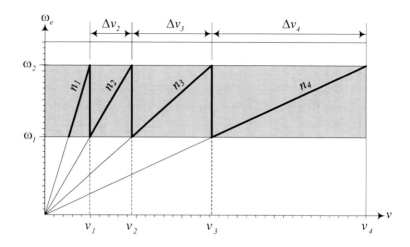

FIGURE 4.13. A gear-speed plot for a geometric gearbox design.

The engine working range is defined by two speeds (ω_1, ω_2)

$$\{(\omega_1, \omega_2), \omega_1 < \omega_M < \omega_2\} \qquad (4.95)$$

When the engine reaches the maximum speed ω_2 in the gear number i with ratio n_i, we gear up to n_{i+1} to jump the engine speed down to ω_1. The engine's speed jump is kept constant for any gear change from n_i to n_{i+1}. Employing the speed equation (4.48), we have

$$\begin{aligned} \triangle\omega &= \omega_2 - \omega_1 = \frac{n_{i-1}\, n_d}{R_w} v_x - \frac{n_i\, n_d}{R_w} v_x \\ &= (n_{i-1} - n_i) \frac{n_d}{R_w} v_x \qquad (4.96) \end{aligned}$$

and therefore,

$$\begin{aligned} \frac{\omega_2 - \omega_1}{\omega_1} &= \frac{n_{i-1} - n_i}{n_i} \\ \frac{\omega_2}{\omega_1} - 1 &= \frac{n_{i-1}}{n_i} - 1 \\ \frac{\omega_2}{\omega_1} &= \frac{n_{i-1}}{n_i} = c_g \qquad (4.97) \end{aligned}$$

Let us indicate the maximum vehicle speed in gear n_i, when $\omega_e = \omega_2$, by v_i and in gear n_{i-1} by v_{i-1}, then,

$$\omega_2 = \frac{n_i\, n_d}{R_w} v_i = \frac{n_{i-1}\, n_d}{R_w} v_{i-1} \qquad (4.98)$$

and therefore, the maximum speed in gear i to the maximum speed in gear $i-1$ is equal to the inverse of the gear ratios

$$c_g = \frac{n_{i-1}}{n_i} = \frac{v_i}{v_{i-1}} \qquad (4.99)$$

The change in vehicle speed between gear n_{i-1} and n_i is indicated by

$$\triangle v_i = v_i - v_{i-1} \tag{4.100}$$

and is called *speed span*.

Having the step jump c_g and knowing the maximum speed v_i of the vehicle in gear n_i, are enough to find the maximum velocity of the car in the other gears

$$v_i = c_g v_{i-1} \tag{4.101}$$

$$v_{i-1} = \frac{1}{c_g} v_i \tag{4.102}$$

$$v_{i+1} = c_g v_i \tag{4.103}$$

The speed span of vehicle with a geometric gearbox increases by gearing up. ∎

Example 151 *A gearbox with three gears.*
Consider an $m = 860$ kg car having an engine with $\eta = \eta_d \eta_g = 0.84$ and the power-speed relation of

$$P_e = 100 - \frac{100}{398^2} (\omega_e - 398)^2 \text{ kW} \tag{4.104}$$

where ω_e is in $[\text{rad}/\text{s}]$. We define the working range for the engine as

$$272 \,\text{rad}/\text{s} \,(\approx 2600 \,\text{rpm}) \le \omega_e \le 524 \,\text{rad}/\text{s} \,(\approx 5000 \,\text{rpm}) \tag{4.105}$$

when the power is $100 \,\text{kW} \ge P_e \ge 90 \,\text{kW}$. The power performance curve (4.104) is illustrated in Figure 4.14 and the working range is shaded.
The differential of the vehicle uses $n_d = 4$, and the effective tire radius is $R_w = 0.326$ m. We would like to design a three-gear geometric gearbox to have the minimum time required to reach the speed $v_x = 100 \,\text{km}/\text{h} \approx 27.78 \,\text{m}/\text{s} \approx 62 \,\text{mi}/\text{h}$. We assume that the total resistance force is constant, and the engine cannot accelerate the car at $v_x = 180 \,\text{km}/\text{h} = 50 \,\text{m}/\text{s} \approx 112 \,\text{mi}/\text{h}$ anymore. Assume that every gear change takes 0.47 s and we need $t_0 = 2.58$ s to adjust the engine speed to the car speed in the first gear.
Using the speed equation (4.48), the relationship between vehicle and engine speeds is

$$v_x = \frac{R_w}{n_d n_i} \omega_e = \frac{0.326}{4 n_i} \omega_e \tag{4.106}$$

At the maximum speed $v_x = 50 \,\text{m}/\text{s}$, the engine is rotating at the upper limit of the working range $\omega_e = 524 \,\text{rad}/\text{s}$ and the gearbox is operating in the third gear. Therefore, Equation (4.106) provides us with

$$n_3 = \frac{0.326}{4} \frac{\omega_e}{v_x} = \frac{0.326}{4} \frac{524}{50} = 0.85412 \tag{4.107}$$

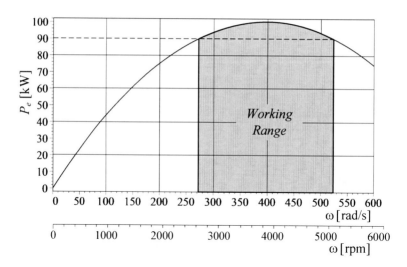

FIGURE 4.14. The power performance curve (4.104) and its working range.

The speed equation

$$v_x = \frac{0.326}{4 \times 0.85412} \omega_e \qquad (4.108)$$

is applied as long as the gearbox is operating in the third gear $n_i = n_3$, and ω_e is in the working range. By decreasing ω_e and sweeping down over the working range, the speed of the car will reduce. At the lower range $\omega_e = 272 \, \text{rad/s}$, the vehicle's speed would be

$$
\begin{aligned}
v_x &= \frac{0.326}{4 \times 0.85412} \times 272 = 25.95 \, \text{m/s} \\
&\approx 93.43 \, \text{km/h} \approx 58 \, \text{mi/h}
\end{aligned}
\qquad (4.109)
$$

At this speed we should gear down to n_2, so the engine jumps back to the higher range $\omega_e = 524 \, \text{rad/s}$. This provides us with

$$n_2 = \frac{0.326}{4} \frac{\omega_e}{v_x} = \frac{0.326}{4} \frac{524}{25.95} = 1.6457 \qquad (4.110)$$

Therefore, the engine and vehicle speed relationship in the second gear would be

$$v_x = \frac{0.326}{4 \times 1.6457} \omega_e \qquad (4.111)$$

that is applicable as long as $n_i = n_2$, and ω_e is in the working range. Sweeping down the engine's angular velocity reduces the vehicle speed to

$$
\begin{aligned}
v_x &= \frac{0.326}{4 \times 1.6457} \times 272 = 13.47 \, \text{m/s} \\
&\approx 48.49 \, \text{km/h} \approx 30.1 \, \text{mi/h}
\end{aligned}
\qquad (4.112)
$$

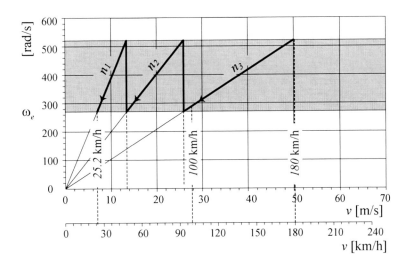

FIGURE 4.15. The gear-speed plot for a three-gear gearbox.

At this speed we should gear down to n_1 and jump the engine again to the higher range $\omega_e = 524\,\mathrm{rad/s}$. This provides us

$$n_1 = \frac{0.326}{4}\frac{\omega_e}{v_x} = \frac{0.326}{4}\frac{524}{13.47} = 3.1705 \qquad (4.113)$$

and therefore, the speed equation for the first gear is

$$v_x = \frac{0.326}{4 \times 3.1705}\omega_e \qquad (4.114)$$

At the lower range of the engine's speed in the first gear $n_i = n_1$, the speed of the vehicle is

$$v_x = \frac{0.326}{4 \times 3.1705} \times 272 = 7\,\mathrm{m/s} \approx 25.2\,\mathrm{km/h} \approx 15.6\,\mathrm{mi/h} \qquad (4.115)$$

Therefore, the three-gear gearbox uses the following gear ratios:

$$n_1 = 3.1705 \qquad n_2 = 1.6457 \qquad n_3 = 0.85412 \qquad (4.116)$$

*The speed equations for the three gears are shown in the **gear-speed** plot of Figure 4.15. The figure also shows the gear switching points and how the vehicle speed is reducing from $v_x = 50\,\mathrm{m/s}$ to $v_x = 7\,\mathrm{m/s}$.*

To evaluate the required time to reach the desired speed, we need to find the traction force F_x from the traction equation and integrate.

$$
\begin{aligned}
F_x &= \eta\frac{n_i n_d}{R_w}\frac{P_e}{\omega_e} = \frac{\eta}{\omega_e}\frac{n_i n_d}{R_w}\left(100 - \frac{100}{398^2}\left(\omega_e - 398\right)^2\right)\\
&= \frac{25}{39\,601}\frac{\eta}{R_w^2}n_d n_i\left(796 R_w - n_d n_i v_x\right)\,\mathrm{kN} \qquad (4.117)
\end{aligned}
$$

At the maximum speed, the gearbox is in the third gear and the traction force F_x is equal to the total resistance force F_R, which is assumed to be constant.

$$F_x = F_R = \frac{\eta P_e}{v_x} = \frac{0.84 \times 90}{50} = 1.512\,\text{kN} \qquad (4.118)$$

Therefore, the traction force in the first gear is

$$
\begin{aligned}
F_x &= \frac{25}{39601}\frac{\eta}{R_w^2} n_d n_1 \left(796 R_w - n_d n_1 v_x\right) \\
&= \frac{25}{39601}\frac{0.84}{0.326^2} \times 4 \times 3.1705\left(796 \times 0.326 - 4 \times 3.1705 v_x\right) \\
&= 16.421 - 0.80252 v_x\,\text{kN} \qquad (4.119)
\end{aligned}
$$

Based on Newton's equation of motion

$$F_x - F_R = m\frac{dv_x}{dt} \qquad (4.120)$$

we can evaluate the required time to sweep the velocity from zero to $v_x = 13.47\,\text{m/s}$

$$
\begin{aligned}
t_1 &= m \int_0^{13.47} \frac{1}{F_x - F_R}\,dv_x \\
&= 860 \int_0^{13.47} \frac{10^{-3}}{16.421 - 0.80252 v_x - 1.512}\,dv_x = 1.3837\,\text{s} \qquad (4.121)
\end{aligned}
$$

In second gear, we have

$$
\begin{aligned}
F_x &= \frac{25}{39601}\frac{\eta}{R_w^2} n_d n_2 \left(796 R_w - n_d n_2 v_x\right) \\
&= \frac{25}{39601}\frac{0.84}{0.326^2} \times 4 \times 1.6457\left(796 \times 0.326 - 4 \times 1.6457 v_x\right) \\
&= 8.5235 - 0.21622 v_x\,\text{kN} \qquad (4.122)
\end{aligned}
$$

and therefore, the sweep time in the second gear is

$$
\begin{aligned}
t_2 &= m \int_{13.47}^{25.95} \frac{1}{F_x - F_R}\,dv_x \\
&= 860 \int_{13.47}^{25.95} \frac{10^{-3}}{8.5235 - 0.21622 v_x - 1.512}\,dv_x = 4.2712\,\text{s} \qquad (4.123)
\end{aligned}
$$

Finally, the traction equation in the third gear is

$$
\begin{aligned}
F_x &= \frac{25}{39601}\frac{\eta}{R_w^2} n_d n_3 \left(796 R_w - n_d n_3 v_x\right) \\
&= \frac{25}{39601}\frac{0.84}{0.326^2} \times 4 \times 0.85412\left(796 \times 0.326 - 4 \times 0.85412 v_x\right) \\
&= 4.4237 - 5.8242 \times 10^{-2} v_x\,\text{kN} \qquad (4.124)
\end{aligned}
$$

and the sweep time to $v_x = 27.78\,\mathrm{m/s}$ is

$$
\begin{aligned}
t_3 &= m \int_{25.95}^{27.78} \frac{1}{F_x - F_R} dv_x \\
&= 860 \int_{25.95}^{27.78} \frac{10^{-3}}{4.4237 - 5.8242 \times 10^{-2} v_x - 1.512} dv_x = 1.169 \text{ s} \quad (4.125)
\end{aligned}
$$

Therefore, the total time to reach the speed $v_x = 100\,\mathrm{km/h} \approx 27.78\,\mathrm{m/s}$
is then equal to

$$
\begin{aligned}
t &= t_0 + t_1 + t_2 + t_3 + 3 \times 0.47 \\
&= 2.58 + 1.3837 + 4.2712 + 1.169 + 3 \times 0.47 = 10.814\,\mathrm{s} \quad (4.126)
\end{aligned}
$$

Example 152 *Better performance with a four-speed gearbox.*
A car equipped with a small engine has

$$
m = 860\,\mathrm{kg} \qquad R_w = 0.326\,\mathrm{m} \qquad \eta = 0.84 \qquad n_d = 4 \qquad (4.127)
$$

and the engine operates based on the performance equation of

$$
P_e = 100 - \frac{100}{398^2}\left(\omega_e - 398\right)^2 \mathrm{kW} \qquad (4.128)
$$

where ω_e *is in* $[\mathrm{rad/s}]$. *We chose the engine working range of*

$$
272\,\mathrm{rad/s}\,(\approx 2600\,\mathrm{rpm}) \le \omega_e \le 524\,\mathrm{rad/s}\,(\approx 5000\,\mathrm{rpm}) \qquad (4.129)
$$

when the power is $100\,\mathrm{kW} \ge P_e \ge 90\,\mathrm{kW}$. *We would like to design a gearbox
to minimize the time to reach* $v_x = 100\,\mathrm{km/h} \approx 27.78\,\mathrm{m/s} \approx 62\,\mathrm{mi/h}$.
*The power performance equation (4.128) is illustrated in Figure 4.14 and
the working range is shaded. To make this example comparable to Example
151 we assume that the total resistance force is constant and the engine
cannot accelerate the car at* $v_x = 180\,\mathrm{km/h}$. *Furthermore, we assume that
every gear change takes* $0.47\,\mathrm{s}$ *and a time* $t_0 = 2.58\,\mathrm{s}$ *is needed to completely
engage the engine in the first gear.*
*Let us design a four-gear gearbox and set the third gear such that we
reach the desired speed* $v_x = 27.78\,\mathrm{m/s}$ *at the higher limit of working range*
$\omega_e = 524\,\mathrm{rad/s}$. *The gear-speed plot for such a design is plotted in Figure
4.16.*
*Using the speed equation (4.48), the relationship between vehicle and en-
gine speeds is*

$$
v_x = \frac{R_w}{n_d\,n_i}\omega_e = \frac{0.326}{4\,n_i}\omega_e \qquad (4.130)
$$

At the speed $v_x = 100\,\mathrm{km/h} \approx 27.78\,\mathrm{m/s}$, *the engine's speed is at the upper
limit of the working range* $\omega_e = 524\,\mathrm{rad/s}$ *and the gearbox is operating in
third gear* $n_i = n_3$. *Therefore,*

$$
n_3 = \frac{0.326}{4}\frac{\omega_e}{v_x} = \frac{0.326}{4}\frac{524}{27.78} = 1.5373 \qquad (4.131)
$$

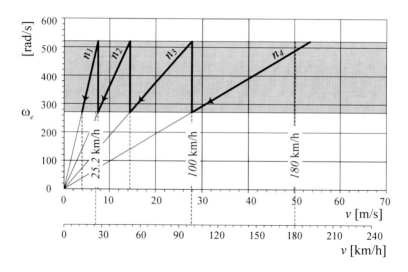

FIGURE 4.16. The gear-speed plot for Example 152.

and the speed equation in the third gear $n_i = n_3$ is

$$v_x = \frac{0.326}{4 \times 1.5373} \omega_e \qquad (4.132)$$

while ω_e is in the working range. By sweeping down to the lower limit of the working range, $\omega_e = 272\,\mathrm{rad/\,s}$, the speed of the car will reduce to

$$
\begin{aligned}
v_x &= \frac{0.326}{4 \times 1.5373} \times 272 = 14.42\,\mathrm{m/\,s}\\
&\approx 51.91\,\mathrm{km/\,h} \approx 32.25\,\mathrm{mi/\,h}
\end{aligned}
\qquad (4.133)
$$

At this speed we should gear down to n_2 and jump to the higher angular speed $\omega_e = 524\,\mathrm{rad/\,s}$. This provides us with

$$n_2 = \frac{0.326}{4} \frac{\omega_e}{v_x} = \frac{0.326}{4} \frac{524}{14.42} = 2.9616 \qquad (4.134)$$

Therefore, the gear-speed relationship in second gear $n_i = n_2$ is

$$v_x = \frac{0.326}{4 \times 2.9616} \omega_e \qquad (4.135)$$

Sweeping down the engine's angular velocity to $\omega_e = 272\,\mathrm{rad/\,s}$, reduces the vehicle speed to

$$
\begin{aligned}
v_x &= \frac{0.326}{4 \times 2.9616} \times 272 = 7.48\,\mathrm{m/\,s}\\
&\approx 26.9\,\mathrm{km/\,h} \approx 16.7\,\mathrm{mi/\,h}
\end{aligned}
\qquad (4.136)
$$

At this speed, we gear down to n_1 and jump again to the higher angular speed $\omega_e = 524\,\mathrm{rad/s}$. This provides us with

$$n_1 = \frac{0.326}{4}\frac{\omega_e}{v_x} = \frac{0.326}{4}\frac{524}{7.48} = 5.7055 \tag{4.137}$$

and therefore, the speed equation for the first gear would be

$$v_x = \frac{0.326}{4 \times 5.7055}\omega_e \tag{4.138}$$

In the first gear, $n_i = n_1$, and the vehicle's speed at the lower range of the engine's speed is

$$
\begin{aligned}
v_x &= \frac{0.326}{4 \times 5.7055} \times 272 = 3.88\,\mathrm{m/s} \\
&\approx 14\,\mathrm{km/h} \approx 8.7\,\mathrm{mi/h}
\end{aligned}
\tag{4.139}
$$

To calculate the fourth gear $n_i = n_4$ we may use the gear-speed equation and set the engine speed to the lower limit $\omega_e = 272\,\mathrm{rad/s}$ while the car is moving at the maximum speed in the third gear.

$$n_4 = \frac{0.326}{4}\frac{\omega_e}{v_x} = \frac{0.326}{4}\frac{272}{27.78} = 0.79798 \tag{4.140}$$

Therefore, the four-gear gearbox uses the following ratios:

$$
\begin{aligned}
n_1 &= 5.7055 & n_2 &= 2.9616 \\
n_3 &= 1.5373 & n_4 &= 0.79798
\end{aligned}
\tag{4.141}
$$

To calculate the required time to reach the desired speed $v_x = 100\,\mathrm{km/h} \approx 27.78\,\mathrm{m/s}$, we need to use the traction equations and find the traction force F_x

$$
\begin{aligned}
F_x &= \eta\frac{n_i n_d}{R_w}\frac{P_e}{\omega_e} = \frac{\eta}{\omega_e}\frac{n_i n_d}{R_w}\left(100 - \frac{100}{398^2}(\omega_e - 398)^2\right) \\
&= \frac{25}{39\,601}\frac{\eta}{R_w^2}n_d n_i\,(796 R_w - n_d n_i v_x)\,\mathrm{kN}
\end{aligned}
\tag{4.142}
$$

At the maximum speed, the gearbox is in the fourth gear and the traction force F_x is equal to the total resistance force F_R, which is assumed to be constant.

$$F_x = F_R = \frac{\eta P_e}{v_x} = \frac{0.84 \times 90}{50} = 1.512\,\mathrm{kN} \tag{4.143}$$

Therefore, the traction force in the first gear is

$$
\begin{aligned}
F_x &= \frac{25}{39601}\frac{\eta}{R_w^2}n_d n_1\,(796 R_w - n_d n_1 v_x) \\
&= \frac{25}{39601}\frac{0.84}{0.326^2} \times 4 \times 5.7055\,(796 \times 0.326 - 4 \times 5.7055 v_x) \\
&= 29.55 - 2.5989 v_x\,\mathrm{kN}
\end{aligned}
\tag{4.144}
$$

Using Newton's equation of motion

$$F_x - F_R = m \frac{dv_x}{dt} \qquad (4.145)$$

we can evaluate the required time to reach the velocity $v_x = 7.48\,\mathrm{m/s}$

$$
\begin{aligned}
t_1 &= m \int_0^{7.48} \frac{1}{F_x - F_R} dv_x \\
&= 860 \int_0^{7.48} \frac{10^{-3}}{29.55 - 2.5989v_x - 1.512} dv_x = 0.39114\,\mathrm{s} \qquad (4.146)
\end{aligned}
$$

In second gear we have

$$
\begin{aligned}
F_x &= \frac{25}{39601} \frac{\eta}{R_w^2} n_d n_2 \left(796 R_w - n_d n_2 v_x\right) \\
&= \frac{25}{39601} \frac{0.84}{0.326^2} \times 4 \times 2.9616 \left(796 \times 0.326 - 4 \times 2.9616 v_x\right) \\
&= 15.339 - 0.70025 v_x\,\mathrm{kN} \qquad (4.147)
\end{aligned}
$$

and therefore, the sweep time in second gear is

$$
\begin{aligned}
t_2 &= m \int_{7.48}^{14.42} \frac{1}{F_x - F_R} dv_x \\
&= 860 \int_{7.48}^{14.42} \frac{10^{-3}}{15.339 - 0.70025 v_x - 1.512} dv_x = 1.0246\,\mathrm{s} \qquad (4.148)
\end{aligned}
$$

The traction equation in the third gear is

$$
\begin{aligned}
F_x &= \frac{25}{39601} \frac{\eta}{R_w^2} n_d n_3 \left(796 R_w - n_d n_3 v_x\right) \\
&= \frac{25}{39601} \frac{0.84}{0.326^2} \times 4 \times 1.5373 \left(796 \times 0.326 - 4 \times 1.5373 v_x\right) \\
&= 7.9621 - 0.18868 v_x\,\mathrm{kN} \qquad (4.149)
\end{aligned}
$$

and the sweep time is

$$
\begin{aligned}
t_3 &= m \int_{14.42}^{27.78} \frac{1}{F_x - F_R} dv_x \\
&= 860 \int_{14.42}^{27.78} \frac{10^{-3}}{7.9621 - 0.18868 v_x - 1.512} dv_x = 5.1359\,\mathrm{s} \qquad (4.150)
\end{aligned}
$$

The total time to reach the speed $v_x = 100\,\mathrm{km/h} \approx 27.78\,\mathrm{m/s}$ *is then equal to*

$$
\begin{aligned}
t &= t_0 + t_1 + t_2 + t_3 + 3 \times 0.07 \\
&= 2.58 + 0.39114 + 1.0246 + 5.1359 + 3 \times 0.47 = 10.542\,\mathrm{s} \qquad (4.151)
\end{aligned}
$$

Example 153 *Changing the working range.*
Consider a car that is equipped with a small engine having the power performance equation of

$$P_e = 100 - \frac{100}{398^2}(\omega_e - 398)^2 \text{ kW} \qquad (4.152)$$

where ω_e is in $[\text{rad/s}]$. The characteristics of the car are

$$m = 860 \text{ kg} \qquad R_w = 0.326 \text{ m} \qquad \eta = 0.84 \qquad n_d = 4 \qquad (4.153)$$

The engine provides a maximum power $P_M = 100 \text{ kW}$ at $\omega_M = 400 \text{ rad/s}$.
The total resistance force is assumed to be constant, and the maximum attainable speed is assumed to be $v_x = 180 \text{ km/h}$. Furthermore, we assume that every gear change takes 0.07s and a minimum time $t_0 = 0.18\text{s}$ is needed to adjust the engine speed to the car speed in the first gear. We would like to design a four-gear gearbox to minimize the time to reach $v_x = 100 \text{ km/h} \approx 27.78 \text{ m/s}$.
To find the best working range for the engine, we set the third gear to reach the desired speed $v_x = 100 \text{ km/h}$ at the upper limit of the working range. Therefore, when we gear up, the fourth gear starts with the lower limit of the working range. If the fourth gear is set such that the car reaches the maximum speed $v_x = 180 \text{ km/h} \approx 50 \text{ m/s}$ at the upper limit of the working range, then the gear-speed equation

$$\omega_e = \frac{n_i\, n_d}{R_w} v_x \qquad (4.154)$$

provides us with

$$\omega_{Max} = \frac{4n_4}{0.326} \times 50 \qquad \omega_{min} = \frac{4n_4}{0.326} \times 27.78 \qquad (4.155)$$

By setting ω_{min} and ω_{Max} to an equal distance from $\omega_M = 400 \text{ rad/s}$,

$$\frac{\omega_{Max} + \omega_{min}}{2} = 400 \qquad (4.156)$$

we find

$$n_4 = 0.83826 \qquad \omega_{min} = 285.73 \text{ rad/s} \qquad \omega_{Max} = 514.27 \text{ rad/s} \quad (4.157)$$

We are designing a gearbox such that the ratio ω_e/v_x is kept constant in each gear. The engine speed jumps up from ω_{min} to ω_{Max} when we gear down from n_4 to n_3 at ω_{min}, hence,

$$\omega_{Max} = \frac{4n_3}{0.326} \times 27.78 = 514.27 \qquad (4.158)$$
$$n_3 = 1.5087 \qquad (4.159)$$

Therefore, the speed of the car in the third gear at the lower limit of the engine speed is

$$v_x = 27.78 \frac{\omega_{min}}{\omega_{Max}} = 27.78 \times \frac{27.78}{50} = 15.435 \, \text{m/s} \qquad (4.160)$$

The engine's speed jumps again to ω_{Max} when we gear it down from n_3 to n_2, hence,

$$\omega_{Max} = \frac{4n_2}{0.326} \times 15.435 = 514.27 \qquad (4.161)$$

$$n_2 = 2.715\,5 \qquad (4.162)$$

Finally the speed of the car in second gear at the lower limit of the engine speed is

$$v_x = 15.435 \frac{\omega_{min}}{\omega_{Max}} = 15.435 \times \frac{27.78}{50} = 8.5757 \, \text{m/s} \qquad (4.163)$$

that provides us with the following gear ratio in the first gear

$$\omega_{Max} = \frac{4n_1}{0.326} \times 8.5757 = 514.27 \qquad (4.164)$$

$$n_1 = 4.8874 \qquad (4.165)$$

The speed of the car in the first gear at the lower limit of the engine speed is then equal to

$$v_x = 8.5757 \frac{\omega_{min}}{\omega_{Max}} = 8.5757 \times \frac{27.78}{50} = 4.7647 \, \text{m/s} \qquad (4.166)$$

Therefore, the four gears of the gearbox have the following ratios:

$$n_1 = 4.8874 \qquad n_2 = 2.7155$$

$$n_3 = 1.5087 \qquad n_4 = 0.83826 \qquad (4.167)$$

and the working range for the engine is

$$285.73 \, \text{rad/s} \, (\approx 2730 \, \text{rpm}) \le \omega_e \le 514.27 \, \text{rad/s} \, (\approx 4911 \, \text{rpm}) \qquad (4.168)$$

The power performance curve (4.152) is illustrated in Figure 4.17 and the working range is shaded. The gear-speed plot of this design is also plotted in Figure 4.18.

Balance of the traction force F_x and the total resistance force F_R at the maximum speed provides us with

$$F_x = F_R = \frac{\eta P_e}{v_x} = \frac{0.84 \times 90}{50} = 1.512 \, \text{kN} \qquad (4.169)$$

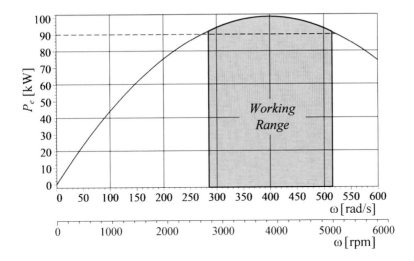

FIGURE 4.17. The power performance curve (4.152) and its working range.

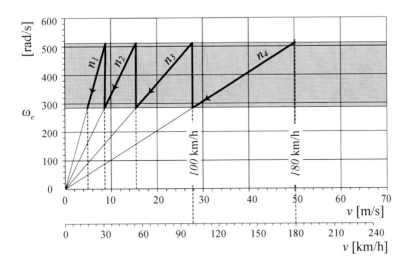

FIGURE 4.18. The gear-speed plot for Example 153.

The traction force in the first gear is

$$
\begin{aligned}
F_x &= \frac{25}{39601}\frac{\eta}{R_w^2}n_d n_1 \left(796 R_w - n_d n_1 v_x\right)\\
&= \frac{25}{39601}\frac{0.84}{0.326^2}\times 4 \times 4.8874\left(796\times 0.326 - 4\times 4.8874 v_x\right)\\
&= 25.313 - 1.907 v_x \ \text{kN}
\end{aligned}
\tag{4.170}
$$

The time in the first gear n_1 can be calculated by integrating Newton's equation of motion

$$
F_x - F_R = m\frac{dv_x}{dt}
\tag{4.171}
$$

and sweep the velocity from $v_x = 0$ to $v_x = 8.5757\,\text{m/s}$

$$
\begin{aligned}
t_1 &= m\int_0^{8.5757}\frac{1}{F_x - F_R}dv_x\\
&= 860\int_0^{8.5757}\frac{10^{-3}}{25.313 - 1.907 v_x - 1.512}dv_x = 0.52398\,\text{s}
\end{aligned}
\tag{4.172}
$$

In the second gear, the traction force is

$$
\begin{aligned}
F_x &= \frac{25}{39601}\frac{\eta}{R_w^2}n_d n_2 \left(796 R_w - n_d n_2 v_x\right)\\
&= \frac{25}{39601}\frac{0.84}{0.326^2}\times 4 \times 2.7155\left(796\times 0.326 - 4\times 2.7155 v_x\right)\\
&= 14.064 - 0.5887 v_x \ \text{kN}
\end{aligned}
\tag{4.173}
$$

and therefore, the sweep time in the second gear is

$$
\begin{aligned}
t_2 &= m\int_{8.5757}^{15.435}\frac{1}{F_x - F_R}dv_x\\
&= 860\int_{8.5757}^{15.435}\frac{10^{-3}}{14.064 - 0.5887 v_x - 1.512}dv_x = 1.1286\,\text{s}
\end{aligned}
\tag{4.174}
$$

In the third gear, the traction force is

$$
\begin{aligned}
F_x &= \frac{25}{39601}\frac{\eta}{R_w^2}n_d n_3 \left(796 R_w - n_d n_3 v_x\right)\\
&= \frac{25}{39601}\frac{0.84}{0.326^2}\times 4 \times 1.5087\left(796\times 0.326 - 4\times 1.5087 v_x\right)\\
&= 7.814 - 0.18172 v_x \ \text{kN}
\end{aligned}
\tag{4.175}
$$

and the third sweep time is

$$
\begin{aligned}
t_3 &= m\int_{15.435}^{27.78}\frac{1}{F_x - F_R}dv_x\\
&= 860\int_{15.435}^{27.78}\frac{10^{-3}}{7.814 - 0.18172 v_x - 1.512}dv_x = 4.8544\,\text{s}
\end{aligned}
\tag{4.176}
$$

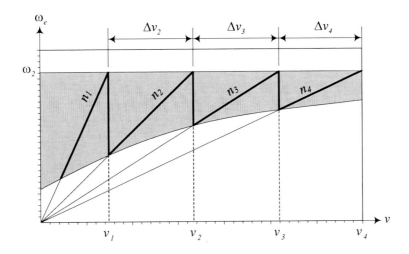

FIGURE 4.19. A gear-speed plot for a progressive gearbox design.

The total time to reach the speed $v_x = 100\,\mathrm{km/h} \approx 27.78\,\mathrm{m/s}$ *is then equal to*

$$
\begin{aligned}
t &= t_0 + t_1 + t_2 + t_3 + 3 \times 0.07 \\
&= 2.58 + 0.52398 + 1.1286 + 4.8544 + 3 \times 0.47 = 10.497\,\mathrm{s} \quad (4.177)
\end{aligned}
$$

4.4.2 ★ *Progressive Ratio Gearbox Design*

When the speed span of a vehicle in any two successive gears is kept constant, we call the gearbox *progressive*. The design condition for a progressive gearbox is

$$
n_{i+1} = \frac{n_i n_{i-1}}{2n_{i-1} - n_i} \quad (4.178)
$$

where n_{i-1}, n_i, and n_{i+1} are the transmission ratios of three successive gears.

Proof. A progressive gearbox has constant vehicle speed span in any gear. Therefore, a progressive gearbox must have a gear-speed plot such as that shown in Figure 4.19.

Indicating the maximum vehicle speed in gear n_i by v_i, in gear n_{i-1} by v_{i-1}, and in gear n_{i+1} by v_{i+1}, we have

$$
\omega_2 = \frac{n_i\, n_d}{R_w}\, v_i = \frac{n_{i-1}\, n_d}{R_w}\, v_{i-1} = \frac{n_{i+1}\, n_d}{R_w}\, v_{i+1} \quad (4.179)
$$

The difference in vehicle speed at maximum engine speed is

$$
\triangle v = v_i - v_{i-1} = v_{i+1} - v_i \quad (4.180)
$$

and therefore,

$$v_{i+1} + v_{i-1} = 2v_i \tag{4.181}$$

$$\frac{v_{i+1}}{v_i} + \frac{v_{i-1}}{v_i} = 2 \tag{4.182}$$

$$\frac{n_i}{n_{i+1}} + \frac{n_i}{n_{i-1}} = 2 \tag{4.183}$$

$$n_{i+1} = \frac{n_i n_{i-1}}{2n_{i-1} - n_i} \tag{4.184}$$

The step jump of a progressive gearbox decreases in higher gears. If the step jump c_{g_i} between n_i and n_{i+1} is

$$\frac{n_i}{n_{i+1}} = c_{g_i} \tag{4.185}$$

then,

$$c_{g_i} = 2 - \frac{1}{c_{g_{i-1}}} \tag{4.186}$$

∎

Example 154 *A progressive gearbox with three gears.*
The power performance of the engine of a car with $\eta = \eta_d \eta_g = 0.84$ is

$$P_e = 100 - \frac{100}{398^2}(\omega_e - 398)^2 \text{ kW} \tag{4.187}$$

where ω_e is in $[\text{rad}/\text{s}]$ and

$$P_M = 100\,\text{kW} \qquad \omega_M = 400\,\text{rad}/\text{s} \tag{4.188}$$

The differential ratio and the equivalent tire radius are

$$n_d = 4 \qquad R_w = 0.326\,\text{m} \tag{4.189}$$

The total resistance force is assumed to be constant and the engine cannot accelerate the car at maximum speed $v_x = 180\,\text{km}/\text{h} = 50\,\text{m}/\text{s} \approx 112\,\text{mi}/\text{h}$ anymore.
Using the speed equation (4.48), the relationship between vehicle and engine speed is

$$v_x = \frac{R_w}{n_d n_i}\omega_e = \frac{0.326}{4\,n_i}\omega_e \tag{4.190}$$

We would like to design a four-gear progressive gearbox with equal speed spans, such that the car achieves the maximum speed in the fourth gear when the engine is at the maximum permissible speed $\omega_2 = \omega_e = 524\,\text{rad}/\text{s} \approx 5000\,\text{rpm}$.

Dividing the maximum velocity of the car into four equal segments determines the speed span at each gear

$$\triangle v = \frac{v_M}{4} = \frac{50}{4} = 12.5 \, \text{m/s} \tag{4.191}$$

Therefore, the maximum car speed for the four gears are

$$
\begin{aligned}
v_1 &= 12.5 & \text{(4.192)} \\
v_2 &= v_1 + \triangle v = 12.5 + 12.5 = 25 \, \text{m/s} & \text{(4.193)} \\
v_3 &= v_2 + \triangle v = 25 + 12.5 = 37.5 \, \text{m/s} & \text{(4.194)} \\
v_4 &= v_3 + \triangle v = 37.5 + 12.5 = 50 \, \text{m/s} & \text{(4.195)}
\end{aligned}
$$

Considering that the maximum car speeds are all achieved at the maximum permissible engine speed, $\omega_e = 524 \, \text{rad/s}$, the speed equation (4.190) determines the gear ratios.

$$
\begin{aligned}
n_4 &= \frac{0.326}{4v_4}\omega_e = \frac{0.326}{4 \times 50}524 = 0.85412 & \text{(4.196)} \\
n_3 &= \frac{0.326}{4v_3}\omega_e = \frac{0.326}{4 \times 37.5}524 = 1.1388 & \text{(4.197)} \\
n_2 &= \frac{0.326}{4v_2}\omega_e = \frac{0.326}{4 \times 25}524 = 1.7082 & \text{(4.198)} \\
n_1 &= \frac{0.326}{4v_1}\omega_e = \frac{0.326}{4 \times 12.5}524 = 3.4165 & \text{(4.199)}
\end{aligned}
$$

The minimum engine speed, ω_1, at which the gear should be changed is not a constant. Assume the gearbox is in n_1 and the engine speed reaches the maximum value of $\omega_e = 524 \, \text{rad/s}$. At this time we need to gear up to n_2 and drop the engine speed. Substituting v_1 in Equation (4.198) determines the engine speed at the first gear up

$$\omega_e = \frac{4v_1 n_2}{0.326} = \frac{4 \times 12.5 \times 1.7082}{0.326} = 261.99 \, \text{rad/s} \tag{4.200}$$

Substituting v_2 in Equation (4.197) determines the engine speed at the second gear up

$$\omega_e = \frac{4v_2 n_3}{0.326} = \frac{4 \times 25 \times 1.1388}{0.326} = 349.33 \, \text{rad/s} \tag{4.201}$$

Substituting v_3 in Equation (4.196) determines the engine speed at the third gear up

$$\omega_e = \frac{4v_3 n_4}{0.326} = \frac{4 \times 37.5 \times 0.85412}{0.326} = 393 \, \text{rad/s} \tag{4.202}$$

Figure 4.20 illustrates the speed equations of the gearbox to show the relations of the engine speed and car velocity in different gears.

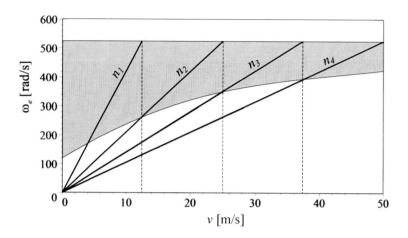

FIGURE 4.20. The relations between the engine speed and car velocity in different gears.

4.5   Summary

The maximum attainable power P_e of an internal combustion engine is a function of the engine angular velocity ω_e. This function must be determined by experiment however, the *power performance* function $P_e = P_e(\omega_e)$, can be estimated by a mathematical function such as

$$P_e = P_1\,\omega_e + P_2\,\omega_e^2 + P_3\,\omega_e^3 \tag{4.203}$$

where,

$$P_1 = \frac{P_M}{\omega_M} \qquad P_2 = \frac{P_M}{\omega_M^2} \qquad P_3 = -\frac{P_M}{\omega_M^3} \tag{4.204}$$

ω_M is the angular velocity, measured in $[\,\mathrm{rad/s}\,]$, at which the engine power reaches the maximum value P_M, measured in $[\,\mathrm{W} = \mathrm{N\,m/s}\,]$.

The engine torque T_e is that provides P_e is

$$T_e = \frac{P_e}{\omega_e} = P_1 + P_2\,\omega_e + P_3\,\omega_e^2 \tag{4.205}$$

An ideal engine is the one that produces a constant power regardless of speed. For the ideal engine, we have

$$P_e = P_0 \qquad T_e = \frac{P_0}{\omega_e} \tag{4.206}$$

We use a gearbox to make the engine approximately work at a constant power close to the P_M. To design a gearbox we use two equations: the *speed equation*

$$\omega_e = \frac{n_i\,n_d}{R_w}\,v_x \tag{4.207}$$

and the *traction equation*

$$T_e = \frac{1}{\eta} \frac{R_w}{n_i n_d} F_x \tag{4.208}$$

These equations state that the forward velocity v_x of a vehicle is proportional to the angular velocity of the engine ω_e, and the tire *traction force* F_x is proportional to the engine torque T_e, where, R_w is the effective tire radius, n_d is the differential transmission ratio, n_i is the gearbox transmission ratio in gear number i, and η is the overall driveline efficiency.

4.6   Key Symbols

$a \equiv \ddot{x}$	acceleration
$a_i, i = 0, \cdots, 6$	coefficients of function $T_e = T_e\left(\omega_e\right)$
a_x	acceleration capacity
AWD	all-wheel-drive
c_g	constant relative gear ratio
C_c	slip factor
d	distance traveled
D	clutch diameter
E	energy
F_x	traction force
FWD	front-wheel-drive
H	thermal value of fuel
m	vehicle mass
$n = \omega_{in}/\omega_{out}$	gear reduction ratio
n_i	gearbox transmission ratio in gear number i
n_d	transmission ratio
n_g	overall transmission ratio
P	power
P_0	ideal engine constant power
P_1, P_2, P_3	coefficients of the power performance function
P_e	maximum attainable power of an engine
$P_e = P_e\left(\omega_e\right)$	power performance function
P_M	maximum power
q	fuel consumption per unit distance
$r = \omega/\omega_n$	frequency ratio
RWD	rear-wheel-drive
T_d	differential input torque
T_e	engine torque
T_M	maximum torque
T_w	wheel torque
$v \equiv \dot{x}$, $\mathbf{v}$	velocity
v_{min}	minimum vehicle speed corresponding to ω_{min}
$\triangle v$	difference in maximum vehicle speed at two different gears
x, y, z, $\mathbf{x}$	displacement
η	overall efficiency
η_c	convertor efficiency
η_e	engine efficiency
η_M	mechanical efficiency
η_t	transmission efficiency
η_t	thermal efficiency
η_T	thermal efficiency

η_V	volumetric efficiency
μ_x	traction coefficient
ρ	oil density
ρ_f	fuel density
ϕ	slope of the road
ω_d	differential input angular velocity
ω_e	engine angular velocity
ω_{min}	minimum engine speed
ω_M	engine angular velocity at maximum power
ω_{Max}	maximum engine speed
ω_p	pump angular velocity
$\omega_r = \omega_{out}/\omega_{in}$	speed ratio

Exercises

1. Power performance.

   Audi $R8^{TM}$ with $m = 1558$ kg, has a $V8$ engine with

   $$P_M = 313\,\text{kW} \approx 420\,\text{hp} \ at \ \omega_M = 7800 \ rpm$$

   and Audi TT Coupe^{TM} with $m = 1430$ kg, has a $V6$ engine with

   $$P_M = 184\,\text{kW} \approx 250\,\text{hp} \ at \ \omega_M = 6300 \ rpm$$

   Determine the power performance equations of their engines and compare the power mass ratio, P_M/m of the cars.

2. Power and torque performance.

   A model of Nissan $NISMO$ 350Z with $m = 1522$ kg, has a $V6$ engine with

   $$
   \begin{aligned}
   P_M &= 228\,\text{kW} \approx 306\,\text{hp} \ at \ \omega_M = 6800 \ \text{rpm} \\
   T_M &= 363\,\text{N m} \approx 268\,\text{lb ft} \ at \ \omega = 4800 \ \text{rpm}
   \end{aligned}
   $$

   Determine the power and torque performance equations, and compare T_M from the torque equation with the above reported number.

3. Power performance and modeling.

   BMW X3^{TM} is offered in four models with four different engines: xDrive20d, xDrive25i, xDrive30i, and xDrive30d. The maximum power and torque of the models and their associated engine speed are

	P_M kW	ω_M rpm	T_M N m	ω rpm	m kg
xDrive20d	130	4000	350	1750	1750
xDrive25i	160	6500	250	2750	1755
xDrive30i	200	6650	315	2750	1765
xDrive30d	160	4000	500	1750	1810

   (a) Employ Equation (4.1) to express the power performance of these vehicles.

   (b) Determine the maximum torque of each vehicle based on the analytic power performance equation. Compare the calculated maximum torques with the actual values and determine which vehicle can better be modeled by analytic equation.

4. Fuel consumption conversion.

A model of Subaru Impreza WRX STI^{TM} with $m = 1521$ kg, has a turbocharged flat-4 engine with

$$P_M = 219 \, \text{kW} \approx 293 \, \text{hp} \; at \; \omega_M = 6000 \; \text{rpm}$$

Fuel consumption of the car is 19 mi/gal in city and 25 mi/gal in highway. Determine the fuel consumption in liter per 100 km.

5. Traction force.

   A model of Mercedes-Benz SLR 722 Edition^{TM} with $m = 1724$ kg, has a supercharged $V8$ engine with

   $$P_M = 485 \, \text{kW} \approx 650 \, \text{hp} \; at \; \omega_M = 6500 \; \text{rpm}$$

   The maximum speed of the car is

   $$v_M = 337 \, \text{km/h} \approx 209 \, \text{mi/h}$$

   Assume the maximum speed happens at the maximum power and use an overall efficiency $\eta = 0.75$ to determine the traction force at the maximum speed.

6. Gearbox speed equation.

   BMW X3^{TM} has four models with four different gearboxes: xDrive20d, xDrive25i, xDrive30i, and xDrive30d. The mass of the vehicles and the gear ratio of their gearbox are given as

	n_1	n_2	n_3	n_4	n_5	n_6	n_r	n_d
xDrive20d	4.17	2.34	1.52	1.14	0.87	0.69	3.4	3.73
xDrive25i	4.07	2.37	1.55	1.16	0.85	0.67	3.20	4.44
xDrive30i	4.07	2.37	1.55	1.16	0.85	0.67	3.20	4.44
xDrive30d	4.17	2.34	1.52	1.14	0.87	0.69	3.40	3.46

   All models are using the same tire and rim: $235/55$ R 17, $8Jx17$, and some other specifications of the vehicles are given in Exercise 3 if needed.

   (a) Employ the speed equation and plot the them on four figures for the vehicles.

   (b) Assuming $\eta = 1$, plot the acceleration capacity of the vehicles on the same figure to compare.

   (c) Assume ω_2 is $1.2\omega_M$ and make a table of different ω_1 at each gear change for all vehicles.

   (d) Examine the gearbox stability condition for all gearboxes.

   (e) ★ Determine the best constant ω_1 and ω_2 for each gearbox such that the gearbox is as close as possible to a geometric design.

7. Car speed and engine speed.

A model of Toyota Camry^{TM} has a 3.5-liter, 6-cylinder engine with

$$P_M = 268\,\text{hp } at \; \omega_M = 6200 \; \text{rpm}$$

The car uses transaxle/front-wheel drive and is equipped with a six-speed ECT-i automatic transmission.

$1st$ gear ratio $= n_1 = 3.300$ $2nd$ gear ratio $= n_2 = 1.900$
$3rd$ gear ratio $= n_3 = 1.420$ $4th$ gear ratio $= n_4 = 1.000$
$5th$ gear ratio $= n_5 = 0.713$ $6th$ gear ratio $= n_6 = 0.609$
reverse gear ratio $= n_r = 4.148$ final drive ratio $= n_d = 3.685$

Determine the speed of the car at each gear, when the engine is running at ω_M, and it is equipped with

(a) $P215/55R17$ tires

(b) $P215/60R16$ tires.

8. Gear-speed equations.

A model of Ford Mondeo^{TM} is equipped with a 2.0-liter engine, which has

$$T_M = 185\,\text{N m } at \; \omega_e = 4500 \; \text{rpm}$$

It has a manual five-speed gearbox.

$1st$ gear ratio $= n_1 = 3.42$ $2nd$ gear ratio $= n_2 = 2.14$
$3rd$ gear ratio $= n_3 = 1.45$ $4th$ gear ratio $= n_4 = 1.03$
$5th$ gear ratio $= n_5 = 0.81$
reverse gear ratio $= n_r = 3.46$ final drive ratio $= n_d = 4.06$

If the tires of the car are $205/55R16$, determine the gear-speed equations for each gear.

9. Final drive and gear ratios.

A model of Renault/Dacia Logan^{TM} with $m = 1115\,\text{kg}$, has a four-cylinder engine with

$$\begin{aligned}
P_M &= 77\,\text{kW} \approx 105\,\text{hp } at \; \omega_M = 5750 \; \text{rpm} \\
T_M &= 148\,\text{N m } at \; \omega_e = 3750 \; \text{rpm} \\
v_M &= 183\,\text{km/ h} \\
Tires &= 185/65R15
\end{aligned}$$

It has a five-speed gearbox. When the engine is running at 1000 rpm the speed of the car at each gear is

$$\begin{aligned}
\text{speed in the } 1st \text{ gear} &= v_{M_1} = 7.25\,\text{km/h}\\
\text{speed in the } 2nd \text{ gear} &= v_{M_2} = 13.18\,\text{km/h}\\
\text{speed in the } 3rd \text{ gear} &= v_{M_3} = 19.37\,\text{km/h}\\
\text{speed in the } 4th \text{ gear} &= v_{M_4} = 26.21\,\text{km/h}\\
\text{speed in the } 5th \text{ gear} &= v_{M_5} = 33.94\,\text{km/h}
\end{aligned}$$

Evaluate the gear ratios $n_i, i = 1, 2, \cdots 5$ for $n_d = 4$.

10. Traction equation.

A model of Jeep Wrangler^{TM} is equipped with a $V6$ engine and has the following specifications.

$$\begin{aligned}
P_M &= 153\,\text{kW} \approx 205\,\text{hp } at \ \omega_M = 5200 \text{ rpm}\\
T_M &= 325\,\text{N m} \approx 240\,\text{lb ft } at \ \omega_e = 4000 \text{ rpm}
\end{aligned}$$

A model of the car may have a six-speed manual transmission with the following gear ratios

$1st$ gear ratio $= n_1 = 4.46$ $2nd$ gear ratio $= n_2 = 2.61$
$3rd$ gear ratio $= n_3 = 1.72$ $4th$ gear ratio $= n_4 = 1.25$
$5th$ gear ratio $= n_5 = 1.00$ $6th$ gear ratio $= n_6 = 0.84$
reverse gear ratio $= n_r = 4.06$ final drive ratio $= n_d = 3.21$

or a four-speed automatic transmission with the following gear ratios.

$1st$ gear ratio $= n_1 = 2.84$ $2nd$ gear ratio $= n_2 = 1.57$
$3rd$ gear ratio $= n_3 = 1.0$ $4th$ gear ratio $= n_4 = 0.69$
reverse gear ratio $= n_r = 2.21$ final drive ratio $= n_d = 4.10$

Assume
$$\eta = 0.8 \qquad Tires = 245/75R16$$

and determine the traction equation for the two models.

11. Acceleration capacity.

Lamborghini Murcielago^{TM} is equipped with a 6.2-liter $V12$ engine and has the following specifications.

$$\begin{aligned}
P_M &= 631\,\text{hp } at \ \omega_M = 8000 \text{ rpm}\\
T_M &= 487\,\text{lb ft } at \ \omega_e = 6000 \text{ rpm}
\end{aligned}$$

$$\begin{aligned}
m &= 3638\,\text{lb}\\
Front\ tire &= P245/35ZR18\\
Rear\ tire &= P335/30ZR18
\end{aligned}$$

The gearbox of the car uses ratios close to the following values.

$$1st \text{ gear ratio} = n_1 = 2.94 \qquad 2nd \text{ gear ratio} = n_2 = 2.056$$
$$3rd \text{ gear ratio} = n_3 = 1.520 \qquad 4th \text{ gear ratio} = n_4 = 1.179$$
$$5th \text{ gear ratio} = n_5 = 1.030 \qquad 6th \text{ gear ratio} = n_6 = 0.914$$
$$\text{reverse gear ratio} = n_r = 2.529 \quad \text{final drive ratio} = n_d = 3.42$$

If $\eta = 0.8$, then

(a) determine the wheel torque function at each gear

(b) determine the acceleration capacity of the car.

12. ★ Gearbox stability.

A model of Jaguar XJ^{TM} is a rear-wheel drive car with a 4.2-liter $V8$ engine. Some of the car's specifications are close to the following values.

$$
\begin{aligned}
m &= 3638\,\text{lb} \qquad l = 119.4\,\text{in} \\
Front \; tire &= P235/50R18 \\
Rear \; tire &= P235/50R18 \\
P_M &= 300\,\text{hp} \; at \; \omega_M = 6000 \;\text{rpm}
\end{aligned}
$$

If gear ratios of the car's gearbox are

$$1st \text{ gear ratio} = n_1 = 4.17 \qquad 2nd \text{ gear ratio} = n_2 = 2.34$$
$$3rd \text{ gear ratio} = n_3 = 1.52 \qquad 4th \text{ gear ratio} = n_4 = 1.14$$
$$5th \text{ gear ratio} = n_5 = 0.87 \qquad 6th \text{ gear ratio} = n_6 = 0.69$$
$$\text{reverse gear ratio} = n_r = 3.4 \quad \text{final drive ratio} = n_d = 2.87$$

Check the gearbox stability condition. In case the relative gear ratio is not constant, determine the new gear ratios using the relative ratio of the first two gears.

13. ★ Geometric gearbox design.

Lamborghini Diablo^{TM} is a rear-wheel drive car that was built in years $1990 - 2000$. The car is equipped with a 5.7-liter $V12$ engine. Some of the car's specifications are given.

$$
\begin{aligned}
P_M &= 492\,\text{hp} \; at \; \omega_M = 7000 \;\text{rpm} \\
T_M &= 580\,\text{N\,m} \approx 428\,\text{lb\,ft} \; at \; \omega_e = 5200 \;\text{rpm} \\
v_M &= 328\,\text{km/h} \approx 203\,\text{mi/h}
\end{aligned}
$$

$$
\begin{aligned}
m &= 1576\,\text{kg} \approx 3474\,\text{lb} \qquad l = 2650\,\text{mm} \approx 104\,\text{in} \\
w_f &= 1540\,\text{mm} \approx 60.6\,\text{in} \qquad w_r = 1640\,\text{mm} \approx 64.6\,\text{in} \\
Front \; tire &= 245/40ZR17 \qquad Rear \; tire = 335/35ZR17
\end{aligned}
$$

The gear ratios of the car's gearbox are close to the following values.

$$
\begin{aligned}
1st \text{ gear ratio} &= n_1 = 2.31 & v_M &= 97.3\,\text{km/h} \approx 60.5\,\text{mi/h} \\
2nd \text{ gear ratio} &= n_2 = 1.52 & v_M &= 147.7\,\text{km/h} \approx 91.8\,\text{mi/h} \\
3rd \text{ gear ratio} &= n_3 = 1.12 & v_M &= 200.2\,\text{km/h} \approx 124\,\text{mi/h} \\
4th \text{ gear ratio} &= n_4 = 0.88 & v_M &= 254.8\,\text{km/h} \approx 158.4\,\text{mi/h} \\
5th \text{ gear ratio} &= n_5 = 0.68 & v_M &= 325\,\text{km/h} \approx 202\,\text{mi/h} \\
\text{reverse gear ratio} &= n_r = 2.12 & v_M &= 105.7\,\text{km/h} \approx 65.7\,\text{mi/h} \\
\text{final drive ratio} &= n_d = 2.41 &
\end{aligned}
$$

Assume $\eta = 0.9$ and

(a) Determine the step jump c_g for each gear change.

(b) Determine the speed span for each gear change.

(c) Determine the engine speed at the maximum car speed for each gear.

(d) Determine the power performance equation and find the engine power at the maximum car speed for each gear.

(e) There is a difference between the car's top speed and the maximum speed in the 5th gear. Find the engine power at the car's top speed. Based on the top speed, determine the overall resistance forces.

14. Manual and auto transmission comparison.

A model of Nissan $U12$ Pintara^{TM} may come with manual or auto transmission. A model with a manual transmission has gear ratios and characteristics close to the following values

$$
\begin{aligned}
1st \text{ gear ratio} = n_1 = 3.285 & \qquad 2nd \text{ gear ratio} = n_2 = 1.850 \\
3rd \text{ gear ratio} = n_3 = 1.272 & \qquad 4th \text{ gear ratio} = n_4 = 0.954 \\
5th \text{ gear ratio} = n_5 = 0.740 & \\
\text{reverse gear ratio} = n_r = 3.428 & \qquad \text{final drive ratio} = n_d = 3.895
\end{aligned}
$$

and the model with an auto transmission has gear ratios close to the following values.

$$
\begin{aligned}
1st \text{ gear ratio} = n_1 = 2.785 & \qquad 2nd \text{ gear ratio} = n_2 = 1.545 \\
3rd \text{ gear ratio} = n_3 = 1.000 & \qquad 4th \text{ gear ratio} = n_4 = 0.694 \\
\text{reverse gear ratio} = n_r = 2.272 & \qquad \text{final drive ratio} = n_d = 3.876
\end{aligned}
$$

Compare the transmissions according to geometric design condition and determine which one has the maximum deviation.

15. ★ Progressive and geometric gearbox design.

An all wheel drive model of Hyundai Santa Fe^{TM} has specifications close to the following numbers.

$$
\begin{aligned}
P_M &= 242\,\text{hp } at \ \omega_M = 6000 \ \text{rpm} \\
T_M &= 226\,\text{lb ft } at \ \omega_e = 4500 \ \text{rpm} \\
m &= 1724\,\text{kg} \approx 4022\,\text{lb} \\
l &= 2700\,\text{mm} \approx 106.3\,\text{in} \\
Tires &= P235/70R16
\end{aligned}
$$

1st gear ratio = n_1 = 3.79   2nd gear ratio = n_2 = 2.06
3rd gear ratio = n_3 = 1.42   4th gear ratio = n_4 = 1.03
5th gear ratio = n_5 = 0.73
reverse gear ratio = n_r = 3.81   final drive ratio = n_d = 3.68

The car can reach a speed $v = 200.2\,\text{km/h} \approx 124\,\text{mi/h}$ at the maximum power P_M in the final gear $n_5 = 0.73$. Accept n_1 and n_5 and redesign the gear ratios based on a progressive and a geometric gearbox.

16. ★ Engine performance estimation.

Consider a RWD vehicle with the following specifications.

$$
\begin{aligned}
m &= 6300\,\text{lb} \\
l &= 153\,\text{in} \\
F_{z_1}/F_{z_2} &= 4410/6000 \\
Tires &= 245/75R16
\end{aligned}
$$

If an experiment shows that

$$
\begin{aligned}
v_M &= 62.6\,\text{mi/h} \quad at \quad 3\% \ \text{slope} \\
v_M &= 52.1\,\text{mi/h} \quad at \quad 6\% \ \text{slope} \\
v_M &= 0 \quad at \quad 33.2\% \ \text{slope}
\end{aligned}
$$

estimate the maximum torque and maximum power of the vehicle. Assume $\eta = 0.85$.

Hint: assume that when the vehicle is stuck on a road with the maximum slope, the engine is working at the maximum torque. However, when the vehicle is moving on a slope at the maximum speed, the engine is working at the maximum power. Slope 3% means the angle of the road with horizon is

$$
\phi = \tan^{-1}\frac{3}{100}
$$

17. ★ Gearbox design.

Consider a RWD vehicle with the following specifications.

$$
\begin{aligned}
P_M &= 141\,\text{kW} \approx 189\,\text{hp } at\ \omega_M = 7800\ \text{rpm} \\
T_M &= 181\,\text{N m} \approx 133\,\text{lb ft } at\ \omega_e = 6800\ \text{rpm} \\
v_M &= 237\,\text{km/ h} \approx 147\,\text{mi/ h} \\
\eta &= 0.90
\end{aligned}
$$

$$
\begin{aligned}
m &= 875\,\text{kg} \qquad\qquad l = 2300\,\text{mm} \\
Front\ tirest &= 195/50R16 \qquad Rear\ tirest = 225/45R17
\end{aligned}
$$

$1st$ gear ratio $= n_1 = 3.116$ $2nd$ gear ratio $= n_2 = 2.050$
$3rd$ gear ratio $= n_3 = 1.481$ $4th$ gear ratio $= n_4 = 1.166$
$5th$ gear ratio $= n_5 = 0.916$ $6th$ gear ratio $= n_6 = 0.815$
reverse gear ratio $= n_r = 3.250$ final drive ratio $= n_d = 4.529$

Based on the maximum velocity at the $6th$ gear n_6, redesign the gear ratios. Use $\pm 20\%$ around the maximum power for the working range.

Part II

Vehicle Kinematics

5

★ Applied Kinematics

Position, velocity, and acceleration are called kinematics information. Rotational position analysis is the key to calculate kinematics of relatively moving rigid bodies. In this chapter, we review kinematics and show applied methods to calculate the relative kinematic information of rigid bodies. A vehicle has many moving sub-systems such as suspensions, and the vehicle can be treated as a moving rigid body in an inertia coordinate frame.

5.1 Rotation About Global Cartesian Axes

Consider an orthogonal Cartesian body coordinate frame $B\,(Oxyz)$ fixed to a rigid body B that is attached to the ground G at the origin point O. The orientation of the rigid body B with respect to the global coordinate

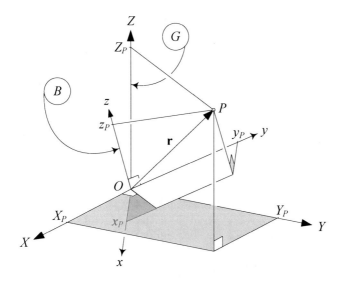

FIGURE 5.1. Rotation of a body coordinate frame B about point O in the global coordinate frame G.

frame $G\,(OXYZ)$ fixed to the ground is known when the orientation of $Oxyz$ with respect to $OXYZ$ is determined. Figure 5.1 illustrates a body coordinate B rotating about point O in the global coordinate frame G.

If the rigid body, B rotates α degrees about the Z-axis of the global

R.N. Jazar, *Vehicle Dynamics: Theory and Application*,
DOI 10.1007/978-1-4614-8544-5_5, © Springer Science+Business Media New York 2014

coordinate frame, then coordinates of any point P of the rigid body in the local and global coordinate frames are related by the equation

$$^G\mathbf{r} = R_{Z,\alpha}\,^B\mathbf{r} \tag{5.1}$$

where

$$R_{Z,\alpha} = \begin{bmatrix} \cos\alpha & -\sin\alpha & 0 \\ \sin\alpha & \cos\alpha & 0 \\ 0 & 0 & 1 \end{bmatrix} \tag{5.2}$$

and

$$^G\mathbf{r} = \begin{bmatrix} X \\ Y \\ Z \end{bmatrix} \qquad ^B\mathbf{r} = \begin{bmatrix} x \\ y \\ z \end{bmatrix} \tag{5.3}$$

Similarly, rotation β degrees about the Y-axis, and γ degrees about the X-axis of the global frame relate the local and global coordinates of point P by the following equations:

$$^G\mathbf{r} = R_{Y,\beta}\,^B\mathbf{r} \tag{5.4}$$
$$^G\mathbf{r} = R_{X,\gamma}\,^B\mathbf{r} \tag{5.5}$$

where

$$R_{Y,\beta} = \begin{bmatrix} \cos\beta & 0 & \sin\beta \\ 0 & 1 & 0 \\ -\sin\beta & 0 & \cos\beta \end{bmatrix} \tag{5.6}$$

$$R_{X,\gamma} = \begin{bmatrix} 1 & 0 & 0 \\ 0 & \cos\gamma & -\sin\gamma \\ 0 & \sin\gamma & \cos\gamma \end{bmatrix} \tag{5.7}$$

Proof. Let $(\hat{\imath}, \hat{\jmath}, \hat{k})$ and $(\hat{I}, \hat{J}, \hat{K})$ be the unit vectors along the coordinate axes of $Oxyz$ and $OXYZ$ respectively. The rigid body has a space fixed point at O, which is the common origin of $Oxyz$ and $OXYZ$. Consider a body fixed point P of a rigid body that turns α rad about the z-axis. Figure 5.2 illustrate the top view of the coordinate frames after the rotation while the dashed lines indicate the point at initial position (X_1, Y_1, Z_1).

The position vector $\mathbf{r}$ of P can be expressed in body and global coordinate frames by

$$^B\mathbf{r} = x\hat{\imath} + y\hat{\jmath} + z\hat{k} \tag{5.8}$$
$$^G\mathbf{r} = X\hat{I} + Y\hat{J} + Z\hat{K} \tag{5.9}$$

where $^B\mathbf{r}$ refers to the position vector $\mathbf{r}$ expressed in the body coordinate frame B, and $^G\mathbf{r}$ refers to the position vector $\mathbf{r}$ expressed in the global coordinate frame G.

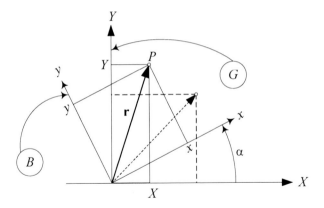

FIGURE 5.2. Position vectors of point P before and after the rotation of the local frame about the Z-axis of the global frame.

Using Equation (5.8) and the definition of the inner product, we may write

$$X = \hat{I} \cdot \mathbf{r} = \hat{I} \cdot x\hat{\imath} + \hat{I} \cdot y\hat{\jmath} + \hat{I} \cdot z\hat{k} \qquad (5.10)$$

$$Y = \hat{J} \cdot \mathbf{r} = \hat{J} \cdot x\hat{\imath} + \hat{J} \cdot y\hat{\jmath} + \hat{J} \cdot z\hat{k} \qquad (5.11)$$

$$Z = \hat{K} \cdot \mathbf{r} = \hat{K} \cdot x\hat{\imath} + \hat{K} \cdot y\hat{\jmath} + \hat{K} \cdot z\hat{k} \qquad (5.12)$$

or equivalently

$$\begin{bmatrix} X \\ Y \\ Z \end{bmatrix} = \begin{bmatrix} \hat{I} \cdot \hat{\imath} & \hat{I} \cdot \hat{\jmath} & \hat{I} \cdot \hat{k} \\ \hat{J} \cdot \hat{\imath} & \hat{J} \cdot \hat{\jmath} & \hat{J} \cdot \hat{k} \\ \hat{K} \cdot \hat{\imath} & \hat{K} \cdot \hat{\jmath} & \hat{K} \cdot \hat{k} \end{bmatrix} \begin{bmatrix} x \\ y \\ z \end{bmatrix} \qquad (5.13)$$

Analyzing Figure 5.2 indicates that

$$\begin{array}{lll} \hat{I} \cdot \hat{\imath} = \cos\alpha & \hat{I} \cdot \hat{\jmath} = -\sin\alpha & \hat{I} \cdot \hat{k} = 0 \\ \hat{J} \cdot \hat{\imath} = \sin\alpha & \hat{J} \cdot \hat{\jmath} = \cos\alpha & \hat{J} \cdot \hat{k} = 0 \\ \hat{K} \cdot \hat{\imath} = 0 & \hat{K} \cdot \hat{\jmath} = 0 & \hat{K} \cdot \hat{k} = 1 \end{array} \qquad (5.14)$$

Combining Equations (5.13) and (5.14) shows that

$$^{G}\mathbf{r} = R_{Z,\alpha} \, ^{B}\mathbf{r} \qquad (5.15)$$

$$\begin{bmatrix} X \\ Y \\ Z \end{bmatrix} = \begin{bmatrix} \cos\alpha & -\sin\alpha & 0 \\ \sin\alpha & \cos\alpha & 0 \\ 0 & 0 & 1 \end{bmatrix} \begin{bmatrix} x \\ y \\ z \end{bmatrix} \qquad (5.16)$$

The elements of the Z-rotation matrix, $R_{Z,\alpha}$, are called the *direction cosines* of $^{B}\mathbf{r}$ with respect to $OXYZ$.

$$R_{Z,\alpha} = \begin{bmatrix} \cos\alpha & -\sin\alpha & 0 \\ \sin\alpha & \cos\alpha & 0 \\ 0 & 0 & 1 \end{bmatrix} \qquad (5.17)$$

Equation (5.15) states that the vector **r** at the second position in the global coordinate frame is equal to $R_{Z,\alpha}$ times the position vector in the local coordinate frame. Hence, we are able to find the global coordinates of a point of a rigid body after rotation about the Z-axis, if we have its local coordinates.

Similarly, rotation β about the Y-axis and rotation γ about the X-axis are expressed by the *Y-rotation matrix* $R_{Y,\beta}$ and the *X-rotation matrix* $R_{X,\gamma}$ respectively.

$$R_{Y,\beta} = \begin{bmatrix} \cos\beta & 0 & \sin\beta \\ 0 & 1 & 0 \\ -\sin\beta & 0 & \cos\beta \end{bmatrix} \qquad (5.18)$$

$$R_{X,\gamma} = \begin{bmatrix} 1 & 0 & 0 \\ 0 & \cos\gamma & -\sin\gamma \\ 0 & \sin\gamma & \cos\gamma \end{bmatrix} \qquad (5.19)$$

The rotation matrices $R_{Z,\alpha}$, $R_{Y,\beta}$, and $R_{X,\gamma}$ are called *basic global rotation matrices*. We usually refer to the first, second, and third rotations about the axes of the global coordinate frame by α, β, and γ respectively.
∎

Example 155 *Successive rotation about global axes.*
The final position of the point $P(1,2,3)$ after a 30 deg rotation about the Z-axis, followed by 30 deg about the X-axis, and then 90 deg about the Y-axis can be found by first multiplying $R_{Z,30}$ by $[1,2,3]^T$ to get the new global position after the first rotation

$$\begin{bmatrix} X_2 \\ Y_2 \\ Z_2 \end{bmatrix} = \begin{bmatrix} \cos\frac{\pi}{6} & -\sin\frac{\pi}{6} & 0 \\ \sin\frac{\pi}{6} & \cos\frac{\pi}{6} & 0 \\ 0 & 0 & 1 \end{bmatrix} \begin{bmatrix} 1 \\ 2 \\ 3 \end{bmatrix} = \begin{bmatrix} -0.134 \\ 2.23 \\ 3 \end{bmatrix} \qquad (5.20)$$

and then multiplying $R_{X,30}$ by $[-0.134, 2.23, 3]^T$ to calculate the position of P after the second rotation

$$\begin{bmatrix} X_3 \\ Y_3 \\ Z_3 \end{bmatrix} = \begin{bmatrix} 1 & 0 & 0 \\ 0 & \cos\frac{\pi}{6} & -\sin\frac{\pi}{6} \\ 0 & \sin\frac{\pi}{6} & \cos\frac{\pi}{6} \end{bmatrix} \begin{bmatrix} -0.134 \\ 2.23 \\ 3 \end{bmatrix} = \begin{bmatrix} -0.134 \\ 0.433 \\ 3.714 \end{bmatrix}$$
$$(5.21)$$

and finally multiplying $R_{Y,90}$ by $[-0.134, 0.433, 3.714]^T$ to find the final position of P after the third rotation.

$$\begin{bmatrix} X_4 \\ Y_4 \\ Z_4 \end{bmatrix} = \begin{bmatrix} \cos\frac{\pi}{2} & 0 & \sin\frac{\pi}{2} \\ 0 & 1 & 0 \\ -\sin\frac{\pi}{2} & 0 & \cos\frac{\pi}{2} \end{bmatrix} \begin{bmatrix} -0.134 \\ 0.433 \\ 3.714 \end{bmatrix} = \begin{bmatrix} 3.714 \\ 0.433 \\ 0.134 \end{bmatrix} \qquad (5.22)$$

Example 156 *Global rotation, local position.*

If a point P is moved to $^G\mathbf{r} = [2,3,2]^T$ after a 60 deg rotation about the Z-axis, its position in the local coordinate is

$$^B\mathbf{r} = R_{Z,60}^{-1}\,{}^G\mathbf{r}$$

$$\begin{bmatrix} x \\ y \\ z \end{bmatrix} = \begin{bmatrix} \cos\frac{\pi}{3} & -\sin\frac{\pi}{3} & 0 \\ \sin\frac{\pi}{3} & \cos\frac{\pi}{3} & 0 \\ 0 & 0 & 1 \end{bmatrix}^{-1} \begin{bmatrix} 2 \\ 3 \\ 2 \end{bmatrix} = \begin{bmatrix} 3.6 \\ -0.23 \\ 2 \end{bmatrix} \tag{5.23}$$

Example 157 *Time-dependent global rotation.*

Assume a rigid body B is continuously turning about the Z-axis at a rate of $0.1\,\mathrm{rad/s}$. The transformation matrix of the body is

$$^G R_B = \begin{bmatrix} \cos 0.1t & -\sin 0.1t & 0 \\ \sin 0.1t & \cos 0.1t & 0 \\ 0 & 0 & 1 \end{bmatrix} \tag{5.24}$$

The rotation

$$\begin{bmatrix} X \\ Y \\ Z \end{bmatrix} = \begin{bmatrix} \cos 0.1t & -\sin 0.1t & 0 \\ \sin 0.1t & \cos 0.1t & 0 \\ 0 & 0 & 1 \end{bmatrix} \begin{bmatrix} x \\ y \\ z \end{bmatrix}$$

$$= \begin{bmatrix} x\cos 0.1t - y\sin 0.1t \\ y\cos 0.1t + x\sin 0.1t \\ z \end{bmatrix} \tag{5.25}$$

will move any point of B on a circle with radius $R = \sqrt{X^2 + Y^2}$ parallel to the (X,Y)-plane:

$$X^2 + Y^2 = (x\cos 0.1t - y\sin 0.1t)^2 + (y\cos 0.1t + x\sin 0.1t)^2$$
$$= x^2 + y^2 = R^2 \tag{5.26}$$

Consider a point P at $^B\mathbf{r} = [\,1\ \ 0\ \ 0\,]^T$. The point will be seen at

$$\begin{bmatrix} X \\ Y \\ Z \end{bmatrix} = \begin{bmatrix} \cos 0.1 & -\sin 0.1 & 0 \\ \sin 0.1 & \cos 0.1 & 0 \\ 0 & 0 & 1 \end{bmatrix} \begin{bmatrix} 1 \\ 0 \\ 0 \end{bmatrix} = \begin{bmatrix} 0.995 \\ 0.0998 \\ 0 \end{bmatrix} \tag{5.27}$$

after $t = 1\,\mathrm{s}$, and at

$$\begin{bmatrix} X \\ Y \\ Z \end{bmatrix} = \begin{bmatrix} \cos 0.6 & -\sin 0.6 & 0 \\ \sin 0.6 & \cos 0.6 & 0 \\ 0 & 0 & 1 \end{bmatrix} \begin{bmatrix} 1 \\ 0 \\ 0 \end{bmatrix} = \begin{bmatrix} 0.825 \\ 0.564 \\ 0 \end{bmatrix} \tag{5.28}$$

after $t = 6\,\mathrm{s}$,

5.2 Successive Rotation About Global Cartesian Axes

The final global position of a point P of a rigid body B with position vector $\mathbf{r}$, after a sequence of rotations R_1, R_2, R_3, ..., R_n about the global axes can be found by

$$^G\mathbf{r} = {}^GR_B\,{}^B\mathbf{r} \tag{5.29}$$

where,

$$^GR_B = R_n \cdots R_3 R_2 R_1 \tag{5.30}$$

and $^G\mathbf{r}$ and $^B\mathbf{r}$ indicate the position vector $\mathbf{r}$ expressed in the global and local coordinate frames, respectively. The transformation matrix ^GR_B is called the *global rotation matrix* which maps the local coordinates to their corresponding global coordinates.

Example 158 *Successive global rotations, global position.*

The point P of a rigid body that is attached to the global frame at O is located at

$$\begin{bmatrix} X \\ Y \\ Z \end{bmatrix} = \begin{bmatrix} 0.0 \\ 0.26 \\ 0.97 \end{bmatrix} \tag{5.31}$$

The rotation matrix to find the new position of the point after a -29 deg rotation about the X-axis, followed by 30 deg about the Z-axis, and again 132 deg about the X-axis is

$$^GR_B = R_{X,132}R_{Z,30}R_{X,-29} = \begin{bmatrix} 0.87 & -0.44 & -0.24 \\ -0.33 & -0.15 & -0.93 \\ 0.37 & 0.89 & -0.27 \end{bmatrix} \tag{5.32}$$

Therefore, its new position is at

$$\begin{bmatrix} X \\ Y \\ Z \end{bmatrix} = \begin{bmatrix} 0.87 & -0.44 & -0.24 \\ -0.33 & -0.15 & -0.93 \\ 0.37 & 0.89 & -0.27 \end{bmatrix} \begin{bmatrix} 0.0 \\ 0.26 \\ 0.97 \end{bmatrix} = \begin{bmatrix} -0.35 \\ -0.94 \\ -0.031 \end{bmatrix} \tag{5.33}$$

Example 159 *Order of rotation, and order of matrix multiplication.*

Changing the order of global rotation matrices is equivalent to changing the order of rotations.

The position of a point P of a rigid body B is located at $^B\mathbf{r} = \begin{bmatrix} 1 & 2 & 3 \end{bmatrix}^T$. Its global position after rotation 30 deg about the X-axis and then 45 deg about the Y-axis is at

$$\begin{aligned} ^G\mathbf{r}_1 &= R_{Y,45}\,R_{X,30}\,{}^B\mathbf{r} \\ &= \begin{bmatrix} 0.53 & -0.84 & 0.13 \\ 0.0 & 0.15 & 0.99 \\ -0.85 & -0.52 & 0.081 \end{bmatrix} \begin{bmatrix} 1 \\ 2 \\ 3 \end{bmatrix} = \begin{bmatrix} -0.76 \\ 3.27 \\ -1.64 \end{bmatrix} \end{aligned} \tag{5.34}$$

and if we change the order of rotations, then its position would be at

$$
\begin{aligned}
{}^{G}\mathbf{r}_2 &= R_{X,30} \, R_{Y,45} \, {}^{B}\mathbf{r} \\
&=
\begin{bmatrix}
0.53 & 0.0 & 0.85 \\
-0.84 & 0.15 & 0.52 \\
-0.13 & -0.99 & 0.081
\end{bmatrix}
\begin{bmatrix}
1 \\
2 \\
3
\end{bmatrix}
=
\begin{bmatrix}
3.08 \\
1.02 \\
-1.86
\end{bmatrix}
\end{aligned}
\tag{5.35}
$$

These two final positions of P are $d = \left| {}^{G}\mathbf{r}_1 - {}^{G}\mathbf{r}_2 \right| = 4.456$ apart.

Example 160 *Global roll-pitch-yaw angles.*
    The rotation about the X-axis of the global coordinate frame is called **roll**, *the rotation about the Y-axis of the global coordinate frame is called* **pitch**, *and the rotation about the Z-axis of the global coordinate frame is called* **yaw**. *The global roll-pitch-yaw rotation matrix is*

$$
\begin{aligned}
{}^{G}R_B &= R_{Z,\gamma} \, R_{Y,\beta} \, R_{X,\alpha} \\
&=
\begin{bmatrix}
c\beta c\gamma & -c\alpha s\gamma + c\gamma s\alpha s\beta & s\alpha s\gamma + c\alpha c\gamma s\beta \\
c\beta s\gamma & c\alpha c\gamma + s\alpha s\beta s\gamma & -c\gamma s\alpha + c\alpha s\beta s\gamma \\
-s\beta & c\beta s\alpha & c\alpha c\beta
\end{bmatrix}
\end{aligned}
\tag{5.36}
$$

Given the roll, pitch, and yaw angles, we can compute the overall rotation matrix using Equation (5.36). Also, we are able to compute the equivalent roll, pitch, and yaw angles when a rotation matrix is given. Suppose that r_{ij} indicates the element of row i and column j of the roll-pitch-yaw rotation matrix (5.36), then the roll angle is

$$
\alpha = \tan^{-1}\left(\frac{r_{32}}{r_{33}} \right)
\tag{5.37}
$$

and the pitch angle is

$$
\beta = -\sin^{-1}\left(r_{31} \right)
\tag{5.38}
$$

and the yaw angle is

$$
\gamma = \tan^{-1}\left(\frac{r_{21}}{r_{11}} \right)
\tag{5.39}
$$

provided that $\cos \beta \neq 0$.

5.3   Rotation About Local Cartesian Axes

Consider a rigid body B with a space-fixed point at point O. The local body coordinate frame $B(Oxyz)$ is coincident with the global coordinate frame $G(OXYZ)$, where the origin of both frames are on the fixed point O. If the body undergoes a rotation φ about the z-axis of its local coordinate

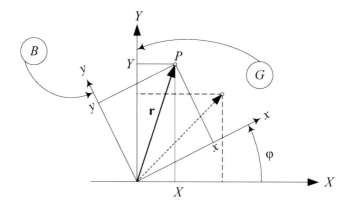

FIGURE 5.3. Position vectors of point P before and after rotation of the local frame about the z-axis of the local frame.

frame, then coordinates of any point of the rigid body in the local and global coordinate frames are related by the equation

$$^B\mathbf{r} = R_{z,\varphi}\,^G\mathbf{r} \tag{5.40}$$

Figure 5.3 depicts the top view of the system. The vectors $^G\mathbf{r}$ and $^B\mathbf{r}$ are the position vectors of the point in the global and local frames respectively

$$^G\mathbf{r} = \begin{bmatrix} X \\ Y \\ Z \end{bmatrix} \qquad ^B\mathbf{r} = \begin{bmatrix} x \\ y \\ z \end{bmatrix} \tag{5.41}$$

and $R_{z,\varphi}$ is the z-rotation matrix

$$R_{z,\varphi} = \begin{bmatrix} \cos\varphi & \sin\varphi & 0 \\ -\sin\varphi & \cos\varphi & 0 \\ 0 & 0 & 1 \end{bmatrix} \tag{5.42}$$

Similarly, rotation θ about the y-axis and rotation ψ about the x-axis are expressed by the y-rotation matrix $R_{y,\theta}$ and the x-rotation matrix $R_{x,\psi}$ respectively.

$$R_{y,\theta} = \begin{bmatrix} \cos\theta & 0 & -\sin\theta \\ 0 & 1 & 0 \\ \sin\theta & 0 & \cos\theta \end{bmatrix} \tag{5.43}$$

$$R_{x,\psi} = \begin{bmatrix} 1 & 0 & 0 \\ 0 & \cos\psi & \sin\psi \\ 0 & -\sin\psi & \cos\psi \end{bmatrix} \tag{5.44}$$

Proof. Vector $\mathbf{r}$ indicates the position of a point P of the rigid body B after the rotation. Using the unit vectors $(\hat{\imath}, \hat{\jmath}, \hat{k})$ along the axes of the local coordinate frame $B(Oxyz)$, and $(\hat{I}, \hat{J}, \hat{K})$ along the axes of the global coordinate frame $G(OXYZ)$, the expression of the position vectors $\mathbf{r}$ in both coordinate frames would be

$$
\begin{aligned}
{}^B\mathbf{r} &= x\hat{\imath} + y\hat{\jmath} + z\hat{k} \tag{5.45} \\
{}^G\mathbf{r} &= X\hat{I} + Y\hat{J} + Z\hat{K} \tag{5.46}
\end{aligned}
$$

The components of ${}^B\mathbf{r}$ can be found if we have the components of ${}^G\mathbf{r}$. Using Equation (5.45) and the definition of the inner product, we may write

$$
\begin{aligned}
x &= \hat{\imath} \cdot \mathbf{r} = \hat{\imath} \cdot X\hat{I} + \hat{\imath} \cdot Y\hat{J} + \hat{\imath} \cdot Z\hat{K} \tag{5.47} \\
y &= \hat{\jmath} \cdot \mathbf{r} = \hat{\jmath} \cdot X\hat{I} + \hat{\jmath} \cdot Y\hat{J} + \hat{\jmath} \cdot Z\hat{K} \tag{5.48} \\
z &= \hat{k} \cdot \mathbf{r} = \hat{k} \cdot X\hat{I} + \hat{k} \cdot Y\hat{J} + \hat{k} \cdot Z\hat{K} \tag{5.49}
\end{aligned}
$$

or equivalently

$$
\begin{bmatrix} x \\ y \\ z \end{bmatrix} = \begin{bmatrix} \hat{\imath} \cdot \hat{I} & \hat{\imath} \cdot \hat{J} & \hat{\imath} \cdot \hat{K} \\ \hat{\jmath} \cdot \hat{I} & \hat{\jmath} \cdot \hat{J} & \hat{\jmath} \cdot \hat{K} \\ \hat{k} \cdot \hat{I} & \hat{k} \cdot \hat{J} & \hat{k} \cdot \hat{K} \end{bmatrix} \begin{bmatrix} X \\ Y \\ Z \end{bmatrix} \tag{5.50}
$$

The elements of the z-rotation matrix $R_{z,\varphi}$ are the *direction cosines* of ${}^G\mathbf{r}_2$ with respect to $Oxyz$. So, the elements of the matrix in Equation (5.50) are

$$
\begin{aligned}
\hat{\imath} \cdot \hat{I} &= \cos\varphi & \hat{\imath} \cdot \hat{J} &= \sin\varphi & \hat{\imath} \cdot \hat{K} &= 0 \\
\hat{\jmath} \cdot \hat{I} &= -\sin\varphi & \hat{\jmath} \cdot \hat{J} &= \cos\varphi & \hat{\jmath} \cdot \hat{K} &= 0 \\
\hat{k} \cdot \hat{I} &= 0 & \hat{k} \cdot \hat{J} &= 0 & \hat{k} \cdot \hat{K} &= 1
\end{aligned} \tag{5.51}
$$

Combining Equations (5.50) and (5.51), we can find the components of ${}^B\mathbf{r}$ by multiplying z-rotation matrix $R_{z,\varphi}$ and vector ${}^G\mathbf{r}$.

$$
\begin{bmatrix} x \\ y \\ z \end{bmatrix} = \begin{bmatrix} \cos\varphi & \sin\varphi & 0 \\ -\sin\varphi & \cos\varphi & 0 \\ 0 & 0 & 1 \end{bmatrix} \begin{bmatrix} X \\ Y \\ Z \end{bmatrix} \tag{5.52}
$$

It can also be shown in a short form:

$$
{}^B\mathbf{r} = R_{z,\varphi}\,{}^G\mathbf{r} \tag{5.53}
$$

where

$$
R_{z,\varphi} = \begin{bmatrix} \cos\varphi & \sin\varphi & 0 \\ -\sin\varphi & \cos\varphi & 0 \\ 0 & 0 & 1 \end{bmatrix} \tag{5.54}
$$

Equation (5.53) says that after rotation about the z-axis of the local coordinate frame, the position vector in the local frame is equal to $R_{z,\varphi}$

times the position vector in the global frame. Hence, after rotation about the z-axis, we are able to find the coordinates of any point of a rigid body in a local coordinate frame, if we have its coordinates in the global frame.

Similarly, rotation θ about the y-axis and rotation ψ about the x-axis are described by the y-rotation matrix $R_{y,\theta}$ and the x-rotation matrix $R_{x,\psi}$ respectively.

$$R_{y,\theta} = \begin{bmatrix} \cos\theta & 0 & -\sin\theta \\ 0 & 1 & 0 \\ \sin\theta & 0 & \cos\theta \end{bmatrix} \tag{5.55}$$

$$R_{x,\psi} = \begin{bmatrix} 1 & 0 & 0 \\ 0 & \cos\psi & \sin\psi \\ 0 & -\sin\psi & \cos\psi \end{bmatrix} \tag{5.56}$$

We usually indicate the first, second, and third rotations about the local axes by φ, θ, and ψ respectively. ∎

Example 161 *Local rotation, local position.*

If a local coordinate frame $B(Oxyz)$ has been rotated 60 deg about the z-axis and a point P in the global coordinate frame $G(OXYZ)$ is at $(4,3,2)$, its coordinates in the local coordinate frame B are

$$\begin{bmatrix} x \\ y \\ z \end{bmatrix} = \begin{bmatrix} \cos\frac{\pi}{3} & \sin\frac{\pi}{3} & 0 \\ -\sin\frac{\pi}{3} & \cos\frac{\pi}{3} & 0 \\ 0 & 0 & 1 \end{bmatrix} \begin{bmatrix} 4 \\ 3 \\ 2 \end{bmatrix} = \begin{bmatrix} 4.60 \\ -1.97 \\ 2.0 \end{bmatrix} \tag{5.57}$$

Example 162 *Local rotation, global position.*

If a local coordinate frame $B(Oxyz)$ has been rotated 60 deg about the z-axis and a point P in the local frame B is at $(4,3,2)$, its position in the global coordinate frame $G(OXYZ)$ is at

$$\begin{bmatrix} X \\ Y \\ Z \end{bmatrix} = \begin{bmatrix} \cos\frac{\pi}{3} & \sin\frac{\pi}{3} & 0 \\ -\sin\frac{\pi}{3} & \cos\frac{\pi}{3} & 0 \\ 0 & 0 & 1 \end{bmatrix}^{-1} \begin{bmatrix} 4 \\ 3 \\ 2 \end{bmatrix} = \begin{bmatrix} -0.60 \\ 4.96 \\ 2.0 \end{bmatrix} \tag{5.58}$$

Example 163 *Successive local rotation, global position.*

Let us turn a rigid body -90 deg about the y-axis and then 90 deg about the x-axis. If a body point P is at $^B\mathbf{r} = \begin{bmatrix} 9.5 & -10.1 & 10.1 \end{bmatrix}^T$, then its position in the global coordinate frame is at

$$^G\mathbf{r}_2 = [R_{x,90}\, R_{y,-90}]^{-1}\, ^B\mathbf{r} = R_{y,-90}^{-1}\, R_{x,90}^{-1}\, ^B\mathbf{r}$$

$$= R_{y,-90}^T\, R_{x,90}^T\, ^B\mathbf{r}_P = \begin{bmatrix} 10.1 \\ -10.1 \\ 9.5 \end{bmatrix} \tag{5.59}$$

The inverse and transpose of the rotation transformation matrices are equal

$$R^{-1} = R^T \tag{5.60}$$

such matrices are called orthogonal matrices.

Example 164 *Global position and postmultiplication of rotation matrix.*

The local position of a point P after rotation is $^B\mathbf{r} = \begin{bmatrix} 1 & 2 & 3 \end{bmatrix}^T$. If the local rotation matrix to transform $^G\mathbf{r}$ to $^B\mathbf{r}$ is given as

$$R_{z,\varphi} = \begin{bmatrix} \cos\varphi & \sin\varphi & 0 \\ -\sin\varphi & \cos\varphi & 0 \\ 0 & 0 & 1 \end{bmatrix} = \begin{bmatrix} \cos\frac{\pi}{6} & \sin\frac{\pi}{6} & 0 \\ -\sin\frac{\pi}{6} & \cos\frac{\pi}{6} & 0 \\ 0 & 0 & 1 \end{bmatrix} \qquad (5.61)$$

then we may find the global position vector $^G\mathbf{r}$ by postmultiplication $^B R_{z,\varphi}$ by the local position vector $^B\mathbf{r}^T$,

$$\begin{aligned} ^G\mathbf{r}^T &= {}^B\mathbf{r}^T R_{z,\varphi} = \begin{bmatrix} 1 & 2 & 3 \end{bmatrix} \begin{bmatrix} \cos\frac{\pi}{6} & \sin\frac{\pi}{6} & 0 \\ -\sin\frac{\pi}{6} & \cos\frac{\pi}{6} & 0 \\ 0 & 0 & 1 \end{bmatrix} \\ &= \begin{bmatrix} -0.13 & 2.23 & 3.0 \end{bmatrix} \end{aligned} \qquad (5.62)$$

instead of premultiplication of $R_{z,\varphi}^{-1}$ by $^B\mathbf{r}$.

$$\begin{aligned} ^G\mathbf{r} &= R_{z,\varphi}^{-1} \, {}^B\mathbf{r} \\ &= \begin{bmatrix} \cos\frac{\pi}{6} & -\sin\frac{\pi}{6} & 0 \\ \sin\frac{\pi}{6} & \cos\frac{\pi}{6} & 0 \\ 0 & 0 & 1 \end{bmatrix} \begin{bmatrix} 1 \\ 2 \\ 3 \end{bmatrix} = \begin{bmatrix} -0.13 \\ 2.23 \\ 3 \end{bmatrix} \end{aligned} \qquad (5.63)$$

5.4 Successive Rotation About Local Cartesian Axes

The final global position of a point P in a rigid body B with position vector $\mathbf{r}$, after some rotations R_1, R_2, R_3, ..., R_n about the local axes, can be found by

$$^B\mathbf{r} = {}^B R_G \, {}^G\mathbf{r} \qquad (5.64)$$

where

$$^B R_G = R_n \cdots R_3 R_2 R_1 \qquad (5.65)$$

The transformation matrix $^B R_G$ is called the *local rotation matrix* and it maps the global coordinates to their corresponding local coordinates.

Example 165 *Successive local rotation, local position.*

A local coordinate frame $B(Oxyz)$ that initially is coincident with a global coordinate frame $G(OXYZ)$ undergoes a rotation $\varphi = 30\deg$ about the z-axis, then $\theta = 30\deg$ about the x-axis, and then $\psi = 30\deg$ about the y-axis. The local coordinates of a point P located at $X = 5$, $Y = 30$, $Z = 10$ can be found by

$$\begin{bmatrix} x \\ y \\ z \end{bmatrix} = {}^B R_G \begin{bmatrix} 5 \\ 30 \\ 10 \end{bmatrix} \qquad (5.66)$$

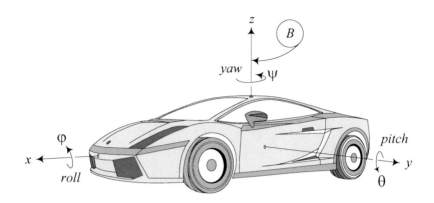

FIGURE 5.4. Local roll-pitch-yaw angles.

The local rotation matrix is

$$^{B}R_{G} = R_{y,30}R_{x,30}R_{z,30} = \begin{bmatrix} 0.63 & 0.65 & -0.43 \\ -0.43 & 0.75 & 0.50 \\ 0.65 & -0.125 & 0.75 \end{bmatrix} \qquad (5.67)$$

and coordinates of P in the local frame are

$$\begin{bmatrix} x \\ y \\ z \end{bmatrix} = \begin{bmatrix} 0.63 & 0.65 & -0.43 \\ -0.43 & 0.75 & 0.50 \\ 0.65 & -0.125 & 0.75 \end{bmatrix} \begin{bmatrix} 5 \\ 30 \\ 10 \end{bmatrix} = \begin{bmatrix} 18.35 \\ 25.35 \\ 7.0 \end{bmatrix} \qquad (5.68)$$

Example 166 *Successive local rotation.*

The rotation matrix for a body point $P(x, y, z)$ after rotation $R_{z,\varphi}$ followed by $R_{x,\theta}$ and $R_{y,\psi}$ is

$$\begin{aligned} ^{B}R_{G} &= R_{y,\psi}R_{x,\theta}R_{z,\varphi} \\ &= \begin{bmatrix} c\varphi c\psi - s\theta s\varphi s\psi & c\psi s\varphi + c\varphi s\theta s\psi & -c\theta s\psi \\ -c\theta s\varphi & c\theta c\varphi & s\theta \\ c\varphi s\psi + s\theta c\psi s\varphi & s\varphi s\psi - c\varphi s\theta c\psi & c\theta c\psi \end{bmatrix} \end{aligned} \qquad (5.69)$$

Example 167 *Local roll-pitch-yaw angles*

*Rotation about the x-axis of the local frame is called **roll** or **bank**, rotation about y-axis of the local frame is called **pitch** or **attitude**, and rotation about the z-axis of the local frame is called **yaw**, **spin**, or **heading**. The local roll-pitch-yaw angles of a car are shown in Figure 5.4.*

The local roll-pitch-yaw rotation matrix is

$$\begin{aligned} ^{B}R_{G} &= R_{z,\psi}R_{y,\theta}R_{x,\varphi} \\ &= \begin{bmatrix} c\theta c\psi & c\varphi s\psi + s\theta c\psi s\varphi & s\varphi s\psi - c\varphi s\theta c\psi \\ -c\theta s\psi & c\varphi c\psi - s\theta s\varphi s\psi & c\psi s\varphi + c\varphi s\theta s\psi \\ s\theta & -c\theta s\varphi & c\theta c\varphi \end{bmatrix} \end{aligned} \qquad (5.70)$$

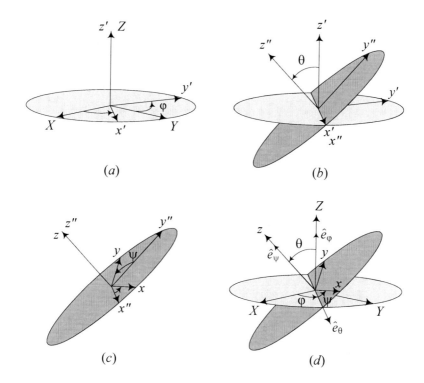

FIGURE 5.5. (a) First Euler angle. (b) Second Euler angle. (c) Third Euler angle.

Example 168 ★ *Euler Angles*

The rotation about the Z-axis of the global coordinate is called preces-sion, the rotation about the x-axis of the local coordinate is called nu-tation, and the rotation about the z-axis of the local coordinate is called spin. The precession-nutation-spin rotation angles are also called Euler an-gles. Rotation matrix based on Euler angles has application in rigid body kinematics. To find the Euler angles rotation matrix to go from the global frame $G(OXYZ)$ to the final body frame $B(Oxyz)$, we employ a body frame $B'(Ox'y'z')$ as shown in Figure 5.5(a) that before the first rotation coin-cides with the global frame. Let there be a rotation φ about the z'-axis. Because Z-axis and z'-axis are coincident, we have

$$^{B'}\mathbf{r} = {}^{B'}R_G\,{}^G\mathbf{r} \tag{5.71}$$

$$^{B'}R_G = R_{z,\varphi} = \begin{bmatrix} \cos\varphi & \sin\varphi & 0 \\ -\sin\varphi & \cos\varphi & 0 \\ 0 & 0 & 1 \end{bmatrix} \tag{5.72}$$

Next we consider the $B'(Ox'y'z')$ frame as a new fixed global frame and in-troduce a new body frame $B''(Ox''y''z'')$. Before the second rotation, the two frames coincide. Then, we execute a θ rotation about x''-axis as shown in

Figure 5.5(b). The transformation between $B'(Ox'y'z')$ *and* $B''(Ox''y''z'')$
is

$$^{B''}\mathbf{r} = \,^{B''}R_{B'}\,^{B'}\mathbf{r} \tag{5.73}$$

$$^{B''}R_{B'} = R_{x,\theta} = \begin{bmatrix} 1 & 0 & 0 \\ 0 & \cos\theta & \sin\theta \\ 0 & -\sin\theta & \cos\theta \end{bmatrix} \tag{5.74}$$

Finally, we consider the $B''(Ox''y''z'')$ *frame as a new fixed global frame and consider the final body frame* $B(Oxyz)$ *to coincide with* B'' *before the third rotation. We now execute a* ψ *rotation about the* z''*-axis as shown in Figure 5.5(c). The transformation between* $B''(Ox''y''z'')$ *and* $B(Oxyz)$ *is*

$$^{B}\mathbf{r} = \,^{B}R_{B''}\,^{B''}\mathbf{r} \tag{5.75}$$

$$^{B}R_{B''} = R_{z,\psi} = \begin{bmatrix} \cos\psi & \sin\psi & 0 \\ -\sin\psi & \cos\psi & 0 \\ 0 & 0 & 1 \end{bmatrix} \tag{5.76}$$

By the rule of composition of rotations, the transformation from $G(OXYZ)$ *to* $B(Oxyz)$ *is*

$$^{B}\mathbf{r} = \,^{B}R_{G}\,^{G}\mathbf{r} \tag{5.77}$$

where

$$
\begin{aligned}
^{B}R_{G} &= R_{z,\psi}R_{x,\theta}R_{z,\varphi} \\
&= \begin{bmatrix} c\varphi c\psi - c\theta s\varphi s\psi & c\psi s\varphi + c\theta c\varphi s\psi & s\theta s\psi \\ -c\varphi s\psi - c\theta c\psi s\varphi & -s\varphi s\psi + c\theta c\varphi c\psi & s\theta c\psi \\ s\theta s\varphi & -c\varphi s\theta & c\theta \end{bmatrix}
\end{aligned} \tag{5.78}
$$

and therefore,

$$
\begin{aligned}
^{B}R_{G}^{-1} &= \,^{B}R_{G}^{T} = \,^{G}R_{B} = [R_{z,\psi}R_{x,\theta}R_{z,\varphi}]^{T} \\
&= \begin{bmatrix} c\varphi c\psi - c\theta s\varphi s\psi & -c\varphi s\psi - c\theta c\psi s\varphi & s\theta s\varphi \\ c\psi s\varphi + c\theta c\varphi s\psi & -s\varphi s\psi + c\theta c\varphi c\psi & -c\varphi s\theta \\ s\theta s\psi & s\theta c\psi & c\theta \end{bmatrix}
\end{aligned} \tag{5.79}
$$

Given the angles of precession φ, nutation θ, and spin ψ, we can compute the overall rotation matrix using Equation (5.78). Also we are able to compute the equivalent precession, nutation, and spin angles when a rotation matrix is given.

If r_{ij} indicates the element of row i and column j of the precession-nutation-spin rotation matrix, then,

$$\theta = \cos^{-1}(r_{33}) \tag{5.80}$$

$$\varphi = -\tan^{-1}\left(\frac{r_{31}}{r_{32}}\right) \tag{5.81}$$

$$\psi = \tan^{-1}\left(\frac{r_{13}}{r_{23}}\right) \tag{5.82}$$

provided that $\sin\theta \neq 0$.

Example 169 ★ *Euler angles of a local rotation matrix.*

   The local rotation matrix after a rotation 30 deg *about the z-axis,* 30 deg *about the x-axis, and* 30 deg *about the y-axis is*

$$
{}^B R_G = R_{y,30} R_{x,30} R_{z,30}
$$
$$
= \begin{bmatrix} 0.63 & 0.65 & -0.43 \\ -0.43 & 0.75 & 0.50 \\ 0.65 & -0.125 & 0.75 \end{bmatrix} \tag{5.83}
$$

and therefore, the local coordinates of a sample point at $X = 5$, $Y = 30$, *and* $Z = 10$ *are*

$$
\begin{bmatrix} x \\ y \\ z \end{bmatrix} = \begin{bmatrix} 0.63 & 0.65 & -0.43 \\ -0.43 & 0.75 & 0.50 \\ 0.65 & -0.125 & 0.75 \end{bmatrix} \begin{bmatrix} 5 \\ 30 \\ 10 \end{bmatrix} = \begin{bmatrix} 18.35 \\ 25.35 \\ 7.0 \end{bmatrix} \tag{5.84}
$$

The Euler angles of the corresponding precession-nutation-spin rotation matrix are

$$
\theta = \cos^{-1}(0.75) = 41.41 \deg \tag{5.85}
$$
$$
\varphi = -\tan^{-1}\left(\frac{0.65}{-0.125}\right) = 79.15 \deg \tag{5.86}
$$
$$
\psi = \tan^{-1}\left(\frac{-0.43}{0.50}\right) = -40.7 \deg \tag{5.87}
$$

We can examine that the precession-nutation-spin rotation matrix for $\varphi = 79.15$ deg, $\theta = 41.41$ deg, *and* $\psi = -40.7$ deg *is also*

$$
{}^B R_G = R_{z,-40.7} R_{x,41.41} R_{z,79.15} \tag{5.88}
$$
$$
= \begin{bmatrix} 0.63 & 0.65 & -0.43 \\ -0.43 & 0.75 & 0.50 \\ 0.65 & -0.125 & 0.75 \end{bmatrix}
$$

Therefore, $R_{y,30} R_{x,30} R_{z,30} = R_{z,\psi} R_{x,\theta} R_{z,\varphi}$ *when* $\varphi = 79.15$ deg, $\theta = 41.41$ deg, *and* $\psi = -40.7$ deg. *In other words, the rigid body attached to the local frame moves to the final configuration by undergoing either three consecutive rotations* $\varphi = 79.15$ deg, $\theta = 41.41$ deg, *and* $\psi = -40.7$ deg *about the* z, x, *and* z *axes respectively, or three consecutive rotations* 30 deg, 30 deg, *and* 30 deg *about the* z, x, *and* y *axes. Hence the triple rotations between the initial and final configuration of a rigid body is not unique.*

Example 170 ★ *Relative rotation matrix of two bodies.*

   Consider a rigid body B_1 *with an orientation matrix* ${}^1 R_G$ *made by Euler angles* $\varphi = 30$ deg, $\theta = -45$ deg, $\psi = 60$ deg, *and another rigid body* B_2

having $\varphi = 10\,\deg$, $\theta = 25\,\deg$, $\psi = -15\,\deg$, with respect to the global frame. To find the relative rotation matrix $^1R_2$ to map the coordinates of the second body frame B_2 to the first body frame B_1, we need to find the individual rotation matrices first.

$$
\begin{aligned}
^1R_G &= R_{z,60}R_{x,-45}R_{z,30} \\
&= \begin{bmatrix} 0.127 & 0.78 & -0.612 \\ -0.927 & -0.127 & -0.354 \\ -0.354 & 0.612 & 0.707 \end{bmatrix}
\end{aligned} \tag{5.89}
$$

$$
\begin{aligned}
^2R_G &= R_{z,10}R_{x,25}R_{z,-15} \\
&= \begin{bmatrix} 0.992 & -6.33\times10^{-2} & -0.109 \\ 0.103 & 0.907 & 0.408 \\ 7.34\times10^{-2} & -0.416 & 0.906 \end{bmatrix}
\end{aligned} \tag{5.90}
$$

The desired rotation matrix $^1R_2$ may be found by

$$^1R_2 = \,^1R_G\,^GR_2 \tag{5.91}$$

which is equal to

$$
\begin{aligned}
^1R_2 &= \,^1R_G\,^2R_G^T \\
&= \begin{bmatrix} 0.992 & 0.103 & 7.34\times10^{-2} \\ -6.33\times10^{-2} & 0.907 & -0.416 \\ -0.109 & 0.408 & 0.906 \end{bmatrix}
\end{aligned} \tag{5.92}
$$

Example 171 ★ *Euler angles rotation matrix for small angles.*

The Euler rotation matrix $^BR_G = R_{z,\psi}R_{x,\theta}R_{z,\varphi}$ for very small Euler angles $\varphi, \theta,$ and ψ is approximated by

$$^BR_G = \begin{bmatrix} 1 & \gamma & 0 \\ -\gamma & 1 & \theta \\ 0 & -\theta & 1 \end{bmatrix} \tag{5.93}$$

where

$$\gamma = \varphi + \psi \tag{5.94}$$

Therefore, when the angles of rotation are small, the angles φ and ψ are indistinguishable.

If $\theta \to 0$ then the Euler rotation matrix $^BR_G = R_{z,\psi}R_{x,\theta}R_{z,\varphi}$ approaches to

$$
\begin{aligned}
^BR_G &= \begin{bmatrix} c\varphi c\psi - s\varphi s\psi & c\psi s\varphi + c\varphi s\psi & 0 \\ -c\varphi s\psi - c\psi s\varphi & -s\varphi s\psi + c\varphi c\psi & 0 \\ 0 & 0 & 1 \end{bmatrix} \\
&= \begin{bmatrix} \cos(\varphi+\psi) & \sin(\varphi+\psi) & 0 \\ -\sin(\varphi+\psi) & \cos(\varphi+\psi) & 0 \\ 0 & 0 & 1 \end{bmatrix}
\end{aligned} \tag{5.95}
$$

and therefore, the angles φ and ψ are indistinguishable even if the value of φ and ψ are finite. Hence, the Euler set of angles in rotation matrix (5.78) is not unique when $\theta = 0$.

Example 172 ★ *Angular velocity vector in terms of Euler frequencies.*
  A Eulerian local frame $E\,(O, \hat{e}_\varphi, \hat{e}_\theta, \hat{e}_\psi)$ can be introduced by defining unit vectors $\hat{e}_\varphi$, $\hat{e}_\theta$, and $\hat{e}_\psi$ as shown in Figure 5.5(d). Although the Eulerian frame is not necessarily orthogonal, it is very useful in rigid body kinematic analysis.
  The angular velocity vector $_G\boldsymbol{\omega}_B$ of the body frame $B(Oxyz)$ with respect to the global frame $G(OXYZ)$ can be written in Euler angles frame E as the sum of three Euler angle rate vectors:

$$\,_G^E\boldsymbol{\omega}_B = \dot{\varphi}\hat{e}_\varphi + \dot{\theta}\hat{e}_\theta + \dot{\psi}\hat{e}_\psi \tag{5.96}$$

where the rate of Euler angles, $\dot{\varphi}$, $\dot{\theta}$, and $\dot{\psi}$ are called Euler frequencies.
  To find $_G\boldsymbol{\omega}_B$ in the body frame we must express the unit vectors $\hat{e}_\varphi$, $\hat{e}_\theta$, and $\hat{e}_\psi$ in the body frame. The unit vector $\hat{e}_\varphi = \begin{bmatrix} 0 & 0 & 1 \end{bmatrix}^T = \hat{K}$ is in the global frame and can be transformed to the body frame after three rotations

$$^B\hat{e}_\varphi = {}^B R_G\,\hat{K} = R_{z,\psi} R_{x,\theta} R_{z,\varphi} \hat{K} = \begin{bmatrix} \sin\theta \sin\psi \\ \sin\theta \cos\psi \\ \cos\theta \end{bmatrix} \tag{5.97}$$

The unit vector $\hat{e}_\theta = \begin{bmatrix} 1 & 0 & 0 \end{bmatrix}^T = \hat{\imath}'$ is in the intermediate frame $Ox'y'z'$ and needs to get two rotations $R_{x,\theta}$ and $R_{z,\psi}$ to be transformed to the body frame

$$^B\hat{e}_\theta = {}^B R_{Ox'y'z'}\,\hat{\imath}' = R_{z,\psi}\, R_{x,\theta}\,\hat{\imath}' = \begin{bmatrix} \cos\psi \\ -\sin\psi \\ 0 \end{bmatrix} \tag{5.98}$$

The unit vector $\hat{e}_\psi$ is already in the body frame, $\hat{e}_\psi = \begin{bmatrix} 0 & 0 & 1 \end{bmatrix}^T = \hat{k}$. Therefore, $_G\boldsymbol{\omega}_B$ is expressed in the body coordinate frame as

$$
\begin{aligned}
\,_G^B\boldsymbol{\omega}_B &= \dot{\varphi} \begin{bmatrix} \sin\theta \sin\psi \\ \sin\theta \cos\psi \\ \cos\theta \end{bmatrix} + \dot{\theta} \begin{bmatrix} \cos\psi \\ -\sin\psi \\ 0 \end{bmatrix} + \dot{\psi} \begin{bmatrix} 0 \\ 0 \\ 1 \end{bmatrix} \\
&= \left(\dot{\varphi}\sin\theta\sin\psi + \dot{\theta}\cos\psi \right)\hat{\imath} + \left(\dot{\varphi}\sin\theta\cos\psi - \dot{\theta}\sin\psi \right)\hat{\jmath} \\
&\quad + \left(\dot{\varphi}\cos\theta + \dot{\psi} \right)\hat{k}
\end{aligned}
\tag{5.99}
$$

and therefore, components of $_G\boldsymbol{\omega}_B$ in body frame $B\,(Oxyz)$ are related to *the Euler angle frame $E\,(O\varphi\theta\psi)$ by the following relationship:*

$$
{}_G^B\boldsymbol{\omega}_B = {}^B R_E\, {}_G^E\boldsymbol{\omega}_B \tag{5.100}
$$

$$
\begin{bmatrix} \omega_x \\ \omega_y \\ \omega_z \end{bmatrix} = \begin{bmatrix} \sin\theta\sin\psi & \cos\psi & 0 \\ \sin\theta\cos\psi & -\sin\psi & 0 \\ \cos\theta & 0 & 1 \end{bmatrix} \begin{bmatrix} \dot\varphi \\ \dot\theta \\ \dot\psi \end{bmatrix} \tag{5.101}
$$

Then, ${}_G\boldsymbol{\omega}_B$ can be expressed in the global frame using an inverse transformation of Euler rotation matrix (5.78)

$$
{}_G^G\boldsymbol{\omega}_B = {}^B R_G^{-1}\, {}_G^B\boldsymbol{\omega}_B = {}^B R_G^{-1} \begin{bmatrix} \dot\varphi\sin\theta\sin\psi + \dot\theta\cos\psi \\ \dot\varphi\sin\theta\cos\psi - \dot\theta\sin\psi \\ \dot\varphi\cos\theta + \dot\psi \end{bmatrix} \tag{5.102}
$$

$$
= \left(\dot\theta\cos\varphi + \dot\psi\sin\theta\sin\varphi\right)\hat I + \left(\dot\theta\sin\varphi - \dot\psi\cos\varphi\sin\theta\right)\hat J
$$
$$
+ \left(\dot\varphi + \dot\psi\cos\theta\right)\hat K \tag{5.103}
$$

and hence, components of ${}_G\boldsymbol{\omega}_B$ in global coordinate frame $G(OXYZ)$ are related to the Euler angle coordinate frame $E\,(O\varphi\theta\psi)$ by the following relationship:

$$
{}_G^G\boldsymbol{\omega}_B = {}^G R_E\, {}_G^E\boldsymbol{\omega}_B \tag{5.104}
$$

$$
\begin{bmatrix} \omega_X \\ \omega_Y \\ \omega_Z \end{bmatrix} = \begin{bmatrix} 0 & \cos\varphi & \sin\theta\sin\varphi \\ 0 & \sin\varphi & -\cos\varphi\sin\theta \\ 1 & 0 & \cos\theta \end{bmatrix} \begin{bmatrix} \dot\varphi \\ \dot\theta \\ \dot\psi \end{bmatrix} \tag{5.105}
$$

Example 173 ★ *Euler frequencies based on a Cartesian angular velocity vector.*

The vector ${}_G^B\boldsymbol{\omega}_B$, that indicates the angular velocity of a rigid body B with respect to the global frame G written in frame B, is related to the Euler frequencies by

$$
{}_G^B\boldsymbol{\omega}_B = {}^B R_E\, {}_G^E\boldsymbol{\omega}_B
$$

$$
{}_G^B\boldsymbol{\omega}_B = \begin{bmatrix} \omega_x \\ \omega_y \\ \omega_z \end{bmatrix} = \begin{bmatrix} \sin\theta\sin\psi & \cos\psi & 0 \\ \sin\theta\cos\psi & -\sin\psi & 0 \\ \cos\theta & 0 & 1 \end{bmatrix} \begin{bmatrix} \dot\varphi \\ \dot\theta \\ \dot\psi \end{bmatrix} \tag{5.106}
$$

The matrix of coefficients is not an orthogonal matrix because,

$$
{}^B R_E^T \neq {}^B R_E^{-1} \tag{5.107}
$$

$$
{}^B R_E^T = \begin{bmatrix} \sin\theta\sin\psi & \sin\theta\cos\psi & \cos\theta \\ \cos\psi & -\sin\psi & 0 \\ 0 & 0 & 1 \end{bmatrix} \tag{5.108}
$$

$$
{}^B R_E^{-1} = \frac{1}{\sin\theta} \begin{bmatrix} \sin\psi & \cos\psi & 0 \\ \sin\theta\cos\psi & -\sin\theta\sin\psi & 0 \\ -\cos\theta\sin\psi & -\cos\theta\cos\psi & 1 \end{bmatrix} \tag{5.109}
$$

It is because the Euler angles coordinate frame $E\,(O\varphi\theta\psi)$ is not an orthogonal frame. For the same reason, the matrix of coefficients that relates the Euler frequencies and the components of $_{G}^{G}\boldsymbol{\omega}_{B}$

$$\begin{matrix} {}_{G}^{G}\boldsymbol{\omega}_{B} & = & {}^{G}R_{E}\,{}_{G}^{E}\boldsymbol{\omega}_{B} \end{matrix} \qquad (5.110)$$

$$\begin{matrix} {}_{G}^{G}\boldsymbol{\omega}_{B} & = & \begin{bmatrix} \omega_{X} \\ \omega_{Y} \\ \omega_{Z} \end{bmatrix} = \begin{bmatrix} 0 & \cos\varphi & \sin\theta\sin\varphi \\ 0 & \sin\varphi & -\cos\varphi\sin\theta \\ 1 & 0 & \cos\theta \end{bmatrix} \begin{bmatrix} \dot{\varphi} \\ \dot{\theta} \\ \dot{\psi} \end{bmatrix} \end{matrix} \qquad (5.111)$$

is not an orthogonal matrix. Therefore, the Euler frequencies based on local and global decomposition of the angular velocity vector $_{G}\boldsymbol{\omega}_{B}$ must solely be found by the inverse of coefficient matrices

$$\begin{matrix} {}_{G}^{E}\boldsymbol{\omega}_{B} & = & {}^{B}R_{E}^{-1}\,{}_{G}^{B}\boldsymbol{\omega}_{B} \end{matrix} \qquad (5.112)$$

$$\begin{bmatrix} \dot{\varphi} \\ \dot{\theta} \\ \dot{\psi} \end{bmatrix} = \frac{1}{\sin\theta} \begin{bmatrix} \sin\psi & \cos\psi & 0 \\ \sin\theta\cos\psi & -\sin\theta\sin\psi & 0 \\ -\cos\theta\sin\psi & -\cos\theta\cos\psi & 1 \end{bmatrix} \begin{bmatrix} \omega_{x} \\ \omega_{y} \\ \omega_{z} \end{bmatrix} \qquad (5.113)$$

and

$$\begin{matrix} {}_{G}^{E}\boldsymbol{\omega}_{B} & = & {}^{G}R_{E}^{-1}\,{}_{G}^{G}\boldsymbol{\omega}_{B} \end{matrix} \qquad (5.114)$$

$$\begin{bmatrix} \dot{\varphi} \\ \dot{\theta} \\ \dot{\psi} \end{bmatrix} = \frac{1}{\sin\theta} \begin{bmatrix} -\cos\theta\sin\varphi & \cos\theta\cos\varphi & 1 \\ \sin\theta\cos\varphi & \sin\theta\sin\varphi & 0 \\ \sin\varphi & -\cos\varphi & 0 \end{bmatrix} \begin{bmatrix} \omega_{X} \\ \omega_{Y} \\ \omega_{Z} \end{bmatrix} \qquad (5.115)$$

Using (5.112) and (5.114), it can be verified that the transformation matrix $^{B}R_{G} = {}^{B}R_{E}\,{}^{G}R_{E}^{-1}$ would be the same as Euler transformation matrix (5.78).

The angular velocity vector can thus be expressed as

$$\begin{matrix} {}_{G}\boldsymbol{\omega}_{B} & = & \begin{bmatrix} \hat{\imath} & \hat{\jmath} & \hat{k} \end{bmatrix} \begin{bmatrix} \omega_{x} \\ \omega_{y} \\ \omega_{z} \end{bmatrix} = \begin{bmatrix} \hat{I} & \hat{J} & \hat{K} \end{bmatrix} \begin{bmatrix} \omega_{X} \\ \omega_{Y} \\ \omega_{Z} \end{bmatrix} \\ \\ & = & \begin{bmatrix} \hat{K} & \hat{e}_{\theta} & \hat{k} \end{bmatrix} \begin{bmatrix} \dot{\varphi} \\ \dot{\theta} \\ \dot{\psi} \end{bmatrix} \end{matrix} \qquad (5.116)$$

5.5 General Transformation

Consider a general situation in which two coordinate frames, $G(OXYZ)$ and $B(Oxyz)$ with a common origin O, are employed to express the components of a vector $\mathbf{r}$. There is always a *transformation matrix* $^{G}R_{B}$ to map the components of $\mathbf{r}$ from the coordinate frame $B(Oxyz)$ to the other coordinate frame $G(OXYZ)$.

$$^G\mathbf{r} = {}^GR_B\,{}^B\mathbf{r} \tag{5.117}$$

In addition, the inverse map, $^B\mathbf{r} = {}^GR_B^{-1}\,{}^G\mathbf{r}$, can be done by ^BR_G

$$^B\mathbf{r} = {}^BR_G\,{}^G\mathbf{r} \tag{5.118}$$

When the coordinate frames B and G are orthogonal, the rotation matrix ^GR_B is called an *orthogonal matrix*. The transpose R^T and inverse R^{-1} of an orthogonal matrix $[R]$ are equal:

$$^BR_G = {}^GR_B^{-1} = {}^GR_B^T \tag{5.119}$$
$$\left|{}^GR_B\right| = \left|{}^BR_G\right| = 1 \tag{5.120}$$

Because of the matrix orthogonality condition, only three of the nine elements of ^GR_B are independent.
Proof. Decomposition of the unit vectors of $G(OXYZ)$ along the axes of $B(Oxyz)$

$$\hat{I} = (\hat{I}\cdot\hat{\imath})\hat{\imath} + (\hat{I}\cdot\hat{\jmath})\hat{\jmath} + (\hat{I}\cdot\hat{k})\hat{k} \tag{5.121}$$
$$\hat{J} = (\hat{J}\cdot\hat{\imath})\hat{\imath} + (\hat{J}\cdot\hat{\jmath})\hat{\jmath} + (\hat{J}\cdot\hat{k})\hat{k} \tag{5.122}$$
$$\hat{K} = (\hat{K}\cdot\hat{\imath})\hat{\imath} + (\hat{K}\cdot\hat{\jmath})\hat{\jmath} + (\hat{K}\cdot\hat{k})\hat{k} \tag{5.123}$$

introduces the transformation matrix ^GR_B to map the local frame to the global frame

$$\begin{bmatrix}\hat{I}\\\hat{J}\\\hat{K}\end{bmatrix} = \begin{bmatrix}\hat{I}\cdot\hat{\imath} & \hat{I}\cdot\hat{\jmath} & \hat{I}\cdot\hat{k}\\\hat{J}\cdot\hat{\imath} & \hat{J}\cdot\hat{\jmath} & \hat{J}\cdot\hat{k}\\\hat{K}\cdot\hat{\imath} & \hat{K}\cdot\hat{\jmath} & \hat{K}\cdot\hat{k}\end{bmatrix}\begin{bmatrix}\hat{\imath}\\\hat{\jmath}\\\hat{k}\end{bmatrix} = {}^GR_B\begin{bmatrix}\hat{\imath}\\\hat{\jmath}\\\hat{k}\end{bmatrix} \tag{5.124}$$

where

$$^GR_B = \begin{bmatrix}\hat{I}\cdot\hat{\imath} & \hat{I}\cdot\hat{\jmath} & \hat{I}\cdot\hat{k}\\\hat{J}\cdot\hat{\imath} & \hat{J}\cdot\hat{\jmath} & \hat{J}\cdot\hat{k}\\\hat{K}\cdot\hat{\imath} & \hat{K}\cdot\hat{\jmath} & \hat{K}\cdot\hat{k}\end{bmatrix}$$
$$= \begin{bmatrix}\cos(\hat{I},\hat{\imath}) & \cos(\hat{I},\hat{\jmath}) & \cos(\hat{I},\hat{k})\\\cos(\hat{J},\hat{\imath}) & \cos(\hat{J},\hat{\jmath}) & \cos(\hat{J},\hat{k})\\\cos(\hat{K},\hat{\imath}) & \cos(\hat{K},\hat{\jmath}) & \cos(\hat{K},\hat{k})\end{bmatrix} \tag{5.125}$$

The elements of ^GR_B are direction cosines of the axes of $G(OXYZ)$ in frame $B(Oxyz)$. This set of nine direction cosines then completely specifies the orientation of the frame $B(Oxyz)$ in the frame $G(OXYZ)$, and can be used to map the local coordinates of any point (x,y,z) to its corresponding global coordinates (X,Y,Z).

Alternatively, using the method of unit vector decomposition to develop the matrix ^BR_G leads to

$$^B\mathbf{r} = {}^BR_G\,{}^G\mathbf{r} = {}^GR_B^{-1}\,{}^G\mathbf{r} \tag{5.126}$$

$$^BR_G = \begin{bmatrix} \hat{\imath}\cdot\hat{I} & \hat{\imath}\cdot\hat{J} & \hat{\imath}\cdot\hat{K} \\ \hat{\jmath}\cdot\hat{I} & \hat{\jmath}\cdot\hat{J} & \hat{\jmath}\cdot\hat{K} \\ \hat{k}\cdot\hat{I} & \hat{k}\cdot\hat{J} & \hat{k}\cdot\hat{K} \end{bmatrix}$$

$$= \begin{bmatrix} \cos(\hat{\imath},\hat{I}) & \cos(\hat{\imath},\hat{J}) & \cos(\hat{\imath},\hat{K}) \\ \cos(\hat{\jmath},\hat{I}) & \cos(\hat{\jmath},\hat{J}) & \cos(\hat{\jmath},\hat{K}) \\ \cos(\hat{k},\hat{I}) & \cos(\hat{k},\hat{J}) & \cos(\hat{k},\hat{K}) \end{bmatrix} \tag{5.127}$$

and shows that the inverse of a transformation matrix is equal to the transpose of the transformation matrix,

$$^GR_B^{-1} = {}^GR_B^T \tag{5.128}$$

A matrix with condition (5.128) is called *orthogonal*. Orthogonality of R comes from the fact that it maps an orthogonal coordinate frame to another orthogonal coordinate frame.

The transformation matrix R has only three *independent* elements. The constraint equations among the elements of R will be found by applying the orthogonality condition (5.128).

$$^GR_B \cdot {}^GR_B^T = I \tag{5.129}$$

$$\begin{bmatrix} r_{11} & r_{12} & r_{13} \\ r_{21} & r_{22} & r_{23} \\ r_{31} & r_{32} & r_{33} \end{bmatrix}\begin{bmatrix} r_{11} & r_{21} & r_{31} \\ r_{12} & r_{22} & r_{32} \\ r_{13} & r_{23} & r_{33} \end{bmatrix} = \begin{bmatrix} 1 & 0 & 0 \\ 0 & 1 & 0 \\ 0 & 0 & 1 \end{bmatrix} \tag{5.130}$$

Therefore, the dot product of any two different rows of ^GR_B is zero, and the dot product of any row of ^GR_B with the same row is one.

$$\begin{aligned} r_{11}^2 + r_{12}^2 + r_{13}^2 &= 1 \\ r_{21}^2 + r_{22}^2 + r_{23}^2 &= 1 \\ r_{31}^2 + r_{32}^2 + r_{33}^2 &= 1 \\ r_{11}r_{21} + r_{12}r_{22} + r_{13}r_{23} &= 0 \\ r_{11}r_{31} + r_{12}r_{32} + r_{13}r_{33} &= 0 \\ r_{21}r_{31} + r_{22}r_{32} + r_{23}r_{33} &= 0 \end{aligned} \tag{5.131}$$

These relations are also true for columns of ^GR_B, and evidently for rows and columns of ^BR_G. The orthogonality condition can be summarized in the following equation:

$$\hat{\mathbf{r}}_{H_i} \cdot \hat{\mathbf{r}}_{H_j} = \hat{\mathbf{r}}_{H_i}^T\,\hat{\mathbf{r}}_{H_j} = \sum_{i=1}^{3} r_{ij}r_{ik} = \delta_{jk} \qquad (j,k=1,2,3) \tag{5.132}$$

where r_{ij} is the element of row i and column j of the transformation matrix R, and δ_{jk} is the *Kronecker's delta*

$$\delta_{jk} = \begin{cases} 1 & j = k \\ 0 & j \neq k \end{cases} \tag{5.133}$$

Equation (5.132) gives six independent relations satisfied by nine direction cosines. As a result, there are only three independent direction cosines. The independent elements of the matrix R cannot be in the same row or column, or any diagonal.

The determinant of a transformation matrix is equal to one,

$$\left| {}^G R_B \right| = 1 \tag{5.134}$$

It is because of Equation (5.129), and noting that

$$\left| {}^G R_B \cdot {}^G R_B^T \right| = \left| {}^G R_B \right| \cdot \left| {}^G R_B^T \right| = \left| {}^G R_B \right| \cdot \left| {}^G R_B \right| = \left| {}^G R_B \right|^2 = 1 \tag{5.135}$$

Using linear algebra and row vectors $\hat{\mathbf{r}}_{H_1}, \hat{\mathbf{r}}_{H_2}$, and $\hat{\mathbf{r}}_{H_3}$ of ${}^G R_B$, we know that

$$\left| {}^G R_B \right| = \hat{\mathbf{r}}_{H_1}^T \cdot (\hat{\mathbf{r}}_{H_2} \times \hat{\mathbf{r}}_{H_3}) \tag{5.136}$$

and because the coordinate system is right handed, we have $\hat{\mathbf{r}}_{H_2} \times \hat{\mathbf{r}}_{H_3} = \hat{\mathbf{r}}_{H_1}$ so $\left| {}^G R_B \right| = \hat{\mathbf{r}}_{H_1}^T \cdot \hat{\mathbf{r}}_{H_1} = 1$. ∎

Example 174 *Elements of transformation matrix.*

The position vector $\mathbf{r}$ of a point P may be expressed in terms of its components with respect to either $G\,(OXYZ)$ or $B\,(Oxyz)$ frames. If ${}^G \mathbf{r} = 100\hat{I} - 50\hat{J} + 150\hat{K}$, and we are looking for components of $\mathbf{r}$ in the $B\,(Oxyz)$ frame, then we have to find the proper rotation matrix ${}^B R_G$ first. Assume that the angle between the x and X axes is 40 deg, and the angle between the y and Y axes is 60 deg.

The row elements of ${}^B R_G$ are the direction cosines of the $B\,(Oxyz)$ axes in the $G\,(OXYZ)$ coordinate frame. The x-axis lies in the XZ plane at 40 deg from the X-axis, and the angle between y and Y is 60 deg. Therefore,

$$
{}^B R_G = \begin{bmatrix} \hat{i} \cdot \hat{I} & \hat{i} \cdot \hat{J} & \hat{i} \cdot \hat{K} \\ \hat{j} \cdot \hat{I} & \hat{j} \cdot \hat{J} & \hat{j} \cdot \hat{K} \\ \hat{k} \cdot \hat{I} & \hat{k} \cdot \hat{J} & \hat{k} \cdot \hat{K} \end{bmatrix} = \begin{bmatrix} \cos 40 & 0 & \sin 40 \\ \hat{j} \cdot \hat{I} & \cos 60 & \hat{j} \cdot \hat{K} \\ \hat{k} \cdot \hat{I} & \hat{k} \cdot \hat{J} & \hat{k} \cdot \hat{K} \end{bmatrix}
$$

$$
= \begin{bmatrix} 0.766 & 0 & 0.643 \\ \hat{j} \cdot \hat{I} & 0.5 & \hat{j} \cdot \hat{K} \\ \hat{k} \cdot \hat{I} & \hat{k} \cdot \hat{J} & \hat{k} \cdot \hat{K} \end{bmatrix} \tag{5.137}
$$

and by using ${}^B R_G \, {}^G R_B = {}^B R_G \, {}^B R_G^T = I$

$$
\begin{bmatrix} 0.766 & 0 & 0.643 \\ r_{21} & 0.5 & r_{23} \\ r_{31} & r_{32} & r_{33} \end{bmatrix} \begin{bmatrix} 0.766 & r_{21} & r_{31} \\ 0 & 0.5 & r_{32} \\ 0.643 & r_{23} & r_{33} \end{bmatrix} = \begin{bmatrix} 1 & 0 & 0 \\ 0 & 1 & 0 \\ 0 & 0 & 1 \end{bmatrix} \tag{5.138}
$$

we obtain a set of six equations to find the missing elements.

$$\begin{aligned}
0.766\,r_{21} + 0.643\,r_{23} &= 0 \\
0.766\,r_{31} + 0.643\,r_{33} &= 0 \\
r_{21}^2 + r_{23}^2 + 0.25 &= 1 \\
r_{21}r_{31} + 0.5r_{32} + r_{23}r_{33} &= 0 \\
r_{31}^2 + r_{32}^2 + r_{33}^2 &= 1
\end{aligned} \tag{5.139}$$

Solving these equations provides the following transformation matrix:

$$^B R_G = \begin{bmatrix} 0.766 & 0 & 0.643 \\ 0.557 & 0.5 & -0.663 \\ -0.322 & 0.866 & 0.383 \end{bmatrix} \tag{5.140}$$

and then we can find the components of $^B\mathbf{r}$.

$$\begin{aligned}
^B\mathbf{r} &= {}^B R_G \, {}^G\mathbf{r} \\
&= \begin{bmatrix} 0.766 & 0 & 0.643 \\ 0.557 & 0.5 & -0.663 \\ -0.322 & 0.866 & 0.383 \end{bmatrix} \begin{bmatrix} 100 \\ -50 \\ 150 \end{bmatrix} = \begin{bmatrix} 173.05 \\ -68.75 \\ -18.05 \end{bmatrix} \tag{5.141}
\end{aligned}$$

Example 175 *Global position, using $^B\mathbf{r}$ and $^B R_G$.*

The position vector $\mathbf{r}$ of a point P may be expressed in either $G\,(OXYZ)$ or $B\,(Oxyz)$ frames. If $^B\mathbf{r} = 100\hat{\imath} - 50\hat{\jmath} + 150\hat{k}$, and $^B R_G$ is the transformation matrix to map $^G\mathbf{r}$ to $^B\mathbf{r}$

$$^B\mathbf{r} = {}^B R_G \, {}^G\mathbf{r} = \begin{bmatrix} 0.766 & 0 & 0.643 \\ 0.557 & 0.5 & -0.663 \\ -0.322 & 0.866 & 0.383 \end{bmatrix} {}^G\mathbf{r} \tag{5.142}$$

then the components of $^G\mathbf{r}$ in $G\,(OXYZ)$ would be

$$\begin{aligned}
^G\mathbf{r} &= {}^G R_B \, {}^B\mathbf{r} = {}^B R_G^T \, {}^B\mathbf{r} \\
&= \begin{bmatrix} 0.766 & 0.557 & -0.322 \\ 0 & 0.5 & 0.866 \\ 0.643 & -0.663 & 0.383 \end{bmatrix} \begin{bmatrix} 100 \\ -50 \\ 150 \end{bmatrix} = \begin{bmatrix} 0.45 \\ 104.9 \\ 154.9 \end{bmatrix} \tag{5.143}
\end{aligned}$$

Example 176 *Two points transformation matrix.*

The global position vector of two points, P_1 and P_2, of a rigid body B are

$$^G\mathbf{r}_{P_1} = \begin{bmatrix} 1.077 \\ 1.365 \\ 2.666 \end{bmatrix} \qquad {}^G\mathbf{r}_{P_2} = \begin{bmatrix} -0.473 \\ 2.239 \\ -0.959 \end{bmatrix} \tag{5.144}$$

The origin of the body $B\,(Oxyz)$ is fixed on the origin of $G\,(OXYZ)$, and the points P_1 and P_2 are lying on the local x-axis and y-axis respectively.

To find ^GR_B, we use the local unit vectors $^G\hat{\imath}$ and $^G\hat{\jmath}$

$$^G\hat{\imath} = \frac{^G\mathbf{r}_{P_1}}{|^G\mathbf{r}_{P_1}|} = \begin{bmatrix} 0.338 \\ 0.429 \\ 0.838 \end{bmatrix} \qquad ^G\hat{\jmath} = \frac{^G\mathbf{r}_{P_2}}{|^G\mathbf{r}_{P_2}|} = \begin{bmatrix} -0.191 \\ 0.902 \\ -0.387 \end{bmatrix} \tag{5.145}$$

to obtain $^G\hat{k}$

$$\begin{aligned}
^G\hat{k} &= \hat{\imath} \times \hat{\jmath} = \tilde{\imath}\,\hat{\jmath} \\
&= \begin{bmatrix} 0 & -0.838 & 0.429 \\ 0.838 & 0 & -0.338 \\ -0.429 & 0.338 & 0 \end{bmatrix} \begin{bmatrix} -0.191 \\ 0.902 \\ -0.387 \end{bmatrix} \\
&= \begin{bmatrix} -0.922 \\ -0.029 \\ 0.387 \end{bmatrix}
\end{aligned} \tag{5.146}$$

where $\tilde{\imath}$ is the skew-symmetric matrix corresponding to $\hat{\imath}$, and $\tilde{\imath}\,\hat{\jmath}$ is an alternative for $\hat{\imath} \times \hat{\jmath}$.

Hence, the transformation matrix using the coordinates of two points $^G\mathbf{r}_{P_1}$ and $^G\mathbf{r}_{P_2}$ would be

$$\begin{aligned}
^GR_B &= \begin{bmatrix} ^G\hat{\imath} & ^G\hat{\jmath} & ^G\hat{k} \end{bmatrix} \\
&= \begin{bmatrix} 0.338 & -0.191 & -0.922 \\ 0.429 & 0.902 & -0.029 \\ 0.838 & -0.387 & 0.387 \end{bmatrix}
\end{aligned} \tag{5.147}$$

Example 177 *Length invariant of a position vector.*

Describing a vector in different frames utilizing rotation matrices does not affect the length and direction properties of the vector. Therefore, the length of a vector is an invariant

$$|\mathbf{r}| = |^G\mathbf{r}| = |^B\mathbf{r}| \tag{5.148}$$

The length invariant property can be shown by

$$\begin{aligned}
|\mathbf{r}|^2 &= {}^G\mathbf{r}^T\,{}^G\mathbf{r} = \begin{bmatrix} ^GR_B\,{}^B\mathbf{r} \end{bmatrix}^T\,{}^GR_B\,{}^B\mathbf{r} \\
&= {}^B\mathbf{r}^T\,{}^GR_B^T\,{}^GR_B\,{}^B\mathbf{r} = {}^B\mathbf{r}^T\,{}^B\mathbf{r}
\end{aligned} \tag{5.149}$$

Example 178 ★ *Skew symmetric matrices for $\hat{\imath}$, $\hat{\jmath}$, and $\hat{k}$.*

The definition of skew symmetric matrix $\tilde{a}$ corresponding to a vector $\mathbf{a}$ is defined by

$$\tilde{a} = \begin{bmatrix} 0 & -a_3 & a_2 \\ a_3 & 0 & -a_1 \\ -a_2 & a_1 & 0 \end{bmatrix} \tag{5.150}$$

Hence,

$$\tilde{\imath} = \begin{bmatrix} 0 & 0 & 0 \\ 0 & 0 & -1 \\ 0 & 1 & 0 \end{bmatrix} \quad \tilde{\jmath} = \begin{bmatrix} 0 & 0 & 1 \\ 0 & 0 & 0 \\ -1 & 0 & 0 \end{bmatrix} \quad \tilde{k} = \begin{bmatrix} 0 & -1 & 0 \\ 1 & 0 & 0 \\ 0 & 0 & 0 \end{bmatrix} \tag{5.151}$$

Example 179 *Inverse of Euler angles rotation matrix.*
Precession-nutation-spin or Euler angle rotation matrix (5.78)

$$
\begin{aligned}
{}^B R_G &= R_{z,\psi} R_{x,\theta} R_{z,\varphi} \\
&= \begin{bmatrix}
c\varphi c\psi - c\theta s\varphi s\psi & c\psi s\varphi + c\theta c\varphi s\psi & s\theta s\psi \\
-c\varphi s\psi - c\theta c\psi s\varphi & -s\varphi s\psi + c\theta c\varphi c\psi & s\theta c\psi \\
s\theta s\varphi & -c\varphi s\theta & c\theta
\end{bmatrix}
\end{aligned} \tag{5.152}
$$

must be inverted to be a transformation matrix to map body coordinates to global coordinates.

$$
\begin{aligned}
{}^G R_B &= {}^B R_G^{-1} = R_{z,\varphi}^T R_{x,\theta}^T R_{z,\psi}^T \\
&= \begin{bmatrix}
c\varphi c\psi - c\theta s\varphi s\psi & -c\varphi s\psi - c\theta c\psi s\varphi & s\theta s\varphi \\
c\psi s\varphi + c\theta c\varphi s\psi & -s\varphi s\psi + c\theta c\varphi c\psi & -c\varphi s\theta \\
s\theta s\psi & s\theta c\psi & c\theta
\end{bmatrix}
\end{aligned} \tag{5.153}
$$

The transformation matrix (5.152) is called a local Euler rotation matrix, and (5.153) is called a global Euler rotation matrix.

Example 180 ★ *Alternative proof for transformation matrix.*
Starting with an identity

$$
\begin{bmatrix} \hat{\imath} & \hat{\jmath} & \hat{k} \end{bmatrix}
\begin{bmatrix} \hat{\imath} \\ \hat{\jmath} \\ \hat{k} \end{bmatrix} = 1 \tag{5.154}
$$

we may write

$$
\begin{bmatrix} \hat{I} \\ \hat{J} \\ \hat{K} \end{bmatrix} =
\begin{bmatrix} \hat{I} \\ \hat{J} \\ \hat{K} \end{bmatrix}
\begin{bmatrix} \hat{\imath} & \hat{\jmath} & \hat{k} \end{bmatrix}
\begin{bmatrix} \hat{\imath} \\ \hat{\jmath} \\ \hat{k} \end{bmatrix} \tag{5.155}
$$

Because the matrix multiplication can be performed in any order, we find

$$
\begin{bmatrix} \hat{I} \\ \hat{J} \\ \hat{K} \end{bmatrix} =
\begin{bmatrix}
\hat{I}\cdot\hat{\imath} & \hat{I}\cdot\hat{\jmath} & \hat{I}\cdot\hat{k} \\
\hat{J}\cdot\hat{\imath} & \hat{J}\cdot\hat{\jmath} & \hat{J}\cdot\hat{k} \\
\hat{K}\cdot\hat{\imath} & \hat{K}\cdot\hat{\jmath} & \hat{K}\cdot\hat{k}
\end{bmatrix}
\begin{bmatrix} \hat{\imath} \\ \hat{\jmath} \\ \hat{k} \end{bmatrix} =
{}^G R_B \begin{bmatrix} \hat{\imath} \\ \hat{\jmath} \\ \hat{k} \end{bmatrix} \tag{5.156}
$$

where,

$$
{}^G R_B = \begin{bmatrix} \hat{I} \\ \hat{J} \\ \hat{K} \end{bmatrix}
\begin{bmatrix} \hat{\imath} & \hat{\jmath} & \hat{k} \end{bmatrix} \tag{5.157}
$$

Following the same method we can show that

$$
{}^B R_G = \begin{bmatrix} \hat{\imath} \\ \hat{\jmath} \\ \hat{k} \end{bmatrix}
\begin{bmatrix} \hat{I} & \hat{J} & \hat{K} \end{bmatrix} \tag{5.158}
$$

5.6 Local and Global Rotations

The global rotation matrix $^G R_B$ is the inverse of the local rotation matrix $^B R_G$ and vice versa,

$$^G R_B = {^B R_G^{-1}} \qquad ^B R_G = {^G R_B^{-1}} \tag{5.159}$$

The premultiplication of the global rotation matrix is equal to postmultiplication of the local rotation matrix.

Proof. Let us show rotation matrices about the global axes by $[Q]$ and rotation matrices about the local axes by $[A]$. Consider a sequence of global rotations and their resultant global rotation matrix $^G Q_B$ to transform a position vector $^B \mathbf{r}$ to $^G \mathbf{r}$:

$$^G \mathbf{r} = {^G Q_B} {^B \mathbf{r}} \tag{5.160}$$

The global position vector $^G \mathbf{r}$ can also be transformed to $^B \mathbf{r}$ using a local rotation matrix $^B A_G$:

$$^B \mathbf{r} = {^B A_G} {^G \mathbf{r}} \tag{5.161}$$

Combining Equations (5.160) and (5.161) yields

$$
\begin{aligned}
^G \mathbf{r} &= {^G Q_B} {^B A_G} {^G \mathbf{r}} & (5.162) \\
^B \mathbf{r} &= {^B A_G} {^G Q_B} {^B \mathbf{r}} & (5.163)
\end{aligned}
$$

and hence,

$$^G Q_B {^B A_G} = {^B A_G} {^G Q_B} = \mathbf{I} \tag{5.164}$$

Therefore, the global and local rotation matrices are inverse of each other:

$$
\begin{aligned}
^G Q_B &= {^B A_G^{-1}} & (5.165) \\
^G Q_B^{-1} &= {^B A_G} & (5.166)
\end{aligned}
$$

Assume that

$$
\begin{aligned}
^G Q_B &= Q_n \cdots Q_3 Q_2 Q_1 & (5.167) \\
^B A_G &= A_n \cdots A_3 A_2 A_1 & (5.168)
\end{aligned}
$$

then

$$
\begin{aligned}
^G Q_B &= {^B A_G^{-1}} = A_1^{-1} A_2^{-1} A_3^{-1} \cdots A_n^{-1} & (5.169) \\
^B A_G &= {^G Q_B^{-1}} = Q_1^{-1} Q_2^{-1} Q_3^{-1} \cdots Q_n^{-1} & (5.170)
\end{aligned}
$$

and Equation (5.164) becomes

$$Q_n \cdots Q_2 Q_1 A_n \cdots A_2 A_1 = A_n \cdots A_2 A_1 Q_n \cdots Q_2 Q_1 = \mathbf{I} \tag{5.171}$$

Therefore,

$$
\begin{aligned}
Q_n \cdots Q_3 Q_2 Q_1 &= A_1^{-1} A_2^{-1} A_3^{-1} \cdots A_n^{-1} & (5.172) \\
A_n \cdots A_3 A_2 A_1 &= Q_1^{-1} Q_2^{-1} Q_3^{-1} \cdots Q_n^{-1} & (5.173)
\end{aligned}
$$

or

$$Q_1^{-1}Q_2^{-1}Q_3^{-1}\cdots Q_n^{-1}Q_n\cdots Q_3Q_2Q_1 = \mathbf{I} \qquad (5.174)$$
$$A_1^{-1}A_2^{-1}A_3^{-1}\cdots A_n^{-1}A_n\cdots A_3A_2A_1 = \mathbf{I} \qquad (5.175)$$

Hence, the effect of order of rotations about the global coordinate axes is equivalent to the effect of the same rotations about the local coordinate axes performed in reverse order:

$$^G Q_B = A_1^{-1}A_2^{-1}A_3^{-1}\cdots A_n^{-1} \qquad (5.176)$$
$$^B A_G = Q_1^{-1}Q_2^{-1}Q_3^{-1}\cdots Q_n^{-1} \qquad (5.177)$$

Therefore the global rotation matrix $^G R_B$ is the inverse of the local rotation matrix $^B R_G$ and vice versa,

$$^G R_B = {}^B R_G^{-1} \qquad ^B R_G = {}^G R_B^{-1} \qquad (5.178)$$

■

Example 181 *Postmultiplication of rotation matrix.*
Assume that the local position of a point P after a rotation is at $^B\mathbf{r} = \begin{bmatrix} 1 & 2 & 3 \end{bmatrix}^T$. *If the local rotation matrix to transform* $^G\mathbf{r}$ *to* $^B\mathbf{r}$ *is given as*

$$^B R_{z,\varphi} = \begin{bmatrix} \cos\varphi & \sin\varphi & 0 \\ -\sin\varphi & \cos\varphi & 0 \\ 0 & 0 & 1 \end{bmatrix} = \begin{bmatrix} \cos 30 & \sin 30 & 0 \\ -\sin 30 & \cos 30 & 0 \\ 0 & 0 & 1 \end{bmatrix} \qquad (5.179)$$

then we may find the global position vector $^G\mathbf{r}$ *by postmultiplying* $^B R_{z,\varphi}$ *and the local position vector* $^B\mathbf{r}^T$,

$$\begin{aligned} ^G\mathbf{r}^T &= {}^B\mathbf{r}^T \, {}^B R_{z,\varphi} = \begin{bmatrix} 1 & 2 & 3 \end{bmatrix} \begin{bmatrix} \cos 30 & \sin 30 & 0 \\ -\sin 30 & \cos 30 & 0 \\ 0 & 0 & 1 \end{bmatrix} \\ &= \begin{bmatrix} -0.13 & 2.23 & 3 \end{bmatrix} \end{aligned} \qquad (5.180)$$

instead of premultiplying of $^B R_{z,\varphi}^{-1}$ *by* $^B\mathbf{r}$:

$$\begin{aligned} ^G\mathbf{r} &= {}^B R_{z,\varphi}^{-1} \, {}^B\mathbf{r} \\ &= \begin{bmatrix} \cos 30 & -\sin 30 & 0 \\ \sin 30 & \cos 30 & 0 \\ 0 & 0 & 1 \end{bmatrix} \begin{bmatrix} 1 \\ 2 \\ 3 \end{bmatrix} = \begin{bmatrix} -0.13 \\ 2.23 \\ 3 \end{bmatrix} \end{aligned} \qquad (5.181)$$

5.7 Axis-angle Rotation

The finial orientation of a rigid body after a finite number of rotations is equivalent to a unique rotation about a unique axis. Determination of the

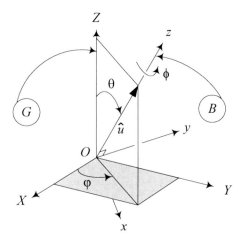

FIGURE 5.6. Axis-angle rotation when the rotation axix $\hat{u}$ is coincident with the local z-axis.

angle and axis is called the *orientation kinematics* of rigid bodies. Assume the body frame $B(Oxyz)$ rotate ϕ about a fixed line in the global frame $G(OXYZ)$. The line is indicated by a unit vector $\hat{u}$ with directional cosines u_1, u_2, u_3,

$$\hat{u} = u_1\hat{I} + u_2\hat{J} + u_3\hat{K} \tag{5.182}$$

$$\sqrt{u_1^2 + u_2^2 + u_3^2} = 1 \tag{5.183}$$

Two parameters are needed to define the axis of rotation that goes through O and one parameter is needed to define the amount of rotation about the axis. This is called the *axis-angle* representation of a rotation. An axis-angle rotation needs three independent parameters to be defined.

The axis-angle transformation matrix ^GR_B that transforms the coordinates of the body frame $B(Oxyz)$ to the associated coordinates in the global frame $G(OXYZ)$,

$$^G\mathbf{r} = {}^GR_B \, {}^B\mathbf{r} \tag{5.184}$$

is

$$^GR_B = R_{\hat{u},\phi} = \mathbf{I}\cos\phi + \hat{u}\hat{u}^T \operatorname{vers}\phi + \tilde{u}\sin\phi \tag{5.185}$$

$$^GR_B = \begin{bmatrix} u_1^2 \operatorname{vers}\phi + c\phi & u_1u_2\operatorname{vers}\phi - u_3s\phi & u_1u_3\operatorname{vers}\phi + u_2s\phi \\ u_1u_2\operatorname{vers}\phi + u_3s\phi & u_2^2\operatorname{vers}\phi + c\phi & u_2u_3\operatorname{vers}\phi - u_1s\phi \\ u_1u_3\operatorname{vers}\phi - u_2s\phi & u_2u_3\operatorname{vers}\phi + u_1s\phi & u_3^2\operatorname{vers}\phi + c\phi \end{bmatrix} \tag{5.186}$$

where

$$\operatorname{vers}\phi = versine\,\phi = 1 - \cos\phi = 2\sin^2\frac{\phi}{2} \tag{5.187}$$

and $\tilde{u}$ is the skew-symmetric matrix associated with the vector $\hat{u}$,

$$\tilde{u} = \begin{bmatrix} 0 & -u_3 & u_2 \\ u_3 & 0 & -u_1 \\ -u_2 & u_1 & 0 \end{bmatrix} \tag{5.188}$$

A matrix $\tilde{u}$ is *skew-symmetric* if

$$\tilde{u}^T = -\tilde{u} \tag{5.189}$$

For any transformation matrix ${}^G R_B$, we may determine the axis $\hat{u}$ and angle ϕ to provide the same matrix by

$$\tilde{u} = \frac{1}{2\sin\phi}\left({}^G R_B - {}^G R_B^T\right) \tag{5.190}$$

$$\cos\phi = \frac{1}{2}\left(\text{tr}\left({}^G R_B\right) - 1\right) \tag{5.191}$$

Equation (5.186) is called the *angle-axis* or *axis-angle rotation matrix* and is the most general transformation matrix for rotation of a body frame B in a global frame G. If the axis of rotation (5.182) coincides with a global coordinate axis Z, Y, or X, then Equation (5.186) reduces to the principal local rotation matrices.

Proof. The rotation ϕ about an axis $\hat{u}$ is equivalent to a sequence of rotations about the axes of a body frame B such that the local frame is first rotated to bring one of its axes, say the z-axis, into coincidence with the rotation axis $\hat{u}$ followed by a rotation ϕ about that local axis, then the reverse of the first sequence of rotations.

Figure 5.6 illustrates an axis of rotation $\hat{u} = u_1\hat{I} + u_2\hat{J} + u_3\hat{K}$, the global frame $G(OXYZ)$, and the rotated body frame $B(Oxyz)$ when the local z-axis is coincident with $\hat{u}$. Assume that the body and global frames were coincident initially. When we apply a sequence of rotations φ about the z-axis and θ about the y-axis on the body frame $B(Oxyz)$, the local z-axis will become coincident with the rotation axis $\hat{u}$. Now, we apply the rotation ϕ about $\hat{u}$ and then perform the sequence of rotations φ and θ backward. Following Equation (5.169), the rotation matrix ${}^G R_B$ to map the coordinates of a point in the body frame to its coordinates in the global frame after rotation ϕ about $\hat{u}$ is

$$\begin{aligned} {}^G R_B &= {}^B R_G^{-1} = {}^B R_G^T = R_{\hat{u},\phi} \\ &= [R_{z,-\varphi}\, R_{y,-\theta}\, R_{z,\phi}\, R_{y,\theta}\, R_{z,\varphi}]^T \\ &= R_{z,\varphi}^T\, R_{y,\theta}^T\, R_{z,\phi}^T\, R_{y,-\theta}^T\, R_{z,-\varphi}^T \end{aligned} \tag{5.192}$$

Substituting equations

$$\begin{aligned} \sin\varphi &= \frac{u_2}{\sqrt{u_1^2+u_2^2}} & \cos\varphi &= \frac{u_1}{\sqrt{u_1^2+u_2^2}} \\ \sin\theta &= \sqrt{u_1^2+u_2^2} & \cos\theta &= u_3 \\ \sin\theta\sin\varphi &= u_2 & \sin\theta\cos\varphi &= u_1 \end{aligned} \tag{5.193}$$

in $^G R_B$ will provide us with the axis-angle rotation matrix

$$^G R_B = R_{\hat{u},\phi} = \tag{5.194}$$
$$\begin{bmatrix} u_1^2 \text{ vers } \phi + c\phi & u_1 u_2 \text{ vers } \phi - u_3 s\phi & u_1 u_3 \text{ vers } \phi + u_2 s\phi \\ u_1 u_2 \text{ vers } \phi + u_3 s\phi & u_2^2 \text{ vers } \phi + c\phi & u_2 u_3 \text{ vers } \phi - u_1 s\phi \\ u_1 u_3 \text{ vers } \phi - u_2 s\phi & u_2 u_3 \text{ vers } \phi + u_1 s\phi & u_3^2 \text{ vers } \phi + c\phi \end{bmatrix}$$

The matrix (5.194) can be decomposed to

$$R_{\hat{u},\phi} = \cos\phi \begin{bmatrix} 1 & 0 & 0 \\ 0 & 1 & 0 \\ 0 & 0 & 1 \end{bmatrix} + (1 - \cos\phi) \begin{bmatrix} u_1 \\ u_2 \\ u_3 \end{bmatrix} \begin{bmatrix} u_1 & u_2 & u_3 \end{bmatrix}$$
$$+ \sin\phi \begin{bmatrix} 0 & -u_3 & u_2 \\ u_3 & 0 & -u_1 \\ -u_2 & u_1 & 0 \end{bmatrix} \tag{5.195}$$

which can be summarized as (5.185). Showing the rotation matrix by its elements $^G R_B = R_{\hat{u},\phi} = [r_{ij}]$, we have

$$r_{ij} = \delta_{ij} \cos\phi + u_i u_j (1 - \cos\phi) - \epsilon_{ijk} u_k \sin\phi \tag{5.196}$$

The *axis-angle rotation equation* (5.185) is also called the *Rodriguez rotation formula*. We can show that the Rodriguez rotation formula may also be expressed by any of the following equivalent forms:

$$R_{\hat{u},\phi} = \mathbf{I} + \tilde{u} \sin\phi + 2\tilde{u}^2 \sin^2 \frac{\phi}{2} \tag{5.197}$$

$$R_{\hat{u},\phi} = \mathbf{I} + 2\tilde{u} \sin\frac{\phi}{2} \left(\mathbf{I} \cos\frac{\phi}{2} + \tilde{u} \sin\frac{\phi}{2} \right) \tag{5.198}$$

$$R_{\hat{u},\phi} = \mathbf{I} + \tilde{u} \sin\phi + \tilde{u}^2 \text{ vers } \phi \tag{5.199}$$

$$R_{\hat{u},\phi} = \left[\mathbf{I} - \hat{u}\hat{u}^T \right] \cos\phi + \tilde{u} \sin\phi + \hat{u}\hat{u}^T \tag{5.200}$$

$$R_{\hat{u},\phi} = \mathbf{I} + \tilde{u}^2 + \tilde{u} \sin\phi - \tilde{u}^2 \cos\phi \tag{5.201}$$

The *inverse of an angle-axis rotation* is

$$^B R_G = {}^G R_B^T = R_{\hat{u},-\phi} = \mathbf{I} \cos\phi + \hat{u}\hat{u}^T \text{ vers } \phi - \tilde{u} \sin\phi \tag{5.202}$$

It means that the orientation of B in G when B is rotated ϕ about $\hat{u}$ is the same as the orientation of G in B when B is rotated $-\phi$ about $\hat{u}$. The rotation $R_{\hat{u},-\phi}$ is also called the *reverse rotation*.

We may examine Equations (5.190) and (5.191) by direct substitution

$$^G R_B - {}^G R_B^T = \begin{bmatrix} 0 & -2u_3 \sin\phi & 2u_2 \sin\phi \\ 2u_3 \sin\phi & 0 & -2u_1 \sin\phi \\ -2u_2 \sin\phi & 2u_1 \sin\phi & 0 \end{bmatrix}$$
$$= 2\sin\phi \begin{bmatrix} 0 & -u_3 & u_2 \\ u_3 & 0 & -u_1 \\ -u_2 & u_1 & 0 \end{bmatrix} = 2\tilde{u} \sin\phi \tag{5.203}$$

and

$$
\begin{aligned}
\operatorname{tr}\left(^{G}R_{B}\right) &= r_{11} + r_{22} + r_{33} \\
&= 3\cos\phi + u_1^2\left(1 - \cos\phi\right) + u_2^2\left(1 - \cos\phi\right) + u_3^2\left(1 - \cos\phi\right) \\
&= 3\cos\phi + u_1^2 + u_2^2 + u_3^2 - \left(u_1^2 + u_2^2 + u_3^2\right)\cos\phi \\
&= 2\cos\phi + 1
\end{aligned} \tag{5.204}
$$

The axis of rotation $\hat{u}$ is also called the Euler axis or the eigenaxis of rotation. ■

Example 182 ★ *Axis-angle rotation when $\hat{u} = \hat{K}$.*
If the local frame $B\left(Oxyz\right)$ rotates about the Z-axis, then

$$
\hat{u} = \hat{K} \tag{5.205}
$$

and the transformation matrix (5.186) reduces to

$$
\begin{aligned}
^{G}R_{B} &= \begin{bmatrix} 0\operatorname{vers}\phi + \cos\phi & 0\operatorname{vers}\phi - 1\sin\phi & 0\operatorname{vers}\phi + 0\sin\phi \\ 0\operatorname{vers}\phi + 1\sin\phi & 0\operatorname{vers}\phi + \cos\phi & 0\operatorname{vers}\phi - 0\sin\phi \\ 0\operatorname{vers}\phi - 0\sin\phi & 0\operatorname{vers}\phi + 0\sin\phi & 1\operatorname{vers}\phi + \cos\phi \end{bmatrix} \\
&= \begin{bmatrix} \cos\phi & -\sin\phi & 0 \\ \sin\phi & \cos\phi & 0 \\ 0 & 0 & 1 \end{bmatrix}
\end{aligned} \tag{5.206}
$$

which is equivalent to the rotation matrix about the Z-axis of the global frame.

Example 183 ★ *Rotation about a rotated local axis.*
If the body coordinate frame $B\left(Oxyz\right)$ rotates φ deg about the global Z-axis, then the x-axis would be along

$$
\begin{aligned}
\hat{u}_x &= {}^{G}R_{Z,\varphi}\,\hat{\imath} \\
&= \begin{bmatrix} \cos\varphi & -\sin\varphi & 0 \\ \sin\varphi & \cos\varphi & 0 \\ 0 & 0 & 1 \end{bmatrix} \begin{bmatrix} 1 \\ 0 \\ 0 \end{bmatrix} = \begin{bmatrix} \cos\varphi \\ \sin\varphi \\ 0 \end{bmatrix}
\end{aligned} \tag{5.207}
$$

Rotation θ about $\hat{u}_x = \left(\cos\varphi\right)\hat{I} + \left(\sin\varphi\right)\hat{J}$ is defined by Rodriguez's formula (5.186)

$$
^{G}R_{\hat{u}_x,\theta} = \begin{bmatrix} \cos^2\varphi\operatorname{vers}\theta + \cos\theta & \cos\varphi\sin\varphi\operatorname{vers}\theta & \sin\varphi\sin\theta \\ \cos\varphi\sin\varphi\operatorname{vers}\theta & \sin^2\varphi\operatorname{vers}\theta + \cos\theta & -\cos\varphi\sin\theta \\ -\sin\varphi\sin\theta & \cos\varphi\sin\theta & \cos\theta \end{bmatrix} \tag{5.208}
$$

Now, rotation φ about the global Z-axis followed by rotation θ about the local x-axis is transformed by

$$
\begin{aligned}
^{G}R_{B} &= {}^{G}R_{\hat{u}_x,\theta}\,{}^{G}R_{Z,\varphi} \\
&= \begin{bmatrix} \cos\varphi & -\cos\theta\sin\varphi & \sin\theta\sin\varphi \\ \sin\varphi & \cos\theta\cos\varphi & -\cos\varphi\sin\theta \\ 0 & \sin\theta & \cos\theta \end{bmatrix}
\end{aligned} \tag{5.209}
$$

that must be equal to $[R_{x,\theta} R_{z,\varphi}]^{-1} = R_{z,\varphi}^T R_{x,\theta}^T.$

Example 184 ★ *Axis and angle of a rotation matrix.*

A body coordinate frame, B, undergoes three Euler rotations $(\varphi, \theta, \psi) = (30, 45, 60)$ deg *with respect to a global frame* G. *The rotation matrix to transform coordinates of* B *to* G *is*

$$
\begin{aligned}
{}^G R_B &= {}^B R_G^T = [R_{z,\psi} R_{x,\theta} R_{z,\varphi}]^T = R_{z,\varphi}^T R_{x,\theta}^T R_{z,\psi}^T \\
&= \begin{bmatrix} 0.126\,83 & -0.926\,78 & 0.353\,55 \\ 0.780\,33 & -0.126\,83 & -0.612\,37 \\ 0.612\,37 & 0.353\,55 & 0.707\,11 \end{bmatrix}
\end{aligned}
\tag{5.210}
$$

The unique angle-axis of rotation for this rotation matrix can then be found by Equations (5.190) and (5.191).

$$
\phi = \cos^{-1}\left(\frac{1}{2} \left(\mathrm{tr}\, \left({}^G R_B \right) - 1 \right) \right) = \cos^{-1}(-0.146\,45) = 98\,\mathrm{deg}
\tag{5.211}
$$

$$
\begin{aligned}
\tilde{u} &= \frac{1}{2\sin\phi} \left({}^G R_B - {}^G R_B^T \right) \\
&= \begin{bmatrix} 0.0 & -0.862\,85 & -0.130\,82 \\ 0.862\,85 & 0.0 & -0.488\,22 \\ 0.130\,82 & 0.488\,22 & 0.0 \end{bmatrix}
\end{aligned}
\tag{5.212}
$$

$$
\hat{u} = \begin{bmatrix} 0.488\,22 \\ -0.130\,82 \\ 0.862\,85 \end{bmatrix}
\tag{5.213}
$$

As a double check, we may verify the angle-axis rotation formula and derive the same rotation matrix.

$$
\begin{aligned}
{}^G R_B &= R_{\hat{u},\phi} = \mathbf{I}\cos\phi + \hat{u}\hat{u}^T \,\mathrm{vers}\,\phi + \tilde{u}\sin\phi \\
&= \begin{bmatrix} 0.126\,82 & -0.926\,77 & 0.353\,54 \\ 0.780\,32 & -0.126\,83 & -0.612\,37 \\ 0.612\,36 & 0.353\,55 & 0.707\,09 \end{bmatrix}
\end{aligned}
\tag{5.214}
$$

5.8 Rigid Body Motion

Consider a rigid body with a local coordinate frame B $(oxyz)$ that is moving in a global coordinate frame $G(OXYZ)$. The rigid body can rotate in G, while point o of B can translate relative to the origin O of G, as is shown in Figure 5.7. The vector ${}^G\mathbf{d}$ indicates the position of the moving origin o relative to the fixed origin O. The coordinates of a body point P in local and global frames are related by

$$
{}^G\mathbf{r}_P = {}^G R_B \, {}^B\mathbf{r}_P + {}^G\mathbf{d}
\tag{5.215}
$$

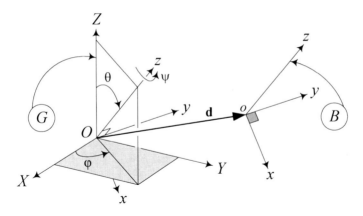

FIGURE 5.7. Rotation and translation of a body frame.

where

$$
{}^G\mathbf{r} = \begin{bmatrix} X_P \\ Y_P \\ Z_P \end{bmatrix} \qquad {}^B\mathbf{r} = \begin{bmatrix} x_P \\ y_P \\ z_P \end{bmatrix} \qquad {}^G\mathbf{d} = \begin{bmatrix} X_o \\ Y_o \\ Z_o \end{bmatrix} \qquad (5.216)
$$

The vector ${}^G\mathbf{d}$ is called the *displacement* or *translation* of B with respect to G, and ${}^G R_B$ is the *rotation matrix* that transforms ${}^B\mathbf{r}$ to ${}^G\mathbf{r}$ when ${}^G\mathbf{d} = 0$. Such a combination of a rotation and translation in Equation (5.215) is called the *rigid body motion* in which the configuration of the body can be expressed by the position of the origin o of B and the orientation of B.

Decomposition of a rigid motion into a rotation and a translation is the practical method for expressing spatial displacement of rigid bodies. We show the translation by a vector and the rotation by the orthogonal Cartesian transformation matrix.

Proof. Consider a body frame B that is initially coincident with a globally fixed frame G. Figure 5.7 illustrates a translated and rotated frame B in G. The most general rotation is represented by the *Rodriguez rotation formula*, (5.185)

$$
{}^G R_B = R_{\hat{u},\phi} = \mathbf{I}\cos\phi + \hat{u}\hat{u}^T \operatorname{vers}\phi + \tilde{u}\sin\phi \qquad (5.217)
$$

All points of the body have the same displacement for the translation ${}^G\mathbf{d}$. Therefore, translation of a rigid body is independent of the local position vector ${}^B\mathbf{r}$. Because of that, we can represent the most general displacement of a rigid body by a rotation, and a translation:

$$
{}^G\mathbf{r} = \left(\mathbf{I}\cos\phi + \hat{u}\hat{u}^T \operatorname{vers}\phi + \tilde{u}\sin\phi\right) {}^B\mathbf{r} + {}^G\mathbf{d} = {}^G R_B {}^B\mathbf{r} + {}^G\mathbf{d} \quad (5.218)
$$

Equation (5.218) indicates that the most general displacement of a rigid body is a rotation about an axis and a translation along an axis. The choice of the point of reference o is arbitrary; however, when this point is chosen and the body coordinate frame is set up, the rotation and translation are determined.

Based on translation and rotation, the configuration of a body can be uniquely determined by six independent parameters: three translation components X_o, Y_o, Z_o and three rotational components. If a body moves in such a way that its rotational components remain constant, the motion is called a *pure translation*; if it moves in such a way that X_o, Y_o, and Z_o remain constant, the motion is called a *pure rotation*. Therefore, a rigid body has three translational and three rotational degrees of freedom. ■

Example 185 *Rotation and translation of a body coordinate frame.*

A body coordinate frame $B(oxyz)$ that is originally coincident with global coordinate frame $G(OXYZ)$ rotates 60 deg about the X-axis and translates to $\begin{bmatrix} 3 & 4 & 5 \end{bmatrix}^T$. The global position of a point at $^B\mathbf{r} = \begin{bmatrix} x & y & z \end{bmatrix}^T$ is

$$
\begin{aligned}
^G\mathbf{r} &= {}^G R_B \, {}^B\mathbf{r} + {}^G\mathbf{d} \\
&= \begin{bmatrix} 1 & 0 & 0 \\ 0 & \cos\dfrac{\pi}{3} & -\sin\dfrac{\pi}{3} \\ 0 & \sin\dfrac{\pi}{3} & \cos\dfrac{\pi}{3} \end{bmatrix} \begin{bmatrix} x \\ y \\ z \end{bmatrix} + \begin{bmatrix} 3 \\ 4 \\ 5 \end{bmatrix} \\
&= (x+3)\,\hat{I} + \left(\frac{1}{2}y - \frac{\sqrt{3}}{2}z + 4 \right) \hat{J} + \left(\frac{1}{2}z + \sqrt{3}y + 5 \right) \hat{K} \quad (5.219)
\end{aligned}
$$

Example 186 *Composition of rigid body motion.*

Consider a rigid motion of body B_1 with respect to body B_2, and then, a rigid motion of body B_2 with respect to frame G such that

$$
\begin{aligned}
^2\mathbf{r} &= {}^2R_1 \, {}^1\mathbf{r} + {}^2\mathbf{d}_1 && (5.220) \\
^G\mathbf{r} &= {}^G R_2 \, {}^2\mathbf{r} + {}^G\mathbf{d}_2 && (5.221)
\end{aligned}
$$

These two motions may be combined to determine a rigid motion that transforms $^1\mathbf{r}$ to $^G\mathbf{r}$:

$$
\begin{aligned}
^G\mathbf{r} &= {}^G R_2 \left({}^2R_1 \, {}^1\mathbf{r} + {}^2\mathbf{d}_1 \right) + {}^G\mathbf{d}_2 \\
&= {}^G R_2 \, {}^2R_1 \, {}^1\mathbf{r} + {}^G R_2 \, {}^2\mathbf{d}_1 + {}^G\mathbf{d}_2 \\
&= {}^G R_1 \, {}^1\mathbf{r} + {}^G\mathbf{d}_1 \quad (5.222)
\end{aligned}
$$

Therefore,

$$
\begin{aligned}
^G R_1 &= {}^G R_2 \, {}^2R_1 && (5.223) \\
^G \mathbf{d}_1 &= {}^G R_2 \, {}^2\mathbf{d}_1 + {}^G\mathbf{d}_2 && (5.224)
\end{aligned}
$$

which shows that the transformation from frame B_1 to frame G can be done by rotation $^G R_1$ and translation $^G\mathbf{d}_1$.

Example 187 *Rotation of a translated rigid body.*
Point P of a rigid body B has the initial position vector $^B\mathbf{r}_P$.

$$^B\mathbf{r}_P = \begin{bmatrix} 1 \\ 2 \\ 3 \end{bmatrix} \tag{5.225}$$

If the body rotates 45 deg about the x-axis, and then translates to $^G\mathbf{d} = \begin{bmatrix} 3 & 2 & 1 \end{bmatrix}^T$, the final position of P would be

$$^G\mathbf{r} = {}^B R_{x,45}^T \, {}^B\mathbf{r}_P + {}^G\mathbf{d} \tag{5.226}$$

$$= \begin{bmatrix} 1 & 0 & 0 \\ 0 & \cos\frac{\pi}{4} & \sin\frac{\pi}{4} \\ 0 & -\sin\frac{\pi}{4} & \cos\frac{\pi}{4} \end{bmatrix}^T \begin{bmatrix} 1 \\ 2 \\ 3 \end{bmatrix} + \begin{bmatrix} 3 \\ 2 \\ 1 \end{bmatrix} = \begin{bmatrix} 4 \\ 1.29 \\ 4.53 \end{bmatrix}$$

5.9 Angular Velocity

Consider a rotating rigid body $B(Oxyz)$ with a fixed point O in a reference frame $G(OXYZ)$. The motion of the body can be expressed by a time varying rotation transformation matrix between the global and body frames to map the instantaneous coordinates of any fixed point in body frame B into their coordinates in the global frame G

$$^G\mathbf{r}(t) = {}^G R_B(t) \, {}^B\mathbf{r} \tag{5.227}$$

The velocity of a body point in the global frame is

$$^G\dot{\mathbf{r}}(t) = {}^G\mathbf{v}(t) = {}^G\dot{R}_B(t) \, {}^B\mathbf{r} = {}_G\tilde{\omega}_B \, {}^G\mathbf{r}(t) = {}_G\boldsymbol{\omega}_B \times {}^G\mathbf{r}(t) \tag{5.228}$$

where ${}_G\boldsymbol{\omega}_B$ is the *angular velocity vector* of B with respect to G. It is equal to a rotation with *angular rate* $\dot{\phi}$ about an *instantaneous axis of rotation* $\hat{u}$.

$$\boldsymbol{\omega} = \begin{bmatrix} \omega_1 \\ \omega_2 \\ \omega_3 \end{bmatrix} = \dot{\phi}\,\hat{u} \tag{5.229}$$

The angular velocity vector is associated with a skew symmetric matrix ${}_G\tilde{\omega}_B$ called the *angular velocity matrix*

$$\tilde{\omega} = \begin{bmatrix} 0 & -\omega_3 & \omega_2 \\ \omega_3 & 0 & -\omega_1 \\ -\omega_2 & \omega_1 & 0 \end{bmatrix} \tag{5.230}$$

where

$$_G\tilde{\omega}_B = {}^G\dot{R}_B \, {}^G R_B^T = \dot{\phi}\,\tilde{u} \tag{5.231}$$

Proof. Consider a rigid body with a fixed point O and an attached frame $B(Oxyz)$. The body frame B is initially coincident with the global frame G. Therefore, the position vector of a body point P is

$$^G\mathbf{r}(t_0) = {}^B\mathbf{r} \tag{5.232}$$

The global time derivative of $^G\mathbf{r}$ is

$$
\begin{aligned}
^G\mathbf{v} &= {}^G\dot{\mathbf{r}} = \frac{^Gd}{dt}\,{}^G\mathbf{r}(t) = \frac{^Gd}{dt}\left[{}^GR_B(t)\,{}^B\mathbf{r}\right] \\
&= \frac{^Gd}{dt}\left[{}^GR_B(t)\,{}^G\mathbf{r}(t_0)\right] = {}^G\dot{R}_B(t)\,{}^B\mathbf{r} \tag{5.233}
\end{aligned}
$$

Eliminating $^B\mathbf{r}$ between (5.227) and (5.233) determines the velocity of the point in global frame

$$^G\mathbf{v} = {}^G\dot{R}_B(t)\,{}^GR_B^T(t)\,{}^G\mathbf{r}(t) \tag{5.234}$$

We denote the coefficient of $^G\mathbf{r}(t)$ by $\tilde{\omega}$

$$_G\tilde{\omega}_B = {}^G\dot{R}_B\,{}^GR_B^T \tag{5.235}$$

and write Equation (5.234) as

$$^G\mathbf{v} = {}_G\tilde{\omega}_B\,{}^G\mathbf{r}(t) \tag{5.236}$$

or as

$$^G\mathbf{v} = {}_G\boldsymbol{\omega}_B \times {}^G\mathbf{r}(t) \tag{5.237}$$

The time derivative of the orthogonality condition, $^GR_B\,{}^GR_B^T = \mathbf{I}$, introduces an important identity

$$^G\dot{R}_B\,{}^GR_B^T + {}^GR_B\,{}^G\dot{R}_B^T = 0 \tag{5.238}$$

which can be utilized to show that $_G\tilde{\omega}_B = [{}^G\dot{R}_B\,{}^GR_B^T]$ is a skew-symmetric matrix because

$$^GR_B\,{}^G\dot{R}_B^T = \left[{}^G\dot{R}_B\,{}^GR_B^T\right]^T \tag{5.239}$$

The vector $_G^G\boldsymbol{\omega}_B$ is called the *instantaneous angular velocity* of the body B relative to the global frame G as seen from the G frame.

Since a vectorial equation can be expressed in any coordinate frame, we may use either of the following expressions for the velocity of a body point in body or global frames

$$
\begin{aligned}
_G^G\mathbf{v}_P &= {}_G^G\boldsymbol{\omega}_B \times {}^G\mathbf{r}_P \tag{5.240} \\
_G^B\mathbf{v}_P &= {}_G^B\boldsymbol{\omega}_B \times {}^B\mathbf{r}_P \tag{5.241}
\end{aligned}
$$

where $_G^G\mathbf{v}_P$ is the global velocity of point P expressed in the global frame and $_G^B\mathbf{v}_P$ is the global velocity of point P expressed in the body frame.

$$_G^G\mathbf{v}_P = {}^G R_B\, _G^B\mathbf{v}_P = {}^G R_B \left(_G^B\omega_B \times {}^B\mathbf{r}_P \right) \tag{5.242}$$

$_G^G\mathbf{v}_P$ and $_G^B\mathbf{v}_P$ can be converted to each other using a rotation matrix

$$
\begin{aligned}
_G^B\mathbf{v}_P &= {}^G R_B^T\, _G^G\mathbf{v}_P = {}^G R_B^T\, _G\tilde{\omega}_B\, _G^G\mathbf{r}_P \\
&= {}^G R_B^T\, {}^G\dot{R}_B\, {}^G R_B^T\, _G^G\mathbf{r}_P = {}^G R_B^T\, {}^G\dot{R}_B\, _G^B\mathbf{r}_P
\end{aligned} \tag{5.243}
$$

showing that

$$_G^B\tilde{\omega}_B = {}^G R_B^T\, {}^G\dot{R}_B \tag{5.244}$$

which is called the *instantaneous angular velocity* of B relative to the global frame G as seen from the B frame. From the definitions of $_G\tilde{\omega}_B$ and $_G^B\tilde{\omega}_B$ we are able to transform the two angular velocity matrices by

$$
\begin{aligned}
_G\tilde{\omega}_B &= {}^G R_B\, _G^B\tilde{\omega}_B\, {}^G R_B^T \tag{5.245} \\
_G^B\tilde{\omega}_B &= {}^G R_B^T\, _G\tilde{\omega}_B\, {}^G R_B \tag{5.246}
\end{aligned}
$$

or equivalently

$$
\begin{aligned}
{}^G\dot{R}_B &= {}_G\tilde{\omega}_B\, {}^G R_B \tag{5.247} \\
{}^G\dot{R}_B &= {}^G R_B\, _G^B\tilde{\omega}_B \tag{5.248} \\
_G\tilde{\omega}_B\, {}^G R_B &= {}^G R_B\, _G^B\tilde{\omega}_B \tag{5.249}
\end{aligned}
$$

The angular velocity of B in G is negative of the angular velocity of G in B if both are expressed in the same coordinate frame.

$$
\begin{aligned}
_G^G\tilde{\omega}_B &= -_B^G\tilde{\omega}_G \tag{5.250} \\
_G^B\tilde{\omega}_B &= -_B^B\tilde{\omega}_G \tag{5.251}
\end{aligned}
$$

$_G\boldsymbol{\omega}_B$ can always be expressed in the form

$$_G\boldsymbol{\omega}_B = \omega\hat{u} \tag{5.252}$$

where $\hat{u}$ is a unit vector parallel to $_G\boldsymbol{\omega}_B$ and indicates the *instantaneous axis of rotation*. ∎

Example 188 *Rotation of a body point about a global axis.*
   Consider a rigid body that is turning about the Z-axis with $\dot{\alpha} = 10 \deg/s$. The global velocity of a point $P(5, 30, 10)$, when the body is turned $\alpha =$

30 deg, *is*

$$
\begin{aligned}
{}^G\mathbf{v}_P &= {}^G\dot{R}_B(t)\,{}^B\mathbf{r}_P \\[2mm]
&= \frac{{}^Gd}{dt}\left(\begin{bmatrix} \cos\alpha & -\sin\alpha & 0 \\ \sin\alpha & \cos\alpha & 0 \\ 0 & 0 & 1 \end{bmatrix}\right)\begin{bmatrix} 5 \\ 30 \\ 10 \end{bmatrix} \\[2mm]
&= \dot{\alpha}\begin{bmatrix} -\sin\alpha & -\cos\alpha & 0 \\ \cos\alpha & -\sin\alpha & 0 \\ 0 & 0 & 0 \end{bmatrix}\begin{bmatrix} 5 \\ 30 \\ 10 \end{bmatrix} \\[2mm]
&= \frac{10\pi}{180}\begin{bmatrix} -\sin\frac{\pi}{6} & -\cos\frac{\pi}{6} & 0 \\ \cos\frac{\pi}{6} & -\sin\frac{\pi}{6} & 0 \\ 0 & 0 & 0 \end{bmatrix}\begin{bmatrix} 5 \\ 30 \\ 10 \end{bmatrix} = \begin{bmatrix} -4.97 \\ -1.86 \\ 0 \end{bmatrix}
\end{aligned}
\tag{5.253}
$$

at this moment, point P is at

$$
\begin{aligned}
{}^G\mathbf{r}_P &= {}^GR_B\,{}^B\mathbf{r}_P \\[2mm]
&= \begin{bmatrix} \cos\frac{\pi}{6} & -\sin\frac{\pi}{6} & 0 \\ \sin\frac{\pi}{6} & \cos\frac{\pi}{6} & 0 \\ 0 & 0 & 1 \end{bmatrix}\begin{bmatrix} 5 \\ 30 \\ 10 \end{bmatrix} = \begin{bmatrix} -10.67 \\ 28.48 \\ 10 \end{bmatrix}
\end{aligned}
\tag{5.254}
$$

Example 189 *Rotation of a global point about a global axis.*

A point P of a rigid body is at ${}^B\mathbf{r}_P = \begin{bmatrix} 5 & 30 & 10 \end{bmatrix}^T$. *When it is turned* $\alpha = 30\deg$ *about the Z-axis, the global position of P is*

$$
\begin{aligned}
{}^G\mathbf{r}_P &= {}^GR_B\,{}^B\mathbf{r}_P \\[2mm]
&= \begin{bmatrix} \cos\frac{\pi}{6} & -\sin\frac{\pi}{6} & 0 \\ \sin\frac{\pi}{6} & \cos\frac{\pi}{6} & 0 \\ 0 & 0 & 1 \end{bmatrix}\begin{bmatrix} 5 \\ 30 \\ 10 \end{bmatrix} = \begin{bmatrix} -10.67 \\ 28.48 \\ 10 \end{bmatrix}
\end{aligned}
\tag{5.255}
$$

If the body is turning with $\dot{\alpha} = 10\deg/\mathrm{s}$, *the global velocity of the point P will be*

$$
\begin{aligned}
{}^G\mathbf{v}_P &= {}^G\dot{R}_B\,{}^GR_B^T\,{}^G\mathbf{r}_P \\[2mm]
&= \frac{10\pi}{180}\begin{bmatrix} -s\frac{\pi}{6} & -c\frac{\pi}{6} & 0 \\ c\frac{\pi}{6} & -s\frac{\pi}{6} & 0 \\ 0 & 0 & 0 \end{bmatrix}\begin{bmatrix} c\frac{\pi}{6} & -s\frac{\pi}{6} & 0 \\ s\frac{\pi}{6} & c\frac{\pi}{6} & 0 \\ 0 & 0 & 1 \end{bmatrix}^T\begin{bmatrix} -10.67 \\ 28.48 \\ 10 \end{bmatrix} \\[2mm]
&= \begin{bmatrix} -4.97 \\ -1.86 \\ 0 \end{bmatrix}
\end{aligned}
\tag{5.256}
$$

Example 190 ★ *Principal angular velocities.*

The principal rotational matrices about the axes X, Y, and Z are

$$R_{X,\gamma} = \begin{bmatrix} 1 & 0 & 0 \\ 0 & \cos\gamma & -\sin\gamma \\ 0 & \sin\gamma & \cos\gamma \end{bmatrix} \qquad R_{Y,\beta} = \begin{bmatrix} \cos\beta & 0 & \sin\beta \\ 0 & 1 & 0 \\ -\sin\beta & 0 & \cos\beta \end{bmatrix}$$

$$R_{Z,\alpha} = \begin{bmatrix} \cos\alpha & -\sin\alpha & 0 \\ \sin\alpha & \cos\alpha & 0 \\ 0 & 0 & 1 \end{bmatrix} \tag{5.257}$$

and hence, their time derivatives are

$$\dot{R}_{X,\gamma} = \dot{\gamma}\begin{bmatrix} 0 & 0 & 0 \\ 0 & -s\gamma & -c\gamma \\ 0 & c\gamma & -s\gamma \end{bmatrix} \qquad \dot{R}_{Y,\beta} = \dot{\beta}\begin{bmatrix} -s\beta & 0 & c\beta \\ 0 & 0 & 0 \\ -c\beta & 0 & -s\beta \end{bmatrix}$$

$$\dot{R}_{Z,\alpha} = \dot{\alpha}\begin{bmatrix} -s\alpha & -c\alpha & 0 \\ c\alpha & -s\alpha & 0 \\ 0 & 0 & 0 \end{bmatrix} \tag{5.258}$$

Therefore, the principal angular velocity matrices about axes X, Y, and Z are

$$_G\tilde{\omega}_X = \dot{R}_{X,\gamma}R_{X,\gamma}^T = \dot{\gamma}\begin{bmatrix} 0 & 0 & 0 \\ 0 & 0 & -1 \\ 0 & 1 & 0 \end{bmatrix} \tag{5.259}$$

$$_G\tilde{\omega}_Y = \dot{R}_{Y,\beta}R_{Y,\beta}^T = \dot{\beta}\begin{bmatrix} 0 & 0 & 1 \\ 0 & 0 & 0 \\ -1 & 0 & 0 \end{bmatrix} \tag{5.260}$$

$$_G\tilde{\omega}_Z = \dot{R}_{Z,\alpha}R_{Z,\alpha}^T = \dot{\alpha}\begin{bmatrix} 0 & -1 & 0 \\ 1 & 0 & 0 \\ 0 & 0 & 0 \end{bmatrix} \tag{5.261}$$

which are equivalent to

$$_G\tilde{\omega}_X = \dot{\gamma}\tilde{I} \qquad _G\tilde{\omega}_Y = \dot{\beta}\tilde{J} \qquad _G\tilde{\omega}_Z = \dot{\alpha}\tilde{K} \tag{5.262}$$

and therefore, the principal angular velocity vectors are

$$\begin{aligned} _G\boldsymbol{\omega}_X &= \omega_X\,\hat{I} = \dot{\gamma}\hat{I} \tag{5.263} \\ _G\boldsymbol{\omega}_Y &= \omega_Y\,\hat{J} = \dot{\beta}\hat{J} \tag{5.264} \\ _G\boldsymbol{\omega}_Z &= \omega_Z\,\hat{K} = \dot{\alpha}\hat{K} \tag{5.265} \end{aligned}$$

Utilizing the same technique, we can find the following principal angular velocity matrices about the local axes:

$$_G^B\tilde{\omega}_x = R_{x,\psi}^T\,\dot{R}_{x,\psi} = \dot{\psi}\begin{bmatrix} 0 & 0 & 0 \\ 0 & 0 & -1 \\ 0 & 1 & 0 \end{bmatrix} = \dot{\psi}\tilde{\imath} \tag{5.266}$$

$$
{}_G^B \tilde{\omega}_y = R_{y,\theta}^T \dot{R}_{y,\theta} = \dot{\theta} \begin{bmatrix} 0 & 0 & 1 \\ 0 & 0 & 0 \\ -1 & 0 & 0 \end{bmatrix} = \dot{\theta}\, \hat{j} \tag{5.267}
$$

$$
{}_G^B \tilde{\omega}_z = R_{z,\varphi}^T \dot{R}_{z,\varphi} = \dot{\varphi} \begin{bmatrix} 0 & -1 & 0 \\ 1 & 0 & 0 \\ 0 & 0 & 0 \end{bmatrix} = \dot{\varphi}\, \hat{k} \tag{5.268}
$$

Example 191 *Decomposition of an angular velocity vector.*
   Every angular velocity vector can be decomposed to three principal angular velocity vectors.

$$
\begin{aligned}
{}_G\boldsymbol{\omega}_B &= \left({}_G\boldsymbol{\omega}_B \cdot \hat{I}\right)\hat{I} + \left({}_G\boldsymbol{\omega}_B \cdot \hat{J}\right)\hat{J} + \left({}_G\boldsymbol{\omega}_B \cdot \hat{K}\right)\hat{K} \tag{5.269}\\
&= \omega_X \hat{I} + \omega_Y \hat{J} + \omega_Z \hat{K} = \dot{\gamma}\hat{I} + \dot{\beta}\hat{J} + \dot{\alpha}\hat{K}\\
&= \boldsymbol{\omega}_X + \boldsymbol{\omega}_Y + \boldsymbol{\omega}_Z
\end{aligned}
$$

Example 192 *Combination of angular velocities.*
   Starting from a combination of rotations

$$
{}^0R_2 = {}^0R_1\,{}^1R_2 \tag{5.270}
$$

and taking a time derivative, we find

$$
{}^0\dot{R}_2 = {}^0\dot{R}_1\,{}^1R_2 + {}^0R_1\,{}^1\dot{R}_2 \tag{5.271}
$$

Now, substituting the derivative of rotation matrices with

$$
{}^0\dot{R}_2 = {}_0\tilde{\omega}_2\,{}^0R_2 \qquad {}^0\dot{R}_1 = {}_0\tilde{\omega}_1\,{}^0R_1 \qquad {}^1\dot{R}_2 = {}_1\tilde{\omega}_2\,{}^1R_2 \tag{5.272}
$$

results in

$$
\begin{aligned}
{}_0\tilde{\omega}_2\,{}^0R_2 &= {}_0\tilde{\omega}_1\,{}^0R_1\,{}^1R_2 + {}^0R_1\,{}_1\tilde{\omega}_2\,{}^1R_2\\
&= {}_0\tilde{\omega}_1\,{}^0R_2 + {}^0R_1\,{}_1\tilde{\omega}_2\,{}^0R_1^T\,{}^0R_1\,{}^1R_2\\
&= {}_0\tilde{\omega}_1\,{}^0R_2 + {}_1^0\tilde{\omega}_2\,{}^0R_2 \tag{5.273}
\end{aligned}
$$

where

$$
{}^0R_1\,{}_1\tilde{\omega}_2\,{}^0R_1^T = {}_1^0\tilde{\omega}_2 \tag{5.274}
$$

Therefore, we find

$$
{}_0\tilde{\omega}_2 = {}_0\tilde{\omega}_1 + {}_1^0\tilde{\omega}_2 \tag{5.275}
$$

which indicates that the angular velocities may be added relatively:

$$
{}_0\boldsymbol{\omega}_2 = {}_0\boldsymbol{\omega}_1 + {}_1^0\boldsymbol{\omega}_2 \tag{5.276}
$$

This result also holds for any number of angular velocities

$$
{}_0\boldsymbol{\omega}_n = {}_0\boldsymbol{\omega}_1 + {}_1^0\boldsymbol{\omega}_2 + {}_2^0\boldsymbol{\omega}_3 + \cdots + {}_{n-1}^0\boldsymbol{\omega}_n = \sum_{i=1}^{n} {}_{i-1}^0\boldsymbol{\omega}_i \tag{5.277}
$$

Example 193 ★ *Angular velocity in terms of Euler frequencies.*

The angular velocity vector can be expressed by Euler frequencies. Therefore,

$$
\begin{aligned}
{}_G^B\boldsymbol{\omega}_B &= \omega_x\hat{\imath} + \omega_y\hat{\jmath} + \omega_z\hat{k} = \dot{\varphi}\hat{e}_\varphi + \dot{\theta}\hat{e}_\theta + \dot{\psi}\hat{e}_\psi \\
&= \dot{\varphi}\begin{bmatrix} \sin\theta\sin\psi \\ \sin\theta\cos\psi \\ \cos\theta \end{bmatrix} + \dot{\theta}\begin{bmatrix} \cos\psi \\ -\sin\psi \\ 0 \end{bmatrix} + \dot{\psi}\begin{bmatrix} 0 \\ 0 \\ 1 \end{bmatrix} \\
&= \begin{bmatrix} \sin\theta\sin\psi & \cos\psi & 0 \\ \sin\theta\cos\psi & -\sin\psi & 0 \\ \cos\theta & 0 & 1 \end{bmatrix}\begin{bmatrix} \dot{\varphi} \\ \dot{\theta} \\ \dot{\psi} \end{bmatrix}
\end{aligned}
\tag{5.278}
$$

and

$$
\begin{aligned}
{}_G^G\boldsymbol{\omega}_B &= {}^BR_G^{-1}\,{}_G^B\boldsymbol{\omega}_B = {}^BR_G^{-1}\begin{bmatrix} \dot{\varphi}\sin\theta\sin\psi + \dot{\theta}\cos\psi \\ \dot{\varphi}\sin\theta\cos\psi - \dot{\theta}\sin\psi \\ \dot{\varphi}\cos\theta + \dot{\psi} \end{bmatrix} \\
&= \begin{bmatrix} 0 & \cos\varphi & \sin\theta\sin\varphi \\ 0 & \sin\varphi & -\cos\varphi\sin\theta \\ 1 & 0 & \cos\theta \end{bmatrix}\begin{bmatrix} \dot{\varphi} \\ \dot{\theta} \\ \dot{\psi} \end{bmatrix}
\end{aligned}
\tag{5.279}
$$

where the inverse of the Euler transformation matrix is

$$
{}^BR_G^{-1} = \begin{bmatrix} c\varphi c\psi - c\theta s\varphi s\psi & -c\varphi s\psi - c\theta c\psi s\varphi & s\theta s\varphi \\ c\psi s\varphi + c\theta c\varphi s\psi & -s\varphi s\psi + c\theta c\varphi c\psi & -c\varphi s\theta \\ s\theta s\psi & s\theta c\psi & c\theta \end{bmatrix}
\tag{5.280}
$$

Example 194 ★ *Angular velocity in terms of rotation frequencies.*

Consider the Euler angles transformation matrix:

$$
{}^BR_G = R_{z,\psi}R_{x,\theta}R_{z,\varphi}
\tag{5.281}
$$

The angular velocity matrix is then equal to

$$
\begin{aligned}
{}_B\tilde{\omega}_G &= {}^B\dot{R}_G\,{}^BR_G^T \\
&= \left(\dot{\varphi}\,R_{z,\psi}R_{x,\theta}\frac{dR_{z,\varphi}}{dt} + \dot{\theta}\,R_{z,\psi}\frac{dR_{x,\theta}}{dt}R_{z,\varphi} + \dot{\psi}\frac{dR_{z,\psi}}{dt}R_{x,\theta}R_{z,\varphi} \right) \\
&\quad\times \left(R_{z,\psi}R_{x,\theta}R_{z,\varphi} \right)^T \\
&= \dot{\varphi}\,R_{z,\psi}R_{x,\theta}\frac{dR_{z,\varphi}}{dt}R_{z,\varphi}^TR_{x,\theta}^TR_{z,\psi}^T + \dot{\theta}\,R_{z,\psi}\frac{dR_{x,\theta}}{dt}R_{x,\theta}^TR_{z,\psi}^T \\
&\quad + \dot{\psi}\frac{dR_{z,\psi}}{dt}R_{z,\psi}^T
\end{aligned}
\tag{5.282}
$$

which, in matrix form, is

$$
{}_B\tilde{\omega}_G = \dot{\varphi}\begin{bmatrix} 0 & \cos\theta & -\sin\theta\cos\psi \\ -\cos\theta & 0 & \sin\theta\sin\psi \\ \sin\theta\cos\psi & -\sin\theta\sin\psi & 0 \end{bmatrix}
$$
$$
+\dot{\theta}\begin{bmatrix} 0 & 0 & \sin\psi \\ 0 & 0 & \cos\psi \\ -\sin\psi & -\cos\psi & 0 \end{bmatrix} + \dot{\psi}\begin{bmatrix} 0 & 1 & 0 \\ -1 & 0 & 0 \\ 0 & 0 & 0 \end{bmatrix} \quad (5.283)
$$

or

$$
{}_B\tilde{\omega}_G = \begin{bmatrix} 0 & \dot{\psi}+\dot{\varphi}c\theta & \dot{\theta}s\psi-\dot{\varphi}s\theta c\psi \\ -\dot{\psi}-\dot{\varphi}c\theta & 0 & \dot{\theta}c\psi+\dot{\varphi}s\theta s\psi \\ -\dot{\theta}s\psi+\dot{\varphi}s\theta c\psi & -\dot{\theta}c\psi-\dot{\varphi}s\theta s\psi & 0 \end{bmatrix} \quad (5.284)
$$

The corresponding angular velocity vector is

$$
{}_B\boldsymbol{\omega}_G = -\begin{bmatrix} \dot{\theta}c\psi+\dot{\varphi}s\theta s\psi \\ -\dot{\theta}s\psi+\dot{\varphi}s\theta c\psi \\ \dot{\psi}+\dot{\varphi}c\theta \end{bmatrix}
$$
$$
= -\begin{bmatrix} \sin\theta\sin\psi & \cos\psi & 0 \\ \sin\theta\cos\psi & -\sin\psi & 0 \\ \cos\theta & 0 & 1 \end{bmatrix}\begin{bmatrix} \dot{\varphi} \\ \dot{\theta} \\ \dot{\psi} \end{bmatrix} \quad (5.285)
$$

However,

$$
{}_B^B\tilde{\omega}_G = -{}_G^B\tilde{\omega}_B \quad (5.286)
$$
$$
{}_B^B\boldsymbol{\omega}_G = -{}_G^B\boldsymbol{\omega}_B \quad (5.287)
$$

and therefore,

$$
{}_G^B\boldsymbol{\omega}_B = \begin{bmatrix} \sin\theta\sin\psi & \cos\psi & 0 \\ \sin\theta\cos\psi & -\sin\psi & 0 \\ \cos\theta & 0 & 1 \end{bmatrix}\begin{bmatrix} \dot{\varphi} \\ \dot{\theta} \\ \dot{\psi} \end{bmatrix} \quad (5.288)
$$

Example 195 ★ *Coordinate transformation of angular velocity.*
    Angular velocity ${}_1\boldsymbol{\omega}_2$ of coordinate frame B_2 with respect to B_1 and expressed in B_1 can be expressed in base coordinate frame B_0 according to

$$
{}^0R_1\,{}_1\tilde{\omega}_2\,{}^0R_1^T = {}_1^0\tilde{\omega}_2 \quad (5.289)
$$

To show this equation, it is enough to apply both sides on an arbitrary vector ${}^0\mathbf{r}$. Therefore, the left-hand side would be

$$
\begin{aligned}
{}^0R_1\,{}_1\tilde{\omega}_2\,{}^0R_1^T\,{}^0\mathbf{r} &= {}^0R_1\,{}_1\tilde{\omega}_2\,{}^1R_0\,{}^0\mathbf{r} = {}^0R_1\,{}_1\tilde{\omega}_2\,{}^1\mathbf{r} \\
&= {}^0R_1\left({}_1\boldsymbol{\omega}_2 \times {}^1\mathbf{r}\right) = {}^0R_1\,{}_1\boldsymbol{\omega}_2 \times {}^0R_1\,{}^1\mathbf{r} \\
&= {}_1^0\boldsymbol{\omega}_2 \times {}^0\mathbf{r} \quad (5.290)
\end{aligned}
$$

which is equal to the right-hand side after applying on the vector $^0\mathbf{r}$

$$^0_1\tilde{\omega}_2\,{}^0\mathbf{r} = {}^0_1\boldsymbol{\omega}_2 \times {}^0\mathbf{r} \tag{5.291}$$

Example 196 ★ *Time derivative of unit vectors.*

Using Equation (5.241) we can define the time derivative of unit vectors of a body coordinate frame $B(\hat{\imath},\hat{\jmath},\hat{k})$, rotating in the global coordinate frame $G(\hat{I},\hat{J},\hat{K})$

$$\frac{{}^G d\hat{\imath}}{dt} = {}^B_G\boldsymbol{\omega}_B \times \hat{\imath} \qquad \frac{{}^G d\hat{\jmath}}{dt} = {}^B_G\boldsymbol{\omega}_B \times \hat{\jmath} \qquad \frac{{}^G d\hat{k}}{dt} = {}^B_G\boldsymbol{\omega}_B \times \hat{k} \tag{5.292}$$

5.10 ★ Time Derivative and Coordinate Frames

The time derivative of a vector depends on the coordinate frame in which we are taking the derivative. The time derivative of a vector $\mathbf{r}$ in the global frame is called the *G-derivative* and is denoted by

$$\frac{{}^G d}{dt}\mathbf{r}$$

while the time derivative of the vector in the body frame is called the *B-derivative* and is denoted by

$$\frac{{}^B d}{dt}\mathbf{r}$$

The left superscript on the derivative symbol indicates the frame in which the derivative is taken, and hence, its unit vectors are considered constant.

Time derivative is called simple if the vector is expressed in the same coordinate frame that we are taking the derivative, because the unit vectors are constant and scalar coefficients are the only time variables. The simple derivatives of $^B\mathbf{r}_P$ in B and $^G\mathbf{r}_P$ in G are

$$\frac{{}^B d}{dt}\,{}^B\mathbf{r}_P = {}^B\dot{\mathbf{r}}_P = {}^B\mathbf{v}_P = \dot{x}\,\hat{\imath} + \dot{y}\,\hat{\jmath} + \dot{z}\,\hat{k} \tag{5.293}$$

$$\frac{{}^G d}{dt}\,{}^G\mathbf{r}_P = {}^G\dot{\mathbf{r}}_P = {}^G\mathbf{v}_P = \dot{X}\,\hat{I} + \dot{Y}\,\hat{J} + \dot{Z}\,\hat{K} \tag{5.294}$$

It is also possible to find the *G-derivative* of $^B\mathbf{r}_P$ and the *B-derivative* of $^G\mathbf{r}_P$. We define the *G-derivative* of a body vector $^B\mathbf{r}_P$ by

$$^B_G\mathbf{v}_P = \frac{{}^G d}{dt}\,{}^B\mathbf{r}_P \tag{5.295}$$

and similarly, a *B-derivative* of a global vector $^G\mathbf{r}_P$ by

$$^G_B\mathbf{v}_P = \frac{{}^B d}{dt}\,{}^G\mathbf{r}_P \tag{5.296}$$

When point P is moving in frame B while B is rotating in G, the G-derivative of ${}^B\mathbf{r}_P(t)$ is defined by

$$\frac{{}^G d}{dt} {}^B\mathbf{r}_P(t) = {}^B\dot{\mathbf{r}}_P + {}^B_G\boldsymbol{\omega}_B \times {}^B\mathbf{r}_P = {}^B_G\dot{\mathbf{r}}_P \qquad (5.297)$$

and the B-derivative of ${}^G\mathbf{r}_P$ is defined by

$$\frac{{}^B d}{dt} {}^G\mathbf{r}_P(t) = {}^G\dot{\mathbf{r}}_P - {}_G\boldsymbol{\omega}_B \times {}^G\mathbf{r}_P = {}^G_B\dot{\mathbf{r}}_P \qquad (5.298)$$

Proof. Let $G(OXYZ)$ with unit vectors $\hat{I}$, $\hat{J}$, and $\hat{K}$ be the global coordinate frame, and let $B(Oxyz)$ with unit vectors $\hat{\imath}$, $\hat{\jmath}$, and $\hat{k}$ be a body coordinate frame. The position vector of a moving point P, can be expressed in the body and global frames

$$
\begin{aligned}
{}^B\mathbf{r}_P(t) &= x(t)\,\hat{\imath} + y(t)\,\hat{\jmath} + z(t)\,\hat{k} & (5.299)\\
{}^G\mathbf{r}_P(t) &= X(t)\,\hat{I} + Y(t)\,\hat{J} + Z(t)\,\hat{K} & (5.300)
\end{aligned}
$$

The time derivative of ${}^B\mathbf{r}_P$ in B and ${}^G\mathbf{r}_P$ in G are

$$
\begin{aligned}
\frac{{}^B d}{dt} {}^B\mathbf{r}_P &= {}^B\dot{\mathbf{r}}_P = {}^B\mathbf{v}_P = \dot{x}\,\hat{\imath} + \dot{y}\,\hat{\jmath} + \dot{z}\,\hat{k} & (5.301)\\
\frac{{}^G d}{dt} {}^G\mathbf{r}_P &= {}^G\dot{\mathbf{r}}_P = {}^G\mathbf{v}_P = \dot{X}\,\hat{I} + \dot{Y}\,\hat{J} + \dot{Z}\,\hat{K} & (5.302)
\end{aligned}
$$

because the unit vectors of B in Equation (5.299) and the unit vectors of G in Equation (5.300) are considered constant.

Using Equation (5.241) for the global velocity of a body fixed point P, expressed in body frame

$$\frac{{}^B}{G}\mathbf{v}_P = {}^B_G\boldsymbol{\omega}_B \times {}^B\mathbf{r}_P = \frac{{}^G d}{dt} {}^B\mathbf{r}_P \qquad (5.303)$$

and definition (5.295), we can find the G-derivative of the position vector ${}^B\mathbf{r}_P$ as

$$
\begin{aligned}
\frac{{}^G d}{dt} {}^B\mathbf{r}_P &= \frac{{}^G d}{dt}\left(x\hat{\imath} + y\hat{\jmath} + z\hat{k} \right)\\
&= \dot{x}\,\hat{\imath} + \dot{y}\,\hat{\jmath} + \dot{z}\,\hat{k} + x\frac{{}^G d\hat{\imath}}{dt} + y\frac{{}^G d\hat{\jmath}}{dt} + z\frac{{}^G d\hat{k}}{dt}\\
&= {}^B\dot{\mathbf{r}}_P + x\,{}^B_G\boldsymbol{\omega}_B \times \hat{\imath} + y\,{}^B_G\boldsymbol{\omega}_B \times \hat{\jmath} + z\,{}^B_G\boldsymbol{\omega}_B \times \hat{k}\\
&= {}^B\dot{\mathbf{r}}_P + {}^B_G\boldsymbol{\omega}_B \times \left(x\hat{\imath} + y\hat{\jmath} + z\hat{k} \right)\\
&= {}^B\dot{\mathbf{r}}_P + {}^B_G\boldsymbol{\omega}_B \times {}^B\mathbf{r}_P = \frac{{}^B d}{dt} {}^B\mathbf{r}_P + {}^B_G\boldsymbol{\omega}_B \times {}^B\mathbf{r}_P \quad (5.304)
\end{aligned}
$$

We achieved this result because the x, y, and z components of ${}^B\mathbf{r}_P$ are scalar. Scalars are invariant with respect to frame transformations. Therefore, if x is a scalar then,

$$\frac{{}^G d}{dt} x = \frac{{}^B d}{dt} x = \dot{x} \tag{5.305}$$

The B-derivative of ${}^G\mathbf{r}_P$ can be found similarly

$$
\begin{aligned}
\frac{{}^B d}{dt} {}^G\mathbf{r}_P &= \frac{{}^B d}{dt} \left(X\hat{I} + Y\hat{J} + Z\hat{K} \right) \\
&= \dot{X}\hat{I} + \dot{Y}\hat{J} + \dot{Z}\hat{K} + X\frac{{}^B d\hat{I}}{dt} + Y\frac{{}^B d\hat{J}}{dt} + Z\frac{{}^B d\hat{K}}{dt} \\
&= {}^G\dot{\mathbf{r}}_P + {}^B_G\boldsymbol{\omega}_G \times {}^G\mathbf{r}_P
\end{aligned}
\tag{5.306}
$$

and therefore,

$$\frac{{}^B d}{dt} {}^G\mathbf{r}_P = {}^G\dot{\mathbf{r}}_P - {}_G\boldsymbol{\omega}_B \times {}^G\mathbf{r}_P \tag{5.307}$$

The angular velocity of B relative to G is a vector quantity and can be expressed in either frames.

$$
\begin{aligned}
{}^G_G\boldsymbol{\omega}_B &= \omega_X\hat{I} + \omega_Y\hat{J} + \omega_Z\hat{K} \tag{5.308} \\
{}^B_G\boldsymbol{\omega}_B &= \omega_x\hat{i} + \omega_y\hat{j} + \omega_z\hat{k} \tag{5.309}
\end{aligned}
$$

■

Example 197 ★ *Time derivative of a moving point in B.*

Consider a local frame B that is rotating in G by $\dot{\alpha}$ about the Z-axis, and a moving point at ${}^B\mathbf{r}_P(t) = t\hat{i}$. Therefore,

$$
\begin{aligned}
{}^G\mathbf{r}_P &= {}^G R_B \, {}^B\mathbf{r}_P = R_{Z,\alpha}(t)\, {}^B\mathbf{r}_P \\
&= \begin{bmatrix} \cos\alpha & -\sin\alpha & 0 \\ \sin\alpha & \cos\alpha & 0 \\ 0 & 0 & 1 \end{bmatrix} \begin{bmatrix} t \\ 0 \\ 0 \end{bmatrix} \\
&= t\cos\alpha\,\hat{I} + t\sin\alpha\,\hat{J}
\end{aligned}
\tag{5.310}
$$

The angular velocity matrix is

$$_G\tilde{\omega}_B = {}^G\dot{R}_B \, {}^G R_B^T = \dot{\alpha}\tilde{K} \tag{5.311}$$

that gives

$$_G\boldsymbol{\omega}_B = \dot{\alpha}\hat{K} \tag{5.312}$$

It can also be verified that

$$_G^B\tilde{\omega}_B = {}^G R_B^T \, {}^G_G\tilde{\omega}_B \, {}^G R_B = \dot{\alpha}\tilde{k} \tag{5.313}$$

and therefore,

$$\substack{B \\ G}\boldsymbol{\omega}_B = \dot\alpha \hat{k} \tag{5.314}$$

Now we can find the following derivatives:

$$\frac{^B d}{dt}\,{}^B\mathbf{r}_P = {}^B\dot{\mathbf{r}}_P = \hat{i} \tag{5.315}$$

$$\frac{^G d}{dt}\,{}^G\mathbf{r}_P = {}^G\dot{\mathbf{r}}_P = (\cos\alpha - t\dot\alpha \sin\alpha)\,\hat{I} + (\sin\alpha + t\dot\alpha \cos\alpha)\,\hat{J} \tag{5.316}$$

For the mixed derivatives, we start with

$$
\begin{aligned}
\frac{^G d}{dt}\,{}^B\mathbf{r}_P &= \frac{^B d}{dt}\,{}^B\mathbf{r}_P + {}_G^B\boldsymbol{\omega}_B \times {}^B\mathbf{r}_P \\[2mm]
&= \begin{bmatrix} 1 \\ 0 \\ 0 \end{bmatrix} + \dot\alpha \begin{bmatrix} 0 \\ 0 \\ 1 \end{bmatrix} \times \begin{bmatrix} t \\ 0 \\ 0 \end{bmatrix} = \begin{bmatrix} 1 \\ t\dot\alpha \\ 0 \end{bmatrix} \\[2mm]
&= \hat{i} + t\dot\alpha\hat{j} = {}_G^B\dot{\mathbf{r}}_P \tag{5.317}
\end{aligned}
$$

which is the global velocity of P expressed in B. We may, however, transform ${}_G^B\dot{\mathbf{r}}_P$ to the global frame and find the global velocity expressed in G.

$$
\begin{aligned}
{}^G\dot{\mathbf{r}}_P &= {}^G R_B\,{}_G^B\dot{\mathbf{r}}_P \\[2mm]
&= \begin{bmatrix} \cos\alpha & -\sin\alpha & 0 \\ \sin\alpha & \cos\alpha & 0 \\ 0 & 0 & 1 \end{bmatrix} \begin{bmatrix} 1 \\ t\dot\alpha \\ 0 \end{bmatrix} = \begin{bmatrix} \cos\alpha - t\dot\alpha \sin\alpha \\ \sin\alpha + t\dot\alpha \cos\alpha \\ 0 \end{bmatrix} \\[2mm]
&= (\cos\alpha - t\dot\alpha \sin\alpha)\,\hat{I} + (\sin\alpha + t\dot\alpha \cos\alpha)\,\hat{J} \tag{5.318}
\end{aligned}
$$

The next derivative is

$$
\begin{aligned}
\frac{^B d}{dt}\,{}^G\mathbf{r}_P &= {}^G\dot{\mathbf{r}}_P - {}_G\boldsymbol{\omega}_B \times {}^G\mathbf{r}_P \\[2mm]
&= \begin{bmatrix} \cos\alpha - t\dot\alpha \sin\alpha \\ \sin\alpha + t\dot\alpha \cos\alpha \\ 0 \end{bmatrix} - \dot\alpha \begin{bmatrix} 0 \\ 0 \\ 1 \end{bmatrix} \times \begin{bmatrix} t\cos\alpha \\ t\sin\alpha \\ 0 \end{bmatrix} \\[2mm]
&= \begin{bmatrix} \cos\alpha \\ \sin\alpha \\ 0 \end{bmatrix} = (\cos\alpha)\,\hat{I} + (\sin\alpha)\,\hat{J} = {}_B^G\dot{\mathbf{r}}_P \tag{5.319}
\end{aligned}
$$

which is the velocity of P relative to B and expressed in G. To express this velocity in B we apply a frame transformation

$$
\begin{aligned}
{}^B\dot{\mathbf{r}}_P &= {}^G R_B^T\,{}_B^G\dot{\mathbf{r}}_P \\[2mm]
&= \begin{bmatrix} \cos\alpha & -\sin\alpha & 0 \\ \sin\alpha & \cos\alpha & 0 \\ 0 & 0 & 1 \end{bmatrix}^T \begin{bmatrix} \cos\alpha \\ \sin\alpha \\ 0 \end{bmatrix} = \begin{bmatrix} 1 \\ 0 \\ 0 \end{bmatrix} = \hat{i} \tag{5.320}
\end{aligned}
$$

Sometimes it is more applied if we transform the vector to the same frame in which we are taking the derivative and then apply the differential operator. Therefore,

$$
\begin{aligned}
\frac{{}^G d}{dt} {}^B \mathbf{r}_P &= \frac{{}^G d}{dt} \left({}^G R_B \, {}^B \mathbf{r}_P \right) \\
&= \frac{{}^G d}{dt} \begin{bmatrix} t \cos \alpha \\ t \sin \alpha \\ 0 \end{bmatrix} = \begin{bmatrix} \cos \alpha - t \dot\alpha \sin \alpha \\ \sin \alpha + t \dot\alpha \cos \alpha \\ 0 \end{bmatrix}
\end{aligned} \tag{5.321}
$$

and

$$
\frac{{}^B d}{dt} {}^G \mathbf{r}_P = \frac{{}^B d}{dt} \left({}^G R_B^T \, {}^G \mathbf{r}_P \right) = \frac{{}^B d}{dt} \begin{bmatrix} t \\ 0 \\ 0 \end{bmatrix} = \begin{bmatrix} 1 \\ 0 \\ 0 \end{bmatrix} \tag{5.322}
$$

Example 198 ★ *Orthogonality of position and velocity vectors.*
If the position vector of a body point in global frame is denoted by $\mathbf{r}$ *then*

$$
\frac{d\mathbf{r}}{dt} \cdot \mathbf{r} = 0 \tag{5.323}
$$

To show this property we may take a derivative from

$$
\mathbf{r} \cdot \mathbf{r} = r^2 \tag{5.324}
$$

and find

$$
\frac{d}{dt} (\mathbf{r} \cdot \mathbf{r}) = \frac{d\mathbf{r}}{dt} \cdot \mathbf{r} + \mathbf{r} \cdot \frac{d\mathbf{r}}{dt} = 2 \frac{d\mathbf{r}}{dt} \cdot \mathbf{r} = 0 \tag{5.325}
$$

Equation (5.323) is correct in every coordinate frame and for every constant length vector, as long as the vector and the derivative are expressed in the same coordinate frame.

Example 199 ★ *Derivative transformation formula.*
The global velocity of a fixed point in the body coordinate frame $B \, (Oxyz)$ *can be found by Equation (5.228). Now consider a point* P *that can move in* $B \, (Oxyz)$. *In this case, the body position vector* ${}^B \mathbf{r}_P$ *is not constant, and therefore, the global velocity of such a point expressed in* B *is*

$$
\frac{{}^G d}{dt} {}^B \mathbf{r}_P = \frac{{}^B d}{dt} {}^B \mathbf{r}_P + {}^B_G \boldsymbol{\omega}_B \times {}^B \mathbf{r}_P = {}^B_G \dot{\mathbf{r}}_P \tag{5.326}
$$

Sometimes the result of Equation (5.326) is utilized to define transformation of the differential operator from a body to a global coordinate frame

$$
\frac{{}^G d}{dt} \square = \frac{{}^B d}{dt} \square + {}^B_G \boldsymbol{\omega}_B \times {}^B \square = {}^B_G \dot{\square} \tag{5.327}
$$

The final result is ${}^B_G \dot{\square}$ *showing the global* (G) *time derivative expressed in body frame* (B). *The vector* $\square$ *might be any vector such as position, velocity, angular velocity, momentum, angular velocity, or even a time-varying force vector.*

Equation (5.327) is called the **derivative transformation formula** *and relates the time derivative of a vector as it would be seen from frame G to its derivative as seen in frame B. The derivative transformation formula (5.327) is general and can be applied to every vector for derivative transformation between every two relatively moving coordinate frames.*

Example 200 ★ *Differential equation for rotation matrix.*

Equation (5.231) for defining the angular velocity matrix may be written as a first-order differential equation

$$\frac{d}{dt}\, {}^{G}R_B - {}^{G}R_B\, {}_{G}\tilde{\omega}_B = 0 \tag{5.328}$$

The solution of the equation confirms the exponential definition of the rotation matrix as

$$ {}^{G}R_B = e^{\tilde{\omega}t} \tag{5.329}$$

or

$$\tilde{\omega}t = \dot{\phi}\,\tilde{u} = \ln\left({}^{G}R_B\right) \tag{5.330}$$

Example 201 ★ *Acceleration of a body point in the global frame.*

The angular acceleration vector of a rigid body $B(Oxyz)$ in the global frame $G(OXYZ)$ is denoted by ${}_{G}\alpha_B$ and is defined as the global time derivative of ${}_{G}\omega_B$.

$$ {}_{G}\alpha_B = \frac{{}^{G}d}{dt}\, {}_{G}\omega_B \tag{5.331}$$

Using this definition, the acceleration of a fixed body point in the global frame is

$$ {}^{G}\mathbf{a}_P = \frac{{}^{G}d}{dt}\left({}_{G}\omega_B \times {}^{G}\mathbf{r}_P\right) = {}_{G}\alpha_B \times {}^{G}\mathbf{r}_P + {}_{G}\omega_B \times \left({}_{G}\omega_B \times {}^{G}\mathbf{r}_P\right) \tag{5.332}$$

Example 202 ★ *Alternative definition of angular velocity vector.*

The angular velocity vector of a rigid body $B(\hat{i}, \hat{j}, \hat{k})$ in global frame $G(\hat{I}, \hat{J}, \hat{K})$ can also be defined by

$$ {}^{B}_{G}\omega_B = \hat{i}(\frac{{}^{G}d\hat{j}}{dt} \cdot \hat{k}) + \hat{j}(\frac{{}^{G}d\hat{k}}{dt} \cdot \hat{i}) + \hat{k}(\frac{{}^{G}d\hat{i}}{dt} \cdot \hat{j}) \tag{5.333}$$

Proof. *Consider a body coordinate frame B moving with a fixed point in the global coordinate frame G. The fixed point of the body is taken as the origin of both coordinate frames, as shown in Figure 5.8. To describe the motion of the body, it is sufficient to describe the motion of the local unit vectors $\hat{i}$, $\hat{j}$, $\hat{k}$. Let $\mathbf{r}_P$ be the position vector of a body point P. Then, ${}^{B}\mathbf{r}_P$ is a vector with constant components.*

$$ {}^{B}\mathbf{r}_P = x\hat{i} + y\hat{j} + z\hat{k} \tag{5.334}$$

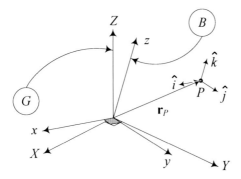

FIGURE 5.8. A body coordinate frame moving with a fixed point in the global coordinate frame.

When the body moves, it is only the unit vectors $\hat{\imath}$, $\hat{\jmath}$, and $\hat{k}$ that vary relative to the global coordinate frame. Therefore, the vector of differential displacement is

$$dr_P = x\, d\hat{\imath} + y\, d\hat{\jmath} + z\, d\hat{k} \tag{5.335}$$

which can also be expressed by

$$d\mathbf{r}_P = (d\mathbf{r}_P \cdot \hat{\imath})\, \hat{\imath} + (d\mathbf{r}_P \cdot \hat{\jmath})\, \hat{\jmath} + \left(d\mathbf{r}_P \cdot \hat{k}\right) \hat{k} \tag{5.336}$$

Substituting (5.335) in the right-hand side of (5.336) yields

$$
\begin{aligned}
d\mathbf{r}_P = {} & \left(x\hat{\imath} \cdot d\hat{\imath} + y\hat{\imath} \cdot d\hat{\jmath} + z\hat{\imath} \cdot d\hat{k} \right) \hat{\imath} \\
& + \left(x\hat{\jmath} \cdot d\hat{\imath} + y\hat{\jmath} \cdot d\hat{\jmath} + z\hat{\jmath} \cdot d\hat{k} \right) \hat{\jmath} \\
& + \left(x\hat{k} \cdot d\hat{\imath} + y\hat{k} \cdot d\hat{\jmath} + z\hat{k} \cdot d\hat{k} \right) \hat{k}
\end{aligned}
\tag{5.337}
$$

Utilizing the unit vectors' relationships

$$\hat{\jmath} \cdot d\hat{\imath} = -\hat{\imath} \cdot d\hat{\jmath} \tag{5.338}$$

$$\hat{k} \cdot d\hat{\jmath} = -\hat{\jmath} \cdot d\hat{k} \tag{5.339}$$

$$\hat{\imath} \cdot d\hat{k} = -\hat{k} \cdot d\hat{\imath} \tag{5.340}$$

$$\hat{\imath} \cdot d\hat{\imath} = \hat{\jmath} \cdot d\hat{\jmath} = \hat{k} \cdot d\hat{k} = 0 \tag{5.341}$$

$$\hat{\imath} \cdot \hat{\jmath} = \hat{\jmath} \cdot \hat{k} = \hat{k} \cdot \hat{\imath} = 0 \tag{5.342}$$

$$\hat{\imath} \cdot \hat{\imath} = \hat{\jmath} \cdot \hat{\jmath} = \hat{k} \cdot \hat{k} = 1 \tag{5.343}$$

the $d\mathbf{r}_P$ reduces to

$$
\begin{aligned}
d\mathbf{r}_P = {} & \left(z\hat{\imath} \cdot d\hat{k} - y\hat{\jmath} \cdot d\hat{\imath} \right) \hat{\imath} + \left(x\hat{\jmath} \cdot d\hat{\imath} - z\hat{k} \cdot d\hat{\jmath} \right) \hat{\jmath} \\
& + \left(y\hat{k} \cdot d\hat{\jmath} - x\hat{\imath} \cdot d\hat{k} \right) \hat{k}
\end{aligned}
\tag{5.344}
$$

This equation can be rearranged to be expressed as a vector product

$$d\mathbf{r}_P = \left((\hat{k} \cdot d\hat{j})\hat{i} + (\hat{i} \cdot d\hat{k})\hat{j} + (\hat{j} \cdot d\hat{i})\hat{k} \right) \times \left(x\hat{i} + y\hat{j} + z\hat{k} \right) \tag{5.345}$$

or

$$_G^B\dot{\mathbf{r}}_P = \left((\hat{k} \cdot \frac{^G d\hat{j}}{dt})\hat{i} + (\hat{i} \cdot \frac{^G d\hat{k}}{dt})\hat{j} + (\hat{j} \cdot \frac{^G d\hat{i}}{dt})\hat{k} \right) \times \left(x\hat{i} + y\hat{j} + z\hat{k} \right)$$

$$\tag{5.346}$$

Comparing this result with

$$\dot{\mathbf{r}}_P = {}_G\boldsymbol{\omega}_B \times \mathbf{r}_P \tag{5.347}$$

shows that

$$_G^B\boldsymbol{\omega}_B = \hat{i}\left(\frac{^G d\hat{j}}{dt} \cdot \hat{k} \right) + \hat{j}\left(\frac{^G d\hat{k}}{dt} \cdot \hat{i} \right) + \hat{k}\left(\frac{^G d\hat{i}}{dt} \cdot \hat{j} \right) \tag{5.348}$$

■

Example 203 ★ *Alternative proof for angular velocity definition (5.333).*

The angular velocity definition presented in Equation (5.333) can also be shown by direct substitution for $^G R_B$ in the angular velocity matrix $_G^B\tilde{\omega}_B$

$$_G^B\tilde{\omega}_B = {}^G R_B^T \, {}^G \dot{R}_B \tag{5.349}$$

Therefore,

$$_G^B\tilde{\omega}_B = \begin{bmatrix} \hat{i} \cdot \hat{I} & \hat{i} \cdot \hat{J} & \hat{i} \cdot \hat{K} \\ \hat{j} \cdot \hat{I} & \hat{j} \cdot \hat{J} & \hat{j} \cdot \hat{K} \\ \hat{k} \cdot \hat{I} & \hat{k} \cdot \hat{J} & \hat{k} \cdot \hat{K} \end{bmatrix} \cdot \frac{^G d}{dt} \begin{bmatrix} \hat{I} \cdot \hat{i} & \hat{I} \cdot \hat{j} & \hat{I} \cdot \hat{k} \\ \hat{J} \cdot \hat{i} & \hat{J} \cdot \hat{j} & \hat{J} \cdot \hat{k} \\ \hat{K} \cdot \hat{i} & \hat{K} \cdot \hat{j} & \hat{K} \cdot \hat{k} \end{bmatrix}$$

$$= \begin{bmatrix} \left(\hat{i} \cdot \frac{^G d\hat{i}}{dt} \right) & \left(\hat{i} \cdot \frac{^G d\hat{j}}{dt} \right) & \left(\hat{i} \cdot \frac{^G d\hat{k}}{dt} \right) \\ \\ \left(\hat{j} \cdot \frac{^G d\hat{i}}{dt} \right) & \left(\hat{j} \cdot \frac{^G d\hat{j}}{dt} \right) & \left(\hat{j} \cdot \frac{^G d\hat{k}}{dt} \right) \\ \\ \left(\hat{k} \cdot \frac{^G d\hat{i}}{dt} \right) & \left(\hat{k} \cdot \frac{^G d\hat{j}}{dt} \right) & \left(\hat{k} \cdot \frac{^G d\hat{k}}{dt} \right) \end{bmatrix} \tag{5.350}$$

which shows that

$$_G^B\boldsymbol{\omega}_B = \begin{bmatrix} \left(\frac{^G d\hat{j}}{dt} \cdot \hat{k} \right) \\ \\ \left(\frac{^G d\hat{k}}{dt} \cdot \hat{i} \right) \\ \\ \left(\frac{^G d\hat{i}}{dt} \cdot \hat{j} \right) \end{bmatrix} \tag{5.351}$$

Example 204 ★ *Second derivative.*

In general, $^G d\mathbf{r}/dt$ *is a variable vector in* $G(OXYZ)$ *and in any other coordinate frame such as* $B\,(oxyz)$. *Therefore, it can be differentiated in either coordinate frames* G *or* B. *However, the order of differentiating is important. In general,*

$$\frac{^B d}{dt}\frac{^G d\mathbf{r}}{dt} \neq \frac{^G d}{dt}\frac{^B d\mathbf{r}}{dt} \qquad (5.352)$$

As an example, consider a rotating body coordinate frame about the Z-*axis, and a variable vector as*

$$^G\mathbf{r} = t\hat{I} \qquad (5.353)$$

Therefore,

$$\frac{^G d\mathbf{r}}{dt} = \,^G\dot{\mathbf{r}} = \hat{I} \qquad (5.354)$$

and hence,

$$
\begin{aligned}
^B\left(\frac{^G d\mathbf{r}}{dt}\right) &= \,^B_G\dot{\mathbf{r}} = R^T_{Z,\varphi}\left[\hat{I}\right] = \begin{bmatrix} \cos\varphi & \sin\varphi & 0 \\ -\sin\varphi & \cos\varphi & 0 \\ 0 & 0 & 1 \end{bmatrix}\begin{bmatrix} 1 \\ 0 \\ 0 \end{bmatrix} \\
&= \cos\varphi\hat{i} - \sin\varphi\hat{j} \qquad (5.355)
\end{aligned}
$$

which provides

$$\frac{^B d}{dt}\frac{^G d\mathbf{r}}{dt} = -\dot{\varphi}\sin\varphi\hat{i} - \dot{\varphi}\cos\varphi\hat{j} \qquad (5.356)$$

and

$$^G\left(\frac{^B d}{dt}\frac{^G d\mathbf{r}}{dt}\right) = -\dot{\varphi}\hat{J} \qquad (5.357)$$

Now

$$^B\mathbf{r} = R^T_{Z,\varphi}\left[t\hat{I}\right] = t\cos\varphi\hat{i} - t\sin\varphi\hat{j} \qquad (5.358)$$

that provides

$$\frac{^B d\mathbf{r}}{dt} = \left(-t\dot{\varphi}\sin\varphi + \cos\varphi\right)\hat{i} - \left(\sin\varphi + t\dot{\varphi}\cos\varphi\right)\hat{j} \qquad (5.359)$$

and

$$
\begin{aligned}
^G\left(\frac{^B d\mathbf{r}}{dt}\right) &= \,^G_B\dot{\mathbf{r}} = R_{Z,\varphi}\left(\left(-t\dot{\varphi}\sin\varphi + \cos\varphi\right)\hat{i} - \left(\sin\varphi + t\dot{\varphi}\cos\varphi\right)\hat{j}\right) \\
&= \begin{bmatrix} \cos\varphi & -\sin\varphi & 0 \\ \sin\varphi & \cos\varphi & 0 \\ 0 & 0 & 1 \end{bmatrix}\begin{bmatrix} -t\dot{\varphi}\sin\varphi + \cos\varphi \\ -\sin\varphi - t\dot{\varphi}\cos\varphi \\ 0 \end{bmatrix} \\
&= \hat{I} - t\dot{\varphi}\hat{J} \qquad (5.360)
\end{aligned}
$$

which shows

$$\frac{^G d}{dt}\frac{^B d\mathbf{r}}{dt} = -\left(\dot{\varphi} + t\ddot{\varphi}\right)\hat{J} \neq \frac{^B d}{dt}\frac{^G d\mathbf{r}}{dt}$$

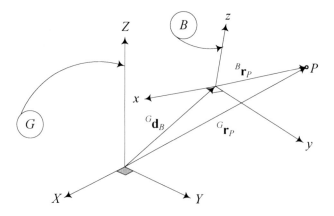

FIGURE 5.9. A rigid body with an attached coordinate frame $B\,(oxyz)$ moving freely in a global coordinate frame $G(OXYZ)$.

5.11   Rigid Body Velocity

Consider a rigid body with an attached local coordinate frame $B\,(oxyz)$ moving freely in a fixed global coordinate frame $G(OXYZ)$, as shown in Figure 5.9. The rigid body can rotate in the global frame, while the origin of the body frame B can translate relative to the origin of G. The coordinates of a body point P in local and global frames are related by the following equation:

$$^{G}\mathbf{r}_P = {}^{G}R_B \,{}^{B}\mathbf{r}_P + {}^{G}\mathbf{d}_B \tag{5.361}$$

where $^{G}\mathbf{d}_B$ indicates the position of the moving origin o relative to the fixed origin O.

The velocity of the point P in G is

$$
\begin{aligned}
^{G}\mathbf{v}_P &= {}^{G}\dot{\mathbf{r}}_P = {}^{G}\dot{R}_B \,{}^{B}\mathbf{r}_P + {}^{G}\dot{\mathbf{d}}_B = {}_{G}\tilde{\omega}_B \,{}^{G}_{B}\mathbf{r}_P + {}^{G}\dot{\mathbf{d}}_B \\
&= {}_{G}\tilde{\omega}_B \left({}^{G}\mathbf{r}_P - {}^{G}\mathbf{d}_B\right) + {}^{G}\dot{\mathbf{d}}_B \\
&= {}_{G}\boldsymbol{\omega}_B \times \left({}^{G}\mathbf{r}_P - {}^{G}\mathbf{d}_B\right) + {}^{G}\dot{\mathbf{d}}_B
\end{aligned} \tag{5.362}
$$

Proof. Direct differentiating shows

$$
\begin{aligned}
^{G}\mathbf{v}_P &= \frac{{}^{G}d}{dt}\,{}^{G}\mathbf{r}_P = {}^{G}\dot{\mathbf{r}}_P = \frac{{}^{G}d}{dt}\left({}^{G}R_B \,{}^{B}\mathbf{r}_P + {}^{G}\mathbf{d}_B\right) \\
&= {}^{G}\dot{R}_B \,{}^{B}\mathbf{r}_P + {}^{G}\dot{\mathbf{d}}_B
\end{aligned} \tag{5.363}
$$

The local position vector $^{B}\mathbf{r}_P$ can be substituted from (5.361) to obtain

$$
\begin{aligned}
^{G}\mathbf{v}_P &= {}^{G}\dot{R}_B \,{}^{G}R_B^{T}\left({}^{G}\mathbf{r}_P - {}^{G}\mathbf{d}_B\right) + {}^{G}\dot{\mathbf{d}}_B \\
&= {}_{G}\tilde{\omega}_B \left({}^{G}\mathbf{r}_P - {}^{G}\mathbf{d}_B\right) + {}^{G}\dot{\mathbf{d}}_B \\
&= {}_{G}\boldsymbol{\omega}_B \times \left({}^{G}\mathbf{r}_P - {}^{G}\mathbf{d}_B\right) + {}^{G}\dot{\mathbf{d}}_B
\end{aligned} \tag{5.364}
$$

It may also be written using relative position vector

$$^G\mathbf{v}_P = {}_G\boldsymbol{\omega}_B \times {}_B^G\mathbf{r}_P + {}^G\dot{\mathbf{d}}_B \qquad (5.365)$$

■

Example 205 *Geometric interpretation of rigid body velocity.*

Consider a body point P of a moving rigid body. The global velocity of point P

$$^G\mathbf{v}_P = {}_G\boldsymbol{\omega}_B \times {}_B^G\mathbf{r}_P + {}^G\dot{\mathbf{d}}_B$$

is a vector addition of rotational and translational velocities, both expressed in the global frame. At the moment, the body frame is assumed to be coincident with the global frame, and the body frame has a velocity $^G\dot{\mathbf{d}}_B$ with respect to the global frame. The translational velocity $^G\dot{\mathbf{d}}_B$ is a common property for every point of the body, but the rotational velocity ${}_G\boldsymbol{\omega}_B \times {}_B^G\mathbf{r}_P$ differs for different points of the body.

Example 206 *Velocity of a moving point in a moving body frame.*

Assume that point P in Figure 5.9 is moving in frame B, indicated by time varying position vector $^B\mathbf{r}_P(t)$. The global velocity of P is a composition of the velocity of P in B, rotation of B relative to G, and velocity of B relative to G.

$$
\begin{aligned}
\frac{^G d}{dt}{}^G\mathbf{r}_P &= \frac{^G d}{dt}\left({}^G\mathbf{d}_B + {}^G R_B \,{}^B\mathbf{r}_P\right) \\
&= \frac{^G d}{dt}{}^G\mathbf{d}_B + \frac{^G d}{dt}\left({}^G R_B \,{}^B\mathbf{r}_P\right) \\
&= {}^G\dot{\mathbf{d}}_B + {}_B^G\dot{\mathbf{r}}_P + {}_G\boldsymbol{\omega}_B \times {}_B^G\mathbf{r}_P \qquad (5.366)
\end{aligned}
$$

Example 207 *Velocity of a body point in multiple coordinate frames.*

Consider three frames, B_0, B_1 and B_2, as shown in Figure 5.10. Let us analyze the velocity of point P. If the point is stationary in a frame, say B_2, then the time derivative of $^2\mathbf{r}_P$ in B_2 is zero. If frame B_2 is moving relative to frame B_1, then, the time derivative of $^1\mathbf{r}_P$ is a combination of the rotational component due to rotation of B_2 relative to B_1 and the velocity of B_2 relative to B_1. In forward velocity kinematics, the velocities must be measured in the base frame B_0. Therefore, the velocity of point P in the base frame is a combination of the velocity of B_2 relative to B_1 and the velocity of B_1 relative to B_0.

The global coordinate of the body point P is

$$^0\mathbf{r}_P = {}^0\mathbf{d}_1 + {}_1^0\mathbf{d}_2 + {}_2^0\mathbf{r}_P = {}^0\mathbf{d}_1 + {}^0R_1 \,{}^1\mathbf{d}_2 + {}^0R_2 \,{}^2\mathbf{r}_P \qquad (5.367)$$

Therefore, the velocity of point P can be found by combining the relative velocities

$$
\begin{aligned}
^0\dot{\mathbf{r}}_P &= {}^0\dot{\mathbf{d}}_1 + \left({}^0\dot{R}_1 \,{}^1\mathbf{d}_2 + {}^0R_1 \,{}^1\dot{\mathbf{d}}_2\right) + {}^0\dot{R}_2 \,{}^2\mathbf{r}_P \\
&= {}^0\dot{\mathbf{d}}_1 + {}_0^0\boldsymbol{\omega}_1 \times {}_1^0\mathbf{d}_2 + {}^0R_1 \,{}^1\dot{\mathbf{d}}_2 + {}_0^0\boldsymbol{\omega}_2 \times {}_2^0\mathbf{r}_P \qquad (5.368)
\end{aligned}
$$

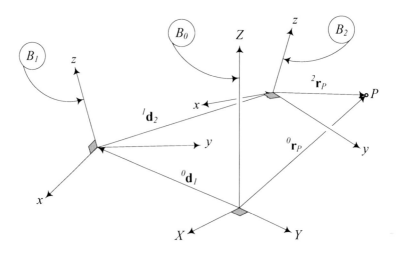

FIGURE 5.10. A rigid body coordinate frame B_2 is moving in a frame B_1 that is moving in the base coordinate frame B_0.

It is usually easier to use the relative velocity method and write

$$ {}_0^0\mathbf{v}_P = {}_0^0\mathbf{v}_1 + {}_1^0\mathbf{v}_2 + {}_2^0\mathbf{v}_P \tag{5.369} $$

because

$$ {}_0^0\mathbf{v}_1 = {}_0^0\dot{\mathbf{d}}_1 \qquad {}_1^0\mathbf{v}_2 = {}_0^0\boldsymbol{\omega}_1 \times {}_1^0\mathbf{d}_2 + {}^0R_1\,{}_1^1\dot{\mathbf{d}}_2 \qquad {}_2^0\mathbf{v}_P = {}_0^0\boldsymbol{\omega}_2 \times {}_2^0\mathbf{r}_P \tag{5.370} $$

and therefore,

$$ {}^0\mathbf{v}_P = {}^0\dot{\mathbf{d}}_1 + {}_0^0\boldsymbol{\omega}_1 \times {}_1^0\mathbf{d}_2 + {}^0R_1\,{}_1^1\dot{\mathbf{d}}_2 + {}_0^0\boldsymbol{\omega}_2 \times {}_2^0\mathbf{r}_P \tag{5.371} $$

Example 208 *Velocity vectors in different coordinate frames.*

To express the velocity vectors in different coordinate frames we need only to premultiply them by a rotation matrix. Hence, considering ${}_j^k\mathbf{v}_i$ as the velocity of the origin of the B_i coordinate frame with respect to the origin of frame B_j expressed in frame B_k, we can write

$$ {}_j^k\mathbf{v}_i = -{}_i^k\mathbf{v}_j \tag{5.372} $$

and

$$ {}_j^k\mathbf{v}_i = {}^kR_m\,{}_j^m\mathbf{v}_i \tag{5.373} $$

and therefore,

$$ \frac{{}^id}{dt}\,{}_i^i\mathbf{r}_P = {}^i\mathbf{v}_P = {}_j^i\mathbf{v}_P + {}_i^i\boldsymbol{\omega}_j \times {}_j^i\mathbf{r}_P \tag{5.374} $$

Example 209 ★ *Zero velocity points.*

To answer whether there is always a point with zero velocity, we may utilize Equation (5.364) and write

$$_G\tilde{\omega}_B \left({}^G\mathbf{r}_0 - {}^G\mathbf{d}_B \right) + {}^G\dot{\mathbf{d}}_B = 0 \tag{5.375}$$

to search for ${}^G\mathbf{r}_0$ *which refers to a point with zero velocity*

$$^G\mathbf{r}_0 = {}^G\mathbf{d}_B - {}_G\tilde{\omega}_B^{-1} \, {}^G\dot{\mathbf{d}}_B \tag{5.376}$$

however, the skew symmetric matrix $_G\tilde{\omega}_B$ *is singular and has no inverse. In other words, there is no general solution for Equation (5.375).*

If we restrict ourselves to planar motions, say XY-plane, then $_G\boldsymbol{\omega}_B = \omega\hat{K}$ and $_G\tilde{\omega}_B^{-1} = 1/\omega$. Hence, in 2D space there is a point at any time with zero velocity at position ${}^G\mathbf{r}_0$ *given by*

$$^G\mathbf{r}_0(t) = {}^G\mathbf{d}_B(t) - \frac{1}{\omega}\, {}^G\dot{\mathbf{d}}_B(t) \tag{5.377}$$

*The zero velocity point is called the **pole** or **instantaneous center of rotation**. The position of the pole is generally a function of time and the path of its motion is called a **centroid**.*

Example 210 ★ *Eulerian and Lagrangian view points.*

*When a variable quantity is measured within the stationary global coordinate frame, it is called absolute or the **Lagrangian** viewpoint. When the variable is measured within a moving body coordinate frame, it is called relative or the **Eulerian** viewpoint.*

In 2D planar motion of a rigid body, there is always a pole of zero velocity at

$$^G\mathbf{r}_0 = {}^G\mathbf{d}_B - \frac{1}{\omega}\, {}^G\dot{\mathbf{d}}_B \tag{5.378}$$

The position of the pole in the body coordinate frame can be found by substituting for ${}^G\mathbf{r}$ from (5.361)

$$^G R_B \, {}^B\mathbf{r}_0 + {}^G\mathbf{d}_B = {}^G\mathbf{d}_B - {}_G\tilde{\omega}_B^{-1}\, {}^G\dot{\mathbf{d}}_B \tag{5.379}$$

and solving for the position of the zero velocity point in the body coordinate frame ${}^B\mathbf{r}_0$.

$$
\begin{aligned}
^B\mathbf{r}_0 &= -{}^G R_B^T \, {}_G\tilde{\omega}_B^{-1}\, {}^G\dot{\mathbf{d}}_B = -{}^G R_B^T \left[{}^G\dot{R}_B \, {}^G R_B^T \right]^{-1} {}^G\dot{\mathbf{d}}_B \\
&= -{}^G R_B^T \left[{}^G R_B \, {}^G\dot{R}_B^{-1} \right] {}^G\dot{\mathbf{d}}_B = -{}^G\dot{R}_B^{-1}\, {}^G\dot{\mathbf{d}}_B
\end{aligned} \tag{5.380}
$$

Therefore, ${}^G\mathbf{r}_0$ indicates the path of motion of the pole in the global frame, while ${}^B\mathbf{r}_0$ indicates the same path in the body frame. The ${}^G\mathbf{r}_0$ refers to the Lagrangian centroid and ${}^B\mathbf{r}_0$ refers to the Eulerian centroid.

5.12   Angular Acceleration

Consider a rotating rigid body $B(Oxyz)$ with a fixed point O in a reference frame $G(OXYZ)$. Equation (5.228), for the velocity vector of a point in a fixed origin body frame,

$$^G\dot{\mathbf{r}}(t) = {}^G\mathbf{v}(t) = {}_G\tilde{\omega}_B \, {}^G\mathbf{r}(t) = {}_G\boldsymbol{\omega}_B \times {}^G\mathbf{r}(t) \qquad (5.381)$$

can be utilized to find the acceleration vector of the body point

$$
\begin{aligned}
^G\ddot{\mathbf{r}} &= \frac{^Gd}{dt}{}^G\dot{\mathbf{r}}(t) = {}_G\boldsymbol{\alpha}_B \times {}^G\mathbf{r} + {}_G\boldsymbol{\omega}_B \times \left({}_G\boldsymbol{\omega}_B \times {}^G\mathbf{r} \right) && (5.382) \\
&= \left(\ddot{\phi}\hat{u} + \dot{\phi}\dot{\hat{u}} \right) \times {}^G\mathbf{r} + \dot{\phi}^2 \, \hat{u} \times \left(\hat{u} \times {}^G\mathbf{r} \right) && (5.383)
\end{aligned}
$$

$_G\boldsymbol{\alpha}_B$ is the *angular acceleration vector* of the body with respect to the G frame.

$$_G\boldsymbol{\alpha}_B = \frac{^Gd}{dt}{}_G\boldsymbol{\omega}_B \qquad (5.384)$$

Proof. Differentiating Equation (5.381) gives

$$
\begin{aligned}
^G\mathbf{a} &= {}^G\dot{\mathbf{v}} = {}^G\ddot{\mathbf{r}} = {}_G\dot{\boldsymbol{\omega}}_B \times {}^G\mathbf{r} + {}_G\boldsymbol{\omega}_B \times {}^G\dot{\mathbf{r}} \\
&= {}_G\boldsymbol{\alpha}_B \times {}^G\mathbf{r} + {}_G\boldsymbol{\omega}_B \times \left({}_G\boldsymbol{\omega}_B \times {}^G\mathbf{r} \right) && (5.385)
\end{aligned}
$$

and because

$$
\begin{aligned}
\boldsymbol{\omega} &= \dot{\phi}\hat{u} && (5.386) \\
\boldsymbol{\alpha} &= \ddot{\phi}\hat{u} + \dot{\phi}\dot{\hat{u}} && (5.387)
\end{aligned}
$$

we derive Equation (5.383). Therefore, the position, velocity, and acceleration vectors of a body point are

$$^B\mathbf{r}_P = x\hat{i} + y\hat{j} + z\hat{k} \qquad (5.388)$$

$$^G\mathbf{v}_P = {}^G\dot{\mathbf{r}}_P = \frac{^Gd}{dt}{}^B\mathbf{r}_P = {}_G\boldsymbol{\omega}_B \times {}^G\mathbf{r} \qquad (5.389)$$

$$
\begin{aligned}
^G\mathbf{a}_P &= {}^G\dot{\mathbf{v}}_P = {}^G\ddot{\mathbf{r}}_P = \frac{^Gd^2}{dt^2}{}^B\mathbf{r}_P = {}_G\boldsymbol{\alpha}_B \times {}^G\mathbf{r} + {}_G\boldsymbol{\omega}_B \times {}^G\dot{\mathbf{r}} \\
&= {}_G\boldsymbol{\alpha}_B \times {}^G\mathbf{r} + {}_G\boldsymbol{\omega}_B \times \left({}_G\boldsymbol{\omega}_B \times {}^G\mathbf{r} \right) && (5.390)
\end{aligned}
$$

The angular acceleration expressed in the body frame is the body derivative of the angular velocity vector. To show this, we use the derivative transport formula (5.327)

$$^B_G\boldsymbol{\alpha}_B = \frac{^Gd}{dt}{}^B_G\boldsymbol{\omega}_B = \frac{^Bd}{dt}{}^B_G\boldsymbol{\omega}_B + {}^B_G\boldsymbol{\omega}_B \times {}^B_G\boldsymbol{\omega}_B = \frac{^Bd}{dt}{}^B_G\boldsymbol{\omega}_B = {}^B_G\dot{\boldsymbol{\omega}}_B \qquad (5.391)$$

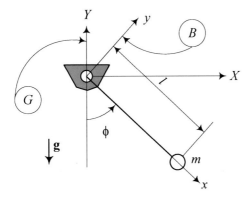

FIGURE 5.11. A simple pendulum.

The angular acceleration of B in G can always be expressed in the form

$$_G\boldsymbol{\alpha}_B = {}_G\alpha_B\,\hat{u}_\alpha \qquad (5.392)$$

where $\hat{u}_\alpha$ is a unit vector parallel to $_G\boldsymbol{\alpha}_B$. The angular velocity and angular acceleration vectors are not parallel in general, and therefore,

$$\hat{u}_\alpha \;\neq\; \hat{u}_\omega \qquad (5.393)$$
$$_G\boldsymbol{\alpha}_B \;\neq\; {}_G\dot{\boldsymbol{\omega}}_B \qquad (5.394)$$

However, the only special case is when the axis of rotation is fixed in both G and B frames. In this case

$$_G\boldsymbol{\alpha}_B = \alpha\,\hat{u} = \dot{\omega}\,\hat{u} = \ddot{\phi}\,\hat{u} \qquad (5.395)$$

■

Example 211 *Velocity and acceleration of a simple pendulum.*
A point mass attached to a massless rod and hanging from a revolute joint is called a simple pendulum. Figure 5.11 illustrates a simple pendulum. A local coordinate frame B is attached to the pendulum that rotates in a global frame G. The position vector of the bob and the angular velocity vector $_G\boldsymbol{\omega}_B$ are

$$^B\mathbf{r} = l\hat{\imath} \qquad ^G\mathbf{r} = {}^GR_B\,{}^B\mathbf{r} = \begin{bmatrix} l\sin\phi \\ -l\cos\phi \\ 0 \end{bmatrix} \qquad (5.396)$$

$$_G^B\boldsymbol{\omega}_B = \dot{\phi}\hat{k} \qquad _G\boldsymbol{\omega}_B = {}^GR_B^{T}\,{}_G^B\boldsymbol{\omega}_B = \dot{\phi}\hat{K} \qquad (5.397)$$

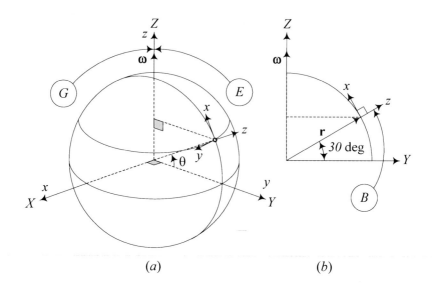

FIGURE 5.12. The motion of a vehicle at 30 deg latitude and heading north on the Earth.

$$
{}^G R_B = \begin{bmatrix} \cos\left(\frac{3}{2}\pi + \phi\right) & -\sin\left(\frac{3}{2}\pi + \phi\right) & 0 \\ \sin\left(\frac{3}{2}\pi + \phi\right) & \cos\left(\frac{3}{2}\pi + \phi\right) & 0 \\ 0 & 0 & 1 \end{bmatrix}
$$

$$
= \begin{bmatrix} \sin\phi & \cos\phi & 0 \\ -\cos\phi & \sin\phi & 0 \\ 0 & 0 & 1 \end{bmatrix} \tag{5.398}
$$

Its velocity is therefore given by

$$
{}^B_G \mathbf{v} = {}^B \dot{\mathbf{r}} + {}^B_G \boldsymbol{\omega}_B \times {}^B_G \mathbf{r} = 0 + \dot{\phi}\hat{k} \times l\hat{i} = l\dot{\phi}\hat{j} \tag{5.399}
$$

$$
{}^G \mathbf{v} = {}^G R_B \, {}^B \mathbf{v} = \begin{bmatrix} l\dot{\phi}\cos\phi \\ l\dot{\phi}\sin\phi \\ 0 \end{bmatrix} \tag{5.400}
$$

The acceleration of the bob is then equal to

$$
{}^B_G \mathbf{a} = {}^B_G \dot{\mathbf{v}} + {}^B_G \boldsymbol{\omega}_B \times {}^B_G \mathbf{v} = l\ddot{\phi}\hat{j} + \dot{\phi}\hat{k} \times l\dot{\phi}\hat{j} = l\ddot{\phi}\hat{j} - l\dot{\phi}^2\hat{i} \tag{5.401}
$$

$$
{}^G \mathbf{a} = {}^G R_B \, {}^B \mathbf{a} = \begin{bmatrix} l\ddot{\phi}\cos\phi - l\dot{\phi}^2\sin\phi \\ l\ddot{\phi}\sin\phi + l\dot{\phi}^2\cos\phi \\ 0 \end{bmatrix} \tag{5.402}
$$

Example 212 *Motion of a vehicle on the Earth.*

Consider the motion of a vehicle on the Earth at latitude 30 deg and heading north, as shown in Figure 5.12. The vehicle has the velocity $v =$

$_{E}^{B}\dot{\mathbf{r}} = 80$ km/ h $= 22.22$ m/ s *and acceleration* $a = {}_{E}^{B}\ddot{\mathbf{r}} = 0.1$ m/ s^2, *both with respect to the road. The radius of the Earth is* R, *and hence, the vehicle's kinematics are*

$$
\begin{aligned}
{}_{E}^{B}\mathbf{r} &= R\hat{k} \text{ m} & {}_{E}^{B}\dot{\mathbf{r}} = 22.22\hat{\imath} \text{ m/ s} & \qquad {}_{E}^{B}\ddot{\mathbf{r}} = 0.1\hat{\imath} \text{ m/ s}^2 & (5.403)\\
\dot{\theta} &= \frac{v}{R} \text{ rad/ s} & \ddot{\theta} = \frac{a}{R} \text{ rad/ s}^2 & & (5.404)
\end{aligned}
$$

There are three coordinate frames involved. A body coordinate frame B *is attached to the vehicle. A global coordinate* G *is set up at the center of the Earth. Another local coordinate frame* E *is rigidly attached to the Earth and turns with the Earth. The frames* E *and* G *are assumed coincident at the moment. The angular velocity of* B *is*

$$
\begin{aligned}
{}_{G}^{B}\boldsymbol{\omega}_{B} &= {}_{G}\boldsymbol{\omega}_{E} + {}_{E}^{G}\boldsymbol{\omega}_{B} = {}^{B}R_{G}\left(\omega_{E}\,\hat{K} + \dot{\theta}\hat{I}\right)\\
&= (\omega_{E}\cos\theta)\,\hat{\imath} + (\omega_{E}\sin\theta)\,\hat{k} + \dot{\theta}\hat{\jmath}\\
&= (\omega_{E}\cos\theta)\,\hat{\imath} + (\omega_{E}\sin\theta)\,\hat{k} + \frac{v}{R}\hat{\jmath} & (5.405)
\end{aligned}
$$

Therefore, the velocity and acceleration of the vehicle are

$$
\begin{aligned}
{}_{G}^{B}\mathbf{v} &= {}^{B}\dot{\mathbf{r}} + {}_{G}^{B}\boldsymbol{\omega}_{B} \times {}_{G}^{B}\mathbf{r} = 0 + {}_{G}^{B}\boldsymbol{\omega}_{B} \times R\hat{k}\\
&= v\hat{\imath} - (R\omega_{E}\cos\theta)\,\hat{\jmath} & (5.406)\\
{}_{G}^{B}\mathbf{a} &= {}_{G}^{B}\dot{\mathbf{v}} + {}_{G}^{B}\boldsymbol{\omega}_{B} \times {}_{G}^{B}\mathbf{v}\\
&= a\hat{\imath} + \left(R\omega_{E}\,\dot{\theta}\sin\theta\right)\hat{\jmath} + \begin{bmatrix} \omega_{E}\cos\theta \\ \dfrac{v}{R} \\ \omega_{E}\sin\theta \end{bmatrix} \times \begin{bmatrix} v \\ -R\omega_{E}\cos\theta \\ 0 \end{bmatrix}\\
&= \begin{bmatrix} a + R\omega_{E}^{2}\,\cos\theta\sin\theta \\ 2R\omega_{E}\,\dot{\theta}\sin\theta \\ -\frac{1}{R}v^{2} - R\omega_{E}^{2}\cos^{2}\theta \end{bmatrix} & (5.407)
\end{aligned}
$$

The term $a\hat{\imath}$ *is the acceleration relative to Earth,* $(2R\omega_{E}\,\dot{\theta}\sin\theta)\hat{\jmath}$ *is the Coriolis acceleration,* $-\frac{v^{2}}{R}\hat{k}$ *is the centrifugal acceleration due to traveling, and* $-(R\omega_{E}^{2}\cos^{2}\theta)$ *is centrifugal acceleration due to Earth's rotation.*

Substituting the numerical values and accepting $R = 6.3677 \times 10^{6}$ m *provides*

$$
\begin{aligned}
{}_{G}^{B}\mathbf{v} &= 22.22\hat{\imath} - 6.3677 \times 10^{6} \left(\frac{2\pi}{24 \times 3600}\frac{366.25}{365.25}\right) \cos\frac{\pi}{6}\hat{\jmath}\\
&= 22.22\hat{\imath} - 402.13\hat{\jmath} \text{ m/ s} & (5.408)\\
{}_{G}^{B}\mathbf{a} &= 1.5662 \times 10^{-2}\hat{\imath} + 1.6203 \times 10^{-3}\hat{\jmath} - 2.5473 \times 10^{-2}\hat{k} \text{ m/ s}^2 & (5.409)
\end{aligned}
$$

Example 213 ★ *Combination of angular accelerations.*

It is shown that the angular velocity of several bodies rotating relative to each other can be related according to (5.277)

$$_0\boldsymbol{\omega}_n = {_0\boldsymbol{\omega}_1} + {_1^0\boldsymbol{\omega}_2} + {_2^0\boldsymbol{\omega}_3} + \cdots + {_{n-1}^0\boldsymbol{\omega}_n} \tag{5.410}$$

The angular accelerations of several relatively rotating rigid bodies follow the same rule:

$$_0\boldsymbol{\alpha}_n = {_0\boldsymbol{\alpha}_1} + {_1^0\boldsymbol{\alpha}_2} + {_2^0\boldsymbol{\alpha}_3} + \cdots + {_{n-1}^0\boldsymbol{\alpha}_n} \tag{5.411}$$

Let us consider a pair of relatively rotating rigid links in a base coordinate frame B_0 with a fixed point at O. The angular velocities of the links are related by

$$_0\boldsymbol{\omega}_2 = {_0\boldsymbol{\omega}_1} + {_1^0\boldsymbol{\omega}_2} \tag{5.412}$$

So, their angular accelerations are

$$_0\boldsymbol{\alpha}_1 = \frac{^0d}{dt} {_0\boldsymbol{\omega}_1} \tag{5.413}$$

$$_0\boldsymbol{\alpha}_2 = \frac{^0d}{dt} {_0\boldsymbol{\omega}_2} = {_0\boldsymbol{\alpha}_1} + {_1^0\boldsymbol{\alpha}_2} \tag{5.414}$$

Example 214 ★ *Angular acceleration and Euler angles.*

The angular velocity $_G^B\boldsymbol{\omega}_B$ in terms of Euler angles is

$$_G^G\boldsymbol{\omega}_B = \begin{bmatrix} \omega_X \\ \omega_Y \\ \omega_Z \end{bmatrix} = \begin{bmatrix} 0 & \cos\varphi & \sin\theta\sin\varphi \\ 0 & \sin\varphi & -\cos\varphi\sin\theta \\ 1 & 0 & \cos\theta \end{bmatrix} \begin{bmatrix} \dot\varphi \\ \dot\theta \\ \dot\psi \end{bmatrix}$$

$$= \begin{bmatrix} \dot\theta\cos\varphi + \dot\psi\sin\theta\sin\varphi \\ \dot\theta\sin\varphi - \dot\psi\cos\varphi\sin\theta \\ \dot\varphi + \dot\psi\cos\theta \end{bmatrix} \tag{5.415}$$

The angular acceleration is then equal to

$$_G^G\boldsymbol{\alpha}_B = \frac{^Gd}{dt} {_G^G\boldsymbol{\omega}_B} \tag{5.416}$$

$$= \begin{bmatrix} \cos\varphi\left(\ddot\theta + \dot\varphi\dot\psi\sin\theta\right) + \sin\varphi\left(\ddot\psi\sin\theta + \dot\theta\dot\psi\cos\theta - \dot\theta\dot\varphi\right) \\ \sin\varphi\left(\ddot\theta + \dot\varphi\dot\psi\sin\theta\right) + \cos\varphi\left(\dot\theta\dot\varphi - \ddot\psi\sin\theta - \dot\theta\dot\psi\cos\theta\right) \\ \ddot\varphi + \ddot\psi\cos\theta - \dot\theta\dot\psi\sin\theta \end{bmatrix}$$

The angular acceleration vector in the body coordinate frame is then equal

to

$$\begin{aligned}
{}^B_G\boldsymbol{\alpha}_B &= {}^G R_B^T {}^G_G\boldsymbol{\alpha}_B \qquad\qquad\qquad\qquad\qquad\qquad (5.417)\\[2mm]
&= \begin{bmatrix} c\varphi c\psi - c\theta s\varphi s\psi & c\psi s\varphi + c\theta c\varphi s\psi & s\theta s\psi \\ -c\varphi s\psi - c\theta c\psi s\varphi & -s\varphi s\psi + c\theta c\varphi c\psi & s\theta c\psi \\ s\theta s\varphi & -c\varphi s\theta & c\theta \end{bmatrix} {}^G_G\boldsymbol{\alpha}_B \\[2mm]
&= \begin{bmatrix} \cos\psi\left(\ddot\theta + \dot\varphi\dot\psi\sin\theta\right) + \sin\psi\left(\ddot\varphi\sin\theta + \dot\theta\dot\varphi\cos\theta - \dot\theta\dot\psi\right) \\ \cos\psi\left(\ddot\varphi\sin\theta + \dot\theta\dot\varphi\cos\theta - \dot\theta\dot\psi\right) - \sin\psi\left(\ddot\theta + \dot\varphi\dot\psi\sin\theta\right) \\ \ddot\varphi\cos\theta - \ddot\psi - \dot\theta\dot\varphi\sin\theta \end{bmatrix}
\end{aligned}$$

5.13 Rigid Body Acceleration

Consider a rigid body with an attached local coordinate frame $B\,(oxyz)$ moving freely in a fixed global coordinate frame $G(OXYZ)$. The rigid body can rotate in the global frame, while the origin of the body frame B can translate relative to the origin of G. The coordinates of a body point P in local and global frames, as shown in Figure 5.13, are related by the equation

$$ {}^G\mathbf{r}_P = {}^G R_B \, {}^B\mathbf{r}_P + {}^G\mathbf{d}_B \qquad\qquad (5.418)$$

where ${}^G\mathbf{d}_B$ indicates the position of the moving origin o relative to the fixed origin O.

The acceleration of point P in G is

$$\begin{aligned}
{}^G\mathbf{a}_P &= {}^G\dot{\mathbf{v}}_P = {}^G\ddot{\mathbf{r}}_P = {}_G\boldsymbol{\alpha}_B \times \left({}^G\mathbf{r}_P - {}^G\mathbf{d}_B\right)\\
&\quad + {}_G\boldsymbol{\omega}_B \times \left({}_G\boldsymbol{\omega}_B \times \left({}^G\mathbf{r}_P - {}^G\mathbf{d}_B\right)\right) + {}^G\ddot{\mathbf{d}}_B \qquad (5.419)
\end{aligned}$$

Proof. The acceleration of point P is a consequence of differentiating the velocity equation (5.364) or (5.365).

$$\begin{aligned}
{}^G\mathbf{a}_P &= \frac{{}^Gd}{dt}{}^G\mathbf{v}_P = {}_G\boldsymbol{\alpha}_B \times {}^G_B\mathbf{r}_P + {}_G\boldsymbol{\omega}_B \times {}^G_B\dot{\mathbf{r}}_P + {}^G\ddot{\mathbf{d}}_B\\
&= {}_G\boldsymbol{\alpha}_B \times {}^G_B\mathbf{r}_P + {}_G\boldsymbol{\omega}_B \times \left({}_G\boldsymbol{\omega}_B \times {}^G_B\mathbf{r}_P\right) + {}^G\ddot{\mathbf{d}}_B\\
&= {}_G\boldsymbol{\alpha}_B \times \left({}^G\mathbf{r}_P - {}^G\mathbf{d}_B\right)\\
&\quad + {}_G\boldsymbol{\omega}_B \times \left({}_G\boldsymbol{\omega}_B \times \left({}^G\mathbf{r}_P - {}^G\mathbf{d}_B\right)\right) + {}^G\ddot{\mathbf{d}}_B \qquad (5.420)
\end{aligned}$$

The term ${}_G\boldsymbol{\omega}_B \times \left({}_G\boldsymbol{\omega}_B \times {}^G_B\mathbf{r}_P\right)$ is called *centripetal acceleration* and is independent of the angular acceleration. The term ${}_G\boldsymbol{\alpha}_B \times {}^G_B\mathbf{r}_P$ is called *tangential acceleration* and is perpendicular to ${}^G_B\mathbf{r}_P$. ∎

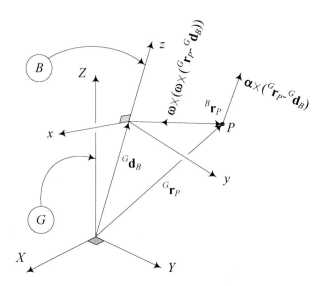

FIGURE 5.13. A rigid body with coordinate frame $B\,(oxyz)$ moving freely in a fixed global coordinate frame $G(OXYZ)$.

Example 215 *Acceleration of joint 2 of a 2R planar manipulator.*

A 2R planar manipulator is illustrated in Figure 5.14. The elbow joint has a circular motion about the base joint. Knowing that

$$_0\boldsymbol{\omega}_1 = \dot{\theta}_1 \, {}^0\hat{k}_0 \tag{5.421}$$

we can write

$$
\begin{aligned}
_0\boldsymbol{\alpha}_1 &= {}_0\dot{\boldsymbol{\omega}}_1 = \ddot{\theta}_1 \, {}^0\hat{k}_0 & (5.422)\\
_0\dot{\boldsymbol{\omega}}_1 \times {}^0\mathbf{r}_1 &= \ddot{\theta}_1 \, {}^0\hat{k}_0 \times {}^0\mathbf{r}_1 = \ddot{\theta}_1 \, R_{Z,\theta+90} \, {}^0\mathbf{r}_1 & (5.423)\\
_0\boldsymbol{\omega}_1 \times \left({}_0\boldsymbol{\omega}_1 \times {}^0\mathbf{r}_1\right) &= -\dot{\theta}_1^2 \, {}^0\mathbf{r}_1 & (5.424)
\end{aligned}
$$

and calculate the acceleration of the elbow joint

$$^0\ddot{\mathbf{r}}_1 = \ddot{\theta}_1 \, R_{Z,\theta+90} \, {}^0\mathbf{r}_1 - \dot{\theta}_1^2 \, {}^0\mathbf{r}_1 \tag{5.425}$$

Example 216 *Acceleration of a moving point in a moving body frame.*
Assume the point P in Figure 5.13 is indicated by a time varying local position vector ${}^B\mathbf{r}_P(t)$. Then, the velocity and acceleration of P can be found by applying the derivative transformation formula (5.327).

$$
\begin{aligned}
{}^G\mathbf{v}_P &= {}^G\dot{\mathbf{d}}_B + {}^B\dot{\mathbf{r}}_P + {}_G^B\boldsymbol{\omega}_B \times {}^B\mathbf{r}_P \\
&= {}^G\dot{\mathbf{d}}_B + {}^B\mathbf{v}_P + {}_G^B\boldsymbol{\omega}_B \times {}^B\mathbf{r}_P & (5.426)
\end{aligned}
$$

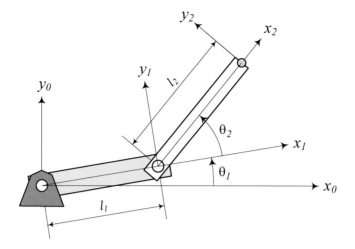

FIGURE 5.14. A 2R planar manipulator.

$$
\begin{aligned}
{}^G\mathbf{a}_P &= {}^G\ddot{\mathbf{d}}_B + {}^B\ddot{\mathbf{r}}_P + {}^B_G\boldsymbol{\omega}_B \times {}^B\dot{\mathbf{r}}_P + {}^B_G\dot{\boldsymbol{\omega}}_B \times {}^B\mathbf{r}_P \\
&\quad + {}^B_G\boldsymbol{\omega}_B \times \left({}^B\dot{\mathbf{r}}_P + {}^B_G\boldsymbol{\omega}_B \times {}^B\mathbf{r}_P \right) \\
&= {}^G\ddot{\mathbf{d}}_B + {}^B\mathbf{a}_P + 2{}^B_G\boldsymbol{\omega}_B \times {}^B\mathbf{v}_P + {}^B_G\dot{\boldsymbol{\omega}}_B \times {}^B\mathbf{r}_P \\
&\quad + {}^B_G\boldsymbol{\omega}_B \times \left({}^B_G\boldsymbol{\omega}_B \times {}^B\mathbf{r}_P \right) \qquad (5.427)
\end{aligned}
$$

It is also possible to take derivative from Equation (5.361) with the assumption ${}^B\dot{\mathbf{r}}_P \neq 0$ *and find the acceleration of P.*

$$
{}^G\mathbf{r}_P = {}^GR_B \, {}^B\mathbf{r}_P + {}^G\mathbf{d}_B \qquad (5.428)
$$

$$
\begin{aligned}
{}^G\dot{\mathbf{r}}_P &= {}^G\dot{R}_B \, {}^B\mathbf{r}_P + {}^GR_B \, {}^B\dot{\mathbf{r}}_P + {}^G\dot{\mathbf{d}}_B \\
&= {}_G\boldsymbol{\omega}_B \times {}^GR_B \, {}^B\mathbf{r}_P + {}^GR_B \, {}^B\dot{\mathbf{r}}_P + {}^G\dot{\mathbf{d}}_B \qquad (5.429)
\end{aligned}
$$

$$
\begin{aligned}
{}^G\ddot{\mathbf{r}}_P &= {}_G\dot{\boldsymbol{\omega}}_B \times {}^GR_B \, {}^B\mathbf{r}_P + {}_G\boldsymbol{\omega}_B \times {}^G\dot{R}_B \, {}^B\mathbf{r}_P + {}_G\boldsymbol{\omega}_B \times {}^GR_B \, {}^B\dot{\mathbf{r}}_P \\
&\quad + {}^G\dot{R}_B \, {}^B\dot{\mathbf{r}}_P + {}^GR_B \, {}^B\ddot{\mathbf{r}}_P + {}^G\ddot{\mathbf{d}}_B \\
&= {}_G\dot{\boldsymbol{\omega}}_B \times {}^G_B\mathbf{r}_P + {}_G\boldsymbol{\omega}_B \times \left({}_G\boldsymbol{\omega}_B \times {}^G\mathbf{r}_P \right) + 2 {}_G\boldsymbol{\omega}_B \times {}^G_B\dot{\mathbf{r}}_P \\
&\quad + {}^G_B\ddot{\mathbf{r}}_P + {}^G\ddot{\mathbf{d}}_B \qquad (5.430)
\end{aligned}
$$

The third term on the right-hand side is called the **Coriolis acceleration.** *The Coriolis acceleration is perpendicular to both* ${}_G\boldsymbol{\omega}_B$ *and* ${}^B\dot{\mathbf{r}}_P$.

Example 217 ★ *Acceleration of a body point.*

Consider a rigid body is moving and rotating in a global frame. The acceleration of a body point can be found by taking twice the time derivative of its position vector

$$^G\mathbf{r}_P = {}^GR_B \, {}^B\mathbf{r}_P + {}^G\mathbf{d}_B \tag{5.431}$$

$$^G\dot{\mathbf{r}}_P = {}^G\dot{R}_B \, {}^B\mathbf{r}_P + {}^G\dot{\mathbf{d}}_B \tag{5.432}$$

$$^G\ddot{\mathbf{r}}_P = {}^G\ddot{R}_B \, {}^B\mathbf{r}_P + {}^G\ddot{\mathbf{d}}_B$$

$$= {}^G\ddot{R}_B \, {}^GR_B^T \left({}^G\mathbf{r}_P - {}^G\mathbf{d}_B \right) + {}^G\ddot{\mathbf{d}}_B \tag{5.433}$$

Differentiating the angular velocity matrix

$$_G\tilde{\omega}_B = {}^G\dot{R}_B \, {}^GR_B^T \tag{5.434}$$

yields

$$_G\dot{\tilde{\omega}}_B = \frac{^Gd}{dt} \, _G\tilde{\omega}_B = {}^G\ddot{R}_B \, {}^GR_B^T + {}^G\dot{R}_B \, {}^G\dot{R}_B^T$$

$$= {}^G\ddot{R}_B \, {}^GR_B^T + {}_G\tilde{\omega}_B \, {}_G\tilde{\omega}_B^T \tag{5.435}$$

and therefore,

$$^G\ddot{R}_B \, {}^GR_B^T = {}_G\dot{\tilde{\omega}}_B - {}_G\tilde{\omega}_B \, {}_G\tilde{\omega}_B^T \tag{5.436}$$

The acceleration vector of the body point becomes

$$^G\ddot{\mathbf{r}}_P = \left({}_G\dot{\tilde{\omega}}_B - {}_G\tilde{\omega}_B \, {}_G\tilde{\omega}_B^T \right) \left({}^G\mathbf{r}_P - {}^G\mathbf{d}_B \right) + {}^G\ddot{\mathbf{d}}_B \tag{5.437}$$

where

$$_G\dot{\tilde{\omega}}_B = {}_G\tilde{\alpha}_B = \begin{bmatrix} 0 & -\dot{\omega}_3 & \dot{\omega}_2 \\ \dot{\omega}_3 & 0 & -\dot{\omega}_1 \\ -\dot{\omega}_2 & \dot{\omega}_1 & 0 \end{bmatrix} \tag{5.438}$$

and

$$_G\tilde{\omega}_B \, {}_G\tilde{\omega}_B^T = \begin{bmatrix} \omega_2^2 + \omega_3^2 & -\omega_1\omega_2 & -\omega_1\omega_3 \\ -\omega_1\omega_2 & \omega_1^2 + \omega_3^2 & -\omega_2\omega_3 \\ -\omega_1\omega_3 & -\omega_2\omega_3 & \omega_1^2 + \omega_2^2 \end{bmatrix} \tag{5.439}$$

5.14 ★ Screw Motion

Based on the *Chasles theorem*, any rigid body motion can be expressed by a single translation along an axis combined with a unique rotation about the axis. Such a motion is called *screw*. Consider the *screw* motion illustrated in Figure 5.15. Point P rotates about the screw axis indicated by $\hat{u}$ and simultaneously translates along the same axis. Hence, any point on the *screw axis* moves along the axis, while any point off the axis moves along a *helix*.

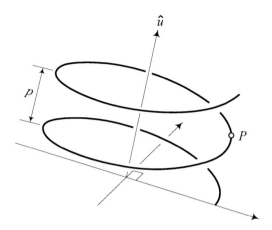

FIGURE 5.15. A screw motion is the translation along a line combined with a rotation about the line.

The angular rotation of the rigid body about the screw is called *twist*. The *Pitch* of a screw, p, is the ratio of *translation*, h, to *rotation*, ϕ.

$$p = \frac{h}{\phi} \tag{5.440}$$

So, pitch is the rectilinear distance through which the rigid body translates parallel to the axis of screw for a unit rotation. If $p > 0$, then the screw is *right-handed*, and if $p < 0$, it is *left-handed*.

A screw is shown by $\check{s}(h, \phi, \hat{u}, \mathbf{s})$ and is indicated by a unit vector $\hat{u}$, a *location* vector $\mathbf{s}$, a twist angle ϕ, and a translation h (or pitch p). The location vector $\mathbf{s}$ indicates the global position of a point on the screw axis. The twist angle ϕ, the twist axis $\hat{u}$, and the translation h (or pitch p) are called *screw parameters*.

The screw is another transformation method to express the motion of a rigid body. A linear displacement along an axis combined with an angular displacement about the same axis arises in steering kinematics of vehicles. If $^B\mathbf{r}_P$ indicates the position vector of a body point, its position vector in the global frame after a screw motion is

$$^G\mathbf{r}_P = \check{s}(h, \phi, \hat{u}, \mathbf{s})\, ^B\mathbf{r}_P \tag{5.441}$$

that is equivalent to a translation $^G\mathbf{d}_B$ along with a rotation $^G R_B$.

$$^G\mathbf{r}_P = {}^G R_B\, ^B\mathbf{r}_P + {}^G\mathbf{d}_B \tag{5.442}$$

We may introduce a 4×4 matrix $[T]$, that is called the *homogeneous matrix*,

$$^G T_B = \begin{bmatrix} {}^G R_B & {}^G\mathbf{d} \\ 0 & 1 \end{bmatrix} \tag{5.443}$$

and combine the translation and rotation to express the motion with only a matrix multiplication

$$^G\mathbf{r}_P = {^GT_B}\,^B\mathbf{r}_P \tag{5.444}$$

where, $^G\mathbf{r}_P$ and $^B\mathbf{r}_P$ are expanded with an extra zero element to be consistent with the 4×4 matrix $[T]$.

$$^G\mathbf{r}_P = \begin{bmatrix} X \\ Y \\ Z \\ 0 \end{bmatrix} \qquad ^B\mathbf{r}_P = \begin{bmatrix} x \\ y \\ z \\ 0 \end{bmatrix} \tag{5.445}$$

Homogeneous matrix representation can be used for screw transformations to combine the screw rotation and screw translation about the screw axis.

If $\hat{u}$ passes through the origin of the coordinate frame, then $\mathbf{s} = 0$ and the screw motion is called *central screw* $\check{s}(h, \phi, \hat{u})$. For a central screw we have

$$^G\check{s}_B(h, \phi, \hat{u}) = D_{\hat{u},h}\, R_{\hat{u},\phi} \tag{5.446}$$

where,

$$D_{\hat{u},h} = \begin{bmatrix} 1 & 0 & 0 & hu_1 \\ 0 & 1 & 0 & hu_2 \\ 0 & 0 & 1 & hu_3 \\ 0 & 0 & 0 & 1 \end{bmatrix} \tag{5.447}$$

$$R_{\hat{u},\phi} = \begin{bmatrix} u_1^2 \operatorname{vers}\phi + c\phi & u_1 u_2 \operatorname{vers}\phi - u_3 s\phi & u_1 u_3 \operatorname{vers}\phi + u_2 s\phi & 0 \\ u_1 u_2 \operatorname{vers}\phi + u_3 s\phi & u_2^2 \operatorname{vers}\phi + c\phi & u_2 u_3 \operatorname{vers}\phi - u_1 s\phi & 0 \\ u_1 u_3 \operatorname{vers}\phi - u_2 s\phi & u_2 u_3 \operatorname{vers}\phi + u_1 s\phi & u_3^2 \operatorname{vers}\phi + c\phi & 0 \\ 0 & 0 & 0 & 1 \end{bmatrix} \tag{5.448}$$

and hence,

$$^G\check{s}_B(h, \phi, \hat{u}) = \begin{bmatrix} ^GR_B & ^G\mathbf{d} \\ 0 & 1 \end{bmatrix} =$$

$$\begin{bmatrix} u_1^2 \operatorname{vers}\phi + c\phi & u_1 u_2 \operatorname{vers}\phi - u_3 s\phi & u_1 u_3 \operatorname{vers}\phi + u_2 s\phi & hu_1 \\ u_1 u_2 \operatorname{vers}\phi + u_3 s\phi & u_2^2 \operatorname{vers}\phi + c\phi & u_2 u_3 \operatorname{vers}\phi - u_1 s\phi & hu_2 \\ u_1 u_3 \operatorname{vers}\phi - u_2 s\phi & u_2 u_3 \operatorname{vers}\phi + u_1 s\phi & u_3^2 \operatorname{vers}\phi + c\phi & hu_3 \\ 0 & 0 & 0 & 1 \end{bmatrix} \tag{5.449}$$

As a result, a central screw transformation matrix includes the pure or fundamental translations and rotations as special cases because a pure translation corresponds to $\phi = 0$, and a pure rotation corresponds to $h = 0$ (or $p = \infty$).

When the screw is not central and $\hat{u}$ is not passing through the origin, a screw motion to move $\mathbf{p}$ to $\mathbf{p}''$ is denoted by

$$\begin{aligned} \mathbf{p}'' &= (\mathbf{p} - \mathbf{s}) \cos\phi + (1 - \cos\phi)(\hat{u} \cdot (\mathbf{p} - \mathbf{s}))\,\hat{u} \\ &\quad + (\hat{u} \times (\mathbf{p} - \mathbf{s})) \sin\phi + \mathbf{s} + h\hat{u} \end{aligned} \tag{5.450}$$

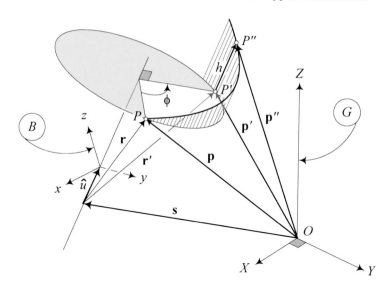

FIGURE 5.16. Screw motion of a rigid body.

or

$$\mathbf{p}'' = {}^{G}R_B\,(\mathbf{p} - \mathbf{s}) + \mathbf{s} + h\hat{u} = {}^{G}R_B\,\mathbf{p} + \mathbf{s} - {}^{G}R_B\,\mathbf{s} + h\hat{u} \qquad (5.451)$$

and therefore,

$$\mathbf{p}'' = \check{s}(h, \phi, \hat{u}, \mathbf{s})\mathbf{p} = [T]\,\mathbf{p} \qquad (5.452)$$

where

$$[T] = \begin{bmatrix} {}^{G}R_B & {}^{G}\mathbf{s} - {}^{G}R_B\,{}^{G}\mathbf{s} + h\hat{u} \\ 0 & 1 \end{bmatrix} = \begin{bmatrix} {}^{G}R_B & {}^{G}\mathbf{d} \\ 0 & 1 \end{bmatrix} \qquad (5.453)$$

The vector ${}^{G}\mathbf{s}$, called *location vector*, is the global position of the body frame before screw motion. The vectors $\mathbf{p}''$ and $\mathbf{p}$ are global positions of a point P after and before screw, as shown in Figure 5.16.

The screw axis is indicated by the unit vector $\hat{u}$. Now a body point P moves from its first position to its second position P' by a rotation about $\hat{u}$. Then it moves to P'' by a translation h parallel to $\hat{u}$. The initial position of P is pointed by $\mathbf{p}$ and its final position is pointed by $\mathbf{p}''$.

A screw motion is a four variable function $\check{s}(h, \phi, \hat{u}, \mathbf{s})$. A screw has a line of action $\hat{u}$ at ${}^{G}\mathbf{s}$, a twist ϕ, and a translation h.

The instantaneous screw axis was first used by Mozzi (1730 − 1813) in 1763 although Chasles (1793 − 1880) is credited with this discovery.

Proof. The angle-axis rotation formula (5.185) relates $\mathbf{r}'$ and $\mathbf{r}$, which are position vectors of P after and before rotation ϕ about $\hat{u}$ when $\mathbf{s} = 0$, $h = 0$.

$$\mathbf{r}' = \mathbf{r}\cos\phi + (1 - \cos\phi)\,(\hat{u}\cdot\mathbf{r})\,\hat{u} + (\hat{u}\times\mathbf{r})\sin\phi \qquad (5.454)$$

However, when the screw axis does not pass through the origin of $G(OXYZ)$, then $\mathbf{r}'$ and $\mathbf{r}$ must accordingly be substituted with:

$$\mathbf{r} = \mathbf{p} - \mathbf{s} \tag{5.455}$$
$$\mathbf{r}' = \mathbf{p}'' - \mathbf{s} - h\hat{u} \tag{5.456}$$

where $\mathbf{r}'$ is a vector after rotation and hence in G coordinate frame, and $\mathbf{r}$ is a vector before rotation and hence in B coordinate frame.

Therefore, the relationship between the new and old positions of the body point P after a screw motion is

$$\begin{aligned}
\mathbf{p}'' = &\ (\mathbf{p} - \mathbf{s}) \cos \phi + (1 - \cos \phi)\,(\hat{u} \cdot (\mathbf{p} - \mathbf{s}))\,\hat{u} \\
&+ (\hat{u} \times (\mathbf{p} - \mathbf{s})) \sin \phi + (\mathbf{s} + h\hat{u})
\end{aligned} \tag{5.457}$$

Equation (5.457) is the *Rodriguez formula* for the most general rigid body motion. Defining new notations $^G\mathbf{p} = \mathbf{p}''$ and $^B\mathbf{p} = \mathbf{p}$ and also noting that $\mathbf{s}$ indicates a point on the rotation axis and therefore rotation does not affect $\mathbf{s}$, we may factor out $^B\mathbf{p}$ and write the Rodriguez formula in the following form

$$\begin{aligned}
^G\mathbf{p} = &\ \left(\mathbf{I}\cos\phi + \hat{u}\hat{u}^T\,(1 - \cos\phi) + \tilde{u}\,\sin\phi\right)\,{}^B\mathbf{p} \\
&- \left(\mathbf{I}\cos\phi + \hat{u}\hat{u}^T\,(1 - \cos\phi) + \tilde{u}\,\sin\phi\right)\,{}^G\mathbf{s} + {}^G\mathbf{s} + h\hat{u} \tag{5.458}
\end{aligned}$$

which can be rearranged to show that a screw can be represented by a homogeneous transformation

$$\begin{aligned}
^G\mathbf{p} &= {}^GR_B\,{}^B\mathbf{p} + {}^G\mathbf{s} - {}^GR_B\,{}^G\mathbf{s} + h\hat{u} \\
&= {}^GR_B\,{}^B\mathbf{p} + {}^G\mathbf{d} = {}^GT_B\,{}^B\mathbf{p} \tag{5.459}
\end{aligned}$$

$$\begin{aligned}
^GT_B &= {}^G\check{s}_B(h, \phi, \hat{u}, \mathbf{s}) \tag{5.460} \\
&= \begin{bmatrix} {}^GR_B & {}^G\mathbf{s} - {}^GR_B\,{}^G\mathbf{s} + h\hat{u} \\ 0 & 1 \end{bmatrix} = \begin{bmatrix} {}^GR_B & {}^G\mathbf{d} \\ 0 & 1 \end{bmatrix}
\end{aligned}$$

where,

$$^GR_B = \mathbf{I}\cos\phi + \hat{u}\hat{u}^T\,(1 - \cos\phi) + \tilde{u}\,\sin\phi \tag{5.461}$$
$$^G\mathbf{d} = \left((\mathbf{I} - \hat{u}\hat{u}^T)\,(1 - \cos\phi) - \tilde{u}\sin\phi\right)\,{}^G\mathbf{s} + h\hat{u} \tag{5.462}$$

Direct substitution shows that:

$$^GR_B = \begin{bmatrix} u_1^2 \operatorname{vers}\phi + c\phi & u_1 u_2 \operatorname{vers}\phi - u_3 s\phi & u_1 u_3 \operatorname{vers}\phi + u_2 s\phi \\ u_1 u_2 \operatorname{vers}\phi + u_3 s\phi & u_2^2 \operatorname{vers}\phi + c\phi & u_2 u_3 \operatorname{vers}\phi - u_1 s\phi \\ u_1 u_3 \operatorname{vers}\phi - u_2 s\phi & u_2 u_3 \operatorname{vers}\phi + u_1 s\phi & u_3^2 \operatorname{vers}\phi + c\phi \end{bmatrix} \tag{5.463}$$

$$
{}^G \mathbf{d} = \begin{bmatrix} hu_1 + \left(\left(1 - u_1^2\right) s_1 - u_1 \left(s_2 u_2 + s_3 u_3\right)\right) \text{vers}\,\phi + \left(s_2 u_3 - s_3 u_2\right) s\phi \\ hu_2 + \left(\left(1 - u_2^2\right) s_2 - u_2 \left(s_3 u_3 + s_1 u_1\right)\right) \text{vers}\,\phi + \left(s_3 u_1 - s_1 u_3\right) s\phi \\ hu_3 + \left(\left(1 - u_3^2\right) s_3 - u_3 \left(s_1 u_1 + s_2 u_2\right)\right) \text{vers}\,\phi + \left(s_1 u_2 - s_2 u_1\right) s\phi \end{bmatrix}
$$
(5.464)

This representation of a rigid motion requires six independent parameters, namely one for rotation angle ϕ, one for translation h, two for screw axis $\hat{u}$, and two for location vector ${}^G \mathbf{s}$. It is because three components of $\hat{u}$ are related to each other according to

$$
\hat{u}^T \hat{u} = 1
$$
(5.465)

and the location vector ${}^G \mathbf{s}$ can locate any arbitrary point on the screw axis. It is convenient to choose the point where it has the minimum distance from O to make ${}^G \mathbf{s}$ perpendicular to $\hat{u}$. Let us indicate the *shortest location vector* by ${}^G \mathbf{s_0}$, then there is a constraint among the components of the location vector

$$
{}^G \mathbf{s}_0^T \, \hat{u} = 0
$$
(5.466)

If $\mathbf{s} = 0$ then the screw axis passes through the origin of G and (5.460) reduces to (5.449).

The screw parameters ϕ and h, together with the screw axis $\hat{u}$ and location vector ${}^G \mathbf{s}$, completely define a rigid motion of $B(oxyz)$ in $G(OXYZ)$. Having the screw parameters and screw axis, we can find the elements of the transformation matrix by Equations (5.463) and (5.464). So, given the transformation matrix ${}^G T_B$, we can find the screw angle and axis by

$$
\begin{aligned}
\cos\phi &= \frac{1}{2}\left(\text{tr}\left({}^G R_B\right) - 1\right) = \frac{1}{2}\left(\text{tr}\left({}^G T_B\right) - 2\right) \\
&= \frac{1}{2}\left(r_{11} + r_{22} + r_{33} - 1\right)
\end{aligned}
$$
(5.467)

$$
\hat{u} = \frac{1}{2\sin\phi}\left({}^G R_B - {}^G R_B^T\right)
$$
(5.468)

hence,

$$
\hat{u} = \frac{1}{2\sin\phi}\begin{bmatrix} r_{32} - r_{23} \\ r_{13} - r_{31} \\ r_{21} - r_{12} \end{bmatrix}
$$
(5.469)

To find all the required screw parameters, we must also find h and coordinates of one point on the screw axis. Because the points on the screw axis are invariant under the rotation, we must have

$$
\begin{bmatrix} r_{11} & r_{12} & r_{13} & r_{14} \\ r_{21} & r_{22} & r_{23} & r_{24} \\ r_{31} & r_{32} & r_{33} & r_{34} \\ 0 & 0 & 0 & 1 \end{bmatrix} \begin{bmatrix} X \\ Y \\ Z \\ 1 \end{bmatrix} = \begin{bmatrix} 1 & 0 & 0 & hu_1 \\ 0 & 1 & 0 & hu_2 \\ 0 & 0 & 1 & hu_3 \\ 0 & 0 & 0 & 1 \end{bmatrix} \begin{bmatrix} X \\ Y \\ Z \\ 1 \end{bmatrix}
$$
(5.470)

where (X, Y, Z) are coordinates of points on the screw axis.

We may find the intersection point of the screw line with YZ-plane, as a sample point, by setting $X_s = 0$ and searching for $\mathbf{s} = \begin{bmatrix} 0 & Y_s & Z_s \end{bmatrix}^T$. Therefore,

$$
\begin{bmatrix}
r_{11} - 1 & r_{12} & r_{13} & r_{14} - hu_1 \\
r_{21} & r_{22} - 1 & r_{23} & r_{24} - hu_2 \\
r_{31} & r_{32} & r_{33} - 1 & r_{34} - hu_3 \\
0 & 0 & 0 & 0
\end{bmatrix}
\begin{bmatrix}
0 \\ Y_s \\ Z_s \\ 1
\end{bmatrix}
=
\begin{bmatrix}
0 \\ 0 \\ 0 \\ 0
\end{bmatrix}
\tag{5.471}
$$

which generates three equations to be solved for Y_s, Z_s, and h.

$$
\begin{bmatrix}
h \\ Y_s \\ Z_s
\end{bmatrix}
=
\begin{bmatrix}
u_1 & -r_{12} & -r_{13} \\
u_2 & 1 - r_{22} & -r_{23} \\
u_3 & -r_{32} & 1 - r_{33}
\end{bmatrix}^{-1}
\begin{bmatrix}
r_{14} \\ r_{24} \\ r_{34}
\end{bmatrix}
\tag{5.472}
$$

Now we can find the shortest location vector $^G\mathbf{s}_0$ by

$$
^G\mathbf{s}_0 = \mathbf{s} - (\mathbf{s} \cdot \hat{u})\hat{u}
\tag{5.473}
$$

∎

Example 218 ★ *Central screw transformation of a base unit vector.*
Consider two initially coincident frames $G(OXYZ)$ and $B(oxyz)$. The body performs a screw motion along the Y-axis for $h = 2$ and $\phi = 90$ deg. The position of a body point at $\begin{bmatrix} 1 & 0 & 0 & 1 \end{bmatrix}^T$ can be found by applying the central screw transformation.

$$
\check{s}(h, \phi, \hat{u}) = \check{s}(2, \frac{\pi}{2}, \hat{J}) = D(2\hat{J}) R(\hat{J}, \frac{\pi}{2})
\tag{5.474}
$$

$$
=
\begin{bmatrix}
1 & 0 & 0 & 0 \\
0 & 1 & 0 & 2 \\
0 & 0 & 1 & 0 \\
0 & 0 & 0 & 1
\end{bmatrix}
\begin{bmatrix}
0 & 0 & 1 & 0 \\
0 & 1 & 0 & 0 \\
-1 & 0 & 0 & 0 \\
0 & 0 & 0 & 1
\end{bmatrix}
=
\begin{bmatrix}
0 & 0 & 1 & 0 \\
0 & 1 & 0 & 2 \\
-1 & 0 & 0 & 0 \\
0 & 0 & 0 & 1
\end{bmatrix}
$$

Therefore,

$$
^G\hat{\imath} = \check{s}(2, \frac{\pi}{2}, \hat{J}) \, ^B\hat{\imath}
\tag{5.475}
$$

$$
=
\begin{bmatrix}
0 & 0 & 1 & 0 \\
0 & 1 & 0 & 2 \\
-1 & 0 & 0 & 0 \\
0 & 0 & 0 & 1
\end{bmatrix}
\begin{bmatrix}
1 \\ 0 \\ 0 \\ 1
\end{bmatrix}
=
\begin{bmatrix}
0 \\ 2 \\ -1 \\ 1
\end{bmatrix}
$$

The pitch of this screw is

$$
p = \frac{h}{\phi} = \frac{2}{\frac{\pi}{2}} = \frac{4}{\pi} = 1.2732 \qquad unit/rad
\tag{5.476}
$$

Example 219 ★ *Screw transformation of a point.*
Consider two initially parallel coordinate frames $G(OXYZ)$ and $B(oxyz)$. The body performs a screw motion along $X = 2$ and parallel to the Y-axis for $h = 2$ and $\phi = 90$ deg. Therefore, the body coordinate frame is at location $\mathbf{s} = \begin{bmatrix} 2 & 0 & 0 \end{bmatrix}^T$. The position of a body point at $^B\mathbf{r} = \begin{bmatrix} 3 & 0 & 0 & 1 \end{bmatrix}^T$ can be found by applying the screw transformation, which is

$$^G T_B = \begin{bmatrix} ^GR_B & \mathbf{s} - {}^GR_B\mathbf{s} + h\hat{u} \\ 0 & 1 \end{bmatrix} = \begin{bmatrix} 0 & 0 & 1 & 2 \\ 0 & 1 & 0 & 2 \\ -1 & 0 & 0 & 2 \\ 0 & 0 & 0 & 1 \end{bmatrix} \tag{5.477}$$

because,

$$^GR_B = \begin{bmatrix} 0 & 0 & 1 \\ 0 & 1 & 0 \\ -1 & 0 & 0 \end{bmatrix} \quad \mathbf{s} = \begin{bmatrix} 2 \\ 0 \\ 0 \end{bmatrix} \quad \hat{u} = \begin{bmatrix} 0 \\ 1 \\ 0 \end{bmatrix} \tag{5.478}$$

Therefore, the position vector of $^G\mathbf{r}$ would then be

$$\begin{aligned} ^G\mathbf{r} &= {}^G T_B\,{}^B\mathbf{r} \\ &= \begin{bmatrix} 0 & 0 & 1 & 2 \\ 0 & 1 & 0 & 2 \\ -1 & 0 & 0 & 2 \\ 0 & 0 & 0 & 1 \end{bmatrix} \begin{bmatrix} 3 \\ 0 \\ 0 \\ 1 \end{bmatrix} = \begin{bmatrix} 2 \\ 2 \\ -1 \\ 1 \end{bmatrix} \end{aligned} \tag{5.479}$$

Example 220 ★ *Rotation of a vector.*
Transformation equation $^G\mathbf{r} = {}^GR_B\,{}^B\mathbf{r}$ and Rodriguez rotation formula (5.185) describe the rotation of any vector fixed in a rigid body. However, the vector can conveniently be described in terms of two points fixed in the body to derive the screw equation.

A reference point P_1 with position vector $\mathbf{r}_1$ at the tail, and a point P_2 with position vector $\mathbf{r}_2$ at the head, define a vector in the rigid body. Then the transformation equation between body and global frames can be written as

$$^G(\mathbf{r}_2 - \mathbf{r}_1) = {}^GR_B\,{}^B(\mathbf{r}_2 - \mathbf{r}_1) \tag{5.480}$$

Assume the original and final positions of the reference point P_1 are along the rotation axis. Equation (5.480) can then be rearranged in a form suitable for calculating coordinates of the new position of point P_2 in a transformation matrix form

$$\begin{aligned} ^G\mathbf{r}_2 &= {}^GR_B\,{}^B(\mathbf{r}_2 - \mathbf{r}_1) + {}^G\mathbf{r}_1 = {}^GR_B\,{}^B\mathbf{r}_2 + {}^G\mathbf{r}_1 - {}^GR_B\,{}^B\mathbf{r}_1 \\ &= {}^G T_B\,{}^B\mathbf{r}_2 \end{aligned} \tag{5.481}$$

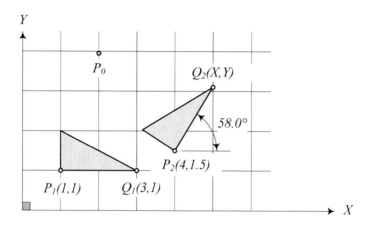

FIGURE 5.17. Motion in a plane.

where

$$
{}^{G}T_{B} = \begin{bmatrix} {}^{G}R_{B} & {}^{G}\mathbf{r}_{1} - {}^{G}R_{B} \, {}^{B}\mathbf{r}_{1} \\ 0 & 1 \end{bmatrix}
\tag{5.482}
$$

It is compatible with screw motion (5.460) for $h = 0$.

Example 221 ★ *Special cases for screw determination.*

There are two special cases for screws. The first one occurs when $r_{11} = r_{22} = r_{33} = 1$, then, $\phi = 0$ and the motion is a pure translation h parallel to $\hat{u}$, where,

$$
\hat{u} = \frac{r_{14} - s_1}{h} \hat{I} + \frac{r_{24} - s_2}{h} \hat{J} + \frac{r_{34} - s_3}{h} \hat{K}
\tag{5.483}
$$

Since there is no unique screw axis in this case, we cannot locate any specific point on the screw axis.

The second special case occurs when $\phi = 180 \deg$. In this case

$$
\hat{u} = \begin{bmatrix} \sqrt{\frac{1}{2}(r_{11}+1)} \\ \sqrt{\frac{1}{2}(r_{22}+1)} \\ \sqrt{\frac{1}{2}(r_{33}+1)} \end{bmatrix}
\tag{5.484}
$$

however, h and (X, Y, Z) can again be calculated from (5.472).

Example 222 ★ *Rotation and translation in a plane.*

Assume that a plane is displaced from position 1 to position 2 according to Figure 5.17. New coordinates of Q_2 are

$$\mathbf{r}_{Q_2} = {}^2R_1 \left(\mathbf{r}_{Q_1} - \mathbf{r}_{P_1} \right) + \mathbf{r}_{P_2} \tag{5.485}$$

$$= \begin{bmatrix} \cos 58 & -\sin 58 & 0 \\ \sin 58 & \cos 58 & 0 \\ 0 & 0 & 1 \end{bmatrix} \left(\begin{bmatrix} 3 \\ 1 \\ 0 \end{bmatrix} - \begin{bmatrix} 1 \\ 1 \\ 0 \end{bmatrix} \right) + \begin{bmatrix} 4 \\ 1.5 \\ 0 \end{bmatrix}$$

$$= \begin{bmatrix} 1.06 \\ 1.696 \\ 0 \end{bmatrix} + \begin{bmatrix} 4 \\ 1.5 \\ 0 \end{bmatrix} = \begin{bmatrix} 5.06 \\ 3.196 \\ 0.0 \end{bmatrix}$$

or equivalently

$$\mathbf{r}_{Q_2} = {}^2T_1 \, \mathbf{r}_{Q_1} = \begin{bmatrix} {}^2R_1 & \mathbf{r}_{P_2} - {}^2R_1 \, \mathbf{r}_{P_1} \\ 0 & 1 \end{bmatrix} \mathbf{r}_{Q_1} \tag{5.486}$$

$$= \begin{bmatrix} \cos 58 & -\sin 58 & 0 & 4.318 \\ \sin 58 & \cos 58 & 0 & 0.122 \\ 0 & 0 & 1 & 0 \\ 0 & 0 & 0 & 1 \end{bmatrix} \begin{bmatrix} 3 \\ 1 \\ 0 \\ 1 \end{bmatrix} = \begin{bmatrix} 5.06 \\ 3.196 \\ 0 \\ 1 \end{bmatrix}$$

Example 223 ★ *Pole of planar motion.*

In the planar motion of a rigid body, going from position 1 to position 2, there is always one point in the plane of motion that does not change its position. Hence, the body can be considered as having rotated about this point, which is known as the finite rotation pole. The transformation matrix can be used to locate the pole. Figure 5.17 depicts a planar motion of a triangle. To locate the pole of motion $P_0(X_0, Y_0)$ we need the transformation of the motion. Using the data given in Figure 5.17 we have

$$\begin{aligned} {}^2T_1 &= \begin{bmatrix} {}^2R_1 & \mathbf{r}_{P_2} - {}^2R_1 \, \mathbf{r}_{P_1} \\ 0 & 1 \end{bmatrix} \\ &= \begin{bmatrix} c\alpha & -s\alpha & 0 & -c\alpha + s\alpha + 4 \\ s\alpha & c\alpha & 0 & -c\alpha - s\alpha + 3.5 \\ 0 & 0 & 1 & 0 \\ 0 & 0 & 0 & 1 \end{bmatrix} \end{aligned} \tag{5.487}$$

The pole would be conserved under the transformation. Therefore,

$$\mathbf{r}_{P_0} = {}^2T_1 \, \mathbf{r}_{P_0} \tag{5.488}$$

$$\begin{bmatrix} X_0 \\ Y_0 \\ 0 \\ 1 \end{bmatrix} = \begin{bmatrix} \cos \alpha & -\sin \alpha & 0 & -\cos \alpha + \sin \alpha + 4 \\ \sin \alpha & \cos \alpha & 0 & -\cos \alpha - \sin \alpha + 1.5 \\ 0 & 0 & 1 & 0 \\ 0 & 0 & 0 & 1 \end{bmatrix} \begin{bmatrix} X_0 \\ Y_0 \\ 0 \\ 1 \end{bmatrix}$$

which for $\alpha = 58 \deg$ *provides*

$$X_0 = -1.5 \sin \alpha + 1 - 4 \cos \alpha = 2.049 \tag{5.489}$$

$$Y_0 = 4 \sin \alpha + 1 - 1.5 \cos \alpha = 3.956 \tag{5.490}$$

Example 224 ★ *Determination of screw parameters.*

We are able to determine screw parameters when we have the original and final position of three non-colinear points of a rigid body. Assume $\mathbf{p}_0$, $\mathbf{q}_0$, *and* $\mathbf{r}_0$ *denote the position of points* P, Q, *and* R *before the screw motion, and* $\mathbf{p}_1$, $\mathbf{q}_1$, *and* $\mathbf{r}_1$ *denote their positions after the screw motion.*

To determine screw parameters, ϕ, $\hat{u}$, h, *and* $\mathbf{s}$, *we should solve the following three simultaneous Rodriguez equations:*

$$\mathbf{p}_1 - \mathbf{p}_0 = \tan\frac{\phi}{2}\hat{u} \times (\mathbf{p}_1 + \mathbf{p}_0 - 2\mathbf{s}) + h\hat{u} \qquad (5.491)$$

$$\mathbf{q}_1 - \mathbf{q}_0 = \tan\frac{\phi}{2}\hat{u} \times (\mathbf{q}_1 + \mathbf{q}_0 - 2\mathbf{s}) + h\hat{u} \qquad (5.492)$$

$$\mathbf{r}_1 - \mathbf{r}_0 = \tan\frac{\phi}{2}\hat{u} \times (\mathbf{r}_1 + \mathbf{r}_0 - 2\mathbf{s}) + h\hat{u} \qquad (5.493)$$

We start with subtracting Equation (5.493) from (5.491) and (5.492).

$$(\mathbf{p}_1 - \mathbf{p}_0) - (\mathbf{r}_1 - \mathbf{r}_0) = \tan\frac{\phi}{2}\hat{u} \times [(\mathbf{p}_1 + \mathbf{p}_0) - (\mathbf{r}_1 - \mathbf{r}_0)] \quad (5.494)$$

$$(\mathbf{q}_1 - \mathbf{q}_0) - (\mathbf{r}_1 - \mathbf{r}_0) = \tan\frac{\phi}{2}\hat{u} \times [(\mathbf{q}_1 + \mathbf{q}_0) - (\mathbf{r}_1 - \mathbf{r}_0)] \quad (5.495)$$

Now multiplying both sides of (5.494) by $[(\mathbf{q}_1 - \mathbf{q}_0) - (\mathbf{r}_1 - \mathbf{r}_0)]$ *which is perpendicular to* $\hat{u}$

$$[(\mathbf{q}_1 - \mathbf{q}_0) - (\mathbf{r}_1 - \mathbf{r}_0)] \times [(\mathbf{p}_1 - \mathbf{p}_0) - (\mathbf{r}_1 - \mathbf{r}_0)]$$
$$= \tan\frac{\phi}{2}[(\mathbf{q}_1 - \mathbf{q}_0) - (\mathbf{r}_1 - \mathbf{r}_0)] \times \{\hat{u} \times [(\mathbf{p}_1 + \mathbf{p}_0) - (\mathbf{r}_1 - \mathbf{r}_0)]\} \qquad (5.496)$$

gives us

$$[(\mathbf{q}_1 - \mathbf{q}_0) - (\mathbf{r}_1 - \mathbf{r}_0)] \times [(\mathbf{p}_1 + \mathbf{p}_0) - (\mathbf{r}_1 - \mathbf{r}_0)]$$
$$= \tan\frac{\phi}{2}\{[(\mathbf{q}_1 - \mathbf{q}_0) - (\mathbf{r}_1 - \mathbf{r}_0)] \cdot [(\mathbf{p}_1 + \mathbf{p}_0) - (\mathbf{r}_1 - \mathbf{r}_0)]\}\,\hat{u} \qquad (5.497)$$

and therefore, the rotation angle can be found by equating $\tan\frac{\phi}{2}$ *with the norm of the right-hand side of the following equation:*

$$\tan\frac{\phi}{2}\hat{u} = \frac{[(\mathbf{q}_1 - \mathbf{q}_0) - (\mathbf{r}_1 - \mathbf{r}_0)] \times (\mathbf{p}_1 + \mathbf{p}_0) - (\mathbf{r}_1 - \mathbf{r}_0)}{[(\mathbf{q}_1 - \mathbf{q}_0) - (\mathbf{r}_1 - \mathbf{r}_0)] \cdot (\mathbf{p}_1 + \mathbf{p}_0) - (\mathbf{r}_1 - \mathbf{r}_0)} \qquad (5.498)$$

To find $\mathbf{s}$, *we may start with the cross product of* $\hat{u}$ *with Equation (5.491).*

$$\hat{u} \times (\mathbf{p}_1 - \mathbf{p}_0) = \hat{u} \times \left[\tan\frac{\phi}{2}\hat{u} \times (\mathbf{p}_1 + \mathbf{p}_0 - 2\mathbf{s}) + h\hat{u}\right] \qquad (5.499)$$

$$= \tan\frac{\phi}{2}\{[\hat{u} \cdot (\mathbf{p}_1 + \mathbf{p}_0)]\,\hat{u} - (\mathbf{p}_1 + \mathbf{p}_0) + 2[\mathbf{s} - (\hat{u} \cdot \mathbf{s})\,\hat{u}]\}$$

Note that $\mathbf{s} - (\hat{u} \cdot \mathbf{s})\,\hat{u}$ *is the component of* $\mathbf{s}$ *perpendicular to* $\hat{u}$, *where* $\mathbf{s}$ *is a vector from the origin of the global frame* $G(OXYZ)$ *to an arbitrary* *point on the screw axis. This perpendicular component indicates a vector with the shortest distance between O and $\hat{u}$. Let's assume $\mathbf{s}_0$ is the name of the shortest $\mathbf{s}$. Therefore,*

$$
\begin{aligned}
\mathbf{s}_0 &= \mathbf{s} - (\hat{u} \cdot \mathbf{s})\,\hat{u} \\
&= \frac{1}{2}\left[\frac{\hat{u} \times \mathbf{p}_1 - \mathbf{p}_0}{\tan\frac{\phi}{2}} - [\hat{u} \cdot (\mathbf{p}_1 + \mathbf{p}_0)]\,\hat{u} + \mathbf{p}_1 + \mathbf{p}_0\right]
\end{aligned}
\tag{5.500}
$$

The last parameter of the screw is the pitch h, which can be found from any one of the Equations (5.491), (5.492), or (5.493).

$$
h = \hat{u} \cdot (\mathbf{p}_1 - \mathbf{p}_0) = \hat{u} \cdot (\mathbf{q}_1 - \mathbf{q}_0) = \hat{u} \cdot (\mathbf{r}_1 - \mathbf{r}_0)
\tag{5.501}
$$

Example 225 ★ *Alternative derivation of screw transformation.*

Assume the screw axis does not pass through the origin of G. If $^G\mathbf{s}$ is the position vector of some point on the axis $\hat{u}$, then we can derive the matrix representation of screw $\check{s}(h, \phi, \hat{u}, \mathbf{s})$ by translating the screw axis back to the origin, performing the central screw motion, and translating the line back to its original position.

$$
\begin{aligned}
\check{s}(h, \phi, \hat{u}, \mathbf{s}) &= D(^G\mathbf{s})\,\check{s}(h, \phi, \hat{u})\,D(-^G\mathbf{s}) \\
&= D(^G\mathbf{s})\,D(h\hat{u})\,R(\hat{u}, \phi)\,D(-^G\mathbf{s}) \\
&= \begin{bmatrix} \mathbf{I} & ^G\mathbf{s} \\ 0 & 1 \end{bmatrix}\begin{bmatrix} ^GR_B & h\hat{u} \\ 0 & 1 \end{bmatrix}\begin{bmatrix} \mathbf{I} & -^G\mathbf{s} \\ 0 & 1 \end{bmatrix} \\
&= \begin{bmatrix} ^GR_B & ^G\mathbf{s} - ^GR_B\,^G\mathbf{s} + h\hat{u} \\ 0 & 1 \end{bmatrix}
\end{aligned}
\tag{5.502}
$$

Example 226 ★ *Rotation about an off-center axis.*

Rotation of a rigid body about an axis indicated by $\hat{u}$ and passing through a point at $^G\mathbf{s}$, where $^G\mathbf{s} \times \hat{u} \neq 0$ is a rotation about an off-center axis. The transformation matrix associated with an off-center rotation can be obtained from the screw transformation by setting $h = 0$. Therefore, an off-center rotation transformation is

$$
^GT_B = \begin{bmatrix} ^GR_B & ^G\mathbf{s} - ^GR_B\,^G\mathbf{s} \\ 0 & 1 \end{bmatrix}
\tag{5.503}
$$

Example 227 ★ *Principal central screw.*

There are three principal central screws, namely the x-screw, y-screw, and z-screw, which are

$$
\check{s}(h_Z, \alpha, \hat{K}) = \begin{bmatrix} \cos\alpha & -\sin\alpha & 0 & 0 \\ \sin\alpha & \cos\alpha & 0 & 0 \\ 0 & 0 & 1 & p_Z\,\alpha \\ 0 & 0 & 0 & 1 \end{bmatrix}
\tag{5.504}
$$

$$\check{s}(h_Y, \beta, \hat{J}) = \begin{bmatrix} \cos\beta & 0 & \sin\beta & 0 \\ 0 & 1 & 0 & p_Y\,\beta \\ -\sin\beta & 0 & \cos\beta & 0 \\ 0 & 0 & 0 & 1 \end{bmatrix} \qquad (5.505)$$

$$\check{s}(h_X, \gamma, \hat{I}) = \begin{bmatrix} 1 & 0 & 0 & p_X\,\gamma \\ 0 & \cos\gamma & -\sin\gamma & 0 \\ 0 & \sin\gamma & \cos\gamma & 0 \\ 0 & 0 & 0 & 1 \end{bmatrix} \qquad (5.506)$$

Example 228 ★ *Proof of Chasles theorem.*

Let $[T]$ be an arbitrary spatial displacement, and decompose it into a rotation R about $\hat{u}$ and a translation D.

$$[T] = [D][R] \qquad (5.507)$$

We may also decompose the translation $[D]$ into two components $[D_\parallel]$ and $[D_\perp]$, parallel and perpendicular to $\hat{u}$, respectively.

$$[T] = [D_\parallel][D_\perp][R] \qquad (5.508)$$

Now $[D_\perp][R]$ is a planar motion, and is therefore equivalent to some rotation $[R'] = [D_\perp][R]$ about an axis parallel to the rotation axis $\hat{u}$. This yields the decomposition $[T] = [D_\parallel][R']$. This decomposition completes the proof, since the axis of $[D_\parallel]$ can be taken equal to $\hat{u}$.

Example 229 ★ *Every rigid motion is a screw.*

To show that any proper rigid motion can be considered as a screw motion, we must show that a homogeneous transformation matrix

$$^{G}T_B = \begin{bmatrix} ^{G}R_B & ^{G}\mathbf{d} \\ 0 & 1 \end{bmatrix} \qquad (5.509)$$

can be written in the form

$$^{G}T_B = \begin{bmatrix} ^{G}R_B & (\mathbf{I} - {}^{G}R_B)\mathbf{s} + h\hat{u} \\ 0 & 1 \end{bmatrix} \qquad (5.510)$$

This problem is then equivalent to the following equation to find h and $\hat{u}$.

$$^{G}\mathbf{d} = (\mathbf{I} - {}^{G}R_B)\mathbf{s} + h\hat{u} \qquad (5.511)$$

The matrix $[\mathbf{I} - {}^{G}R_B]$ is singular because $^{G}R_B$ always has 1 as an eigenvalue. This eigenvalue corresponds to $\hat{u}$ as an eigenvector. Therefore,

$$[\mathbf{I} - {}^{G}R_B]\hat{u} = [\mathbf{I} - {}^{G}R_B^{T}]\hat{u} = 0 \qquad (5.512)$$

and an inner product shows that

$$\begin{aligned} \hat{u} \cdot {}^{G}\mathbf{d} &= \hat{u} \cdot \left[\mathbf{I} - {}^{G}R_B\right]\mathbf{s} + \hat{u} \cdot h\hat{u} \\ &= \left[\mathbf{I} - {}^{G}R_B\right]\hat{u} \cdot \mathbf{s} + \hat{u} \cdot h\hat{u} \end{aligned} \qquad (5.513)$$

which leads to

$$h = \hat{u} \cdot {}^G\mathbf{d} \qquad (5.514)$$

Now we may use h to find **s**

$$\mathbf{s} = \left[\mathbf{I} - {}^GR_B\right]^{-1} ({}^G\mathbf{d} - h\hat{u}). \qquad (5.515)$$

5.15 Summary

To analyze the relative motion of rigid bodies, we instal a body coordinate frame at the mass center of each body. The relative motion of the bodies can be expressed by the relative motion of the frames.

Coordinates of a point in two Cartesian coordinate frames with a common origin are convertible based on nine directional cosines of the three axes of a frame in the other. The conversion of coordinates in the two frames can be cast in matrix transformation

$$ {}^G\mathbf{r} = {}^GR_B\,{}^B\mathbf{r} \qquad (5.516)$$

$$\begin{bmatrix} X_2 \\ Y_2 \\ Z_2 \end{bmatrix} = \begin{bmatrix} \hat{I}\cdot\hat{\imath} & \hat{I}\cdot\hat{\jmath} & \hat{I}\cdot\hat{k} \\ \hat{J}\cdot\hat{\imath} & \hat{J}\cdot\hat{\jmath} & \hat{J}\cdot\hat{k} \\ \hat{K}\cdot\hat{\imath} & \hat{K}\cdot\hat{\jmath} & \hat{K}\cdot\hat{k} \end{bmatrix} \begin{bmatrix} x_2 \\ y_2 \\ z_2 \end{bmatrix} \qquad (5.517)$$

where,

$$ {}^GR_B = \begin{bmatrix} \cos(\hat{I},\hat{\imath}) & \cos(\hat{I},\hat{\jmath}) & \cos(\hat{I},\hat{k}) \\ \cos(\hat{J},\hat{\imath}) & \cos(\hat{J},\hat{\jmath}) & \cos(\hat{J},\hat{k}) \\ \cos(\hat{K},\hat{\imath}) & \cos(\hat{K},\hat{\jmath}) & \cos(\hat{K},\hat{k}) \end{bmatrix} \qquad (5.518)$$

The transformation matrix ${}^GR_B$ is orthogonal and therefore its inverse is equal to its transpose.

$$ {}^GR_B^{-1} = {}^GR_B^T \qquad (5.519)$$

When a body coordinate frame B and a global frame G have a common origin and frame B rotates continuously with respect to frame G, the rotation matrix ${}^GR_B$ is time dependent.

$$ {}^G\mathbf{r}(t) = {}^GR_B(t)\,{}^B\mathbf{r} \qquad (5.520)$$

Then, the global velocity of a point in B is

$$ {}^G\dot{\mathbf{r}}(t) = {}^G\mathbf{v}(t) = {}^G\dot{R}_B(t)\,{}^B\mathbf{r} = {}_G\tilde{\omega}_B\,{}^G\mathbf{r}(t) \qquad (5.521)$$

where ${}_G\tilde{\omega}_B$ is the skew symmetric angular velocity matrix

$$ {}_G\tilde{\omega}_B = {}^G\dot{R}_B\,{}^GR_B^T \qquad (5.522)$$

$$ {}_G\tilde{\omega}_B = \begin{bmatrix} 0 & -\omega_3 & \omega_2 \\ \omega_3 & 0 & -\omega_1 \\ -\omega_2 & \omega_1 & 0 \end{bmatrix} \qquad (5.523)$$

The matrix $_G\tilde{\omega}_B$ is associated with the angular velocity vector $_G\omega_B = \dot{\phi}\,\hat{u}$, which is equal to an angular rate $\dot{\phi}$ about the instantaneous axis of rotation $\hat{u}$.

Angular velocities of connected rigid bodies may be added relatively to find the angular velocity of the nth body in the base coordinate frame

$$_0\omega_n = {}_0\omega_1 + {}_1^0\omega_2 + {}_2^0\omega_3 + \cdots + {}_{n-1}^0\omega_n = \sum_{i=1}^{n} {}_{i-1}^0\omega_i \qquad (5.524)$$

Relative time derivatives between the global and a coordinate frames attached to a moving rigid body must be taken according to the following rules.

$$\frac{{}^Bd}{dt}{}^B\mathbf{r}_P = {}^B\dot{\mathbf{r}}_P = {}^B\mathbf{v}_P = \dot{x}\,\hat{\imath} + \dot{y}\,\hat{\jmath} + \dot{z}\,\hat{k} \qquad (5.525)$$

$$\frac{{}^Gd}{dt}{}^G\mathbf{r}_P = {}^G\dot{\mathbf{r}}_P = {}^G\mathbf{v}_P = \dot{X}\,\hat{I} + \dot{Y}\,\hat{J} + \dot{Z}\,\hat{K} \qquad (5.526)$$

$$\frac{{}^Gd}{dt}{}^B\mathbf{r}_P(t) = {}^B\dot{\mathbf{r}}_P + {}_G^B\omega_B \times {}^B\mathbf{r}_P = {}_G^B\dot{\mathbf{r}}_P \qquad (5.527)$$

$$\frac{{}^Bd}{dt}{}^G\mathbf{r}_P(t) = {}^G\dot{\mathbf{r}}_P - {}_G\omega_B \times {}^G\mathbf{r}_P = {}_B^G\dot{\mathbf{r}}_P \qquad (5.528)$$

The global velocity of a point P in a moving frame B at

$$^G\mathbf{r}_P = {}^G R_B\,{}^B\mathbf{r}_P + {}^G\mathbf{d}_B \qquad (5.529)$$

is

$$
\begin{aligned}
{}^G\mathbf{v}_P &= {}^G\dot{\mathbf{r}}_P = {}_G\tilde{\omega}_B \left({}^G\mathbf{r}_P - {}^G\mathbf{d}_B\right) + {}^G\dot{\mathbf{d}}_B \\
&= {}_G\omega_B \times \left({}^G\mathbf{r}_P - {}^G\mathbf{d}_B\right) + {}^G\dot{\mathbf{d}}_B \qquad (5.530)
\end{aligned}
$$

When a body coordinate frame B and a global frame G have a common origin, the global acceleration of a point P in frame B is

$$^G\ddot{\mathbf{r}} = \frac{{}^Gd}{dt}{}^G\mathbf{v}_P = {}_G\alpha_B \times {}^G\mathbf{r} + {}_G\omega_B \times \left({}_G\omega_B \times {}^G\mathbf{r}\right) \qquad (5.531)$$

where, $_G\alpha_B$ is the angular acceleration of B with respect to G

$$_G\alpha_B = \frac{{}^Gd}{dt}{}_G\omega_B \qquad (5.532)$$

However, when the body coordinate frame B has a rigid motion with respect to G, then

$$
\begin{aligned}
{}^G\mathbf{a}_P &= \frac{{}^Gd}{dt}{}^G\mathbf{v}_P = {}_G\alpha_B \times \left({}^G\mathbf{r}_P - {}^G\mathbf{d}_B\right) \\
&\quad + {}_G\omega_B \times \left({}_G\omega_B \times \left({}^G\mathbf{r}_P - {}^G\mathbf{d}_B\right)\right) + {}^G\ddot{\mathbf{d}}_B \qquad (5.533)
\end{aligned}
$$

where $^G\mathbf{d}_B$ indicates the position of the origin of B with respect to the origin of G.

Angular accelerations of two connected rigid bodies are related according to

$$_0\boldsymbol{\alpha}_2 = {}_0\boldsymbol{\alpha}_1 + {}_1^0\boldsymbol{\alpha}_2 + {}_0\boldsymbol{\omega}_1 \times {}_1^0\boldsymbol{\omega}_2 \qquad (5.534)$$

5.16   Key Symbols

$B, \, Oxyz$	body Cartesian coordinate frame
${}^G\mathbf{d}_B$	position vector of body coordinate frame B in G
${}^0_1\mathbf{d}_2$	position of frame B_2 respect to B_1 expressed in B_0
$\hat{e}_\varphi, \hat{e}_\theta, \hat{e}_\psi$	Euler angle coordinate frame unit vectors
$G, \, OXYZ$	global Cartesian coordinate frame
$\hat{\imath}, \hat{\jmath}, \hat{k}$	body coordinate frame unit vectors
$\tilde{\imath}, \tilde{\jmath}, \tilde{k}$	skew symmetric matrix associated to $\hat{\imath}, \hat{\jmath}, \hat{k}$
$\hat{I}, \hat{J}, \hat{K}$	global coordinate frame unit vectors
$p = h/\phi$	pitch of screw
P	point
${}^G\mathbf{r}$	position vector in global coordinate frame
${}^B\mathbf{r}$	position vector in body coordinate frame
$\hat{\mathbf{r}}_{H_1}, \hat{\mathbf{r}}_{H_2}, \hat{\mathbf{r}}_{H_3}$	row vectors of a rotation matrix
R_Z	rotation matrix about the global Z-axis
R_Y	rotation matrix about the global Y-axis
R_X	rotation matrix about the global X-axis
$\dot{R}$	time derivative of a rotation matrix R
${}^G R_B$	rotation matrix from local frame to global frame
R^T	transpose of a rotation matrix
R^{-1}	inverse of a rotation matrix
R_z	rotation matrix about the body z-axis
R_y	rotation matrix about the body y-axis
R_x	rotation matrix about the body x-axis
${}^{B_1} R_{B_2}$	rotation matrix from coordinate frame B_1 to B_2
${}^B R_G$	rotation matrix from global to local coordinate frame
$\check{s}(h, \phi, \hat{u}, \mathbf{s})$	screw motion
t	time
$\hat{u}, \phi$	axis and angle of rotation
$\hat{u}_\alpha$	instant angular acceleration axis
$\hat{u}_\omega$	instant angular velocity axis
x, y, z	body coordinates of a point
X, Y, Z	body coordinates of a point
$x, y, z, \, \mathbf{x}$	displacement
${}^G\boldsymbol{\alpha}_B$	angular acceleration of body B expressed in G
δ_{jk}	Kronecker's delta
ϵ_{ijk}	permutation symbol
$\dot{\varphi}, \dot{\theta}, \dot{\psi}$	Euler frequencies
${}^G\boldsymbol{\omega}_B$	angular velocity of rigid body B expressed in G
$\tilde{\omega}$	skew symmetric matrix associated to ω

Exercises

1. Body point and global rotations.

   The point P is at $\mathbf{r}_P = (1, 2, 1)$ in a body coordinate $B(Oxyz)$. Find the final global position of P after a rotation of 30 deg about the X-axis, followed by a 45 deg rotation about the Z-axis.

2. Body point after global rotation.

   Find the position of a point P in the local coordinate frame if it is moved to $^G\mathbf{r}_P = [1, 3, 2]^T$ after

   (a) a 60 deg rotation about Z-axis.

   (b) a 60 deg rotation about X-axis.

   (c) a 60 deg rotation about Z-axis followed by a 60 deg rotation about X-axis.

   (d) ★ Is it possible to combine the rotations in part (c) and do only one rotation about the bisector of X and Z-axes?

3. Invariant of a vector.

   A point was at $^B\mathbf{r}_P = [1, 2, z]^T$. After a rotation of 60 deg about the X-axis, followed by a 30 deg rotation about the Z-axis, it is at

   $$^G\mathbf{r}_P = \begin{bmatrix} X \\ Y \\ 2.933 \end{bmatrix}$$

   Find z, X, and Y.

4. ★ Constant length vector.

   Show that the length of a vector will not change by rotation.

   $$\left|^G\mathbf{r}\right| = \left|^G R_B \, ^B\mathbf{r}\right|$$

   Show that the distance between two body points will not change by rotation.

   $$\left|^B\mathbf{p}_1 - ^B\mathbf{p}_2\right| = \left|^G R_B \, ^B\mathbf{p}_1 - ^G R_B \, ^B\mathbf{p}_2\right|$$

5. Global roll-pitch-yaw rotation angles.

   Calculate the role, pitch, and yaw angles for the following rotation matrix:

   $$^B R_G = \begin{bmatrix} 0.53 & -0.84 & 0.13 \\ 0.0 & 0.15 & 0.99 \\ -0.85 & -0.52 & 0.081 \end{bmatrix}$$

6. Body point, local rotation.

   What is the global coordinates of a body point at $^B\mathbf{r}_P = [2, 2, 3]^T$, after a rotation of 60 deg about the x-axis?

7. Two local rotations.

   Find the global coordinates of a body point at $^B\mathbf{r}_P = [2, 2, 3]^T$ after a rotation of 60 deg about the x-axis followed by 60 deg about the z-axis.

8. Combination of local and global rotations.

   Find the final global position of a body point at $^B\mathbf{r}_P = [10, 10, -10]^T$ after a rotation of 45 deg about the x-axis followed by 60 deg about the Z-axis.

9. Combination of global and local rotations.

   Find the final global position of a body point at $^B\mathbf{r}_P = [10, 10, -10]^T$ after a rotation of 45 deg about the X-axis followed by 60 deg about the z-axis.

10. ★ Euler angles from rotation matrix.

    Find the Euler angles for the following rotation matrix:

    $$^B R_G = \begin{bmatrix} 0.53 & -0.84 & 0.13 \\ 0.0 & 0.15 & 0.99 \\ -0.85 & -0.52 & 0.081 \end{bmatrix}$$

11. ★ Equivalent Euler angles to two rotations.

    Find the Euler angles corresponding to the rotation matrix $^B R_G = A_{y,45} A_{x,30}$.

12. ★ Equivalent Euler angles to three rotations.

    Find the Euler angles corresponding to the rotation matrix $^B R_G = A_{z,60} A_{y,45} A_{x,30}$.

13. ★ Local and global positions, Euler angles.

    Find the conditions between the Euler angles to transform $^G\mathbf{r}_P = [1, 1, 0]^T$ to $^B\mathbf{r}_P = [0, 1, 1]^T$.

14. Elements of rotation matrix.

    The elements of rotation matrix $^G R_B$ are

    $$^G R_B = \begin{bmatrix} \cos(\hat{I}, \hat{\imath}) & \cos(\hat{I}, \hat{\jmath}) & \cos(\hat{I}, \hat{k}) \\ \cos(\hat{J}, \hat{\imath}) & \cos(\hat{J}, \hat{\jmath}) & \cos(\hat{J}, \hat{k}) \\ \cos(\hat{K}, \hat{\imath}) & \cos(\hat{K}, \hat{\jmath}) & \cos(\hat{K}, \hat{k}) \end{bmatrix}$$

    Find $^G R_B$ if $^G\mathbf{r}_{P_1} = (0.7071, -1.2247, 1.4142)$ is a point on the x-axis, and $^G\mathbf{r}_{P_2} = (2.7803, 0.38049, -1.0607)$ is a point on the y-axis.

15. Local position, global velocity.

A body is turning about the Z-axis at a constant angular rate $\dot{\alpha} = 2\,\mathrm{rad/sec}$. Find the global velocity of a point at

$$
{}^{B}\mathbf{r} = \begin{bmatrix} 5 \\ 30 \\ 10 \end{bmatrix}
$$

16. Global position, constant angular velocity.

A body is turning about the Z-axis at a constant angular rate $\dot{\alpha} = 2\,\mathrm{rad/s}$. Find the global position of a point at

$$
{}^{B}\mathbf{r} = \begin{bmatrix} 5 \\ 30 \\ 10 \end{bmatrix}
$$

after $t = 3\,\mathrm{sec}$ if the body and global coordinate frames were coincident at $t = 0\,\mathrm{sec}$.

17. Turning about x-axis.

Find the angular velocity matrix when the body coordinate frame is turning $35\,\mathrm{deg/sec}$ at $45\,\mathrm{deg}$ about the x-axis.

18. Combined rotation and angular velocity.

Find the rotation matrix for a body frame after $30\,\mathrm{deg}$ rotation about the Z-axis, followed by $30\,\mathrm{deg}$ about the X-axis, and then $90\,\mathrm{deg}$ about the Y-axis. Then calculate the angular velocity of the body if it is turning with $\dot{\alpha} = 20\,\mathrm{deg/sec}$, $\dot{\beta} = -40\,\mathrm{deg/sec}$, and $\dot{\gamma} = 55\,\mathrm{deg/sec}$ about the Z, Y, and X axes respectively.

19. Angular velocity, expressed in body frame.

The point P is at $\mathbf{r}_P = (1, 2, 1)$ in a body coordinate $B(Oxyz)$. Find ${}^{B}_{G}\tilde{\omega}_B$ when the body frame is turned $30\,\mathrm{deg}$ about the X-axis at a rate $\dot{\gamma} = 75\,\mathrm{deg/sec}$, followed by $45\,\mathrm{deg}$ about the Z-axis at a rate $\dot{\alpha} = 25\,\mathrm{deg/sec}$.

20. Global roll-pitch-yaw angular velocity.

Calculate the angular velocity for a global roll-pitch-yaw rotation of $\alpha = 30\,\mathrm{deg}$, $\beta = 30\,\mathrm{deg}$, and $\gamma = 30\,\mathrm{deg}$ with $\dot{\alpha} = 20\,\mathrm{deg/sec}$, $\dot{\beta} = -20\,\mathrm{deg/sec}$, and $\dot{\gamma} = 20\,\mathrm{deg/sec}$.

21. Roll-pitch-yaw angular velocity.

Find ${}^{B}_{G}\tilde{\omega}_B$ and ${}_{G}\tilde{\omega}_B$ for the role, pitch, and yaw rates equal to $\dot{\alpha} = 20\,\mathrm{deg/sec}$, $\dot{\beta} = -20\,\mathrm{deg/sec}$, and $\dot{\gamma} = 20\,\mathrm{deg/sec}$ respectively, and having the following rotation matrix:

$$
{}^B R_G = \begin{bmatrix} 0.53 & -0.84 & 0.13 \\ 0.0 & 0.15 & 0.99 \\ -0.85 & -0.52 & 0.081 \end{bmatrix}
$$

22. ★ Differentiating in local and global frames.

Consider a local point at ${}^B r_P = t\hat{\imath} + \hat{\jmath}$. The local frame B is rotating in G by $\dot{\alpha}$ about the Z-axis. Calculate $\frac{{}^B d}{dt} {}^B r_P$, $\frac{{}^G d}{dt} {}^G r_P$, $\frac{{}^B d}{dt} {}^G r_P$, and $\frac{{}^G d}{dt} {}^B r_P$.

23. ★ Transformation of angular velocity exponents.

Show that
$$
{}^B_G \tilde{\omega}^n_B = {}^G R^T_B \, {}_G \tilde{\omega}^n_B \, {}^G R_B
$$

24. Local position, global acceleration.

A body is turning about the Z-axis at a constant angular acceleration $\ddot{\alpha} = 2\,\mathrm{rad/sec^2}$. Find the global velocity of a point, when $\dot{\alpha} = 2\,\mathrm{rad/sec}$, $\alpha = \pi/3\,\mathrm{rad}$ and

$$
{}^B \mathbf{r} = \begin{bmatrix} 5 \\ 30 \\ 10 \end{bmatrix}
$$

25. Global position, constant angular acceleration.

A body is turning about the Z-axis at a constant angular acceleration $\ddot{\alpha} = 2\,\mathrm{rad/sec^2}$. Find the global position of a point at

$$
{}^B \mathbf{r} = \begin{bmatrix} 5 \\ 30 \\ 10 \end{bmatrix}
$$

after $t = 3\,\mathrm{sec}$ if the body and global coordinate frames were coincident at $t = 0\,\mathrm{sec}$.

26. Turning about x-axis.

Find the angular acceleration matrix when the body coordinate frame is turning $-5\,\mathrm{deg/sec^2}$, $35\,\mathrm{deg/sec}$ at $45\,\mathrm{deg}$ about the x-axis.

27. Angular acceleration and Euler angles.

Calculate the angular velocity and acceleration vectors in body and global coordinate frames if the Euler angles and their derivatives are:

$$
\begin{array}{lll}
\varphi = .25\,\mathrm{rad} & \dot{\varphi} = 2.5\,\mathrm{rad/sec} & \ddot{\varphi} = 25\,\mathrm{rad/sec^2} \\
\theta = -.25\,\mathrm{rad} & \dot{\theta} = -4.5\,\mathrm{rad/sec} & \ddot{\theta} = 35\,\mathrm{rad/sec^2} \\
\psi = .5\,\mathrm{rad} & \dot{\psi} = 3\,\mathrm{rad/sec} & \ddot{\psi} = 25\,\mathrm{rad/sec^2}
\end{array}
$$

28. Combined rotation and angular acceleration.

Find the rotation matrix for a body frame after 30 deg rotation about the Z-axis, followed by 30 deg about the X-axis, and then 90 deg about the Y-axis. Then calculate the angular velocity of the body if it is turning with $\dot{\alpha} = 20\,\deg/\sec$, $\dot{\beta} = -40\,\deg/\sec$, and $\dot{\gamma} = 55\,\deg/\sec$ about the Z, Y, and X axes respectively. Finally, calculate the angular acceleration of the body if it is turning with $\ddot{\alpha} = 2\,\deg/\sec^2$, $\ddot{\beta} = 4\,\deg/\sec^2$, and $\ddot{\gamma} = -6\,\deg/\sec^2$ about the Z, Y, and X axes.

6

Applied Mechanisms

The mechanisms that are used in vehicle subsystems are mostly made of four-bar linkages. Double *A*-arm for independent suspension, and trapezoidal steering are two subsystems examples in vehicle . In this chapter, we review the analysis and design methods for such mechanisms.

6.1 Four-Bar Linkage

An individual rigid member that can have relative motion with respect to all other members is called a *link*. A link may also be called a *bar*, *body*, *arm*, or a *member*. Any two or more links connected together, such that no relative motion can occur among them, are considered a single link.

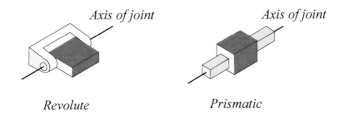

Axis of joint   *Axis of joint*

Revolute   *Prismatic*

FIGURE 6.1. A revolute and a prismatic joint.

Two links are connected by a *joint* where their relative motion can be expressed by a single coordinate. Joints are typically *revolute* (rotary) or *prismatic* (translatory). Figure 6.1 illustrates the revolute and prismatic joints. A *revolute joint* (R), is like a hinge that allows relative rotation between the two connected links. A *prismatic joint* (P), allows a relative translation between the two connected links.

Relative rotation or translation, between two connected links by a revolute or prismatic joint, occurs about a line called *joint axis*. The value of the single variable describing the relative position of two connected links at a joint is called the *joint coordinate* or *joint variable*. It is an *angle* for a revolute joint, or a *distance* for a prismatic joint.

A set of connected links is called a *mechanism*. A *linkage* is made by attaching, and fixing, one link of a mechanism to the ground. The fixed link is called the *ground link*. There are two types of linkages, *closed loop* or *parallel*, and *open loop* or *serial*. In vehicle subsystems we usually use

R.N. Jazar, *Vehicle Dynamics: Theory and Application*,   319
DOI 10.1007/978-1-4614-8544-5_6, © Springer Science+Business Media New York 2014

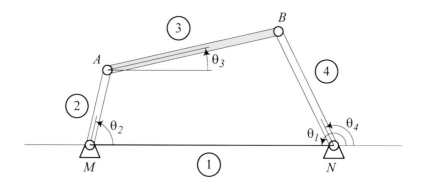

FIGURE 6.2. A four-bar linkage.

closed-loop linkages. Open-loop linkages are used in robotic systems where an actuator controls the joint variable at each joint.

A four-bar linkage is shown in Figure 6.2. Link number 1 is the ground link MN. The ground link is the base and used as a reference link. We measure all the variables with respect to the ground link. Link number $2 \equiv MA$ is usually the *input link* which is controlled by the *input angle* θ_2. Link number $4 \equiv NB$ is usually the *output link* with angular position θ_4, and link number $3 \equiv AB$ is the *coupler link* with angular position θ_3 that connects the input and output links.

The angular position of the output and coupler links, θ_4 and θ_3, are functions of the links' length and the value of the *input variable* θ_2. The angles θ_4 and θ_3 can be calculated by the functions

$$\theta_4 = 2\tan^{-1}\left(\frac{-B \pm \sqrt{B^2 - 4AC}}{2A}\right) \tag{6.1}$$

$$\theta_3 = 2\tan^{-1}\left(\frac{-E \pm \sqrt{E^2 - 4DF}}{2D}\right) \tag{6.2}$$

where,

$$A = J_3 - J_1 + (1 - J_2)\cos\theta_2 \tag{6.3}$$
$$B = -2\sin\theta_2 \tag{6.4}$$
$$C = J_1 + J_3 - (1 + J_2)\cos\theta_2 \tag{6.5}$$
$$D = J_5 - J_1 + (1 + J_4)\cos\theta_2 \tag{6.6}$$
$$E = -2\sin\theta_2 \tag{6.7}$$
$$F = J_5 + J_1 - (1 - J_4)\cos\theta_2 \tag{6.8}$$

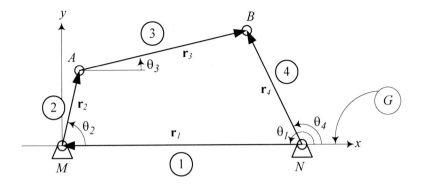

FIGURE 6.3. Expressing a four-bar linkage with a vector loop.

and

$$J_1 = \frac{d}{a} \tag{6.9}$$

$$J_2 = \frac{d}{c} \tag{6.10}$$

$$J_3 = \frac{a^2 - b^2 + c^2 + d^2}{2ac} \tag{6.11}$$

$$J_4 = \frac{d}{b} \tag{6.12}$$

$$J_5 = \frac{c^2 - d^2 - a^2 - b^2}{2ab} \tag{6.13}$$

Proof. We may show a four-bar linkage by a vector loop as is shown in Figure 6.3. The direction of each vector is arbitrary. However, the angle of each vector should be measured with respect to the positive direction of the x-axis. The vector expression of each link is shown in Table 6.1.

Table 6.1 - Vector representation of the four-bar linkage shown in Figure 6.3.

Link	Name	Vector	Length	Angle	Variable
1	Ground	$\mathbf{r}_1$	d	$\theta_1 = 180\,\mathrm{deg}$	–
2	Input	$\mathbf{r}_2$	a	θ_2	θ_2
3	Coupler	$\mathbf{r}_3$	b	θ_3	θ_3
4	Output	$\mathbf{r}_4$	c	θ_4	θ_4

The vector loop in the global coordinate frame G is

$$^{G}\mathbf{r}_4 + {}^{G}\mathbf{r}_1 + {}^{G}\mathbf{r}_2 - {}^{G}\mathbf{r}_3 = \mathbf{0} \tag{6.14}$$

where,

$$
\begin{aligned}
{}^{G}\mathbf{r}_1 &= -d\,\hat{\imath} & (6.15)\\
{}^{G}\mathbf{r}_2 &= a\,(\cos\theta_2\,\hat{\imath} + \sin\theta_2\,\hat{\jmath}) & (6.16)\\
{}^{G}\mathbf{r}_3 &= b\,(\cos\theta_3\,\hat{\imath} + \sin\theta_3\,\hat{\jmath}) & (6.17)\\
{}^{G}\mathbf{r}_4 &= c\,(\cos\theta_4\,\hat{\imath} + \sin\theta_4\,\hat{\jmath}) & (6.18)
\end{aligned}
$$

and the left superscript G reminds that the vectors are expressed in the global coordinate frame attached to the ground link. Substituting the Cartesian expressions for the planar vectors in Equation (6.14) yields

$$
\begin{aligned}
-d\,\hat{\imath} + a\,(\cos\theta_2\,\hat{\imath} + \sin\theta_2\,\hat{\jmath}) + b\,(\cos\theta_3\,\hat{\imath} + \sin\theta_3\,\hat{\jmath}) \\
- c\,(\cos\theta_4\,\hat{\imath} + \sin\theta_4\,\hat{\jmath}) = 0 \qquad (6.19)
\end{aligned}
$$

We may decompose Equation (6.19) into sin and cos components.

$$
\begin{aligned}
a\sin\theta_2 + b\sin\theta_3 - c\sin\theta_4 &= 0 & (6.20)\\
-d + a\cos\theta_2 + b\cos\theta_3 - c\cos\theta_4 &= 0 & (6.21)
\end{aligned}
$$

To derive the relationship between the input angle θ_2 and the output angle θ_4, the coupler angle θ_3 must be eliminated between Equations (6.20) and (6.21). Transferring the terms not containing θ_3 to the other side of the equations, and squaring both sides, provides the following equations:

$$
\begin{aligned}
(b\sin\theta_3)^2 &= (-a\sin\theta_2 + c\sin\theta_4)^2 & (6.22)\\
(b\cos\theta_3)^2 &= (-a\cos\theta_2 + c\cos\theta_4 + d)^2 & (6.23)
\end{aligned}
$$

By adding Equations (6.22) and (6.23), and simplifying, we derive the following equation:

$$
J_1\cos\theta_4 - J_2\cos\theta_2 + J_3 = \cos(\theta_4 - \theta_2) \qquad (6.24)
$$

where

$$
J_1 = \frac{d}{a} \qquad J_2 = \frac{d}{c} \qquad J_3 = \frac{a^2 - b^2 + c^2 + d^2}{2ac} \qquad (6.25)
$$

Equation (6.24) is called the *Freudenstein's equation*. The Freudenstein's equation may be expanded by using trigonometry

$$
\sin\theta_4 = \frac{2\tan\dfrac{\theta_4}{2}}{1 + \tan^2\dfrac{\theta_4}{2}} \qquad \cos\theta_4 = \frac{1 - \tan^2\dfrac{\theta_4}{2}}{1 + \tan^2\dfrac{\theta_4}{2}} \qquad (6.26)
$$

to provide a more practical equation

$$
A\tan^2\frac{\theta_4}{2} + B\tan\frac{\theta_4}{2} + C = 0 \qquad (6.27)
$$

where A, B, and C are functions of the input variable.

$$A = J_3 - J_1 + (1 - J_2)\cos\theta_2 \tag{6.28}$$
$$B = -2\sin\theta_2 \tag{6.29}$$
$$C = J_1 + J_3 - (1 + J_2)\cos\theta_2 \tag{6.30}$$

Equation (6.27) is quadratic in $\tan(\theta_4/2)$ and can be used to find the output angle θ_4 as a function of the input angle θ_2.

$$\theta_4 = 2\tan^{-1}\left(\frac{-B \pm \sqrt{B^2 - 4AC}}{2A}\right) \tag{6.31}$$

To find the relationship between the input angle θ_2 and the coupler angle θ_3, the output angle θ_4 must be eliminated between Equations (6.20) and (6.21). Transferring the terms not containing θ_4 to the right-hand side of the equations, and squaring both sides, provides

$$(c\sin\theta_4)^2 = (a\sin\theta_2 + b\sin\theta_3)^2 \tag{6.32}$$
$$(c\cos\theta_4)^2 = (a\cos\theta_2 + b\cos\theta_3 - d)^2 \tag{6.33}$$

By adding Equations (6.32) and (6.33), and simplifying, we derive the equation:

$$J_1\cos\theta_3 + J_4\cos\theta_2 + J_5 = \cos(\theta_3 - \theta_2) \tag{6.34}$$

where

$$J_4 = \frac{d}{b} \qquad J_5 = \frac{c^2 - d^2 - a^2 - b^2}{2ab} \tag{6.35}$$

Equation (6.34) may be expanded and transformed to

$$D\tan^2\frac{\theta_3}{2} + E\tan\frac{\theta_3}{2} + F = 0 \tag{6.36}$$

where D, E, and F are functions of the input variable.

$$D = J_5 - J_1 + (1 + J_4)\cos\theta_2 \tag{6.37}$$
$$E = -2\sin\theta_2 \tag{6.38}$$
$$F = J_5 + J_1 - (1 - J_4)\cos\theta_2 \tag{6.39}$$

Equation (6.36) is quadratic in $\tan(\theta_3/2)$ and can be solved to find the coupler angle θ_3.

$$\theta_3 = 2\tan^{-1}\left(\frac{-E \pm \sqrt{E^2 - 4DF}}{2D}\right) \tag{6.40}$$

Equations (6.31) and (6.40) calculate the output and coupler angles θ_4 and θ_3 as two functions of the input angle θ_2, provided that the lengths a, b, c, and d are given. ∎

Example 230 *Two possible configurations for a four-bar linkage.*

At any angle θ_2, and for suitable values of a, b, c, and d, Equations (6.1) and (6.2) provide us with two values for the output and coupler angles, θ_4 and θ_3. Both solutions are possible and provide two different configurations for each input angle θ_2.

A suitable set of (a, b, c, d) make the four-bar linkage a closed loop. A suitable set of numbers makes the radicals in Equations (6.1) and (6.2) real.

As an example, consider a linkage with

$$a = 1 \qquad b = 2 \qquad c = 2.5 \qquad d = 3 \qquad (6.41)$$

The $J_i, i = 1, 2, 3, 4, 5$ are functions of the links' length and are equal to

$$
\begin{aligned}
J_1 &= \frac{d}{a} = 3 \\[2mm]
J_2 &= \frac{d}{c} = \frac{3}{2.5} = 1.2 \\[2mm]
J_3 &= \frac{a^2 - b^2 + c^2 + d^2}{2ac} = 2.45 \\[2mm]
J_4 &= \frac{d}{b} = 1.5 \\[2mm]
J_5 &= \frac{c^2 - d^2 - a^2 - b^2}{2ab} = -1.9375 \qquad (6.42)
\end{aligned}
$$

The coefficients of the quadratic equations are then calculated.

$$
\begin{aligned}
A &= -0.6914213562 & B &= -1.414213562 \\
C &= 3.894365082 & D &= -3.169733048 \qquad (6.43) \\
E &= -1.414213562 & F &= 1.416053390
\end{aligned}
$$

Using the minus sign of (6.1) and (6.2), the output and coupler angles at $\theta_2 = \pi/4 \, \mathrm{rad} = 45 \deg$ are

$$
\begin{aligned}
\theta_4 &\approx 2 \, \mathrm{rad} \approx 114.73 \deg \\
\theta_3 &\approx 0.897 \, \mathrm{rad} \approx 51.42 \deg \qquad (6.44)
\end{aligned}
$$

and using the plus sign, they are

$$
\begin{aligned}
\theta_4 &\approx -2.6 \, \mathrm{rad} \approx -149 \deg \\
\theta_3 &\approx -1.495 \, \mathrm{rad} \approx -85.7 \deg \qquad (6.45)
\end{aligned}
$$

Figure 6.4 depicts the two possible configurations of the linkage for $\theta_2 = 45 \deg$. The configuration in Figure 6.4(a) is called **convex, non-crossed,** *or* **elbow-up,** *and the configuration in Figure 6.4(b) is called* **concave, crossed,** *or* **elbow-down.**

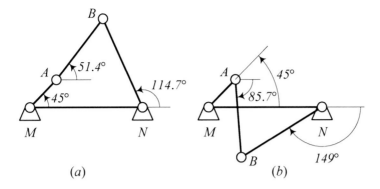

FIGURE 6.4. Two possible configuration of a four-bar linkage having the same input angle θ_2.

Example 231 *Velocity analysis of a four-bar linkage.*
The velocity analysis of a four-bar linkage is possible by taking a time derivative of Equations (6.20) and (6.21),

$$\frac{d}{dt}\left(a \sin \theta_2 + b \sin \theta_3 - c \sin \theta_4\right)$$

$$= a\omega_2 \cos \theta_2 + b\omega_3 \cos \theta_3 - c\omega_4 \cos \theta_4 = 0 \qquad (6.46)$$

$$\frac{d}{dt}\left(-d + a \cos \theta_2 + b \cos \theta_3 - c \cos \theta_4\right)$$

$$= -a\omega_2 \sin \theta_2 - b\omega_3 \sin \theta_3 + c\omega_4 \sin \theta_4 = 0 \qquad (6.47)$$

where

$$\omega_2 = \dot{\theta}_2 \qquad \omega_3 = \dot{\theta}_3 \qquad \omega_4 = \dot{\theta}_4 \qquad (6.48)$$

Assuming θ_2 and ω_2 are given, and θ_3, θ_4 are known from Equations (6.1) and (6.2), we solve Equations (6.46) and (6.47), for ω_3 and ω_4.

$$\omega_4 = \frac{a \sin \left(\theta_2 - \theta_3\right)}{c \sin \left(\theta_4 - \theta_3\right)}\omega_2 \qquad (6.49)$$

$$\omega_3 = \frac{a \sin \left(\theta_2 - \theta_4\right)}{b \sin \left(\theta_4 - \theta_3\right)}\omega_2 \qquad (6.50)$$

Example 232 *Velocity of moving joints for a four-bar linkage.*
Having the coordinates θ_2, θ_3, θ_4 and velocities ω_2, ω_3, ω_4 enables us to calculate the absolute and relative velocities of points A and B shown in Figure 6.3. The absolute velocity of a point refers to the its velocity relative to the ground link, and relative velocity of a point refers to its velocity relative of another moving point.

The absolute velocity of points A and B are

$$
{}^{G}\mathbf{v}_A = {}_{G}\boldsymbol{\omega}_2 \times {}^{G}\mathbf{r}_2
$$

$$
= \begin{bmatrix} 0 \\ 0 \\ \omega_2 \end{bmatrix} \times \begin{bmatrix} a\cos\theta_2 \\ a\sin\theta_2 \\ 0 \end{bmatrix} = \begin{bmatrix} -a\omega_2\sin\theta_2 \\ a\omega_2\cos\theta_2 \\ 0 \end{bmatrix} \tag{6.51}
$$

$$
{}^{G}\mathbf{v}_B = {}_{G}\boldsymbol{\omega}_4 \times {}^{G}\mathbf{r}_4
$$

$$
= \begin{bmatrix} 0 \\ 0 \\ \omega_4 \end{bmatrix} \times \begin{bmatrix} c\cos\theta_4 \\ c\sin\theta_4 \\ 0 \end{bmatrix} = \begin{bmatrix} -c\omega_4\sin\theta_4 \\ c\omega_4\cos\theta_4 \\ 0 \end{bmatrix} \tag{6.52}
$$

and the velocity of point B with respect to point A is

$$
{}^{G}\mathbf{v}_{B/A} = {}^{G}\mathbf{v}_B - {}^{G}\mathbf{v}_A = \begin{bmatrix} -c\omega_4\sin\theta_4 \\ c\omega_4\cos\theta_4 \\ 0 \end{bmatrix} - \begin{bmatrix} -a\omega_2\sin\theta_2 \\ a\omega_2\cos\theta_2 \\ 0 \end{bmatrix}
$$

$$
= \begin{bmatrix} a\omega_2\sin\theta_2 - c\omega_4\sin\theta_4 \\ c\omega_4\cos\theta_4 - a\omega_2\cos\theta_2 \\ 0 \end{bmatrix} \tag{6.53}
$$

The velocity of point B with respect to A can also be found as

$$
{}^{G}\mathbf{v}_{B/A} = {}^{G}R_2\,{}^{2}\mathbf{v}_B = {}^{G}R_2\,{}^{2}\mathbf{v}_B = {}^{G}R_2\left({}_{2}\boldsymbol{\omega}_3 \times {}^{2}\mathbf{r}_3\right) = {}_{G}\boldsymbol{\omega}_3 \times {}^{G}\mathbf{r}_3
$$

$$
= \begin{bmatrix} 0 \\ 0 \\ \omega_3 \end{bmatrix} \times \begin{bmatrix} b\cos\theta_3 \\ b\sin\theta_3 \\ 0 \end{bmatrix} = \begin{bmatrix} -b\omega_3\sin\theta_3 \\ b\omega_3\cos\theta_3 \\ 0 \end{bmatrix} \tag{6.54}
$$

Equations (6.53) and (6.54) are both correct and convertible to each other.

Example 233 *Acceleration analysis of a four-bar linkage.*

   The acceleration analysis of a four-bar linkage is possible by taking a time derivative from Equations (6.46) and (6.47),

$$
\frac{d}{dt}\left(a\,\omega_2\cos\theta_2 + b\,\omega_3\cos\theta_3 - c\,\omega_4\cos\theta_4\right)
$$

$$
= a\alpha_2\cos\theta_2 + b\alpha_3\cos\theta_3 - c\alpha_4\cos\theta_4
$$
$$
-a\omega_2^2\sin\theta_2 - b\omega_3^2\sin\theta_3 + c\omega_4^2\sin\theta_4 = 0 \tag{6.55}
$$

$$
\frac{d}{dt}\left(-a\,\omega_2\sin\theta_2 - b\,\omega_3\sin\theta_3 + c\,\omega_4\sin\theta_4\right)
$$

$$
= -a\alpha_2\sin\theta_2 - b\alpha_3\sin\theta_3 + c\alpha_4\sin\theta_4
$$
$$
-a\omega_2^2\cos\theta_2 - b\omega_3^2\cos\theta_3 + c\omega_4^2\cos\theta_4 = 0 \tag{6.56}
$$

where

$$
\alpha_2 = \dot{\omega}_2 \qquad \alpha_3 = \dot{\omega}_3 \qquad \alpha_4 = \dot{\omega}_4 \tag{6.57}
$$

Assuming θ_2, ω_2, and α_2 are given as the kinematics of the input link, θ_3, θ_4 are known from Equations (6.1) and (6.2), and ω_3, ω_4 are known from Equations (6.49) and (6.50), we solve Equations (6.55) and (6.56), for α_3 and α_4.

$$\alpha_4 = \frac{C_3 C_4 - C_1 C_6}{C_1 C_5 - C_2 C_4} \tag{6.58}$$

$$\alpha_3 = \frac{C_3 C_5 - C_2 C_6}{C_1 C_5 - C_2 C_4} \tag{6.59}$$

where

$$
\begin{aligned}
C_1 &= c \sin\theta_4 \qquad\qquad C_2 = b \sin\theta_3 \\
C_3 &= a\alpha_2 \sin\theta_2 + a\omega_2^2 \cos\theta_2 + b\omega_3^2 \cos\theta_3 - c\omega_4^2 \cos\theta_4 \\
C_4 &= c \cos\theta_4 \qquad\qquad C_5 = b \cos\theta_3 \\
C_6 &= a\alpha_2 \cos\theta_2 - a\omega_2^2 \sin\theta_2 - b\omega_3^2 \sin\theta_3 + c\omega_4^2 \sin\theta_4 \qquad (6.60)
\end{aligned}
$$

Example 234 *Acceleration of moving joints for a four-bar linkage.*

Having the angular kinematics of a four-bar linkage θ_2, θ_3, θ_4, ω_2, ω_3, ω_4, α_2, α_3, and α_4 is necessary and enough to calculate the absolute and relative accelerations of points A and B shown in Figure 6.3.

The absolute acceleration of points A and B are

$$
\begin{aligned}
{}^{G}\mathbf{a}_A &= {}_{G}\boldsymbol{\alpha}_2 \times {}^{G}\mathbf{r}_2 + {}_{G}\boldsymbol{\omega}_2 \times \left({}_{G}\boldsymbol{\omega}_2 \times {}^{G}\mathbf{r}_2\right) \\
&= \begin{bmatrix} -a\alpha_2 \sin\theta_2 - a\omega_2^2 \cos\theta_2 \\ a\alpha_2 \cos\theta_2 - a\omega_2^2 \sin\theta_2 \\ 0 \end{bmatrix} \tag{6.61}
\end{aligned}
$$

$$
\begin{aligned}
{}^{G}\mathbf{a}_B &= {}_{G}\boldsymbol{\alpha}_4 \times {}^{G}\mathbf{r}_4 + {}_{G}\boldsymbol{\omega}_4 \times \left({}_{G}\boldsymbol{\omega}_4 \times {}^{G}\mathbf{r}_4\right) \\
&= \begin{bmatrix} -c\alpha_4 \sin\theta_4 - c\omega_4^2 \cos\theta_4 \\ c\alpha_4 \cos\theta_4 - c\omega_4^2 \sin\theta_4 \\ 0 \end{bmatrix} \tag{6.62}
\end{aligned}
$$

where

$$
{}^{G}\mathbf{r}_2 = \begin{bmatrix} a\cos\theta_2 \\ a\sin\theta_2 \\ 0 \end{bmatrix} \qquad
{}^{G}\mathbf{r}_4 = \begin{bmatrix} c\cos\theta_4 \\ c\sin\theta_4 \\ 0 \end{bmatrix} \tag{6.63}
$$

$$
{}_{G}\boldsymbol{\omega}_2 = \begin{bmatrix} 0 \\ 0 \\ \omega_2 \end{bmatrix} \qquad
{}_{G}\boldsymbol{\omega}_4 = \begin{bmatrix} 0 \\ 0 \\ \omega_4 \end{bmatrix} \tag{6.64}
$$

$$
{}_{G}\boldsymbol{\alpha}_2 = \begin{bmatrix} 0 \\ 0 \\ \alpha_2 \end{bmatrix} \qquad
{}_{G}\boldsymbol{\alpha}_4 = \begin{bmatrix} 0 \\ 0 \\ \alpha_4 \end{bmatrix} \tag{6.65}
$$

The acceleration of point B with respect to point A is

$$
\begin{aligned}
{}^{G}\mathbf{a}_{B/A} &= {}_{G}\boldsymbol{\alpha}_{3} \times {}^{G}\mathbf{r}_{3} + {}_{G}\boldsymbol{\omega}_{3} \times \left({}_{G}\boldsymbol{\omega}_{3} \times {}^{G}\mathbf{r}_{3} \right) \\
&= \begin{bmatrix} -b\alpha_{3}\sin\theta_{3} - b\omega_{3}^{2}\cos\theta_{3} \\ b\alpha_{3}\cos\theta_{3} - b\omega_{3}^{2}\sin\theta_{3} \\ 0 \end{bmatrix} \qquad (6.66)
\end{aligned}
$$

where

$$
{}^{G}\mathbf{r}_{3} = \begin{bmatrix} b\cos\theta_{3} \\ b\sin\theta_{3} \\ 0 \end{bmatrix} \qquad {}_{G}\boldsymbol{\omega}_{3} = \begin{bmatrix} 0 \\ 0 \\ \omega_{3} \end{bmatrix} \qquad {}_{G}\boldsymbol{\alpha}_{3} = \begin{bmatrix} 0 \\ 0 \\ \alpha_{3} \end{bmatrix} \qquad (6.67)
$$

Example 235 *Grashoff criterion.*

The ability of a four-bar linkage to have a rotary link is determined by the Grashoff criterion. Assume the four links have the lengths s, l, p, and q, where

$$
\begin{aligned}
l &= \quad longest\ link \\
s &= \quad shortest\ link \\
p, q &= \quad the\ other\ two\ links
\end{aligned}
$$

then, the **Grashoff criterion** states that the linkage has a rotary link if

$$
l + s < p + q \qquad (6.68)
$$

Different types of a Grashoff mechanism are:

1. *Shortest link is the input link, then the mechanism is a crank-rocker.*
2. *Shortest link is the ground link, and the mechanism is a crank-crank.*
3. *At all other conditions, the mechanism is a rocker-rocker.*

A crank-crank mechanism is also called a drag-link.

Example 236 *Limit positions for a four-bar linkage.*

When the output link of a four-bar linkage stops while the input link can turn, we call the linkage is at a **limit position**. It occurs when the angle between the input and coupler links is either $180\,\mathrm{deg}$ or $360\,\mathrm{deg}$. Limit positions of a four-bar linkage, if there are any, must be determined by the designer to make sure the linkage is designed properly. A sample of limit positions for a four-bar linkage are shown in Figure 6.5.

We show the limit angles of the output link by $\theta_{4_{L1}}$, $\theta_{4_{L2}}$, and the corresponding input angles by $\theta_{2_{L1}}$, $\theta_{2_{L2}}$. They can be calculated as

$$
\theta_{2_{L1}} = \cos^{-1}\left[\frac{(a+b)^{2} + d^{2} - c^{2}}{2d(a+b)}\right] \qquad (6.69)
$$

$$
\theta_{4_{L1}} = \cos^{-1}\left[\frac{(a+b)^{2} - d^{2} - c^{2}}{2cd}\right] \qquad (6.70)
$$

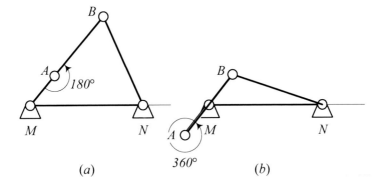

FIGURE 6.5. Limit position for a four-bar linkage.

$$\theta_{2_{L2}} = \cos^{-1}\left[\frac{(b-a)^2 + d^2 - c^2}{2d\,(b-a)}\right] \tag{6.71}$$

$$\theta_{4_{L2}} = \cos^{-1}\left[\frac{(b-a)^2 - d^2 - c^2}{2cd}\right] \tag{6.72}$$

The sweep angle of the output link would be

$$\phi = \theta_{4_{L2}} - \theta_{4_{L1}} \tag{6.73}$$

Example 237 *Dead positions for a four-bar linkage.*

*When the input link of a four-bar linkage locks, we say the linkage is at a **dead position**. It happens when the angle between the output and coupler links is either 180 deg or 360 deg. Dead positions of a four-bar linkage, if there are any, must be determined by the designer to make sure the linkage is never stuck in a dead position. A dead position for a four-bar linkage is shown in Figure 6.6.*

We show the dead angle of the output link by $\theta_{4_{D1}}$, $\theta_{4_{D2}}$, and the corresponding input angles by $\theta_{2_{D1}}$, $\theta_{2_{D2}}$. They can be calculated as

$$\theta_{2_{D1}} = \cos^{-1}\left[\frac{a^2 + d^2 - (b+c)^2}{2ad}\right] \tag{6.74}$$

$$\theta_{4_{D1}} = \cos^{-1}\left[\frac{a^2 - d^2 - (b+c)^2}{2\,(b+c)\,d}\right] \tag{6.75}$$

$$\theta_{2_{D2}} = \cos^{-1}\left[\frac{a^2 + d^2 - (b-c)^2}{2ad}\right] \tag{6.76}$$

$$\theta_{4_{D2}} = \cos^{-1}\left[\frac{a^2 - d^2 - (b-c)^2}{2ad}\right] \tag{6.77}$$

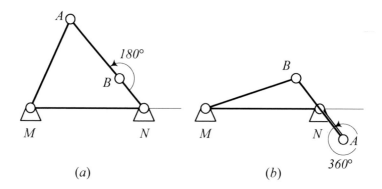

FIGURE 6.6. Dead position for a four-bar linkage.

Example 238 ★ *Designing a four-bar linkage using Freudenstein's equation.*

Designing a mechanism can be thought of as determining the required lengths of the links to accomplish a specific task.

Freudenstein's equation (6.24)

$$J_1 \cos \theta_4 - J_2 \cos \theta_2 + J_3 = \cos (\theta_4 - \theta_2) \tag{6.78}$$

$$J_1 = \frac{d}{a} \qquad J_2 = \frac{d}{c} \qquad J_3 = \frac{a^2 - b^2 + c^2 + d^2}{2ac} \tag{6.79}$$

determines the input-output relationship of a four-bar linkage. This equation can be utilized to design a four-bar linkage for three associated input-output angles.

Figure 6.7 illustrates the four popular windshield wiper systems. Double-arm parallel method is the most popular wiping system that serves most passenger cars. The double-arm opposing method has been using since last century, however, it was never very popular. The single-arm simple method is not very efficient, so the controlled single-arm is designed to maximize the wiped area.

*Wipers are used on windshields, and headlights. Figure 6.8 illustrates a sample of double-arm parallel windshield wiper mechanism. A four-bar linkage makes the main mechanism match the angular positions of the left and right wipers. A **dyad** or a two-link connects the driving motor to the main four-bar linkage and converts the rotational output of the motor into the back-and-forth motion of the wipers.*

The input and output links of the main four-bar linkage at three different positions are shown in Figure 6.9. We show the beginning and the end angles for the input link by θ_{21} and θ_{23}, and for the output link by θ_{41} and θ_{43} respectively. To design the mechanism we must match the angular positions of the left and right blades at the beginning and at the end positions. Let us add another match point approximately in the middle of the total sweep

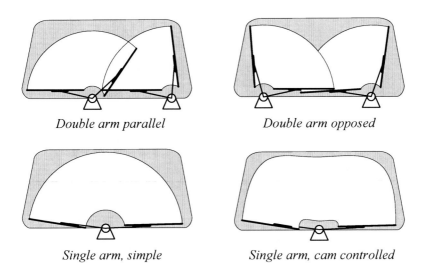

FIGURE 6.7. Four popular windshield wiper systems.

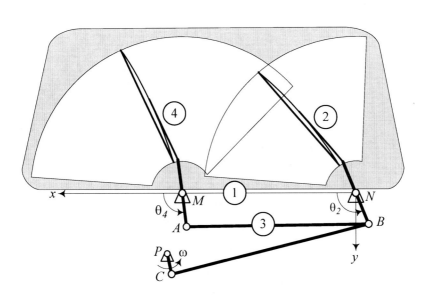

FIGURE 6.8. A sample of double-arm parallel windshield wiper mechanism.

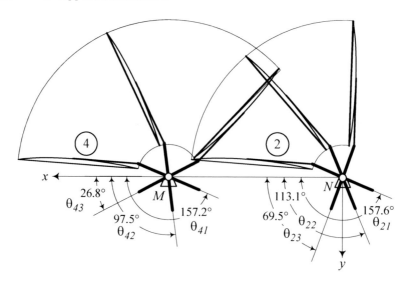

FIGURE 6.9. The input and output links of the main four-bar linkage of a windshield wiper at three different positions.

angles and design a four-bar linkage to match the angles indicated in Table 6.2.

Table 6.2 - Matching angles for a four-bar linkage of the double-arm parallel mechanism shown in Figure 6.9.

Matching	Input angle	Output angle
1	$\theta_{21} = 157.6\deg \approx 2.75\,\mathrm{rad}$	$\theta_{41} = 157.2\deg \approx 2.74\,\mathrm{rad}$
2	$\theta_{22} = 113.1\deg \approx 1.97\,\mathrm{rad}$	$\theta_{42} = 97.5\deg \approx 1.7\,\mathrm{rad}$
3	$\theta_{23} = 69.5\deg \approx 1.213\,\mathrm{rad}$	$\theta_{43} = 26.8\deg \approx 0.468\,\mathrm{rad}$

Substituting the input and output angles in Freudenstein's equation (6.24)

$$
\begin{aligned}
J_1 \cos\theta_{41} - J_2 \cos\theta_{21} + J_3 &= \cos(\theta_{41} - \theta_{21}) \\
J_1 \cos\theta_{42} - J_2 \cos\theta_{22} + J_3 &= \cos(\theta_{42} - \theta_{22}) \\
J_1 \cos\theta_{43} - J_2 \cos\theta_{23} + J_3 &= \cos(\theta_{43} - \theta_{23})
\end{aligned}
\qquad (6.80)
$$

provides us with equations

$$
\begin{aligned}
J_1 \cos 2.74 - J_2 \cos 2.75 + J_3 &= \cos(2.74 - 2.75) \\
J_1 \cos 1.7 - J_2 \cos 1.97 + J_3 &= \cos(1.7 - 1.97) \\
J_1 \cos 0.468 - J_2 \cos 1.213 + J_3 &= \cos(0.468 - 1.213)
\end{aligned}
\qquad (6.81)
$$

The set of equations (6.81) is linear for the unknowns J_1, J_2, and J_3

$$
\begin{bmatrix}
-0.92044 & 0.9243 & 1 \\
-0.12884 & 0.38868 & 1 \\
0.89247 & -0.35021 & 1
\end{bmatrix}
\begin{bmatrix}
J_1 \\
J_2 \\
J_3
\end{bmatrix}
=
\begin{bmatrix}
0.99995 \\
0.96377 \\
0.73509
\end{bmatrix}
\qquad (6.82)
$$

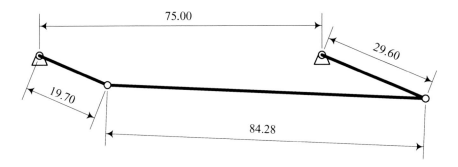

FIGURE 6.10. The main four-bar linkage of the windshield wiper at the initial position measured in [cm].

with the following solution:

$$\begin{bmatrix} J_1 \\ J_2 \\ J_3 \end{bmatrix} = \begin{bmatrix} 2.5284 \\ 3.8043 \\ -0.18911 \end{bmatrix} \qquad (6.83)$$

The three factors J_1, J_2, J_3 should be used to find the four links' length.

$$J_1 = \frac{d}{a} \qquad J_2 = \frac{d}{c} \qquad J_3 = \frac{a^2 - b^2 + c^2 + d^2}{2ac} \qquad (6.84)$$

So, we may set the length of one of the links, based on the physical situation and solve the equations for the other three lengths. Traditionally, we use $a = 1$ and find the remaining lengths. Then, the designed mechanism can be magnified or shrunk to fit the required geometry. In this example, we find

$$a = 1 \qquad b = 2.8436 \qquad c = 0.66462 \qquad d = 2.5284 \qquad (6.85)$$

Assuming a distance $d = 75\,\mathrm{cm} \approx 29.5\,\mathrm{in}$ for a real passenger car, between the left and right fixed joints M and N, we find the following dimensions:

$$\begin{aligned} a &= 296\,\mathrm{mm} \qquad b = 843\,\mathrm{mm} \\ c &= 197\,\mathrm{mm} \qquad d = 750\,\mathrm{mm} \end{aligned} \qquad (6.86)$$

Such a mechanism is shown in Figure 6.10 at the initial position.

Example 239 ★ *Equal sweep angles for input and output links.*
   Let us place the second matching point of the windshield wiper mechanism in Example 238 exactly in the middle of the total sweep angles

$$\begin{aligned} \theta_{22} &= \frac{157.6 + 69.5}{2} = 113.55\,\mathrm{deg} \approx 1.982\,\mathrm{rad} \\ \theta_{42} &= \frac{157.2 + 26.8}{2} = 92\,\mathrm{deg} \approx 1.605\,\mathrm{rad} \end{aligned} \qquad (6.87)$$

The first and second sweep angles for such matching points would be equal. Having equal sweep angles makes the motion of the wipers more uniform, although it cannot guarantee that the angular speed ratio of the left and right blades remains constant.

The matching points for the main four-bar linkage of the windshield wiper with equal sweep angles are indicated in Table 6.3.

Table 6.3 - *Equal sweep angle matching points for the four-bar linkage of the double-arm parallel mechanism shown in Figure 6.9.*

Matching	Input angle	Output angle
1	$\theta_{21} = 157.6 \deg \approx 2.75 \, \text{rad}$	$\theta_{41} = 157.2 \deg \approx 2.74 \, \text{rad}$
2	$\theta_{22} = 113.55 \deg \approx 1.982 \, \text{rad}$	$\theta_{42} = 92 \deg \approx 1.605 \, \text{rad}$
3	$\theta_{23} = 69.5 \deg \approx 1.213 \, \text{rad}$	$\theta_{43} = 26.8 \deg \approx 0.468 \, \text{rad}$

Substituting the angles in Freudenstein's equation (6.24) provides us with three equations:

$$
\begin{aligned}
J_1 \cos 2.74 - J_2 \cos 2.75 + J_3 &= \cos(2.74 - 2.75) \\
J_1 \cos 1.605 - J_2 \cos 1.982 + J_3 &= \cos(1.605 - 1.982) \\
J_1 \cos 0.468 - J_2 \cos 1.213 + J_3 &= \cos(0.468 - 1.213)
\end{aligned}
\tag{6.88}
$$

The set of equations can be written in a matrix form for the three unknowns J_1, J_2, and J_3

$$
\begin{bmatrix}
-0.920\,44 & 0.924\,3 & 1 \\
-.0332 & .3993 & 1 \\
0.892\,47 & -0.350\,21 & 1
\end{bmatrix}
\begin{bmatrix}
J_1 \\
J_2 \\
J_3
\end{bmatrix}
=
\begin{bmatrix}
0.99995 \\
.929589 \\
0.73509
\end{bmatrix}
\tag{6.89}
$$

with the solution.

$$
\begin{bmatrix}
J_1 \\
J_2 \\
J_3
\end{bmatrix}
=
\begin{bmatrix}
0.276 \\
0.6 \\
0.699
\end{bmatrix}
\tag{6.90}
$$

Using $a = 1$ and the three factors J_1, J_2, and J_3

$$
J_1 = \frac{d}{a} \qquad J_2 = \frac{d}{c} \qquad J_3 = \frac{a^2 - b^2 + c^2 + d^2}{2ac}
\tag{6.91}
$$

we can find the links' length.

$$
a = 1 \qquad b = 0.803 \qquad c = 0.46 \qquad d = 0.276
\tag{6.92}
$$

Assuming a distance $d = 75 \, \text{cm} \approx 29.5 \, \text{in}$ between the left and right fixed joint M and N, we find the following dimensions for a real passenger car:

$$
\begin{aligned}
a &= 2717 \, \text{mm} \qquad b = 2182 \, \text{mm} \\
c &= 1250 \, \text{mm} \qquad d = 750 \, \text{mm}
\end{aligned}
\tag{6.93}
$$

These dimensions do not show a practical design because the links may be longer than the width of the vehicle.

It shows that the designed mechanism is highly dependent on the second match point. So, it might be possible to design a desirable mechanism by choosing a suitable second match point.

Example 240 ★ *Second match point and link's length.*

To examine how the design of the windshield wiper mechanism in Example 238 is dependent on the second match point, let us set

$$\theta_{22} = \frac{157.6 + 69.5}{2} = 113.55 \deg \approx 1.982\,\text{rad} \tag{6.94}$$

and make θ_{42} a variable. The three matching points for the main four-bar linkage of the windshield wiper are indicated in Table 6.4.

Table 6.4 - Variable second match point for the four-bar linkage of the double-arm parallel mechanism shown in Figure 6.9.

Matching	Input angle	Output angle
1	$\theta_{21} = 157.6 \deg \approx 2.75\,\text{rad}$	$\theta_{41} = 157.2 \deg \approx 2.74\,\text{rad}$
2	$\theta_{22} = 113.55 \deg \approx 1.982\,\text{rad}$	θ_{42}
3	$\theta_{23} = 69.5 \deg \approx 1.213\,\text{rad}$	$\theta_{43} = 26.8 \deg \approx 0.468\,\text{rad}$

The Freudenstein's equation (6.24) provides us with

$$J_1 \cos 2.74 - J_2 \cos 2.75 + J_3 = \cos(2.74 - 2.75)$$
$$J_1 \cos \theta_{42} - J_2 \cos 1.982 + J_3 = \cos(\theta_{42} - 1.982)$$
$$J_1 \cos 0.468 - J_2 \cos 1.213 + J_3 = \cos(0.468 - 1.213) \tag{6.95}$$

The set of equations gives the following solutions:

$$J_1 = \frac{79.657 \cos(\theta_{42} - 1.9815) - 70.96}{79.657 \cos \theta_{42} + 13.828}$$
$$J_2 = \frac{93.642 \cos(\theta_{42} - 1.981) + 13.681 \cos \theta_{42} - 81.045}{65.832 \cos \theta_{42} + 11.428}$$
$$J_3 = \frac{32.357 - 25.959 \cos(\theta_{42} - 1.981) + 53.184 \cos(\theta_{42})}{11.428 + 65.83 \cos(\theta_{42})} \tag{6.96}$$

Having $d = 75\,\text{cm} \approx 29.5\,\text{in}$ between the left and right fixed joint M and N as the ground link, and using the factors J_1, J_2, and J_3

$$J_1 = \frac{d}{a} \qquad J_2 = \frac{d}{c} \qquad J_3 = \frac{a^2 - b^2 + c^2 + d^2}{2ac} \tag{6.97}$$

we can find the length of the other links a, b, and c as functions of θ_{42}. Figure 6.11 illustrates how the angle θ_{42} affects the lengths of the links.

To fit the mechanism under the hood in a small space, we need to have the lengths a and c much shorter than the ground d. Based on Figure 6.11,

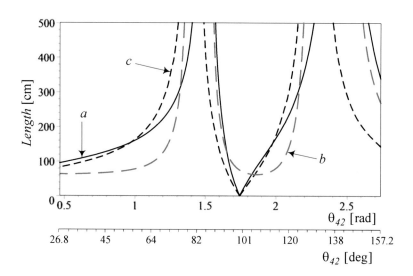

FIGURE 6.11. The length of links a, b, and c as functions of θ_{42}.

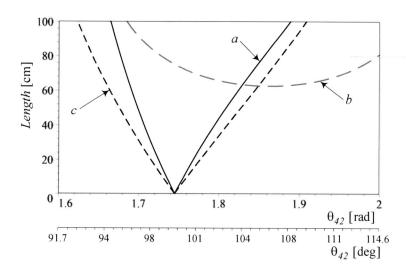

FIGURE 6.12. Maginification of the plot for the length of links a, b, and c as functions of θ_{42}, around the optimal design.

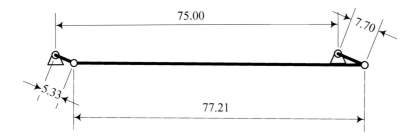

FIGURE 6.13. The finalized main four-bar linkage of the windshield wiper at the initial position measured in [cm].

a possible solution would be around $\theta_{42} = 100$ deg. Figure 6.12 illustrates a magnified view around $\theta_{42} = 100$ deg.

To have the length of a and c less than $100\,\mathrm{mm} \approx 3.94\,\mathrm{in}$ we pick $\theta_{42} = 99.52\,\mathrm{deg} \approx 1.737\,\mathrm{rad}$. Then the factors J_1, J_2, and J_3 are

$$
\begin{bmatrix} J_1 \\ J_2 \\ J_3 \end{bmatrix} = \begin{bmatrix} 9.740208376 \\ 14.06262379 \\ -3.032892944 \end{bmatrix} \tag{6.98}
$$

and the links for $d = 75\,\mathrm{cm} \approx 29.5\,\mathrm{in}$ are

$$
\begin{aligned}
a &= 77\,\mathrm{mm} & b &= 772\,\mathrm{mm} \\
c &= 53.3\,\mathrm{mm} & d &= 750\,\mathrm{mm}
\end{aligned} \tag{6.99}
$$

These numbers show a compact and reasonable mechanism. Figure 6.13 illustrates the finalized four-bar linkage of the windshield wiper at the initial position.

Example 241 ★ *Designing a dyad to attach a motor.*

The main four-bar linkage of a windshield wiper is a rocker-rocker mechanism because both the input and the output links must oscillate between two specific limits. To run the wipers and lock them at the limits, a two-link dyad can be designed. First we set the point of installing a rotary motor according to the physical conditions. Let point P, as shown in Figure 6.14, be the point at which we install the electric motor to run the mechanism. The next step would be to select a point on the input link to attach the second link of the dyad. Although joint B is usually the best choice, we select a point on the extension of the input link, indicated by D. There must be a dyad between joints D and P with lengths p and q. When the mechanism is at the initial position, joint D is at the longest distance form the motor P, and when it is at the final position, joint D is at the shortest distance form the motor P. Let us show the longest distance between P and D by l

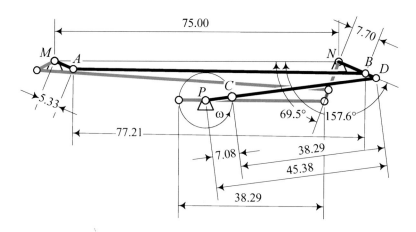

FIGURE 6.14. The final design of the windshield mechanism at the initial and final positions.

and the shortest distance by s.

$$l = \text{longest distance between } P \text{ and } D$$
$$s = \text{shortest distance between } P \text{ and } D$$
$$p, q = \text{dyad lengths between } P \text{ and } D$$

When P and D are at the maximum distance, the two-link dyad must be along each other, and when P and D are at the minimum distance, the two-link dyad must be on top of each other. Therefore,

$$l = q + p \qquad s = q - p \tag{6.100}$$

where p is the shortest link, and q is the longest link of the dyad. Solving Equations (6.100) for p and q provides

$$p = \frac{l - s}{2} \qquad q = \frac{l + s}{2} \tag{6.101}$$

In this example we measure

$$l = 453.8 \, \text{mm} \qquad s = 312.1 \, \text{mm} \tag{6.102}$$

and calculate for p and q

$$p = 70.8 \, \text{mm} \qquad q = 382.9 \, \text{mm} \tag{6.103}$$

The final design of the windshield mechanism and the running motor is shown in Figure 6.14 at the initial and final positions. The shorter link of the running dyad, p, must be attached to the motor at P, and the larger

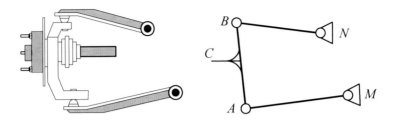

FIGURE 6.15. Double A arm suspension in a four-bar linkage mechanism.

link, q, connects joint D to the shorter link to C. The motor will turn the shorter link, PC continuously at an angular speed ω, while the longer link, CD, will run the mechanism and protect the wiper links from going beyond the initial and final angles.

Example 242 *Application of four-bar linkage in a vehicle.*

*The **double A-arm** suspension is a very popular mechanism for independent suspension of road vehicles. Figure 6.15 illustrates a double A-arm suspension and its equivalent kinematic model. We attach the wheel to a coupler point at C. The double A-arm is also called **double wishbone** suspension.*

6.2 Slider-Crank Mechanism

A *slider-crank mechanism* is shown in Figure 6.16. A slider-crank mechanism is a four-bar linkage. Link number 1 is the ground, which is the base and reference link. Link number $2 \equiv MA$ is usually the input link, which is controlled by the input angle θ_2. Link number 4 is the slider link that is usually considered as the output link. The output variable is the horizontal distance s between the slider and a fixed point on the ground, which is usually the revolute joint at M. If the slider slides on a flat surface, we define the horizon by a straight line parallel to the flat surface and passing through M. The link number $3 \equiv AB$ is the coupler link with angular position θ_3, which connects the input link to the output slider. This mechanism is called the slider-crank because in most applications, the input link is a crank link that rotates 360 deg, and the output is a slider.

The position of the output slider, s, and the angular position of the coupler link, θ_3, are functions of the link's length and the value of the input variable θ_2. The functions are

$$s = \frac{-G \pm \sqrt{G^2 - 4H}}{2} \tag{6.104}$$

$$\theta_3 = \sin^{-1}\left(\frac{e - a\sin\theta_2}{-b}\right) \tag{6.105}$$

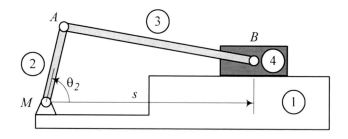

FIGURE 6.16. A slider-crank mechanism.

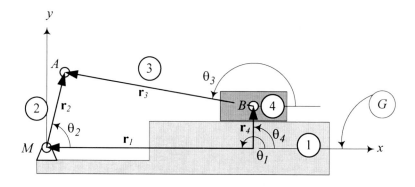

FIGURE 6.17. Expressing a slider-crank mechanism by a vector loop.

where

$$G = -2a \cos\theta_2 \tag{6.106}$$

$$H = a^2 + e^2 - b^2 - 2ae \sin\theta_2 \tag{6.107}$$

Proof. We show the slider-crank mechanism by a vector loop, as shown in Figure 6.17. The direction of each vector is arbitrary, however the angles should be associated to the vector's direction and be measured from the positive direction of the x-axis. The links and their expression vectors are shown in Table 6.5.

Table 6.5 - Vector representation of the slider-crank mechanism shown in Figure 6.17.

Link	Vector	Length	Angle	Variable
1	$^G\mathbf{r}_1$	s	$\theta_1 = 180$ deg	s
2	$^G\mathbf{r}_2$	a	θ_2	θ_2
3	$^G\mathbf{r}_3$	b	θ_3	θ_3
4	$^G\mathbf{r}_4$	e	$\theta_4 = 90$ deg	—

The equation of the vector loop is

$$^{G}\mathbf{r}_1 + {}^{G}\mathbf{r}_2 - {}^{G}\mathbf{r}_3 - {}^{G}\mathbf{r}_4 = \mathbf{0} \tag{6.108}$$

We may decompose the vector equation (6.108) into sin and cos components to generate two independent equations.

$$a \sin\theta_2 - b \sin\theta_3 - e = 0 \tag{6.109}$$
$$a \cos\theta_2 - b \cos\theta_3 - s = 0 \tag{6.110}$$

To derive the relationship between the input angle θ_2 and the output position s, the coupler angle θ_3 must be eliminated between Equations (6.109) and (6.110). Transferring the terms containing θ_3 to the other side of the equations, and squaring both sides, we get

$$(b \sin\theta_3)^2 = (a \sin\theta_2 - e)^2 \tag{6.111}$$
$$(b \cos\theta_3)^2 = (a \cos\theta_2 - s)^2 \tag{6.112}$$

By adding Equations (6.111) and (6.112), we derive the following equation:

$$s^2 - 2as \cos\theta_2 + a^2 + e^2 - b^2 - 2ae \sin\theta_2 = 0 \tag{6.113}$$

or

$$s^2 + Gs + H = 0 \tag{6.114}$$

where

$$G = -2a \cos\theta_2 \tag{6.115}$$
$$H = a^2 + e^2 - b^2 - 2ae \sin\theta_2 \tag{6.116}$$

Equation (6.114) is quadratic in s and provides two solutions:

$$s = \frac{-G \pm \sqrt{G^2 - 4H}}{2} \tag{6.117}$$

To find the relationship between the input angle θ_2 and the coupler angle θ_3, we can use Equations (6.109) or (6.110) to solve for θ_3.

$$\theta_3 = \sin^{-1}\left(\frac{e - a \sin\theta_2}{-b}\right) \tag{6.118}$$

$$\theta_3 = \cos^{-1}\left(\frac{s - a \cos\theta_2}{-b}\right) \tag{6.119}$$

Equations (6.117) and (6.118) calculate the output and coupler variables s and θ_3 as functions of the input angle θ_2, provided that the lengths a, b, and e are given. ∎

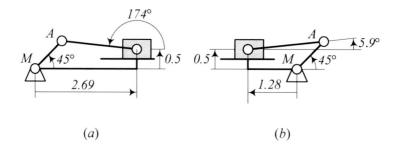

(a)   (b)

FIGURE 6.18. Two possible configurations of a slider-crank mechanism having the same input angle θ_2.

Example 243 *Two possible configurations for a slider-crank mechanism.*
At any angle θ_2, and for suitable values of a, b, and e, Equation (6.117) provides two values for the output variable s. Both solutions are possible and provide two different configurations for the mechanism. A suitable set of (a, b, e) is the numbers that make the radical in Equations (6.117) real.
As an example, consider a slider-crank mechanism at $\theta_2 = \pi/4\,\mathrm{rad} = 45\,\mathrm{deg}$ with the lengths

$$a = 1 \qquad b = 2 \qquad e = 0.5 \tag{6.120}$$

To solve the possible configurations, we start by calculating the coefficients of the quadratic equation (6.114)

$$
\begin{aligned}
G &= -2a\cos\theta_2 = -1.4142 & (6.121)\\
H &= a^2 + e^2 - b^2 - 2ae\sin\theta_2 = -3.4571 & (6.122)
\end{aligned}
$$

Employing Equation (6.117) provides

$$s = \frac{-G \pm \sqrt{G^2 - 4H}}{2} = \begin{cases} 2.696 \\ -1.282 \end{cases} \tag{6.123}$$

The corresponding coupler angle θ_3 can be calculated form either Equation (6.118) or (6.119).

$$
\begin{aligned}
\theta_3 &= \sin^{-1}\left(\frac{e - a\sin\theta_2}{-b}\right) = \cos^{-1}\left(\frac{s - a\cos\theta_2}{-b}\right)\\
&\approx \begin{cases} 3.037\,\mathrm{rad} \approx 174\,\mathrm{deg} \\ 0.103\,\mathrm{rad} \approx 5.9\,\mathrm{deg} \end{cases}
\end{aligned}
\tag{6.124}
$$

Figure 6.18 depicts the two possible configurations of the mechanism for $\theta_2 = 45\,\mathrm{deg}$.

Example 244 *Velocity analysis of a slider-crank mechanism.*

The velocity analysis of a slider-crank mechanism is possible by taking a time derivative of Equations (6.109) and (6.110),

$$\frac{d}{dt}\left(a\sin\theta_2 - b\sin\theta_3 - e\right) = a\,\omega_2\cos\theta_2 - b\,\omega_3\cos\theta_3 = 0 \qquad (6.125)$$

$$\frac{d}{dt}\left(a\cos\theta_2 - b\cos\theta_3 - s\right) = -a\,\omega_2\sin\theta_2 + b\,\omega_3\sin\theta_3 - \dot{s} = 0 \qquad (6.126)$$

where

$$\omega_2 = \dot{\theta}_2 \qquad \omega_3 = \dot{\theta}_3 \qquad (6.127)$$

Assuming θ_2 and ω_2 are given values, and s, θ_3 are known from Equations (6.104) and (6.105), we may solve Equations (6.125) and (6.126) for $\dot{s}$ and ω_3.

$$\dot{s} = \frac{\sin\left(\theta_3 - \theta_2\right)}{\cos\theta_3}a\omega_2 \qquad (6.128)$$

$$\omega_3 = \frac{\cos\theta_2}{\cos\theta_3}\frac{a}{b}\omega_2 \qquad (6.129)$$

Example 245 *Velocity of moving joints for a slider-crank mechanism.*

Having the coordinates θ_2, θ_3, s and velocities ω_2, ω_3, $\dot{s}$ enables us to calculate the absolute and relative velocities of points A and B shown in Figure 6.17. The absolute velocities of points A and B are

$$
\begin{aligned}
{}^{G}\mathbf{v}_A &= {}^{G}\boldsymbol{\omega}_2 \times {}^{G}\mathbf{r}_2 \\
&= \begin{bmatrix} 0 \\ 0 \\ \omega_2 \end{bmatrix} \times \begin{bmatrix} a\cos\theta_2 \\ a\sin\theta_2 \\ 0 \end{bmatrix} = \begin{bmatrix} -a\omega_2\sin\theta_2 \\ a\omega_2\cos\theta_2 \\ 0 \end{bmatrix} \qquad (6.130)
\end{aligned}
$$

$$
{}^{G}\mathbf{v}_B = \dot{s}\,\hat{\imath} = \begin{bmatrix} \dfrac{\sin\left(\theta_3 - \theta_2\right)}{\cos\theta_3}a\omega_2 \\ 0 \\ 0 \end{bmatrix} \qquad (6.131)
$$

and the velocity of point B with respect to point A is

$$
\begin{aligned}
{}^{G}\mathbf{v}_{B/A} &= {}^{G}\mathbf{v}_B - {}^{G}\mathbf{v}_A \\
&= \begin{bmatrix} \dfrac{\sin\left(\theta_3 - \theta_2\right)}{\cos\theta_3}a\omega_2 \\ 0 \\ 0 \end{bmatrix} - \begin{bmatrix} -a\omega_2\sin\theta_2 \\ a\omega_2\cos\theta_2 \\ 0 \end{bmatrix} \\
&= \begin{bmatrix} a\omega_2\sin\theta_2 + a\dfrac{\omega_2}{\cos\theta_3}\sin\left(\theta_3 - \theta_2\right) \\ -a\omega_2\cos\theta_2 \\ 0 \end{bmatrix} \qquad (6.132)
\end{aligned}
$$

The velocity of point B with respect to A can also be found as

$$
{}^G\mathbf{v}_{B/A} = {}^GR_2\,{}^2\mathbf{v}_B = {}^GR_2\,{}^2\mathbf{v}_B = {}^GR_2\left({}_2\boldsymbol{\omega}_3 \times {}^2\mathbf{r}_3\right) = {}_G\boldsymbol{\omega}_3 \times {}^G\mathbf{r}_3
$$

$$
= \begin{bmatrix} 0 \\ 0 \\ \omega_3 \end{bmatrix} \times \begin{bmatrix} b\cos\theta_3 \\ b\sin\theta_3 \\ 0 \end{bmatrix} = \begin{bmatrix} -b\omega_3\sin\theta_3 \\ b\omega_3\cos\theta_3 \\ 0 \end{bmatrix} \tag{6.133}
$$

Equations (6.132) and (6.133) are both correct and convertible to each other.

Example 246 *Acceleration analysis of a slider-crank mechanism.*

The acceleration analysis of a slider-crank mechanism is possible by taking another time derivative from Equations (6.125) and (6.126),

$$
\frac{d}{dt}\left(a\,\omega_2\cos\theta_2 - b\,\omega_3\cos\theta_3\right)
$$

$$
= a\alpha_2\cos\theta_2 - b\alpha_3\cos\theta_3 - a\omega_2^2\sin\theta_2 + b\omega_3^2\sin\theta_3 = 0 \tag{6.134}
$$

$$
\frac{d}{dt}\left(-a\,\omega_2\sin\theta_2 + b\,\omega_3\sin\theta_3 - \dot{s}\right)
$$

$$
= -a\alpha_2\sin\theta_2 - b\alpha_3\sin\theta_3 + a\omega_2^2\cos\theta_2 + b\omega_3^2\cos\theta_3 - \ddot{s} = 0 \tag{6.135}
$$

where,

$$
\alpha_2 = \dot{\omega}_2 \qquad \alpha_3 = \dot{\omega}_3 \tag{6.136}
$$

Assuming θ_2, ω_2, and α_2 are given values as the kinematics of the input link, s, θ_3, are known from Equations (6.104) and (6.105), and $\dot{s}$, ω_3 are known from Equations (6.128) and (6.129), we may solve Equations (6.134) and (6.135) for $\ddot{s}$ and α_3.

$$
\ddot{s} = \frac{-a\alpha_2\sin(\theta_2+\theta_3) + b\omega_3^2\cos 2\theta_3 + a\omega_2^2\cos(\theta_2-\theta_3)}{\cos\theta_3} \tag{6.137}
$$

$$
\alpha_3 = \frac{a\alpha_2\cos\theta_2 - a\omega_2^2\sin\theta_2 + b\omega_3^2\sin\theta_3}{b\cos\theta_3} \tag{6.138}
$$

Example 247 *Acceleration of moving joints of a slider-crank mechanism.*

Having the angular kinematics of a slider-crank mechanism θ_2, θ_3, s, ω_2, ω_3, $\dot{s}$, α_2, α_3, and $\ddot{s}$ are necessary and enough to calculate the absolute and relative accelerations of points A and B, shown in Figure 6.17.

The absolute acceleration of points A and B are

$$
{}^G\mathbf{a}_A = {}_G\boldsymbol{\alpha}_2 \times {}^G\mathbf{r}_2 + {}_G\boldsymbol{\omega}_2 \times \left({}_G\boldsymbol{\omega}_2 \times {}^G\mathbf{r}_2\right)
$$

$$
= \begin{bmatrix} -a\alpha_2\sin\theta_2 - a\omega_2^2\cos\theta_2 \\ a\alpha_2\cos\theta_2 - a\omega_2^2\sin\theta_2 \\ 0 \end{bmatrix} \tag{6.139}
$$

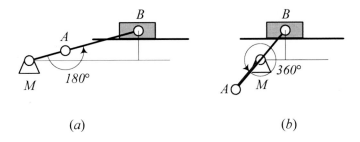

FIGURE 6.19. Limit position for a slider-crank mechanism.

$$^G\mathbf{a}_B = \ddot{s}\,\hat{\imath}$$

$$= \begin{bmatrix} \dfrac{-a\alpha_2 \sin(\theta_2 + \theta_3) + b\omega_3^2 \cos 2\theta_3 + a\omega_2^2 \cos(\theta_2 - \theta_3)}{\cos\theta_3} \\ 0 \\ 0 \end{bmatrix} \quad (6.140)$$

The acceleration of point B with respect to point A is

$$^G\mathbf{a}_{B/A} = {}_G\boldsymbol{\alpha}_3 \times {}^G\mathbf{r}_3 + {}_G\boldsymbol{\omega}_3 \times \left({}_G\boldsymbol{\omega}_3 \times {}^G\mathbf{r}_3 \right)$$

$$= \begin{bmatrix} -b\alpha_3 \sin\theta_3 - b\omega_3^2 \cos\theta_3 \\ b\alpha_3 \cos\theta_3 - b\omega_3^2 \sin\theta_3 \\ 0 \end{bmatrix} \quad (6.141)$$

Example 248 *Limit positions for a slider-crank mechanism.*

*When the output slider of a slider-crank mechanism stops while the input link can turn, we say the slider is at a **limit position**. It occurs when the angle between the input and coupler links is either* 180 deg *or* 360 deg. *Limit positions of a slider-crank mechanism are usually dictated by the design requirements. A limit position for a slider-crank mechanism is shown in Figure 6.19.*

We show the limit angle of the input link by $\theta_{2_{L1}}$, $\theta_{2_{L2}}$, *and the corresponding horizontal distance of the slider by* s_{Max}, s_{min}. *They can be calculated by the equations:*

$$\theta_{2_{L1}} = \sin^{-1}\left[\dfrac{e}{b+a}\right] \quad (6.142)$$

$$s_{Max} = \sqrt{(b+a)^2 - e^2} \quad (6.143)$$

$$\theta_{2_{L2}} = \pi + \sin^{-1}\left[\dfrac{e}{b-a}\right] \quad (6.144)$$

$$s_{min} = \sqrt{(b-a)^2 - e^2} \quad (6.145)$$

The length of stroke that the slider travels repeatedly would be

$$s = s_{Max} - s_{min} = \sqrt{(b+a)^2 - e^2} - \sqrt{(b-a)^2 - e^2} \qquad (6.146)$$

Example 249 ★ *Quick return slider-crank mechanism.*

Consider a slider-crank with a rotating input link at a constant angular velocity ω_2. The required time for the slider to move from s_{min} to s_{Max} is

$$t_1 = \frac{\theta_{2_{L2}} - \theta_{2_{L1}}}{\omega_2} = \frac{1}{\omega_2}\left(\pi + \sin^{-1}\left[\frac{e}{b-a}\right] - \sin^{-1}\left[\frac{e}{b+a}\right]\right) \qquad (6.147)$$

and the required time for returning from s_{Max} to s_{min} is

$$t_2 = \frac{\theta_{2_{L1}} - \theta_{2_{L2}}}{\omega_2} = \frac{1}{\omega_2}\left(\sin^{-1}\left[\frac{e}{b+a}\right] - \pi - \sin^{-1}\left[\frac{e}{b-a}\right]\right) \qquad (6.148)$$

If $e = 0$, then

$$\theta_{2_{L1}} = 0 \qquad \theta_{2_{L2}} = 180\,\text{deg} \qquad (6.149)$$

and therefore,

$$t_1 = t_2 = \frac{\pi}{\omega_2} \qquad (6.150)$$

However, when $e < 0$ then,

$$t_2 < t_1 \qquad (6.151)$$

*and the slider returns to s_{min} faster. Such a mechanism is called a **quick return** mechanism.*

6.3   Inverted Slider-Crank Mechanism

An *inverted slider-crank mechanism* is shown in Figure 6.20. It is a four-link mechanism. Link number 1 is the ground link, which is the base and reference link. Link number $2 \equiv MA$ is usually the input link, which is controlled by the input angle θ_2. Link number 4 is the slider link and is usually considered as the output link. The slider link has a revolute joint with the ground and a prismatic joint with the coupler link $3 \equiv AB$. The output variable can be the angle of the slider with the horizon, or the length AB. The link number $3 \equiv AB$ is the coupler link with angular position θ_3.

If we attach the coupler link of a slider-crank mechanism to the ground, an inverted slider-crank mechanism is made. Changing the grounded link produces a new mechanism that is called an *inversion* of the previous mechanism. Hence, the inverted slider-crank is an inversion of a slider-crank mechanism.

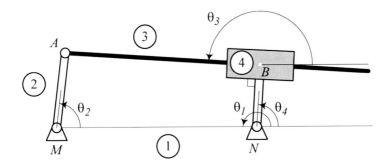

FIGURE 6.20. An inverted slider-crank mechanism.

The angular position of the output slider θ_4 and the length of the coupler link b are functions of the length of the links and the value of the input variable θ_2. These variables are:

$$b = \pm\sqrt{a^2 + d^2 - e^2 - 2ad\cos\theta_2} \qquad (6.152)$$

$$\theta_4 = \theta_3 + \frac{\pi}{2} = 2\tan^{-1}\left(\frac{-H \pm \sqrt{H^2 - 4GI}}{2G}\right) \qquad (6.153)$$

where

$$G = d - e - a\cos\theta_2 \qquad (6.154)$$

$$H = 2a\sin\theta_2 \qquad (6.155)$$

$$I = a\cos\theta_2 - d - e \qquad (6.156)$$

Proof. We show the inverted slider-crank mechanism by a vector loop as shown in Figure 6.21. The direction of each vector is arbitrary, however, the angles should be associated with the vector's direction and be measured from positive direction of the x-axis. The links and their expression vectors are shown in Table 6.6.

Table 6.6 - Vector representation of the inverted slider-crank mechanism shown in Figure 6.21.

Link	Vector	Length	Angle	Variable
1	$^G\mathbf{r}_1$	d	$\theta_1 = 180\deg$	d
2	$^G\mathbf{r}_2$	a	θ_2	θ_2
3	$^G\mathbf{r}_3$	b	θ_3	θ_3 or θ_4
4	$^G\mathbf{r}_4$	e	$\theta_4 = \theta_3 + 90\deg$	$-$

The vector loop is

$$^G\mathbf{r}_1 + {}^G\mathbf{r}_2 - {}^G\mathbf{r}_3 - {}^G\mathbf{r}_4 = \mathbf{0} \qquad (6.157)$$

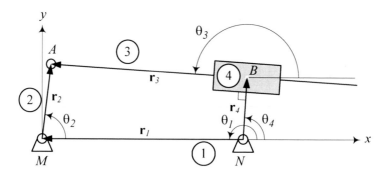

FIGURE 6.21. Kinematic model of an inverted slider-crank mechanism.

which can be decomposed into sin and cos components.

$$a \sin \theta_2 - b \sin \left(\theta_4 - \frac{\pi}{2} \right) - e \sin \theta_4 = 0 \tag{6.158}$$

$$-d + a \cos \theta_2 - b \sin \left(\theta_4 - \frac{\pi}{2} \right) - e \cos \theta_4 = 0 \tag{6.159}$$

To derive the relationship between the input angle θ_2 and the output θ_4, we eliminate b between Equations (6.158) and (6.159) and find

$$(a \cos \theta_2 - d) \cos \theta_4 + a \sin \theta_2 \sin \theta_4 - e = 0 \tag{6.160}$$

To have a better expression suitable for computer programming, we may use trigonometric equations

$$\sin \theta_4 = \frac{2 \tan \frac{\theta_4}{2}}{1 + \tan^2 \frac{\theta_4}{2}} \qquad \cos \theta_4 = \frac{1 - \tan^2 \frac{\theta_4}{2}}{1 + \tan^2 \frac{\theta_4}{2}} \tag{6.161}$$

to transform Equation (6.160) to a more useful equation

$$G \tan^2 \frac{\theta_4}{2} + H \tan \frac{\theta_4}{2} + I = 0 \tag{6.162}$$

where, G, H, and I are functions of the input variable.

$$G = d - e - a \cos \theta_2 \tag{6.163}$$
$$H = 2a \sin \theta_2 \tag{6.164}$$
$$I = a \cos \theta_2 - d - e \tag{6.165}$$

Equation (6.162) is quadratic in $\tan (\theta_4/2)$ and can be used to find the output angle θ_4.

$$\theta_4 = 2 \tan^{-1} \left(\frac{-H \pm \sqrt{H^2 - 4GI}}{2G} \right) \tag{6.166}$$

To find the relationship between the input angle θ_2 and the coupler length b, we may solve Equations (6.158) and (6.159) for $\sin\theta_4$ and $\cos\theta_4$

$$\sin\theta_4 = \frac{ab\cos\theta_2 - ae\sin\theta_2 + bd}{b^2 + e^2} \tag{6.167}$$

$$\cos\theta_4 = -\frac{ab\sin\theta_2 - ae\cos\theta_2 + ed}{b^2 + e^2} \tag{6.168}$$

or substitute (6.166) in (6.160) and solve for b. By squaring and adding Equations (6.167) and (6.168), we find the following equation:

$$a^2 - b^2 + d^2 - e^2 - 2ad\cos\theta_2 = 0 \tag{6.169}$$

which must be solved for b.

$$b = \pm\sqrt{a^2 + d^2 - e^2 - 2ad\cos\theta_2} \tag{6.170}$$

∎

Example 250 *Two possible configurations for an inverted slider-crank mechanism.*

At any angle θ_2, and for suitable values of a, d, and e, Equations (6.152) and (6.153) provide two values for the output b and coupler angles θ_4. Both solutions are possible and provide two different configurations for the mechanism. A suitable set of (a, d, e) are the numbers that make the radicals in Equations (6.152) and (6.153) real.

Consider an inverted slider-crank mechanism at $\theta_2 = \pi/4\,\mathrm{rad} = 45\,\deg$ with the lengths

$$a = 1 \qquad e = 0.5 \qquad d = 3 \tag{6.171}$$

The parameters of Equation (6.152) would be

$$\begin{aligned} G &= d - e - a\cos\theta_2 = 1.7929 \\ H &= 2a\sin\theta_2 = 1.4142 \\ I &= a\cos\theta_2 - d - e = -2.7929 \end{aligned} \tag{6.172}$$

Now, Equation (6.162) gives two real values for θ_4

$$\theta_4 \approx \begin{cases} 1.48\,\mathrm{rad} \approx 84.8\,\deg \\ -2.08\,\mathrm{rad} \approx -120\,\deg \end{cases} \tag{6.173}$$

Using $\theta_4 = 1.48\,\mathrm{rad}$, Equations (6.167) provides us with

$$b \approx 2.33 \tag{6.174}$$

and when $\theta_4 = -1.732\,\mathrm{rad}$ we get

$$b = -2.28 \tag{6.175}$$

Figure 6.22 depicts the two configurations of the mechanism for $\theta_2 = 45\,\deg$.

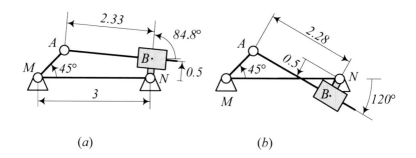

(a)   (b)

FIGURE 6.22. Two configurations of an inverted slider crank mechanism for $\theta_2 = 45\deg$.

Example 251 *Velocity analysis of an inverted slider-crank mechanism.*
The velocity analysis of a slider-crank mechanism can be found by taking a time derivative of Equations (6.158) and (6.159),

$$\frac{d}{dt}\left(a\sin\theta_2 + b\cos\theta_4 - e\sin\theta_4\right)$$

$$= a\,\omega_2\cos\theta_2 - b\,\omega_4\sin\theta_4 + \dot{b}\cos\theta_4 - e\,\omega_4\cos\theta_4 = 0 \quad (6.176)$$

$$\frac{d}{dt}\left(a\cos\theta_2 + b\cos\theta_4 - e\cos\theta_4 - d\right)$$

$$= -a\,\omega_2\sin\theta_2 - b\,\omega_4\sin\theta_4 + \dot{b}\cos\theta_4 + e\,\omega_4\sin\theta_4 = 0 \quad (6.177)$$

where

$$\omega_2 = \dot{\theta}_2 \qquad \omega_4 = \omega_3 = \dot{\theta}_4 \qquad (6.178)$$

Assuming θ_2 and ω_2 are given, and b, θ_4 are known from Equations (6.152) and (6.153), we can solve Equations (6.176) and (6.177) for $\dot{b}$ and ω_4.

$$\dot{b} = \frac{a}{b}\omega_2\left[b\cos\left(\theta_4 - \theta_2\right) - e\sin\left(\theta_4 - \theta_2\right)\right] \qquad (6.179)$$

$$\omega_4 = \omega_3 = \frac{a}{b}\omega_2\sin\left(\theta_2 - \theta_4\right) \qquad (6.180)$$

Example 252 *Velocity of moving joints of inverted slider-crank mechanism.*
Having the coordinates θ_2, θ_4, b and velocities ω_2, ω_4, $\dot{b}$ enables us to calculate the absolute and relative velocities of points A and B shown in Figure 6.21. The absolute and relative velocities of points A and B are

$$
{}^G\mathbf{v}_A = {}_G\boldsymbol{\omega}_2 \times {}^G\mathbf{r}_2
$$

$$
= \begin{bmatrix} 0 \\ 0 \\ \omega_2 \end{bmatrix} \times \begin{bmatrix} a\cos\theta_2 \\ a\sin\theta_2 \\ 0 \end{bmatrix} = \begin{bmatrix} -a\omega_2\sin\theta_2 \\ a\omega_2\cos\theta_2 \\ 0 \end{bmatrix} \qquad (6.181)
$$

$$
{}^G\mathbf{v}_{B_4} = {}_G\boldsymbol{\omega}_4 \times {}^G\mathbf{r}_4
$$

$$
= \begin{bmatrix} 0 \\ 0 \\ \omega_4 \end{bmatrix} \times \begin{bmatrix} e\cos\theta_4 \\ e\sin\theta_4 \\ 0 \end{bmatrix} = \begin{bmatrix} -e\omega_4\sin\theta_4 \\ e\omega_4\cos\theta_4 \\ 0 \end{bmatrix} \qquad (6.182)
$$

$$
{}^G\mathbf{v}_{B_3/A} = {}_G\boldsymbol{\omega}_3 \times \left(-{}^G\mathbf{r}_3\right)
$$

$$
= \begin{bmatrix} 0 \\ 0 \\ \omega_4 \end{bmatrix} \times \begin{bmatrix} -b\cos\theta_4 \\ -b\sin\theta_4 \\ 0 \end{bmatrix} = \begin{bmatrix} b\omega_4\sin\theta_4 \\ -b\omega_4\cos\theta_4 \\ 0 \end{bmatrix} \qquad (6.183)
$$

$$
{}^G\mathbf{v}_{B_3} = {}^G\mathbf{v}_{B_3/A} + {}^G\mathbf{v}_A
$$

$$
= \begin{bmatrix} b\omega_4\sin\theta_4 \\ -b\omega_4\cos\theta_4 \\ 0 \end{bmatrix} + \begin{bmatrix} -a\omega_2\sin\theta_2 \\ a\omega_2\cos\theta_2 \\ 0 \end{bmatrix}
$$

$$
= \begin{bmatrix} b\omega_4\sin\theta_4 - a\omega_2\sin\theta_2 \\ a\omega_2\cos\theta_2 - b\omega_4\cos\theta_4 \\ 0 \end{bmatrix} \qquad (6.184)
$$

$$
{}^G\mathbf{v}_{B_3/B_4} = {}^G\mathbf{v}_{B_3} - {}^G\mathbf{v}_{B_4}
$$

$$
= \begin{bmatrix} b\omega_4\sin\theta_4 - a\omega_2\sin\theta_2 \\ a\omega_2\cos\theta_2 - b\omega_4\cos\theta_4 \\ 0 \end{bmatrix} - \begin{bmatrix} -e\omega_4\sin\theta_4 \\ e\omega_4\cos\theta_4 \\ 0 \end{bmatrix}
$$

$$
= \begin{bmatrix} \omega_4 e\sin\theta_4 - a\omega_2\sin\theta_2 + b\omega_4\sin\theta_4 \\ a\omega_2\cos\theta_2 - \omega_4 e\cos\theta_4 - b\omega_4\cos\theta_4 \\ 0 \end{bmatrix} \qquad (6.185)
$$

Example 253 *Acceleration analysis of inverted slider-crank mechanism.*
The acceleration analysis of an inverted slider-crank mechanism can be found by taking another time derivative from Equations (6.176) and (6.177),

$$
\frac{d}{dt}\left(a\omega_2\cos\theta_2 - b\omega_4\sin\theta_4 + \dot{b}\cos\theta_4 - e\omega_4\cos\theta_4\right)
$$
$$
= a\alpha_2\cos\theta_2 - a\omega_2^2\sin\theta_2 - b\alpha_4\sin\theta_4 + b\omega_4^2\cos\theta_4
$$
$$
+\ddot{b}\cos\theta_4 - \dot{b}\omega_4\sin\theta_4 - e\alpha_4\cos\theta_4 + e\omega_4^2\sin\theta_4 = 0 \qquad (6.186)
$$

$$
\frac{d}{dt}\left(-a\omega_2\sin\theta_2 - b\omega_4\sin\theta_4 + \dot{b}\cos\theta_4 + e\omega_4\sin\theta_4\right)
$$
$$
= -a\alpha_2\sin\theta_2 + a\omega_2^2\cos\theta_2 - b\alpha_4\sin\theta_4 - b\omega_4^2\cos\theta_4
$$
$$
+\ddot{b}\cos\theta_4 - \dot{b}\omega_4\sin\theta_4 + e\alpha_4\sin\theta_4 + e\omega_4^2\cos\theta_4 = 0 \qquad (6.187)
$$

where

$$
\alpha_2 = \dot{\omega}_2 \qquad \alpha_4 = \alpha_3 = \dot{\omega}_4 = \dot{\omega}_3 \qquad (6.188)
$$

Assuming θ_2, ω_2, and α_4 are given values as the kinematics of the input link, b, θ_4, are known from Equations (6.152) and (6.153), and also $\dot{b}$, ω_4 are known from Equations (6.179) and (6.180), we may solve Equations (6.186) and (6.187), for $\ddot{b}$ and α_4

$$\ddot{b} = \frac{C_7 C_{12} - C_9 C_{10}}{C_7 C_{11} - C_8 C_{10}} \tag{6.189}$$

$$\alpha_4 = \frac{C_9 C_{11} - C_8 C_{12}}{C_7 C_{11} - C_8 C_{10}} \tag{6.190}$$

where,

$$
\begin{aligned}
C_7 &= \sin\theta_4 \\
C_8 &= b\cos\theta_4 + e\sin\theta_4 \\
C_9 &= a\alpha_2 \sin\theta_2 + a\omega_2^2 \cos\theta_2 - 2\dot{b}\omega_4 \cos\theta_4 \\
&\quad + b\omega_4^2 \sin\theta_4 - e\omega_4^2 \cos\theta_4 \\
C_{10} &= \cos\theta_4 \\
C_{11} &= -b\sin\theta_4 + e\cos\theta_4 \\
C_{12} &= a\alpha_2 \cos\theta_2 - a\omega_2^2 \sin\theta_2 + 2\dot{b}\omega_4 \sin\theta_4 \\
&\quad + b\omega_4^2 \cos\theta_4 + e\omega_4^2 \sin\theta_4 \tag{6.191}
\end{aligned}
$$

Example 254 *Application of inverted slider mechanism in vehicles.*

The McPherson strut suspension is a very popular mechanism for independent front suspension of street cars. Figure 6.23 illustrates a McPherson strut suspension and its equivalent kinematic model. We attach the wheel to a coupler point at C.

The piston rod of the shock absorber serves as a kingpin axis at the top of the strut. At the bottom, the shock absorber pivots on a ball joint on a single lower arm. The McPherson strut, also called the Chapman strut, was invented by Earl McPherson in the 1940s. It was first introduced on the 1949 Ford Vedette, and also adopted in the 1951 Ford Consul, and then become one of the dominating suspension systems because of its compactness and low cost.

6.4   Instant Center of Rotation

In the general planar motion of a rigid body, at a given instant, the velocity of any point of the body can be expressed as a rotation about an axis perpendicular to the plane. The axis intersects the plane at a point called the *instantaneous center of rotation* of the body with respect to the ground. The instantaneous center of rotation is also called *instant center, centro,* and *pole*.

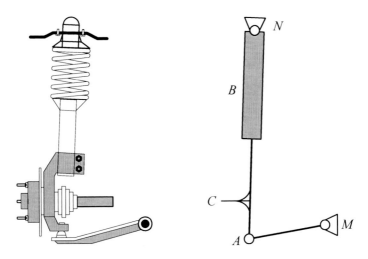

FIGURE 6.23. The McPherson strut suspension is an inverted slider mechanism.

If the directions of the velocities of two different body points A and B are known, the instant center of rotation I is at the intersection of the lines perpendicular to the velocity vectors $\mathbf{v}_A$ and $\mathbf{v}_B$. Such a situation is shown in Figure 6.24(a).

In the exceptional case where the points A and B are such that the velocity vectors $\mathbf{v}_A$ and $\mathbf{v}_B$ are perpendicular to the line AB then, the instantaneous center of rotation I is at the intersection of AB with the line joining the extremities of the velocity vectors. Such a situation is shown in Figure 6.24(b).

There is an instant center of rotation between every two relatively moving links. The instant center is a point common to both bodies that has the same velocity in each body coordinate frame.

The three instant centers , I_{12}, I_{23}, and I_{13} between three links numbered 1, 2, and 3 lie on a straight line. This statement is called the *Kennedy theorem* for three instant centers.

Proof. Consider the two ground connected and relatively moving bodies in Figure 6.25. The ground is link number 1. The links number 2 and 3 are pivoted to the ground at points M and N, and are rotating with angular velocities ω_2 and ω_3 with respect to the ground. The two links are contacted at point C.

The revolute joint at M is the instant center I_{12} between link 2 and 1, and the revolute joint at N is the instant center I_{13} between links 3 and 1. The velocity of any point of the link 2 with respect to the ground is perpendicular to the connection line of the point to M. Similarly, The velocity of any point of the link 3 with respect to the ground is perpendicular to the connection line of the point to N. Therefore, the velocity of the contact

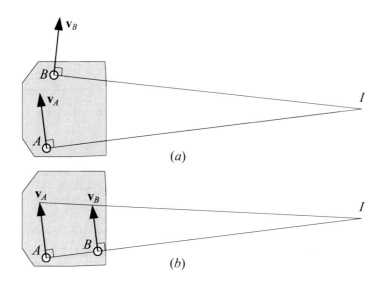

FIGURE 6.24. Determination of the instantaneous center of rotation I for a moving rigid body.

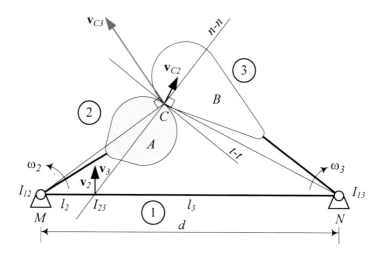

FIGURE 6.25. A 3-link mechanism with the ground as link number 1, and two moving links, numbers 2 and 3.

point C as a point of link 2 is $\mathbf{v}_{C_2}$, perpendicular to the radius MC, and the velocity of C as a point of link 3 is $\mathbf{v}_{C_3}$, perpendicular to the radius NC.

The instant center of rotation I_{23} must be a point with the same ground velocity in both bodies. Let us draw the normal line $n-n$, and tangential line $t-t$ to the curves of links 2 and 3, at the contact point C. The normal components of $\mathbf{v}_{C_2}$ and $\mathbf{v}_{C_3}$ must be equal to keep the bodies in touch. So is the velocity of any other point on $n-n$. The only difference between the velocities $\mathbf{v}_{C_2}$ and $\mathbf{v}_{C_3}$ is in their tangent components. Therefore, the instant center of rotation I_{23} must be on the normal line $n-n$ at a position where the velocity of the point on both bodies are equal on the line $t-t$. The intersection of the normal line $n-n$ with the center line MN is the only possible point for the instant center of rotation I_{13} at which the velocity on both bodies coincide. As their components on $n-n$ are equal, they must be equal.

Let us define

$$I_{12}I_{23} = l_2 \tag{6.192}$$
$$I_{13}I_{23} = l_3 \tag{6.193}$$

then, because the velocities of the two bodies must be equal at the common instant center of rotation, we have

$$l_2\omega_2 = l_3\omega_3 \tag{6.194}$$

or

$$\frac{\omega_2}{\omega_3} = \frac{l_3}{l_2} = \frac{1}{1 + \dfrac{d}{l_2}} \tag{6.195}$$

where, d is the length of the ground link MN. ∎

Example 255 *Number of instant centers.*

There is one instant center between every two relatively moving bodies. So, there are three instant centers between three bodies. The number N of instant centers between n relatively moving bodies is

$$N = \frac{n(n-1)}{2} \tag{6.196}$$

Thus, a four-bar linkage has six instant centers, I_{12}, I_{13}, I_{14}, I_{23}, I_{24}, I_{34}. The symbol I_{ij} indicates the instant center of rotation between links i and j. Because two links have only one instant center, we have

$$I_{ij} = I_{ji} \tag{6.197}$$

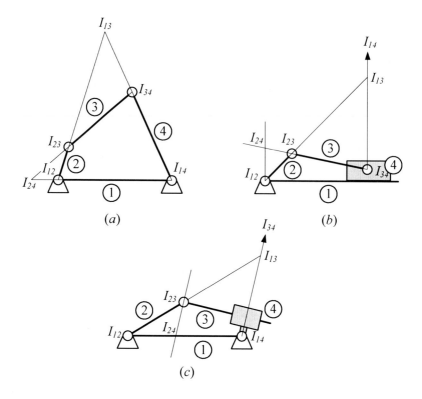

FIGURE 6.26. The instant centers of rotation for a four-bar linkage, a slider-crank, and an inverted slider-crank mechanism.

The four instant centers of rotation for a four-bar linkage, a slider-crank, and an inverted slider-crank mechanisms are shown in Figure 6.26. The instant center of rotation for two links that slide on each other is at infinity, on a line normal to the common tangent. So, I_{14} in Figure 6.26(b) is on a line perpendicular to the ground at infinity, and I_{34} in Figure 6.26(c) is on the perpendicular to the link 3 at infinity.

Figure 6.27 depicts the 15 instant centers for a six-link mechanism.

Example 256 *Application of instant center of rotation in vehicles.*

Figure 6.28 illustrates a double A-arm suspension and its equivalent four-bar linkage kinematic model. The wheel will be fastened to the coupler link AB, witch connects the upper A-arm BN to the lower A-arm AN. The A-arms are connected to the body with two revolute joints at N and M. The body of the vehicle acts as the ground link for the suspension mechanism.

Points N and M are, respectively, the instant centers of rotation for the upper and lower arms with respect to the body. The intersection point of the extension line for the upper and lower A-arms indicates the instant center of rotation for the coupler with respect to the body. When the suspension

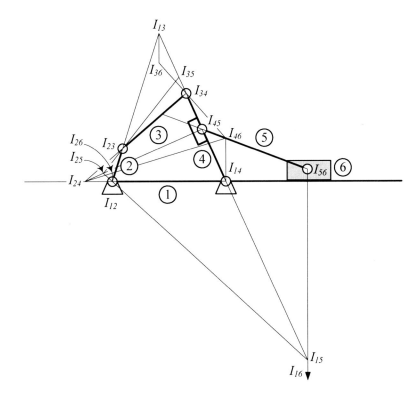

FIGURE 6.27. Fifteen instant center of rotations for a 6-link mechanism.

*moves, the wheel will rotate about point I with respect to the body. Point I is called the **roll center** of the wheel and body.*

Example 257 *The instant centers of rotation may not be stationary.*

When a mechanism moves, the instant centers of rotation may move, if they are not at a fixed joint with the ground. Figure 6.29 illustrates a four-bar linkage at a few different positions and shows the instant centers of rotation for the coupler with respect to the ground I_{13}. Point I_{13} will move when the linkage moves, and traces a path shown in the figure.

For a given set of the lengths a, b, c, d, the coordinates of points M, A, B, N are as given in Table 6.7.

Table 6.7 - Coordinates of the joints of a four-bar linkage.

point	x-coordinate	y-coordinate
M	0	0
A	$a\cos\theta_2$	$a\sin\theta_2$
B	$d + c\cos\theta_4$	$c\sin\theta_4$
N	d	0

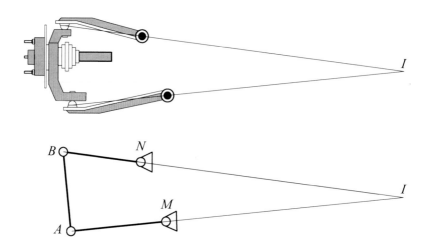

FIGURE 6.28. The roll center of a double A-arm suspension and its equivalent kinematic model.

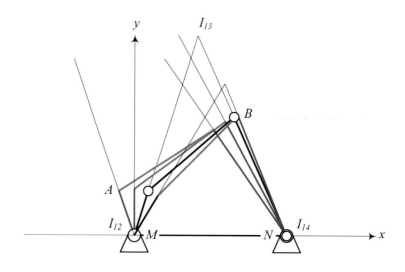

FIGURE 6.29. Path of motion for the instant center of rotation I_{13}.

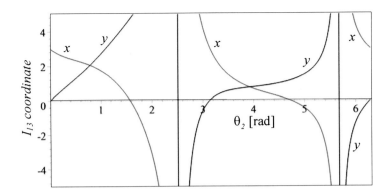

FIGURE 6.30. The coordinates (x, y) of the instant center of rotation I_{13} as functions of θ_2.

The equation of the line MA is

$$y = \frac{y_A}{x_A} x = (\tan \theta_2)\, x \qquad (6.198)$$

and the equation of NB is

$$y = \frac{y_B}{x_B - d}(x - d) = (\tan \theta_4)(x - d) \qquad (6.199)$$

The instant center of rotation I_{23} of the coupler link AB with respect to the ground MN is at the intersection of the lines MA and NB.

$$x = d\frac{\tan \theta_4}{\tan \theta_4 - \tan \theta_2} \qquad (6.200)$$

$$y = d\frac{\tan \theta_4 \tan \theta_2}{\tan \theta_4 - \tan \theta_2} \qquad (6.201)$$

Figure 6.30 depicts the coordinates (x, y) of the instant center of rotation I_{13} for

$$a = 1 \qquad b = 2 \qquad c = 2.5 \qquad d = 3 \qquad (6.202)$$

The position of the center of rotation I_{23} goes to infinity whenever $\theta_4 = \theta_2$ and the input and output links become parallel. The Freudenstein's equation (6.24)

$$J_1 \cos \theta_4 - J_2 \cos \theta_2 + J_3 = \cos(\theta_4 - \theta_2) \qquad (6.203)$$

under the parallel condition, yields

$$\cos \theta_2 = \frac{1 - J_3}{J_1 - J_2} = \frac{(a - c)^2 - b^2 + d^2}{2d\,(a - c)} \qquad (6.204)$$

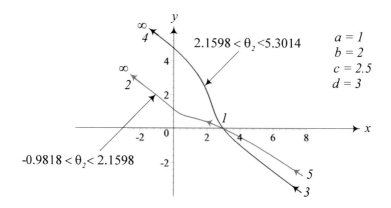

FIGURE 6.31. The path of motion of the instant center of rotation I_{23} when $0 \leq \theta_2 \leq 2\pi$.

The values of $\theta_2 = \theta_4$ for the sample data (6.202) are:

$$\theta_2 = \cos^{-1} \frac{(a-c)^2 - b^2 + d^2}{2d\,(a-c)} = \begin{cases} 2.1598\,\text{rad} \\ 5.3014\,\text{rad} \end{cases} \qquad (6.205)$$

Figure 6.31 shows the path of motion of the instant center of rotation I_{23} when $0 \leq \theta_2 \leq 2\pi$. Starting from $\theta_2 = 0$, I_{23} moves from $(3,0)$, indicated by 1, to point 2 and goes to infinity when $\theta_2 \to 2.1598$. When θ_2 goes beyond 2.1598, I_{23} appears at point 3 and moves to point 4 and infinity when $\theta_2 \to 5.3014$. The instant center I_{23} again appears at point 5 and moves to $(3,0)$ when $5.3014 < \theta_2 \leq 2\pi$.

Example 258 *Sliding a slender on the wall.*

Figure 6.32 illustrates a slender bar AB sliding at points A and C. We have the velocity axes of the points A and C, and therefore, we can find the instant center of rotation I.

The coordinates of point I are a function of the parameter θ:

$$\begin{aligned} x_I &= h \cot \theta & (6.206) \\ y_I &= h + x_I \cot \theta = h\left(1 + \cot^2 \theta\right) & (6.207) \end{aligned}$$

Eliminating θ between x and y, generates the path of motion for I.

$$y_I = h\left(1 + \frac{x_I^2}{h^2}\right) \qquad (6.208)$$

Example 259 ★ *Plane motion of a rigid body.*

The plane motion of a rigid body is such that all points of the body move only in parallel planes. So, to study the motion of the body, it is enough to examine the motion of points in one plane.

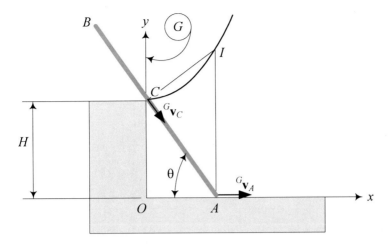

FIGURE 6.32. A slender bar AB sliding at points A and C.

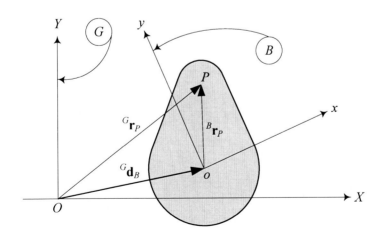

FIGURE 6.33. A rigid body in a planar motion.

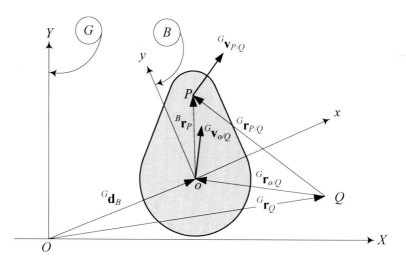

FIGURE 6.34. Instant center of rotation Q, for planar motion of a rigid body.

Figure 6.33 illustrates a rigid body in a planar motion and the corresponding coordinate frames. The position and velocity of a body point P are:

$$^G\mathbf{r}_P = {}^G\mathbf{d}_B + {}^GR_B {}^B\mathbf{r}_P = {}^G\mathbf{d}_B + {}^G_B\mathbf{r}_P \qquad (6.209)$$

$$^G\mathbf{v}_P = {}^G\dot{\mathbf{d}}_B + {}_G\boldsymbol{\omega}_B \times \left({}^G\mathbf{r}_P - {}^G\mathbf{d}_B\right) = {}^G\dot{\mathbf{d}}_B + {}_G\boldsymbol{\omega}_B \times {}^G_B\mathbf{r}_P \qquad (6.210)$$

where, $^G\mathbf{d}_B$ indicates the position of the moving origin o relative to the fixed origin O. The term $^G\dot{\mathbf{d}}_B$ is the velocity of point o and, ${}_G\boldsymbol{\omega}_B \times {}^G_B\mathbf{r}_P$ is the velocity of point P relative to o, both in G.

$$^G\mathbf{v}_{P/o} = {}_G\boldsymbol{\omega}_B \times {}^G_B\mathbf{r}_P \qquad (6.211)$$

Although it is not a correct view, it might sometimes help if we interpret $^G\dot{\mathbf{d}}_B$ as the translational velocity and ${}_G\boldsymbol{\omega}_B \times {}^G_B\mathbf{r}_P$ as the rotational velocity components of $^G\mathbf{v}_P$. Then, the velocity of any point P of the rigid body is a superposition of the velocity $^G\dot{\mathbf{d}}_B$ of an arbitrary point o and the angular velocity ${}_G\boldsymbol{\omega}_B \times {}^G_B\mathbf{r}_P$ of the points P about o.

The relative velocity vector $^G\mathbf{v}_{P/o}$ is perpendicular to the relative position vector ${}^G_B\mathbf{r}_P$. Employing the same concept we interpret that the velocity of points P and o with respect to another point Q are perpendicular to ${}^G_B\mathbf{r}_{P/Q}$ and ${}^G_B\mathbf{r}_{o/Q}$ respectively. We may search for a point Q, as the instantaneous center of rotation, at which the velocity is zero. Points o, P, and Q are shown in Figure 6.34.

Assuming a position vector $^G\mathbf{r}_{o/Q}$ for the instant center Q, we define

$$^G\mathbf{r}_{o/Q} = a_Q\, {}^G\dot{\mathbf{d}}_B + b_Q\, {}_G\boldsymbol{\omega}_B \times {}^G\dot{\mathbf{d}}_B \qquad (6.212)$$

then, following (6.210), the velocity of point Q can be expressed by

$$
\begin{aligned}
{}^G\mathbf{v}_Q &= {}^G\dot{\mathbf{d}}_B + {}_G\boldsymbol{\omega}_B \times {}^G\mathbf{r}_{Q/o} = {}^G\dot{\mathbf{d}}_B - {}_G\boldsymbol{\omega}_B \times {}^G\mathbf{r}_{o/Q} \\
&= {}^G\dot{\mathbf{d}}_B - {}_G\boldsymbol{\omega}_B \times \left(a_Q\,{}^G\dot{\mathbf{d}}_B + b_Q\,{}_G\boldsymbol{\omega}_B \times {}^G\dot{\mathbf{d}}_B\right) \\
&= {}^G\dot{\mathbf{d}}_B - a_Q\,{}_G\boldsymbol{\omega}_B \times {}^G\dot{\mathbf{d}}_B - b_Q\,{}_G\boldsymbol{\omega}_B \times \left({}_G\boldsymbol{\omega}_B \times {}^G\dot{\mathbf{d}}_B\right) \\
&= 0 \qquad\qquad\qquad\qquad\qquad\qquad\qquad\qquad (6.213)
\end{aligned}
$$

Now, using the equations

$$
{}_G\boldsymbol{\omega}_B = \omega\hat{K} \tag{6.214}
$$

$$
{}_G\boldsymbol{\omega}_B \times \left({}_G\boldsymbol{\omega}_B \times {}^G\dot{\mathbf{d}}_B\right) = \left({}_G\boldsymbol{\omega}_B \cdot {}^G\dot{\mathbf{d}}_B\right){}_G\boldsymbol{\omega}_B - \omega^2\,{}^G\dot{\mathbf{d}}_B \tag{6.215}
$$

$$
{}_G\boldsymbol{\omega}_B \cdot {}^G\dot{\mathbf{d}}_B = 0 \tag{6.216}
$$

we find

$$
\left(1 + b_Q\,\omega^2\right){}^G\dot{\mathbf{d}}_B - a_Q\,{}_G\boldsymbol{\omega}_B \times {}^G\dot{\mathbf{d}}_B = 0 \tag{6.217}
$$

Because ${}^G\dot{\mathbf{d}}_B$ *and* ${}_G\boldsymbol{\omega}_B \times {}^G\dot{\mathbf{d}}_B$ *must be perpendicular, Equation (6.217) provides*

$$
1 + b_Q\,\omega^2 = 0 \qquad a_Q = 0 \tag{6.218}
$$

and therefore,

$$
{}^G\mathbf{r}_{Q/o} = \frac{1}{\omega^2}\left({}_G\boldsymbol{\omega}_B \times {}^G\dot{\mathbf{d}}_B\right) \tag{6.219}
$$

Example 260 ★ *Instantaneous center of acceleration.*

For planar motions of rigid bodies, it is possible to find a body point with zero acceleration. Such a point may be called the instantaneous center of acceleration. When a rigid body is in a planar motion, we can express the acceleration of a body point P, such as shown in Figure 6.34, as

$$
\begin{aligned}
{}^G\mathbf{a}_P &= {}^G\ddot{\mathbf{d}}_B + {}_G\boldsymbol{\alpha}_B \times \left({}^G\mathbf{r}_P - {}^G\mathbf{d}_B\right) \\
&\quad + {}_G\boldsymbol{\omega}_B \times \left({}_G\boldsymbol{\omega}_B \times \left({}^G\mathbf{r}_P - {}^G\mathbf{d}_B\right)\right) \\
&= {}^G\ddot{\mathbf{d}}_B + {}_G\boldsymbol{\alpha}_B \times {}^G_B\mathbf{r}_P + {}_G\boldsymbol{\omega}_B \times \left({}_G\boldsymbol{\omega}_B \times {}^G_B\mathbf{r}_P\right) \tag{6.220}
\end{aligned}
$$

The term ${}_G\boldsymbol{\omega}_B \times \left({}_G\boldsymbol{\omega}_B \times {}^G_B\mathbf{r}_P\right)$ *is the centripetal acceleration, and the term* ${}_G\boldsymbol{\alpha}_B \times {}^G_B\mathbf{r}_P$ *is the tangential acceleration. Because the motion is planar, the angular velocity vector is always parallel to* $\hat{k}$ *and* $\hat{K}$ *unit vectors.*

$$
{}_G\boldsymbol{\omega}_B = \omega\hat{K} \qquad {}_G\boldsymbol{\alpha}_B = \alpha\hat{K} \tag{6.221}
$$

Therefore, the velocity ${}^G\mathbf{v}_P$ *and acceleration* ${}^G\mathbf{a}_P$ *can be simplified to*

$$
{}^G\mathbf{a}_P = {}^G\ddot{\mathbf{d}}_B + {}_G\boldsymbol{\alpha}_B \times {}^G_B\mathbf{r}_P - \omega^2\,{}^G_B\mathbf{r}_P \tag{6.222}
$$

We now look for a zero acceleration point S and express its position vector by

$$^G\mathbf{r}_{S/o} = a_S\,{}^G\ddot{\mathbf{d}}_B + b_S\,{}_G\boldsymbol{\alpha}_B \times {}^G\ddot{\mathbf{d}}_B \tag{6.223}$$

and based on (6.222) we have,

$$
\begin{aligned}
{}^G\mathbf{a}_S &= {}^G\ddot{\mathbf{d}}_B + {}_G\boldsymbol{\alpha}_B \times {}_B^G\mathbf{r}_S - \omega^2\,{}_B^G\mathbf{r}_S \\
&= {}^G\ddot{\mathbf{d}}_B + {}_G\boldsymbol{\alpha}_B \times {}^G\mathbf{r}_{S/o} - \omega^2\,{}^G\mathbf{r}_{S/o} \\
&= {}^G\ddot{\mathbf{d}}_B + {}_G\boldsymbol{\alpha}_B \times \left(a_S\,{}^G\ddot{\mathbf{d}}_B + b_S\,{}_G\boldsymbol{\alpha}_B \times {}^G\ddot{\mathbf{d}}_B\right) \\
&\quad -\omega^2\left(a_S\,{}^G\ddot{\mathbf{d}}_B + b_S\,{}_G\boldsymbol{\alpha}_B \times {}^G\ddot{\mathbf{d}}_B\right) \\
&= {}^G\ddot{\mathbf{d}}_B + a_S\,{}_G\boldsymbol{\alpha}_B \times {}^G\ddot{\mathbf{d}}_B + b_S\,{}_G\boldsymbol{\alpha}_B \times \left({}_G\boldsymbol{\alpha}_B \times {}^G\ddot{\mathbf{d}}_B\right) \\
&\quad -a_S\,\omega^2\,{}^G\ddot{\mathbf{d}}_B - b_S\,\omega^2\,{}_G\boldsymbol{\alpha}_B \times {}^G\ddot{\mathbf{d}}_B = 0 \tag{6.224}
\end{aligned}
$$

Simplifying the above equation yields

$$\left(1 - a_S\,\omega^2 - b_S\,\alpha^2\right){}^G\ddot{\mathbf{d}}_B + \left(a_S - b_S\,\omega^2\right){}_G\boldsymbol{\alpha}_B \times {}^G\ddot{\mathbf{d}}_B = 0 \tag{6.225}$$

and because ${}^G\ddot{\mathbf{d}}_B$ and ${}_G\boldsymbol{\alpha}_B \times {}^G\ddot{\mathbf{d}}_B$ are perpendicular, we must have

$$1 - a_S\,\omega^2 - b_S\,\alpha^2 = 0 \qquad a_S - b_S\,\omega^2 = 0 \tag{6.226}$$

and hence,

$$a_S = \frac{\omega^2}{\omega^2 + \alpha^2} \qquad b_S = \frac{1}{\omega^2 + \alpha^2} \tag{6.227}$$

The position vector of the instant center of acceleration is then equal to

$$^G\mathbf{r}_{S/o} = \frac{1}{\omega^2 + \alpha^2}\left(\omega^2\,{}^G\ddot{\mathbf{d}}_B + {}_G\boldsymbol{\alpha}_B \times {}^G\ddot{\mathbf{d}}_B\right) \tag{6.228}$$

6.5   Coupler Point Curve

To provide flexibility between wheels and the vehicle chassis, wheels are attached to chassis by a mechanism. The most common suspension mechanisms are double A-arm and inverted slider-crank. The wheel of the vehicle will then be attached to a point of the coupler link of the mechanism. The relative motion of the wheel and chassis will be calculated by analysis of the displacement of the coupler point.

6.5.1   Coupler Point Curve for Four-Bar Linkages

Figure 6.35 illustrates a four-bar linkage $MNAB$ and a coupler point at C. When the mechanism moves, the coupler point C will move on a path.

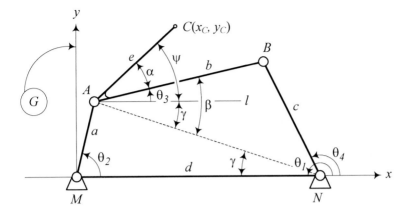

FIGURE 6.35. A four-bar linkage $MNAB$ and a coupler point at C.

The path of the coupler point is called the *coupler point curve*. Considering θ_2 as the input variable of the mechanism, the parametric coordinates of the coupler point curve (x_C, y_C) are

$$x_C = a\cos\theta_2 + e\cos(\beta - \gamma + \alpha) \tag{6.229}$$
$$y_C = a\sin\theta_2 + e\sin(\beta - \gamma + \alpha) \tag{6.230}$$

where,

$$\gamma = \tan^{-1}\frac{a\sin\theta_2}{d - a\cos\theta_2} \tag{6.231}$$

$$\beta = \tan^{-1}\frac{\sqrt{4b^2 f^2 - (b^2 + f^2 - c^2)^2}}{b^2 + f^2 - c^2} \tag{6.232}$$

$$f = \sqrt{a^2 + d^2 - 2ad\cos\theta_2} \tag{6.233}$$

Proof. The position of the coupler point C in Figure 6.35 is defined by the polar coordinates length e and angle α with respect to the coupler link AB, and by (x_C, y_C) in the Cartesian coordinate frame G attached to the ground. The length of the links are indicated by $MA = a$, $AB = b$, $NB = c$, and $MN = d$. We show the angle $\angle ANM$ by γ and $\angle BAN$ by β. Let us draw a line l through A and parallel to the ground link MN, then,

$$\angle NAl = \angle ANM = \gamma \tag{6.234}$$
$$\angle CAl = \psi \tag{6.235}$$
$$\psi = \beta - \gamma + \alpha \tag{6.236}$$

The global coordinates of point C are then

$$x_C = a\cos\theta_2 + e\cos\psi \tag{6.237}$$
$$y_C = a\sin\theta_2 + e\sin\psi \tag{6.238}$$

where, ψ comes from Equation (6.236). The angle β can be calculated from the cosine law in $\triangle BAN$,

$$\cos\beta = \frac{b^2 + f^2 - c^2}{2bf} \tag{6.239}$$

where, $f = AN$. Applying the cosine law in $\triangle AMN$ shows that f is equal to

$$f = \sqrt{a^2 + d^2 - 2ad\cos\theta_2} \tag{6.240}$$

given by Equation (6.233).

For computer calculation ease, it is better to find β from the trigonometric equation

$$\tan^2\beta = \sec^2\beta - 1 \tag{6.241}$$

and substitute $\sec\beta$ from Equation (6.239).

$$\beta = \tan^{-1}\frac{\sqrt{4b^2f^2 - (b^2 + f^2 - c^2)^2}}{b^2 + f^2 - c^2} \tag{6.242}$$

The angle γ can be found from a tan equation based on the vertical distance of point A from the ground link MN.

$$\gamma = \tan^{-1}\frac{a\sin\theta_2}{d - a\cos\theta_2} \tag{6.243}$$

Therefore, the coordinates x_C and y_C can be calculated as two parametric functions of θ_2 for a given set of a, b, c, d, e, and α. ∎

Example 261 *A poorly designed double A-arm suspension mechanism.*

Figure 6.36 illustrates a double A-arm suspension mechanism and its equivalent four-bar linkage kinematic model. Points M and N are fixed joints on the body, and points A and B are moving joints attached to the wheel supporting coupler link. Point C is on the spindle and supposed to be the wheel center. When the wheel moves up and down, the wheel center moves on a the couple point curve shown in the figure. The wheel center of proper suspension mechanism is supposed to move vertically for the working range suspension travel. However, the wheel center of the suspension of Figure 6.36 moves on a high curvature path and generates an undesired camber.

A small motion of the kinematic model of suspension is shown in Figure 6.37, and the actual suspension and wheel configurations are shown in the figure.

6.5.2 Coupler Point Curve for a Slider-Crank Mechanism

Figure 6.38 illustrates a slider-crank mechanism and a coupler point at C. When the mechanism moves, coupler point C will move on a coupler point

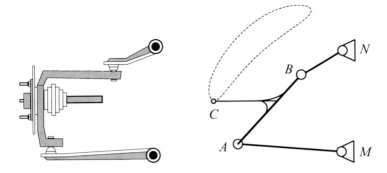

FIGURE 6.36. A double A arm suspension mechanism and its equivalent four-bar linkage kinematic model.

curve with the following parametric equation:

$$x_C = a \cos \theta_2 + c \cos (\alpha - \gamma) \qquad (6.244)$$
$$y_C = a \sin \theta_2 + c \sin (\alpha - \gamma) \qquad (6.245)$$

The angle θ_2 is the input angle and acts as the parameter of the equations. The angle γ can be calculated from

$$\gamma = \sin^{-1} \frac{a \sin \theta_2 - e}{b} \qquad (6.246)$$

Proof. We attach a planar Cartesian coordinate frame to the ground link at M. The x-axis is parallel to the ground indicated by the sliding surface, as shown in Figure 6.38. Drawing a line l through A and parallel to the ground shows that

$$\beta = \alpha - \gamma \qquad (6.247)$$

where γ is the angle between the coupler link and the ground.
   The coordinates (x_C, y_C) for the coupler point C are

$$x_C = a \cos \theta_2 + c \cos \beta \qquad (6.248)$$
$$y_C = a \sin \theta_2 + c \sin \beta \qquad (6.249)$$

To calculate the angle γ, we examine $\triangle AEB$ and find

$$\sin \gamma = \frac{AE}{AB} = \frac{a \sin \theta_2 - e}{b} \qquad (6.250)$$

that finalizes the proof of Equation (6.246). Therefore, the coordinates x_C and y_C can be calculated as two parametric functions of θ_2 for a given set of a, b, c, e, and α. ∎

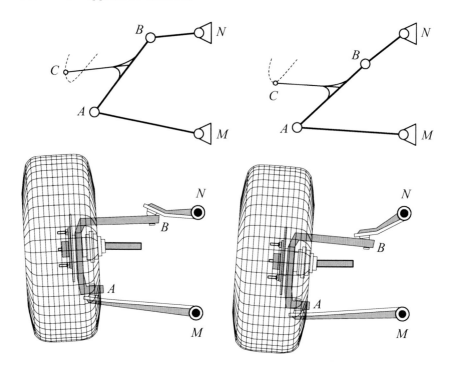

FIGURE 6.37. A small motion of the kinematic model and the actual suspension configurations.

Example 262 *A centric and symmetric slider-crank mechanism.*

Point $C(x_C, y_C)$ is the coupler point of a centric and symmetric slider-crank mechanism shown in Figure 6.39. It is centric because $e = 0$, and is symmetric because $a = b$, and therefore, $\theta_2 = \theta_4$. Point C is on the coupler link AB and is at a distance kb from A, where $0 < k < 1$.

The coordinates of point C are

$$
\begin{aligned}
x_C &= a\cos\theta_2 + ka\cos\theta_2 = a(1+k)\cos\theta_2 & (6.251)\\
y_C &= a\sin\theta_2 - ka\sin\theta_2 = a(1-k)\sin\theta_2 & (6.252)
\end{aligned}
$$

and therefore,

$$
\cos\theta_2 = \frac{x_C}{a(1+k)} \qquad \sin\theta_2 = \frac{y_C}{a(1-k)} \qquad (6.253)
$$

Using $\cos^2\theta_2 + \sin^2\theta_2 = 1$, we can show that the coupler point C will move on an ellipse.

$$
\frac{x_C^2}{a^2(1+k)^2} + \frac{y_C^2}{a^2(1-k)^2} = 1 \qquad (6.254)
$$

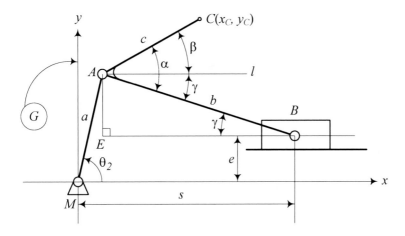

FIGURE 6.38. A slider-crank mechanism and a coupler point at C.

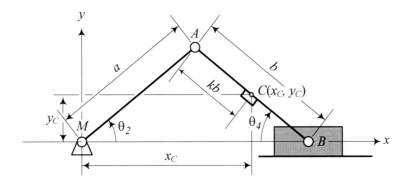

FIGURE 6.39. A centric and symmetric slider-crank mechanism.

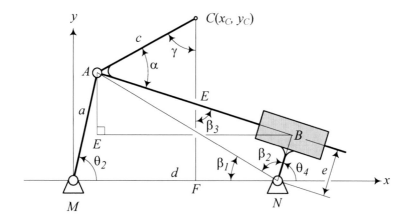

FIGURE 6.40. An inverted slider-crank mechanism and a coupler point at C.

6.5.3 Coupler Point Curve for Inverted Slider-Crank Mechanism

Figure 6.40 illustrates an inverted slider-crank mechanism and a coupler point at C. When the mechanism moves, the coupler point C will move on a curve with the following parametric equation:

$$x_C = a \cos \theta_2 + c \cos (\pi - \alpha - \theta_4) \qquad (6.255)$$
$$y_C = a \sin \theta_2 + c \sin (\pi - \alpha - \theta_4) \qquad (6.256)$$

The angle θ_2 is the input angle and acts as a parameter, and θ_4 is the angle of the output link, given by Equation (6.153).

$$\theta_4 = 2 \tan^{-1} \left(\frac{-H \pm \sqrt{H^2 - 4GI}}{2G} \right) \qquad (6.257)$$

$$G = d - e - a \cos \theta_2 \qquad (6.258)$$
$$H = 2a \sin \theta_2 \qquad (6.259)$$
$$I = a \cos \theta_2 - d - e \qquad (6.260)$$

Proof. We attach a planar Cartesian coordinate frame to the ground link at MN. Drawing a vertical line through C defines the variable angle $\gamma = \angle ACF$ as shown in Figure 6.40. We also define three angles $\beta_1 = \angle ANM$, $\beta_2 = \angle ANB$, and $\beta_3 = \angle BEF$ to simplify the calculations. From the triangle $\triangle ACE$, we find

$$\gamma = \pi - \beta_3 - \alpha \qquad (6.261)$$

and from quadrilateral $\square EFNB$, we find

$$\beta_3 + \angle EFN + \beta_2 + \beta_1 + \angle EBN = 2\pi \qquad (6.262)$$

However,

$$\angle EBN = \frac{\pi}{2} \qquad (6.263)$$

$$\angle EFN = \frac{\pi}{2} \qquad (6.264)$$

and therefore,

$$\beta_3 + \beta_2 + \beta_1 = \pi \qquad (6.265)$$

The output angle θ_4 is

$$\theta_4 = \pi - (\beta_2 + \beta_1) \qquad (6.266)$$

and thus,

$$\theta_4 = \beta_3 \qquad (6.267)$$

Now the angle γ may be written as

$$\gamma = \pi - \theta_4 - \alpha \qquad (6.268)$$

where θ_4 is the output angle, found in Equation (6.153).

Therefore, the coordinates x_C and y_C can be calculated as two parametric functions of θ_2 for a given set of a, d, c, e, and α. ∎

6.6 ★ Universal Joint

The *universal joint* shown in Figure 6.41 is a mechanism to connect rotating shafts that intersect at an angle φ. The universal joint is also known as *Hook's coupling*, *Hook joint*, *Cardan joint*, or *Yoke joint*.

Figure 6.42 illustrates a practical universal joint. There are four links in a universal joint: link number 1 is the ground, which has a revolute joint with the input link 2 and the output link 4. The input and the output links are connected with a cross-link 3. The universal joint is a three-dimensional four-bar linkage for which the cross-link acts as a coupler link.

The driver and driven shafts make a complete revolution at the same time, but the velocity ratio is not constant throughout the revolution. The angular velocity of the output shaft 4 relative to the input shaft 2 is called *speed ratio* Ω and is a function of the angular position of the input shaft θ, and the angle between the shafts φ.

$$\Omega = \frac{\omega_4}{\omega_2} = \frac{\cos\varphi}{1 - \sin^2\varphi\cos^2\theta} \qquad (6.269)$$

Proof. A universal joint may appear in many shapes, however, regardless of how it is constructed, it has essentially the form shown in Figure 6.42. Each connecting shaft ends in a U-shaped yoke. The yokes are connected by a rigid cross-link. The ends of the cross-link are set in bearings in the

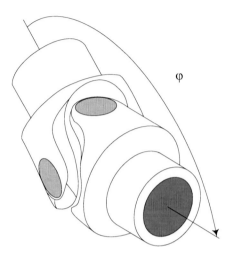

FIGURE 6.41. A universal joint.

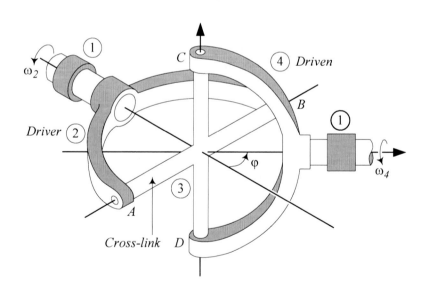

FIGURE 6.42. A universal joint with four links: link 1 is the ground, link 2 is the input, link 4 is the output, and the cross-link 3 is a coupler link.

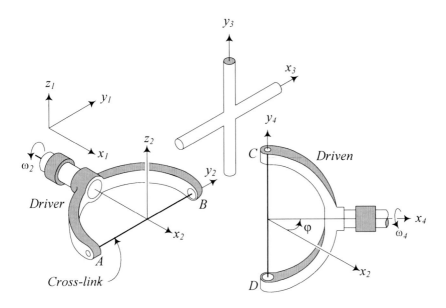

FIGURE 6.43. A separate illustration of the input, output, and the cross links for a universal joint.

yokes. When the driver yoke turns, the cross-link rotates relative to the yoke about its axis AB. The cross-link also rotates relative to the driven yoke about the axis CD.

Although the driver and driven shafts make a complete revolution together, their velocity ratio is not constant throughout the revolution. A disassembled illustration of the input, output, and the cross links is shown in Figure 6.43.

The angular velocity of the cross-link may be shown by

$$_1\omega_3 = {}_1\omega_2 + {}_2^1\omega_3 = {}_1\omega_4 + {}_4^1\omega_3 \tag{6.270}$$

where, $_1\omega_2$ is the angular velocity of the driver yoke about the x_2-axis and $_2^1\omega_3$ is the angular velocity of the cross-link about the axis AB relative to the driver yoke expressed in the ground coordinate frame.

Figure 6.44 shows that the unit vectors $\hat{j}_2$ and $\hat{j}_3$ are along the arms of the cross link, and the unit vectors $\hat{i}_2$ and $\hat{i}_4$ are along the shafts. Having the angular velocity vectors,

$$_1\omega_2 = \begin{bmatrix} \omega_{21}\,\hat{i}_1 \\ \hat{j}_1 \\ \hat{k}_1 \end{bmatrix} = \begin{bmatrix} \omega_{21} \\ 0 \\ 0 \end{bmatrix} \tag{6.271}$$

$$_1\omega_4 = \begin{bmatrix} \omega_{41}\,\hat{i}_4 \\ \hat{j}_4 \\ \hat{k}_4 \end{bmatrix} = \begin{bmatrix} \omega_{41} \\ 0 \\ 0 \end{bmatrix} \tag{6.272}$$

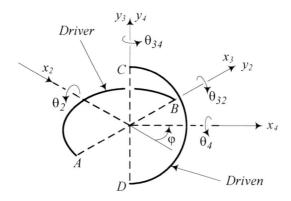

FIGURE 6.44. A kinematic model for a universal joint.

$$_2\boldsymbol{\omega}_3 = \begin{bmatrix} \hat{\imath}_2 \\ \omega_{32}\,\hat{\jmath}_2 \\ \hat{k}_2 \end{bmatrix} \qquad {}^3_2\boldsymbol{\omega}_3 = \begin{bmatrix} \omega_{32}\,\hat{\imath}_3 \\ \hat{\jmath}_3 \\ \hat{k}_3 \end{bmatrix} \qquad (6.273)$$

$$_4\boldsymbol{\omega}_3 = \begin{bmatrix} \hat{\imath}_4 \\ \omega_{34}\,\hat{\jmath}_4 \\ \hat{k}_4 \end{bmatrix} \qquad {}^3_4\boldsymbol{\omega}_3 = \begin{bmatrix} \hat{\imath}_3 \\ \omega_{34}\,\hat{\jmath}_3 \\ \hat{k}_3 \end{bmatrix} \qquad (6.274)$$

we can simplify Equation (6.270) to

$$\omega_{32}\,\hat{\imath}_3 + \omega_{21}\,\hat{\imath}_2 = \omega_{41}\,\hat{\imath}_4 + \omega_{34}\,\hat{\jmath}_3 \qquad (6.275)$$

However, because the cross-link coordinate frame is right-handed, we have

$$\hat{\imath}_3 \times \hat{\jmath}_3 = \hat{k}_3 \qquad (6.276)$$

and therefore,

$$(\omega_{32}\,\hat{\imath}_3 + \omega_{21}\,\hat{\imath}_2) \cdot \hat{k}_3 = (\omega_{41}\,\hat{\imath}_4 + \omega_{34}\,\hat{\jmath}_3) \cdot \hat{k}_3 \qquad (6.277)$$

that yields

$$\omega_{21}\,\hat{\imath}_2 \cdot \hat{k}_3 = \omega_{41}\,\hat{\imath}_4 \cdot \hat{k}_3 \qquad (6.278)$$

Now the required equation for the speed ratio $\Omega = \omega_{41}/\omega_{21}$ is

$$\Omega = \frac{\omega_{41}}{\omega_{21}} = \frac{\hat{\imath}_3 \times \hat{\jmath}_3 \cdot \hat{\imath}_2}{\hat{\imath}_3 \times \hat{\jmath}_3 \cdot \hat{\imath}_4} \qquad (6.279)$$

The unit vector $\hat{\jmath}_3$ is perpendicular to $\hat{\imath}_3$ and $\hat{\imath}_4$, we may write

$$\hat{\jmath}_3 = a\hat{\imath}_4 \times \hat{\imath}_3 \qquad (6.280)$$

where a is a coefficient. Now

$$\hat{\imath}_3 \times \hat{\jmath}_3 = \hat{\imath}_3 \times (a\hat{\imath}_4 \times \hat{\imath}_3) = a\left[\hat{\imath}_4 - (\hat{\imath}_3 \cdot \hat{\imath}_4)\,\hat{\imath}_3\right] \qquad (6.281)$$

and because
$$\hat{i}_3 \cdot \hat{i}_2 = 0 \tag{6.282}$$

we find

$$
\begin{aligned}
\Omega &= \frac{\omega_{41}}{\omega_{21}} = \frac{\hat{i}_2 \cdot a \left[\hat{i}_4 - (\hat{i}_3 \cdot \hat{i}_4)\,\hat{i}_3\right]}{\hat{i}_3 \cdot a \left[\hat{i}_4 - (\hat{i}_3 \cdot \hat{i}_4)\,\hat{i}_3\right]} \\
&= \frac{\hat{i}_2 \cdot \hat{i}_4}{1 - (\hat{i}_3 \cdot \hat{i}_4)^2} = \frac{\cos\varphi}{1 - (\hat{i}_3 \cdot \hat{i}_4)^2}
\end{aligned}
\tag{6.283}
$$

If we show the angular position of the input yoke by θ, then

$$
\begin{aligned}
\hat{i}_3 &= \cos\theta\,\hat{j}_1 + \sin\theta\,\hat{k}_1 & \tag{6.284} \\
\hat{i}_4 &= \cos\varphi\,\hat{i}_1 - \sin\varphi\,\hat{j}_1 & \tag{6.285}
\end{aligned}
$$

and the final equation for the speed ratio is found as

$$\Omega = \frac{\omega_{41}}{\omega_{21}} = \frac{\cos\varphi}{1 - \sin^2\varphi\cos^2\theta} \tag{6.286}$$

This formula shows that although both shafts complete one revolution at the same time, the ratio of their angular speeds varies with the angle of rotation $\theta(t)$ of the driver and is also a function of the shaft angle φ. Thus, even if the angular speed ω_{21} of the drive shaft is constant, the angular speed ω_{41} of a driven shaft will not be uniform. ∎

Example 263 ★ *Graphical illustration of the universal joint speed ratio* Ω.

Figure 6.45 depicts a three-dimensional plot for Ω *as a function of* θ *and* φ. *The* Ω-surface *is plotted for one revolution of the drive shaft and every possible angle between the two shafts.*

$$-\pi < \theta < \pi \qquad -\pi < \varphi < \pi \tag{6.287}$$

A two-dimensional view for Ω *is depicted in Figure 6.46. When* $\varphi \lessgtr$ 10 deg *there is not much fluctuation in speed ratio, however, when the angle between the two shafts is more than* 10 deg *then the speed ratio* Ω *cannot be assumed constant. The universal joint get stuck when* $\varphi = 90$ deg, *because theoretically*

$$\lim_{\varphi \to 90} \Omega = indefinite \tag{6.288}$$

The behavior of Ω *as a function of* θ *and* φ *can be better viewed in a polar coordinate, as shown in Figure 6.47.*

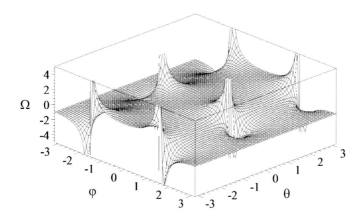

FIGURE 6.45. A three-dimensional plot for the speed ratio of a universal joint, Ω as a function of the input angle θ and the angle between input and output shafts φ.

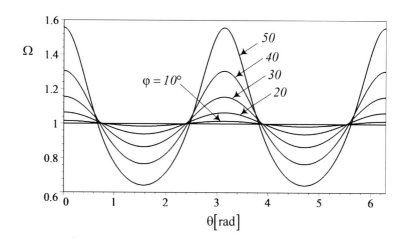

FIGURE 6.46. A two-dimensional view of Ω as a function of the input angle θ and the angle between input and output shafts φ.

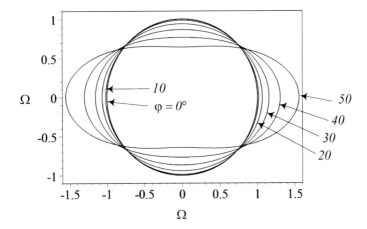

FIGURE 6.47. The behavior of speed ratio Ω as a function of θ and φ in a polar coordinate.

Example 264 ★ *Maximum and minimum of ω_{41} in one revolution.*
    The maximum value of Ω is

$$\Omega_M = \frac{1}{\cos\varphi} \tag{6.289}$$

that occurs at

$$\theta = 0, \pi \tag{6.290}$$

and the minimum value of Ω is

$$\Omega_m = \cos\varphi \tag{6.291}$$

that occurs at

$$\theta = \frac{\pi}{2}, \frac{3\pi}{2} \tag{6.292}$$

Example 265 ★ *Double universal joint.*
    To eliminate the non-uniform speed ratio between the input and output shafts, connected by a universal joint, we can connect a second joint to make the intermediate shaft have a variable speed ratio with respect to both the input and the output shafts in such a way that the overall speed ratio between the input and output shafts remains equal to one. Figure 6.48 depicts the two possible arrangements to cancel the alternative speed ratio of the universal joints between two shafts in an angle or offset.

Example 266 ★ *Alternative proof for universal joint equation.*
    Consider a universal joint of Figure 6.42. Looking along the axis of the input shaft, we see points A and B moving on a circle and points C and D moving on an ellipse as shown in Figure 6.49(a). This is because A and B

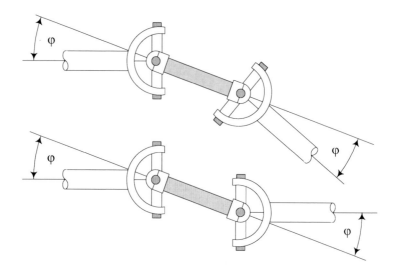

FIGURE 6.48. Two possible arrangements to cancel the alternative speed ratio of the universal joints between two shafts in an angle or offset.

trace a circle in a normal plane, and C and D trace a circle in a rotated plane by the angle φ. Assume the universal joint starts rotating when the axis CD of the cross link is at the intersection of the planes of motion CD and AB, as shown in Figure 6.49(a). If the axis AB turns an angle θ, then the projection of the axis CD will turn on the ellipse, as can be seen in Figure 6.49(b). However, the angle of rotation CD is θ_4 different than θ when we look at the axis CD along the output shaft.

Looking along the input shaft, the axis AB starts from $A_1 B_1$ and moves to $A_2 B_2$ after rotation θ. From the same viewpoint, the axis CD starts from $C_1 D_1$ and moves to $C_2 D_2$, however, CD would be at $C_2' D_2'$, if it were looking along the output shaft. The geometric relationship between the angles are

$$\frac{C_2' R}{OR} = \tan \theta_4 \qquad \frac{C_2 R}{OR} = \tan \theta \qquad \frac{C_2 R}{C_2' R} = \cos \varphi \qquad (6.293)$$

Therefore,

$$\tan \theta = \tan \theta_4 \cos \varphi \qquad (6.294)$$

which after differentiation becomes

$$\frac{\omega_2}{\csc^2 \theta} = \frac{\omega_4}{\csc^2 \theta_4} \cos \varphi \qquad (6.295)$$

Eliminating θ_4 between (6.294) and (6.295), we find the relationship between the input and output shafts' angular velocities.

$$\omega_4 = \frac{\cos \varphi}{\sin^2 \theta + \cos^2 \theta \cos^2 \varphi} \omega_2 \qquad (6.296)$$

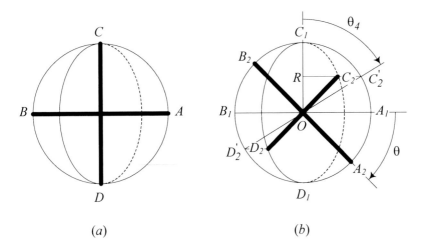

FIGURE 6.49. Rotation of the cross link from a viewpoint along the input shaft.

The speed ratio would then be the same as (6.269).

$$\Omega = \frac{\omega_4}{\omega_2} = \frac{\cos\varphi}{\sin^2\theta + \cos^2\theta\cos^2\varphi} = \frac{\cos\varphi}{1 - \sin^2\varphi\cos^2\theta} \qquad (6.297)$$

Example 267 ★ *History of the universal joint.*
    The need to transmit a rotary motion from one shaft to another, which are intersecting at an angle, was a problem for installing clocktowers in the 1300s. The transmission of the rotation to the hands should be displaced because of tower construction. Cardano (1501−1576) in 1550, Hooke (1635 − 1703) in 1663, and Schott (1608 − 1666) in 1664 used the joint for transferring rotary motion. Hooke was the first who said that the rotary motion between the input and output shafts is not uniform. However, Monge (1746−1818) was the first person who published the mathematical principles of the joint in 1794, and later Poncelet (1788 − 1867) in 1822.
    As an alternative proof for the universal joint speed ratio equation, Poncelet (1788−1867) in 1824 used spherical trigonometry to find the universal joint speed ratio equation, and Jazar in 2011 employed the coordinate transformation method to derive the equation.
    The universal joint can be used to transfer torque at larger angles than flexible couplings. One universal joint may be used to transmit power up to $\varphi = 15$ deg depending on the application. Universal joints are available in a wide variety of torque capacities.

6.7 Summary

Every movable component of a vehicle, such as the doors, hoods, wind-shield wipers, axles, wheels, and suspensions, are connected to the vehicle body using some mechanisms. The four-bar linkage and inverted slider-crank mechanism are the two common mechanisms that we use to connect wheels of independent suspensions to the vehicle chassis. There are analytic equations for determining configuration of the all links of a mechanism with respect to one of the links that is assumed to be fixed and called the ground link.

The wheels are installed on a spindle, which is rigidly attached to the coupler link of the mechanisms. The center of the wheel will move on a coupler point curve. Analytic expressions determine the path of motion of the center as well as the angle of the wheel with respect to the body.

6.8   Key Symbols

$a \equiv \ddot{x}$	acceleration
a, b	coefficients of equation
$a, b, c, \cdots$	length of the links of a linkage
$\mathbf{a}$	acceleration vector
$A, B, \cdots$	coefficients of quadratic equations
b	relative position of an inverted slider
$\dot{b}$	relative speed of a slider
$\ddot{b}$	relative acceleration of a slider
$C_1, C_2, \cdots$	link acceleration parameters of linkages
$\mathbf{d}$	position vector of a moving frame
$f = 1/T$	cyclic frequency [Hz]
g	gravitational acceleration
$\hat{\imath}, \hat{\jmath}, \hat{k}$	unit vectors of Cartesian coordinate frames
I	instant center of rotation
$J_1, J_2, \cdots$	link position parameters of linkages
k	a parameter between zero and one, $0 < k < 1$
l	length, length of the longest link
n	number of links
N	number of instant center of rotations
p, q	length of the middle links
Q	instantaneous center of velocity
$\mathbf{r}$	joint relative position vector
s	displacement position of an slider
s	length of the shortest link
S	instantaneous center of acceleration
$\dot{s}$	speed of a slider
$\ddot{s}$	acceleration of a slider
t	time
T	period
$x, y, z, \ \mathbf{x}$	displacement
x, y, z	Cartesian coordinates
x_C, y_C	coupler point coordinates
$\mathbf{v}$	velocity vector
$\boldsymbol{\alpha}$	angular acceleration vector
α_i	angular acceleration of link number i
θ_i	angular position of link number i
θ	angular position of input and output axles of a universal joint
φ	angle between the input and output axles of a universal joint
$\boldsymbol{\omega}$	angular velocity vector
ω_i	angular velocity of link number i
Ω	angular velocity ratio

Exercises

1. Two possible configurations for a four-bar linkage.

   Consider a four-bar linkage with the following links.

   $$a = 10\,\text{cm} \qquad b = 25\,\text{cm} \qquad c = 30\,\text{cm} \qquad d = 25\,\text{cm}$$

   If $\theta_2 = 30\,\text{deg}$ what would be the angles θ_3 and θ_4 for a convex configuration?

2. Angular velocity of a four-bar linkage output link.

   Consider a four-bar linkage with the following links.

   $$a = 10\,\text{cm} \qquad b = 25\,\text{cm} \qquad c = 30\,\text{cm} \qquad d = 25\,\text{cm}$$

   Determine the angular velocity of the output link ω_4 at $\theta_2 = 30\,\text{deg}$ if $\omega_2 = 2\pi\,\text{rad/s}$.

3. Angular acceleration of a four-bar linkage output link.

   Consider a four-bar linkage with the following links.

   $$a = 10\,\text{cm} \qquad b = 25\,\text{cm} \qquad c = 30\,\text{cm} \qquad d = 25\,\text{cm}$$

   Determine the angular acceleration of the output link α_4 at $\theta_2 = 30\,\text{deg}$ if $\alpha_2 = 0.2\pi\,\text{rad/s}^2$ and $\omega_2 = 2\pi\,\text{rad/s}$.

4. Grashoff criterion.

   Consider a four-bar linkage with the following links.

   $$a = 10\,\text{cm} \qquad b = 25\,\text{cm} \qquad c = 30\,\text{cm}$$

   Determine the limit values of the length d to satisfy the Grashoff criterion.

5. Limit and dead positions.

   Consider a four-bar linkage with the following links.

   $$a = 10\,\text{cm} \qquad b = 25\,\text{cm} \qquad c = 30\,\text{cm} \qquad d = 25\,\text{cm}$$

   Determine if there is any limit or dead positions for the linkage.

6. ★ Limit position determination.

   Explain how we may be able to determine the limit positions of a four-bar linkage by the following condition.

   $$\frac{d\theta}{dt} = 0$$

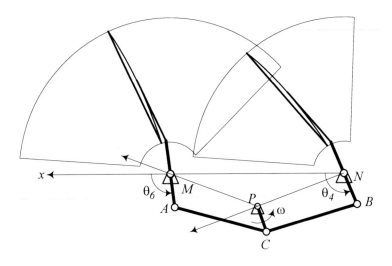

FIGURE 6.50. A six-link windshield wiper mechanism.

7. ★ Design a windshield wiper mechanism.

Figure 6.50 illustrates a six-link windshield wiper mechanism by combining two four-bar linkages. The link PC is attached to a driver motor. Assume $PC = 1$ and $MP = NP$. Design the mechanism such that the links MA and NB are matching the required angles given in the following table.

Matching	NB angle	MA angle
1	$\theta_{41} = 157.6 \deg \approx 2.75 \, \mathrm{rad}$	$\theta_{61} = 157.2 \deg \approx 2.74 \, \mathrm{rad}$
3	$\theta_{43} = 69.5 \deg \approx 1.213 \, \mathrm{rad}$	$\theta_{63} = 26.8 \deg \approx 0.468 \, \mathrm{rad}$

8. Two possible configurations for a slider-crank mechanism.

Consider a slider-crank mechanism with the following links.

$$a = 10 \, \mathrm{cm} \qquad b = 45 \, \mathrm{cm} \qquad e = 0$$

If $\theta_2 = 30 \deg$ what would be angle θ_3 and position of the slides s for a convex configuration?

9. Angular velocity and acceleration of the slider of a slider-crank mechanism.

Consider a slider-crank mechanism with the following links.

$$a = 10 \, \mathrm{cm} \qquad b = 45 \, \mathrm{cm} \qquad e = 0$$

Determine the angular velocity and acceleration of the slider at $\theta_2 = 30 \deg$ if $\omega_2 = 2\pi \, \mathrm{rad/s}$ and $\alpha_2 = 0.2\pi \, \mathrm{rad/s^2}$.

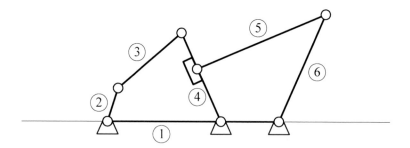

FIGURE 6.51. A 6-bar linkage.

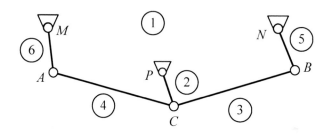

FIGURE 6.52. A 6-bar linkage.

10. Quick return time.

Consider a slider-crank mechanism with the following links.

$$a = 10\,\text{cm} \qquad b = 45\,\text{cm} \qquad e = 3\,\text{cm}$$

Determine the difference time between go and return half cycle of the slider motion if $\omega_2 = 2\pi\,\text{rad/s}$.

11. Two possible configurations for an inverted slider-crank mechanism.

Consider an inverted slider-crank mechanism with the following links.

$$a = 10\,\text{cm} \qquad d = 45\,\text{cm} \qquad e = 5\,\text{cm}$$

If $\theta_2 = 30\,\text{deg}$, what would be the angle θ_3 and position of the slides b?

12. Instant center of rotation.

Find the instant center of rotations for the 6-bar linkage shown in

(a) Figure 6.51.
(b) Figure 6.52.

13. A coupler point of a four-bar linkage.

Consider a four-bar linkage with the following links

$$a = 10\,\text{cm} \qquad b = 25\,\text{cm} \qquad c = 30\,\text{cm} \qquad d = 25\,\text{cm}$$

and a coupler point with the following parameters.

$$e = 10\,\text{cm} \qquad \alpha = 30\,\text{deg}$$

Determine the coordinates of the coupler point if $\theta_2 = 30\,\text{deg}$.

14. A coupler point of a slider-crank mechanism.

Consider a slider-crank mechanism with the following parameters.

$$a \;=\; 10\,\text{cm} \qquad b = 45\,\text{cm} \qquad e = 3\,\text{cm}$$
$$c \;=\; 10\,\text{cm} \qquad \alpha = 30\,\text{deg}$$

Determine the coordinates of the coupler point.

15. A coupler point of an inverted slider-crank mechanism if $\theta_2 = 30\,\text{deg}$.

Consider an inverted slider-crank mechanism with the following parameters.

$$a \;=\; 10\,\text{cm} \qquad d = 45\,\text{cm} \qquad e = 5\,\text{cm}$$
$$c \;=\; 10\,\text{cm} \qquad \alpha = 30\,\text{deg}$$

Determine the coordinates of the coupler point if $\theta_2 = 30\,\text{deg}$.

7

Steering Dynamics

To maneuver a vehicle we need a steering mechanism to turn steerable wheels. Steering dynamics which we review in this chapter, introduces the requirements and challenges to have a steering system to guide a vehicle on non-straight paths.

7.1 Kinematic Steering

Consider a front-wheel-steering (FWS) vehicle that is turning to the left, as is shown in Figure 7.1. When the vehicle is moving very slowly, there is a kinematic condition between the inner and outer wheels that allows them to turn slip-free. The kinematic condition is called the *Ackerman condition* and is expressed by

$$\cot \delta_o - \cot \delta_i = \frac{w}{l} \tag{7.1}$$

where, δ_i is the steer angle of the *inner wheel*, and δ_o is the steer angle of the *outer wheel*. The inner and outer wheels are defined based on the turning center O.

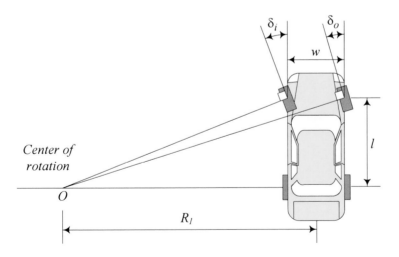

FIGURE 7.1. A front-wheel-steering vehicle and the Ackerman condition.

The distance between the steer axes of the steerable wheels is called the *track* and is shown by w. The distance between the front and real axles

R.N. Jazar, *Vehicle Dynamics: Theory and Application*,
DOI 10.1007/978-1-4614-8544-5_7, © Springer Science+Business Media New York 2014

is called the *wheelbase* and is shown by l. Track w and wheelbase l are considered as kinematic width and length of the vehicle.

We may also use the following more general equation for the kinematic condition between the steer angles of a FWS vehicle

$$\cot \delta_{fr} - \cot \delta_{fl} = \frac{w_f}{l} \tag{7.2}$$

where, δ_{fl} and δ_{fr} are the steer angles of the front left and front right wheels. In this equation the steer angle is measured from the x-axis and is positive if it is about positive z-axis.

The mass center of a steered vehicle will turn on a circle with radius R,

$$R = \sqrt{a_2^2 + l^2 \cot^2 \delta} \tag{7.3}$$

where δ is the cot-average of the inner and outer steer angles.

$$\cot \delta = \frac{\cot \delta_o + \cot \delta_i}{2} \tag{7.4}$$

The angle δ is the equivalent steer angle of a bicycle having the same wheelbase l and radius of rotation R.

Proof. To have all wheels turning freely on a curved road, all the tire axes must intersect at a common point. This criteria is the Ackerman condition. The tire axis is the normal line to the tire-plane at the center of the tire.

Figure 7.2 illustrates a vehicle turning left about the turning center O. Therefore, the inner wheels are the left wheels that are closer to the center of rotation. The inner and outer steer angles δ_i and δ_o may be calculated from the triangles $\triangle OAD$ and $\triangle OBC$ as:

$$\tan \delta_i = \frac{l}{R_1 - \dfrac{w}{2}} \tag{7.5}$$

$$\tan \delta_o = \frac{l}{R_1 + \dfrac{w}{2}} \tag{7.6}$$

Eliminating R_1

$$R_1 = \frac{1}{2}w + \frac{l}{\tan \delta_i} = -\frac{1}{2}w + \frac{l}{\tan \delta_o} \tag{7.7}$$

provides us with the Ackerman condition (7.1), which is a direct relationship between δ_i and δ_o.

$$\cot \delta_o - \cot \delta_i = \frac{w}{l} \tag{7.8}$$

To find the vehicle's turning radius R, we define an equivalent bicycle model, as shown in Figure 7.3. The radius of rotation R is perpendicular

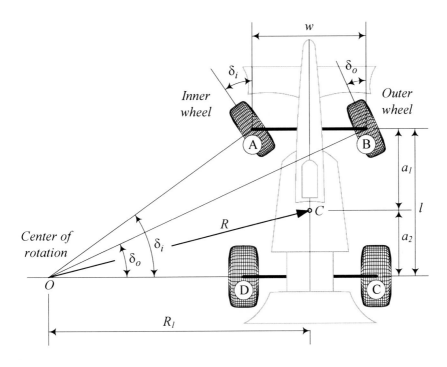

FIGURE 7.2. A front-wheel-steering vehicle and steer angles of the inner and outer wheels.

to the vehicle's velocity vector **v** at the mass center C. Using the geometry shown in the bicycle model, we have

$$R^2 = a_2^2 + R_1^2 \tag{7.9}$$

$$\cot \delta = \frac{R_1}{l} = \frac{1}{2}(\cot \delta_i + \cot \delta_o) \tag{7.10}$$

and therefore,

$$R = \sqrt{a_2^2 + l^2 \cot^2 \delta} \tag{7.11}$$

The Ackerman condition is needed when the speed of the vehicle is very small, and slip angles are very close to zero. In these conditions, there would be no lateral force and no centrifugal force to balance each other. The Ackerman steering condition is also called the *kinematic steering condition*, because it is only a static condition at zero velocity.

A device that provides steering according to the Ackerman condition (7.1) is called *Ackerman steering*, *Ackerman mechanism*, or *Ackerman geometry*. There is no practical four-bar linkage steering mechanism that can provide the Ackerman condition perfectly. However, we may design a multi-bar linkages to work close to the condition and be exact at a few angles.

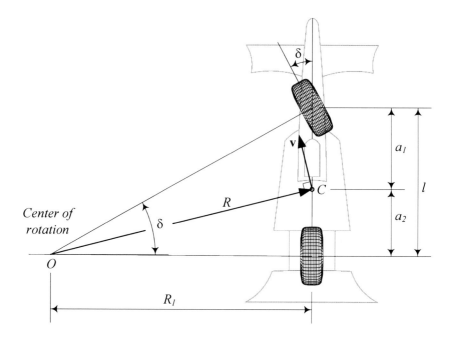

FIGURE 7.3. Equivalent bicycle model for a front-wheel-steering vehicle.

Figure 7.4 illustrates the Ackerman condition for different values of w/l. The difference between the inner and outer steer angles decreases by decreasing w/l. ∎

Example 268 *Steering angles and radii.*
Consider a vehicle with the following dimensions and inner steer angle:

$$l = 103.1\,\text{in} \approx 2.619\,\text{m} \qquad w = 61.6\,\text{in} \approx 1.565\,\text{m}$$
$$a_2 = 60\,\text{in} \approx 1.524\,\text{m} \qquad \delta_i = 12\,\text{deg} \approx 0.209\,\text{rad} \qquad (7.12)$$

The kinematic steering characteristics of the vehicle would be

$$\delta_o = \cot^{-1}\left(\frac{w}{l} + \cot\delta_i\right) = 0.186\,\text{rad} \approx 10.661\,\text{deg} \qquad (7.13)$$

$$\delta = \cot^{-1}\left(\frac{\cot\delta_o + \cot\delta_i}{2}\right) = 0.19684\,\text{rad} \approx 11.278\,\text{deg} \qquad (7.14)$$

$$R_1 = l\cot\delta_i + \frac{1}{2}w = 516.9\,\text{in} \approx 13.129\,\text{m} \qquad (7.15)$$

$$R = \sqrt{a_2^2 + l^2\cot^2\delta} = 520.46\,\text{in} \approx 13.219\,\text{m} \qquad (7.16)$$

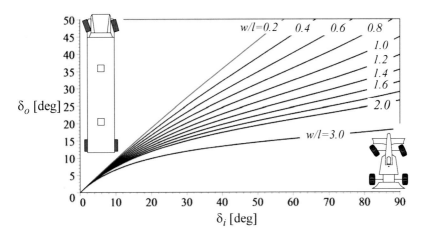

FIGURE 7.4. Effect of w/l on the Ackerman condition for front-wheel-steering vehicles.

Example 269 w *is the front track.*

Most cars have different tracks in front and rear. The track w in the kinematic condition (7.1) refers to the front track w_f. The rear track has no effect on the kinematic condition of a front-wheel-steering vehicle. The rear track w_r of a FWS vehicle can be zero with the same kinematic steering condition (7.1).

Example 270 *Width requirement.*

The kinematic steering condition can be used to calculate the space requirement of a vehicle during a turn. Consider the front wheels of a two-axle vehicle, steered according to the Ackerman geometry as shown in Figure 7.5.

The outer point of the front of the vehicle will run on the maximum radius R_{Max}, whereas a point on the inner side of the vehicle at the location of the rear axle will run on the minimum radius R_{min}. The front outer point has an overhang distance g from the front axle. The maximum radius R_{Max} is

$$R_{Max} = \sqrt{(R_{min} + w)^2 + (l + g)^2} \qquad (7.17)$$

Therefore, the required space for turning is a ring with a width $\triangle R$

$$\triangle R = R_{Max} - R_{min} = \sqrt{(R_{min} + w)^2 + (l + g)^2} - R_{min} \qquad (7.18)$$

The required space $\triangle R$ can be calculated based on the steer angle by substituting R_{min}

$$R_{min} = R_1 - \frac{1}{2}w = \frac{l}{\tan \delta_i} = \frac{l}{\tan \delta_o} - w \qquad (7.19)$$

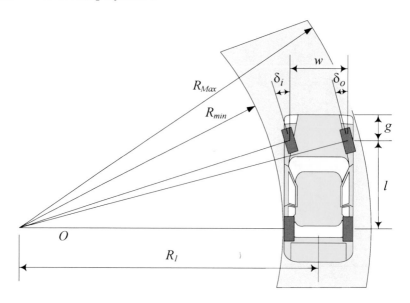

FIGURE 7.5. The required space for a turning two-axle vehicle.

and getting

$$\triangle R = \sqrt{\left(\frac{l}{\tan \delta_i} + 2w\right)^2 + (l+g)^2} - \frac{l}{\tan \delta_i} \tag{7.20}$$

$$= \sqrt{\left(\frac{l}{\tan \delta_o} + w\right)^2 + (l+g)^2} - \frac{l}{\tan \delta_o} + w \tag{7.21}$$

In this example the width of the car w_v and the track w are assumed to be equal. The width of vehicles are always greater than their track.

$$w_v > w \tag{7.22}$$

The requirement for more space becomes more critical when the vehicle is longer and w/l increases. Figure 7.6 compares the required space $\triangle R$ for a bus and a long truck.

Example 271 $\triangle R$ *for a semitrailer.*

Figure 7.7 illustrates a semitrailer in a turn. Analysis of the tractor provides us with the radius R_1 as a function of the steer angle and front width of the tractor.

$$R_1 = \frac{1}{2}w + \frac{a_1}{\tan \delta_i} = -\frac{1}{2}w + \frac{a_1}{\tan \delta_o} \tag{7.23}$$

Therefore, the maximum radius, which belongs to the tractor, is

$$R_{Max} = \sqrt{\left(R_1 + \frac{w}{2}\right)^2 + a_3^2} \tag{7.24}$$

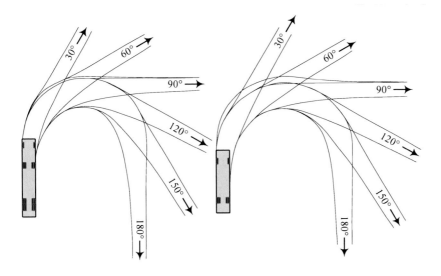

FIGURE 7.6. The required space $\triangle R$ for a bus and a long truck.

The minimum radius R_{min} belongs to the trailer and is

$$R_{min} = \sqrt{R_1 - a_2^2} - \frac{1}{2}w_r \qquad (7.25)$$

Therefore, the required space for turning is a ring with a width $\triangle R$.

$$\triangle R = R_{Max} - R_{min} = \sqrt{\left(R_1 + \frac{w}{2}\right)^2 + a_3^2} - \sqrt{R_1 - a_2^2} - \frac{w_r}{2} \qquad (7.26)$$

Example 272 *Trapezoidal steering mechanism.*

*Figure 7.8 illustrates a symmetric four-bar linkage called **trapezoidal steering mechanism**, used for more than 100 years as a steering connection. The mechanism has two characteristic parameters: angle β and offset arm length d. A steered position of the trapezoidal mechanism is shown in Figure 7.9 to illustrate the inner and outer steer angles δ_i and δ_o.*

The relationship between the inner and outer steer angles of a trapezoidal steering mechanism is given by an implicit function:

$$\sin(\beta + \delta_i) + \sin(\beta - \delta_o)$$
$$= \frac{w}{d} - \sqrt{\left(\frac{w}{d} - 2\sin\beta\right)^2 - (\cos(\beta - \delta_o) - \cos(\beta + \delta_i))^2} \qquad (7.27)$$

To prove this equation, we examine Figure 7.10. In the triangle $\triangle ABC$ we can write

$$(w - 2d\sin\beta)^2 = (w - d\sin(\beta - \delta_o) - d\sin(\beta + \delta_i))^2$$
$$+ (d\cos(\beta - \delta_o) - d\cos(\beta + \delta_i))^2 \qquad (7.28)$$

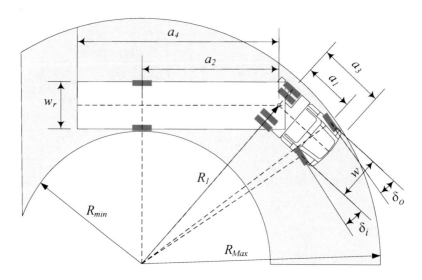

FIGURE 7.7. A semitrailer in a turn.

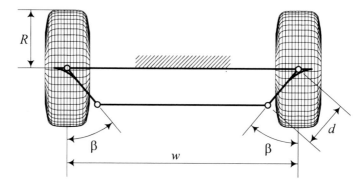

FIGURE 7.8. A trapezoidal steering mechanism.

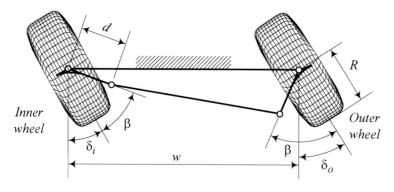

FIGURE 7.9. Steered configuration of a trapezoidal steering mechanism.

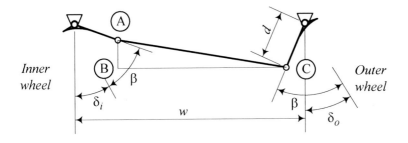

FIGURE 7.10. Trapezoidal steering triangle ABC.

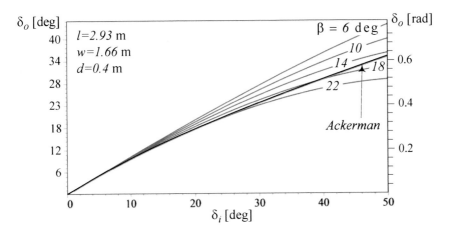

FIGURE 7.11. Behavior of a trapezoidal steering mechanism, compared to the associated Ackerman mechanism $d = 0.4\,\mathrm{m}$.

and derive Equation (7.27) with some manipulation.

The functionality of a trapezoidal steering mechanism, compared to the associated Ackerman condition, is shown in Figure 7.11 and 7.12 for $l = 2.93\,\mathrm{m} \approx 9.61\,\mathrm{ft}$, $w = 1.66\,\mathrm{m} \approx 5.45\,\mathrm{ft}$ *and respectively for* $d = 0.4\,\mathrm{m} \approx 1.3\,\mathrm{ft}$ *and* $d = 0.2\,\mathrm{m} \approx 0.65\,\mathrm{ft}$. *The horizontal axis shows the inner steer angle and the vertical axis shows the outer steer angle. It shows that for given* l *and* w, *a mechanism with* $18\,\mathrm{deg} \lesssim \beta \lesssim 22\,\mathrm{deg}$ *is the best simulator of the Ackerman mechanism if* $\delta_i < 50\,\mathrm{deg}$.

To examine the trapezoidal steering mechanism and compare it with the Ackerman condition, we define an error parameter $e = \delta_{D_o} - \delta_{A_o}$. The error e is the difference between the outer steer angles calculated by the trapezoidal mechanism and the Ackerman condition at the same inner steer angle δ_i.

$$e = \Delta\delta_o = \delta_{D_o} - \delta_{A_o} \qquad (7.29)$$

Figure 7.13 and 7.14 depicts the error e *for the same sample steering mechanisms using the angle* β *as a parameter.*

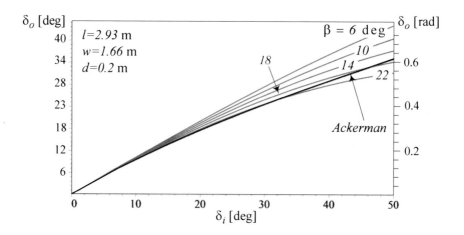

FIGURE 7.12. Behavior of a trapezoidal steering mechanism, compared to the associated Ackerman mechanism $d = 0.2\,\mathrm{m}$.

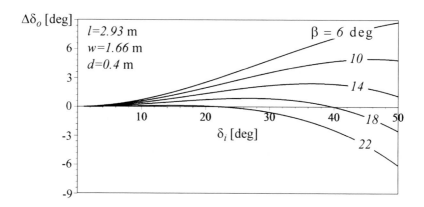

FIGURE 7.13. The error parameter $e = \delta_{D_o} - \delta_{A_o}$ for a sample trapezoidal steering mechanism $d = 0.4\,\mathrm{m}$.

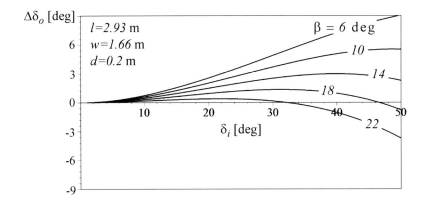

FIGURE 7.14. The error parameter $e = \delta_{D_o} - \delta_{A_o}$ for a sample trapezoidal steering mechanism $d = 0.2\,\mathrm{m}$.

Example 273 ★ *Locked rear axle.*

In a design of simple vehicles, we sometimes eliminate the differential and use a locked rear axle in which no relative rotation between the left and right wheels is possible. Such a simple design is usually used in toy cars, or small off-road vehicles such as a mini Baja.

Consider the vehicle shown in Figure 7.2. In a slow left turn, the speed of the inner rear wheel must be

$$v_{ri} = \left(R_1 - \frac{w}{2}\right) r = R_w \omega_{ri} \qquad (7.30)$$

and the speed of the outer rear wheel must be

$$v_{ro} = \left(R_1 + \frac{w}{2}\right) r = R_w \omega_{ro} \qquad (7.31)$$

where, r is the yaw velocity of the vehicle, R_w is rear wheels radius, and ω_{ri}, ω_{ro} should respectively be the angular velocities of the rear inner and outer wheels about their common axle. If the rear axle is locked, we have

$$\omega_{ri} = \omega_{ro} = \omega \qquad (7.32)$$

however,

$$\left(R_1 - \frac{w}{2}\right) \neq \left(R_1 + \frac{w}{2}\right) \qquad (7.33)$$

which shows it is impossible to have a locked axle for a nonzero w.

Turning reduces the load on the inner wheels and makes the rear inner wheel of a locked axle overcome the friction force and spin. Hence, the traction of the inner wheel drops to the maximum friction force under a reduced load. However, the load on the outer wheels increases and hence, the higher friction limit of the outer wheel helps to have higher traction force on the outer rear wheel.

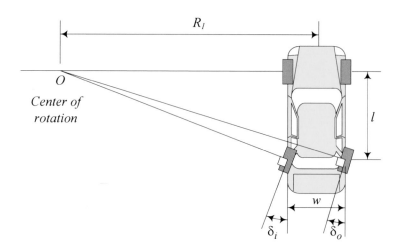

FIGURE 7.15. A rear-wheel-steering vehicle.

Eliminating the differential and using a locked drive axle is an impractical design for street cars. However, it can be an acceptable design for small and light cars moving on dirt or slippery surfaces. It reduces the cost and simplifies the design significantly.

In a conventional two-wheel-drive motor vehicle, the rear wheels are driven using a differential, and the vehicle is steered by changing the direction of the front wheels. With an ideal differential, equal torque is delivered to each drive wheel. The rotational speed of the drive wheels are determined by the differential and the tire-road characteristics. However, a vehicle using a differential has disadvantages when one wheel has lower traction. Differences in traction characteristics of each of the drive wheels may come from different tire-road characteristics or weight distribution. Because a differential delivers equal torque, the wheel with greater tractive ability can deliver only the same amount of torque as the wheel with the lower traction. The steering behavior of a vehicle with a differential is relatively stable under changing tire-road conditions. However, the total thrust may be reduced when the traction conditions are different for drive wheel.

Example 274 ★ *Rear-wheel-steering.*

Rear-wheel-steering is used where high maneuverability is a necessity on a low-speed vehicle, such as forklifts. Rear-wheel-steering is not used on street vehicles because it is unstable at high speeds. The center of rotation for a rear-wheel-steering vehicle is always a point on the extension of the front axle.

Figure 7.15 illustrates a rear-wheel-steering vehicle. The kinematic steer-

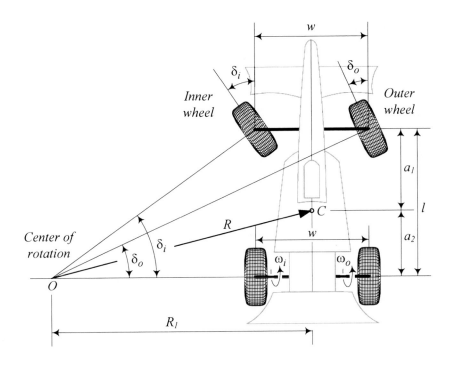

FIGURE 7.16. Kinematic condition of a FWS vehicle using the angular velocity of the inner and outer wheels.

ing condition (7.1) remains the same for a rear-wheel steering vehicle.

$$\cot \delta_o - \cot \delta_i = \frac{w}{l} \qquad (7.34)$$

Example 275 ★ *Alternative kinematic steer angles equation.*

Consider a rear-wheel-drive vehicle with front steerable wheels as shown in Figure 7.16. Assume that the front and rear tracks of the vehicle are equal and the drive wheels are turning without slip. If we show the angular velocities of the inner and outer drive wheels by ω_i and ω_o, respectively, the kinematic steer angles of the front wheels can be expressed by

$$\delta_i = \tan^{-1}\left(\frac{l}{w}\left(\frac{\omega_o}{\omega_i} - 1\right)\right) \qquad (7.35)$$

$$\delta_o = \tan^{-1}\left(\frac{l}{w}\left(1 - \frac{\omega_i}{\omega_o}\right)\right) \qquad (7.36)$$

To prove these equations, we may start from the non-slipping condition for the drive wheels:

$$\frac{R_w \,\omega_o}{R_1 + \dfrac{w}{2}} = \frac{R_w \,\omega_i}{R_1 - \dfrac{w}{2}} \qquad (7.37)$$

Equation (7.37) can be rearranged to

$$\frac{\omega_o}{\omega_i} = \frac{R_1 + \dfrac{w}{2}}{R_1 - \dfrac{w}{2}} \tag{7.38}$$

and substituted in Equations (7.35) and (7.36) to reduce them to Equations (7.5) and (7.6).

The equality (7.37) is the yaw rate of the vehicle about the z-axis, which is also the vehicle's angular velocity about the center of rotation.

$$r = \frac{R_w \, \omega_o}{R_1 + \dfrac{w}{2}} = \frac{R_w \, \omega_i}{R_1 - \dfrac{w}{2}} \tag{7.39}$$

Example 276 ★ *Unequal front and rear tracks.*

We usually design vehicles with different tracks in the front and rear. It is a common design for race cars, which are usually equipped with wider and larger rear tires to increase traction and stability. Such a vehicle is illustrated in Figure 7.17. For street cars we usually use the same tires in the front and rear, however, it is common to have a few centimeters of larger rear track.

The yaw rate or angular velocity of the vehicle is

$$r = \frac{R_w \, \omega_o}{R_1 + \dfrac{w_r}{2}} = \frac{R_w \, \omega_i}{R_1 - \dfrac{w_r}{2}} \tag{7.40}$$

and the kinematic steer angles of the front wheels are

$$\delta_i = \tan^{-1} \frac{2l \, (\omega_o + \omega_i)}{w_f \, (\omega_o - \omega_i) + w_r \, (\omega_o + \omega_i)} \tag{7.41}$$

$$\delta_o = \tan^{-1} \frac{2l \, (\omega_o - \omega_i)}{w_f \, (\omega_o - \omega_i) + w_r \, (\omega_o + \omega_i)} \tag{7.42}$$

To show these equations, we should find R_1 from Equation (7.40)

$$R_1 = \frac{w_r}{2} \frac{\omega_o + \omega_i}{\omega_o - \omega_i} \tag{7.43}$$

and substitute it in the following geometric equations.

$$\tan \delta_i = \frac{l}{R_1 - \dfrac{w_f}{2}} \tag{7.44}$$

$$\tan \delta_o = \frac{l}{R_1 + \dfrac{w_f}{2}} \tag{7.45}$$

In the above equations, w_f is the front track, w_r is the rear track, and R_w is the wheel effective radius.

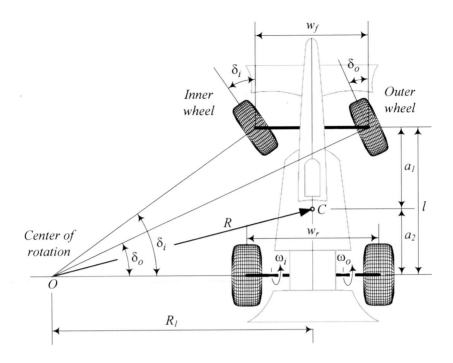

FIGURE 7.17. Kinematic steering condition for a vehicle with different tracks in the front and in the back.

Example 277 ★ *Independent rear-wheel-drive.*

For some special-purpose vehicles, such as moon rovers and autonomous mobile robots, we may attach each drive wheel to an independently controlled motor to apply any desired angular velocity. The steerable wheels of such vehicles are usually designed to be able to turn more than 90 deg to the left and right. Such a vehicle is highly maneuverable at low speeds.

Figure 7.18 illustrates the advantages of such a high steerable vehicle and its possible turnings. Figures 7.18(a)-(c) illustrate forward maneuvering. The arrows next to the rear wheels, illustrate the magnitude of the angular velocity of the wheel, and the arrows on the front wheels illustrate the direction of their motion. The maneuvering in backward motion is illustrated in Figures 7.18(d)-(f). Having such a vehicle allows us to turn the vehicle about any point on the rear axle including the inner points. In Figure 7.18(g) the vehicle is turning about the center of the rear right wheel, and in Figure 7.18(h) about the center of the rear left wheel. Figure 7.18(i) illustrates a rotation about the center point of the rear axle.

In any of the above scenarios, the steer angle of the front wheels should be determined using a proper equation, such as (7.44) and (7.45). The ratio of the outer to inner angular velocities of the drive wheels ω_o/ω_i may be

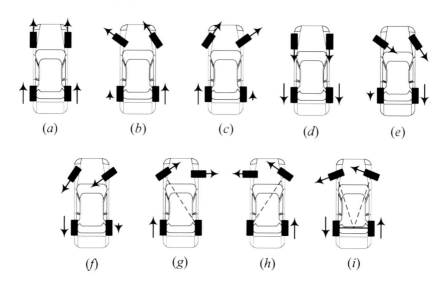

FIGURE 7.18. A highly steerable vehicle.

determined using either the outer or inner steer angles.

$$\frac{\omega_o}{\omega_i} = \frac{\delta_o (w_f - w_r) - 2l}{\delta_o (w_f + w_r) - 2l} \qquad (7.46)$$

$$\frac{\omega_o}{\omega_i} = \frac{\delta_i (w_f + w_r) + 2l}{\delta_i (w_f - w_r) + 2l} \qquad (7.47)$$

Example 278 ★ *Race car steering.*

The Ackerman or kinematic steering is a correct condition when the turning speed of the vehicle is very slow. When the vehicle turns very fast, significant lateral acceleration is needed and therefore, the wheels operate at high slip angles. Furthermore, the loads on the inner wheels will be much lower than the outer wheels. Tire performance curves show that by increasing the wheel load, less slip angle is required to reach the peak of the lateral force. Under these conditions the inner front wheel of a kinematic steering vehicle might be at a higher slip angle than required for maximum lateral force. Therefore, the inner wheel of a vehicle in a high speed turn must operate at a lower steer angle than kinematic steering and the outer wheel must operate at a higher steer angle than kinematic steering. These requirements reduces the difference between steer angles of the inner and outer wheels.

For race cars, it is common to use parallel or reverse steering. Ackerman, parallel, and reverse Ackerman steering are illustrated in Figure 7.19.

The correct steer angle is a function of the instant wheel load, road condition, speed, and tire characteristics. Furthermore, the vehicle must also be able to turn at a low speed under an Ackerman steering condition. Hence, there is no ideal steering mechanism unless we control the steer angle of

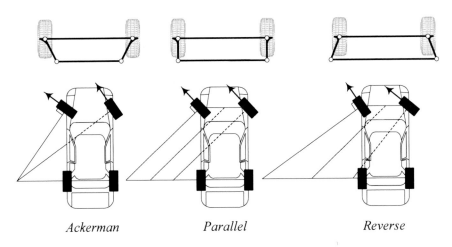

| Ackerman | Parallel | Reverse |

FIGURE 7.19. By increasing the speed at a turn, parallel or reverse steering is needed instead of Ackerman steering.

each steerable wheel independently using a steer by wire and smart system.

Example 279 ★ *Speed dependent steering system.*
There is a speed adjustment idea that says it is better to have a harder steering system at high speeds that needs higher torques. This idea can be applied in power steering systems to make them speed dependent, such that the steering be heavily assisted at low speeds and lightly assisted at high speeds. The idea is supported by this fact that the drivers might need large steering for parking, and small steering when traveling at high speeds.

Example 280 ★ *Ackerman condition history.*
Correct steering geometry was a major problem in the early days of carriages, horse-drawn vehicles, and cars. Four- or six-wheel cars and carriages always left rubber marks behind. This is why there were so many three-wheeled cars and carriages in the past. The problem was making a mechanism to give the inner wheel a smaller turning radius than the outside wheel when the vehicle was driven in a circle.

The required geometric condition for a front-wheel-steering four-wheel-carriage was introduced in 1816 by George Langensperger in Munich, Germany. Langensperger's mechanism is illustrated in Figure 7.20.

Rudolf Ackerman met Langensperger and saw his invention and acted as Langensperger's patent agent in London to introduce the invention to British carriage builders. Car manufacturers have been adopting and improving the Ackerman geometry for their steering mechanisms since 1881.

The basic design of vehicle steering systems has changed little since the invention of the steering mechanism. The driver's steering input is transmitted by a shaft through some type of gear reduction mechanism to generate

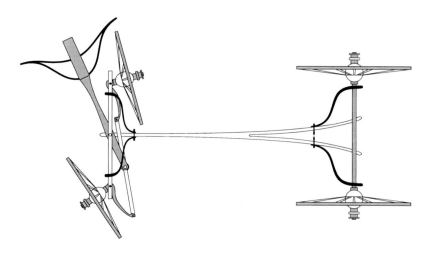

FIGURE 7.20. Langensperger invention for the steering geometry condition.

steering motion at the front wheels.

7.2   Vehicles with More Than Two Axles

If a vehicle has more than two axles, all the axles, except one, must be steerable to provide slip-free turning at zero velocity. When an n-axle vehicle has only one non-steerable axle, there are $n - 1$ geometric steering conditions. A three-axle vehicle with two steerable axles is shown in Figure 7.22.

To indicate the geometry of a multi-axle vehicle, we start from the front axle and measure the longitudinal distance a_i between axle i and the mass center C. Hence, a_1 is the longitudinal distance between the front axle and C, and a_2 is the distance between the second axle and C, and so on. Furthermore, we number the wheels in a clockwise rotation starting from the front left wheel as number 1. In the United States, Europe, and other left-hand drive (LHD) system countries, the front left wheel is the driver's wheel, while in Britain and other right-hand drive (RHD) system countries the front left wheel is the passenger wheel

For the three-axle vehicle shown in Figure 7.22, there are two independent Ackerman conditions:

$$\cot \delta_2 - \cot \delta_1 = \frac{w}{a_1 + a_3} \tag{7.48}$$

$$\cot \delta_3 - \cot \delta_4 = \frac{w}{a_2 + a_3} \tag{7.49}$$

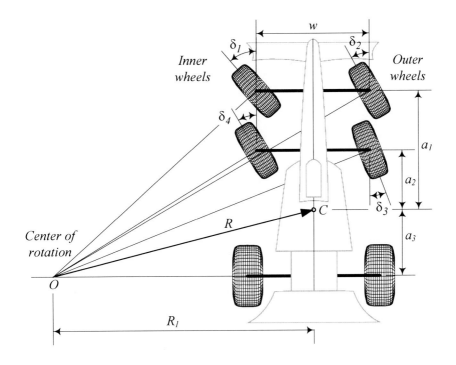

FIGURE 7.21. Steering of a three-axle vehicle.

Example 281 *A six-wheel vehicle with one steerable axle.*

When a multi-axle vehicle has only one steerable axle, slip-free rotation is impossible for the non-steering wheels. The kinematic length or wheel-base of the vehicle is not clearly defined, and it is not possible to define a kinematic steering condition. Strong wear occurs for the tires, especially at low speeds and large steer angles. Hence, such a combination is not recommended. However, in case of a long three-axle vehicle with two non-steerable axles close to each other, an approximated analysis is possible for low-speed steering.

*Figure 7.22 illustrates a six-wheel vehicle with only one steerable axle in front. We design the steering mechanism such that the center of rotation O is on a lateral line, called the **midline**, between the couple rear axles. The kinematic length of the vehicle, l, is the distance between the front axle and the midline. For this design we have*

$$\cot \delta_o - \cot \delta_i = \frac{w}{l} \tag{7.50}$$

and

$$l = a_1 + a_2 + \frac{a_3 - a_2}{2} \tag{7.51}$$

$$R_1 = l \cot \delta_o - \frac{w}{2} = l \cot \delta_i + \frac{w}{2} \tag{7.52}$$

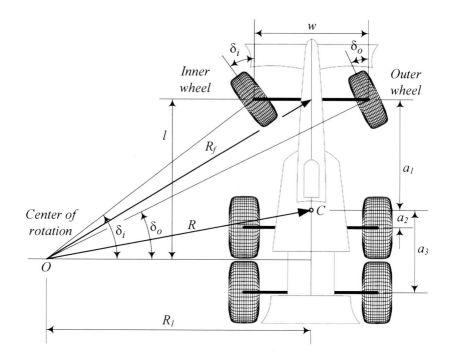

FIGURE 7.22. A six-wheel vehicle with one steerable axle in front.

The center of the front axle and the mass center of the vehicle are turning about O by radii R_f and R.

$$R_f = \frac{R_1}{\cos\left(\tan^{-1}\dfrac{l}{R_1}\right)} \qquad R = \frac{R_1}{\cos\left(\tan^{-1}\dfrac{a_3 - a_2}{2R_1}\right)} \qquad (7.53)$$

If the radius of rotation is large compared to the wheelbase, we may approximate Equations (7.52) and (7.53) to:

$$R_1 = \frac{l}{2}\left(\cot\delta_o + \cot\delta_i\right) \qquad (7.54)$$

$$R_f \approx \frac{R_1}{\cos\left(\dfrac{l}{R_1}\right)} \qquad R \approx \frac{R_1}{\cos\left(\dfrac{a_3 - a_2}{2R_1}\right)} \qquad (7.55)$$

   To avoid strong wear of tires of heavy trucks, it is possible to lift an axle when the vehicle is not carrying heavy loads. For such a vehicle, we may design the steering mechanism to follow an Ackerman condition based on a wheelbase for the non-lifted axle. However, when this vehicle is carrying a heavy load and using all the axles, the liftable axle encounters huge wear at large steer angles.

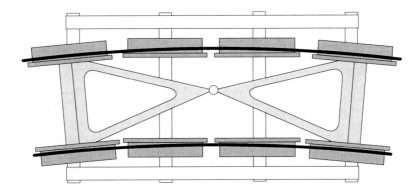

FIGURE 7.23. A self-steering axle mechanism for locomotive wagons.

*Another option for multi-axle vehicles is to use **self-steering** wheels that can adjust themselves to minimize sideslip. Such wheels cannot provide lateral force, and hence, cannot help in maneuvers very much. Self-steering wheels may be installed on buggies and trailers. Such a self-steering axle mechanism for locomotive wagons is shown in Figure 7.23.*

7.3 ★ Vehicle with Trailer

If a four-wheel vehicle has a trailer with one axle, it is possible to derive a kinematic condition for slip-free steering. Figure 7.24 illustrates a vehicle with a one-axle trailer. The mass center of the vehicle is turning on a circle with radius R, while the trailer is turning on a circle with radius R_t.

$$R_t = \sqrt{\left(l\cot\delta_i + \frac{1}{2}w\right)^2 + b_1^2 - b_2^2} \qquad (7.56)$$

$$R_t = \sqrt{\left(l\cot\delta_o - \frac{1}{2}w\right)^2 + b_1^2 - b_2^2} \qquad (7.57)$$

At a steady-state condition, the angle between the trailer and the vehicle is

$$\theta = \begin{cases} 2\tan^{-1}\left[\dfrac{1}{b_1 - b_2}\left(R_t \mp \sqrt{R_t^2 - b_1^2 + b_2^2}\right)\right] & b_1 - b_2 \neq 0 \\[2ex] 2\tan^{-1}\dfrac{1}{2R_t}(b_1 + b_2) & b_1 - b_2 = 0 \end{cases} \qquad (7.58)$$

Proof. Using the right triangle $\triangle OAB$ in Figure 7.24, we may write the trailer's radius of rotation as

$$R_t = \sqrt{R_1^2 + b_1^2 - b_2^2} \qquad (7.59)$$

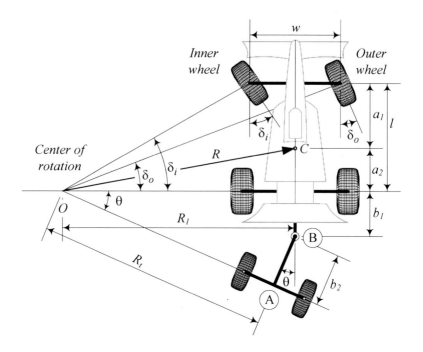

FIGURE 7.24. A vehicle with a one-axle trailer.

because the length $\overline{OB}$ is

$$\overline{OB}^2 = R_t^2 + b_2^2 = R_1^2 + b_1^2 \tag{7.60}$$

Substituting R_1 from Equation (7.7) shows that the trailer's radius of rotation is related to the vehicle's geometry and steer angles by

$$R_t = \sqrt{\left(l \cot \delta_i + \frac{1}{2}w\right)^2 + b_1^2 - b_2^2} \tag{7.61}$$

$$R_t = \sqrt{\left(l \cot \delta_o - \frac{1}{2}w\right)^2 + b_1^2 - b_2^2} \tag{7.62}$$

$$R_t = \sqrt{R^2 - a_2^2 + b_1^2 - b_2^2} \tag{7.63}$$

Using the equation

$$R_t \sin \theta = b_1 + b_2 \cos \theta \tag{7.64}$$

and employing trigonometry, we calculate the angle θ between the trailer and the vehicle as (7.58).

The minus sign in (7.58), in case $b_1 - b_2 \neq 0$, is the usual case in forward motion, and the plus sign is a solution associated with a backward motion. Both possible θ for a set of configuration (R_t, b_1, b_2) are shown in Figure

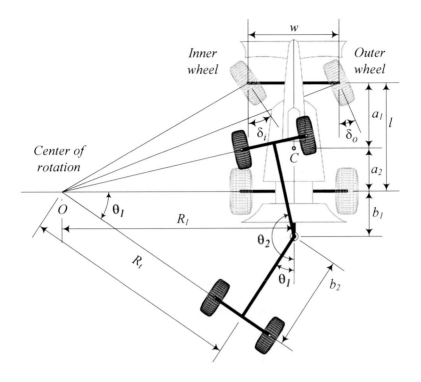

FIGURE 7.25. Two possible angle θ for a set of (R_t, b_1, b_2).

7.25. The unstable and undesirable θ_2 is called a **jackknifing** configuration. The second configuration usually interferes with physical boundaries of vehicles and is not achievable without crash.   ■

Example 282 ★ *Two possible trailer-vehicle angles.*

Consider a four-wheel vehicle that is pulling a one-axle trailer with the following dimensions:

$$
\begin{aligned}
l &= 104\,\text{in} \approx 2.641\,\text{m} & w &= 64\,\text{in} \approx 1.626\,\text{m} \\
b_1 &= 24\,\text{in} \approx 0.61\,\text{m} & b_2 &= 90\,\text{in} \approx 2.286\,\text{m} \\
a_2 &= 50\,\text{in} \approx 1.27\,\text{m} & \delta_i &= 12\,\text{deg} \approx 0.209\,\text{rad} \quad (7.65)
\end{aligned}
$$

The kinematic steering characteristics of the vehicle would be

$$
\delta_o = \cot^{-1}\left(\frac{w}{l} + \cot\delta_i\right) \approx 0.185\,\text{rad} \approx 10.626\,\text{deg} \tag{7.66}
$$

$$
R_t = \sqrt{\left(l\cot\delta_i + \frac{1}{2}w\right)^2 + b_1^2 - b_2^2} = 515.09\,\text{in} \approx 13.083\,\text{m} \tag{7.67}
$$

$$
R_1 = l\cot\delta_i + \frac{1}{2}w = 522.34\,\text{in} \approx 13.268\,\text{m} \tag{7.68}
$$

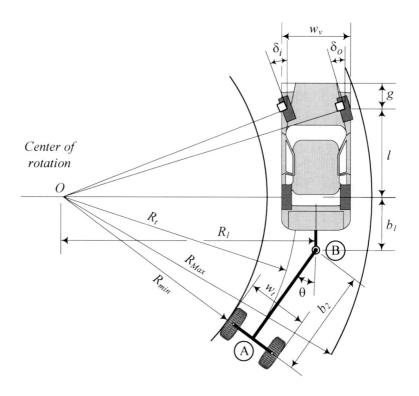

FIGURE 7.26. A two-axle vehicle with a trailer is steered according to the Ackerman condition.

$$\delta = \cot^{-1}\left(\frac{\cot\delta_o + \cot\delta_i}{2}\right) = 0.20274\,\text{rad} \approx 11.616\,\text{deg} \quad (7.69)$$

$$R = \sqrt{a_2^2 + l^2\cot^2\delta} = 508.39\,\text{in} \approx 12.913\,\text{m} \quad (7.70)$$

$$\theta = 2\tan^{-1}\left[\frac{1}{b_1 - b_2}\left(R_t \mp \sqrt{R_t^2 - b_1^2 + b_2^2}\right)\right]$$

$$= \begin{cases} \theta_1 = 0.21890\,\text{rad} \approx 12.542\,\text{deg} \\ \theta_2 = -3.0145\,\text{rad} \approx -172.72\,\text{deg} \end{cases} \quad (7.71)$$

Example 283 ★ *Space requirement.*

The kinematic steering condition can be used to calculate the space requirement of a vehicle with a trailer during a turn. Consider that the front wheels of a two-axle vehicle with a trailer are steered according to the Ackerman geometry, as shown in Figure 7.26.

The outer point of the front of the vehicle will run on the maximum radius R_{Max}, whereas a point on the inner side of the wheel at the trailer's rear *axle will run on the minimum radius R_{min}. The maximum radius R_{Max} is*

$$R_{Max} = \sqrt{\left(R_1 + \frac{w_v}{2}\right)^2 + (l+g)^2} \tag{7.72}$$

where

$$R_1 = \sqrt{\left(R_{min} + \frac{w_t}{2}\right)^2 + b_2^2 - b_1^2} \tag{7.73}$$

and the width of the vehicle is shown by w_v.

The required space for turning the vehicle and trailer is a ring with a width $\triangle R$, which is a function of the vehicle and trailer geometry.

$$\triangle R = R_{Max} - R_{min} \tag{7.74}$$

The required space $\triangle R$ can be calculated based on the steer angle by substituting R_{min}

$$
\begin{aligned}
R_{min} &= R_t - \frac{1}{2}w_t = \sqrt{\left(l\cot\delta_i + \frac{1}{2}w\right)^2 + b_1^2 - b_2^2} - \frac{1}{2}w_t \\
&= \sqrt{\left(l\cot\delta_o - \frac{1}{2}w\right)^2 + b_1^2 - b_2^2} - \frac{1}{2}w_t \\
&= \sqrt{R^2 - a_2^2 + b_1^2 - b_2^2} - \frac{1}{2}w_t \tag{7.75}
\end{aligned}
$$

7.4 Steering Mechanisms

A steering system begins with the *steering wheel* or *steering handle*. The driver's steering input is transmitted by a shaft through a gear reduction system, usually rack-and-pinion or recirculating ball bearings. The steering gear output goes to steerable wheels to generate motion through a *steering mechanism*. The lever, which transmits motion from the steering gear to the steering linkage, is called *Pitman arm*.

The direction of each wheel is controlled by one steering arm. The steering arm is attached to the steerable wheel hub by a keyway, locking taper, and a hub. In some vehicles, it is an integral part of a one-piece hub and steering knuckle.

To achieve good maneuverability, a minimum steering angle of approximately 35 deg must be provided at the front wheels of passenger cars.

A sample parallelogram steering mechanism and its components are shown in Figure 7.27. The parallelogram steering linkage is common on independent front-wheel vehicles. There are many varieties of steering mechanisms each with some advantages and disadvantages.

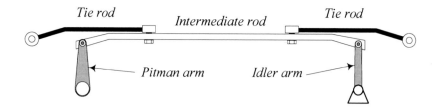

FIGURE 7.27. A sample parallelogram steering linkage and its components.

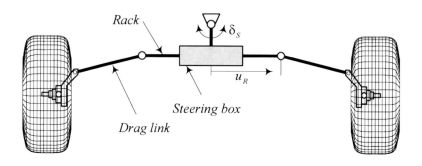

FIGURE 7.28. A rack-and-pinion steering system.

Example 284 *Steering ratio.*

The **steering ratio** *is the rotation angle of a steering wheel divided by the steer angle of the steerable wheels. The steering ratio of street cars is around* 10 : 1 *and the steering ratio of race cars is about* 5 : 1. *Generally speaking, the steering ratio of ground vehicles varies between* 1 : 1 *to* 20 : 1.

The steering ratio of Ackerman steering is different for inner and outer wheels. Furthermore, it has a nonlinear behavior and is a function of the wheel angle.

Example 285 *Rack-and-pinion steering.*

*Rack-and-pinion is the most common steering system of passenger cars. Figure 7.28 illustrates a sample rack-and-pinion steering system. The rack is either in front or behind the steering axle. The driver's rotary steering command δ_S is transformed by a steering box to translation $u_R = u_R(\delta_S)$ of the racks, and then by the **drag links** to the wheel steering $\delta_i = \delta_i(u_R)$, $\delta_o = \delta_o(u_R)$. The overall steering ratio depends on the ratio of the steering box and on the kinematics of the steering linkage. The drag link is also called the **tie rod**.*

Example 286 *Lever arm steering system.*

*Figure 7.29 illustrates a steering linkage that sometimes is called a **lever arm** steering system. Using a lever arm steering system, large steering angles at the wheels are possible. This steering system is used on trucks*

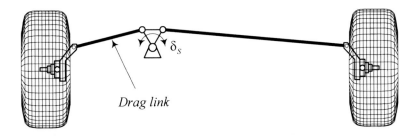

FIGURE 7.29. A lever arm steering system.

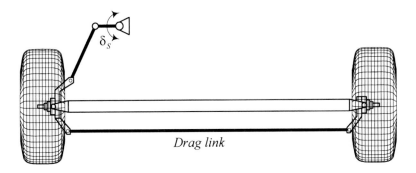

FIGURE 7.30. A drag link steering system.

with large wheel bases and independent wheel suspensions at the front axle. The steering box and triangle can also be placed outside of the axle's center.

Example 287 *Drag link steering system.*

It is sometimes better to send the steering command to only one wheel and connect the other one to the first wheel by a drag link, as shown in Figure 7.30. Such steering linkages are usually used for trucks and busses with a front solid axle. The rotations of the steering wheel are transformed by a steering box to the motion of the steering arm and then to the rotation of the left wheel. A drag link transmits the rotation of the left wheel to the right wheel.

Figure 7.31 shows a sample for connecting a steering mechanism to the Pitman arm of the left wheel and using a trapezoidal linkage to connect the right wheel to the left wheel.

Example 288 *Multi-link steering mechanism.*

In busses and big trucks, the driver may sit more than $2\,\mathrm{m} \approx 7\,\mathrm{ft}$ in front of the front axle. These vehicles need large steering angles at the front wheels to achieve good maneuverability. So a more sophisticated multi-link steering mechanism is needed. A sample multi-link steering mechanism is shown in Figure 7.32.

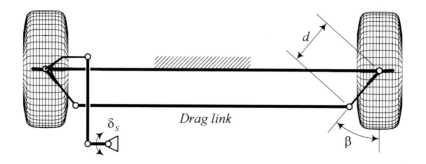

FIGURE 7.31. Connection of the Pitman arm to a trapezoidal steering mechanism.

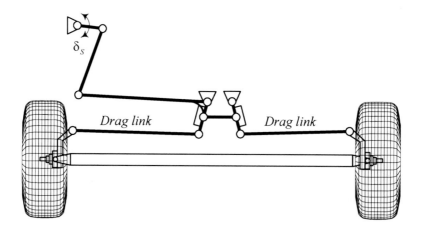

FIGURE 7.32. A multi-link steering mechanism.

The rotations of the steering wheel are transformed by the steering box to a steering lever arm. The lever arm is connected to a distributing linkage, which turns the left and right wheels by a long tire rod.

Example 289 ★ *Reverse efficiency.*

The ability of the steering mechanism to feedback the road inputs to the driver is called **reverse efficiency***. Feeling the applied steering torque or aligning moment helps the driver to make smoother turn.*

Rack-and-pinion and recirculating ball steering gears have a feedback of the wheels steering torque to the driver. However, worm and sector steering gears have very weak feedback. Low feedback may be desirable for off-road vehicles, to reduce the driver's fatigue.

Because of safety, the steering torque feedback should be proportional to the speed of the vehicle. In this way, the required torque to steer the vehicle is higher at higher speeds. Such steering prevents a sharp and high steer angle. A steering damper with a damping coefficient increasing with speed is the mechanism that provides such behavior. A steering damper can also reduce shimmy vibrations.

Example 290 ★ *Power steering.*

Power steering has been developed in the 1950s when a hydraulic power steering assist was first introduced. Since then, power assist has gradually become a standard component in automotive steering systems. Using hydraulic pressure, supplied by an engine-driven pump, amplifies the driver-applied torque at the steering wheel. As a result, the steering effort is reduced.

In recent years, electric torque amplifiers were introduced in automotive steering systems as a substitute for hydraulic amplifiers. Electrical steering eliminates the need for the hydraulic pump. Electric power steering is more efficient than conventional power steering, because the electric power steering motor needs to provide assistance only when the steering wheel is turned, whereas the hydraulic pump runs constantly. The assist level is also tunable by vehicle type, road speed, and driver preference.

Example 291 *Bump steering.*

The steer angle caused by the vertical motion of the wheel with respect to the body is called **bump steering***. Bump steering is usually an undesirable phenomenon and is a function of the suspension and steering mechanisms. If the vehicle has a bump steering character, then the wheel steers when it runs over a bump or when the vehicle rolls in a turn. As a result, the vehicle will travel in a path not selected by the driver.*

Bump steering occurs when the end of the tie rod is not at the instant center of rotation of the suspension mechanism. Hence, in a suspension deflection, the suspension and steering mechanisms will rotate about different centers.

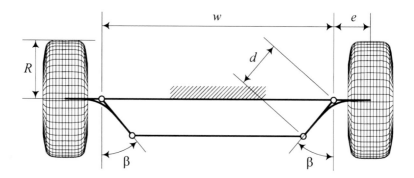

FIGURE 7.33. An offset design for wheel attachment to an steering mechanism.

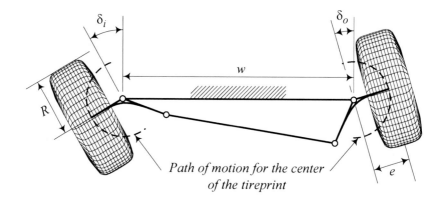

FIGURE 7.34. Offset attachment of steerable wheels to a trapezoidal steering mechanism.

Example 292 ★ *Offset steering axis.*

   Theoretically, the steering axis of each steerable wheel must vertically go through the center of the wheel at the tire-plane to minimize the required steering torque. Figure 7.31 is an example of matching the center of a wheel with the steering axis. However, it is possible to attach the wheels to the steering mechanism, using an offset design, as shown in Figure 7.33.

   Figure 7.34 depicts a steered trapezoidal mechanism with offset wheels attachment. The path of motion for the center of the tireprint for an offset design is a circle with offset arm radius e. Such a design is not recommended for street vehicles, especially because of the huge steering torque of stationary vehicles. However, the steering torque reduces dramatically to an acceptable value when the vehicle is moving. Furthermore, an offset design sometimes makes more room to attach other devices, and simplifies manufacturing. So, it may be used for small off-road vehicles, such as a mini Baja, and toy vehicles.

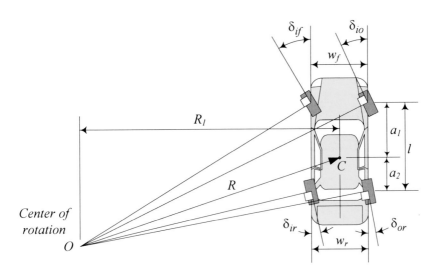

FIGURE 7.35. A positive four-wheel steering vehicle.

7.5 ★ Four wheel steering.

At very low speeds, the kinematic steering condition that the perpendicular lines to each tire meet at one point, must be applied. The intersection point is the *turning center* of the vehicle.

Figure 7.35 illustrates a *positive* four-wheel steering vehicle, and Figure 7.36 illustrates a *negative* $4WS$ vehicle. In a *positive* $4WS$ configuration the front and rear wheels steer in the same direction, and in a *negative* $4WS$ configuration the front and rear wheels steer opposite to each other. The kinematic condition between the steer angles of a $4WS$ vehicle is

$$\cot \delta_{of} - \cot \delta_{if} = \frac{w_f}{l} - \frac{w_r}{l} \frac{\cot \delta_{of} - \cot \delta_{if}}{\cot \delta_{or} - \cot \delta_{ir}} \tag{7.76}$$

where, w_f and w_r are the front and rear tracks, δ_{if} and δ_{of} are the steer angles of the front inner and outer wheels, δ_{ir} and δ_{or} are the steer angles of the rear inner and outer wheels, and l is the wheelbase of the vehicle. We may also use the following more general equation for the kinematic condition between the steer angles of a $4WS$ vehicle

$$\cot \delta_{fr} - \cot \delta_{fl} = \frac{w_f}{l} - \frac{w_r}{l} \frac{\cot \delta_{fr} - \cot \delta_{fl}}{\cot \delta_{rr} - \cot \delta_{rl}} \tag{7.77}$$

where, δ_{fl} and δ_{fr} are the steer angles of the front left and front right wheels, and δ_{rl} and δ_{rr} are the steer angles of the rear left and rear right wheels.

If we define the steer angles according to the sign convention shown in Figure 7.37 then, Equation (7.77) expresses the kinematic condition for

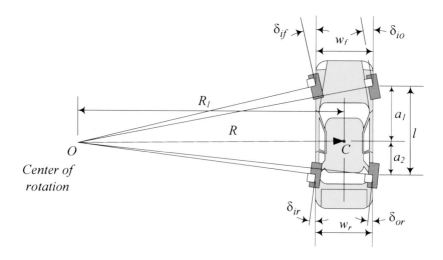

FIGURE 7.36. A negative four-wheel steering vehicle.

both, positive and negative $4WS$ systems. Employing the wheel coordinate frame (x_w, y_w, z_w), we define the steer angle as the angle between the vehicle x-axis and the wheel x_w-axis, measured about the z-axis. Therefore, a steer angle is positive when the wheel is turned to the left, and it is negative when the wheel is turned to the right.

Proof. The slip-free condition for wheels of a $4WS$ in a turn requires that the normal lines to the center of each tire-plane intersect at a common point. This is the kinematic steering condition.

Figure 7.38 illustrates a positive $4WS$ vehicle in a left turn. The inner wheels are the left wheels that are closer to the turning center O. The longitudinal distance between point O and the axles of the car are indicated by c_1, and c_2 measured in the body coordinate frame.

The front inner and outer steer angles δ_{if}, δ_{of} may be calculated from the triangles $\triangle OAE$ and $\triangle OBF$, while the rear inner and outer steer angles δ_{ir}, δ_{or} may be calculated from the triangles $\triangle ODG$ and $\triangle OCH$ as follows.

$$\tan \delta_{if} = \frac{c_1}{R_1 - \dfrac{w_f}{2}} \tag{7.78}$$

$$\tan \delta_{of} = \frac{c_1}{R_1 + \dfrac{w_f}{2}} \tag{7.79}$$

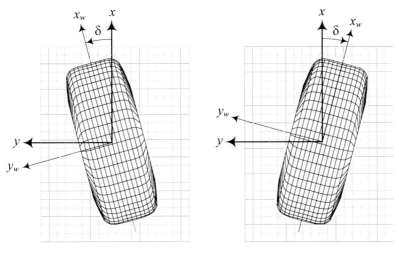

Positive steer angle   Negative steer angle

FIGURE 7.37. Sign convention for steer angles.

$$\tan \delta_{ir} = \frac{c_2}{R_1 - \dfrac{w_r}{2}} \tag{7.80}$$

$$\tan \delta_{or} = \frac{c_2}{R_1 + \dfrac{w_r}{2}} \tag{7.81}$$

Eliminating R_1

$$R_1 = \frac{1}{2}w_f + \frac{c_1}{\tan \delta_{if}} = -\frac{1}{2}w_f + \frac{c_1}{\tan \delta_{of}} \tag{7.82}$$

between (7.78) and (7.79) provides the kinematic condition between the front steering angles δ_{if} and δ_{of}.

$$\cot \delta_{of} - \cot \delta_{if} = \frac{w_f}{c_1} \tag{7.83}$$

Similarly, we may eliminate R_1

$$R_1 = \frac{1}{2}w_r + \frac{c_2}{\tan \delta_{ir}} = -\frac{1}{2}w_r + \frac{c_2}{\tan \delta_{or}} \tag{7.84}$$

between (7.80) and (7.81) to provide the kinematic condition between the rear steering angles δ_{ir} and δ_{or}.

$$\cot \delta_{or} - \cot \delta_{ir} = \frac{w_r}{c_2} \tag{7.85}$$

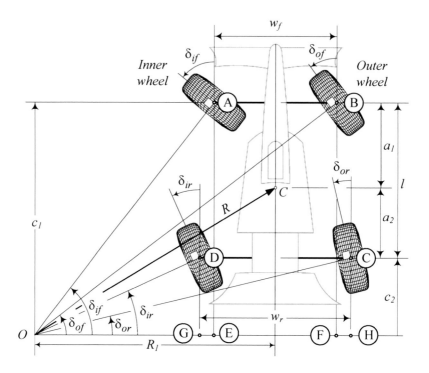

FIGURE 7.38. Illustration of a positive four-wheel steering vehicle in a left turn.

Using the constraint

$$c_1 - c_2 = l \qquad (7.86)$$

we may combine Equations (7.83) and (7.85)

$$\frac{w_f}{\cot \delta_{of} - \cot \delta_{if}} - \frac{w_r}{\cot \delta_{or} - \cot \delta_{ir}} = l \qquad (7.87)$$

to find the kinematic condition (7.76) between the steer angles of the front and rear wheels for a positive $4WS$ vehicle.

Figure 7.39 illustrates a negative $4WS$ vehicle in a left turn. The inner wheels are the left wheels that are closer to the turning center O. The front inner and outer steer angles δ_{if}, δ_{of} can be calculated from the triangles $\triangle OAE$ and $\triangle OBF$, while the rear inner and outer steer angles δ_{ir}, δ_{or} may be calculated from the triangles $\triangle ODG$ and $\triangle OCH$ as follows.

$$\tan \delta_{if} = \frac{c_1}{R_1 - \dfrac{w_f}{2}} \qquad (7.88)$$

$$\tan \delta_{of} = \frac{c_1}{R_1 + \dfrac{w_f}{2}} \qquad (7.89)$$

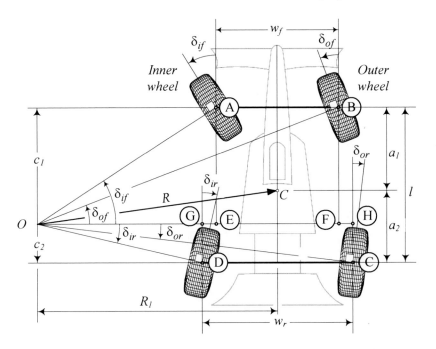

FIGURE 7.39. Illustration of a negative four-wheel steering vehicle in a left turn.

$$- \tan \delta_{ir} = \frac{-c_2}{R_1 - \dfrac{w_r}{2}} \qquad (7.90)$$

$$- \tan \delta_{or} = \frac{-c_2}{R_1 + \dfrac{w_r}{2}} \qquad (7.91)$$

Eliminating R_1

$$R_1 = \frac{1}{2} w_f + \frac{c_1}{\tan \delta_{if}} = -\frac{1}{2} w_f + \frac{c_1}{\tan \delta_{of}} \qquad (7.92)$$

between (7.88) and (7.89) provides us with the kinematic condition between the front steering angles δ_{if} and δ_{of}.

$$\cot \delta_{of} - \cot \delta_{if} = \frac{w_f}{c_1} \qquad (7.93)$$

Similarly, we may eliminate R_1

$$R_1 = \frac{1}{2} w_r + \frac{c_2}{\tan \delta_{ir}} = -\frac{1}{2} w_r + \frac{c_2}{\tan \delta_{or}} \qquad (7.94)$$

between (7.90) and (7.91) to provide the kinematic condition between the rear steering angles δ_{ir} and δ_{or}.

$$\cot \delta_{or} - \cot \delta_{ir} = \frac{w_r}{c_2} \qquad (7.95)$$

Using the constraint

$$c_1 - c_2 = l \qquad (7.96)$$

we may combine Equations (7.93) and (7.95)

$$\frac{w_f}{\cot \delta_{of} - \cot \delta_{if}} - \frac{w_r}{\cot \delta_{or} - \cot \delta_{ir}} = l \qquad (7.97)$$

to find the kinematic condition (7.76) between the steer angles of the front and rear wheels for a negative $4WS$ vehicle.

Using the sign convention of Figure 7.37, we may re-examine Figures 7.38 and 7.39. When the steer angle of the front wheels are positive then, the steer angle of the rear wheels are negative in a negative $4WS$ system, and are positive in a positive $4WS$ system. Therefore, Equation (7.77)

$$\cot \delta_{fr} - \cot \delta_{fl} = \frac{w_f}{l} - \frac{w_r}{l} \frac{\cot \delta_{fr} - \cot \delta_{fl}}{\cot \delta_{rr} - \cot \delta_{rl}} \qquad (7.98)$$

expresses the kinematic condition for both, positive and negative $4WS$ systems. Similarly, the following equations can uniquely determine c_1 and c_2 regardless of the positive or negative $4WS$ system.

$$c_1 = \frac{w_f}{\cot \delta_{fr} - \cot \delta_{fl}} \qquad (7.99)$$

$$c_2 = \frac{w_r}{\cot \delta_{rr} - \cot \delta_{rl}} \qquad (7.100)$$

Four-wheel steering or all wheel steering AWS may be applied on vehicles to improve steering response, increase the stability at high speed maneuvering, or decrease turning radius at low speeds. A negative $4WS$ has shorter turning radius R than a front-wheel steering (FWS) vehicle.

For a FWS vehicle, the perpendicular to the front wheels meet at a point on the extension of the rear axle. However, for a $4WS$ vehicle, the intersection point can be any point in the xy plane. The point is the *turning center* of the car and its position depends on the steer angles of the wheels. Positive steering is also called *same steer*, and a negative steering is also called *counter steer*. ∎

Example 293 ★ *Steering angles relationship.*
Consider a car with the following dimensions.

$$l = 2.8\,\text{m} \qquad w_f = 1.35\,\text{m} \qquad w_r = 1.4\,\text{m} \qquad (7.101)$$

The set of equations (7.78)-(7.81) which are the same as (7.88)-(7.91) must be used to find the kinematic steer angles of the wheels. Assume one of the angles, such as

$$\delta_{if} = 15\,\text{deg} \qquad (7.102)$$

is a known as an input steer angle. To find the other steer angles, we need to know the position of the turning center O. The position of the turning center can be determined if we have one of the three parameters c_1, c_2, R_1. To clarify this fact, let us assume that the car is turning left and we know the value of δ_{if}. Therefore, the perpendicular line to the front left wheel is known. The turning center can be any point on this line. When we pick a point, the other wheels can be adjusted accordingly.

The steer angles for a 4WS system is a set of four equations, each with two variables.

$$\delta_{if} = \delta_{if}(c_1, R_1) \qquad \delta_{of} = \delta_{of}(c_1, R_1) \qquad (7.103)$$
$$\delta_{ir} = \delta_{ir}(c_2, R_1) \qquad \delta_{or} = \delta_{or}(c_2, R_1) \qquad (7.104)$$

If c_1 and R_1 are known, we will be able to determine the steer angles δ_{if}, δ_{of}, δ_{ir}, and δ_{or} uniquely. However, a practical situation is when we have one of the steer angles, such as δ_{if}, and we need to determine the required steer angle of the other wheels, δ_{of}, δ_{ir}, δ_{or}. It can be done if we know c_1 or R_1.

The turning center is the curvature center of the path of motion. If the path of motion is known, then at any point of the road, the turning center can be found in the vehicle coordinate frame.

In this example, let us assume

$$R_1 = 50\,\text{m} \qquad (7.105)$$

therefore, from Equation (7.78), we have

$$c_1 = \left(R_1 - \frac{w_f}{2}\right)\tan\delta_{if} = \left(50 - \frac{1.35}{2}\right)\tan\frac{\pi}{12} = 13.217\,\text{m} \qquad (7.106)$$

Because $c_1 > l$ and $\delta_{if} > 0$ the vehicle is in a positive 4WS configuration and the turning center is behind the rear axle of the car.

$$c_2 = c_1 - l = 13.217 - 2.8 = 10.417\,\text{m} \qquad (7.107)$$

Now, employing Equations (7.79)-(7.81) provides us the other steer angles.

$$\delta_{of} = \tan^{-1}\frac{c_1}{R_1 + \frac{w_f}{2}} = \tan^{-1}\frac{13.217}{50 + \frac{1.35}{2}}$$
$$= 0.25513\,\text{rad} \approx 14.618\,\text{deg} \qquad (7.108)$$

$$\delta_{ir} = \tan^{-1}\frac{c_2}{R_1 - \frac{w_r}{2}} = \tan^{-1}\frac{10.417}{50 - \frac{1.4}{2}}$$
$$= 0.20824\,\text{rad} \approx 11.931\,\text{deg} \qquad (7.109)$$

$$\delta_{or} = \tan^{-1}\frac{c_2}{R_1 + \dfrac{w_r}{2}} = \tan^{-1}\frac{10.417}{50 + \dfrac{1.4}{2}}$$

$$= 0.20264\,\text{rad} \approx 11.61\,\text{deg} \tag{7.110}$$

Example 294 ★ *Position of the turning center.*

The turning center of a vehicle, in the vehicle body coordinate frame, is at a point with coordinates (x_O, y_O). The coordinates of the turning center are

$$x_O = -a_2 - c_2 = -a_2 - \frac{w_r}{\cot\delta_{or} - \cot\delta_{ir}} \tag{7.111}$$

$$y_O = R_1 = \frac{l + \dfrac{1}{2}\left(w_f \tan\delta_{if} - w_r \tan\delta_{ir}\right)}{\tan\delta_{if} - \tan\delta_{ir}} \tag{7.112}$$

Equation (7.112) is found by substituting c_1 and c_2 from (7.92) and (7.94) in (7.96), and defining y_O in terms of δ_{if} and δ_{ir}. It is also possible to define y_O in terms of δ_{of} and δ_{or}.

Equations (7.111) and (7.112) can be used to define the coordinates of the turning center for both positive and negative 4WS systems.

As an example, let us examine a car with

$$l = 2.8\,\text{m} \qquad w_f = 1.35\,\text{m} \qquad w_r = 1.4\,\text{m} \qquad a_1 = a_2 \tag{7.113}$$

$$\begin{aligned}\delta_{if} &= 0.26180\,\text{rad} \approx 15\,\text{deg}\\ \delta_{of} &= 0.25513\,\text{rad} \approx 14.618\,\text{deg}\\ \delta_{ir} &= 0.20824\,\text{rad} \approx 11.931\,\text{deg}\\ \delta_{or} &= 0.20264\,\text{rad} \approx 11.61\,\text{deg}\end{aligned} \tag{7.114}$$

and find the position of the turning center.

$$\begin{aligned}x_O &= -a_2 - \frac{w_r}{\cot\delta_{or} - \cot\delta_{ir}}\\ &= -\frac{2.8}{2} - \frac{1.4}{\cot 0.20264 - \cot 0.20824} = -11.802\,\text{m}\end{aligned} \tag{7.115}$$

$$\begin{aligned}y_O &= \frac{l + \dfrac{1}{2}\left(w_f \tan\delta_{if} - w_r \tan\delta_{ir}\right)}{\tan\delta_{if} - \tan\delta_{ir}}\\[2mm] &= \frac{2.8 + \dfrac{1}{2}\left(1.35\tan 0.26180 - 1.4\tan 0.20824\right)}{\tan 0.26180 - \tan 0.20824} = 50.011\,\text{m}\end{aligned} \tag{7.116}$$

The position of turning center for a FWS vehicle is at

$$x_O = -a_2 \qquad y_O = \frac{1}{2}w_f + \frac{l}{\tan\delta_{if}} \tag{7.117}$$

and for a RWS vehicle is at

$$x_O = a_1 \qquad y_O = \frac{1}{2} w_r + \frac{l}{\tan \delta_{ir}} \tag{7.118}$$

Example 295 ★ *Curvature radius.*
  Consider a road as a path of motion that is expressed mathematically by a function $Y = f(X)$, in a global coordinate frame. The radius of curvature R_κ of such a road at point X is

$$R_\kappa = \frac{\left(1 + Y'^2\right)^{3/2}}{Y''} \tag{7.119}$$

where

$$Y' = \frac{dY}{dX} \qquad Y'' = \frac{d^2Y}{dX^2} \tag{7.120}$$

Consider a road with a given equation

$$Y = \frac{X^2}{200} \qquad Y' = \frac{X}{100} \qquad Y'' = \frac{1}{100} \tag{7.121}$$

where both X and Y are measured in meter $[\text{m}]$. The curvature radius of the road is.

$$R_\kappa = \frac{\left(1 + Y'^2\right)^{3/2}}{Y''} = 100 \left(\frac{1}{10000} X^2 + 1\right)^{3/2} \tag{7.122}$$

At $X = 30\,\text{m}$, we have

$$Y = \frac{9}{2}\,\text{m} \qquad Y' = \frac{3}{10} \qquad Y'' = \frac{1}{100}\,\text{m}^{-1} \tag{7.123}$$

and therefore,

$$R_\kappa = 113.80\,\text{m} \tag{7.124}$$

Example 296 ★ *Symmetric four-wheel steering system.*
  Figure 7.40 illustrates a symmetric 4WS vehicle that the front and rear wheels steer opposite to each other equally. The kinematic steering condition for a symmetric steering is simplified to

$$\cot \delta_o - \cot \delta_i = \frac{w_f}{l} + \frac{w_r}{l} \tag{7.125}$$

and c_1 and c_2 are reduced to

$$c_1 = c_2 = -\frac{1}{2}l \tag{7.126}$$

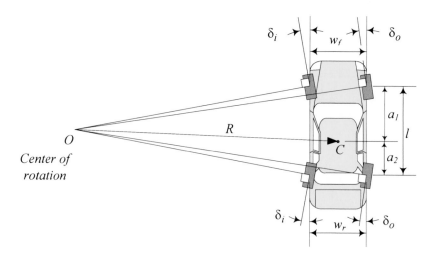

FIGURE 7.40. A symmetric four-wheel steering vehicle.

Example 297 ★ c_2/c_1 *ratio.*

*Longitudinal distance of the turning center of a vehicle from the front axle is c_1 and from the rear axle is c_2. We show the ratio of these distances by c_s and call it the 4WS **factor**.*

$$c_s = \frac{c_2}{c_1} = \frac{w_r}{w_f} \frac{\cot \delta_{fr} - \cot \delta_{fl}}{\cot \delta_{rr} - \cot \delta_{rl}} \tag{7.127}$$

c_s *is negative for a negative 4WS vehicle and is positive for a positive 4WS vehicle. When $c_s = 0$, the car is FWS, and when $c_s = -\infty$, the car is RWS. A symmetric 4WS system has $c_s = -\frac{1}{2}$.*

Example 298 ★ *Steering length l_s.*

*For a 4WS vehicle, we may define a **steering length** l_s as*

$$
\begin{aligned}
l_s &= \frac{c_1 + c_2}{l} = \frac{l}{c_1} + 2c_s \\
&= \frac{1}{l} \left(\frac{w_f}{\cot \delta_{fr} - \cot \delta_{fl}} + \frac{w_r}{\cot \delta_{rr} - \cot \delta_{rl}} \right)
\end{aligned}
\tag{7.128}
$$

Steering length l_s is 1 for a FWS car, zero for a symmetric steering car, and -1 for a RWS car. When a car has a negative 4WS system then, $-1 < l_s < 1$, and when the car has a positive 4WS system then, $l_s > 1$ or $l_s < -1$. The case $l_s > 1$ occurs when the turning center is behind the car, and the case $l_s < -1$ occurs when the turning center is ahead of the car.

A comparison of FWS and 4WS is that the turning center of a FWS car is always on the extension of the rear axel, and its steering length l_s is always equal to 1. However, the turning center of a 4WS car can be:

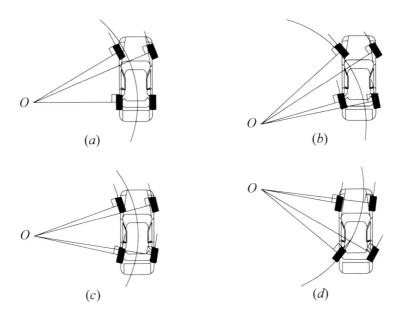

FIGURE 7.41. A comparison among the different steering lengths.

1. *ahead of the front axle, if* $l_s < -1$
2. *between the front and rear axles, if* $-1 < l_s < 1$
3. *behind the rear axle, if* $l_s > 1$

Figure 7.41 illustrates the comparison among the different steering lengths. A FWS car is shown in Figure 7.41(a), while the 4WS systems with $l_s < -1$, $-1 < l_s < 1$, and $l_s < 1$ are shown in Figures 7.41(b)-(d) respectively.

Example 299 ★ *FWS and Ackerman condition.*
    By substituting
$$\delta_{rr} \to 0 \qquad \delta_{rl} \to 0$$
in the 4WS kinematic condition (7.77) the equation reduces to the Ackerman condition (7.1) for the FWS vehicles.

$$\cot \delta_{fr} - \cot \delta_{fl} = \frac{w}{l} \qquad (7.129)$$

Writing the Ackerman condition as this equation frees us from checking the inner and outer wheels.

Example 300 ★ *Turning radius.*
    To find the vehicle's turning radius R, we may define equivalent bicycle models as shown in Figure 7.42 and 7.43 for positive and negative 4WS vehicles. The radius of rotation R is perpendicular to the vehicle's velocity vector **v** *at the mass center C.*

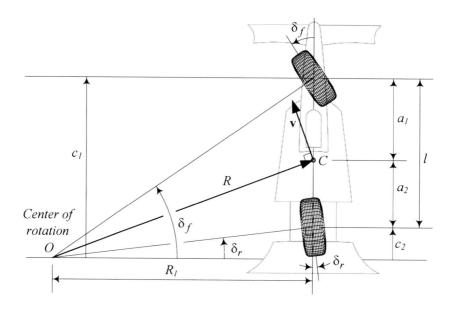

FIGURE 7.42. Bicycle model for a positive $4WS$ vehicle.

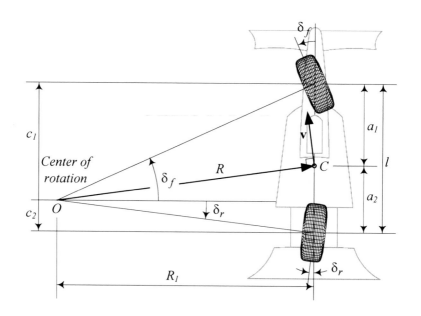

FIGURE 7.43. Bicycle model for a negative $4WS$ vehicle.

Let us examine the positive $4WS$ configuration of Figure 7.42. Using the geometry shown in the bicycle model, we have

$$R^2 = (a_2 + c_2)^2 + R_1^2 \tag{7.130}$$

$$\cot \delta_f = \frac{R_1}{c_1} = \frac{1}{2}\left(\cot \delta_{if} + \cot \delta_{of}\right) \tag{7.131}$$

and therefore,

$$R = \sqrt{(a_2 + c_2)^2 + c_1^2 \cot^2 \delta_f} \tag{7.132}$$

Examining Figure 7.43 shows that the turning radius of a negative $4WS$ vehicle can be determined from the same equation (7.132).

Example 301 ★ *Passive and active four-wheel steering.*

*The negative $4WS$ is not recommended at high speeds because of high yaw rates, and the positive steering is not recommended at low speeds because of increasing radius of turning. Therefore, to maximize the advantages of a $4WS$ system, we need a smart system to allow the vehicle to change the mode of steering according to the speed of the vehicle and adjust the steer angles for different purposes. A **smart steering** is also called **active steering** system.*

An active system may provide a negative steering at low speeds and a positive steering at high speeds. In a negative steering, the rear wheels are steered in the opposite directions as the front wheels to turn in a significantly smaller radius, while in positive steering, the rear wheels are steered in the same direction as the front wheels to increase the lateral force.

When the $4WS$ system is passive, there is a constant proportional ratio between the front and rear steer angles which is equivalent to have a constant c_s. A passive steering may be applied in vehicles to compensate some undesirable vehicle tendencies. As an example, in a FWS system, the rear wheels tend to steer slightly to the outside of a turn. Such tendency can reduce stability.

Example 302 ★ *Autodriver.*

Consider a car at the global position (X, Y) that is moving on a road, as shown in Figure 7.44. Point C indicates the center of curvature of the road at the car's position. The center of curvature of the road is supposed to be the turning center of the car at the instant of consideration.

There is a global coordinate frame G attached to the ground, and a vehicle coordinate frame B attached to the car at its mass center C. The z and Z axes are parallel and the angle ψ indicates the angle between the X and x axes. If (X_C, Y_C) are the coordinates of C in the global coordinate frame G

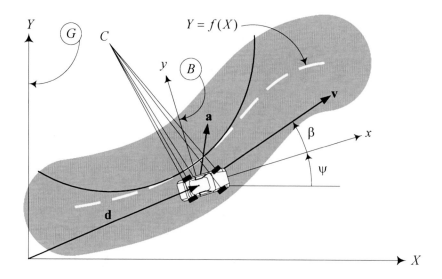

FIGURE 7.44. Illustration of a car that is moving on a road at the point that O is the center of curvature.

then, the coordinates of C in B would be

$$
{}^B\mathbf{r}_C = R_{z,\psi}\left({}^G\mathbf{r}_C - {}^G\mathbf{d}\right) \tag{7.133}
$$

$$
\begin{bmatrix} x_C \\ y_C \\ 0 \end{bmatrix} = \begin{bmatrix} \cos\psi & \sin\psi & 0 \\ -\sin\psi & \cos\psi & 0 \\ 0 & 0 & 1 \end{bmatrix}\left(\begin{bmatrix} X_C \\ Y_C \\ 0 \end{bmatrix} - \begin{bmatrix} X \\ Y \\ 0 \end{bmatrix}\right)
$$

$$
= \begin{bmatrix} (X_C - X)\cos\psi + (Y_C - Y)\sin\psi \\ (Y_C - Y)\cos\psi - (X_C - X)\sin\psi \\ 0 \end{bmatrix} \tag{7.134}
$$

Having coordinates of C in the vehicle coordinate frame is enough to determine R_1, c_1, and c_2.

$$
R_1 = y_C = (Y_C - Y)\cos\psi - (X_C - X)\sin\psi \tag{7.135}
$$

$$
c_2 = -a_2 - x_C = -(X_C - X)\cos\psi - (Y_C - Y)\sin\psi - a_2 \tag{7.136}
$$

$$
c_1 = c_2 + l = -(X_C - X)\cos\psi - (Y_C - Y)\sin\psi + a_1 \tag{7.137}
$$

Then, the required steer angles of the wheels can be uniquely determined by Equations (7.78)-(7.81).

*It is possible to define a road by a mathematical function $Y = f(X)$ in a global coordinate frame. At any point X of the road, the position of the vehicle and the position of the turning center in the vehicle coordinate frame can be determined. The required steer angles can accordingly be set to keep the vehicle on the road and run the vehicle in the correct direction. This principle may be used to design an **autodriver**.*

As an example, let us consider a car that is moving tangent to a road with a given equation

$$Y = \frac{X^2}{200} \tag{7.138}$$

where both X and Y are measured in meter $[\text{m}]$. At $X = 30\,\text{m}$, we have $Y = 4.5\,\text{m}$ and $Y' = 0.3$, $Y'' = 0.01$, and therefore

$$\psi = \arctan \frac{dY}{dX} = \arctan 0.3 = 0.29146\,\text{rad} \approx 16.7\,\text{deg} \tag{7.139}$$

The curvature radius at $(30, 4.5)$, from Example 295, is

$$R_\kappa = 113.80\,\text{m} \tag{7.140}$$

The tangent line to the road is

$$Y - 4.5 = 0.3\,(X - 30) \tag{7.141}$$

and therefore the perpendicular line to the road is

$$Y - 4.5 = -\frac{10}{3}\,(X - 30) \tag{7.142}$$

Having $R_\kappa = 113.80\,\text{m}$,

$$R_\kappa = \frac{\left(1 + Y'^2\right)^{3/2}}{Y''} = \frac{\left(1 + 0.3^2\right)^{3/2}}{0.01} = 113.8\,\text{m} \tag{7.143}$$

we have

$$(X - X_C)^2 + (Y - Y_C)^2 = R_\kappa^2 \tag{7.144}$$

and we can find the global coordinates of the curvature center (X_C, Y_C) at the proper intersection of the line (7.142) and circle (7.144).

$$X_C = -2.7002\,\text{m} \qquad Y_C = 113.5\,\text{m} \tag{7.145}$$

The coordinates of the turning center in the body frame would then be

$$\begin{bmatrix} x_C \\ y_C \\ 0 \end{bmatrix} = \begin{bmatrix} \cos\psi & \sin\psi & 0 \\ -\sin\psi & \cos\psi & 0 \\ 0 & 0 & 1 \end{bmatrix} \left(\begin{bmatrix} X_C \\ Y_C \\ 0 \end{bmatrix} - \begin{bmatrix} X \\ Y \\ 0 \end{bmatrix} \right) = \begin{bmatrix} 0 \\ 113.8 \\ 0 \end{bmatrix} \tag{7.146}$$

Example 303 ★ *Curvature equation.*

Consider a vehicle that is moving on a path $Y = f(X)$ with velocity $\mathbf{v}$ and acceleration $\mathbf{a}$. The curvature $\kappa = 1/R$ of the path that the vehicle is moving on is

$$\kappa = \frac{1}{R} = \frac{a_n}{v^2} \tag{7.147}$$

where, a_n is the normal component of the acceleration **a**. *The normal component a_n is toward the rotation center and is equal to*

$$
\begin{aligned}
a_n &= \left| \frac{\mathbf{v}}{v} \times \mathbf{a} \right| = \frac{1}{v} |\mathbf{v} \times \mathbf{a}| \\
&= \frac{1}{v} (a_Y v_X - a_X v_Y) = \frac{\ddot{Y}\dot{X} - \ddot{X}\dot{Y}}{\sqrt{\dot{X}^2 + \dot{Y}^2}}
\end{aligned}
\tag{7.148}
$$

and therefore,

$$
\kappa = \frac{\ddot{Y}\dot{X} - \ddot{X}\dot{Y}}{\left(\dot{X}^2 + \dot{Y}^2 \right)^{3/2}} = \frac{\ddot{Y}\dot{X} - \ddot{X}\dot{Y}}{\dot{X}^3} \frac{1}{\left(1 + \dfrac{\dot{Y}^2}{\dot{X}^2} \right)^{3/2}}
\tag{7.149}
$$

However,

$$
Y' = \frac{dY}{dX} = \frac{\dot{Y}}{\dot{X}}
\tag{7.150}
$$

$$
Y'' = \frac{d^2 Y}{dX^2} = \frac{d}{dx}\left(\frac{\dot{Y}}{\dot{X}} \right) = \frac{d}{dt}\left(\frac{\dot{Y}}{\dot{X}} \right) \frac{1}{\dot{X}} = \frac{\ddot{Y}\dot{X} - \ddot{X}\dot{Y}}{\dot{X}^3}
\tag{7.151}
$$

and we find the equation for the curvature of the path and radius of the curvature based on the equation of the path.

$$
\kappa = \frac{Y''}{(1 + Y'^2)^{3/2}}
\tag{7.152}
$$

$$
R_\kappa = \frac{1}{\kappa} = \frac{(1 + Y'^2)^{3/2}}{Y''} = \frac{\left(\dot{X}^2 + \dot{Y}^2 \right)^{3/2}}{\ddot{Y}\dot{X} - \ddot{X}\dot{Y}}
\tag{7.153}
$$

As an example, let us consider a road with a given equation

$$
Y = \frac{X^2}{200}
\tag{7.154}
$$

At a point with $X = 30$ m, we have

$$
Y = \frac{9}{2}\,\text{m} \qquad Y' = \frac{3}{10} \qquad Y'' = \frac{1}{100}\,\text{m}^{-1}
\tag{7.155}
$$

and therefore,

$$
\kappa = 8.7874 \times 10^{-3}\,\text{m}^{-1} \qquad R_\kappa = 113.80\,\text{m}
\tag{7.156}
$$

Example 304 ★ *Center of curvature in global coordinate frame.*
   Assume a planar road being expressed by a parametric equation in a global frame as

$$
X = X(t) \qquad Y = Y(t)
\tag{7.157}
$$

The perpendicular line to the road at a point (X_0, Y_0) is:

$$Y = Y_0 - (X - X_0)/\tan\theta \tag{7.158}$$

$$\tan\theta = \frac{dY}{dX} = \frac{dY/dt}{dX/dt} = \frac{\dot{Y}}{\dot{X}} \tag{7.159}$$

The curvature center (X_C, Y_C) is on the perpendicular line and is at a distance R_κ from the point $(X, Y) = (X_0, Y_0)$. Therefore,

$$X_C = X - \frac{\dot{Y}\left(\dot{X}^2 + \dot{Y}^2\right)}{\ddot{Y}\dot{X} - \ddot{X}\dot{Y}} \qquad Y_C = Y + \frac{\dot{X}\left(\dot{X}^2 + \dot{Y}^2\right)}{\ddot{Y}\dot{X} - \ddot{X}\dot{Y}} \tag{7.160}$$

Employing Equation (7.146), the curvature center in the body coordinate frame of a moving vehicle that its x-axis makes the angle ψ with the global X-axis is

$$\begin{aligned}
x_C &= (X_C - X)\cos\psi + (Y_C - Y)\sin\psi \\
&= \left(\dot{X}\sin\psi - \dot{Y}\cos\psi\right)\frac{\dot{X}^2 + \dot{Y}^2}{\ddot{Y}\dot{X} - \ddot{X}\dot{Y}}
\end{aligned} \tag{7.161}$$

$$\begin{aligned}
y_C &= (Y_C - Y)\cos\psi - (X_C - X)\sin\psi \\
&= \left(\dot{X}\cos\psi + \dot{Y}\sin\psi\right)\frac{\dot{X}^2 + \dot{Y}^2}{\ddot{Y}\dot{X} - \ddot{X}\dot{Y}}
\end{aligned} \tag{7.162}$$

As and example, the global coordinate frame of a parabolic path

$$X = t \qquad Y = \frac{t^2}{200} \tag{7.163}$$

would be

$$X_C = X - \frac{\dot{X}^2 + \dot{Y}^2}{\ddot{Y}\dot{X} - \ddot{X}\dot{Y}}\dot{Y} = 2.5244 \times 10^{-29}t - 0.0001t^3 \tag{7.164}$$

$$Y_C = Y + \frac{\dot{X}^2 + \dot{Y}^2}{\ddot{Y}\dot{X} - \ddot{X}\dot{Y}}\dot{X} = 0.015\,t^2 + 100 \tag{7.165}$$

Example 305 ★ *An elliptic path and curvature center.*
  Consider an elliptic path with equations

$$X = a\cos t \qquad Y = b\sin t \tag{7.166}$$

$$a = 100\,\text{m} \qquad b = 65\,\text{m} \tag{7.167}$$

The curvature center of the road in the global coordinate frame is at

$$X_C = X - \frac{\dot{Y}\left(\dot{X}^2 + \dot{Y}^2\right)}{\ddot{Y}\dot{X} - \ddot{X}\dot{Y}} = \frac{a^2 - b^2}{a}\cos^3 t = \frac{231}{4}\cos^3 t \tag{7.168}$$

$$Y_C = Y + \frac{\dot{X}\left(\dot{X}^2 + \dot{Y}^2\right)}{\ddot{Y}\dot{X} - \ddot{X}\dot{Y}} = -\frac{a^2 - b^2}{b}\sin^3 t = \frac{-1155}{13}\sin^3 t \tag{7.169}$$

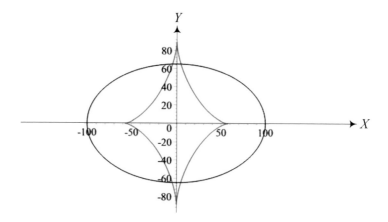

FIGURE 7.45. An elliptic path and its curvature center.

Therefore, the curvature center in the vehicle coordinate frame would be

$$
\begin{aligned}
x_C &= \left(\dot{X} \sin \psi - \dot{Y} \cos \psi \right) \frac{\dot{X}^2 + \dot{Y}^2}{\ddot{Y} \dot{X} - \ddot{X} \dot{Y}} \\
&= \left(\frac{a^2 - b^2}{a} \cos^3 t - a \cos t \right) \cos \psi \\
&\quad - \left(\frac{a^2 - b^2}{b} \sin^3 t + b \sin t \right) \sin \psi
\end{aligned}
\tag{7.170}
$$

$$
\begin{aligned}
y_C &= \left(\dot{X} \cos \psi + \dot{Y} \sin \psi \right) \frac{\dot{X}^2 + \dot{Y}^2}{\ddot{Y} \dot{X} - \ddot{X} \dot{Y}} \\
&= - \left(\frac{a^2 - b^2}{a} \sin^3 t + b \sin t \right) \cos \psi \\
&\quad - \left(\frac{a^2 - b^2}{a} \cos^3 t - a \cos t \right) \sin \psi
\end{aligned}
\tag{7.171}
$$

Figure 7.45 illustrates the elliptic path and its curvature center.

7.6 ★ Road Design

Roads are made by continuously connecting straight and circular paths by proper transition turning sections. Having a continues and well behaved curvature is a necessary criterion in road design. The *clothoid spiral* is the best smooth transition connecting curve in road design which is expressed

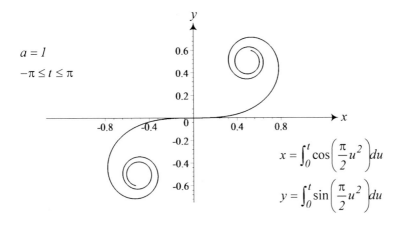

FIGURE 7.46. The clothoide curve for $a = 1$ and $-\pi \leq t \leq \pi$.

by parametric equations called *Fresnel Integrals*:

$$X(t) = a \int_0^t \cos\left(\frac{\pi}{2}u^2\right) du \qquad (7.172)$$

$$Y(t) = a \int_0^t \sin\left(\frac{\pi}{2}u^2\right) du \qquad (7.173)$$

The curvature of the clothoid curve varies linearly with arc length and this linearity makes clothoid the smoothest driving transition curve. Figure 7.46 illustrates the clothoid curve for the *scaling parameter* $a = 1$ and *variable* $-\pi \leq t \leq \pi$. The scaling parameter a is only a magnification factor that shrinks or magnifies the curve. The range of t determines the variation of curvature within the clothoid, as well as the initial and final tangent angles of the clothoid curve.

The *arc length*, s, of a clothoid for a given value of t is

$$s = at \qquad (7.174)$$

If the variable t indicates time then, a would be the speed of motion along the path. The *curvature* κ and *radius of curvature* R of a clothoid at a given t is

$$\kappa = \frac{\pi t}{a} \qquad (7.175)$$

$$R = \frac{1}{\kappa} = \frac{a}{\pi t} \qquad (7.176)$$

The *tangent angle* θ of a clothoid for a given value t is

$$\theta = \frac{\pi}{2}t^2 \qquad (7.177)$$

Having a road with linearly increasing curvature is equivalent to entering the path with a steering wheel at the neutral position and turning the steering wheel with a constant angular velocity. This is a desirable and natural driving action.

Proof. Arc length s of a parametric planar curve $X = X(t)$, $Y = Y(t)$ between t_1 and t_2 is calculated by

$$s = \int_{t_1}^{t_2} \sqrt{\left(\frac{dX}{dt}\right)^2 + \left(\frac{dY}{dt}\right)^2}\, dt \tag{7.178}$$

Substituting the clothoid spiral parametric expression (7.172)-(7.173), we have

$$
\begin{aligned}
s &= \int_{t_1}^{t_2} \sqrt{\left(\frac{dX}{du}\right)^2 + \left(\frac{dY}{du}\right)^2}\, dt \\
&= a \int_0^t \sqrt{\cos^2\left(\frac{\pi}{2}u^2\right) + \sin^2\left(\frac{\pi}{2}u^2\right)}\, dt = a \int_0^t dt = at
\end{aligned} \tag{7.179}
$$

Curvature κ of a planar curve $X = X(s)$, $Y = Y(s)$ that is expressed parametrically by its arc length s is

$$\kappa = \sqrt{\left(\frac{d^2 X}{ds^2}\right)^2 + \left(\frac{d^2 Y}{ds^2}\right)^2} \tag{7.180}$$

Using the result of (7.179), we can replace the variable t with arc length s

$$t = \frac{s}{a} \tag{7.181}$$

and defined the clothoid spiral parametric equations (7.172)-(7.173) as

$$X(s) = a \int_0^{s/a} \cos\left(\frac{\pi}{2}u^2\right) du \tag{7.182}$$

$$Y(s) = a \int_0^{s/a} \sin\left(\frac{\pi}{2}u^2\right) du \tag{7.183}$$

Therefore, the curvature of clothoid spiral is

$$\kappa = \frac{\pi s}{a^2} \sqrt{\cos^2\left(\frac{\pi}{2}\frac{s^2}{a^2}\right) + \sin^2\left(\frac{\pi}{2}\frac{s^2}{a^2}\right)} = \frac{\pi s}{a^2} = \frac{\pi t}{a} \tag{7.184}$$

The slope $\tan\theta$ of the tangent to clothoid spiral at a point t is

$$\tan\theta = \frac{dY}{dX} = \frac{dY/dt}{dX/dt} = \frac{a\sin\left(\frac{\pi}{2}t^2\right)}{a\cos\left(\frac{\pi}{2}t^2\right)} = \tan\left(\frac{\pi}{2}t^2\right) \tag{7.185}$$

and therefore, the slope angle θ of the tangent line is

$$\theta = \frac{\pi}{2}t^2 = \frac{\pi}{2}\frac{s^2}{a^2} \tag{7.186}$$

The clothoid curve approaches the point $(a/2, a/2)$ at infinity

$$X(t) = \lim_{t\to\infty}\left(a\int_0^t \cos\left(\frac{\pi}{2}u^2\right)du\right) = \frac{a}{2} \tag{7.187}$$

$$Y(t) = \lim_{t\to\infty}\left(a\int_0^t \sin\left(\frac{\pi}{2}u^2\right)du\right) = \frac{a}{2} \tag{7.188}$$

Combining the equations for arc length and curvature, we can express the curvature κ of a clothoid curve such that it varies linearly with its arc-length s. The curvature of such a curve is

$$\kappa(s) = \frac{\pi}{a^2}s = ks \tag{7.189}$$

where, s is the arc length, k is the *sharpness* or the rate of change of curvature.

Figure 7.47 illustrates a design graph of the relationship between the clothoid and parameters of magnification factor a, curvature κ, and slope θ. The higher magnification factor a the larger the clothoid. The clothoid curves of different a are intersected by the constant slope lines of θ. The curves for constant curvature k intersect both, the constant a and constant θ curves.

The clothoid transition equation is a proper solution for any required change in any parameter of a road. As an example the change of the bank angle from a flat straight road to a tilted road on a circular path needs a transition bank angle. ∎

Example 306 *History of clothoid.*
The clothoid spiral is also called Cornu spiral, referring to Alfred Cornu (1841 – 1902), a French physicist who rediscovered the clothoid spiral. It may also be called Euler spiral, as Leonard Euler (1707 – 1783) was the the first codiscoverer of the curve with Jacques Bernoulli (1654 – 1705) who formulated the clothoid spiral on deformations of elastic members. It is also called Fresnel spiral credited to Augustin-Jean Fresnel spiral (1788 – 1827) who independently rediscovered the curve in his work on the fringes of diffraction of light through a slot. In vehicle dynamics and road design industry it may also be called the transition spiral to refer to the road connections corners.

The clothoid spiral is also called Cornu spiral, referring to Alfred Cornu (1841 – 1902), a French physicist who rediscovered the clothoid spiral. It may also be called Euler spiral, as Leonard Euler (1707 – 1783) was the the first codiscoverer of the curve Jacques Bernoulli (1654 – 1705) who formulated

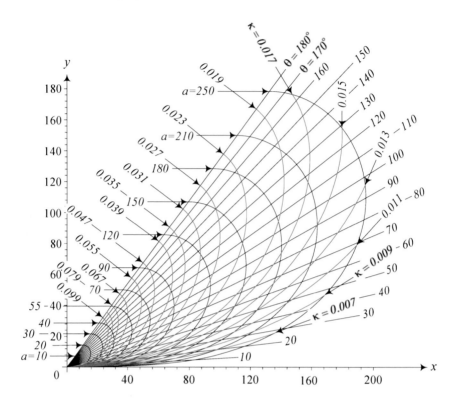

FIGURE 7.47. A design graph of the relationship between the clothoid and parameters of maginification factor a, curvature κ, and slope θ.

the clothoid spiral on deformations of elastic members. It is also called Augustin-Jean Fresnel spiral (1788 − 1827) who independently rediscovered the curve in his work on the fringes of diffraction of light through a slot. In vehicle dynamics and road design industry it may also be called the transition spiral to refer to the road connections corners.

In the 19th century it became clear that we need a track shape with gradually varying curvature. Although circles were being used for most of the path, a correct transition curve was needed to gradually change the curvature from one path to the other. Arthur Talbot (1857−1942) in 1880 derived the same integrals as Bernoulli and Fresnel and introduced the railway transition spirals. Because of this the clothoid spiral is also called Talbot curve. Talbot curve has bee used in railways and road construction.

It is said that Clotho was one of the three Fates who spun the thread of human life, by winding it around the spindle. At the beginning of the 20th century, the Italian mathematician Ernesto Cesàro (1859−1906), from this poetic reference gave the name "clothoid" to the curve with a double spiral shape.

Example 307 ★ *Derivative of an definite integral.*
Differentiation of an definite integral is based on the Leibniz formula

$$\frac{d}{dt} \int_{a(t)}^{b(t)} f(u,t)\, du = \int_{a(t)}^{b(t)} \frac{df}{dt}\, du + f(b(t),t)\frac{db}{dt} - f(a(t),t)\frac{da}{dt} \quad (7.190)$$

Taking derivative of the clothoid spiral parametric expression in calculating the arc length s of (7.178) is based on the Leibniz formula.

$$\frac{dX}{dt} = a\frac{d}{dt}\int_0^t \cos\left(\frac{\pi}{2}u^2\right) du = a\cos\left(\frac{\pi}{2}t^2\right) \quad (7.191)$$

$$\frac{dY}{dt} = a\frac{d}{dt}\int_0^t \sin\left(\frac{\pi}{2}u^2\right) du = a\sin\left(\frac{\pi}{2}t^2\right) \quad (7.192)$$

The calculation of the curvature κ of (7.180) is also based on the Leibniz formula.

$$\frac{dX}{ds} = a\frac{d}{dt}\int_0^{s/a} \cos\left(\frac{\pi}{2}u^2\right) du = \cos\left(\frac{\pi}{2}\frac{s^2}{a^2}\right) \quad (7.193)$$

$$\frac{dY}{ds} = a\frac{d}{dt}\int_0^{s/a} \sin\left(\frac{\pi}{2}u^2\right) du = \sin\left(\frac{\pi}{2}\frac{s^2}{a^2}\right) \quad (7.194)$$

$$\frac{d^2X}{ds^2} = \frac{d}{ds}\cos\left(\frac{\pi}{2}\frac{s^2}{a^2}\right) = -\frac{\pi s}{a^2}\sin\left(\frac{\pi}{2}\frac{s^2}{a^2}\right) \quad (7.195)$$

$$\frac{d^2Y}{ds^2} = \frac{d}{ds}\sin\left(\frac{\pi}{2}\frac{s^2}{a^2}\right) = \frac{\pi s}{a^2}\cos\left(\frac{\pi}{2}\frac{s^2}{a^2}\right) \quad (7.196)$$

Example 308 *A connecting road with given a and κ.*
Let us set $a = 200$ and plot a clothoid road starting from $(0,0)$ and end up at a point with a given curvature of $\kappa = 0.01$. Using

$$t = \frac{\kappa a}{\pi} \quad (7.197)$$

we can define the parametric equations of the transition road (7.172)-(7.173) as

$$X(\kappa) = a\int_0^{\kappa a/\pi} \cos\left(\frac{\pi}{2}u^2\right) du \quad (7.198)$$

$$Y(\kappa) = a\int_0^{\kappa a/\pi} \sin\left(\frac{\pi}{2}u^2\right) du \quad (7.199)$$

The coordinates of the clothoid road at $\kappa = 0.01$ and $a = 200$ are

$$X_0 = 122.2596310 \qquad Y_0 = 26.24682756 \quad (7.200)$$

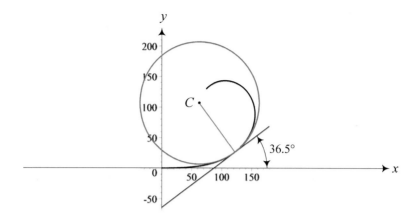

FIGURE 7.48. The tangent line, normal line, and the tangent circle to the clothoid at the point where, $\kappa = 0.01$ for a given $a = 200$. The clothoid is plotted up to $\kappa = 0.025$.

The slope of the road at the point is

$$\theta = \frac{1}{2\pi}a^2\kappa^2 = 0.6366197722\,\text{rad} = 36.475\,\text{deg} \qquad (7.201)$$

and therefore the tangent line to the road is

$$Y = -64.14007833 + 0.7393029502X \qquad (7.202)$$

and the normal line to the road is

$$Y = 191.6183183 - 1.352625469X \qquad (7.203)$$

Having the radius of the tangent curvature circle, $R = 1/\kappa = 100$, we are able to find the coordinates of the center C of the tangent circle on the line (7.203).

$$X_C = 62.81155414 \qquad Y_C = 106.6578104 \qquad (7.204)$$

Figure 7.48 illustrates the tangent line, normal line, and the tangent circle to the clothoid at the point where, $\kappa = 0.01$. The clothoid is plotted up to $\kappa = 0.025$.

Example 309 ★ *Connecting a straight road to a circle.*

Assume that we need to define a clothoid road to begin with zero curvature and meet a given circular curve. Let us consider the road to be on the X-axis and the circle of $R = 100\,\text{m}$ at center $C\,(62.811, 106.658)$.

$$(X - 62.811)^2 + (Y - 106.658)^2 = 100^2 \qquad (7.205)$$

Therefore, the transition road must begin with $\kappa = 0$ on the X-axis and touch the circle at a point when its curvature is $\kappa = 1/100$. Because of

$$\kappa = \frac{\pi s}{a^2} = \frac{\pi t}{a} \tag{7.206}$$

we can define the parametric equations of the transition road (7.172)-(7.173) by κ

$$X(\kappa) = a \int_0^{\kappa a/\pi} \cos\left(\frac{\pi}{2}u^2\right) du$$

$$Y(\kappa) = a \int_0^{\kappa a/\pi} \sin\left(\frac{\pi}{2}u^2\right) du \tag{7.207}$$

Having $\kappa = 0.01$ at the destination point, we find the coordinates of the end of the clothoid as functions of a.

$$X(a) = a \int_0^{0.01a/\pi} \cos\left(\frac{\pi}{2}u^2\right) du$$

$$Y(a) = a \int_0^{0.01a/\pi} \sin\left(\frac{\pi}{2}u^2\right) du \tag{7.208}$$

We need to find the magnifying factor, a, such that the clothoid (7.207) touches the circle (7.205) with the same slope. The slope of the circle at (X, Y) is

$$Y' = \tan\theta = -\frac{(X - 62.811)}{(Y - 106.658)} \tag{7.209}$$

and the slope angle of the clothoid is

$$\theta = \frac{\pi}{2}t^2 = \frac{1}{2\pi}a^2\kappa^2 = 1.5915 \times 10^{-5}a^2 \tag{7.210}$$

To make the clothoid have the same slope, we derive an equation that relates the magnification factor, a, to the components of the final point of the clothoid.

$$\arctan\frac{-(X - 62.811)}{Y - 106.658} = 1.5915 \times 10^{-5}a^2 \tag{7.211}$$

Equations (7.211), (7.208) and (7.205) provide us with an equation to find

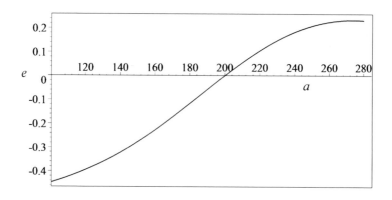

FIGURE 7.49. The error $e = \arctan \dfrac{-(X - 62.811)}{Y - 106.658} - 1.5915 \times 10^{-5} a^2$ as a function of a.

a. *Let us define and plot an error equation versus* a *in Figure 7.49*

$$
\begin{aligned}
e &= \arctan \frac{-(X - 62.811)}{Y - 106.658} - 1.5915 \times 10^{-5} a^2 \\
&= \arctan \frac{-(X - 62.811)}{\sqrt{100^2 - (X - 62.811)^2}} - 1.5915 \times 10^{-5} a^2 \\
&= \arctan \frac{-\left(a \int_0^{0.01a/\pi} \cos\left(\frac{\pi}{2}u^2\right) du - 62.811\right)}{\sqrt{100^2 - \left(a \int_0^{0.01a/\pi} \cos\left(\frac{\pi}{2}u^2\right) du - 62.811\right)^2}} \\
&\quad - 1.5915 \times 10^{-5} a^2
\end{aligned} \tag{7.212}
$$

and solve the equation for a.

$$
a = 200 \tag{7.213}
$$

The clothoid equation are

$$
\begin{aligned}
X(\kappa) &= a \int_0^{\kappa a/\pi} \cos\left(\frac{\pi}{2}u^2\right) du \\
Y(\kappa) &= a \int_0^{\kappa a/\pi} \sin\left(\frac{\pi}{2}u^2\right) du
\end{aligned} \tag{7.214}
$$

where at $\kappa = 0.01$ *reaches to*

$$
X_0 = 122.2596310 \qquad Y_0 = 26.24682756 \tag{7.215}
$$

The slope of the road at the point, the tangent line to the road and the normal line to the road are

$$\theta = \frac{1}{2\pi}a^2\kappa^2 = 0.6366197722\,\text{rad} = 36.475\,\text{deg} \qquad (7.216)$$

$$Y = -64.14007833 + 0.7393029502X \qquad (7.217)$$

$$Y = 191.6183183 - 1.352625469X \qquad (7.218)$$

Figure 7.48 illustrates the clothoid, tangent line, normal line, and the tangent circle to the clothoid at the point where, $\kappa = 0.01$. The clothoid is plotted up to $\kappa = 0.025$.

Example 310 ★ *Connecting a straight road to another circle.*
   Assume that we need to determine a clothoid road to begin with zero curvature on the X-axis and meet a given circular curve of $R = 80\,\text{m}$ at center $C\,(100, 100)$.

$$(X - 100)^2 + (Y - 100)^2 = 80^2 \qquad (7.219)$$

The expression of the transition road (7.172)-(7.173) by κ is:

$$X(\kappa) = a \int_0^{\kappa a/\pi} \cos\left(\frac{\pi}{2}u^2\right) du$$

$$Y(\kappa) = a \int_0^{\kappa a/\pi} \sin\left(\frac{\pi}{2}u^2\right) du \qquad (7.220)$$

Using $\kappa = 1/R = 0.0125$ at the destination point determines the coordinates of the end of the clothoid as functions of a.

$$X(a) = a \int_0^{0.0125a/\pi} \cos\left(\frac{\pi}{2}u^2\right) du$$

$$Y(a) = a \int_0^{0.0125a/\pi} \sin\left(\frac{\pi}{2}u^2\right) du \qquad (7.221)$$

The slope of the tangent to the circle (7.219) at a point (X, Y) is

$$Y' = \tan\theta = -\frac{(X - 100)}{Y - 100} \qquad (7.222)$$

and the slope angle of the clothoid as a function of a is

$$\theta = \frac{\pi}{2}t^2 = \frac{1}{2\pi}a^2\kappa^2 = 2.4868 \times 10^{-5}a^2 \qquad (7.223)$$

The clothoid should have the same slope, therefore,

$$\arctan\frac{-(X - 100)}{Y - 100} = 2.4868 \times 10^{-5}a^2 \qquad (7.224)$$

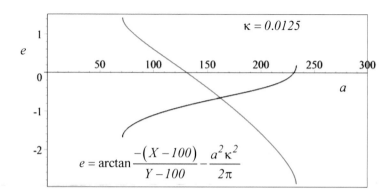

FIGURE 7.50. Plot of $e = \arctan \dfrac{-(X-100)}{\pm\sqrt{80^2 - (X-100)^2}} - 2.4868 \times 10^{-5}a^2$ versus a.

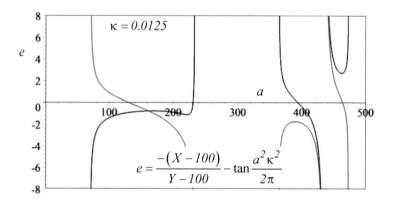

FIGURE 7.51. Plot of $e = \dfrac{-(X-100)}{\pm\sqrt{80^2 - (X-100)^2}} - \tan\left(2.4868 \times 10^{-5}a^2\right)$ versus a.

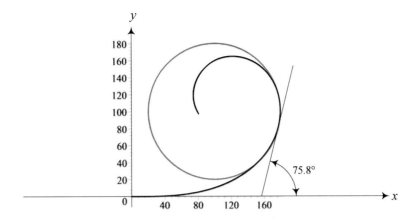

FIGURE 7.52. The transition road starting on the X-axis and goes to a circle of radius $R = 80\,\mathrm{m}$ at center $C\,(100\,\mathrm{m}, 100\,\mathrm{m})$.

Equations (7.224) and (7.219) along with (7.221) provide us with an equation to find a. However, substituting $Y = Y(X)$ and replacing tan *and* arctan *generate four equations to be solved for possible a. To visualize the possible solutions, let us define two error equations (7.225)-(7.226).*

$$e = \arctan \frac{-(X - 100)}{\pm\sqrt{80^2 - (X - 100)^2}} - 2.4868 \times 10^{-5} a^2 \quad (7.225)$$

$$e = \frac{-(X - 100)}{\pm\sqrt{80^2 - (X - 100)^2}} - \tan\left(2.4868 \times 10^{-5} a^2\right) \quad (7.226)$$

Figure 7.50 depicts Equation (7.225) and Figure 7.51 shows Equation (7.226). Equation (7.225) provides the solutions of

$$a = 230.7098693 \qquad a = 130.8889343 \qquad (7.227)$$

and Equation (7.226) provide the solutions of

$$
\begin{aligned}
a &= 230.7098693 & a &= 130.8889343 \\
a &= 394.0940573 & a &= 463.5589702
\end{aligned}
\qquad (7.228)
$$

The correct answer is a = 230.7098693 and Figure 7.52 depicts the circle and the proper clothoid. Using a, we define the clothoid equation

$$
\begin{aligned}
X(\kappa) &= a \int_0^{\kappa a/\pi} \cos\left(\frac{\pi}{2} u^2\right) du \\
Y(\kappa) &= a \int_0^{\kappa a/\pi} \sin\left(\frac{\pi}{2} u^2\right) du
\end{aligned}
\qquad (7.229)
$$

which at $\kappa = 0.0125$ reaches

$$X_0 = 177.5691613 \qquad Y_0 = 82.38074640 \tag{7.230}$$

at angle

$$\theta = \frac{1}{2\pi} a^2 \kappa^2 = 1.3236 \, \text{rad} \approx 75.84 \, \text{deg} \tag{7.231}$$

Example 311 *Using the design chart.*

Assume we are asked to find a clothoid transition road to connect a straight road to a circle of radius $R = 58.824$ m. Having R is equivalent to have the destination curvature $\kappa = 1/R = 0.017$. The desired circle must be tangent to a clothoid with a given a at the point that the clothoid is intersecting the curve of $\kappa = 0.017$.

The clothoid for $a = 250$ m hits the curve of $\kappa = 0.017$ at a point for which we have

$$\begin{aligned} X &= 147.3884878 \, \text{m} & Y &= 176.4421850 \, \text{m} & (7.232) \\ \theta &= 164.7102491 \, \text{deg} & s &= 338.2042540 \, \text{m} & (7.233) \end{aligned}$$

The clothoid for $a = 210$ m hits the curve of $\kappa = 0.017$ at a point for which we have

$$\begin{aligned} X &= 157.4739501 \, \text{m} & Y &= 119.7133227 \, \text{m} & (7.234) \\ \theta &= 116.2195518 \, \text{deg} & s &= 238.6369216 \, \text{m} & (7.235) \end{aligned}$$

The clothoid for $a = 180$ m hits the curve of $\kappa = 0.017$ at a point for which we have

$$\begin{aligned} X &= 140.1918463 \, \text{m} & Y &= 74.21681673 \, \text{m} & (7.236) \\ \theta &= 85.38579313 \, \text{deg} & s &= 175.3250853 \, \text{m} & (7.237) \end{aligned}$$

The clothoid for $a = 150$ m hits the curve of $\kappa = 0.017$ at a point for which we have

$$\begin{aligned} X &= 109.3442240 \, \text{m} & Y &= 38.89541829 \, \text{m} & (7.238) \\ \theta &= 59.29568967 \, \text{deg} & s &= 121.7535314 \, \text{m} & (7.239) \end{aligned}$$

The clothoid for $a = 120$ m hits the curve of $\kappa = 0.017$ at a point for which we have

$$\begin{aligned} X &= 74.57259185 \, \text{m} & Y &= 16.67204291 \, \text{m} & (7.240) \\ \theta &= 37.94924139 \, \text{deg} & s &= 77.92226012 \, \text{m} & (7.241) \end{aligned}$$

Figure 7.53 illustrates these solutions. The number of solutions is practically infinite and the best solution depends on safety, cost, and physical constraints of the field.

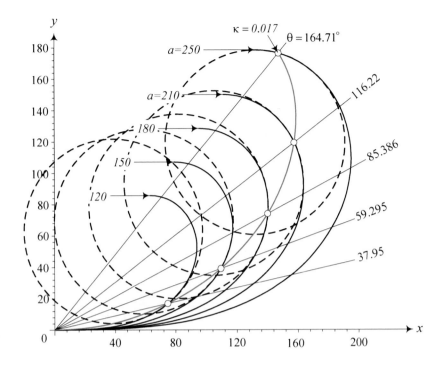

FIGURE 7.53. A few clothoid transition road to connect a straight road to a circle of radius $R = 58.824\,\mathrm{m}$.

Example 312 ★ *Clothoid spiral as an optimal curve problem.*

Clothoid spiral is the shortest curve connecting two given points with given initial and final tangent angles and curvatures. The angle and curvature are varying. The problem can be formulated as follows:

Given two points (X_1, Y_1) and (X_2, Y_2) and two angles θ_1 and θ_2, find a curve (clothoid segment) which satisfies:

$$X(0) = X_1 \qquad Y(0) = Y_1 \qquad \tan\theta_1 = \frac{dY(0)}{dX(0)} = \frac{dY(0)/dt}{dX(0)/dt} \qquad (7.242)$$

$$X(s) = X_2 \qquad Y(s) = Y_2 \qquad \tan\theta_2 = \frac{dY(s)}{dX(s)} = \frac{dY(s)/dt}{dX(s)/dt} \qquad (7.243)$$

with minimal arc length s.

Example 313 ★ *Clothoid shift to meet a given circle.*

It is not generally possible to design a clothoid starting at the origin and meet a given circle at an arbitrary center and radius. However, it is possible to start the clothoid from other points on the x-axis to meet the given circle.

Assume we need to design a clothoid starting on the x-axis to meet a given circle

$$(x - x_C)^2 + (y - y_C)^2 = R^2 \tag{7.244}$$

where (x_C, y_C) indicates the coordinates of the center of the circle, and R is the radius of the circle. Substituting t in terms of a and R

$$t = \frac{a}{\pi R} = \frac{a\kappa}{\pi} \tag{7.245}$$

we define the equation of clothoid to have the same radius of curvature as the circle at the end point.

$$
\begin{aligned}
x\left(\kappa\right) &= a \int_0^{a/(R\pi)} \cos\left(\frac{\pi}{2} u^2\right) du \\
y\left(\kappa\right) &= a \int_0^{a/(R\pi)} \sin\left(\frac{\pi}{2} u^2\right) du
\end{aligned}
\tag{7.246}
$$

Let us assume there is a y at which the slope of the clothoid

$$\tan\theta = \tan\left(\frac{\pi t^2}{2}\right) = \tan\left(\frac{a^2}{2\pi R^2}\right) \tag{7.247}$$

and the circle

$$\tan\theta = -\frac{x - x_C}{y - y_C} \tag{7.248}$$

are equal.

$$\tan\left(\frac{a^2}{2\pi R^2}\right) = -\frac{x - x_C}{y - y_C} \tag{7.249}$$

Searching for a match point in the right half circle

$$y - y_C = \sqrt{R^2 - (x - x_C)^2} \tag{7.250}$$

makes the slope equation to be a function of a

$$\tan\left(\frac{a^2}{2\pi R^2}\right) \sqrt{R^2 - (x - x_C)^2} + \frac{x - x_C}{y - y_C} = 0 \tag{7.251}$$

or

$$
\tan\left(\frac{a^2}{2\pi R^2}\right) \sqrt{R^2 - \left(a \int_0^{a/(R\pi)} \cos\left(\frac{\pi}{2} u^2\right) du - x_C\right)^2}
$$
$$
+ a \int_0^{a/(R\pi)} \cos\left(\frac{\pi}{2} u^2\right) du - x_C = 0 \tag{7.252}
$$

Solution of this equation provides us with an a for which the clothoid ends at a point with the same curvature as the circle. At the same y of the end

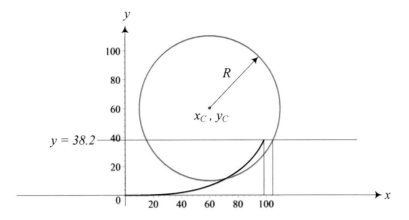

FIGURE 7.54. A clothoid starting at origin ends at a point with the same slope and curvature as the given circle, at the same y.

point, the slope of the clothoid is also equal to the slope of the circle. A proper shift of the clothoid on the x-axis will match the clothoid and the circle.

As an example, let us assume that the circle is

$$(x - 60)^2 + (y - 60)^2 = 50^2 \tag{7.253}$$

and therefore, the slope equation will be

$$\tan\left(\frac{a^2}{2\pi 50^2}\right) \sqrt{50^2 - \left(a \int_0^{a/(50\pi)} \cos\left(\frac{\pi}{2}u^2\right) du - 60\right)^2}$$

$$+ a \int_0^{a/(50\pi)} \cos\left(\frac{\pi}{2}u^2\right) du - 60 = 0 \tag{7.254}$$

Numerical solution of the equation is

$$a = 132.6477323 \tag{7.255}$$

The plot of the clothoid and the circle at this moment are shown in Figure 7.54.

For the calculated a, the value of t at the end point of the clothoid is

$$t = \frac{a}{\pi R} = \frac{2.652954646}{\pi} = 0.84446 \tag{7.256}$$

and therefore, the coordinates of the end point are

$$x(\kappa) = 132.6 \int_0^{0.84} \cos\left(\frac{\pi}{2}u^2\right) du = 98.75389126$$

$$y(\kappa) = 132.6 \int_0^{0.84} \sin\left(\frac{\pi}{2}u^2\right) du = 38.22304651 \tag{7.257}$$

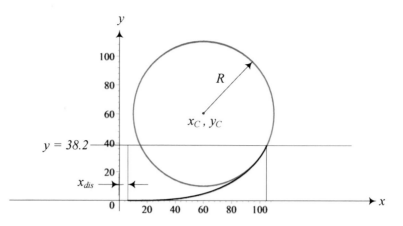

FIGURE 7.55. A shifted clothoid starting on a point on the x-axis ends at a point on the given circle with the same slope and curvature.

At the point, the radius of curvature of the clothoid is

$$R = \frac{1}{\kappa} = \frac{a}{\pi t} = \frac{132.6477323}{0.84446\pi} = 50 \qquad (7.258)$$

and the slope is

$$\theta = \frac{\pi}{2}t^2 = \frac{\pi}{2}0.84446^2 = 1.1202 \, \text{rad} \qquad (7.259)$$

The x-coordinate of the circle at the same $y = 38.22304651$,

$$38.22304651 - 60 = \sqrt{50^2 - (x - 60)^2} \qquad (7.260)$$

is

$$x_{circle} = 105.0084914 \qquad (7.261)$$

If we shift the clothoid by the difference between x_{circle} and $x_{clothoid}$

$$\begin{aligned} x_{dis} &= x_{circle} - x_{clothoid} \\ &= 105.0084914 - 98.75389126 = 6.2546 \end{aligned} \qquad (7.262)$$

then the clothoid and circle meet at a point on the circle with all requirements to have a smooth transition. Figure 7.55 illustrates the result.

Example 314 ★ *Complex expression and proof of curvature.*
    Let us define a curve in complex plane as

$$C(s) = a \int_0^{s/a} e^{i\pi u^2/2} du \qquad (7.263)$$

The derivative of the curve is an equation with absolute value of a.

$$\frac{dC(s)}{ds} = e^{i\pi s^2/(2a^2)} \tag{7.264}$$

The curvature of the curve is

$$\kappa = \left| \frac{d^2C(s)}{ds^2} \right| = \left| i \frac{\pi s}{a^2} e^{i\pi s^2/(2a^2)} \right| = \frac{\pi s}{a^2} \tag{7.265}$$

Example 315 ★ *Parametric form of a straight road.*
   The equation of a straight road that connects two points $P_1(x_1, y_1, z_1)$ and $P_2(x_2, y_2, z_3)$ is

$$\frac{x - x_1}{x_2 - x_1} = \frac{y - y_1}{y_2 - y_1} = \frac{z - z_1}{z_2 - z_1} \tag{7.266}$$

This line may also be expressed by the following parametric equations.

$$
\begin{aligned}
x &= x_1 + (x_2 - x_1)t \\
y &= y_1 + (y_2 - y_1)t \\
z &= z_1 + (z_2 - z_1)t
\end{aligned} \tag{7.267}
$$

Example 316 ★ *Arc length of a planar road.*
   A planar road in the (x, y)-plane

$$y = f(x) \tag{7.268}$$

can be expressed vectorially by

$$\mathbf{r} = x\hat{\imath} + y(x)\hat{\jmath} \tag{7.269}$$

The displacement element on the curve

$$\frac{d\mathbf{r}}{dx} = \hat{\imath} + \frac{dy}{dx}\hat{\jmath} \tag{7.270}$$

provides us with

$$\left(\frac{ds}{dx}\right)^2 = \frac{d\mathbf{r}}{dx} \cdot \frac{d\mathbf{r}}{dx} = 1 + \left(\frac{dy}{dx}\right)^2 \tag{7.271}$$

Therefore, the arc length of the curve between $x = x_1$ and $x = x_2$ is

$$s = \int_{x_1}^{x_2} \sqrt{1 + \left(\frac{dy}{dx}\right)^2}\, dx \tag{7.272}$$

In case the curve is given parametrically,

$$x = x(t) \qquad y = y(t) \tag{7.273}$$

we have

$$\left(\frac{ds}{dt}\right)^2 = \frac{d\mathbf{r}}{dt} \cdot \frac{d\mathbf{r}}{dt} = \left(\frac{dx}{dt}\right)^2 + \left(\frac{dy}{dt}\right)^2 \tag{7.274}$$

and hence,

$$s = \int_{t_1}^{t_2} \left|\frac{d\mathbf{r}}{dt}\right| = \int_{t_1}^{t_2} \sqrt{\left(\frac{dx}{dt}\right)^2 + \left(\frac{dy}{dt}\right)^2} \, dt \tag{7.275}$$

Let us show a circle with radius R by its polar expression using the angle θ as a parameter, as an example:

$$x = R\cos\theta \qquad y = R\sin\theta \tag{7.276}$$

The arc length between $\theta = 0$ and $\theta = \pi/2$ would then be one-fourth the perimeter of the circle. Therefore, the equation for calculating the perimeter of a circle with radius R is

$$
\begin{aligned}
s &= 4\int_0^{\pi/2} \sqrt{\left(\frac{dx}{d\theta}\right)^2 + \left(\frac{dy}{d\theta}\right)^2} \, d\theta = R\int_0^{\pi/2} \sqrt{\sin^2\theta + \cos^2\theta} \, d\theta \\
&= 4R\int_0^{\pi/2} d\theta = 2\pi R
\end{aligned}
\tag{7.277}
$$

Using Equation (7.272), we can define the equation of the road as

$$y = \int_{x_1}^{x_2} \sqrt{\left(\frac{ds}{dx}\right)^2 - 1} \, dx \tag{7.278}$$

Example 317 *A figure 8 as an approximately correct road.*
Sometimes, matching slopes, instead of matching curvatures, can be used to design an approximately correct road. Let us make a closed road in the shape of a symmetric figure eight with two 180 deg circular paths. Assuming

$$a = 200 \tag{7.279}$$

the equations of the clothoid road starting from the origin are:

$$X(t) = 200\int_0^t \cos\left(\frac{\pi}{2}u^2\right) du \tag{7.280}$$

$$Y(t) = 200\int_0^t \sin\left(\frac{\pi}{2}u^2\right) du \tag{7.281}$$

The slope (7.177) of the curve would be parallel to the symmetric line $Y = X$ when

$$\theta = \frac{\pi}{4} \qquad t = \sqrt{\frac{2\theta}{\pi}} = \frac{\sqrt{2}}{2} = t_0 \tag{7.282}$$

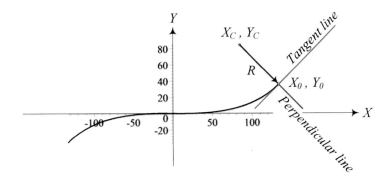

FIGURE 7.56. A clothoid between two point at which the tangent lines are parallel to $Y = X$.

At $t = t_0$ the clothoid is at

$$X_0 \;\; = \;\; 200 \int_0^{\sqrt{2}/2} \cos\left(\frac{\pi}{2}u^2\right) du = 132.943 \qquad (7.283)$$

$$Y_0 \;\; = \;\; 200 \int_0^{\sqrt{2}/2} \sin\left(\frac{\pi}{2}u^2\right) du = 35.424 \qquad (7.284)$$

where the tangent and perpendicular lines respectively are

$$Y \;\; = \;\; 35.424 + (X - 132.943)\tan\theta = -97.519 + X \qquad (7.285)$$
$$Y \;\; = \;\; 35.424 - (X - 132.943)/\tan\theta = 168.37 - X \qquad (7.286)$$

as are shown in Figure 7.56 for the clothoid from $t = -t_0$ to $t = t_0$.
The perpendicular line hits the symmetric line $Y = X$ at

$$X_C = Y_C = 84.184 \qquad (7.287)$$

which would be the center of a circular path with

$$R = \sqrt{(X_0 - X_C)^2 + (Y_0 - Y_C)^2} = 68.956 \qquad (7.288)$$

to connect (X_0, Y_0) to its mirror point with respect to $Y = X$ at

$$X_1 = 35.424 \qquad Y_1 = 132.943 \qquad (7.289)$$

The mirror clothoid

$$X \;\; = \;\; 200 \int_{-\sqrt{2}/2}^{\sqrt{2}/2} \sin\left(\frac{\pi}{2}u^2\right) du \qquad (7.290)$$

$$Y \;\; = \;\; 200 \int_{-\sqrt{2}/2}^{\sqrt{2}/2} \cos\left(\frac{\pi}{2}u^2\right) du \qquad (7.291)$$

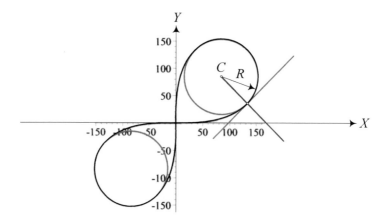

FIGURE 7.57. A slope match symmetric figure 8 road based on two clothoid and two circular parts.

will complete the figure 8 road as is shown in Figure 7.57.

Therefore, the parametric equations of the road, beginning from the origin and moving in the X-direction are as follows. The parameter t is not continuous and not with the same dimension in all equations.

$$X(t) = 200 \int_0^t \cos\left(\frac{\pi}{2}u^2\right) du \qquad 0 \le t \le \frac{\sqrt{2}}{2} \qquad (7.292)$$

$$Y(t) = 200 \int_0^t \sin\left(\frac{\pi}{2}u^2\right) du \qquad 0 \le t \le \frac{\sqrt{2}}{2} \qquad (7.293)$$

$$X(t) = X_C + R\cos t \qquad -\frac{\pi}{4} \le t \le 2.3562 \qquad (7.294)$$

$$Y(t) = Y_C + R\sin t \qquad -\frac{\pi}{4} \le t \le 2.3562 \qquad (7.295)$$

$$X(t) = 200 \int_0^t \sin\left(\frac{\pi}{2}u^2\right) du \qquad \frac{\sqrt{2}}{2} \le t \le -\frac{\sqrt{2}}{2} \qquad (7.296)$$

$$Y(t) = 200 \int_0^t \cos\left(\frac{\pi}{2}u^2\right) du \qquad \frac{\sqrt{2}}{2} \le t \le -\frac{\sqrt{2}}{2} \qquad (7.297)$$

$$X(t) = X_C + R\cos t \qquad 2.3562 \le t \le 5.4978 \qquad (7.298)$$
$$Y(t) = Y_C + R\sin t \qquad 2.3562 \le t \le 5.4978 \qquad (7.299)$$

$$X(t) = 200 \int_0^t \cos\left(\frac{\pi}{2}u^2\right) du \qquad -\frac{\sqrt{2}}{2} \le t \le 0 \qquad (7.300)$$

$$Y(t) = 200 \int_0^t \sin\left(\frac{\pi}{2}u^2\right) du \qquad -\frac{\sqrt{2}}{2} \le t \le 0 \qquad (7.301)$$

Using s as the road length, we may define the equations with a smooth and continuos parameter.

$$X(t) = 200 \int_0^{s/200} \cos\left(\frac{\pi}{2}u^2\right) du \qquad (7.302)$$

$$Y(t) = 200 \int_0^{s/200} \sin\left(\frac{\pi}{2}u^2\right) du \qquad (7.303)$$

$$0 \le s \le 100\sqrt{2} \qquad (7.304)$$

$$X(t) = X_C + R\cos\left(\frac{s - 100\sqrt{2}}{R} - \frac{\pi}{4}\right) \qquad (7.305)$$

$$Y(t) = Y_C + R\sin\left(\frac{s - 100\sqrt{2}}{R} - \frac{\pi}{4}\right) \qquad (7.306)$$

$$100\sqrt{2} \le s \le 358.05 \qquad (7.307)$$

$$X(t) = 200 \int_0^{(499.47-s)/200} \sin\left(\frac{\pi}{2}u^2\right) du \qquad (7.308)$$

$$Y(t) = 200 \int_0^{(499.47-s)/200} \cos\left(\frac{\pi}{2}u^2\right) du \qquad (7.309)$$

$$358.05 \le s \le 640.89 \qquad (7.310)$$

$$X(t) = -X_C + R\cos\left(\frac{640.89 - s}{R} - \frac{\pi}{4}\right) \qquad (7.311)$$

$$Y(t) = -Y_C + R\sin\left(\frac{640.89 - s}{R} - \frac{\pi}{4}\right) \qquad (7.312)$$

$$640.89 \le s \le 857.52 \qquad (7.313)$$

$$X(t) = 200 \int_0^{(s-857.52-100\sqrt{2})/200} \cos\left(\frac{\pi}{2}u^2\right) du \qquad (7.314)$$

$$Y(t) = 200 \int_0^{(s-857.52-100\sqrt{2})/200} \sin\left(\frac{\pi}{2}u^2\right) du \qquad (7.315)$$

$$857.52 \le s \le 998.94 \qquad (7.316)$$

The variation of X and Y for $0 \le s \le 998.94$ are depicted in Figure 7.58.

Example 318 *A figure 8 correct road.*

   Let us design a closed road in the shape of a symmetric figure eight with a curvature transition between the clothoids and the circular paths. Assuming

$$a = 200 \qquad (7.317)$$

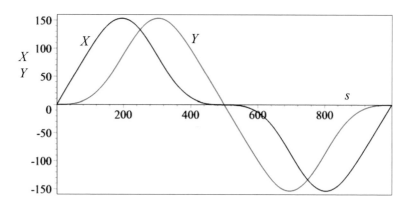

FIGURE 7.58. The variation of X and Y for $0 \leq s \leq 998.94$ of the figure 8 road.

the equations of the clothoid road starting from the origin are:

$$X(t) = 200 \int_0^t \cos\left(\frac{\pi}{2}u^2\right) du \tag{7.318}$$

$$Y(t) = 200 \int_0^t \sin\left(\frac{\pi}{2}u^2\right) du \tag{7.319}$$

The slope (7.177) at t of the curve is

$$\theta = \frac{\pi}{2}t^2 \tag{7.320}$$

where the tangent and perpendicular lines respectively are:

$$Y = Y(t) + (X - X(t))\tan\theta \tag{7.321}$$
$$Y = Y(t) - (X - X(t))/\tan\theta \tag{7.322}$$

The radius of curvature of the clothoid at t is

$$R = \frac{1}{\kappa} = \frac{a}{\pi t} \tag{7.323}$$

If a clothoid point (X,Y) exist at which the radius of curvature is equal to the distance of the point from the line $Y = X$ on the perpendicular line, then we can have a circular path starting at the point. The intersection of the clothoid point (X,Y) and the symmetric line $Y = X$ is the point $Y_C = X_C$ where

$$Y_C = X_C = \frac{X(t) + Y(t)\tan\dfrac{\pi t^2}{2}}{\tan\dfrac{\pi t^2}{2} + 1} \tag{7.324}$$

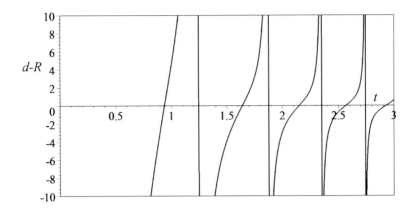

FIGURE 7.59. Plot of the equation $d - R = d - \dfrac{a}{\pi t} = 0$ to find a t at which the curvature of the clothoid matches with a symmetric circle.

The distance of the clothoid and the point (X_C, Y_C) on the perpendicular line is

$$
\begin{aligned}
d &= \sqrt{(Y(t) - Y_C)^2 + (X(t) - X_C)^2} \\
&= \sqrt{\left(\frac{Y(t) - X(t)}{\tan\frac{\pi t^2}{2} + 1}\right)^2 + \left(\frac{(X(t) - Y(t))\tan\frac{\pi t^2}{2}}{\tan\frac{\pi t^2}{2} + 1}\right)^2} \quad (7.325)
\end{aligned}
$$

Equating d and R provides us with an equation to be solved for the t at which the clothoid terminates and a circle with the same curvature starts.

$$
d - \frac{a}{\pi t} = 0 \quad (7.326)
$$

As is shown in Figure 7.59, the equation has multiple solutions, and the first solution is at

$$
t = t_0 = 0.9371211755 \quad (7.327)
$$

At $t = t_0$ the clothoid and its kinematics are

$$
X_0 = 200 \int_0^{0.9371211755} \cos\left(\frac{\pi}{2}u^2\right) du = 154.77 \quad (7.328)
$$

$$
Y_0 = 200 \int_0^{0.9371211755} \sin\left(\frac{\pi}{2}u^2\right) du = 75.154 \quad (7.329)
$$

$$
R = \frac{1}{\kappa} = 67.93355959 \quad (7.330)
$$

$$
\theta = 1.379467204 \,\text{rad} = 79.038 \,\text{deg} \quad (7.331)
$$

The intersection of the perpendicular to the clothoid at (X_0, Y_0) with $Y = X$ is at

$$X_C = 88.0724138 < X_0 \qquad (7.332)$$
$$Y_C = 88.0724138 > Y_0 \qquad (7.333)$$

We can check the distance of (X_0, Y_0) and (X_C, Y_C) to be equal to R.

Therefore, using the parametric s as the road length, the equations of the road, beginning from the origin and moving in the X-direction can be expressed as

$$X(t) = 200 \int_0^{s/a} \cos\left(\frac{\pi}{2} u^2\right) du \qquad 0 \le s \le s_0 \qquad (7.334)$$

$$Y(t) = 200 \int_0^{s/a} \sin\left(\frac{\pi}{2} u^2\right) du \qquad 0 \le s \le s_0 \qquad (7.335)$$

$$s_0 = a t_0 = 187.42 \qquad (7.336)$$

$$X(t) = X_C + R\cos\left(\frac{s - s_0}{R} - \theta_0\right) \qquad s_0 \le s \le s_1 \qquad (7.337)$$

$$Y(t) = Y_C + R\sin\left(\frac{s - s_0}{R} - \theta_0\right) \qquad s_0 \le s \le s_1 \qquad (7.338)$$

$$\theta_0 = \arctan\frac{Y_C - Y_0}{X_0 - X_C} = 0.19132\,\mathrm{rad} = 10.962\,\mathrm{deg} \qquad (7.339)$$

$$s_1 = s_0 + R\left(\frac{\pi}{2} + 2\theta_0\right) = 320.12 \qquad (7.340)$$

$$X(t) = 200 \int_0^{(s_1+s_0-s)/a} \sin\left(\frac{\pi}{2} u^2\right) du \qquad s_1 \le s \le s_2 \quad (7.341)$$

$$Y(t) = 200 \int_0^{(s_1+s_0-s)/a} \cos\left(\frac{\pi}{2} u^2\right) du \qquad s_1 \le s \le s_2 \quad (7.342)$$

$$s_2 = s_1 + 2s_0 = 694.96 \qquad (7.343)$$

$$X(t) = -X_C + R\cos\left(\frac{s_2 - s}{R} - \left(\frac{\pi}{2} - \theta_0\right)\right) \qquad s_2 \le s \le s_3 \quad (7.344)$$

$$Y(t) = -Y_C + R\sin\left(\frac{s_2 - s}{R} - \left(\frac{\pi}{2} - \theta_0\right)\right) \qquad s_2 \le s \le s_3 \quad (7.345)$$

$$s_3 = s_2 + R\left(\frac{\pi}{2} + 2\theta_0\right) = 827.66 \qquad (7.346)$$

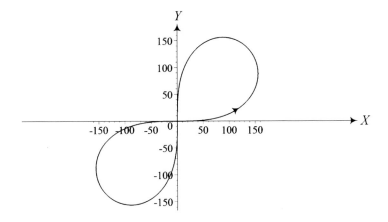

FIGURE 7.60. A curvature match symmetric figure 8 road based on two clothoid and two circular parts.

$$X(t) = 200 \int_0^{(s-s_3-s_0)/200} \cos\left(\frac{\pi}{2}u^2\right) du \qquad s_3 \le s \le s_4 \,(7.347)$$

$$Y(t) = 200 \int_0^{(s-s_3-s_0)/200} \sin\left(\frac{\pi}{2}u^2\right) du \qquad s_3 \le s \le s_4 \,(7.348)$$

$$s_4 = s_3 + s_0 = 1015.1 \qquad\qquad (7.349)$$

The road is shown in Figure 7.60.

Example 319 ★ *Spatial road.*
If the position vector $^G\mathbf{r}_P$ of a moving car is such that each component is a function of a variable t,

$$^G\mathbf{r} = {}^G\mathbf{r}(t) = x(t)\,\hat{\imath} + y(t)\,\hat{\jmath} + z(t)\,\hat{k} \qquad (7.350)$$

then the end point of the position vector indicates a curve C in G. The curve $^G\mathbf{r} = {}^G\mathbf{r}(t)$ reduces to a point on C if we fix the parameter t. The functions

$$x = x(t) \qquad y = y(t) \qquad z = z(t) \qquad (7.351)$$

are the parametric equations of the curve. When the parameter t is the arc length s, the infinitesimal arc distance ds on the curve is

$$ds^2 = d\mathbf{r} \cdot d\mathbf{r} \qquad (7.352)$$

The arc length s of a curve is defined as the limit of the diagonal of a rectangular box as the length of the sides uniformly approaches zero.
When the space curve is a straight line that passes through point $P(x_0, y_0, z_0)$

where $x_0 = x(t_0)$, $y_0 = y(t_0)$, $z_0 = z(t_0)$, its equation can be shown by

$$\frac{x - x_0}{\alpha} = \frac{y - y_0}{\beta} = \frac{z - z_0}{\gamma} \tag{7.353}$$

$$\alpha^2 + \beta^2 + \gamma^2 = 1 \tag{7.354}$$

where α, β, and γ are the directional cosines of the line. The equation of the tangent line to the space curve (7.351) at a point $P(x_0, y_0, z_0)$ is

$$\frac{x - x_0}{dx/dq} = \frac{y - y_0}{dy/dq} = \frac{z - z_0}{dz/dq} \tag{7.355}$$

$$\left(\frac{dx}{dq}\right)^2 + \left(\frac{dy}{dq}\right)^2 + \left(\frac{dz}{dq}\right)^2 = 1 \tag{7.356}$$

To show this, let us consider a position vector $^G\mathbf{r} = {}^G\mathbf{r}(s)$ that describes a space curve using the length parameter s:

$$^G\mathbf{r} = {}^G\mathbf{r}(s) = x(s)\,\hat{\imath} + y(s)\,\hat{\jmath} + z(s)\,\hat{k} \tag{7.357}$$

The arc length s is measured from a fixed point on the curve. By a very small change ds, the position vector will move to a very close point such that the increment in the position vector would be

$$d\mathbf{r} = dx(s)\,\hat{\imath} + dy(s)\,\hat{\jmath} + dz(s)\,\hat{k} \tag{7.358}$$

The lengths of $d\mathbf{r}$ and ds are equal for infinitesimal displacement:

$$ds = \sqrt{dx^2 + dy^2 + dz^2} \tag{7.359}$$

The arc length has a better expression in the square form:

$$ds^2 = dx^2 + dy^2 + dz^2 = d\mathbf{r} \cdot d\mathbf{r} \tag{7.360}$$

If the parameter of the space curve is q instead of s, the increment arc length would be

$$\left(\frac{ds}{dt}\right)^2 = \frac{d\mathbf{r}}{dt} \cdot \frac{d\mathbf{r}}{dt} \tag{7.361}$$

Therefore, the arc length between two points on the curve can be found by integration:

$$s = \int_{t_1}^{t_2} \sqrt{\frac{d\mathbf{r}}{dt} \cdot \frac{d\mathbf{r}}{dt}}\, dt = \int_{t_1}^{t_2} \sqrt{\left(\frac{dx}{dt}\right)^2 + \left(\frac{dy}{dt}\right)^2 + \left(\frac{dz}{dt}\right)^2}\, dt \tag{7.362}$$

Let us expand the parametric equations of the curve (7.351) at a point $P(x_0, y_0, z_0)$,

$$
\begin{aligned}
x &= x_0 + \frac{dx}{dt}\Delta t + \frac{1}{2}\frac{d^2 x}{dt^2}\Delta t^2 + \cdots \\
y &= y_0 + \frac{dy}{dt}\Delta t + \frac{1}{2}\frac{d^2 y}{dt^2}\Delta t^2 + \cdots \\
z &= z_0 + \frac{dz}{dt}\Delta t + \frac{1}{2}\frac{d^2 z}{dt^2}\Delta t^2 + \cdots
\end{aligned}
\tag{7.363}
$$

and ignore the nonlinear terms to find the tangent line to the curve at the point:

$$\frac{x - x_0}{dx/dt} = \frac{y - y_0}{dy/dt} = \frac{z - z_0}{dz/dt} = \triangle t \tag{7.364}$$

Example 320 ★ *Length of a spatial road.*
Consider a spatial closed road with the following parametric equations:

$$
\begin{aligned}
x &= (a + b\sin\theta)\cos\theta \\
y &= (a + b\sin\theta)\sin\theta \\
z &= b + b\cos\theta
\end{aligned} \tag{7.365}
$$

The total length of the road can be found by the integral of ds for θ from 0 to 2π:

$$
\begin{aligned}
s &= \int_{\theta_1}^{\theta_2} \sqrt{\frac{d\mathbf{r}}{d\theta} \cdot \frac{d\mathbf{r}}{d\theta}}\, d\theta = \int_{\theta_1}^{\theta_2} \sqrt{\left(\frac{\partial x}{\partial\theta}\right)^2 + \left(\frac{\partial x}{\partial\theta}\right)^2 + \left(\frac{\partial x}{\partial\theta}\right)^2}\, d\theta \\
&= \int_0^{2\pi} \frac{\sqrt{2}}{2}\sqrt{2a^2 + 3b^2 - b^2\cos 2\theta + 4ab\sin\theta}\, d\theta
\end{aligned} \tag{7.366}
$$

7.7 ★ Steering Mechanism Optimization

Optimization of a steering mechanism means to design a system that works as closely as possible to a desired function. Assume the Ackerman kinematic condition is the desired function for a steering system. Comparing the function of the designed steering mechanism to the Ackerman condition, we define an error function e to compare the two systems. An example for the e function can be the difference between the outer steer angles of the designed mechanism δ_{D_o} and the Ackerman δ_{A_o} for the same inner angle δ_i.

The error function may be the absolute value of the maximum difference,

$$e = \max|\delta_{D_o} - \delta_{A_o}| \tag{7.367}$$

or the root mean square (RMS) of the difference between the two functions

$$e = \sqrt{\frac{1}{\delta_2 - \delta_1}\int_{\delta_1}^{\delta_2}(\delta_{D_o} - \delta_{A_o})^2\, d\delta_i} \tag{7.368}$$

or simply

$$e = \sqrt{\int_{\delta_1}^{\delta_2}(\delta_{D_o} - \delta_{A_o})^2\, d\delta_i} \tag{7.369}$$

in a specific range of the inner steer angle $\delta_1 < \delta_i < \delta_2$.

The error e, would be a function of a set of parameters. Minimization of the error function for a parameter, over the working range of the steer angle δ_i, generates the optimized value of the parameter.

The RMS function (7.368) is defined for continuous variables δ_{D_o} and δ_{A_o}. However, depending on the designed mechanism, it is not always possible to find a closed-form equation for e. In such a case, the error function cannot be defined explicitly, and hence, the error function should be evaluated for n different values of the inner steer angle δ_i numerically. The RMS error function for a set of discrete values of e is define by

$$e = \sqrt{\frac{1}{n} \sum_{i=1}^{n} (\delta_{D_o} - \delta_{A_o})^2} \qquad (7.370)$$

where n is the number of subdivision of δ_i in the working range $\delta_1 \leq \delta_i \leq \delta_2$. The error function (7.368) or (7.370) must be evaluated for different values of a parameter. Then a plot for $e = e(parameter)$ can show the trend of variation of e as a function of the parameter. If there is a minimum for e, then the optimal value for the parameter can be found. Otherwise, the trend of the e function can show the direction for minimum searching.

Example 321 ★ *Optimized trapezoidal steering mechanism.*

The inner-outer angles relationship for a trapezoidal steering mechanism, shown in Figure 7.8 is

$$\sin (\beta + \delta_i) + \sin (\beta - \delta_o)$$
$$= \frac{w}{d} - \sqrt{\left(\frac{w}{d} - 2\sin\beta\right)^2 - (\cos(\beta - \delta_o) - \cos(\beta + \delta_i))^2} \quad (7.371)$$

Comparing Equation (7.371) with the Ackerman condition,

$$\cot \delta_o - \cot \delta_i = \frac{w}{l} \qquad (7.372)$$

we define an error function

$$e = \sqrt{\frac{1}{n} \sum_{i=1}^{n} (\delta_{D_o} - \delta_{A_o})^2} \qquad (7.373)$$

and search for its minimum to optimize the trapezoidal steering mechanism.

Consider a vehicle with the dimensions

$$w = 1.66\,\mathrm{m} \qquad l = 2.93\,\mathrm{m} \qquad (7.374)$$

Let's optimize a trapezoidal steering mechanism for

$$d = 0.4\,\mathrm{m} \qquad (7.375)$$

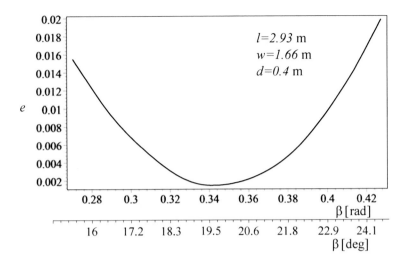

FIGURE 7.61. Error function $e = e(\beta)$ for a specific trapezoidal steering mechanism, with a minimum at $\beta \approx 19.5\,\mathrm{deg}$.

and use β as a parameter.

A plot of comparison between such a mechanism and the Ackerman condition, for a set of different β, is shown in Figure 7.11, and their difference $\Delta\delta_o = \delta_{D_o} - \delta_{A_o}$ is shown in Figure 7.13.

We may set a value for β, say $\beta = 6\,\mathrm{deg}$, and evaluate δ_{D_o} and δ_{A_o} at $n = 100$ different values of δ_i for a working range such as $-40\,\mathrm{deg} \le \delta_i \le 40\,\mathrm{deg}$. Then, we calculate the associated error function e

$$e = \sqrt{\frac{1}{100}\sum_{i=1}^{100}(\delta_{D_o} - \delta_{A_o})^2} \qquad (7.376)$$

for the specific β. Now we conduct our calculation again for new values of β, such as $\beta = 10\,\mathrm{deg}, 11\,\mathrm{deg}, \cdots$. Figure 7.61 depicts the function $e = e(\beta)$ with a minimum at $\beta \approx 19.5\,\mathrm{deg}$.

The geometry of the optimal trapezoidal steering mechanism is shown in Figure 7.62(a). The two side arms intersect at point G on their extensions. For an optimal mechanism, the intersection of point G may be in front or behind of the rear axle. However, it is recommended to put the intersection point at the center of the rear axle and design a near optimal trapezoidal steering mechanism. Using the recommendation, it is possible to eliminate the optimization process and get a good enough design. Such an estimated design is shown in Figure 7.62(b). The angle β for the optimal design is $\beta = 19.5\,\mathrm{deg}$, and for the estimated design is $\beta \approx 15.8\,\mathrm{deg}$.

$$\beta \approx \arctan\frac{w/2}{l} = \arctan\frac{1.66/2}{2.93} = 0.27604\,\mathrm{rad} = 15.816\,\mathrm{deg} \qquad (7.377)$$

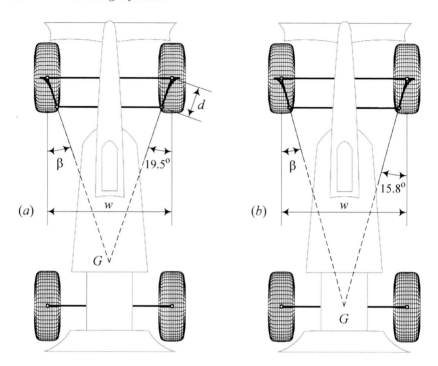

FIGURE 7.62. The geometry of the optimal trapezoidal steering mechanism and the estimated design.

Example 322 ★ *No exact Ackerman mechanism.*

It is not possible to make a simple steering linkage to work exactly based on the Ackerman steering condition. However, it is possible to optimize various steering linkages, for a working range, to function close to the Ackerman condition, and be exact at a few points. An isosceles trapezoidal linkage is not as exact as the Ackerman steering at every arbitrary turning radius, however, it is simple enough to be mass produced, and exact enough to be installed in street cars.

The trapezoidal linkage cannot act as Ackerman mechanism because the Ackerman condition is a function of the wheel base l, while the trapezoidal linkage is not a function of l.

Example 323 ★ *Optimization of a multi-link steering mechanism.*

Assume that we wish to design a multi-link steering mechanism for a vehicle with the following dimensions.

$$w = 2.4\,\text{m} \qquad l = 4.8\,\text{m} \qquad a_2 = 0.45l = 2.16\,\text{m} \qquad (7.378)$$

Due to space constraints, the positions of some joints of the mechanism are fixed as shown in Figure 7.63. However, we may vary the length x to

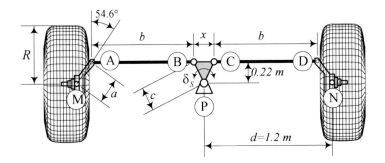

FIGURE 7.63. A multi-link steering mechanism that must be optimized by vary-ing x.

design the best mechanism to function as close as possible to the Ackerman condition.

$$\cot \delta_2 - \cot \delta_1 = \frac{w}{l} = \frac{1}{2} \qquad (7.379)$$

The steering wheel input δ_s turns the triangle PBC which turns both the left and the right wheels.

The vehicle must be able to turn in a minimum circle with radius R_m.

$$R_m = 10\,\mathrm{m} \qquad (7.380)$$

The minimum turning radius determines the maximum steer angle δ_M

$$
\begin{aligned}
R_m &= \sqrt{a_2^2 + l^2 \cot^2 \delta_M} \\
10 &= \sqrt{(0.45 \times 4.8)^2 + 4.8^2 \cot^2 \delta_M} \\
\delta_M &= 0.4569\,\mathrm{rad} \approx 26.178\,\mathrm{deg} \qquad (7.381)
\end{aligned}
$$

where δ_M is the cot-average of the inner and outer steer angles. Having R and δ_M is enough to determine δ_o and δ_i.

$$
\begin{aligned}
R_1 &= l \cot \delta_M = 4.8 \cot 0.4569 = 9.7642\,\mathrm{m} \qquad (7.382) \\
\delta_i &= \tan^{-1} \frac{l}{R_1 - \dfrac{w}{2}} = 0.51085\,\mathrm{rad} \approx 29.270\,\mathrm{deg} \qquad (7.383) \\
\delta_o &= \tan^{-1} \frac{l}{R_1 + \dfrac{w}{2}} = 0.41265\,\mathrm{rad} \approx 23.643\,\mathrm{deg} \qquad (7.384)
\end{aligned}
$$

Because the mechanism is symmetric, each wheel of the steering mechanism in Figure 7.63 must be able to turn at least 29.27 deg. To be safe, we try to optimize the mechanism for a wider range, say $\delta = \pm 35$ deg, however,

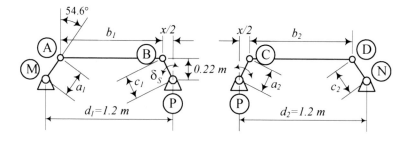

FIGURE 7.64. The multi-link steering is a 6-link mechanism that may be treeted as two combined 4-bar linkages.

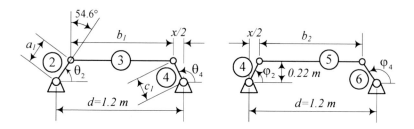

FIGURE 7.65. The input and output angles of the two 4-bar linkages.

vehicles in normal driving conditions barely receive steer angles beyond $\delta = \pm 15$ deg. Therefore, we try to optimize the steering mechanism for this practical range.

The multi-link steering mechanism is a six-link Watt linkage. We divide the mechanism into two four-bar linkages. The linkage 1 is on the left and the linkage 2 is on the right, as shown in Figure 7.64. We may assume that MA is the input link of the left linkage and PB is its output link. Link PB is rigidly attached to PC, which is the input of the right linkage. The output of the right linkage is ND. To find the inner-outer steer angles relationship, we need to find the angle of ND as a function of the angle of MA. In the following analysis, a_1 and a_2 are the length of the input links of the mechanism 1 and 2.

Figure 7.65 illustrates the link numbers, and the input-output angles of the four-bar linkages. The steer angles can be calculated based on the angle of these two links.

$$\delta_1 = \theta_2 - (90 - 54.6) \deg \tag{7.385}$$
$$\delta_2 = \varphi_4 - (90 + 54.6) \deg \tag{7.386}$$

The lengths of the links for the mechanisms are collected in Table 7.1.

Table 7.1 - *Link numbers, and the input-output angles*
for the multi-link steering mechanism

	Left linkage	
Link	Length	angle
1	$d_1 = 1.2$	180
2	$a_1 = 0.22/\cos 54.6 = 0.37978$	θ_2
3	$b_1 = 1.2 - 0.22\tan 54.6 - \frac{x}{2}$ $= 0.89043 - \frac{x}{2}$	θ_3
4	$c_1 = \sqrt{0.22^2 + x^2/4}$	θ_4

	Right linkage	
Link	Length	angle
1	$d_2 = 1.2$	180
4	$a_2 = \sqrt{0.22^2 + x^2/4}$	$\varphi_2 = \theta_4 - 2\tan^{-1}\frac{x}{0.44}$
5	$b_2 = 0.89043 - \frac{x}{2}$	φ_3
6	$c_2 = 0.22/\cos 54.6 = 0.37978$	φ_4

Equation (6.1) that is repeated below, provides the angle θ_4 as a function of θ_2.

$$\theta_4 = 2\tan^{-1}\left(\frac{-B_1 \pm \sqrt{B_1^2 - 4A_1C_1}}{2A_1}\right) \tag{7.387}$$

$$\begin{aligned} A_1 &= J_3 - J_1 + (1 - J_2)\cos\theta_2 & B_1 &= -2\sin\theta_2 \\ C_1 &= J_1 + J_3 - (1 + J_2)\cos\theta_2 \end{aligned} \tag{7.388}$$

$$\begin{aligned} J_{11} &= \frac{d_1}{a_1} & J_{12} &= \frac{d_1}{c_1} & J_{13} &= \frac{a_1^2 - b_1^2 + c_1^2 + d_1^2}{2a_1c_1} \\ J_{14} &= \frac{d_1}{b_1} & J_{15} &= \frac{c_1^2 - d_1^2 - a_1^2 - b_1^2}{2a_1b_1} \end{aligned} \tag{7.389}$$

The same equation (7.387) can be used to connect the input-output angles of the right four-bar linkage.

$$\varphi_4 = 2\tan^{-1}\left(\frac{-B_2 \pm \sqrt{B_2^2 - 4A_2C_2}}{2A_2}\right) \tag{7.390}$$

$$\begin{aligned} A_2 &= J_{23} - J_{21} + (1 - J_{22})\cos\varphi_2 & B_2 &= -2\sin\varphi_2 \\ C_2 &= J_{21} + J_{23} - (1 + J_{22})\cos\varphi_2 \end{aligned} \tag{7.391}$$

$$\begin{aligned} J_{21} &= \frac{d_2}{a_2} & J_{22} &= \frac{d_2}{c_2} & J_{23} &= \frac{a_2^2 - b_2^2 + c_2^2 + d_2^2}{2a_2c_2} \\ J_{24} &= \frac{d_2}{b_2} & J_{25} &= \frac{c_2^2 - d_2^2 - a_2^2 - b_2^2}{2a_2b_2} \end{aligned} \tag{7.392}$$

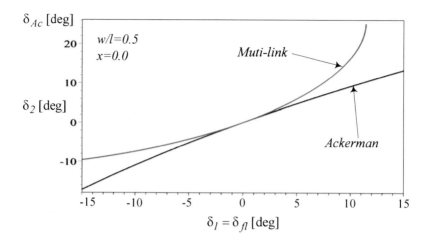

FIGURE 7.66. Steer angles δ_2 and δ_{Ac} versus δ_1.

Starting with a guess value for x, we are able to calculate the length of the links. Using Equations (7.387) and (7.390), along with (7.385) and (7.386), we calculate δ_2 for a given value of δ_1.

Let us start with $x = 0$, then

$$a_1 = 0.37978\,\text{m} \qquad b_1 = .89043\,\text{m} \qquad c_1 = 0.22\,\text{m} \qquad (7.393)$$
$$a_2 = 0.22\,\text{m} \qquad b_2 = 0.89043\,\text{m} \qquad c_2 = 0.37978\,\text{m} \qquad (7.394)$$

Using Equations (7.385) and (7.387), we may calculate the output of the first four-bar linkage, θ_4, for a range of the left steer angle $-15\deg < \delta_1 < 15\deg$. The following constraint, provides the numerical values for φ_2 to be used as the input of the right four-bar linkage.

$$\varphi_2 = \theta_4 - 2\tan^{-1}\frac{x}{0.44} \qquad (7.395)$$

Then, using Equations (7.390) and (7.387), we can calculate the steer angle δ_2 for the right wheel.

Figure 7.66 depicts the numerical values of the steer angles δ_2 and δ_{Ac} versus δ_1. The angle δ_{Ac} is the steer angle of the right wheel based on the Ackerman equation (7.379).

Having δ_2 and δ_{Ac}, we calculate the difference Δ

$$\Delta = \delta_2 - \delta_{Ac} \qquad (7.396)$$

for n different values of δ_1 in the working range angle $-15\deg < \delta_1 < 15\deg$. Based on the n numbers for Δ, we may find the error e.

$$e = \sqrt{\frac{\Delta^2}{n}} \qquad (7.397)$$

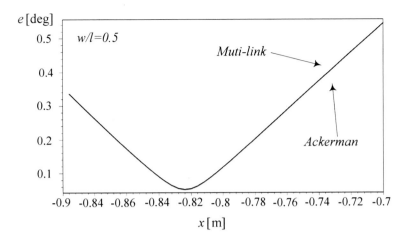

FIGURE 7.67. Illustration of the error function $e = e(x)$.

Changing the value of x and recalculating e evaluates the function $e = e(x)$.

Figure 7.67 illustrates the result of the calculation. It shows that the error is minimum for $x \approx -0.824$ m, which is the best length for the base of the triangle PBC.

The behavior of the multi-link steering mechanism for different values of x, is shown in Figure 7.68. The Ackerman condition is also plotted to compare with the optimal multi-link mechanism. The optimality of $x = -0.824$ m may be clearer in Figure 7.69 that shows the difference $\Delta = \delta_2 - \delta_{Ac}$ for different values of x. However, because the calculation has been done for the range $-15 \deg < \delta_1 < 15 \deg$, there is no guarantee that the mechanism exists or is optimal out of this range.

The optimal multi-link steering mechanism along with the length of its links is shown in Figure 7.70. The mechanism and the meaning of negative value for x are shown in Figure 7.71 where the mechanism is in a positive turning position.

7.8 ★ Trailer-Truck Kinematics

Consider a car pulling a one-axle trailer, as shown in Figure 7.72. We may normalize the dimensions such that the length of the trailer is 1. The positions of the car at the hinge point and the trailer at the center of its axle are shown by vectors **r** and **s**.

Assuming **r** is a given differentiable function of time t, we would like to examine the steering behavior of the trailer by calculating **s**, and predict jackknifing. When the car is moving forward, we say the car and trailer are

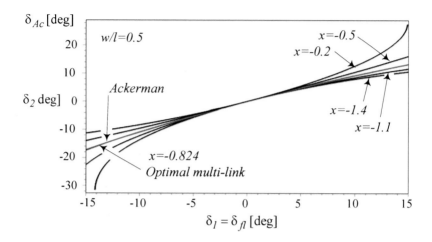

FIGURE 7.68. The behavior of the multi-link steering mechanism for different values of x.

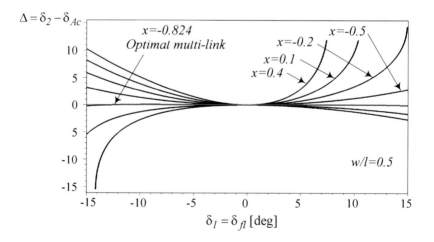

FIGURE 7.69. Illustration of the difference $\Delta = \delta_2 - \delta_{Ac}$ for different values of x.

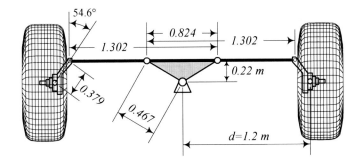

FIGURE 7.70. The optimal multi-link steering mechanism along with the length of its links.

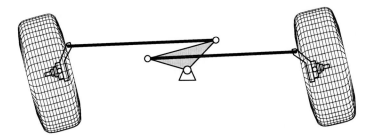

FIGURE 7.71. The optimal multi-link steering mechanism in a positive turning position.

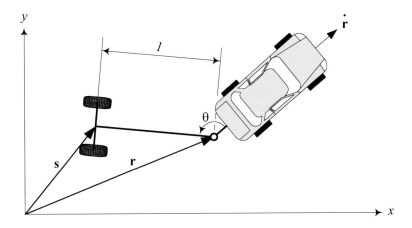

FIGURE 7.72. A car pulling a one-axle trailer.

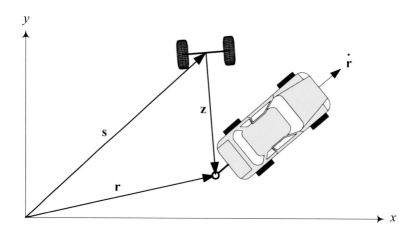

FIGURE 7.73. A jackknifed configuration of a car pulling a one-axle trailer.

jackknifed if

$$\dot{\mathbf{r}} \cdot \mathbf{z} < 0 \qquad (7.398)$$

where

$$\mathbf{z} = \mathbf{r} - \mathbf{s} \qquad (7.399)$$

A jackknifed configuration is shown in Figure 7.73, while Figure 7.72 is showing an *unjackknifed* configuration.

Mathematically, we want to know if the truck-trailer will jackknife for a given path of motion $\mathbf{r} = \mathbf{r}(t)$ and what conditions we must impose on $\mathbf{r}(t)$ to prevent jackknifing.

The velocity of the trailer can be expressed by

$$\dot{\mathbf{s}} = c\,(\mathbf{r} - \mathbf{s}) \qquad (7.400)$$

where

$$c = \dot{\mathbf{r}} \cdot \mathbf{z} \qquad (7.401)$$

and the unjackknifing condition is

$$c > 0 \qquad (7.402)$$

Assume the twice continuously differentiable function $\mathbf{r}$ is the path of the car motion. If $|\mathbf{z}| = 1$, and $\mathbf{r}$ has a radius of curvature $R(t) > 1$, and

$$\dot{\mathbf{r}}(0) \cdot \mathbf{z}(0) > 0 \qquad (7.403)$$

then

$$\dot{\mathbf{r}}(t) \cdot \mathbf{z}(t) > 0 \qquad (7.404)$$

for all $t > 0$.

Therefore, if the car is moving forward and the car-trailer combination is not originally jackknifed, then it will remain unjackknifed.

Proof. Let us set the normalized constant trailer length as 1 to have $\mathbf{z}$ as a unit vector

$$|\mathbf{z}| = |\mathbf{r} - \mathbf{s}| = 1 \tag{7.405}$$

and

$$(\mathbf{r} - \mathbf{s}) \cdot (\mathbf{r} - \mathbf{s}) = 1 \tag{7.406}$$

The nonslip condition of the wheels of the trailer constrains the trailer velocity vector $\dot{\mathbf{s}}$ to be directed along the trailer axis $\mathbf{z}$ and therefore,

$$\dot{\mathbf{s}} = c(\mathbf{r} - \mathbf{s}) = c\mathbf{z} \tag{7.407}$$

Differentiating (7.406) yields

$$2(\dot{\mathbf{r}} - \dot{\mathbf{s}}) \cdot (\mathbf{r} - \mathbf{s}) = 0 \tag{7.408}$$
$$\dot{\mathbf{r}} \cdot (\mathbf{r} - \mathbf{s}) = \dot{\mathbf{s}} \cdot (\mathbf{r} - \mathbf{s}) \tag{7.409}$$

and therefore,

$$\dot{\mathbf{r}} \cdot (\mathbf{r} - \mathbf{s}) = c(\mathbf{r} - \mathbf{s}) \cdot (\mathbf{r} - \mathbf{s}) \tag{7.410}$$
$$c = \dot{\mathbf{r}} \cdot (\mathbf{r} - \mathbf{s}) = \dot{\mathbf{r}} \cdot \mathbf{z} \tag{7.411}$$

Having c enables us to express the trailer velocity vector $\dot{\mathbf{s}}$ of Equation (7.407) as

$$\dot{\mathbf{s}} = [\dot{\mathbf{r}} \cdot (\mathbf{r} - \mathbf{s})] (\mathbf{r} - \mathbf{s}) = (\dot{\mathbf{r}} \cdot \mathbf{z}) \mathbf{z} \tag{7.412}$$

There are three situations

1. When $c > 0$, the velocity vector of the trailer $\dot{\mathbf{s}}$ is along the trailer axis, $\mathbf{z}$. The trailer follows the car and the system is stable.

2. When $c = 0$, the velocity vector of the trailer $\dot{\mathbf{s}}$ is zero. In this case, the trailer spins about the center of its axle and the system is neutral-stable.

3. When $c < 0$, the velocity vector of the trailer $\dot{\mathbf{s}}$ is along the negative trailer axis, $-\mathbf{z}$. The trailer does not follow the car and the system is unstable.

Using a Cartesian coordinate expression, we may show the car and trailer position vectors by

$$\mathbf{r} = \begin{bmatrix} x_c \\ y_c \end{bmatrix} \tag{7.413}$$

$$\mathbf{s} = \begin{bmatrix} x_t \\ y_t \end{bmatrix} \tag{7.414}$$

and therefore,

$$\dot{\mathbf{s}} = \begin{bmatrix} \dot{x}_t \\ \dot{y}_t \end{bmatrix} = [\dot{\mathbf{r}} \cdot (\mathbf{r} - \mathbf{s})] (\mathbf{r} - \mathbf{s})$$

$$= \begin{bmatrix} \dot{x}_c (x_c - x_t)^2 + (x_c - x_t)(y_c - y_t) \dot{y}_c \\ \dot{x}_c (x_c - x_t)(y_c - y_t) + (y_c - y_t)^2 \dot{y}_c \end{bmatrix} \qquad (7.415)$$

$$c = (x_c - x_t) \dot{x}_c + (y_c - y_t) \dot{y}_c$$

$$= \dot{x}_c x_c + \dot{y}_c y_c - (\dot{x}_c x_t + \dot{y}_c y_t) \qquad (7.416)$$

Let us define a function $f(t) = \dot{\mathbf{r}} \cdot \mathbf{z}$ and assume that conclusion (7.404) is wrong while assumption (7.403) is correct. Then there exists a time $t_1 > 0$ such that $f(t_1) = 0$ and $f'(t_1) \leq 0$. Using $|\mathbf{z}| = 1$ and $\dot{\mathbf{r}} \neq 0$, we have $\dot{\mathbf{r}}(t_1) \cdot \mathbf{z}(t_1) = 0$ and therefore, $\dot{\mathbf{r}}(t_1)$ is perpendicular to $\mathbf{z}(t_1)$. The derivative $f'(t)$ would be

$$f'(t) = \ddot{\mathbf{r}} \cdot \mathbf{z} + \dot{\mathbf{r}} \cdot \dot{\mathbf{z}} = \ddot{\mathbf{r}} \cdot \mathbf{z} + \dot{\mathbf{r}} \cdot (\dot{\mathbf{r}} - \dot{\mathbf{s}})$$

$$= \ddot{\mathbf{r}} \cdot \mathbf{z} + |\dot{\mathbf{r}}|^2 - \dot{\mathbf{r}} \cdot \dot{\mathbf{s}} = \ddot{\mathbf{r}} \cdot \mathbf{z} + |\dot{\mathbf{r}}|^2 - \dot{\mathbf{r}} \cdot ((\dot{\mathbf{r}} \cdot \mathbf{z}) \mathbf{z})$$

$$= \ddot{\mathbf{r}} \cdot \mathbf{z} + |\dot{\mathbf{r}}|^2 - (\dot{\mathbf{r}} \cdot \mathbf{z})^2 = \ddot{\mathbf{r}} \cdot \mathbf{z} + |\dot{\mathbf{r}}|^2 - f^2(t) \qquad (7.417)$$

and therefore,

$$f'(t_1) = \ddot{\mathbf{r}} \cdot \mathbf{z} + |\dot{\mathbf{r}}|^2 \qquad (7.418)$$

The acceleration $\ddot{\mathbf{r}}$ in a normal-tangential coordinate frame $(\hat{u}_n, \hat{u}_t)$ is

$$\ddot{\mathbf{r}} = \frac{d |\dot{\mathbf{r}}|}{dt} \hat{u}_t + \kappa |\dot{\mathbf{r}}|^2 \hat{u}_n \qquad (7.419)$$

$$\kappa = \frac{1}{R} \qquad (7.420)$$

where $\hat{u}_n$ and $\hat{u}_t$ are the unit normal and tangential vectors. $\hat{u}_t$ is parallel to $\dot{\mathbf{r}}(t_1)$, and $\hat{u}_n$ is parallel to $\mathbf{z}(t_1)$. Hence,

$$\ddot{\mathbf{r}} \cdot \mathbf{z} = \pm \kappa(t_1) |\dot{\mathbf{r}}(t_1)|^2 \qquad (7.421)$$

and

$$f'(t_1) = |\dot{\mathbf{r}}(t_1)|^2 \pm \kappa(t_1) |\dot{\mathbf{r}}(t_1)|^2 = [1 \pm \kappa(t_1)] |\dot{\mathbf{r}}(t_1)|^2 \qquad (7.422)$$

where κ is the curvature of the path. Because $\kappa(t_1) = 1/R(t) > 0$, we conclude that $f'(t_1) > 0$, and it is not possible to have $f'(t_1) \leq 0$. ∎

Example 324 ★ *Straight motion of the car with constant velocity.*

Consider a car moving forward in a straight line with a constant velocity. We may use a normalization and set the speed of the car as 1 moving in positive x direction starting from $x = 0$. Using a two-dimensional vector expression we have

$$\mathbf{r} = \begin{bmatrix} x_c \\ y_c \end{bmatrix} = \begin{bmatrix} t \\ 0 \end{bmatrix} \qquad (7.423)$$

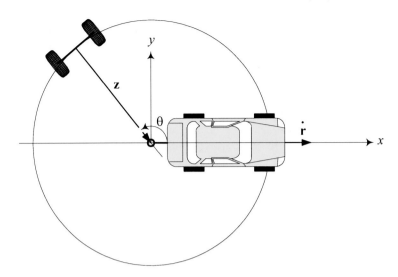

FIGURE 7.74. The initial position of a one-axle trailer pulled by a car moving forward in a straight line with a constant velocity is a circle about the hinge point.

Because of (7.405), we have

$$\mathbf{z}(0) = \mathbf{r}(0) - \mathbf{s}(0) = -\mathbf{s}(0) \tag{7.424}$$

and therefore, the initial position of the trailer must lie on a unit circle as shown in Figure 7.74.

Using two dimensional vectors, we may express $\mathbf{z}(0)$ as a function of θ

$$\mathbf{z}(0) = -\mathbf{s}(0) = \begin{bmatrix} x_t(0) \\ y_t(0) \end{bmatrix} = \begin{bmatrix} \cos\theta \\ \sin\theta \end{bmatrix} \tag{7.425}$$

and simplify Equation (7.415) as

$$\dot{\mathbf{s}} = \begin{bmatrix} \dot{x}_t \\ \dot{y}_t \end{bmatrix} = \begin{bmatrix} (t - x_t)^2 \\ -y_t(t - x_t) \end{bmatrix} \tag{7.426}$$

Equation (7.426) is a set of two coupled first-order ordinary differential equations with the solution

$$\mathbf{s} = \begin{bmatrix} x_t \\ y_t \end{bmatrix} = \begin{bmatrix} t + \dfrac{e^{-2t} - C_1}{e^{-2t} + C_1} \\ \dfrac{C_2 e^{-t}}{e^{-2t} + C_1} \end{bmatrix} \tag{7.427}$$

Applying the initial conditions (7.425) yields

$$C_1 = \frac{\cos\theta - 1}{\cos\theta + 1} \qquad C_2 = \frac{2\sin\theta}{\cos\theta + 1} \tag{7.428}$$

If $\theta \neq k\pi$, then the solution depends on time, and when time goes to infinity, the solution leads to the following limits asymptotically:

$$\lim_{t\to\infty} x_t = t - 1 \qquad \lim_{t\to\infty} y_t = 0 \qquad (7.429)$$

When the car is moving with a constant velocity, this solution shows that the trailer will approach the position of straight forward moving, following the car.

We may also consider that the car is reversing, the solution shows that, except for the unstable initial condition $\theta = \pi$, all solutions eventually approach the jackknifed configuration.

If $\theta = 0$, then

$$C_1 = 0 \qquad C_2 = 0 \qquad x_t = t + 1 \qquad y_t = 0 \qquad (7.430)$$

and the trailer moves in an unstable configuration. Any deviation from $\theta = 0$ leads to the stable limit solution (7.429).

If $\theta = \pi$, then

$$C_1 = \infty \qquad C_2 = \infty \qquad x_t = t - 1 \qquad y_t = 0 \qquad (7.431)$$

and the trailer follows the car in an stable configuration. Any deviation from $\theta = 0$ will disappear after a while.

Example 325 ★ *Straight car motion with different initial θ.*

Consider a car moving on the x-axis with constant speed. The car is pulling a trailer, which is initially at θ_0 as shown in Figure 7.74. Using a normalized length, we assume the distance between the center of the trailer axle and the hinge is the length of trailer, and is equal to 1.

If we show the global position of the car at hinge by $\mathbf{r} = \begin{bmatrix} x_c & y_c \end{bmatrix}^T$ and the global position of the trailer by $\mathbf{s} = \begin{bmatrix} x_t & y_t \end{bmatrix}^T$, then the position of the trailer is a function of the car's motion. When the position of the car is given by a time-dependent vector function

$$\mathbf{r} = \begin{bmatrix} x_c(t) \\ y_c(t) \end{bmatrix} \qquad (7.432)$$

the trailer position can be found by solving two coupled differential equations.

$$\dot{x}_t = (x_c - x_t)^2 \dot{x}_c + (x_c - x_t)(y_c - y_t)\dot{y}_c \qquad (7.433)$$
$$\dot{y}_t = (x_c - x_t)(y_c - y_t)\dot{x}_c + (y_c - y_t)^2 \dot{y}_c \qquad (7.434)$$

For a constantly uniform car motion $\mathbf{r} = \begin{bmatrix} t & 0 \end{bmatrix}^T$, Equations (7.433) and (7.434) reduce to

$$\dot{x}_t = (t - x_t)^2 \qquad (7.435)$$
$$\dot{y}_t = -y_t(t - x_t) \qquad (7.436)$$

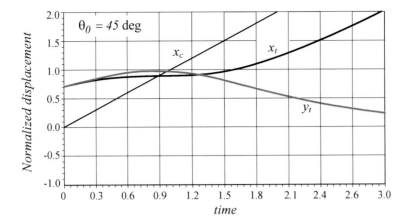

FIGURE 7.75. Trailer kinematics for a $\theta = 45$ deg starting position.

The first equation (7.435) is independent of the second equation (7.436) and can be solved independently.

$$x_t = \frac{C_1 e^{2t}(t-1) - t - 1}{C_1 e^{2t} - 1} = t + \frac{e^{-2t} - C_1}{e^{-2t} + C_1} \qquad (7.437)$$

Substituting Equation (7.437) in (7.436) generates an independent differential equation for y_t:

$$\dot{y}_t = \frac{e^{-2t} - C_1}{e^{-2t} + C_1} y_t \qquad (7.438)$$

with the solution

$$y_t = \frac{C_2 e^{-t}}{e^{-2t} + C_1} \qquad (7.439)$$

When the trailer starts from $\mathbf{s} = \begin{bmatrix} \cos\theta & \sin\theta \end{bmatrix}^T$, the constants of integral would be equal to Equations (7.428) and therefore,

$$x_t = t + \frac{e^{-2t}(\cos\theta + 1) - \cos\theta + 1}{e^{-2t}(\cos\theta + 1) + \cos\theta - 1} \qquad (7.440)$$

$$y_t = \frac{2e^{-t}\sin\theta}{e^{-2t}(\cos\theta + 1) + \cos\theta - 1} \qquad (7.441)$$

Figures 7.75, 7.76, and 7.77 illustrate the behavior of the trailer starting from $\theta = 45$ deg, $\theta = 90$ deg, $\theta = 135$ deg.

Example 326 ★ *Circular motion of the car with constant velocity.*
   Consider a car pulling a trailer as Figure 7.72 shows. The car is traveling along a circle of radius $R > 1$, based on a normalized length in which the length of the trailer is 1. In a circular motion with a normalized angular

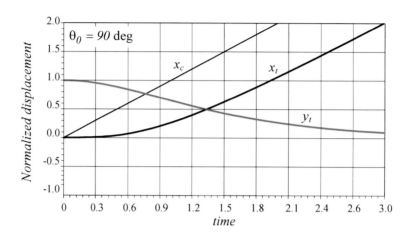

FIGURE 7.76. Trailer kinematics for a $\theta = 90\,\mathrm{deg}$ starting position.

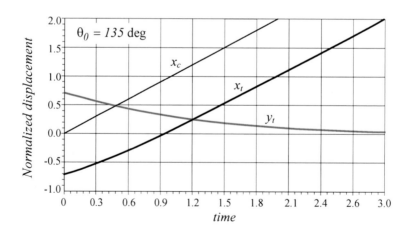

FIGURE 7.77. Trailer kinematics for a $\theta = 135\,\mathrm{deg}$ starting position.

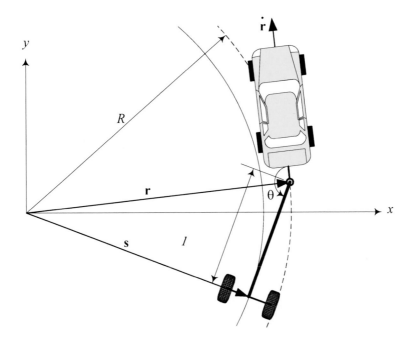

FIGURE 7.78. Steady state configuration of a car-trailer combination.

velocity $\omega = 1$ and period $T = 2\pi$, the position of the car is given by a time-dependent vector function:

$$\mathbf{r} = \begin{bmatrix} x_c(t) \\ y_c(t) \end{bmatrix} = \begin{bmatrix} R\cos(t) \\ R\sin(t) \end{bmatrix} \tag{7.442}$$

The initial position of the trailer must lie on a unit circle with a center at $\mathbf{r}(0) = \begin{bmatrix} x_c(0) & y_c(0) \end{bmatrix}^T.$

$$\mathbf{s}(0) = \begin{bmatrix} x_t(0) \\ y_t(0) \end{bmatrix} = \begin{bmatrix} x_c(0) \\ y_c(0) \end{bmatrix} + \begin{bmatrix} \cos\theta \\ \sin\theta \end{bmatrix} \tag{7.443}$$

The car-trailer combination approaches a steady-state configuration as shown in Figure 7.78.
    Substituting

$$\dot{\mathbf{r}} = \begin{bmatrix} -R\sin(t) \\ R\cos(t) \end{bmatrix} \tag{7.444}$$

and the initial conditions (7.443) in (7.415) will generate two differential equations for the trailer position.

$$\dot{x}_t = R(R\cos t - x_t)(x_t \sin t - y_t \cos t) \tag{7.445}$$
$$\dot{y}_t = R(R\sin t - y_t)(x_t \sin t - y_t \cos t) \tag{7.446}$$

Assuming $\mathbf{r}(0) = \begin{bmatrix} 0 & 0 \end{bmatrix}^T$ *the steady-state solutions of these equations are*

$$x_t = R_c \cos(t - \alpha) \tag{7.447}$$
$$y_t = R_c \sin(t - \alpha) \tag{7.448}$$

where R_c is the trailer's radius of rotation, and α is the angular position of the trailer behind the car.

$$R_c = \sqrt{R^2 - 1} \qquad \sin \alpha = \frac{1}{R} \qquad \cos \alpha = \frac{R_c}{R} \tag{7.449}$$

We may check the solution by employing two new variables, u and v, such that

$$u = x_t \sin t - y_t \cos t \tag{7.450}$$
$$v = x_t \cos t + y_t \sin t \tag{7.451}$$

and

$$\dot{u} = v \tag{7.452}$$

Using the new variables we find

$$x_t = u \sin t + v \cos t \tag{7.453}$$
$$y_t = -u \cos t + v \sin t \tag{7.454}$$

$$\dot{x}_t = Ru(R \cos t - u \sin t - v \cos t) \tag{7.455}$$
$$\dot{y}_t = Ru(R \sin t + u \cos(t) - v \sin t) \tag{7.456}$$

Direct differentiating from (7.450), (7.451), (7.453), and (7.454) shows that

$$\dot{u} = x_t \cos t + y_t \sin t + \dot{x}_t (\sin t) - \dot{y}_t (\cos t) \tag{7.457}$$
$$\dot{v} = -x_t \sin t + y_t \cos t + \dot{x}_t \cos t + \dot{y}_t \sin t \tag{7.458}$$

$$\dot{x}_t = \dot{u} \sin t + \dot{v} \cos t + u \cos t - v \sin t \tag{7.459}$$
$$\dot{y}_t = \dot{u} \cos t - \dot{v} \sin t - u \sin t - v \cos t \tag{7.460}$$

and therefore, the problem can be expressed in a new set of equations.

$$\dot{u} = v - Ru^2 \tag{7.461}$$
$$\dot{v} = u \left(R^2 - Rv - 1 \right) \tag{7.462}$$

In the steady-state condition the time differentials must be zero, and therefore, the steady-state solutions would be the answers to the following algebraic equations:

$$v - Ru^2 = 0 \tag{7.463}$$
$$u \left(R^2 - Rv - 1 \right) = 0 \tag{7.464}$$

There are three sets of solutions.

$$\{u = 0, v = 0\} \tag{7.465}$$

$$\{u = \frac{R_c}{R}, v = \frac{R_c^2}{R}\} \tag{7.466}$$

$$\{u = -\frac{R_c}{R}, v = \frac{R_c^2}{R}\} \tag{7.467}$$

The first solution is associated with $\mathbf{s} = 0$,

$$x_t = 0 \qquad y_t = 0 \tag{7.468}$$

which shows that the center of the trailer's axle remains at the origin while the car is turning on a circle $R = 1$. *This is a stable motion.*

The second solution is associated with

$$x_t = \frac{R_c}{R} \sin t + \frac{R_c^2}{R} \cos t \qquad y_t = -\frac{R_c}{R} \cos t + \frac{R_c^2}{R} \sin t \tag{7.469}$$

which are equivalent to (7.447) and (7.448).

To examine the stability of the second solution, we may substitute a perturbed solution

$$u = \frac{R_c}{R} + p \qquad v = \frac{R_c^2}{R} + q \tag{7.470}$$

in the linearized equations of motion (7.461) and (7.462) at the second set of solutions

$$\dot{u} = v - 2R_c u \qquad \dot{v} = -R_c v \tag{7.471}$$

Therefore, we derive two equations for the perturbed functions p and q.

$$\dot{p} = q - 2R_c p \qquad \dot{q} = R_c q \tag{7.472}$$

The set of linear perturbed equations can be set in a matrix form

$$\begin{bmatrix} \dot{p} \\ \dot{q} \end{bmatrix} = \begin{bmatrix} -2R_c & 1 \\ 0 & -R_c \end{bmatrix} \begin{bmatrix} p \\ q \end{bmatrix} \tag{7.473}$$

The stability of Equation (7.473) is determined by the eigenvalues λ_i *of the coefficient matrix, which are*

$$\lambda_1 = -R_c \qquad \lambda_2 = -2R_c \tag{7.474}$$

Because both eigenvalues λ_1 *and* λ_2 *are negative, the solution of the perturbed equations symptomatically goes to zero. Therefore, the second set of solutions (7.466) is stable and it absorbs any near path that starts close to it.*

The third solution is associated with

$$x_t = -\frac{R_c}{R} \sin t + \frac{R_c^2}{R} \cos t \qquad y_t = \frac{R_c}{R} \cos t + \frac{R_c^2}{R} \sin t \tag{7.475}$$

The linearized equations of motion at the third set of solutions (7.467) are

$$\dot{u} = v + 2R_c u \qquad \dot{v} = R_c v \tag{7.476}$$

The perturbed equations would then be

$$\begin{bmatrix} \dot{p} \\ \dot{q} \end{bmatrix} = \begin{bmatrix} 2R_c & 1 \\ 0 & R_c \end{bmatrix} \begin{bmatrix} p \\ q \end{bmatrix} \tag{7.477}$$

with two positive eigenvalues

$$\lambda_1 = R_c \qquad \lambda_2 = 2R_c \tag{7.478}$$

Positive eigenvalues show that the solution of the perturbed equations diverges and goes to infinity. Therefore, the third set of solutions (7.467) is unstable and repels any near path that starts close to it.

Example 327 ★ *Slalom maneuver.*

    To analyze the behavior of a trailer, we need to have the motion of the car as a given function of time

$$\mathbf{r} = \begin{bmatrix} x_c(t) \\ y_c(t) \end{bmatrix} \tag{7.479}$$

and substitute the function into equations

$$\dot{x}_t = (x_c - x_t)^2 \dot{x}_c + (x_c - x_t)(y_c - y_t)\dot{y}_c \tag{7.480}$$
$$\dot{y}_t = (x_c - x_t)(y_c - y_t)\dot{x}_c + (y_c - y_t)^2 \dot{y}_c \tag{7.481}$$

and solve the equations to determine the the the trailer position

$$\mathbf{s} = \begin{bmatrix} x_t \\ y_t \end{bmatrix} \tag{7.482}$$

Slalom is a usual test to examine the handling behavior of vehicles. Let us assume that a car with a trailer is moving on a steady state sinusoidal path about the x-axis

$$\mathbf{r} = \begin{bmatrix} vt \\ \sin t \end{bmatrix} \tag{7.483}$$

The behavior of the car should be found by solving two coupled differential equation

$$\begin{aligned} \dot{x}_t &= (vt - x_t)^2 v + (vt - x_t)(\sin t - y_t)\cos t \\ &= (vt - x_t)((vt - x_t)v + (\sin t - y_t)\cos t) \end{aligned} \tag{7.484}$$
$$\begin{aligned} \dot{y}_t &= (vt - x_t)(\sin t - y_t)v + (\sin t - y_t)^2 \cos t \\ &= (\sin t - y_t)((vt - x_t)v + (\sin t - y_t)\cos t) \end{aligned} \tag{7.485}$$

In steady state condition, we have

$$\dot{x}_t = 0 \qquad \dot{y}_t = 0 \tag{7.486}$$

and therefore the possible solutions for the trailer motion is a similar motion to the car:

$$\mathbf{s} = \left[\begin{array}{c} vt \\ \sin t \end{array} \right] \tag{7.487}$$

7.9 Summary

Steering is required to guide a vehicle to a desired direction. When a vehicle turns, the wheels closer to the center of rotation are called the inner wheels, and the wheels further from the center of rotation are called the outer wheels. If the speed of a vehicle is very slow, there is a kinematic condition between the inner and outer steerable wheels, called the Ackerman condition.

Street cars are four-wheel vehicles and usually have front-wheel-steering. The kinematic condition between the inner and outer steered wheels is

$$\cot \delta_o - \cot \delta_i = \frac{w}{l} \tag{7.488}$$

where δ_i is the steer angle of the inner wheel, δ_o is the steer angle of the outer wheel, w is the track, and l is the wheelbase of the vehicle. Track w and wheel base l are considered the kinematic width and length of the vehicle.

The mass center of a steered vehicle will turn on a circle with radius R,

$$R = \sqrt{a_2^2 + l^2 \cot^2 \delta} \tag{7.489}$$

where δ is the cot-average of the inner and outer steer angles.

$$\cot \delta = \frac{\cot \delta_o + \cot \delta_i}{2} \tag{7.490}$$

The angle δ is the equivalent steer angle of a bicycle having the same wheelbase l and radius of rotation R.

7.10   Key Symbols

$4WS$	four-wheel-steering
a, b, c, d	lengths of the links of a four-bar linkage
a_i	distance of the axle number i from the mass center
A, B, C	input angle parameters of a four-bar linkage
AWS	all-wheel-steering
b_1	distance of the hinge point from rear axle
b_2	distance of trailer axle from the hinge point
c	stability index of a trailer motion
c_1	longitudinal distance of turn center and front axle of a $4WS$ car
c_2	longitudinal distance of turn center and rear axle of a $4WS$ car
c_s	$4WS$ factor
C	mass center, curvature center
$C_1, C_2, \cdots$	constants of integration
d	arm length in trapezoidal steering mechanism
e	error
e	length of the offset arm
FWS	front-wheel-steering
g	overhang distance
J	link parameters of a four-bar linkage
l	wheelbase
l_s	steering length
n	number of increments
O	center of rotation in a turn, curvature center
p	perturbation in u
q	perturbation in v
r	yaw velocity of a turning vehicle
$\mathbf{r}$	position vector of a car at the hinge
R	radius of rotation at mass center
R_1	radius of rotation at the center of the rear axle for FWS
R_1	horizontal distance of O and the center of axles
R_c	trailer's radius of rotation
R_t	radius of rotation at the center of the trailer axle
R_w	radius of the rear wheel
R_κ	curvature radius
RWS	rear wheel steering
$\mathbf{s}$	position vector of a trailer at the axle center
t	time
u	temporary variable in car-trailer analysis
u_R	steering rack translation
$\hat{u}$	unit vector
$v \equiv \dot{x}, \mathbf{v}$	vehicle velocity, temporary variable in car-trailer analysis

v_{ri}	speed of the inner rear wheel
v_{ro}	speed of the outer rear wheel
w	track
w_f	front track
w_r	rear track
$x, y, z, \mathbf{x}$	displacement
$\mathbf{z} = \mathbf{r} - \mathbf{s}$	position vector of a trailer relative to the car
β	arm angle in trapezoidal steering mechanism
δ	cot-average of the inner and outer steer angles
$\delta_1 = \delta_{fl}$	front left wheel steer angle
$\delta_2 = \delta_{fr}$	front right wheel steer angle
δ_{Ac}	steer angle based on Ackerman condition
δ_{fl}	front left wheel steer angle
δ_{fr}	front right wheel steer angle
δ_i	inner wheel
δ_{rl}	rear left wheel steer angle
δ_{rr}	rear right wheel steer angle
δ_o	outer wheel
δ_S	steer command
$\Delta = \delta_2 - \delta_{Ac}$	steer angle difference
θ	angle between trailer and vehicle longitudinal axes
κ	curvature of a road
λ	eigenvalue
ω	angular velocity
$\omega_i = \omega_{ri}$	angular velocity of the rear inner wheel
$\omega_o = \omega_{ro}$	angular velocity of the rear outer wheel

Exercises

1. Bicycle model and radius of rotation.

   Mercedes-Benz $GL450^{TM}$ has the following dimensions.

   $$l = 121.1\,\text{in} \qquad w_f = 65.0\,\text{in} \qquad w_r = 65.1\,\text{in} \qquad R = 39.7\,\text{ft}$$

   Assume $a_1 = a_2$ and use an average track to determine the maximum steer angle δ for a bicycle model of the car.

2. Radius of rotation.

   Consider a two-axle truck that is offered in different wheelbases.

   $$l = 109\,\text{in} \qquad l = 132.5\,\text{in} \qquad l = 150.0\,\text{in} \qquad l = 176.0\,\text{in}$$

   If the front track of the vehicles is

   $$w = 70\,\text{in}$$

   and $a_1 = a_2$, calculate the radius of rotations for $\delta = 30\,\text{deg}$.

3. Required space.

   Consider a two-axle vehicle with the following dimensions.

   $$l = 4\,\text{m} \qquad w = 1.3\,\text{m} \qquad g = 1.2\,\text{m}$$

   Determine R_{min}, R_{Max}, and ΔR for $\delta = 30\,\text{deg}$.

4. Rear wheel steering lift truck.

   A battery powered lift truck has the following dimensions.

   $$l = 55\,\text{in} \qquad w = 30\,\text{in}$$

   Calculate the radius of rotations if $\delta = 55\,\text{deg}$ and $a_1 = a_2$.

5. Wheel angular velocity.

   Consider a two-axle vehicle with

   $$l = 2.7\,\text{m} \qquad w = 1.36\,\text{m}$$

   What is the angular velocity ratio of w_o/w_i as a function of the steer angle δ?

6. A three-axle vehicle.

   A three-axle vehicle is shown in Figure 7.21. Find the relationship between δ_2 and δ_3, and also between δ_1 and δ_6.

7. A three-axle truck.

Consider a three-axle truck that has only one steerable axle in front. The dimensions of the truck are

$$a_1 = 5300\,\text{mm} \qquad a_2 = 300\,\text{mm} \qquad a_3 = 1500\,\text{mm}$$
$$w = 1800\,\text{mm}$$

Determine maximum steer angles of the front wheels if the truck is supposed to be able to turn with $R = 11$ m.

8. ★ Three-wheel vehicle steering condition.

Figure 7.79 illustrates a three-wheel vehicle with all wheels steerable. Determine the kinematic steering condition.

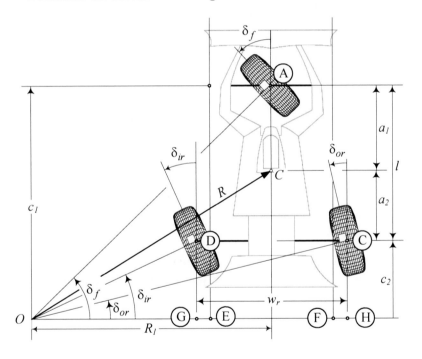

FIGURE 7.79. A three-wheel vehicle with all wheels steerable.

(a) Determine the kinematic steering condition between the three wheels.

(b) Determine R, c_1, c_2, and R_1.

(c) Determine the required condition on the kinematic steering of FWS vehicles to be able to simplify them to the three-wheel steering condition.

9. Optimization of trapezoidal steering mechanism.

   Use the ideal kinematic steering condition

   $$\cot \delta_{A_o} - \cot \delta_i = \frac{w}{l}$$

   and the equation of the trapezoidal steering mechanism

   $$\begin{aligned}
   (w - 2d \sin \beta)^2 &= (w - d \sin (\beta - \delta_{D_o}) - d \sin (\beta + \delta_i))^2 \\
   &\quad + (d \cos (\beta - \delta_{D_o}) - d \cos (\beta + \delta_i))^2
   \end{aligned}$$

   to define the function $e(\beta) = \delta_{D_o} - \delta_{A_o}$. For a set of given w, l, d, δ_i determine the minimum of e using derivative method. Is there any solution? Is the answer (if any) unique?

10. A vehicle with a one-axle trailer.

    Determine the steady state angle between the trailer and vehicle with the following dimensions.

    $$\begin{aligned}
    a_1 &= 1000\,\text{mm} & a_2 &= 1300\,\text{mm} \\
    b_1 &= 1200\,\text{mm} & b_2 &= 1800\,\text{mm} \\
    w_t &= 1100\,\text{mm} & w_v &= 1500\,\text{mm} \\
    g &= 800\,\text{mm} & \delta_i &= 12\,\text{deg}
    \end{aligned}$$

    What is the rotation radius of the trailer R_t, and the vehicle R?

    Determine minimum radius $R_{\min}$, maximum R_{Max}, and difference radius ΔR?

11. Best d for trapezoidal steering mechanism.

    A trapezoidal steering mechanism is shown in Figure 7.80. Assume

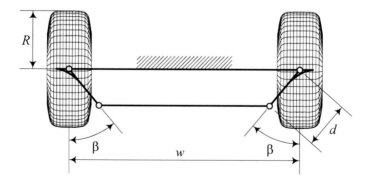

FIGURE 7.80. A trapezoidal steering mechanism.

$$w = 1.66 \, \text{m} \qquad l = 2.93 \, \text{m} \qquad \beta \approx 19.5 \, \text{deg}$$

and determine the best d by minimization of the RMS error function e

$$e = \sqrt{\frac{1}{n} \sum_{i=1}^{n} (\delta_{D_o} - \delta_{A_o})^2}$$

where n is the number of subdivision of δ_i in the working range $-30 \le \delta_i \le 30$.

12. ★ Coordinates of the turning center.

Determine the coordinates of the turning center for the vehicle in Exercise 16 if $\delta_{fl} = 5 \, \text{deg}$ and $c_1 = 1300 \, \text{mm}$.

13. ★ Steer angle for given curvature center.

Consider a car with

$$l = 2.8 \, \text{m} \qquad w_f = 1.35 \, \text{m} \qquad w_r = 1.4 \, \text{m} \qquad a_1 = a_2$$

and the coordrinates of the curvature center in body coordinate frame

$$x_O = 60.336 \qquad y_O = 118.3$$

Determine the required steer angles of a $4WS$ vehicle.

14. ★ Motion of the turning center.

Consider a stationary $4WS$ vehicle as is shown in Figure 7.81.

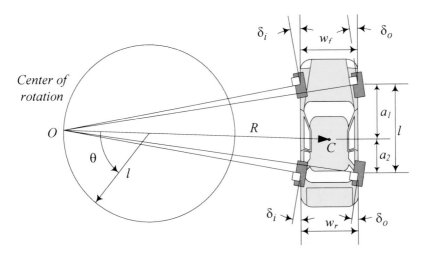

FIGURE 7.81. A stationary $4WS$ vehicle with a moving turning center.

Assume the inner front and rear steer angles are equal at the moment and

$$R = 3l$$

(a) Determine the turning center of rotation in the vehicle body coordinate frame.

(b) Determine the steer angles of the wheels at the moment.

(c) The kinematic turning center of the vehicle is moving on a circle with radius equal to the wheelbase of the car, l. Determine the steer angles as functions of θ.

15. Local curvature center.

Calculate the coordinates of the curvature center in the body coordinate frame of a car that is moving tangent to the road in Example 304.

16. ★ Turning radius of a $4WS$ vehicle.

Consider a FWS vehicle with the following dimensions.

$$l = 2300\,\text{mm} \qquad \frac{a_1}{a_2} = \frac{38}{62}$$
$$w_f = 1457\,\text{mm} \qquad w_r = 1607\,\text{mm}$$

Determine the turning radius of the vehicle for $\delta_{fl} = 5$ deg.

What should be the steer angles of the rear wheels to decrease 10% of the turning radius, if we make the vehicle $4WS$?

17. ★ Different front and rear tracks.

Lotus 2-Eleven^{TM} is a RWD sportscar with the following specifications.

$$l = 2300\,\text{mm} \qquad w_f = 1457\,\text{mm} \qquad w_r = 1607\,\text{mm}$$
$$\text{Front tire} = 195/50R16 \qquad \text{Rear tire} = 225/45R17$$
$$\frac{F_{z_1}}{F_{z_2}} = \frac{38}{62}$$

Determine the angular velocity ratio of ω_o/ω_i, R, δ_i, and δ_o for $\delta = 5$ deg.

18. ★ Coordinates of turning center.

Determine the coordinates of turning center of a $4WS$ vehicle in terms of outer steer angles δ_{of} and δ_{or}.

19. ★ Turning radius.

Determine the turning radius of a $4WS$ vehicle in terms of δ_r.

$$\cot \delta_r = \frac{1}{2} \left(\cot \delta_{ir} + \cot \delta_{or} \right)$$

20. Curvature radius.

Consider a road of motion as

$$Y = \sin \frac{2\pi}{100} X$$

(a) Determine curvature of the road as a function of X.

(b) Determine radius curvature of the road as a function of X.

(c) Determine the path of the curvature center.

21. ★ A three-axle car.

Consider a three-axle off-road pick-up car. Assume

$$a_1 = 1100\,\text{mm} \qquad a_2 = 1240\,\text{mm} \qquad a_3 = 1500\,\text{mm}$$
$$w = 1457\,\text{mm}$$

and determine δ_o, R_1, R_f, and R if $\delta_i = 10\,\text{deg}$.

22. ★ A $3D$ road.

Consider a spatial road with the parametric equations

$$x = (a + b \sin \theta) \cos \theta$$
$$y = (a + b \sin \theta) \sin \theta$$
$$z = b + b \cos \theta$$

(a) Show that the curvature of the road as a function of the parameter θ is

$$\kappa = \frac{2}{\sqrt{4a^2 + 6b^2 - 2b^2 \cos 2\theta + 8ab \sin \theta}}$$

(b) Determine the path of the road curvature center for $a = 200$, $b = 150$.

23. ★ Steering mechanism optimization.

Find the optimum length x for the multi-link steering mechanism shown in Figure 7.82 to operate as close as possible to the kinematic steering condition. The vehicle has a track $w = 2.64\,\text{m}$ and a wheelbase $l = 3.84\,\text{m}$, and must be able to turn the front wheels within a working range equal to $-22\,\text{deg} \le \delta \le 22\,\text{deg}$. Use $a_1 = 380\,\text{mm}$.

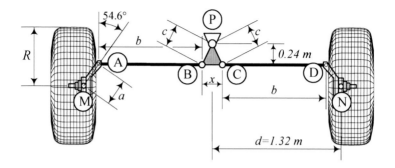

FIGURE 7.82. A multi-link steering mechanism tha must be optimized by varying x.

24. A car on an ellipse.

Consider a $4WS$ vehicle that is moving on an elliptic path.

$$X = a\cos\theta \qquad Y = b\sin\theta$$

(a) Determine the global coordinates of the curvature center.

(b) Determine the coordinates of the curvature center in the vehicle body coordinate.

(c) Determine the steer angle of the wheels if the car is always moving tangent to the road.

25. ★ Road connection.

Design a clothoid connection road to connect the X-axis to the shown line in Figure 7.83. How many design are possible?

26. ★ Curvature of a road.

The curvature of a planar road can be calculated by

$$\kappa = \left| \frac{d^2y/dx^2}{\left(1 + (dy/dx)^2\right)^{3/2}} \right| = \frac{x'y'' - y'x''}{(x'^2 + y'^2)^{3/2}}$$

(a) Consider a road with equation

$$X\left(t\right) = \int_0^t \cos\left(\frac{u^2}{2}\right) du \qquad Y\left(t\right) = \int_0^t \sin\left(\frac{u^2}{2}\right) du$$

and verify that the curvature of the road is

$$\kappa = t$$

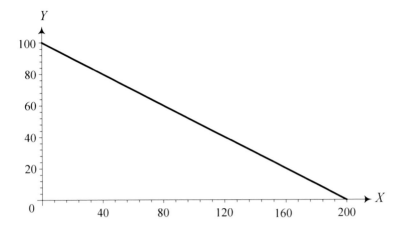

FIGURE 7.83. A road connecting (0.100) to $(200, 0)$.

(b) Consider a road with equation

$$X(t) = \int_0^t \cos\left(\frac{u^3}{3}\right) du \qquad Y(t) = \int_0^t \sin\left(\frac{u^3}{3}\right) du$$

and verify that the curvature of the road is

$$\kappa = t^2$$

27. ★ A loop road.

Figure 7.84 illustrates two straight roads tangentially connected to two circular paths.

(a) Design proper clothoid transitions to connect the straight roads to the circular paths.

(b) Determine and plot the loci of the curvature center of the designed loop.

28. Steering mechanism design.

Figure 7.85 illustrates a steering mechanism. Suppose the middle box can control the length of x and y.

(a) Determine x and y as functions of l, w, d, β such that the mechanism follows the Ackerman kinematic condition.

(b) Plot x, y, $x - y$, and x/y for a symmetric range of inner steering angle $\delta_i = \pm 35\,\text{deg}$ if

$$w = 1.66\,\text{m} \qquad l = 2.93\,\text{m} \qquad d = 0.4 \qquad \beta = 19.5\,\text{deg}$$

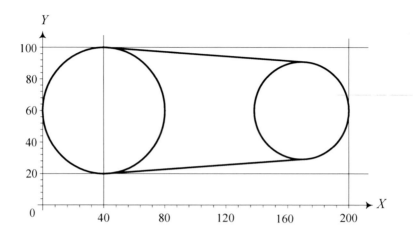

FIGURE 7.84. A loop road by two straight paths tangentially connected to two circular paths.

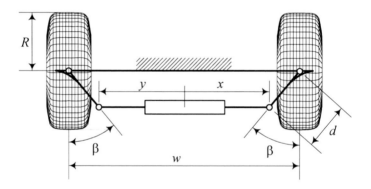

FIGURE 7.85. A steering mechanism with controlled length of x and y.

29. ★ Road design.

    Assume we are asked to find a clothoid transition road to connect a straight road to a circle of radius $R = 58.824\,\text{m}$. Assume the clothoid begins from $(0,0)$ and determine the center of the circular path and the steer angles of a $4WS$ car as a function of s for

    (a) $a = 250\,\text{m}$

    (b) $a = 210\,\text{m}$

    (c) $a = 180\,\text{m}$

    (d) $a = 150\,\text{m}$

    (e) $a = 120\,\text{m}$

30. Clothoid.

Show that

$$\frac{dX}{dt} = a\frac{dX}{ds} \qquad \frac{dY}{dt} = a\frac{dY}{ds}$$

31. Figure 8 road.

The problem in example 317 is that the curvature of the road at the ends of clothoids are not equal to the curvature of the connecting circular paths. Find the curvatures at the connection points.

8

Suspension Mechanisms

The suspension is what links the wheels to the vehicle chassis and allows relative motion. This chapter covers the suspension mechanisms, and discusses the possible relative motions between the wheel and the vehicle chassis. The wheels, through the suspension linkage, must propel, steer, and stop the vehicle, and support the associated forces.

8.1 Solid Axle Suspension

The simplest way to attach a pair of wheels to a chassis is to mount them at two ends of a *solid axle*, such as the one that is shown in Figure 8.1.

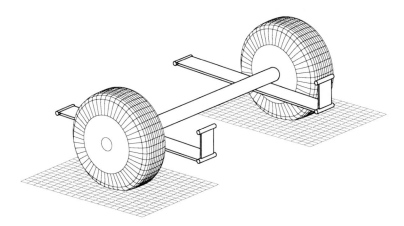

FIGURE 8.1. A solid axle with leaf spring suspension.

The solid axle must be attached to the body such that an up and down motion in the z-direction, as well as a roll rotation about the x-axis, are possible. Therefore,, no forward and lateral translation, and also no rotation about the axle and the z-axis, are allowed. There are many combinations of links and springs that can provide the kinematic and dynamic requirements. The simplest design is to clamp the axle to the middle of two leaf springs with their ends tied and shackled to the chassis as shown schematically in Figure 8.1. A side view of a multi-leaf spring and solid axle is shown in Figure 8.2. A suspension with a solid axle between the left and right wheels is called *dependent suspension*.

R.N. Jazar, *Vehicle Dynamics: Theory and Application*,
DOI 10.1007/978-1-4614-8544-5_8, © Springer Science+Business Media New York 2014

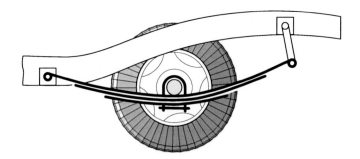

FIGURE 8.2. A side view of a multi-leaf spring and solid axle suspension.

The performance of a solid axle with leaf springs suspension can be improved by adding a linkage to guide the axle kinematically and provide dynamic support to carry the non z-direction forces.

The solid axle with leaf spring combination came to vehicle industry from horse-drawn vehicles.

Example 328 *Hotchkiss drive.*

*When a live solid axle is connected to the chassis with nothing but two leaf springs, it is called the **Hotchkiss drive**, which is the name of the car that used it first. Figure 8.2 illustrates a Hotchkiss suspension. The main problems of Hotchkiss drive are locating the axle under lateral and longitudinal forces, and having a low mass ratio $\varepsilon = m_s/m_u$, where m_s is the sprung mass and m_u is the unsprung mass.*

Sprung mass refers to all masses that are supported by the spring, such as vehicle body. Unsprung mass refers to all masses that are attached to and not supported by the spring, such as wheel, axle, or brakes.

Example 329 *Leaf spring suspension and flexibility problem.*

The solid axle suspension systems with longitudinal leaf springs have many drawbacks. The main problem lies in the fact that springs themselves should act as locating members. Springs are supposed to flex under load, but their flexibility is needed in only one direction. However, it is the nature of leaf springs to twist and bend laterally and hence, flex also and twist in planes other than the tireplane. Leaf springs are not suited for taking up the driving and braking traction forces. These forces tend to push the springs into an S-shaped profile, as shown in Figure 8.3. The driving and braking flexibility of leaf springs may generate a negative caster and increase instability.

Long springs provide better ride. However, long sprigs exaggerate their bending and twisting under different load conditions.

Example 330 *Leaf spring suspension and flexibility solution.*

To reduce the effect of a horizontal force and S-shaped profile appearance in a solid axle with leaf springs, the axle may be attached to the chassis by a

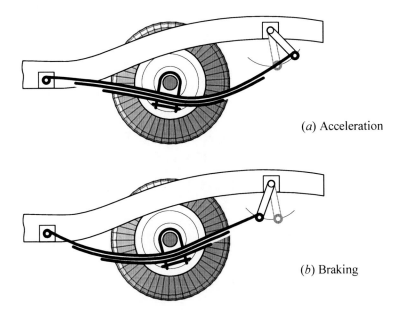

(a) Acceleration

(b) Braking

FIGURE 8.3. A driving and braking trust, force leaf springs into an S shaped profile.

longitudinal bar as Figure 8.4(a) shows. Such a bar is called an **anti-tramp** *bar, and the suspension is the simplest cure for longitudinal problems of a Hotchkiss drive.*

A solid axle with an anti-tramp bar may kinematically be approximated by a four-bar linkage, as shown in Figure 8.4(b). Although an anti-tramp bar may control the shape of the leaf spring, it introduces a twisting angle problem when the axle is moving up and down, as shown in Figure 8.5. Twisting the axle and the wheel about the axle is called caster.

The solid axle is frequently used to help keeping the wheels perpendicular to the road.

Example 331 *Leaf spring location problem.*

The front wheels need room to steer left and right. Therefore, leaf springs cannot be attached close to the wheel hubs, and must be placed closer to the middle of the axle. That gives a narrow spring-base, which means that a small side force can sway or tilt the body relative to the axle through a considerable roll angle. This is uncomfortable for the vehicle passengers, and may also produce unwanted steering.

The solid axle positively prevents the camber change by body roll. The wheels remain almost upright and hence, do not roll on a side. However, a solid axle shifts laterally from its static plane and its center does not remain on the vehicle's longitudinal axis under a lateral force.

A solid axle produces bump-camber when single-wheel bump occurs. If

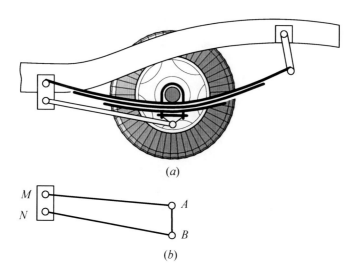

(a)

(b)

FIGURE 8.4. (a) Adding an anti-tramp bar to guide a solid axle. (b) Equivalent kinematic model.

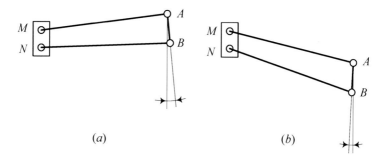

(a)   (b)

FIGURE 8.5. An anti-tramp bar introduces a twistng angle problem. (a) The wheel moves up and (b) The wheel moves down.

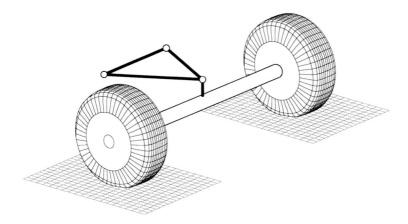

FIGURE 8.6. A solid axle suspension with a triangulated linkage.

the right wheel goes over a bump, the axle is raised at its right end, and that tilts the left wheel hub, putting the left wheel at a camber angle for the duration of deflection.

Example 332 *Triangular linkage.*

A triangulated linkage, as shown in Figure 8.6, may be attached to a solid axle to provide lateral resistance in turning maneuvers, and twist resistance during acceleration and braking.

Example 333 *Panhard arm and lateral displacement.*

High spring rate is a problem of leaf springs. Reducing their stiffness by narrowing them and using fewer leaves, reduces the lateral stiffness and increases the directional stability of the suspension significantly. A Panhard arm is a bar that attaches a solid axle suspension to the chassis laterally. Figure 8.7 illustrates a solid axle and a Panhard arm to limit the lateral displacement of the axle. Figure 8.8 shows a triangular linkage and a Panhard arm combination for guiding a solid axle.

A double triangle mechanism, as shown in Figure 8.9, is an alternative design to guide the axle and support it laterally.

Example 334 *Straight line linkages.*

There are many mechanisms that provide a straight line motion to be used in suspension design. The simplest mechanisms are four-bar linkages with a coupler point moving straight. Some of the most applied and famous linkages are shown in Figure 8.10. By having proper lengths, the Watt, Robert, Chebyshev, and Evance linkages can make the coupler point C move on a straight line vertically. Such a mechanism and straight motion can be used to guide a solid axle.

Two Watt suspension mechanisms with a Panhard arm are shown in Figures 8.11 and 8.12.

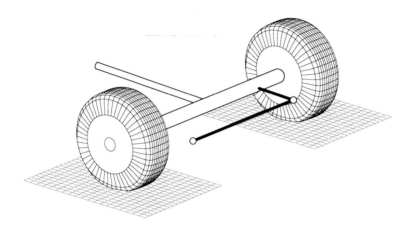

FIGURE 8.7. A solid axle and a Panhard arm to guide the axle.

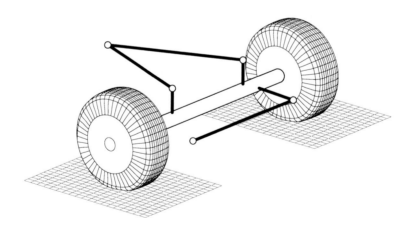

FIGURE 8.8. A triangle mechanism and a Panhard arm to guide a solid axle.

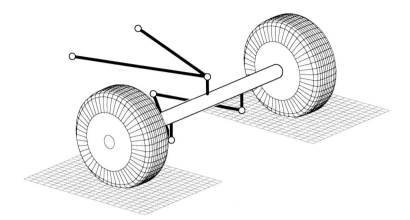

FIGURE 8.9. Double triangle suspension mechanism.

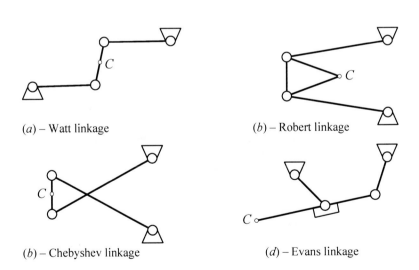

(a) – Watt linkage

(b) – Robert linkage

(b) – Chebyshev linkage

(d) – Evans linkage

FIGURE 8.10. Some linkages with straight line motion.

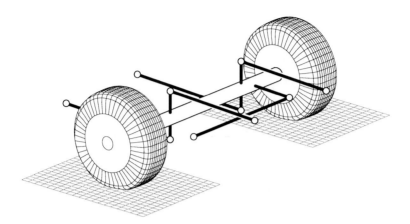

FIGURE 8.11. A Watt suspension mechanisms with a Panhard arm.

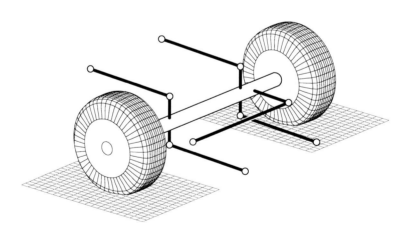

FIGURE 8.12. A Watt suspension mechanisms with a Panhard arm.

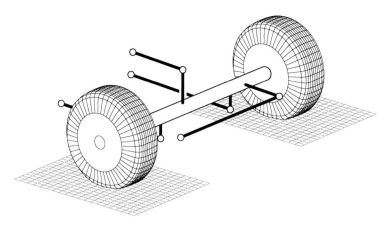

FIGURE 8.13. A Robert suspension mechanism with a Panhard arm.

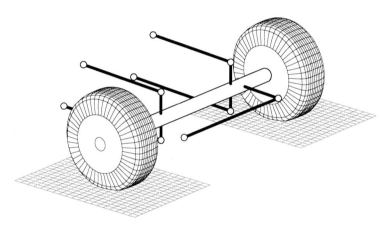

FIGURE 8.14. A Robert suspension mechanism with a Panhard arm.

Figures 8.13, 8.14, and 8.15 illustrate three combinations of Robert suspension linkages equipped with a Panhard arm.

Example 335 *Solid axle suspension and unsprung mass problem.*

Solid axles are usually heavier than the same size independent suspensions. A solid axle is a part of unsprung members, and hence, the unsprung mass is increased where using solid axle suspension. A heavy unsprung mass ruins both, the ride and handling quality of a vehicle. Lightening the solid axle makes it weaker and increases the most dangerous problem in vehicles: axle breakage. The solid axle must be strong enough to make sure it will not break under any loading conditions at any age. As a rough estimate, 90% of the leaf spring mass may also be counted as unsprung mass, which makes the problem worse.

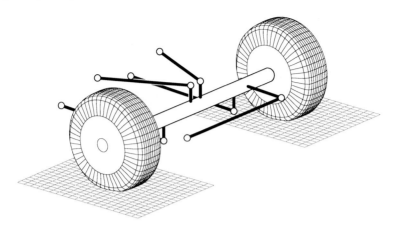

FIGURE 8.15. A Robert suspension mechanism with a Panhard arm.

The unsprung mass problem is worse in front, and it is the main reason that they are no longer being used in street cars. However, front solid axles are still common on trucks and buses. These are heavy vehicles and solid axle suspension does not reduce the mass ratio $\varepsilon = m_s/m_u$ very much.

*When a vehicle is rear-wheel-drive and a solid axle suspension is used in the back, the suspension is called **live axle**. A live axle is a casing that contains a differential, and two drive shafts. The drive shafts are connected to the wheel hubs. A live axle can be three to four times heavier than a dead I-beam axle. It is called live axle because of rotating gears and shafts inside the axle.*

Example 336 *Solid axle and coil spring.*

To decrease the unsprung mass and increase vertical flexibility of solid axle suspensions, it is possible to equip them with coil springs instead of leaf springs. A sample of a solid axle suspension with coil spring is shown in Figure 8.16. The suspension mechanism is made of four longitudinal bars between the axle and chassis. The springs may have some lateral or longitudinal angle to introduce some lateral or longitudinal compliance as well.

Example 337 *De Dion axle.*

When a solid axle is a dead axle with no driving wheels, the connecting beam between the left and right wheels may have different shapes to do different jobs, usually to give the wheels independent flexibility. We may also modify the shape of a live axle to attach the differential to the chassis and reduce the unsprung mass.

De Dion design is a modification of a beam axle that may be used as a dead axle or to attach the differential to the chassis and transfer the driving power to the drive wheels by employing universal joints and split shafts. Figure 8.17 illustrates a De Dion suspension.

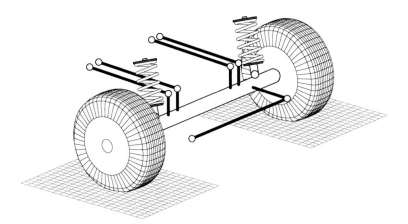

FIGURE 8.16. A solid axle suspension with coil springs.

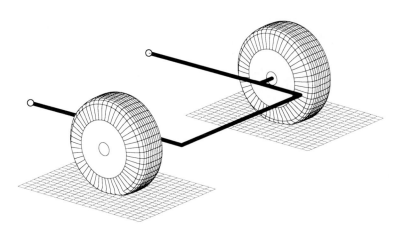

FIGURE 8.17. A De Dion suspension.

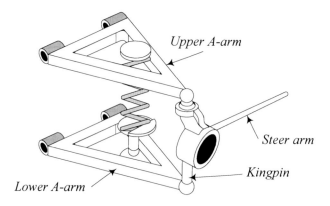

FIGURE 8.18. A double *A*-arm suspension.

8.2 Independent Suspension

Independent suspensions are introduced to allow a wheel to move up and down without affecting the opposite wheel. There are many forms and designs of independent suspensions. However, *double A-arm* and *McPherson strut* suspensions are the simplest and the most common designs. Figure 8.18 illustrates a sample of a double *A*-arm and Figure 8.19 shows a McPherson suspension.

Kinematically, a double *A*-arm suspension mechanism is a four-bar linkage with the chassis as the ground link, and coupler as the wheel carrying link. A McPherson suspension is an inverted slider mechanism that has the chassis as the ground link and the coupler as the wheel carrying link. A double *A*-arm and a McPherson suspension mechanism are schematically shown in Figures 8.20 and 8.21 respectively.

Double *A*-arm, is also called *double wishbone*, or *short/long arm* suspension. McPherson also may be written as MacPherson.

Example 338 *Double A-arm suspension and spring position.*

Consider a double A-arm suspension mechanism. The coil spring may be between the lower arm and the chassis, as shown in Figure 8.18. It is also possible to install the spring between the upper arm and the chassis, or between the upper and lower arms. In either case, the lower or the upper arm, which supports the spring, is made stronger and the other arm acts as a connecting arm.

Example 339 *Multi-link suspension mechanism.*

When the two side bars of an A-arm are attached to each other with a joint, as shown in Figure 8.22, then the double A-arm is called a multi-link mechanism. Such a multi-link suspension is a six-bar mechanism that may have a better coupler motion than a double A-arm mechanism. However, multi-link suspensions are more expensive, less reliable, and more compli-

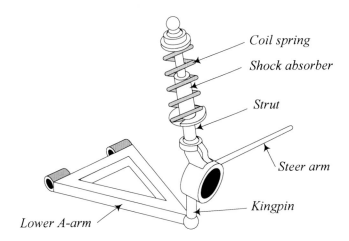

FIGURE 8.19. A McPherson suspension.

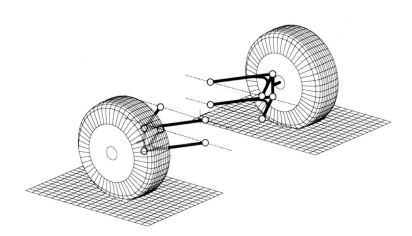

FIGURE 8.20. A double A-arm suspension mechanism on the left and right wheels.

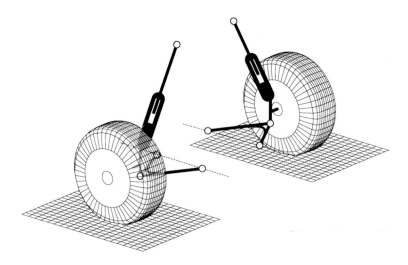

FIGURE 8.21. A McPherson suspension mechanism on the left and right wheels.

cated compare to a double A-arm four-bar linkage. There are vehicles with more than six-link suspension with possibly better kinematic performance.

Example 340 *Swing arm suspension.*

An independent suspension may be as simple as a triangle shown in Figure 8.23. The base of the triangle is jointed to the chassis and the wheel to the tip point. The base of the triangle is aligned with the longitudinal axis of the vehicle. Such a suspension mechanism is called a **swing axle** or **swing arm**.

Compared to the other suspension mechanisms, a swing arm suspension has the maximum camber angle variation.

Example 341 *Trailing arm suspension.*

Figure 8.24 illustrates a trailing arm suspension that is a longitudinal arm with a lateral axis of rotation. The camber angle of the wheel, supported by a trailing arm, will not change during the up and down motion.

Trailing arm suspension has been successfully using in a variety of front-wheel-drive vehicles, to suspend their rear wheels.

Example 342 *Semi-trailing arm*

Semi-trailing arm suspension, as shown in Figure 8.25, is a compromise between the swing arm and trailing arm suspensions. The joint axis may have any angle, however an angle not too far from 45 deg is more applied. Such suspensions have acceptable camber angle change, while they can handle both, the lateral and longitudinal forces. Semi-trailing design has successfully applied to a series of rear-wheel-drive cars for several decades.

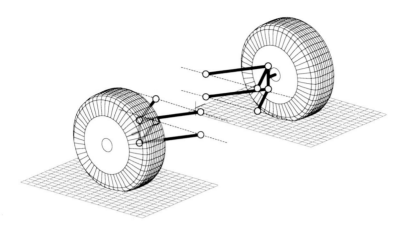

FIGURE 8.22. A multi-link suspension mechanism.

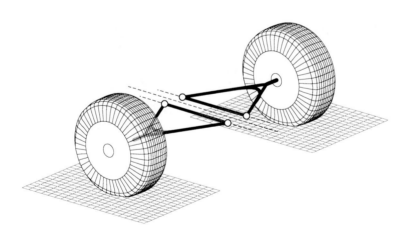

FIGURE 8.23. A swing arm suspension.

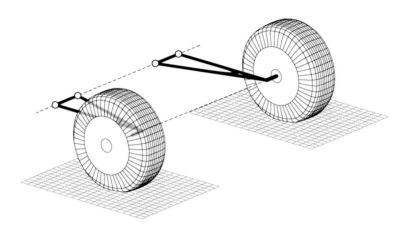

FIGURE 8.24. A trailing arm suspension.

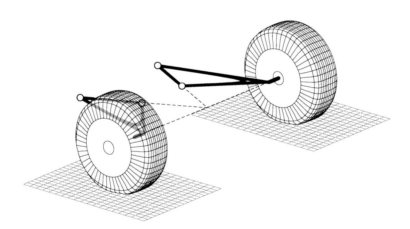

FIGURE 8.25. A semi-trailing arm suspension.

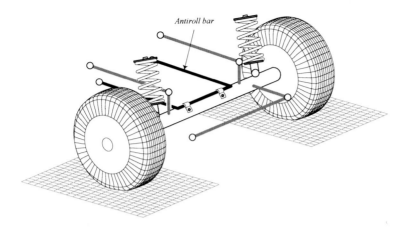

FIGURE 8.26. An antiroll bar attached to a solid axle with coil springs.

Example 343 *Antiroll bar and roll stiffness.*

Coil springs are used in vehicles because they are less stiff with better ride comfort compared to leaf springs. The roll stiffness of the vehicle with coil springs is usually less than vehicles with leaf springs. To increase the roll stiffness of such suspensions, an **antiroll bar** should be used. Leaf springs with reduced layers, uni-leaf, trapezoidal, or nonuniform thickness may also need an antiroll bar to compensate their reduced roll stiffness. The antiroll bar is also called a **stabilizer**. Figure 8.26 illustrates an antiroll bar attached to a solid axle with coil springs.

Example 344 *Need for longitudinal compliance.*

When a vehicle goes over a bump, the first action is a displacement that tends to push the wheel backward relative to the rest of the vehicle. So, the lifting force has a longitudinal component, which will be felt inside the vehicle even if the suspension system has some horizontal compliance.

There are situations in which the horizontal component of the force is even higher than the vertical component. Leaf springs can somewhat absorb this horizontal force by flattening out and stretching the distance from the forward spring anchor and the axle. Such a stretch is usually less than $1/2\,\mathrm{in} \approx 1\,\mathrm{cm}$.

8.3 Roll Center and Roll Axis

The *roll axis* is the instantaneous line about which the body of a vehicle rolls. Roll axis is found by connecting the *roll center* of the front and rear suspensions of the vehicle. Assume we cut a vehicle laterally to disconnect the front and rear halves of the vehicle. Then, the roll center of the front or rear suspension is the instantaneous center of rotation of the body with respect to the ground.

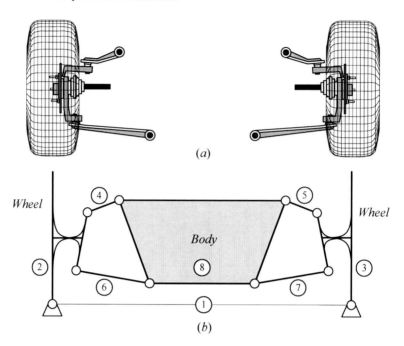

FIGURE 8.27. (*a*) An example of a double A-arm front suspensions. (*b*) Kinematically equivalent mechanism for the front half of the double A-arm suspension.

Figure 8.27(*a*) illustrates a sample of suspension of a car with a double A-arm mechanism. To find the roll center of the body with respect to the ground, we analyze the two-dimensional kinematically equivalent mechanism shown in Figure 8.27(*b*). The center of tireprint is the instant center of rotation of the wheel with respect to the ground, so the wheels are jointed links to the ground at their center of tireprints. In this figure, the relatively moving bodies are numbered staring from 1 for the ground and ending with number 8 for the chassis.

The instant center I_{18} is the roll center of the body with respect to the ground. To find I_{18}, we apply the Kennedy theorem and find the intersection of the line $\overline{I_{12}I_{28}}$ and $\overline{I_{13}I_{38}}$ as shown in Figure 8.28.

The point I_{28} and I_{38} are the instant center of rotation for the wheels with respect to the body. The instant center of rotation of a wheel with respect to the body is called *suspension roll center* or simply *suspension center*. So, to find the roll center of the front or rear half of a car, we should determine the suspension roll centers, and find the intersection of the lines connecting the suspension roll centers to the center of their associated tireprints.

The *Kennedy theorem* states that the instant center of every three relatively moving objects are colinear.

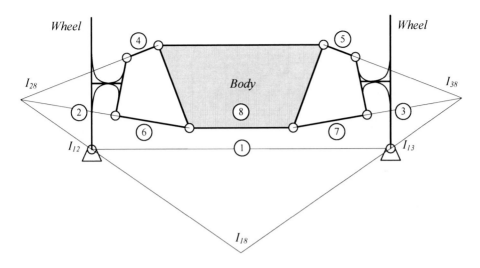

FIGURE 8.28. The roll center I_{18} is at the intersection of lines $\overline{I_{12}I_{28}}$ and $\overline{I_{13}I_{38}}$.

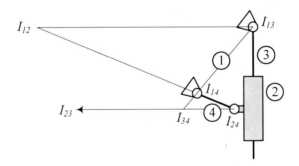

FIGURE 8.29. Instant center or rotation for an example of inverted slider crank mechanism.

Example 345 *McPherson suspension roll center.*

A McPherson suspension is an inverted slider crank mechanism. The instant centers of an example of inverted slider crank mechanism are shown in Figure 8.29. The point I_{12} is the suspension roll center, which is the instant center of rotation for the wheel link number 2 with respect to the chassis link number 1.

A car with a McPherson suspension is shown in Figure 8.30(a). The kinematic equivalent mechanism is depicted in Figure 8.30(b). Suspension roll centers along with the body roll center are shown in Figure 8.31. To find the roll center of the front or rear half of a car, we determine each suspension roll center and then find the intersection of the lines connecting the suspension roll centers to the center of the associated tireprint.

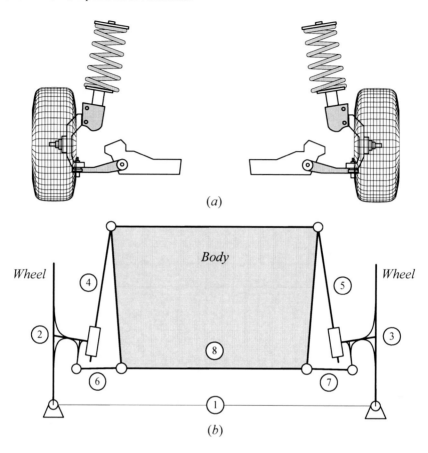

FIGURE 8.30. (a) McPherson suspension system. (b) Kinematics model of the McPherson suspension.

Example 346 *Roll center of double A-arm suspension.*

The roll center of an independent suspension such as a double A-arm can be internal or external. The kinematic model of a double A-arm suspension for the front left wheel of a car is illustrated in Figure 8.32. The suspension roll center in Figure 8.32(a) is internal, and in Figure 8.32(b) is external. An internal suspension roll center is toward the vehicle body, while an external suspension roll center goes away from the vehicle body.

A suspension roll center may be on, above, or below the road surface, as shown in Figure 8.33(a)-(c) for an external suspension roll center. When the suspension roll center is on the ground, above the ground, or below the ground, the vehicle roll center would be on the ground, below the ground, and above the ground, respectively.

Example 347 *Roll axis, roll height, roll torque.*

Both the front and rear suspensions of a car have roll centers. They are

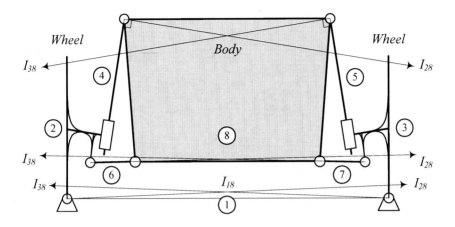

FIGURE 8.31. Roll center of a car with a McPherson suspension system.

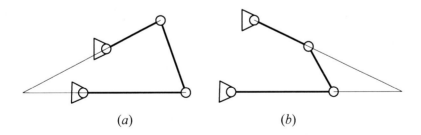

FIGURE 8.32. The kinematic model of the suspension of a front left wheel: (a) an internal roll center, and (b) an external roll center.

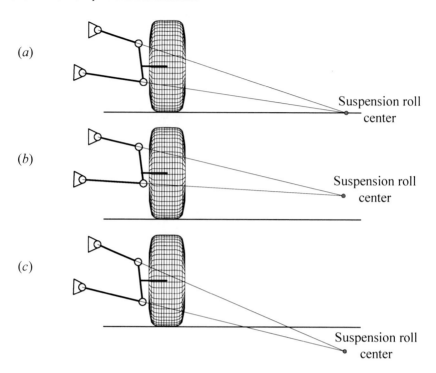

FIGURE 8.33. A suspension roll center at (a) on, (b) above, and (c) below the road surface.

*in general at different heights. The line connecting the front and rear roll centers is the **roll axis** of the car about which the car will roll. The vertical distance of the mass center and the roll axis is called the **roll height,** h_r. Figure 8.34 illustrates the roll axis and roll height of a car. When the car is turning in a curved path, the centripetal force $f_y = mv^2/R$ is the effective lateral force at the mass center that generates a **roll torque** M_x about the roll center.*

$$M_x = \frac{mv^2}{R} h_r \tag{8.1}$$

It is common that the mass center locate above the roll axis. As a result the car will roll outward of the turning path. However, we can technically design the suspensions of a car such that the roll axis pass above the mass center. Such a car will roll inward in a turning path and acts similar to a boat. If the roll axis passes through the mass center, the car will not roll on a curved path.

The roll torque, and hence, the roll angle of a car increases by the roll height h_r, and square of the speed v^2. Therefore, doubling the speed needs to have four times shorter roll height for the same roll angle.

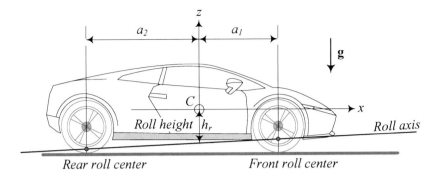

FIGURE 8.34. The roll axis is the line connecting the front and rear suspension roll centers.

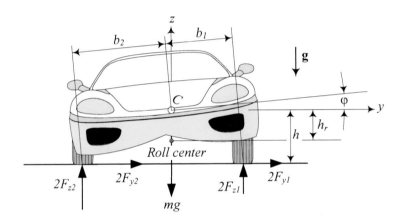

FIGURE 8.35. A car in a turning path about the z-axis.

Example 348 *Vertical tireprint forces.*

*The roll torque makes the car body roll until it is balanced by the moment of the forces of the deflected springs. The resultant moment of the spring forces for one radian or one degree roll of the body is called the **roll stiffness** k_φ.*

$$M_x = k_\varphi \varphi \qquad (8.2)$$

Any roll angle causes a proportional deflection in the main springs of the car and as a result decreases the vertical force of the tireprint of the inner tires and increases the vertical force of the outer tires. Consider a car in a turning path about the z-axis as is shown in Figure 8.35.

Assuming similar springs with stiffness k and a horizontal roll axis, the roll angle φ provides us with

$$M_r = 2k\left(b_1 + b_2\right)\varphi \qquad (8.3)$$

indicating that the roll stiffness of the car is:

$$k_\varphi = 2k\,(b_1 + b_2) \tag{8.4}$$

Combining with Equation (8.1), yields:

$$\varphi = \frac{mv^2}{R}\frac{h_r}{k_\varphi} = \frac{mv^2}{R}\frac{h_r}{2k\,(b_1 + b_2)} \tag{8.5}$$

To decrease the roll angle, we need to increase the roll stiffness k_φ and decrease the roll height h_r. As the spring rates are generally designed based on vertical ride comfort, adding anti-roll bar is a common approach to increase k_φ if needed. To design the roll stiffness of a car, the roll natural frequency of the car must also be considered.

If the spring rate of the front and rear suspensions are different then, the roll stiffness of the car would be

$$k_\varphi = k_f\,(b_1 + b_2) + k_r\,(b_1 + b_2) \tag{8.6}$$

where, k_r is the stiffness of the front suspension and k_r is the stiffness of the rear suspension.

Assume that the wheel load on the left and right sides of the car are equal at the rest position,

$$F_{z_1} = \frac{1}{2}mg\frac{b_2}{w} \qquad F_{z_2} = \frac{1}{2}mg\frac{b_1}{w} \tag{8.7}$$

$$w = b_1 + b_2 \tag{8.8}$$

we may calculate the wheel loads for a rolled car as

$$F_{z_1} = \frac{1}{2}mg\frac{b_2}{w} - kb_1\varphi \qquad F_{z_2} = \frac{1}{2}mg\frac{b_1}{w} + kb_2\varphi \tag{8.9}$$

The lateral force of a tire is proportional to its vertical load. Therefore, there is a critical roll angle φ_c at which the vertical force and hence, the lateral force of the tire disappear. This is the first step of instability and roll over of a car in a turn. As a result, the roll angle must be limited to have all tires on the road.

$$-\frac{1}{2}\frac{mg}{kw}\frac{b_1}{b_2} \le \varphi \le \frac{1}{2}\frac{mg}{kw}\frac{b_2}{b_1} \tag{8.10}$$

As a reasonable approximation, we may assume $b_1 \approx b_2$ and calculate the critical roll angle.

$$\varphi_c \approx \pm\frac{1}{2}\frac{mg}{kw} \tag{8.11}$$

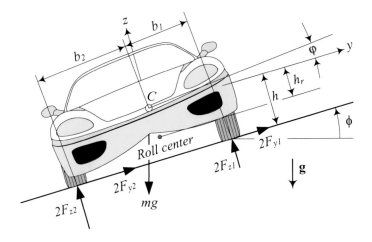

FIGURE 8.36. A car with a roll angle φ on a road with the bank angle ϕ.

Consider a car is turning on a banked road such as shown in Figure 8.36. If the bank angle is ϕ while the roll angle is φ, the vertical force under the each left and right tires are:

$$F_{z_1} = \frac{1}{2}\frac{mg}{w}\left(b_2 \cos\phi - h \sin\phi\right) - kb_1\varphi \tag{8.12}$$

$$F_{z_2} = \frac{1}{2}\frac{mg}{w}\left(b_1 \cos\phi + h \sin\phi\right) + kb_1\varphi \tag{8.13}$$

Example 349 ★ *Camber variation of double A-arm suspension.*

When a wheel moves up and down with respect to the vehicle body, depending on the suspension mechanism, the wheel may camber. Figure 8.37 illustrates the kinematic models a double A-arm suspension mechanism.

The mechanism is equivalent to a four-bar linkage with the ground link as the vehicle chassis. The wheel is attached to a coupler point C of the mechanism. We set a local suspension coordinate frame (x, y) with the x-axis indicating the ground link MN. The x-axis makes a constant angle θ_0 with the vertical direction. The suspension mechanism has a length a for the upper A-arm, b for the coupler link, c for the lower A-arm, and d for the ground link. The configuration of the suspension is determined by the angles θ_2, θ_3, and θ_4, all measured from the positive direction of the x-axis. When the suspension is at its equilibrium position, the links of the double A-arm suspension make initial angles θ_{20} θ_{30}, and θ_{40} with the x-axis. The equilibrium position of a suspension is called the **rest position**.

To determine the camber angle during the fluctuation of the wheel, we should determine the variation of the coupler angle θ_3, as a function of vertical motion z of the coupler point C.

Using θ_2 as a parameter, we can find the coordinates (x_C, y_C) of the

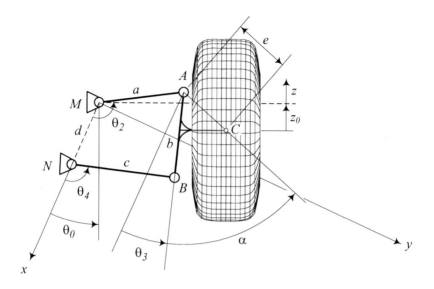

FIGURE 8.37. The kinematic model for a double A-arm suspension mechanism.

coupler point C in the suspension coordinate frame (x, y) as

$$
\begin{aligned}
x_C &= a \cos \theta_2 + e \cos (p - q + \alpha) & (8.14) \\
y_C &= a \sin \theta_2 + e \sin (p - q + \alpha) & (8.15)
\end{aligned}
$$

where,

$$
q = \tan^{-1} \frac{a \sin \theta_2}{d - a \cos \theta_2} \tag{8.16}
$$

$$
p = \tan^{-1} \frac{\sqrt{4b^2 f^2 - (b^2 + f^2 - c^2)^2}}{b^2 + f^2 - c^2} \tag{8.17}
$$

$$
f = \sqrt{a^2 + d^2 - 2ad \cos \theta_2} \tag{8.18}
$$

The position vector of the coupler point is $\mathbf{u}_C$

$$
\mathbf{u}_C = x_C \hat{\imath} + y_C \hat{\jmath} \tag{8.19}
$$

and the unit vector in the tire displacement z-direction is

$$
\hat{u}_z = - \cos \theta_0 \hat{\imath} - \sin \theta_0 \hat{\jmath} \tag{8.20}
$$

Therefore, the displacement z in terms of x_C and y_C is:

$$
z = \mathbf{u}_C \cdot \hat{u}_z = -x_C \cos \theta_0 - y_C \sin \theta_0 \tag{8.21}
$$

The initial coordinates of the coupler point C and the initial value of z are:

$$x_{C0} = a \cos \theta_{20} + e \cos (p_0 - q_0 + \alpha) \tag{8.22}$$

$$y_{C0} = a \sin \theta_{20} + e \sin (p_0 - q_0 + \alpha) \tag{8.23}$$

$$z_0 = -x_{C0} \cos \theta_0 - y_{C0} \sin \theta_0 \tag{8.24}$$

and hence, the vertical displacement of the wheel center can be calculated by

$$h = z - z_0 \tag{8.25}$$

The initial angle of the coupler link with the vertical direction is $\theta_0 - \theta_{30}$. Therefore, the camber angle of the wheel would be

$$\gamma = (\theta_0 - \theta_3) - (\theta_0 - \theta_{30}) = \theta_{30} - \theta_3 \tag{8.26}$$

The angle of the coupler link with respect to the x-direction is equal to

$$\theta_3 = 2 \tan^{-1} \left(\frac{-E \pm \sqrt{E^2 - 4DF}}{2D} \right) \tag{8.27}$$

where,

$$D = J_5 - J_1 + (1 + J_4) \cos \theta_2 \tag{8.28}$$

$$E = -2 \sin \theta_2 \tag{8.29}$$

$$F = J_5 + J_1 - (1 - J_4) \cos \theta_2 \tag{8.30}$$

and

$$J_1 = \frac{d}{a} \qquad J_2 = \frac{d}{c} \qquad J_3 = \frac{a^2 - b^2 + c^2 + d^2}{2ac}$$

$$J_4 = \frac{d}{b} \qquad J_5 = \frac{c^2 - d^2 - a^2 - b^2}{2ab} \tag{8.31}$$

Substituting (8.27) and (8.26), and then, eliminating θ_2 between (8.26) and (8.21) provides us with the relationship between the vertical motion of the wheel, z, and the camber angle γ.

Example 350 ★ *Camber angle and wheel fluctuations.*
Consider the double A-arm suspension that is shown in Figure 8.37. The dimensions of the equivalent kinematic model are:

$$a = 22.4\,\text{cm} \qquad b = 22.1\,\text{cm} \qquad c = 27.3\,\text{cm}$$

$$d = 17.4\,\text{cm} \qquad \theta_0 = 24.3\,\text{deg} \tag{8.32}$$

The coupler point C is at:

$$e = 14.8\,\text{cm} \qquad \alpha = 54.8\,\text{deg} \tag{8.33}$$

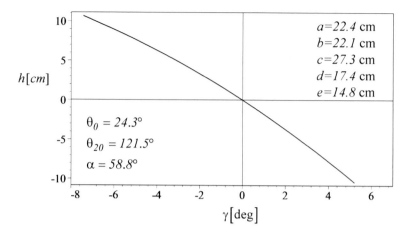

FIGURE 8.38. The vertical displacement of the wheel center h as a function of the camber angle γ.

If the angle θ_2 at the rest position is at

$$\theta_{20} = 121.5 \deg \tag{8.34}$$

then the initial angle of the other links are:

$$\theta_{30} = 18.36 \deg \qquad \theta_{40} = 107.32 \deg \tag{8.35}$$

At the rest position, the coupler point is at:

$$x_{C0} = -7.417 \, \text{cm} \qquad y_{C0} = 33.26 \, \text{cm} \qquad z_0 = -6.929 \, \text{cm} \tag{8.36}$$

We may calculate h and γ by varying the parameter θ_2. Figure 8.38 illustrates h as a function of the camber angle γ. The vertical tire displacement h versus θ_2, θ_3, θ_3 are respectively shown in Figures 8.39, 8.40, 8.41. Figure 8.42 depicts the angle change of the upper arm $\theta_2 - \theta_{20}$ versus the camber angle γ.

For this suspension mechanism, the wheel gains a positive camber when the wheel moves up, and gains a negative camber when the tire moves down. The mechanism is shown in Figure 8.43, when the wheel is at the rest position and has a positive or a negative displacement.

8.4 ★ Car Tire Relative Angles

There are four major wheel alignment parameters that affect the dynamics of a vehicle: *toe*, *camber*, *caster*, and *thrust angle*.

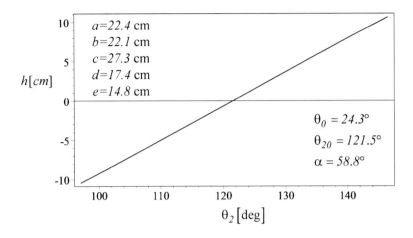

FIGURE 8.39. The vertical displacement of the wheel center h versus θ_2.

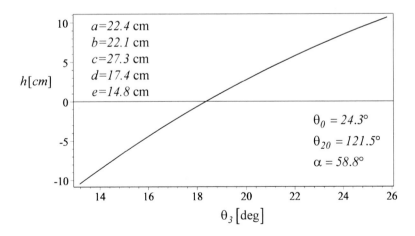

FIGURE 8.40. The vertical displacement of the wheel center h versus θ_3.

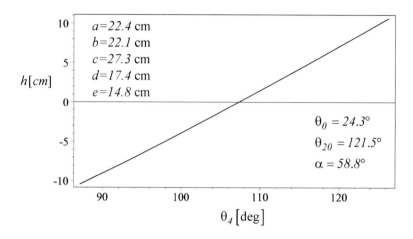

FIGURE 8.41. The vertical displacement of the wheel center h versus θ_4.

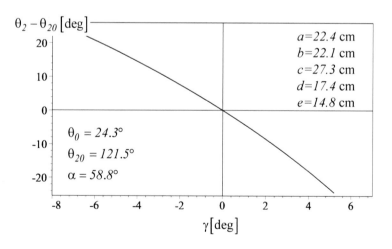

FIGURE 8.42. The angle change of the upper arm $\theta_2 - \theta_{20}$ versus the camber angle γ.

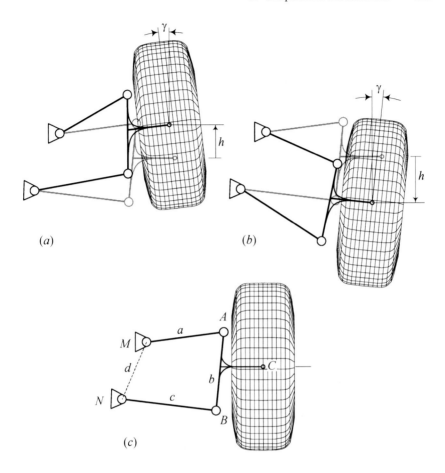

FIGURE 8.43. A double A-arm suspension mechanism when the wheel is at: (a) a positive displacement, (b) a negative displacement, and (c) rest position.

8.4.1 ★ Toe

When a pair of wheels is set such that their leading edges are pointed toward each other, the wheel pair is said to have *toe-in*. If the leading edges point away from each other, the pair is said to have *toe-out*. Toe-in and toe-out of front wheel configurations of a car are illustrated in Figure 8.44.

The amount of toe can be expressed in degrees of the angle to which the wheels are not parallel. However, it is more common to express the toe-in and toe-out as the difference between the track widths as measured at the leading and trailing edges of the tires. Toe settings affect three major performances: tire wear, straight-line stability, and corner entry handling.

For minimum tire wear and power loss, the wheels on a given axle of a car should be neutral and point directly ahead when the car is running in a straight line. Excessive toe-in causes accelerated wear at the outboard

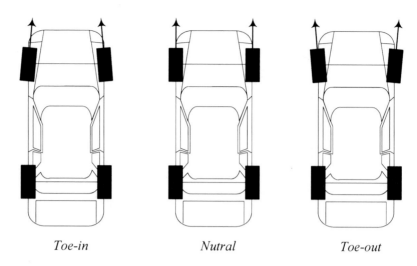

Toe-in   Nutral   Toe-out

FIGURE 8.44. Toe-in, nutral, and toe-out configuration on the front wheels of a car.

edges of the tires, while too much toe-out causes wear at the inboard edges.

Toe-in increases the directional stability of the vehicle, and toe-out increases the steering response. Hence, a toe-in setting makes the steering function lazy, while a toe-out makes the vehicle unstable. As a result, most street cars are set to have a few angles toe-in and race cars are set to have a few angles toe-out in their front wheels.

With four wheel independent suspensions, the toe may also be set at the rear of the car. Toe settings at the rear have the same effect on wear, directional stability, and turn-in response as they do on the front. However, we usually do not set a rear-drive race car toed out in the rear, because of excessive instability.

When driving torque is applied to the wheels, they pull themselves forward and try to create toe-in. On the other hand, a non-driven wheel or a braking wheel will tend to toe-out.

Example 351 *Toe-in and directional stability.*

Toe settings have an impact on directional stability. When the steering wheel is centered, toe-in causes the wheels to tend to move along paths that intersect each other in front of the vehicle. However, if the wheels' angle are equal, the wheels' lateral forces are in balance and no turn results. Toe-in setup can increase the directional stability caused by little steering fluctuations and keep the car moving straight. Steering fluctuations may be a result of road disturbances, wind, bank angle, bump steering and so on. Little steering disturbance, to the left of example, increases the directional angle and the lateral force of the left wheel and decreases the directional angle and the lateral force of the right wheel. Therefore a rightward resultant lateral force will be generated that opposes the disturbance.

If a car is set up with toe-out, the front wheels are aligned so that slight disturbances cause the wheel pair to assume rolling directions that approach a turn. Therefore, toe-out encourages the initiation of a turn, while toe-in discourages it. Toe-out makes the steering quicker. So, it may be used in vehicles for a faster response. The toe setting on a particular car becomes a trade-off between the straight-line stability afforded by toe-in and the quick steering response by toe-out. Toe-out is not desirable for street cars, however, race car drivers are willing to drive a car with a little directional instability, for sharper turn-in to the corners. So street cars are generally set up with toe-in, while race cars are often set up with toe-out.

Example 352 *Toe-in and toe-out in the front and rear axles.*

Front toe-in: slower steering response, more straight-line stability, greater wear at the outboard edges of the tires.

Front toe-zero: medium steering response, minimum power loss, minimum tire wear.

Front toe-out: quicker steering response, less straight-line stability, greater wear at the inboard edges of the tires.

Rear toe-in: straight-line stability, traction out of the corner, more steerability, higher top speed.

8.4.2 ★ *Caster Angle*

Caster is the angle to which the steering axis is tilted forward or rearward from vertical as viewed from the side in the direction of the y-axes. The steering axis is the line about which the wheel will turn when steered. Assume the wheel is straight to have the body frame and the wheel frame coincident. If the steering axis is turned about the wheel y_w-axis then the wheel has *positive caster*. If the steering axis is turned about the wheel $-y_w$-axis, then the wheel has *negative caster*. Positive and negative caster configurations on the front wheel of a car are shown in Figure 8.45.

Negative caster aids in centering the steering wheel after a turn and makes the front tires straighten quicker. Most street cars are made with 4 deg to 6 deg negative caster. Negative caster tends to straighten the wheel when the vehicle is traveling forward, and thus is used to enhance straight-line stability. It is caused by the position of the backward friction force in the tireprint and the steering axis point on the ground, about which the wheel steers. If the tireprint is behind the point, then we have a stable system similar to a hanging pendulum under the gravitational force being lower than the hanging fulcrum.

Example 353 *Negative caster of shopping carts.*

The steering axis of a shopping cart wheel is set to be in front of where the wheel contacts the ground. As the cart is pushed forward, the steering

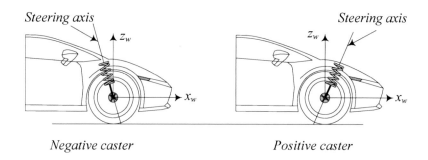

Negative caster   Positive caster

FIGURE 8.45. A positive and negative caster configuration on front wheel of a car.

axis pulls the wheel along. Because the wheel drags along the ground, it falls directly in line behind the steering axis. The force that causes the wheel to follow the steering axis is proportional to the distance between the steering axis and the wheel-to-ground contact point, if the caster is small. This distance is referred to as trail. The cars' steering axis intersects the ground at a point in front of the tireprint, and thus the same effect as seen in the shopping cart casters is achieved.

While greater caster angles improves straight-line stability, they cause an increase in steering effort.

Example 354 *Characteristics of caster in front axle.*

Zero castor provides: easy steering into the corner, low steering out of the corner, low straight-line stability.

Negative caster provides: lazy steering into the corner, easy steering out of the corner, more straight-line directional stability, high tireprint area during turn, good steering feel.

When a negative castered wheel rotates about the steering axis, the wheel gains camber. This camber is generally favorable for cornering.

8.4.3 ★ Camber

Camber is the lateral angle of a wheel about the x-axis relative to the vertical line to the road. Figure 8.46 illustrates the wheel number 1 of a vehicle. If the wheel plane is turned about the x-axis and leans toward the chassis, it has a positive camber. The wheel has a negative camber if its plane is turned about $-x$-axis.

The cornering force that a tire can develop is a function of its angles relative to the road surface. The wheel camber has a significant effect on the lateral force development and road holding of a car. A tire develops its maximum lateral force at a small camber angle. This fact is due to the contribution of camber thrust, which is an additional lateral force generated by elastic deformation as the tread rubber pulls through the tire/road

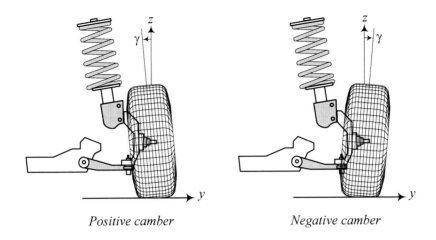

Positive camber   *Negative camber*

FIGURE 8.46. A positive and negative camber on the front wheel of a car.

interface.

To optimize a tire's performance in a turn, the suspension should provide a slight camber angle in the direction of rotation. As the body rolls in a turn, the suspension deflects vertically. The wheel is connected to the chassis by suspension mechanism, which must rotate to allow for the wheel deflection. Therefore, the wheel can be subject to some camber changes as the suspension moves up and down. So, the more the wheel deflects from its static position, the more difficult it is to maintain an ideal camber angle. To provide a good ride comfort in passenger cars, a relatively large wheel travel and soft roll stiffness are needed. As a result, suspension design to provide a good camber presents a design challenge, while the small wheel travel and high roll stiffness inherent in racing cars reduces the problem.

Example 355 *Castor versus camber.*

Camber doesn't improve turn-in as the positive caster does. Camber is not generally good for tire wear. Camber in one wheel does not improve directional stability. Camber adversely affects braking and acceleration efforts.

Example 356 *Camber versus sideslip.*

Cars usually use sideslip angle to generate the required lateral force for turning. Bikes usually use camber angle to generate the required lateral force to turn. Because a tire needs camber angle ten times of the sideslip angle to produce the same amount of lateral force, roll of bikes in a turn is an observable motion, while sideslip angle is hard to see.

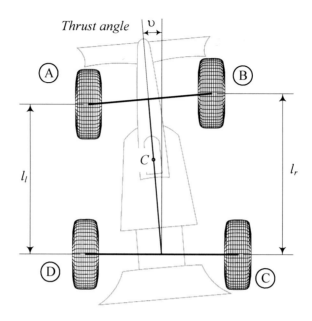

FIGURE 8.47. Thrust angle.

8.4.4 ★ Thrust Angle

The *thrust angle* v is the angle between vehicle's centerline and perpendicular to the rear axle. It compares the direction that the rear axle is aimed with the centerline of the vehicle. A nonzero angle configuration is shown in Figure 8.47.

Zero angle confirms that the rear axle is parallel to the front axle, and the wheelbase on both sides of the vehicle are the same. A reason for nonzero thrust angle would have unequal toe-in or toe-out on both sides of the axle.

Example 357 *Torque reaction.*

There are two kinds of torque reactions in rear-wheel-drive vehicles: 1. the reaction of the axle housing to rotate in the opposite direction of the crown wheel rotation, and 2. the reaction of axle housing to spin about its own center, opposite to the direction of pinion's rotation.

The first reaction leads to a lifting force in the differential causing a wind-up in springs. The second reaction leads to a lifting force on the right wheels.

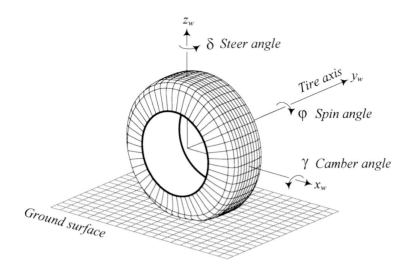

FIGURE 8.48. Six degrees of freedom of a wheel with respect to a vehicle body.

8.5 ★ Suspension Requirements and Coordinate Frames

The suspension mechanism should allow some relative motions between the wheel and the vehicle body. The relative motions are needed to pass the road irregularities and steering. Therefore, to function properly, a suspension mechanism should have some kinematic and dynamics requirements.

8.5.1 Kinematic Requirements

To express the motions of a wheel, we attach a wheel coordinate system $W (ox_w y_w z_w)$ to the center of the wheel. A wheel, as a rigid body, has *six* degrees-of-freedom (DOF) with respect to the vehicle body: three translations and three rotations, as shown in Figure 8.48.

The axes x_w, y_w, and z_w indicate the direction of forward, lateral, and vertical translations and rotations. In the position shown in the figure, the rotation about the x_w-axis is the *camber* angle, about the y_w-axis is the *spin*, and about the z_w-axis is the *steer* angle.

Consider a non-steerable wheel. Translation in z_w-direction and spin about the y_w-axis are the only *two* DOF allowed. So, we need to take four DOF. If the wheel is steerable, then translation in the z_w-direction, spin about the y_w-axis, and steer rotation about the z_w-axis are the three DOF allowed and we must take three DOF of a steerable wheel.

Kinematically, non-steerable and steerable wheels should be supported

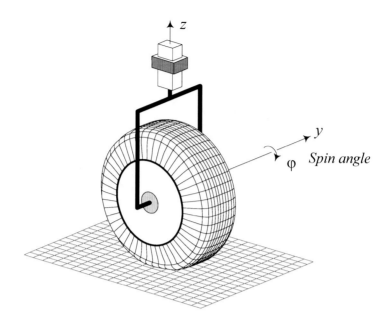

FIGURE 8.49. A non-steerable wheel must have two DOF.

as shown in Figures 8.49 and 8.50 respectively. Providing the required freedom, as well as taking the non-allowed DOF, are the *kinematic requirements* of a suspension mechanism.

8.5.2 Dynamic Requirements

Wheels should be able to propel, steer, and stop the vehicle. So, the suspension system must transmit the driving traction and deceleration braking forces between the vehicle body and the ground. The suspension members must also resist lateral forces acting on the vehicle. Hence, the wheel suspension system must make the wheel rigid for the taken DOF. However, there must also be some compliance members to limit the untaken DOF. The most important compliant members are spring and dampers to provide returning and resistance forces in the z-direction.

8.5.3 Wheel, wheel-body, and tire Coordinate Frames

Three coordinate frames are employed to express the orientation of a tire and wheel with respect to the vehicle: the wheel frame W, wheel-body frame C, and tire frame T. The wheel coordinate frame $W(x_w, y_w, z_w)$ is attached to the center of the wheel. It follows every translation and rotation of the wheel except the spin. Hence, the x_w and z_w axes are always in the

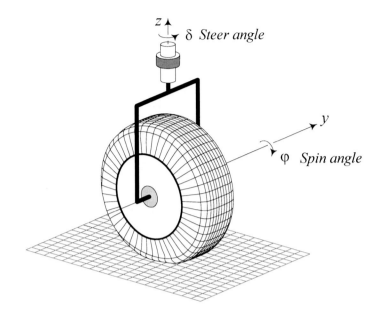

FIGURE 8.50. A steerable wheel must have three DOF.

tire-plane, while the y_w-axis is always parallel to the spin axis. A wheel coordinate frame is shown in Figure 8.48.

When the wheel is straight and upright and the W frame is parallel to the vehicle coordinate frame, we attach a wheel-body coordinate frame $C(x_c, y_c, z_c)$ at the center of the wheel parallel to the vehicle coordinate axes. The wheel-body frame C is motionless with respect to the vehicle coordinate and does not follow any motion of the wheel.

The tire coordinate frame $T(x_t, y_t, z_t)$ is set at the center of the tireprint. The z_t-axis is always perpendicular to the ground. The x_t-axis is along the intersection line of the tire-plane and the ground. The tire frame does not follows the spin and camber rotations of the tire however, it follows the steer angle rotation about the z_c-axis.

Figure 8.51 illustrates a tire and a wheel coordinate frames.

Example 358 *Visualization of the wheel, tire, and wheel-body frames.*

Figure 8.52 illustrates the relative configuration of a wheel-body frame C, a tire frame T, and a wheel frame W. If the steering axis is along the z_c-axis then, the rotation of the wheel about the z_c-axis is the steer angle δ. Rotation about the x_t-axis is the camber angle γ.

Generally speaking, the steering axis may have any angle and may go through any point of the ground plane.

Example 359 *Wheel to tire coordinate frame transformation.*

If $^T\mathbf{d}_W$ indicates the T-expression of the position vector of the wheel

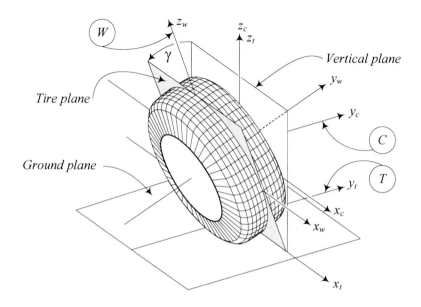

FIGURE 8.51. Illustration of tire and wheel coordinate frames.

frame origin relative to the tire frame origin, then having the coordinates of a point P in the wheel frame, we can find its coordinates in the tire frame by:

$$^T\mathbf{r}_P = {}^T R_W \, {}^W\mathbf{r}_P + {}^T\mathbf{d}_W \tag{8.37}$$

If $^W\mathbf{r}_P$ indicates the position vector of a point P in the wheel frame,

$$^W\mathbf{r}_P = \begin{bmatrix} x_P \\ y_P \\ z_P \end{bmatrix} \tag{8.38}$$

then the coordinates of the point P in the tire frame $^T\mathbf{r}_P$ are

$$\begin{aligned}
^T\mathbf{r}_P &= {}^T R_W \, {}^W\mathbf{r}_P + {}^T\mathbf{d} = {}^T R_W \, {}^W\mathbf{r}_P + {}^T R_W \, {}^W_T\mathbf{d}_W \\
&= \begin{bmatrix} x_P \\ y_P \cos\gamma - R_w \sin\gamma - z_P \sin\gamma \\ R_w \cos\gamma + z_P \cos\gamma + y_P \sin\gamma \end{bmatrix}
\end{aligned} \tag{8.39}$$

where, $^W_T\mathbf{d}_W$ is the W-expression of the position vector of the wheel frame in the tire frame, R_w is the radius of the tire, and $^T R_W$ is the rotation

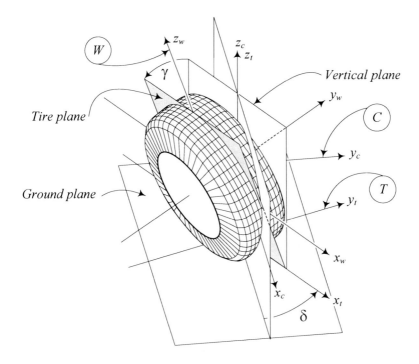

FIGURE 8.52. Illustration of tire, wheel, and body coordinate frames.

matrix to transfer from the wheel frame W to the tire frame T.

$$
{}^{T}R_{W} = \begin{bmatrix} 1 & 0 & 0 \\ 0 & \cos\gamma & -\sin\gamma \\ 0 & \sin\gamma & \cos\gamma \end{bmatrix} \tag{8.40}
$$

$$
{}^{W}_{T}\mathbf{d}_{W} = \begin{bmatrix} 0 \\ 0 \\ R_{w} \end{bmatrix} \tag{8.41}
$$

As an example, the center of the wheel ${}^{W}\mathbf{r}_{P} = {}^{W}\mathbf{r}_{o} = \mathbf{0}$ is the origin of the wheel frame W, that is at ${}^{T}\mathbf{r}_{o}$ in the tire coordinate frame T.

$$
{}^{T}\mathbf{r}_{o} = {}^{T}\mathbf{d}_{W} = {}^{T}R_{W}\,{}^{W}_{T}\mathbf{d}_{W} = \begin{bmatrix} 0 \\ -R_{w}\sin\gamma \\ R_{w}\cos\gamma \end{bmatrix} \tag{8.42}
$$

Example 360 ★ *Tire to wheel coordinate frame transformation.*

If $\mathbf{r}_{P}$ indicates the position vector of a point P in the tire coordinate frame,

$$
{}^{T}\mathbf{r}_{P} = \begin{bmatrix} x_{P} \\ y_{P} \\ z_{P} \end{bmatrix} \tag{8.43}
$$

then the position vector $^W\mathbf{r}_P$ of the point P in the wheel coordinate frame is

$$^W\mathbf{r}_P = {}^W R_T {}^T\mathbf{r}_P - {}^W_T \mathbf{d}_W \tag{8.44}$$

$$= \begin{bmatrix} x_P \\ y_P \cos\gamma + z_P \sin\gamma \\ z_P \cos\gamma - R_w - y_P \sin\gamma \end{bmatrix}$$

because

$$^W R_T = \begin{bmatrix} 1 & 0 & 0 \\ 0 & \cos\gamma & \sin\gamma \\ 0 & -\sin\gamma & \cos\gamma \end{bmatrix} \qquad ^W\mathbf{d}_T = \begin{bmatrix} 0 \\ 0 \\ R_w \end{bmatrix} \tag{8.45}$$

and we may multiply both sides of Equation (8.37) by $^T R_W^T$ to get

$$^T R_W^T {}^T\mathbf{r}_P = {}^W\mathbf{r}_P + {}^T R_W^T {}^T\mathbf{d}_W = {}^W\mathbf{r}_P + {}^W_T\mathbf{d}_W \tag{8.46}$$

$$^W\mathbf{r}_P = {}^W R_T {}^T\mathbf{r}_P - {}^W_T\mathbf{d}_W \tag{8.47}$$

As an example, the center of tireprint in the wheel frame is at

$$^W\mathbf{r}_P = \begin{bmatrix} 1 & 0 & 0 \\ 0 & \cos\gamma & -\sin\gamma \\ 0 & \sin\gamma & \cos\gamma \end{bmatrix}^T \begin{bmatrix} 0 \\ 0 \\ 0 \end{bmatrix} - \begin{bmatrix} 0 \\ 0 \\ R_w \end{bmatrix} = \begin{bmatrix} 0 \\ 0 \\ -R_w \end{bmatrix} \tag{8.48}$$

Example 361 ★ *Wheel to tire homogeneous transformation matrices.*
  The transformation from the wheel to tire coordinate frame may also be expressed by a 4×4 homogeneous transformation matrix $^T T_W$,

$$^T\mathbf{r}_P = {}^T T_W {}^W\mathbf{r}_P = \begin{bmatrix} ^T R_W & ^T\mathbf{d}_W \\ 0 & 1 \end{bmatrix} {}^W\mathbf{r}_P \tag{8.49}$$

where

$$^T T_W = \begin{bmatrix} 1 & 0 & 0 & 0 \\ 0 & \cos\gamma & -\sin\gamma & -R_w \sin\gamma \\ 0 & \sin\gamma & \cos\gamma & R_w \cos\gamma \\ 0 & 0 & 0 & 1 \end{bmatrix} \tag{8.50}$$

The corresponding homogeneous transformation matrix $^W T_T$ from the tire to wheel frame would be

$$^W T_T = \begin{bmatrix} ^W R_T & ^W\mathbf{d}_T \\ 0 & 1 \end{bmatrix} = \begin{bmatrix} 1 & 0 & 0 & 0 \\ 0 & \cos\gamma & -\sin\gamma & 0 \\ 0 & \sin\gamma & \cos\gamma & -R_w \\ 0 & 0 & 0 & 1 \end{bmatrix} \tag{8.51}$$

It can be checked that $^WT_T = {}^TT_W^{-1}$, using the inverse of a homogeneous transformation matrix rule.

$$^TT_W^{-1} = \begin{bmatrix} {}^TR_W & {}^T\mathbf{d}_W \\ 0 & 1 \end{bmatrix}^{-1} = \begin{bmatrix} {}^TR_W^T & -{}^TR_W^T \, {}^T\mathbf{d}_W \\ 0 & 1 \end{bmatrix}$$

$$= \begin{bmatrix} {}^WR_T & -{}^WR_T \, {}^T\mathbf{d}_W \\ 0 & 1 \end{bmatrix} \tag{8.52}$$

Example 362 ★ *Tire to wheel-body frame transformation.*
The origin of the tire frame is at $^C\mathbf{d}_T$ in the wheel-body frame.

$$^C\mathbf{d}_T = \begin{bmatrix} 0 \\ 0 \\ -R_w \end{bmatrix} \tag{8.53}$$

The tire frame can steer about the z_c-axis with respect to the wheel-body frame. The associated transformation matrix is

$$^CR_T = \begin{bmatrix} \cos\delta & -\sin\delta & 0 \\ \sin\delta & \cos\delta & 0 \\ 0 & 0 & 1 \end{bmatrix} \tag{8.54}$$

Therefore, the transformation between the tire and wheel-body frames can be expressed by

$$^C\mathbf{r} = {}^CR_T \, {}^T\mathbf{r} + {}^C\mathbf{d}_T \tag{8.55}$$

or equivalently, by a homogeneous transformation matrix ^CT_T.

$$^CT_T = \begin{bmatrix} {}^CR_T & {}^C\mathbf{d}_T \\ 0 & 1 \end{bmatrix} = \begin{bmatrix} \cos\delta & -\sin\delta & 0 & 0 \\ \sin\delta & \cos\delta & 0 & 0 \\ 0 & 0 & 1 & -R_w \\ 0 & 0 & 0 & 1 \end{bmatrix} \tag{8.56}$$

As an example, the wheel-body coordinates of the point P on the tread of a negatively steered tire at the position shown in Figure 8.53, are:

$$^C\mathbf{r} = {}^CT_T \, {}^T\mathbf{r}_P$$

$$= \begin{bmatrix} \cos-\delta & -\sin-\delta & 0 & 0 \\ \sin-\delta & \cos-\delta & 0 & 0 \\ 0 & 0 & 1 & -R_w \\ 0 & 0 & 0 & 1 \end{bmatrix} \begin{bmatrix} R_w \\ 0 \\ R_w \\ 1 \end{bmatrix} = \begin{bmatrix} R_w\cos\delta \\ -R_w\sin\delta \\ 0 \\ 1 \end{bmatrix} \tag{8.57}$$

The homogeneous transformation matrix for tire to wheel-body frame ^TT_C is:

$$^TT_C = {}^CT_T^{-1} = \begin{bmatrix} {}^CR_T & {}^C\mathbf{d}_T \\ 0 & 1 \end{bmatrix}^{-1} = \begin{bmatrix} {}^CR_T^T & -{}^CR_T^T \, {}^C\mathbf{d}_T \\ 0 & 1 \end{bmatrix}$$

$$= \begin{bmatrix} {}^CR_T^T & -{}^T_C\mathbf{d}_T \\ 0 & 1 \end{bmatrix} = \begin{bmatrix} \cos\delta & \sin\delta & 0 & 0 \\ -\sin\delta & \cos\delta & 0 & 0 \\ 0 & 0 & 1 & R_w \\ 0 & 0 & 0 & 1 \end{bmatrix} \tag{8.58}$$

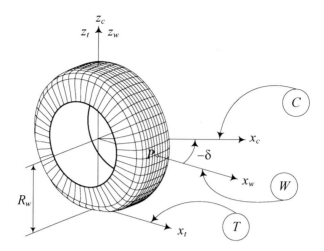

FIGURE 8.53. The tire, wheel, and wheel body frames of a steered wheel.

Example 363 ★ *Cycloid.*

Assume that the wheel in Figure 8.53 is turning with angular velocity ω and has no slip on the ground. If the point P is at the center of the tireprint when $t = 0$,

$$
{}^{M}\mathbf{r}_P = \begin{bmatrix} 0 \\ 0 \\ -R_w \end{bmatrix} \tag{8.59}
$$

then we can find its position in the wheel frame at a time t by employing another coordinate frame M. The frame M is called the rim frame and is stuck to the wheel at its center. Because of spin, the M frame turns about the y_w-axis, and therefore, the rotation matrix to go from the rim frame to the wheel frame is:

$$
{}^{W}R_M = \begin{bmatrix} \cos\omega t & 0 & \sin\omega t \\ 0 & 1 & 0 \\ -\sin\omega t & 0 & \cos\omega t \end{bmatrix} \tag{8.60}
$$

So the coordinates of P in the wheel frame are:

$$
{}^{W}\mathbf{r}_P = {}^{W}R_M {}^{M}\mathbf{r}_P = \begin{bmatrix} -R_w \sin t\omega \\ 0 \\ -R_w \cos t\omega \end{bmatrix} \tag{8.61}
$$

The center of the wheel is moving with speed $v_x = R_w \omega$ and it is at ${}^{G}\mathbf{r} = \begin{bmatrix} v_x t & 0 & R_w \end{bmatrix}$ in the global coordinate frame G on the ground. Hence, the coordinates of point P in the global frame G, would be

$$
{}^{G}\mathbf{r}_P = {}^{W}\mathbf{r}_P + \begin{bmatrix} v_x t \\ 0 \\ R_w \end{bmatrix} = \begin{bmatrix} R_w (\omega t - \sin t\omega) \\ 0 \\ R_w (1 - \cos t\omega) \end{bmatrix} \tag{8.62}
$$

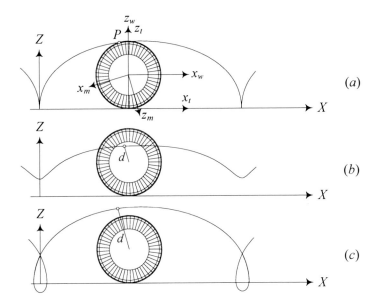

FIGURE 8.54. A cycloid (a), curtate cycloid (b), and prolate cycloid (c).

*The path of motion of point P in the (X, Z)-plane can be found by elim-inating t between X and Z coordinates. However, it is easier to expressed the path by using ωt as a parameter. Such a path is called **cycloid**.*

*In general case, the interested point P can be at any distance from the center of the rim frame. If the point is at a distance $d \neq R_w$, then its path of motion is called the **trochoid**. A trochoid is called a **curtate cycloid** when $d < R_w$ and a **prolate cycloid** when $d > R_w$. Figure 8.54(a)-(c) illustrate a cycloid, curtate cycloid, and prolate cycloid respectively.*

Example 364 ★ *Wheel to wheel-body frame transformation.*

The homogeneous transformation matrix ${}^{C}T_W$ to go from the wheel frame to the wheel-body frame will be found by combined transformation.

$$
{}^{C}T_W = {}^{C}T_T \, {}^{T}T_W \tag{8.63}
$$

$$
= \begin{bmatrix} c\delta & -s\delta & 0 & 0 \\ s\delta & c\delta & 0 & 0 \\ 0 & 0 & 1 & -R_w \\ 0 & 0 & 0 & 1 \end{bmatrix} \begin{bmatrix} 1 & 0 & 0 & 0 \\ 0 & c\gamma & -s\gamma & -R_w \sin\gamma \\ 0 & s\gamma & c\gamma & R_w \cos\gamma \\ 0 & 0 & 0 & 1 \end{bmatrix}
$$

$$
= \begin{bmatrix} \cos\delta & -\cos\gamma\sin\delta & \sin\gamma\sin\delta & R_w \sin\gamma\sin\delta \\ \sin\delta & \cos\gamma\cos\delta & -\cos\delta\sin\gamma & -R_w \cos\delta\sin\gamma \\ 0 & \sin\gamma & \cos\gamma & R_w \cos\gamma - R_w \\ 0 & 0 & 0 & 1 \end{bmatrix}
$$

If $\mathbf{r}_P$ indicates the position vector of a point P in the wheel coordinate *frame,*

$$^W\mathbf{r}_P = \begin{bmatrix} x_P \\ y_P \\ z_P \end{bmatrix} \tag{8.64}$$

then the homogeneous position vector $^C\mathbf{r}_P$ of the point P in the wheel-body coordinate frame would be:

$$
\begin{aligned}
^C\mathbf{r}_P &= {}^C T_W \, ^W\mathbf{r}_P \\
&= \begin{bmatrix} x_P \cos\delta - y_P \cos\gamma \sin\delta + (R_w + z_P)\sin\gamma\sin\delta \\ x_P \sin\delta + y_P \cos\gamma\cos\delta - (R_w + z_P)\cos\delta\sin\gamma \\ -R_w + (R_w + z_P)\cos\gamma + y_P\sin\gamma \\ 1 \end{bmatrix}
\end{aligned} \tag{8.65}
$$

The position of the wheel center $^W\mathbf{r} = \mathbf{0}$, for a cambered and steered wheel is at

$$^C\mathbf{r} = {}^C T_W \, ^W\mathbf{r} = \begin{bmatrix} R_w \sin\gamma \sin\delta \\ -R_w \cos\delta \sin\gamma \\ -R_w(1-\cos\gamma) \\ 1 \end{bmatrix} \tag{8.66}$$

The $z_c = R_w(\cos\gamma - 1)$ indicates how much the center of the wheel comes down when the wheel cambers.

If the wheel is not steerable, then $\delta = 0$ and the transformation matrix $^C T_W$ reduces to

$$^C T_W = \begin{bmatrix} 1 & 0 & 0 & 0 \\ 0 & \cos\gamma & -\sin\gamma & -R_w\sin\gamma \\ 0 & \sin\gamma & \cos\gamma & R_w(\cos\gamma - 1) \\ 0 & 0 & 0 & 1 \end{bmatrix} \tag{8.67}$$

that shows

$$
\begin{aligned}
^C\mathbf{r}_P &= {}^C T_W \, ^W\mathbf{r}_P \\
&= \begin{bmatrix} x_P \\ y_P \cos\gamma - R_w\sin\gamma - z_P\sin\gamma \\ z_P\cos\gamma + y_P\sin\gamma + R_w(\cos\gamma - 1) \\ 1 \end{bmatrix}
\end{aligned} \tag{8.68}
$$

Example 365 ★ *Tire to vehicle coordinate frame transformation.*

Figure 8.55 illustrates the first and fourth tires of a 4-wheel vehicle. There is a body coordinate frame $B(x,y,z)$ attached to the mass center C of the vehicle. There are also two tire coordinate frames $T_1(x_{t_1}, y_{t_1}, z_{t_1})$ and $T_4(x_{t_4}, y_{t_4}, z_{t_4})$ attached to the tires 1 and 4 at the center of their tireprints.

The origin of the tire coordinate frame T_1 is at $^B\mathbf{d}_1$

$$^B\mathbf{d}_{T_1} = \begin{bmatrix} a_1 \\ b_1 \\ -h \end{bmatrix} \tag{8.69}$$

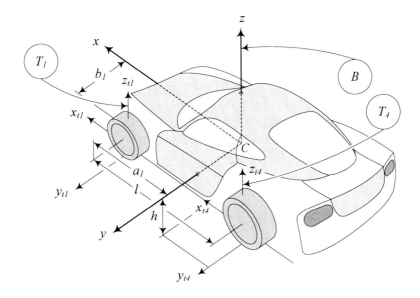

FIGURE 8.55. The coordinate frames of the first and fourth tires of a 4-wheel vehicle with respect to the body frame.

where, a_1 is the longitudinal distance between C and the front axle, b_1 is the lateral distance between C and the tireprint of the tire 1, and h is the height of C from the ground level. If P is a point in the tire frame at $^{T_1}\mathbf{r}_P$

$$^{T_1}\mathbf{r}_P = \begin{bmatrix} x_P \\ y_P \\ z_P \end{bmatrix} \tag{8.70}$$

then its coordinates in the body frame are

$$\begin{aligned}
^{B}\mathbf{r}_P &= {}^{B}R_{T_1}{}^{T_1}\mathbf{r}_P + {}^{B}\mathbf{d}_{T_1} \\
&= \begin{bmatrix} x_P \cos \delta_1 - y_P \sin \delta_1 + a_1 \\ y_P \cos \delta_1 + x_P \sin \delta_1 + b_1 \\ z_P - h \end{bmatrix}
\end{aligned} \tag{8.71}$$

The transformation matrix $^{B}R_{T_1}$ is a result of steering about the z_1-axis.

$$^{B}R_{T_1} = \begin{bmatrix} \cos \delta_1 & -\sin \delta_1 & 0 \\ \sin \delta_1 & \cos \delta_1 & 0 \\ 0 & 0 & 1 \end{bmatrix} \tag{8.72}$$

Employing Equation (8.37), we may examine a wheel point P at $^{W}\mathbf{r}_P$

$$^{W}\mathbf{r}_P = \begin{bmatrix} x_P \\ y_P \\ z_P \end{bmatrix} \tag{8.73}$$

and find the body coordinates of the point

$$
\begin{aligned}
{}^B\mathbf{r}_P &= {}^B R_{T_1} {}^{T_1}\mathbf{r}_P + {}^B\mathbf{d}_{T_1} \\
&= {}^B R_{T_1} \left({}^{T_1} R_W {}^W\mathbf{r}_P + {}^{T_1}\mathbf{d}_W\right) + {}^B\mathbf{d}_{T_1} \\
&= {}^B R_{T_1} {}^{T_1} R_W {}^W\mathbf{r}_P + {}^B R_{T_1} {}^{T_1}\mathbf{d}_W + {}^B\mathbf{d}_{T_1} \\
&= {}^B R_W {}^W\mathbf{r}_P + {}^B R_{T_1} {}^{T_1}\mathbf{d}_W + {}^B\mathbf{d}_{T_1}
\end{aligned} \tag{8.74}
$$

$$
{}^B\mathbf{r}_P = \begin{bmatrix}
x_P \cos\delta_1 - y_P \cos\gamma \sin\delta_1 + (R_w + z_P)\sin\gamma \sin\delta_1 + a_1 \\
x_P \sin\delta_1 + y_P \cos\gamma \cos\delta_1 - (R_w + z_P)\cos\delta_1 \sin\gamma + b_1 \\
(R_w + z_P)\cos\gamma + y_P \sin\gamma - h
\end{bmatrix} \tag{8.75}
$$

where,

$$
\begin{aligned}
{}^B R_W &= {}^B R_{T_1} {}^{T_1} R_W \\
&= \begin{bmatrix}
\cos\delta_1 & -\cos\gamma \sin\delta_1 & \sin\gamma \sin\delta_1 \\
\sin\delta_1 & \cos\gamma \cos\delta_1 & -\cos\delta_1 \sin\gamma \\
0 & \sin\gamma & \cos\gamma
\end{bmatrix}
\end{aligned} \tag{8.76}
$$

$$
{}^{T_1}\mathbf{d}_W = \begin{bmatrix}
0 \\
-R_w \sin\gamma \\
R_w \cos\gamma
\end{bmatrix} \tag{8.77}
$$

Example 366 ★ *Wheel-body to vehicle transformation.*

The wheel-body coordinate frames is always parallel to the vehicle frame. The origin of the wheel-body coordinate frame of the wheel number 1 is at

$$
{}^B\mathbf{d}_{W_1} = \begin{bmatrix}
a_1 \\
b_1 \\
-h + R_w
\end{bmatrix} \tag{8.78}
$$

Hence the transformation between the two frames is only a displacement.

$$
{}^B\mathbf{r} = {}^B I_{W_1} {}^{W_1}\mathbf{r} + {}^B\mathbf{d}_{W_1} \tag{8.79}
$$

8.6 ★ Caster Theory

The *steer axis* of a wheel may have any angle and be at any location with respect to the wheel-body coordinate frame. The wheel-body frame $C(x_c, y_c, z_c)$ is a frame at the center of the wheel at its rest position, and is parallel to the vehicle coordinate frame. The frame C does not follow any motion of the wheel. The steer axis is the *kingpin axis* of rotation.

Figure 8.56 illustrates the front and side views of a wheel and its steering axis. The steering axis has angle φ with (y_c, z_c) plane, and angle θ with (x_c, z_c) plane. The angles φ and θ are measured about the y_c and x_c axes,

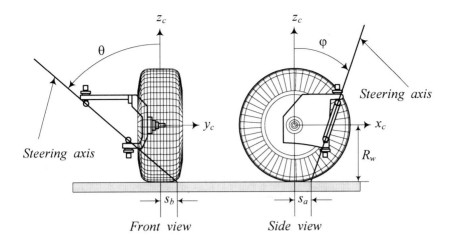

FIGURE 8.56. The front and side views of a wheel and its steering axis.

respectively. The angle φ is the *caster angle* of the wheel, while the angle θ is the *lean angle*. The steering axis of the wheel, as shown in Figure 8.56, is at positive caster and lean angles. The steering axis intersects the ground plane at a point that has coordinates $(s_a, s_b, -R_w)$ in the wheel-body coordinate frame.

If we indicate the steering axis by the unit vector $\hat{u}$, then the components of $\hat{u}$ are functions of the caster and lean angles.

$$^C\hat{u} = \begin{bmatrix} u_1 \\ u_2 \\ u_3 \end{bmatrix} = \frac{1}{\sqrt{\cos^2\varphi + \cos^2\theta \sin^2\varphi}} \begin{bmatrix} \cos\theta \sin\varphi \\ -\cos\varphi \sin\theta \\ \cos\theta \cos\varphi \end{bmatrix} \qquad (8.80)$$

The position vector of the point that $\hat{u}$ intersects the ground plane, is called the *location vector* **s** that in the wheel-body frame has the following coordinates:

$$^C\mathbf{s} = \begin{bmatrix} s_a \\ s_b \\ -R_w \end{bmatrix} \qquad (8.81)$$

We express the rotation of the wheel about the steering axis $\hat{u}$ by a zero pitch screw motion $\check{s}$.

$$\begin{aligned} ^C T_W &= {}^C \check{s}_W(0, \delta, \hat{u}, \mathbf{s}) \\ &= \begin{bmatrix} ^C R_W & {}^C\mathbf{s} - {}^C R_W \, {}^C\mathbf{s} \\ 0 & 1 \end{bmatrix} = \begin{bmatrix} ^C R_W & {}^C\mathbf{d}_W \\ 0 & 1 \end{bmatrix} \end{aligned} \qquad (8.82)$$

Proof. The steering axis is at the intersection of the *caster plane* π_C and the *lean plane* π_L, both expressed in the wheel-body coordinate frame. The two planes can be indicated by their normal unit vectors $\hat{n}_1$ and $\hat{n}_2$.

$$
{}^C\hat{n}_1 \quad = \quad \begin{bmatrix} 0 \\ \cos\theta \\ \sin\theta \end{bmatrix} \tag{8.83}
$$

$$
{}^C\hat{n}_2 \quad = \quad \begin{bmatrix} -\cos\varphi \\ 0 \\ \sin\varphi \end{bmatrix} \tag{8.84}
$$

The unit vector $\hat{u}$ on the intersection of the caster and lean planes can be found by

$$
\hat{u} = \frac{\hat{n}_1 \times \hat{n}_2}{|\hat{n}_1 \times \hat{n}_2|} \tag{8.85}
$$

where,

$$
\hat{n}_1 \times \hat{n}_2 \quad = \quad \begin{bmatrix} \cos\theta\sin\varphi \\ -\cos\varphi\sin\theta \\ \cos\theta\cos\varphi \end{bmatrix} \tag{8.86}
$$

$$
|\hat{n}_1 \times \hat{n}_2| \quad = \quad \sqrt{\cos^2\varphi + \cos^2\theta\sin^2\varphi} \tag{8.87}
$$

and therefore,

$$
{}^C\hat{u} = \begin{bmatrix} u_1 \\ u_2 \\ u_3 \end{bmatrix} = \begin{bmatrix} \dfrac{\cos\theta\sin\varphi}{\sqrt{\cos^2\varphi + \cos^2\theta\sin^2\varphi}} \\ \dfrac{-\cos\varphi\sin\theta}{\sqrt{\cos^2\varphi + \cos^2\theta\sin^2\varphi}} \\ \dfrac{\cos\theta\cos\varphi}{\sqrt{\cos^2\varphi + \cos^2\theta\sin^2\varphi}} \end{bmatrix} \tag{8.88}
$$

Steering axis does not follow any motion of the wheel except the wheel hop in the z-direction. Therefore, it is assumed that the steering axis is a fixed line with respect to the vehicle, and the steer angle δ is the rotation angle about $\hat{u}$.

The intersection point of the steering axis and the ground plane defines the *location vector* **s**.

$$
{}^C\mathbf{s} = \begin{bmatrix} s_a \\ s_b \\ -R_w \end{bmatrix} \tag{8.89}
$$

The components s_a and s_b are called the *forward* and *lateral locations* respectively.

Using the axis-angle rotation $(\hat{u}, \delta)$, and the location vector **s**, we can define the steering of the wheel by a screw motion $\check{s}$ with zero pitch. Employing Equations (5.460)-(5.464), we find the transformation screw for wheel frame W to wheel-body frame C.

$$\begin{aligned}
{}^C T_W &= {}^C \breve{s}_W(0, \delta, \hat{u}, \mathbf{s}) & (8.90)\\
&= \begin{bmatrix} {}^C R_W & {}^C \mathbf{s} - {}^C R_W {}^C \mathbf{s} \\ 0 & 1 \end{bmatrix} = \begin{bmatrix} {}^C R_W & {}^C \mathbf{d} \\ 0 & 1 \end{bmatrix}
\end{aligned}$$

$$\begin{aligned}
{}^C R_W &= \mathbf{I} \cos \delta + \hat{u}\hat{u}^T \operatorname{vers} \delta + \tilde{u} \sin \delta & (8.91)\\
{}^C \mathbf{d}_W &= \left((\mathbf{I} - \hat{u}\hat{u}^T) \operatorname{vers} \delta - \tilde{u} \sin \delta \right) {}^C \mathbf{s} & (8.92)
\end{aligned}$$

$$\tilde{u} = \begin{bmatrix} 0 & -u_3 & u_2 \\ u_3 & 0 & -u_1 \\ -u_2 & u_1 & 0 \end{bmatrix} \qquad (8.93)$$

$$\operatorname{vers} \delta = 1 - \cos \delta \qquad (8.94)$$

Direct substitution shows that ${}^C R_W$ and ${}^C \mathbf{d}_W$ are:

$$
{}^C R_W = \begin{bmatrix}
u_1^2 \operatorname{vers} \delta + c\delta & u_1 u_2 \operatorname{vers} \delta - u_3 s\delta & u_1 u_3 \operatorname{vers} \delta + u_2 s\delta \\
u_1 u_2 \operatorname{vers} \delta + u_3 s\delta & u_2^2 \operatorname{vers} \delta + c\delta & u_2 u_3 \operatorname{vers} \delta - u_1 s\delta \\
u_1 u_3 \operatorname{vers} \delta - u_2 s\delta & u_2 u_3 \operatorname{vers} \delta + u_1 s\delta & u_3^2 \operatorname{vers} \delta + c\delta
\end{bmatrix}
$$
$$(8.95)$$

$$
{}^C \mathbf{d}_W = \begin{bmatrix}
(s_1 - u_1(s_3 u_3 + s_2 u_2 + s_1 u_1)) \operatorname{vers} \delta + (s_2 u_3 - s_3 u_2) \sin \delta \\
(s_2 - u_2(s_3 u_3 + s_2 u_2 + s_1 u_1)) \operatorname{vers} \delta + (s_3 u_1 - s_1 u_3) \sin \delta \\
(s_3 - u_3(s_3 u_3 + s_2 u_2 + s_1 u_1)) \operatorname{vers} \delta + (s_1 u_2 - s_2 u_1) \sin \delta
\end{bmatrix}
$$
$$(8.96)$$

The vector ${}^C \mathbf{d}_W$ indicates the position of the wheel center with respect to the wheel-body frame. The matrix ${}^C T_W$ is the homogeneous transformation from wheel frame W to wheel-body frame C, when the wheel is steered by the angle δ about the steering axis $\hat{u}$. ∎

Example 367 ★ *Zero steer angle.*

To examine the screw transformation, we may check the zero steering. Substituting $\delta = 0$ simplifies the rotation matrix ${}^C R_W$ and the position vector ${}^C \mathbf{d}_W$ to $\mathbf{I}$ and $\mathbf{0}$ respectively

$$
{}^C R_W = \begin{bmatrix} 1 & 0 & 0 \\ 0 & 1 & 0 \\ 0 & 0 & 1 \end{bmatrix} \qquad {}^C \mathbf{d}_W = \begin{bmatrix} 0 \\ 0 \\ 0 \end{bmatrix} \qquad (8.97)
$$

indicating that at zero steering, the wheel frame W and wheel-body frame C are coincident.

Example 368 ★ *Steer angle transformation for zero lean and caster.*

Consider a wheel with a steer axis coincident with z_w. Such a wheel has no lean or caster angle. When the wheel is steered by the angle δ, we can find the coordinates of a wheel point P in the wheel-body coordinate frame

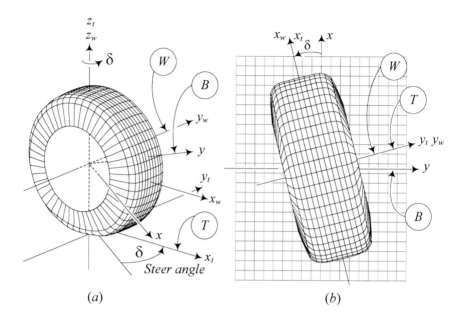

FIGURE 8.57. A steered wheel by the angle δ and with a steer axis coincident with z_w. (a) A 3D illustration of the wheel. (b) Top view of the wheel.

using transformation method. Figure 8.57(a) illustrates a 3D view, and Figure 8.57(b) a top view of such a wheel.

Assume $^W\mathbf{r}_P = [x_w, y_w, z_w]^T$ is the position vector of a wheel point, then its position vector in the wheel-body coordinate frame C is

$$
\begin{aligned}
^C\mathbf{r}_P &= {}^C R_W {}^W\mathbf{r}_P = R_{z,\delta} {}^W\mathbf{r}_P \qquad (8.98) \\
&= \begin{bmatrix} \cos\delta & -\sin\delta & 0 \\ \sin\delta & \cos\delta & 0 \\ 0 & 0 & 1 \end{bmatrix} \begin{bmatrix} x_w \\ y_w \\ z_w \end{bmatrix} = \begin{bmatrix} x_w\cos\delta - y_w\sin\delta \\ y_w\cos\delta + x_w\sin\delta \\ z_w \end{bmatrix}
\end{aligned}
$$

We assumed that the wheel-body coordinate is installed at the center of the wheel and is parallel to the vehicle coordinate frame. Therefore, the transformation from the frame W to the frame C is a rotation δ about the wheel-body z-axis. There would be no camber angle when the lean and caster angles are zero and steer axis is on the z_w-axis.

Example 369 ★ *Zero lean, zero lateral location.*

The case of zero lean, $\theta = 0$, and zero lateral location, $s_b = 0$, is impor-tant in caster dynamics of cars bicycle model. The screw transformation for this case will be simplified to

$$
^C\hat{u} = \begin{bmatrix} u_1 \\ u_2 \\ u_3 \end{bmatrix} = \begin{bmatrix} \sin\varphi \\ 0 \\ \cos\varphi \end{bmatrix} \qquad ^C\mathbf{s} = \begin{bmatrix} s_a \\ 0 \\ -R_w \end{bmatrix} \qquad (8.99)
$$

$$
{}^C R_W = \begin{bmatrix} \sin^2 \varphi \operatorname{vers} \delta + \cos \delta & -\cos \varphi \sin \delta & \sin \varphi \cos \varphi \operatorname{vers} \delta \\ \cos \varphi \sin \delta & \cos \delta & -\sin \varphi \sin \delta \\ \sin \varphi \cos \varphi \operatorname{vers} \delta & \sin \varphi \sin \delta & \cos^2 \varphi \operatorname{vers} \delta + \cos \delta \end{bmatrix}
$$

$$(8.100)$$

$$
{}^C \mathbf{d} = \begin{bmatrix} \cos \varphi \, (s_a \cos \varphi + R_w \sin \varphi) \operatorname{vers} \delta \\ -(s_a \cos \varphi + R_w \sin \varphi) \sin \delta \\ -\dfrac{1}{2} \, (R_w - R_w \cos 2\varphi + s_a \sin 2\varphi) \operatorname{vers} \delta \end{bmatrix}
$$

$$(8.101)$$

Example 370 ★ *Position of the tireprint.*

The center of tireprint in the wheel coordinate frame is at $\mathbf{r}_T$

$$
{}^W \mathbf{r}_T = \begin{bmatrix} 0 \\ 0 \\ -R_w \end{bmatrix}
$$

$$(8.102)$$

If we assume the width of the tire is zero and the wheel is steered, the center of tireprint would be at

$$
{}^C \mathbf{r}_T = {}^C T_W \, {}^W \mathbf{r}_T = \begin{bmatrix} x_T \\ y_T \\ z_T \end{bmatrix}
$$

$$(8.103)$$

where

$$
x_T = \left(1 - u_1^2\right)(1 - \cos \delta) \, s_a + (u_3 \sin \delta - u_1 u_2 \, (1 - \cos \delta)) \, s_b \quad (8.104)
$$

$$
y_T = -(u_3 \sin \delta + u_1 u_2 \, (1 - \cos \delta)) \, s_a + \left(1 - u_2^2\right)(1 - \cos \delta) \, s_b \quad (8.105)
$$

$$
z_T = (u_2 \sin \delta - u_1 u_3 \, (1 - \cos \delta)) \, s_a
$$
$$
- (u_1 \sin \delta + u_2 u_3 \, (1 - \cos \delta)) \, s_b - R_w \quad (8.106)
$$

or

$$
x_T = s_b \left(\frac{\cos \theta \cos \varphi \sin \delta}{\sqrt{\cos^2 \theta \sin^2 \varphi + \cos^2 \varphi}} + \frac{1}{4} \frac{\sin 2\theta \sin 2\varphi \, (1 - \cos \delta)}{\cos^2 \theta \sin^2 \varphi + \cos^2 \varphi} \right)
$$
$$
+ s_a \left(1 - \frac{\cos^2 \theta \sin^2 \varphi}{\cos^2 \theta \sin^2 \varphi + \cos^2 \varphi} \right)(1 - \cos \delta) \quad (8.107)
$$

$$
y_T = -s_a \left(\frac{\cos \theta \cos \varphi \sin \delta}{\sqrt{\cos^2 \theta \sin^2 \varphi + \cos^2 \varphi}} - \frac{1}{4} \frac{\sin 2\theta \sin 2\varphi \, (1 - \cos \delta)}{\cos^2 \theta \sin^2 \varphi + \cos^2 \varphi} \right)
$$
$$
+ s_b \left(1 - \frac{\cos^2 \varphi \sin^2 \theta}{\cos^2 \theta \sin^2 \varphi + \cos^2 \varphi} \right)(1 - \cos \delta) \quad (8.108)
$$

$$
z_T = -R_w - \frac{s_b \cos \theta \sin \varphi + s_a \cos \varphi \sin \theta}{\sqrt{\cos^2 \theta \sin^2 \varphi + \cos^2 \varphi}} \sin \delta
$$
$$
+ \frac{1}{2} \frac{s_b \cos^2 \varphi \sin 2\theta - s_a \cos^2 \theta \sin 2\varphi}{\cos^2 \theta \sin^2 \varphi + \cos^2 \varphi}(1 - \cos \delta) \quad (8.109)
$$

Example 371 ★ *Wheel center drop.*

The z_T coordinate in (8.106) or (8.109) indicates the height that the center of the tireprint will move vertically with respect to the wheel-body frame when the wheel is steering. If the steer angle is zero, $\delta = 0$, then z_T is at

$$z_T = -R_w \tag{8.110}$$

Because the center of tireprint must be on the ground, $H = -R_w - z_T$ indicates the height that the center of the wheel will drop during steering.

$$
\begin{aligned}
H &= -R_w - z_T \\
&= \frac{s_b \cos\theta \sin\varphi + s_a \cos\varphi \sin\theta}{\sqrt{\cos^2\theta \sin^2\varphi + \cos^2\varphi}} \sin\delta \\
&\quad - \frac{1}{2}\frac{s_b \cos^2\varphi \sin 2\theta - s_a \cos^2\theta \sin 2\varphi}{\cos^2\theta \sin^2\varphi + \cos^2\varphi}(1 - \cos\delta)
\end{aligned}
\tag{8.111}
$$

The z_T coordinate of the tireprint may be simplified for different designs:
1. If the lean angle is zero, $\theta = 0$, then z_T is at

$$z_T = -R_w - \frac{1}{2}s_a \sin 2\varphi (1 - \cos\delta) - s_b \sin\varphi \sin\delta \tag{8.112}$$

2. If the lean angle and lateral location are zero, $\theta = 0$, $s_b = 0$, then z_T is at

$$z_T = -R_w - \frac{1}{2}s_a \sin 2\varphi (1 - \cos\delta) \tag{8.113}$$

In this case, the wheel center drop may be expressed by a dimensionless equation.

$$\frac{H}{s_a} = \frac{1}{2}\sin 2\varphi (1 - \cos\delta) \tag{8.114}$$

Figure 8.58 illustrates H/s_a for the caster angle $\varphi = 5\deg$, $0\deg$, $-5\deg$, $-10\deg$, $-15\deg$, $-20\deg$, and the steer angle δ in the range $-10\deg < \delta < 10\deg$. For street cars, we set the steering axis with a positive longitudinal location $s_a > 0$, and a few degrees negative caster angle $\varphi < 0$. In this case the wheel center drops as is shown in the figure.
3. If the caster angle is zero, $\varphi = 0$, then z_T is at

$$z_T = -R_w + \frac{1}{2}s_b \sin 2\theta (1 - \cos\delta) - s_a \sin\theta \sin\delta \tag{8.115}$$

4. If the caster angle and lateral location are zero, $\varphi = 0$, $s_b = 0$, then z_T is at

$$z_T = -R_w - s_a \sin\theta \sin\delta \tag{8.116}$$

In this case, the wheel center drop may be expressed by a dimensionless equation.

$$\frac{H}{s_a} = -\sin\theta \sin\delta \tag{8.117}$$

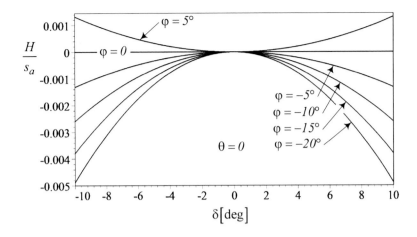

FIGURE 8.58. H/s_a for the caster angle $\varphi = 5\deg$ 0, $-5\deg$, $-10\deg$, $-15\deg$, $-20\deg$ and the steer angle in the range $-10\deg < \delta < 10\deg$.

Figure 8.59 illustrates H/s_a for the lean angle $\theta = 5\deg$, 0, $-5\deg$, $-10\deg$, $-15\deg$, $-20\deg$ and the steer angle δ in the range $-10\deg < \delta < 10\deg$. The steering axis of street cars is usually set with a positive longitudinal location $s_a > 0$, and a few degrees positive lean angle $\theta > 0$. In this case the wheel center lowers when the wheel number 1 turns to the right, and elevates when the wheel turns to the left.

Comparison of Figures 8.58 and 8.59 shows that the lean angle has much more affect on the wheel center drop than the caster angle.

    5. *If the lateral location is zero, $s_b = 0$, then z_T is at*

$$z_T = -R_w - s_a \frac{\cos\varphi \sin\theta}{\sqrt{\cos^2\theta \sin^2\varphi + \cos^2\varphi}} \sin\delta$$
$$-\frac{1}{2} s_a \frac{\cos^2\theta \sin 2\varphi}{\cos^2\theta \sin^2\varphi + \cos^2\varphi} (1 - \cos\delta) \qquad (8.118)$$

and the wheel center drop, H, may be expressed by a dimensionless equation.

$$\frac{H}{s_a} = -\frac{1}{2} \frac{\cos^2\theta \sin^2\varphi (1 - \cos\delta)}{\cos^2\theta \sin^2\varphi + \cos^2\varphi} - \frac{\cos\varphi \sin\theta \sin\delta}{\sqrt{\cos^2\theta \sin^2\varphi + \cos^2\varphi}} \qquad (8.119)$$

Example 372 ★ *Position of the wheel center.*

    As given by Equation (8.96), the wheel center is at $^C\mathbf{d}_W$ with respect to the wheel-body frame.

$$^C\mathbf{d}_W = \begin{bmatrix} x_W \\ y_W \\ z_W \end{bmatrix} \qquad (8.120)$$

Substituting for $\hat{u}$ and $\mathbf{s}$ from (8.80) and (8.81) in (8.96) provides us with

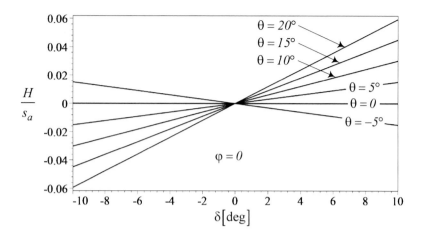

FIGURE 8.59. H/s_a for the lean angle $\theta = 5\deg 0$, $-5\deg$, $-10\deg$, $-15\deg$, $-20\deg$ and the steer angle in the range $-10\deg < \delta < 10\deg$.

the coordinates of the wheel center in the wheel-body frame as

$$
\begin{aligned}
x_W &= (s_a - u_1(-R_w u_3 + s_b u_2 + s_a u_1))(1 - \cos \delta) \\
&\quad + (s_b u_3 + R_w u_2) \sin \delta
\end{aligned}
\tag{8.121}
$$

$$
\begin{aligned}
y_W &= (s_b - u_2(-R_w u_3 + s_b u_2 + s_a u_1))(1 - \cos \delta) \\
&\quad - (R_w u_1 + s_a u_3) \sin \delta
\end{aligned}
\tag{8.122}
$$

$$
\begin{aligned}
z_W &= (-R_w - u_3(-R_w u_3 + s_b u_2 + s_a u_1))(1 - \cos \delta) \\
&\quad + (s_a u_2 - s_b u_1) \sin \delta
\end{aligned}
\tag{8.123}
$$

or

$$
\begin{aligned}
x_W &= s_a(1 - \cos \delta) \\
&\quad + \frac{\left(\frac{1}{2} R_w \sin 2\varphi - s_a \sin^2 \varphi\right) \cos^2 \theta + \frac{1}{4} s_b \sin 2\theta \sin 2\varphi}{\cos^2 \varphi + \cos^2 \theta \sin^2 \varphi}(1 - \cos \delta) \\
&\quad + \frac{(s_b \cos \theta - R_w \sin \theta)}{\sqrt{\cos^2 \varphi + \cos^2 \theta \sin^2 \varphi}} \cos \varphi \sin \delta
\end{aligned}
\tag{8.124}
$$

$$
\begin{aligned}
y_W &= s_b(1 - \cos \delta) \\
&\quad - \frac{\frac{1}{2}(R_w \sin 2\theta + s_b \sin^2 \theta) \cos^2 \varphi - \frac{1}{4} s_a \sin 2\theta \sin 2\varphi}{\cos^2 \varphi + \cos^2 \theta \sin^2 \varphi}(1 - \cos \delta) \\
&\quad - \frac{R_w \sin \varphi + s_a \cos \varphi}{\sqrt{\cos^2 \varphi + \cos^2 \theta \sin^2 \varphi}} \cos \theta \sin \delta
\end{aligned}
\tag{8.125}
$$

$$z_W = -R_w\left(1 - \cos\delta\right)$$

$$+\frac{\left(R_w\cos^2\theta + \frac{1}{2}s_b\sin 2\theta\right)\cos^2\varphi - \frac{1}{2}s_a\cos^2\theta\sin 2\varphi}{\cos^2\varphi + \cos^2\theta\sin^2\varphi}\left(1 - \cos\delta\right)$$

$$-\frac{s_a\cos\varphi\sin\theta + s_b\cos\theta\sin\varphi}{\sqrt{\cos^2\varphi + \cos^2\theta\sin^2\varphi}}\sin\delta \qquad (8.126)$$

The z_W coordinate indicates how the center of the wheel moves vertically with respect to the wheel-body frame, when the wheel is steering. It shows that $z_W = 0$, as long as $\delta = 0$.

The z_W coordinate of the wheel center may be simplified for different designs:

1. If the lean angle is zero, $\theta = 0$, then z_W is at

$$z_W = -R_w\left(1 - \cos^2\varphi\right)\left(1 - \cos\delta\right) - s_b\sin\varphi\sin\delta$$
$$-\frac{1}{2}s_a\sin 2\varphi\left(1 - \cos\delta\right) \qquad (8.127)$$

2. If the lean angle and lateral location are zero, $\theta = 0$, $s_b = 0$, then z_W is at

$$z_W = -R_w\left(1 - \cos^2\varphi\right)\left(1 - \cos\delta\right) - \frac{1}{2}s_a\sin 2\varphi\left(1 - \cos\delta\right) \qquad (8.128)$$

3. If the caster angle is zero, $\varphi = 0$, then z_W is at

$$z_W = -R_w\left(1 - \cos^2\theta\right)\left(1 - \cos\delta\right) - s_a\sin\theta\sin\delta$$
$$+\frac{1}{2}s_b\sin 2\theta\left(1 - \cos\delta\right) \qquad (8.129)$$

4. If the caster angle and lateral location are zero, $\varphi = 0$, $s_b = 0$, then z_W is at

$$z_W = -R_w\left(1 - \cos^2\theta\right)\left(1 - \cos\delta\right) - s_a\sin\theta\sin\delta \qquad (8.130)$$

5. If the lateral location is zero, $s_b = 0$, then z_T is at

$$z_W = -R_w\left(1 - \cos\delta\right) - \frac{s_a\cos\varphi\sin\theta}{\sqrt{\cos^2\varphi + \cos^2\theta\sin^2\varphi}}\sin\delta$$

$$+\frac{R_w\cos^2\theta\cos^2\varphi - \frac{1}{2}s_a\cos^2\theta\sin 2\varphi}{\cos^2\varphi + \cos^2\theta\sin^2\varphi}\left(1 - \cos\delta\right) \qquad (8.131)$$

In each case of the above designs, the height of the wheel center with respect to the ground level can be found by adding H to z_W. The equations for calculating H are given in Example 372.

Example 373 ★ *Camber theory.*

Having a non-zero lean and caster angles causes a camber angle γ for a steered wheel. To find the camber angle of an steered wheel, we may determine the angle between the camber line and the vertical direction z_c. The **camber line** *is the line connecting the wheel center and the center of tireprint.*

The coordinates of the center of tireprint (x_T, y_T, z_T) are given in Equations (8.107)-(8.109), and the coordinates of the wheel center (x_W, y_W, z_W) are given in Equations (8.124)-(8.126). The line connecting (x_T, y_T, z_T) to (x_W, y_W, z_W) may be indicated by the unit vector $\hat{l}_c$

$$\hat{l}_c = \frac{(x_W - x_T)\,\hat{I} + (y_W - y_T)\,\hat{J} + (z_W - z_T)\,\hat{K}}{\sqrt{(x_W - x_T)^2 + (y_W - y_T)^2 + (z_W - z_T)^2}} \tag{8.132}$$

in which $\hat{I}$, $\hat{J}$, $\hat{K}$, are the unit vectors of the wheel-body coordinate frame C.

The camber angle is the angle between $\hat{l}_c$ and $\hat{K}$, which can be found by the inner vector product.

$$\begin{aligned}
\gamma &= \arccos\left(\hat{l}_c \cdot \hat{K}\right) \\
&= \arccos\frac{(z_W - z_T)}{\sqrt{(x_W - x_T)^2 + (y_W - y_T)^2 + (z_W - z_T)^2}}
\end{aligned} \tag{8.133}$$

As an special case, let us determine the camber angle when the lean angle and lateral location are zero, $\theta = 0$, $s_b = 0$. In this case, we have

$$x_T = s_a\left(1 - \sin^2\varphi\right)(\cos\delta - 1) \tag{8.134}$$

$$y_T = -s_a\cos\varphi\sin\delta \tag{8.135}$$

$$z_T = z_T = -R_w - \frac{1}{2}s_a\sin 2\varphi\left(1 - \cos\delta\right) \tag{8.136}$$

$$x_W = \left(s_a + \frac{1}{2}R_w\sin 2\varphi - s_a\sin^2\varphi\right)(1 - \cos\delta) \tag{8.137}$$

$$y_W = s_b\left(1 - \cos\delta\right) - R_w\sin\varphi + s_a\cos\varphi\sin\delta \tag{8.138}$$

$$z_W = \left(R_w\left(\cos^2\varphi - 1\right) - \frac{1}{2}s_a\sin 2\varphi\right)(1 - \cos\delta). \tag{8.139}$$

8.7   Summary

There are two general types of suspensions: *dependent*, in which the left and right wheels on an axle are rigidly connected, and *independent*, in which the left and right wheels are disconnected. Solid axle is the most common dependent suspension, while McPherson and double A-arm are the most common independent suspensions.

The *roll axis* is the instantaneous line about which the body of a vehicle rolls. Roll axis is found by connecting the *roll center* of the front and rear suspensions of the vehicle. The instant center of rotation of a wheel with respect to the body is called the *suspension roll center*. To find the roll center of the front or rear half of a car, we should determine the suspension roll centers, and find the intersection of the lines connecting the suspension roll centers to the center of their associated tireprints.

Three coordinate frames are employed to express the orientation of a tire and wheel with respect to the vehicle body: the wheel frame W, wheel-body frame C, and tire frame T. A wheel coordinate frame $W(x_w, y_w, z_w)$ is attached to the center of a wheel. It follows every translation and rotation of the wheel except the spin. Hence, the x_w and z_w axes are always in the tire-plane, while the y_w-axis is always along the spin axis. When the wheel is straight and the W frame is parallel to the vehicle coordinate frame, we attach a wheel-body coordinate frame $C(x_c, y_c, z_c)$ at the center of the wheel parallel to the vehicle coordinate axes. The wheel-body frame C is motionless with respect to the vehicle coordinate and does not follow any motion of the wheel. The tire coordinate frame $T(x_t, y_t, z_t)$ is set at the center of the tireprint. The z_t-axis is always perpendicular to the ground. The x_t-axis is along the intersection line of the tire-plane and the ground. The tire frame does not follow the spin and camber rotations of the tire; however, it follows the steer angle rotation about the z_c-axis.

We define the orientation and position of a steering axis by the caster angle φ, lean angle θ, and the intersection point of the axis with the ground surface at (s_a, s_b) with respect to the center of tireprint. Because of these parameters, a steered wheel will camber and generates a lateral force. This is called the *caster theory*. The camber angle γ of a steered wheel for $\theta = 0$, and $s_b = 0$ is:

$$
\begin{aligned}
\gamma &= \cos^{-1}\left(\hat{i}_c \cdot \hat{K}\right) \\
&= \cos^{-1} \frac{(z_W - z_T)}{\sqrt{(x_W - x_T)^2 + (y_W - y_T)^2 + (z_W - z_T)^2}}
\end{aligned}
\tag{8.140}
$$

where

$$
\begin{aligned}
x_T &= s_a\left(1 - \sin^2\varphi\right)(\cos\delta - 1) \tag{8.141} \\
y_T &= -s_a\cos\varphi\sin\delta \tag{8.142} \\
z_T &= z_T = -R_w - \frac{1}{2}s_a\sin 2\varphi\left(1 - \cos\delta\right) \tag{8.143}
\end{aligned}
$$

$$x_W = \left(s_a + \frac{1}{2}R_w \sin 2\varphi - s_a \sin^2 \varphi\right)(1 - \cos \delta) \qquad (8.144)$$

$$y_W = s_b (1 - \cos \delta) - R_w \sin \varphi + s_a \cos \varphi \sin \delta \qquad (8.145)$$

$$z_W = \left(R_w \left(\cos^2 \varphi - 1\right) - \frac{1}{2}s_a \sin 2\varphi\right)(1 - \cos \delta) \qquad (8.146)$$

8.8   Key Symbols

a, b, c, d	lengths of the links of a four-bar linkage
a_i	distance of the axle number i from the mass center
$A, B, \cdots$	coefficients in equation for calculating θ_3
b_1, b_2	distance of left and right wheels from mass center
$B(x, y, z)$	vehicle coordinate frame
C	mass center, coupler point
$C(x_c, y_c, z_c)$	wheel-body coordinate frame
$^C_T \mathbf{d}_W$	C expression of the position of W with respect to T
d	radial distance from center of a wheel
e, α	polar coordinates of a coupler point
g	overhang, gravitational acceleration
$\mathbf{g}$	gravitational acceleration vector
h	height of mass center
$h = z - z_0$	vertical displacement of the wheel center
h_r	roll height
H	wheel center drop
I_{ij}	instant center of rotation between link i and link j
$\overline{I_{ij} I_{mn}}$	a line connecting I_{ij} and I_{mn}
$\hat{I}, \hat{J}, \hat{K}$	unit vectors of the wheel-body frame C
$\mathbf{I}$	identity matrix
$J_1, J_2, \cdots$	length function for calculating θ_3
k	stiffness, speing stiffness
k_f	spring stiffness of the front wheels
k_i	spring stiffness of the wheel number i
k_r	spring stiffness of the rear wheel number i
k_φ	roll stiffness
l	wheelbase
$\hat{l}_c$	unit vector on the line (x_T, y_T, z_T) to (x_W, y_W, z_W)
m	mass
m_s	sprung mass
m_u	unsprung mass
M	moment
$\hat{n}_1$	normal unit vectors to π_L
$\hat{n}_2$	normal unit vectors to π_C
P	point
q, p, f	parameters for calculating couple point coordinate
$\mathbf{r}$	position vector
R	curvature radius of the road, radius of rotation
R_w	tire radius
$^T R_W$	rotation matrix to go from W frame to T frame
$\mathbf{s}$	position vector of the steer axis
s_a	forward location of the steer axis

s_b	lateral location of the steer axis
$\check{s}_W (0, \delta, \hat{u}, \mathbf{s})$	zero pitch screw about the steer axis
$T (x_t, y_t, z_t)$	tire coordinate system
$^T T_W$	homogeneous transformation to go from W to T
u_1, u_2, u_3	components of $\hat{u}$
$\hat{u}$	steer axis unit vector
$\tilde{u}$	skew symmetric matrix associated to $\hat{u}$
$\mathbf{u}_C$	position vector of the coupler point
$\hat{u}_z$	unit vector in the z-direction
v, v_x	forward speed
$\mathrm{vers}\, \delta$	$1 - \cos \delta$
w	track
$W (x_w y_w z_w)$	wheel coordinate system
x, y	suspension coordinate frame
x_C, y_C	coordinate of a couple point
x_T, y_T, z_T	wheel-body coordinates of the origin of T frame
x_W, y_W, z_W	wheel-body coordinates of the origin of W frame
$W (x_w y_w z_w)$	wheel coordinate system
z	vertical position of the wheel center
z_0	initial vertical position of the wheel center
α	angle of a coupler point with upper A-arm
γ	camber angle
δ	steer angle
$\varepsilon = m_s/m_u$	sprung to unsprung mass ratio
θ	lean angle
θ_0	angle between the ground link and the z-direction
θ_i	angular position of link number i
θ_2	angular position of the upper A-arm
θ_3	angular position of the coupler link
θ_4	angular position of link lower A-arm
θ_{i0}	initial angular position of θ_i
$\boldsymbol{\pi}_C$	caster plane
$\boldsymbol{\pi}_L$	lean plane
υ	thrust angle
φ	roll angle, caster angle, spin angle
ϕ	bank angle
ω	angular velocity
φ_c	critical roll angle

Exercises

1. Roll center.

   Determine the roll center of the kinematic models of vehicles shown in Figures 8.60 to 8.63.

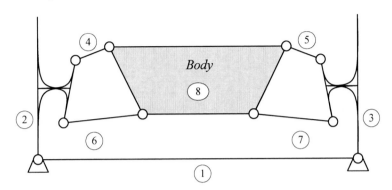

FIGURE 8.60.

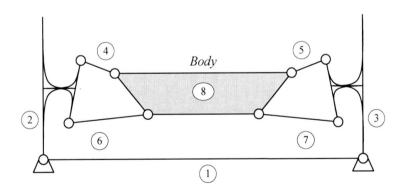

FIGURE 8.61.

2. Upper A-arm and roll center.

   Design the upper A-arm for the suspensions that are shown in Figures 8.64 to 8.66, such that the roll center of the vehicle is at point P.

3. Lower arm and roll center.

   Design the lower arm for the McPherson suspensions that are shown in Figures 8.67 to 8.69, such that the roll center of the vehicle is at point P.

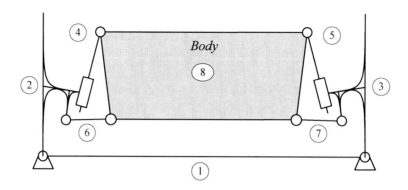

FIGURE 8.62.

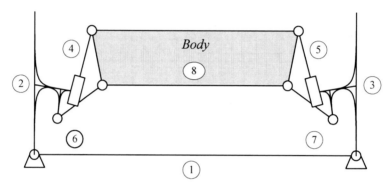

FIGURE 8.63.

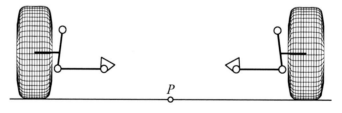

FIGURE 8.64.

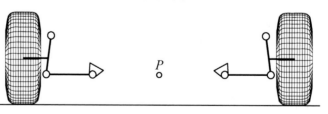

FIGURE 8.65.

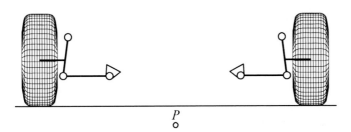

FIGURE 8.66.

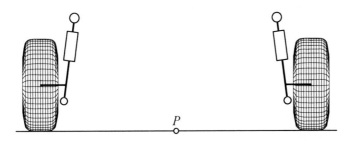

FIGURE 8.67.

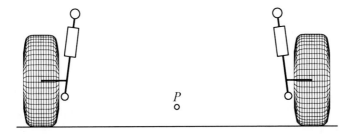

FIGURE 8.68.

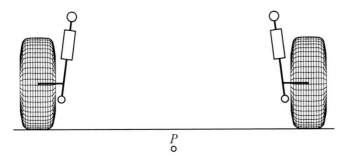

FIGURE 8.69.

4. ★ Position of the roll center and mass center.

Figure 8.70 illustrates the wheels and mass center C of a vehicle. Design a double A-arm suspension such that the roll center of the

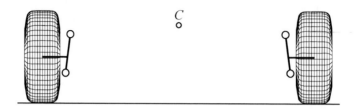

FIGURE 8.70.

vehicle is

(a) above C.

(b) on C.

(c) below C.

(d) Is it possible to make street cars with a roll center on or above C?

(e) What would be the advantages or disadvantages of a roll center on or above C.

5. Asymmetric position of the roll center.

Design double A-arm suspensions for the vehicle shown in 8.71, such that the roll center of the vehicle is at point P. What would be the advantages or disadvantages of an asymmetric roll center?

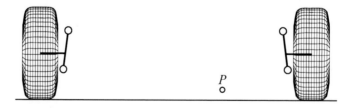

FIGURE 8.71.

6. ★ Camber angle variation.

Consider a double A-arm suspension such that is shown in Figure 8.72. Assume that the dimensions of the equivalent kinematic model

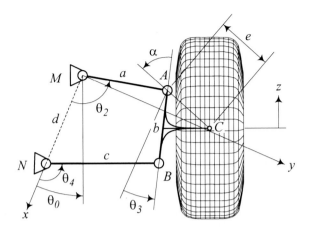

FIGURE 8.72. A double A-arm suspension kinematics.

are:

a = 22.57 cm $b = 18.88$ cm $c = 29.8$ cm $d = 24.8$ cm

θ_0 = 23.5 deg

and the coupler point C is at:

$$e = 14.8 \text{ cm} \qquad \alpha = 56.2 \text{ deg}$$

Draw a graph to show the variation of the camber angle, when the wheel is moving up and down.

7. ★ Steer axis unit vector.

Determine the C expression of the unit vector $\hat{u}$ on the steer axis, for a caster angle $\varphi = 15$ deg, and a lean angle $\theta = 8$ deg.

8. ★ Location vector and steer axis.

Determine the location vector $\mathbf{s}$, if the steer axis is going through the wheel center. The caster and lean angles are $\varphi = 10$ deg and $\theta = 0$ deg.

9. ★ Homogeneous transformation matrix $^C T_W$.

Determine $^C T_W$ for $\varphi = 8$ deg, $\theta = 12$ deg, and the location vector $^C \mathbf{s}$

$$^C \mathbf{s} = \begin{bmatrix} 3.8 \text{ cm} & 1.8 \text{ cm} & -R_w \end{bmatrix}^T$$

(a) The vehicle uses a tire $235/35ZR19$.

(b) The vehicle uses a tire $P215/65R15\ 96H$.

10. ★ Wheel drop.

Find the coordinates of the tireprint for

$$\varphi = 10\,\text{deg} \qquad \theta = 10\,\text{deg} \qquad {}^{C}\mathbf{s} = \begin{bmatrix} 3.8\,\text{cm} & 1.8\,\text{cm} & 38\,\text{cm} \end{bmatrix}$$

if $\delta = 18\,\text{deg}$. How much is the wheel drop H.

11. ★ Wheel drop and steer angle.

Draw a plot to show the wheel drop H at different steer angle δ for the given data in Exercise 10.

12. ★ Camber and steering.

Draw a plot to show the camber angle γ at different steer angle δ for the following characteristics:

$$\varphi = 10\,\text{deg} \qquad \theta = 0\,\text{deg} \qquad {}^{C}\mathbf{s} = \begin{bmatrix} 3.8\,\text{cm} & 0\,\text{cm} & 38\,\text{cm} \end{bmatrix}$$

13. ★ Parallelogram suspension.

If the double A-arm suspension is geometrically made by a parallelogram then, the wheel will always remain perpendicular to the road. This is the best design to have uniform tire wear. A parallelogram suspension however, has a variable track when the load of the car changes or the wheel moves up and down. The maximum track of a vehicle with such a suspension occurs when the upper and lower arms of the suspension are parallel to the ground such as the one illustrated in Figure 8.73.

Calculate and show the track changes of the suspension for $z = \pm 25\,\text{cm}$.

14. ★ Track change suspension.

If the rest position a parallelogram suspension is as shown in Figure 8.74 then, the track of the car will increase by increasing the load of the car. Having a wider track, increases the stability of the car, so such a suspension works well. Calculate and show the track changes of the suspension for $z = \pm 25\,\text{cm}$ if the initial value of $\theta_2 = \theta_4 = 75\,\text{deg}$.

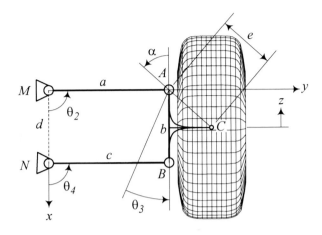

FIGURE 8.73. A parallelogram double A-arm suspension.

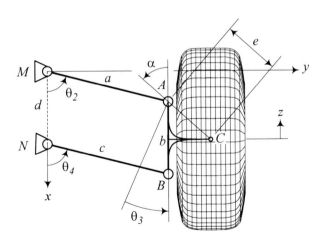

FIGURE 8.74. A tilted parallelogram double A-arm suspension.

Part III

Vehicle Dynamics

9

★ Applied Dynamics

Dynamics of a rigid vehicle may be considered as the motion of a rigid body with respect to a fixed global coordinate frame. The principles of Dynamics as well as Newton and Euler equations of motion that describe the translational and rotational motion of the rigid body are reviewed in this chapter.

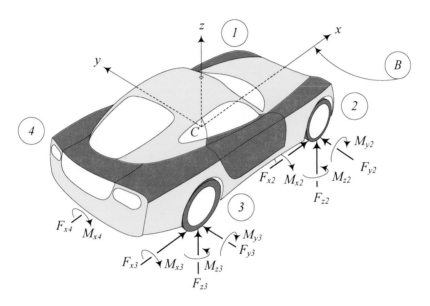

FIGURE 9.1. The force system of a vehicle is the applied forces and moments at the tireprints.

9.1 Elements of Dynamics

In this section, we will review the definition of the elements that are being used in Dynamics.

9.1.1 Force and Moment

In Newtonian dynamics, the forces acting on a system of connected rigid bodied can be divided into *internal* and *external forces*. Internal forces are

R.N. Jazar, *Vehicle Dynamics: Theory and Application*,
DOI 10.1007/978-1-4614-8544-5_9, © Springer Science+Business Media New York 2014

acting between connected bodies, and external forces are acting from out-side of the system. An external force can be a *contact force*, such as traction force at tireprint of a driving wheel, or a *body force*, such as gravitational force on vehicle's body.

The external forces and moments are called *load*, and a set of forces and moments acting on a rigid body, such as forces and moments on the vehicle shown in Figure 9.1, is called a *force system*. The *resultant* or *total force* **F** is the vectorial sum of all the external forces acting on a body, and the *resultant* or *total moment* **M** is the vectorial sum of all the moments of the external forces.

$$\mathbf{F} = \sum_i \mathbf{F}_i \tag{9.1}$$

$$\mathbf{M} = \sum_i \mathbf{M}_i \tag{9.2}$$

Consider a force **F** acting on a point P at $\mathbf{r}_P$. The *moment of the force* about a directional line l with unit vector $\hat{u}$ passing through the origin is

$$\mathbf{M}_l = \hat{u} \cdot (\mathbf{r}_P \times \mathbf{F}) \tag{9.3}$$

The moment of the force **F**, about a point Q at $\mathbf{r}_Q$ is

$$\mathbf{M}_Q = (\mathbf{r}_P - \mathbf{r}_Q) \times \mathbf{F} \tag{9.4}$$

so, the moment of **F** about the origin is

$$\mathbf{M} = \mathbf{r}_P \times \mathbf{F} \tag{9.5}$$

The moment of a force may also be called *torque* or *moment*.

The effect of a force system is equivalent to the effect of the resultant force and resultant moment of the force system. Any two force systems are equivalent if their resultant forces and resultant moments are equal. If the resultant force of a force system is zero, the resultant moment of the force system is independent of the origin of the coordinate frame. Such a resultant moment is called *couple*.

When a force system is reduced to a resultant $\mathbf{F}_P$ and $\mathbf{M}_P$ with respect to a reference point P, we may change the reference point to another point Q and find the new resultants as

$$\mathbf{F}_Q = \mathbf{F}_P \tag{9.6}$$

$$\mathbf{M}_Q = \mathbf{M}_P + (\mathbf{r}_P - \mathbf{r}_Q) \times \mathbf{F}_P = \mathbf{M}_P + {}_Q\mathbf{r}_P \times \mathbf{F}_P \tag{9.7}$$

9.1.2 Momentum

The *momentum* of a moving rigid body is a vector quantity equal to the total mass of the body times the translational velocity of the mass center of the body.

$$\mathbf{p} = m\mathbf{v} \tag{9.8}$$

The momentum **p** is also called *translational momentum* or *linear momentum*.

Consider a rigid body with momentum **p**. The *moment of momentum*, **L**, about a directional line l with directional unit vector $\hat{u}$ passing through the origin is

$$\mathbf{L}_l = \hat{u} \cdot (\mathbf{r}_C \times \mathbf{p}) \qquad (9.9)$$

where $\mathbf{r}_C$ is the position vector of the mass center C. The moment of momentum about the origin is

$$\mathbf{L} = \mathbf{r}_C \times \mathbf{p} \qquad (9.10)$$

The moment of momentum **L** is also called *angular momentum*.

9.1.3 Vectors

A vector expresses any physical quantity that can be represented by a directed section of a line with a start point, such as O, and an end point, such as P. A vector can be shown by an ordered pair of points with an arrow, such as $\overrightarrow{OP}$. Therefore, a sign $\overrightarrow{PP}$ indicates a zero vector at point P. A vector may have up to five characteristics: *length, axis, end point, direction, physical quantity*. However, the length and direction are necessary for every vector.

1. *Length.* The length of a vector corresponds to the magnitude of the physical quantity that the vector is representing.

2. *Axis.* The straight line on which the vector sits, is the axis of the vector. The vector axis is also called the *line of action*.

3. *End point.* A start or an end point, called the *affecting point*, indicates the point at which the vector is applied.

4. *Direction.* The direction of a vector indicates at what direction on the axis the vector is pointing.

5. *Physical quantity.* Any vector represents a physical quantity. If a physical quantity can be represented by a vector, it is called a *vectorial physical quantity*. The magnitude of the quantity is proportional to the length of the vector. Although a vector may be dimensionless, a vector that represents no physical quantity is meaningless, .

Depending on the physical quantity and application, there are seven types of vectors: *vecpoint, vecline, vecface, vecfree, vecpoline, vecpoface, pecporee*.

1. *Vecpoint*. When all of the vector characteristics; length, axis, end point, direction, and physical quantity; are specified, the vector is called a *bounded vector*, *point vector*, or *vecpoint*. Such vecpoint is fixed at a point with no movability.

2. *Vecline*. If the affecting point of a vector are not fixed on the axis, the vector is called a *sliding vector*, *line vector*, or *vecline*. A vecline is free to slide on its axis.

3. *Vecface*. If the affecting point of a vector can move on a surface while the vector displaces parallel to itself, the vector is called a *surface vector* or *vecface*. If the surface is a plane, then the vector is a *plane vector* or *veclane*.

4. *Vecfree*. If the axis of a vector is not fixed, the vector is called a *free vector*, *direction vector,* or *vecfree*. A vecfree can move parallel to itself to any point of a specified space while keeping its direction.

5. *Vecpoline*. If the start point of a vector is fixed while the end point can slide on a line, the vector is a *point-line vector* or *vecpoline*. A vecpoline has a constraint variable length and orientation. However, if the start and end points of a vecpoline are on the sliding line, its orientation is constant.

6. *Vecpoface*. If the start point of a vector is fixed, while the end point can slide on a surface, the vector is a *point-surface vector* or *vecpoface*. A vecpoface has a constraint variable length and orientation. If the surface is a plane, the vector is called a *point-plane vector* or *vecpolane*. The start point of a vecpoface may also be on the sliding surface.

7. *Vecporee*. When the start point of a vector is fixed and the end point can move anywhere in a specified space, the vector is called a *point-free vector* or *vecporee*. A vecporee has a variable length and orientation.

Two vectors are comparable only if they represent the same physical quantity and are expressed in the same coordinate frame. Two vectors are equal if they are comparable and are the same type and have the same characteristics. Two vectors are equivalent if they are comparable and the same type and can be substituted with each other.

Vectors can only be added if they are coaxial. In case the vectors are not coaxial, the decomposed expression of vectors must be used to add the vectors.

Force is a sliding vector and couple and moment are free vectors.

9.1.4 Equation of Motion

The application of a force system is emphasized by *Newton's second and third laws of motion*. The second law of motion, also called the *Newton's equation of motion*, states that the global rate of change of *linear momentum* is proportional to the global *applied force*.

$$^G\mathbf{F} = \frac{^Gd}{dt}\,{}^G\mathbf{p} = \frac{^Gd}{dt}\left(m\,{}^G\mathbf{v}\right) \tag{9.11}$$

The third Newton's law of motion states that the action and reaction forces acting between two bodies are equal and opposite.

The second law of motion can be expanded to include rotational motions. Hence, the second law of motion also states that the global rate of change of *angular momentum* is proportional to the global *applied moment*.

$$^G\mathbf{M} = \frac{^Gd}{dt}\,{}^G\mathbf{L} \tag{9.12}$$

Proof. Differentiating from moment of momentum (9.10) shows that

$$\frac{^Gd}{dt}\,{}^G\mathbf{L} = \frac{^Gd}{dt}\left(\mathbf{r}_C \times \mathbf{p}\right) = \left(\frac{^Gd\mathbf{r}_C}{dt} \times \mathbf{p} + \mathbf{r}_C \times \frac{^Gd\mathbf{p}}{dt}\right)$$

$$= {}^G\mathbf{r}_C \times \frac{^Gd\mathbf{p}}{dt} = {}^G\mathbf{r}_C \times {}^G\mathbf{F} = {}^G\mathbf{M} \tag{9.13}$$

∎

9.1.5 Work and Energy

Kinetic energy K of a moving body point P with mass m at a position $^G\mathbf{r}_P$, and having a velocity $^G\mathbf{v}_P$, is

$$K = \frac{1}{2}m\,{}^G v_P^2 = \frac{1}{2}m\left({}^G\dot{\mathbf{d}}_B + {}^B\mathbf{v}_P + {}^B_G\boldsymbol{\omega}_B \times {}^B\mathbf{r}_P\right)^2 \tag{9.14}$$

where, G indicates the global coordinate frame in which the velocity vector $\mathbf{v}_P$ is expressed. The work done by the applied force $^G\mathbf{F}$ on m in moving from point 1 to point 2 on a path, indicated by a vector $^G\mathbf{r}$, is

$$_1W_2 = \int_1^2 {}^G\mathbf{F} \cdot d\,{}^G\mathbf{r} \tag{9.15}$$

However,

$$\int_1^2 {}^G\mathbf{F} \cdot d\,{}^G\mathbf{r} = m\int_1^2 \frac{^Gd}{dt}\,{}^G\mathbf{v} \cdot {}^G\mathbf{v}\,dt = \frac{1}{2}m\int_1^2 \frac{d}{dt}v^2\,dt$$

$$= \frac{1}{2}m\left(v_2^2 - v_1^2\right) = K_2 - K_1 \tag{9.16}$$

that shows $_1W_2$ is equal to the difference of the kinetic energy between terminal and initial points.

$$_1W_2 = K_2 - K_1 \tag{9.17}$$

Equation (9.17) is called *principle of work and energy*. If there is a scalar *potential field function* $V = V(x, y, z)$ such that

$$\mathbf{F} = -\nabla V = -\frac{dV}{d\mathbf{r}} = -\left(\frac{\partial V}{\partial x}\hat{\imath} + \frac{\partial V}{\partial y}\hat{\jmath} + \frac{\partial V}{\partial z}\hat{k}\right) \tag{9.18}$$

then the principle of work and energy simplifies to the principle of *conservation of energy*,

$$K_1 + V_1 = K_2 + V_2 \tag{9.19}$$

The value of the potential field function $V = V(x, y, z)$ is the *potential energy* of the system.

Example 374 *Position of mass center.*
    The position of the mass center of a rigid body in a coordinate frame is indicated by $^B\mathbf{r}_C$ *and is usually measured in a body coordinate frame* B.

$$^B\mathbf{r}_C = \frac{1}{m}\int_B {}^B\mathbf{r}\,dm \tag{9.20}$$

$$\begin{bmatrix} x_C \\ y_C \\ z_C \end{bmatrix} = \begin{bmatrix} \frac{1}{m}\int_B x\,dm \\ \frac{1}{m}\int_B y\,dm \\ \frac{1}{m}\int_B z\,dm \end{bmatrix} \tag{9.21}$$

Applying the mass center integral on the symmetric and uniform L-section rigid body with $\rho = 1$ *shown in Figure 9.2 provides us with the position of mass center* C *of the section. The* x *position of* C *is*

$$x_C = \frac{1}{m}\int_B x\,dm = \frac{1}{A}\int_B x\,dA = -\frac{b^2 + ab - a^2}{4ab + 2a^2} \tag{9.22}$$

and because of symmetry, we have

$$y_C = -x_C = \frac{b^2 + ab - a^2}{4ab + 2a^2} \tag{9.23}$$

When $a = b$, *the position of* C *reduces to*

$$y_C = -x_C = \frac{1}{2}b \tag{9.24}$$

Example 375 ★ *Every force system is equivalent to a wrench.*
    The Poinsot theorem states: Every force system is equivalent to a single force, plus a moment parallel to the force. Let $\mathbf{F}$ *and* $\mathbf{M}$ *be the resultant force*

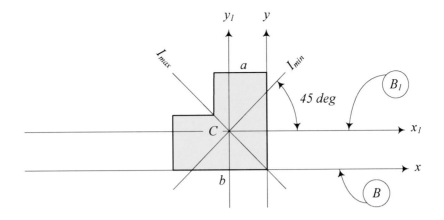

FIGURE 9.2. Principal coordinate frame for a symmetric L-section.

and moment of a force system. We decompose the moment into parallel and perpendicular components, $\mathbf{M}_{\parallel}$ *and* $\mathbf{M}_{\perp}$*, to the force axis. The force* $\mathbf{F}$ *and the perpendicular moment* $\mathbf{M}_{\perp}$ *can be replaced by a single force* $\mathbf{F}'$ *parallel to* $\mathbf{F}$*. Therefore, the force system is reduced to a force* $\mathbf{F}'$ *and a moment* $\mathbf{M}_{\parallel}$ *parallel to each other. A force and a moment about the force axis is called a* ***wrench****.*

The Poinsot theorem is similar to the Chasles theorem that states: Every rigid body motion is equivalent to a screw, which is a translation plus a rotation about the axis of translation.

There is no simple relationship between screw and wrench similar to Newton equation of motion. If there was such relation as the second time derivative of screw is proportional to the applied wrench then we would have an equation of motion as a combination of Newton and Euler equation.

Example 376 ★ *Motion of a moving point in a moving body frame.*

The velocity and acceleration of a moving point P *as shown in Figure 5.9 are found in Example 216.*

$$
{}^{G}\mathbf{v}_P = {}^{G}\dot{\mathbf{d}}_B + {}^{G}R_B \left({}^{B}\mathbf{v}_P + {}_{G}^{B}\boldsymbol{\omega}_B \times {}^{B}\mathbf{r}_P \right) \tag{9.25}
$$

$$
{}^{G}\mathbf{a}_P = {}^{G}\ddot{\mathbf{d}}_B + {}^{G}R_B \left({}^{B}\mathbf{a}_P + 2\,{}_{G}^{B}\boldsymbol{\omega}_B \times {}^{B}\mathbf{v}_P + {}_{G}^{B}\dot{\boldsymbol{\omega}}_B \times {}^{B}\mathbf{r}_P \right)
$$
$$
+ {}^{G}R_B \left({}_{G}^{B}\boldsymbol{\omega}_B \times \left({}_{G}^{B}\boldsymbol{\omega}_B \times {}^{B}\mathbf{r}_P \right) \right) \tag{9.26}
$$

Therefore, the equation of motion for the point mass P *is*

$$
{}^{G}\mathbf{F} = m\,{}^{G}\mathbf{a}_P
$$
$$
= m \left({}^{G}\ddot{\mathbf{d}}_B + {}^{G}R_B \left({}^{B}\mathbf{a}_P + 2\,{}_{G}^{B}\boldsymbol{\omega}_B \times {}^{B}\mathbf{v}_P + {}_{G}^{B}\dot{\boldsymbol{\omega}}_B \times {}^{B}\mathbf{r}_P \right) \right)
$$
$$
+ m\,{}^{G}R_B \left({}_{G}^{B}\boldsymbol{\omega}_B \times \left({}_{G}^{B}\boldsymbol{\omega}_B \times {}^{B}\mathbf{r}_P \right) \right) \tag{9.27}
$$

Example 377 *Newton's equation in a rotating frame.*

Consider a spherical rigid body, such as Earth, with a fixed point that is rotating with a constant angular velocity ω. The equation of motion for a moving point vehicle P on the rigid body is found by setting $^G\ddot{\mathbf{d}}_B = {}^B_G\dot{\boldsymbol{\omega}}_B = 0$ in the equation of motion of a moving point in a moving body frame (9.27)

$$^B\mathbf{F} = m\,{}^B\mathbf{a}_P + m\,{}^B_G\boldsymbol{\omega}_B \times \left({}^B_G\boldsymbol{\omega}_B \times {}^B\mathbf{r}_P\right) + 2m\,{}^B_G\boldsymbol{\omega}_B \times {}^B\dot{\mathbf{r}}_P \quad (9.28)$$
$$\neq m\,{}^B\mathbf{a}_P$$

which shows that the Newton's equation of motion $\mathbf{F} = m\,\mathbf{a}$ must be modified for rotating frames.

Example 378 *Coriolis force.*

The equation of motion of a moving vehicle point on the surface of the Earth is

$$^B\mathbf{F} = m\,{}^B\mathbf{a}_P + m\,{}^B_G\boldsymbol{\omega}_B \times \left({}^B_G\boldsymbol{\omega}_B \times {}^B\mathbf{r}_P\right) + 2m\,{}^B_G\boldsymbol{\omega}_B \times {}^B\mathbf{v}_P \quad (9.29)$$

which can be rearranged to

$$^B\mathbf{F} - m\,{}^B_G\boldsymbol{\omega}_B \times \left({}^B_G\boldsymbol{\omega}_B \times {}^B\mathbf{r}_P\right) - 2m\,{}^B_G\boldsymbol{\omega}_B \times {}^B\mathbf{v}_P = m\,{}^B\mathbf{a}_P \quad (9.30)$$

Equation (9.30) is the equation of motion for an observer in the rotating frame, which in this case is an observer on the Earth. The left-hand side of this equation is called the **effective force** $\mathbf{F}_{eff}$,

$$\mathbf{F}_{eff} = {}^B\mathbf{F} - m\,{}^B_G\boldsymbol{\omega}_B \times \left({}^B_G\boldsymbol{\omega}_B \times {}^B\mathbf{r}_P\right) - 2m\,{}^B_G\boldsymbol{\omega}_B \times {}^B\mathbf{v}_P \quad (9.31)$$

because it seems that the particle is moving in the body coordinate B under the influence of this force.

The second term is negative of the centrifugal force and pointing outward. The maximum value of this force on the Earth is on the equator which is about 0.3% of the acceleration of gravity.

$$r\omega^2 = 6378.388 \times 10^3 \times \left(\frac{2\pi}{24 \times 3600} \frac{366.25}{365.25}\right)^2$$
$$= 3.3917 \times 10^{-2}\,\mathrm{m/s^2} \quad (9.32)$$

If we add the variation of the gravitational acceleration because of a change of radius from $R = 6356912\,\mathrm{m}$ at the pole to $R = 6378388\,\mathrm{m}$ on the equator, then the variation of the acceleration of gravity becomes 0.53%. So, generally speaking, a sportsman such as a pole-vaulter who has practiced in the north pole can show a better record in a competition held on the equator.

The third term of the effective force is called the **Coriolis force** or **Coriolis effect**, F_C, which is perpendicular to both ω and $^B\mathbf{v}_P$. For a mass m moving on the north hemisphere at a latitude θ towards the equator, we *should provide a lateral eastward force equal to the Coriolis effect to force the mass, to keep its direction relative to the ground.*

$$F_C = 2m\,_G^B\boldsymbol{\omega}_B \times {}^B\mathbf{v}_m = 1.4584 \times 10^{-4}\,{}^B\mathbf{p}_m \cos\theta \ \ \mathrm{kg\,m/s^2} \qquad (9.33)$$

The Coriolis effect is the reason why the west side of railways, roads, and rivers wear faster than east side. The lack of providing the Coriolis force is the reason for turning the direction of winds, projectiles, flood, and falling objects.

Example 379 *Work, force, and kinetic energy in a unidirectional motion.*

A vehicle with mass $m = 1200\,\mathrm{kg}$ has an initial kinetic energy $K = 6000\,\mathrm{J}$. The mass is under a constant force $\mathbf{F} = F\hat{I} = 4000\hat{I}$ and moves from $X(0) = 0$ to $X(t_f) = 1000\,\mathrm{m}$ at a terminal time t_f. The work done by the force during this motion is

$$W = \int_{\mathbf{r}(0)}^{\mathbf{r}(t_f)} \mathbf{F} \cdot d\mathbf{r} = \int_0^{1000} 4000\,dX = 4 \times 10^6\,\mathrm{N\,m} = 4\,\mathrm{MJ} \qquad (9.34)$$

The kinetic energy at the terminal time is

$$K(t_f) = W + K(0) = 4006000\,\mathrm{J} \qquad (9.35)$$

which shows that the terminal speed of the mass is

$$v_2 = \sqrt{\frac{2K(t_f)}{m}} \approx 81.7\,\mathrm{m/s} \approx 22.694\,\mathrm{km/h} \qquad (9.36)$$

Example 380 *Direct and inverse dynamics dynamics.*
   When the applied force is time varying and is a known function, then,

$$\mathbf{F}(t) = m\,\ddot{\mathbf{r}} \qquad (9.37)$$

The general solution for the equation of motion can be found by integration.

$$\dot{\mathbf{r}}(t) = \dot{\mathbf{r}}(t_0) + \frac{1}{m}\int_{t_0}^{t} \mathbf{F}(t)dt \qquad (9.38)$$

$$\mathbf{r}(t) = \mathbf{r}(t_0) + \dot{\mathbf{r}}(t_0)(t - t_0) + \frac{1}{m}\int_{t_0}^{t}\int_{t_0}^{t} \mathbf{F}(t)dt\,dt \qquad (9.39)$$

This kind of problem is called direct or forward dynamics.
   If the path of motion $\mathbf{r}(t)$ is a known function of time, then the required force to move the system on the path would be found by differentiation.

$$\mathbf{F}(t) = \frac{d^2}{dt^2}\left(m\,\ddot{\mathbf{r}}\right) \qquad (9.40)$$

This kind of problem is called indirect or inverse dynamics.

Example 381 ★ *Force function in equation of motion.*

By definition, force is qualitatively whatever that changes the motion, and quantitatively, is whatever that is equal to mass times acceleration. Mathematically, the equation of motion provides us with a vectorial second-order differential equation

$$m\ddot{\mathbf{r}} = \mathbf{F}\left(\dot{\mathbf{r}}, \mathbf{r}, t\right) \tag{9.41}$$

It is assumed that the force function is generally only a function of time t, position $\mathbf{r}$, and velocity $\dot{\mathbf{r}}$. In other words, the Newton equation of motion is correct as long as we can show that the force is only a function of $\dot{\mathbf{r}}, \mathbf{r}, t$.

If there is an applied force that depends on the acceleration, jerk, or other variables that cannot be reduced to $\dot{\mathbf{r}}, \mathbf{r}, t$, the system is not Newtonian. There is no known equation of motion for non-Newtonian dynamic systems, because

$$\mathbf{F}\left(\mathbf{r}, \dot{\mathbf{r}}, \ddot{\mathbf{r}}, \dddot{\mathbf{r}}, \cdots, t\right) \neq m\ddot{\mathbf{r}} \tag{9.42}$$

In Newtonian mechanics, we assume that force can only be a function of $\dot{\mathbf{r}}, \mathbf{r}, t$. In real world, force may be a function of everything; however, we always ignore any other variables than $\dot{\mathbf{r}}, \mathbf{r}, t$.

Because Equation (9.41) is a linear equation for force $\mathbf{F}$, it accepts the force superposition principle. When a mass m is affected by several forces $\mathbf{F}_1$, $\mathbf{F}_2$, $\mathbf{F}_3$, $\cdots$, we may calculate their summation vectorially

$$\mathbf{F} = \mathbf{F}_1 + \mathbf{F}_2 + \mathbf{F}_3 + \cdots \tag{9.43}$$

and apply the resultant force on m. Therefore, if a force $\mathbf{F}_1$ provides us with acceleration $\ddot{\mathbf{r}}_1$, and $\mathbf{F}_2$ provides us with $\ddot{\mathbf{r}}_2$,

$$m\ddot{\mathbf{r}}_1 = \mathbf{F}_1 \qquad m\ddot{\mathbf{r}}_2 = \mathbf{F}_2 \tag{9.44}$$

then the resultant force $\mathbf{F}_3 = \mathbf{F}_1 + \mathbf{F}_2$ provides us with the acceleration $\ddot{\mathbf{r}}_3$ such that

$$\ddot{\mathbf{r}}_3 = \ddot{\mathbf{r}}_1 + \ddot{\mathbf{r}}_2 \tag{9.45}$$

To see that the Newton equation of motion is not correct when the force is not only a function of $\dot{\mathbf{r}}, \mathbf{r}, t$, let us assume that a particle with mass m is under two acceleration dependent forces $F_1(\ddot{x})$ and $F_2(\ddot{x})$ on x-axis.

$$m\ddot{x}_1 = F_1(\ddot{x}_1) \qquad m\ddot{x}_2 = F_2(\ddot{x}_2) \tag{9.46}$$

The acceleration of m under the action of both forces would be $\ddot{x}_3$

$$m\ddot{x}_3 = F_1(\ddot{x}_3) + F_2(\ddot{x}_3) \tag{9.47}$$

however, we must have

$$\ddot{x}_3 = \ddot{x}_1 + \ddot{x}_2 \tag{9.48}$$

while we have:

$$\begin{aligned} m\left(\ddot{x}_1 + \ddot{x}_2\right) &= F_1(\ddot{x}_1 + \ddot{x}_2) + F_2(\ddot{x}_1 + \ddot{x}_2) \\ &\neq F_1(\ddot{x}_1) + F_2(\ddot{x}_2) \end{aligned} \tag{9.49}$$

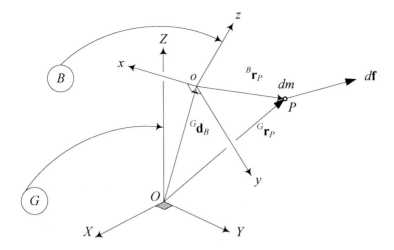

FIGURE 9.3. A body point mass moving with velocity $^G\mathbf{v}_P$ and acted on by force $d\mathbf{f}$.

9.2 Rigid Body Translational Dynamics

Figure 9.3 depicts a moving body B in a global coordinate frame G. Assume that the body frame is attached at the mass center of the body. Point P indicates an infinitesimal sphere of the body, which has a very small mass dm. The point mass dm is acted on by an infinitesimal force $d\mathbf{f}$ and has a global velocity $^G\mathbf{v}_P$.

According to Newton's law of motion we have

$$d\mathbf{f} = {}^G\mathbf{a}_P \, dm \tag{9.50}$$

However, the equation of motion for the whole body in the global coordinate frame is

$${}^G\mathbf{F} = m \, {}^G\mathbf{a}_B \tag{9.51}$$

which can be expressed in the body coordinate frame as

$${}^B\mathbf{F} = m \, {}^B_G\mathbf{a}_B + m \, {}^B_G\boldsymbol{\omega}_B \times {}^B\mathbf{v}_B \tag{9.52}$$

$$\begin{bmatrix} F_x \\ F_y \\ F_z \end{bmatrix} = \begin{bmatrix} ma_x + m\left(\omega_y v_z - \omega_z v_y\right) \\ ma_y - m\left(\omega_x v_z - \omega_z v_x\right) \\ ma_z + m\left(\omega_x v_y - \omega_y v_x\right) \end{bmatrix} \tag{9.53}$$

In these equations, $\mathbf{a}_B$ is the acceleration vector of the body mass center C in the global frame, m is the total mass of the body, and $\mathbf{F}$ is the resultant of the external forces acted on the body at C.

Proof. A body coordinate frame at the mass center is called a *central frame*. If frame B is central, then the *center of mass*, C, is defined such that

$$\int_B {}^B\mathbf{r}_{dm}\, dm = 0 \tag{9.54}$$

The global position vector of dm is related to its local position vector by

$$^G\mathbf{r}_{dm} = {}^G\mathbf{d}_B + {}^G R_B \, {}^B\mathbf{r}_{dm} \tag{9.55}$$

where $^G\mathbf{d}_B$ is the global position vector of the central body frame, and therefore,

$$\begin{aligned}\int_B {}^G\mathbf{r}_{dm}\, dm &= \int_B {}^G\mathbf{d}_B\, dm + {}^G R_B \int_m {}^B\mathbf{r}_{dm}\, dm \\ &= \int_B {}^G\mathbf{d}_B\, dm = {}^G\mathbf{d}_B \int_B dm = m\, {}^G\mathbf{d}_B \end{aligned} \tag{9.56}$$

A time derivative of both sides shows that

$$m\, {}^G\dot{\mathbf{d}}_B = m\, {}^G\mathbf{v}_B = \int_B {}^G\dot{\mathbf{r}}_{dm}\, dm = \int_B {}^G\mathbf{v}_{dm}\, dm \tag{9.57}$$

and another derivative is

$$m\, {}^G\dot{\mathbf{v}}_B = m\, {}^G\mathbf{a}_B = \int_B {}^G\dot{\mathbf{v}}_{dm}\, dm \tag{9.58}$$

However, we have $d\mathbf{f} = {}^G\dot{\mathbf{v}}_P\, dm$ and therefore,

$$m\, {}^G\mathbf{a}_B = \int_B d\mathbf{f} \tag{9.59}$$

The integral on the right-hand side is the resultant of all the forces acting on the body. The internal forces cancel one another out, so the net result is the vector sum of all the externally applied forces, $\mathbf{F}$, and therefore,

$$^G\mathbf{F} = m\, {}^G\mathbf{a}_B = m\, {}^G\dot{\mathbf{v}}_B \tag{9.60}$$

In the body coordinate frame we have

$$\begin{aligned}^B\mathbf{F} &= {}^B R_G\, {}^G\mathbf{F} = m\, {}^B R_G\, {}^G\mathbf{a}_B = m\, {}^B_G\mathbf{a}_B \\ &= m\, {}^B\mathbf{a}_B + m\, {}^B_G\boldsymbol{\omega}_B \times {}^B\mathbf{v}_B \end{aligned} \tag{9.61}$$

The expanded form of the Newton's equation in the body coordinate frame is then equal to

$$^B\mathbf{F} = m\, {}^B\mathbf{a}_B + m\, {}^B_G\boldsymbol{\omega}_B \times {}^B\mathbf{v}_B \tag{9.62}$$

$$\begin{bmatrix} F_x \\ F_y \\ F_z \end{bmatrix} = m \begin{bmatrix} a_x \\ a_y \\ a_z \end{bmatrix} + m \begin{bmatrix} \omega_x \\ \omega_y \\ \omega_z \end{bmatrix} \times \begin{bmatrix} v_x \\ v_y \\ v_z \end{bmatrix}$$

$$= \begin{bmatrix} ma_x + m\,(\omega_y v_z - \omega_z v_y) \\ ma_y - m\,(\omega_x v_z - \omega_z v_x) \\ ma_z + m\,(\omega_x v_y - \omega_y v_x) \end{bmatrix} \tag{9.63}$$

■

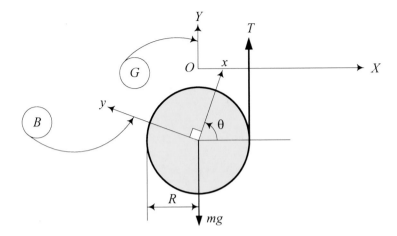

FIGURE 9.4. A ribbon of negligible weight and thickness that is wound tightly around a uniform massive disc.

Example 382 *A wound ribbon.*

Figure 9.4 illustrates a ribbon of negligible weight and thickness that is wound tightly around a uniform massive disc of radius R and mass m. The ribbon is fastened to a rigid support, and the disc is released to roll down vertically. There are two forces acting on the disc during the motion, its weight mg and the tension of the ribbon T. The translational equation of motion of the disc is expressed easier in the global coordinate frame:

$$\sum F_Y = -mg + T = m\ddot{Y} \tag{9.64}$$

The rotational equation of motion is simpler if expressed in the body coordinate frame:

$$\sum M_z = TR = {}^{B}I \, {}^{B}_{G}\dot{\boldsymbol{\omega}}_B + {}^{B}_{G}\boldsymbol{\omega}_B \times {}^{B}I \, {}^{B}_{G}\boldsymbol{\omega}_B = I\ddot{\theta} \tag{9.65}$$

There is a constraint between the coordinates Y and θ:

$$Y = Y_0 - R\theta \tag{9.66}$$

To solve the motion, let us eliminate T between (9.64) and (9.65) to obtain

$$m\ddot{Y} = -mg + \frac{I}{R}\ddot{\theta} \tag{9.67}$$

and use the constraint to eliminate $\ddot{Y}$:

$$\ddot{\theta} = \frac{mg}{\left(\dfrac{I}{R} + mR\right)} \tag{9.68}$$

Now, we can find T, $\ddot{Y}$ and Y:

$$T = \frac{I}{R}\ddot{\theta} = \frac{I}{mR^2 + I}mg \tag{9.69}$$

$$\ddot{Y} = -\frac{mR^2}{mR^2 + I}g \qquad Y = -\frac{mR^2}{mR^2 + I}gt^2 + \dot{Y}(0)\,t + Y(0) \tag{9.70}$$

For a point mass with $I = 0$, the falling acceleration is the same as the free fall of a particle. However, being a rigid body and having $I \neq 0$, the falling acceleration of the disc will be less. This is because the kinetic energy of the disc splits between rotation and translation.

9.3   Rigid Body Rotational Dynamics

The rigid body rotational equation of motion is the *Euler equation*

$$
\begin{aligned}
{}^B\mathbf{M} &= \frac{{}^G d}{dt}\,{}^B\mathbf{L} = {}^B\dot{\mathbf{L}} + {}^B_G\boldsymbol{\omega}_B \times {}^B\mathbf{L} \\
&= {}^BI\,{}^B_G\dot{\boldsymbol{\omega}}_B + {}^B_G\boldsymbol{\omega}_B \times \left({}^BI\,{}^B_G\boldsymbol{\omega}_B\right)
\end{aligned}
\tag{9.71}
$$

where $\mathbf{L}$ is the *angular momentum*

$$
{}^B\mathbf{L} = {}^BI\,{}^B_G\boldsymbol{\omega}_B \tag{9.72}
$$

and I is the *mass moment* of the rigid body.

$$
I = \begin{bmatrix} I_{xx} & I_{xy} & I_{xz} \\ I_{yx} & I_{yy} & I_{yz} \\ I_{zx} & I_{zy} & I_{zz} \end{bmatrix} \tag{9.73}
$$

The elements of I are functions of the mass distribution of the rigid body and can be defined by

$$
I_{ij} = \int_B \left(r_i^2 \delta_{mn} - x_{im}x_{jn}\right) dm \qquad i,j = 1,2,3 \tag{9.74}
$$

where δ_{ij} is Kronecker's delta.

$$
\delta_{mn} = \begin{cases} 1 & if \quad m = n \\ 0 & if \quad m \neq n \end{cases} \tag{9.75}
$$

The expanded form of the Euler equation (9.71) is

$$
\begin{aligned}
M_x &= I_{xx}\dot{\omega}_x + I_{xy}\dot{\omega}_y + I_{xz}\dot{\omega}_z - (I_{yy} - I_{zz})\,\omega_y\omega_z \\
&\quad - I_{yz}\left(\omega_z^2 - \omega_y^2\right) - \omega_x\left(\omega_z I_{xy} - \omega_y I_{xz}\right)
\end{aligned}
\tag{9.76}
$$

$$M_y = I_{yx}\dot{\omega}_x + I_{yy}\dot{\omega}_y + I_{yz}\dot{\omega}_z - (I_{zz} - I_{xx})\,\omega_z\omega_x$$
$$-I_{xz}\left(\omega_x^2 - \omega_z^2\right) - \omega_y\left(\omega_x I_{yz} - \omega_z I_{xy}\right) \tag{9.77}$$

$$M_z = I_{zx}\dot{\omega}_x + I_{zy}\dot{\omega}_y + I_{zz}\dot{\omega}_z - (I_{xx} - I_{yy})\,\omega_x\omega_y$$
$$-I_{xy}\left(\omega_y^2 - \omega_x^2\right) - \omega_z\left(\omega_y I_{xz} - \omega_x I_{yz}\right) \tag{9.78}$$

which can be reduced to

$$
\begin{aligned}
M_1 &= I_1\dot{\omega}_1 - (I_2 - I_2)\,\omega_2\omega_3 \\
M_2 &= I_2\dot{\omega}_2 - (I_3 - I_1)\,\omega_3\omega_1 \\
M_3 &= I_3\dot{\omega}_3 - (I_1 - I_2)\,\omega_1\omega_2
\end{aligned}
\tag{9.79}
$$

in a special Cartesian coordinate frame called the *principal coordinate frame*. The principal coordinate frame is denoted by numbers 123 to indicate the first, second, and third *principal axes*. The parameters I_{ij}, $i \neq j$ are zero in the principal frame. The body and principal coordinate frame sit at the mass center C.

Kinetic energy of a rotating rigid body is

$$
\begin{aligned}
K &= \frac{1}{2}\left(I_{xx}\omega_x^2 + I_{yy}\omega_y^2 + I_{zz}\omega_z^2\right) - I_{xy}\omega_x\omega_y - I_{yz}\omega_y\omega_z - I_{zx}\omega_z\omega_x \\
&= \frac{1}{2}\boldsymbol{\omega} \cdot \mathbf{L} = \frac{1}{2}\boldsymbol{\omega}^T I \,\boldsymbol{\omega}
\end{aligned}
\tag{9.80}
$$

that in the principal coordinate frame reduces to

$$K = \frac{1}{2}\left(I_1\omega_1^2 + I_2\omega_2^2 + I_3\omega_3^2\right) \tag{9.81}$$

Proof. Let m_i be the mass of the ith particle of a rigid body B, which is made of n particles and

$$\mathbf{r}_i = {}^B\mathbf{r}_i = \begin{bmatrix} x_i & y_i & z_i \end{bmatrix}^T \tag{9.82}$$

is the Cartesian position vector of m_i in a central body fixed coordinate frame $Oxyz$. Assume that

$$\boldsymbol{\omega} = {}^B_G\boldsymbol{\omega}_B = \begin{bmatrix} \omega_x & \omega_y & \omega_z \end{bmatrix}^T \tag{9.83}$$

is the angular velocity of the rigid body with respect to the ground, expressed in the body coordinate frame.

The angular momentum of m_i is

$$
\begin{aligned}
\mathbf{L}_i &= \mathbf{r}_i \times m_i\dot{\mathbf{r}}_i = m_i\left[\mathbf{r}_i \times (\boldsymbol{\omega} \times \mathbf{r}_i)\right] \\
&= m_i\left[(\mathbf{r}_i \cdot \mathbf{r}_i)\,\boldsymbol{\omega} - (\mathbf{r}_i \cdot \boldsymbol{\omega})\,\mathbf{r}_i\right] \\
&= m_i r_i^2\boldsymbol{\omega} - m_i\left(\mathbf{r}_i \cdot \boldsymbol{\omega}\right)\mathbf{r}_i
\end{aligned}
\tag{9.84}
$$

Hence, the angular momentum of the rigid body would be

$$\mathbf{L} = \boldsymbol{\omega} \sum_{i=1}^{n} m_i r_i^2 - \sum_{i=1}^{n} m_i \left(\mathbf{r}_i \cdot \boldsymbol{\omega}\right) \mathbf{r}_i \tag{9.85}$$

Substitution for $\mathbf{r}_i$ and $\boldsymbol{\omega}$ provides us with

$$
\begin{aligned}
\mathbf{L} &= \left(\omega_x \hat{\imath} + \omega_y \hat{\jmath} + \omega_z \hat{k}\right) \sum_{i=1}^{n} m_i \left(x_i^2 + y_i^2 + z_i^2\right) \\
&\quad - \sum_{i=1}^{n} m_i \left(x_i \omega_x + y_i \omega_y + z_i \omega_z\right) \cdot \left(x_i \hat{\imath} + y_i \hat{\jmath} + z_i \hat{k}\right) \tag{9.86}
\end{aligned}
$$

and therefore,

$$
\begin{aligned}
\mathbf{L} &= \sum_{i=1}^{n} m_i \left(x_i^2 + y_i^2 + z_i^2\right) \omega_x \hat{\imath} + \sum_{i=1}^{n} m_i \left(x_i^2 + y_i^2 + z_i^2\right) \omega_y \hat{\jmath} \\
&\quad + \sum_{i=1}^{n} m_i \left(x_i^2 + y_i^2 + z_i^2\right) \omega_z \hat{k} \\
&\quad - \sum_{i=1}^{n} m_i \left(x_i \omega_x + y_i \omega_y + z_i \omega_z\right) x_i \hat{\imath} - \sum_{i=1}^{n} m_i \left(x_i \omega_x + y_i \omega_y + z_i \omega_z\right) y_i \hat{\jmath} \\
&\quad - \sum_{i=1}^{n} m_i \left(x_i \omega_x + y_i \omega_y + z_i \omega_z\right) z_i \hat{k} \tag{9.87}
\end{aligned}
$$

or

$$
\begin{aligned}
\mathbf{L} &= \sum_{i=1}^{n} m_i \left[\left(x_i^2 + y_i^2 + z_i^2\right) \omega_x - \left(x_i \omega_x + y_i \omega_y + z_i \omega_z\right) x_i\right] \hat{\imath} \\
&\quad + \sum_{i=1}^{n} m_i \left[\left(x_i^2 + y_i^2 + z_i^2\right) \omega_y - \left(x_i \omega_x + y_i \omega_y + z_i \omega_z\right) y_i\right] \hat{\jmath} \\
&\quad + \sum_{i=1}^{n} m_i \left[\left(x_i^2 + y_i^2 + z_i^2\right) \omega_z - \left(x_i \omega_x + y_i \omega_y + z_i \omega_z\right) z_i\right] \hat{k} \tag{9.88}
\end{aligned}
$$

which can be rearranged as

$$
\mathbf{L} = \sum_{i=1}^{n} \left[m_i \left(y_i^2 + z_i^2 \right) \right] \omega_x \hat{\imath} + \sum_{i=1}^{n} \left[m_i \left(z_i^2 + x_i^2 \right) \right] \omega_y \hat{\jmath}
$$

$$
+ \sum_{i=1}^{n} \left[m_i \left(x_i^2 + y_i^2 \right) \right] \omega_z \hat{k}
$$

$$
- \left(\sum_{i=1}^{n} (m_i x_i y_i) \, \omega_y + \sum_{i=1}^{n} (m_i x_i z_i) \, \omega_z \right) \hat{\imath}
$$

$$
- \left(\sum_{i=1}^{n} (m_i y_i z_i) \, \omega_z + \sum_{i=1}^{n} (m_i y_i x_i) \, \omega_x \right) \hat{\jmath}
$$

$$
- \left(\sum_{i=1}^{n} (m_i z_i x_i) \, \omega_x + \sum_{i=1}^{n} (m_i z_i y_i) \, \omega_y \right) \hat{k} \tag{9.89}
$$

By introducing the mass moment matrix I with the following elements,

$$
I_{xx} = \sum_{i=1}^{n} \left[m_i \left(y_i^2 + z_i^2 \right) \right] \tag{9.90}
$$

$$
I_{yy} = \sum_{i=1}^{n} \left[m_i \left(z_i^2 + x_i^2 \right) \right] \tag{9.91}
$$

$$
I_{zz} = \sum_{i=1}^{n} \left[m_i \left(x_i^2 + y_i^2 \right) \right] \tag{9.92}
$$

$$
I_{xy} = I_{yx} = - \sum_{i=1}^{n} (m_i x_i y_i) \tag{9.93}
$$

$$
I_{yz} = I_{zy} = - \sum_{i=1}^{n} (m_i y_i z_i) \tag{9.94}
$$

$$
I_{zx} = I_{xz} = - \sum_{i=1}^{n} (m_i z_i x_i) \tag{9.95}
$$

we can write the angular momentum $\mathbf{L}$ in a concise form

$$
L_x = I_{xx}\omega_x + I_{xy}\omega_y + I_{xz}\omega_z \tag{9.96}
$$
$$
L_y = I_{yx}\omega_x + I_{yy}\omega_y + I_{yz}\omega_z \tag{9.97}
$$
$$
L_z = I_{zx}\omega_x + I_{zy}\omega_y + I_{zz}\omega_z \tag{9.98}
$$

or in a matrix form

$$
\begin{bmatrix} L_x \\ L_y \\ L_z \end{bmatrix} = \begin{bmatrix} I_{xx} & I_{xy} & I_{xz} \\ I_{yx} & I_{yy} & I_{yz} \\ I_{zx} & I_{zy} & I_{zz} \end{bmatrix} \begin{bmatrix} \omega_x \\ \omega_y \\ \omega_z \end{bmatrix} \tag{9.99}
$$

$$
\mathbf{L} = I \cdot \boldsymbol{\omega} \tag{9.100}
$$

For a rigid body that is a continuous solid, the summations must be replaced by integrations over the volume of the body as in Equation (9.74).

The Euler equation of motion for a rigid body is

$$^{B}\mathbf{M} = \frac{^{G}d}{dt}\,^{B}\mathbf{L} \tag{9.101}$$

where $^{B}\mathbf{M}$ is the resultant of the external moments applied on the rigid body. The angular momentum vector $^{B}\mathbf{L}$ is defined in the body coordinate frame B. Hence, its time derivative in the global coordinate frame is

$$\frac{^{G}d\,^{B}\mathbf{L}}{dt} = {}^{B}\dot{\mathbf{L}} + {}^{B}_{G}\boldsymbol{\omega}_{B} \times {}^{B}\mathbf{L} \tag{9.102}$$

Therefore,

$$^{B}\mathbf{M} = \frac{d\mathbf{L}}{dt} = \dot{\mathbf{L}} + \boldsymbol{\omega} \times \mathbf{L} = I\dot{\boldsymbol{\omega}} + \boldsymbol{\omega} \times (I\boldsymbol{\omega}) \tag{9.103}$$

or in expanded form

$$
\begin{aligned}
^{B}\mathbf{M} =\ & \left(I_{xx}\dot{\omega}_{x} + I_{xy}\dot{\omega}_{y} + I_{xz}\dot{\omega}_{z}\right)\hat{\imath} + \omega_{y}\left(I_{xz}\omega_{x} + I_{yz}\omega_{y} + I_{zz}\omega_{z}\right)\hat{\imath} \\
& -\omega_{z}\left(I_{xy}\omega_{x} + I_{yy}\omega_{y} + I_{yz}\omega_{z}\right)\hat{\imath} \\
& +\left(I_{yx}\dot{\omega}_{x} + I_{yy}\dot{\omega}_{y} + I_{yz}\dot{\omega}_{z}\right)\hat{\jmath} + \omega_{z}\left(I_{xx}\omega_{x} + I_{xy}\omega_{y} + I_{xz}\omega_{z}\right)\hat{\jmath} \\
& -\omega_{x}\left(I_{xz}\omega_{x} + I_{yz}\omega_{y} + I_{zz}\omega_{z}\right)\hat{\jmath} \\
& +\left(I_{zx}\dot{\omega}_{x} + I_{zy}\dot{\omega}_{y} + I_{zz}\dot{\omega}_{z}\right)\hat{k} + \omega_{x}\left(I_{xy}\omega_{x} + I_{yy}\omega_{y} + I_{yz}\omega_{z}\right)\hat{k} \\
& -\omega_{y}\left(I_{xx}\omega_{x} + I_{xy}\omega_{y} + I_{xz}\omega_{z}\right)\hat{k}
\end{aligned} \tag{9.104}
$$

and therefore, the most general form of the Euler equations of motion for a rigid body in a body frame attached to C are

$$
\begin{aligned}
M_{x} =\ & I_{xx}\dot{\omega}_{x} + I_{xy}\dot{\omega}_{y} + I_{xz}\dot{\omega}_{z} - (I_{yy} - I_{zz})\,\omega_{y}\omega_{z} \\
& -I_{yz}\left(\omega_{z}^{2} - \omega_{y}^{2}\right) - \omega_{x}\left(\omega_{z}I_{xy} - \omega_{y}I_{xz}\right) \tag{9.105}
\end{aligned}
$$

$$
\begin{aligned}
M_{y} =\ & I_{yx}\dot{\omega}_{x} + I_{yy}\dot{\omega}_{y} + I_{yz}\dot{\omega}_{z} - (I_{zz} - I_{xx})\,\omega_{z}\omega_{x} \\
& -I_{xz}\left(\omega_{x}^{2} - \omega_{z}^{2}\right) - \omega_{y}\left(\omega_{x}I_{yz} - \omega_{z}I_{xy}\right) \tag{9.106}
\end{aligned}
$$

$$
\begin{aligned}
M_{z} =\ & I_{zz}\dot{\omega}_{x} + I_{zy}\dot{\omega}_{y} + I_{zz}\dot{\omega}_{z} - (I_{xx} - I_{yy})\,\omega_{x}\omega_{y} \\
& -I_{xy}\left(\omega_{y}^{2} - \omega_{x}^{2}\right) - \omega_{z}\left(\omega_{y}I_{xz} - \omega_{x}I_{yz}\right) \tag{9.107}
\end{aligned}
$$

Assume that we are able to rotate the body frame about its origin to find an orientation that makes $I_{ij} = 0$, for $i \neq j$. In such a coordinate frame, which is the principal frame, the Euler equations reduce to

$$
\begin{aligned}
M_{1} &= I_{1}\dot{\omega}_{1} - (I_{2} - I_{2})\,\omega_{2}\omega_{3} \tag{9.108} \\
M_{2} &= I_{2}\dot{\omega}_{2} - (I_{3} - I_{1})\,\omega_{3}\omega_{1} \tag{9.109} \\
M_{3} &= I_{3}\dot{\omega}_{3} - (I_{1} - I_{2})\,\omega_{1}\omega_{2} \tag{9.110}
\end{aligned}
$$

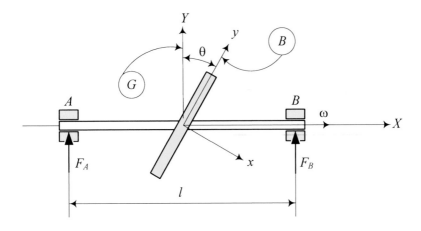

FIGURE 9.5. A disc with mass m and radius r, mounted on a massless turning shaft.

The kinetic energy of a rigid body may be found by the integral of the kinetic energy of a mass element dm, over the whole body.

$$
\begin{aligned}
K &= \frac{1}{2} \int_B \dot{\mathbf{v}}^2 dm = \frac{1}{2} \int_B (\boldsymbol{\omega} \times \mathbf{r}) \cdot (\boldsymbol{\omega} \times \mathbf{r}) \, dm \\
&= \frac{\omega_x^2}{2} \int_B \left(y^2 + z^2 \right) dm + \frac{\omega_y^2}{2} \int_B \left(z^2 + x^2 \right) dm + \frac{\omega_z^2}{2} \int_B \left(x^2 + y^2 \right) dm \\
&\quad - \omega_x \omega_y \int_B xy \, dm - \omega_y \omega_z \int_B yz \, dm - \omega_z \omega_x \int_B zx \, dm \\
&= \frac{1}{2} \left(I_{xx} \omega_x^2 + I_{yy} \omega_y^2 + I_{zz} \omega_z^2 \right) \\
&\quad - I_{xy} \omega_x \omega_y - I_{yz} \omega_y \omega_z - I_{zx} \omega_z \omega_x
\end{aligned} \tag{9.111}
$$

The kinetic energy can be rearranged to a matrix multiplication form

$$
K = \frac{1}{2} \boldsymbol{\omega}^T I \, \boldsymbol{\omega} = \frac{1}{2} \boldsymbol{\omega} \cdot \mathbf{L} \tag{9.112}
$$

When the body frame is principal, the kinetic energy will simplify to

$$
K = \frac{1}{2} \left(I_1 \omega_1^2 + I_2 \omega_2^2 + I_3 \omega_3^2 \right) \tag{9.113}
$$

∎

Example 383 *A tilted disc on a massless shaft.*
   Figure 9.5 illustrates a disc with mass m and radius r, mounted on a massless shaft. The shaft is turning with a constant angular speed ω. The disc is attached to the shaft at an angle θ. Because of θ, the bearings at A and B must support a rotating force.

*To analyze the system, we attach a principal body coordinate frame at
the disc center as shown in the figure. The angular velocity vector in the
body frame is*

$$\,^{B}_{G}\boldsymbol{\omega}_B = \omega\cos\theta\,\hat{\imath} + \omega\sin\theta\,\hat{\jmath} \tag{9.114}$$

and the mass moment matrix is

$$\,^{B}I = \frac{1}{4}\begin{bmatrix} 2mr^2 & 0 & 0 \\ 0 & mr^2 & 0 \\ 0 & 0 & mr^2 \end{bmatrix} \tag{9.115}$$

*Substituting (9.114) and (9.115) in (9.108)-(9.110), with $1 \equiv x$, $2 \equiv y$,
$3 \equiv z$, yields*

$$M_x = 0 \qquad M_y = 0 \qquad M_z = \frac{mr^2}{4}\omega\cos\theta\sin\theta \tag{9.116}$$

Therefore, the bearing reaction forces F_A and F_B are

$$F_A = -F_B = -\frac{M_z}{l} = -\frac{mr^2}{4l}\omega\cos\theta\sin\theta \tag{9.117}$$

Example 384 *Steady rotation of a freely rotating rigid body.*
   The Newton-Euler equations of motion for a rigid body are

$$\,^{G}\mathbf{F} = m\,^{G}\dot{\mathbf{v}} \tag{9.118}$$

$$\,^{B}\mathbf{M} = I\,^{B}_{G}\dot{\boldsymbol{\omega}}_B + \,^{B}_{G}\boldsymbol{\omega}_B \times \,^{B}\mathbf{L} \tag{9.119}$$

*Consider a situation in which the resultant applied force and moment on
the body are zero.*

$$\,^{G}\mathbf{F} = \,^{B}\mathbf{F} = 0 \qquad \,^{G}\mathbf{M} = \,^{B}\mathbf{M} = 0 \tag{9.120}$$

*Based on the Newton's equation, the velocity of the mass center will be
constant in the global coordinate frame. However, the Euler equation reduces
to*

$$\dot{\omega}_1 = \frac{I_2 - I_3}{I_1}\omega_2\omega_3 \tag{9.121}$$

$$\dot{\omega}_2 = \frac{I_3 - I_1}{I_{22}}\omega_3\omega_1 \tag{9.122}$$

$$\dot{\omega}_3 = \frac{I_1 - I_2}{I_3}\omega_1\omega_2 \tag{9.123}$$

that show the angular velocity can be constant if

$$I_1 = I_2 = I_3 \tag{9.124}$$

*or if two principal moments of inertia, say I_1 and I_2, are zero and the third
angular velocity, in this case ω_3, is initially zero, or if the angular velocity
vector is initially parallel to a principal axis.*

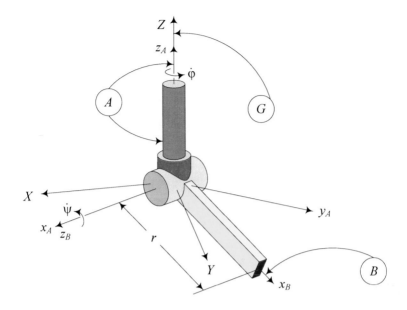

FIGURE 9.6. A two-link manipulator.

Example 385 *Angular momentum of a two-link manipulator.*

A two-link manipulator is shown in Figure 9.6. Link A rotates with angular velocity $\dot\varphi$ about the z-axis of its local coordinate frame. Link B is attached to link A and has angular velocity $\dot\psi$ with respect to A about the x_A-axis. We assume that A and G were coincident at $\varphi = 0$, therefore, the rotation matrix between A and G is

$$
{}^G R_A = \begin{bmatrix} \cos\varphi(t) & -\sin\varphi(t) & 0 \\ \sin\varphi(t) & \cos\varphi(t) & 0 \\ 0 & 0 & 1 \end{bmatrix} \tag{9.125}
$$

Frame B is related to frame A by Euler angles $\varphi = 90 \deg$, $\theta = 90 \deg$, and ψ, hence,

$$
{}^A R_B = \begin{bmatrix} c\pi c\psi - c\pi s\pi s\psi & -c\pi s\psi - c\pi c\psi s\pi & s\pi s\pi \\ c\psi s\pi + c\pi c\pi s\psi & -s\pi s\psi + c\pi c\pi c\psi & -c\pi s\pi \\ s\pi s\psi & s\pi c\psi & c\pi \end{bmatrix}
$$
$$
= \begin{bmatrix} -\cos\psi & \sin\psi & 0 \\ \sin\psi & \cos\psi & 0 \\ 0 & 0 & -1 \end{bmatrix} \tag{9.126}
$$

and therefore,

$$
{}^G R_B = {}^G R_A \, {}^A R_B \tag{9.127}
$$

$$
= \begin{bmatrix} -\cos\varphi\cos\psi - \sin\varphi\sin\psi & \cos\varphi\sin\psi - \cos\psi\sin\varphi & 0 \\ \cos\varphi\sin\psi - \cos\psi\sin\varphi & \cos\varphi\cos\psi + \sin\varphi\sin\psi & 0 \\ 0 & 0 & -1 \end{bmatrix}
$$

The angular velocity of A in G, and B in A are

$$
{}_G\boldsymbol{\omega}_A = \dot{\varphi}\hat{K} \qquad {}_A\boldsymbol{\omega}_B = \dot{\psi}\hat{\imath}_A \tag{9.128}
$$

Moment of inertia matrices for the arms A and B can be defined as

$$
{}^A I_A = \begin{bmatrix} I_{A1} & 0 & 0 \\ 0 & I_{A2} & 0 \\ 0 & 0 & I_{A3} \end{bmatrix} \qquad {}^B I_B = \begin{bmatrix} I_{B1} & 0 & 0 \\ 0 & I_{B2} & 0 \\ 0 & 0 & I_{B3} \end{bmatrix} \tag{9.129}
$$

These moments of inertia must be transformed to the global frame

$$
{}^G I_A = {}^G R_B \, {}^A I_A \, {}^G R_A^T \qquad {}^G I_B = {}^G R_B \, {}^B I_B \, {}^G R_B^T \tag{9.130}
$$

The total angular momentum of the manipulator is

$$
{}^G\mathbf{L} = {}^G\mathbf{L}_A + {}^G\mathbf{L}_B \tag{9.131}
$$

where

$$
{}^G\mathbf{L}_A = {}^G I_A \, {}_G\boldsymbol{\omega}_A \tag{9.132}
$$

$$
{}^G\mathbf{L}_B = {}^G I_B \, {}_G\boldsymbol{\omega}_B = {}^G I_B \left({}^G_A\boldsymbol{\omega}_B + {}_G\boldsymbol{\omega}_A \right) \tag{9.133}
$$

Example 386 ★ *Poinsot's construction.*

Consider a freely rotating rigid body with an attached principal coordinate frame. Having $\mathbf{M} = 0$ provides a motion under constant angular momentum and constant kinetic energy

$$
\mathbf{L} = I\,\boldsymbol{\omega} = cte \tag{9.134}
$$

$$
K = \frac{1}{2}\boldsymbol{\omega}^T I\,\boldsymbol{\omega} = cte \tag{9.135}
$$

Because the length of the angular momentum $\mathbf{L}$ *is constant, the equation*

$$
L^2 = \mathbf{L} \cdot \mathbf{L} = L_x^2 + L_y^2 + L_z^2 = I_1^2\omega_1^2 + I_2^2\omega_2^2 + I_3^2\omega_3^2 \tag{9.136}
$$

introduces an ellipsoid in the $(\omega_1, \omega_2, \omega_3)$ *coordinate frame, called the **momentum ellipsoid**. The tip of all possible angular velocity vectors must lie on the surface of the momentum ellipsoid. The kinetic energy also defines an **energy ellipsoid** in the same coordinate frame so that the tip of the angular velocity vectors must also lie on its surface.*

$$
K = \frac{1}{2}\left(I_1\omega_1^2 + I_2\omega_2^2 + I_3\omega_3^2 \right) \tag{9.137}
$$

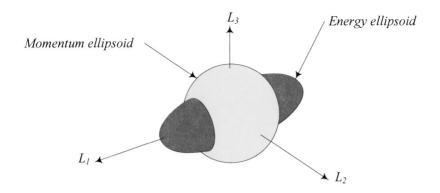

FIGURE 9.7. Intersection of the momentum and energy ellipsoids.

In other words, the dynamics of moment-free motion of a rigid body requires that the corresponding angular velocity $\boldsymbol{\omega}(t)$ satisfy both Equations (9.136) and (9.137) and therefore lie on the intersection of the momentum and energy ellipsoids.

For clarity, we may define the ellipsoids in the (L_x, L_y, L_z) coordinate system as

$$L_x^2 + L_y^2 + L_z^2 = L^2 \tag{9.138}$$

$$\frac{L_x^2}{2I_1 K} + \frac{L_y^2}{2I_2 K} + \frac{L_z^2}{2I_3 K} = 1 \tag{9.139}$$

Equation (9.138) is a sphere and Equation (9.139) is an ellipsoid with $\sqrt{2I_i K}$ as semi-axes. To have a meaningful motion, these two shapes must intersect. The intersection forms a trajectory for the tip point of $\mathbf{L}$, as shown in Figure 9.7.

It can be deduced that for a certain value of angular momentum there are maximum and minimum limit values for acceptable kinetic energy. Assuming

$$I_1 > I_3 > I_3 \tag{9.140}$$

the limits of possible kinetic energy are

$$K_{\min} = \frac{L^2}{2I_1} \qquad K_{\max} = \frac{L^2}{2I_3} \tag{9.141}$$

and the corresponding motions are turning about the axes I_1 and I_3 respectively.

Example 387 ★ *Alternative derivation of Euler equations of motion.*

Assume that the moment of the small force $d\mathbf{f}$ is shown by $d\mathbf{m}$ and a mass element is shown by dm, then,

$$d\mathbf{m} = {}^G\mathbf{r}_{dm} \times d\mathbf{f} = {}^G\mathbf{r}_{dm} \times {}^G\dot{\mathbf{v}}_{dm}\, dm \tag{9.142}$$

The global angular momentum dl of dm is equal to

$$dl = {}^G\mathbf{r}_{dm} \times {}^G\mathbf{v}_{dm}\,dm \tag{9.143}$$

and according to (9.12) we have

$$d\mathbf{m} = \frac{{}^Gd}{dt}dl \tag{9.144}$$

$${}^G\mathbf{r}_{dm} \times d\mathbf{f} = \frac{{}^Gd}{dt}\left({}^G\mathbf{r}_{dm} \times {}^G\mathbf{v}_{dm}\,dm\right) \tag{9.145}$$

Integrating over the body results in

$$\int_B {}^G\mathbf{r}_{dm} \times d\mathbf{f} = \int_B \frac{{}^Gd}{dt}\left({}^G\mathbf{r}_{dm} \times {}^G\mathbf{v}_{dm}\,dm\right)$$

$$= \frac{{}^Gd}{dt}\int_B \left({}^G\mathbf{r}_{dm} \times {}^G\mathbf{v}_{dm}\,dm\right) \tag{9.146}$$

However, we have

$${}^G\mathbf{r}_{dm} = {}^G\mathbf{d}_B + {}^G R_B\,{}^B\mathbf{r}_{dm} \tag{9.147}$$

where ${}^G\mathbf{d}_B$ is the global position vector of the central body frame, can simplify the left-hand side of the integral to

$$\int_B {}^G\mathbf{r}_{dm} \times d\mathbf{f} = \int_B \left({}^G\mathbf{d}_B + {}^G R_B\,{}^B\mathbf{r}_{dm}\right) \times d\mathbf{f}$$

$$= \int_B {}^G\mathbf{d}_B \times d\mathbf{f} + \int_B {}^G_B\mathbf{r}_{dm} \times d\mathbf{f}$$

$$= {}^G\mathbf{d}_B \times {}^G\mathbf{F} + {}^G\mathbf{M}_C \tag{9.148}$$

where $\mathbf{M}_C$ is the resultant external moment about the body mass center C. The right-hand side of Equation (9.146) is

$$\frac{{}^Gd}{dt}\int_B \left({}^G\mathbf{r}_{dm} \times {}^G\mathbf{v}_{dm}\,dm\right)$$

$$= \frac{{}^Gd}{dt}\int_B \left(\left({}^G\mathbf{d}_B + {}^G R_B\,{}^B\mathbf{r}_{dm}\right) \times {}^G\mathbf{v}_{dm}\,dm\right)$$

$$= \frac{{}^Gd}{dt}\int_B \left({}^G\mathbf{d}_B \times {}^G\mathbf{v}_{dm}\right)dm + \frac{{}^Gd}{dt}\int_B \left({}^G_B\mathbf{r}_{dm} \times {}^G\mathbf{v}_{dm}\right)dm$$

$$= \frac{{}^Gd}{dt}\left({}^G\mathbf{d}_B \times \int_B {}^G\mathbf{v}_{dm}\,dm\right) + \frac{{}^Gd}{dt}\mathbf{L}_C$$

$$= {}^G\dot{\mathbf{d}}_B \times \int_B {}^G\mathbf{v}_{dm}\,dm + {}^G\mathbf{d}_B \times \int_B {}^G\dot{\mathbf{v}}_{dm}\,dm + \frac{d}{dt}\mathbf{L}_C \tag{9.149}$$

We use $\mathbf{L}_C$ for moment of momentum about the body mass center. Because the body frame is at the mass center, we have

$$\int_B {}^G\mathbf{r}_{dm}\, dm = m\,{}^G\mathbf{d}_B = m\,{}^G\mathbf{r}_C \tag{9.150}$$

$$\int_B {}^G\mathbf{v}_{dm}\, dm = m\,{}^G\dot{\mathbf{d}}_B = m\,{}^G\mathbf{v}_C \tag{9.151}$$

$$\int_B {}^G\dot{\mathbf{v}}_{dm}\, dm = m\,{}^G\ddot{\mathbf{d}}_B = m\,{}^G\mathbf{a}_C \tag{9.152}$$

and therefore,

$$\frac{{}^Gd}{dt}\int_B \left({}^G\mathbf{r}_{dm} \times {}^G\mathbf{v}_{dm}\, dm\right) = {}^G\mathbf{d}_B \times {}^G\mathbf{F} + \frac{{}^Gd}{dt}\,{}^G\mathbf{L}_C \tag{9.153}$$

Substituting (9.148) and (9.153) in (9.146) provides us with the Euler equation of motion in the global frame, indicating that the resultant of externally applied moments about C is equal to the global derivative of angular momentum about C.

$$^G\mathbf{M}_C = \frac{{}^Gd}{dt}\,{}^G\mathbf{L}_C \tag{9.154}$$

The Euler equation in the body coordinate can be found by transforming (9.154)

$$\begin{aligned}
{}^B\mathbf{M}_C &= {}^GR_B^T\,{}^G\mathbf{M}_C = {}^GR_B^T\frac{{}^Gd}{dt}\mathbf{L}_C = \frac{{}^Gd}{dt}\,{}^GR_B^T\,\mathbf{L}_C = \frac{{}^Gd}{dt}\,{}^B\mathbf{L}_C \\
&= {}^B\dot{\mathbf{L}}_C + {}^B_G\boldsymbol{\omega}_B \times {}^B\mathbf{L}_C
\end{aligned} \tag{9.155}$$

9.4   Mass Moment Matrix

In analyzing the motion of rigid bodies, two types of integrals arise that belong to the geometry of the body. The first type defines the center of mass and is important when the translation motion of the body is considered. The second is the *mass moment* that appears when the rotational motion of the body is considered. The mass moment is also called *moment of inertia*, *centrifugal moments*, or *deviation moments*. Every rigid body has a 3×3 moment of inertia matrix I, which is denoted by

$$I = \begin{bmatrix} I_{xx} & I_{xy} & I_{xz} \\ I_{yx} & I_{yy} & I_{yz} \\ I_{zx} & I_{zy} & I_{zz} \end{bmatrix} \tag{9.156}$$

The diagonal elements I_{ij}, $i = j$ are called *polar moments of inertia*

$$I_{xx} = I_x = \int_B \left(y^2 + z^2\right) dm \tag{9.157}$$

$$I_{yy} = I_y = \int_B \left(z^2 + x^2 \right) dm \qquad (9.158)$$

$$I_{zz} = I_z = \int_B \left(x^2 + y^2 \right) dm \qquad (9.159)$$

and the off-diagonal elements I_{ij}, $i \neq j$ are called *products of inertia*

$$I_{xy} = I_{yx} = -\int_B xy\, dm \qquad (9.160)$$

$$I_{yz} = I_{zy} = -\int_B yz\, dm \qquad (9.161)$$

$$I_{zx} = I_{xz} = -\int_B zx\, dm \qquad (9.162)$$

The elements of I for a rigid body, made of discrete point masses, are defined in Equation (9.74).

The elements of I are calculated about a body coordinate frame attached to the mass center C of the body. Therefore, I is a frame-dependent quantity and must be written like $^B I$ to show the frame it is computed in.

$$
\begin{aligned}
^B I &= \int_B \begin{bmatrix} y^2 + z^2 & -xy & -zx \\ -xy & z^2 + x^2 & -yz \\ -zx & -yz & x^2 + y^2 \end{bmatrix} dm & (9.163) \\
&= \int_B \left(r^2 \mathbf{I} - \mathbf{r}\mathbf{r}^T \right) dm & (9.164) \\
&= \int_B -\tilde{r}\tilde{r}\, dm & (9.165)
\end{aligned}
$$

Mass moments can be transformed from a coordinate frame B_1 to another coordinate frame B_2, both installed at the mass center of the body, according to the rule of the *rotated-axes theorem*

$$^{B_2} I = \, ^{B_2} R_{B_1} \, ^{B_1} I \, ^{B_2} R_{B_1}^T \qquad (9.166)$$

Transformation of the mass moment from a central frame B_1 located at $^{B_2} \mathbf{r}_C$ to another frame B_2, which is parallel to B_1, is, according to the rule of *parallel-axes theorem*,

$$^{B_2} I = \, ^{B_1} I + m \tilde{r}_C \tilde{r}_C^T \qquad (9.167)$$

If the local coordinate frame $Oxyz$ is located such that the products of inertia vanish, the local coordinate frame is called the *principal coordinate frame* and the associated moments of inertia are called *principal moments of inertia.* Principal axes and principal moments of inertia can be found by solving the following equation for I:

$$\begin{vmatrix} I_{xx} - I & I_{xy} & I_{xz} \\ I_{yx} & I_{yy} - I & I_{yz} \\ I_{zx} & I_{zy} & I_{zz} - I \end{vmatrix} = 0 \tag{9.168}$$

$$\det\left([I_{ij}] - I\,[\delta_{ij}]\right) = 0 \tag{9.169}$$

Since Equation (9.169) is a cubic equation in I, we obtain three eigenvalues

$$I_1 = I_x \qquad I_2 = I_y \qquad I_3 = I_z \tag{9.170}$$

that are the principal moments of inertia.

Proof. Consider two coordinate frames with a common origin at the mass center of a rigid body as shown in Figure (a). The angular velocity and angular momentum of a rigid body transform from the frame B_1 to the frame B_2 by vector transformation rule

$$^{B_2}\boldsymbol{\omega} = {}^{B_2}R_{B_1}\,{}^{B_1}\boldsymbol{\omega} \tag{9.171}$$

$$^{B_2}\mathbf{L} = {}^{B_2}R_{B_1}\,{}^{B_1}\mathbf{L} \tag{9.172}$$

However, $\mathbf{L}$ and $\boldsymbol{\omega}$ are related according to Equation (9.72)

$$^{B_1}\mathbf{L} = {}^{B_1}I\,{}^{B_1}\boldsymbol{\omega} \tag{9.173}$$

and therefore,

$$^{B_2}\mathbf{L} = {}^{B_2}R_{B_1}\,{}^{B_1}I\,{}^{B_2}R_{B_1}^T\,{}^{B_2}\boldsymbol{\omega} = {}^{B_2}I\,{}^{B_2}\boldsymbol{\omega} \tag{9.174}$$

which shows how to transfer the moment of inertia from the coordinate frame B_1 to a rotated frame B_2

$$^{B_2}I = {}^{B_2}R_{B_1}\,{}^{B_1}I\,{}^{B_2}R_{B_1}^T \tag{9.175}$$

Now consider a central frame B_1, shown in Figure 9.8(b), at $^{B_2}\mathbf{r}_C$, which rotates about the origin of a fixed frame B_2 such that their axes remain parallel. The angular velocity and angular momentum of the rigid body transform from frame B_1 to frame B_2 by

$$^{B_2}\boldsymbol{\omega} = {}^{B_1}\boldsymbol{\omega} \tag{9.176}$$

$$^{B_2}\mathbf{L} = {}^{B_1}\mathbf{L} + (\mathbf{r}_C \times m\mathbf{v}_C) \tag{9.177}$$

Therefore,

$$\begin{aligned} ^{B_2}\mathbf{L} &= {}^{B_1}\mathbf{L} + m\,{}^{B_2}\mathbf{r}_C \times \left({}^{B_2}\boldsymbol{\omega} \times {}^{B_2}\mathbf{r}_C\right) \\ &= {}^{B_1}\mathbf{L} + \left(m\,{}^{B_2}\tilde{r}_C\,{}^{B_2}\tilde{r}_C^T\right){}^{B_2}\boldsymbol{\omega} \\ &= \left({}^{B_1}I + m\,{}^{B_2}\tilde{r}_C\,{}^{B_2}\tilde{r}_C^T\right){}^{B_2}\boldsymbol{\omega} \end{aligned} \tag{9.178}$$

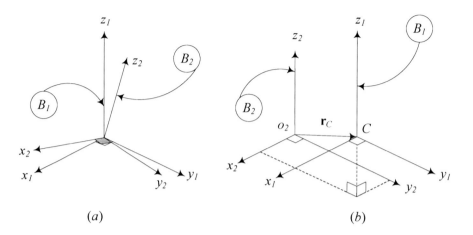

FIGURE 9.8. (a) Two coordinate frames with a common origin at the mass center of a rigid body. (b) A central coordinate frame B_1 and a translated frame B_2.

which shows how to transfer the mass moment from frame B_1 to a parallel frame B_2

$$^{B_2}I = {}^{B_1}I + m\tilde{r}_C \tilde{r}_C^T \qquad (9.179)$$

The parallel-axes theorem is also called the *Huygens-Steiner theorem*.

Referring to Equation (9.175) for transformation of the mass moment to a rotated frame, we can always find a frame in which ^{B_2}I is diagonal. In such a frame, we have

$$^{B_2}R_{B_1} {}^{B_1}I = {}^{B_2}I {}^{B_2}R_{B_1} \qquad (9.180)$$

or

$$
\begin{bmatrix} r_{11} & r_{12} & r_{13} \\ r_{21} & r_{22} & r_{23} \\ r_{31} & r_{32} & r_{33} \end{bmatrix}
\begin{bmatrix} I_{xx} & I_{xy} & I_{xz} \\ I_{yx} & I_{yy} & I_{yz} \\ I_{zx} & I_{zy} & I_{zz} \end{bmatrix}
$$
$$
= \begin{bmatrix} I_1 & 0 & 0 \\ 0 & I_2 & 0 \\ 0 & 0 & I_3 \end{bmatrix}
\begin{bmatrix} r_{11} & r_{12} & r_{13} \\ r_{21} & r_{22} & r_{23} \\ r_{31} & r_{32} & r_{33} \end{bmatrix} \qquad (9.181)
$$

which shows that I_1, I_2, and I_3 are eigenvalues of ^{B_1}I. These eigenvalues can be found by solving the following equation for λ:

$$
\begin{vmatrix} I_{xx} - \lambda & I_{xy} & I_{xz} \\ I_{yx} & I_{yy} - \lambda & I_{yz} \\ I_{zx} & I_{zy} & I_{zz} - \lambda \end{vmatrix} = 0 \qquad (9.182)
$$

The eigenvalues I_1, I_2, and I_3 are *principal mass moments*, and their associated eigenvectors are called *principal directions*. The coordinate frame made by the eigenvectors is the *principal body coordinate frame*. In the principal coordinate frame, the rigid body angular momentum is

$$
\begin{bmatrix} L_1 \\ L_2 \\ L_3 \end{bmatrix} = \begin{bmatrix} I_1 & 0 & 0 \\ 0 & I_2 & 0 \\ 0 & 0 & I_3 \end{bmatrix} \begin{bmatrix} \omega_1 \\ \omega_2 \\ \omega_3 \end{bmatrix} \tag{9.183}
$$

∎

Example 388 *Principal moments of inertia.*
   Consider the inertia matrix I

$$
I = \begin{bmatrix} 20 & -2 & 0 \\ -2 & 30 & 0 \\ 0 & 0 & 40 \end{bmatrix} \tag{9.184}
$$

We set up the determinant (9.169)

$$
\begin{vmatrix} 20 - \lambda & -2 & 0 \\ -2 & 30 - \lambda & 0 \\ 0 & 0 & 40 - \lambda \end{vmatrix} = 0 \tag{9.185}
$$

which leads to the following characteristic equation.

$$
(20 - \lambda)(30 - \lambda)(40 - \lambda) - 4(40 - \lambda) = 0 \tag{9.186}
$$

   Three roots of Equation (9.186) are

$$
I_1 = 30.385 \qquad I_2 = 19.615 \qquad I_3 = 40 \tag{9.187}
$$

and therefore, the principal moment of inertia matrix is

$$
I = \begin{bmatrix} 30.385 & 0 & 0 \\ 0 & 19.615 & 0 \\ 0 & 0 & 40 \end{bmatrix} \tag{9.188}
$$

Example 389 *Principal coordinate frame.*
   Consider the inertia matrix I

$$
I = \begin{bmatrix} 20 & -2 & 0 \\ -2 & 30 & 0 \\ 0 & 0 & 40 \end{bmatrix} \tag{9.189}
$$

the direction of a principal axis x_i is established by solving

$$
\begin{bmatrix} I_{xx} - I_i & I_{xy} & I_{xz} \\ I_{yx} & I_{yy} - I_i & I_{yz} \\ I_{zx} & I_{zy} & I_{zz} - I_i \end{bmatrix} \begin{bmatrix} \cos \alpha_i \\ \cos \beta_i \\ \cos \gamma_i \end{bmatrix} = \begin{bmatrix} 0 \\ 0 \\ 0 \end{bmatrix} \tag{9.190}
$$

for direction cosines, which must also satisfy

$$\cos^2 \alpha_i + \cos^2 \beta_i + \cos^2 \gamma_i = 1 \tag{9.191}$$

For the first principal moment of inertia $I_1 = 30.385$ *we have*

$$\begin{bmatrix} 20 - 30.385 & -2 & 0 \\ -2 & 30 - 30.385 & 0 \\ 0 & 0 & 40 - 30.385 \end{bmatrix} \begin{bmatrix} \cos \alpha_1 \\ \cos \beta_1 \\ \cos \gamma_1 \end{bmatrix} = \begin{bmatrix} 0 \\ 0 \\ 0 \end{bmatrix} \tag{9.192}$$

or

$$-10.385 \cos \alpha_1 - 2 \cos \beta_1 + 0 = 0 \tag{9.193}$$
$$-2 \cos \alpha_1 - 0.385 \cos \beta_1 + 0 = 0 \tag{9.194}$$
$$0 + 0 + 9.615 \cos \gamma_1 = 0 \tag{9.195}$$

and we obtain

$$\alpha_1 = 79.1 \deg \qquad \beta_1 = 169.1 \deg \qquad \gamma_1 = 90.0 \deg \tag{9.196}$$

Using $I_2 = 19.615$ *for the second principal axis*

$$\begin{bmatrix} 20 - 19.62 & -2 & 0 \\ -2 & 30 - 19.62 & 0 \\ 0 & 0 & 40 - 19.62 \end{bmatrix} \begin{bmatrix} \cos \alpha_2 \\ \cos \beta_2 \\ \cos \gamma_2 \end{bmatrix} = \begin{bmatrix} 0 \\ 0 \\ 0 \end{bmatrix} \tag{9.197}$$

we obtain

$$\alpha_2 = 10.9 \deg \qquad \beta_2 = 79.1 \deg \qquad \gamma_2 = 90.0 \deg \tag{9.198}$$

The third principal axis is for $I_3 = 40$

$$\begin{bmatrix} 20 - 40 & -2 & 0 \\ -2 & 30 - 40 & 0 \\ 0 & 0 & 40 - 40 \end{bmatrix} \begin{bmatrix} \cos \alpha_3 \\ \cos \beta_3 \\ \cos \gamma_3 \end{bmatrix} = \begin{bmatrix} 0 \\ 0 \\ 0 \end{bmatrix} \tag{9.199}$$

which leads to

$$\alpha_3 = 90.0 \deg \qquad \beta_3 = 90.0 \deg \qquad \gamma_3 = 0.0 \deg \tag{9.200}$$

Example 390 *Mass moment of a rigid rectangular bar.*
    Consider a homogeneous rectangular link with mass m, *length* l, *width* w, *and height* h, *as shown in Figure 9.9(a).*
    The local central coordinate frame is attached to the link at its mass center. The moments of inertia matrix of the link can be found by the integral method. We begin with calculating I_{xx}

$$\begin{aligned} I_{xx} &= \int_B \left(y^2 + z^2 \right) dm = \int_v \left(y^2 + z^2 \right) \rho \, dv = \frac{m}{lwh} \int_v \left(y^2 + z^2 \right) dv \\ &= \frac{m}{lwh} \int_{-h/2}^{h/2} \int_{-w/2}^{w/2} \int_{-l/2}^{l/2} \left(y^2 + z^2 \right) dx \, dy \, dz = \frac{m}{12} \left(w^2 + h^2 \right) \end{aligned} \tag{9.201}$$

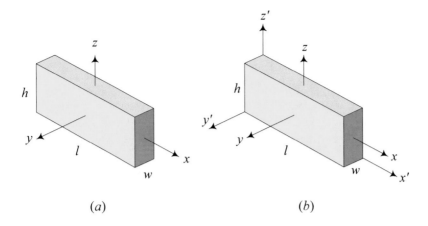

FIGURE 9.9. (a) A homogeneous rigid rectangular link. (b) A homogeneous rigid rectangular link in the principal and non principal frames.

The I_{yy} and I_{zz} can be calculated similarly

$$I_{yy} = \frac{m}{12}\left(h^2 + l^2\right) \qquad I_{zz} = \frac{m}{12}\left(l^2 + w^2\right) \qquad (9.202)$$

The coordinate frame is central and therefore, the products of inertia must be zero. To show this, we examine I_{xy}.

$$
\begin{aligned}
I_{xy} &= I_{yx} = -\int_B xy\,dm = \int_v xy\rho\,dv \\
&= \frac{m}{lwh}\int_{-h/2}^{h/2}\int_{-w/2}^{w/2}\int_{-l/2}^{l/2} xy\,dx\,dy\,dz = 0
\end{aligned}
\qquad (9.203)
$$

Therefore, the mass moment for the rigid rectangular bar in its central frame is

$$
I = \begin{bmatrix} \frac{m}{12}\left(w^2 + h^2\right) & 0 & 0 \\ 0 & \frac{m}{12}\left(h^2 + l^2\right) & 0 \\ 0 & 0 & \frac{m}{12}\left(l^2 + w^2\right) \end{bmatrix}
\qquad (9.204)
$$

Example 391 *Translation of the inertia matrix.*

The moment of inertia matrix of the rigid body shown in Figure 9.9(b), in the principal frame $B(oxyz)$ is given in Equation (9.204). The mass moment matrix in the non-principal frame $B'(ox'y'z')$ can be found by applying the parallel-axes transformation formula (9.179).

$$ {}^{B'}I = {}^{B}I + m \, {}^{B'}\tilde{r}_C \, {}^{B'}\tilde{r}_C^T \qquad (9.205) $$

The mass center is at

$$ {}^{B'}\mathbf{r}_C = \frac{1}{2}\begin{bmatrix} l \\ w \\ h \end{bmatrix} \qquad (9.206) $$

and therefore,

$$B'\tilde{r}_C = \frac{1}{2} \begin{bmatrix} 0 & -h & w \\ h & 0 & -l \\ -w & l & 0 \end{bmatrix} \tag{9.207}$$

that provides

$$B'I = \begin{bmatrix} \frac{1}{3}h^2 m + \frac{1}{3}mw^2 & -\frac{1}{4}lmw & -\frac{1}{4}hlm \\ -\frac{1}{4}lmw & \frac{1}{3}h^2 m + \frac{1}{3}l^2 m & -\frac{1}{4}hmw \\ -\frac{1}{4}hlm & -\frac{1}{4}hmw & \frac{1}{3}l^2 m + \frac{1}{3}mw^2 \end{bmatrix} \tag{9.208}$$

Example 392 *Principal rotation matrix.*

Consider a body mass moment matrix as

$$I = \begin{bmatrix} 2/3 & -1/2 & -1/2 \\ -1/2 & 5/3 & -1/4 \\ -1/2 & -1/4 & 5/3 \end{bmatrix} \tag{9.209}$$

The eigenvalues and eigenvectors of I are

$$I_1 = 0.2413 \qquad \mathbf{w}_1 = \begin{bmatrix} 2.351 & 1 & 1 \end{bmatrix}^T \tag{9.210}$$

$$I_2 = 1.8421 \qquad \mathbf{w}_2 = \begin{bmatrix} -0.851 & 1 & 1 \end{bmatrix}^T \tag{9.211}$$

$$I_3 = 1.9167 \qquad \mathbf{w}_3 = \begin{bmatrix} 0 & -1 & 1 \end{bmatrix}^T \tag{9.212}$$

The normalized eigenvector matrix W is equal to the transpose of the required transformation matrix to make the inertia matrix diagonal

$$\begin{aligned} W &= \begin{bmatrix} \mathbf{w}_1 & \mathbf{w}_2 & \mathbf{w}_3 \end{bmatrix} = {}^2R_1^T \\ &= \begin{bmatrix} 0.856\,9 & -0.515\,6 & 0.0 \\ 0.364\,48 & 0.605\,88 & -0.707\,11 \\ 0.364\,48 & 0.605\,88 & 0.707\,11 \end{bmatrix} \end{aligned} \tag{9.213}$$

We may verify that

$$\begin{aligned} {}^2I &\approx {}^2R_1 \, {}^1I \, {}^2R_1^T = W^T \, {}^1I \, W \\ &= \begin{bmatrix} 0.2413 & -1 \times 10^{-4} & 0.0 \\ -1 \times 10^{-4} & 1.842\,1 & -1 \times 10^{-19} \\ 0.0 & 0.0 & 1.916\,7 \end{bmatrix} \end{aligned} \tag{9.214}$$

Example 393 ★ *Relative diagonal mass moments.*

Using the definitions for mass moments (9.157), (9.158), and (9.159) it is seen that the inertia matrix is symmetric, and

$$\int_B \left(x^2 + y^2 + z^2 \right) dm = \frac{1}{2} \left(I_{xx} + I_{yy} + I_{zz} \right) \tag{9.215}$$

and also

$$I_{xx} + I_{yy} \geq I_{zz} \qquad I_{yy} + I_{zz} \geq I_{xx} \qquad I_{zz} + I_{xx} \geq I_{yy} \tag{9.216}$$

Noting that

$$(y - z)^2 \geq 0 \tag{9.217}$$

it is evident that

$$(y^2 + z^2) \geq 2yz \tag{9.218}$$

and therefore

$$I_{xx} \geq 2I_{yz} \tag{9.219}$$

and similarly

$$I_{yy} \geq 2I_{zx} \qquad I_{zz} \geq 2I_{xy} \tag{9.220}$$

Example 394 ★ *Coefficients of the characteristic equation.*
   The determinant (9.182)

$$\begin{vmatrix} I_{xx} - \lambda & I_{xy} & I_{xz} \\ I_{yx} & I_{yy} - \lambda & I_{yz} \\ I_{zx} & I_{zy} & I_{zz} - \lambda \end{vmatrix} = 0 \tag{9.221}$$

for calculating the principal moments of inertia, leads to a third-degree equation for λ, called the **characteristic equation**.

$$\lambda^3 - a_1 \lambda^2 + a_2 \lambda - a_3 = 0 \tag{9.222}$$

The coefficients of the characteristic equation are called the **principal invariants** *of $[I]$. The coefficients of the characteristic equation can directly be found from the following equations:*

$$a_1 = I_{xx} + I_{yy} + I_{zz} = \mathrm{tr}\,[I] \tag{9.223}$$

$$\begin{aligned} a_2 &= I_{xx}I_{yy} + I_{yy}I_{zz} + I_{zz}I_{xx} - I_{xy}^2 - I_{yz}^2 - I_{zx}^2 \\ &= \begin{vmatrix} I_{xx} & I_{xy} \\ I_{yx} & I_{yy} \end{vmatrix} + \begin{vmatrix} I_{yy} & I_{yz} \\ I_{zy} & I_{zz} \end{vmatrix} + \begin{vmatrix} I_{xx} & I_{xz} \\ I_{zx} & I_{zz} \end{vmatrix} \\ &= \frac{1}{2} \left(a_1^2 - \mathrm{tr}\,[I^2] \right) \end{aligned} \tag{9.224}$$

$$\begin{aligned} a_3 &= I_{xx}I_{yy}I_{zz} + I_{xy}I_{yz}I_{zx} + I_{zy}I_{yx}I_{xz} \\ &\quad - \left(I_{xx}I_{yz}I_{zy} + I_{yy}I_{zx}I_{xz} + I_{zz}I_{xy}I_{yx} \right) \\ &= I_{xx}I_{yy}I_{zz} + 2I_{xy}I_{yz}I_{zx} - \left(I_{xx}I_{yz}^2 + I_{yy}I_{zx}^2 + I_{zz}I_{xy}^2 \right) \\ &= \det [I] \end{aligned} \tag{9.225}$$

Example 395 ★ *The principal mass moments are coordinate invariants.*

   The roots of the inertia characteristic equation are the principal mass moments. They are all real but not necessarily different. The principal mass *moments are extreme values. That is, the principal mass moments deter-mine the smallest and the largest values of I_{ii}. Since the smallest and largest values of I_{ii} do not depend on the choice of the body coordinate frame, the solution of the characteristic equation is not dependent of the coordinate frame.*

In other words, if I_1, I_2, and I_3 are the principal moments of inertia for ${}^{B_1}I$, the principal moments of inertia for ${}^{B_2}I$ are also I_1, I_2, and I_3, where

$$^{B_2}I = {}^{B_2}R_{B_1}\,{}^{B_1}I\,{}^{B_2}R_{B_1}^T \tag{9.226}$$

We conclude that I_1, I_2, and I_3 are coordinate invariants of the matrix $[I]$, and therefore any quantity that depends on I_1, I_2, and I_3 is also coordinate invariant. The matrix $[I]$ has only three independent invariants and every other invariant can be expressed in terms of I_1, I_2, and I_3.

Since I_1, I_2, and I_3 are the solutions of the characteristic equation of $[I]$ given in (9.222), we may write the determinant (9.182) in the form

$$(\lambda - I_1)(\lambda - I_2)(\lambda - I_3) = 0 \tag{9.227}$$

The expanded form of this equation is

$$\lambda^3 - (I_1 + I_2 + I_3)\lambda^2 + (I_1I_2 + I_2I_3 + I_3I_1)a_2\lambda - I_1I_2I_3 = 0 \tag{9.228}$$

By comparing (9.228) and (9.222) we conclude that

$$
\begin{aligned}
a_1 &= I_{xx} + I_{yy} + I_{zz} = I_1 + I_2 + I_3 & (9.229)\\
a_2 &= I_{xx}I_{yy} + I_{yy}I_{zz} + I_{zz}I_{xx} - I_{xy}^2 - I_{yz}^2 - I_{zx}^2 \\
&= I_1I_2 + I_2I_3 + I_3I_1 & (9.230)\\
a_3 &= I_{xx}I_{yy}I_{zz} + 2I_{xy}I_{yz}I_{zx} - \left(I_{xx}I_{yz}^2 + I_{yy}I_{zx}^2 + I_{zz}I_{xy}^2\right) \\
&= I_1I_2I_3 & (9.231)
\end{aligned}
$$

Being able to express the coefficients a_1, a_2, and a_3 as functions of I_1, I_2, and I_3 determines that the coefficients of the characteristic equation are also coordinate-invariant.

Example 396 ★ *Short notation for the elements of inertia matrix.*

Taking advantage of the Kronecker's delta (5.133) we may write the elements of the moment of inertia matrix I_{ij} in short notation forms

$$I_{ij} = \int_B \left((x_1^2 + x_2^2 + x_3^2)\,\delta_{ij} - x_ix_j\right) dm \tag{9.232}$$

$$I_{ij} = \int_B \left(r^2\delta_{ij} - x_ix_j\right) dm \tag{9.233}$$

$$I_{ij} = \int_B \left(\sum_{k=1}^3 x_kx_k\delta_{ij} - x_ix_j\right) dm \tag{9.234}$$

where we adopted the following notations:

$$x_1 = x \qquad x_2 = y \qquad x_3 = z \qquad (9.235)$$

Example 397 ★ *Mass moment with respect to plane, line, and point.*
   The mass moments of a system of particles may be defined with respect to a plane, a line, or a point as the sum of the products of the mass of the particles into the square of the perpendicular distance from the particle to the plane, line, or point. For a continuous body, the sum would be definite integral over the volume of the body.
   The mass moments with respect to the xy, yz, and zx-plane are

$$I_{z^2} = \int_B z^2 dm \qquad I_{y^2} = \int_B y^2 dm \qquad I_{x^2} = \int_B x^2 dm \qquad (9.236)$$

The mass moments with respect to the x, y, and z axes are

$$I_x = \int_B \left(y^2 + z^2\right) dm \qquad (9.237)$$

$$I_y = \int_B \left(z^2 + x^2\right) dm \qquad (9.238)$$

$$I_z = \int_B \left(x^2 + y^2\right) dm \qquad (9.239)$$

and therefore,

$$I_x = I_{y^2} + I_{z^2} \qquad I_y = I_{z^2} + I_{x^2} \qquad I_z = I_{x^2} + I_{y^2} \qquad (9.240)$$

The mass moment with respect to the origin is

$$I_o = \int_B \left(x^2 + y^2 + z^2\right) dm = I_{x^2} + I_{y^2} + I_{z^2} = \frac{1}{2}\left(I_x + I_y + I_z\right) \quad (9.241)$$

Because the choice of the coordinate frame is arbitrary, we can say that the mass moments with respect to a line is the sum of the mass moments with respect to any two mutually orthogonal planes that pass through the line. The mass moments with respect to a point have similar meaning for three mutually orthogonal planes intersecting at the point.

9.5   Lagrange's Form of Newton's Equations of Motion

Newton's equation of motion can be transformed to

$$\frac{d}{dt}\left(\frac{\partial K}{\partial \dot{q}_r}\right) - \frac{\partial K}{\partial q_r} = F_r \qquad r = 1, 2, \cdots n \qquad (9.242)$$

where

$$F_r = \sum_{i=1}^{n} \left(F_{ix}\frac{\partial f_i}{\partial q_1} + F_{iy}\frac{\partial g_i}{\partial q_2} + F_{iz}\frac{\partial h_i}{\partial q_n} \right) \tag{9.243}$$

Equation (9.242) is called the *Lagrange equation of motion*, where K is the kinetic energy of the n degree-of-freedom (DOF) system, q_r, $r = 1, 2, \cdots, n$ are the generalized coordinates of the system, $\mathbf{F} = \begin{bmatrix} F_{ix} & F_{iy} & F_{iz} \end{bmatrix}^T$ is the external force acting on the ith particle of the system, and F_r is the generalized force associated to q_r.

Proof. Let m_i be the mass of one of the particles of a system and let (x_i, y_i, z_i) be its Cartesian coordinates in a globally fixed coordinate frame. Assume that the coordinates of every individual particle are functions of another set of coordinates $q_1, q_2, q_3, \cdots, q_n$, and possibly time t.

$$x_i = f_i(q_1, q_2, q_3, \cdots, q_n, t) \tag{9.244}$$
$$y_i = g_i(q_1, q_2, q_3, \cdots, q_n, t) \tag{9.245}$$
$$z_i = h_i(q_1, q_2, q_3, \cdots, q_n, t) \tag{9.246}$$

If F_{xi}, F_{yi}, F_{zi} are components of the total force acting on the particle m_i, then the Newton equations of motion for the particle would be

$$F_{xi} = m_i\ddot{x}_i \tag{9.247}$$
$$F_{yi} = m_i\ddot{y}_i \tag{9.248}$$
$$F_{zi} = m_i\ddot{z}_i \tag{9.249}$$

We respectively multiply both sides of these equations by

$$\frac{\partial f_i}{\partial q_r} \quad \frac{\partial g_i}{\partial q_r} \quad \frac{\partial h_i}{\partial q_r} \tag{9.250}$$

respectively, and add them up for all the particles to have

$$\sum_{i=1}^{n} m_i \left(\ddot{x}_i\frac{\partial f_i}{\partial q_r} + \ddot{y}_i\frac{\partial g_i}{\partial q_r} + \ddot{z}_i\frac{\partial h_i}{\partial q_r} \right) = \sum_{i=1}^{n} \left(F_{xi}\frac{\partial f_i}{\partial q_r} + F_{yi}\frac{\partial g_i}{\partial q_r} + F_{zi}\frac{\partial h_i}{\partial q_r} \right) \tag{9.251}$$

where n is the total number of particles.

Taking a time derivative of Equation (9.244),

$$\dot{x}_i = \frac{\partial f_i}{\partial q_1}\dot{q}_1 + \frac{\partial f_i}{\partial q_2}\dot{q}_2 + \frac{\partial f_i}{\partial q_3}\dot{q}_3 + \cdots + \frac{\partial f_i}{\partial q_n}\dot{q}_n + \frac{\partial f_i}{\partial t} \tag{9.252}$$

we find

$$\frac{\partial \dot{x}_i}{\partial \dot{q}_r} = \frac{\partial}{\partial \dot{q}_r}\left(\frac{\partial f_i}{\partial q_1}\dot{q}_1 + \frac{\partial f_i}{\partial q_2}\dot{q}_2 + \cdots + \frac{\partial f_i}{\partial q_n}\dot{q}_n + \frac{\partial f_i}{\partial t} \right) = \frac{\partial f_i}{\partial q_r} \tag{9.253}$$

and therefore,

$$\ddot{x}_i \frac{\partial f_i}{\partial q_r} = \ddot{x}_i \frac{\partial \dot{x}_i}{\partial \dot{q}_r} = \frac{d}{dt}\left(\dot{x}_i \frac{\partial \dot{x}_i}{\partial \dot{q}_r}\right) - \dot{x}_i \frac{d}{dt}\left(\frac{\partial \dot{x}_i}{\partial \dot{q}_r}\right) \qquad (9.254)$$

However,

$$\begin{aligned}
\dot{x}_i \frac{d}{dt}\left(\frac{\partial \dot{x}_i}{\partial \dot{q}_r}\right) &= \dot{x}_i \frac{d}{dt}\left(\frac{\partial f_i}{\partial q_r}\right) \\
&= \dot{x}_i \left(\frac{\partial^2 f_i}{\partial q_1 \partial q_r}\dot{q}_1 + \cdots + \frac{\partial^2 f_i}{\partial q_n \partial q_r}\dot{q}_n + \frac{\partial^2 f_i}{\partial t \partial q_r}\right) \\
&= \dot{x}_i \frac{\partial}{\partial q_r}\left(\frac{\partial f_i}{\partial q_1}\dot{q}_1 + \frac{\partial f_i}{\partial q_2}\dot{q}_2 + \cdots + \frac{\partial f_i}{\partial q_n}\dot{q}_n + \frac{\partial f_i}{\partial t}\right) \\
&= \dot{x}_i \frac{\partial \dot{x}_i}{\partial q_r} \qquad (9.255)
\end{aligned}$$

and we have

$$\ddot{x}_i \frac{\partial \dot{x}_i}{\partial \dot{q}_r} = \frac{d}{dt}\left(\dot{x}_i \frac{\partial \dot{x}_i}{\partial \dot{q}_r}\right) - \dot{x}_i \frac{\partial \dot{x}_i}{\partial q_r} \qquad (9.256)$$

which is equal to

$$\ddot{x}_i \frac{\dot{x}_i}{\dot{q}_r} = \frac{d}{dt}\left[\frac{\partial}{\partial \dot{q}_r}\left(\frac{1}{2}\dot{x}_i^2\right)\right] - \frac{\partial}{\partial q_r}\left(\frac{1}{2}\dot{x}_i^2\right) \qquad (9.257)$$

Now substituting (9.254) and (9.257) in the left-hand side of (9.251) leads to

$$\begin{aligned}
\sum_{i=1}^{n} m_i &\left(\ddot{x}_i \frac{\partial f_i}{\partial q_r} + \ddot{y}_i \frac{\partial g_i}{\partial q_r} + \ddot{z}_i \frac{\partial h_i}{\partial q_r}\right) \\
&= \sum_{i=1}^{n} m_i \frac{d}{dt}\left[\frac{\partial}{\partial \dot{q}_r}\left(\frac{1}{2}\dot{x}_i^2 + \frac{1}{2}\dot{y}_i^2 + \frac{1}{2}\dot{z}_i^2\right)\right] \\
&\quad - \sum_{i=1}^{n} m_i \frac{\partial}{\partial q_r}\left(\frac{1}{2}\dot{x}_i^2 + \frac{1}{2}\dot{y}_i^2 + \frac{1}{2}\dot{z}_i^2\right) \\
&= \frac{1}{2}\sum_{i=1}^{n} m_i \frac{d}{dt}\left[\frac{\partial}{\partial \dot{q}_r}\left(\dot{x}_i^2 + \dot{y}_i^2 + \dot{z}_i^2\right)\right] \\
&\quad - \frac{1}{2}\sum_{i=1}^{n} m_i \frac{\partial}{\partial q_r}\left(\dot{x}_i^2 + \dot{y}_i^2 + \dot{z}_i^2\right) \qquad (9.258)
\end{aligned}$$

where

$$\frac{1}{2}\sum_{i=1}^{n} m_i \left(\dot{x}_i^2 + \dot{y}_i^2 + \dot{z}_i^2\right) = K \qquad (9.259)$$

is the *kinetic energy* of the system. Therefore, the Newton equations of motion (9.247), (9.248), and (9.249) are converted to

$$\frac{d}{dt}\left(\frac{\partial K}{\partial \dot{q}_r}\right) - \frac{\partial K}{\partial q_r} = \sum_{i=1}^{n}\left(F_{xi}\frac{\partial f_i}{\partial q_r} + F_{yi}\frac{\partial g_i}{\partial q_r} + F_{zi}\frac{\partial h_i}{\partial q_r}\right) \qquad (9.260)$$

Because of (9.244), (9.245), and (9.246), the kinetic energy is a function of $q_1, q_2, q_3, \cdots, q_n$ and time t. The left-hand side of Equation (9.260) includes the kinetic energy of the whole system and the right-hand side is a generalized force and shows the effect of changing coordinates from x_i to q_j on the external forces. Let us assume that the coordinate q_r alters to $q_r + \delta q_r$ while the other coordinates $q_1, q_2, q_3, \cdots, q_{r-1}, q_{r+1}, \cdots, q_n$ and time t are unaltered. So, the coordinates of m_i are changed to

$$x_i + \frac{\partial f_i}{\partial q_r}\delta q_r \qquad y_i + \frac{\partial g_i}{\partial q_r}\delta q_r \qquad z_i + \frac{\partial h_i}{\partial q_r}\delta q_r \qquad (9.261)$$

Such a displacement is called virtual displacement. The work done in this virtual displacement by all forces acting on the particles of the system is

$$\delta W = \sum_{i=1}^{n}\left(F_{xi}\frac{\partial f_i}{\partial q_r} + F_{yi}\frac{\partial g_i}{\partial q_r} + F_{zi}\frac{\partial h_i}{\partial q_r}\right)\delta q_r \qquad (9.262)$$

Because the work done by internal forces appears in opposite pairs, only the work done by external forces remains in Equation (9.262). Let us denote the virtual work by

$$\delta W = F_r\left(q_1, q_2, q_3, \cdots, q_n, t\right)\delta q_r \qquad (9.263)$$

Then we have

$$\frac{d}{dt}\left(\frac{\partial K}{\partial \dot{q}_r}\right) - \frac{\partial K}{\partial q_r} = F_r \qquad (9.264)$$

where

$$F_r = \sum_{i=1}^{n}\left(F_{xi}\frac{\partial f_i}{\partial q_r} + F_{yi}\frac{\partial g_i}{\partial q_r} + F_{zi}\frac{\partial h_i}{\partial q_r}\right) \qquad (9.265)$$

Equation (9.264) is the Lagrange form of equations of motion. This equation is true for all values of r from 1 to n. We thus have n second-order ordinary differential equations in which $q_1, q_2, q_3, \cdots, q_n$ are the dependent variables and t is the independent variable. The coordinates $q_1, q_2, q_3, \cdots, q_n$ are called *generalized coordinates* and they can be any measurable parameters to provide the configuration of the system. The number of equations and the number of dependent variables are equal, therefore, the equations are theoretically sufficient to determine the motion of all m_i. ■

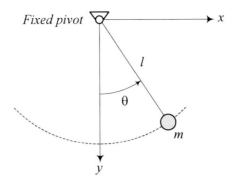

FIGURE 9.10. A simple pendulum.

Example 398 *Equation of motion for a simple pendulum.*

A pendulum is shown in Figure 9.10. Using x and y for the Cartesian position of m, and using $\theta = q$ as the generalized coordinate, we have

$$x = f(\theta) = l \sin \theta \tag{9.266}$$

$$y = g(\theta) = l \cos \theta \tag{9.267}$$

$$K = \frac{1}{2} m \left(\dot{x}^2 + \dot{y}^2 \right) = \frac{1}{2} m l^2 \dot{\theta}^2 \tag{9.268}$$

and therefore,

$$\frac{d}{dt} \left(\frac{\partial K}{\partial \dot{\theta}} \right) - \frac{\partial K}{\partial \theta} = \frac{d}{dt} (ml^2 \dot{\theta}) = ml^2 \ddot{\theta} \tag{9.269}$$

The external force components, acting on m, are

$$F_x = 0 \qquad F_y = mg \tag{9.270}$$

and therefore,

$$F_\theta = F_x \frac{\partial f}{\partial \theta} + F_y \frac{\partial g}{\partial \theta} = -mgl \sin \theta \tag{9.271}$$

Hence, the equation of motion for the pendulum is

$$ml^2 \ddot{\theta} = -mgl \sin \theta \tag{9.272}$$

Example 399 *A pendulum attached to an oscillating mass.*

Figure 9.11 illustrates a vibrating mass with a hanging pendulum. The pendulum can act as a vibration absorber if designed properly. This system has two degrees of freedom and therefore, needs two generalized coordinates.

Adopting x and θ as the generalized coordinate and starting with coordinate relationships

$$x_M = f_M = x \qquad\qquad y_M = g_M = 0 \tag{9.273}$$

$$x_m = f_m = x + l \sin \theta \qquad y_m = g_m = l \cos \theta \tag{9.274}$$

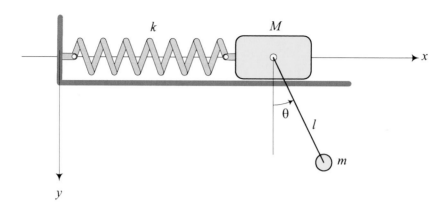

FIGURE 9.11. A vibrating mass with a hanging pendulum.

we may find the kinetic energy in terms of x and θ.

$$K = \frac{1}{2}M\left(\dot{x}_M^2 + \dot{y}_M^2\right) + \frac{1}{2}m\left(\dot{x}_m^2 + \dot{y}_m^2\right)$$

$$= \frac{1}{2}M\dot{x}^2 + \frac{1}{2}m\left(\dot{x}^2 + l^2\dot{\theta}^2 + 2l\dot{x}\dot{\theta}\cos\theta\right) \qquad (9.275)$$

The left-hand side of the Lagrange equations are

$$\frac{d}{dt}\left(\frac{\partial K}{\partial \dot{x}}\right) - \frac{\partial K}{\partial x} = (M+m)\ddot{x} + ml\ddot{\theta}\cos\theta - ml\dot{\theta}^2\sin\theta \qquad (9.276)$$

$$\frac{d}{dt}\left(\frac{\partial K}{\partial \dot{\theta}}\right) - \frac{\partial K}{\partial \theta} = ml^2\ddot{\theta} + ml\ddot{x}\cos\theta \qquad (9.277)$$

The external forces acting on M and m are

$$F_{x_M} = -kx \qquad F_{y_M} = 0 \qquad F_{x_m} = 0 \qquad F_{y_m} = mg \qquad (9.278)$$

Therefore, the generalized forces are

$$F_x = F_{x_M}\frac{\partial f_M}{\partial x} + F_{y_M}\frac{\partial g_M}{\partial x} + F_{x_m}\frac{\partial f_m}{\partial x} + F_{y_m}\frac{\partial g_m}{\partial x}$$

$$= -kx \qquad (9.279)$$

$$F_\theta = F_{x_M}\frac{\partial f_M}{\partial \theta} + F_{y_M}\frac{\partial g_M}{\partial \theta} + F_{x_m}\frac{\partial f_m}{\partial \theta} + F_{y_m}\frac{\partial g_m}{\partial \theta}$$

$$= -mgl\sin\theta \qquad (9.280)$$

and finally the Lagrange equations of motion are

$$(M+m)\ddot{x} + ml\ddot{\theta}\cos\theta - ml\dot{\theta}^2\sin\theta = -kx \qquad (9.281)$$

$$ml^2\ddot{\theta} + ml\ddot{x}\cos\theta = -mgl\sin\theta \qquad (9.282)$$

Example 400 *Kinetic energy of the Earth.*

Earth is approximately a rotating rigid body about a fixed axis. The two motions of the Earth are called **revolution** about the sun, and **rotation** about an axis approximately fixed in the Earth. The kinetic energy of the Earth due to its rotation is

$$
\begin{aligned}
K_1 &= \frac{1}{2} I \omega_1^2 \\
&= \frac{1}{2} \frac{2}{5} \left(5.9742 \times 10^{24}\right) \left(\frac{6356912 + 6378388}{2}\right)^2 \left(\frac{2\pi}{24 \times 3600} \frac{366.25}{365.25}\right)^2 \\
&= 2.5762 \times 10^{29} \text{ J}
\end{aligned}
\tag{9.283}
$$

where $I = 2MR^2/5$ is the mass moment of the Earth, and ω_1 is the angular speed about the Earth's axis. The kinetic energy of the Earth due to its revolution is

$$
\begin{aligned}
K_2 &= \frac{1}{2} M r^2 \omega_2^2 \\
&= \frac{1}{2} \left(5.9742 \times 10^{24}\right) \left(1.49475 \times 10^{11}\right)^2 \left(\frac{2\pi}{24 \times 3600} \frac{1}{365.25}\right)^2 \\
&= 2.6457 \times 10^{33} \text{ J}
\end{aligned}
\tag{9.284}
$$

where r is the distance from the sun, and ω_2 is the angular speed about the sun. The total kinetic energy of the Earth is $K = K_1 + K_2$. However, the ratio of the revolutionary to rotational kinetic energies is

$$
\frac{K_2}{K_1} = \frac{2.6457 \times 10^{33}}{2.5762 \times 10^{29}} \approx 10000
\tag{9.285}
$$

Therefore, rotation of the Earth contribute only 0.1% of the total kinetic energy of the Earth.

Example 401 ★ *Explicit form of Lagrange equations.*

Assume the coordinates of every particle are functions of the coordinates $q_1, q_2, q_3, \cdots, q_n$ but not the time t. The kinetic energy of the system made of n massive particles can be written as

$$
K = \frac{1}{2} \sum_{i=1}^{n} m_i \left(\dot{x}_i^2 + \dot{y}_i^2 + \dot{z}_i^2\right) = \frac{1}{2} \sum_{j=1}^{n} \sum_{k=1}^{n} a_{jk} \dot{q}_j \dot{q}_k
\tag{9.286}
$$

where the coefficients a_{jk} are functions of $q_1, q_2, q_3, \cdots, q_n$ and

$$
a_{jk} = a_{kj}
\tag{9.287}
$$

The Lagrange equations of motion

$$
\frac{d}{dt}\left(\frac{\partial K}{\partial \dot{q}_r}\right) - \frac{\partial K}{\partial q_r} = F_r \qquad r = 1, 2, \cdots n
\tag{9.288}
$$

are then equal to

$$\frac{d}{dt}\sum_{m=1}^{n}a_{mr}\dot{q}_m - \frac{1}{2}\sum_{j=1}^{n}\sum_{k=1}^{n}\frac{\partial a_{jk}}{\partial q_r}\dot{q}_j\dot{q}_k = F_r \tag{9.289}$$

or

$$\sum_{m=1}^{n}a_{mr}\ddot{q}_m + \sum_{k=1}^{n}\sum_{n=1}^{n}\Gamma^r_{k,n}\dot{q}_k\dot{q}_n = F_r \tag{9.290}$$

where $\Gamma^i_{j,k}$ *is called the* **Christoffel operator**

$$\Gamma^i_{j,k} = \frac{1}{2}\left(\frac{\partial a_{ij}}{\partial q_k} + \frac{\partial a_{ik}}{\partial q_j} - \frac{\partial a_{kj}}{\partial q_i}\right) \tag{9.291}$$

9.6 Lagrangian Mechanics

Assume for some forces $\mathbf{F} = \begin{bmatrix} F_{ix} & F_{iy} & F_{iz} \end{bmatrix}^T$ there is a function V, called *potential energy*, such that the force is derivable from V

$$\mathbf{F} = -\nabla V \tag{9.292}$$

Such a force is called *potential* or *conservative force*. Then, the Lagrange equation of motion can be written as

$$\frac{d}{dt}\left(\frac{\partial \mathcal{L}}{\partial \dot{q}_r}\right) - \frac{\partial \mathcal{L}}{\partial q_r} = Q_r \qquad r = 1, 2, \cdots n \tag{9.293}$$

where

$$\mathcal{L} = K - V \tag{9.294}$$

is the *Lagrangean* of the system and Q_r is the nonpotential generalized force.

Proof. Assume the external forces $\mathbf{F} = \begin{bmatrix} F_{xi} & F_{yi} & F_{zi} \end{bmatrix}^T$ acting on the system are conservative.

$$\mathbf{F} = -\nabla V \tag{9.295}$$

The work done by these forces in an arbitrary virtual displacement δq_1, $\delta q_2, \delta q_3, \cdots, \delta q_n$ is

$$\partial W = -\frac{\partial V}{\partial q_1}\delta q_1 - \frac{\partial V}{\partial q_2}\delta q_2 - \cdots \frac{\partial V}{\partial q_n}\delta q_n \tag{9.296}$$

then the Lagrange equation becomes

$$\frac{d}{dt}\left(\frac{\partial K}{\partial \dot{q}_r}\right) - \frac{\partial K}{\partial q_r} = -\frac{\partial V}{\partial q_1} \qquad r = 1, 2, \cdots n. \tag{9.297}$$

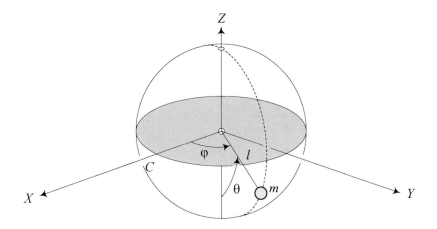

FIGURE 9.12. A spherical pendulum.

Introducing the Lagrangean function $\mathcal{L} = K - V$ converts the Lagrange equation to

$$\frac{d}{dt}\left(\frac{\partial \mathcal{L}}{\partial \dot{q}_r}\right) - \frac{\partial \mathcal{L}}{\partial q_r} = 0 \qquad r = 1, 2, \cdots n \qquad (9.298)$$

for a conservative system. The Lagrangean $\mathcal{L}$ is also called *kinetic potential*.

If a force is not conservative, then the virtual work done by the force is

$$\delta W = \sum_{i=1}^{n}\left(F_{xi}\frac{\partial f_i}{\partial q_r} + F_{yi}\frac{\partial g_i}{\partial q_r} + F_{zi}\frac{\partial h_i}{\partial q_r}\right)\delta q_r = Q_r\,\delta q_r \qquad (9.299)$$

and the equation of motion would be

$$\frac{d}{dt}\left(\frac{\partial \mathcal{L}}{\partial \dot{q}_r}\right) - \frac{\partial \mathcal{L}}{\partial q_r} = Q_r \qquad r = 1, 2, \cdots n \qquad (9.300)$$

where Q_r is the nonpotential generalized force doing work in a virtual displacement of the rth generalized coordinate q_r. ■

Example 402 *Spherical pendulum.*
    A pendulum analogy is utilized in modeling of many dynamical problems. Figure 9.12 illustrates a spherical pendulum with mass m and length l. The angles φ and θ may be used as describing coordinates of the system.
    The Cartesian coordinates of the mass m as a function of the generalized coordinates are

$$\begin{bmatrix} X \\ Y \\ Z \end{bmatrix} = \begin{bmatrix} r\cos\varphi\sin\theta \\ r\sin\theta\sin\varphi \\ -r\cos\theta \end{bmatrix} \qquad (9.301)$$

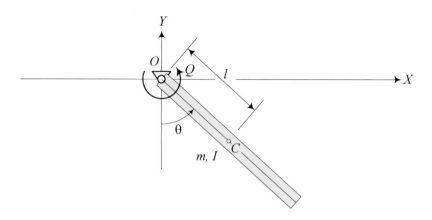

FIGURE 9.13. A controlled compound pendulum.

Having the constraint

$$\dot{X}^2 + \dot{Y}^2 + \dot{Z}^2 - r^2 = 0 \qquad (9.302)$$

makes the three coordinates r, θ, φ dependent with only two coordinates independent. Taking θ, φ as the two generalized coordinates, the kinetic and potential energies of the pendulum would be

$$K = \frac{1}{2}m\left(l^2\dot{\theta}^2 + l^2\dot{\varphi}^2 \sin^2\theta\right) \qquad (9.303)$$

$$V = -mgl\cos\theta \qquad (9.304)$$

The kinetic potential function of this system is then equal to

$$\mathcal{L} = \frac{1}{2}m\left(l^2\dot{\theta}^2 + l^2\dot{\varphi}^2 \sin^2\theta\right) + mgl\cos\theta \qquad (9.305)$$

which leads to the following equations of motion:

$$\ddot{\theta} - \dot{\varphi}^2 \sin\theta\cos\theta + \frac{g}{l}\sin\theta = 0 \qquad (9.306)$$

$$\ddot{\varphi}\sin^2\theta + 2\dot{\varphi}\dot{\theta}\sin\theta\cos\theta = 0 \qquad (9.307)$$

Example 403 *Controlled compound pendulum.*

A massive arm is attached to a ceiling at a pin joint O as illustrated in Figure 9.13. Assume that there is viscous friction in the joint where an ideal motor can apply a torque Q to move the arm. The rotor of an ideal motor has no mass moment by assumption.

The kinetic and potential energies of the manipulator are

$$K = \frac{1}{2}I\dot{\theta}^2 = \frac{1}{2}\left(I_C + ml^2\right)\dot{\theta}^2 \qquad (9.308)$$

$$V = -mgl\cos\theta \qquad (9.309)$$

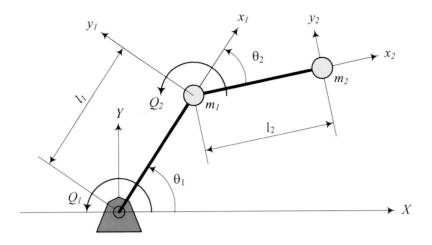

FIGURE 9.14. A model for a 2R planar manipulator.

where m is the mass and I is the mass moment of the pendulum about O. The Lagrangean of the manipulator is

$$\mathcal{L} = K - V = \frac{1}{2} I \dot{\theta}^2 + mgl \cos \theta \qquad (9.310)$$

and therefore, the equation of motion of the pendulum is

$$M = \frac{d}{dt} \left(\frac{\partial \mathcal{L}}{\partial \dot{\theta}} \right) - \frac{\partial \mathcal{L}}{\partial \theta} = I \ddot{\theta} + mgl \sin \theta \qquad (9.311)$$

The generalized force M is the contribution of the motor torque Q and the viscous friction torque $-c\dot{\theta}$. Hence, the equation of motion of the manipulator is

$$Q = I \ddot{\theta} + c \dot{\theta} + mgl \sin \theta \qquad (9.312)$$

Example 404 *An ideal 2R planar manipulator dynamics.*
An ideal model of a 2R planar manipulator is illustrated in Figure 9.14. It is called ideal because we assume the links are massless and there is no friction in the system. There is a motor on the ground that applies a torque Q_1 on the link one. The second link has the mass m_1 and runs the second link and the load m_2 at the endpoint. We take the absolute angle θ_1 and the relative angle θ_2 as the generalized coordinates to express the configuration of the manipulator.
The global positions of m_1 and m_2 are

$$\begin{bmatrix} X_1 \\ Y_2 \end{bmatrix} = \begin{bmatrix} l_1 \cos \theta_1 \\ l_1 \sin \theta_1 \end{bmatrix} \qquad (9.313)$$

$$\begin{bmatrix} X_2 \\ Y_2 \end{bmatrix} = \begin{bmatrix} l_1 \cos \theta_1 + l_2 \cos (\theta_1 + \theta_2) \\ l_1 \sin \theta_1 + l_2 \sin (\theta_1 + \theta_2) \end{bmatrix} \qquad (9.314)$$

and therefore, the global velocity of the masses are

$$\begin{bmatrix} \dot{X}_1 \\ \dot{Y}_1 \end{bmatrix} = \begin{bmatrix} -l_1\dot{\theta}_1 \sin\theta_1 \\ l_1\dot{\theta}_1 \cos\theta_1 \end{bmatrix} \tag{9.315}$$

$$\begin{bmatrix} \dot{X}_2 \\ \dot{Y}_2 \end{bmatrix} = \begin{bmatrix} -l_1\dot{\theta}_1 \sin\theta_1 - l_2\left(\dot{\theta}_1 + \dot{\theta}_2\right)\sin(\theta_1 + \theta_2) \\ l_1\dot{\theta}_1 \cos\theta_1 + l_2\left(\dot{\theta}_1 + \dot{\theta}_2\right)\cos(\theta_1 + \theta_2) \end{bmatrix} \tag{9.316}$$

The kinetic energy of this manipulator is made of kinetic energy of the masses and is equal to

$$\begin{aligned} K &= K_1 + K_2 = \frac{1}{2}m_1\left(\dot{X}_1^2 + \dot{Y}_1^2\right) + \frac{1}{2}m_2\left(\dot{X}_2^2 + \dot{Y}_2^2\right) \\ &= \frac{1}{2}m_1 l_1^2\dot{\theta}_1^2 \\ &\quad + \frac{1}{2}m_2\left(l_1^2\dot{\theta}_1^2 + l_2^2\left(\dot{\theta}_1 + \dot{\theta}_2\right)^2 + 2l_1 l_2\dot{\theta}_1\left(\dot{\theta}_1 + \dot{\theta}_2\right)\cos\theta_2\right) \end{aligned} \tag{9.317}$$

The potential energy of the manipulator is

$$\begin{aligned} V &= V_1 + V_2 = m_1 g Y_1 + m_2 g Y_2 \\ &= m_1 g l_1 \sin\theta_1 + m_2 g\left(l_1 \sin\theta_1 + l_2 \sin(\theta_1 + \theta_2)\right) \end{aligned} \tag{9.318}$$

The Lagrangean is then obtained from Equations (9.317) and (9.318)

$$\begin{aligned} \mathcal{L} &= K - V \qquad\qquad\qquad\qquad\qquad\qquad (9.319) \\ &= \frac{1}{2}m_1 l_1^2\dot{\theta}_1^2 \\ &\quad + \frac{1}{2}m_2\left(l_1^2\dot{\theta}_1^2 + l_2^2\left(\dot{\theta}_1 + \dot{\theta}_2\right)^2 + 2l_1 l_2\dot{\theta}_1\left(\dot{\theta}_1 + \dot{\theta}_2\right)\cos\theta_2\right) \\ &\quad - \left(m_1 g l_1 \sin\theta_1 + m_2 g\left(l_1 \sin\theta_1 + l_2 \sin(\theta_1 + \theta_2)\right)\right) \end{aligned}$$

which provides the required partial derivatives as follows:

$$\frac{\partial\mathcal{L}}{\partial\theta_1} = -(m_1 + m_2)g l_1 \cos\theta_1 - m_2 g l_2 \cos(\theta_1 + \theta_2) \tag{9.320}$$

$$\begin{aligned} \frac{\partial\mathcal{L}}{\partial\dot{\theta}_1} &= (m_1 + m_2)l_1^2\dot{\theta}_1 + m_2 l_2^2\left(\dot{\theta}_1 + \dot{\theta}_2\right) \\ &\quad + m_2 l_1 l_2\left(2\dot{\theta}_1 + \dot{\theta}_2\right)\cos\theta_2 \end{aligned} \tag{9.321}$$

$$\begin{aligned} \frac{d}{dt}\left(\frac{\partial\mathcal{L}}{\partial\dot{\theta}_1}\right) &= (m_1 + m_2)l_1^2\ddot{\theta}_1 + m_2 l_2^2\left(\ddot{\theta}_1 + \ddot{\theta}_2\right) \\ &\quad + m_2 l_1 l_2\left(2\ddot{\theta}_1 + \ddot{\theta}_2\right)\cos\theta_2 \\ &\quad - m_2 l_1 l_2\dot{\theta}_2\left(2\dot{\theta}_1 + \dot{\theta}_2\right)\sin\theta_2 \end{aligned} \tag{9.322}$$

$$\frac{\partial \mathcal{L}}{\partial \theta_2} = -m_2 l_1 l_2 \dot{\theta}_1 \left(\dot{\theta}_1 + \dot{\theta}_2\right) \sin \theta_2 - m_2 g l_2 \cos \left(\theta_1 + \theta_2\right) \quad (9.323)$$

$$\frac{\partial \mathcal{L}}{\partial \dot{\theta}_2} = m_2 l_2^2 \left(\dot{\theta}_1 + \dot{\theta}_2\right) + m_2 l_1 l_2 \dot{\theta}_1 \cos \theta_2 \quad (9.324)$$

$$\frac{d}{dt}\left(\frac{\partial \mathcal{L}}{\partial \dot{\theta}_2}\right) = m_2 l_2^2 \left(\ddot{\theta}_1 + \ddot{\theta}_2\right) + m_2 l_1 l_2 \ddot{\theta}_1 \cos \theta_2 - m_2 l_1 l_2 \dot{\theta}_1 \dot{\theta}_2 \sin \theta_2 \quad (9.325)$$

Therefore, the equations of motion for the 2R manipulator are

$$\begin{aligned} Q_1 &= \frac{d}{dt}\left(\frac{\partial \mathcal{L}}{\partial \dot{\theta}_1}\right) - \frac{\partial \mathcal{L}}{\partial \theta_1} \\ &= (m_1 + m_2) l_1^2 \ddot{\theta}_1 + m_2 l_2^2 \left(\ddot{\theta}_1 + \ddot{\theta}_2\right) \\ &\quad + m_2 l_1 l_2 \left(2\ddot{\theta}_1 + \ddot{\theta}_2\right) \cos \theta_2 - m_2 l_1 l_2 \dot{\theta}_2 \left(2\dot{\theta}_1 + \dot{\theta}_2\right) \sin \theta_2 \\ &\quad + (m_1 + m_2) g l_1 \cos \theta_1 + m_2 g l_2 \cos \left(\theta_1 + \theta_2\right) \quad (9.326) \end{aligned}$$

$$\begin{aligned} Q_2 &= \frac{d}{dt}\left(\frac{\partial \mathcal{L}}{\partial \dot{\theta}_2}\right) - \frac{\partial \mathcal{L}}{\partial \theta_2} \\ &= m_2 l_2^2 \left(\ddot{\theta}_1 + \ddot{\theta}_2\right) + m_2 l_1 l_2 \ddot{\theta}_1 \cos \theta_2 - m_2 l_1 l_2 \dot{\theta}_1 \dot{\theta}_2 \sin \theta_2 \\ &\quad + m_2 l_1 l_2 \dot{\theta}_1 \left(\dot{\theta}_1 + \dot{\theta}_2\right) \sin \theta_2 + m_2 g l_2 \cos \left(\theta_1 + \theta_2\right) \quad (9.327) \end{aligned}$$

The generalized forces Q_1 and Q_2 are the required forces to drive the generalized coordinates. In this case, Q_1 is the torque at the base motor and Q_2 is the torque of the motor at m_1.

The equations of motion can be rearranged to have a more systematic form

$$\begin{aligned} Q_1 &= \left((m_1 + m_2) l_1^2 + m_2 l_2 (l_2 + 2l_1 \cos \theta_2)\right) \ddot{\theta}_1 \\ &\quad + m_2 l_2 (l_2 + l_1 \cos \theta_2) \ddot{\theta}_2 \\ &\quad - 2m_2 l_1 l_2 \sin \theta_2 \, \dot{\theta}_1 \dot{\theta}_2 - m_2 l_1 l_2 \sin \theta_2 \, \dot{\theta}_2^2 \\ &\quad + (m_1 + m_2) g l_1 \cos \theta_1 + m_2 g l_2 \cos \left(\theta_1 + \theta_2\right) \quad (9.328) \end{aligned}$$

$$\begin{aligned} Q_2 &= m_2 l_2 (l_2 + l_1 \cos \theta_2) \ddot{\theta}_1 + m_2 l_2^2 \ddot{\theta}_2 + m_2 l_1 l_2 \sin \theta_2 \, \dot{\theta}_1^2 \\ &\quad + m_2 g l_2 \cos \left(\theta_1 + \theta_2\right) \quad (9.329) \end{aligned}$$

Example 405 *Mechanical energy.*

If a system of masses m_i are moving in a potential force field

$$\mathbf{F}_{m_i} = -\nabla_i V \quad (9.330)$$

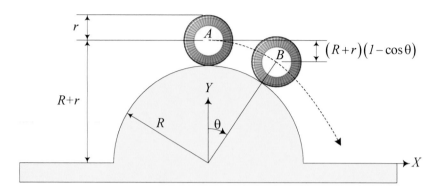

FIGURE 9.15. A wheel turning, without slip, over a cylindrical hill.

their Newton equations of motion will be

$$m_i \ddot{\mathbf{r}}_i = -\nabla_i V \qquad i = 1, 2, \cdots n \tag{9.331}$$

The inner product of equations of motion with $\dot{\mathbf{r}}_i$ and adding the equations

$$\sum_{i=1}^{n} m_i \dot{\mathbf{r}}_i \cdot \ddot{\mathbf{r}}_i = -\sum_{i=1}^{n} \dot{\mathbf{r}}_i \cdot \nabla_i V \tag{9.332}$$

and then, integrating over time

$$\frac{1}{2} \sum_{i=1}^{n} m_i \dot{\mathbf{r}}_i \cdot \dot{\mathbf{r}}_i = -\int \sum_{i=1}^{n} \mathbf{r}_i \cdot \nabla_i V \tag{9.333}$$

shows that

$$K = -\int \sum_{i=1}^{n} \left(\frac{\partial V}{\partial x_i} x_i + \frac{\partial V}{\partial y_i} y_i + \frac{\partial V}{\partial z_i} z_i \right) = -V + E \tag{9.334}$$

*where E is the constant of integration. E is called the **mechanical energy** of the system and is equal to kinetic plus potential energies.*

$$E = K + V \tag{9.335}$$

Example 406 *Falling wheel.*

Figure 9.15 illustrates a wheel rolling over a cylindrical hill. We may use the conservation of mechanical energy to find the angle at which the wheel leaves the hill.

At the initial instant of time, the wheel is at point A. We assume the initial kinetic and potential, and hence, the mechanical energies are zero. When the wheel is turning over the hill, its angular velocity, ω, is

$$\omega = \frac{v}{r} \tag{9.336}$$

where v is the speed at the center of the wheel. At any other point B, the wheel achieves some kinetic energy and loses some potential energy. At a certain angle, where the normal component of the weight cannot provide more centripetal force,

$$mg \cos \theta = \frac{mv^2}{R+r} \tag{9.337}$$

the wheel separates from the surface. Employing the conservation of energy, we have

$$E_A = E_B \tag{9.338}$$
$$K_A + V_A = K_B + V_B \tag{9.339}$$

The kinetic and potential energy at the separation point B are

$$K_B = \frac{1}{2}mv^2 + \frac{1}{2}I_C\omega^2 \tag{9.340}$$
$$V_B = -mg(R+r)(1-\cos\theta) \tag{9.341}$$

where I_C is the mass moment of the wheel about its center. Therefore,

$$\frac{1}{2}mv^2 + \frac{1}{2}I_C\omega^2 = mg(R+r)(1-\cos\theta) \tag{9.342}$$

and substituting (9.336) and (9.337) provides

$$\left(1 + \frac{I_C}{mr^2}\right)(R+r)g\cos\theta = 2g(R+r)(1-\cos\theta) \tag{9.343}$$

and therefore, the separation angle is

$$\theta = \cos^{-1}\frac{2mr^2}{I_C + 3mr^2} \tag{9.344}$$

Let us examine the equation for a disc wheel with

$$I_C = \frac{1}{2}mr^2 \tag{9.345}$$

and find the separation angle.

$$\theta = \cos^{-1}\frac{4}{7} \approx 0.96\,\text{rad} \approx 55.15\,\text{deg} \tag{9.346}$$

Example 407 *Turning wheel over a step.*

Figure 9.16 illustrates a wheel of radius R rolling with speed v to go over a step with height $H < R$.

We may use the principle of energy conservation and find the speed of the wheel after getting across the step. Employing the conservation of energy,

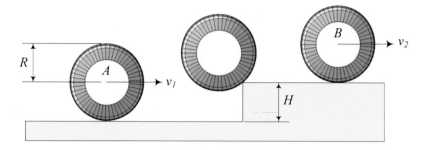

FIGURE 9.16. A rolling wheel moving up a step.

we have

$$E_A = E_B \tag{9.347}$$

$$K_A + V_A = K_B + V_B \tag{9.348}$$

$$\frac{1}{2}mv_1^2 + \frac{1}{2}I_C\omega_1^2 + 0 = \frac{1}{2}mv_2^2 + \frac{1}{2}I_C\omega_2^2 + mgH \tag{9.349}$$

$$\left(m + \frac{I_C}{R^2}\right)v_1^2 = \left(m + \frac{I_C}{R^2}\right)v_2^2 + 2mgH \tag{9.350}$$

and therefore,

$$v_2 = \sqrt{v_1^2 - \frac{2gH}{1 + \dfrac{I_C}{mR^2}}} \tag{9.351}$$

The condition for having a real v_2 is

$$v_1 > \sqrt{\frac{2gH}{1 + \dfrac{I_C}{mR^2}}} \tag{9.352}$$

The second speed (9.351) and the condition (9.352) for a solid disc are

$$v_2 = \sqrt{v_1^2 - \frac{4}{3}Hg} \tag{9.353}$$

$$v_1 > \sqrt{\frac{4}{3}Hg} \tag{9.354}$$

because we assumed that

$$I_C = \frac{1}{2}mR^2 \tag{9.355}$$

Example 408 *Trebuchet.*

   *A **trebuchet**, shown schematically in Figure 9.17, is a shooting weapon of war powered by a falling massive counterweight m_1. A beam AB is pivoted to the chassis with two unequal sections a and b.*

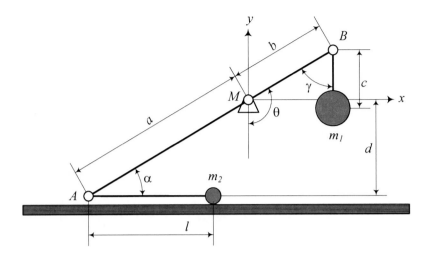

FIGURE 9.17. A trebuchet at starting position.

The figure shows a trebuchet at its initial configuration. The origin of a global coordinate frame is set at the pivot point. The counterweight m_1 is at (x_1, y_1) and is hinged at the shorter arm of the beam at a distance c from the end B. The mass of the projectile is m_2 and it is at the end of a massless sling with a length l attached to the end of the longer arm of the beam. The three independent variable angles α, θ, γ describe the motion of the device. We consider the parameters a, b, c, d, l, m_1, m_2 constant, and determine the equations of motion by the Lagrange method.

Figure 9.18 illustrates the trebuchet when it is in motion. The position coordinates of masses m_1 and m_2 are

$$x_1 = b\sin\theta - c\sin(\theta + \gamma) \tag{9.356}$$
$$y_1 = -b\cos\theta + c\cos(\theta + \gamma) \tag{9.357}$$

and

$$x_2 = -a\sin\theta - l\sin(-\theta + \alpha) \tag{9.358}$$
$$y_2 = -a\cos\theta - l\cos(-\theta + \alpha) \tag{9.359}$$

Taking a time derivative provides the velocity components

$$\dot{x}_1 = b\dot{\theta}\cos\theta - c\left(\dot{\theta} + \dot{\gamma}\right)\cos(\theta + \gamma) \tag{9.360}$$

$$\dot{y}_1 = b\dot{\theta}\sin\theta - c\left(\dot{\theta} + \dot{\gamma}\right)\sin(\theta + \gamma) \tag{9.361}$$

$$\dot{x}_2 = l(c - \dot{\alpha})\cos(\alpha - \theta) - a\dot{\theta}\cos(\theta) \tag{9.362}$$
$$\dot{y}_2 = a\dot{\theta}\sin\theta - l\left(\dot{\theta} - \dot{\alpha}\right)\sin(\alpha - \theta) \tag{9.363}$$

which shows that the kinetic energy of the system is

$$
\begin{aligned}
K &= \frac{1}{2}m_1 v_1^2 + \frac{1}{2}m_2 v_2^2 = \frac{1}{2}m_1\left(\dot{x}_1^2 + \dot{y}_1^2\right) + \frac{1}{2}m_2\left(\dot{x}_2^2 + \dot{y}_2^2\right) \\
&= \frac{1}{2}m_1\left(\left(b^2 + c^2\right)\dot{\theta}^2 + c^2\dot{\gamma}^2 + 2c^2\dot{\theta}\dot{\gamma}\right) \\
&\quad - m_1 bc\dot{\theta}\left(\dot{\theta} + \dot{\gamma}\right)\cos\gamma + \frac{1}{2}m_2\left(\left(a^2 + l^2\right)\dot{\theta}^2 + l^2\dot{\alpha}^2 - 2l^2\dot{\theta}\dot{\alpha}\right) \\
&\quad - m_2 al\dot{\theta}\left(\dot{\theta} - \dot{\alpha}\right)\cos\left(2\theta - \alpha\right)
\end{aligned} \tag{9.364}
$$

The potential energy of the system can be calculated by y-position of the masses.

$$
\begin{aligned}
V &= m_1 g y_1 + m_2 g y_2 \\
&= m_1 g\left(-b\cos\theta + c\cos\left(\theta + \gamma\right)\right) \\
&\quad + m_2 g\left(-a\cos\theta - l\cos\left(-\theta + \alpha\right)\right)
\end{aligned} \tag{9.365}
$$

Having the energies K and V, we can set up the Lagrangean $\mathcal{L}$.

$$
\mathcal{L} = K - V \tag{9.366}
$$

Using the Lagrangean, we are able to find the three equations of motion.

$$
\frac{d}{dt}\left(\frac{\partial \mathcal{L}}{\partial \dot{\theta}}\right) - \frac{\partial \mathcal{L}}{\partial \theta} = 0 \tag{9.367}
$$

$$
\frac{d}{dt}\left(\frac{\partial \mathcal{L}}{\partial \dot{\alpha}}\right) - \frac{\partial \mathcal{L}}{\partial \alpha} = 0 \tag{9.368}
$$

$$
\frac{d}{dt}\left(\frac{\partial \mathcal{L}}{\partial \dot{\gamma}}\right) - \frac{\partial \mathcal{L}}{\partial \gamma} = 0 \tag{9.369}
$$

The trebuchet appeared in 500 to 400 B.C. in China and was developed by Persian armies around 300 B.C. It was used by the Arabs against the Romans during 600 to 1200 A.D. The trebuchet may also be called the **manjaniq, catapults,** or **onager**. The Persian word "Manganic" is the root manjaniq as well as the words "mechanic" and "machine".

9.7 Summary

The translational and rotational equations of motion for a rigid body, expressed in the global coordinate frame G, are

$$
{}^G\mathbf{F} = \frac{{}^G d}{dt}{}^G\mathbf{p} \tag{9.370}
$$

$$
{}^G\mathbf{M} = \frac{{}^G d}{dt}{}^G\mathbf{L} \tag{9.371}
$$

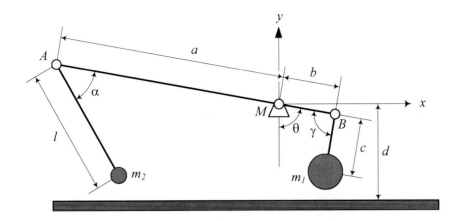

FIGURE 9.18. A trebuchet in motion.

where $^G\mathbf{F}$ and $^G\mathbf{M}$ indicate the resultant of the external forces and moments applied on the rigid body, measured at the mass center C.

The vector $^G\mathbf{p}$ is the momentum and $^G\mathbf{L}$ is the moment of momentum for the rigid body at C

$$\mathbf{p} = m\,\mathbf{v} \tag{9.372}$$

$$\mathbf{L} = \mathbf{r}_C \times \mathbf{p} \tag{9.373}$$

The expression of the equations of motion in the body coordinate frame are

$$^B\mathbf{F} = {}^G\dot{\mathbf{p}} + {}^B_G\boldsymbol{\omega}_B \times {}^B\mathbf{p} = m\,{}^B\mathbf{a}_B + m\,{}^B_G\boldsymbol{\omega}_B \times {}^B\mathbf{v}_B \tag{9.374}$$

$$^B\mathbf{M} = {}^B\dot{\mathbf{L}} + {}^B_G\boldsymbol{\omega}_B \times {}^B\mathbf{L} = {}^B I\,{}^B_G\dot{\boldsymbol{\omega}}_B + {}^B_G\boldsymbol{\omega}_B \times \left({}^B I\,{}^B_G\boldsymbol{\omega}_B\right) \tag{9.375}$$

where I is the mass moment of the rigid body.

$$I = \begin{bmatrix} I_{xx} & I_{xy} & I_{xz} \\ I_{yx} & I_{yy} & I_{yz} \\ I_{zx} & I_{zy} & I_{zz} \end{bmatrix} \tag{9.376}$$

The elements of I are functions of the mass distribution of the rigid body and are defined by

$$I_{ij} = \int_B \left(r_i^2 \delta_{mn} - x_{im} x_{jn}\right) dm \qquad i,j = 1,2,3 \tag{9.377}$$

where δ_{ij} is Kronecker's delta.

Every rigid body has a principal body coordinate frame in which the mass moment matrix is diagonal

$$^B I = \begin{bmatrix} I_1 & 0 & 0 \\ 0 & I_2 & 0 \\ 0 & 0 & I_3 \end{bmatrix} \tag{9.378}$$

The rotational equation of motion in the principal coordinate frame simplifies to

$$\begin{aligned}
M_1 &= I_1 \dot{\omega}_1 - (I_2 - I_2) \omega_2 \omega_3 \\
M_2 &= I_2 \dot{\omega}_2 - (I_3 - I_1) \omega_3 \omega_1 \\
M_3 &= I_3 \dot{\omega}_3 - (I_1 - I_2) \omega_1 \omega_2
\end{aligned} \qquad (9.379)$$

The equations of motion for a mechanical system having n DOF can also be found by the Lagrange equation

$$\frac{d}{dt} \left(\frac{\partial \mathcal{L}}{\partial \dot{q}_r} \right) - \frac{\partial \mathcal{L}}{\partial q_r} = Q_r \qquad r = 1, 2, \cdots n \qquad (9.380)$$

$$\mathcal{L} = K - V \qquad (9.381)$$

where $\mathcal{L}$ is the *Lagrangean* of the system, K is the kinetic energy, V is the potential energy, and Q_r is the nonpotential generalized force.

$$Q_r = \sum_{i=1}^{n} \left(Q_{ix} \frac{\partial f_i}{\partial q_1} + Q_{iy} \frac{\partial g_i}{\partial q_2} + Q_{iz} \frac{\partial h_i}{\partial q_n} \right) \qquad (9.382)$$

The parameters q_r, $r = 1, 2, \cdots, n$ are the generalized coordinates of the system, $\mathbf{Q} = \begin{bmatrix} Q_{ix} & Q_{iy} & Q_{iz} \end{bmatrix}^T$ is the external force acting on the ith particle of the system, and Q_r is the generalized force associated to q_r. When (x_i, y_i, z_i) are Cartesian coordinates in a globally fixed coordinate frame for the particle m_i, then its coordinates may be functions of another set of generalized coordinates $q_1, q_2, q_3, \cdots, q_n$ and possibly time t.

$$\begin{aligned}
x_i &= f_i(q_1, q_2, q_3, \cdots, q_n, t) & (9.383) \\
y_i &= g_i(q_1, q_2, q_3, \cdots, q_n, t) & (9.384) \\
z_i &= h_i(q_1, q_2, q_3, \cdots, q_n, t) & (9.385)
\end{aligned}$$

9.8   Key Symbols

a, b, w, h	length
$\mathbf{a}$	acceleration
C	mass center
$\mathbf{d}$	position vector of the body coordinate frame
$d\mathbf{f}$	infinitesimal force
dm	infinitesimal mass
$d\mathbf{m}$	infinitesimal moment
E	mechanical energy
$\mathbf{F}$	force
F_C	Coriolis force
g	gravitational acceleration
H	height
I	mass moment matrix
I_1, I_2, I_3	principal moment of inertia
K	kinetic energy
l	directional line
$\mathbf{L}$	moment of momentum
$\mathcal{L} = K - V$	Lagrangean
m	mass
$\mathbf{M}$	moment
$\mathbf{p}$	momentum
P, Q	points in rigid body
Q	torque
r	radius of disc
$\mathbf{r}$	position vector
R	radius
R	rotation matrix
t	time
T	tension force
$\hat{u}$	unit vector to show the directional line
$v \equiv \dot{x}, \mathbf{v}$	velocity
V	potential energy
$\mathbf{w}$	eigenvector
W	work, eigenvector matrix
$x, y, z, \mathbf{x}$	displacement
δ_{ij}	Kronecker's delta
$\Gamma^i_{j,k}$	Christoffel operator
λ	eigenvalue
φ, θ, ψ	Euler angles
$\omega, \boldsymbol{\omega}$	angular velocity
$\|$	parallel
$\perp$	orthogonal
∇	gradient

Exercises

1. Kinetic energy of a rigid link.

   Consider a straight and uniform bar as a rigid bar. The bar has a mass m. Show that the kinetic energy of the bar can be expressed as

   $$K = \frac{1}{6} m \left(\mathbf{v}_1 \cdot \mathbf{v}_1 + \mathbf{v}_1 \cdot \mathbf{v}_2 + \mathbf{v}_2 \cdot \mathbf{v}_2 \right)$$

   where $\mathbf{v}_1$ and $\mathbf{v}_2$ are the velocity vectors of the endpoints of the bar.

2. Discrete particles.

   There are three particles $m_1 = 10\,\text{kg}$, $m_2 = 20\,\text{kg}$, $m_3 = 30\,\text{kg}$, at

   $$\mathbf{r}_1 = \begin{bmatrix} 1 \\ -1 \\ 1 \end{bmatrix} \quad \mathbf{r}_1 = \begin{bmatrix} -1 \\ -3 \\ 2 \end{bmatrix} \quad \mathbf{r}_1 = \begin{bmatrix} 2 \\ -1 \\ -3 \end{bmatrix}$$

   Their velocities are

   $$\mathbf{v}_1 = \begin{bmatrix} 2 \\ 1 \\ 1 \end{bmatrix} \quad \mathbf{v}_1 = \begin{bmatrix} -1 \\ 0 \\ 2 \end{bmatrix} \quad \mathbf{v}_1 = \begin{bmatrix} 3 \\ -2 \\ -1 \end{bmatrix}$$

   Find the position and velocity of the system at C. Calculate the system's momentum and moment of momentum. Calculate the system's kinetic energy and determine the rotational and translational parts of the kinetic energy.

3. Newton's equation of motion in the body frame.

   Show that Newton's equation of motion in the body frame is

   $$\begin{bmatrix} F_x \\ F_y \\ F_z \end{bmatrix} = m \begin{bmatrix} a_x \\ a_y \\ a_z \end{bmatrix} + \begin{bmatrix} 0 & -\omega_z & \omega_y \\ \omega_z & 0 & -\omega_x \\ -\omega_y & \omega_x & 0 \end{bmatrix} \begin{bmatrix} v_x \\ v_y \\ v_z \end{bmatrix}$$

4. Work on a curved path.

   A particle of mass m is moving on a circular path given by

   $$^G\mathbf{r}_P = \cos\theta\, \hat{I} + \sin\theta\, \hat{J} + 4\,\hat{K}$$

   Calculate the work done by a force $^G\mathbf{F}$ when the particle moves from $\theta = 0$ to $\theta = \frac{\pi}{2}$.

   (a)

   $$^G\mathbf{F} = \frac{z^2 - y^2}{(x+y)^2}\, \hat{I} + \frac{y^2 - x^2}{(x+y)^2}\, \hat{J} + \frac{x^2 - y^2}{(x+z)^2}\, \hat{K}$$

(b)
$$^G\mathbf{F} = \frac{z^2 - y^2}{(x+y)^2}\,\hat{I} + \frac{2y}{x+y}\,\hat{J} + \frac{x^2 - y^2}{(x+z)^2}\,\hat{K}$$

5. Principal moments of inertia.

Find the principal moments of inertia and directions for the following inertia matrices:

(a)
$$[I] = \begin{bmatrix} 3 & 2 & 2 \\ 2 & 2 & 0 \\ 2 & 0 & 4 \end{bmatrix}$$

(b)
$$[I] = \begin{bmatrix} 3 & 2 & 4 \\ 2 & 0 & 2 \\ 4 & 2 & 3 \end{bmatrix}$$

(c)
$$[I] = \begin{bmatrix} 100 & 20\sqrt{3} & 0 \\ 20\sqrt{3} & 60 & 0 \\ 0 & 0 & 10 \end{bmatrix}$$

6. Rotated moment of inertia matrix.

A principal moment of inertia matrix ^{B_2}I is given as
$$[I] = \begin{bmatrix} 3 & 0 & 0 \\ 0 & 5 & 0 \\ 0 & 0 & 4 \end{bmatrix}$$

The principal frame was achieved by rotating the initial body coordinate frame 30 deg about the x-axis, followed by 45 deg about the z-axis. Find the initial moment of inertia matrix ^{B_1}I.

7. Rotation of moment of inertia matrix.

Find the required rotation matrix that transforms the moment of inertia matrix $[I]$ to an diagonal matrix.
$$[I] = \begin{bmatrix} 3 & 2 & 2 \\ 2 & 2 & 0.1 \\ 2 & 0.1 & 4 \end{bmatrix}$$

8. ★ Cubic equations.

The solution of a cubic equation
$$ax^3 + bx^2 + cx + d = 0$$

where $a \neq 0$, can be found in a systematic way.

Transform the equation to a new form with discriminant $4p^3 + q^2$,

$$y^3 + 3py + q = 0$$

using the transformation $x = y - \frac{b}{3a}$, where,

$$p = \frac{3ac - b^2}{9a^2} \qquad q = \frac{2b^3 - 9abc + 27a^2 d}{27a^3}$$

The solutions are then

$$
\begin{aligned}
y_1 &= \sqrt[3]{\alpha} - \sqrt[3]{\beta} \\
y_2 &= e^{\frac{2\pi i}{3}} \sqrt[3]{\alpha} - e^{\frac{4\pi i}{3}} \sqrt[3]{\beta} \qquad y_3 = e^{\frac{4\pi i}{3}} \sqrt[3]{\alpha} - e^{\frac{2\pi i}{3}} \sqrt[3]{\beta}
\end{aligned}
$$

where,

$$\alpha = \frac{-q + \sqrt{q^2 + 4p^3}}{2} \qquad \beta = \frac{-q + \sqrt{q^2 + 4p^3}}{2}$$

For real values of p and q, if the discriminant is positive, then one root is real, and two roots are complex conjugates. If the discriminant is zero, then there are three real roots, of which at least two are equal. If the discriminant is negative, then there are three unequal real roots.

Apply this theory for the characteristic equation of the matrix $[I]$ and show that the principal moments of inertia are real.

9. **Kinematics of a moving car on the Earth.**

   The location of a vehicle on the Earth is described by its longitude φ from a fixed meridian, say, the Greenwich meridian, and its latitude θ from the equator, as shown in Figure 9.19. We attach a coordinate frame B at the center of the Earth with the x-axis on the equator's plane and the y-axis pointing to the vehicle. There are also two coordinate frames E and G where E is attached to the Earth and G is the global coordinate frame. Show that the angular velocity of B and the velocity of the vehicle are

$$
\begin{aligned}
{}^{B}_{G}\omega_B &= \dot{\theta}\,\hat{\imath}_B + (\omega_E + \dot{\varphi}) \sin\theta\,\hat{\jmath}_B + (\omega_E + \dot{\varphi}) \cos\theta\,\hat{k} \\
{}^{B}_{G}\mathbf{v}_P &= -r\,(\omega_E + \dot{\varphi}) \cos\theta\,\hat{\imath}_B + r\dot{\theta}\,\hat{k}
\end{aligned}
$$

   Calculate the acceleration of the vehicle.

10. **Global differential of angular momentum.**

    Convert the moment of inertia ${}^{B}I$ and the angular velocity ${}^{B}_{G}\omega_B$ to the global coordinate frame and then find the differential of angular momentum. It is an alternative method to show that

$$\frac{{}^{G}d}{dt}\,{}^{B}\mathbf{L} = \frac{{}^{G}d}{dt}\left({}^{B}I\,{}^{B}_{G}\omega_B\right) = {}^{B}\dot{\mathbf{L}} + {}^{B}_{G}\omega_B \times {}^{B}\mathbf{L} = I\dot{\omega} + \omega \times (I\omega)$$

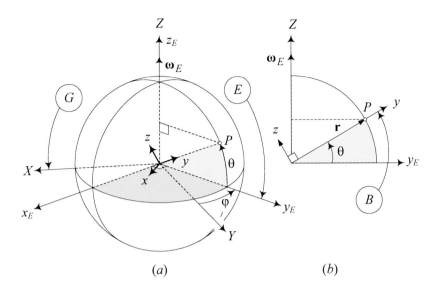

FIGURE 9.19. The location on the Earth is defined by longitude φ and latitude θ.

11. Lagrange method and nonlinear vibrating system.

    Use the Lagrange method and find the equation of motion for the pendulum shown in Figure 9.20. The stiffness of the linear spring is k.

12. Forced vibration of a pendulum.

    Figure 9.21 illustrates a simple pendulum having a length l and a bob with mass m. Find the equation of motion if

    (a) the pivot O has a dictated motion in X direction

    $$X_O = a \sin \omega t$$

    (b) the pivot O has a dictated motion in Y direction

    $$Y_O = b \sin \omega t$$

    (c) the pivot O has a uniform motion on a circle

    $$\mathbf{r}_O = R \cos \omega t \; \hat{I} + R \sin \omega t \; \hat{J}$$

13. Equations of motion from Lagrangean.

    Consider a physical system with a Lagrangean as

    $$\mathcal{L} = \frac{1}{2} m \left(a\dot{x} + b\dot{y} \right)^2 - \frac{1}{2} k \left(ax + by \right)^2$$

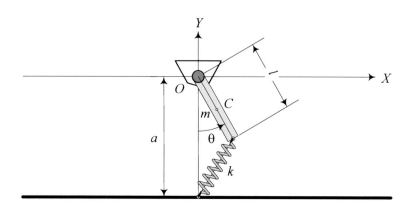

FIGURE 9.20. A compound pendulum attached with a linear spring at the tip point.

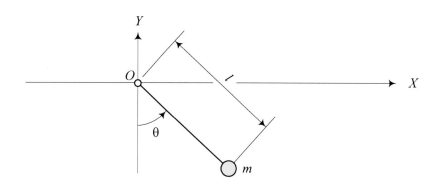

FIGURE 9.21. A pendulum with a vibrating pivot.

and find the equations of motion. The coefficients m, k, a, and b are constant.

14. Lagrangean from equation of motion.

Find the Lagrangean associated to the following equations of motions:

(a)
$$mr^2\ddot{\theta} + k_1 l_1 \theta + k_2 l_2 \theta + mgl = 0$$

(b)
$$\ddot{r} - r\dot{\theta}^2 = 0 \qquad r^2\ddot{\theta} + 2r\dot{r}\dot{\theta} = 0$$

15. Trebuchet.

Derive the equations of motion for the trebuchet shown in Figure 9.17.

16. Simplified trebuchet.

Three simplified models of a trebuchet are shown in Figures 9.22 to 9.24. Derive and compare their equations of motion.

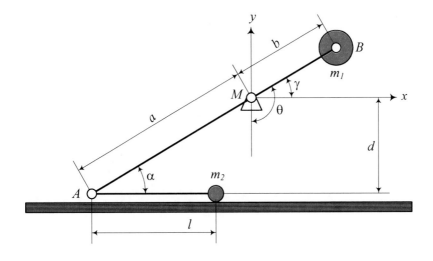

FIGURE 9.22. A simplified models of a trebuchet.

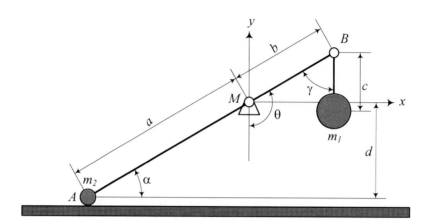

FIGURE 9.23. A simplified models of a trebuchet.

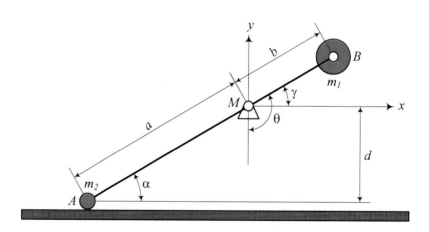

FIGURE 9.24. A simplified models of a trebuchet.

10

Vehicle Planar Dynamics

In this chapter we develop a dynamic model for a rigid vehicle in a planar motion. The planar model is applicable whenever the forward, lateral and yaw velocities are important and are enough to examine the behavior of a vehicle.

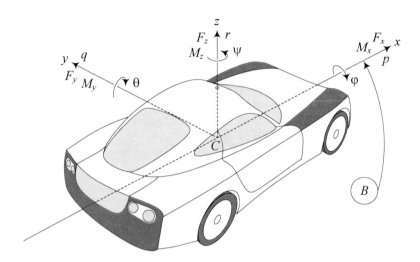

FIGURE 10.1. Vehicle body coordinate frame $B(Cxyz)$.

10.1 Vehicle Coordinate Frame

The equations of motion in vehicle dynamics are usually expressed in a set of vehicle coordinate frame $B(Cxyz)$, attached to the vehicle at the mass center C, as shown in Figure 10.1. The x-axis is a longitudinal axis passing through C and directed forward. The y-axis goes laterally to the left from the driver's viewpoint. The z-axis makes the coordinate system a right-hand triad. When the car is parked on a flat horizontal road, the z-axis is perpendicular to the ground, opposite to the gravitational acceleration **g**.

To show the vehicle orientation, we use three angles: *roll angle* φ about the x-axis, *pitch angle* θ about the y-axis, and *yaw angle* ψ about the z-axis. Because the rate of the orientation angles are important in vehicle

R.N. Jazar, *Vehicle Dynamics: Theory and Application*, DOI 10.1007/978-1-4614-8544-5_10, © Springer Science+Business Media New York 2014

dynamics, we usually show them by a special character and call them *roll rate*, *pitch rate*, and *yaw rate* respectively.

$$\dot{\varphi} = p \tag{10.1}$$
$$\dot{\theta} = q \tag{10.2}$$
$$\dot{\psi} = r \tag{10.3}$$

The resultant of external forces and moments, that the vehicle receives from the ground and environment, makes the *vehicle force system* $(\mathbf{F}, \mathbf{M})$. This force system will be expressed in the body coordinate frame.

$$^B\mathbf{F} = F_x\hat{\imath} + F_y\hat{\jmath} + F_z\hat{k} \tag{10.4}$$
$$^B\mathbf{M} = M_x\hat{\imath} + M_y\hat{\jmath} + M_z\hat{k} \tag{10.5}$$

The individual components of the 3D vehicle force system are shown in Figure 10.1. These components have special names and importance.

1. *Longitudinal force F_x.* It is a force acting along the x-axis. The resultant $F_x > 0$ if the vehicle is accelerating, and $F_x < 0$ if the vehicle is braking. Longitudinal force is also called *forward force*, or *traction force*.

2. *Lateral force F_y.* It is an orthogonal force to both F_x and F_z. The resultant $F_y > 0$ if it is leftward from the driver's viewpoint. Lateral force is usually a result of steering and is the main reason to generate a yaw moment and turn a vehicle.

3. *Normal force F_z.* It is a vertical force, normal to the ground plane. The resultant $F_z > 0$ if it is upward. Normal force is also called *vertical force* or *vehicle load.*

4. *Roll moment M_x.* It is a longitudinal moment about the x-axis. The resultant $M_x > 0$ if the vehicle tends to turn about the x-axis. The roll moment is also called the *bank moment, tilting torque,* or *overturning moment.*

5. *Pitch moment M_y.* It is a lateral moment about the y-axis. The resultant $M_y > 0$ if the vehicle tends to turn about the y-axis and move the head down.

6. *Yaw moment M_z.* It is an upward moment about the z-axis. The resultant $M_z > 0$ if the tire tends to turn about the z-axis. The yaw moment is also called the *aligning moment.*

The position and orientation of the vehicle coordinate frame $B(Cxyz)$ is measured with respect to a grounded fixed coordinate frame $G(OXYZ)$.

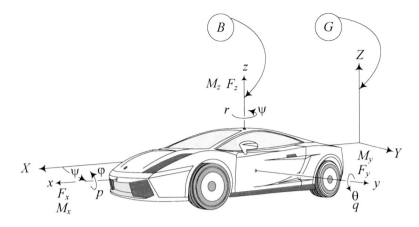

FIGURE 10.2. Illustration of a moving vehicle, indicated by its body coordinate frame B in a global coordinate frame G.

The vehicle coordinate frame is called the *body frame* or *vehicle frame*, and the grounded frame is called the *global coordinate frame*. Analysis of the vehicle motion is equivalent to express the position and orientation of $B(Cxyz)$ in $G(OXYZ)$. Figure 10.2 shows how a moving vehicle is indicated by a body frame B in a global frame G.

The angle between the X and x axes measured from X to x about Z is the *yaw angle* ψ and is called the *heading angle*. The velocity vector $\mathbf{v}$ of the vehicle makes an angle β with the body x-axis, measured from x to $\mathbf{v}$ about Z, which is called *sideslip angle* or *attitude angle*. The vehicle's velocity vector $\mathbf{v}$ makes an angle $\beta + \psi$ with the global X-axis, measured from X to $\mathbf{v}$ about Z that is called the *cruise angle*. A positive configuration of these angles are shown in the top view of a moving vehicle in Figure 10.3.

There are many situations in which we need to number the wheels of a vehicle. We start numbering from the front left wheel as number 1, and then the front right wheel would be number 2. Numbering increases sequentially on the right wheels going to the back of the vehicle up to the rear right wheel. Then, we go to the left of the vehicle and continue numbering the wheels from the rear left toward the front. Each wheel is indicated by a position vector $\mathbf{r}_i$

$$^{B}\mathbf{r}_i = x_i i + y_i j + z_i k \tag{10.6}$$

expressed in the body coordinate frame B. Numbering of a four wheel vehicle is shown in Figure 10.3.

Example 409 *Wheel numbers and their position vectors.*

Figure 10.4 depicts a six-wheel passenger car. The wheel numbers are indicated next to each wheel. The front left wheel is wheel number 1, and the front right wheel is number 2. Moving to the back on the right side,

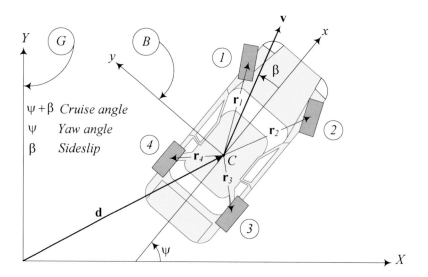

FIGURE 10.3. Top view of a moving vehicle to show the yaw angle ψ between the X and x axes, the sideslip angle β between the x-axis and the velocity vector **v**, and the crouse angle $\beta + \psi$ between the X-axis and the velocity vector **v**.

we count the wheels numbered 3 and 4. The back left wheel gets number 5, and then moving forward on the left side, the only unnumbered wheel is the wheel number 6.

If the global position vector of the car's mass center is given by

$$
{}^{G}\mathbf{d} = \begin{bmatrix} X_C \\ Y_C \end{bmatrix}
\tag{10.7}
$$

and the body position vectors of the wheels are

$$
{}^{B}\mathbf{r}_1 = \begin{bmatrix} a_1 \\ w/2 \end{bmatrix} \qquad
{}^{B}\mathbf{r}_2 = \begin{bmatrix} a_1 \\ -w/2 \end{bmatrix}
\tag{10.8}
$$

$$
{}^{B}\mathbf{r}_3 = \begin{bmatrix} -a_2 \\ -w/2 \end{bmatrix} \qquad
{}^{B}\mathbf{r}_4 = \begin{bmatrix} -a_3 \\ -w/2 \end{bmatrix}
\tag{10.9}
$$

$$
{}^{B}\mathbf{r}_5 = \begin{bmatrix} -a_3 \\ w/2 \end{bmatrix} \qquad
{}^{B}\mathbf{r}_6 = \begin{bmatrix} -a_2 \\ w/2 \end{bmatrix}
\tag{10.10}
$$

then the global position of the wheels are

$$
{}^{G}\mathbf{r}_1 = {}^{G}\mathbf{d} + {}^{G}R_B\,{}^{B}\mathbf{r}_1 = \begin{bmatrix} X_C - \dfrac{1}{2}w\sin\psi + a_1\cos\psi \\[2mm] Y_C + \dfrac{1}{2}w\cos\psi + a_1\sin\psi \end{bmatrix}
\tag{10.11}
$$

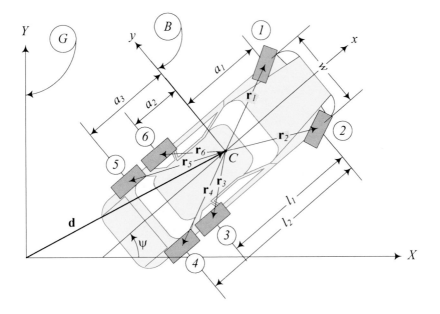

FIGURE 10.4. A six-wheel passenger car and its wheel numbering.

$$^{G}\mathbf{r}_2 = {}^{G}\mathbf{d} + {}^{G}R_B\,{}^{B}\mathbf{r}_2 = \begin{bmatrix} X_C + \frac{1}{2}w\sin\psi + a_1\cos\psi \\ Y_C - \frac{1}{2}w\cos\psi + a_1\sin\psi \end{bmatrix} \tag{10.12}$$

$$^{G}\mathbf{r}_3 = {}^{G}\mathbf{d} + {}^{G}R_B\,{}^{B}\mathbf{r}_3 = \begin{bmatrix} X_C + \frac{1}{2}w\sin\psi - a_2\cos\psi \\ Y_C - \frac{1}{2}w\cos\psi - a_2\sin\psi \end{bmatrix} \tag{10.13}$$

$$^{G}\mathbf{r}_4 = {}^{G}\mathbf{d} + {}^{G}R_B\,{}^{B}\mathbf{r}_4 = \begin{bmatrix} X_C + \frac{1}{2}w\sin\psi - a_3\cos\psi \\ Y_C - \frac{1}{2}w\cos\psi - a_3\sin\psi \end{bmatrix} \tag{10.14}$$

$$^{G}\mathbf{r}_5 = {}^{G}\mathbf{d} + {}^{G}R_B\,{}^{B}\mathbf{r}_5 = \begin{bmatrix} X_C - \frac{1}{2}w\sin\psi - a_3\cos\psi \\ Y_C + \frac{1}{2}w\cos\psi - a_3\sin\psi \end{bmatrix} \tag{10.15}$$

$$^{G}\mathbf{r}_6 = {}^{G}\mathbf{d} + {}^{G}R_B\,{}^{B}\mathbf{r}_6 = \begin{bmatrix} X_C - \frac{1}{2}w\sin\psi - a_2\cos\psi \\ Y_C + \frac{1}{2}w\cos\psi - a_2\sin\psi \end{bmatrix} \tag{10.16}$$

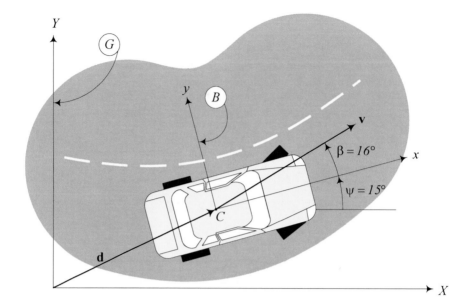

FIGURE 10.5. A car moving on a road with sideslip angle β and heading angle ψ.

where the rotation matrix between the global G and body coordinate B is

$$^{G}R_{B} = \begin{bmatrix} \cos\psi & -\sin\psi \\ \sin\psi & \cos\psi \end{bmatrix} \qquad (10.17)$$

Example 410 *Cruise angle, attitude angle, and heading angle.*
  Figure 10.5 illustrates a car moving on a road with the angles

$$\psi = 15\,\mathrm{deg} \qquad \beta = 16\,\mathrm{deg} \qquad (10.18)$$

The heading angle of the car is

$$Heading\ angle = \psi = 15\,\mathrm{deg} \qquad (10.19)$$

which is the angle between a reference X-axis on the road and the car's longitudinal x-axis. The attitude angle of the car is

$$Attitude\ angle = \beta = 16\,\mathrm{deg} \qquad (10.20)$$

which is the angle between the body longitudinal x-axis and the direction of the car's motion. The cruise angle of the car is

$$Cruise\ angle = \beta + \psi = 31\,\mathrm{deg} \qquad (10.21)$$

which is the angle between the reference X-axis on the road and the car's direction of motion.

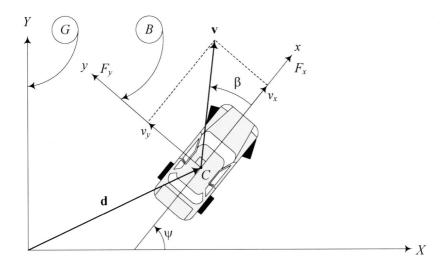

FIGURE 10.6. A rigid vehicle in a planar motion.

10.2 Rigid Vehicle Newton-Euler Dynamics

A rigid vehicle is assumed to act similar to a flat box moving on a horizontal surface. A rigid vehicle has a planar motion with three degrees of freedom that are: translation in the x and y directions, and a rotation about the z-axis. The *Newton-Euler equations of motion* for a rigid vehicle in the body coordinate frame B, attached to the vehicle at its mass center C, are:

$$F_x = m\,\dot{v}_x - m\omega_z\,v_y \tag{10.22}$$

$$F_y = m\,\dot{v}_y + m\omega_z\,v_x \tag{10.23}$$

$$M_z = \dot{\omega}_z\,I_z \tag{10.24}$$

Proof. Figure 10.6 illustrates a rigid vehicle in a planar motion. A global coordinate frame G is attached to the ground and a body coordinate frame B is attached to the vehicle at the mass center C. The Z and z axes are parallel, and the orientation of the frame B is indicated by the heading angle ψ between the X and x axes. The global position vector of the mass center is denoted by $^G\mathbf{d}$.

The velocity vector of the vehicle, expressed in the body frame, is

$$^B\mathbf{v}_C = \begin{bmatrix} v_x \\ v_y \\ 0 \end{bmatrix} \tag{10.25}$$

where v_x is the forward component and v_y is the lateral component of $\mathbf{v}$.

The rigid body equations of motion in the body coordinate frame are:

$$
\begin{aligned}
{}^B\mathbf{F} &= {}^B R_G \, {}^G\mathbf{F} = {}^B R_G \left(m \, {}^G\mathbf{a}_B \right) = m \, {}^B_G\mathbf{a}_B \\
&= m \, {}^B\dot{\mathbf{v}}_B + m \, {}^B_G\boldsymbol{\omega}_B \times {}^B\mathbf{v}_B
\end{aligned}
\tag{10.26}
$$

$$
\begin{aligned}
{}^B\mathbf{M} &= \frac{{}^G d}{dt} \, {}^B\mathbf{L} = {}^B_G\dot{\mathbf{L}}_B = {}^B\dot{\mathbf{L}} + {}^B_G\boldsymbol{\omega}_B \times {}^B\mathbf{L} \\
&= {}^B I \, {}^B_G\dot{\boldsymbol{\omega}}_B + {}^B_G\boldsymbol{\omega}_B \times \left({}^B I \, {}^B_G\boldsymbol{\omega}_B \right)
\end{aligned}
\tag{10.27}
$$

The force, moment, and kinematic vectors for the rigid vehicle are:

$$
{}^B\mathbf{F}_C = \begin{bmatrix} F_x \\ F_y \\ 0 \end{bmatrix}
\qquad
{}^B\mathbf{M}_C = \begin{bmatrix} 0 \\ 0 \\ M_z \end{bmatrix}
\tag{10.28}
$$

$$
{}^B\dot{\mathbf{v}}_C = \begin{bmatrix} \dot{v}_x \\ \dot{v}_y \\ 0 \end{bmatrix}
\qquad
{}^B_G\boldsymbol{\omega}_B = \begin{bmatrix} 0 \\ 0 \\ \omega_z \end{bmatrix}
\tag{10.29}
$$

$$
{}^B_G\dot{\boldsymbol{\omega}}_B = \begin{bmatrix} 0 \\ 0 \\ \dot{\omega}_z \end{bmatrix}
\tag{10.30}
$$

We may assume that the body coordinate is the principal coordinate frame of the vehicle to have a diagonal moment of inertia matrix.

$$
{}^B I = \begin{bmatrix} I_1 & 0 & 0 \\ 0 & I_2 & 0 \\ 0 & 0 & I_3 \end{bmatrix}
\tag{10.31}
$$

Substituting the above vectors and matrices in the equations of motion (10.26)-(10.27) provides the following equations:

$$
\begin{aligned}
{}^B\mathbf{F} &= m \, {}^B\dot{\mathbf{v}}_B + m \, {}^B_G\boldsymbol{\omega}_B \times {}^B\mathbf{v}_B \\
&= m \begin{bmatrix} \dot{v}_x \\ \dot{v}_y \\ 0 \end{bmatrix} + m \begin{bmatrix} 0 \\ 0 \\ \omega_z \end{bmatrix} \times \begin{bmatrix} v_x \\ v_y \\ 0 \end{bmatrix} = \begin{bmatrix} m\dot{v}_x - m\omega_z v_y \\ m\dot{v}_y + m\omega_z v_x \\ 0 \end{bmatrix}
\end{aligned}
\tag{10.32}
$$

$$
\begin{aligned}
{}^B\mathbf{M} &= {}^B I \, {}^B_G\dot{\boldsymbol{\omega}}_B + {}^B_G\boldsymbol{\omega}_B \times \left({}^B I \, {}^B_G\boldsymbol{\omega}_B \right) \\
&= \begin{bmatrix} I_1 & 0 & 0 \\ 0 & I_2 & 0 \\ 0 & 0 & I_3 \end{bmatrix} \begin{bmatrix} 0 \\ 0 \\ \dot{\omega}_z \end{bmatrix} \\
&\quad + \begin{bmatrix} 0 \\ 0 \\ \omega_z \end{bmatrix} \times \left(\begin{bmatrix} I_1 & 0 & 0 \\ 0 & I_2 & 0 \\ 0 & 0 & I_3 \end{bmatrix} \begin{bmatrix} 0 \\ 0 \\ \omega_z \end{bmatrix} \right) = \begin{bmatrix} 0 \\ 0 \\ I_3 \dot{\omega}_z \end{bmatrix}
\end{aligned}
\tag{10.33}
$$

The first two Newton equations (10.32) and the third Euler equation (10.33) are the only nonzero equations that make up the set of equations of motion (10.22)-(10.24) for the planar rigid vehicle. ∎

Example 411 *Rigid vehicle and Lagrange method.*
The kinetic energy of a rigid vehicle in a planar motion is,

$$
\begin{aligned}
K &= \frac{1}{2} {}^G\mathbf{v}_B^T \, m \, {}^G\mathbf{v}_B + \frac{1}{2} {}^G\boldsymbol{\omega}_B^T \, {}^G I \, {}^G\boldsymbol{\omega}_B \\
&= \frac{1}{2} \begin{bmatrix} v_X \\ v_Y \\ 0 \end{bmatrix}^T m \begin{bmatrix} v_X \\ v_Y \\ 0 \end{bmatrix} + \frac{1}{2} \begin{bmatrix} 0 \\ 0 \\ \omega_Z \end{bmatrix}^T {}^G I \begin{bmatrix} 0 \\ 0 \\ \omega_Z \end{bmatrix} \\
&= \frac{1}{2} m v_X^2 + \frac{1}{2} m v_Y^2 + \frac{1}{2} I_3 \omega_Z^2 = \frac{1}{2} m \left(\dot{X}^2 + \dot{Y}^2 \right) + \frac{1}{2} I_z \dot{\psi}^2 \quad (10.34)
\end{aligned}
$$

where,

$$
\begin{aligned}
{}^G I &= {}^G R_B \, {}^B I \, {}^G R_B^T \\
&= \begin{bmatrix} \cos\psi & -\sin\psi & 0 \\ \sin\psi & \cos\psi & 0 \\ 0 & 0 & 1 \end{bmatrix} \begin{bmatrix} I_1 & 0 & 0 \\ 0 & I_2 & 0 \\ 0 & 0 & I_3 \end{bmatrix} \begin{bmatrix} \cos\psi & -\sin\psi & 0 \\ \sin\psi & \cos\psi & 0 \\ 0 & 0 & 1 \end{bmatrix}^T \\
&= \begin{bmatrix} I_1 \cos^2\psi + I_2 \sin^2\psi & (I_1 - I_2)\sin\psi\cos\psi & 0 \\ (I_1 - I_2)\sin\psi\cos\psi & I_2 \cos^2\psi + I_1 \sin^2\psi & 0 \\ 0 & 0 & I_3 \end{bmatrix} \quad (10.35)
\end{aligned}
$$

and

$$
{}^G\mathbf{v}_B = \begin{bmatrix} v_X \\ v_Y \\ 0 \end{bmatrix} = \begin{bmatrix} \dot{X} \\ \dot{Y} \\ 0 \end{bmatrix} \quad (10.36)
$$

$$
{}^G\boldsymbol{\omega}_B = \begin{bmatrix} 0 \\ 0 \\ \omega_Z \end{bmatrix} = \begin{bmatrix} 0 \\ 0 \\ r \end{bmatrix} = \begin{bmatrix} 0 \\ 0 \\ \dot{\psi} \end{bmatrix} \quad (10.37)
$$

The resultant external force system are:

$$
{}^G\mathbf{F}_C = \begin{bmatrix} F_X \\ F_Y \\ 0 \end{bmatrix} \qquad {}^G\mathbf{M}_C = \begin{bmatrix} 0 \\ 0 \\ M_Z \end{bmatrix} \quad (10.38)
$$

Applying the Lagrange method

$$
\frac{d}{dt}\left(\frac{\partial K}{\partial \dot{q}_i} \right) - \frac{\partial K}{\partial q_i} = F_i \qquad i = 1, 2, \cdots n \quad (10.39)
$$

and using the coordinates X, Y, and ψ for q_i, generates the following equations of motion in the global coordinate frame:

$$m\dot{v}_x = m\frac{d}{dt}\dot{X} = F_X \qquad (10.40)$$

$$m\dot{v}_y = m\frac{d}{dt}\dot{Y} = F_Y \qquad (10.41)$$

$$I_z\dot{\omega}_z = I_z\frac{d}{dt}\dot{\psi} = M_Z \qquad (10.42)$$

Example 412 *Transforming to the body coordinate frame.*
We may find the rigid vehicle's equations of motion in the body coordinate frame by expressing the global equations of motion (10.40)-(10.42) in the vehicle's body coordinate B, using the transformation matrix ^GR_B.

$$^GR_B = \begin{bmatrix} \cos\psi & -\sin\psi & 0 \\ \sin\psi & \cos\psi & 0 \\ 0 & 0 & 1 \end{bmatrix} \qquad (10.43)$$

The body expression of the velocity vector is equal to

$$^G_B\mathbf{v}_C = {}^GR_B\,{}^B\mathbf{v}_C \qquad (10.44)$$

$$\begin{bmatrix} v_X \\ v_Y \\ 0 \end{bmatrix} = \begin{bmatrix} \cos\psi & -\sin\psi & 0 \\ \sin\psi & \cos\psi & 0 \\ 0 & 0 & 1 \end{bmatrix} \begin{bmatrix} v_x \\ v_y \\ 0 \end{bmatrix}$$

$$= \begin{bmatrix} v_x\cos\psi - v_y\sin\psi \\ v_y\cos\psi + v_x\sin\psi \\ 0 \end{bmatrix} \qquad (10.45)$$

and therefore, the global acceleration components are

$$\begin{bmatrix} \dot{v}_X \\ \dot{v}_Y \\ 0 \end{bmatrix} = \begin{bmatrix} \left(\dot{v}_x - \dot{\psi}\,v_y\right)\cos\psi - \left(\dot{v}_y + \dot{\psi}\,v_x\right)\sin\psi \\ \left(\dot{v}_y + \dot{\psi}\,v_x\right)\cos\psi + \left(\dot{v}_x - \dot{\psi}\,v_y\right)\sin\psi \\ 0 \end{bmatrix} \qquad (10.46)$$

The global Newton's equation of motion is

$$^G\mathbf{F}_C = m\,{}^G\dot{\mathbf{v}}_C \qquad (10.47)$$

and the force vector transformation is

$$^G\mathbf{F}_C = {}^GR_B\,{}^B\mathbf{F}_C \qquad (10.48)$$

therefore, the body coordinate expression for the equations of motion is

$$^B\mathbf{F}_C = {}^GR_B^T\,{}^G\mathbf{F}_C = m\,{}^GR_B^T\,{}^G\dot{\mathbf{v}}_C \qquad (10.49)$$

Substituting the associated vectors generates the Newton equations of motion in the body coordinate frame.

$$
\begin{bmatrix} F_x \\ F_y \\ 0 \end{bmatrix} = m\,{}^G R_B^T \begin{bmatrix} \left(\dot{v}_x - \dot{\psi}\,v_y\right)\cos\psi - \left(\dot{v}_y + \dot{\psi}\,v_x\right)\sin\psi \\ \left(\dot{v}_y + \dot{\psi}\,v_x\right)\cos\psi + \left(\dot{v}_x - \dot{\psi}\,v_y\right)\sin\psi \\ 0 \end{bmatrix}
$$

$$
= m \begin{bmatrix} \dot{v}_x - \dot{\psi}\,v_y \\ \dot{v}_y + \dot{\psi}\,v_x \\ 0 \end{bmatrix} \tag{10.50}
$$

Applying the same procedure for moment transformation,

$$
{}^G\mathbf{M}_C = {}^G R_B\,{}^B\mathbf{M}_C \tag{10.51}
$$

$$
\begin{bmatrix} 0 \\ 0 \\ M_Z \end{bmatrix} = \begin{bmatrix} \cos\psi & -\sin\psi & 0 \\ \sin\psi & \cos\psi & 0 \\ 0 & 0 & 1 \end{bmatrix} \begin{bmatrix} 0 \\ 0 \\ M_z \end{bmatrix} = \begin{bmatrix} 0 \\ 0 \\ M_z \end{bmatrix} \tag{10.52}
$$

we find the Euler equation in the body coordinate frame.

$$
M_z = \dot{\omega}_z\,I_z \tag{10.53}
$$

Example 413 *Vehicle path.*
When we find the translational and rotational velocities of a rigid vehicle, v_X, v_Y, r, we may find the path of motion for the vehicle by integration.

$$
\psi = \int \dot{\psi}\,dt = \psi_0 + \int r\,dt \tag{10.54}
$$

$$
X = \int \dot{X}\,dt = \int (v_x\cos\psi - v_y\sin\psi)\,dt \tag{10.55}
$$

$$
Y = \int \dot{Y}\,dt = \int (v_x\sin\psi + v_y\cos\psi)\,dt \tag{10.56}
$$

Example 414 ★ *Equations of motion using principal method.*
The equations of motion for a rigid vehicle in a planar motion may also be found by principle of differential calculus. Consider a vehicle at time $t = 0$ that has a lateral velocity v_y, a yaw rate r, and a forward velocity v_x. The longitudinal x-axis makes angle ψ with a fixed X-axis as shown in Figure 10.7. Point $P(x,y)$ indicates a general point of the vehicle. The velocity components of point P are

$$
v_{Px} = v_x - y\,r \qquad v_{Py} = v_y + x\,r \tag{10.57}
$$

because

$$
{}^B_G\mathbf{v}_P = {}^B\mathbf{v}_C + {}^B_G\boldsymbol{\omega}_B \times {}^B\mathbf{r}_P
$$

$$
= \begin{bmatrix} v_x \\ v_y \\ 0 \end{bmatrix} + \begin{bmatrix} 0 \\ 0 \\ r \end{bmatrix} \times \begin{bmatrix} x \\ y \\ 0 \end{bmatrix} \tag{10.58}
$$

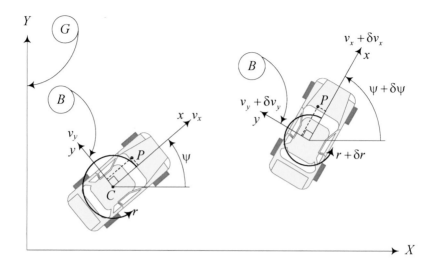

FIGURE 10.7. A vehicle at time $t = 0$ and $t = dt$ moving with a lateral velocity v_y, a yaw rate r, and a forward velocity v_x at a heading angle ψ.

After an increment of time, at $t = dt$, the vehicle has moved to a new position. The velocity components of point P at the second position are

$$v'_{Px} = (v_x + dv_x) - y(r + dr) \tag{10.59}$$
$$v'_{Py} = (v_y + dv_y) + x(r + dr) \tag{10.60}$$

However,

$$v_{Px} + dv_{Px} = v'_{Px} \cos d\psi - v'_{Py} \sin d\psi \tag{10.61}$$
$$v_{Py} + dv_{Py} = v'_{Px} \sin d\psi + v'_{Py} \cos d\psi \tag{10.62}$$

and therefore,

$$
\begin{aligned}
dv_{Px} = {} & [(v_x + dv_x) - y(r + dr)] \cos d\psi \\
& - [(v_y + dv_y) + x(r + dr)] \sin d\psi - (v_x - y\,r) \quad (10.63)
\end{aligned}
$$
$$
\begin{aligned}
dv_{Py} = {} & [(v_x + dv_x) - y(r + dr)] \sin d\psi \\
& + [(v_y + dv_y) + x(r + dr)] \cos d\psi - (v_y + x\,r) \quad (10.64)
\end{aligned}
$$

We simplify Equations (10.63) and (10.64) and divide them by dt.

$$
\begin{aligned}
\frac{dv_{Px}}{dt} = {} & \frac{1}{dt} \left([dv_x - y\,dr] \cos d\psi\right) \\
& - \frac{1}{dt} \left([(v_y + dv_y) + x(r + dr)] \sin d\psi\right) \quad (10.65)
\end{aligned}
$$
$$
\begin{aligned}
\frac{dv_{Py}}{dt} = {} & \frac{1}{dt} \left([dv_y + x\,dr] \cos d\psi\right) \\
& + \frac{1}{dt} \left([(v_x + dv_x) - y(r + dr)] \sin d\psi\right) \quad (10.66)
\end{aligned}
$$

When $dt \to 0$, then $\sin d\psi \to \psi$ and $\cos d\psi \to 1$, and we may substitute $\dot{\psi} = r$ to get the acceleration components of point P.

$$\dot{v}_{Px} = a_{Px} = \dot{v}_x - v_y r - y\dot{r} + x r^2 \qquad (10.67)$$
$$\dot{v}_{Py} = a_{Py} = \dot{v}_y + v_x r + x\dot{r} - y r^2 \qquad (10.68)$$

Let us assume point P has a small mass dm. Multiplying dm by the acceleration components of point P and integrating over the whole rigid vehicle must be equal to the applied external force system.

$$F_x = \int_m a_{Px}\, dm \qquad (10.69)$$

$$F_y = \int_m a_{Py}\, dm \qquad (10.70)$$

$$M_z = \int_m (x\, a_{Py} - y\, a_{Px})\, dm \qquad (10.71)$$

Substituting for accelerations and assuming the body coordinate frame is the principal frame at the mass center C, we find

$$
\begin{aligned}
F_x &= \int_m \left(\dot{v}_x - v_y r - y\dot{r} + x r^2\right) dm \\
&= m\left(\dot{v}_x - v_y r\right) - \dot{r}\int_m y\, dm + r^2 \int_m x\, dm \\
&= m\left(\dot{v}_x - v_y r\right) \qquad (10.72)
\end{aligned}
$$

$$
\begin{aligned}
F_y &= \int_m \left(\dot{v}_y + v_x r + x\dot{r} - y r^2\right) dm \\
&= m\left(\dot{v}_y + v_x r\right) + \dot{r}\int_m x\, dm - r^2 \int_m y\, dm \\
&= m\left(\dot{v}_y + v_x r\right) \qquad (10.73)
\end{aligned}
$$

$$
\begin{aligned}
M_z &= \int_m \left(x\left(\dot{v}_y + v_x r + x\dot{r} - y r^2\right) - y\left(\dot{v}_x - v_y r - y\dot{r} + x r^2\right)\right) dm \\
&= \dot{r}\int_m \left(x^2 + y^2\right) dm + \left(\dot{v}_y + v_x r\right)\int_m x\, dm \\
&\quad - \left(\dot{v}_x - v_y r\right)\int_m y\, dm - 2r^2 \int_m xy\, dm = I_z\, \dot{r} \qquad (10.74)
\end{aligned}
$$

because for a principal coordinate frame we have

$$\int_m x\, dm = 0 \qquad \int_m y\, dm = 0 \qquad \int_m xy\, dm = 0 \qquad (10.75)$$

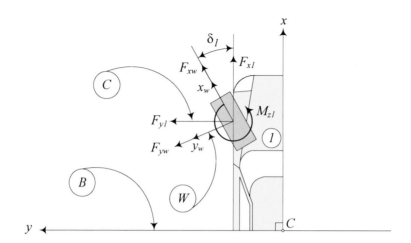

FIGURE 10.8. The force system at the tireprint of tire number 1.

10.3 Force System Acting on a Rigid Vehicle

To determine the force system on a rigid vehicle, first we define the force system at the tireprint of a wheel. Then, we transform and apply the tire force system on the body of the vehicle.

10.3.1 Tire Force and Body Force Systems

Figure 10.8 depicts wheel number 1 of a vehicle. The components of the force system in the C-frame on xy-plane, because of the forces at the tireprint of the wheel number i, are

$$F_{x_i} = F_{x_{w_i}} \cos \delta_i - F_{y_{w_i}} \sin \delta_i \qquad (10.76)$$
$$F_{y_i} = F_{y_{w_i}} \cos \delta_i + F_{x_{w_i}} \sin \delta_i \qquad (10.77)$$
$$M_{z_i} = M_{z_{w_i}} \qquad (10.78)$$

Therefore, the total planar force system on the rigid vehicle in the body coordinate frame is

$$^B F_x = \sum_i F_{x_i} = \sum_i F_{x_w} \cos \delta_i - \sum_i F_{y_w} \sin \delta_i \qquad (10.79)$$
$$^B F_y = \sum_i F_{y_i} = \sum_i F_{y_w} \cos \delta_i + \sum_i F_{x_w} \sin \delta_i \qquad (10.80)$$
$$^B M_z = \sum_i M_{z_i} + \sum_i x_i F_{y_i} - \sum_i y_i F_{x_i} \qquad (10.81)$$

Proof. The wheel coordinate frame $W(x_w, y_w, z_w)$, shown by or W, is a local coordinate frame attached to the center of the wheel. The tire coordinate frame $T\,(x_t, y_t, z_t)$ is set at the center of the tireprint. Let us assume that the force system in the wheel frame is

$$
{}^W\mathbf{F}_w = \begin{bmatrix} F_{x_w} & F_{y_w} & F_{y_w} \end{bmatrix}^T \tag{10.82}
$$

$$
{}^W\mathbf{M}_w = \begin{bmatrix} M_{x_w} & M_{y_w} & M_{y_w} \end{bmatrix}^T \tag{10.83}
$$

which is because of the force system at the tireprint

$$
{}^T\mathbf{F}_w = \begin{bmatrix} F_{x_{w_i}} & F_{y_{w_i}} & F_{z_{w_i}} \end{bmatrix}^T \tag{10.84}
$$

$$
{}^T\mathbf{M}_w = \begin{bmatrix} M_{x_{w_i}} & M_{y_{w_i}} & M_{z_{w_i}} \end{bmatrix}^T \tag{10.85}
$$

The tire force in the x_w-direction, $F_{x_{w_i}}$, is a combination of the longitudinal force $F_{x_{w_1}}$, defined by (3.119) or (4.49), and the tire roll resistance F_{r_1} defined in (3.88). The tire force in the y_w-direction, $F_{y_{w_i}}$, is a combination of the lateral force $F_{y_{w_1}}$ defined by (3.153) and (3.177), and the tire roll resistance F_{r_1} defined in (3.88). The tire moment in the z_w-direction, $M_{z_{w_i}}$, is a resultant of the aligning moment $M_{z_{w_1}}$ defined by (3.156) and (3.184).

The rotation matrix between the wheel frame W and the wheel-body coordinate frame C, parallel to the vehicle coordinate frame B, is

$$
{}^C R_W = \begin{bmatrix} \cos \delta_1 & -\sin \delta_1 \\ \sin \delta_1 & \cos \delta_1 \end{bmatrix} \tag{10.86}
$$

and therefore, the force system at the tireprint of the wheel, parallel to the vehicle coordinate frame, is

$$
{}^C\mathbf{F}_w = {}^C R_W\, {}^W\mathbf{F}_w \tag{10.87}
$$

$$
\begin{bmatrix} F_{x_1} \\ F_{y_1} \end{bmatrix} = \begin{bmatrix} \cos \delta_1 & -\sin \delta_1 \\ \sin \delta_1 & \cos \delta_1 \end{bmatrix} \begin{bmatrix} F_{x_w} \\ F_{y_w} \end{bmatrix}
$$

$$
= \begin{bmatrix} F_{x_w} \cos \delta_1 - F_{y_w} \sin \delta_1 \\ F_{y_w} \cos \delta_1 + F_{x_w} \sin \delta_1 \end{bmatrix} \tag{10.88}
$$

$$
{}^C\mathbf{M}_w = {}^C R_W\, {}^W\mathbf{M}_w \tag{10.89}
$$

$$
M_{z_1} = M_{z_w} \tag{10.90}
$$

Transforming the force system of each tire to the body coordinate frame B, located at the body mass center C, generates the total force system applied on the vehicle

$$
{}^B\mathbf{F} = \sum_i F_{x_i}\,\hat{\imath} + \sum_i F_{y_i}\,\hat{\jmath} \tag{10.91}
$$

$$
{}^B\mathbf{M} = \sum_i M_{z_i}\,\hat{k} + \sum_i {}^B\mathbf{r}_i \times {}^B\mathbf{F}_{w_i} \tag{10.92}
$$

where, $^B\mathbf{r}_i$ is the position vector of the wheel number i.

$$^B\mathbf{r}_i = x_i\hat{\imath} + y_i\hat{\jmath} + z_i\hat{k} \tag{10.93}$$

Expanding Equations (10.91) and (10.92) provides us with the total planar force system.

$$^BF_x = \sum_i F_{x_w}\cos\delta_i - \sum_i F_{y_w}\sin\delta_i \tag{10.94}$$

$$^BF_y = \sum_i F_{y_w}\cos\delta_i + \sum_i F_{x_w}\sin\delta_i \tag{10.95}$$

$$^BM_z = \sum_i M_{z_i} + \sum_i x_i F_{y_i} - \sum_i y_i F_{x_i} \tag{10.96}$$

∎

Example 415 *Difference between tireprint and wheel frames.*

To express the orientation of a wheel and the force system, three coordinate frames are needed: the wheel frame W, wheel-body frame C, and tire frame T. The wheel coordinate frame $W\,(x_w, y_w, z_w)$ is attached to the center of the wheel and follows every motion of the wheel except the wheel spin. Therefore, the x_w and z_w axes are always in the tire-plane, while the y_w-axis is always on the spin axis.

We also attach a wheel-body coordinate frame $C\,(x_c, y_c, z_c)$ at the center of the wheel parallel and fixed with respect to the vehicle coordinate axes $B\,(x, y, z)$. The wheel-body frame C is motionless with respect to the vehicle coordinate and does not follow any motion of the wheel. When the wheel is upright straight, the W-frame coincides with C-frame and becomes parallel to the vehicle coordinate frame. The W-frame makes the steer angle δ and camber angle γ with respect to the C-frame.

The tire coordinate frame $T\,(x_t, y_t, z_t)$ is set at the center of the tireprint with the z_t-axis perpendicular to the ground and parallel to the z-axis. The x_t-axis is along the intersection line of the tire-plane and the ground. The tire frame follows the steer angle rotation about the z_c-axis but it does not follows the spin and camber rotations of the tire.

To determine the difference between the T and W frames, let us use $^T\mathbf{d}_W$ to indicates the T-expression of the position vector of the wheel frame origin relative to the tire frame origin. Having the coordinates of a point P in the wheel frame, we can find its coordinates in the tire frame by:

$$^T\mathbf{r}_P = {}^TR_W\,{}^W\mathbf{r}_P + {}^T\mathbf{d}_W \tag{10.97}$$

If $^W\mathbf{r}_P$ indicates the position vector of a point P in the wheel frame,

$$^W\mathbf{r}_P = \begin{bmatrix} x_P & y_P & z_P \end{bmatrix}^T \tag{10.98}$$

then the coordinates of the point P in the tire frame $^T\mathbf{r}_P$ are

$$
\begin{aligned}
^T\mathbf{r}_P &= {}^TR_W{}^W\mathbf{r}_P + {}^T\mathbf{d} = {}^TR_W{}^W\mathbf{r}_P + {}^TR_W{}^W_T\mathbf{d}_W \\[4pt]
&= \begin{bmatrix} x_P \\ y_P \cos\gamma - R_w \sin\gamma - z_P \sin\gamma \\ R_w \cos\gamma + z_P \cos\gamma + y_P \sin\gamma \end{bmatrix}
\end{aligned}
\tag{10.99}
$$

because

$$
^TR_W = \begin{bmatrix} 1 & 0 & 0 \\ 0 & \cos\gamma & -\sin\gamma \\ 0 & \sin\gamma & \cos\gamma \end{bmatrix} \qquad {}^W_T\mathbf{d}_W = \begin{bmatrix} 0 \\ 0 \\ R_w \end{bmatrix}
\tag{10.100}
$$

where, $^W_T\mathbf{d}_W$ is the W-expression of the position vector of the wheel frame in the tire frame, R_w is the radius of the wheel, and ^TR_W is the transformation matrix from the W to T.

The center of the wheel $^W\mathbf{r}_P = {}^W\mathbf{r}_o = \mathbf{0}$ is the origin of the wheel frame W, that will be at $^T\mathbf{r}_o$ in the tire coordinate frame T.

$$
^T\mathbf{r}_o = {}^T\mathbf{d}_W = {}^TR_W{}^W_T\mathbf{d}_W = \begin{bmatrix} 0 \\ -R_w \sin\gamma \\ R_w \cos\gamma \end{bmatrix}
\tag{10.101}
$$

If the camber angle is zero, $\gamma = 0$, then

$$
^T\mathbf{r}_o = \begin{bmatrix} 0 \\ 0 \\ R_w \end{bmatrix} = {}^W_T\mathbf{d}_W \qquad \gamma = 0
\tag{10.102}
$$

If the force system at the tireprint is $^T\mathbf{F}_w$ and $^T\mathbf{M}_w$ then the force system in the W-frame at the center of the wheel would be

$$
^W\mathbf{F}_w = {}^WR_T{}^T\mathbf{F}_w = {}^TR_W^T{}^T\mathbf{F}_w
\tag{10.103}
$$

$$
\begin{bmatrix} F_{x_{w_i}} \\ F_{y_{w_i}} \\ F_{z_{w_i}} \end{bmatrix} = \begin{bmatrix} 1 & 0 & 0 \\ 0 & \cos\gamma & -\sin\gamma \\ 0 & \sin\gamma & \cos\gamma \end{bmatrix}^T \begin{bmatrix} F_{x_{w_i}} \\ F_{y_{w_i}} \\ F_{z_{w_i}} \end{bmatrix}
\tag{10.104}
$$

$$
\begin{aligned}
^W\mathbf{M}_w &= {}^TR_W^T\left({}^T\mathbf{M}_w - {}^T\mathbf{r}_o \times {}^T\mathbf{F}_w\right) \\[4pt]
&= \begin{bmatrix} R_w F_{y_{w_i}} \cos\gamma + R_w F_{z_{w_i}} \sin\gamma \\ M_{z_{w_i}} \sin\gamma - R_w F_{x_{w_i}} \\ M_{z_{w_i}} \cos\gamma \end{bmatrix}
\end{aligned}
\tag{10.105}
$$

where

$$
^T\mathbf{r}_o = \begin{bmatrix} 0 \\ -R_w \sin\gamma \\ R_w \cos\gamma \end{bmatrix} \qquad {}^T\mathbf{M}_w = \begin{bmatrix} 0 \\ 0 \\ M_{z_{w_i}} \end{bmatrix}
\tag{10.106}
$$

The wheel force system at zero camber, $\gamma = 0$, reduces to

$$
{}^W\mathbf{F}_w = \begin{bmatrix} F_{x_{w_i}} \\ F_{y_{w_i}}\cos\gamma + F_{z_{w_i}}\sin\gamma \\ F_{z_{w_i}}\cos\gamma - F_{y_{w_i}}\sin\gamma \end{bmatrix} \qquad {}^W\mathbf{M}_w = \begin{bmatrix} R_w F_{y_{w_i}} \\ -R_w F_{x_{w_i}} \\ M_{z_{w_i}} \end{bmatrix} \qquad (10.107)
$$

10.3.2 Tire Lateral Force

Figure 10.9(a) illustrates a tire, moving along the velocity vector $\mathbf{v}$ at a sideslip angle α. The tire is steered by the steer angle δ. If the angle between the vehicle x-axis and the velocity vector $\mathbf{v}$ is shown by β, then

$$
\alpha = \beta - \delta \qquad (10.108)
$$

The angle β is called the *wheel sideslip angle* while α is the *tire sideslip angle*. If the word "sideslip" is used individually, it always refer to the tire sideslip angle α. The lateral force, generated by a tire, is dependent on sideslip angle α that is proportional to the sideslip for small α.

$$
F_y = -C_\alpha\,\alpha = -C_\alpha\,(\beta - \delta) \qquad (10.109)
$$

Proof. A wheel coordinate frame $W(x_w, y_w)$ is attached to the wheel at the center of wheel as shown in Figure 10.9(a). The orientation of the wheel frame is measured with respect to the wheel-body coordinate frame $C(x_c, y_c)$, parallel to the vehicle frame $B(x, y)$. The angle between the x and x_w axes is the wheel steer angle δ, measured about the z_w-axis. The wheel is moving along the tire velocity vector $\mathbf{v}$. The angle between the x_w-axis and $\mathbf{v}$ is the *tire sideslip angle* α, and the angle between the body x-axis and $\mathbf{v}$ is called the *wheel sideslip angle* β. The angles α, β, and δ in Figure 10.9(a) are positive. The figure shows that

$$
\alpha = \beta - \delta \qquad (10.110)
$$

Practically, when a steered wheel is moving forward, the relationship between the angles α, β, and δ are such that the velocity vector sits between the x and x_w axes. A practical situation is shown in Figure 10.9(b). A steer angle will turn the heading of the wheel by a δ angle. However, because of tire flexibility, the velocity vector of the wheel is lazier than the heading and turns by a β angle, where $\beta < \delta$. As a result, a positive steer angle δ generates a negative sideslip angle α. Analysis of Figure 10.9(b) and using the definition for positive direction of the angles, show that under a practical situation we have the same relation (10.108).

According to (3.154), the existence of a sideslip angle is sufficient to generate a lateral force F_y, which is proportional to α when the angle is small.

$$
F_y = -C_\alpha\,\alpha = -C_\alpha\,(\beta - \delta) \qquad (10.111)
$$

∎

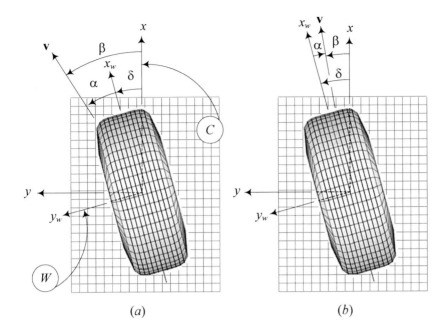

FIGURE 10.9. Angular orientation of a moving tire along the velocity vector **v** at a sideslip angle α and a steer angle δ. (a) $\alpha > 0$. (b) $\alpha < 0$.

Example 416 *Extreme velocity cases of a wheel.*

Consider a wheel as is shown in Figure 10.9(b) which has a spinning angular velocity of $\omega \neq 0$ on a frictionless ground. Therefore, the velocity of the wheel center would be zero, $\mathbf{v} = 0$. The sideslip angle of such a wheel would be zero, $\alpha = 0$.

Now consider the wheel which has a zero spinning angular velocity $\omega = 0$ and a nonzero translational velocity $\mathbf{v} \neq 0$. The sideslip angle of such a wheel would be as is shown in Figure 10.9(b). The sideslip angle of a wheel is not a function of the spinning angular velocity of a wheel.

10.3.3 Two-wheel Model and Body Force Components

Figure 10.10 illustrates the forces in the xy-plane acting at the wheel center of a front-wheel-steering four-wheel vehicle. When we ignore the roll motion of the vehicle, the body z and global Z axes are parallel and consequently the xy-plane remains parallel to the road's XY-plane. Therefore, we may use a *two-wheel model* for the vehicle. Figure 10.11 illustrates a two-wheel model for a vehicle with no roll motion. The two-wheel model is also called a *bicycle model*, although a two-wheel model does not act similar to a traditional bicycle.

The force system applied on a bicycle model of vehicle, in which only the

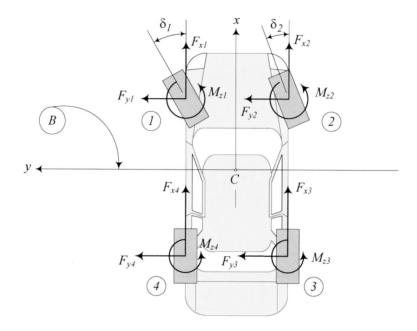

FIGURE 10.10. A front-wheel-steering four-wheel vehicle and the forces in the xy-plane acting at the wheel center to be considered equal to the trireprint's forces.

front wheel is steerable, and the steer angle δ is assumed to be small, are:

$$F_x \approx F_{x_f} + F_{x_r} \tag{10.112}$$

$$F_y \approx F_{y_f} + F_{y_r} \tag{10.113}$$

$$M_z \approx a_1 F_{y_f} - a_2 F_{y_r} \tag{10.114}$$

where, $\left(F_{x_f}, F_{y_f}\right)$ and $\left(F_{x_r}, F_{y_r}\right)$ are the planar forces on the tireprint of the front and rear wheels and we consider them at the wheel center.

The vehicle lateral force F_y and moment M_z depend only on the front and rear wheels' lateral forces F_{y_f} and F_{y_r}, which are functions of the tires sideslip angles α_f and α_r. They can be approximated by:

$$F_y = \left(-\frac{a_1}{v_x}C_{\alpha f} + \frac{a_2}{v_x}C_{\alpha r}\right)r - (C_{\alpha f} + C_{\alpha r})\beta + C_{\alpha f}\delta \tag{10.115}$$

$$M_z = \left(-\frac{a_1^2}{v_x}C_{\alpha f} - \frac{a_2^2}{v_x}C_{\alpha r}\right)r - (a_1 C_{\alpha f} - a_2 C_{\alpha r})\beta + a_1 C_{\alpha f}\delta \tag{10.116}$$

where $C_{\alpha f}$ and $C_{\alpha r}$ are sum of the sideslip coefficients of the left and right tires in front and rear, respectively.

$$C_{\alpha f} = C_{\alpha f_L} + C_{\alpha f_R} \tag{10.117}$$

$$C_{\alpha r} = C_{\alpha r_L} + C_{\alpha r_R} \tag{10.118}$$

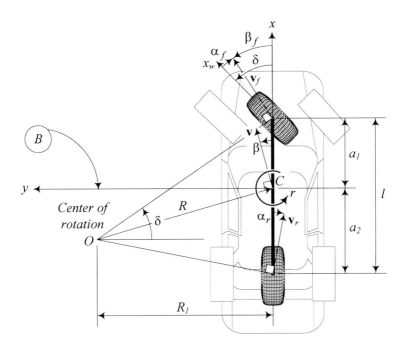

FIGURE 10.11. A two-wheel model for a vehicle moving with no roll.

Proof. For the two-wheel vehicle, we use the cot-average (7.4) of the outer and inner steer angles as the only steer angle δ.

$$\cot \delta = \frac{\cot \delta_o + \cot \delta_i}{2} \qquad (10.119)$$

Furthermore, we define a single sideslip coefficient $C_{\alpha f}$ and $C_{\alpha r}$ as (10.117) and (10.118) for the front and rear wheels.

Employing Equations (10.79)-(10.81) and ignoring the aligning moments M_{z_i}, the applied forces on the two-wheel vehicle are:

$$
\begin{aligned}
F_x &= F_{x_f} \cos \delta + F_{x_r} - F_{y_f} \sin \delta & (10.120) \\
F_y &= F_{y_f} \cos \delta + F_{y_r} + F_{x_f} \sin \delta & (10.121) \\
M_z &= a_1 F_{y_f} \cos \delta + a_1 F_{x_f} \sin \delta - a_2 F_{y_r} & (10.122)
\end{aligned}
$$

The force equations can be approximated by the following equations, if we assume δ small.

$$
\begin{aligned}
F_x &\approx F_{x_f} + F_{x_r} & (10.123) \\
F_y &\approx F_{y_f} + F_{y_r} & (10.124) \\
M_z &\approx a_1 F_{y_f} - a_2 F_{y_r} & (10.125)
\end{aligned}
$$

Assume the wheel number i of a rigid vehicle is located at (x_i, y_i) in the body coordinate frame. The velocity of the wheel number i is

$$
{}^B\mathbf{v}_i = {}^B\mathbf{v} + {}^B\dot{\psi} \times {}^B\mathbf{r}_i \tag{10.126}
$$

in which ${}^B\mathbf{r}_i$ is the position vector of the wheel number i, ${}^B\mathbf{v}$ is the velocity vector of the vehicle at its mass center C, and ${}^B\dot{\psi}$ is the yaw rate of the vehicle.

$$
{}^B\dot{\psi} = \omega_z \hat{k} = r\hat{k} \tag{10.127}
$$

Expanding Equation (10.126) provides the following velocity vector for the wheel number i expressed in the vehicle coordinate frame at C.

$$
\begin{bmatrix} v_{x_i} \\ v_{y_i} \\ 0 \end{bmatrix} = \begin{bmatrix} v_x \\ v_y \\ 0 \end{bmatrix} + \begin{bmatrix} 0 \\ 0 \\ r \end{bmatrix} \times \begin{bmatrix} x_i \\ y_i \\ 0 \end{bmatrix} = \begin{bmatrix} v_x - y_i r \\ v_y + x_i r \\ 0 \end{bmatrix} \tag{10.128}
$$

The wheel sideslip β_i for the wheel i, is the angle between the vehicle body x-axis and the wheel velocity vector $\mathbf{v}_i$.

$$
\beta_i = \arctan\left(\frac{v_{y_i}}{v_{x_i}}\right) = \arctan\left(\frac{v_y + x_i r}{v_x - y_i r}\right) \tag{10.129}
$$

If the wheel number i has a steer angle δ_i then, its tire sideslip angle α_i, that generates a lateral force F_{y_w} on the tire, is

$$
\alpha_i = \beta_i - \delta_i = \arctan\left(\frac{v_y + x_i r}{v_x - y_i r}\right) - \delta_i \tag{10.130}
$$

The wheel sideslip angles β_i for the front and rear wheels of a two-wheel vehicle model, β_f and β_r, are

$$
\beta_f = \arctan\left(\frac{v_{y_f}}{v_{x_f}}\right) = \arctan\left(\frac{v_y + a_1 r}{v_x}\right) \tag{10.131}
$$

$$
\beta_r = \arctan\left(\frac{v_{y_r}}{v_{x_r}}\right) = \arctan\left(\frac{v_y - a_2 r}{v_x}\right) \tag{10.132}
$$

and the vehicle sideslip angle β is

$$
\beta = \arctan\left(\frac{v_y}{v_x}\right) \tag{10.133}
$$

Assuming small angles for wheel and vehicle sideslips β_f, β_r, and β, the tire sideslip angles for the front and rear wheels, α_f and α_r, may be approximated as

$$
\alpha_f = \beta_f - \delta = \frac{1}{v_x}(v_y + a_1 r) - \delta = \beta + \frac{a_1}{v_x}r - \delta \tag{10.134}
$$

$$
\alpha_r = \beta_r = \frac{1}{v_x}(v_y - a_2 r) = \beta - \frac{a_2}{v_x}r \tag{10.135}
$$

For small sideslip angles, the associated lateral forces are

$$F_{yf} = -C_{\alpha f}\,\alpha_f \qquad (10.136)$$
$$F_{yr} = -C_{\alpha r}\,\alpha_r \qquad (10.137)$$

and therefore, the second and third equations of motion (10.113) and (10.114) can be written as

$$
\begin{aligned}
F_y &= F_{yf} + F_{yr} = -C_{\alpha f}\,\alpha_f - C_{\alpha r}\,\alpha_r \\
&= -C_{\alpha f}\left(\beta + \frac{a_1}{v_x}r - \delta\right) - C_{\alpha r}\left(\beta - \frac{a_2}{v_x}r\right) \qquad (10.138) \\
M_z &= a_1 F_{yf} - a_2 F_{yr} = -a_1 C_{\alpha f}\,\alpha_f + a_2 C_{\alpha r}\,\alpha_r \\
&= -a_1 C_{\alpha f}\left(\beta + \frac{a_1}{v_x}r - \delta\right) + a_2 C_{\alpha r}\left(\beta - \frac{a_2}{v_x}r\right) \qquad (10.139)
\end{aligned}
$$

which can be rearranged to the following force system

$$
F_y = \left(\frac{a_2}{v_x}C_{\alpha r} - \frac{a_1}{v_x}C_{\alpha f}\right)r - (C_{\alpha f} + C_{\alpha r})\beta + C_{\alpha f}\delta \qquad (10.140)
$$
$$
M_z = \left(-\frac{a_1^2}{v_x}C_{\alpha f} - \frac{a_2^2}{v_x}C_{\alpha r}\right)r + (a_2 C_{\alpha r} - a_1 C_{\alpha f})\beta + a_1 C_{\alpha f}\delta \qquad (10.141)
$$

The parameters $C_{\alpha f}$, $C_{\alpha r}$ are the sideslip coefficient for the front and rear tires, r is the yaw rate, δ is the front wheel steer angle, and β is the sideslip angle of the vehicle.

These equations are dependent on three parameters, r, β, δ, and may be written as

$$
\begin{aligned}
F_y &= F_y\,(r, \beta, \delta) = \frac{\partial F_y}{\partial r}r + \frac{\partial F_y}{\partial \beta}\beta + \frac{\partial F_y}{\partial \delta}\delta \\
&= C_r\,r + C_\beta\,\beta + C_\delta\,\delta \qquad (10.142)
\end{aligned}
$$

$$
\begin{aligned}
M_z &= M_z\,(r, \beta, \delta) = \frac{\partial M_z}{\partial r}r + \frac{\partial M_z}{\partial \beta}\beta + \frac{\partial M_z}{\partial \delta}\delta \\
&= D_r\,r + D_\beta\,\beta + D_\delta\,\delta \qquad (10.143)
\end{aligned}
$$

where the *force system coefficients* are

$$
C_r = \frac{\partial F_y}{\partial r} = -\frac{a_1}{v_x}C_{\alpha f} + \frac{a_2}{v_x}C_{\alpha r} \qquad (10.144)
$$
$$
C_\beta = \frac{\partial F_y}{\partial \beta} = -C_{\alpha f} - C_{\alpha r} \qquad (10.145)
$$
$$
C_\delta = \frac{\partial F_y}{\partial \delta} = C_{\alpha f} \qquad (10.146)
$$

$$D_r = \frac{\partial M_z}{\partial r} = -\frac{a_1^2}{v_x}C_{\alpha f} - \frac{a_2^2}{v_x}C_{\alpha r} \qquad (10.147)$$

$$D_\beta = \frac{\partial M_z}{\partial \beta} = a_2 C_{\alpha r} - a_1 C_{\alpha f} \qquad (10.148)$$

$$D_\delta = \frac{\partial M_z}{\partial \delta} = a_1 C_{\alpha f} \qquad (10.149)$$

The coefficients C_r, C_β, C_δ, D_r, D_β, D_δ are slopes of the curves for lateral force F_y and yaw moment M_z as a function of r, β, and δ respectively. ∎

Example 417 *Physical significance of the force system coefficients.*

Assuming a steady-state condition and constant values for r, β, δ, $C_{\alpha f}$, and $C_{\alpha r}$, the lateral force F_y and the yaw moment M_z can be written as a superposition of three independent forces proportional to r, β, and δ.

$$F_y = C_r\, r + C_\beta\, \beta + C_\delta\, \delta \qquad (10.150)$$
$$M_z = D_r\, r + D_\beta\, \beta + D_\delta\, \delta \qquad (10.151)$$

C_r *indicates the proportionality between the lateral force F_y and the yaw rate r. The value of C_r decreases by increasing the forward velocity of the vehicle, v_x.*

C_β *indicates the proportionality between the lateral force F_y and the vehicle sideslip angle β. C_β is always negative and indicates the lateral stiffness for the whole vehicle. C_β acts similar to the sideslip coefficient of tires C_α.*

C_δ *indicates the proportionality between the lateral force F_y and the steer angle δ. C_δ is always positive and generates greater lateral force by increasing the steer angle.*

D_r *indicates the proportionality between the yaw moment M_z and the yaw rate r. D_r is a negative number and is called the yaw damping coefficient because it is proportional to the angular velocity in the M_z-direction. The value of D_r decreases with the forward velocity of the vehicle, v_x.*

D_β *indicates the proportionality between the yaw moment M_z and the vehicle sideslip angle β. D_β indicates the under/oversteer behavior and hence, indicates the directional stability of a vehicle. A negative D_β tries to align the vehicle with the velocity vector.*

D_δ *indicates the proportionality between the yaw moment M_z and the steer angle δ. Because δ is the input command to control the maneuvering of a vehicle, D_δ is called the control moment coefficient. D_δ is a positive number and increases with a_1 and $C_{\alpha f}$.*

Example 418 *Kinematic steering of a two-wheel vehicle.*

For the two-wheel vehicle shown in Figure 10.12, we use the cot-average (7.4) of the outer and inner steer angles as the input steer angle,

$$\cot \delta = \frac{\cot \delta_o + \cot \delta_i}{2} \qquad (10.152)$$

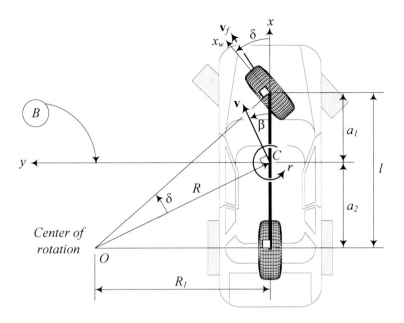

FIGURE 10.12. A two-wheel model for a vehicle moving with no roll.

where,

$$\tan \delta_i = \frac{l}{R_1 - \dfrac{w}{2}} \qquad \tan \delta_o = \frac{l}{R_1 + \dfrac{w}{2}} \qquad (10.153)$$

The radius of rotation R for the two-wheel vehicle is given by (7.3)

$$R = \sqrt{a_2^2 + l^2 \cot^2 \delta} \qquad (10.154)$$

and is measured at the mass center of the steered vehicle.

Example 419 *Four-wheel rigid vehicle.*
  The force on each wheel of a planar model of vehicles is

$$\mathbf{F}_1 = \begin{bmatrix} F_{x_1} \cos \delta_1 - F_{y_1} \sin \delta_1 \\ F_{x_1} \sin \delta_1 + F_{y_1} \cos \delta_1 \\ 0 \end{bmatrix} \qquad \mathbf{F}_3 = \begin{bmatrix} F_{x_3} \\ F_{y_3} \\ 0 \end{bmatrix} \qquad (10.155)$$

$$\mathbf{F}_2 = \begin{bmatrix} F_{x_2} \cos \delta_2 - F_{y_2} \sin \delta_2 \\ F_{x_2} \sin \delta_2 + F_{y_2} \cos \delta_2 \\ 0 \end{bmatrix} \qquad \mathbf{F}_4 = \begin{bmatrix} F_{x_4} \\ F_{y_4} \\ 0 \end{bmatrix} \qquad (10.156)$$

The position vector of the wheels are

$$\mathbf{r}_1 = \begin{bmatrix} a_1 \\ b_1 \\ 0 \end{bmatrix} \qquad \mathbf{r}_3 = \begin{bmatrix} -a_2 \\ b_1 \\ 0 \end{bmatrix} \tag{10.157}$$

$$\mathbf{r}_2 = \begin{bmatrix} a_1 \\ -b_2 \\ 0 \end{bmatrix} \qquad \mathbf{r}_4 = \begin{bmatrix} -a_2 \\ -b_2 \\ 0 \end{bmatrix} \tag{10.158}$$

and therefore, the force system on a planar model of vehicles is

$$\mathbf{F} = \sum \mathbf{F}_i \tag{10.159}$$

$$= \begin{bmatrix} F_{x_1}\cos\delta_1 + F_{x_2}\cos\delta_2 - F_{y_1}\sin\delta_1 - F_{y_2}\sin\delta_2 + F_{x_3} + F_{x_4} \\ F_{y_1}\cos\delta_1 + F_{y_2}\cos\delta_2 + F_{x_1}\sin\delta_1 + F_{x_2}\sin\delta_2 + F_{y_3} + F_{y_4} \\ 0 \end{bmatrix}$$

$$\mathbf{M} = \sum \mathbf{r}_i \times \mathbf{F}_i = \begin{bmatrix} 0 \\ 0 \\ M_z \end{bmatrix} \tag{10.160}$$

$$\begin{aligned}
M_z = \; & a_1 F_{y_1}\cos\delta_1 + a_1 F_{y_2}\cos\delta_2 - a_2 F_{y_3} - a_2 F_{y_4} \\
& + b_1 F_{y_1}\sin\delta_1 - b_2 F_{y_2}\sin\delta_2 + b_2 F_{x_2}\cos\delta_2 - b_1 F_{x_1}\cos\delta_1 \\
& + a_1 F_{x_1}\sin\delta_1 + a_1 F_{x_2}\sin\delta_2 + b_2 F_{x_4} - b_1 F_{x_3}
\end{aligned} \tag{10.161}$$

Let us assume a planar dynamic model of vehicles with four wheels such that its equation can be approximated by

$$F_x \approx F_{x_1} + F_{x_2} + F_{x_3} + F_{x_4} \tag{10.162}$$

$$F_y \approx F_{y_1} + F_{y_2} + F_{y_3} + F_{y_4} \tag{10.163}$$

$$M_z \approx a_1 F_{y_1} + a_1 F_{y_2} - a_2 F_{y_3} - a_2 F_{y_4} \tag{10.164}$$

in which, we assumed small steer angles and ignored the yew moment caused by imbalance longitudinal forces, F_{x_i}. Knowing that

$$v_{x_i} = v_x - y_i\, r \qquad v_{y_i} = v_y + x_i\, r \tag{10.165}$$

and tire angles as

$$\beta_i = \arctan\left(\frac{v_{y_i}}{v_{x_i}}\right) = \arctan\left(\frac{v_y + x_i\, r}{v_x - y_i\, r}\right) \tag{10.166}$$

$$\alpha_i = \beta_i - \delta_i = \arctan\left(\frac{v_y + x_i\, r}{v_x - y_i\, r}\right) - \delta_i \tag{10.167}$$

we find that

$$\beta_1 = \arctan\left(\frac{v_{y_1}}{v_{x_1}}\right) = \arctan\left(\frac{v_y + a_1\, r}{v_x - b_1\, r}\right) \tag{10.168}$$

$$\beta_2 = \arctan\left(\frac{v_{y_2}}{v_{x_2}}\right) = \arctan\left(\frac{v_y + a_1\, r}{v_x + b_2\, r}\right) \tag{10.169}$$

$$\beta_3 = \arctan\left(\frac{v_{y_3}}{v_{x_3}}\right) = \arctan\left(\frac{v_y - a_2\, r}{v_x + b_2\, r}\right) \qquad (10.170)$$

$$\beta_4 = \arctan\left(\frac{v_{y_4}}{v_{x_4}}\right) = \arctan\left(\frac{v_y - a_2\, r}{v_x - b_1\, r}\right) \qquad (10.171)$$

Employing the vehicle sideslip angle $\beta = \arctan(v_y/v_x)$ and assuming small angles for β_f, β_r, and β, the tire sideslip angles would be:

$$\alpha_1 = \beta_1 - \delta_1 = \frac{v_y + a_1\, r}{v_x - b_1\, r} - \delta_1 = \frac{\beta + \dfrac{a_1}{v_x}r}{1 - \dfrac{b_1}{v_x}r} - \delta_1 \qquad (10.172)$$

$$\alpha_2 = \beta_2 - \delta_2 = \frac{v_y + a_1\, r}{v_x + b_2\, r} - \delta_2 = \frac{\beta + \dfrac{a_1}{v_x}r}{1 + \dfrac{b_2}{v_x}r} - \delta_2 \qquad (10.173)$$

$$\alpha_3 = \beta_3 = \frac{v_y - a_2\, r}{v_x + b_2\, r} = \frac{\beta - \dfrac{a_2}{v_x}r}{1 + \dfrac{b_2}{v_x}r} \qquad (10.174)$$

$$\alpha_4 = \beta_4 = \frac{v_y - a_2\, r}{v_x - b_1\, r} = \frac{\beta - \dfrac{a_2}{v_x}r}{1 - \dfrac{b_1}{v_x}r} \qquad (10.175)$$

Modeling the lateral forces as

$$F_{y_i} = -C_{\alpha_i}\,\alpha_i \qquad (10.176)$$

and assuming that the left and right wheels have the same sideslip coefficients

$$C_{\alpha_1} = C_{\alpha_2} \qquad C_{\alpha_3} = C_{\alpha_4} \qquad (10.177)$$

we find the second and third equations of motion as

$$F_y = F_{y_1} + F_{y_2} + F_{y_3} + F_{y_4} = -C_{\alpha_1}\alpha_1 - C_{\alpha_1}\alpha_2 - C_{\alpha_4}\alpha_3 - C_{\alpha_4}\alpha_4$$

$$= -C_{\alpha_1}\left(\frac{\beta + \dfrac{a_1}{v_x}r}{1 - \dfrac{b_1}{v_x}r} - \delta_1\right) - C_{\alpha_1}\left(\frac{\beta + \dfrac{a_1}{v_x}r}{1 + \dfrac{b_2}{v_x}r} - \delta_2\right)$$

$$- C_{\alpha_4}\frac{\beta - \dfrac{a_2}{v_x}r}{1 + \dfrac{b_2}{v_x}r} - C_{\alpha_4}\frac{\beta - \dfrac{a_2}{v_x}r}{1 - \dfrac{b_1}{v_x}r} \qquad (10.178)$$

$$
\begin{aligned}
M_z &= a_1 F_{y_1} + a_1 F_{y_2} - a_2 F_{y_3} - a_2 F_{y_4} \\
&= -a_1 C_{\alpha_1} \alpha_1 - a_1 C_{\alpha_1} \alpha_2 + a_2 C_{\alpha_4} \alpha_3 + a_2 C_{\alpha_4} \alpha_4 \\
&= -a_1 C_{\alpha_1} \left(\dfrac{\beta + \dfrac{a_1}{v_x} r}{1 - \dfrac{b_1}{v_x} r} - \delta_1 \right) - a_1 C_{\alpha_1} \left(\dfrac{\beta + \dfrac{a_1}{v_x} r}{1 + \dfrac{b_2}{v_x} r} - \delta_2 \right)
\end{aligned}
$$

$$
+ a_2 C_{\alpha_4} \dfrac{\beta - \dfrac{a_2}{v_x} r}{1 + \dfrac{b_2}{v_x} r} + a_2 C_{\alpha_4} \dfrac{\beta - \dfrac{a_2}{v_x} r}{1 - \dfrac{b_1}{v_x} r} \tag{10.179}
$$

These equations are nonlinear and therefore, they cannot be arranged at the bicycle model.

Example 420 *Linearized equations of four-wheel rigid vehicle*

Using the Taylor series expansion formula of a two variable function $f(\beta, r)$

$$
df = \frac{\partial f}{\partial \beta} \beta + \frac{\partial f}{\partial r} r + \frac{1}{2} \frac{\partial^2 f}{\partial \beta^2} \beta^2 + \frac{1}{2} \frac{\partial^2 f}{\partial \beta \partial r} \beta r + \frac{1}{2} \frac{\partial^2 f}{\partial r \partial \beta} r \beta + \frac{1}{2} \frac{\partial^2 f}{\partial r^2} r^2 + \cdots \tag{10.180}
$$

we have

$$
\dfrac{\beta + \dfrac{a_1}{v_x} r}{1 - \dfrac{b_1}{v_x} r} \approx \beta + \dfrac{a_1}{v_x} r \qquad\qquad \dfrac{\beta + \dfrac{a_1}{v_x} r}{1 + \dfrac{b_2}{v_x} r} \approx \beta + \dfrac{a_1}{v_x} r \tag{10.181}
$$

$$
\dfrac{\beta - \dfrac{a_2}{v_x} r}{1 + \dfrac{b_2}{v_x} r} \approx \beta - \dfrac{a_2}{v_x} r \qquad\qquad \dfrac{\beta - \dfrac{a_2}{v_x} r}{1 - \dfrac{b_1}{v_x} r} \approx \beta - \dfrac{a_2}{v_x} r \tag{10.182}
$$

and therefore, Equations (10.178) and (10.179) become

$$
\begin{aligned}
F_y &= -C_{\alpha_1} \left(\beta + \frac{a_1}{v_x} r - \delta_1 \right) - C_{\alpha_1} \left(\beta + \frac{a_1}{v_x} r - \delta_2 \right) \\
&\quad - C_{\alpha_4} \left(\beta - \frac{a_2}{v_x} r \right) - C_{\alpha_4} \left(\beta - \frac{a_2}{v_x} r \right)
\end{aligned} \tag{10.183}
$$

$$
\begin{aligned}
M_z &= -a_1 C_{\alpha_1} \left(\beta + \frac{a_1}{v_x} r - \delta_1 \right) - a_1 C_{\alpha_1} \left(\beta + \frac{a_1}{v_x} r - \delta_2 \right) \\
&\quad + a_2 C_{\alpha_4} \left(\beta - \frac{a_2}{v_x} r \right) + a_2 C_{\alpha_4} \left(\beta - \frac{a_2}{v_x} r \right)
\end{aligned} \tag{10.184}
$$

Rearranged to the force system yields

$$F_y = 2\left(\frac{a_2}{v_x}C_{\alpha_4} - \frac{a_1}{v_x}C_{\alpha_1}\right)r - 2\left(C_{\alpha_1} + C_{\alpha_3}\right)\beta$$
$$+ C_{\alpha_1}\left(\delta_1 + \delta_2\right) \tag{10.185}$$

$$M_z = -2\left(\frac{a_1^2}{v_x}C_{\alpha_1} + \frac{a_2^2}{v_x}C_{\alpha_4}\right)r + 2\left(a_2 C_{\alpha_4} - a_1 C_{\alpha_1}\right)\beta$$
$$+ a_1 C_{\alpha_1}\left(\delta_1 + \delta_2\right) \tag{10.186}$$

These equations are comparaible with (10.115) and (10.116).

This analysis indicates that adopting the bicycle model of vehicles is equivalent to accepting a linearized equation of α_i as function of r and β.

10.4 Two-wheel Rigid Vehicle Dynamics

We combine the planar equations of motion (10.22)-(10.24) with the force expressions (10.112)-(10.114) to express the motion of a two-wheel rigid vehicle with no roll, by the following set of equations:

$$\dot{v}_x = \frac{1}{m}F_x + r\,v_y = \frac{1}{m}\left(F_{x_f} + F_{x_r}\right) + r\,v_y \tag{10.187}$$

$$\begin{bmatrix} \dot{v}_y \\ \dot{r} \end{bmatrix} = \begin{bmatrix} \dfrac{C_\beta}{mv_x} & \dfrac{C_r}{m} - v_x \\[2ex] \dfrac{D_\beta}{I_z v_x} & \dfrac{D_r}{I_z} \end{bmatrix} \begin{bmatrix} v_y \\ r \end{bmatrix} + \begin{bmatrix} \dfrac{C_\delta}{m} \\[2ex] \dfrac{D_\delta}{I_z} \end{bmatrix}\delta$$

$$= \begin{bmatrix} -\dfrac{C_{\alpha f} + C_{\alpha r}}{mv_x} & \dfrac{-a_1 C_{\alpha f} + a_2 C_{\alpha r}}{mv_x} - v_x \\[2ex] -\dfrac{a_1 C_{\alpha f} - a_2 C_{\alpha r}}{I_z v_x} & -\dfrac{a_1^2 C_{\alpha f} + a_2^2 C_{\alpha r}}{I_z v_x} \end{bmatrix}\begin{bmatrix} v_y \\ r \end{bmatrix}$$

$$+ \begin{bmatrix} \dfrac{C_{\alpha f}}{m} \\[2ex] \dfrac{a_1 C_{\alpha f}}{I_z} \end{bmatrix}\delta \tag{10.188}$$

These sets of equations are linear when a vehicle is moving at a constant forward speed. Having $\dot{v}_x = 0$, the set of Equations (10.188) become independent of the first Equation (10.187). Then, the lateral velocity v_y and yaw rate r of the vehicle will change according to the two coupled Equations (10.188).

Assuming the steer angle δ is the input command, the lateral velocity v_y and the yaw rate r may be assumed as the output. Hence, we may consider Equation (10.188) as a linear control system to write them as

$$\dot{\mathbf{q}} = [A]\,\mathbf{q} + \mathbf{u} \tag{10.189}$$

in which $[A]$ is a coefficient matrix, $\mathbf{q}$ is the vector of control variables, and $\mathbf{u}$ is the vector of inputs.

$$[A] = \begin{bmatrix} -\dfrac{C_{\alpha f} + C_{\alpha r}}{m v_x} & \dfrac{-a_1 C_{\alpha f} + a_2 C_{\alpha r}}{m v_x} - v_x \\[4mm] -\dfrac{a_1 C_{\alpha f} - a_2 C_{\alpha r}}{I_z v_x} & -\dfrac{a_1^2 C_{\alpha f} + a_2^2 C_{\alpha r}}{I_z v_x} \end{bmatrix} \tag{10.190}$$

$$\mathbf{q} = \begin{bmatrix} v_y \\ r \end{bmatrix} \tag{10.191}$$

$$\mathbf{u} = \begin{bmatrix} \dfrac{C_{\alpha f}}{m} \\[3mm] \dfrac{a_1 C_{\alpha f}}{I_z} \end{bmatrix} \delta \tag{10.192}$$

Proof. The Newton-Euler equations of motion for a rigid vehicle in the local coordinate frame B, attached to the vehicle at its mass center C, are given in Equations (10.22)–(10.24) as

$$\begin{aligned} F_x &= m\dot{v}_x - m r\, v_y \tag{10.193} \\ F_y &= m\dot{v}_y + m r\, v_x \tag{10.194} \\ M_z &= \dot{r}\, I_z \tag{10.195} \end{aligned}$$

The approximate force system applied on a two-wheel rigid vehicle is found in Equations (10.112)–(10.114)

$$\begin{aligned} F_x &\approx F_{x_f} + F_{x_r} \tag{10.196} \\ F_y &\approx F_{y_f} + F_{y_r} \tag{10.197} \\ M_z &\approx a_1 F_{y_f} - a_2 F_{y_r} \tag{10.198} \end{aligned}$$

and in terms of tire characteristics, in (10.115) and (10.116).

$$\begin{aligned} F_x &= \frac{T_w}{R_w} \tag{10.199} \\ F_y &= \left(-\frac{a_1}{v_x}C_{\alpha f} + \frac{a_2}{v_x}C_{\alpha r}\right)r - (C_{\alpha f} + C_{\alpha r})\beta + C_{\alpha f}\delta \\ &= C_r r + C_\beta \beta + C_\delta \delta \tag{10.200} \\ M_z &= \left(-\frac{a_1^2}{v_x}C_{\alpha f} - \frac{a_2^2}{v_x}C_{\alpha r}\right)r - (a_1 C_{\alpha f} - a_2 C_{\alpha r})\beta + a_1 C_{\alpha f}\delta \\ &= D_r r + D_\beta \beta + D_\delta \delta \tag{10.201} \end{aligned}$$

Substituting (10.199)–(10.201) in (10.193)–(10.195) produces the following equations of motion:

$$m \dot{v}_x - mr\, v_y = F_x \qquad (10.202)$$

$$m \dot{v}_y + mr\, v_x = C_r\, r + C_\beta\, \beta + C_\delta\, \delta \qquad (10.203)$$

$$\dot{r}\, I_z = D_r\, r + D_\beta\, \beta + D_\delta\, \delta \qquad (10.204)$$

Solving the equations for $\dot{v}_x$, $\dot{v}_y$, and $\dot{r}$ yields

$$\dot{v}_x = \frac{F_x}{m} + r\, v_y \qquad (10.205)$$

$$
\begin{aligned}
\dot{v}_y &= \frac{1}{m}\left(C_r\, r + C_\beta\, \beta + C_\delta\, \delta\right) - r\, v_x \\
&= \frac{1}{m}\left(-\frac{a_1}{v_x}C_{\alpha f} + \frac{a_2}{v_x}C_{\alpha r}\right) r \\
&\quad - \frac{1}{m}\left(C_{\alpha f} + C_{\alpha r}\right)\beta + \frac{1}{m}C_{\alpha f}\delta - r\, v_x \qquad (10.206)
\end{aligned}
$$

$$
\begin{aligned}
\dot{r} &= \frac{1}{I_z}\left(D_r\, r + D_\beta\, \beta + D_\delta\, \delta\right) \\
&= \frac{1}{I_z}\left(-\frac{a_1^2}{v_x}C_{\alpha f} - \frac{a_2^2}{v_x}C_{\alpha r}\right) r \\
&\quad - \frac{1}{I_z}\left(a_1 C_{\alpha f} - a_2 C_{\alpha r}\right)\beta + \frac{1}{I_z}a_1 C_{\alpha f}\delta \qquad (10.207)
\end{aligned}
$$

The vehicle sideslip angle β can be substituted by vehicle velocity components

$$\beta = \frac{v_y}{v_x} \qquad (10.208)$$

So, we can find the equations of motion in terms of the three variables, v_x, v_y, r.

$$\dot{v}_x = \frac{F_x}{m} + r\, v_y \qquad (10.209)$$

$$
\begin{aligned}
\dot{v}_y &= \frac{1}{m}\left(C_r\, r + \frac{C_\beta}{v_x} v_y + C_\delta\, \delta\right) - r\, v_x \\
&= \frac{1}{m v_x}\left(-a_1 C_{\alpha f} + a_2 C_{\alpha r}\right) r \\
&\quad - \frac{1}{m v_x}\left(C_{\alpha f} + C_{\alpha r}\right) v_y + \frac{1}{m}C_{\alpha f}\delta - r\, v_x \qquad (10.210)
\end{aligned}
$$

$$
\begin{aligned}
\dot{r} &= \frac{1}{I_z}\left(D_r\, r + \frac{D_\beta}{v_x} v_y + D_\delta\, \delta\right) \\
&= \frac{1}{I_z v_x}\left(-a_1^2 C_{\alpha f} - a_2^2 C_{\alpha r}\right) r \\
&\quad - \frac{1}{I_z v_x}\left(a_1 C_{\alpha f} - a_2 C_{\alpha r}\right) v_y + \frac{1}{I_z}a_1 C_{\alpha f}\delta \qquad (10.211)
\end{aligned}
$$

These equations are coupled. The first equation (10.209) depends on the
yaw rate r and the lateral velocity v_y, which are the output of the second
and third equations, (10.210) and (10.211). Also Equations (10.210) and
(10.211) depend on v_x which is the output of the first equation. However,
if we assume the vehicle is moving with a constant forward speed,

$$v_x = const \tag{10.212}$$

then Equation (10.209) becomes an algebraic equation and then Equations
(10.210) and (10.211) become independent with (10.209). So, the second
and third equations can be treated independent of the first one.

Equations (10.210) and (10.211) can also be considered as two coupled
differential equations describing the behavior of a dynamic system. The
dynamic system receives the steering angle δ as the input, and uses v_x as
a parameter to generate two outputs, v_y and r.

$$\begin{bmatrix} \dot{v}_y \\ \dot{r} \end{bmatrix} = \begin{bmatrix} -\dfrac{C_{\alpha f}+C_{\alpha r}}{mv_x} & \dfrac{-a_1 C_{\alpha f}+a_2 C_{\alpha r}}{mv_x}-v_x \\ -\dfrac{a_1 C_{\alpha f}-a_2 C_{\alpha r}}{I_z v_x} & -\dfrac{a_1^2 C_{\alpha f}+a_2^2 C_{\alpha r}}{I_z v_x} \end{bmatrix} \begin{bmatrix} v_y \\ r \end{bmatrix} + \begin{bmatrix} \dfrac{C_{\alpha f}}{m} \\ \dfrac{a_1 C_{\alpha f}}{I_z} \end{bmatrix} \delta \tag{10.213}$$

We can rearrange Equation (10.213) in the following form to show the
input-output relationship:

$$\dot{\mathbf{q}} = [A]\,\mathbf{q} + \mathbf{u} \tag{10.214}$$

The vector $\mathbf{q}$ is called the *control variables vector*, and $\mathbf{u}$ is called the *inputs
vector*. The matrix $[A]$ is the control variable coefficients matrix.

Employing the force system coefficients C_r, C_β, C_δ, D_r, D_β, and D_δ from
(10.144)-(10.149) for a front-wheel steering vehicle, we may write the set
of Equations (10.213) as

$$\begin{bmatrix} \dot{v}_y \\ \dot{r} \end{bmatrix} = \begin{bmatrix} \dfrac{C_\beta}{mv_x} & \dfrac{C_r}{m}-v_x \\ \dfrac{D_\beta}{I_z v_x} & \dfrac{D_r}{I_z} \end{bmatrix} \begin{bmatrix} v_y \\ r \end{bmatrix} + \begin{bmatrix} \dfrac{C_\delta}{m} \\ \dfrac{D_\delta}{I_z} \end{bmatrix} \delta \tag{10.215}$$

■

Example 421 *Equations of motion based on kinematic angles.*
*The equations of motion (10.188) can be expressed based on only the
angles β, r, and δ, by employing (10.208).*

Taking a time derivative from Equation (10.208) for constant v_x

$$\dot\beta = \frac{\dot v_y}{v_x} \tag{10.216}$$

and substituting it in Equations (10.206) shows that we can transform the equation for $\dot\beta$ to:

$$\begin{aligned}
v_x\dot\beta &= \frac{1}{m}\left(-\frac{a_1}{v_x}C_{\alpha f} + \frac{a_2}{v_x}C_{\alpha r}\right)r \\
&\quad -\frac{1}{m}(C_{\alpha f}+C_{\alpha r})\beta + \frac{1}{m}C_{\alpha f}\delta - r\,v_x
\end{aligned} \tag{10.217}$$

Therefore, the set of equations of motion can be expressed in terms of the vehicle's angular variables β, r, and δ.

$$\begin{bmatrix}\dot\beta\\\dot r\end{bmatrix} = \begin{bmatrix} -\dfrac{C_{\alpha f}+C_{\alpha r}}{mv_x} & \dfrac{-a_1 C_{\alpha f}+a_2 C_{\alpha r}}{mv_x^2}-1 \\[3mm] -\dfrac{a_1 C_{\alpha f}-a_2 C_{\alpha r}}{I_z} & -\dfrac{a_1^2 C_{\alpha f}+a_2^2 C_{\alpha r}}{I_z v_x}\end{bmatrix}\begin{bmatrix}\beta\\r\end{bmatrix}$$
$$+\begin{bmatrix}\dfrac{C_{\alpha f}}{mv_x}\\[3mm]\dfrac{a_1 C_{\alpha f}}{I_z}\end{bmatrix}\delta \tag{10.218}$$

Employing the force system coefficients C_r, C_β, C_δ, D_r, D_β, and D_δ for a front-wheel steering vehicle, we can write the set of Equations (10.218) as

$$\begin{bmatrix}\dot\beta\\\dot r\end{bmatrix}=\begin{bmatrix}\dfrac{C_\beta}{mv_x}&\dfrac{C_r}{mv_x}-1\\[3mm]\dfrac{D_\beta}{I_z}&\dfrac{D_r}{I_z}\end{bmatrix}\begin{bmatrix}\beta\\r\end{bmatrix}+\begin{bmatrix}\dfrac{C_\delta}{mv_x}\\[3mm]\dfrac{D_\delta}{I_z}\end{bmatrix}\delta \tag{10.219}$$

Example 422 *Four-wheel-steering vehicles.*

Consider a vehicle with steerable wheels in front and rear. Let us indicate the steer angle in front and rear by δ_f and δ_r respectively. To find the equations of motion of the bicycle planar model, we begin with Equation (10.108) for the relationship between α, β, and δ

$$\alpha = \beta - \delta \tag{10.220}$$

and apply the equation to the front and rear wheels.

$$\alpha_f = \beta_f - \delta_f = \frac{1}{v_x}(v_y+a_1 r)-\delta_f = \beta + \frac{a_1 r}{v_x}-\delta_f \tag{10.221}$$
$$\alpha_r = \beta_r - \delta_r = \frac{1}{v_x}(v_y-a_2 r)-\delta_r = \beta - \frac{a_2 r}{v_x}-\delta_r \tag{10.222}$$

When the sideslip angles are small, the associated lateral forces are

$$F_{yf} = -C_{\alpha f}\,\alpha_f \qquad F_{yr} = -C_{\alpha r}\,\alpha_r \qquad (10.223)$$

Substituting these equations in the second and third equations of motion (10.113) and (10.114) yields:

$$
\begin{aligned}
F_y &= F_{yf} + F_{yr} = \left(-\frac{a_1}{v_x}C_{\alpha f} + \frac{a_2}{v_x}C_{\alpha r}\right) r \\
&\quad - (C_{\alpha f} + C_{\alpha r})\beta + C_{\alpha f}\delta_f + C_{\alpha r}\delta_r & (10.224) \\
M_z &= a_1 F_{yf} - a_2 F_{yr} = \left(-\frac{a_1^2}{v_x}C_{\alpha f} - \frac{a_2^2}{v_x}C_{\alpha r}\right) r \\
&\quad - (a_1 C_{\alpha f} - a_2 C_{\alpha r})\beta + a_1 C_{\alpha f}\delta_f - a_2 C_{\alpha r}\delta_r & (10.225)
\end{aligned}
$$

The Newton-Euler equations of motion for a rigid vehicle are given in Equations (10.22)–(10.24) as

$$
\begin{aligned}
F_x &= m\,\dot{v}_x - mr\,v_y & (10.226) \\
F_y &= m\,\dot{v}_y + mr\,v_x & (10.227) \\
M_z &= \dot{r}\,I_z & (10.228)
\end{aligned}
$$

and therefore, the equations of motion for a four-wheel-steering vehicle are

$$m\,\dot{v}_x - mr\,v_y = F_x \qquad (10.229)$$

$$
\begin{aligned}
m\,\dot{v}_y + mr\,v_x &= \left(-\frac{a_1}{v_x}C_{\alpha f} + \frac{a_2}{v_x}C_{\alpha r}\right) r \\
&\quad - (C_{\alpha f} + C_{\alpha r})\beta + C_{\alpha f}\delta_f + C_{\alpha r}\delta_r & (10.230)
\end{aligned}
$$

$$
\begin{aligned}
\dot{r}\,I_z &= \left(-\frac{a_1^2}{v_x}C_{\alpha f} - \frac{a_2^2}{v_x}C_{\alpha r}\right) r \\
&\quad - (a_1 C_{\alpha f} - a_2 C_{\alpha r})\beta + a_1 C_{\alpha f}\delta_f - a_2 C_{\alpha r}\delta_r & (10.231)
\end{aligned}
$$

Solving these equations for $\dot{v}_x$, $\dot{v}_y$, and $\dot{r}$ provide us with three first order coupled equations.

$$\dot{v}_x = \frac{F_x}{m} + r\,v_y \qquad (10.232)$$

$$
\begin{aligned}
\dot{v}_y &= \frac{1}{m}\left(-\frac{a_1}{v_x}C_{\alpha f} + \frac{a_2}{v_x}C_{\alpha r}\right) r \\
&\quad - \frac{1}{m}(C_{\alpha f} + C_{\alpha r})\beta + \frac{1}{m}C_{\alpha f}\delta_f + \frac{1}{m}C_{\alpha r}\delta_r - r\,v_x & (10.233)
\end{aligned}
$$

$$
\begin{aligned}
\dot{r} &= \frac{1}{I_z}\left(-\frac{a_1^2}{v_x}C_{\alpha f} - \frac{a_2^2}{v_x}C_{\alpha r}\right) r \\
&\quad - \frac{1}{I_z}(a_1 C_{\alpha f} - a_2 C_{\alpha r})\beta + \frac{1}{I_z}a_1 C_{\alpha f}\delta_f - \frac{1}{I_z}a_2 C_{\alpha r}\delta_r & (10.234)
\end{aligned}
$$

The three variables v_x, v_y, β, are related by the vehicle sideslip angle

$$\beta = \frac{v_y}{v_x} \tag{10.235}$$

Using this relation, we can replace β by v_y/v_x and transform the equations to the following set of three coupled first order ordinary differential equations:

$$\dot{v}_x = \frac{F_x}{m} + r\,v_y \tag{10.236}$$

$$\dot{v}_y = \frac{1}{mv_x}\left(-a_1 C_{\alpha f} + a_2 C_{\alpha r}\right) r$$
$$\quad - \frac{1}{mv_x}\left(C_{\alpha f} + C_{\alpha r}\right) v_y + \frac{1}{m}C_{\alpha f}\delta_f + \frac{1}{m}C_{\alpha r}\delta_r - r\,v_x \tag{10.237}$$

$$\dot{r} = \frac{1}{I_z v_x}\left(-a_1^2 C_{\alpha f} - a_2^2 C_{\alpha r}\right) r$$
$$\quad - \frac{1}{I_z v_x}\left(a_1 C_{\alpha f} - a_2 C_{\alpha r}\right) v_y + \frac{1}{I_z}a_1 C_{\alpha f}\delta_f - \frac{1}{I_z}a_2 C_{\alpha r}\delta_r \tag{10.238}$$

The second and third equations may be cast in a matrix form for $\begin{bmatrix} v_y & r \end{bmatrix}^T$:

$$\begin{bmatrix} \dot{v}_y \\ \dot{r} \end{bmatrix} = \begin{bmatrix} -\dfrac{C_{\alpha f} + C_{\alpha r}}{mv_x} & \dfrac{-a_1 C_{\alpha f} + a_2 C_{\alpha r}}{mv_x} - v_x \\ -\dfrac{a_1 C_{\alpha f} - a_2 C_{\alpha r}}{I_z v_x} & -\dfrac{a_1^2 C_{\alpha f} + a_2^2 C_{\alpha r}}{I_z v_x} \end{bmatrix} \begin{bmatrix} v_y \\ r \end{bmatrix}$$
$$\quad + \begin{bmatrix} \dfrac{1}{m}C_{\alpha f} & \dfrac{1}{m}C_{\alpha r} \\ \dfrac{1}{I_z}a_1 C_{\alpha f} & -\dfrac{1}{I_z}a_2 C_{\alpha r} \end{bmatrix} \begin{bmatrix} \delta_f \\ \delta_r \end{bmatrix} \tag{10.239}$$

or to the following form for $\begin{bmatrix} \beta & r \end{bmatrix}^T$:

$$\begin{bmatrix} \dot{\beta} \\ \dot{r} \end{bmatrix} = \begin{bmatrix} -\dfrac{C_{\alpha f} + C_{\alpha r}}{mv_x} & \dfrac{-a_1 C_{\alpha f} + a_2 C_{\alpha r}}{mv_x^2} - 1 \\ -\dfrac{a_1 C_{\alpha f} - a_2 C_{\alpha r}}{I_z} & -\dfrac{a_1^2 C_{\alpha f} + a_2^2 C_{\alpha r}}{I_z v_x} \end{bmatrix} \begin{bmatrix} \beta \\ r \end{bmatrix}$$
$$\quad + \begin{bmatrix} \dfrac{1}{mv_x}C_{\alpha f} & \dfrac{1}{mv_x}C_{\alpha r} \\ \dfrac{1}{I_z}a_1 C_{\alpha f} & -\dfrac{1}{I_z}a_2 C_{\alpha r} \end{bmatrix} \begin{bmatrix} \delta_f \\ \delta_r \end{bmatrix} \tag{10.240}$$

For computerization of the equations of motion, it is better to write them as

$$\begin{bmatrix} \dot{v}_y \\ \dot{r} \end{bmatrix} = \begin{bmatrix} \dfrac{C_\beta}{mv_x} & \dfrac{C_r}{m} - v_x \\ \dfrac{D_\beta}{I_z v_x} & \dfrac{D_r}{I_z} \end{bmatrix} \begin{bmatrix} v_y \\ r \end{bmatrix} + \begin{bmatrix} \dfrac{C_{\delta_f}}{m} & \dfrac{C_{\delta_r}}{m} \\ \dfrac{D_{\delta_f}}{I_z} & \dfrac{D_{\delta_r}}{I_z} \end{bmatrix} \begin{bmatrix} \delta_f \\ \delta_r \end{bmatrix}$$

(10.241)

or

$$\begin{bmatrix} \dot{\beta} \\ \dot{r} \end{bmatrix} = \begin{bmatrix} \dfrac{C_\beta}{mv_x} & \dfrac{C_r}{mv_x} - 1 \\ \dfrac{D_\beta}{I_z} & \dfrac{D_r}{I_z} \end{bmatrix} \begin{bmatrix} \beta \\ r \end{bmatrix} + \begin{bmatrix} \dfrac{C_{\delta_f}}{mv_x} & \dfrac{C_{\delta_r}}{mv_x} \\ \dfrac{D_{\delta_f}}{I_z} & \dfrac{D_{\delta_r}}{I_z} \end{bmatrix} \begin{bmatrix} \delta_f \\ \delta_r \end{bmatrix}$$

(10.242)

where

$$C_r = \frac{\partial F_y}{\partial r} = -\frac{a_1}{v_x} C_{\alpha f} + \frac{a_2}{v_x} C_{\alpha r}$$

(10.243)

$$C_\beta = \frac{\partial F_y}{\partial \beta} = -(C_{\alpha f} + C_{\alpha r})$$

(10.244)

$$C_{\delta_f} = \frac{\partial F_y}{\partial \delta_f} = C_{\alpha f}$$

(10.245)

$$C_{\delta_r} = \frac{\partial F_y}{\partial \delta_r} = C_{\alpha r}$$

(10.246)

$$D_r = \frac{\partial M_z}{\partial r} = -\frac{a_1^2}{v_x} C_{\alpha f} - \frac{a_2^2}{v_x} C_{\alpha r}$$

(10.247)

$$D_\beta = \frac{\partial M_z}{\partial \beta} = -(a_1 C_{\alpha f} - a_2 C_{\alpha r})$$

(10.248)

$$D_{\delta_f} = \frac{\partial M_z}{\partial \delta_f} = a_1 C_{\alpha r}$$

(10.249)

$$D_{\delta_r} = \frac{\partial M_z}{\partial \delta_r} = -a_2 C_{\alpha r}$$

(10.250)

Equation (10.239) may be rearranged in the following form to show the input-output relationship:

$$\dot{\mathbf{q}} = [A]\,\mathbf{q} + [B]\,\mathbf{u}$$

(10.251)

The vector **q** *is called the control variables vector,*

$$\mathbf{q} = \begin{bmatrix} v_y \\ r \end{bmatrix}$$

(10.252)

and **u** *is called the inputs vector.*

$$\mathbf{u} = \begin{bmatrix} \delta_f \\ \delta_r \end{bmatrix} \qquad (10.253)$$

The matrix $[A]$ *is the control variable coefficients matrix and the matrix* $[B]$ *is the input coefficient matrix.*

To double check, we may substitute $\delta_r = 0$, *and* $\delta_f = \delta$ *to reduce Equations* (10.239) *to* (10.213) *for a front-wheel-steering vehicle.*

Example 423 *Rear-wheel-steering vehicle.*

Rear wheel steering is frequently employed in lift trucks and construction vehicles. The equations of motion for rear steering vehicles are similar to those in front steering. To find the equations of motion, we substitute $\delta_f = 0$ *in Equations* (10.239) *to find these equations*

$$\begin{bmatrix} \dot{v}_y \\ \dot{r} \end{bmatrix} = \begin{bmatrix} -\dfrac{C_{\alpha f} + C_{\alpha r}}{mv_x} & \dfrac{-a_1 C_{\alpha f} + a_2 C_{\alpha r}}{mv_x} - v_x \\ -\dfrac{a_1 C_{\alpha f} - a_2 C_{\alpha r}}{I_z v_x} & -\dfrac{a_1^2 C_{\alpha f} + a_2^2 C_{\alpha r}}{I_z v_x} \end{bmatrix} \begin{bmatrix} v_y \\ r \end{bmatrix}$$

$$+ \begin{bmatrix} \dfrac{1}{m} C_{\alpha r} \\ -\dfrac{1}{I_z} a_2 C_{\alpha r} \end{bmatrix} \delta_r \qquad (10.254)$$

These equations are valid as long as the angles are very small. However, most of the rear steering construction vehicles work at a high steer angle. Therefore, these equations cannot predict the behavior of the construction vehicles very well.

Example 424 ★ *Better model for two-wheel vehicles.*

Because of the steer angle, a reaction moment appears at the tireprints of the front and rear wheels, which act as external moments M_f *and* M_r *on the wheels. Therefore, the total steering reaction moment on the front and rear wheels are*

$$M_f \approx 2D_{\delta_f} M_z \qquad M_r \approx 2D_{\delta_r} M_z \qquad (10.255)$$

where

$$D_{\delta_f} = \frac{dM_z}{d\delta_f} \qquad D_{\delta_r} = \frac{dM_z}{d\delta_r} \qquad (10.256)$$

Figure 10.13 illustrates a two-wheel vehicle model with the reaction moments M_f *and* M_r. *The force system on the vehicle is*

$$F_x \approx F_{x_f} + F_{x_r} \qquad (10.257)$$
$$F_y \approx F_{y_f} + F_{y_r} \qquad (10.258)$$
$$M_z \approx a_1 F_{y_f} - a_2 F_{y_r} + M_f + M_r \qquad (10.259)$$

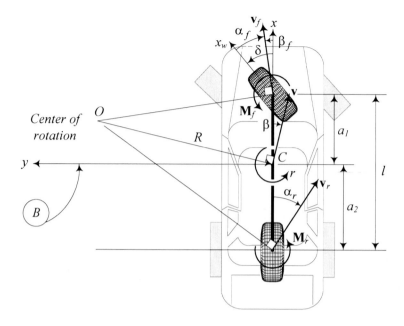

FIGURE 10.13. A two-wheel vehicle and its force system, including the steer moment reactions.

Example 425 ★ *The race car* 180 deg *quick turn from reverse.*

You have seen that stunt car drivers can turn a 180 deg *quickly when the car is moving backward. Here is how they do this. The driver moves backward when the car is in reverse gear. To make a fast* 180 deg *turn without stopping, the driver may follow these steps:* 1.*The driver should push the gas pedal to gain enough speed,* 2.*free the gas pedal and put the gear in neutral,* 3.*cut the steering wheel sharply around* 90 deg, 4.*change the gear to drive, and* 5.*push the gas pedal and return the steering wheel to* 0 deg *after the car has completed the* 180 deg *turn.*

The backward speed before step 2 *may be around* 20 m/s ≈ 70 km/h ≈ 45 mi/h. *Steps* 2 *to* 4 *should be done fast and almost simultaneously. Figure* 10.14(a) *illustrates the* 180 deg *fast turning maneuver from reverse.*

This example should never be performed by the reader of this book due to high safety risk.

Example 426 ★ *The race car* 180 deg *quick turn from forward.*

You have seen that stunt car drivers can turn a 180 deg *quickly when the car is moving forward. Here is how they do this. The driver moves forward when the car is in drive or a forward gear. To make a fast* 180 deg *turn without stopping, the driver may follow these steps:* 1.*The driver should push the gas pedal to gain enough speed,* 2.*free the gas pedal and put the gear in neutral,* 3.*cut the steering wheel sharply around* 90 deg *while pulling*

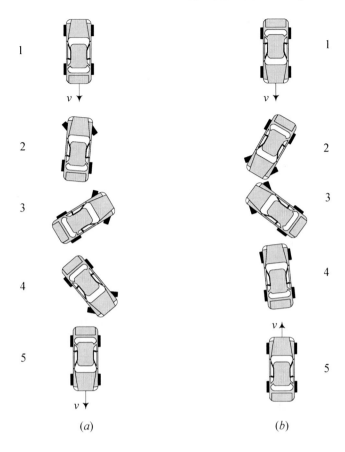

FIGURE 10.14. (a). A 180 deg, fast turning maneuver from reverse. (b). A 180 deg, fast turning maneuver from forward.

hard the hand brake, 4.*while the rear of the car swings around, return the steering wheel to* 0 deg *and put the gear into drive, and* 5.*push the gas pedal after the car has completed the* 180 deg *turn. Figure 10.14(b) illustrates this maneuver.*

The forward speed before step 2 *may be around* $20\,\mathrm{m/s} \approx 70\,\mathrm{km/h} \approx 45\,\mathrm{mi/h}$. *Steps* 2 *to* 4 *should be done fast and almost simultaneously. The* 180 deg *fast turning from forward is more difficult than backward and can be done because the hand brakes are connected to the rear wheels. It can be done better when the rear of a car is lighter than the front to slide easier. Road condition, nonuniform friction, slippery surface can cause flipping the car and spinning out of control.*

This example should never be performed by the reader of this book due to high safety risk.

10.5 Steady-State Turning

The turning of a front-wheel-steering two-wheel rigid vehicle at its steady-state condition is governed by the following equations:

$$F_x = -mr\,v_y \tag{10.260}$$
$$C_r\,r + C_\beta\,\beta + C_\delta\,\delta = mr\,v_x \tag{10.261}$$
$$D_r\,r + D_\beta\,\beta + D_\delta\,\delta = 0 \tag{10.262}$$

or equivalently, by

$$F_x = -\frac{m}{R}v_x\,v_y \tag{10.263}$$
$$C_\beta\,\beta + \left(C_r\,v_x - m\,v_x^2\right)\frac{1}{R} = -C_\delta\,\delta \tag{10.264}$$
$$D_\beta\,\beta + D_r\,v_x\frac{1}{R} = -D_\delta\,\delta \tag{10.265}$$

The first equation determines the required forward force to keep v_x constant when the car is on a circular path. The second and third equations show the steady-state values of the output variables, vehicle slip angle β, and path curvature κ,

$$\kappa = \frac{1}{R} = \frac{r}{v_x} \tag{10.266}$$

for a constant steering input δ at a constant forward speed v_x. The steady-state output-input relationships of the vehicle are defined by the following responses:

1. *Curvature response, S_κ*

$$S_\kappa = \frac{\kappa}{\delta} = \frac{1}{R\delta} = \frac{C_\delta D_\beta - C_\beta D_\delta}{v_x\left(D_r C_\beta - C_r D_\beta + m v_x D_\beta\right)} \tag{10.267}$$

2. *Sideslip response, S_β*

$$S_\beta = \frac{\beta}{\delta} = \frac{D_\delta\left(C_r - m v_x\right) - D_r C_\delta}{D_r C_\beta - C_r D_\beta + m v_x D_\beta} \tag{10.268}$$

3. *Yaw rate response, S_r*

$$S_r = \frac{r}{\delta} = \frac{\kappa}{\delta}v_x = S_\kappa v_x = \frac{C_\delta D_\beta - C_\beta D_\delta}{D_r C_\beta - C_r D_\beta + m v_x D_\beta} \tag{10.269}$$

4. *Centripetal acceleration response, S_a*

$$S_a = \frac{v_x^2/R}{\delta} = \frac{\kappa}{\delta}v_x^2 = S_\kappa v_x^2 = \frac{\left(C_\delta D_\beta - C_\beta D_\delta\right)v_x}{D_r C_\beta - C_r D_\beta + m v_x D_\beta} \tag{10.270}$$

5. *Lateral velocity response, S_y.*

$$S_y = \frac{v_y}{\delta} = S_\beta v_x = \frac{D_\delta \left(C_r - mv_x\right) - D_r C_\delta}{D_r C_\beta - C_r D_\beta + mv_x D_\beta} v_x \qquad (10.271)$$

Proof. In steady-state conditions, all the variables are constant, and hence, their derivatives are zero. Therefore, the equations of motion (10.202)–(10.204) reduce to

$$F_x = -mr\, v_y \qquad (10.272)$$
$$F_y = mr\, v_x \qquad (10.273)$$
$$M_z = 0 \qquad (10.274)$$

where the lateral force F_y and yaw moment M_z from (10.142) and (10.143) are

$$F_y = C_r r + C_\beta \beta + C_\delta \delta \qquad (10.275)$$
$$M_z = D_r r + D_\beta \beta + D_\delta \delta \qquad (10.276)$$

Therefore, the equations describing the steady-state turning of a two-wheel rigid vehicle are

$$F_x = -mr\, v_y \qquad (10.277)$$
$$C_r r + C_\beta \beta + C_\delta \delta = mr\, v_x \qquad (10.278)$$
$$D_r r + D_\beta \beta + D_\delta \delta = 0 \qquad (10.279)$$

At steady-state turning, the mass center of the vehicle moves on a circle with radius R at a forward speed v_x and angular velocity r, so v_x and r are approximately related by

$$v_x \approx R\, r \qquad (10.280)$$

Substituting (10.280) in Equations (10.277)-(10.279), we may write the equations as (10.263)-(10.265). Equation (10.263) may be used to calculate the required traction force to keep the motion steady. However, Equations (10.264) and (10.265) can be used to determine the steady-state responses of the vehicle. We use the curvature definition (10.266) and write the equations in matrix form

$$\begin{bmatrix} C_\beta & C_r v_x - m v_x^2 \\ D_\beta & D_r v_x \end{bmatrix} \begin{bmatrix} \beta \\ \kappa \end{bmatrix} = \begin{bmatrix} -C_\delta \\ -D_\delta \end{bmatrix} \delta \qquad (10.281)$$

Solving the equations for β and κ shows that

$$\begin{bmatrix} \beta \\ \kappa \end{bmatrix} = \begin{bmatrix} C_\beta & C_r v_x - m v_x^2 \\ D_\beta & D_r v_x \end{bmatrix}^{-1} \begin{bmatrix} -C_\delta \\ -D_\delta \end{bmatrix} \delta$$

$$= \begin{bmatrix} \dfrac{D_\delta \left(C_r - mv_x\right) - D_r C_\delta}{D_r C_\beta - C_r D_\beta + mv_x D_\beta} \\[3ex] \dfrac{C_\delta D_\beta - C_\beta D_\delta}{v_x \left(D_r C_\beta - C_r D_\beta + mv_x D_\beta\right)} \end{bmatrix} \delta \qquad (10.282)$$

Based on the solutions of (10.282) and by using Equation (10.280), we are able to define different output-input relationships as (10.267)-(10.270).
    Using

$$\beta = \frac{v_y}{v_x} \tag{10.283}$$

Equations (10.278)-(10.279) will be

$$C_r \, r + \frac{C_\beta}{v_x} v_y + C_\delta \, \delta \;=\; m r \, v_x \tag{10.284}$$

$$D_r \, r + \frac{D_\beta}{v_x} v_y + D_\delta \, \delta \;=\; 0 \tag{10.285}$$

rearrangement in matrix form

$$\begin{bmatrix} \dfrac{C_\beta}{v_x} & C_r \, v_x - m \, v_x^2 \\[2mm] \dfrac{D_\beta}{v_x} & D_r \, v_x \end{bmatrix} \begin{bmatrix} v_y \\ \kappa \end{bmatrix} = \begin{bmatrix} -C_\delta \\ -D_\delta \end{bmatrix} \delta \tag{10.286}$$

yields:

$$\begin{bmatrix} v_y \\ \kappa \end{bmatrix} = \begin{bmatrix} \dfrac{D_\delta \left(C_r - m v_x \right) - D_r C_\delta}{D_r C_\beta - C_r D_\beta + m v_x D_\beta} v_x \\[4mm] \dfrac{C_\delta D_\beta - C_\beta D_\delta}{v_x \left(D_r C_\beta - C_r D_\beta + m v_x D_\beta \right)} \end{bmatrix} \delta \tag{10.287}$$

Then the lateral velocity response S_y in (10.271) would be an outcome of this result. ∎

Example 427 *Force system coefficients for a car.*
    Consider a front-wheel-steering, four-wheel-car with the following characteristics.

$$\begin{aligned} C_{\alpha f_L} &= C_{\alpha f_R} = 500 \,\mathrm{N/\,deg} \approx 28648 \,\mathrm{N/\,rad} \\ &\approx 112.4 \,\mathrm{lb/\,deg} \approx 6440 \,\mathrm{lb/\,rad} \end{aligned} \tag{10.288}$$

$$\begin{aligned} C_{\alpha r_L} &= C_{\alpha r_R} = 460 \,\mathrm{N/\,deg} \approx 26356 \,\mathrm{N/\,rad} \\ &\approx 103.4 \,\mathrm{lb/\,deg} \approx 5924.4 \,\mathrm{lb/\,rad} \end{aligned} \tag{10.289}$$

$$\begin{aligned} mg &= 9000 \,\mathrm{N} \approx 2023 \,\mathrm{lb} \qquad m = 917 \,\mathrm{kg} \approx 62.8 \,\mathrm{slug} \\ I_z &= 1128 \,\mathrm{kg\, m^2} \approx 832 \,\mathrm{slug\, ft^2} \\ a_1 &= 91 \,\mathrm{cm} \approx 2.98 \,\mathrm{ft} \qquad a_2 = 164 \,\mathrm{cm} \approx 5.38 \,\mathrm{ft} \end{aligned} \tag{10.290}$$

The sideslip coefficient of an equivalent bicycle model are

$$C_{\alpha f} = C_{\alpha f_L} + C_{\alpha f_R} = 57296 \,\mathrm{N/\,rad} \tag{10.291}$$

$$C_{\alpha r} = C_{\alpha r_L} + C_{\alpha r_R} = 52712 \,\mathrm{N/\,rad} \tag{10.292}$$

The force system coefficients C_r, C_β, C_δ, D_r, D_β, D_δ are then equal to the following in which v_x is measured in $[\mathrm{m/s}]$.

$$C_r = -\frac{a_1}{v_x}C_{\alpha f} + \frac{a_2}{v_x}C_{\alpha r} = \frac{34308}{v_x}\,\mathrm{N\,s/rad} \tag{10.293}$$

$$C_\beta = -(C_{\alpha f} + C_{\alpha r}) = -110008\,\mathrm{N/rad} \tag{10.294}$$

$$C_\delta = C_{\alpha f} = 57296\,\mathrm{N/rad} \tag{10.295}$$

$$D_r = -\frac{a_1^2}{v_x}C_{\alpha f} - \frac{a_2^2}{v_x}C_{\alpha r} = -\frac{189221}{v_x}\,\mathrm{N\,m\,s/rad} \tag{10.296}$$

$$D_\beta = -(a_1 C_{\alpha f} - a_2 C_{\alpha r}) = 34308\,\mathrm{N\,m/rad} \tag{10.297}$$

$$D_\delta = a_1 C_{\alpha f} = 52139\,\mathrm{N\,m/rad} \tag{10.298}$$

The coefficients C_r and D_r are the only force coefficients that are functions of the forward speed v_x. As an example C_r and D_r at

$$v_x = 10\,\mathrm{m/s} = 36\,\mathrm{km/h} \approx 32.81\,\mathrm{ft/s} \approx 22.37\,\mathrm{mi/h} \tag{10.299}$$

are

$$C_r = 3430.8\,\mathrm{N\,s/rad} \qquad D_r = -18922\,\mathrm{N\,m\,s/rad} \tag{10.300}$$

and at

$$v_x = 30\,\mathrm{m/s} = 108\,\mathrm{km/h} \approx 98.43\,\mathrm{ft/s} \approx 67.11\,\mathrm{mi/h} \tag{10.301}$$

are

$$C_r = 1143.6\,\mathrm{N\,s/rad} \qquad D_r = -6307.3\,\mathrm{N\,m\,s/rad} \tag{10.302}$$

Example 428 *Steady state responses of an understeer vehicle.*

The steady-state responses are function of the forward velocity of a vehicle. To visualize how these steady-state parameters vary when the speed of a vehicle changes, we calculate S_κ, S_β, S_r, S_a, S_y for a vehicle with the following characteristics.

$$C_{\alpha f_L} = C_{\alpha f_R} \approx 3000\,\mathrm{N/rad} \qquad C_{\alpha r_L} = C_{\alpha r_R} \approx 3000\,\mathrm{N/rad} \tag{10.303}$$

$$m = 1000\,\mathrm{kg} \qquad I_z = 1650\,\mathrm{kg\,m^2} \tag{10.304}$$

$$a_1 = 1.0\,\mathrm{m} \qquad a_2 = 1.5\,\mathrm{m} \tag{10.305}$$

$$K = \frac{m}{l^2}\left(\frac{a_2}{C_{\alpha f}} - \frac{a_1}{C_{\alpha r}}\right) = 1.33 \times 10^{-2} \tag{10.306}$$

$$S_\kappa = \frac{30}{75 + v_x^2} \qquad S_\beta = \frac{45 - 2v_x^2}{75 + v_x^2} \qquad S_r = S_\kappa v_x = \frac{30v_x}{75 + v_x^2}$$

$$S_a = S_\kappa v_x^2 = \frac{30v_x^2}{75 + v_x^2} \qquad\qquad S_y = \frac{45 - 2v_x^2}{75 + v_x^2}v_x \tag{10.307}$$

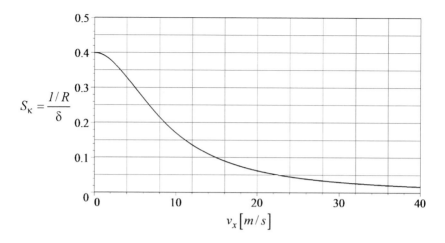

FIGURE 10.15. Curvature response, S_κ, as a function of forward velocity v_x.

The parameter K is called the stability factor and determines if a vehicle is understeer or oversteer. A positive K indicates a stable understeer vehicle and considered desirable. Figures 10.15-10.19 illustrate the steady-states variations by increasing the forward velocity.

At very slow speed, $v_x \approx 0$, the steady state responses are at

$$S_\kappa = 0.4 \qquad S_\beta = 0.6 \qquad S_r = 0 \qquad S_a = 0 \qquad (10.308)$$

By increasing the speed of the car, S_κ decreases and approaches zero, which means the radius of rotation increases for constant δ.

The value of S_β also decreases by increasing v_x and approaches a constant negative value, $S_\beta \to -2$. Therefore, the center of rotation starts at a point on the rear axis when $\beta > 0$, and moves away from the car and moves forward where $\beta < 0$.

S_r has a strange behavior which begins from zero and increases with v_x for a while to a maximum at $v_x = \sqrt{75}$. Then it decreases and asymptotically approaches zero.

S_a shows the centripetal acceleration of the car while it is turning on a circle with a constant speed. The value of S_a starts with zero at $v_x = 0$ and increases rapidly to approach $S_a \to 30$ asymptotically when $v_x \to \infty$.

The lateral velocity response S_y indicates the speed of the vehicle at its mass center in the y-direction. It starts at zero and increases to a maximum before decreasing monotonically to negative values.

Example 429 *Where is S_κ before steady state?*

For a given vehicle, the geometric parameters a_1, a_2, m, $C_{\alpha f}$, $C_{\alpha r}$ are set. When the steer angle δ and forward velocity v_x are constant, it take some time for the car to achieve the steady state value of S_κ. Now let us assume δ remains unchanged and, we change v_x to a new constant value. The vehicle

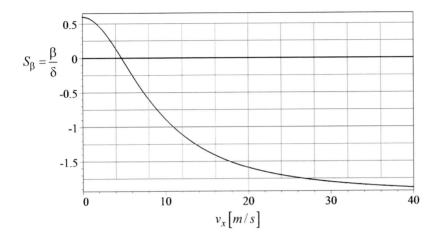

FIGURE 10.16. Sideslip response, S_β, as a function of forward velocity v_x.

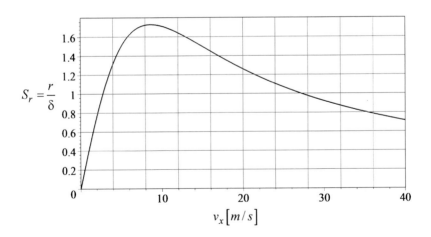

FIGURE 10.17. Yaw rate response, S_r, as a function of forward velocity v_x.

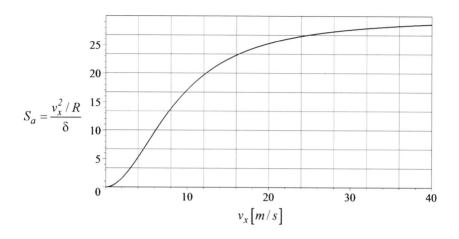

FIGURE 10.18. Lateral acceleration response, S_a, as a function of forward velocity v_x.

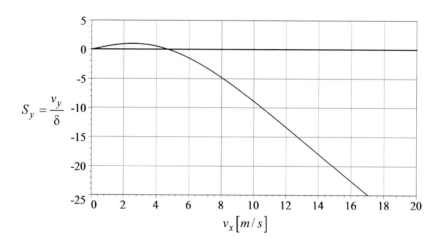

FIGURE 10.19. Lateral velocity response, S_y, as a function of forward velocity v_x.

will move from the previous point on S_κ to a new value. However, moving from one point to the other point will not be done by following the S_κ-curve, as this curve is valid only at steady-state conditions, while moving between two steady-state points is a transition.

When a set of parameters and variables is fixed, approached from above or below of the S_κ-curve to the set point on S_κ depends the initial conditions of the vehicle.

Example 430 *Curvature response behavior S_κ.*

Substituting the force coefficients (10.144)-(10.149) in the curvature response, S_κ

$$S_\kappa = \frac{\kappa}{\delta} = \frac{1}{R\delta} = \frac{C_\delta D_\beta - C_\beta D_\delta}{v_x\,(D_r C_\beta - D_\beta\,(C_r - mv_x))} \tag{10.309}$$

yields:

$$S_\kappa = \frac{lC_{\alpha f}C_{\alpha r}}{m\,(C_{\alpha r}a_2 - C_{\alpha f}a_1)\,v_x^2 + l^2 C_{\alpha f}C_{\alpha r}} \tag{10.310}$$

For a given set of parameters a_1, a_2, m, $C_{\alpha f}$, $C_{\alpha r}$, the curvature response begins at

$$\lim_{v_x=0} S_\kappa = \frac{1}{a_1 + a_2} = \frac{1}{l} \tag{10.311}$$

and ends up at

$$\lim_{v_x=\infty} S_\kappa = 0 \tag{10.312}$$

If $C_{\alpha f} = C_{\alpha r} = C_\alpha$ and $a_1 = a_2 = l/2$ then S_κ simplifies to be constant and independent of v_x.

$$S_\kappa = \frac{1}{l} \tag{10.313}$$

and therefore, the radius of rotation is also constant and velocity independent.

$$R = \frac{l}{\delta} \tag{10.314}$$

If $C_{\alpha f} = C_{\alpha r} = C_\alpha$ and $a_1 \neq a_2$ then S_κ simplifies to

$$S_\kappa = \frac{lC_\alpha}{l^2 C_\alpha + mv_x^2\,(a_2 - a_1)} \tag{10.315}$$

and therefore, the radius of rotation is also velocity dependent

$$R = \frac{l^2 C_\alpha + mv_x^2\,(a_2 - a_1)}{lC_\alpha \delta} \tag{10.316}$$

At a given speed, depending on the position of the mass center, R can be between two extreme values when $a_1 = 0$ and $a_2 = 0$

$$\frac{lC_\alpha + mv_x^2}{lC_\alpha} < \frac{R\delta}{l} < \frac{lC_\alpha - mv_x^2}{lC_\alpha} \tag{10.317}$$

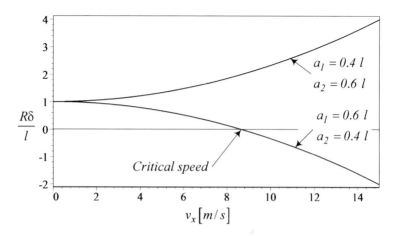

FIGURE 10.20. The limits of steady state $R\delta/l$ for a vehicle and the existence of a critical speed for $a_1 < a_2$.

In practical conditions, we have $0.4l < a_1 < 0.6l$ and $0.6l > a_2 > 0.4l$. Figure 10.20 illustrates the limits of R depending on the mass center position, for the following characteristics.

$$C_{\alpha f} = 6000\,\mathrm{N/rad} \qquad C_{\alpha r} = 6000\,\mathrm{N/rad} \qquad (10.318)$$
$$m = 1000\,\mathrm{kg} \qquad I_z = 1650\,\mathrm{kg\,m^2} \qquad (10.319)$$
$$a_1 = 1.0\,\mathrm{m} \qquad a_2 = 1.5\,\mathrm{m} \qquad (10.320)$$

When $a_1 < a_2$ and the front of the car is heavier than the back, the steady-state R increases monotonically by increasing v_x. However, when $a_1 > a_2$ and the back of the car is heavier than the front then, R decreases monotonically by increasing v_x starting from a positive value. Therefore, for a heavier back and positive δ, a car begins to turn to left at low speed. By increasing v_x, the steady-state value of R decreases until it becomes theoretically zero for a critical speed v_c, at which the vehicle would be unstable and uncontrollable.

$$v_c = \sqrt{\frac{C_\alpha}{m\,(a_1 - a_2)}}\; l \qquad a_1 > a_2 \qquad (10.321)$$

Example 431 *Sideslip response behavior S_β.*

Substituting the force coefficients (10.144)-(10.149) in the sideslip response, S_β

$$S_\beta = \frac{\beta}{\delta} = \frac{D_\delta\,(C_r - mv_x) - D_r C_\delta}{D_r C_\beta - C_r D_\beta + mv_x D_\beta} \qquad (10.322)$$

yields:

$$S_\beta = \frac{a_2 l C_{\alpha f} C_{\alpha r} - m a_1 v_x^2 C_{\alpha f}}{m\,(C_{\alpha r} a_2 - C_{\alpha f} a_1)\, v_x^2 + l^2 C_{\alpha f} C_{\alpha r}} \qquad (10.323)$$

For a given set of parameters a_1, a_2, m, $C_{\alpha f}$, $C_{\alpha r}$, the sideslip response begins at

$$\lim_{v_x=0} S_\beta = \frac{a_2}{a_1 + a_2} = \frac{a_2}{l} \tag{10.324}$$

and ends up at

$$\lim_{v_x=\infty} S_\beta = \frac{a_1 C_{\alpha f}}{a_1 C_{\alpha f} - a_2 C_{\alpha r}} \tag{10.325}$$

If $C_{\alpha f} = C_{\alpha r} = C_\alpha$ and $a_1 = a_2 = l/2$ then S_β simplifies to

$$S_\beta = \frac{l C_\alpha - m v_x^2}{2 l C_\alpha} \tag{10.326}$$

If $C_{\alpha f} = C_{\alpha r} = C_\alpha$ and $a_1 \neq a_2$ then S_β simplifies to

$$S_\beta = \frac{a_2 l C_\alpha - m a_1 v_x^2}{m (a_2 - a_1) v_x^2 + l^2 C_\alpha} \tag{10.327}$$

S_β is proportional to β which indicates the angle of the vehicle velocity vector measured from the body longitudinal x-axis. At very low speed, $\beta > 0$ and the front wheel is turning on a bigger circle than the rear. By increasing v_x, the vehicle sideslip angle decreases until it becomes zero at a critical speed

$$v_c = \sqrt{\frac{a_2 l C_{\alpha r}}{m a_1}} \tag{10.328}$$

When $v_x = v_c$, the vehicle is turning such that the x-axis is perpendicular to the radius of rotation R at the mass center. When $v_x > v_c$, then $\beta < 0$ and the rear wheel is turning on a bigger circle than the front.

Example 432 *Yaw rate response behavior S_r.*

Substituting the force coefficients (10.144)-(10.149) in the yaw rate response, S_r

$$S_r = \frac{r}{\delta} = \frac{\kappa}{\delta} v_x = S_\kappa v_x = \frac{C_\delta D_\beta - C_\beta D_\delta}{D_r C_\beta - C_r D_\beta + m v_x D_\beta} \tag{10.329}$$

yields:

$$S_r = \frac{v_x l C_{\alpha f} C_{\alpha r}}{m (C_{\alpha r} a_2 - C_{\alpha f} a_1) v_x^2 + l^2 C_{\alpha f} C_{\alpha r}} \tag{10.330}$$

For a given set of parameters a_1, a_2, m, $C_{\alpha f}$, $C_{\alpha r}$, the yaw rate response begins at

$$\lim_{v_x=0} S_r = 0 \qquad \frac{d S_r}{d x_x} > 0 \tag{10.331}$$

and ends up at

$$\lim_{v_x=\infty} S_r = 0 \qquad \frac{d S_r}{d x_x} < 0 \tag{10.332}$$

Therefore, there must be at least a maximum for S_r at a critical speed. Taking a derivative

$$\frac{d S_r}{d x_x} = l C_{fa} C_{ra} \frac{l^2 C_{fa} C_{ra} - m \left(C_{ar} a_2 - C_{af} a_1 \right) v_x^2}{\left(l^2 C_{fa} C_{ra} + m \left(C_{ar} a_2 - C_{af} a_1 \right) v_x^2 \right)^2} \tag{10.333}$$

determines that S_r is at maximum when v_x is at

$$v_c = \sqrt{\frac{l^2 C_{af} C_{ar}}{m \left(C_{ar} a_2 - C_{af} a_1 \right)}} \tag{10.334}$$

Therefore, the maximum yaw rate of a vehicle at $v_x = v_c$ is

$$r = \frac{\delta}{2} \sqrt{\frac{C_{af} C_{ar}}{m \left(C_{ar} a_2 - C_{af} a_1 \right)}} \tag{10.335}$$

If $C_{af} = C_{ar} = C_\alpha$ and $a_1 = a_2 = l/2$ then S_r simplifies to

$$S_r = \frac{v_x}{l} \tag{10.336}$$

If $C_{af} = C_{ar} = C_\alpha$ and $a_1 \neq a_2$ then S_r simplifies to

$$S_r = \frac{l v_x C_\alpha}{l^2 C_\alpha + m \left(a_2 - a_1 \right) v_x^2} \tag{10.337}$$

Example 433 *Centripetal acceleration response behavior S_a.*
Examining Figure 10.21, we know that

$$\begin{aligned} v_x &= v_f \cos \beta_f = v \cos \beta = v_r \cos \alpha_r & (10.338) \\ R_1 &= R \cos \beta & (10.339) \end{aligned}$$

Therefore, the centripetal acceleration of the vehicle is v^2/R or equivalently v_x^2/R_1 is not proportional to the centripetal acceleration response S_a.

$$\frac{v^2}{R} = \frac{v_x^2}{R \cos^2 \beta} = \frac{S_a \delta}{\cos^2 \beta} \tag{10.340}$$

However, considering small β indicates that

$$S_a = \frac{1}{\delta} \frac{v_x^2}{R} = \frac{1}{\delta} \frac{v^2 \cos^2 \beta}{R} \approx \frac{1}{\delta} \frac{v^2}{R} \tag{10.341}$$

Substituting the force coefficients (10.144)-(10.149) in the centripetal acceleration response, S_a

$$S_a = \frac{v_x^2/R}{\delta} = \frac{\kappa}{\delta} v_x^2 = S_\kappa v_x^2 = \frac{\left(C_\delta D_\beta - C_\beta D_\delta \right) v_x}{D_r C_\beta - C_r D_\beta + m v_x D_\beta} \tag{10.342}$$

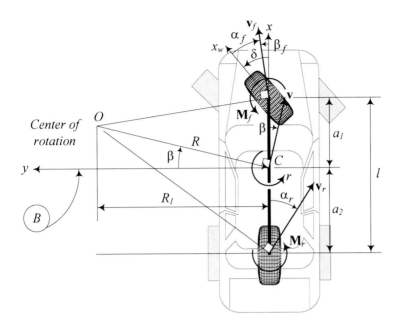

FIGURE 10.21. Kinematics of a moving vehicle at steady-state conditions.

yields:

$$S_a = \frac{v_x^2 l C_{\alpha f} C_{\alpha r}}{m \left(C_{\alpha r} a_2 - C_{\alpha f} a_1 \right) v_x^2 + l^2 C_{\alpha f} C_{\alpha r}} \qquad (10.343)$$

For a given set of parameters a_1, a_2, m, $C_{\alpha f}$, $C_{\alpha r}$, *the centripetal acceleration response begins at*

$$\lim_{v_x = 0} S_a = 0 \qquad (10.344)$$

and ends up at

$$\lim_{v_x = \infty} S_a = \frac{l C_{\alpha f} C_{\alpha r}}{m \left(C_{\alpha r} a_2 - C_{\alpha f} a_1 \right)} \qquad (10.345)$$

If $C_{\alpha f} = C_{\alpha r} = C_\alpha$ *and* $a_1 = a_2 = l/2$ *then* S_a *simplifies to*

$$S_a = \frac{v_x^2}{l} \qquad (10.346)$$

If $C_{\alpha f} = C_{\alpha r} = C_\alpha$ *and* $a_1 \neq a_2$ *then* S_a *simplifies to*

$$S_a = \frac{l v_x^2 C_\alpha}{l^2 C_\alpha + m \left(a_2 - a_1 \right) v_x^2} \qquad (10.347)$$

The lateral acceleration response $S_a = \left(v_x^2 / R \right) / \delta$ *is actually the centripetal acceleration of the vehicle of its mass center. It is the exact lateral acceleration only when* $\beta = 0$ *and* R *is perpendicular to the x-axis. It is the only* *time that the centripetal acceleration has no longitudinal component. However, assuming a vehicle as a massive point turning about a central point with radius R justifies to use the term of lateral acceleration.*

Example 434 *Lateral velocity response behavior S_y.*

Substituting the force coefficients (10.144)-(10.149) in the lateral velocity response, S_y

$$S_y = \frac{v_y}{\delta} = S_\beta v_x = \frac{D_\delta (C_r - mv_x) - D_r C_\delta}{D_r C_\beta - C_r D_\beta + mv_x D_\beta} v_x \qquad (10.348)$$

yields:

$$S_y = \frac{a_2 l C_{\alpha f} C_{\alpha r} - ma_1 v_x^2 C_{\alpha f}}{m (C_{\alpha r} a_2 - C_{\alpha f} a_1) v_x^2 + l^2 C_{\alpha f} C_{\alpha r}} v_x \qquad (10.349)$$

For a given set of parameters a_1, a_2, m, $C_{\alpha f}$, $C_{\alpha r}$, the lateral velocity response begins at

$$\lim_{v_x=0} S_y = 0 \qquad (10.350)$$

and ends up at

$$\lim_{v_x=\infty} S_y = \begin{cases} -\infty & a_1 C_{f\alpha} - a_2 C_{r\alpha} < 0 \\ \infty & a_1 C_{f\alpha} - a_2 C_{r\alpha} > 0 \end{cases} \qquad (10.351)$$

If $C_{\alpha f} = C_{\alpha r} = C_\alpha$ and $a_1 = a_2 = l/2$ then S_y simplifies to

$$S_y = \frac{l C_\alpha - mv_x^2}{2 l C_\alpha} v_x \qquad (10.352)$$

If $C_{\alpha f} = C_{\alpha r} = C_\alpha$ and $a_1 \neq a_2$ then S_y simplifies to

$$S_y = \frac{a_2 l C_\alpha - ma_1 v_x^2}{m (a_2 - a_1) v_x^2 + l^2 C_\alpha} v_x \qquad (10.353)$$

S_y is proportional to β which indicates the angle of the vehicle velocity vector measured from the body longitudinal x-axis. At very low speed, $\beta > 0$ and the front wheel is turning on a bigger circle than the rear to make v_y positive. By increasing v_x, the vehicle sideslip angle decreases until it becomes zero at a critical speed at which the lateral velocity also becomes zero.

$$v_c = \sqrt{\frac{a_2 l C_{\alpha r}}{ma_1}} \qquad (10.354)$$

When $v_x = v_c$, the vehicle is turning such that the x-axis is perpendicular to the radius of rotation R at the mass center. Therefore, the lateral component of the velocity vector at the mass center becomes zero. When $v_x > v_c$, then $\beta < 0$ and the rear wheel is turning on a bigger circle than the front. At this moment, the velocity vector of C has a negative y-component.

As a result, v_y is not a good indicator to show if the vehicle is turning in y-direction or not. v_y is only the y-component of the vehicle velocity vector $\mathbf{v}$ at the mass center C. The velocity vector, and therefore, v_y, at other points of the longitudinal axis of the vehicle are different. The common property of $\mathbf{v}$ at any point of the x-axis is that all of them must have the same v_x. The best indicator of the rotation of a vehicle is the yaw rate r, which shows how fast and in what direction the vehicle is turning.

Example 435 *Steady-state center of rotation.*

Having the steady-state responses $S_k = 1/R/\delta$ and $S_\beta = \beta/\delta$, we are able to determine the position of the rotation center of a vehicle in the body coordinate frame. The coordinate of the center of rotation (x_O, y_O) with respect to the mass center are

$$x_O = -R\sin\beta = -\frac{1}{S_k\delta}\sin(S_\beta\delta) \tag{10.355}$$

$$y_O = R\cos\beta = \frac{1}{S_k\delta}\cos(S_\beta\delta) \tag{10.356}$$

Consider a vehicle with the following characteristics

$$C_{\alpha f_L} = C_{\alpha f_R} \approx 3000\,\mathrm{N/\,rad} \qquad C_{\alpha r_L} = C_{\alpha r_R} \approx 3000\,\mathrm{N/\,rad} \tag{10.357}$$

$$m = 1000\,\mathrm{kg} \qquad I_z = 1650\,\mathrm{kg\,m^2} \tag{10.358}$$

$$a_1 = 1.0\,\mathrm{m} \qquad a_2 = 1.5\,\mathrm{m} \tag{10.359}$$

$$K = \frac{m}{l^2}\left(\frac{a_2}{C_{\alpha f}} - \frac{a_1}{C_{\alpha r}}\right) = 1.33 \times 10^{-2} \tag{10.360}$$

$$S_\kappa = \frac{30}{75 + v_x^2} \qquad S_\beta = \frac{45 - 2v_x^2}{75 + v_x^2} \tag{10.361}$$

The variation of x_O, y_O when v_x changes are shown in Figure 10.22.

$$x_O = -R\sin\beta = -\frac{75 + v_x^2}{30\delta}\sin\left(\frac{45 - 2v_x^2}{75 + v_x^2}\delta\right) \tag{10.362}$$

$$y_O = R\cos\beta = \frac{75 + v_x^2}{30\delta}\cos\left(\frac{45 - 2v_x^2}{75 + v_x^2}\delta\right) \tag{10.363}$$

x_O starts from $x_O = -a_2$ on the real axel at zero velocity and moves forward. y_O starts from $y_O = l\cot\delta$ on the rear axel at zero velocity and moves away from the vehicle.

Figure 10.23 illustrates the loci of the steady-state rotation center at different speeds and a constant steer angle for the given vehicle. For this understeer vehicle, by increasing the forward speed, the rotation center moves away and forward with respect to the vehicle. This example shows that kinematic steering condition determine the rotation center only when the speed

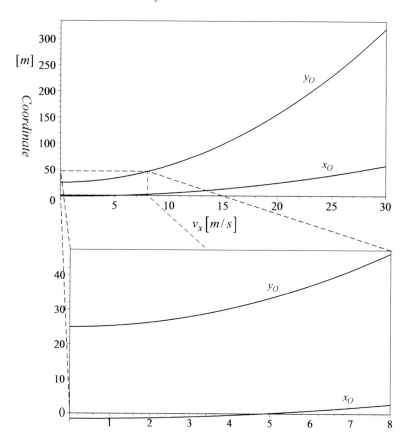

FIGURE 10.22. The coordinate of the center of rotation (x_O, y_O) with respect to the mass center of a vehicle for different speed.

of the care is zero. By increasing the speed of, the actual steady state rotation center will be determined by dynamic parameters of the vehicle and not the kinematic steering condition.

Figure 10.24 illustrates three steady-state conditions of the vehicle with the circular path of motions and velocity vectors. The three conditions are for v_x to be lower, equal, and higher than the critical speed at which $\beta = 0$.

Example 436 *Under steering, over steering, neutral steering.*
    Curvature response S_κ indicates how the radius of turning will change

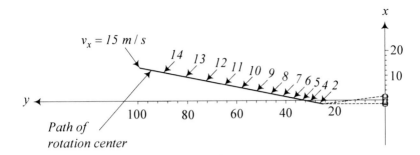

FIGURE 10.23. The loci of the steady-state rotation center at different speeds and a constant steer angle for a given vehicle.

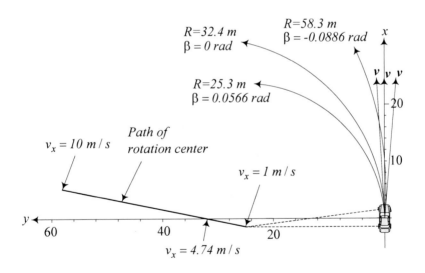

FIGURE 10.24. The three steady-state conditions of a vehicle for v_x to be lower, equal, and higher than the critical speed at which $\beta = 0$.

with a change in steer angle. S_κ can be expressed as

$$S_\kappa = \frac{\kappa}{\delta} = \frac{1/R}{\delta} = \frac{1}{l}\frac{1}{1 + Kv_x^2} \tag{10.364}$$

$$K = \frac{m}{l^2}\left(\frac{a_2}{C_{\alpha f}} - \frac{a_1}{C_{\alpha r}}\right) \tag{10.365}$$

*where K is called the **stability factor**. It determines that the vehicle is*

$$
\begin{array}{lll}
Understeer & if & K > 0 \\
Neutral & if & K = 0 \\
Oversteer & if & K < 0
\end{array} \tag{10.366}
$$

To find K we may rewrite S_κ as

$$
\begin{aligned}
S_\kappa &= \frac{\kappa}{\delta} = \frac{1/R}{\delta} = \frac{C_\delta D_\beta - C_\beta D_\delta}{v_x\left(D_r C_\beta - C_r D_\beta + mv_x D_\beta\right)} \\
&= \frac{1}{v_x\left(\dfrac{D_r C_\beta - C_r D_\beta}{C_\delta D_\beta - C_\beta D_\delta} + \dfrac{mv_x D_\beta}{C_\delta D_\beta - C_\beta D_\delta}\right)} \\
&= \frac{1}{l + \dfrac{mv_x^2 D_\beta}{C_\delta D_\beta - C_\beta D_\delta}} = \frac{1}{l}\frac{1}{1 + \dfrac{m}{l}\dfrac{D_\beta}{C_\delta D_\beta - C_\beta D_\delta}v_x^2} \\
&= \frac{1}{l}\frac{1}{1 + Kv_x^2}
\end{aligned} \tag{10.367}
$$

Therefore,

$$K = \frac{m}{l}\frac{D_\beta}{C_\delta D_\beta - C_\beta D_\delta} \tag{10.368}$$

which, after substituting the force system coefficients (10.144)-(10.149) will be equal to

$$K = \frac{m}{l^2}\left(\frac{a_2}{C_{\alpha f}} - \frac{a_1}{C_{\alpha r}}\right) \tag{10.369}$$

The sign of stability factor K determines if S_κ is an increasing or decreasing function of velocity v_x. The sign of K depends on the relative weight of $a_2/C_{\alpha f}$ and $a_1/C_{\alpha r}$, which are dependent on the position of mass center a_1, a_2, and sideslip coefficients of the front and rear tires $C_{\alpha f}$, $C_{\alpha r}$.
If $K > 0$ then

$$\frac{a_2}{C_{\alpha f}} > \frac{a_1}{C_{\alpha r}} \tag{10.370}$$

and $S_\kappa = \kappa/\delta$ and $dS_\kappa/dv_x < 0$. Hence, the curvature of the path $\kappa = 1/R$ decreases with speed for a constant δ. Decreasing κ indicates that the radius of the steady-state circle, R, increases by increasing speed v_x. A positive stability factor is desirable. A vehicle with $K > 0$ is stable and is called

understeer. For an understeer vehicle, we need to increase the steering angle if we increase the speed of the vehicle to keep the same turning circle.

If $K < 0$ then

$$\frac{a_2}{C_{\alpha f}} < \frac{a_1}{C_{\alpha r}} \qquad (10.371)$$

and $S_\kappa = \kappa/\delta$ and $dS_\kappa/dv_x > 0$. Hence, the curvature of the path $\kappa = 1/R$ increases with speed for a constant δ. Increasing κ indicates that the radius of the steady-state circle, R, decreases by increasing speed v_x. A negative stability factor is undesirable. A vehicle with $K < 0$ is unstable and is called **oversteer.** For an oversteer vehicle, we need to decrease the steering angle when we increase the speed of the vehicle, to keep the same turning circle.

If $K = 0$ then

$$\frac{a_2}{C_{\alpha f}} = \frac{a_1}{C_{\alpha r}} \qquad (10.372)$$

then $S_\kappa = \kappa/\delta$ is not a function of v_x because $dS_\kappa/dv_x = 0$. Hence, the curvature of the path, $\kappa = 1/R$ remains constant for a constant δ regardless of v_x. Having a constant κ indicates that the radius of the steady-state circle, R, will not change by changing the speed v_x. A zero stability factor is neutral and a vehicle with $K = 0$ is on the border of stability and is called a **neutral steer.** When driving a neutral steer vehicle, we do not need to change the steering angle if we increase or decrease the speed of the vehicle, to keep the same turning circle.

As an example, consider a car with the following characteristics:

$$
\begin{aligned}
C_{\alpha f} &= \ 57296 \,\text{N/rad} & C_{\alpha r} &= 52712 \,\text{N/rad} & (10.373) \\
m &= \ 917 \,\text{kg} \approx 62.8 \,\text{slug} & & & (10.374) \\
a_2 &= \ 164 \,\text{cm} \approx 5.38 \,\text{ft} & a_1 &= 91 \,\text{cm} \approx 2.98 \,\text{ft} & (10.375)
\end{aligned}
$$

This car has a stability factor K and a curvature response S_κ equal to

$$K = \frac{m}{l^2}\left(\frac{a_2}{C_{\alpha f}} - \frac{a_1}{C_{\alpha r}}\right) = 1.602 \times 10^{-3} \qquad (10.376)$$

$$S_\kappa = \frac{1}{l}\frac{1}{1 + Kv_x^2} = \frac{0.39216}{1 + 1.602 \times 10^{-3} v_x^2} \qquad (10.377)$$

Now assume we fill the trunk and change the car's characteristics to a new set.

$$
\begin{aligned}
m &= \ 1400 \,\text{kg} \approx 95.9 \,\text{slug} & & & (10.378) \\
a_1 &= \ 125 \,\text{cm} \approx 4.1 \,\text{ft} & a_2 &= 130 \,\text{cm} \approx 4.26 \,\text{ft} & (10.379)
\end{aligned}
$$

The new stability factor K and curvature response S_κ are

$$K = -2.21 \times 10^{-4} \qquad (10.380)$$

$$S_\kappa = -\frac{0.392\,16}{2.21 \times 10^{-4} v_x^2 - 1} \qquad (10.381)$$

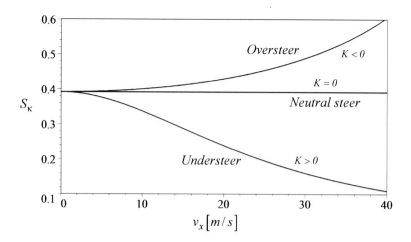

FIGURE 10.25. Comarison of the curvature response S_κ for a car with $K = 1.602 \times 10^{-3}$, $K = -2.21 \times 10^{-4}$, and $K = 0$.

Figure 10.25 compares the curvature response S_κ for the two situations and a neutral steering. We assumed that increasing weight did not change the tire characteristics, and we kept the same sideslip coefficients.

Example 437 *Critical speed v_c.*

If $K < 0$ then S_κ increases by increasing v_x. Therefore, the steering angle must be decreased to maintain a constant radius path. At a critical speed v_c

$$v_c = \sqrt{-\frac{1}{K}} \tag{10.382}$$

the denominator of S_κ becomes zero and then

$$S_\kappa \to \infty \tag{10.383}$$

At the critical speed, any decrease in steering angle cannot keep the path. When $v_x = v_c$, the curvature κ is not a function of steering angle δ, and any radius of rotation is theoretically possible for a constant δ. The critical speed makes the system unstable. Controlling an oversteer vehicle gets harder by $v_x \to v_c$ and becomes uncontrollable when $v_x = v_c$.

The critical speed of an oversteer car with characteristics

$$
\begin{aligned}
C_{\alpha f} &= 57296 \,\text{N/rad} & C_{\alpha r} &= 52712 \,\text{N/rad} & (10.384) \\
m &= 1400 \,\text{kg} \approx 95.9 \,\text{slug} & & & (10.385) \\
a_1 &= 125 \,\text{cm} \approx 4.1 \,\text{ft} & a_2 &= 130 \,\text{cm} \approx 4.26 \,\text{ft} & (10.386)
\end{aligned}
$$

is

$$v_c = \sqrt{-\frac{1}{K}} = 67.33 \,\text{m/s} \tag{10.387}$$

because

$$K = \frac{m}{l^2}\left(\frac{a_2}{C_{\alpha f}} - \frac{a_1}{C_{\alpha r}}\right) = -2.2059 \times 10^{-4} \tag{10.388}$$

Example 438 *Neutral steer point.*

The **neutral steer point** of the bicycle model of a vehicle is the point along the longitudinal axis at which the mass center allows neutral steering. To find the neutral steer point P_N we define a distance a_N from the front axle to have $K = 0$

$$\frac{l - a_N}{C_{\alpha f}} - \frac{a_N}{C_{\alpha r}} = 0 \tag{10.389}$$

therefore,

$$a_N = \frac{C_{\alpha r}}{C_{\alpha f} + C_{\alpha r}}l \tag{10.390}$$

The **neutral distance** d_N

$$d_N = a_N - a_1 \tag{10.391}$$

indicates how much the mass center can move to have neutral steering.

For example, the neutral steer point P_N for a car with characteristics

$$C_{\alpha f} = 57296\,\text{N/rad} \qquad C_{\alpha r} = 52712\,\text{N/rad} \tag{10.392}$$
$$a_1 = 91\,\text{cm} \approx 2.98\,\text{ft} \qquad a_2 = 164\,\text{cm} \approx 5.38\,\text{ft} \tag{10.393}$$

is at

$$a_N = 1.2219\,\text{m} \tag{10.394}$$

Therefore, the mass center can move a distance d_N forward and still have an understeer car.

$$d_N = a_N - a_1 \approx 31.2\,\text{cm} \tag{10.395}$$

Example 439 *Steady state responses of an oversteer vehicle.*

To compare the steady-state responses of under and oversteer vehicles, we examine how the steady-state parameters S_κ, S_β, S_r, S_a, S_y vary when the speed of an oversteer vehicle changes. The following characteristics are similar to Example 428 except the position of the mass center.

$$C_{\alpha f_L} = C_{\alpha f_R} \approx 3000\,\text{N/rad} \qquad C_{\alpha r_L} = C_{\alpha r_R} \approx 3000\,\text{N/rad} \tag{10.396}$$

$$m = 1000\,\text{kg} \qquad I_z = 1650\,\text{kg m}^2 \tag{10.397}$$
$$a_1 = 1.28\,\text{m} \qquad a_2 = 1.22\,\text{m} \tag{10.398}$$
$$K = \frac{m}{l^2}\left(\frac{a_2}{C_{\alpha f}} - \frac{a_1}{C_{\alpha r}}\right) = -1.33 \times 10^{-2} \tag{10.399}$$

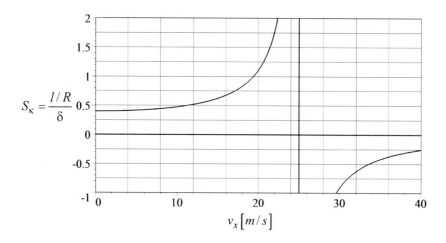

FIGURE 10.26. Curvature response, S_κ, of an oversteer vehicle as a function of forward velocity v_x.

$$S_\kappa = \frac{250}{625 - v_x^2} \qquad\qquad S_\beta = \frac{350 - 21.33v_x^2}{75 - v_x^2}$$

$$S_r = S_\kappa v_x = \frac{250v_x}{625 - v_x^2} \qquad S_a = S_\kappa v_x^2 = \frac{250v_x^2}{625 - v_x^2}$$

$$S_y = \frac{350 - 21.33v_x^2}{75 - v_x^2} v_x \qquad\qquad\qquad (10.400)$$

Figures 10.26-10.30 illustrate the steady-states variations by increasing the forward velocity.

By increasing the speed of the car, S_κ increases and symptomatically approaches infinity at a critical speed, which means the radius of rotation decreases for constant δ. The value of S_β monotonically decreases by increasing v_x.

S_r begins from zero and increases with v_x. S_a shows the centripetal acceleration of the car while it is turning on a circle with a constant speed. The value of S_a starts from zero at $v_x = 0$ and increases rapidly to approach $S_a \to \infty$. The lateral velocity response S_y indicates the speed of the vehicle at its mass center in the y-direction. It starts at zero and increases to a maximum before decreasing monotonically to negative values.

Example 440 ★ *Constant lateral force and steady-state response.*

Consider a situation in which there is a constant lateral force F_y on a vehicle and there is no steering angle. At steady-state conditions, the following equations describe the motion of the vehicle.

$$F_y = C_r r + C_\beta \beta = mr\, v_x \qquad\qquad (10.401)$$
$$M_z = D_r r + D_\beta \beta = 0 \qquad\qquad (10.402)$$

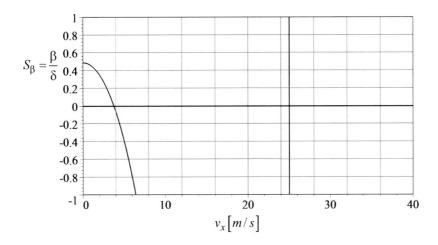

FIGURE 10.27. Sideslip response, S_β, of an oversteer vehicle as a function of forward velocity v_x.

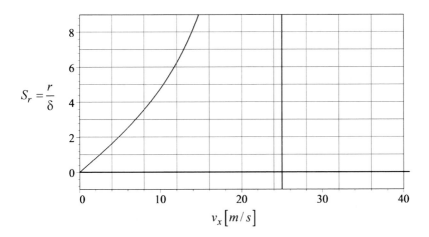

FIGURE 10.28. Yaw rate response, S_r, of an oversteer vehicle as a function of forward velocity v_x.

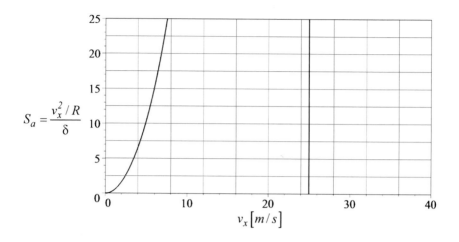

FIGURE 10.29. Lateral acceleration response, S_a, of an oversteer vehicle as a function of forward velocity v_x.

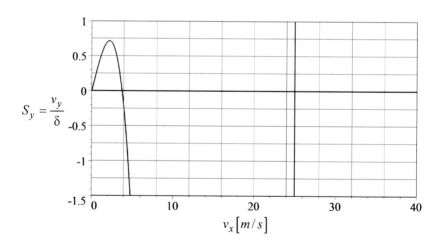

FIGURE 10.30. Lateral velocity response, S_y, of an oversteer vehicle as a function of forward velocity v_x.

Equation (10.401) and (10.402) may be used to define the following steady-state responses

$$S_{y_1} = \frac{\kappa}{F_y} = \frac{1/R}{F_y} = \frac{D_\beta}{v_x\,(C_r D_\beta - C_\beta D_r)} \qquad (10.403)$$

$$S_{y_2} = \frac{r}{F_y} = \frac{D_\beta}{C_r D_\beta - C_\beta D_r} \qquad (10.404)$$

A constant lateral force can be a result of driving straight on a banked road, or having a side wind. A nonzero lateral rotation response S_{y_1} indicates that the vehicle will turn with $\delta = 0$ and $F_y \neq 0$. The lateral rotation response S_{y_1} may be transformed to the following equation to be a function of the stability factor K:

$$S_{y_1} = \frac{1}{v_x\left(C_r - C_\beta \dfrac{D_r}{D_\beta}\right)} = \frac{a_1 C_{\alpha f} - a_2 C_{\alpha r}}{v_x C_{\alpha f} C_{\alpha r} l^2}$$

$$= -\frac{1}{v_x l^2}\left(\frac{a_2}{C_{\alpha f}} - \frac{a_1}{C_{\alpha r}}\right) = -\frac{1}{m v_x} K \qquad (10.405)$$

To drain the rain and water from roads, we build the straight roads with a little bank angle from the center to the shoulder. Consider a moving car on a straight banked road. There is a lateral gravitational force because of the bank angle θ

$$F_y = -mg\sin\theta \approx -mg\theta \qquad (10.406)$$

to pull the car downhill. If the car is an understeer and $K > 0$ then the car will turn downhill, while an oversteer car with $K < 0$ will turn uphill.

Similarly, if there is a side wind in the y-direction and the car is moving on a flat straight road then, an understeer car, $K > 0$, will turn about the z-axis, while an oversteer car, $K < 0$, will turn about the $-z$-axis.

Example 441 ★ *SAE steering definition.*
SAE steer definitions for under and oversteer behaviors are as follows:
US. A vehicle is understeer if the ratio of the steering wheel angle gradient to the overall steering ratio is greater than the Ackerman steer gradient.
OS. A vehicle is oversteer if the ratio of the steering wheel angle gradient to the overall steering ratio is less than the Ackerman steer gradient.
AS. Ackerman steering gradient is

$$S_A = \frac{l}{v_x^2} = \frac{d\,(l/R)}{d\,(v_x^2/R)} \qquad (10.407)$$

Example 442 ★ *Exact steady-state responses*
The steady-state turning of a two-wheel rigid vehicle is expressed by

$$F_x = -mr\,v_y \qquad (10.408)$$
$$C_r\,r + C_\beta\,\beta + C_\delta\,\delta = mr\,v_x \qquad (10.409)$$
$$D_r\,r + D_\beta\,\beta + D_\delta\,\delta = 0 \qquad (10.410)$$

At steady-state turning, the mass center of the vehicle moves on a circle with radius R at a forward speed v_x and angular velocity r, so v_x and r are related by

$$v_x = Rr\cos\beta \qquad v = Rr \qquad R_1 = R\cos\beta \tag{10.411}$$

Substituting (10.280) in Equations (10.277)-(10.279), yields

$$C_r\frac{v_x}{R_1} + C_\beta\,\beta + C_\delta\,\delta = m\frac{v_x^2}{R_1} \tag{10.412}$$

$$D_r\frac{v_x}{R_1} + D_\beta\,\beta + D_\delta\,\delta = 0 \tag{10.413}$$

Let us define a new curvature κ_1 based on R_1

$$\kappa_1 = \frac{1}{R_a} \tag{10.414}$$

and write the equations in matrix form

$$\begin{bmatrix} C_\beta & C_r\,v_x - m\,v_x^2 \\ D_\beta & D_r\,v_x \end{bmatrix}\begin{bmatrix} \beta \\ \kappa_1 \end{bmatrix} = \begin{bmatrix} -C_\delta \\ -D_\delta \end{bmatrix}\delta \tag{10.415}$$

Solving the equations for β and κ_1 shows that

$$\begin{bmatrix} \beta \\ \kappa_1 \end{bmatrix} = \begin{bmatrix} \dfrac{D_\delta\,(C_r - mv_x) - D_r C_\delta}{D_r C_\beta - C_r D_\beta + mv_x D_\beta} \\[3mm] \dfrac{C_\delta D_\beta - C_\beta D_\delta}{v_x\,(D_r C_\beta - C_r D_\beta + mv_x D_\beta)} \end{bmatrix}\delta \tag{10.416}$$

Now we are able to define a more accurate steady-state responses.

1. Curvature response, S_κ $R_1 = R\cos\beta$

$$\begin{aligned} S_\kappa &= \frac{\kappa}{\delta} = \frac{1}{R\delta} = \frac{1}{R_1\delta}\cos\beta \\ &= \frac{C_\delta D_\beta - C_\beta D_\delta}{v_x\,(D_r C_\beta - C_r D_\beta + mv_x D_\beta)}\cos\beta \end{aligned} \tag{10.417}$$

2. Sideslip response, S_β

$$S_\beta = \frac{\beta}{\delta} = \frac{D_\delta\,(C_r - mv_x) - D_r C_\delta}{D_r C_\beta - C_r D_\beta + mv_x D_\beta} \tag{10.418}$$

3. Yaw rate response, S_r

$$S_r = \frac{r}{\delta} = \frac{\kappa}{\delta}v_x = S_\kappa v_x = \frac{C_\delta D_\beta - C_\beta D_\delta}{D_r C_\beta - C_r D_\beta + mv_x D_\beta}\cos\beta \tag{10.419}$$

4. *Centripetal acceleration response, S_a*

$$S_a = \frac{v_x^2/R}{\delta} = \frac{\kappa}{\delta} v_x^2 = S_\kappa v_x^2 = \frac{(C_\delta D_\beta - C_\beta D_\delta) v_x}{D_r C_\beta - C_r D_\beta + m v_x D_\beta} \cos \beta \qquad (10.420)$$

5. *Lateral velocity response, S_y.*

$$S_y = \frac{v_y}{\delta} = S_\beta v_x = \frac{D_\delta (C_r - m v_x) - D_r C_\delta}{D_r C_\beta - C_r D_\beta + m v_x D_\beta} v_x \qquad (10.421)$$

10.6 ★ Linearized Model for a Two-Wheel Vehicle

When the vehicle sideslip angle β is very small, the equations of motion of bicycle model reduce to the following set of equations:

$$F_x = m\dot{v} - mrv\beta \qquad (10.422)$$
$$F_y = mv\left(r + \dot{\beta}\right) + m\beta\dot{v} \qquad (10.423)$$
$$M_z = I_z\dot{r} \qquad (10.424)$$

$$
\begin{aligned}
F_y &= (-C_{\alpha f} - C_{\alpha r})\beta + \frac{1}{v}(a_2 C_{\alpha r} - a_1 C_{\alpha f})r \\
&\quad + \delta_f C_{\alpha f} + \delta_r C_{\alpha r} \qquad (10.425)
\end{aligned}
$$

$$
\begin{aligned}
M_z &= (a_2 C_{\alpha r} - a_1 C_{\alpha f})\beta - \frac{1}{v}(a_1^2 C_{\alpha f} + a_2^2 C_{\alpha r})r \\
&\quad + a_1 C_{\alpha f}\delta_f - a_2 C_{\alpha r}\delta_r \qquad (10.426)
\end{aligned}
$$

Although these equations are not linear if $\mathbf{v}$ is not constant, because of the assumption $\beta \ll 1$, they are called linearized equations of motion.

When the speed of the vehicle is constant, then the equations are

$$F_x = -mrv\beta \qquad (10.427)$$
$$F_y = mv\left(r + \dot{\beta}\right) \qquad (10.428)$$
$$M_z = I_z\dot{r} \qquad (10.429)$$

Proof. For a small sideslip, β, we may assume that,

$$v_x = v\cos\beta \approx v \qquad (10.430)$$
$$v_y = v\sin\beta \approx v\beta \qquad (10.431)$$

Therefore, the equations of motion (10.193)–(10.195) will be simplified to

$$F_x = m\dot{v}_x - mrv_y = m\dot{v} - mrv\beta \qquad (10.432)$$
$$F_y = m\dot{v}_y + mrv_x = m\left(\dot{v}\beta + v\dot{\beta}\right) + mrv \qquad (10.433)$$
$$M_z = \dot{r}I_z \qquad (10.434)$$

Substitute $\dot{v} = 0$ for a constant velocity, these equations will be equal to (10.427)-(10.429).

The tire sideslip angles α_f and α_r can also be simplified to

$$\alpha_f = \beta_f - \delta_f = \frac{1}{v_x}(v_y + a_1 r) - \delta_f = \beta + \frac{a_1 r}{v} - \delta_f \qquad (10.435)$$

$$\alpha_r = \beta_r - \delta_r = \frac{1}{v_x}(v_y - a_2 r) - \delta_r = \beta - \frac{a_2 r}{v} - \delta_r \qquad (10.436)$$

The front and rear tires lateral forces are

$$F_{yf} = -C_{\alpha f}\alpha_f \qquad (10.437)$$
$$F_{yr} = -C_{\alpha r}\alpha_r \qquad (10.438)$$

Substituting these equations in (10.432)-(10.434) and using the definitions

$$F_x \approx F_{x_f} + F_{x_r} \qquad (10.439)$$
$$F_y \approx F_{y_f} + F_{y_r} \qquad (10.440)$$
$$M_z \approx a_1 F_{y_f} - a_2 F_{y_r} \qquad (10.441)$$

results in the force system

$$\begin{aligned} F_y &= F_{yf} + F_{yr} \\ &= (-C_{\alpha f} - C_{\alpha r})\beta + \frac{1}{v}(a_2 C_{\alpha r} - a_1 C_{\alpha f})r \\ &\quad + \delta_f C_{\alpha f} + \delta_r C_{\alpha r} \end{aligned} \qquad (10.442)$$

$$\begin{aligned} M_z &= a_1 F_{yf} - a_2 F_{yr} \\ &= (a_2 C_{\alpha r} - a_1 C_{\alpha f})\beta - \frac{1}{v}\left(a_1^2 C_{\alpha f} + a_2^2 C_{\alpha r}\right)r \\ &\quad + a_1 C_{\alpha f}\delta_f - a_2 C_{\alpha r}\delta_r \end{aligned} \qquad (10.443)$$

Factorization will transform the force system to (10.425)-(10.426). ∎

Example 443 ★ *Front-wheel-steering and constant velocity.*

In most cases, the front wheel is the only steerable wheel, hence $\delta_f = \delta$ and $\delta_r = 0$. This simplifies the equations of motion for a front steering vehicle at constant velocity to

$$F_x = -mrv\beta \qquad (10.444)$$

$$\begin{aligned} mv\dot{\beta} &= (-C_{\alpha f} - C_{\alpha r})\beta \\ &\quad + \left(-\frac{1}{v}(a_1 C_{\alpha f} - a_2 C_{\alpha r}) - mv\right)r + C_{\alpha f}\delta \end{aligned} \qquad (10.445)$$

$$\begin{aligned} I_z \dot{r} &= (a_2 C_{\alpha r} - a_1 C_{\alpha f})\beta \\ &\quad + \left(-\frac{1}{v}\left(C_{\alpha f}a_1^2 + C_{\alpha r}a_2^2\right)\right)r + (a_1 C_{\alpha f})\delta \end{aligned} \qquad (10.446)$$

The second and third equations may be written in a matrix form for simpler calculation.

$$\begin{bmatrix} \dot{\beta} \\ \dot{r} \end{bmatrix} = \begin{bmatrix} \dfrac{-(C_{\alpha f} + C_{\alpha r})}{mv} & \dfrac{a_2 C_{\alpha r} - a_1 C_{\alpha f}}{mv^2} - 1 \\ \dfrac{a_2 C_{\alpha r} - a_1 C_{\alpha f}}{I_z} & \dfrac{-(C_{\alpha f} a_1^2 + C_{\alpha r} a_2^2)}{v I_z} \end{bmatrix} \begin{bmatrix} \beta \\ r \end{bmatrix}$$

$$+ \begin{bmatrix} \dfrac{C_{\alpha f}}{mv} \\ \dfrac{a_1 C_{\alpha f}}{I_z} \end{bmatrix} \delta \qquad (10.447)$$

Example 444 ★ *Steady state conditions and a linearized system.*

The equations of motion for a front steering, two-wheel model of four-wheel vehicles are given in equation (10.447) for linearized angle β. At a steady-state condition we have

$$\begin{bmatrix} \dot{\beta} \\ \dot{r} \end{bmatrix} = 0 \qquad (10.448)$$

and therefore,

$$\begin{bmatrix} \beta \\ r \end{bmatrix} = \begin{bmatrix} \dfrac{-C_{\alpha f} - C_{\alpha r}}{mv} & \dfrac{a_2 C_{\alpha r} - a_1 C_{\alpha f}}{mv^2} - 1 \\ \dfrac{a_2 C_{\alpha r} - a_1 C_{\alpha f}}{I_z} & \dfrac{-C_{\alpha f} a_1^2 - C_{\alpha r} a_2^2}{v I_z} \end{bmatrix}^{-1} \begin{bmatrix} \dfrac{C_{\alpha f}}{mv} \delta \\ \dfrac{a_1 C_{\alpha f}}{I_z} \delta \end{bmatrix}$$

$$= \begin{bmatrix} \dfrac{-\left(a_2^2 C_{\alpha r} + a_1 a_2 C_{\alpha r} - mv^2 a_1\right) C_{\alpha f} \delta}{C_{\alpha f} C_{\alpha r} (a_1 + a_2)^2 - mv^2 (a_1 C_{\alpha f} - a_2 C_{\alpha r})} \\ \dfrac{-(a_1 + a_2) v C_{\alpha f} C_{\alpha r} \delta}{C_{\alpha f} C_{\alpha r} (a_1 + a_2)^2 - mv^2 (a_1 C_{\alpha f} - a_2 C_{\alpha r})} \end{bmatrix} \qquad (10.449)$$

Using Equation (10.449) we can define the following steady-state responses:

2. Sideslip response, S_β

$$S_\beta = \frac{\beta}{\delta} = \frac{-\left(a_2^2 C_{\alpha r} + a_1 a_2 C_{\alpha r} - mv^2 a_1\right) C_{\alpha f}}{C_{\alpha f} C_{\alpha r} l^2 - mv^2 (a_1 C_{\alpha f} - a_2 C_{\alpha r})} \qquad (10.450)$$

3. Yaw rate response, S_r

$$S_r = \frac{r}{\delta} = \frac{-C_{\alpha f} C_{\alpha r} l v}{C_{\alpha f} C_{\alpha r} l^2 - mv^2 (a_1 C_{\alpha f} - a_2 C_{\alpha r})} \qquad (10.451)$$

At steady-state conditions we have

$$r = \frac{v}{R} \qquad (10.452)$$

where R is the radius of the circular path of the vehicle. Employing Equation (10.452) we are able to define two more steady-state responses as follows:

1.*Curvature response, S_κ*

$$
\begin{aligned}
S_\kappa &= \frac{\kappa}{\delta} = \frac{1}{R\delta} = \frac{r}{v\delta} = \frac{1}{v}S_r \\
&= \frac{-lC_{\alpha f}C_{\alpha r}}{C_{\alpha f}C_{\alpha r}l^2 - mv^2\left(a_1 C_{\alpha f} - a_2 C_{\alpha r}\right)}
\end{aligned}
\tag{10.453}
$$

4.*Lateral acceleration response, S_a*

$$
\begin{aligned}
S_a &= \frac{v^2/R}{\delta} = \frac{\kappa}{\delta}v^2 = S_\kappa v^2 \\
&= \frac{-lC_{\alpha f}C_{\alpha r}v^2}{C_{\alpha f}C_{\alpha r}l^2 - mv^2\left(a_1 C_{\alpha f} - a_2 C_{\alpha r}\right)}
\end{aligned}
\tag{10.454}
$$

The above steady-state responses are comparable to the steady-states numbered 1 to 4 given in Equations (10.267)-(10.270) for a more general case.

Example 445 ★ *Understeer and oversteer for a linearized model.*

Employing the curvature response S_κ in Equation (10.453) we may define

$$
\begin{aligned}
S_\kappa &= \frac{\kappa}{\delta} = \frac{1/R}{\delta} = \frac{-lC_{\alpha f}C_{\alpha r}}{C_{\alpha f}C_{\alpha r}l^2 - mv^2\left(a_1 C_{\alpha f} - a_2 C_{\alpha r}\right)} \\
&= \frac{1}{l + \dfrac{mv^2\left(a_1 C_{\alpha f} - a_2 C_{\alpha r}\right)}{-lC_{\alpha f}C_{\alpha r}}} \\
&= \frac{1}{l}\frac{1}{1 + \dfrac{mv^2\left(a_1 C_{\alpha f} - a_2 C_{\alpha r}\right)}{-l^2 C_{\alpha f}C_{\alpha r}}} = \frac{1}{l}\frac{1}{1 + Kv_x^2}
\end{aligned}
\tag{10.455}
$$

where

$$
K = \frac{m}{l^2}\left(\frac{a_2}{C_{\alpha f}} - \frac{a_1}{C_{\alpha r}}\right)
\tag{10.456}
$$

This K is the same as the stability factor given in Equation (10.369). Therefore, the stability factor remains the same for the linearized equations.

Example 446 ★ *Source of nonlinearities.*

*There are three main reasons for nonlinearity in rigid vehicle equations of motion: product of variables, trigonometric functions, and nonlinear characteristic of forces. When the steer angle δ and sideslip angles α_i and β are very small, the forces act linear, trigonometric function are approximately linear, and product of variables are too small. As a result, it is reasonable to ignore all kinds of nonlinearities. This is called **low angles condition** driving, and is correct for low turn and normal speed driving.*

10.7 ★ Transient Response

To examine the transient response of a vehicle and examine how the vehicle will respond to a steering input, the following set of coupled ordinary differential equations must be solved.

$$\dot{v}_x = \frac{1}{m}F_x + r\,v_y \tag{10.457}$$

$$\dot{v}_y = \frac{C_\beta}{mv_x}v_y - \left(v_x - \frac{C_r}{m}\right)r + \frac{C_\delta}{m}\delta(t) \tag{10.458}$$

$$\dot{r} = \frac{D_\beta}{I_z v_x}v_y + \frac{D_r}{I_z}r + \frac{D_\delta}{I_z}\delta(t) \tag{10.459}$$

The answers to this set of equations to a given time dependent steer angle are

$$v_x = v_x(t) \tag{10.460}$$
$$v_y = v_y(t) \tag{10.461}$$
$$r = r(t) \tag{10.462}$$

Such a solution is called *time response* or *transient response*.

Assuming a constant forward velocity, the first equation (10.457) simplifies to an algebraic equation

$$F_x = -mr\,v_y \tag{10.463}$$

$$\begin{bmatrix} \dot{v}_y \\ \dot{r} \end{bmatrix} = \begin{bmatrix} \dfrac{C_\beta}{mv_x} & \dfrac{C_r}{m} - v_x \\ \dfrac{D_\beta}{I_z v_x} & \dfrac{D_r}{I_z} \end{bmatrix} \begin{bmatrix} v_y \\ r \end{bmatrix} + \begin{bmatrix} \dfrac{C_\delta}{m} \\ \dfrac{D_\delta}{I_z} \end{bmatrix} \delta(t) \tag{10.464}$$

and Equations (10.464) become independent from the first one. The set of Equations (10.464) can be written in the form

$$\dot{\mathbf{q}} = [A]\,\mathbf{q} + \mathbf{u} \tag{10.465}$$

in which $[A]$ is a constant coefficient matrix, $\mathbf{q}$ is the vector of control variables, and $\mathbf{u}$ is the vector of inputs.

$$[A] = \begin{bmatrix} -\dfrac{C_{\alpha f} + C_{\alpha r}}{mv_x} & \dfrac{-a_1 C_{\alpha f} + a_2 C_{\alpha r}}{mv_x} - v_x \\ -\dfrac{a_1 C_{\alpha f} - a_2 C_{\alpha r}}{I_z v_x} & -\dfrac{a_1^2 C_{\alpha f} + a_2^2 C_{\alpha r}}{I_z v_x} \end{bmatrix} \tag{10.466}$$

$$= \begin{bmatrix} \dfrac{C_\beta}{mv_x} & \dfrac{C_r}{m} - v_x \\ \dfrac{D_\beta}{I_z v_x} & \dfrac{D_r}{I_z} \end{bmatrix} \tag{10.467}$$

$$\mathbf{q} \;=\; \begin{bmatrix} v_y \\ r \end{bmatrix} \tag{10.468}$$

$$\mathbf{u} \;=\; \begin{bmatrix} \dfrac{C_{\alpha f}}{m} \\[2mm] \dfrac{a_1 C_{\alpha f}}{I_z} \end{bmatrix} \delta\left(t\right) = \begin{bmatrix} \dfrac{C_\delta}{m} \\[2mm] \dfrac{D_\delta}{I_z} \end{bmatrix} \delta\left(t\right) \tag{10.469}$$

To solve the inverse dynamic problem and find the vehicle response, the steering function $\delta\left(t\right)$ must be given.

Example 447 ★ *Forward and inverse dynamic problems.*

Two types of dynamic problems may be defined: 1.*direct or forward and,* 2.*indirect or inverse. In forward dynamics a set of desired functions* $v_x = v_x\left(t\right)$, $v_y = v_y\left(t\right)$, $r = r\left(t\right)$ *are given and the required* $\delta\left(t\right)$ *is asked for. In inverse dynamics, an input function* $\delta = \delta\left(t\right)$ *is given and the output functions* $v_x = v_x\left(t\right)$, $v_y = v_y\left(t\right)$, $r = r\left(t\right)$ *are asked for.*

The forward dynamic problem needs differentiation and the inverse dynamic problem needs integration. Generally speaking, solving an inverse dynamic problem is more complicated than a forward dynamic problem.

Example 448 ★ *Analytic solution to a step steer input.*

Consider a vehicle with the following characteristics

$$
\begin{aligned}
C_{\alpha f} &= 60000\,\mathrm{N/\,rad} & C_{\alpha r} &= 60000\,\mathrm{N/\,rad} \\
m &= 1000\,\mathrm{kg} & I_z &= 1650\,\mathrm{kg\,m^2} \\
a_1 &= 1.0\,\mathrm{m} & a_2 &= 1.5\,\mathrm{m} \\
v_x &= 20\,\mathrm{m/\,s} & &
\end{aligned}
\tag{10.470}
$$

The force system coefficients for the vehicle from (10.144)-(10.149) are

$$
\begin{aligned}
C_r &= 1500\,\mathrm{N\,s/\,rad} & C_\beta &= -120000\,\mathrm{N/\,rad} \\
C_\delta &= 60000\,\mathrm{N/\,rad} & D_r &= -9750\,\mathrm{N\,m\,s/\,rad} \\
D_\beta &= 30000\,\mathrm{N\,m/\,rad} & D_\delta &= 60000\,\mathrm{N\,m/\,rad}
\end{aligned}
\tag{10.471}
$$

Let us assume the steering input is

$$\delta\left(t\right) = \begin{cases} 0.1\,\mathrm{rad} \approx 5.73\,\mathrm{deg} & t > 0 \\ 0 & t \le 0 \end{cases} \tag{10.472}$$

The steady state responses can be used to determine the steady-state values of the vehicle's variable kinematics.

$$R = \frac{1}{S_\kappa \delta} = \frac{D_r C_\beta - C_r D_\beta + mv_x D_\beta}{C_\delta D_\beta - C_\beta D_\delta} \frac{v_x}{\delta} = 38.33\,\text{m} \qquad (10.473)$$

$$\beta = S_\beta \delta = \frac{D_\delta (C_r - mv_x) - D_r C_\delta}{D_r C_\beta - C_r D_\beta + mv_x D_\beta} \delta = -0.0304\,\text{rad} \qquad (10.474)$$

$$r = S_r \delta = \frac{C_\delta D_\beta - C_\beta D_\delta}{D_r C_\beta - C_r D_\beta + mv_x D_\beta} \delta = 0.522\,\text{rad/s} \qquad (10.475)$$

$$\frac{v_x^2}{R} = S_a \delta = \frac{C_\delta D_\beta - C_\beta D_\delta}{D_r C_\beta - C_r D_\beta + mv_x D_\beta} v_x \delta = 10.43\,\text{m/s}^2 \qquad (10.476)$$

$$v_y = S_y \delta = \frac{D_\delta (C_r - mv_x) - D_r C_\delta}{D_r C_\beta - C_r D_\beta + mv_x D_\beta} v_x \delta = -0.608\,\text{m/s} \qquad (10.477)$$

The equations of motion for a zero initial condition

$$\mathbf{q}_0 = \begin{bmatrix} v_y(0) \\ r(0) \end{bmatrix} = \begin{bmatrix} 0 \\ 0 \end{bmatrix} \qquad (10.478)$$

are

$$\dot{v}_y + 6v_y + 18.5r = 60\delta(t) = 6 \qquad (10.479)$$
$$\dot{r} - 0.909v_y + 5.909r = 36.363\delta(t) = 3.636 \qquad (10.480)$$

The solution of the equations of motion are

$$\begin{bmatrix} v_y(t) \\ r(t) \end{bmatrix} = \begin{bmatrix} -0.609 + e^{-5.95t}(2.347\sin 4.1t + 0.609\cos 4.1t) \\ 0.522 + e^{-5.95t}(0.129\sin 4.1t - 0.522\cos 4.1t) \end{bmatrix} \qquad (10.481)$$

To examine the response of the vehicle to the sudden change in steer angle while going straight at a constant speed, we plot the kinematic variables of the vehicle. Figures 10.31 and 10.32 depict the solutions $v_y(t)$ and $r(t)$ respectively.

The steering input is positive and therefore, the vehicle must turn left, in a positive direction of the y-axis. The yaw rate in Figure 10.32 is positive and correctly shows that the vehicle is turning about the z-axis.

We can also find the lateral velocity of the front and rear wheels, by having v_y and r.

$$v_{y_f} = v_y + a_1 r \qquad v_{y_r} = v_y - a_2 r \qquad (10.482)$$

The lateral speed of the front and rear wheels are shown in Figures 10.33 and 10.34. The vehicle sideslip angle $\beta = v_y/v_x$ and the radius of rotation $R = v_x/r$ are also shown in Figures 10.35 and 10.36.

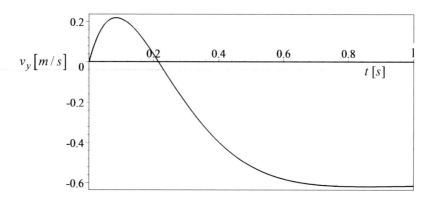

FIGURE 10.31. Lateral velocity response to a sudden change in steer angle.

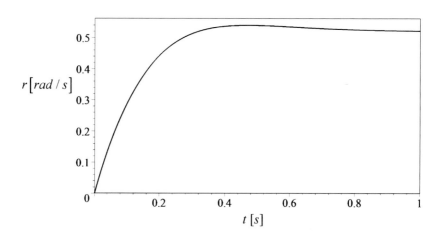

FIGURE 10.32. Yaw velocity response to a sudden change in steer angle.

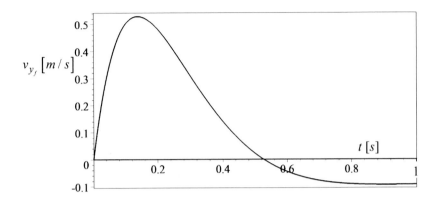

FIGURE 10.33. Lateral velocity response of the front wheel to a sudden change in steer angle.

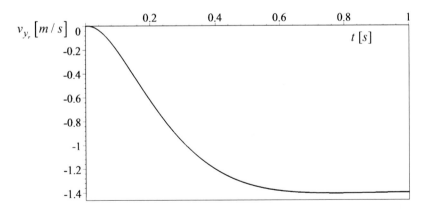

FIGURE 10.34. Lateral velocity response of the rear wheel to a sudden change in steer angle.

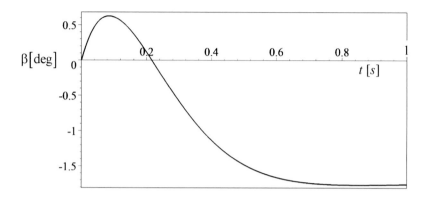

FIGURE 10.35. Sideslip response to a sudden change in steer angle.

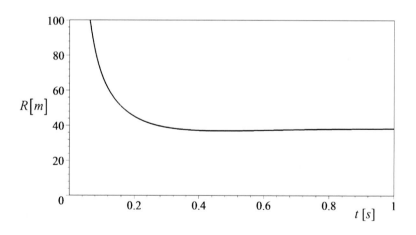

FIGURE 10.36. Radius of rotation response to a sudden change in steer angle.

Example 449 ★ *Time series and free response.*
The response of a vehicle to zero steer angle

$$\delta\left(t\right)=0 \tag{10.483}$$

at constant speed is called the free response. The equation of motion under free dynamics is

$$\dot{\mathbf{q}}=[A]\,\mathbf{q} \tag{10.484}$$

To solve the equations, let us assume

$$[A]=\begin{bmatrix} a & b \\ c & d \end{bmatrix} \tag{10.485}$$

and therefore, the equations of motion are

$$\begin{bmatrix} \dot{v}_y \\ \dot{r} \end{bmatrix}=\begin{bmatrix} a & b \\ c & d \end{bmatrix}\begin{bmatrix} v_y \\ r \end{bmatrix} \tag{10.486}$$

Because the equations are linear, the solutions are exponential functions

$$v_y = Ae^{\lambda t} \qquad r = Be^{\lambda t} \tag{10.487}$$

Substituting the solutions in equations of motion

$$\begin{bmatrix} A\lambda e^{\lambda t} \\ B\lambda e^{\lambda t} \end{bmatrix}=\begin{bmatrix} a & b \\ c & d \end{bmatrix}\begin{bmatrix} Ae^{\lambda t} \\ Be^{\lambda t} \end{bmatrix} \tag{10.488}$$

shows that

$$\begin{bmatrix} a-\lambda & b \\ c & d-\lambda \end{bmatrix}\begin{bmatrix} Ae^{\lambda t} \\ Be^{\lambda t} \end{bmatrix}=0 \tag{10.489}$$

Therefore, the condition for functions (10.487) to be the solution of Equation (10.486) is that the exponent λ to be the eigenvalue of $[A]$. To find λ, we may expand the determinant of the coefficient matrix

$$\det\begin{bmatrix} a-\lambda & b \\ c & d-\lambda \end{bmatrix}=\lambda^2-\left(a+d\right)\lambda+\left(ad-bc\right) \tag{10.490}$$

and find the characteristic equation

$$\lambda^2-\left(a+d\right)\lambda+\left(ad-bc\right)=0 \tag{10.491}$$

The solution of the characteristic equations is

$$\lambda_{1,2}=\frac{1}{2}\left(a+d\right)\pm\frac{1}{2}\sqrt{\left(a-d\right)^2+4bc} \tag{10.492}$$

Having the eigenvalues $\lambda_{1,2}$ provides us with the general solution for the free dynamics of a bicycle vehicle:

$$v_y = A_1 e^{\lambda_1 t} + A_2 e^{\lambda_2 t} \tag{10.493}$$
$$r = B_1 e^{\lambda_1 t} + B_2 e^{\lambda_2 t} \tag{10.494}$$

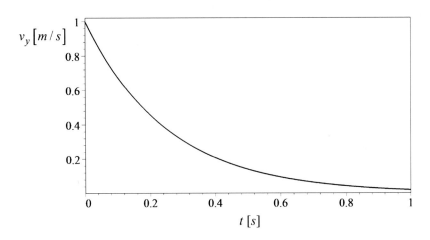

FIGURE 10.37. Lateral velocity response in example 449.

The coefficients A_1, A_2, B_1, and B_2, must be found from initial conditions:
As an example, consider a vehicle with

$$
\begin{aligned}
C_{\alpha f} &= 57296\,\mathrm{N/rad} & C_{\alpha r} &= 52712\,\mathrm{N/rad} \\
m &= 1400\,\mathrm{kg} \approx 95.9\,\mathrm{slug} & I_z &= 1128\,\mathrm{kg\,m}^2 \approx 832\,\mathrm{slug\,ft}^2 \\
a_1 &= 125\,\mathrm{cm} \approx 4.1\,\mathrm{ft} & a_2 &= 130\,\mathrm{cm} \approx 4.26\,\mathrm{ft} \\
v_x &= 20\,\mathrm{m/s} & & \hspace{2em}(10.495)
\end{aligned}
$$

which starts from

$$
\mathbf{q}_0 = \begin{bmatrix} v_y(0) \\ r(0) \end{bmatrix} = \begin{bmatrix} 1 \\ 0 \end{bmatrix} \hspace{2em}(10.496)
$$

Substituting these values provide us with

$$
\begin{bmatrix} \dot{v}_y \\ \dot{r} \end{bmatrix} = \begin{bmatrix} -3.929 & -31.051 \\ -13.716 & -79170.337 \end{bmatrix} \begin{bmatrix} v_y \\ r \end{bmatrix} \hspace{2em}(10.497)
$$

and their solutions are

$$
\begin{aligned}
v_y &= -0.173 \times 10^{-3} \left(e^{-3.92t} - e^{-79170.34t} \right) & (10.498) \\
r &= e^{-3.92t} + 0.68 \times 10 \times 10^{-7} e^{-79170.34t} & (10.499)
\end{aligned}
$$

Figures 10.37 and 10.38 illustrate the time responses. Figure 10.39 is a magnification of Figure 10.38 to show that r does not jump to a negative point but decreases rapidly and then approaches zero gradually.

Example 450 ★ *Matrix exponentiation.*
The exponential function $e^{[A]t}$ is called matrix exponentiation. This function is defined as a matrix time series.

$$
e^{[A]t} = I + [A]\,t + \frac{[A]^2}{2!}t^2 + \frac{[A]^3}{3!}t^3 + \cdots \hspace{2em}(10.500)
$$

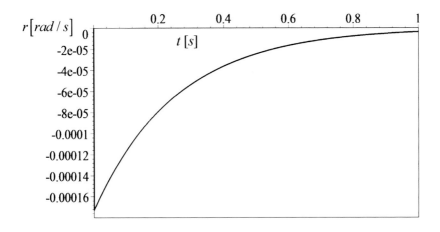

FIGURE 10.38. Yaw velocity response in example 449.

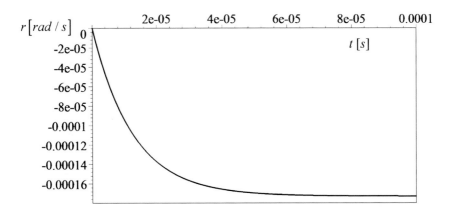

FIGURE 10.39. Yaw velocity response in example 449.

This series always converges. As an example assume

$$[A] = \begin{bmatrix} 0.1 & 0.2 \\ -0.3 & 0.4 \end{bmatrix} \tag{10.501}$$

then

$$
e^{[A]t} \approx \begin{bmatrix} 1 & 0 \\ 0 & 1 \end{bmatrix} + \begin{bmatrix} 0.1 & 0.2 \\ -0.3 & 0.4 \end{bmatrix} t + \frac{1}{2} \begin{bmatrix} 0.1 & 0.2 \\ -0.3 & 0.4 \end{bmatrix}^2 t^2 + \cdots
$$

$$
\approx \begin{bmatrix} 1 + 0.1t - 0.025\,t^2 + \cdots & 0.2t + 0.05t^2 + \cdots \\ -0.3t - 0.075\,t^2 + \cdots & 1 + 0.4t + 0.05t^2 + \cdots \end{bmatrix} \tag{10.502}
$$

Example 451 ★ *Time series and free response.*
  The response of a vehicle to zero steer angle

$$\delta(t) = 0 \tag{10.503}$$

at constant speed is the free response. The equation of motion under free dynamics is

$$\dot{\mathbf{q}} = [A]\,\mathbf{q} \tag{10.504}$$

The solution of this differential equation with the general initial conditions

$$\mathbf{q}(0) = \mathbf{q}_0 \tag{10.505}$$

is

$$\mathbf{q}(t) = e^{[A]t}\mathbf{q}_0 \tag{10.506}$$

If the eigenvalues of $[A]$ are negative then $\mathbf{q}(t) \to \mathbf{0}$ for $\forall \mathbf{q}_0$.
  The free response in a series can be expressed as

$$
\begin{aligned}
\mathbf{q}(t) &= e^{[A]t}\mathbf{q}_0 \\
&= \left(I + [A]t + \frac{[A]^2}{2!}t^2 + \frac{[A]^3}{3!}t^3 + \cdots \right)\mathbf{q}_0
\end{aligned} \tag{10.507}
$$

For example, consider a vehicle with the following characteristics:

$$
\begin{aligned}
C_{\alpha f} &= 57296\,\mathrm{N/\,rad} & C_{\alpha r} &= 52712\,\mathrm{N/\,rad} \\
m &= 1400\,\mathrm{kg} \approx 95.9\,\mathrm{slug} & I_z &= 1128\,\mathrm{kg\,m^2} \approx 832\,\mathrm{slug\,ft^2} \\
a_1 &= 125\,\mathrm{cm} \approx 4.1\,\mathrm{ft} & a_2 &= 130\,\mathrm{cm} \approx 4.26\,\mathrm{ft} \\
v_x &= 20\,\mathrm{m/\,s}
\end{aligned} \tag{10.508}
$$

which start from

$$\mathbf{q}_0 = \begin{bmatrix} v_y(0) \\ r(0) \end{bmatrix} = \begin{bmatrix} 1 \\ 0 \end{bmatrix} \tag{10.509}$$

Employing the vehicle's characteristics, we have

$$[A] = \begin{bmatrix} -3.929 & -31.051 \\ -13.716 & -79170.337 \end{bmatrix} \tag{10.510}$$

and therefore, the time response of the vehicle is

$$
\begin{bmatrix} v_y(t) \\ r(t) \end{bmatrix} = \begin{bmatrix} 1 & 0 \\ 0 & 1 \end{bmatrix} \begin{bmatrix} 1 \\ 0 \end{bmatrix} + \begin{bmatrix} -3.929 & -31.051 \\ -13.716 & -79170.337 \end{bmatrix} t \begin{bmatrix} 1 \\ 0 \end{bmatrix}
$$

$$
+ \frac{1}{2} \begin{bmatrix} -3.929 & -31.051 \\ -13.716 & -79170.337 \end{bmatrix}^2 t^2 \begin{bmatrix} 1 \\ 0 \end{bmatrix}
$$

$$
+ \frac{1}{6} \begin{bmatrix} -3.929 & -31.051 \\ -13.716 & -79170.337 \end{bmatrix}^3 t^3 \begin{bmatrix} 1 \\ 0 \end{bmatrix} \cdots \qquad (10.511)
$$

Accepting an approximate solution up to cubic degree provides the following approximate solution:

$$
\begin{bmatrix} v_y(t) \\ r(t) \end{bmatrix} \approx \begin{bmatrix} -5.620\,3 \times 10^6 t^3 + 220.67 t^2 - 3.929 t + 1 \\ -1.432\,9 \times 10^{10} t^3 + 5.4298 \times 10^5 t^2 - 13.716 t \end{bmatrix}
$$
$$
(10.512)
$$

Example 452 ★ *Response of an understeer vehicle to a step input.*

The response of dynamic systems to a step input is a traditional method to examine the behavior of dynamic systems. A step input for vehicle dynamics is a sudden change in steer angle from zero to a nonzero constant value.

Consider a vehicle with the following characteristics:

$$
\begin{aligned}
C_{\alpha f} &= 57296 \, \text{N/rad} & C_{\alpha r} &= 52712 \, \text{N/rad} \\
m &= 917 \, \text{kg} \approx 62.8 \, \text{slug} & I_z &= 1128 \, \text{kg m}^2 \approx 832 \, \text{slug ft}^2 \\
a_1 &= 0.91 \, \text{m} \approx 2.98 \, \text{ft} & a_2 &= 1.64 \, \text{m} \approx 5.38 \, \text{ft} \\
v_x &= 20 \, \text{m/s} & & \qquad (10.513)
\end{aligned}
$$

and a sudden change in the steering input to a constant value

$$
\delta(t) = \begin{cases} 0.1 \, \text{rad} \approx 5.7296 \, \text{deg} & t > 0 \\ 0 & t \le 0 \end{cases} \qquad (10.514)
$$

The equations of motion for a zero initial condition

$$
\mathbf{q}_0 = \begin{bmatrix} v_y(0) \\ r(0) \end{bmatrix} = \begin{bmatrix} 0 \\ 0 \end{bmatrix} \qquad (10.515)
$$

are

$$
\begin{aligned}
\dot{v}_y + 5.9983 v_y + 18.129 r &= 62.482 \delta(t) = 6.2482 & (10.516) \\
\dot{r} - 1.521 v_y + 8.387 r &= 46.2228 \delta(t) = 4.6223 & (10.517)
\end{aligned}
$$

The force system coefficients for the vehicle from (10.144)-(10.149) are

$$
\begin{aligned}
C_r &= -\frac{a_1}{v_x} C_{\alpha f} + \frac{a_2}{v_x} C_{\alpha r} = 1715.416 \, \text{N s/rad} & (10.518) \\
C_\beta &= -(C_{\alpha f} + C_{\alpha r}) = -110008 \, \text{N/rad} & (10.519) \\
C_\delta &= C_{\alpha f} = 57296 \, \text{N/rad} & (10.520)
\end{aligned}
$$

$$D_r = -\frac{a_1^2}{v_x}C_{\alpha f} - \frac{a_2^2}{v_x}C_{\alpha r} = -9461.05 \,\mathrm{N\,m\,s/\,rad} \qquad (10.521)$$

$$D_\beta = -(a_1 C_{\alpha f} - a_2 C_{\alpha r}) = 34308.32 \,\mathrm{N\,m/\,rad} \qquad (10.522)$$

$$D_\delta = a_1 C_{\alpha f} = 52139.36 \,\mathrm{N\,m/\,rad} \qquad (10.523)$$

Equations (10.267)-(10.270) indicate that the steady-state response of the vehicle, when $t \to \infty$, are

$$S_\kappa = \frac{\kappa}{\delta} = \frac{1}{R\delta} = 0.2390051454 \qquad (10.524)$$

$$S_\beta = \frac{\beta}{\delta} = -0.2015419091 \qquad (10.525)$$

$$S_r = \frac{r}{\delta} = \frac{\kappa}{\delta}v_x = S_\kappa v_x = 4.780102908 \qquad (10.526)$$

$$S_a = \frac{v_x^2/R}{\delta} = \frac{\kappa}{\delta}v_x^2 = S_\kappa v_x^2 = 95.60205816 \qquad (10.527)$$

$$S_y = \frac{v_y}{\delta} = S_\beta v_x = -4.030838182 \qquad (10.528)$$

Therefore, the steady-state characteristics of the vehicle with $\delta = 0.1$ must be

$$R = 41.84 \,\mathrm{m} \qquad (10.529)$$

$$\beta = -0.02015 \,\mathrm{rad} \approx -1.1545 \,\mathrm{deg} \qquad (10.530)$$

$$r = 0.478 \,\mathrm{rad/\,s} \qquad (10.531)$$

$$\frac{v_x^2}{R} = 9.56 \,\mathrm{m/\,s^2} \qquad (10.532)$$

Substituting the input function (10.535) and solving the equations provides the following solutions:

$$\begin{bmatrix} v_y(t) \\ r(t) \end{bmatrix} = \begin{bmatrix} -0.4 + e^{-7.193t}(1.789\sin 5.113t + 0.403\cos 5.113t) \\ 0.478 + e^{-7.193t}(0.232\sin 5.113t + 0.478\cos 5.113t) \end{bmatrix} \qquad (10.533)$$

Figures 10.40 and 10.41 depict the solutions.

Having $v_y(t)$ and $r(t)$ are enough to calculate the other kinematic variables as well as the required forward force F_x to maintain the constant speed.

$$F_x = -mr\,v_y \qquad (10.534)$$

Figures 10.42 and 10.43 show the kinematics variables of the vehicle, and Figure 10.44 depicts how the require F_x is changing as a function of time.

Example 453 ★ *Response of an oversteer vehicle to a step input.*

Let's assume the steering input is

$$\delta(t) = \begin{cases} 0.1 \,\mathrm{rad} \approx 5.7296 \,\mathrm{deg} & t > 0 \\ 0 & t \le 0 \end{cases} \qquad (10.535)$$

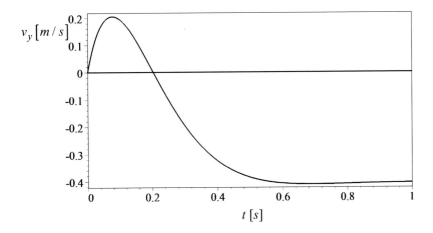

FIGURE 10.40. Lateral velocity response in example 452.

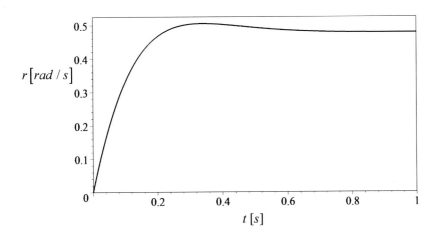

FIGURE 10.41. Yaw rate response in example 452.

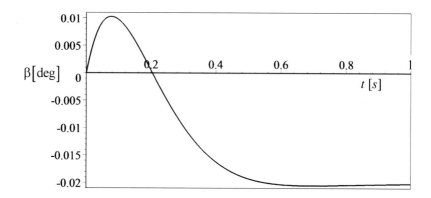

FIGURE 10.42. Sideslip angle response in example 452.

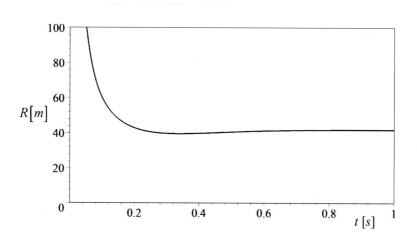

FIGURE 10.43. Radius of rotation response in example 452.

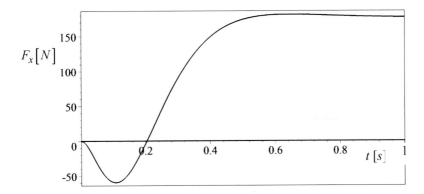

FIGURE 10.44. The required forward force F_x to keep the speed constant, in example 452.

and the vehicle characteristics are

$$
\begin{array}{ll}
C_{\alpha f} = 57296\,\mathrm{N/rad} & C_{\alpha r} = 52712\,\mathrm{N/rad} \\
m = 1400\,\mathrm{kg} \approx 95.9\,\mathrm{slug} & I_z = 1128\,\mathrm{kg\,m^2} \approx 832\,\mathrm{slug\,ft^2} \\
a_1 = 1.25\,\mathrm{m} \approx 4.1\,\mathrm{ft} & a_2 = 1.30\,\mathrm{m} \approx 4.26\,\mathrm{ft} \\
v_x = 20\,\mathrm{m/s} & (10.536)
\end{array}
$$

The equations of motion for a zero initial conditions

$$
\mathbf{q}_0 = \left[\begin{array}{c} v_y\,(0) \\ r\,(0) \end{array} \right] = \left[\begin{array}{c} 0 \\ 0 \end{array} \right] \tag{10.537}
$$

are

$$
\begin{aligned}
\dot{v}_y + 3.928857143 v_y + 20.11051429 r &= 40.92571429 \delta\,(t) \\
&= 4.092571429 \qquad (10.538) \\
\dot{r} + 0.1371631206 v_y + 7.91703369 r &= 63.4929078 \delta\,(t) \\
&= 6.34929078 \qquad (10.539)
\end{aligned}
$$

Substituting the input function (10.535) and solving the equations provides the following solutions:

$$
\left[\begin{array}{c} v_y\,(t) \\ r\,(t) \end{array} \right] = \left[\begin{array}{c} 6.3 e^{-3.328t} - 2.943 e^{-8.518t} - 3.361 \\ -0.188 e^{-3.328t} - 0.672 e^{-8.518t} + 0.86 \end{array} \right] \tag{10.540}
$$

Figures 10.45 and 10.46 depict the solutions.

Example 454 ★ *Standard steer inputs.*

Step and sinusoidal excitation inputs are the most general input to examine the behavior of a vehicle. Furthermore, some other transient inputs

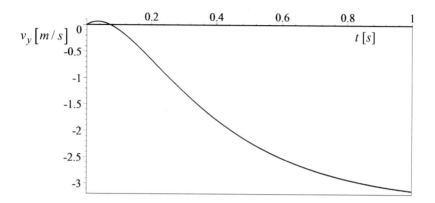

FIGURE 10.45. Lateral velocity response for example 453.

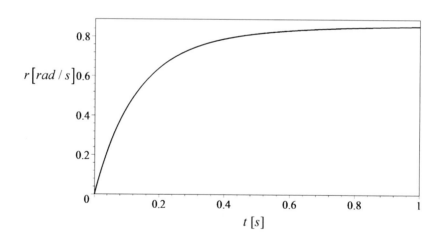

FIGURE 10.46. Yaw velocity response for example 453.

may also be used to analyze the dynamic behavior of a vehicle. Single sine steering, linearly increasing steering, and half sine lane change steering are the most common transient steering inputs.

Example 455 ★ *Position of the rotation center.*
   The position of the center of rotation O in the vehicle body coordinate is at

$$x_O = -R\sin\beta = -\frac{1}{S_k\delta}\sin(S_\beta\delta) \tag{10.541}$$

$$y_O = R\cos\beta = \frac{1}{S_k\delta}\cos(S_\beta\delta) \tag{10.542}$$

because β is positive when it is about positive direction of the z-axis. Figure 10.21 illustrates a two-wheel vehicle model, the vehicle body coordinate frame, and the center of rotation O.
   At steady-state conditions the radius of rotation R can be found from the curvature response S_κ, and the vehicle sideslip angle β can be found from the sideslip response S_β.

$$R = \frac{1}{\delta S_\kappa} = \frac{v_x\left(D_rC_\beta - C_rD_\beta + mv_xD_\beta\right)}{(C_\delta D_\beta - C_\beta D_\delta)\delta} \tag{10.543}$$

$$\beta = \delta S_\beta = \frac{D_\delta\left(C_r - mv_x\right) - D_rC_\delta}{D_rC_\beta - C_rD_\beta + mv_xD_\beta}\delta \tag{10.544}$$

Therefore, the position of the actual center of rotation O is at

$$x_O = -\frac{v_x\left(D_rC_\beta - C_rD_\beta + mv_xD_\beta\right)}{(C_\delta D_\beta - C_\beta D_\delta)\delta}\sin\left(\frac{D_\delta\left(C_r - mv_x\right) - D_rC_\delta}{D_rC_\beta - C_rD_\beta + mv_xD_\beta}\delta\right) \tag{10.545}$$

$$y_O = \frac{v_x\left(D_rC_\beta - C_rD_\beta + mv_xD_\beta\right)}{(C_\delta D_\beta - C_\beta D_\delta)\delta}\cos\left(\frac{D_\delta\left(C_r - mv_x\right) - D_rC_\delta}{D_rC_\beta - C_rD_\beta + mv_xD_\beta}\delta\right) \tag{10.546}$$

Assuming a small β, we may find the position of O approximately.

$$x \approx -\frac{D_\delta\left(C_r - mv_x\right) - D_rC_\delta}{C_\delta D_\beta - C_\beta D_\delta}v_x \tag{10.547}$$

$$y \approx \frac{D_rC_\beta - C_rD_\beta + mv_xD_\beta}{(C_\delta D_\beta - C_\beta D_\delta)\delta}v_x \tag{10.548}$$

Example 456 ★ *Variable v_x.*
   When v_x is variable, the equations of motion are no longer linear and therefore, a numerical integration is needed. Let us assume that a vehicle has the characteristics of

$$\begin{aligned}
C_{\alpha f} &= 60000\,\text{N/rad} & C_{\alpha r} &= 60000\,\text{N/rad} \\
m &= 1000\,\text{kg} & I_z &= 1650\,\text{kg m}^2 \\
a_1 &= 1.0\,\text{m} & a_2 &= 1.5\,\text{m}
\end{aligned} \tag{10.549}$$

Let us assume that the vehicle changes its forward velocity from zero to a maximum of $v_x = 20\,\mathrm{m/s}$ and remains constant while the steering is kept constant at

$$\delta\left(t\right) = 0.1\,\mathrm{rad} \approx 5.73\,\mathrm{deg} \tag{10.550}$$

The force system coefficients for the vehicle from (10.144)-(10.149) are

$$C_r = \frac{30000}{v_x}\,\mathrm{N\,s/\,rad} \qquad C_\beta = -120000\,\mathrm{N/\,rad}$$

$$C_\delta = 60000\,\mathrm{N/\,rad} \qquad D_r = -\frac{195000}{v_x}\,\mathrm{N\,m\,s/\,rad}$$

$$D_\beta = 30000\,\mathrm{N\,m/\,rad} \qquad D_\delta = 60000\,\mathrm{N\,m/\,rad} \tag{10.551}$$

We may assume that v_x varies linearly with time and reaches the maximum speed $v_x = 20\,\mathrm{m/s}$ at $t = t_0$.

$$v_x = \begin{cases} \dfrac{20}{t_0}t\ \mathrm{m/s} & 0 < t < t_0 \\ 20\ \mathrm{m/s} & t_0 < t \end{cases} \tag{10.552}$$

The equations of motion for the vehicle are

$$\dot{v}_y = -\frac{120}{v_x}v_y - r\left(v_x - \frac{30}{v_x}\right) + 60\delta\left(t\right) \tag{10.553}$$

$$\dot{r} = \frac{18.182}{v_x}v_y - 118.18\frac{r}{v_x} + 36.364\delta\left(t\right) \tag{10.554}$$

The graphical illustration of the solutions are shown in Figures 10.47 and 10.48 for

$$t_0 = 20\,\mathrm{s} \tag{10.555}$$

Example 457 ★ *Rotation center between two steady states.*
  Consider a vehicle with the following characteristics

$$C_{\alpha f} = 60000\,\mathrm{N/\,rad} \qquad C_{\alpha r} = 60000\,\mathrm{N/\,rad}$$

$$m = 1000\,\mathrm{kg} \qquad I_z = 1650\,\mathrm{kg\,m^2}$$

$$a_1 = 1.0\,\mathrm{m} \qquad a_2 = 1.5\,\mathrm{m} \tag{10.556}$$

$$K = \frac{m}{l^2}\left(\frac{a_2}{C_{\alpha f}} - \frac{a_1}{C_{\alpha r}}\right) = 1.33 \times 10^{-3}$$

The steady-state responses $S_k = 1/R/\delta$ and $S_\beta = \beta/\delta$ are:

$$S_\kappa = \frac{300}{750 + v_x^2} \qquad S_\beta = \frac{450 - 2v_x^2}{750 + v_x^2} \tag{10.557}$$

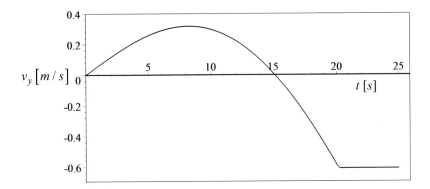

FIGURE 10.47. Lateral velocity of a variable forward velocity vehicle dynamics.

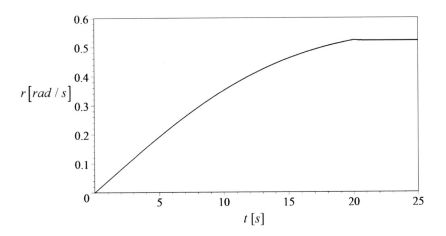

FIGURE 10.48. Yaw rate of a variable forward velocity vehicle dynamics.

The coordinate of the center of rotation (x_O, y_O) in the body coordinate frame at the vehicle mass center are

$$x_O = -R \sin \beta = -\frac{1}{S_k \delta} \sin (S_\beta \delta) \qquad (10.558)$$

$$y_O = R \cos \beta = \frac{1}{S_k \delta} \cos (S_\beta \delta) \qquad (10.559)$$

Let us assume that the car is moving with

$$\delta(t) = 0.1 \,\text{rad} \approx 5.73 \,\text{deg} \qquad (10.560)$$

at constant forward speed of

$$v_x = 3 \,\text{m/s} \qquad (10.561)$$

At the steady-state condition, the vehicle would have

$$
\begin{aligned}
S_\kappa &= 0.3952569170 & S_\beta &= 0.5691699605 \\
S_r &= 1.185770751 & S_y &= 1.707509882
\end{aligned}
\qquad (10.562)
$$

$$
\begin{aligned}
R &= \frac{1}{S_\kappa \delta} = 25.3 \,\text{m} & \beta &= S_\beta \delta = 0.056917 \,\text{rad} \\
r &= S_r \delta = 0.118577 \,\text{rad/s} & v_y &= 0.170751 \,\text{m/s}
\end{aligned}
\qquad (10.563)
$$

We change v_x form $v_x = 3$ m/s at $t = 0$ to the maximum speed $v_x = 20$ m/s at $t = t_0$, and keep it constant. The mathematical expression of v_x with constant acceleration is:

$$v_x = 3 + \frac{20 - 3}{t_0} t \, H(t_0 - t) + (20 - 3) \, H(t - t_0) \ \text{m/s} \qquad (10.564)$$

where $H(t - \tau)$ is the Heaviside function.

$$H(t - \tau) = \begin{cases} 0 & t < \tau \\ 1 & \tau < t \end{cases} \qquad (10.565)$$

Assuming a very high t_0 and very low acceleration, the transition from $v_x = 3$ m/s to $v_x = 20$ m/s will be close to the intermediate steady-state conditions. To have a low velocity sweep, let us change the speed in

$$t_0 = 120 \,\text{s} \qquad (10.566)$$

With such a long time for velocity change, the kinematics of the vehicle will change according to their steady-state values. Figure 10.49 depicts how the forward velocity varies, and Figure 10.50 shows the variation of the lateral velocity v_y and Figure 10.51 depicts the yaw rate r. The loci of the rotation center (x_O, y_O) is shown in Figure 10.52.

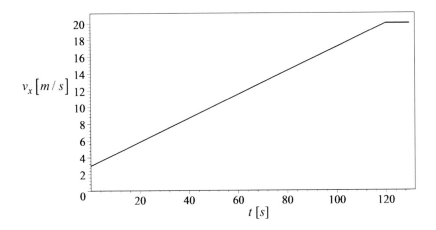

FIGURE 10.49. The forward velocity v_x of a vehicle with a constant low acceleration between $v_x = 3\,\mathrm{m/s}$ and $v_x = 20\,\mathrm{m/s}$.

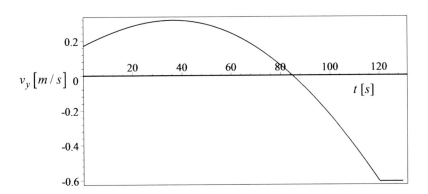

FIGURE 10.50. The lateral velocity v_y of a vehicle with a constant low acceleration between $v_x = 3\,\mathrm{m/s}$ and $v_x = 20\,\mathrm{m/s}$.

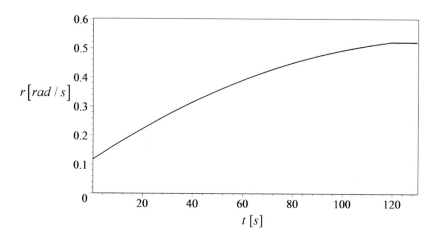

FIGURE 10.51. The yaw rate r of a vehicle with a constant low acceleration between $v_x = 3\,\text{m}/\text{s}$ and $v_x = 20\,\text{m}/\text{s}$.

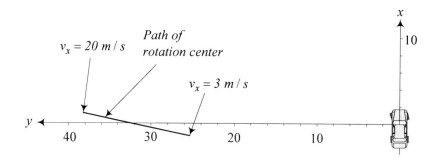

FIGURE 10.52. The loci of the rotation center of a vehicle with a constant low acceleration between $v_x = 3\,\text{m}/\text{s}$ and $v_x = 20\,\text{m}/\text{s}$.

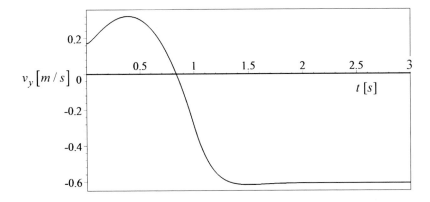

FIGURE 10.53. The lateral velocity v_y of a vehicle with a constant very high acceleration between $v_x = 3\,\mathrm{m/\,s}$ and $v_x = 20\,\mathrm{m/\,s}$.

If t_0 is short, the acceleration is high, and the dynamics of the vehicle will deviate from the almost steady state conditions. To have a fast velocity sweep, let us change the speed in

$$t_0 = 1\,\mathrm{s} \tag{10.567}$$

Figure 10.53 shows the variation of the lateral velocity v_y and Figure 10.54 shows the yaw rate r. The coordinate of the rotation center (x_O, y_O) are shown in Figure 10.55. The loci of the path of rotation center (x_O, y_O) is shown in Figure 10.56 to be compared with the loci at very low acceleration maneuver.

Example 458 ★ *Vehicles act close to steady-state conditions.*

Interestingly, vehicles act on their steady-state condition with a good approximation. In the previous example 457, a vehicle was supposed to change its speed while it is moving at steady state conditions with $v_x = 3\,\mathrm{m/\,s}$ and $\delta(t) = 0.1\,\mathrm{rad}$ to $v_x = 20\,\mathrm{m/\,s}$.

When the acceleration of the vehicle is very low $a = 0.14167\,\mathrm{m/\,s^2}$, it takes 120 s for the vehicle to change its speed. The vehicle is always at its steady state conditions while the speed changes slowly. Figure 10.50-10.52 show how the dynamics of the vehicle changes. At the moment the acceleration becomes zero, the dynamics are at their steady-state values and they stay their since after.

When the acceleration of the vehicle is 120 times higher at a very high value of $a = 17\,\mathrm{m/\,s^2}$, it takes only 1 s for the vehicle to change its speed. After cutting the acceleration, it takes less than 1 s for the vehicle to achieve its steady-state conditions, as are shown in Figures 10.53-10.55. Figure 10.57 illustrates very well how the rotation center of the vehicle deviates from the steady-state in a high acceleration. It also shows how far its rotation center is from its steady-state when acceleration stops.

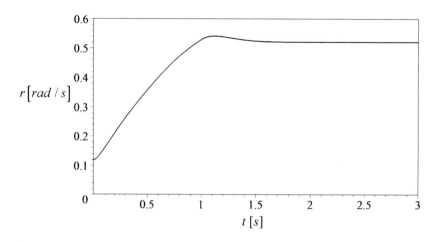

FIGURE 10.54. The yaw rate r of a vehicle with a constant very high acceleration between $v_x = 3\,\text{m/s}$ and $v_x = 20\,\text{m/s}$.

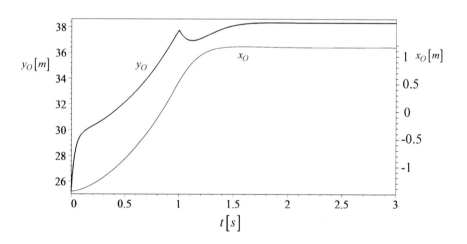

FIGURE 10.55. The coordinate of the rotation center of a vehicle with a constant very high acceleration between $v_x = 3\,\text{m/s}$ and $v_x = 20\,\text{m/s}$.

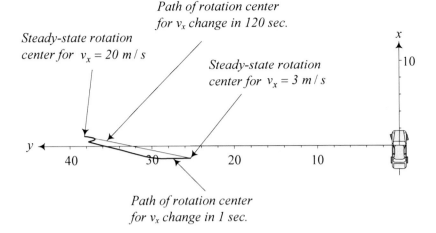

FIGURE 10.56. The loci of the rotation center of a vehicle with a constant very high acceleration between $v_x = 3\,\mathrm{m/s}$ and $v_x = 20\,\mathrm{m/s}$.

As a result of this example, we know that vehicles act based on their steady-state dynamics in normal operations.

Example 459 ★ *Global position of a vehicle.*
A vehicle with the following characteristics

$$
\begin{aligned}
C_{\alpha f} &= 60000\,\mathrm{N/rad} & C_{\alpha r} &= 60000\,\mathrm{N/rad} \\
m &= 1000\,\mathrm{kg} & I_z &= 1650\,\mathrm{kg\,m^2} \\
a_1 &= 1.0\,\mathrm{m} & a_2 &= 1.5\,\mathrm{m} \\
v_x &= 20\,\mathrm{m/s} &&
\end{aligned}
\tag{10.568}
$$

is moving straight with a constant forward velocity of

$$
v_x = 20\,\mathrm{m/s}
\tag{10.569}
$$

At $t = 0$, when the car is at the origin of a global coordinate frame, suddenly we change the steer angle to

$$
\delta(t) = \begin{cases} 0.1\,\mathrm{rad} \approx 5.73\,\mathrm{deg} & t > 0 \\ 0 & t \le 0 \end{cases}
\tag{10.570}
$$

Calculating the force system coefficients for the vehicle from (10.144)-(10.149)

$$
\begin{aligned}
C_r &= 1500\,\mathrm{N\,s/rad} & C_\beta &= -120000\,\mathrm{N/rad} \\
C_\delta &= 60000\,\mathrm{N/rad} & D_r &= -9750\,\mathrm{N\,m\,s/rad} \\
D_\beta &= 30000\,\mathrm{N\,m/rad} & D_\delta &= 60000\,\mathrm{N\,m/rad}
\end{aligned}
\tag{10.571}
$$

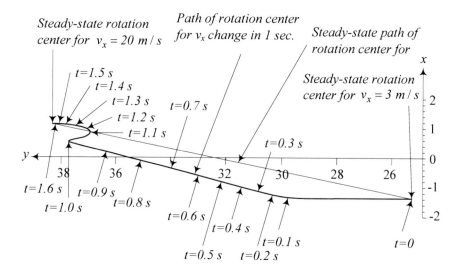

FIGURE 10.57. The loci of the rotation center of a vehicle with a constant very high acceleration between $v_x = 3\,\mathrm{m/s}$ and $v_x = 20\,\mathrm{m/s}$, compared to the steady-state loci.

we determine the steady-state values of the vehicle's variable kinematics.

$$R = \frac{1}{S_\kappa \delta} = 38.33\,\mathrm{m} \qquad \beta = S_\beta \delta = -0.0304\,\mathrm{rad}$$

$$r = S_r \delta = 0.522\,\mathrm{rad/s} \qquad v_y = S_y \delta = -0.608\,\mathrm{m/s} \qquad (10.572)$$

The solution of the equations of motion

$$\dot{v}_y + 6v_y + 18.5r = 60\delta(t) = 6 \qquad (10.573)$$
$$\dot{r} - 0.909v_y + 5.909r = 36.363\delta(t) = 3.636 \qquad (10.574)$$

for a zero initial conditions are

$$\begin{bmatrix} v_y(t) \\ r(t) \end{bmatrix} = \begin{bmatrix} -0.609 + e^{-5.95t}\,(2.347\sin 4.1t + 0.609\cos 4.1t) \\ 0.522 + e^{-5.95t}\,(0.129\sin 4.1t - 0.522\cos 4.1t) \end{bmatrix}$$
$$(10.575)$$

The angular orientation of the body coordinate frame with respect to the fixed global coordinate frame is

$$\psi = \int_0^t r(t)\,dt \qquad (10.576)$$

and the velocity vector of the vehicle in the global frame is

$$^B\mathbf{v} = \begin{bmatrix} v_x \\ v_y(t) \end{bmatrix} \qquad (10.577)$$

Therefore, the velocity vector of the vehicle in the global frame is

$$
{}^G\mathbf{v} = [R]\,{}^B\mathbf{v} = \begin{bmatrix} \cos\psi & -\sin\psi \\ \sin\psi & \cos\psi \end{bmatrix} \begin{bmatrix} v_x \\ v_y(t) \end{bmatrix}
$$

$$
= \begin{bmatrix} v_x\cos\psi - v_y\sin\psi \\ v_y\cos\psi + v_x\sin\psi \end{bmatrix} = \begin{bmatrix} v_X \\ v_Y \end{bmatrix} \tag{10.578}
$$

The global coordinate of the mass center of the vehicle would be

$$
X = \int_0^t v_X\,dt = \int_0^t (v_x\cos\psi - v_y\sin\psi)\,dt \tag{10.579}
$$

$$
Y = \int_0^t v_Y\,dt = \int_0^t (v_y\cos\psi + v_x\sin\psi)\,dt \tag{10.580}
$$

When the steer angle is constant, the vehicle will eventually be turning on a constant circular path. The position of the steady-state rotation center in the body coordinate frame is at

$$
x_O = -R\sin\beta = 1.1651\,\mathrm{m} \tag{10.581}
$$

$$
y_O = R\cos\beta = 38.312\,\mathrm{m} \tag{10.582}
$$

Therefore the global coordinate of the rotation center would be

$$
\begin{bmatrix} X_O \\ Y_O \end{bmatrix} = {}^G_G\mathbf{r}_B + {}^G R_B \begin{bmatrix} x_O \\ y_O \end{bmatrix} \tag{10.583}
$$

where ${}^G_G\mathbf{r}_B$ is the G-expression of the position vector of the origin of the B-frame with respect to the origin of the G-frame at any point on the steady-state circular path. Figure 10.58 illustrates the path of motion of the vehicle and the steady-state rotation center.

Example 460 ★ *Second-order equations.*

The coupled equations of motion (10.188) may be modified to a second-order differential equation of only one variable. To do this, let us rewrite the equations.

$$
\dot{v}_y = \frac{1}{mv_x}\left(v_y C_\beta + rC_r v_x - mrv_x^2 + \delta v_x C_\delta\right) \tag{10.584}
$$

$$
\dot{r} = \frac{1}{I_z v_x}\left(v_y D_\beta + rD_r v_x + \delta v_x D_\delta\right) \tag{10.585}
$$

Assuming a constant forward speed

$$
v_x = const \tag{10.586}
$$

and taking a derivative from Equation (10.585) yields:

$$
\ddot{r} = \frac{1}{I_z v_x}\left(\dot{v}_y D_\beta + \dot{r}D_r v_x\right) \tag{10.587}
$$

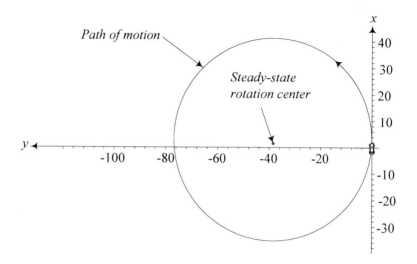

FIGURE 10.58. The path of motion of a vehicle with a step steer angle that is moving with constant forward speed.

We substitute Equation (10.584) in (10.587)

$$\ddot{r} = \frac{1}{I_z v_x}\left(\frac{1}{m v_x}\left(v_y C_\beta + r C_r v_x - mr v_x^2 + \delta v_x C_\delta\right)\right) D_\beta$$
$$+ \frac{1}{I_z v_x}\dot{r} D_r v_x + \dot{\delta} v_x D_\delta \qquad (10.588)$$

and then substitute for v_y from (10.585) and get the following equation:

$$m I_z v_x \ddot{r} - \left(I_z C_\beta + m D_r v_x\right)\dot{r} + \left(D_r C_\beta - C_r F_\beta + m v_x D_\beta\right) r$$
$$= -\left(\delta C_\delta D_\beta - \delta C_\beta D_\delta + m\dot{\delta} v_x D_\delta\right) \qquad (10.589)$$

This equation is similar to the equation of motion for a force vibration single DOF system

$$m_{eq}\ddot{r} + c_{eq}\dot{r} + k_{eq}r = f_{eq}(t) \qquad (10.590)$$

where the equivalent mass m_{eq}, damping c_{eq}, stiffness k_{eq}, and force $f_{eq}(t)$ are:

$$m_{eq} = m I_z v_x \qquad (10.591)$$
$$c_{eq} = -\left(I_z C_\beta + m D_r v_x\right) \qquad (10.592)$$
$$k_{eq} = D_r C_\beta - C_r F_\beta + m v_x D_\beta \qquad (10.593)$$
$$f_e(t) = -m\frac{d\delta(t)}{dt}v_x D_\delta - (C_\delta D_\beta - C_\beta D_\delta)\delta(t) \qquad (10.594)$$

We may use Equation (10.589) and determine the behavior of a vehicle similar to analysis of a vibrating system. The response of the equation to a step steering input may be expressed by rise time, peak time, overshoot, and settling time.

10.8 Summary

A vehicle is effectively modeled as a rigid bicycle in a planar motion by ignoring the roll of the vehicle. Such a vehicle has three DOF in a body coordinate frame attached to the vehicle at C: forward motion, lateral motion, and yaw motion. If the vehicle is front wheel steering, the dynamic equations of such a vehicle are expressed in (v_x, v_y, r) variables in the following set of three coupled first order ordinary differential equations.

$$\dot{v}_x = \frac{F_x}{m} + r\,v_y \tag{10.595}$$

$$\dot{v}_y = \frac{1}{mv_x}\left(-a_1 C_{\alpha f} + a_2 C_{\alpha r}\right) r$$
$$-\frac{1}{mv_x}\left(C_{\alpha f} + C_{\alpha r}\right) v_y + \frac{1}{m}C_{\alpha f}\delta - r\,v_x \tag{10.596}$$

$$\dot{r} = \frac{1}{I_z v_x}\left(-a_1^2 C_{\alpha f} - a_2^2 C_{\alpha r}\right) r$$
$$-\frac{1}{I_z v_x}\left(a_1 C_{\alpha f} - a_2 C_{\alpha r}\right) v_y + \frac{1}{I_z}a_1 C_{\alpha f}\delta \tag{10.597}$$

The second and third equations may be written in a matrix form for $\begin{bmatrix} v_y & r \end{bmatrix}^T$

$$\begin{bmatrix} \dot{v}_y \\ \dot{r} \end{bmatrix} = \begin{bmatrix} -\dfrac{C_{\alpha f}+C_{\alpha r}}{mv_x} & \dfrac{-a_1 C_{\alpha f}+a_2 C_{\alpha r}}{mv_x}-v_x \\ -\dfrac{a_1 C_{\alpha f}-a_2 C_{\alpha r}}{I_z v_x} & -\dfrac{a_1^2 C_{\alpha f}+a_2^2 C_{\alpha r}}{I_z v_x} \end{bmatrix}\begin{bmatrix} v_y \\ r \end{bmatrix}$$
$$+\begin{bmatrix} \dfrac{C_{\alpha f}}{m} \\ \dfrac{a_1 C_{\alpha f}}{I_z} \end{bmatrix}\delta \tag{10.598}$$

or in a matrix form for $\begin{bmatrix} \beta & r \end{bmatrix}^T$.

$$
\begin{bmatrix} \dot{\beta} \\ \dot{r} \end{bmatrix} = \begin{bmatrix} -\dfrac{C_{\alpha f} + C_{\alpha r}}{mv_x} & \dfrac{-a_1 C_{\alpha f} + a_2 C_{\alpha r}}{mv_x^2} - 1 \\[3ex] -\dfrac{a_1 C_{\alpha f} - a_2 C_{\alpha r}}{I_z} & -\dfrac{a_1^2 C_{\alpha f} + a_2^2 C_{\alpha r}}{I_z v_x} \end{bmatrix} \begin{bmatrix} \beta \\ r \end{bmatrix}
$$

$$
+ \begin{bmatrix} \dfrac{C_{\alpha f}}{mv_x} \\[3ex] \dfrac{a_1 C_{\alpha f}}{I_z} \end{bmatrix} \delta \tag{10.599}
$$

10.9   Key Symbols

$a \equiv \ddot{x}$	acceleration
a_i	distance of the axle number i from the mass center
a_N	
$[A]$	force coefficient matrix
b_1	distance of the hinge point from rear axle
b_2	distance of trailer axle from the hinge point
$B(Cxyz)$	vehicle coordinate frame
C	mass center
C_α	tire sideslip coefficient
$C_{\alpha f}$	front sideslip coefficient
$C_{\alpha f_L}$	front left sideslip coefficient
$C_{\alpha f_R}$	front right sideslip coefficient
$C_{\alpha r}$	rear sideslip coefficient
$C_{\alpha r_L}$	rear left sideslip coefficient
$C_{\alpha r_R}$	rear right sideslip coefficient
$C_r, \cdots, D_\delta$	force system coefficients
C_r	proportionality coefficient between F_y and r
C_β	proportionality coefficient between F_y and β
C_δ	proportionality coefficient between F_y and δ
D_r	proportionality coefficient between M_z and r
D_β	proportionality coefficient between M_z and β
D_δ	proportionality coefficient between M_z and δ
$\mathbf{d}$	frame position vector
d_N	neutral distance
dm	mass element
F_i	generalized force
F_x	longitudinal force, forward force, traction force
F_y	lateral force
F_{yf}	front lateral force
F_{yr}	rear lateral force
F_z	normal force, vertical force, vehicle load
$\mathbf{F}, \mathbf{M}$	vehicle force system
$\mathbf{g}, g$	gravitational acceleration
$G(OXYZ)$	global coordinate frame
I	mass moment
K	stability factor
$\mathbf{L}$	moment of momentum
$\mathcal{L}$	Lagrangean
m	mass
M_x	roll moment, bank moment, tilting torque
M_y	pitch moment
M_z	yaw moment, aligning moment

$p = \dot{\varphi}$	roll rate
$\mathbf{p}$	momentum
P_N	neutral steer point
$q = \dot{\theta}$	pitch rate
$\mathbf{q}$	control variable vector
q_i	generalized coordinate
$r = \dot{\psi}$	yaw rate
$\mathbf{r}$	position vector
R	radius of rotation
R_w	tire radius
$^G R_B$	rotation matrix to go from B frame to G frame
$S_\kappa = \kappa/\delta$	curvature response
$S_\beta = \beta/\delta$	sideslip response
$S_r = r/\delta$	yaw rate response
$S_a = v_x^2/R/\delta$	centripetal acceleration response
S_y	lateral velocity responses
S_{y_1}, S_{y_2}	steady-state responses
$S_A = 1/v_x^2$	Ackerman steering gradient
t	time
T	tire coordinate frame
T_w	wheel torque
$\mathbf{u}$	input vector
$v \equiv \dot{x}, \ \mathbf{v}$	velocity
w	wheelbase
$x, y, z, \ \mathbf{x}$	displacement
α	sideslip angle
β	global sideslip angle
β	vehicle sideslip angle, attitude angle
β	attitude angle
$\beta + \psi$	cruise angle
δ	steer angle
δ_f	front steer angle
δ_r	rear steer angle
θ	pitch angle
$\dot{\theta} = q$	pitch rate
$\kappa = 1/R$	curvature
λ	eigenvalue
φ	roll angle
$\dot{\varphi} = p$	roll rate
ψ	yaw angle
$\dot{\psi} = r$	yaw rate
ψ	heading angle
$\boldsymbol{\omega}$	angular velocity
$\dot{\boldsymbol{\omega}}$	angular acceleration

Exercises

1. Force system coefficients.

   Consider a front-wheel-steering car with the following characteristics

   $$C_{\alpha f_L} = C_{\alpha f_R} = 500 \,\mathrm{N/\,deg} \qquad C_{\alpha r_L} = C_{\alpha r_R} = 460 \,\mathrm{N/\,deg}$$
   $$m = 1245 \,\mathrm{kg} \qquad I_z = 1328 \,\mathrm{kg\,m^2}$$
   $$a_1 = 110 \,\mathrm{cm} \qquad a_2 = 132 \,\mathrm{cm}$$

   (a) Determine the force system coefficients C_r, C_β, C_δ, D_r, D_β, D_δ.

   (b) Plot C_r, D_r as functions of v_x for $0 < v_x < 60 \,\mathrm{m/\,s}$.

2. Dimension of the force characteristic coefficients.

   What are the dimensions of the force system coefficients, C_r, C_β, C_δ, D_r, D_β, D_δ?

3. Force system and two-wheel model of a car.

   Consider a front-wheel-steering car with

   $$C_{\alpha r_L} = C_{\alpha r_R} = C_{\alpha f_L} = C_{\alpha f_R} = 500 \,\mathrm{N/\,deg}$$
   $$a_1 = 110 \,\mathrm{cm} \qquad a_2 = 132 \,\mathrm{cm}$$
   $$m = 1205 \,\mathrm{kg} \qquad I_z = 1300 \,\mathrm{kg\,m^2}$$

   (a) determine the force system that applies on the two-wheel model of the car.

   $$F_y = C_r \, r + C_\beta \, \beta + C_\delta \, \delta$$
   $$M_z = D_r \, r + D_\beta \, \beta + D_\delta \, \delta$$

   (b) Write the equations of motion of the car as

   $$F_x = m \, \dot{v}_x - m r \, v_y$$
   $$F_y = m \, \dot{v}_y + m r \, v_x$$
   $$M_z = \dot{r} \, I_z$$

   (c) Derive the force system coefficient that the velocity is measured in km/h instead of m/s.

4. Equations of motion for a front-wheel-steering car.

   Consider a front-wheel-steering car with

   $$C_{\alpha r_L} = C_{\alpha r_R} = C_{\alpha f_L} = C_{\alpha f_R} = 500 \,\mathrm{N/\,deg}$$

$$a_1 = 110 \, \text{cm} \qquad a_2 = 132 \, \text{cm}$$
$$m = 1245 \, \text{kg} \qquad I_z = 1328 \, \text{kg} \, \text{m}^2$$
$$v_x = 40 \, \text{m/s}$$

and develop the equations of motion

$$\dot{\mathbf{q}} = [A]\,\mathbf{q} + \mathbf{u}$$

5. ★ Nonlinear tire behavior.

Let us assume that the sideslip coefficient of the tires of a vehicle are nonlinear such that its lateral force generation capacity drops at high α.

$$C_\alpha = C_1 - C_2 \alpha^2$$

(a) Develop the force system for the vehicle.

(b) Derive the equations of motion for a front wheel steerable planar bicycle car model.

6. No approximation for β.

Do not use the approximations of

$$\beta_f = \arctan\left(\frac{v_{y_f}}{v_{x_f}}\right) \approx \frac{v_{y_f}}{v_{x_f}}$$

$$\beta_r = \arctan\left(\frac{v_{y_r}}{v_{x_r}}\right) \approx \frac{v_{y_r}}{v_{x_r}}$$

$$\beta = \arctan\left(\frac{v_y}{v_x}\right) \approx \frac{v_y}{v_x}$$

(a) Accept the linear functions of

$$F_{y_f} = -C_{\alpha f}\,\alpha_f \qquad F_{y_r} = -C_{\alpha r}\,\alpha_r$$

and develop the force system for the planar bicycle car model.

(b) Assume that

$$F_{y_f} = -C_{1f}\,\alpha_f + C_{2f}\,\alpha_f^3 \qquad F_{y_r} = -C_{1r}\,\alpha_r + C_{2r}\,\alpha_r^3$$

and develop the force system for the planar bicycle car model.

7. Equations of motion in different variables.

Consider a car with the following characteristics

$$C_{\alpha r_L} = C_{\alpha r_R} = C_{\alpha f_L} = C_{\alpha f_R} = 500 \, \text{N/deg}$$
$$a_1 = 100 \, \text{cm} \qquad a_2 = 120 \, \text{cm}$$
$$m = 1000 \, \text{kg} \qquad I_z = 1008 \, \text{kg} \, \text{m}^2$$
$$v_x = 40 \, \text{m/s}$$

and develop the equations of motion

(a) in terms of $(\dot{v}_x, \dot{v}_y, \dot{r})$, if the car is front-wheel steering.

(b) in terms of $(\dot{v}_x, \dot{v}_y, \dot{r})$, if the car is four-wheel steering.

(c) in terms of $\left(\dot{v}_x, \dot{\beta}, \dot{r}\right)$, if the car is front-wheel steering.

(d) in terms of $\left(\dot{v}_x, \dot{\beta}, \dot{r}\right)$, if the car is four-wheel steering.

8. Steady state response parameters.

   Consider a car with the following characteristics

$$
\begin{aligned}
C_{\alpha f_L} &= C_{\alpha f_R} = 500\,\mathrm{N/deg} & C_{\alpha r_L} &= C_{\alpha r_R} = 520\,\mathrm{N/deg} \\
m &= 1245\,\mathrm{kg} & I_z &= 1328\,\mathrm{kg\,m^2} \\
a_1 &= 110\,\mathrm{cm} & a_2 &= 132\,\mathrm{cm}
\end{aligned}
$$

   (a) Determine the steady-state responses S_κ, S_β, S_r, S_a, S_y.

   (b) Plot the steady-state responses as functions of v_x.

   (c) Determine their values at $v_x = 40\,\mathrm{m/s}$.

9. Steady state motion parameters.

   Consider a car with the following characteristics

$$
\begin{aligned}
C_{\alpha f_L} &= C_{\alpha f_R} = 600\,\mathrm{N/deg} & C_{\alpha r_L} &= C_{\alpha r_R} = 550\,\mathrm{N/deg} \\
m &= 1245\,\mathrm{kg} & I_z &= 1128\,\mathrm{kg\,m^2} \\
a_1 &= 120\,\mathrm{cm} & a_2 &= 138\,\mathrm{cm} \\
v_x &= 20\,\mathrm{m/s} & \delta &= 3\,\mathrm{deg}
\end{aligned}
$$

   (a) Determine the steady-state values of r, R, β, and v_x^2/R.

   (b) Find the coordinates of the rotation center.

10. ★ Inertia and steady-state parameters.

    Consider a car that is made up of a uniform solid box with dimensions $260\,\mathrm{cm} \times 140\,\mathrm{cm} \times 40\,\mathrm{cm}$ in $x \times z \times y$ dimensions. If the density of the box is $\rho = 1000\,\mathrm{kg/m^3}$, and the other characteristics are

$$
\begin{aligned}
C_{\alpha f_L} &= C_{\alpha f_R} = 600\,\mathrm{N/deg} & C_{\alpha r_L} &= C_{\alpha r_R} = 550\,\mathrm{N/deg} \\
a_1 &= a_2 = \frac{l}{2} = 1.25\,\mathrm{m}
\end{aligned}
$$

    then,

    (a) Determine m, I_z.

    (b) Determine the steady-state responses S_κ, S_β, S_r, and S_a as functions of v_x.

(c) Determine the velocity v_x at which the car has a radius of turning equal to

$$R = 35\,\mathrm{m}$$

when

$$\delta = 4\,\mathrm{deg}$$

(d) Determine the steady-state parameters r, R, β, v_x^2/R and center of rotation at that speed.

(e) Set the speed of the car at

$$v_x = 20\,\mathrm{m/s}$$

and plot the steady-state responses S_κ, S_β, S_r, and S_a for variable ρ.

11. Stability factor and understeer behavior.

Examine the stability factor K and

(a) determine the condition to have an understeer car, if $a_1 = a_2$.

(b) determine the condition to have an understeer car, if $C_{af} = C_{ar}$.

(c) show that if we use the tires in front and rear such that $C_{af} = C_{ar}$, then the front of the car must be heavier to have an understeer car?

(d) show that if we have a car with $a_1 = a_2$, then we must use tires such that $C_{ar} > C_{af}$ to have an understeer car?

12. Stability factor and mass of the car.

Find a_1 and a_2 in terms of F_{z_1}, F_{z_2}, and mg to rewrite the stability factor K to see the effect of a car's mass distribution.

13. Stability factor and car behavior.

Examine the stability factor of a car with the parameters

$$
\begin{aligned}
C_{\alpha f_L} &= C_{\alpha f_R} = 500\,\mathrm{N/deg} & C_{\alpha r_L} &= C_{\alpha r_R} = 460\,\mathrm{N/deg} \\
m &= 1245\,\mathrm{kg} & I_z &= 1328\,\mathrm{kg\,m^2} \\
a_1 &= 110\,\mathrm{cm} & a_2 &= 132\,\mathrm{cm} \\
v_x &= 30\,\mathrm{m/s}
\end{aligned}
$$

and

(a) determine if the car is understeer, neutral, or oversteer?

(b) determine the neutral distance d_N.

14. ★ Critical speeds of a car.

Compare the critical speeds of curvature, sideslip, and yaw rate responses and order them if possible. Determine the conditions at which they are equal if there is any.

15. Critical speed of a car.

Consider a car with the characteristics

$$
\begin{aligned}
C_{\alpha f_L} &= C_{\alpha f_R} = 700\,\mathrm{N/deg} & C_{\alpha r_L} &= C_{\alpha r_R} = 520\,\mathrm{N/deg} \\
m &= 1245\,\mathrm{kg} & I_z &= 1328\,\mathrm{kg\,m^2} \\
a_1 &= 118\,\mathrm{cm} & a_2 &= 122\,\mathrm{cm}
\end{aligned}
$$

(a) determine if the car is understeer, neutral, or oversteer?

(b) in case of an oversteer situation, determine the neutral distance d_N and the critical speed v_c of the car.

16. Steady-state R for oversteer vehicles.

Consider a car with

$$
\begin{aligned}
C_{\alpha f_L} &= C_{\alpha f_R} = 500\,\mathrm{N/deg} & C_{\alpha r_L} &= C_{\alpha r_R} = 460\,\mathrm{N/deg} \\
m &= 1245\,\mathrm{kg} & I_z &= 1328\,\mathrm{kg\,m^2} \\
a_1 &= 110\,\mathrm{cm} & a_2 &= 132\,\mathrm{cm}
\end{aligned}
$$

(a) Determine and plot R versus v_x.

(b) Switch a_1 and a_2 such that

$$
a_2 = 110\,\mathrm{cm} \qquad a_1 = 132\,\mathrm{cm}
$$

and determine and plot R versus v_x.

17. ★ Step input response at different speed.

Consider a car with the characteristics

$$
\begin{aligned}
C_{\alpha f_L} &= C_{\alpha f_R} = 600\,\mathrm{N/deg} & C_{\alpha r_L} &= C_{\alpha r_R} = 750\,\mathrm{N/deg} \\
m &= 1245\,\mathrm{kg} & I_z &= 1328\,\mathrm{kg\,m^2} \\
a_1 &= 110\,\mathrm{cm} & a_2 &= 132\,\mathrm{cm}
\end{aligned}
$$

and a step input

$$
\delta\left(t\right) = \begin{cases} 5\,\mathrm{deg} & t > 0 \\ 0 & t \le 0 \end{cases}
$$

Determine the time response of the car at the following forward speeds and plot r as a function of time.

(a) $v_x = 10\,\mathrm{m/s}$

(b) $v_x = 20\,\mathrm{m/s}$

(c) $v_x = 30\,\mathrm{m/s}$

(d) $v_x = 40\,\mathrm{m/s}$

(e) Determine r_{Max} in each case and examine if r_{Max} is linearly proportional with v_x.

(f) Find r_{Max} as a function of v_x analytically and describe the care behavior based on the plot of r_{Max}.

18. Steady-state rotation center function.

    Consider a car with

$$
\begin{aligned}
C_{\alpha f_L} &= C_{\alpha f_R} \approx 30000\,\mathrm{N/rad} \qquad C_{\alpha r_L} = C_{\alpha r_R} \approx 30000\,\mathrm{N/rad} \\
m &= 1000\,\mathrm{kg} \qquad I_z = 1650\,\mathrm{kg\,m^2} \\
a_1 &= 1.0\,\mathrm{m} \qquad a_2 = 1.5\,\mathrm{m} \\
K &= \frac{m}{l^2}\left(\frac{a_2}{C_{\alpha f}} - \frac{a_1}{C_{\alpha r}}\right) = 1.33 \times 10^{-2} > 0
\end{aligned}
$$

(a) Determine the coordinate of rotation center (x_O, y_O).

(b) Check if it is possible to eliminate v_x to find $y_O = f(x_O)$.

(c) Approximate $y_O = f(x_O)$ by a linear function.

19. Stability factor analysis.

    Consider a car with

$$
\begin{aligned}
C_{\alpha f} &= 57296\,\mathrm{N/rad} \qquad C_{\alpha r} = 52712\,\mathrm{N/rad} \\
m &= 917\,\mathrm{kg} \qquad I_z = 1128\,\mathrm{kg\,m^2} \\
l &= 2.55\,\mathrm{m} \qquad a_1 = x \qquad a_2 = l - x
\end{aligned}
$$

(a) Plot K as a function of x.

(b) ★ Plot a three dimensional surface of S_κ as a function of x and v_x.

20. ★ Mass center and velocity.

    Let us imagine a car such that the position of its mass center shifts based on its forward speed for $0 < v_x < 100\,\mathrm{m/s}$.

$$
\begin{aligned}
C_{\alpha f} &= 57296\,\mathrm{N/rad} \qquad C_{\alpha r} = 52712\,\mathrm{N/rad} \\
m &= 917\,\mathrm{kg} \qquad I_z = 1128\,\mathrm{kg\,m^2} \\
l &= 2.55\,\mathrm{m} \qquad a_1 = 1 + \frac{v_x}{100} \qquad a_2 = l - 1 - \frac{v_x}{100}
\end{aligned}
$$

(a) Plot K as a function of x.

(b) Plot S_κ as a function of x.

21. ★ Step input response for different steer angle.

Consider a car with the characteristics

$$
\begin{aligned}
C_{\alpha f_L} &= C_{\alpha f_R} = 600\,\text{N/deg} \qquad C_{\alpha r_L} = C_{\alpha r_R} = 750\,\text{N/deg} \\
m &= 1245\,\text{kg} \qquad I_z = 1328\,\text{kg m}^2 \\
a_1 &= 110\,\text{cm} \qquad a_2 = 132\,\text{cm} \\
v_x &= 20\,\text{m/s}
\end{aligned}
$$

Determine the time response of the car to a step input

$$
\delta(t) = \begin{cases} \delta & t > 0 \\ 0 & t \le 0 \end{cases}
$$

and plot r versus v_x when

(a) $\delta = 2\,\text{deg}$, $\delta = 3\,\text{deg}$, $\delta = 4\,\text{deg}$, $\delta = 5\,\text{deg}$, $\delta = 6\,\text{deg}$, $\delta = 7\,\text{deg}$, $\delta = 8\,\text{deg}$, $\delta = 9\,\text{deg}$, $\delta = 10\,\text{deg}$.

(b) Determine r_{Max} in each case and examine if r_{Max} is linearly proportional with v_x.

(c) Find r_{Max} as a function of v_x analytically and describe the care behavior based on the plot of r_{Max}.

22. ★ Eigenvalues and free response.

Consider a car with the characteristics

$$
\begin{aligned}
C_{\alpha f_L} &= C_{\alpha f_R} = 600\,\text{N/deg} \qquad C_{\alpha r_L} = C_{\alpha r_R} = 750\,\text{N/deg} \\
m &= 1245\,\text{kg} \qquad I_z = 1328\,\text{kg m}^2 \\
a_1 &= 110\,\text{cm} \qquad a_2 = 132\,\text{cm} \\
v_x &= 20\,\text{m/s}
\end{aligned}
$$

(a) Determine the eigenvalues of the coefficient matrix $[A]$ and find out if the car is stable at zero steer angle.

(b) In either case, determine the weight distribution ratio, a_1/a_2, such that the car is neutral stable.

(c) Recommend a condition for the weight distribution ratio, a_1/a_2, such that the car is stable.

23. ★ Time response to different steer functions.

Consider a car with the characteristics

$$
\begin{aligned}
C_{\alpha f_L} &= C_{\alpha f_R} = 600\,\text{N/deg} \qquad C_{\alpha r_L} = C_{\alpha r_R} = 750\,\text{N/deg} \\
m &= 1245\,\text{kg} \qquad I_z = 1328\,\text{kg m}^2 \\
a_1 &= 110\,\text{cm} \qquad a_2 = 132\,\text{cm} \\
v_x &= 20\,\text{m/s}
\end{aligned}
$$

Determine the time response of the car to

(a) $\delta(t) = \sin 0.1t$ for $0 < t < 10\pi$ and $\delta(t) = 0$ for $t \leq 0$ and $t \geq 10\pi$.

(b) $\delta(t) = \sin 0.5t$ for $0 < t < 2\pi$ and $\delta(t) = 0$ for $t \leq 0$ and $t \geq 2\pi$.

(c) $\delta(t) = \sin t$ for $0 < t < \pi$ and $\delta(t) = 0$ for $t \leq 0$ and $t \geq \pi$.

24. ★ Transient between two steady-state conditions.

Consider the front-wheel-steering bicycle planar model of a vehicle with the following parameters.

$$
\begin{aligned}
C_{\alpha f} &= 57296\,\mathrm{N/rad} & C_{\alpha r} &= 52712\,\mathrm{N/rad} \\
m &= 917\,\mathrm{kg} & I_z &= 1128\,\mathrm{kg\,m^2} \\
a_1 &= 0.91\,\mathrm{m} & a_2 &= 1.64\,\mathrm{m}
\end{aligned}
$$

(a) Determine the steady state values of S_κ for $v_x = 10\,\mathrm{m/s}$ and $v_x = 30\,\mathrm{m/s}$.

(b) Determine the transient behavior of the vehicle between the steady state condition assuming the speed of the car changes suddenly.

25. Transient path of rotation center.

Consider a vehicle with the following characteristics

$$
\begin{aligned}
C_{\alpha f_L} &= C_{\alpha f_R} \approx 3000\,\mathrm{N/rad} & C_{\alpha r_L} &= C_{\alpha r_R} \approx 3000\,\mathrm{N/rad} \\
m &= 1000\,\mathrm{kg} & I_z &= 1650\,\mathrm{kg\,m^2} \\
a_1 &= 1.0\,\mathrm{m} & a_2 &= 1.5\,\mathrm{m}
\end{aligned}
$$

(a) Determine the steady-state responses S_κ, S_β, S_r, S_a for $v_x = 1\,\mathrm{m/s}$ and $v_x = 10\,\mathrm{m/s}$.

(b) Determine the critical speed at which $\beta = 0$.

(c) Calculate the coordinates of the rotation center for $v_x = 1\,\mathrm{m/s}$, the critical speed, and $v_x = 10\,\mathrm{m/s}$.

(d) ★ Assume the vehicle is at steady state condition for $v_x = 1\,\mathrm{m/s}$ and suddenly the speed changes to $v_x = 10\,\mathrm{m/s}$ and remains constant. Determine the path of the rotation center between the two steady-state conditions.

26. Linear model dynamics.

Consider a vehicle with the following characteristics

$$
\begin{aligned}
C_{\alpha f_L} &= C_{\alpha f_R} \approx 3000\,\mathrm{N/rad} & C_{\alpha r_L} &= C_{\alpha r_R} \approx 3000\,\mathrm{N/rad} \\
m &= 1000\,\mathrm{kg} & I_z &= 1650\,\mathrm{kg\,m^2} \\
a_1 &= 1.0\,\mathrm{m} & a_2 &= 1.5\,\mathrm{m}
\end{aligned}
$$

and compare the transient behavior of the vehicle using the planar model and the linearized model for a step steer change.

$$\delta\left(t\right) = \begin{cases} 5\deg & t > 0 \\ 0 & t \leq 0 \end{cases}$$

11

★ Vehicle Roll Dynamics

In this chapter, we develop a dynamic model for a rigid bicycle vehicle having forward, lateral, yaw, and roll motions. The model of a rollable rigid vehicle is more exact and more effective compared to the rigid bicycle vehicle planar model. Using this model, we are able to analyze the roll behavior of a vehicle as well as its maneuvering.

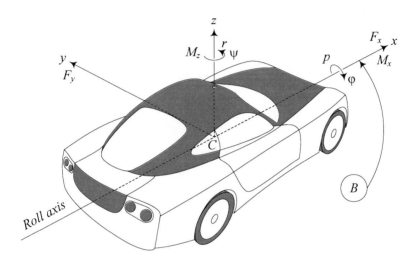

FIGURE 11.1. The DOF of a roll model of rigid vehicles are: x, y, φ, ψ.

11.1 ★ Vehicle Coordinate and DOF

Figure 11.1 illustrates a vehicle with a body coordinate $B(Cxyz)$ at the mass center C. The x-axis is a longitudinal axis passing through C and directed forward. The y-axis goes laterally to the left from the driver's viewpoint. The z-axis makes the coordinate system a right-hand triad. When the car is parked on a flat horizontal road, the z-axis is perpendicular to the ground, opposite to the gravitational acceleration $\mathbf{g}$. The equations of motion of the vehicle should be expressed in $B(Cxyz)$.

Angular orientation and angular velocity of a vehicle are expressed by three angles: *roll* φ, *pitch* θ, *yaw* ψ, and their rates: *roll rate* p, *pitch rate*

R.N. Jazar, *Vehicle Dynamics: Theory and Application*,
DOI 10.1007/978-1-4614-8544-5_11, © Springer Science+Business Media New York 2014

q, *yaw rate* r.

$$p = \dot{\varphi} \qquad (11.1)$$
$$q = \dot{\theta} \qquad (11.2)$$
$$r = \dot{\psi} \qquad (11.3)$$

The vehicle force system $(\mathbf{F}, \mathbf{M})$ is the resultant of external forces and moments that the vehicle receives from the ground and environment. The force system may be expressed in the body coordinate frame as:

$$^B\mathbf{F} = F_x\hat{\imath} + F_y\hat{\jmath} + F_z\hat{k} \qquad (11.4)$$
$$^B\mathbf{M} = M_x\hat{\imath} + M_y\hat{\jmath} + M_z\hat{k} \qquad (11.5)$$

The roll model vehicle dynamics can be expressed by four kinematic variables: the forward motion x, the lateral motion y, the roll angle φ, and the yaw angle ψ. In this model, we do not consider vertical movement z, and pitch motion θ.

11.2 ★ Equations of Motion

A rigid vehicle with roll and yaw motions has a motion with four degrees of freedom, which are translation in x and y directions, and rotation about the x and z axes. The *Newton-Euler equations of motion* for such a rolling rigid vehicle in the body coordinate frame B are:

$$F_x = m\dot{v}_x - mr\,v_y \qquad (11.6)$$
$$F_y = m\dot{v}_y + mr\,v_x \qquad (11.7)$$
$$M_z = I_z\dot{\omega}_z = I_z\dot{r} \qquad (11.8)$$
$$M_x = I_x\dot{\omega}_x = I_x\dot{p} \qquad (11.9)$$

Proof. Consider the vehicle shown in Figure 11.2. A global coordinate frame G is fixed on the ground, and a local coordinate frame B is attached to the vehicle at the mass center C. The orientation of the frame B can be expressed by the heading angle ψ between the x and X axes, and the roll angle φ between the z and Z axes. The global position vector of the mass center is denoted by $^G\mathbf{d}$.

The rigid body equations of motion in the body coordinate frame are:

$$
\begin{aligned}
^B\mathbf{F} &= {}^B R_G\, {}^G\mathbf{F} = {}^B R_G\left(m\,{}^G\mathbf{a}_B\right) = m\,{}^B_G\mathbf{a}_B \\
&= m\,{}^B\dot{\mathbf{v}}_B + m\,{}^B_G\boldsymbol{\omega}_B \times {}^B\mathbf{v}_B
\end{aligned} \qquad (11.10)
$$

$$
\begin{aligned}
^B\mathbf{M} &= \frac{^G d}{dt}\,{}^B\mathbf{L} = {}^B_G\dot{\mathbf{L}}_B = {}^B\dot{\mathbf{L}} + {}^B_G\boldsymbol{\omega}_B \times {}^B\mathbf{L} \\
&= {}^B I\,{}^B_G\dot{\boldsymbol{\omega}}_B + {}^B_G\boldsymbol{\omega}_B \times \left({}^B I\,{}^B_G\boldsymbol{\omega}_B\right)
\end{aligned} \qquad (11.11)
$$

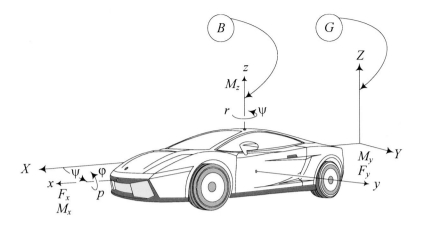

FIGURE 11.2. A vehicle with roll, and yaw rotations.

The velocity vector of the vehicle, expressed in the body frame, is

$$^{B}\mathbf{v}_C = \begin{bmatrix} v_x \\ v_y \\ 0 \end{bmatrix} \tag{11.12}$$

where v_x is the forward component and v_y is the lateral component of $\mathbf{v}$. The other kinematic vectors for the rigid vehicle are:

$$^{B}\dot{\mathbf{v}}_C = \begin{bmatrix} \dot{v}_x \\ \dot{v}_y \\ 0 \end{bmatrix} \tag{11.13}$$

$$^{B}_{G}\boldsymbol{\omega}_B = \begin{bmatrix} \omega_x \\ 0 \\ \omega_z \end{bmatrix} = \begin{bmatrix} p \\ 0 \\ r \end{bmatrix} \tag{11.14}$$

$$^{B}_{G}\dot{\boldsymbol{\omega}}_B = \begin{bmatrix} \dot{\omega}_x \\ 0 \\ \dot{\omega}_z \end{bmatrix} = \begin{bmatrix} \dot{p} \\ 0 \\ \dot{r} \end{bmatrix} \tag{11.15}$$

We may assume that the body coordinate is the principal coordinate frame of the vehicle to have a diagonal moment of inertia matrix.

$$^{B}I = \begin{bmatrix} I_x & 0 & 0 \\ 0 & I_y & 0 \\ 0 & 0 & I_z \end{bmatrix} = \begin{bmatrix} I_1 & 0 & 0 \\ 0 & I_2 & 0 \\ 0 & 0 & I_3 \end{bmatrix} \tag{11.16}$$

Substituting the above vectors and matrices in the equations of motion

(11.10) and (11.11) provides us with the following equations:

$$^B\mathbf{F} = m \, ^B\dot{\mathbf{v}}_B + m \, ^B_G\boldsymbol{\omega}_B \times \, ^B\mathbf{v}_B \tag{11.17}$$

$$\begin{bmatrix} F_x \\ F_y \\ F_z \end{bmatrix} = m \begin{bmatrix} \dot{v}_x \\ \dot{v}_y \\ 0 \end{bmatrix} + m \begin{bmatrix} \omega_x \\ 0 \\ \omega_z \end{bmatrix} \times \begin{bmatrix} v_x \\ v_y \\ 0 \end{bmatrix}$$

$$= \begin{bmatrix} m\dot{v}_x - m\omega_z v_y \\ m\dot{v}_y + m\omega_z v_x \\ m\omega_x v_y \end{bmatrix} \tag{11.18}$$

$$^B\mathbf{M} = \, ^BI \, ^B_G\dot{\boldsymbol{\omega}}_B + \, ^B_G\boldsymbol{\omega}_B \times \left(^BI \, ^B_G\boldsymbol{\omega}_B \right) \tag{11.19}$$

$$\begin{bmatrix} M_x \\ M_y \\ M_z \end{bmatrix} = \begin{bmatrix} I_1 & 0 & 0 \\ 0 & I_2 & 0 \\ 0 & 0 & I_3 \end{bmatrix} \begin{bmatrix} \dot{\omega}_x \\ 0 \\ \dot{\omega}_z \end{bmatrix}$$

$$+ \begin{bmatrix} \omega_x \\ 0 \\ \omega_z \end{bmatrix} \times \left(\begin{bmatrix} I_1 & 0 & 0 \\ 0 & I_2 & 0 \\ 0 & 0 & I_3 \end{bmatrix} \begin{bmatrix} \omega_x \\ 0 \\ \omega_z \end{bmatrix} \right)$$

$$= \begin{bmatrix} I_1\dot{\omega}_x \\ I_1\omega_x\omega_z - I_3\omega_x\omega_z \\ I_3\dot{\omega}_z \end{bmatrix} \tag{11.20}$$

The first two Newton equations (11.18) are the translational equations of motion in the x and y directions.

$$\begin{bmatrix} F_x \\ F_y \end{bmatrix} = \begin{bmatrix} m\dot{v}_x - m\omega_z v_y \\ m\dot{v}_y + m\omega_z v_x \end{bmatrix} \tag{11.21}$$

and the first and third Euler equations (11.20) are the rotational equations of motion in the x and z directions.

The third Newton's equation

$$m\omega_x v_y = F_z \tag{11.22}$$

provides the compatibility condition to keep the vehicle on the road. However, because of ignorance gravity and any motion in y-direction or rotation about y-axis, the equation in z-direction is not complete.

The first and third Euler equations (11.20) are the equations of motion about the x and z axes.

$$\begin{bmatrix} M_x \\ M_z \end{bmatrix} = \begin{bmatrix} I_1\dot{\omega}_x \\ I_3\dot{\omega}_z \end{bmatrix} \tag{11.23}$$

The second Euler equation

$$I_1\omega_x\omega_z - I_3\omega_x\omega_z = M_y \tag{11.24}$$

is another compatibility condition that provides the required pitch moment condition to keep the vehicle on the road. ∎

Example 461 ★ *Motion of a six DOF vehicle.*

Consider a vehicle that moves in space. Such a vehicle has six DOF. To develop the equations of motion of such a vehicle, we need to define the kinematic characteristics as follows:

$$
{}^B\mathbf{v}_C = \begin{bmatrix} v_x \\ v_y \\ v_z \end{bmatrix} \qquad {}^B\dot{\mathbf{v}}_C = \begin{bmatrix} \dot{v}_x \\ \dot{v}_y \\ \dot{v}_z \end{bmatrix} \tag{11.25}
$$

$$
{}^B_G\boldsymbol{\omega}_B = \begin{bmatrix} \omega_x \\ \omega_y \\ \omega_z \end{bmatrix} \qquad {}^B_G\dot{\boldsymbol{\omega}}_B = \begin{bmatrix} \dot{\omega}_x \\ \dot{\omega}_y \\ \dot{\omega}_z \end{bmatrix} \tag{11.26}
$$

The acceleration vector of the vehicle in the body coordinate is

$$
{}^B\mathbf{a} = {}^B\dot{\mathbf{v}}_B + {}^B_G\boldsymbol{\omega}_B \times {}^B\mathbf{v}_B = \begin{bmatrix} \dot{v}_x + \omega_y v_z - \omega_z v_y \\ \dot{v}_y + \omega_z v_x - \omega_x v_z \\ \dot{v}_z + \omega_x v_y - \omega_y v_x \end{bmatrix} \tag{11.27}
$$

and therefore, the Newton equations of motion for the vehicle are

$$
\begin{bmatrix} F_x \\ F_y \\ F_z \end{bmatrix} = m \begin{bmatrix} \dot{v}_x + \omega_y v_z - \omega_z v_y \\ \dot{v}_y + \omega_z v_x - \omega_x v_z \\ \dot{v}_z + \omega_x v_y - \omega_y v_x \end{bmatrix} \tag{11.28}
$$

To find the Euler equations of motion,

$$
{}^B\mathbf{M} = {}^BI\,{}^B_G\dot{\boldsymbol{\omega}}_B + {}^B_G\boldsymbol{\omega}_B \times \left({}^BI\,{}^B_G\boldsymbol{\omega}_B\right) \tag{11.29}
$$

we need to define the mass moment matrix and perform the required matrix calculations. Assume the body coordinate system is the principal coordinate frame. So,

$$
{}^BI\,{}^B_G\dot{\boldsymbol{\omega}}_B + {}^B_G\boldsymbol{\omega}_B \times \left({}^BI\,{}^B_G\boldsymbol{\omega}_B\right)
$$

$$
= \begin{bmatrix} I_1 & 0 & 0 \\ 0 & I_2 & 0 \\ 0 & 0 & I_3 \end{bmatrix} \begin{bmatrix} \dot{\omega}_x \\ \dot{\omega}_y \\ \dot{\omega}_z \end{bmatrix} + \begin{bmatrix} \omega_x \\ \omega_y \\ \omega_z \end{bmatrix} \times \left(\begin{bmatrix} I_1 & 0 & 0 \\ 0 & I_2 & 0 \\ 0 & 0 & I_3 \end{bmatrix} \begin{bmatrix} \omega_x \\ \omega_y \\ \omega_z \end{bmatrix} \right)
$$

$$
= \begin{bmatrix} \dot{\omega}_x I_1 - \omega_y \omega_z I_2 + \omega_y \omega_z I_3 \\ \dot{\omega}_y I_2 + \omega_x \omega_z I_1 - \omega_x \omega_z I_3 \\ \dot{\omega}_z I_3 - \omega_x \omega_y I_1 + \omega_x \omega_y I_2 \end{bmatrix} \tag{11.30}
$$

and therefore, the Euler equations of motion for the vehicle are

$$
\begin{bmatrix} M_x \\ M_y \\ M_z \end{bmatrix} = \begin{bmatrix} \dot{\omega}_x I_1 - \omega_y \omega_z I_2 + \omega_y \omega_z I_3 \\ \dot{\omega}_y I_2 + \omega_x \omega_z I_1 - \omega_x \omega_z I_3 \\ \dot{\omega}_z I_3 - \omega_x \omega_y I_1 + \omega_x \omega_y I_2 \end{bmatrix} \tag{11.31}
$$

Equations (11.28) and (11.31) are the equations of motion of a car with pitch degree of freedom.

Example 462 ★ *Roll rigid vehicle from general motion.*

We may derive the equations of motion for a roll rigid vehicle from the general equations of motion for a vehicle with six DOF (11.28) and (11.31).

Consider a bicycle model of a four-wheel vehicle moving on a road. Because the vehicle cannot move in z-direction and cannot turn about the y-axis, we have

$$v_z = 0 \qquad \dot{v}_z = 0 \qquad \omega_y = 0 \qquad \dot{\omega}_y = 0 \qquad (11.32)$$

Furthermore, the resultant force in the z-direction and moment in y-direction for such a bicycle must be zero,

$$F_z = 0 \qquad M_y = 0 \qquad (11.33)$$

Substitution Equations (11.32)-(11.33) in (11.28) and (11.31) results in the force system.

$$\begin{bmatrix} F_x \\ F_y \\ F_z \end{bmatrix} = m \begin{bmatrix} \dot{v}_x - \omega_z v_y \\ \dot{v}_y + \omega_z v_x \\ \omega_x v_y \end{bmatrix} \qquad (11.34)$$

$$\begin{bmatrix} M_x \\ M_y \\ M_z \end{bmatrix} = \begin{bmatrix} \dot{\omega}_x I_1 \\ \omega_x \omega_z I_1 - \omega_x \omega_z I_3 \\ \dot{\omega}_z I_3 \end{bmatrix} \qquad (11.35)$$

11.3 ★ Vehicle Force System

To determine the force system on a rigid vehicle, we define the force system at the tireprint of a wheel. The lateral force at the tireprint depends on the sideslip angle. Then, we transform and apply the tire force system on the rollable model of the vehicle.

11.3.1 ★ Tire and Body Force Systems

Figure 11.3 depicts the wheel number 1 of a vehicle. The components of the applied force system in the C-frame, because of the forces at the tireprint of the wheel number i are:

$$F_{x_i} = F_{x_{w_i}} \cos \delta_i - F_{y_{w_i}} \sin \delta_i \qquad (11.36)$$
$$F_{y_i} = F_{y_{w_i}} \cos \delta_i + F_{x_{w_i}} \sin \delta_i \qquad (11.37)$$
$$M_{x_i} = M_{x_{w_i}} \qquad (11.38)$$
$$M_{z_i} = M_{z_{w_i}} \qquad (11.39)$$

where (x_i, y_i, z_i) are body coordinates of the wheel number i. In this analy-

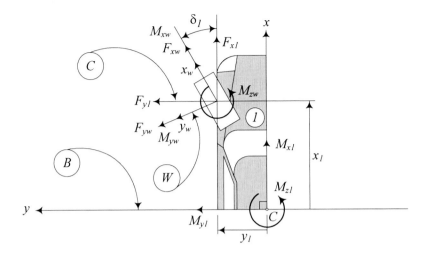

FIGURE 11.3. The force system at the tireprint of tire number 1, and their resultant force system at C.

sis, we ignored the components of the tire moment at the tireprint, $M_{y_{w_i}}$ to simplify the equations.

The total important force system on the rigid vehicle in the body coordinate frame to analyze the roll model of rigid vehicle is

$$^B F_x = \sum_i F_{x_i} = \sum_i F_{x_w} \cos \delta_i - \sum_i F_{y_w} \sin \delta_i \qquad (11.40)$$

$$^B F_y = \sum_i F_{y_i} = \sum_i F_{y_w} \cos \delta_i + \sum_i F_{x_w} \sin \delta_i \qquad (11.41)$$

$$^B M_x = \sum_i M_{x_i} + \sum_i y_i F_{z_i} - \sum_i z_i F_{y_i} \qquad (11.42)$$

$$^B M_z = \sum_i M_{z_i} + \sum_i x_i F_{y_i} - \sum_i y_i F_{x_i} \qquad (11.43)$$

Proof. To simplify the roll model of vehicle dynamics, we ignore the difference between the tire frame at the center of tireprint and wheel frame at the wheel center for small roll angles. The tireprint, the force system generated at the tireprint of the wheel in the wheel frame W is

$$^W \mathbf{F}_w = \begin{bmatrix} F_{x_w} & F_{y_w} & F_{z_w} \end{bmatrix}^T \qquad (11.44)$$

$$^W \mathbf{M}_w = \begin{bmatrix} M_{x_w} & M_{y_w} & M_{z_w} \end{bmatrix}^T \qquad (11.45)$$

The rotation matrix between the wheel frame W and the wheel-frame coordinate frame C, parallel to the vehicle body coordinate frame B, is

$$
{}^{C}R_{W} = \begin{bmatrix} \cos\delta_1 & -\sin\delta_1 & 0 \\ \sin\delta_1 & \cos\delta_1 & 0 \\ 0 & 0 & 1 \end{bmatrix}
\tag{11.46}
$$

and therefore, the force system at the tireprint of the wheel, parallel to the vehicle coordinate frame, is

$$
\begin{aligned}
{}^{C}\mathbf{F}_{w} &= {}^{C}R_{W}\,{}^{W}\mathbf{F}_{w} \\
\begin{bmatrix} F_{x_1} \\ F_{y_1} \\ F_{z_1} \end{bmatrix} &= \begin{bmatrix} \cos\delta_1 & -\sin\delta_1 & 0 \\ \sin\delta_1 & \cos\delta_1 & 0 \\ 0 & 0 & 1 \end{bmatrix} \begin{bmatrix} F_{x_w} \\ F_{y_w} \\ F_{z_w} \end{bmatrix} \\
&= \begin{bmatrix} F_{x_w}\cos\delta_1 - F_{y_w}\sin\delta_1 \\ F_{y_w}\cos\delta_1 + F_{x_w}\sin\delta_1 \\ F_{z_w} \end{bmatrix}
\end{aligned}
\tag{11.47}
$$
$$
\tag{11.48}
$$

$$
\begin{aligned}
{}^{B_1}\mathbf{M}_{w} &= {}^{B_1}R_{B_w}\,{}^{B_w}\mathbf{M}_{w} \\
\begin{bmatrix} M_{x_1} \\ M_{y_1} \\ M_{z_1} \end{bmatrix} &= \begin{bmatrix} \cos\delta_1 & -\sin\delta_1 & 0 \\ \sin\delta_1 & \cos\delta_1 & 0 \\ 0 & 0 & 1 \end{bmatrix} \begin{bmatrix} M_{x_w} \\ M_{y_w} \\ M_{z_w} \end{bmatrix} \\
&= \begin{bmatrix} M_{x_w}\cos\delta_1 - M_{y_w}\sin\delta_1 \\ M_{y_w}\cos\delta_1 + M_{x_w}\sin\delta_1 \\ M_{z_w} \end{bmatrix}
\end{aligned}
\tag{11.49}
$$
$$
\tag{11.50}
$$

Ignoring M_{y_i}, we transform the force system of each tire to the body coordinate frame B, located at the vehicle mass center C, to generate the total force system applied on the vehicle

$$
{}^{B}\mathbf{F} = \sum_{i} {}^{B_i}\mathbf{F}_{w} = \sum_{i} F_{x_i}\,\hat{\imath} + \sum_{i} F_{y_i}\,\hat{\jmath}
\tag{11.51}
$$

$$
\begin{aligned}
{}^{B}\mathbf{M} &= \sum_{i} {}^{B_i}\mathbf{M}_{w} \\
&= \sum_{i} M_{x_i}\,\hat{\imath} + \sum_{i} M_{z_i}\,\hat{k} + \sum_{i} {}^{B}\mathbf{r}_i \times {}^{B}\mathbf{F}_{w_i}
\end{aligned}
\tag{11.52}
$$

where ${}^{B}\mathbf{r}_i$ is the position vector of the wheel number i.

$$
{}^{B}\mathbf{r}_i = x_i\hat{\imath} + y_i\hat{\jmath} + z_i\hat{k}
\tag{11.53}
$$

In deriving Equation (11.51), we have used the equation $\sum_{i} F_{z_i} - mg = 0$. Expanding Equations (11.51) and (11.52) provides the total vehicle force system.

$$^B F_x = \sum_i F_{x_w} \cos \delta_i - \sum_i F_{y_w} \sin \delta_i \qquad (11.54)$$

$$^B F_y = \sum_i F_{y_w} \cos \delta_i + \sum_i F_{x_w} \sin \delta_i \qquad (11.55)$$

$$^B M_x = \sum_i M_{x_i} + \sum_i y_i F_{z_i} - \sum_i z_i F_{y_i} \qquad (11.56)$$

$$^B M_z = \sum_i M_{z_i} + \sum_i x_i F_{y_i} - \sum_i y_i F_{x_i} \qquad (11.57)$$

For a two-wheel vehicle model we have

$$
\begin{aligned}
x_1 &= a_1 \qquad x_2 = -a_2 \\
y_1 &= y_2 = 0 \\
z_1 &= R_f \qquad z_2 = R_w
\end{aligned} \qquad (11.58)
$$

For such a vehicle, after ignoring the effect of tire reduces R_f and R_w, the force system reduces to

$$^B F_x = F_{x_1} \cos \delta_1 + F_{x_2} \cos \delta_2 - F_{y_1} \sin \delta_1 - F_{y_2} \sin \delta_2 \qquad (11.59)$$

$$^B F_y = F_{y_1} \cos \delta_1 + F_{y_2} \cos \delta_2 + F_{x_1} \sin \delta_1 + F_{x_2} \sin \delta_2 \qquad (11.60)$$

$$^B M_x = M_{x_1} + M_{x_2} \qquad (11.61)$$

$$^B M_z = M_{z_1} + M_{z_2} + a_1 F_{y_1} - a_2 F_{y_2} \qquad (11.62)$$

■

11.3.2 ★ Tire Lateral Force

If the steer angle of the steering mechanism is denoted by δ, then the actual steer angle δ_a of a rollable vehicle is

$$\delta_a = \delta + \delta_\varphi \qquad (11.63)$$

where, δ_φ is the *roll-steering* angle.

$$\delta_\varphi = C_{\delta\varphi}\varphi \qquad (11.64)$$

The roll-steering angle δ_φ is proportional to the roll angle φ and the coefficient $C_{\delta\varphi}$ is called the *roll-steering* coefficient. The roll-steering happens because of the suspension mechanisms that generate some steer angle when deflected. The *tire sideslip angle* of each tire of a rollable vehicle is

$$\alpha_i = \beta_i - \delta_a = \beta_i - \delta_i - \delta_\varphi \qquad (11.65)$$

where β_i is the angle between the velocity vector $\mathbf{v}$ and the vehicle body x-axis, and is called the *wheel sideslip angle*.

The generated lateral force by such a wheel for a small sideslip angle, is

$$
\begin{aligned}
F_y &= -C_\alpha \, \alpha_i - C_\varphi \varphi_i \\
&= -C_\alpha \, (\beta_i - \delta_i - C_{\delta\varphi}\varphi_i) - C_\varphi \varphi
\end{aligned}
\tag{11.66}
$$

and C_φ is the *tire camber thrust coefficient*, because of the vehicle's roll. The wheel slip angle β_i of a bicycle vehicle model can be approximated by

$$
\beta_i = \frac{v_y + x_i \, r - C_{\beta_i} p}{v_x}
\tag{11.67}
$$

to find the tire lateral force F_y in terms of the vehicle kinematic variables.

$$
F_y = -x_i \frac{C_\alpha}{v_x} r + \frac{C_\alpha C_\beta}{v_x} p - C_\alpha \beta + (C_\alpha C_{\delta\varphi} - C_\varphi) \, \varphi + C_\alpha \delta_i
\tag{11.68}
$$

$$
\beta = \frac{v_y}{v_x}
\tag{11.69}
$$

C_{β_i} is *wheel slip coefficient*.

Proof. When a vehicle rolls, there are some new reactions in the tires that introduce new dynamic terms in the behavior of the tire. The most important reactions are:

1−Tire camber thrust $F_{y\varphi}$, which is a lateral force because of the vehicle roll. Tire camber thrust is assumed to be proportional to the vehicle roll angle φ.

$$
F_{y\varphi} = -C_\varphi \varphi
\tag{11.70}
$$

$$
C_\varphi = \frac{dF_y}{d\varphi}
\tag{11.71}
$$

2−Wheel roll steering angle δ_φ, which is the wheel steer angle because of the vehicle roll. Most suspension mechanisms provide some steer angle when the vehicle rolls and the mechanism deflects. The wheel roll steering is assumed to be proportional to the vehicle roll angle φ.

$$
\delta_\varphi = C_{\delta\varphi} \varphi
\tag{11.72}
$$

$$
C_{\delta\varphi} = \frac{d\delta}{d\varphi}
\tag{11.73}
$$

Therefore, the actual steer angle δ_a of such a tire is

$$
\delta_a = \delta + \delta_\varphi
\tag{11.74}
$$

Assume the wheel number i of a rigid vehicle is located at

$$
{}^B\mathbf{r}_i = \begin{bmatrix} x_i & y_i & z_i \end{bmatrix}^T
\tag{11.75}
$$

The velocity of the wheel number i is

$$^B\mathbf{v}_i = {}^B\mathbf{v} + {}^B\boldsymbol{\omega} \times {}^B\mathbf{r}_i \tag{11.76}$$

in which $^B\mathbf{v}$ is the velocity vector of the vehicle at its mass center C, and $^B\boldsymbol{\omega}$ is the angular velocity of the vehicle.

$$^B\boldsymbol{\omega} = \dot{\varphi}\hat{i} + \dot{\psi}\hat{k} = p\hat{i} + r\hat{k} \tag{11.77}$$

Expanding Equation (11.76) provides the following velocity vector for the wheel number i expressed in the vehicle coordinate frame at B.

$$\begin{bmatrix} v_{x_i} \\ v_{y_i} \\ v_{z_i} \end{bmatrix} = \begin{bmatrix} v_x \\ v_y \\ 0 \end{bmatrix} + \begin{bmatrix} \dot{\varphi} \\ 0 \\ \dot{\psi} \end{bmatrix} \times \begin{bmatrix} x_i \\ y_i \\ z_i \end{bmatrix} = \begin{bmatrix} v_x - \dot{\psi} y_i \\ v_y - \dot{\varphi} z_i + \dot{\psi} x_i \\ \dot{\varphi} y_i \end{bmatrix} \tag{11.78}$$

Consider a bicycle model for the rollable vehicle to have

$$y_i = 0 \tag{11.79}$$
$$x_1 = a_1 \tag{11.80}$$
$$x_2 = -a_2 \tag{11.81}$$

The wheel slip angle β_i for the wheel i, is defined as the angle between the wheel velocity vector $\mathbf{v}_i$ and the vehicle body x-axis. When the roll angle is very small, β_i is

$$\beta_i = \tan^{-1}\left(\frac{v_{y_i}}{v_{x_i}}\right) \approx \frac{v_{y_i}}{v_{x_i}} \approx \frac{v_y - \dot{\varphi} z_i + \dot{\psi} x_i}{v_x} \tag{11.82}$$

If the tire number i has a steer angle δ_i then, its sideslip angle α_i, that generates a lateral force F_{y_w} on the tire, is

$$\alpha_i = \beta_i - \delta_i \approx \frac{v_y - \dot{\varphi} z_i + \dot{\psi} x_i}{v_x} - \delta_i + \delta_{\varphi_i} \tag{11.83}$$

The wheel slip angle β_i for the front and rear wheels of a two-wheel vehicle, β_f and β_r, are

$$\beta_f = \tan^{-1}\left(\frac{v_{y_f}}{v_{x_f}}\right) \approx \frac{v_{y_f}}{v_{x_f}} \approx \frac{v_y + a_1 r - z_f p}{v_x} \tag{11.84}$$

$$\beta_r = \tan^{-1}\left(\frac{v_{y_r}}{v_{x_r}}\right) \approx \frac{v_{y_r}}{v_{x_r}} \approx \frac{v_y - a_2 r - z_r p}{v_x} \tag{11.85}$$

and the *vehicle slip angle* β is

$$\beta = \tan^{-1}\left(\frac{v_y}{v_x}\right) \approx \frac{v_y}{v_x} \tag{11.86}$$

The z_i coordinate of the wheels is not constant however, its variation is very small. To show the effect of z_i, we substitute it by coefficient C_{β_i} called the *tire roll rate coefficient*, and define coefficients C_{β_f} and C_{β_r} to express the change in β_i because of roll rate p.

$$\beta_i = C_{\beta_i} p \tag{11.87}$$

$$C_{\beta_i} = \frac{d\beta_i}{dp} \tag{11.88}$$

Therefore,

$$\beta_f = \tan^{-1}\left(\frac{v_y + a_1 r - C_{\beta_f} p}{v_x}\right) \tag{11.89}$$

$$\beta_r = \tan^{-1}\left(\frac{v_y - a_2 r - C_{\beta_r} p}{v_x}\right) \tag{11.90}$$

Assuming small angles for slip angles β_f, β, and β_r then, the tire sideslip angles for the front and rear wheels, α_f and α_r, may be approximated as

$$\alpha_f = \frac{1}{v_x}(v_y + a_1 r - z_f p) - \delta - \delta_{\varphi_f}$$
$$= \beta + a_1 \frac{r}{v_x} - C_{\beta_f} \frac{p}{v_x} - \delta - C_{\delta\varphi_f} \varphi \tag{11.91}$$

$$\alpha_r = \frac{1}{v_x}(v_y - a_2 r - z_r p) - \delta_{\varphi_r}$$
$$= \beta - a_2 \frac{r}{v_x} - C_{\beta_r} \frac{p}{v_x} - C_{\delta\varphi_r} \varphi \tag{11.92}$$

■

11.3.3 ★ *Body Force Components on a Two-wheel Model*

Figure 11.4 illustrates the top view of a car and the force systems acting at the wheel center of a front-wheel-steering four-wheel vehicle. When we consider the roll motion of the vehicle, the xy-plane does not remain parallel to the road's XY-plane, however, we may still use a *two-wheel model* for the vehicle.

Figure 11.5 illustrates the force system and Figure 11.6 illustrates the kinematics of a two-wheel model for a vehicle with roll and yaw rotations. The rolling two-wheel model is also called the *bicycle model*.

The force system applied on the bicycle vehicle, with only the front wheel steerable, is

$$F_x = \sum_{i=1}^{2}(F_{x_i} \cos\delta - F_{y_i} \sin\delta) \tag{11.93}$$

$$F_y = \sum_{i=1}^{2} F_{y_i} \tag{11.94}$$

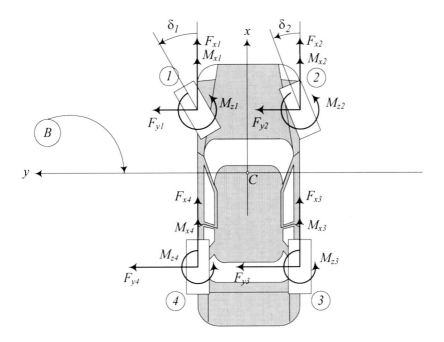

FIGURE 11.4. Top view of a car and the forces system acting at the tireprints.

$$M_x = M_{x_f} + M_{x_r} - wc_f\,\dot\varphi - wk_f\,\varphi \tag{11.95}$$
$$M_z = a_1 F_{y_f} - a_2 F_{y_r} \tag{11.96}$$

where $\left(F_{x_f}, F_{x_r}\right)$ and $\left(F_{y_f}, F_{y_r}\right)$ are the planar forces on the front and rear wheels. The force system may be approximated by the following equations, if the steer angle δ is assumed small:

$$F_x \approx F_{x_f} + F_{x_r} \tag{11.97}$$
$$F_y \approx F_{y_f} + F_{y_r} \tag{11.98}$$
$$M_x \approx C_{T_f} F_{y_f} + C_{T_r} F_{y_r} - k_\varphi\varphi - c_\varphi\dot\varphi \tag{11.99}$$
$$M_z \approx a_1 F_{y_f} - a_2 F_{y_r} \tag{11.100}$$

The vehicle's lateral force F_y and moment M_z depend only on the front and rear wheels' lateral forces F_{y_f} and F_{y_r}, which are functions of the tires' sideslip angles α_f and α_r. They can be approximated by the following equations:

$$
\begin{aligned}
F_y =\ & \left(\frac{a_2}{v_x}C_{\alpha r} - \frac{a_1}{v_x}C_{\alpha f}\right) r + \left(\frac{C_{\alpha f}C_{\beta_f}}{v_x} + \frac{C_{\alpha r}C_{\beta_r}}{v_x}\right) p \\
& + \left(-C_{\alpha f} - C_{\alpha r}\right)\beta + \left(C_{\alpha f}C_{\delta\varphi_f} - C_{\varphi_r} - C_{\varphi_f} + C_{\alpha r}C_{\delta\varphi_r}\right)\varphi \\
& + C_{\alpha f}\delta
\end{aligned}
\tag{11.101}
$$

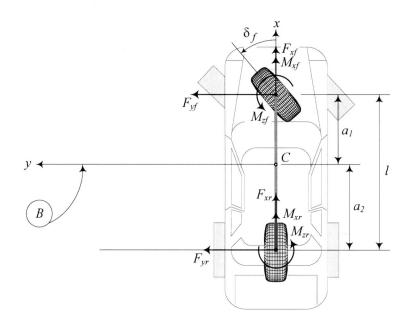

FIGURE 11.5. A two-wheel model for a vehicle with roll and yaw rotations.

$$M_x = \left(\frac{a_2}{v_x} C_{T_r} C_{\alpha r} - \frac{a_1}{v_x} C_{T_f} C_{\alpha f} \right) r$$
$$+ \left(\frac{1}{v_x} C_{\beta_f} C_{T_f} C_{\alpha f} + \frac{1}{v_x} C_{\beta_r} C_{T_r} C_{\alpha r} - c_\varphi \right) p$$
$$+ \left(-C_{T_f} \left(C_{\varphi_f} - C_{\alpha f} C_{\delta \varphi_f} \right) - C_{T_r} \left(C_{\varphi_r} - C_{\alpha r} C_{\delta \varphi_r} \right) - k_\varphi \right) \varphi$$
$$+ C_{T_f} C_{\alpha f} \delta \tag{11.102}$$

$$M_z = \left(-\frac{a_1^2}{v_x} C_{\alpha f} - \frac{a_2^2}{v_x} C_{\alpha r} \right) r + \left(\frac{a_1}{v_x} C_{\beta_f} C_{\alpha f} - \frac{a_2}{v_x} C_{\beta_r} C_{\alpha r} \right) p$$
$$+ \left(a_2 C_{\alpha r} - a_1 C_{\alpha f} \right) \beta$$
$$+ \left(a_2 \left(C_{\varphi_r} - C_{\alpha r} C_{\delta \varphi_r} \right) - a_1 \left(C_{\varphi_f} - C_{\alpha f} C_{\delta \varphi_f} \right) \right) \varphi$$
$$+ a_1 C_{\alpha f} \delta \tag{11.103}$$

where $C_{\alpha f} = C_{\alpha f_L} + C_{\alpha f_R}$ and $C_{\alpha r} = C_{\alpha r_L} + C_{\alpha r_R}$ are equal to the sideslip coefficients of the left and right wheels in front and rear, respectively.

$$C_{\alpha f} = C_{\alpha f_L} + C_{\alpha f_R} \tag{11.104}$$
$$C_{\alpha r} = C_{\alpha r_L} + C_{\alpha r_R} \tag{11.105}$$

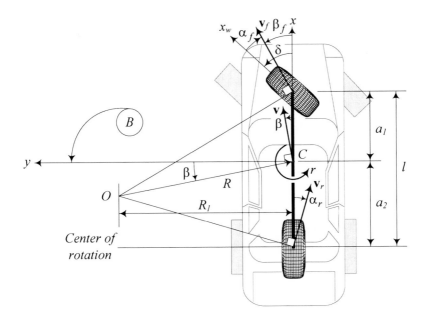

FIGURE 11.6. Kinematics of a two-wheel model for a vehicle with roll and yaw rotations.

Proof. For a two-wheel vehicle, we use the cot-average (7.4) of the outer and inner steer angles as the only steer angle δ.

$$\cot \delta = \frac{\cot \delta_o + \cot \delta_i}{2} \qquad (11.106)$$

Furthermore, we define a single sideslip coefficient $C_{\alpha f}$ and $C_{\alpha r}$ for the front and rear wheels. The coefficient $C_{\alpha f}$ and $C_{\alpha r}$ are equal to sum of the left and right wheels' sideslip coefficients.

Employing Equations (11.40)-(11.43) the forward and lateral forces on the rollable bicycle would be

$$F_x = F_{x_f} \cos \delta + F_{x_r} - F_{y_f} \sin \delta \qquad (11.107)$$
$$F_y = F_{y_f} + F_{y_r} \qquad (11.108)$$

The yaw moment equation does not interact with the vehicle roll. We may also ignore the moments M_{z_i} and assume that the forward forces on the front left and right wheels are equal, as well as the forward forces on the rear left and right wheels. So, the terms $\sum_i y_i F_{x_i}$ cancel each other and the yaw moment reduces to

$$M_z = a_1 F_{y_f} - a_2 F_{y_r} \qquad (11.109)$$

The vehicle roll moment M_x is a summation of the slip and camber moments on the front and rear wheels, M_{x_f}, M_{x_r}, and the moment because of change in normal loads of the left and right wheels $y_i F_{z_i}$. Let's assume that the slip and camber moments are proportional to the wheels' lateral force and write them as

$$M_{x_f} = C_{T_f} F_{y_f} \tag{11.110}$$
$$M_{x_r} = C_{T_r} F_{y_r} \tag{11.111}$$

where C_{T_f} and C_{T_r} are the overall *torque coefficient* of the front and rear wheels respectively.

$$C_{T_f} = \frac{dM_{x_f}}{dF_{y_f}} \tag{11.112}$$

$$C_{T_r} = \frac{dM_{x_r}}{dF_{y_r}} \tag{11.113}$$

Roll moment because of change in normal force of the left and right wheels is a result of force change in springs and dampers. These unbalanced forces generate a roll stiffness moment that is proportional to the vehicle's roll angle

$$M_{x_k} = -k_\varphi \varphi \tag{11.114}$$
$$M_{x_c} = -c_\varphi \dot{\varphi} \tag{11.115}$$

where k_φ and c_φ are the roll stiffness and roll damping of the vehicle.

$$k_\varphi = wk = w(k_f + k_r) \tag{11.116}$$
$$c_\varphi = wc = w(c_f + c_r) \tag{11.117}$$

w is the track of the vehicle and k and c are sum of the front and rear springs' stuffiness and shock absorbers damping. The coefficients k_φ and c_φ are called the *roll stiffness* and *roll damping*, respectively.

$$k = k_f + k_r \tag{11.118}$$
$$c = c_f + c_r \tag{11.119}$$

Therefore, the applied roll moment on the vehicle can be summarized as

$$\begin{aligned} M_x &= M_{x_f} + M_{x_r} + M_{x_c} + M_{x_k} \\ &= C_{T_f} F_{y_f} + C_{T_r} F_{y_r} - w(c_f + c_r)\dot{\varphi} - w(k_f + k_r)\varphi \end{aligned} \tag{11.120}$$

If we assume a small steer angle δ, the vehicle force system can be approximated by the following equations:

$$F_x \approx F_{x_f} + F_{x_r} \tag{11.121}$$
$$F_y \approx F_{y_f} + F_{y_r} \tag{11.122}$$
$$M_x \approx C_{T_f} F_{y_f} + C_{T_r} F_{y_r} - k_\varphi \varphi - c_\varphi \dot{\varphi} \tag{11.123}$$
$$M_z \approx a_1 F_{y_f} - a_2 F_{y_r} \tag{11.124}$$

Substituting for the lateral forces from (11.68), and expanding Equations (11.121)-(11.124) provides the following force system.

$$F_x = F_{x_f} + F_{x_r} \tag{11.125}$$

$$
\begin{aligned}
F_y &= F_{y_f} + F_{y_r} = -C_{\alpha f}\,\alpha_f - C_{\varphi_f}\varphi - C_{\alpha r}\,\alpha_r - C_{\varphi_r}\varphi \\
&= -C_{\alpha f}\left(\beta + a_1\frac{r}{v_x} - C_{\beta_f}\frac{p}{v_x} - \delta - C_{\delta\varphi_f}\varphi\right) - C_{\varphi_f}\varphi \\
&\quad -C_{\alpha r}\left(\beta - a_2\frac{r}{v_x} - C_{\beta_r}\frac{p}{v_x} - C_{\delta\varphi_r}\varphi\right) - C_{\varphi_r}\varphi \\
&= \left(\frac{a_2}{v_x}C_{\alpha r} - \frac{a_1}{v_x}C_{\alpha f}\right)r + \left(\frac{C_{\alpha f}C_{\beta_f}}{v_x} + \frac{C_{\alpha r}C_{\beta_r}}{v_x}\right)p \\
&\quad + (-C_{\alpha f} - C_{\alpha r})\,\beta + \left(C_{\alpha f}C_{\delta\varphi_f} - C_{\varphi_r} - C_{\varphi_f} + C_{\alpha r}C_{\delta\varphi_r}\right)\varphi \\
&\quad + C_{\alpha f}\delta \tag{11.126}
\end{aligned}
$$

$$
\begin{aligned}
M_x &= C_{T_f}F_{y_f} + C_{T_r}F_{y_r} - k_\varphi\varphi - c_\varphi p \\
&= -C_{T_f}\left(C_{\alpha f}\left(\beta + a_1\frac{r}{v_x} - C_{\beta_f}\frac{p}{v_x} - \delta - C_{\delta\varphi_f}\varphi\right) + C_{\varphi_f}\varphi\right) \\
&\quad -C_{T_r}\left(C_{\alpha r}\left(\beta - a_2\frac{r}{v_x} - C_{\beta_r}\frac{p}{v_x} - C_{\delta\varphi_r}\varphi\right) + C_{\varphi_r}\varphi\right) \\
&\quad -k_\varphi\varphi - c_\varphi p \\
&= \left(\frac{a_2}{v_x}C_{T_r}C_{\alpha r} - \frac{a_1}{v_x}C_{T_f}C_{\alpha f}\right)r \\
&\quad + \left(\frac{1}{v_x}C_{\beta_f}C_{T_f}C_{\alpha f} + \frac{1}{v_x}C_{\beta_r}C_{T_r}C_{\alpha r} - c_\varphi\right)p \\
&\quad + (-C_{T_f}C_{\alpha f} - C_{T_r}C_{\alpha r})\,\beta \\
&\quad + \left(-C_{T_f}\left(C_{\varphi_f} - C_{\alpha f}C_{\delta\varphi_f}\right) - C_{T_r}\left(C_{\varphi_r} - C_{\alpha r}C_{\delta\varphi_r}\right) - k_\varphi\right)\varphi \\
&\quad + C_{T_f}C_{\alpha f}\delta \tag{11.127}
\end{aligned}
$$

$$
\begin{aligned}
M_z &= a_1 F_{y_f} - a_2 F_{y_r} \\
&= -a_1\left(C_{\alpha f}\left(\beta + a_1\frac{r}{v_x} - C_{\beta_f}\frac{p}{v_x} - \delta - C_{\delta\varphi_f}\varphi\right) + C_{\varphi_f}\varphi\right) \\
&\quad + a_2\left(C_{\alpha r}\left(\beta - a_2\frac{r}{v_x} - C_{\beta_r}\frac{p}{v_x} - C_{\delta\varphi_r}\varphi\right) + C_{\varphi_r}\varphi\right) \\
&= \left(-\frac{a_1^2}{v_x}C_{\alpha f} - \frac{a_2^2}{v_x}C_{\alpha r}\right)r + \left(\frac{a_1}{v_x}C_{\beta_f}C_{\alpha f} - \frac{a_2}{v_x}C_{\beta_r}C_{\alpha r}\right)p \\
&\quad + (a_2 C_{\alpha r} - a_1 C_{\alpha f})\,\beta \\
&\quad + \left(a_2\left(C_{\varphi_r} - C_{\alpha r}C_{\delta\varphi_r}\right) - a_1\left(C_{\varphi_f} - C_{\alpha f}C_{\delta\varphi_f}\right)\right)\varphi \\
&\quad + a_1 C_{\alpha f}\delta \tag{11.128}
\end{aligned}
$$

The parameters $C_{\alpha f}$ and $C_{\alpha r}$ are the sideslip stiffness for the front and rear wheels, r is the yaw rate, p is the roll rate, φ is the roll angle, δ is the steer angle, and β is the slip angle of the vehicle.

These equations are dependent on five parameters: $r, p, \beta, \varphi, \delta$, and may be written as

$$
\begin{aligned}
F_y &= F_y\left(r,p,\beta,\varphi,\delta\right)\\
&= \frac{\partial F_y}{\partial r}r + \frac{\partial F_y}{\partial p}p + \frac{\partial F_y}{\partial \beta}\beta + \frac{\partial F_y}{\partial \varphi}\varphi + \frac{\partial F_y}{\partial \delta}\delta\\
&= C_r\, r + C_p\, p + C_\beta\, \beta + C_\varphi\, \varphi + C_\delta\, \delta
\end{aligned}
\tag{11.129}
$$

$$
\begin{aligned}
M_x &= M_x\left(r,p,\beta,\varphi,\delta\right)\\
&= \frac{\partial M_x}{\partial r}r + \frac{\partial M_x}{\partial p}p + \frac{\partial M_x}{\partial \beta}\beta + \frac{\partial M_x}{\partial \varphi}\varphi + \frac{\partial M_x}{\partial \delta}\delta\\
&= E_r\, r + E_p\, p + E_\beta\, \beta + E_\varphi\, \varphi + E_\delta\, \delta
\end{aligned}
\tag{11.130}
$$

$$
\begin{aligned}
M_z &= M_z\left(r,p,\beta,\varphi,\delta\right)\\
&= \frac{\partial M_z}{\partial r}r + \frac{\partial M_z}{\partial p}p + \frac{\partial M_z}{\partial \beta}\beta + \frac{\partial M_z}{\partial \varphi}\varphi + \frac{\partial M_z}{\partial \delta}\delta\\
&= D_r\, r + D_p\, p + D_\beta\, \beta + D_\varphi\, \varphi + D_\delta\, \delta
\end{aligned}
\tag{11.131}
$$

where the *force system coefficients* are

$$
C_r = \frac{\partial F_y}{\partial r} = -\frac{a_1}{v_x}C_{\alpha f} + \frac{a_2}{v_x}C_{\alpha r}
\tag{11.132}
$$

$$
C_p = \frac{\partial F_y}{\partial p} = \frac{C_{\alpha f}C_{\beta_f}}{v_x} + \frac{C_{\alpha r}C_{\beta_r}}{v_x}
\tag{11.133}
$$

$$
C_\beta = \frac{\partial F_y}{\partial \beta} = -\left(C_{\alpha f} + C_{\alpha r}\right)
\tag{11.134}
$$

$$
C_\varphi = \frac{\partial F_y}{\partial \varphi} = C_{\alpha r}C_{\delta\varphi_r} + C_{\alpha f}C_{\delta\varphi_f} - C_{\varphi_f} - C_{\varphi_r}
\tag{11.135}
$$

$$
C_\delta = \frac{\partial F_y}{\partial \delta} = C_{\alpha f}
\tag{11.136}
$$

$$E_r = \frac{\partial M_x}{\partial r} = -\frac{a_1}{v_x}C_{T_f}C_{\alpha f} + \frac{a_2}{v_x}C_{T_r}C_{\alpha r} \tag{11.137}$$

$$E_p = \frac{\partial M_x}{\partial p} = \frac{1}{v_x}C_{\beta_f}C_{T_f}C_{\alpha f} + \frac{1}{v_x}C_{\beta_r}C_{T_r}C_{\alpha r} - c_\varphi \tag{11.138}$$

$$E_\beta = \frac{\partial M_x}{\partial \beta} = -C_{T_f}C_{\alpha f} - C_{T_r}C_{\alpha r} \tag{11.139}$$

$$E_\varphi = \frac{\partial M_x}{\partial \varphi} = -C_{T_f}\left(C_{\varphi_f} - C_{\alpha f}C_{\delta\varphi_f}\right) - k_\varphi$$
$$\quad -C_{T_r}\left(C_{\varphi_r} - C_{\alpha r}C_{\delta\varphi_r}\right) \tag{11.140}$$

$$E_\delta = \frac{\partial M_x}{\partial \delta} = C_{T_f}C_{\alpha f} \tag{11.141}$$

$$D_r = \frac{\partial M_z}{\partial r} = -\frac{a_1^2}{v_x}C_{\alpha f} - \frac{a_2^2}{v_x}C_{\alpha r} \tag{11.142}$$

$$D_p = \frac{\partial M_z}{\partial p} = \frac{a_1}{v_x}C_{\beta_f}C_{\alpha f} - \frac{a_2}{v_x}C_{\beta_r}C_{\alpha r} \tag{11.143}$$

$$D_\beta = \frac{\partial M_z}{\partial \beta} = -\left(a_1 C_{\alpha f} - a_2 C_{\alpha r}\right) \tag{11.144}$$

$$D_\varphi = \frac{\partial M_z}{\partial \varphi} = -a_1\left(C_{\varphi_f} - C_{\alpha f}C_{\delta\varphi_f}\right) + a_2\left(C_{\varphi_r} - C_{\alpha r}C_{\delta\varphi_r}\right) \tag{11.145}$$

$$D_\delta = \frac{\partial M_z}{\partial \delta} = a_1 C_{\alpha f} \tag{11.146}$$

The force system coefficients are slopes of the curves for lateral force F_y, roll moment M_x, and yaw moment M_z as a function of r, p, β, φ, and δ respectively. ■

11.4 ★ Two-wheel Rigid Vehicle Dynamics

We may combine the equations of motion (11.6)-(11.9) along with (11.97)-(11.103) for a two-wheel rollable rigid vehicle, and express its motion by the following set of equations:

$$\dot{v}_x = \frac{1}{m}F_x + r\,v_y = \frac{1}{m}\left(F_{x_f} + F_{x_r}\right) + r\,v_y \tag{11.147}$$

$$
\begin{bmatrix} \dot{v}_y \\ \dot{p} \\ \dot{\varphi} \\ \dot{r} \end{bmatrix}
=
\begin{bmatrix}
\dfrac{C_\beta}{mv_x} & \dfrac{C_p}{m} & \dfrac{C_\varphi}{m} & \dfrac{C_r}{m} - v_x \\[2mm]
\dfrac{E_\beta}{I_x v_x} & \dfrac{E_p}{I_x} & \dfrac{E_\varphi}{I_x} & \dfrac{E_r}{I_x} \\[1mm]
0 & 1 & 0 & 0 \\[1mm]
\dfrac{D_\beta}{I_z v_x} & \dfrac{D_p}{I_z} & \dfrac{D_\varphi}{I_z} & \dfrac{D_r}{I_z}
\end{bmatrix}
\begin{bmatrix} v_y \\ p \\ \varphi \\ r \end{bmatrix}
+
\begin{bmatrix}
\dfrac{C_\delta}{m} \\[2mm]
\dfrac{E_\delta}{I_x} \\[1mm]
0 \\[1mm]
\dfrac{D_\delta}{I_z}
\end{bmatrix}
\delta \tag{11.148}
$$

These sets of equations are very useful to analyze vehicle motions, especially when they move at a constant forward speed.

Assuming $\dot{v}_x = 0$, the first equation (11.147) becomes an independent algebraic equation, while the lateral velocity v_y, roll rate p, roll angle φ, and yaw rate r of the vehicle will change according to the four coupled equations (11.148). Considering the steer angle δ is the input command, the other variables v_y, p, φ, and r may be assumed as the outputs. Hence, we may consider Equation (11.148) as a linear control system, and write the equations as

$$\dot{\mathbf{q}} = [A]\,\mathbf{q} + \mathbf{u} \tag{11.149}$$

in which $[A]$ is the coefficient matrix, $\mathbf{q}$ is the vector of control variables, and $\mathbf{u}$ is the vector of inputs.

$$[A] = \begin{bmatrix} \dfrac{C_\beta}{mv_x} & \dfrac{C_p}{m} & \dfrac{C_\varphi}{m} & \dfrac{C_r}{m} - v_x \\[2mm] \dfrac{E_\beta}{I_x v_x} & \dfrac{E_p}{I_x} & \dfrac{E_\varphi}{I_x} & \dfrac{E_r}{I_x} \\[2mm] 0 & 1 & 0 & 0 \\[2mm] \dfrac{D_\beta}{I_z v_x} & \dfrac{D_p}{I_z} & \dfrac{D_\varphi}{I_z} & \dfrac{D_r}{I_z} \end{bmatrix} \tag{11.150}$$

$$\mathbf{q} = \begin{bmatrix} v_y & p & \varphi & r \end{bmatrix}^T \tag{11.151}$$

$$\mathbf{u} = \begin{bmatrix} \dfrac{C_\delta}{m} & \dfrac{E_\delta}{I_x} & 0 & \dfrac{D_\delta}{I_z} \end{bmatrix}^T \delta \tag{11.152}$$

Proof. The Newton-Euler equations of motion for a rigid vehicle in the local coordinate frame B, attached to the vehicle at its mass center C, are given in Equations (11.6)-(11.9) as

$$\begin{aligned} F_x &= m\,\dot{v}_x - mr\,v_y & (11.153) \\ F_y &= m\,\dot{v}_y + mr\,v_x & (11.154) \\ M_z &= I_z\dot{\omega}_z = I_z\dot{r} & (11.155) \\ M_x &= I_x\dot{\omega}_x = I_x\dot{p} & (11.156) \end{aligned}$$

The approximate force system applied on a two-wheel vehicle is found in Equations (11.97)-(11.100)

$$\begin{aligned} F_x &\approx F_{x_f} + F_{x_r} & (11.157) \\ F_y &\approx F_{y_f} + F_{y_r} & (11.158) \\ M_x &\approx C_{T_f} F_{y_f} + C_{T_r} F_{y_r} - k_\varphi \varphi - c_\varphi \dot{\varphi} & (11.159) \\ M_z &\approx a_1 F_{y_f} - a_2 F_{y_r} & (11.160) \end{aligned}$$

and in terms of tire characteristics, in (11.101)-(11.103). These equations could be summarized in (11.129)-(11.131) as follows:

$$F_y = C_r r + C_p p + C_\beta \beta + C_\varphi \varphi + C_\delta \delta \tag{11.161}$$
$$M_x = E_r r + E_p p + E_\beta \beta + E_\varphi \varphi + E_\delta \delta \tag{11.162}$$
$$M_z = D_r r + D_p p + D_\beta \beta + D_\varphi \varphi + D_\delta \delta \tag{11.163}$$

Substituting (11.161)–(11.163) in (11.153)–(11.156) produces the following set of equations of motion:

$$m \dot{v}_x - mr\, v_y = F_x \tag{11.164}$$
$$m \dot{v}_y + mr\, v_x = C_r r + C_p p + C_\beta \beta + C_\varphi \varphi + C_\delta \delta \tag{11.165}$$
$$I_x \dot{p} = E_r r + E_p p + E_\beta \beta + E_\varphi \varphi + E_\delta \delta \tag{11.166}$$
$$\dot{r} I_z = D_r r + D_p p + D_\beta \beta + D_\varphi \varphi + D_\delta \delta \tag{11.167}$$

Employing

$$\beta = \frac{v_y}{v_x} \tag{11.168}$$

we are able to transform these equations to a set of differential equations for v_x, v_y, p, and r.

$$\dot{v}_x = \frac{F_x}{m} + r\, v_y \tag{11.169}$$
$$\dot{v}_y = \left(\frac{C_r}{m} - v_x\right) r + \frac{C_p}{m} p + \frac{C_\beta}{m}\frac{v_y}{v_x} + \frac{C_\varphi}{m}\varphi + \frac{C_\delta}{m}\delta \tag{11.170}$$
$$\dot{p} = \frac{1}{I_x}\left(E_r r + E_p p + E_\beta \frac{v_y}{v_x} + E_\varphi \varphi + E_\delta \delta\right) \tag{11.171}$$
$$\dot{r} = \frac{1}{I_z}\left(D_r r + D_p p + D_\beta \frac{v_y}{v_x} + D_\varphi \varphi + D_\delta \delta\right) \tag{11.172}$$

The first equation (11.169) depends on the yaw rate r and the lateral velocity v_y, which are the outputs of the other equations, (11.170)-(11.172). However, if we assume the vehicle is moving with a constant forward speed,

$$v_x = const. \tag{11.173}$$

then Equations (11.170)-(11.172) become independent with (11.169), and may be treated independent of the first equation.

Equations (11.170)-(11.172) may be considered as three coupled differential equations describing the behavior of a dynamic system. The dynamic system receives the steering angle δ as an input, and uses v_x as a parameter to generates four outputs: v_y, p, φ, and r.

$$
\begin{bmatrix} \dot{v}_y \\ \dot{p} \\ \dot{\varphi} \\ \dot{r} \end{bmatrix} =
\begin{bmatrix}
\dfrac{C_\beta}{mv_x} & \dfrac{C_p}{m} & \dfrac{C_\varphi}{m} & \dfrac{C_r}{m} - v_x \\
\dfrac{E_\beta}{I_x v_x} & \dfrac{E_p}{I_x} & \dfrac{E_\varphi}{I_x} & \dfrac{E_r}{I_x} \\
0 & 1 & 0 & 0 \\
\dfrac{D_\beta}{I_z v_x} & \dfrac{D_p}{I_z} & \dfrac{D_\varphi}{I_z} & \dfrac{D_r}{I_z}
\end{bmatrix}
\begin{bmatrix} v_y \\ p \\ \varphi \\ r \end{bmatrix} +
\begin{bmatrix} \dfrac{C_\delta}{m} \\ \dfrac{E_\delta}{I_x} \\ 0 \\ \dfrac{D_\delta}{I_z} \end{bmatrix} \delta \quad (11.174)
$$

Equation (11.174) may be rearranged to show the input-output relationship.

$$\dot{\mathbf{q}} = [A]\,\mathbf{q} + \mathbf{u} \tag{11.175}$$

The vector $\mathbf{q}$ is called the *control variables vector*, and $\mathbf{u}$ is called the *inputs vector*. The matrix $[A]$ is the control variable coefficients matrix. ∎

Example 463 *Equations of motion based on kinematic angles.*
The equations of motion (11.174) can be expressed based on the angles β, p, φ, r, *and* δ, *by employing (11.168).*
Taking a derivative from Equation (11.168) for constant v_x

$$\dot{\beta} = \frac{\dot{v}_y}{v_x} \tag{11.176}$$

and substituting in Equations (11.165) shows that we can transform the equation for $\dot{\beta}$.

$$mv_x\dot{\beta} + mr\,v_x = C_r r + C_p p + C_\beta \beta + C_\varphi \varphi + C_\delta \delta \tag{11.177}$$

Therefore, the set of equations of motion can be expressed in terms of the vehicle's angular variables.

$$
\begin{bmatrix} \dot{\beta} \\ \dot{p} \\ \dot{\varphi} \\ \dot{r} \end{bmatrix} =
\begin{bmatrix}
\dfrac{C_\beta}{mv_x} & \dfrac{C_p}{mv_x} & \dfrac{C_\varphi}{mv_x} & \dfrac{C_r}{mv_x} - 1 \\
\dfrac{E_\beta}{I_x} & \dfrac{E_p}{I_x} & \dfrac{E_\varphi}{I_x} & \dfrac{E_r}{I_x} \\
0 & 1 & 0 & 0 \\
\dfrac{D_\beta}{I_z} & \dfrac{D_p}{I_z} & \dfrac{D_\varphi}{I_z} & \dfrac{D_r}{I_z}
\end{bmatrix}
\begin{bmatrix} \beta \\ p \\ \varphi \\ r \end{bmatrix} +
\begin{bmatrix} \dfrac{C_\delta}{mv_x} \\ \dfrac{E_\delta}{I_x} \\ 0 \\ \dfrac{D_\delta}{I_z} \end{bmatrix} \delta
$$

$$(11.178)$$

11.5 ★ Steady-State Motion

Turning the front steering of a two-wheel, rollable rigid vehicle at steady-state condition is governed by

$$F_x = -mr\,v_y \tag{11.179}$$

$$C_r\,r + C_\beta\,\beta + C_\varphi\,\varphi + C_\delta\,\delta = mr\,v_x \tag{11.180}$$

$$E_r\,r + E_\beta\,\beta + E_\varphi\,\varphi + E_\delta\,\delta = 0 \tag{11.181}$$

$$D_r\,r + D_\beta\,\beta + D_\varphi\,\varphi + D_\delta\,\delta = 0 \tag{11.182}$$

or equivalently, by the following equations:

$$F_x = -\frac{m}{R}v_x\,v_y \tag{11.183}$$

$$\left(C_r\,v_x - m\,v_x^2\right)\frac{1}{R} + C_\beta\,\beta + C_\varphi\,\varphi = -C_\delta\,\delta \tag{11.184}$$

$$E_r\,v_x\frac{1}{R} + E_\beta\,\beta + E_\varphi\,\varphi = -E_\delta\,\delta \tag{11.185}$$

$$D_r\,v_x\frac{1}{R} + D_\beta\,\beta + D_\varphi\,\varphi = -D_\delta\,\delta \tag{11.186}$$

The first equation determines the required forward force to keep v_x constant. The next three equations show the steady-state values of the output variables, which are: path curvature κ,

$$\kappa = \frac{1}{R} = \frac{r}{v_x} \tag{11.187}$$

vehicle slip angle β, vehicle roll rate p, and vehicle roll angle φ for a constant steering input δ at a constant forward speed v_x. The output-input relationships are defined by the following responses:

1. *Curvature response*, S_κ

$$S_\kappa = \frac{\kappa}{\delta} = \frac{1}{R\delta} = -\frac{Z_1}{v_x Z_0} \tag{11.188}$$

2. *Slip response*, S_β

$$S_\beta = \frac{\beta}{\delta} = \frac{Z_2}{Z_0} \tag{11.189}$$

3. *Yaw rate response*, S_r

$$S_r = \frac{r}{\delta} = \frac{\kappa}{\delta}v_x = S_\kappa v_x = -\frac{Z_1}{Z_0} \tag{11.190}$$

4. *Centripetal acceleration response*, S_a

$$S_a = \frac{v_x^2/R}{\delta} = \frac{\kappa}{\delta}v_x^2 = S_\kappa v_x^2 = \frac{v_x Z_1}{Z_0} \tag{11.191}$$

5. *Lateral velocity response, S_y*

$$S_y = \frac{v_y}{\delta} = S_\beta v_x = \frac{v_x Z_2}{Z_0} \tag{11.192}$$

6. *Roll angle response, S_φ*

$$S_\varphi = \frac{\varphi}{\delta} = -\frac{Z_3}{Z_0} \tag{11.193}$$

$$
\begin{aligned}
Z_0 &= E_\beta \left(D_r C_\varphi - C_r D_\varphi + m v_x D_\varphi \right) \\
&\quad + E_\varphi \left(C_r D_\beta - D_r C_\beta - m v_x D_\beta \right) + E_r \left(C_\beta D_\varphi - D_\beta C_\varphi \right) \quad (11.194) \\
Z_1 &= E_\beta \left(C_\varphi D_\delta - v_x C_\delta D_\varphi \right) - E_\varphi \left(C_\beta D_\delta - v_x C_\delta D_\beta \right) \\
&\quad + E_\delta \left(C_\beta D_\varphi - D_\beta C_\varphi \right) \quad (11.195) \\
Z_2 &= E_\varphi \left(m v_x D_\delta - C_r D_\delta + D_r v_x C_\delta \right) + E_r \left(C_\varphi D_\delta - v_x C_\delta D_\varphi \right) \\
&\quad - E_\delta \left(D_r C_\varphi - C_r D_\varphi + m v_x D_\varphi \right) \quad (11.196) \\
Z_3 &= E_\beta \left(m v_x D_\delta - C_r D_\delta + D_r v_x C_\delta \right) + E_r \left(C_\beta D_\delta - v_x C_\delta D_\beta \right) \\
&\quad - E_\delta \left(D_r C_\beta - C_r D_\beta + m v_x D_\beta \right) \quad (11.197)
\end{aligned}
$$

Proof. In steady-state conditions, all the variables are constant and hence, their derivatives are zero. Therefore, the equations of motion (11.153)–(11.156) reduce to

$$
\begin{aligned}
F_x &= -mr\, v_y & (11.198) \\
F_y &= mr\, v_x & (11.199) \\
M_x &= 0 & (11.200) \\
M_z &= 0 & (11.201)
\end{aligned}
$$

where the lateral force F_y, roll moment M_x, and yaw moment M_z from (11.161)–(11.163) would be

$$
\begin{aligned}
F_y &= C_r r + C_\beta \beta + C_\varphi \varphi + C_\delta \delta & (11.202) \\
M_x &= E_r r + E_\beta \beta + E_\varphi \varphi + E_\delta \delta & (11.203) \\
M_z &= D_r r + D_\beta \beta + D_\varphi \varphi + D_\delta \delta & (11.204)
\end{aligned}
$$

Therefore, the equations describing the steady-state turning of a two-wheel rigid vehicle are equal to

$$
\begin{aligned}
F_x &= -mr\, v_y & (11.205) \\
C_r r + C_\beta \beta + C_\varphi \varphi + C_\delta \delta &= mr\, v_x & (11.206) \\
E_r r + E_\beta \beta + E_\varphi \varphi + E_\delta \delta &= 0 & (11.207) \\
D_r r + D_\beta \beta + D_\varphi \varphi + D_\delta \delta &= 0 & (11.208)
\end{aligned}
$$

Equation (11.205) will be used to calculate the required traction force to keep the motion steady. However, Equations (11.206)-(11.208) can be used to determine the steady-state responses of the vehicle.

$$C_r \frac{v_x}{R} + C_\beta \beta + C_\varphi \varphi + C_\delta \delta = m \frac{v_x}{R} v_x \qquad (11.209)$$

$$E_r \frac{v_x}{R} + E_\beta \beta + E_\varphi \varphi + E_\delta \delta = 0 \qquad (11.210)$$

$$D_r \frac{v_x}{R} + D_\beta \beta + D_\varphi \varphi + D_\delta \delta = 0 \qquad (11.211)$$

At steady-state turning, the vehicle will move on a circle with radius R at a speed v_x and angular velocity r, so

$$v_x \approx R\,r \qquad (11.212)$$

By substituting (11.212) in Equations (11.206)-(11.208) and employing the curvature definition (11.187), we may write the equations in matrix form

$$\begin{bmatrix} C_\beta & C_r\,v_x - m\,v_x^2 & C_\varphi \\ E_\beta & E_r\,v_x & E_\varphi \\ D_\beta & D_r\,v_x & D_\varphi \end{bmatrix} \begin{bmatrix} \beta \\ \kappa \\ \varphi \end{bmatrix} = \begin{bmatrix} -C_\delta \\ -E_\delta \\ -D_\delta \end{bmatrix} \delta \qquad (11.213)$$

Solving the equations for β, κ, and φ and using

$$\beta \approx \frac{v_y}{v_x} \qquad (11.214)$$

enables us to define different output-input relationships as (11.188)-(11.193). ∎

Example 464 *Force system coefficients for a car.*
Consider a front steering, four-wheel car with the following characteristics:

$$C_{\alpha f_L} = C_{\alpha f_R} \approx 28648\,\text{N/ rad} \qquad C_{\alpha r_L} = C_{\alpha r_R} \approx 26356\,\text{N/ rad} \quad (11.215)$$

$$
\begin{aligned}
m &= 917\,\text{kg} & I_x &= 300\,\text{kg m}^2 & I_z &= 1128\,\text{kg m}^2 \\
a_1 &= 0.91\,\text{m} & a_2 &= 1.64\,\text{m} \\
k_\varphi &= 20000\,\text{N/ rad} & c_\varphi &= 1000\,\text{N s/ rad}
\end{aligned}
\qquad (11.216)
$$

$$
\begin{aligned}
C_{\beta_f} &= 0.01 & C_{\beta_r} &= 0.01 & C_{T_f} &= 0.02 \\
C_{T_r} &= 0.02 & C_{\delta\varphi_f} &= 0.01 & C_{\delta\varphi_r} &= 0 \\
C_{\varphi_f} &= 2 & C_{\varphi_r} &= 1
\end{aligned}
\qquad (11.217)
$$

$$v_x = 40\,\text{m/ s} \qquad \delta = 0.1\,\text{rad} \qquad (11.218)$$

The sideslip coefficients of an equivalent bicycle model are

$$C_{\alpha f} = C_{\alpha f_L} + C_{\alpha f_R} = 57296\,\text{N}/\text{rad} \tag{11.219}$$
$$C_{\alpha r} = C_{\alpha r_L} + C_{\alpha r_R} = 52712\,\text{N}/\text{rad} \tag{11.220}$$

The force system coefficients are equal to the following if v_x is measured in $[\text{m}/\text{s}]$:

$$
\begin{aligned}
C_r &= 857.708 & C_p &= 27.502 & C_\beta &= -110008 \\
C_\varphi &= 569.96 & C_\delta &= 57296
\end{aligned}
\tag{11.221}
$$

$$
\begin{aligned}
E_r &= 17.15416 & E_p &= -999.44996 & E_\beta &= -2200.16 \\
E_\varphi &= -19988.6008 & E_\delta &= 1145.92
\end{aligned}
\tag{11.222}
$$

$$
\begin{aligned}
D_r &= -4730.52532 & D_p &= -8.57708 & D_\beta &= 34308.32 \\
D_\varphi &= 521.2136 & D_\delta &= 52139.36
\end{aligned}
\tag{11.223}
$$

The Z_i parameters and the steady-state responses of the vehicle are as follows:

$$
\begin{aligned}
Z_0 &= 0.3493155774 \times 10^{14} \\
Z_1 &= -0.1683862164 \times 10^{16} \\
Z_2 &= 0.1793365332 \times 10^{15} \\
Z_3 &= 0.1629203858 \times 10^{14}
\end{aligned}
\tag{11.224}
$$

$$
\begin{aligned}
S_\kappa &= \frac{\kappa}{\delta} = \frac{1/R}{\delta} = 1.205115283 \\
S_\beta &= \frac{\beta}{\delta} = 5.133940334 \\
S_r &= \frac{r}{\delta} = 48.20461133 \\
S_a &= \frac{v_x^2/R}{\delta} = -1928.184453 \\
S_\varphi &= \frac{\varphi}{\delta} = -0.4663988563
\end{aligned}
\tag{11.225}
$$

Having the steady-state responses, we are able to calculate the steady-state characteristic of the motion.

$$
\begin{aligned}
R &= 8.298\,\text{m} & \tag{11.226} \\
\beta &= 0.5138\,\text{rad} \approx 29.415\,\text{deg} & \tag{11.227} \\
r &= 4.82\,\text{rad}/\text{s} & \tag{11.228}
\end{aligned}
$$

Example 465 ★ *Camber thrust.*

When a vehicle rolls, the wheels of almost all types of suspensions take up a camber angle in the same sense as the roll. The wheel camber is always less than the roll angle.

At steady-state conditions, camber at the front wheels increases the understeer characteristic of the vehicle, while camber at the rear increases the oversteer characteristic of the vehicle. Most road vehicles are made such that in a turn, the rear wheels remain upright and the front wheels camber. These vehicles have an increasing understeer behavior with roll and are more stable.

Example 466 ★ *Roll steer.*

Positive roll steer means the wheel steers about the z-axis when the vehicle rolls about the x-axis. So, when the vehicle turns to the right, a positive roll steer wheel will steer to the left.

Positive roll steer at the front wheels increases the understeer characteristic of the vehicle, while roll steer at the rear increases the oversteer characteristic of the vehicle. Most road vehicles' suspension is made such that in a turn the front wheels have positive roll steering. These vehicles have an increasing understeer behavior with roll and are more stable.

11.6 ★ Time Response

The equations of motion must analytically or numerically be integrated to analyze the time response of a vehicle and examine how the vehicle will respond to a steering input. The equations of motion are a set of coupled ordinary differential equations as expressed here.

$$\dot{v}_x = \frac{1}{m} F_x + r\, v_y \tag{11.229}$$

$$
\begin{bmatrix} \dot{v}_y \\ \dot{p} \\ \dot{\varphi} \\ \dot{r} \end{bmatrix}
=
\begin{bmatrix}
\dfrac{C_\beta}{mv_x} & \dfrac{C_p}{m} & \dfrac{C_\varphi}{m} & \dfrac{C_r}{m} - v_x \\[2mm]
\dfrac{E_\beta}{I_x v_x} & \dfrac{E_p}{I_x} & \dfrac{E_\varphi}{I_x} & \dfrac{E_r}{I_x} \\[2mm]
0 & 1 & 0 & 0 \\[2mm]
\dfrac{D_\beta}{I_z v_x} & \dfrac{D_p}{I_z} & \dfrac{D_\varphi}{I_z} & \dfrac{D_r}{I_z}
\end{bmatrix}
\begin{bmatrix} v_y \\ p \\ \varphi \\ r \end{bmatrix}
+
\begin{bmatrix}
\dfrac{C_\delta}{m} \\[2mm]
\dfrac{E_\delta}{I_x} \\[2mm]
0 \\[2mm]
\dfrac{D_\delta}{I_z}
\end{bmatrix}
\delta(t)
$$

$$\tag{11.230}$$

Their answers to a given time-dependent steer angle $\delta(t)$ are

$$v_x = v_x(t) \qquad (11.231)$$
$$v_y = v_y(t) \qquad (11.232)$$
$$p = p(t) \qquad (11.233)$$
$$\varphi = \varphi(t) \qquad (11.234)$$
$$r = r(t) \qquad (11.235)$$

Such a solution is called the *time response* or *transient response* of the vehicle.

Assuming a constant forward velocity, the first equation (11.229) simplifies to

$$F_x = -mr\,v_y \qquad (11.236)$$

and Equation (11.230) becomes independent from the first one. The set of Equations (11.230) can be written in the form

$$\dot{\mathbf{q}} = [A]\,\mathbf{q} + \mathbf{u} \qquad (11.237)$$

in which $[A]$ is a constant coefficient matrix, $\mathbf{q}$ is the vector of control variables, and $\mathbf{u}$ is the vector of inputs.

To solve the inverse dynamics problem and find the vehicle response, the steering function $\delta(t)$ must be given.

Example 467 ★ *Free dynamics and free response.*

The response of a vehicle to zero steer angle $\delta(t) = 0$ at constant speed is called **free response**. *The equation of motion under a* **free dynamics** *is*

$$\dot{\mathbf{q}} = [A]\,\mathbf{q} \qquad (11.238)$$

To solve the equation, let us assume

$$[A] = \begin{bmatrix} a_{11} & a_{12} & a_{13} & a_{14} \\ a_{21} & a_{22} & a_{23} & a_{24} \\ a_{31} & a_{32} & a_{33} & a_{34} \\ a_{41} & a_{42} & a_{43} & a_{44} \end{bmatrix} \qquad (11.239)$$

and therefore, the equations of motion are

$$\begin{bmatrix} \dot{v}_y \\ \dot{p} \\ \dot{\varphi} \\ \dot{r} \end{bmatrix} = \begin{bmatrix} a_{11} & a_{12} & a_{13} & a_{14} \\ a_{21} & a_{22} & a_{23} & a_{24} \\ a_{31} & a_{32} & a_{33} & a_{34} \\ a_{41} & a_{42} & a_{43} & a_{44} \end{bmatrix} \begin{bmatrix} v_y \\ p \\ \varphi \\ r \end{bmatrix} \qquad (11.240)$$

Because the equations are linear, the solutions are exponential functions

$$v_y = A_1 e^{\lambda t} \qquad (11.241)$$
$$p = A_2 e^{\lambda t} \qquad (11.242)$$
$$\varphi = A_3 e^{\lambda t} \qquad (11.243)$$
$$r = A_4 e^{\lambda t} \qquad (11.244)$$

*Substituting the solutions shows that the condition for functions (11.241)-
(11.244) to be the solution of the equations (11.240) is that the exponent λ
is the eigenvalue of $[A]$. To find λ, we may expand the determinant of the
above coefficient matrix and find the characteristic equation*

$$\det [A] = 0 \tag{11.245}$$

*Having the eigenvalues $\lambda_{1,2,3,4}$ provides the following general solution for
the free dynamics of the vehicle:*

$$
\begin{aligned}
v_y &= A_{11}e^{\lambda_1 t} + A_{12}e^{\lambda_2 t} + A_{13}e^{\lambda_3 t} + A_{14}e^{\lambda_4 t} & (11.246) \\
p &= A_{21}e^{\lambda_1 t} + A_{22}e^{\lambda_2 t} + A_{23}e^{\lambda_3 t} + A_{24}e^{\lambda_4 t} & (11.247) \\
\varphi &= A_{31}e^{\lambda_1 t} + A_{32}e^{\lambda_2 t} + A_{33}e^{\lambda_3 t} + A_{34}e^{\lambda_4 t} & (11.248) \\
r &= A_{41}e^{\lambda_1 t} + A_{42}e^{\lambda_2 t} + A_{43}e^{\lambda_3 t} + A_{44}e^{\lambda_4 t} & (11.249)
\end{aligned}
$$

*The coefficients A_{ij} must be found from initial conditions. The vehicle is
stable as long as the eigenvalues have negative real part.*

*As an example, consider a vehicle with the characteristics given in (11.215)-
(11.217), and the following steer angle and forward velocity.*

$$
\begin{aligned}
v_x &= 40\,\text{m/s} & (11.250) \\
\delta &= 0.1\,\text{rad} & (11.251)
\end{aligned}
$$

*Substituting those values provide the following equations of motion for free
dynamics.*

$$
\begin{bmatrix} \dot{v}_y \\ \dot{p} \\ \dot{\varphi} \\ \dot{r} \end{bmatrix} =
\begin{bmatrix}
-3 & 0.03 & 0.621 & -39.06 \\
-0.18335 & -3.3315 & -66.63 & 0.05718 \\
0 & 1 & 0 & 0 \\
0.76038 & -0.0076 & 0.4621 & -4.194
\end{bmatrix}
\begin{bmatrix} v_y \\ p \\ \varphi \\ r \end{bmatrix}
\tag{11.252}
$$

The eigenvalues of the coefficient matrix are

$$
\begin{aligned}
\lambda_1 &= -3.593983044 + 5.409888448i \\
\lambda_2 &= -3.593983044 - 5.409888448i \\
\lambda_3 &= -1.668194745 + 7.995670961i \\
\lambda_4 &= -1.668194745 - 7.995670961i \tag{11.253}
\end{aligned}
$$

*which shows the vehicle is stable because all of the eigenvalues have real
negative parts.*

*Substituting the eigenvalues in Equations (11.246)-(11.249) provides the
solution with unknown coefficients. Let us examine the free dynamics behavior of the vehicle for a nonzero initial condition.*

$$
\mathbf{q}_0 =
\begin{bmatrix} v_y(0) \\ p(0) \\ \varphi(0) \\ r(0) \end{bmatrix} =
\begin{bmatrix} 0 \\ 0.1 \\ 0 \\ 0 \end{bmatrix}
\tag{11.254}
$$

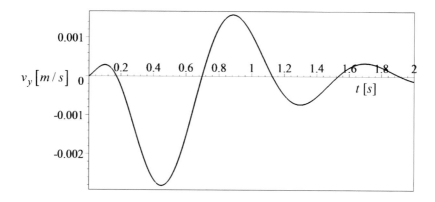

FIGURE 11.7. Lateral velocity response of a vehicle free dynamics.

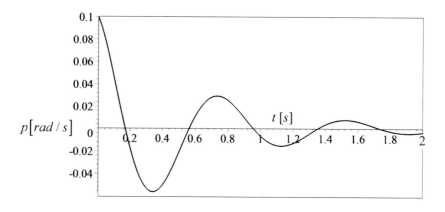

FIGURE 11.8. Roll rate response of a vehicle free dynamics.

Figures 11.7 to 11.10 illustrate the time responses of the vehicle.

Example 468 ★ *Response to a step input.*

The response of a dynamic systems to a step input is a standard test to examine the behavior of the systems. Step input in vehicle dynamics is a sudden change in steer angle from zero to a nonzero constant value.

Consider a vehicle that is moving straight at a constant forward speed

$$v_x = 40 \, \text{m/s} \tag{11.255}$$

with the characteristics given in (11.215)-(11.217) and a sudden change in the steer angle to a constant value

$$\delta(t) = \begin{cases} 0.2 \, \text{rad} \approx 11.459 \, \text{deg} & t > 0 \\ 0 & t \le 0 \end{cases} \tag{11.256}$$

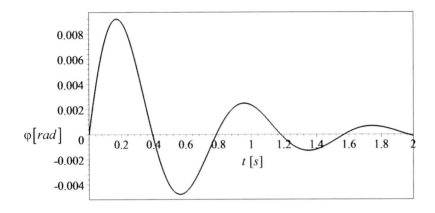

FIGURE 11.9. Roll angle response of a vehicle free dynamics.

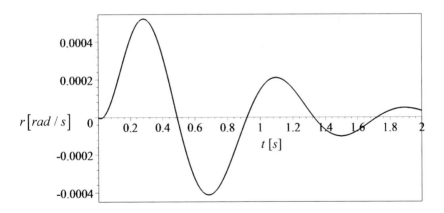

FIGURE 11.10.

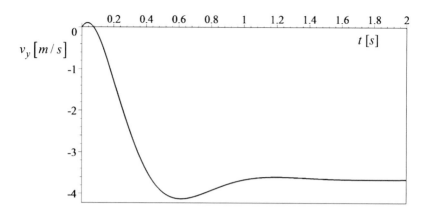

FIGURE 11.11. Lateral velocity response of a vehicle to a step steer angle.

The equations of motion for a non-zero initial conditions

$$\mathbf{q}_0 = \begin{bmatrix} v_y(0) \\ p(0) \\ \varphi(0) \\ r(0) \end{bmatrix} = \begin{bmatrix} 0 \\ 0.1 \\ 0 \\ 0 \end{bmatrix} \tag{11.257}$$

are

$$
\begin{array}{rcl}
\dot{v}_y + 6.91 v_y + 1.62 p - 10.01\varphi + 16.99 r & = & 62\delta(t) \quad (11.258) \\
\dot{p} - 5.6 v_y + 4.06 p + 99.91\varphi + .192 r & = & -69.33\delta(t) \quad (11.259) \\
\dot{\varphi} - p & = & 0 \quad (11.260) \\
\dot{r} - 1.81 v_y + 0.3 p - 5.19\varphi + 6.46 r & = & 32.11\delta(t) \quad (11.261)
\end{array}
$$

Figures 11.11 to 11.14 depict the solutions.

Having $v_y(t)$, $p(t)$, $\varphi(t)$, and $r(t)$ are enough to calculate any other kinematic variables as well as the required forward force F_x to maintain the constant forward speed.

$$F_x = -mr\,v_y \tag{11.262}$$

Figure 11.15 illustrates the required forward force $F_x(t)$.

Example 469 ★ *Lane-change maneuver.*

Passing and lane-change maneuvers are two other standard tests to examine a vehicle's dynamic responses. Lane-change can be expressed by a half-sine or a sine-squared function for steering input. Two examples of

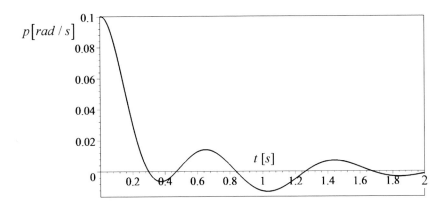

FIGURE 11.12. Roll rate response of a vehicle to a step steer angle.

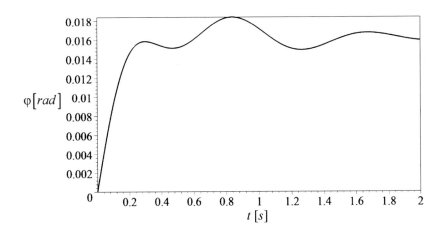

FIGURE 11.13. Roll angle response of a vehicle to a step steer angle.

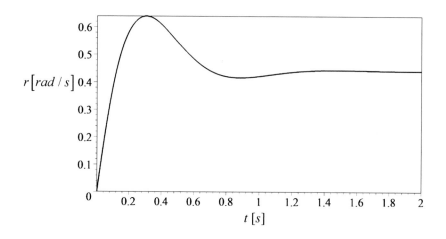

FIGURE 11.14. Yaw rate response of a vehicle to a step steer angle.

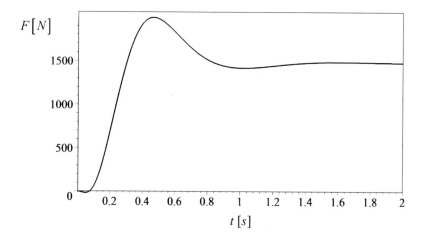

FIGURE 11.15. The required forward force F_x of a vehicle to a step steer angle to maintain the speed constant.

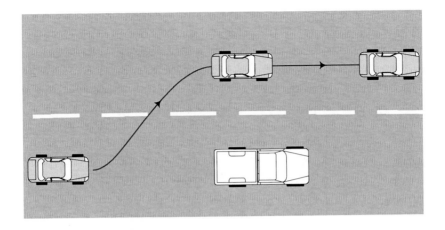

FIGURE 11.16. A lane-change maneuver.

such functions are

$$\delta(t) = \begin{cases} \delta_0 \sin \omega t & t_1 < t < \dfrac{\pi}{\omega} \\ 0 & \dfrac{\pi}{\omega} < t < t_1 \end{cases} \text{ rad} \qquad (11.263)$$

$$\delta(t) = \begin{cases} \delta_0 \sin^2 \omega t & t_1 < t < \dfrac{\pi}{\omega} \\ 0 & \dfrac{\pi}{\omega} < t < t_1 \end{cases} \text{ rad} \qquad (11.264)$$

$$\omega = \frac{\pi L}{v_x} \qquad (11.265)$$

where L is the moving length during the lane-change and v_x is the forward speed of the vehicle. The path of a lane-change car would be similar to Figure 11.16.

Let us examine a vehicle with the characteristics given in (11.215)-(11.217) and a change in half-sine steering input $\delta(t)$.

$$\delta(t) = \begin{cases} 0.2 \sin \dfrac{\pi L}{v_x} t & 0 < t < \dfrac{v_x}{L} \\ 0 & \dfrac{v_x}{L} < t < 0 \end{cases} \text{ rad} \qquad (11.266)$$

$$L = 100 \,\text{m} \qquad v_x = 40 \,\text{m/s} \qquad (11.267)$$

The equations of motion for zero initial conditions are as given in (11.258)-(11.261).

Figures 11.17 to 11.20 show the time responses of the vehicle for the steering function (11.266).

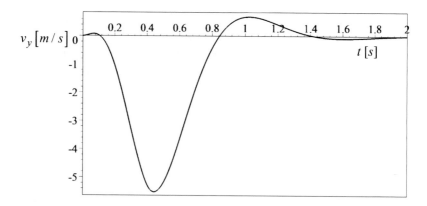

FIGURE 11.17. Lateral velocity response for the steering function (11.266).

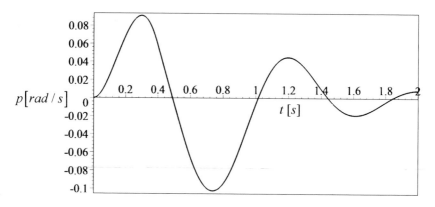

FIGURE 11.18. Roll rate response for the steering function (11.266).

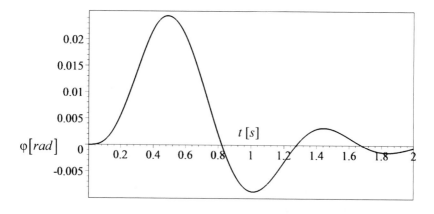

FIGURE 11.19. Roll angle response for the steering function (11.266).

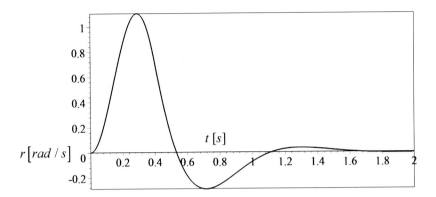

FIGURE 11.20. Yaw rate response for the steering function (11.266).

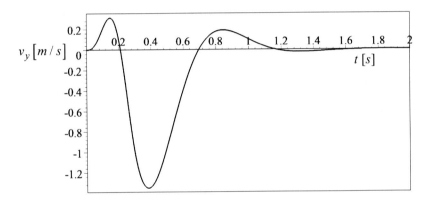

FIGURE 11.21. Lateral velocity response for the steering function (11.268).

Example 470 ★ *Lane-change with a sine-square steer function.*

A good driver should change the steer angle as smoothly as possible to minimize undesired roll angle and roll fluctuation. A sine-square steer function

$$\delta(t) = \begin{cases} \delta_0 \sin^2 \omega t & t_1 < t < \dfrac{\pi}{\omega} \\ 0 & \dfrac{\pi}{\omega} < t < \dfrac{1}{\omega} \end{cases} \text{ rad} \tag{11.268}$$

$$\omega = \frac{\pi L}{v_x} \qquad L = 100 \,\text{m} \qquad v_x = 40 \,\text{m/s} \tag{11.269}$$

which is introduced in Equation (11.264), makes for smoother lane-change steering. The responses of the vehicle in Example 469 to the steering (11.268) are illustrated in Figures 11.21 to 11.24.

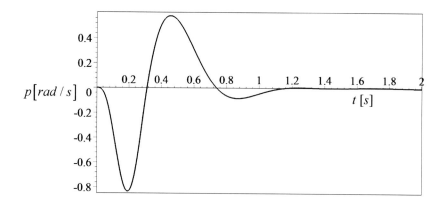

FIGURE 11.22. Roll rate response for the steering function (11.268).

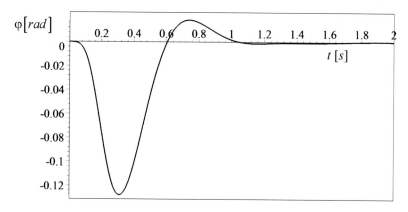

FIGURE 11.23. Roll angle response for the steering function (11.268).

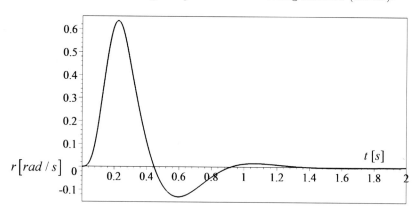

FIGURE 11.24. Yaw rate response for the steering function (11.268).

Example 471 ★ *Vehicle driving and classical feedback control.*

Driving a car is similar to a problem in feedback control. The driver compares desired direction, speed, and acceleration with actual direction, speed, and acceleration. The driver uses the cars's indicator and measuring devices, as well as human sensors to sense the actual direction, speed, and acceleration. When the actual data differs from the desired values, the driver uses the control devices such as gas pedal, brake, steering, and gear selection to improve the actual.

Example 472 ★ *Hatchback, notchback, and station models of a platform.*

It is common in vehicle manufacturing companies to install different bodies on the same chassis and platform to make different models easier. Consider a hatchback, notchback, and station models of a car that use the same platform. To compare the dynamic behavior of the three models, we may examine their response to a step steer function at a constant forward speed.

$$\delta\left(t\right) = 0.1\,\text{rad} \qquad v_x = 40\,\text{m/s} \qquad\qquad (11.270)$$

The common characteristics of the cars are

$$
\begin{aligned}
C_{\alpha f_L} &= C_{\alpha f_R} \approx 26000\,\text{N/rad} & (11.271)\\
C_{\alpha r_L} &= C_{\alpha r_R} \approx 32000\,\text{N/rad} & (11.272)
\end{aligned}
$$

$$
\begin{aligned}
l &= 2.345\,\text{m}\\
C_{\beta_f} &= -0.4 & C_{\beta_r} &= -0.1 & C_{T_f} &= -0.4\\
C_{T_r} &= -0.4 & C_{\delta\varphi_f} &= 0.01 & C_{\delta\varphi_r} &= 0.01\\
C_{\varphi_f} &= 3200 & C_{\varphi_r} &= 0\\
k_\varphi &= 26612\,\text{N m/rad} & c_\varphi &= 1700\,\text{N m s/rad} & & (11.273)
\end{aligned}
$$

For the hatchback, we use

$$
\begin{aligned}
m &= 838.7\,\text{kg} & I_x &= 300\,\text{kg m}^2 & I_z &= 1391\,\text{kg m}^2\\
a_1 &= 0.859\,\text{m} & a_2 &= 1.486\,\text{m} & & & (11.274)
\end{aligned}
$$

and for the notchback we have

$$
\begin{aligned}
m &= 845.4\,\text{kg} & I_x &= 350\,\text{kg m}^2 & I_z &= 1490\,\text{kg m}^2\\
a_1 &= 0.909\,\text{m} & a_2 &= 1.436\,\text{m} & & & (11.275)
\end{aligned}
$$

and finally for the station model we use the following data.

$$
\begin{aligned}
m &= 859\,\text{kg} & I_x &= 400\,\text{kg m}^2 & I_z &= 1680\,\text{kg m}^2\\
a_1 &= 0.945\,\text{m} & a_2 &= 1.4\,\text{m} & & & (11.276)
\end{aligned}
$$

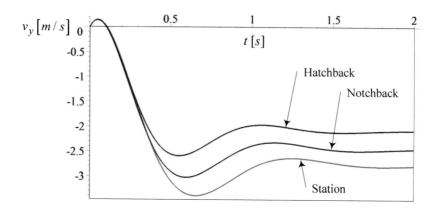

FIGURE 11.25. Lateral velocity response for hatchback, notchback, and station models of a car to a step steer angle.

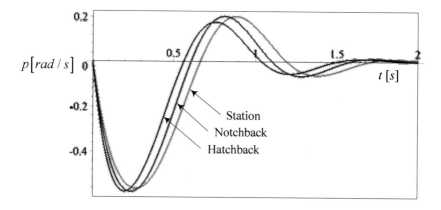

FIGURE 11.26. Roll rate response for hatchback, notchback, and station models of a car to a step steer angle.

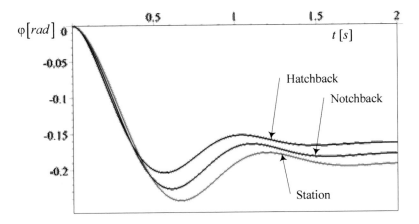

FIGURE 11.27. Roll angle response for hatchback, notchback, and station models of a car to a step steer angle.

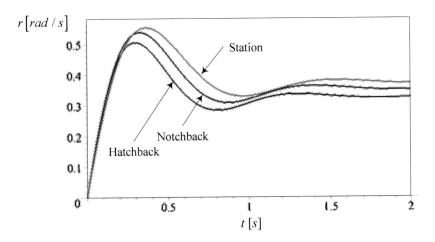

FIGURE 11.28. Yaw rate response for hatchback, notchback, and station models of a car to a step steer angle.

Figure 11.25 compares the lateral velocity responses of the three models. It shows that the steady-state value of the lateral velocity of the station model is higher than the others.

Figure 11.26 compares the roll rate responses of the three models. The hatchback model has the least longitudinal mass moment and hence shows the fastest roll rate response. It will also reach zero steady-state value faster than the other models.

Figure 11.27 compares the roll angle responses of the three models. The station model has the largest roll angle because of the highest longitudinal moment of inertia. It will also reach to the steady-state roll angle later than the other models.

Figure 11.28 compares the yaw rate responses of the three models. The station model has the highest yaw rate because of the highest vertical mass moment. It will also reach to the steady-state yaw rate later than the other models.

11.7 Summary

The best bicycle applied dynamic model for vehicle motion shows the yaw and roll DOF as well as the x and y motions. Such a model is called the rigid vehicle roll model and can be expressed by the following five differential equations.

$$\dot{v}_x = \frac{1}{m}F_x + r\,v_y$$

$$\begin{bmatrix} \dot{v}_y \\ \dot{p} \\ \dot{\varphi} \\ \dot{r} \end{bmatrix} = \begin{bmatrix} \frac{C_\beta}{mv_x} & \frac{C_p}{m} & \frac{C_\varphi}{m} & \frac{C_r}{m}-v_x \\ \frac{E_\beta}{I_x v_x} & \frac{E_p}{I_x} & \frac{E_\varphi}{I_x} & \frac{E_r}{I_x} \\ 0 & 1 & 0 & 0 \\ \frac{D_\beta}{I_z v_x} & \frac{D_p}{I_z} & \frac{D_\varphi}{I_z} & \frac{D_r}{I_z} \end{bmatrix} \begin{bmatrix} v_y \\ p \\ \varphi \\ r \end{bmatrix} + \begin{bmatrix} \frac{C_\delta}{m} \\ \frac{E_\delta}{I_x} \\ 0 \\ \frac{D_\delta}{I_z} \end{bmatrix}\delta$$

The vehicle receives the steering angle δ as the input to generate five outputs v_x, v_y, p, φ, and r. However, keeping the forward speed constant, $v_x = cnst.$, and using it as a parameter, can uncouple the first equation from the others. As a result, a constant forward speed assumption is used in many of the vehicle dynamic examinations of vehicles.

11.8   Key Symbols

$a, \ddot{x}, \mathbf{a}, \dot{\mathbf{v}}$	acceleration
a_{fwd}	front wheel drive acceleration
a_{rwd}	rear wheel drive acceleration
$a_1 = x_1$	distance of first axle from mass center
$a_2 = -x_2$	distance of second axle from mass center
A, B, C	constant parameters
$[A]$	control variable coefficient matrix
b_1	distance of left wheels from mass center
b_2	distance of right wheels from mass center
c	damping
c_f	damping of front suspension
c_r	damping of rear suspension
$c_\varphi = dM_{x_k}/d\dot{\varphi}$	vehicle roll damping
C	mass center of vehicle
$C_{\alpha_i} = dF_y/d\alpha_i$	tire sideslip coefficient
$C_{\beta_i} = v_x \, d\beta_i/dp$	tire roll rate coefficient
$C_{\delta_{\varphi_i}} = d\delta/d\varphi_i$	tire roll steering coefficient
$C_{\varphi_i} = dF_y/d\varphi_i$	tire camber thrust coefficient
$C_{T_i} = dM_x/dF_y$	tire torque coefficient
$C_r = \partial F_y/\partial r$	vehicle yaw-rate lateral-force coefficient
$C_p = \partial F_y/\partial p$	vehicle roll-rate lateral-force coefficient
$C_\beta = \partial F_y/\partial\beta$	vehicle slip-angle lateral-force coefficient
$C_\varphi = \partial F_y/\partial\varphi$	vehicle yaw-angle lateral-force coefficient
$C_\delta = \partial F_y/\partial\delta$	vehicle steer-angle lateral-force coefficient
$D_r = \partial M_z/\partial r$	vehicle yaw-rate yaw-moment coefficient
$D_p = \partial M_z/\partial p$	vehicle roll-rate yaw-moment coefficient
$D_\beta = \partial M_z/\partial\beta$	vehicle slip-angle yaw-moment coefficient
$D_\varphi = \partial M_z/\partial\varphi$	vehicle yaw-angle yaw-moment coefficient
$D_\delta = \partial M_z/\partial\delta$	vehicle steer-angle yaw-moment coefficient
$E_r = \partial M_x/\partial r$	vehicle yaw-rate roll-moment coefficient
$E_p = \partial M_x/\partial p$	vehicle roll-rate roll-moment coefficient
$E_\beta = \partial M_x/\partial\beta$	vehicle slip-angle roll-moment coefficient
$E_\varphi = \partial M_x/\partial\varphi$	vehicle yaw-angle roll-moment coefficient
$E_\delta = \partial M_x/\partial\delta$	vehicle steer-angle roll-moment coefficient
$F, \mathbf{F}$	force
F_x	traction or brake force under a wheel
F_{x_1}	traction or brake force under front wheels
F_{x_2}	traction or brake force under rear wheels
F_{x_t}	horizontal force at hinge
F_z	normal force under a wheel
F_{z_1}	normal force under front wheels
F_{z_2}	normal force under rear wheels
F_{z_3}	normal force under trailer wheels

F_{z_t}	normal force at hinge
g, $\mathbf{g}$	gravitational acceleration
h	height of C
H	height
I	mass moment of inertia
I_1, I_2, I_3	principal mass moment of inertia
k	stiffness
k_f	stiffness of front suspension
k_r	stiffness of rear suspension
$k_\varphi = dM_{x_k}/d\varphi$	vehicle roll stiffness
l	wheel base
L	road wave length
$\mathbf{L}$	angular momentum
m	car mass
M, $\mathbf{M}$	moment
$p = \dot{\varphi}$	roll rate
$\mathbf{p}$	translational momentum
$q = \dot{\theta}$	pitch rate
$\mathbf{q}$	control variable vector
$\mathbf{q}_0$	initial condition vector
$\mathbf{u}$	control input vector
$r = \dot{\psi}$	yaw rate
$\mathbf{r}$	position vector
R	tire radius
R	rotation matrix
R_f	front tire radius
R_H	radius of curvature
R_r	rear tire radius
$S_a = v_x^2/R/\delta$	lateral acceleration response
$S_r = r/\delta$	yaw rate response
$S_\beta = \beta/\delta$	sideslip response
$S_\kappa = \kappa/\delta$	curvature response
$S_\varphi = \varphi/\delta$	roll angle response
t	time
$v, \dot{x}, \mathbf{v}$	velocity
v_c	critical velocity
v_x	forward velocity
v_c	lateral velocity
w	track
z_i	deflection of axil number i
x, y, z	vehicle coordinate axes
x_i, y_i, z_i	coordinates of wheel number i in B
x_w, y_w, z_w	axes of a wheel coordinate frame
X, Y, Z	global coordinate axes
Z_0, Z_1, Z_2, Z_3	steady-state response parameters

α	tire sideslip angle between $\mathbf{v}_w$ and x_w-axis
$\beta = v_y/v_x$	vehicle slip angle between $\mathbf{v}$ and x-axis
β_f	front tire slip angle
β_i	tire slip angle between $\mathbf{v}$ and x-axis
β_r	rear tire slip angle
δ	vehicle steer angle
δ_0	a constant steer angle value
δ_w	tire steer angle between x_w-axis and x-axis
δ_1, δ_f	steer angle of front wheels
δ_2, δ_r	steer angle of rear wheels
δ_a	actual steer angle
δ_φ	roll-steer angle
η	$atan2(a,b)$
θ	pitch angle
$\kappa = 1/R$	path curvature
λ	eigenvalue
μ	friction coefficient
ϕ	slope angle
ϕ_M	maximum slope angle
φ	roll angle
ψ	yaw angle
ω	angular frequency
$\omega, \boldsymbol{\omega}$	angular velocity
$\dot{\omega}, \dot{\boldsymbol{\omega}}$	angular acceleration

Subscriptions

dyn	dynamic
f	front
fwd	front-wheel-drive
i	wheel number
L	left
M	maximum
r	rear
R	right
rwd	rear-wheel-drive
st	statics
w	wheel

Exercises

1. ★ Force system coefficients.

   (a) Consider a front-wheel-steering car with the following charac-
       teristics and

$$
\begin{aligned}
C_{\alpha f_L} &= C_{\alpha f_R} = 600 \, \text{N/deg} \qquad C_{\alpha r_L} = C_{\alpha r_R} = 560 \, \text{N/deg} \\
m &= 1245 \, \text{kg} \qquad I_x = 300 \, \text{kg m}^2 \qquad I_z = 1328 \, \text{kg m}^2 \\
a_1 &= 110 \, \text{cm} \qquad a_2 = 132 \, \text{cm} \\
k_\varphi &= 26612 \, \text{N/rad} \qquad c_\varphi = 1700 \, \text{N s/rad} \\
v_x &= 30 \, \text{m/s}
\end{aligned}
$$

$$
\begin{aligned}
C_{\beta_f} &= -0.4 \qquad C_{\beta_r} = -0.1 \qquad C_{T_f} = -0.4 \\
C_{T_r} &= -0.2 \qquad C_{\delta \varphi_f} = 0.01 \qquad C_{\delta \varphi_r} = 0.01 \\
C_{\varphi_f} &= -3200 \qquad C_{\varphi_r} = -300
\end{aligned}
$$

   (b) Determine the force system coefficients C_r, C_p, C_β, C_φ, C_δ, E_r,
       E_p, E_β, E_φ, E_δ, D_r, D_p, D_β, D_φ, and D_δ.

   (c) Derive the equations of motion of the bicycle roll model of the
       vehicle.

2. ★ Force system and two-wheel model of a car.

   Consider a front-wheel-steering car with the following characteristics

$$
\begin{aligned}
C_{\alpha r_L} &= C_{\alpha r_R} = C_{\alpha f_L} = C_{\alpha f_R} = 500 \, \text{N/deg} \\
I_x &= 300 \, \text{kg m}^2 \qquad I_z = 1328 \, \text{kg m}^2 \\
a_1 &= 110 \, \text{cm} \qquad a_2 = 132 \, \text{cm} \\
m &= 1205 \, \text{kg}
\end{aligned}
$$

$$
\begin{aligned}
k_\varphi &= 26612 \, \text{N/rad} \qquad c_\varphi = 1700 \, \text{N s/rad} \\
v_x &= 30 \, \text{m/s}
\end{aligned}
$$

$$
\begin{aligned}
C_{\beta_f} &= -0.4 \qquad C_{\beta_r} = -0.1 \qquad C_{T_f} = -0.4 \\
C_{T_r} &= -0.2
\end{aligned}
$$

$$
\begin{aligned}
C_{\delta \varphi_f} &= 0.01 \qquad C_{\delta \varphi_r} = 0.01 \\
C_{\varphi_f} &= -3200 \qquad C_{\varphi_r} = -300
\end{aligned}
$$

and determine the force system that applies on the two-wheel model of the car.

$$\begin{aligned}
F_y &= C_r\, r + C_p\, p + C_\beta\, \beta + C_\varphi\, \varphi + C_\delta\, \delta \\
M_x &= E_r\, r + E_p\, p + E_\beta\, \beta + E_\varphi\, \varphi + E_\delta\, \delta \\
M_z &= D_r\, r + D_p\, p + D_\beta\, \beta + D_\varphi\, \varphi + D_\delta\, \delta
\end{aligned}$$

Then, write the equations of motion of the car as

$$\begin{aligned}
F_x &= m\,\dot{v}_x - mr\,v_y \\
F_y &= m\,\dot{v}_y + mr\,v_x \\
M_z &= I_z\dot{r} \\
M_x &= I_x\dot{p}
\end{aligned}$$

3. ★ Equations of motion for a front-wheel-steering car.

Consider a front-wheel-steering car with the following characteristics

$$\begin{aligned}
a_1 &= 110\,\text{cm} \qquad a_2 = 132\,\text{cm} \\
m &= 1245\,\text{kg} \qquad I_z = 1328\,\text{kg m}^2 \qquad I_x = 300\,\text{kg m}^2 \\
v_x &= 40\,\text{m/s}
\end{aligned}$$

$$\begin{aligned}
C_{\alpha r_L} &= C_{\alpha r_R} = C_{\alpha f_L} = C_{\alpha f_R} = 500\,\text{N/deg} \\
k_\varphi &= 26612\,\text{N/rad} \qquad c_\varphi = 1700\,\text{N s/rad}
\end{aligned}$$

$$\begin{aligned}
C_{\beta_f} &= -0.4 \qquad C_{\beta_r} = -0.1 \\
C_{T_f} &= -0.4 \qquad C_{T_r} = -0.2
\end{aligned}$$

$$\begin{aligned}
C_{\delta\varphi_f} &= 0.01 \qquad C_{\delta\varphi_r} = 0.01 \\
C_{\varphi_f} &= -3200 \qquad C_{\varphi_r} = -300
\end{aligned}$$

and develop the equations of motion

$$\dot{\mathbf{q}} = [A]\,\mathbf{q} + \mathbf{u}$$

4. ★ Equations of motion in different variables.

Consider a car with the following characteristics

$$\begin{aligned}
C_{\alpha r_L} &= C_{\alpha r_R} = C_{\alpha f_L} = C_{\alpha f_R} = 500\,\text{N/deg} \\
a_1 &= 100\,\text{cm} \qquad a_2 = 120\,\text{cm} \\
m &= 1000\,\text{kg} \qquad I_z = 1008\,\text{kg m}^2 \\
v_x &= 40\,\text{m/s}
\end{aligned}$$

$$I_x = 300 \, \text{kg m}^2$$
$$k_\varphi = 26612 \, \text{N/rad} \qquad c_\varphi = 1700 \, \text{N s/rad}$$

$$C_{\beta_f} = -0.4 \qquad C_{\beta_r} = -0.1$$
$$C_{T_f} = -0.4 \qquad C_{T_r} = -0.2$$

$$C_{\delta\varphi_f} = 0.01 \qquad C_{\delta\varphi_r} = 0.01$$
$$C_{\varphi_f} = -3200 \qquad C_{\varphi_r} = -300$$

and develop the equations of motion

(a) in terms of $(\dot{v}_x, \dot{v}_y, \dot{p}, \dot{\varphi}, \dot{r})$, if the car is front-wheel steering.

(b) in terms of $\left(\dot{v}_x, \dot{\beta}, \dot{p}, \dot{\varphi}, \dot{r}\right)$, if the car is front-wheel steering.

5. ★ Steady state response parameters.

Consider a car with the following characteristics

$$C_{\alpha f_L} = C_{\alpha f_R} = 500 \, \text{N/deg} \qquad C_{\alpha r_L} = C_{\alpha r_R} = 520 \, \text{N/deg}$$

$$m = 1245 \, \text{kg} \qquad I_z = 1328 \, \text{kg m}^2$$
$$a_1 = 110 \, \text{cm} \qquad a_2 = 132 \, \text{cm}$$
$$v_x = 40 \, \text{m/s}$$

$$I_x = 300 \, \text{kg m}^2$$
$$k_\varphi = 26612 \, \text{N/rad} \qquad c_\varphi = 1700 \, \text{N s/rad}$$

$$C_{\beta_f} = -0.4 \qquad C_{\beta_r} = -0.1$$
$$C_{T_f} = -0.4 \qquad C_{T_r} = -0.2$$

$$C_{\delta\varphi_f} = 0.01 \qquad C_{\delta\varphi_r} = 0.01$$
$$C_{\varphi_f} = -3200 \qquad C_{\varphi_r} = -300$$

and determine the steady-state curvature response S_κ, sideslip response S_β, yaw rate response, S_r, roll angle response, S_φ, and lateral acceleration response S_a.

6. ★ Steady state motion parameters.

Consider a car with the following characteristics

$$C_{\alpha f_L} = C_{\alpha f_R} = 600 \, \text{N/deg} \qquad C_{\alpha r_L} = C_{\alpha r_R} = 550 \, \text{N/deg}$$

$$m = 1245\,\text{kg} \qquad I_z = 1128\,\text{kg m}^2 \qquad I_x = 300\,\text{kg m}^2$$

$$\begin{aligned} a_1 &= 120\,\text{cm} & a_2 &= 138\,\text{cm} \\ v_x &= 20\,\text{m/s} & \delta &= 3\,\text{deg} \end{aligned}$$

$$k_\varphi = 26612\,\text{N/rad} \qquad c_\varphi = 1700\,\text{N s/rad}$$

$$\begin{aligned} C_{\beta_f} &= -0.4 & C_{\beta_r} &= -0.1 \\ C_{T_f} &= -0.4 & C_{T_r} &= -0.2 \end{aligned}$$

$$\begin{aligned} C_{\delta\varphi_f} &= 0.01 & C_{\delta\varphi_r} &= 0.01 \\ C_{\varphi_f} &= -3200 & C_{\varphi_r} &= -300 \end{aligned}$$

and determine the steady state values of r, R, β, φ, and v_x^2/R.

7. ★ Mass moment and steady state parameters.

Consider a car that is made up of a uniform solid box with dimensions $260\,\text{cm} \times 140\,\text{cm} \times 40\,\text{cm}$. If the density of the box is $\rho = 1000\,\text{kg/m}^3$, and the other characteristics are

$$C_{\alpha f_L} = C_{\alpha f_R} = 600\,\text{N/deg} \qquad C_{\alpha r_L} = C_{\alpha r_R} = 550\,\text{N/deg}$$
$$a_1 = a_2 = \frac{l}{2}$$

$$k_\varphi = 26612\,\text{N/rad} \qquad c_\varphi = 1700\,\text{N s/rad}$$

$$\begin{aligned} C_{\beta_f} &= -0.4 & C_{\beta_r} &= -0.1 \\ C_{T_f} &= -0.4 & C_{T_r} &= -0.2 \end{aligned}$$

$$\begin{aligned} C_{\delta\varphi_f} &= 0.01 & C_{\delta\varphi_r} &= 0.01 \\ C_{\varphi_f} &= -3200 & C_{\varphi_r} &= -300 \end{aligned}$$

then,

(a) determine m, I_z.

(b) determine the steady-state responses S_κ, S_β, S_r, and S_a as functions of v_x.

(c) determine the velocity v_x at which the car has a radius of turning equal to
$$R = 35\,\text{m}$$
when
$$\delta = 4\,\text{deg}.$$

(d) determine the steady state parameters r, R, β, φ, and v_x^2/R at that speed.

(e) set the speed of the car at

$$v_x = 20\,\mathrm{m/s}$$

and plot the steady-state responses S_κ, S_β, S_r, and S_a for variable R.

8. ★ Stability factor and understeer behavior.

Define a stability factor K for the vehicle roll model.

9. ★ Stability factor and mass of the car.

Find a_1 and a_2 in terms of F_{z_1}, F_{z_2}, and mg to rewrite the stability factor K to see the effect of a car's mass distribution.

10. ★ Stability factor and car behavior.

Examine the stability factor of a car with the parameters

$$C_{\alpha f_L} = C_{\alpha f_R} = 500\,\mathrm{N/deg} \qquad C_{\alpha r_L} = C_{\alpha r_R} = 460\,\mathrm{N/deg}$$

$$
\begin{aligned}
m &= 1245\,\mathrm{kg} & I_z &= 1328\,\mathrm{kg\,m^2} & I_x &= 300\,\mathrm{kg\,m^2} \\
a_1 &= 110\,\mathrm{cm} & a_2 &= 132\,\mathrm{cm} \\
v_x &= 30\,\mathrm{m/s}
\end{aligned}
$$

$$k_\varphi = 26612\,\mathrm{N/rad} \qquad c_\varphi = 1700\,\mathrm{N\,s/rad}$$

$$
\begin{aligned}
C_{\beta_f} &= -0.4 & C_{\beta_r} &= -0.1 \\
C_{T_f} &= -0.4 & C_{T_r} &= -0.2
\end{aligned}
$$

$$
\begin{aligned}
C_{\delta\varphi_f} &= 0.01 & C_{\delta\varphi_r} &= 0.01 \\
C_{\varphi_f} &= -3200 & C_{\varphi_r} &= -300
\end{aligned}
$$

and determine if the car is understeer, neutral, or oversteer?

11. ★ Critical speed of a car.

Consider a car with the characteristics

$$
\begin{aligned}
C_{\alpha f_L} &= C_{\alpha f_R} = 700\,\mathrm{N/deg} \\
C_{\alpha r_L} &= C_{\alpha r_R} = 520\,\mathrm{N/deg}
\end{aligned}
$$

$$
\begin{aligned}
m &= 1245\,\mathrm{kg} & I_z &= 1328\,\mathrm{kg\,m^2} & I_x &= 300\,\mathrm{kg\,m^2} \\
a_1 &= 118\,\mathrm{cm} & a_2 &= 122\,\mathrm{cm}
\end{aligned}
$$

$$k_\varphi = 26612 \, \text{N/rad} \qquad c_\varphi = 1700 \, \text{N s/rad}$$

$$
\begin{aligned}
C_{\beta_f} &= -0.4 & C_{\beta_r} &= -0.1 \\
C_{T_f} &= -0.4 & C_{T_r} &= -0.2
\end{aligned}
$$

$$
\begin{aligned}
C_{\delta\varphi_f} &= 0.01 & C_{\delta\varphi_r} &= 0.01 \\
C_{\varphi_f} &= -3200 & C_{\varphi_r} &= -300
\end{aligned}
$$

(a) Define a critical speed for oversteer condition.

(b) Determine if the car is understeer, neutral, or oversteer?

12. ★ Step input response at different speed.

Consider a car with the characteristics

$$C_{\alpha f_L} = C_{\alpha f_R} = 600 \, \text{N/deg} \qquad C_{\alpha r_L} = C_{\alpha r_R} = 750 \, \text{N/deg}$$

$$
\begin{aligned}
m &= 1245 \, \text{kg} & I_z &= 1328 \, \text{kg m}^2 & I_x &= 300 \, \text{kg m}^2 \\
a_1 &= 110 \, \text{cm} & a_2 &= 132 \, \text{cm}
\end{aligned}
$$

$$k_\varphi = 26612 \, \text{N/rad} \qquad c_\varphi = 1700 \, \text{N s/rad}$$

$$
\begin{aligned}
C_{\beta_f} &= -0.4 & C_{\beta_r} &= -0.1 \\
C_{T_f} &= -0.4 & C_{T_r} &= -0.2
\end{aligned}
$$

$$
\begin{aligned}
C_{\delta\varphi_f} &= 0.01 & C_{\delta\varphi_r} &= 0.01 \\
C_{\varphi_f} &= -3200 & C_{\varphi_r} &= -300
\end{aligned}
$$

and a step input

$$
\delta(t) = \begin{cases} 5 \deg & t > 0 \\ 0 & t \leq 0 \end{cases}
$$

Determine the time response of the car at

(a) $v_x = 10 \, \text{m/s}$.

(b) $v_x = 20 \, \text{m/s}$.

(c) $v_x = 30 \, \text{m/s}$.

(d) $v_x = 40 \, \text{m/s}$.

13. ★ Step input response for different steer angle.

Consider a car with the characteristics

$$
\begin{aligned}
C_{\alpha f_L} &= C_{\alpha f_R} = 600 \,\text{N}/\deg \qquad C_{\alpha r_L} = C_{\alpha r_R} = 750 \,\text{N}/\deg \\
m &= 1245 \,\text{kg} \qquad I_z = 1328 \,\text{kg m}^2 \qquad I_x = 300 \,\text{kg m}^2 \\
a_1 &= 110 \,\text{cm} \qquad a_2 = 132 \,\text{cm} \\
v_x &= 20 \,\text{m/s} \\
k_\varphi &= 26612 \,\text{N/rad} \qquad c_\varphi = 1700 \,\text{N s/rad} \\
C_{\beta_f} &= -0.4 \qquad C_{\beta_r} = -0.1 \qquad C_{T_f} = -0.4 \qquad C_{T_r} = -0.2 \\
C_{\delta\varphi_f} &= 0.01 \qquad C_{\delta\varphi_r} = 0.01 \qquad C_{\varphi_f} = -3200 \qquad C_{\varphi_r} = -300
\end{aligned}
$$

Determine the time response of the car to a step input

$$
\delta(t) = \begin{cases} \delta & t > 0 \\ 0 & t \le 0 \end{cases}
$$

when

(a) $\delta = 3 \deg$.

(b) $\delta = 5 \deg$.

(c) $\delta = 10 \deg$.

14. ★ Eigenvalues and free response.

Consider a car with the characteristics

$$
C_{\alpha f_L} = C_{\alpha f_R} = 600 \,\text{N}/\deg \qquad C_{\alpha r_L} = C_{\alpha r_R} = 750 \,\text{N}/\deg
$$

$$
\begin{aligned}
m &= 1245 \,\text{kg} \qquad I_z = 1328 \,\text{kg m}^2 \qquad I_x = 300 \,\text{kg m}^2 \\
a_1 &= 110 \,\text{cm} \qquad a_2 = 132 \,\text{cm} \qquad v_x = 20 \,\text{m/s}
\end{aligned}
$$

$$
k_\varphi = 26612 \,\text{N/rad} \qquad c_\varphi = 1700 \,\text{N s/rad}
$$

$$
\begin{aligned}
C_{\beta_f} &= -0.4 \qquad C_{\beta_r} = -0.1 \\
C_{T_f} &= -0.4 \qquad C_{T_r} = -0.2
\end{aligned}
$$

$$
\begin{aligned}
C_{\delta\varphi_f} &= 0.01 \qquad C_{\delta\varphi_r} = 0.01 \\
C_{\varphi_f} &= -3200 \qquad C_{\varphi_r} = -300
\end{aligned}
$$

(a) Determine the eigenvalues of the coefficient matrix $[A]$ and find out if the car is stable at zero steer angle.

(b) In either case, determine the weight distribution ratio, a_1/a_2, such that the car is neutral stable.

(c) Recommend a condition for the weight distribution ratio, a_1/a_2, such that the car is stable.

15. ★ Time response to different steer functions.

Consider a car with the characteristics

$$C_{\alpha f_L} = C_{\alpha f_R} = 600 \,\text{N}/\deg \qquad C_{\alpha r_L} = C_{\alpha r_R} = 750 \,\text{N}/\deg$$

$$
\begin{aligned}
m &= 1245 \,\text{kg} & I_z &= 1328 \,\text{kg m}^2 & I_x &= 300 \,\text{kg m}^2 \\
a_1 &= 110 \,\text{cm} & a_2 &= 132 \,\text{cm} & v_x &= 20 \,\text{m/s}
\end{aligned}
$$

$$k_\varphi = 26612 \,\text{N}/\text{rad} \qquad c_\varphi = 1700 \,\text{N s}/\text{rad}$$

$$
\begin{aligned}
C_{\beta_f} &= -0.4 & C_{\beta_r} &= -0.1 \\
C_{T_f} &= -0.4 & C_{T_r} &= -0.2
\end{aligned}
$$

$$
\begin{aligned}
C_{\delta\varphi_f} &= 0.01 & C_{\delta\varphi_r} &= 0.01 \\
C_{\varphi_f} &= -3200 & C_{\varphi_r} &= -300
\end{aligned}
$$

and a step input

$$\delta(t) = \begin{cases} 5\,\text{deg} & t > 0 \\ 0 & t \le 0 \end{cases}$$

Determine the time response of the car to

(a) $\delta(t) = \sin 0.1t$ for $0 < t < 10\pi$ and $\delta(t) = 0$ for $t \le 0$ and $t \ge 10\pi$.

(b) $\delta(t) = \sin 0.5t$ for $0 < t < 2\pi$ and $\delta(t) = 0$ for $t \le 0$ and $t \ge 2\pi$.

(c) $\delta(t) = \sin t$ for $0 < t < \pi$ and $\delta(t) = 0$ for $t \le 0$ and $t \ge \pi$.

16. ★ Research exercise 1.

Consider a bicycle model of a car such that tires are always upright and remain perpendicular to the road surface. Develop the equations of motion for the roll model of the car.

17. ★ Research exercise 2.

Employ the tire frame T at the tireprint, wheel frame W at the wheel center, and wheel-body frame C that is at the point corresponding to the wheel center at zero δ and zero γ, and remains parallel to the vehicle frame B. Develop the W, C, and B expressions of the generated forces at the tireprint in T frame, and develop a better set of equations for the roll model of a car.

18. ★ Research exercise 3.

Use the caster theory to find the associated camber angle γ for a steer angle δ, when the caster angle φ and lean angle θ are given. Then provide a better set of equations for the roll model of vehicles.

Part IV

Vehicle Vibration

12

Applied Vibrations

Vibration is an inevitable phenomena in vehicle dynamics. In this chapter, we review the principles of vibrations, analysis methods, and their applications, along with the frequency and time responses of vibrating systems. Special attention is devoted to frequency response analysis, because most of the optimization methods for vehicle suspensions and vehicle vibrating components are based on frequency responses.

12.1 Mechanical Vibration Elements

Mechanical vibrations is a result of continuous transformation of kinetic energy K to potential energy V, back and forth. When the potential energy is at its maximum, the kinetic energy is zero and vice versa. When a periodic fluctuations of kinetic energy appears as a periodic motion of a massive body, we call the energy transformation *mechanical vibrations*.

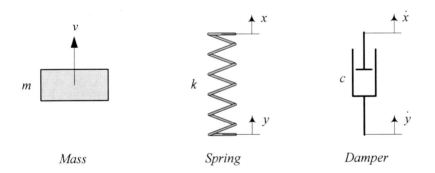

Mass   *Spring*   *Damper*

FIGURE 12.1. A mass m, spring k, and damper c.

The mechanical element that stores kinetic energy is called *mass*, and the mechanical element that stores potential energy, is called *spring*. If the total value of mechanical energy $E = K + V$ decreases during a vibration, there is a mechanical element that dissipates energy. The dissipative element is called *damper*. A mass, spring, and damper are usually illustrated by symbols in Figure 12.1.

The amount of stored kinetic energy in a mass m is proportional to the square of its velocity, v^2. The velocity $v \equiv \dot{x}$ may be a function of position

R.N. Jazar, *Vehicle Dynamics: Theory and Application*,
DOI 10.1007/978-1-4614-8544-5_12, © Springer Science+Business Media New York 2014

and time.

$$K = \frac{1}{2}mv^2 \tag{12.1}$$

The required force f_m to move a mass m is proportional to its acceleration $a \equiv \ddot{x}$.

$$f_m = ma \tag{12.2}$$

A spring is characterized by its *stiffness* k. A force f_k to generate a deflection in spring is proportional to relative displacement of its ends. The stiffness k may be a function of position and time.

$$f_k = -kz = -k(x - y) \tag{12.3}$$

If k is time independent then, the value of stored potential energy in the spring is equal to the work done by the spring force f_k during the spring deflection.

$$V = -\int f_k \, dz = -\int -kz \, dz \tag{12.4}$$

The spring potential energy is then a function of the spring's length change. If the stiffness of a spring, k, is not a function of displacements, it is called *linear spring*. Then, its potential energy is

$$V = \frac{1}{2}kz^2 \tag{12.5}$$

Damping of a damper is measured by the value of mechanical energy loss in one cycle. Equivalently, a damper may be defined by the required force f_c to generate a motion in the damper. If f_c is proportional to the relative velocity of its ends, it is a *linear damper* with a constant damping c.

$$f_c = -c\dot{z} = -c(\dot{x} - \dot{y}) \tag{12.6}$$

Such a damping is also called *viscous damping*.

A vibrating motion x is characterized by *period* T, which is the required time for one complete cycle of vibration, starting from and ending at $(\dot{x} = 0, \ddot{x} < 0)$. *Frequency* f is the number of cycles in one T.

$$f = \frac{1}{T} \tag{12.7}$$

In theoretical vibrations, we usually work with *angular frequency* $\omega \, [\text{rad/ s}]$, and in applied vibrations we use *cyclic frequency* $f \, [\text{Hz}]$.

$$\omega = 2\pi f \tag{12.8}$$

When there is no applied external force or *excitation* on a vibrating system, any possible motion of the system is called *free vibration*. A free vibrating system will oscillate if any one of the kinematic states x, $\dot{x}$, or $\ddot{x}$ is

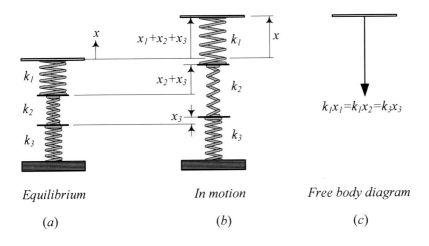

FIGURE 12.2. Three serial springs

not zero. If we apply any external excitation, a possible motion of the system is called *forced vibration*. There are four types of applied excitations: *harmonic, periodic, transient, random*. The harmonic and transient excitations are more applied and more predictable than the periodic and random types. When the excitation is a sinusoidal function of time, it is called *harmonic excitation* and when the excitation disappears after a while or stays steady, it is *transient excitation*. A *random excitation* has no short term pattern, however, we may define some long term averages to characterize a random excitation.

We use f to indicate a harmonically-variable force with amplitude F to be consistent with a harmonic motion x with amplitude X. We also use f for cyclic frequency, however, f is a force unless it is indicated that it is a frequency.

Example 473 *Serial springs and dampers.*

Serial springs have the same force, and a resultant displacement equal to the sum of individual displacements. Figure 12.2 illustrates three serial springs attached to a massless plate and the ground.

The equilibrium position of the springs is the un-stretched configuration in Figure 12.2(a). Applying a displacement x as shown in Figure 12.2(b) generates the free body diagram as shown in Figure 12.2(c). Each spring makes a force $f_i = -k_i x_i$ where x_i is the length change in spring number i. The total displacement of the springs, x, is the sum of their individual displacements, $x = \sum x_i$.

$$x = x_1 + x_2 + x_3 \qquad (12.9)$$

We may substitute a set of serial springs with only one equivalent spring, having a stiffness k_{eq}, that produces the same displacement x under the *same force f_k.*

$$f_k = -k_1 x_1 = -k_2 x_2 = -k_3 x_3 = -k_{eq} x \qquad (12.10)$$

Substituting (12.10) in (12.9)

$$\frac{f_s}{k_{eq}} = \frac{f_s}{k_1} + \frac{f_s}{k_2} + \frac{f_s}{k_3} \qquad (12.11)$$

shows that the inverse of the equivalent stiffness of the serial springs, $1/k_{eq}$, is the sum of their inverse stiffness, $\sum 1/k_i$.

$$\frac{1}{k_{eq}} = \frac{1}{k_1} + \frac{1}{k_2} + \frac{1}{k_3} \qquad (12.12)$$

We assume that velocity $\dot{x}$ has no effect on the force of a linear spring.

Serial dampers have the same force, f_c, and a resultant velocity $\dot{x}$ equal to the sum of individual velocities, $\sum \dot{x}_i$. We may substitute a set of serial dampers with only one equivalent damping c_{eq} that produces the same velocity $\dot{x}$ under the same force f_c. For three parallel dampers, the velocity and force balance

$$\dot{x} = \dot{x}_1 + \dot{x}_2 + \dot{x}_3 \qquad (12.13)$$
$$f_c = -c_1 \dot{x} = -c_2 \dot{x} = -c_3 \dot{x} = -c_{eq} \dot{x} \qquad (12.14)$$

shows that the equivalent damping is

$$\frac{1}{c_{eq}} = \frac{1}{c_1} + \frac{1}{c_2} + \frac{1}{c_3} \qquad (12.15)$$

We assume that displacement x has no effect on the force of a linear damper.

Example 474 *Parallel springs and dampers.*

Parallel springs have the same displacement x, with a resultant force, f_k, equal to sum of the individual forces $\sum f_i$. Figure 12.3 illustrates three parallel springs between a massless plate and the ground.

The equilibrium position of the springs is the un-stretched configuration shown in Figure 12.3(a). Applying a displacement x to all the springs in Figure 12.3(b) generates the free body diagram shown in Figure 12.3(c). Each spring makes a force $-kx$ opposite to the direction of displacement. The resultant force of the springs is

$$f_k = -k_1 x - k_2 x - k_3 x \qquad (12.16)$$

We may substitute parallel springs with only one equivalent stiffness k_{eq} that produces the same force f_k under the same displacement.

$$f_k = -k_{eq} x \qquad (12.17)$$

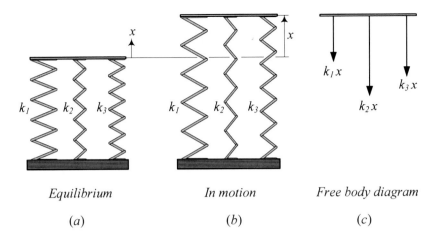

Equilibrium	*In motion*	*Free body diagram*
(a)	*(b)*	*(c)*

FIGURE 12.3. Three parallel springs.

Therefore, the equivalent stiffness of the parallel springs is sum of their stiffness.

$$k_{eq} = k_1 + k_2 + k_3 \qquad (12.18)$$

Parallel dampers have the same speed $\dot{x}$, and a resultant force f_c equal to the sum of individual forces. We may substitute parallel dampers with only one equivalent damping c_{eq} that produces the same force f_c under the same velocity. Consider three parallel dampers such as is shown in Figure 12.4. Their force balance and equivalent damping would be

$$\begin{aligned} f_c &= -c_1\dot{x} - c_2\dot{x} - c_3\dot{x} & (12.19) \\ f_c &= -c_{eq}\dot{x} & (12.20) \\ c_{eq} &= c_1 + c_2 + c_3 & (12.21) \end{aligned}$$

Example 475 *Flexible frame.*

Figure 12.5 depicts a mass m hanging from a frame. The frame is flexible, so it can be modeled by some springs attached to each other, as shown in Figure 12.6(a). If we assume that each beam is simply supported, then the equivalent stiffness for a lateral deflection of each beam at their midspan is

$$k_5 = \frac{48E_5I_5}{l_5^3} \qquad k_4 = \frac{48E_4I_4}{l_4^3} \qquad k_3 = \frac{48E_3I_3}{l_3^3} \qquad (12.22)$$

When the mass is vibrating, the elongation of each spring would be similar to Figure 12.6(b). Assume we separate the mass and springs, and then apply a force f at the end of spring k_1 as shown in Figure 12.6(c). Because the

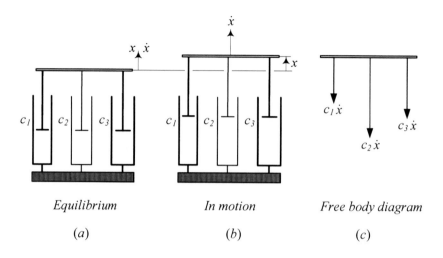

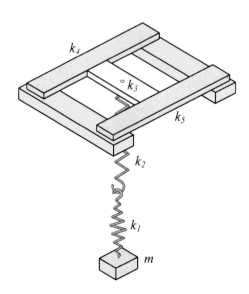

FIGURE 12.4. Three parallel dampers

FIGURE 12.5. A mass m hanging from a flexible frame.

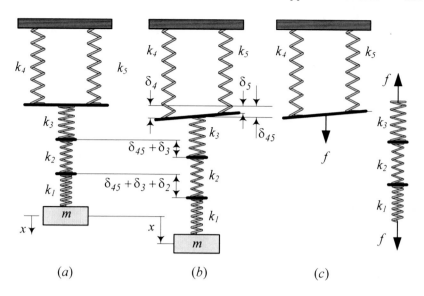

FIGURE 12.6. Equivalent springs model for the flexible frame.

springs k_1, k_2, and k_3 have the same force, and their resultant displacement is the sum of individual displacements, they are in series.

The springs k_4 and k_5 are neither in series nor parallel. To find their equivalent, let us assume that springs k_4 and k_5 support a force equal to $f/2$. Therefore,

$$\delta_4 = \frac{f}{2k_4} \qquad \delta_5 = \frac{f}{2k_5} \tag{12.23}$$

and the displacement at midspan of the lateral beam is

$$\delta_{45} = \frac{\delta_4 + \delta_5}{2} \tag{12.24}$$

Assuming

$$\delta_{45} = \frac{f}{k_{45}} \tag{12.25}$$

we can define an equivalent stiffness k_{45} for k_4 and k_5 as

$$\frac{1}{k_{45}} = \frac{1}{2}\left(\frac{1}{2k_4} + \frac{1}{2k_5}\right) = \frac{1}{4}\left(\frac{1}{k_4} + \frac{1}{k_5}\right) \tag{12.26}$$

Now the equivalent spring k_{45} is in series with the series of k_1, k_2, and k_3. Hence the overall equivalent spring k_{eq} is

$$\begin{aligned}
\frac{1}{k_{eq}} &= \frac{1}{k_1} + \frac{1}{k_2} + \frac{1}{k_3} + \frac{1}{k_{45}} \\
&= \frac{1}{k_1} + \frac{1}{k_2} + \frac{1}{k_3} + \frac{1}{4k_4} + \frac{1}{4k_5}
\end{aligned} \tag{12.27}$$

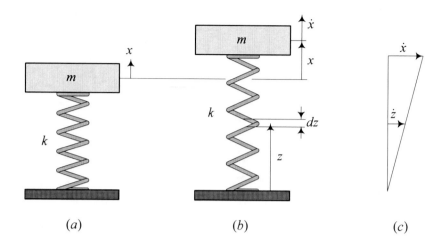

FIGURE 12.7. A vibrating system with a massive spring.

Example 476 ★ *Massive spring.*

In modeling of vibrating systems we ignore the mass of springs and dampers. This assumption is valid as long as the mass of springs and dampers are much smaller than the mass of the body they support. However, when the mass of spring m_s or damper m_d is comparable with the mass of body m, we may define a new system with an equivalent mass m_{eq}

$$m_{eq} = m + \frac{1}{3}m_s \qquad (12.28)$$

which is supported by massless spring and damper.

Consider a vibrating system with a massive spring as shown in Figure 12.7(a). The spring has a mass m_s, and a length l, when the system is at equilibrium. The mass of spring is uniformly distributed along its length, so, we may define a length density as

$$\rho = \frac{m}{l} \qquad (12.29)$$

To show (12.28), we seek for a system with a mass m_{eq} and a massless spring, which can keep the same amount of kinetic energy as the original system. Figure 12.7(b) illustrates the system when the mass m is at position x and has a velocity $\dot{x}$. The spring is between the mass and the ground. So, the base of spring has no velocity, while the other end has the same velocity as m. Let us define a coordinate z that goes from the grounded base of the spring to the end point. An element of spring at z has a length dz and a mass dm.

$$dm = \rho\, dz \qquad (12.30)$$

Assuming a linear velocity distribution of the elements of spring, as shown in Figure 12.7(c), we find the velocity $\dot{z}$ of dm as

$$\dot{z} = \frac{z}{l}\dot{x} \tag{12.31}$$

The kinetic energy of the system is a summation of kinetic energy of the mass m and kinetic energy of the spring.

$$
\begin{aligned}
K &= \frac{1}{2}m\dot{x}^2 + \frac{1}{2}\int_0^l (dm\,\dot{z}^2) = \frac{1}{2}m\dot{x}^2 + \frac{1}{2}\int_0^l \rho\left(\frac{z}{l}\dot{x}\right)^2 dz \\
&= \frac{1}{2}m\dot{x}^2 + \frac{1}{2}\frac{\rho}{l^2}\dot{x}^2 \int_0^l z^2\,dz = \frac{1}{2}m\dot{x}^2 + \frac{1}{2}\frac{\rho}{l^2}\dot{x}^2\left(\frac{1}{3}l^3\right) \\
&= \frac{1}{2}m\dot{x}^2 + \frac{1}{2}\left(\frac{1}{3}\rho l\right)\dot{x}^2 = \frac{1}{2}\left(m + \frac{1}{3}m_s\right)\dot{x}^2 \\
&= \frac{1}{2}m_{eq}\dot{x}^2 \tag{12.32}
\end{aligned}
$$

Therefore, an equivalent system should have a massless spring and a mass $m_{eq} = m + \frac{1}{3}m_s$ to keep the same amount of kinetic energy.

12.2 Newton's Method and Vibrations

Every vibrating system can be modeled as a combination of masses m_i, dampers c_i, and springs k_i. Such a model is called a *discrete* or *lumped* model of the system. A one degree of freedom (DOF) vibrating system, with the following equation of motion, is shown in Figure 12.8.

$$ma = -cv - kx + f(x, v, t) \tag{12.33}$$

To apply Newton's method and find the equations of motion, we assume all the masses m_i are out of the equilibrium at positions x_i with velocities $\dot{x}_i$. Such a situation is shown in Figure 12.8(*b*) for a one DOF system. The free body diagram that is shown in Figure 12.8(*c*) illustrates the applied forces and then, Newton's equation (9.11)

$$ {}^G\mathbf{F} = \frac{{}^Gd}{dt}\,{}^G\mathbf{p} = \frac{{}^Gd}{dt}\left(m\,{}^G\mathbf{v}\right) \tag{12.34}$$

generates the equations of motion.

The *equilibrium position* of a vibrating system is where the potential energy of the system, V, is extremum.

$$\frac{\partial V}{\partial x} = 0 \tag{12.35}$$

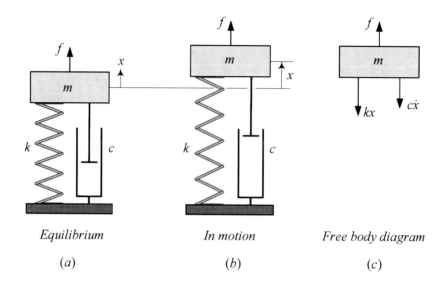

FIGURE 12.8. A one DOF vibrating systems.

We usually set $V = 0$ at the equilibrium position. Linear systems with constant stiffness have only one equilibrium or infinity equilibria, while nonlinear systems may have multiple equilibria. An equilibrium is *stable* if

$$\frac{\partial^2 V}{\partial x^2} > 0 \qquad (12.36)$$

and is *unstable* if

$$\frac{\partial^2 V}{\partial x^2} < 0 \qquad (12.37)$$

The arrangement and the number of employed elements can be used to classify discrete vibrating systems. The number of masses, times the DOF of each mass, makes the total DOF of the vibrating system n. The final set of equations would be n second-order differential equations to be solved for n generalized coordinates. When each mass has one DOF, then the system's DOF is equal to the number of masses. The DOF may also be defined as the minimum number of independent coordinates that defines the configuration of a system.

A one, two, and three DOF model for analysis of vertical vibrations of a vehicle are shown in Figure 12.9(a)-(c). The system in Figure 12.9(a) is called the *quarter car* model, which m_s represents a quarter mass of the body, and m_u represents a wheel. The parameters k_u and c_u are models for tire stiffness and damping. Similarly, k_s and c_u are models for the main suspension of the vehicle. Figure 12.9(c) is called the 1/8 car model which does not show the wheel of the car, and Figure 12.9(b) is a quarter car with a driver m_d and the driver's seat modeled as k_d and c_d.

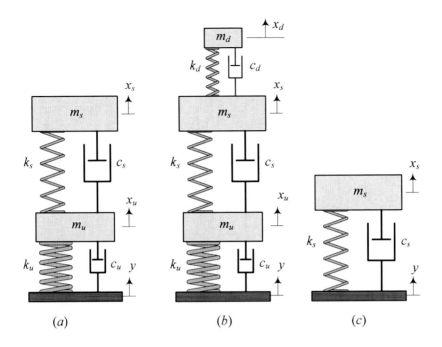

FIGURE 12.9. Two, three, and one DOF models for vertical vibrations of vehicles.

Example 477 1/8 *car model.*

Figure 12.9(c) and 12.10(a) show the simplest model for vertical vibrations of a vehicle. This model is sometimes called 1/8 *car model. The mass* m_s *represents one quarter of the car's body, which is mounted on a suspension made of a spring* k_s *and a damper* c_s. *When* m_s *is vibrating at a position such as in Figure 12.10(b), its free body diagram is as Figure 12.10(c) shows.*

Applying Newton's method, the equation of motion would be

$$m_s\ddot{x} = -k_s\left(x_s - y\right) - c_s\left(\dot{x}_s - \dot{y}\right) \qquad (12.38)$$

which can be simplified to the following equation, when we separate the input y and output x variables.

$$m_s\ddot{x} + c_s\dot{x}_s + k_s x_s = k_s y + c_s\dot{y} \qquad (12.39)$$

Example 478 *Equivalent mass and spring.*

Figure 12.11(a) illustrates a pendulum made by a point mass m attached to a massless bar with length l. The coordinate θ *shows the angular position of the bar. The equation of motion for the pendulum can be found by using the Euler equation and employing the free-body-diagram shown in Figure 12.11(b).*

$$ml^2\ddot{\theta} = -mgl\sin\theta \qquad (12.40)$$

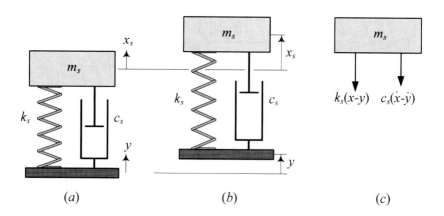

FIGURE 12.10. A 1/8 car model and its free body diagram.

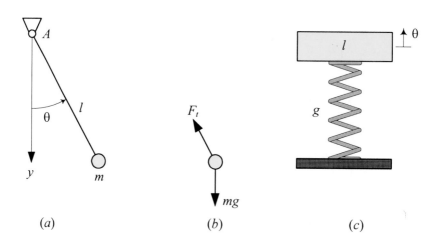

FIGURE 12.11. Equivalent mass-spring vibrator for a pendulum.

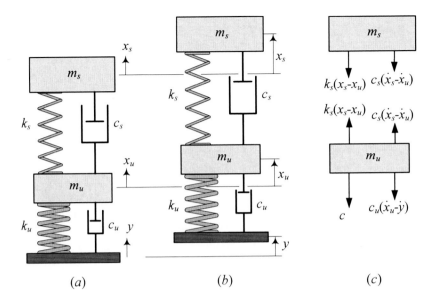

FIGURE 12.12. A 1/4 car model and its free body diagram.

Simplifying the equation of motion and assuming a very small swing angle yields:

$$l\ddot{\theta} + g\theta = 0 \tag{12.41}$$

This equation is equivalent to an equation of motion for a mass-spring system made by a mass $m \equiv l$, and a spring with stiffness $k \equiv g$. The displacement of the mass would be $x \equiv \theta$. Figure 12.11(c) depicts such an equivalent mass-spring system.

Example 479 *Force proportionality.*
   The equation of motion for a vibrating system is a balance between four different forces. A force proportional to displacement, $-kx$, a force proportional to velocity, $-cv$, a force proportional to acceleration, ma, and an applied external force $f(x, v, t)$, which can be a function of displacement, velocity, and time. Based on Newton's method, the force proportional to acceleration, ma, is always equal to the sum of all the other forces.

$$ma = -cv - kx + f(x, v, t) \tag{12.42}$$

Example 480 *A two-DOF base excited system.*
   Figure 12.12(a)-(c) illustrate the equilibrium, motion, and free body diagram of the two-DOF system shown in 12.9(a). The free body diagram is plotted based on the assumption

$$x_s > x_u > y \tag{12.43}$$

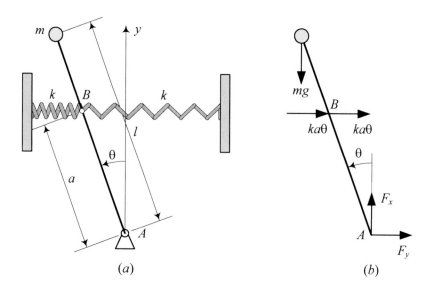

FIGURE 12.13. An inverted pendulum with a tip mass m and two supportive springs.

Applying Newton's method provides two equations of motion as follows

$$m_s \ddot{x}_s = -k_s (x_s - x_u) - c_s (\dot{x}_s - \dot{x}_u) \tag{12.44}$$

$$m_u \ddot{x}_u = k_s (x_s - x_u) + c_s (\dot{x}_s - \dot{x}_u)$$
$$-k_u (x_u - y) - c_u (\dot{x}_u - \dot{y}) \tag{12.45}$$

The assumption (12.43) is not necessarily fulfilled. We can find the same Equations (12.44) and (12.45) using any other assumption, such as $x_s < x_u > y$, $x_s > x_u < y$, or $x_s < x_u < y$. However, having an assumption helps to make a consistent free body diagram.

We usually rearrange the equations of motion for a linear system in a matrix form to take advantage of matrix calculus.

$$[M]\,\dot{\mathbf{x}} + [c]\,\dot{\mathbf{x}} + [k]\,\mathbf{x} = \mathbf{F} \tag{12.46}$$

Rearrangement of Equations (12.44) and (12.45) results in the following set of equations:

$$\begin{bmatrix} m_s & 0 \\ 0 & m_u \end{bmatrix} \begin{bmatrix} \ddot{x}_s \\ \ddot{x}_u \end{bmatrix} + \begin{bmatrix} c_s & -c_s \\ -c_s & c_s + c_u \end{bmatrix} \begin{bmatrix} \dot{x}_s \\ \dot{x}_u \end{bmatrix} +$$
$$\begin{bmatrix} k_s & -k_s \\ -k_s & k_s + k_u \end{bmatrix} \begin{bmatrix} x_s \\ x_u \end{bmatrix} = \begin{bmatrix} 0 \\ k_u y + c_u \dot{y} \end{bmatrix} \tag{12.47}$$

Example 481 ★ *Inverted pendulum.*

Figure 12.13(a) illustrates an inverted pendulum with a tip mass m and a length l. The pendulum is supported by two identical springs attached *to point B at a distance a < l from the pivot A. A free body diagram of the pendulum is shown in Figure 12.13(b). The equation of motion may be found by taking a moment about A.*

$$\sum M_A = I_A \ddot{\theta} \tag{12.48}$$

$$mg\,(l\sin\theta) - 2ka\theta\,(a\cos\theta) = ml^2\ddot{\theta} \tag{12.49}$$

To derive Equation (12.49) we assumed that the springs are long enough to remain almost straight when the pendulum oscillates. Rearrangement and assuming a very small θ shows that the nonlinear equation of motion (12.49) can be approximated by

$$ml^2\ddot{\theta} + \left(mgl - 2ka^2\right)\theta = 0 \tag{12.50}$$

which is equivalent to a linear oscillator

$$m_{eq}\ddot{\theta} + k_{eq}\theta = 0 \tag{12.51}$$

with an equivalent mass m_{eq} and and equivalent k_{eq}.

$$m_{eq} = ml^2 \qquad k_{eq} = mgl - 2ka^2 \tag{12.52}$$

The potential energy of the inverted pendulum can be expressed as

$$V = -mgl\,(1 - \cos\theta) + ka^2\theta^2 \tag{12.53}$$

which has a zero value at $\theta = 0$. The potential energy V is approximately equal to the following equation if θ is very small

$$V \approx -\frac{1}{2}mgl\theta^2 + ka^2\theta^2 \tag{12.54}$$

because

$$\cos\theta \approx 1 - \frac{1}{2}\theta^2 + O\left(\theta^4\right) \tag{12.55}$$

To find the equilibrium positions of the system, we should solve the following equation for any possible θ.

$$\frac{\partial V}{\partial x} = -2mgl\theta + 2ka^2\theta = 0 \tag{12.56}$$

The solution of the equation is

$$\theta = 0 \tag{12.57}$$

that shows the upright vertical position is the only equilibrium of the inverted pendulum as long as θ is very small. However, if

$$mgl = ka^2 \tag{12.58}$$

then any θ around $\theta = 0$ can be an equilibrium position and hence, the inverted pendulum would have infinity equilibria.

A second derivative of the potential energy

$$\frac{\partial^2 V}{\partial x^2} = -2mgl + 2ka^2 \qquad (12.59)$$

indicates that the equilibrium position $\theta = 0$ is stable if

$$ka^2 > mgl \qquad (12.60)$$

A stable equilibrium pulls the system back, if it deviates from the equilibrium, while an unstable equilibrium repels the system. Vibration happens when the equilibrium is stable.

12.3 Frequency Response of Vibrating Systems

Frequency response is the *steady-state* solution of equations of motion, when the system is *harmonically excited*. Steady-state response refers to a constant amplitude oscillation at a given frequency, after the effect of initial conditions dies out. A harmonic excitation is any combination of *sinusoidal functions* that applies on a vibrating system. If the system is linear, then a harmonic excitation generates a harmonic response with a frequency-dependent amplitude. In frequency response analysis, we are looking for the steady-state amplitude of oscillation as a function of the excitation frequency.

A vast amount of vibrating systems in vehicle dynamics can be modeled by a one DOF system. Consider a one DOF mass-spring-damper system. There are four types of one DOF harmonically excited systems:

1. base excitation,
2. eccentric excitation,
3. eccentric base excitation,
4. forced excitation.

These four systems are shown in Figure 12.14 symbolically.

Base excitation is the most practical model for vertical vibration of vehicles. *Eccentric excitation* is a model for every type of rotary motor on a suspension, such as engine on engine mounts. *Eccentric base excitation* is a model for vibration of any equipment mounted on an engine. *Forced excitation* has almost no practical application, however, it is the simplest model for forced vibrations, with good pedagogical use.

For simplicity, we first examine the frequency response of a harmonically forced vibrating system.

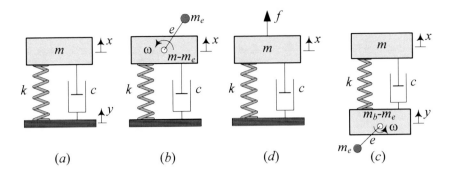

FIGURE 12.14. The four practical types of one DOF harmonically excited systems: a.base excitation, b.eccentric excitation, c.eccentric base excitation, d.forced excitation.

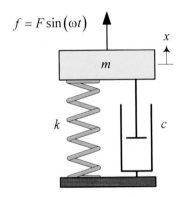

FIGURE 12.15. A harmonically forced excitated, single-DOF system.

12.3.1   Forced Excitation

Figure 12.15 illustrates a one DOF vibrating mass m supported by spring k and a damper c. The absolute motion of m with respect to its equilibrium position is measured by the coordinate x. A sinusoidal excitation force

$$f = F \sin \omega t \tag{12.61}$$

is applied on m and makes the system vibrate.

The equation of motion for the system is

$$m\ddot{x} + c\dot{x} + kx = F \sin \omega t \tag{12.62}$$

which generates a frequency response equal to either of the following functions:

$$
\begin{aligned}
x &= A_1 \sin \omega t + B_1 \cos \omega t & (12.63)\\
&= X \sin (\omega t - \varphi_x) & (12.64)
\end{aligned}
$$

The steady-state response has an *amplitude X*

$$\frac{X}{F/k} = \frac{1}{\sqrt{\left(1 - r^2\right)^2 + \left(2\xi r\right)^2}} \tag{12.65}$$

and a *phase* φ_x

$$\varphi_x = \tan^{-1} \frac{2\xi r}{1 - r^2} \tag{12.66}$$

where we use the *frequency ratio r*, *natural frequency* ω_n, and *damping ratio* ξ.

$$r = \frac{\omega}{\omega_n} \tag{12.67}$$

$$\xi = \frac{c}{2\sqrt{km}} \tag{12.68}$$

$$\omega_n = \sqrt{\frac{k}{m}} \tag{12.69}$$

Phase φ_x indicates the *angular lag* of the response x with respect to the excitation f. Because of the importance of the function $X = X\left(\omega\right)$, it is common to call such a function the *frequency response* of the system. Furthermore, we may use frequency response to every characteristic of the system that is a function of excitation frequency, such as velocity frequency response $\dot{X} = \dot{X}\left(\omega\right)$ and transmitted force frequency response $f_T = f_T\left(\omega\right)$.

The frequency responses for X and φ_x as a function of r and ξ are plotted in Figures 12.16 and 12.17.

Proof. Applying Newton's method and using the free body diagram of the system, as shown in Figure 12.18, generates the equation of motion (12.62), which is a linear differential equation.

The steady-state solution of the linear equation is the same function as the excitation with an unknown amplitude and phase. Therefore, the solution can be (12.63), or (12.64). The solution should be substituted in the equation of motion to find the amplitude and phase of the response. We examine the solution (12.63) and find the following equation:

$$-m\omega^2 \left(A_1 \sin \omega t + B_1 \cos \omega t\right) + c\omega \left(A_1 \cos \omega t - B_1 \sin \omega t\right)$$
$$+ k \left(A_1 \sin \omega t + B_1 \cos \omega t\right) = F \sin \omega t \tag{12.70}$$

The functions $\sin \omega t$ and $\cos \omega t$ are orthogonal therefore, their coefficient must be balanced on both sides of the equal sign. Balancing the coefficients of $\sin \omega t$ and $\cos \omega t$ provides a set of two equations for A_1 and B_1.

$$\begin{bmatrix} k - m\omega^2 & -c\omega \\ c\omega & k - m\omega^2 \end{bmatrix} \begin{bmatrix} A_1 \\ B_1 \end{bmatrix} = \begin{bmatrix} F \\ 0 \end{bmatrix} \tag{12.71}$$

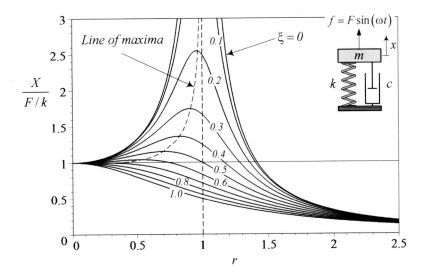

FIGURE 12.16. The position frequency response for $\dfrac{X}{F/k}$.

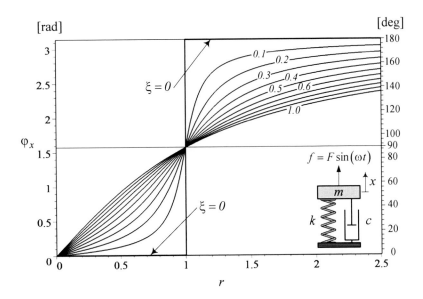

FIGURE 12.17. The frequency response for φ_x.

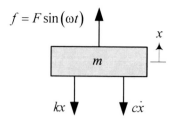

FIGURE 12.18. Free body diagram of the harmonically forced excitated, single-DOF system shown in Figure 12.15.

Solving for coefficients A_1 and B_1

$$
\begin{bmatrix} A_1 \\ B_1 \end{bmatrix} = \begin{bmatrix} k - m\omega^2 & -c\omega \\ c\omega & k - m\omega^2 \end{bmatrix}^{-1} \begin{bmatrix} F \\ 0 \end{bmatrix}
$$

$$
= \begin{bmatrix} \dfrac{k - m\omega^2}{(k - m\omega^2)^2 + c^2\omega^2} F \\[4mm] \dfrac{-c\omega}{(k - m\omega^2)^2 + c^2\omega^2} F \end{bmatrix}
\tag{12.72}
$$

provides us with the steady-state solution (12.63).

Amplitude X and phase φ_x can be found by equating Equations (12.63) and (12.64).

$$
\begin{aligned}
A_1 \sin\omega t + B_1 \cos\omega t &= X \sin(\omega t - \varphi_x) \\
&= X \cos\varphi_x \sin\omega t - X \sin\varphi_x \cos\omega t \tag{12.73}
\end{aligned}
$$

It shows that,

$$
\begin{aligned}
A_1 &= X \cos\varphi_x \tag{12.74}\\
B_1 &= -X \sin\varphi_x \tag{12.75}
\end{aligned}
$$

and therefore,

$$
X = \sqrt{A_1^2 + B_1^2} \tag{12.76}
$$

$$
\tan\varphi_x = \frac{-B_1}{A_1} \tag{12.77}
$$

Substituting A_1 and B_1 from (12.72) yields:

$$
X = \frac{1}{\sqrt{(k - m\omega^2)^2 + c^2\omega^2}} F \tag{12.78}
$$

$$
\tan\varphi_x = \frac{c\omega}{k - m\omega^2} \tag{12.79}
$$

However, we may use more practical expressions (12.65) and (12.66) for amplitude X and phase φ_x by employing r and ξ.

When we apply a constant force $f = F$ on m, a displacement, δ_s, appears.

$$\delta_s = \frac{F}{k} \tag{12.80}$$

If we call δ_s "static amplitude" and X "dynamic amplitude," then, X/δ_s is the ratio of dynamic to static amplitudes. The dynamic amplitude is equal to the static amplitude, $X = \delta_s$, at $r = 0$, and approaches zero, $X \to 0$, when $r \to \infty$. However, X gets a high value when $r \to 1$ and $\omega \to \omega_n$. Theoretically, $X \to \infty$ if $\xi = 0$ when $r = 1$. Frequency domains around the natural frequency is called *resonance zone*. The amplitude of vibration in resonance zone can be reduces by introducing damping. ∎

Example 482 *A forced vibrating system*
Consider a mass-spring-damper system with

$$m = 2\,\text{kg} \qquad k = 100000\,\text{N/m} \qquad c = 100\,\text{N s/m} \tag{12.81}$$

The natural frequency and damping ratio of the system are

$$\omega_n = \sqrt{\frac{k}{m}} = \sqrt{\frac{100000}{2}} = 223.61\,\text{rad/s} \approx 35.6\,\text{Hz} \tag{12.82}$$

$$\xi = \frac{c}{2\sqrt{km}} = \frac{100}{2\sqrt{100000 \times 2}} = 0.1118 \tag{12.83}$$

If a harmonic force f

$$f = 100 \sin 100t \tag{12.84}$$

is applied on m, then the steady-state amplitude of vibrations of the mass, X, would be

$$X = \frac{F/k}{\sqrt{(1-r^2)^2 + (2\xi r)^2}} = 1.24 \times 10^{-3}\,\text{m} \tag{12.85}$$

because

$$r = \frac{\omega}{\omega_n} = 0.44721 \tag{12.86}$$

The phase φ_x of the vibration is

$$\varphi_x = \tan^{-1}\frac{2\xi r}{1-r^2} = 0.124\,\text{rad} \approx 7.12\,\text{deg} \tag{12.87}$$

Therefore, the steady-state vibrations of the mass m can be expressed by the following function.

$$x = 1.24 \times 10^{-3} \sin\left(100t - 0.124\right) \tag{12.88}$$

The value of X and φ_x may also be found from Figures 12.16 and 12.17 approximately.

Example 483 *Velocity and acceleration frequency responses.*
  When we calculate the position frequency response

$$x = A_1 \sin \omega t + B_1 \cos \omega t = X \sin (\omega t - \varphi_x) \qquad (12.89)$$

we are able to calculate the velocity and acceleration frequency responses by derivative.

$$\begin{aligned}
\dot{x} &= A_1 \omega \cos \omega t - B_1 \omega \sin \omega t = X\omega \cos (\omega t - \varphi_x) \\
&= \dot{X} \cos (\omega t - \varphi_x) \qquad (12.90)
\end{aligned}$$

$$\begin{aligned}
\ddot{x} &= -A_1 \omega^2 \sin \omega t - B_1 \omega^2 \cos \omega t = -X\omega^2 \sin (\omega t - \varphi_x) \\
&= \ddot{X} \sin (\omega t - \varphi_x) \qquad (12.91)
\end{aligned}$$

The amplitude of velocity and acceleration frequency responses are shown by $\dot{X}$, $\ddot{X}$

$$\dot{X} = \frac{\omega}{\sqrt{(k - m\omega^2)^2 + c^2\omega^2}} F \qquad (12.92)$$

$$\ddot{X} = \frac{\omega^2}{\sqrt{(k - m\omega^2)^2 + c^2\omega^2}} F \qquad (12.93)$$

which can be written as

$$\frac{\dot{X}}{F/\sqrt{km}} = \frac{r}{\sqrt{(1 - r^2)^2 + (2\xi r)^2}} \qquad (12.94)$$

$$\frac{\ddot{X}}{F/m} = \frac{r^2}{\sqrt{(1 - r^2)^2 + (2\xi r)^2}} \qquad (12.95)$$

The velocity and acceleration frequency responses (12.94) and (12.95) are plotted in Figures 12.19 and 12.20.

Example 484 *Transmitted force to the base.*
  A forced excited system, such as the one in Figure 12.15, transmits a force, f_T, to the ground. The transmitted force is equal to the sum of forces in spring and damper.

$$f_T = f_k + f_c = kx + c\dot{x} \qquad (12.96)$$

Substituting x from (12.63), and A_1, B_1 from (12.72), shows that the frequency response of the transmitted force is

$$\begin{aligned}
f_T &= k (A_1 \sin \omega t + B_1 \cos \omega t) + c\omega (A_1 \cos \omega t - B_1 \sin \omega t) \\
&= (kA_1 - c\omega B_1) \sin t\omega + (kB_1 + c\omega A_1) \cos t\omega \\
&= F_T \sin (\omega t - \varphi_{F_T}) \qquad (12.97)
\end{aligned}$$

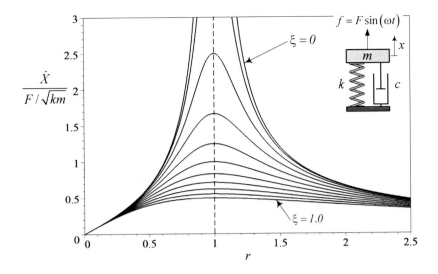

FIGURE 12.19. The velocity frequency response for $\dfrac{\dot{X}}{F/\sqrt{km}}$.

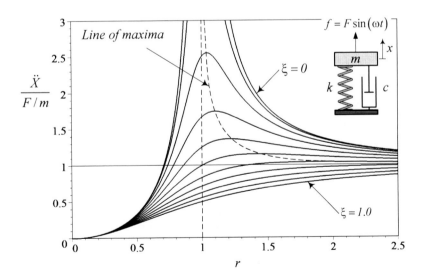

FIGURE 12.20. The acceleration frequency response for $\dfrac{\ddot{X}}{F/m}$.

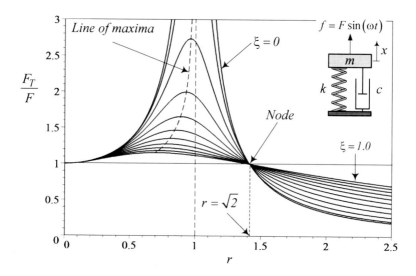

FIGURE 12.21. The frequency response for $\dfrac{F_T}{F}$.

The amplitude F_T and phase φ_{F_T} of f_T are

$$\frac{F_T}{F} = \frac{\sqrt{k + c^2\omega^2}}{\sqrt{(k - m\omega^2)^2 + c^2\omega^2}} \tag{12.98}$$

$$= \frac{\sqrt{1 + (2\xi r)^2}}{\sqrt{(1 - r^2)^2 + (2\xi r)^2}} \tag{12.99}$$

$$\tan\varphi_{F_T} = \frac{c\omega}{k - m\omega^2} = \frac{2\xi r}{1 - r^2} \tag{12.100}$$

because

$$F_T = \sqrt{(kA_1 - c\omega B_1)^2 + (kB_1 + c\omega A_1)^2} \tag{12.101}$$

$$\tan\varphi_{F_T} = \frac{-(kB_1 + c\omega A_1)}{kA_1 - c\omega B_1} \tag{12.102}$$

The transmitted force frequency response F_T/F is plotted in Figure 12.21, and because φ_{F_T} is the same as Equation (12.66), a graph for φ_{F_T} is the same as the one in Figure 12.17.

Example 485 *Alternative method to find transmitted force f_T.*
It is possible to use the equation of motion and substitute x from (12.63),

to find the transmitted force frequency response of f_T as

$$
\begin{aligned}
f_T &= F \sin \omega t - m\ddot{x} = F \sin \omega t + m\omega^2 \left(A_1 \sin \omega t + B_1 \cos \omega t \right) \\
&= \left(mA_1\omega^2 + F \right) \sin t\omega + m\omega^2 B_1 \cos t\omega \\
&= F_T \sin \left(\omega t - \varphi_x \right)
\end{aligned}
\tag{12.103}
$$

Amplitude F_T and phase φ_{F_T} would be the same as (12.99) and (12.100), because

$$
\begin{aligned}
F_T &= \sqrt{\left(mA_1\omega^2 + F \right)^2 + \left(m\omega^2 B_1 \right)^2} \\
&= \frac{\sqrt{k + c^2\omega^2}}{\sqrt{\left(k - m\omega^2 \right)^2 + c^2\omega^2}}
\end{aligned}
\tag{12.104}
$$

$$
\varphi_{F_T} = \arctan \frac{-m\omega^2 B_1}{m\omega^2 A_1 + F} = \arctan \frac{c\omega}{k - m\omega^2}
\tag{12.105}
$$

Example 486 *No mechanical harmonically forced vibration.*
   In mechanics, there is no way to apply a periodic force on an object without attaching a mechanical device and applying a displacement. Hence, the forced vibrating system shown in Figure 12.15 has no practical application in mechanics. However, it is possible to make m from a ferromagnetic material to apply an alternative or periodic magnetic force.

Example 487 ★ *Orthogonality of functions $\sin \omega t$ and $\cos \omega t$.*
   Two functions $f(t)$ and $g(t)$ are orthogonal in $[a, b]$ if

$$
\int_a^b f(t) \, g(t) \, dt = 0
\tag{12.106}
$$

The functions $\sin \omega t$ and $\cos \omega t$ are orthogonal in a period $T = [0, 2\pi/\omega]$.

$$
\int_0^{2\pi/\omega} \sin \omega t \cos \omega t \, dt = 0
\tag{12.107}
$$

Example 488 ★ *Beating in linear systems.*
   Consider a displacement $x(t)$ that is produced by two harmonic forces f_1 and f_2.

$$
f_1 = F_1 \cos \omega_1 t \qquad f_2 = F_2 \cos \omega_2 t
\tag{12.108}
$$

Assume that the steady-state response to f_1 is

$$
x_1(t) = X_1 \cos \left(\omega_1 t + \phi_1 \right)
\tag{12.109}
$$

and response to f_2 is

$$
x_2(t) = X_2 \cos \left(\omega_2 t + \phi_2 \right)
\tag{12.110}
$$

then, because of the linearity of the system, the response to $f_1 + f_2$ would be $x(t) = x_1(t) + x_2(t)$.

$$
\begin{aligned}
x(t) &= x_1(t) + x_2(t) \\
&= X_1 \cos(\omega_1 t + \phi_1) + X_2 \cos(\omega_2 t + \phi_2) \quad (12.111)
\end{aligned}
$$

It is convenient to express $x(t)$ in an alternative method

$$
\begin{aligned}
x(t) = \; &\frac{1}{2}(X_1 + X_2)(\cos(\omega_1 t + \phi_1) + \cos(\omega_2 t + \phi_2)) \\
&+ \frac{1}{2}(X_1 - X_2)(\cos(\omega_1 t + \phi_1) - \cos(\omega_2 t + \phi_2)) \quad (12.112)
\end{aligned}
$$

and convert the sums to a product.

$$
\begin{aligned}
x(t) = \; &(X_1 + X_2)\cos\left(\frac{\omega_1 + \omega_2}{2}t - \frac{\phi_1 + \phi_2}{2}\right) \\
&\times \cos\left(\frac{\omega_1 - \omega_2}{2}t - \frac{\phi_1 - \phi_2}{2}\right) \\
&- (X_1 - X_2)\sin\left(\frac{\omega_1 + \omega_2}{2}t - \frac{\phi_1 + \phi_2}{2}\right) \\
&\times \sin\left(\frac{\omega_1 - \omega_2}{2}t - \frac{\phi_1 - \phi_2}{2}\right) \quad (12.113)
\end{aligned}
$$

This equation may be expressed better as

$$
\begin{aligned}
x(t) = \; &(X_1 + X_2)\cos(\Omega_1 t - \Phi_1)\cos(\Omega_2 t - \Phi_2) \\
&- (X_1 - X_2)\sin(\Omega_1 t - \Phi_1)\sin(\Omega_2 t - \Phi_2) \quad (12.114)
\end{aligned}
$$

if we use the following notations.

$$
\Omega_1 = \frac{\omega_1 + \omega_2}{2} \qquad \Omega_2 = \frac{\omega_1 - \omega_2}{2} \quad (12.115)
$$

$$
\Phi_1 = \frac{\phi_1 + \phi_2}{2} \qquad \Phi_2 = \frac{\phi_1 - \phi_2}{2} \quad (12.116)
$$

Figure 12.22 illustrates a sample plot of $x(t)$ for

$$
\begin{aligned}
X_1 &= 1 \qquad X_2 = 0.8 \qquad \omega_1 = 10 \\
\omega_2 &= 11 \qquad \phi_1 = 0 \qquad \phi_2 = 0 \quad (12.117)
\end{aligned}
$$

The displacement $x(t)$ indicates an oscillation between $X_1 + X_2$ and $X_1 - X_2$, with the higher frequency Ω_1 inside an envelope that oscillates at the lower frequency Ω_2. This behavior is called **beating**.

When $X_1 = X_2 = X$ then,

$$
x(t) = 2X\cos(\Omega_1 t - \Phi_1)\cos(\Omega_2 t - \Phi_2) \quad (12.118)
$$

which becomes zero at every half period $T = 2\pi/\Omega_2$.

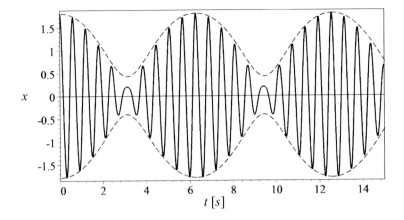

FIGURE 12.22. Beating phenomena.

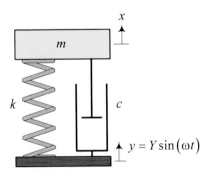

FIGURE 12.23. A harmonically base excited single DOF system.

12.3.2 Base Excitation

Figure 12.23 illustrates a one-DOF base excited vibrating system with a mass m supported by a spring k and a damper c. Base excited system is a good model for vehicle suspension system or any equipment that is mounted on a vibrating base. The absolute motion of m with respect to its equilibrium position is measured by the coordinate x. A sinusoidal excitation motion

$$y = Y \sin \omega t \tag{12.119}$$

is applied to the base of the suspension and makes the system vibrate.

The equation of motion for the system can be expressed by either one of the following equations for the absolute displacement x

$$m\ddot{x} + c\dot{x} + kx = cY\omega \cos \omega t + kY \sin \omega t \tag{12.120}$$

$$\ddot{x} + 2\xi\omega_n \dot{x} + \omega_n^2 x = 2\xi\omega_n \omega Y \cos \omega t + \omega_n^2 Y \sin \omega t \tag{12.121}$$

or either one of the following equations for the relative displacement z.

$$m \ddot{z} + c \dot{z} + kz = m\omega^2 Y \sin \omega t \qquad (12.122)$$

$$\ddot{z} + 2\xi \omega_n \dot{z} + \omega_n^2 z = \omega^2 Y \sin \omega t \qquad (12.123)$$

$$z = x - y \qquad (12.124)$$

The equations of motion generate the following absolute and relative frequency responses.

$$x = A_2 \sin \omega t + B_2 \cos \omega t \qquad (12.125)$$

$$= X \sin (\omega t - \varphi_x) \qquad (12.126)$$

$$z = A_3 \sin \omega t + B_3 \cos \omega t \qquad (12.127)$$

$$= Z \sin (\omega t - \varphi_z) \qquad (12.128)$$

The frequency response of x has an amplitude X, and the frequency response of z has an amplitude Z

$$\frac{X}{Y} = \frac{\sqrt{1 + (2\xi r)^2}}{\sqrt{(1 - r^2)^2 + (2\xi r)^2}} \qquad (12.129)$$

$$\frac{Z}{Y} = \frac{r^2}{\sqrt{(1 - r^2)^2 + (2\xi r)^2}} \qquad (12.130)$$

with the following phases φ_x and φ_z for x and z.

$$\varphi_x = \tan^{-1} \frac{2\xi r^3}{1 - r^2 + (2\xi r)^2} \qquad (12.131)$$

$$\varphi_z = \tan^{-1} \frac{2\xi r}{1 - r^2} \qquad (12.132)$$

The phase φ_x indicates the *angular lag* of the response x with respect to the excitation y. The frequency responses for X, Z, and φ_x as a function of r and ξ are plotted in Figures 12.24, 12.25, and 12.26.

Proof. Newton's method and the free body diagram of the system, as shown in Figure 12.27, generate the equation of motion

$$m \ddot{x} = -c (\dot{x} - \dot{y}) - k (x - y) \qquad (12.133)$$

which, after substituting (12.119), makes the equation of motion (12.120). Equation (12.120) can be transformed to (12.121) by dividing over m and using the definitions (12.67)-(12.69) for natural frequency and damping ratio.

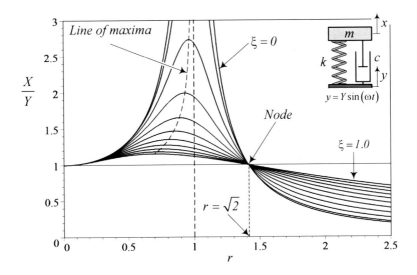

FIGURE 12.24. The position frequency response for $\dfrac{X}{Y}$.

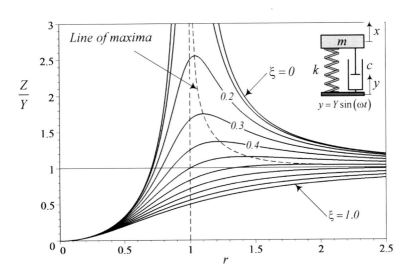

FIGURE 12.25. The frequency response for $\dfrac{Z}{Y}$.

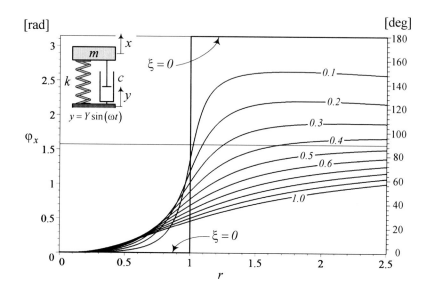

FIGURE 12.26. The frequency response for φ_x.

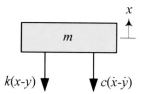

FIGURE 12.27. A harmonically based excitated single-DOF system.

A practical response for a base excited system is the relative displacement

$$z = x - y \tag{12.134}$$

Relative displacement is important because for every mechanical device mounted on a suspension such as vehicle body, we need to control the maximum or minimum distance between the base and the device. Taking derivatives from (12.134)

$$\ddot{z} = \ddot{x} - \ddot{y} \tag{12.135}$$

and substituting in (12.133)

$$m\left(\ddot{z} + \ddot{y}\right) = -c\,\dot{z} - kz \tag{12.136}$$

can be transformed to Equations (12.122) and (12.123).

The steady-state solution of Equation (12.120) can be (12.125), or (12.126). To find the amplitude and phase of the response, we substitute the solution (12.125) in the equation of motion.

$$-m\omega^2 \left(A_2 \sin \omega t + B_2 \cos \omega t\right) + c\omega \left(A_2 \cos \omega t - B_2 \sin \omega t\right)$$
$$+k\left(A_2 \sin \omega t + B_2 \cos \omega t\right) = cY\omega \cos \omega t + kY \sin \omega t \tag{12.137}$$

The coefficients of the functions $\sin \omega t$ and $\cos \omega t$ must balance on both sides of the equation.

$$kA_2 - mA_2\omega^2 - cB_2\omega = Yk \tag{12.138}$$
$$kB_2 - m\omega^2 B_2 + c\omega A_2 = Yc\omega \tag{12.139}$$

Therefore, we find two algebraic equations to calculate A_2 and B_2.

$$\begin{bmatrix} k - m\omega^2 & -c\omega \\ c\omega & k - m\omega^2 \end{bmatrix} \begin{bmatrix} A_2 \\ B_2 \end{bmatrix} = \begin{bmatrix} Yk \\ Yc\omega \end{bmatrix} \tag{12.140}$$

Solving for the coefficients A_2 and B_2

$$\begin{bmatrix} A_2 \\ B_2 \end{bmatrix} = \begin{bmatrix} k - m\omega^2 & -c\omega \\ c\omega & k - m\omega^2 \end{bmatrix}^{-1} \begin{bmatrix} Yk \\ Yc\omega \end{bmatrix}$$
$$= \begin{bmatrix} \dfrac{k\left(k - m\omega^2\right) + c^2\omega^2}{\left(k - m\omega^2\right)^2 + c^2\omega^2} Y \\ \dfrac{c\omega\left(k - m\omega^2\right) - ck\omega}{\left(k - m\omega^2\right)^2 + c^2\omega^2} Y \end{bmatrix} \tag{12.141}$$

provides us with the steady-state solution (12.125).

The amplitude X and phase φ_x can be found by

$$X = \sqrt{A_2^2 + B_2^2} \tag{12.142}$$

$$\tan \varphi_x = \frac{-B_2}{A_2} \tag{12.143}$$

which, after substituting A_2 and B_2 from (12.141), results in the following solutions:

$$X = \frac{\sqrt{k^2 + c^2\omega^2}}{\sqrt{(k - m\omega^2)^2 + c^2\omega^2}} Y \tag{12.144}$$

$$\tan\varphi_x = \frac{-cm\omega^3}{k(k - m\omega^2) + c^2\omega^2} \tag{12.145}$$

A more practical expressions for X and φ_x are Equations (12.129) and (12.131), which can be found by employing r and ξ.

To find the relative displacement frequency response (12.130), we substitute Equation (12.127) in (12.122).

$$-m\omega^2 \left(A_3 \sin\omega t + B_3 \cos\omega t\right) + c\omega \left(A_3 \cos\omega t - B_3 \sin\omega t\right)$$
$$+k\left(A_3 \sin\omega t + B_3 \cos\omega t\right) = m\omega^2 Y \sin\omega t \tag{12.146}$$

Balancing the coefficients of the functions $\sin\omega t$ and $\cos\omega t$

$$kA_2 - mA_2\omega^2 - cB_2\omega = m\omega^2 Y \tag{12.147}$$
$$kB_2 - m\omega^2 B_2 + c\omega A_2 = 0 \tag{12.148}$$

provides two algebraic equations to find A_3 and B_3.

$$\begin{bmatrix} k - m\omega^2 & -c\omega \\ c\omega & k - m\omega^2 \end{bmatrix} \begin{bmatrix} A_3 \\ B_3 \end{bmatrix} = \begin{bmatrix} m\omega^2 Y \\ 0 \end{bmatrix} \tag{12.149}$$

Solving for the coefficients A_3 and B_3

$$\begin{bmatrix} A_3 \\ B_3 \end{bmatrix} = \begin{bmatrix} k - m\omega^2 & -c\omega \\ c\omega & k - m\omega^2 \end{bmatrix}^{-1} \begin{bmatrix} m\omega^2 Y \\ 0 \end{bmatrix}$$

$$= \begin{bmatrix} \dfrac{m\omega^2 \left(k - m\omega^2\right)}{(k - m\omega^2)^2 + c^2\omega^2} Y \\ -\dfrac{mc\omega^3}{(k - m\omega^2)^2 + c^2\omega^2} Y \end{bmatrix} \tag{12.150}$$

provides the steady-state solution (12.127). The amplitude Z and phase φ_z can be found by

$$Z = \sqrt{A_3^2 + B_3^2} \tag{12.151}$$

$$\tan\varphi_z = \frac{-B_3}{A_3} \tag{12.152}$$

which, after substituting A_3 and B_3 from (12.150) yields:

$$Z = \frac{m\omega^2}{\sqrt{(k - m\omega^2)^2 + c^2\omega^2}} Y \tag{12.153}$$

$$\tan\varphi_z = \frac{c\omega}{k - m\omega^2} \tag{12.154}$$

A more practical expression for Z and φ_z are Equations (12.130) and (12.132). ∎

Example 489 *A base excited system.*
Consider a mass-spring-damper system with

$$m = 2\,\text{kg} \qquad k = 100000\,\text{N/m} \qquad c = 100\,\text{N s/m} \qquad (12.155)$$

If a harmonic base excitation y

$$y = 0.002 \sin 350t \qquad (12.156)$$

is applied on the system, then the absolute and relative steady-state amplitude of vibrations of the mass, X and Z would be

$$X = \frac{Y\sqrt{1 + (2\xi r)^2}}{\sqrt{(1 - r^2)^2 + (2\xi r)^2}} = 1.9573 \times 10^{-3}\,\text{m} \qquad (12.157)$$

$$Z = \frac{Y r^2}{\sqrt{(1 - r^2)^2 + (2\xi r)^2}} = 9.589 \times 10^{-4}\,\text{m} \qquad (12.158)$$

because

$$\omega_n = \sqrt{\frac{k}{m}} = 223.61\,\text{rad/s} \approx 35.6\,\text{Hz} \qquad (12.159)$$

$$\xi = \frac{c}{2\sqrt{km}} = 0.1118 \qquad (12.160)$$

$$r = \frac{\omega}{\omega_n} = 1.5652 \qquad (12.161)$$

The phases φ_x and φ_z for x and z are

$$\varphi_x = \tan^{-1} \frac{2\xi r^3}{1 - r^2 + (2\xi r)^2} = 0.489\,\text{rad} \approx 28.02\,\text{deg} \qquad (12.162)$$

$$\varphi_z = \tan^{-1} \frac{2\xi r}{1 - r^2} = 1.8585\,\text{rad} \approx 106.48\,\text{deg} \qquad (12.163)$$

Therefore, the steady-state vibrations of the mass m can be expressed by the following functions.

$$x = 1.9573 \times 10^{-3} \sin (350t - 0.489) \qquad (12.164)$$
$$z = 9.589 \times 10^{-4} \sin (350t - 1.8585) \qquad (12.165)$$

Example 490 *Comparison between frequency responses.*
A comparison shows that Equation (12.130) is equal to Equation (12.94), and therefore the relative frequency response $\frac{Z}{Y}$ for a base excited system,

is the same as acceleration frequency response $\frac{\ddot{X}}{F/m}$ for a forces excited system. Also a graph for φ_z would be the same as Figure 12.17.

Comparing Equations (12.129) and (12.99) indicates that the amplitude frequency response of a base excited system, $\frac{X}{Y}$, is the same as the transmitted force frequency response of a harmonically force excited system $\frac{F_T}{F}$. However, the phase of these two responses are different.

Example 491 *Absolute velocity and acceleration of a base excited system.*

Having the position frequency response of a base excited system

$$x = A_2 \sin \omega t + B_2 \cos \omega t = X \sin(\omega t - \varphi_x) \tag{12.166}$$

we are able to calculate the velocity and acceleration frequency responses.

$$\begin{aligned} \dot{x} &= A_2 \omega \cos \omega t - B_2 \omega \sin \omega t = X\omega \cos(\omega t - \varphi_x) \\ &= \dot{X} \cos(\omega t - \varphi_x) \end{aligned} \tag{12.167}$$

$$\begin{aligned} \ddot{x} &= -A_2 \omega^2 \sin \omega t - B_2 \omega^2 \cos \omega t = -X\omega^2 \sin(\omega t - \varphi_x) \\ &= \ddot{X} \sin(\omega t - \varphi_x) \end{aligned} \tag{12.168}$$

The amplitude of velocity and acceleration frequency responses, $\dot{X}$, $\ddot{X}$ are

$$\dot{X} = \frac{\omega \sqrt{k^2 + c^2\omega^2}}{\sqrt{(k - m\omega^2)^2 + c^2\omega^2}} Y \tag{12.169}$$

$$\ddot{X} = \frac{\omega^2 \sqrt{k^2 + c^2\omega^2}}{\sqrt{(k - m\omega^2)^2 + c^2\omega^2}} Y \tag{12.170}$$

which can be written as

$$\frac{\dot{X}}{\omega_n Y} = \frac{r\sqrt{1 + (2\xi r)^2}}{\sqrt{(1 - r^2)^2 + (2\xi r)^2}} \tag{12.171}$$

$$\frac{\ddot{X}}{\omega_n^2 Y} = \frac{r^2 \sqrt{1 + (2\xi r)^2}}{\sqrt{(1 - r^2)^2 + (2\xi r)^2}}. \tag{12.172}$$

The velocity and acceleration frequency responses (12.171) and (12.172) are plotted in Figures 12.28 and 12.29.

*There is a point in both figures, called the **switching point** or **node**, at which the behavior of $\dot{X}$ and $\ddot{X}$ as a function of ξ switches. Before the node, $\dot{X}$ and $\ddot{X}$ increase by increasing ξ, while they decrease after the node. To find the node, we may find the intersection between frequency response*

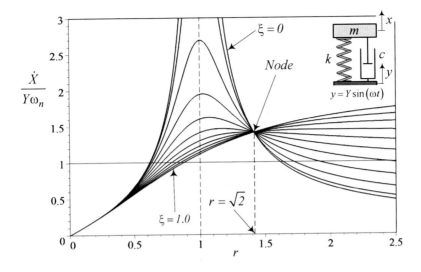

FIGURE 12.28. The velocity frequency response of base excited systems.

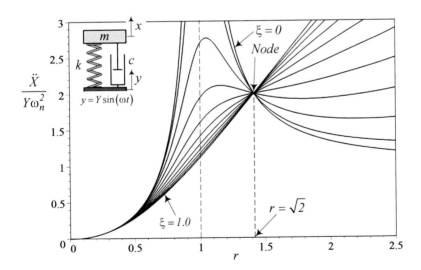

FIGURE 12.29. The acceleration frequency response of base excited systems.

curves for $\xi = 0$ and $\xi = \infty$. We apply this method to the acceleration frequency response.

$$\lim_{\xi \to 0} \frac{\ddot{X}}{\omega_n^2 Y} = \pm \frac{r^2}{(1 - r^2)} \tag{12.173}$$

$$\lim_{\xi \to \infty} \frac{\ddot{X}}{\omega_n^2 Y} = \pm r^2 \tag{12.174}$$

Therefore, the frequency ratio r at the intersection of these two limits is the solution of the equation

$$r^2 \left(r^2 - 2 \right) = 0 \tag{12.175}$$

The nodal frequency response is then equal to

$$r = \sqrt{2} \tag{12.176}$$

The value of acceleration frequency response at the node is not a function of ξ.

$$\lim_{r \to \sqrt{2}} \frac{\ddot{X}}{\omega_n^2 Y} = \frac{2\sqrt{8\xi^2 + 1}}{\sqrt{8\xi^2 + 1}} = 2 \tag{12.177}$$

Applying the same method for the velocity frequency response results in the same nodal frequency ratio $r = \sqrt{2}$. However, the value of the frequency response at the node is different.

$$\lim_{r \to \sqrt{2}} \frac{\dot{X}}{\omega_n Y} = \sqrt{2}\frac{\sqrt{8\xi^2 + 1}}{\sqrt{8\xi^2 + 1}} = \sqrt{2} \tag{12.178}$$

Example 492 *Relative velocity and acceleration of a base excited system.*

We may use the relative displacement frequency response of a base excited system

$$z = A_3 \sin \omega t + B_3 \cos \omega t = Z \sin (\omega t - \varphi_z) \tag{12.179}$$

and calculate the relative velocity and acceleration frequency responses.

$$\begin{aligned} \dot{z} &= A_3 \omega \cos \omega t - B_3 \omega \sin \omega t = Z\omega \cos (\omega t - \varphi_z) \\ &= \dot{Z} \cos (\omega t - \varphi_z) \end{aligned} \tag{12.180}$$

$$\begin{aligned} \ddot{z} &= -A_3 \omega^2 \sin \omega t - B_3 \omega^2 \cos \omega t = -Z\omega^2 \sin (\omega t - \varphi_z) \\ &= \ddot{Z} \sin (\omega t - \varphi_z) \end{aligned} \tag{12.181}$$

The amplitude of velocity and acceleration frequency responses, $\dot{Z}$, $\ddot{Z}$

$$\dot{Z} = \frac{m\omega^3}{\sqrt{(k - m\omega^2)^2 + c^2\omega^2}} Y \tag{12.182}$$

$$\ddot{Z} = \frac{m\omega^4}{\sqrt{(k - m\omega^2)^2 + c^2\omega^2}} Y \tag{12.183}$$

can be written as

$$\frac{\dot{Z}}{\omega_n Y} = \frac{r^3}{\sqrt{(1 - r^2)^2 + (2\xi r)^2}} \tag{12.184}$$

$$\frac{\ddot{Z}}{\omega_n^2 Y} = \frac{r^4}{\sqrt{(1 - r^2)^2 + (2\xi r)^2}} \tag{12.185}$$

Example 493 *Transmitted force to the base of a base excited system.*

The transmitted force f_T to the ground by a base excited system, such as is shown in Figure 12.23, is equal to the sum of forces in the spring and damper.

$$f_T = f_k + f_c = k(x - y) + c(\dot{x} - \dot{y}) \tag{12.186}$$

which based on the equation of motion (12.133) is also equal to

$$f_T = -m\ddot{x} \tag{12.187}$$

Substituting $\ddot{x}$ from (12.168) and (12.172) shows that the frequency response of the transmitted force can be written as

$$\frac{F_T}{kY} = \frac{\omega^2 \sqrt{k^2 + c^2 \omega^2}}{\sqrt{(k - m\omega^2)^2 + c^2 \omega^2}} = \frac{r^2 \sqrt{1 + (2\xi r)^2}}{\sqrt{(1 - r^2)^2 + (2\xi r)^2}} \tag{12.188}$$

The frequency response of $\frac{F_T}{kY}$ is the same as is shown in Figure 12.29.

Example 494 ★ *Line of maxima in X/Y.*

The peak value of the absolute displacement frequency response X/Y happens at different r depending on ξ. To find this relationship, we take a derivative of X/Y, given in Equation (12.129), with respect to r and solve the equation.

$$\frac{d}{dr} \frac{X}{Y} = \frac{2r\left(1 - r^2 - 2r^4 \xi^2\right)}{\sqrt{1 + 4r^2\xi^2}\left((1 - r^2)^2 + (2\xi r)^2\right)^{\frac{3}{2}}} = 0 \tag{12.189}$$

Let us indicate the peak amplitude by $X_{\max}$ and the associated frequency by $r_{\max}$. The value of $r_{\max}^2$ is

$$r_{\max}^2 = \frac{1}{4\xi^2}\left(-1 \pm \sqrt{1 + 8\xi^2}\right) \tag{12.190}$$

which is only a function of ξ.

Substituting the positive sign of (12.190) in (12.129) determines the peak amplitude $X_{\max}$.

$$\frac{X_{\max}}{Y} = \frac{2\sqrt{2}\xi^2 \sqrt[4]{8\xi^2 + 1}}{\sqrt{8\xi^2 + (8\xi^4 - 4\xi^2 - 1)\sqrt{8\xi^2 + 1} + 1}} \tag{12.191}$$

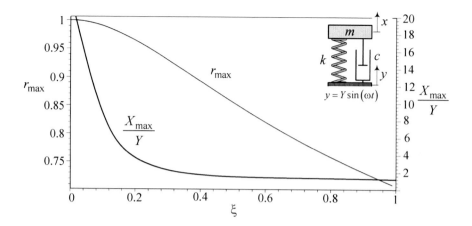

FIGURE 12.30. The peak amplitude $X_{\max}$ and the associated frequency $r_{\max}$, as a function of ξ.

Figure 12.30 shows $X_{\max}$ and $r_{\max}$ as a function of ξ.

Example 495 ★ *Line of maxima in Z/Y.*

The peak value of the relative displacement frequency response Z/Y happens at $r > 1$ depending on ξ. To find this relationship, we take a derivative of Z/Y, given in Equation (12.130), with respect to r and solve the equation.

$$\frac{d}{dr}\frac{Z}{Y} = \frac{2r\left(1 - r^2 - 2r^4\xi^2\right)}{\left((1 - r^2)^2 + (2\xi r)^2\right)^{\frac{3}{2}}} = 0 \tag{12.192}$$

Let us indicate the peak amplitude by $Z_{\max}$ and the associated frequency by $r_{\max}$. The value of $r_{\max}^2$ is

$$r_{\max}^2 = \frac{1}{\sqrt{1 - 2\xi^2}} \tag{12.193}$$

which has a real value for

$$\xi < \frac{\sqrt{2}}{2} \tag{12.194}$$

Substituting (12.193) in (12.130) determines the peak amplitude $Z_{\max}$.

$$\frac{Z_{\max}}{Y} = \frac{1}{2\xi\sqrt{1 - 2\xi^2}} \tag{12.195}$$

As an example, the maximum amplitude of a system with

$$
\begin{aligned}
m &= 2\,\text{kg} & k &= 100000\,\text{N/m} & c &= 100\,\text{N s/m} \\
\omega_n &= 223.61\,\text{rad/s} & \xi &= 0.1118 & Y &= 0.002\,\text{m} \quad (12.196)
\end{aligned}
$$

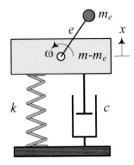

FIGURE 12.31. An eccentric excitated single-DOF system.

is

$$Z_{\max} = \frac{Y}{2\xi\sqrt{1 - 2\xi^2}} = 9.0585 \times 10^{-3}\,\text{m} \qquad (12.197)$$

that occurs at

$$r_{\max} = \frac{1}{\sqrt[4]{1 - 2\xi^2}} = 1.0063 \qquad (12.198)$$

12.3.3   Eccentric Excitation

Figure 12.31 illustrates a one-DOF eccentric excited vibrating system with a mass m supported by a suspension made of a spring k and a damper c. There is an unbalance mass m_e at a distance e that is rotating with an angular velocity ω. An eccentric excited vibrating system is a good model for vibration analysis of the engine of a vehicle, or any rotary motor that is mounted on a stationary base with a flexible suspension.

The absolute motion of m with respect to its equilibrium position is measured by the coordinate x. When the lateral motion of m is protected, a harmonic excitation force

$$f_x = m_e e\omega^2 \sin\omega t \qquad (12.199)$$

is applied on m and makes the system vibrate. The distance e is called the *eccentricity* and m_e is called the *eccentric mass*.

The equation of motion for the system is

$$m\ddot{x} + c\dot{x} + kx = m_e e\omega^2 \sin\omega t \qquad (12.200)$$

or

$$\ddot{x} + 2\xi\omega_n \dot{x} + \omega_n^2 x = \varepsilon e\omega^2 \sin\omega t \qquad (12.201)$$

$$\varepsilon = \frac{m_e}{m} \qquad (12.202)$$

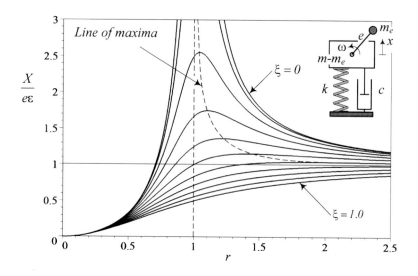

FIGURE 12.32. The position frequency response for $\dfrac{X}{e\varepsilon}$.

The absolute displacement responses of the system is

$$x = A_4 \sin \omega t + B_4 \cos \omega t \tag{12.203}$$
$$= X \sin (\omega t - \varphi_e) \tag{12.204}$$

which has an amplitude X, and phases φ_e

$$\frac{X}{e\varepsilon} = \frac{r^2}{\sqrt{(1 - r^2)^2 + (2\xi r)^2}} \tag{12.205}$$

$$\varphi_e = \tan^{-1} \frac{2\xi r}{1 - r^2} \tag{12.206}$$

Phase φ_e indicates the angular lag of the response x with respect to the excitation $m_e e \omega^2 \sin \omega t$. The frequency responses for X and φ_e as a function of r and ξ are plotted in Figures 12.32 and 12.33.

Proof. Employing the free body diagram of the system, as shown in Figure 12.34, and applying Newton's method in the x-direction generate the equation of motion

$$m\ddot{x} = -c\dot{x} - kx + m_e e \omega^2 \sin \omega t \tag{12.207}$$

Equation (12.200) can be transformed to (12.201) by dividing over m and using the following definitions for natural frequency, damping ratio,

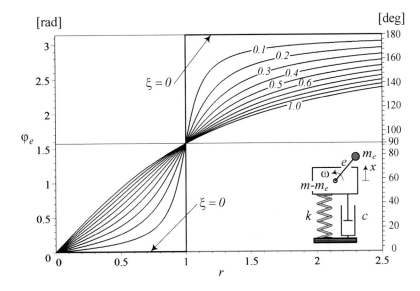

FIGURE 12.33. The frequency response for φ_e.

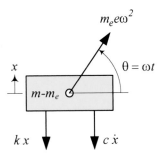

FIGURE 12.34. Free body diagram of an eccentric excitated single-DOF system.

and frequency ratio.

$$\omega_n = \sqrt{\frac{k}{m}} \tag{12.208}$$

$$\xi = \frac{c}{2\sqrt{km}} \tag{12.209}$$

$$r = \frac{\omega}{\omega_n} \tag{12.210}$$

The parameter $\varepsilon = \frac{m_e}{m}$ is called the *mass ratio* and indicates the ratio between the eccentric mass m_e and the total mass m.

The steady-state solution of Equations (12.200) can be (12.203), or (12.204). To find the amplitude and phase of the response, we substitute the solution (12.203) in the equation of motion.

$$-m\omega^2 \left(A_4 \sin \omega t + B_4 \cos \omega t\right) + c\omega \left(A_4 \cos \omega t - B_4 \sin \omega t\right)$$
$$+k \left(A_4 \sin \omega t + B_4 \cos \omega t\right) = m_e e\omega^2 \sin \omega t \tag{12.211}$$

The coefficients of the functions $\sin \omega t$ and $\cos \omega t$ must balance on both sides of the equation.

$$kA_4 - mA_4\omega^2 - cB_4\omega = m_e e\omega^2 \tag{12.212}$$
$$kB_4 - m\omega^2 B_4 + c\omega A_4 = 0 \tag{12.213}$$

Therefore, we find two algebraic equations to calculate A_4 and B_4.

$$\begin{bmatrix} k - \omega^2 m & -c\omega \\ c\omega & k - \omega^2 m \end{bmatrix} \begin{bmatrix} A_4 \\ B_4 \end{bmatrix} = \begin{bmatrix} e\omega^2 m_e \\ 0 \end{bmatrix} \tag{12.214}$$

Solving for the coefficients A_4 and B_4

$$\begin{bmatrix} A_4 \\ B_4 \end{bmatrix} = \begin{bmatrix} k - \omega^2 m & -c\omega \\ c\omega & k - \omega^2 m \end{bmatrix}^{-1} \begin{bmatrix} e\omega^2 m_e \\ 0 \end{bmatrix}$$

$$= \begin{bmatrix} \dfrac{k - m\omega^2 - \omega^2 m_e}{(k - \omega^2 m)^2 + c^2\omega^2} e\omega^2 m_e \\[4mm] \dfrac{-c\omega}{(k - \omega^2 m)^2 + c^2\omega^2} e\omega^2 m_e \end{bmatrix} \tag{12.215}$$

provides us with the steady-state solution (12.203).

The amplitude X and phase φ_e are

$$X = \sqrt{A_4^2 + B_4^2} \tag{12.216}$$

$$\tan \varphi_e = \frac{-B_4}{A_4} \tag{12.217}$$

which, after substituting A_4 and B_4 from (12.215), results in the following solutions.

$$X = \frac{\omega^2 e m_e}{\sqrt{(k - m\omega^2)^2 + c^2\omega^2}} \tag{12.218}$$

$$\tan\varphi_e = \frac{c\omega}{k - m\omega^2} \tag{12.219}$$

A more practical expression for X and φ_e is Equation (12.205) and (12.206), which can be found by employing r and ξ. ∎

Example 496 *An eccentric excited system.*
Consider an engine with a mass m

$$m = 110\,\text{kg} \tag{12.220}$$

that is supported by four engine mounts, each with the following equivalent stiffness and damping.

$$k = 100000\,\text{N}/\text{m} \qquad c = 1000\,\text{N s}/\text{m} \tag{12.221}$$

The engine is running at

$$\omega = 5000\ rpm \approx 523.60\,\text{rad}/\text{s} \approx 83.333\,\text{Hz} \tag{12.222}$$

with the following eccentric parameters.

$$m_e = 0.001\,\text{kg} \qquad e = 0.12\,\text{m} \tag{12.223}$$

The natural frequency ω_n, damping ratio ξ, and mass ratio ε of the system, and frequency ratio r are

$$\omega_n = \sqrt{\frac{k}{m}} = \sqrt{\frac{400000}{110}} = 60.302\,\text{rad}/\text{s} \approx 9.6\,\text{Hz} \tag{12.224}$$

$$\xi = \frac{c}{2\sqrt{km}} = 0.30151 \tag{12.225}$$

$$\varepsilon = \frac{m_e}{m} = \frac{0.001}{110} = 9.0909 \times 10^{-6} \tag{12.226}$$

$$r = \frac{\omega}{\omega_n} = \frac{523.60}{60.302} = 8.683 \tag{12.227}$$

The engine's amplitude of vibration is

$$X = \frac{r^2 e\varepsilon}{\sqrt{(1 - r^2)^2 + (2\xi r)^2}} = 1.1028 \times 10^{-6}\,\text{m} \tag{12.228}$$

However, if the speed of the engine is at the natural frequency of the system,

$$\omega = 576.0\ rpm \approx 60.302\,\text{rad}/\text{s} \approx 9.6\,\text{Hz} \tag{12.229}$$

then the amplitude of the engine's vibration increases to

$$X = \frac{r^2 e \varepsilon}{\sqrt{(1 - r^2)^2 + (2\xi r)^2}} = 1.8091 \times 10^{-6}\,\text{m} \qquad (12.230)$$

Example 497 *Eccentric exciting systems.*

All rotating machines such as engines, turbines, generators, and turning machines can have imperfections in their rotating components or have irregular mass distribution, which creates dynamic imbalances. When the unbalanced components rotate, an eccentric load applies to the structure. The load can be decomposed into two perpendicular harmonic forces in the plane of rotation in lateral and normal directions of the suspension. If the lateral force component is balanced by a reaction, the normal component provides a harmonically variable force with an amplitude depending on the eccentricity $m_e e$. Unbalanced rotating machines are a common source of excitation.

Example 498 *Velocity and acceleration of an eccentric excited system.*

Using the position frequency response of an eccentric excited system

$$x = A_4 \sin \omega t + B_4 \cos \omega t = X \sin (\omega t - \varphi_e) \qquad (12.231)$$

we can find the velocity and acceleration frequency responses.

$$
\begin{aligned}
\dot{x} &= A_4 \omega \cos \omega t - B_4 \omega \sin \omega t = X \omega \cos (\omega t - \varphi_e) \\
&= \dot{X} \cos (\omega t - \varphi_e)
\end{aligned}
\qquad (12.232)
$$

$$
\begin{aligned}
\ddot{x} &= -A_4 \omega^2 \sin \omega t - B_4 \omega^2 \cos \omega t = -X \omega^2 \sin (\omega t - \varphi_e) \\
&= \ddot{X} \sin (\omega t - \varphi_e)
\end{aligned}
\qquad (12.233)
$$

The amplitude of velocity and acceleration frequency responses, $\dot{X}$, $\ddot{X}$ are

$$\frac{\dot{X}}{e\varepsilon} = \frac{\omega^3 e m_e}{\sqrt{(k - m\omega^2)^2 + c^2 \omega^2}} \qquad (12.234)$$

$$\frac{\ddot{X}}{e\varepsilon} = \frac{\omega^4 e m_e}{\sqrt{(k - m\omega^2)^2 + c^2 \omega^2}} \qquad (12.235)$$

which can be written as

$$\frac{\dot{X}}{e\varepsilon \omega_n} = \frac{r^3}{\sqrt{(1 - r^2)^2 + (2\xi r)^2}} \qquad (12.236)$$

$$\frac{\ddot{X}}{e\varepsilon \omega_n^2} = \frac{r^4}{\sqrt{(1 - r^2)^2 + (2\xi r)^2}} \qquad (12.237)$$

Example 499 *Transmitted force to the base of an eccentric excited system.*

The transmitted force

$$f_T = F_T \sin(\omega t - \varphi_T) \tag{12.238}$$

to the ground by an eccentric excited system is equal to the sum of forces in the spring and damper.

$$f_T = f_k + f_c = kx + c\dot{x} \tag{12.239}$$

Substituting x *and* $\dot{x}$ *from (12.203) shows that*

$$f_T = (kA_4 - c\omega B_4)\sin t\omega + (kB_4 + c\omega A_4)\cos t\omega \tag{12.240}$$

therefore, the amplitude of the transmitted force is

$$\begin{aligned}
F_T &= \sqrt{(kA_4 - c\omega B_4)^2 + (kB_4 + c\omega A_4)^2} \\
&= e\omega^2 m_e \sqrt{\frac{c^2\omega^2 + k^2}{(k - m\omega^2)^2 + c^2\omega^2}}
\end{aligned} \tag{12.241}$$

The frequency response of the transmitted force can be simplified to the following applied equation.

$$\frac{F_T}{e\omega^2 m_e} = \frac{\sqrt{1 + (2\xi r)^2}}{\sqrt{(1 - r^2)^2 + (2\xi r)^2}} \tag{12.242}$$

12.3.4 ★ Eccentric Base Excitation

Figure 12.35 illustrates a one-DOF eccentric base excited vibrating system with a mass m suspended by a spring k and a damper c on a base with mass m_b. The base has an unbalance mass m_e at a distance e that is rotating with angular velocity ω. The eccentric base excited system is a good model for vibration analysis of different equipment that are attached to the engine of a vehicle, or any equipment mounted on a rotary motor.

Using the relative motion of m with respect to the base

$$z = x - y \tag{12.243}$$

we may develop the equation of motion as

$$\frac{mm_b}{m_b + m}\ddot{z} + c\dot{z} + kz = \frac{mm_e}{m_b + m}e\omega^2 \sin \omega t \tag{12.244}$$

or

$$\ddot{z} + 2\xi\omega_n \dot{z} + \omega_n^2 z = \varepsilon e\omega^2 \sin \omega t \tag{12.245}$$

$$\varepsilon = \frac{m_e}{m_b} \tag{12.246}$$

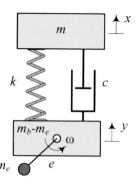

FIGURE 12.35. An eccentric base excited single-DOF system.

The relative displacement response of the system is

$$z = A_5 \sin \omega t + B_5 \cos \omega t \qquad (12.247)$$
$$= Z \sin (\omega t - \varphi_b) \qquad (12.248)$$

which has an amplitude Z and phases φ_b.

$$\frac{Z}{e\varepsilon} = \frac{r^2}{\sqrt{(1 - r^2)^2 + (2\xi r)^2}} \qquad (12.249)$$

$$\varphi_b = \tan^{-1} \frac{2\xi r}{1 - r^2} \qquad (12.250)$$

The frequency responses for Z, and φ_b as a function of r and ξ are plotted in Figures 12.36 and 12.37.

Proof. The free body diagram shown in Figure 12.38, along with Newton's method in the x-direction, may be used to find the equation of motion.

$$m\ddot{x} = -c(\dot{x} - \dot{y}) - k(x - y) \qquad (12.251)$$
$$m_b\ddot{y} = c(\dot{x} - \dot{y}) + k(x - y) - m_e e\omega^2 \sin \omega t \qquad (12.252)$$

Using $z = x - y$, and

$$\ddot{z} = \ddot{x} - \ddot{y} \qquad (12.253)$$

we may combine Equations (12.251) and (12.252) to find the equation of relative motion.

$$\frac{mm_b}{m_b + m}\ddot{z} + c\dot{z} + kz = \frac{mm_e}{m_b + m}e\omega^2 \sin \omega t \qquad (12.254)$$

Equation (12.254) can be transformed to (12.245) if we divide it by $\frac{mm_b}{m_b+m}$

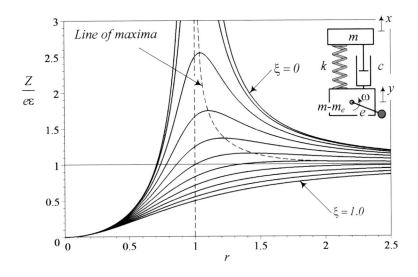

FIGURE 12.36. The position frequency response for $\dfrac{Z}{e\varepsilon}$.

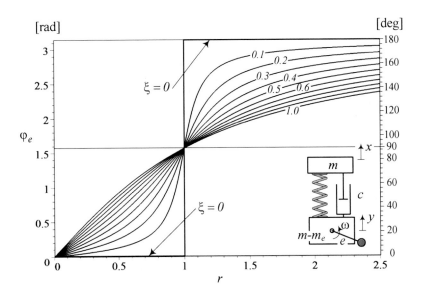

FIGURE 12.37. The frequency response for φ_b.

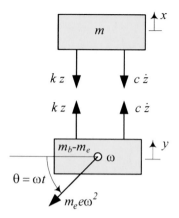

FIGURE 12.38. Free body diagram of an eccentric base excitated single DOF system.

and use the following definitions:

$$\xi = \cfrac{c}{2\sqrt{k\cfrac{mm_b}{m_b + m}}} \qquad (12.255)$$

$$\omega_n = \sqrt{k\cfrac{m_b + m}{mm_b}} \qquad (12.256)$$

The parameter $\varepsilon = \frac{m_e}{m_b}$ is called the *mass ratio* and indicates the ratio between the eccentric mass m_e and the total base mass m_b.

The steady-state solution of Equation (12.245) can be (12.247). To find the amplitude and phase of the response, we substitute the solution in the equation of motion.

$$-\omega^2 (A_5 \sin \omega t + B_5 \cos \omega t) + 2\xi\omega_n\omega (A_5 \cos \omega t - B_5 \sin \omega t)$$
$$+\omega_n^2 (A_5 \sin \omega t + B_5 \cos \omega t) = \varepsilon e\omega^2 \sin \omega t \qquad (12.257)$$

The coefficients of the functions $\sin \omega t$ and $\cos \omega t$ must balance on both sides of the equation.

$$\omega_n^2 A_5 - \omega^2 A_5 - 2\xi\omega\omega_n B_5 = \varepsilon\omega^2 e \qquad (12.258)$$
$$2\xi A_5 \omega\omega_n - B_5\omega^2 + B_5\omega_n^2 = 0 \qquad (12.259)$$

Therefore, we find two algebraic equations to calculate A_5 and B_5.

$$\begin{bmatrix} \omega_n^2 - \omega^2 & -2\xi\omega\omega_n \\ 2\xi\omega\omega_n & \omega_n^2 - \omega^2 \end{bmatrix} \begin{bmatrix} A_5 \\ B_5 \end{bmatrix} = \begin{bmatrix} \varepsilon\omega^2 e \\ 0 \end{bmatrix} \qquad (12.260)$$

Solving for the coefficients A_5 and B_5

$$\begin{bmatrix} A_5 \\ B_5 \end{bmatrix} = \begin{bmatrix} \omega_n^2 - \omega^2 & -2\xi\omega\omega_n \\ 2\xi\omega\omega_n & \omega_n^2 - \omega^2 \end{bmatrix}^{-1} \begin{bmatrix} \varepsilon\omega^2 e \\ 0 \end{bmatrix}$$

$$= \begin{bmatrix} \dfrac{\omega_n^2 - \omega^2}{(\omega_n^2 - \omega^2)^2 + (2\xi\omega\omega_n)^2} \varepsilon\omega^2 e \\[4mm] \dfrac{-2\xi\omega\omega_n}{(\omega_n^2 - \omega^2)^2 + (2\xi\omega\omega_n)^2} \varepsilon\omega^2 e \end{bmatrix} \qquad (12.261)$$

provides the steady-state solution (12.245).

The amplitude Z and phase φ_b can be found by

$$X = \sqrt{A_5^2 + B_5^2} \qquad (12.262)$$

$$\tan\varphi_b = \frac{-B_5}{A_5} \qquad (12.263)$$

which, after substituting A_5 and B_5 from (12.261) yields:

$$Z = \frac{\omega^2 e\varepsilon}{\sqrt{(\omega_n^2 - \omega^2)^2 + (2\xi\omega\omega_n)^2}} \qquad (12.264)$$

$$\tan\varphi_b = \frac{2\xi\omega\omega_n}{\omega_n^2 - \omega^2} \qquad (12.265)$$

Equations (12.264) and (12.265) can be simplified to more practical expressions (12.249) and (12.250) by employing $r = \frac{\omega}{\omega_n}$. ∎

Example 500 ★ *A base eccentric excited system.*
Consider an engine with a mass m_b

$$m_b = 110\,\text{kg} \qquad (12.266)$$

and an air intake device with a mass

$$m = 2\,\text{kg} \qquad (12.267)$$

that is mounted on the engine using an elastic mounts, with the following equivalent stiffness and damping.

$$k = 10000\,\text{N/m} \qquad c = 100\,\text{N s/m} \qquad (12.268)$$

The engine is running at

$$\omega = 576.0\ rpm \approx 60.302\,\text{rad/s} \approx 9.6\,\text{Hz} \qquad (12.269)$$

with the following eccentric parameters.

$$m_e = 0.001\,\text{kg} \qquad e = 0.12\,\text{m} \qquad (12.270)$$

The natural frequency ω_n, damping ratio ξ, and mass ratio ε of the system, and frequency ratio r are

$$\omega_n = \sqrt{k\frac{m_b + m}{mm_b}} = 100\,\text{rad/s} \approx 15.9\,\text{Hz} \qquad (12.271)$$

$$\xi = \frac{c}{2\sqrt{k\dfrac{mm_b}{m_b + m}}} = 0.49995 \qquad (12.272)$$

$$\varepsilon = \frac{m_e}{m_b} = 9.0909 \times 10^{-6} \qquad (12.273)$$

$$r = \frac{\omega}{\omega_n} = 0.60302 \qquad (12.274)$$

The relative amplitude of the device's vibration is

$$Z = \frac{e\varepsilon r^2}{\sqrt{(1 - r^2)^2 + (2\xi r)^2}} = 4.525 \times 10^{-7}\,\text{m} \qquad (12.275)$$

Example 501 ★ *Absolute displacement of the upper mass.*
   Equation (12.251)

$$\ddot{x} = -\frac{c}{m}(\dot{x} - \dot{y}) - \frac{k}{m}(x - y) = -\frac{c}{m}\dot{z} - \frac{k}{m}z \qquad (12.276)$$

along with the solution (12.247) may be used to calculate the displacement frequency response of the upper mass m in the eccentric base excited system shown in Figure 12.35. Assuming a steady-state displacement

$$x = A_6 \sin\omega t + B_6 \cos\omega t = X \sin(\omega t - \varphi_{bx}) \qquad (12.277)$$

we have

$$
\begin{aligned}
-\omega^2 (A_6 \sin\omega t + B_6 \cos\omega t) &= -\frac{c}{m}\dot{z} - \frac{k}{m}z \\
&= -\frac{c}{m}\omega (A_5 \cos\omega t - B_5 \sin\omega t) \\
&\quad - \frac{k}{m}(A_5 \sin\omega t + B_5 \cos\omega t) \\
&= \left(\frac{c}{m}\omega B_5 - \frac{k}{m}A_5\right)\sin t\omega \\
&\quad + \left(-\frac{k}{m}B_5 - \frac{c}{m}\omega A_5\right)\cos t\omega \quad (12.278)
\end{aligned}
$$

and therefore,

$$-\omega^2 A_6 = \frac{c}{m}\omega B_5 - \frac{k}{m}A_5 \qquad (12.279)$$

$$-\omega^2 B_6 = -\frac{k}{m}B_5 - \frac{c}{m}\omega A_5 \qquad (12.280)$$

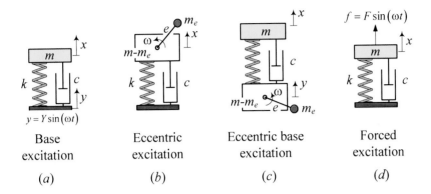

FIGURE 12.39. The four practical types of one DOF harmonically excited systems: a.base excitation, b.eccentric excitation, c.eccentric base excitation, d.forced excitation.

Substituting A_5 and B_5 from (12.261) and using

$$X = \sqrt{A_6^2 + B_6^2} \qquad \tan\varphi_{bx} = \frac{-B_6}{A_6} \qquad (12.281)$$

shows that

$$A_6 = -\frac{2c\xi\omega^2\omega_n + k\left(\omega_n^2 - \omega^2\right)}{\left(\omega_n^2 - \omega^2\right)^2 + \left(2\xi\omega\omega_n\right)^2}\frac{1}{m}\varepsilon e \qquad (12.282)$$

$$B_6 = \frac{-c\left(\omega_n^2 - \omega^2\right) + 2k\xi\omega_n}{\left(\omega_n^2 - \omega^2\right)^2 + \left(2\xi\omega\omega_n\right)^2}\frac{1}{m}\varepsilon\omega e \qquad (12.283)$$

the amplitude X of steady-state vibration of the upper mass in an eccentric base excited system is

$$X = \frac{\sqrt{c^2\omega^2 + k^2}}{\sqrt{\left(\omega_n^2 - \omega^2\right)^2 + \left(2\xi\omega\omega_n\right)^2}}\frac{\varepsilon}{m}e \qquad (12.284)$$

12.3.5 ★ *Classification for the Frequency Responses of One-DOF Forced Vibration Systems*

A harmonically excited one-DOF systems can be one of the four systems shown in Figure 12.39. The dimensionless amplitude of different applied steady-state responses of these systems are equal to one of the following equations (12.285)-(12.292), and the phase of their motion are equal to one of the equations (12.293)-(12.296).

$$S_0 = \frac{1}{\sqrt{(1 - r^2)^2 + (2\xi r)^2}} \qquad (12.285)$$

$$S_1 = \frac{r}{\sqrt{(1 - r^2)^2 + (2\xi r)^2}} \qquad (12.286)$$

$$S_2 = \frac{r^2}{\sqrt{(1 - r^2)^2 + (2\xi r)^2}} \qquad (12.287)$$

$$S_3 = \frac{r^3}{\sqrt{(1 - r^2)^2 + (2\xi r)^2}} \qquad (12.288)$$

$$S_4 = \frac{r^4}{\sqrt{(1 - r^2)^2 + (2\xi r)^2}} \qquad (12.289)$$

$$G_0 = \frac{\sqrt{1 + (2\xi r)^2}}{\sqrt{(1 - r^2)^2 + (2\xi r)^2}} \qquad (12.290)$$

$$G_1 = \frac{r\sqrt{1 + (2\xi r)^2}}{\sqrt{(1 - r^2)^2 + (2\xi r)^2}} \qquad (12.291)$$

$$G_2 = \frac{r^2\sqrt{1 + (2\xi r)^2}}{\sqrt{(1 - r^2)^2 + (2\xi r)^2}} \qquad (12.292)$$

$$\Phi_0 = \tan^{-1}\frac{2\xi r}{1 - r^2} \qquad (12.293)$$

$$\Phi_1 = \tan^{-1}\frac{1 - r^2}{-2\xi r} \qquad (12.294)$$

$$\Phi_2 = \tan^{-1}\frac{-2\xi r}{1 - r^2} \qquad (12.295)$$

$$\Phi_3 = \tan^{-1}\frac{2\xi r^3}{(1 - r^2)^2 + (2\xi r)^2} \qquad (12.296)$$

The function S_0 and G_0 are the main parts in all amplitude frequency responses. To have a sense about the behavior of different responses, we usually plot them as a function of r and using ξ as a parameter. The mass m, stiffness k, and damper c of the system are constant and hence, the excitation frequency ω is the only variable. We combine m, k, c, ω and define two parameters r and ξ to express frequency responses by two variable functions.

To develop a clear classification let us indicate the frequency responses related to the systems shown in Figure 12.39 by adding a subscript and express their different responses as follow:

1. For a *base excitation system*, we usually use the frequency responses of the relative and absolute kinematics Z_B, $\dot{Z}_B$, $\ddot{Z}_B$, X_B, $\dot{X}_B$, $\ddot{X}_B$, along with the transmitted force frequency response F_{T_B}.

2. For an *eccentric excitation system*, we usually use the frequency responses of the absolute kinematics X_E, $\dot{X}_E$, $\ddot{X}_E$, along with the transmitted force frequency response F_{T_E}.

3. For an *eccentric base excitation system*, we usually use the frequency responses of the relative and absolute kinematics Z_R, $\dot{Z}_R$, $\ddot{Z}_R$, X_R, $\dot{X}_R$, $\ddot{X}_R$, Y_R, $\dot{Y}_R$, $\ddot{Y}_R$, along with the transmitted force frequency response F_{T_R}.

4− For a *forced excitation system*, we usually use the frequency responses of the absolute kinematics X_F, $\dot{X}_F$, $\ddot{X}_F$, along with the transmitted force frequency response F_{T_F}

The frequency response of different features of the four systems in Figure 12.39 may be summarized and labeled as follows:

$$S_0 = \frac{X_F}{F/k} \tag{12.297}$$

$$S_1 = \frac{\dot{X}_F}{F/\sqrt{km}} \tag{12.298}$$

$$S_2 = \frac{\ddot{X}_F}{F/m} = \frac{Z_B}{Y} = \frac{X_E}{e\varepsilon_E} = \frac{Z_R}{e\varepsilon_R} \tag{12.299}$$

$$S_3 = \frac{\dot{Z}_B}{\omega_n Y} = \frac{\dot{X}_E}{e\varepsilon_E \omega_n} = \frac{\dot{Z}_R}{e\varepsilon_R \omega_n} \tag{12.300}$$

$$S_4 = \frac{\ddot{Z}_B}{\omega_n^2 Y} = \frac{\ddot{X}_E}{e\varepsilon_E \omega_n^2} = \frac{\ddot{Z}_R}{e\varepsilon_R \omega_n^2} \tag{12.301}$$

$$G_0 = \frac{F_{T_F}}{F} = \frac{X_B}{Y} \tag{12.302}$$

$$G_1 = \frac{\dot{X}_B}{\omega_n Y} \tag{12.303}$$

$$G_2 = \frac{\ddot{X}_B}{\omega_n^2 Y} = \frac{F_{T_B}}{kY} = \frac{F_{T_E}}{e\omega_n^2 m_e} = \frac{F_{T_R}}{e\omega_n^2 m_e}\left(1 + \frac{m_b}{m}\right) \tag{12.304}$$

Figures A.1-A.8 in Appendix A visualize the frequency responses used in analysis and designing of the systems. However, the exact value of the responses should be found from the associated equations.

Proof. The equations of motion for a harmonically forced vibrating single DOF system is always equal to

$$m\ddot{q} + c\dot{q} + kq = f(q, \dot{q}, t) \tag{12.305}$$

where, the variable q is a general coordinate to show the absolute displacement x, or relative displacement $z = x - y$. The forcing term $f(x, \dot{x}, t)$ is a harmonic function which in the general case can be a combination of $\sin \omega t$ and $\cos \omega t$, where ω is the excitation frequency.

$$f(q, \dot{q}, t) = a \sin \omega t + b \cos \omega t \tag{12.306}$$

Depending on the system and the frequency response we are looking for, the coefficients a and b are zero, constant, or proportional to ω, ω^2, ω^3, ω^4, $\cdots$, ω^n. To cover all practical harmonically forced vibrating systems, let us assume

$$a = a_0 + a_1\omega + a_2\omega^2 \tag{12.307}$$
$$b = b_0 + b_1\omega + b_2\omega^2 \tag{12.308}$$

We usually divide the equation of motion (12.305) by m to express it by ξ and ω_n

$$\ddot{q} + 2\xi\omega_n\dot{q} + \omega_n^2 q = \left(A_0 + A_1\omega + A_2\omega^2\right)\sin\omega t$$
$$+ \left(B_0 + B_1\omega + B_2\omega^2\right)\cos\omega t \tag{12.309}$$

where,

$$A_0 + A_1\omega + A_2\omega^2 = \frac{1}{m}\left(a_0 + a_1\omega + a_2\omega^2\right) \tag{12.310}$$

$$B_0 + B_1\omega + B_2\omega^2 = \frac{1}{m}\left(b_0 + b_1\omega + b_2\omega^2\right) \tag{12.311}$$

The solution of the equation of motion would be a harmonic response with unknown coefficients.

$$q = A\sin\omega t + B\cos\omega t \tag{12.312}$$
$$= Q\sin(\omega t - \varphi) \tag{12.313}$$

To find the steady-state amplitude of the response Q and its phase lag φ

$$Q = \sqrt{A^2 + B^2} \tag{12.314}$$

$$\varphi = \tan^{-1}\frac{-B}{A} \tag{12.315}$$

we should substitute the solution in the equation of motion.

$$-\omega^2\left(A\sin\omega t + B\cos\omega t\right) + 2\xi\omega_n\omega\left(A\cos\omega t - B\sin\omega t\right)$$
$$+\omega_n^2\left(A\sin\omega t + B\cos\omega t\right)$$
$$= \left(A_0 + A_1\omega + A_2\omega^2\right)\sin\omega t + \left(B_0 + B_1\omega + B_2\omega^2\right)\cos\omega t \tag{12.316}$$

The coefficients of the functions $\sin \omega t$ and $\cos \omega t$ must balance on both sides of the equation.

$$
\begin{aligned}
\omega_n^2 A - \omega^2 A - 2\xi \omega \omega_n B &= A_0 + A_1 \omega + A_2 \omega^2 & (12.317) \\
2\xi A \omega \omega_n - B\omega^2 + B\omega_n^2 &= B_0 + B_1 \omega + B_2 \omega^2 & (12.318)
\end{aligned}
$$

Therefore, we always find two algebraic equations to calculate A and B.

$$
\begin{bmatrix} \omega_n^2 - \omega^2 & -2\xi \omega \omega_n \\ 2\xi \omega \omega_n & \omega_n^2 - \omega^2 \end{bmatrix} \begin{bmatrix} A \\ B \end{bmatrix} = \begin{bmatrix} A_0 + A_1 \omega + A_2 \omega^2 \\ B_0 + B_1 \omega + B_2 \omega^2 \end{bmatrix} \qquad (12.319)
$$

Solving for the coefficients A and B

$$
\begin{aligned}
\begin{bmatrix} A \\ B \end{bmatrix} &= \begin{bmatrix} \omega_n^2 - \omega^2 & -2\xi \omega \omega_n \\ 2\xi \omega \omega_n & \omega_n^2 - \omega^2 \end{bmatrix}^{-1} \begin{bmatrix} A_0 + A_1 \omega + A_2 \omega^2 \\ B_0 + B_1 \omega + B_2 \omega^2 \end{bmatrix} \\[2mm]
&= \begin{bmatrix} \dfrac{Z_1}{(1-r^2)^2 + (2\xi r)^2} \\[4mm] \dfrac{Z_2}{(1-r^2)^2 + (2\xi r)^2} \end{bmatrix} \qquad (12.320)
\end{aligned}
$$

$$
\begin{aligned}
Z_1 &= 2\xi r \frac{1}{\omega_n^2} \left(B_2 \omega^2 + B_1 \omega + B_0 \right) \\
&\quad + \frac{1}{\omega_n^2} \left(1 - r^2 \right) \left(A_2 \omega^2 + A_1 \omega + A_0 \right) \qquad (12.321)
\end{aligned}
$$

$$
\begin{aligned}
Z_2 &= \frac{1}{\omega_n^2} \left(1 - r^2 \right) \left(B_2 \omega^2 + B_1 \omega + B_0 \right) \\
&\quad - 2\xi r \frac{1}{\omega_n^2} \left(A_2 \omega^2 + A_1 \omega + A_0 \right) \qquad (12.322)
\end{aligned}
$$

provides us with the steady-state solution amplitude Q and phase φ

$$
\begin{aligned}
Q &= \sqrt{A^2 + B^2} & (12.323) \\
\tan \varphi &= \frac{-B}{A} & (12.324)
\end{aligned}
$$

We are able to reproduce any of the steady-state responses S_i and G_i by setting the coefficients A_0, A_1, A_2, B_0, B_1, and B_2 properly. ∎

Example 502 ★ *Base excited frequency responses.*

A one DOF base excited vibrating system is shown in Figure 12.23. The equation of relative motion $z = x - y$ with a harmonic excitation $y = Y \sin \omega t$ is

$$
\ddot{z} + 2\xi \omega_n \dot{z} + \omega_n^2 z = \omega^2 Y \sin \omega t \qquad (12.325)
$$

This equation can be found from Equation (12.309) if

$$A_0 = 0 \quad A_1 = 0 \quad A_2 = Y$$
$$B_0 = 0 \quad B_1 = 0 \quad B_2 = 0 \tag{12.326}$$

So, the frequency response of the system would be

$$Z = Q = \sqrt{A^2 + B^2} = \frac{r^2}{\sqrt{(1 - r^2)^2 + (2\xi r)^2}} Y \tag{12.327}$$

because,

$$\begin{bmatrix} A \\ B \end{bmatrix} = \begin{bmatrix} \dfrac{Z_1}{(1 - r^2)^2 + (2\xi r)^2} \\ \dfrac{Z_2}{(1 - r^2)^2 + (2\xi r)^2} \end{bmatrix} \tag{12.328}$$

$$Z_1 = r^2 \left(1 - r^2\right) Y \tag{12.329}$$
$$Z_2 = 2\xi r^3 Y \tag{12.330}$$

12.4   Time Response of Vibrating Systems

The equations of motion of any linear vibrating systems to examine its transient response is

$$[m]\,\ddot{\mathbf{x}} + [c]\,\dot{\mathbf{x}} + [k]\,\mathbf{x} = \mathbf{F} \tag{12.331}$$
$$\mathbf{x}(0) = \mathbf{x}_0 \tag{12.332}$$
$$\dot{\mathbf{x}}(0) = \dot{\mathbf{x}}_0 \tag{12.333}$$

where the mass matrix $[m]$, stiffness matrix $[k]$, and damping matrix $[c]$ are assumed constant. The time response of the system is the solution $\mathbf{x} = \mathbf{x}(t), t > 0$ for the set of coupled ordinary differential equations. Such a problem is called an *initial-value problem.*

Consider a one-DOF vibrating system

$$m\ddot{x} + c\dot{x} + kx = f\left(x, \dot{x}, t\right) \tag{12.334}$$

with the initial conditions

$$x(0) = x_0 \tag{12.335}$$
$$\dot{x}(0) = \dot{x}_0 \tag{12.336}$$

The coefficients m, c, k are assumed constant, although, they may be functions of time in more general problems. The solution of such a problem, $x = x(t), t > 0$, is unique.

The *order* of an equation is the highest number of derivatives. In mechanical vibrations of lumped models, we work with a set of second-order differential equations. If $x_1(t)$, $x_2(t)$, $\cdots$, $x_n(t)$, are solutions of an n-order equation, then its general solution is

$$x(t) = a_1 x_1(t) + a_2 x_2(t) + \cdots + a_n x_n(t). \tag{12.337}$$

When $f = 0$, the equation is called *homogeneous*,

$$m\ddot{x} + c\dot{x} + kx = 0 \tag{12.338}$$

otherwise it is *non-homogeneous*. The solution of the non-homogeneous equation (12.334) is equal to

$$x(t) = x_h(t) + x_p(t) \tag{12.339}$$

where, $x_h(t)$ is the *homogeneous solution*, and $x_p(t)$ is the *particular solution*. In mechanical vibration, the homogeneous equation is called *free vibration* and its solution is called *free vibration response*. The non-homogeneous equation is called *forced vibration* and its solution is called *forced vibration response*.

An exponential function

$$x = e^{\lambda t} \tag{12.340}$$

satisfies every homogeneous linear differential equation. Therefore, the homogeneous response of the second order equation (12.338) is

$$x_h(t) = a_1 e^{\lambda_1 t} + a_2 e^{\lambda_2 t} \tag{12.341}$$

where the constants a_1 and a_2 depend on the initial conditions. The parameters λ_1 and λ_2 are called *characteristic parameters* or *eigenvalues* of the system The eigenvalues are the solution of an algebraic equation, called a *characteristic equation*, which is the result of substituting the solution (12.340) in Equation (12.338). The characteristic equation is the condition to make the solution (12.340) satisfy the equation of motion (12.338).

A general particular solution of a forced equation is hard to find, however, we know that if the forcing function $f = f(t)$ is a combination of the following functions:

1. a constant, such as $f = a$
2. a polynomial in t, such as $f = a_0 + a_1 t + a_2 t^2 + \cdots + a_n t^n$
3. an exponential function, such as $f = e^{at}$
4. a harmonic function, such as $f = F_1 \sin at + F_2 \cos at$

then the particular solution $x_p(t)$ has the same form as a forcing term.

1. $x_p(t) =$ a constant, such as $x_p(t) = C$
2. $x_p(t) =$ a polynomial of the same degree, such as $x_p(t) = C_0 + C_1 t + C_2 t^2 + \cdots + C_n t^n$

3. $x_p(t) =$ an exponential function, such as $x_p(t) = Ce^{at}$
4. $x_p(t) =$ a harmonic function, such as $x_p(t) = A\sin at + B\cos at$.

If the system is force free, or the forcing term disappears after a while, the solution of the equation is called *time response* or *transient response*. The initial conditions are important in transient response.

When the system has some damping, the effect of initial conditions disappears after a while in both transient and forced vibration responses, and a steady-state response remains. If the forcing term is harmonic, then the steady-state solution is called frequency response.

Example 503 *A homogeneous solution of a second-order linear equation. Consider a system with the following equation of motion:*

$$\ddot{x} + \dot{x} - 2x = 0 \qquad x_0 = 1 \qquad \dot{x}_0 = 7 \tag{12.342}$$

To find the solution, we substitute an exponential solution $x = e^{\lambda t}$ in the equation of motion and find the characteristic equation.

$$\lambda^2 + \lambda - 2 = 0 \tag{12.343}$$

The eigenvalues are

$$\lambda_{1,2} = 1, -2 \tag{12.344}$$

and therefore, the solution is

$$x = a_1 e^t + a_2 e^{-2t} \tag{12.345}$$

Taking a derivative

$$\dot{x} = a_1 e^t - 2a_2 e^{-2t} \tag{12.346}$$

and employing the initial conditions

$$1 = a_1 + a_2 \tag{12.347}$$
$$7 = a_1 - 2a_2 \tag{12.348}$$

provides the constants a_1, a_2, and the solution $x = x(t)$.

$$a_1 = 3 \qquad a_2 = -2 \tag{12.349}$$
$$x = 3e^t - 2e^{-2t} \tag{12.350}$$

Example 504 *Natural frequency.*
Consider a free mass-spring system such as the one shown in Figure 12.40. The system is undamped and free of excitation forces, so its equation of motion is

$$m\ddot{x} + kx = 0 \tag{12.351}$$

To find the solution, let us try a harmonic solution with an unknown frequency Ω.

$$x = A\sin\Omega t + B\cos\Omega t \tag{12.352}$$

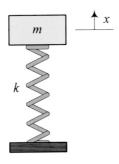

FIGURE 12.40. A mass-spring, single degree-of-freedom vibrating system.

Substituting (12.352) in (12.351) provides

$$-\Omega^2 m \left(A \sin \Omega t + B \cos \Omega t \right) + k \left(A \sin \Omega t + B \cos \Omega t \right) = 0 \qquad (12.353)$$

which can be collected as

$$\left(Bk - Bm\Omega^2 \right) \cos \Omega t + \left(Ak - Am\Omega^2 \right) \sin \Omega t = 0 \qquad (12.354)$$

The coefficients of $\sin \Omega t$ *and* $\cos \Omega t$ *must be zero, and hence,*

$$\Omega = \sqrt{\frac{k}{m}} \qquad (12.355)$$

$$x = A \sin \sqrt{\frac{k}{m}} t + B \cos \sqrt{\frac{k}{m}} t \qquad (12.356)$$

The frequency $\Omega = \sqrt{k/m}$ *is the frequency of vibration of a free and un-damped mass-spring system. It is called **natural frequency** and is shown by a special character* ω_n.

$$\omega_n = \sqrt{\frac{k}{m}} \qquad (12.357)$$

A system has as many natural frequencies as its degrees of freedom.

Example 505 *Free vibration of a single-DOF system.*
   The simplest free vibration equation of motion is

$$m\ddot{x} + c\dot{x} + kx = 0 \qquad (12.358)$$

which is equivalent to

$$\ddot{x} + 2\xi\omega_n \dot{x} + \omega_n^2 x = 0 \qquad (12.359)$$

*The response of a system to free vibration is called **transient response** and depends solely on the initial conditions* $x_0 = x(0)$ *and* $\dot{x}_0 = \dot{x}(0)$.

To determine the solution of the linear equation (12.358), we may search for a solution in an exponential form.

$$x = A\,e^{\lambda t} \qquad\qquad (12.360)$$

Substituting (12.360) in (12.359) provides the characteristic equation

$$\lambda^2 + 2\xi\omega_n\lambda + \omega_n^2 = 0 \qquad\qquad (12.361)$$

to find the eigenvalues $\lambda_{1,2}$.

$$\lambda_{1,2} = -\xi\omega_n \pm \omega_n\sqrt{\xi^2 - 1} \qquad\qquad (12.362)$$

Therefore, the general solution for Equation (12.359) is

$$
\begin{aligned}
x &= A_1\,e^{\lambda_1 t} + A_2\,e^{\lambda_2 t} \\
&= A_1\,e^{\left(-\xi\omega_n + \omega_n\sqrt{\xi^2-1}\right)t} + A_2\,e^{\left(-\xi\omega_n - \omega_n\sqrt{\xi^2-1}\right)t} \\
&= e^{-\xi\omega_n t}\left(A_1\,e^{i\omega_d t} + A_2\,e^{-i\omega_d t}\right) \qquad (12.363)
\end{aligned}
$$

$$\omega_d = \omega_n\sqrt{1-\xi^2} \qquad\qquad (12.364)$$

where ω_d is called **damped natural frequency**.

By using the Euler equation

$$e^{i\alpha} = \cos\alpha + i\sin\alpha \qquad\qquad (12.365)$$

we may modify solution (12.363) to the following forms:

$$
\begin{aligned}
x &= e^{-\xi\omega_n t}\left(B_1\sin\omega_d t + B_2\cos\omega_d t\right) & (12.366) \\
x &= B\,e^{-\xi\omega_n t}\sin\left(\omega_d t + \phi\right) & (12.367)
\end{aligned}
$$

where

$$
\begin{aligned}
B_1 &= i\left(A_1 - A_2\right) & (12.368) \\
B_2 &= A_1 + A_2 & (12.369) \\
B &= \sqrt{B_1^2 + B_2^2} & (12.370) \\
\phi &= \tan^{-1}\frac{B_2}{B_1} & (12.371)
\end{aligned}
$$

Because the displacement x is a real physical quantity, coefficients B_1 and B_2 in Equation (12.366) must also be real. This requires that A_1 and A_2 be complex conjugates. The motion described by Equation (12.367) consists of a harmonic motion of frequency $\omega_d = \omega_n\sqrt{1-\xi^2}$ and a decreasing amplitude $B\,e^{-\xi\omega_n t}$.

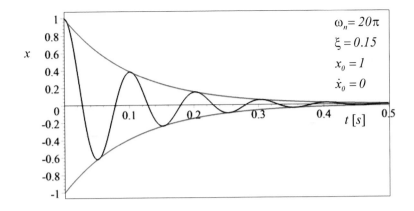

FIGURE 12.41. A sample time response for an under-damped system.

Example 506 *Under-damped, critically-damped, and over-damped systems.*

The time response of a damped one-DOF system is given by Equation (12.363). The solution can be transformed to Equation (12.366) as long as $\xi < 1$.

The value of a damping ratio controls the type of time response of a one-DOF system. Depending on the value of damping, there are three major solution categories:

1. *under-damped,*
2. *critically-damped,*
3. *over-damped.*

*An **under damped** system appears when $\xi < 1$. For such a system, the characteristic parameters (12.362) are complex conjugate*

$$\lambda_{1,2} = -\xi\omega_n \pm i\omega_n\sqrt{1 - \xi^2} \qquad (12.372)$$

and therefore, the general solution (12.363)

$$x = A_1\, e^{\lambda_1 t} + A_2\, e^{\lambda_2 t} \qquad (12.373)$$

can be transformed to (12.366)

$$x = e^{-\xi\omega_n t}\left(B_1 \sin\omega_d t + B_2 \cos\omega_d t\right) \qquad (12.374)$$

An under-damped system has an oscillatory time response with a decaying amplitude as shown in Figure 12.41 for $\xi = 0.15$, $\omega_n = 20\pi\,\text{rad}$, $x_0 = 1$, and $\dot{x}_0 = 0$. The exponential function $Xe^{\pm\xi\omega_n t}$ is an envelope for the curve of response.

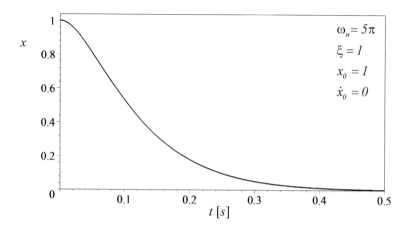

FIGURE 12.42. A sample time response for an critically-damped system.

*A **critically damped** system is when $\xi = 1$. For such a system, the characteristic parameters (12.362) are equal.*

$$\lambda = \lambda_{1,2} = -\omega_n \tag{12.375}$$

When the characteristic values are equal, the time response of the system would be

$$x = A_1 e^{\lambda t} + A_2 t e^{\lambda t} \tag{12.376}$$

which is equal to

$$x = e^{-\xi \omega_n t} (A_1 + A_2 t) \tag{12.377}$$

Figure 12.42 shows a critically-damped response for $\xi = 1$, $\omega_n = 5\pi$ rad, $x_0 = 1$, and $\dot{x}_0 = 0$.

*An **over damped** system is when $\xi > 1$. The characteristic parameters (12.362) for an over-damped system are two real numbers*

$$\lambda_{1,2} = -\xi \omega_n \pm \omega_n \sqrt{\xi^2 - 1} \tag{12.378}$$

and therefore, the exponential solution cannot be converted to harmonic functions.

$$x = A_1 e^{\lambda_1 t} + A_2 e^{\lambda_2 t} \tag{12.379}$$

Starting from any set of initial conditions, the time response of an over-damped system goes to zero exponentially. Figure 12.43 shows an over-damped response for $\xi = 2$, $\omega_n = 10\pi$ rad, $x_0 = 1$, and $\dot{x}_0 = 0$.

Example 507 *Free vibration and initial conditions.*

Consider a one DOF mass-spring-damper in a free vibration. The general motion of the system, given in Equation (12.366), is

$$x = e^{-\xi \omega_n t} (B_1 \sin \omega_d t + B_2 \cos \omega_d t) \tag{12.380}$$

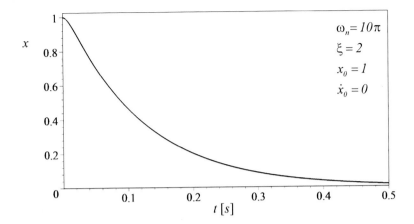

FIGURE 12.43. A sample time response for an over-damped system.

If the initial conditions of the system are

$$x(0) = x_0 \qquad \dot{x}(0) = \dot{x}_0 \qquad (12.381)$$

then,

$$x_0 = B_2 \qquad \dot{x}_0 = -\xi\omega_n B_2 + B_1\omega_d \qquad (12.382)$$

and hence,

$$B_1 = \frac{\dot{x}_0 + \xi\omega_n x_0}{\omega_d} \qquad B_2 = x_0 \qquad (12.383)$$

Substituting B_1 and B_2 in solution (12.380) generates the general solution for free vibration of a single-DOF system.

$$x = e^{-\xi\omega_n t}\left(\frac{\dot{x}_0 + \xi\omega_n x_0}{\omega_d}\sin\omega_d t + x_0\cos\omega_d t\right) \qquad (12.384)$$

The solution can also be written as

$$x = e^{-\xi\omega_n t}\left(x_0\left(\cos\omega_d t + \frac{\xi}{\omega_d}\omega_n\sin\omega_d t\right) + \frac{\dot{x}_0}{\omega_d}\sin\omega_d t\right) \qquad (12.385)$$

If the initial conditions of the system are substituted in solution (12.367)

$$x = Be^{-\xi\omega_n t}\sin(\omega_d t + \phi) \qquad (12.386)$$

then,

$$x_0 = B\sin\phi \qquad \dot{x}_0 = -B\xi\omega_n\sin\phi + B\omega_d\cos\phi \qquad (12.387)$$

To solve for B and ϕ, we may write

$$B = \frac{x_0}{\sin\phi} \qquad \tan\phi = \frac{\omega_d x_0}{\dot{x}_0 + \xi\omega_n x_0} \qquad (12.388)$$

and therefore,

$$B = \frac{1}{\omega_d}\sqrt{(\omega_d x_0)^2 + (\dot{x}_0 + \xi\omega_n x_0)^2} \tag{12.389}$$

Now the solution (12.386) becomes

$$\begin{aligned} x &= \frac{e^{-\xi\omega_n t}}{\omega_d}\sqrt{(\omega_d x_0)^2 + (\dot{x}_0 + \xi\omega_n x_0)^2} \\ &\times \sin\left(\omega_d t + \tan^{-1}\frac{\omega_d x_0}{\dot{x}_0 + \xi\omega_n x_0}\right) \end{aligned} \tag{12.390}$$

Example 508 *Free vibration, initial conditions, and critically damping.*
  If the system is critically damped, then the time response to free vibrations is

$$x = e^{-\xi\omega_n t}\left(A_1 + A_2 t\right) \tag{12.391}$$

Using the initial conditions, $x(0) = x_0$, $\dot{x}(0) = \dot{x}_0$, we can find the coefficients A_1 and A_2 as

$$A_1 = x_0 \qquad A_2 = \dot{x}_0 + \xi\omega_n x_0 \tag{12.392}$$

and therefore the general critically-damped response is

$$x = e^{-\xi\omega_n t}\left(x_0 + (\dot{x}_0 + \xi\omega_n x_0)\,t\right) \tag{12.393}$$

Example 509 *Free vibration, initial conditions, and over damping.*
  If the system is over-damped, then the characteristic parameters $\lambda_{1,2}$ are real and the time response to free vibrations is a real exponential function.

$$x = A_1\,e^{\lambda_1 t} + A_2\,e^{\lambda_2 t} \tag{12.394}$$

Using the initial conditions, $x(0) = x_0$, $\dot{x}(0) = \dot{x}_0$,

$$x_0 = A_1 + A_2 \qquad \dot{x}_0 = \lambda_1 A_1 + \lambda_2 A_2 \tag{12.395}$$

we can find the coefficients A_1 and A_2 as

$$A_1 = \frac{\dot{x}_0 - \lambda_2 x_0}{\lambda_1 - \lambda_2} \qquad A_2 = \frac{\lambda_1 x_0 - \dot{x}_0}{\lambda_1 - \lambda_2} \tag{12.396}$$

Hence, the general over-damped response is

$$x = \frac{\dot{x}_0 - \lambda_2 x_0}{\lambda_1 - \lambda_2}\,e^{\lambda_1 t} + \frac{\lambda_1 x_0 - \dot{x}_0}{\lambda_1 - \lambda_2}\,e^{\lambda_2 t} \tag{12.397}$$

Example 510 *Work done by a harmonic force.*
  The work done by a harmonic force

$$f(t) = F\sin\left(\omega t + \varphi\right) \tag{12.398}$$

acting on a body with a harmonic displacement

$$x(t) = X \sin(\omega t) \tag{12.399}$$

during one period

$$T = \frac{2\pi}{\omega} \tag{12.400}$$

is equal to

$$
\begin{aligned}
W &= \int_0^{2\pi/\omega} f(t)dx = \int_0^{2\pi/\omega} f(t)\frac{dx}{dt}dt \\
&= FX\omega \int_0^{2\pi/\omega} \sin(\omega t + \varphi)\cos(\omega t)\,dt \\
&= FX \int_0^{2\pi} \sin(\omega t + \varphi)\cos(\omega t)\,d(\omega t) \\
&= FX \int_0^{2\pi} \left(\sin\varphi\cos^2\omega t + \cos\varphi\sin\omega t\cos\omega t\right)d(\omega t) \\
&= \pi FX \sin\varphi \tag{12.401}
\end{aligned}
$$

The work W is a function of the phase φ between f and x. When $\varphi = \frac{\pi}{2}$ then the work is maximum

$$W_{Max} = \pi F_0 X_0 \tag{12.402}$$

and when $\varphi = 0$, the work is minimum.

$$W_{min} = 0 \tag{12.403}$$

Example 511 ★ *Response to a step input.*
   Step input is an standard and the most important transient excitation by which we examine and compare vibrating systems. Consider a linear second order system with the following equation of motion.

$$\ddot{x} + 2\xi\omega_n\dot{x} + \omega_n^2 x = f(t) \tag{12.404}$$
$$\xi < 1 \tag{12.405}$$

A step input, is a sudden change of the forcing function $f(t)$ from zero to a constant and steady value. If the value is unity then,

$$f(t) = \begin{cases} 1\,N/kg & t > 0 \\ 0 & t \le 0 \end{cases} \tag{12.406}$$

*This excitation is called **unit step input**, and the response of the system is called the **unit step response**. Linearity of the equation of motion guarantees that the response to a non-unit step input is proportional to the unit step response.*

Consider a force function as

$$f(t) = \begin{cases} F_0 \, \text{N/kg} & t > 0 \\ 0 & t \le 0 \end{cases} \tag{12.407}$$

The general solution of Equation (12.404) for (12.407) is equal to sum of the homogeneous and particular solutions, $x = x_h + x_p$. The homogeneous solution was given by Equation (12.363) in Example 505. The particular solution would be constant $x_p = C$ because the input is constant $f(t) = F_0$. Substituting $x_p = C$ in Equation (12.404) provides us with

$$C = \frac{F_0}{\omega_n^2} \tag{12.408}$$

Therefore, the general solution of Equation (12.404) is

$$\begin{aligned} x &= x_h + x_p \\ &= \frac{F_0}{\omega_n^2} + e^{-\xi \omega_n t} \left(A \cos \omega_d t + B \sin \omega_d t \right) \qquad t \ge 0 \end{aligned} \tag{12.409}$$

$$\omega_d = \omega_n \sqrt{1 - \xi^2} \tag{12.410}$$

The zero initial conditions are the best to explore the natural behavior of the system. Applying a zero initial condition

$$x(0) = 0 \qquad \dot{x}(0) = 0 \tag{12.411}$$

provides two equations for A and B

$$\frac{F_0}{\omega_n^2} + A = 0 \tag{12.412}$$

$$\xi \omega_n A + \omega_d B = 0 \tag{12.413}$$

with the following solutions.

$$A = -\frac{F_0}{\omega_n^2} \qquad B = -\frac{\xi F_0}{\omega_d \omega_n} \tag{12.414}$$

Therefore, the step response is

$$x = \frac{F_0}{\omega_n^2} \left(1 - e^{-\xi \omega_n t} \left(\cos \omega_d t + \frac{\xi \omega_n}{\omega_d} \sin \omega_d t \right) \right) \tag{12.415}$$

Figure 12.44 depicts a step input for the following numerical values.

$$\xi = 0.3 \qquad \omega_n = 1 \qquad F_0 = 1 \tag{12.416}$$

*There are some characteristics for a step response: **rise time** t_r, **peak time** t_P, **peak value** x_P, **overshoot** $S = x_P - \frac{F_0}{\omega_n^2}$, and **settling time** t_s.*

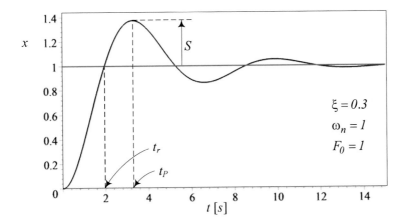

FIGURE 12.44. Response of a one DOF vibrating system to a step input.

Rise time t_r is the first time that the response $x(t)$ reaches the value of the step input $\frac{F_0}{\omega_n^2}$.

$$t_r = \frac{2}{\omega_d} \tan^{-1} \frac{\xi + 1}{\sqrt{1 - \xi^2}} \qquad (12.417)$$

Rise time may also be defined as the inverse of the largest slope of the step response, or as the time it takes to pass from 10% to 90% of the steady-state value.

Peak time t_P is the first time that the response $x(t)$ reaches its maximum value.

$$t_P = \frac{\pi}{\omega_d} \qquad (12.418)$$

Peak value x_P is the value of $x(t)$ when $t = t_P$.

$$x_P = \frac{F_0}{\omega_n^2} \left(1 + e^{-\xi \omega_n \frac{\pi}{\omega_d}} \right) = \frac{F_0}{\omega_n^2} \left(1 + e^{-\xi \frac{\pi}{\sqrt{1 - \xi^2}}} \right) \qquad (12.419)$$

Overshoot S indicates how much the response $x(t)$ exceeds the step input.

$$S = x_P - \frac{F_0}{\omega_n^2} = \frac{F_0}{\omega_n^2} e^{-\xi \frac{\pi}{\sqrt{1 - \xi^2}}} \qquad (12.420)$$

Settling time t_s is, by definition, four times of the time constant of the exponential function $e^{-\xi \omega_n t}$.

$$t_s = \frac{4}{\xi \omega_n} \qquad (12.421)$$

Settling time may also be defines as the required time that the step response $x(t)$ needs to settles within a $\pm p\%$ window of the step input. The value *p = 2 is commonly used.*

$$t_s \approx \frac{\ln\left(p\sqrt{1-\xi^2}\right)}{\xi\omega_n} \tag{12.422}$$

For the given data in (12.416) we find the following characteristic values.

$$
\begin{array}{ccc}
t_r & = & 1.966 \qquad t_P = 3.2933 \qquad t_s = 13.333 \\
x_P & = & 1.3723 \qquad S = 0.3723
\end{array}
\tag{12.423}
$$

12.5 Vibration Application and Measurement

The measurable vibration parameters, such as period T and amplitude X, may be used to identify mechanical characteristics of the vibrating system. In most vibration measurement and test methods, a transient or harmonically steady-state vibration will be examined. Using time and kinematic measurement devices, we measure amplitude and period of response, and use the analytic equations to find the required data.

Example 512 *Damping ratio determination.*
Damping ratio of an under-damped one-DOF system can be found by

$$\xi = \frac{1}{\sqrt{4(n-1)^2\pi^2 + \ln^2\frac{x_1}{x_n}}}\ln\frac{x_1}{x_n} \approx \frac{1}{2(n-1)\pi}\ln\frac{x_1}{x_n} \tag{12.424}$$

which is based on a plot of $x = x(t)$ and peak amplitudes x_i.
To derive this equation, consider the free vibration of an under-damped one-DOF system with the following equation of motion:

$$\ddot{x} + 2\xi\omega_n\dot{x} + \omega_n^2 x = 0. \tag{12.425}$$

The time response of the system is given in Equation (12.366) as

$$x = X e^{-\xi\omega_n t}\cos(\omega_d t + \phi) \tag{12.426}$$

where the constants X and ϕ are dependent on initial conditions.
Figure 12.45 illustrates a sample of the x-response. The peak amplitudes x_i are

$$
\begin{array}{ccc}
x_1 & = & e^{-\xi\omega_n t_1}\left(X\cos(\omega_d t_1 + \phi)\right) \tag{12.427} \\
x_2 & = & e^{-\xi\omega_n t_2}\left(X\cos(\omega_d t_2 + \phi)\right) \tag{12.428}
\end{array}
$$

$$\vdots$$

$$x_n = e^{-\xi\omega_n t_n}\left(X\cos(\omega_d t_n + \phi)\right) \tag{12.429}$$

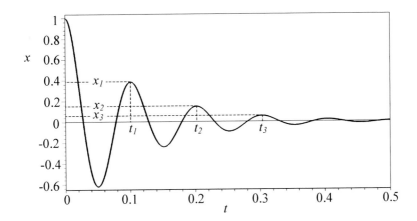

FIGURE 12.45. An x-response for the free vibration of an under-damped one-DOF system.

The ratio of the first two peaks is

$$\frac{x_1}{x_2} = e^{-\xi\omega_n(t_1-t_2)}\frac{\cos(\omega_d t_1 + \phi)}{\cos(\omega_d t_2 + \phi)} \tag{12.430}$$

Because the time difference between t_1 and t_2 is the period of oscillation

$$T_d = t_2 - t_1 = \frac{2\pi}{\omega_d} = \frac{2\pi}{\omega_n\sqrt{1-\xi^2}} \tag{12.431}$$

we may simplify Equation (12.430) to

$$\frac{x_1}{x_2} = e^{\xi\omega_n T_d}\frac{\cos(\omega_d t_1 + \phi)}{\cos(\omega_d(t_1+T_d) + \phi)} = e^{\xi\omega_n T_d}\frac{\cos(\omega_d t_1 + \phi)}{\cos(\omega_d t_1 + 2\pi + \phi)}$$
$$= e^{\xi\omega_n T_d} \tag{12.432}$$

This equation shows that,

$$\ln\frac{x_1}{x_2} = \xi\omega_n T_d = \frac{2\pi\xi}{\sqrt{1-\xi^2}} \tag{12.433}$$

which can be used to evaluate the damping ratio ξ.

$$\xi \approx \frac{1}{\sqrt{4\pi^2 + \ln^2\frac{x_1}{x_2}}}\ln\frac{x_1}{x_2} \tag{12.434}$$

For a better evaluation we may measure the ratio between x_1 and any other x_n, and use the following equation:

$$\xi \approx \frac{1}{\sqrt{4(n-1)^2\pi^2 + \ln^2\frac{x_1}{x_n}}}\ln\frac{x_1}{x_n} \tag{12.435}$$

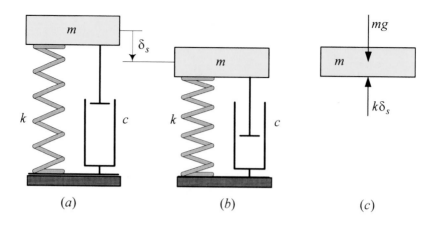

FIGURE 12.46. Static deflection and natural frequency determination.

If $\xi \ll 1$, then $\sqrt{1 - \xi^2} \approx 1$, and we may evaluate ξ from (12.433) with a simpler equation.

$$\xi \approx \frac{1}{2(n-1)\pi} \ln \frac{x_1}{x_n} \tag{12.436}$$

Example 513 *Natural frequency determination.*

Natural frequency of a mass-spring-damper system can be found by measuring the static deflection of the system. Consider a one-DOF system shown in Figure 12.46(a) that barely touches the ground. Assume that the spring has no tension or compression before touching the ground. When the system rests on the ground as shown in Figure 12.46(b), the spring is compressed by a static deflection $\delta_s = mg/k$ because of gravity. We may determine the natural frequency of the system by measuring δ_s

$$\omega_n = \sqrt{\frac{g}{\delta_s}} \tag{12.437}$$

because

$$\delta_s = \frac{mg}{k} = \frac{g}{\omega_n^2} \tag{12.438}$$

Example 514 *Mass moments determination.*

Mass moments are important characteristics of a vehicle that affect its dynamic behavior. The main mass moments I_x, I_y, and I_z can be calculated by an oscillating experiment.

Figure 12.47 illustrates an oscillating platform hung from point A. Assume the platform has a mass M and a mass moment I_0 about the pivot point A. Ignoring the mass of cables, we can write the Euler equation about point A

$$\sum M_y = I_0 \ddot{\theta} = -Mgh_1 \sin\theta \tag{12.439}$$

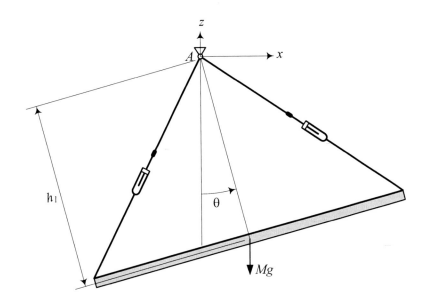

FIGURE 12.47. An oscillating platform hung from point A.

and derive the equation of motion.

$$I_0 \ddot{\theta} + Mgh_1 \sin \theta = 0 \qquad (12.440)$$

If the angle of oscillation θ is very small, then $\sin \theta \approx \theta$ and therefore, Equation (12.440) reduces to a linear equation

$$\ddot{\theta} + \omega_n^2 \theta = 0 \qquad \omega_n = \sqrt{\frac{Mgh_1}{I_0}} \qquad (12.441)$$

where ω_n is the natural frequency of the oscillation.

ω_n can be assumed as the frequency of small oscillation about the point A when the platform is set free after a small deviation from equilibrium position. The natural period of oscillation $T_n = 2\pi/\omega_n$ is what we can measure, and therefore, the mass moment I_0 is equal to

$$I_0 = \frac{1}{4\pi^2} Mgh_1 T_n^2 \qquad (12.442)$$

The natural period T_n may be measured by an average period of a few cycles, or more accurately, by an accelerometer.

Now consider the swing shown in Figure 12.48. A car with mass m at C is on the platform such that C is exactly above the mass center of the platform. Because the location of the mass center C is known, the distance between C and the fulcrum A is also known as h_2.

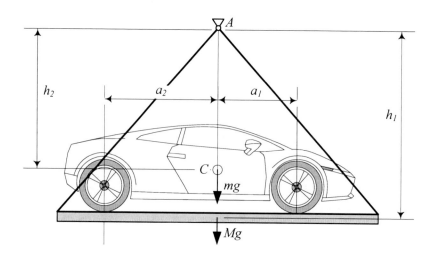

FIGURE 12.48. A car with mass m on an oscillating platform hung from point A.

To find the car's pitch mass moment I_y about C, we apply the Euler equation about point A, when the oscillator is deviated from the equilibrium condition.

$$\sum M_y = I_A \ddot{\theta} \qquad (12.443)$$

$$-Mgh_1 \sin\theta - mgh_2 \sin\theta = I_0 + I_y + mh_2^2 \qquad (12.444)$$

Assuming very small oscillation, we may use $\sin\theta \approx \theta$ and then Equation (12.444) reduces to a linear oscillator

$$\ddot{\theta} + \omega_n^2 \theta = 0 \qquad (12.445)$$

$$\omega_n = \sqrt{\frac{(Mh_1 + mh_2)\,g}{I_0 + I_y + mh_2^2}} \qquad (12.446)$$

Therefore, the pitch mass moment I_y can be calculated by measuring the natural period of oscillation $T_n = 2\pi/\omega_n$ from the following equation.

$$I_y = \frac{1}{4\pi^2}\,(Mh_1 + mh_2)\,gT_n^2 - I_0 - mh_2^2 \qquad (12.447)$$

To determine the roll moment of inertia, we may put the car on the platform as shown in Figure 12.49.

Having I_x and I_y we may put the car on the platform, at an angle α, to find its mass moment about the axis passing through C and parallel to the swing axis. Then, the product mass moment I_{xy} can be calculated by transformation calculus.

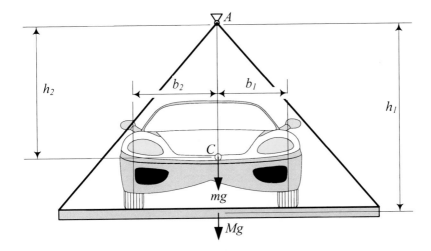

FIGURE 12.49. Roll moment of inertia measurement, using a swinging platform.

Example 515 *Sample data.*
   Tables 12.1 *indicates an example of data for the mass center position,
mass moment, and geometry of street cars, that are close to a Mercedes-
Benz A-Class.*

Table 12.1 - *Sample data close to a Mercedes-Benz A-Class.*

wheelbase	2424 mm
front track	1492 mm
rear track	1426 mm
mass	1245 kg
a_1	1100 mm
a_2	1323 mm
h	580 mm
I_x	335 kg m^2
I_y	1095 kg m^2
I_z	1200 kg m^2

12.6 ★ Vibration Optimization Theory

The first goal in vibration optimization is to reduce the vibration amplitude
of a primary mass to zero, when the system is under a forced vibration.
There are two principal methods for decreasing the vibration amplitude of
a primary mass: *vibration absorber*, and *vibration isolator*.

   When the suspension of a *primary system* is not easy to change, we add
another vibrating system, known as the *vibration absorber* or *secondary*

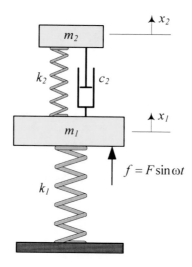

FIGURE 12.50. A secondary vibration absorber system (m_2, c_2, k_2) added to a primary vibrating system (m_1, k_1).

system, to absorb the vibrations of the primary system. The vibration absorber increases the DOF of the system, and is an applied method for vibration reduction in frequency domain. It can work very well in a few specific frequencies, and may be designed to work well in a range of frequencies.

Consider a mass m_1 supported by a suspension made of only a spring k_1, as shown in Figure 12.50. There is a harmonic force $f = F \sin \omega t$ applied on m_1. We add a secondary system (m_2, c_2, k_2) to the primary mass m_1 and make a two-DOF vibrating system. Such a system is sometimes called *Frahm absorber*, or *Frahm damper*.

It is possible to design the suspension of the secondary system (c_2, k_2) to reduce the amplitude of vibration m_1 to zero at any specific excitation frequency ω. However, if the excitation frequency is variable, we can adjust k_2 at the optimal value $k_2^\star$,

$$k_2^\star = \frac{m_1 m_2}{(m_1 + m_2)^2} k_1 \qquad (12.448)$$

and select c_2 within the range

$$2m_2\omega_1\xi_1^\star < c_2 < 2m_2\omega_1\xi_2^\star \qquad (12.449)$$

to minimize the amplitude of m_1 over the whole frequency range. The optimal $\xi_1^\star$ and $\xi_2^\star$ are the positive values of

$$\xi_1^\star = \sqrt{\frac{-B - \sqrt{B^2 - 4AC}}{2A}} \tag{12.450}$$

$$\xi_2^\star = \sqrt{\frac{-B + \sqrt{B^2 - 4AC}}{2A}} \tag{12.451}$$

where

$$A = 16Z_8 - 4r^2\left(4Z_4 + 8Z_5\right) \tag{12.452}$$
$$B = 4Z_9 - 4Z_6 r^2 - Z_7\left(4Z_4 + 8Z_5\right) + 4Z_3 Z_8 \tag{12.453}$$
$$C = Z_3 Z_9 - Z_6 Z_7 \tag{12.454}$$

and

$$Z_3 = 2\left(r^2 - \alpha^2\right) \tag{12.455}$$
$$Z_4 = \left[r^2\left(1 + \varepsilon\right) - 1\right]^2 \tag{12.456}$$
$$Z_5 = r^2\left(1 + \varepsilon\right)\left[r^2\left(1 + \varepsilon\right) - 1\right] \tag{12.457}$$
$$Z_6 = 2\left[\varepsilon\alpha^2 r^2 - \left(r^2 - \alpha^2\right)\left(r^2 - 1\right)\right]$$
$$\times \left[\varepsilon\alpha^2 - \left(r^2 - \alpha^2\right) - \left(r^2 - 1\right)\right] \tag{12.458}$$
$$Z_7 = \left(r^2 - \alpha^2\right)^2 \tag{12.459}$$
$$Z_8 = r^2\left[r^2\left(1 + \varepsilon\right) - 1\right]^2 \tag{12.460}$$
$$Z_9 = \left[\varepsilon\alpha^2 r^2 - \left(r^2 - 1\right)\left(r^2 - \alpha^2\right)\right]^2 \tag{12.461}$$

$$\varepsilon = \frac{m_2}{m_1} \tag{12.462}$$

$$\omega_1 = \sqrt{\frac{k_1}{m_1}} \tag{12.463}$$

$$\omega_2 = \sqrt{\frac{k_2}{m_2}} \tag{12.464}$$

$$\alpha = \frac{\omega_2}{\omega_1} \tag{12.465}$$

$$r = \frac{\omega}{\omega_1} \tag{12.466}$$

$$\xi = \frac{c_2}{2m_2\omega_1} \tag{12.467}$$

$$\mu = \frac{X_1}{F/k_1} \tag{12.468}$$

Proof. The equations of motion for the system shown in Figure 12.50 are:

$$m_1\ddot{x}_1 + c_2\left(\dot{x}_1 - \dot{x}_2\right) + k_1 x_1 + k_2\left(x_1 - x_2\right) = F\sin\omega t \qquad (12.469)$$
$$m_2\ddot{x}_2 - c_2\left(\dot{x}_1 - \dot{x}_2\right) - k_2\left(x_1 - x_2\right) = 0 \qquad (12.470)$$

To find the frequency response of the system, we substitute the following solutions in the equations of motion:

$$x_1 = A_1\cos\omega t + B_1\sin\omega t \qquad (12.471)$$
$$x_2 = A_2\cos\omega t + B_2\sin\omega t \qquad (12.472)$$

Assuming a steady-state condition, we find the following set of equations for A_1, B_1, A_2, B_2

$$\begin{bmatrix} a_{11} & c_2\omega & -k_2 & -c_2\omega \\ -c_2\omega & a_{22} & c_2\omega & -k_2 \\ -k_2 & -c_2\omega & a_{33} & c_2\omega \\ c_2\omega & -k_2 & -c_2\omega & a_{44} \end{bmatrix} \begin{bmatrix} A_1 \\ B_1 \\ A_2 \\ B_2 \end{bmatrix} = \begin{bmatrix} 0 \\ F \\ 0 \\ 0 \end{bmatrix} \qquad (12.473)$$

$$a_{11} = a_{22} = k_1 + k_2 - m_1\omega^2 \qquad (12.474)$$
$$a_{33} = a_{44} = k_2 - m_2\omega^2 \qquad (12.475)$$

The steady-state amplitude X_1 for vibration of the primary mass m_1 is found by

$$X_1 = \sqrt{A_1^2 + B_1^2} \qquad (12.476)$$

and is equal to

$$\left(\frac{X_1}{F}\right)^2 = \frac{\left(k_2 - \omega^2 m_2\right)^2 + \omega^2 c_2^2}{Z_1^2 + \omega^2 c_2^2 Z_2^2} \qquad (12.477)$$

where,

$$Z_1 = \left(k_1 - \omega^2 m_1\right)\left(k_2 - \omega^2 m_2\right) - \omega^2 m_2 k_2 \qquad (12.478)$$
$$Z_2 = k_1 - \omega^2 m_1 - \omega^2 m_2 \qquad (12.479)$$

Introducing the parameters (12.462)-(12.468) we may rearrange the frequency response (12.477) to

$$\mu^2 = \frac{4\xi^2 r^2 + \left(r^2 - \alpha^2\right)^2}{4\xi^2 r^2 \left[r^2\left(1 + \varepsilon\right) - 1\right]^2 + \left[\varepsilon\alpha^2 r^2 - \left(r^2 - 1\right)\left(r^2 - \alpha^2\right)\right]^2} \qquad (12.480)$$

The parameter ε is the *mass ratio* between m_2 and the m_1, ω_1 is the angular natural frequency of the main system, ω_2 is the angular natural frequency of the vibration absorber system, α is the natural frequency ratio, r is the excitation frequency ratio, ξ is the damping ratio, and μ is the amplitude ratio between dynamic amplitude X_1 and the static deflection F/k_1.

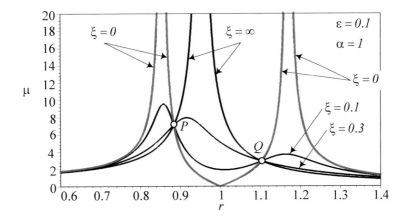

FIGURE 12.51. Behavior of ferquency response μ for a set of parameters and different damping ratios.

Figure 12.51 illustrates the behavior of frequency response μ for

$$\varepsilon = 0.1 \tag{12.481}$$
$$\alpha = 1 \tag{12.482}$$

and

$$\xi = 0 \tag{12.483}$$
$$\xi = 0.2 \tag{12.484}$$
$$\xi = 0.3 \tag{12.485}$$
$$\xi = \infty \tag{12.486}$$

All the curves pass through two nodes P and Q, independent of the damping ratio ξ. To find the parameters that control the position of the nodes, we find the intersection points of the curves for $\xi = 0$ and $\xi = \infty$. Setting $\xi = 0$ and $\xi = \infty$ yields:

$$\mu^2 = \frac{\left(r^2 - \alpha^2\right)^2}{\left[\varepsilon \alpha^2 r^2 - \left(r^2 - 1\right)\left(r^2 - \alpha^2\right)\right]^2} \tag{12.487}$$

$$\mu^2 = \frac{1}{\left[r^2\left(1 + \varepsilon\right) - 1\right]^2} \tag{12.488}$$

When $\xi = 0$, the system is an undamped linear two-DOF system with two natural frequencies. The vibration amplitude of the system approaches infinity $\mu \to \infty$ when the excitation frequency approaches either of the natural frequencies. When $\xi = \infty$, there would be no relative motion between m_1 and m_2 and the system reduces to an undamped linear one-DOF system with one natural frequency

$$\omega_n = \sqrt{\frac{k_1}{m_1 + m_2}} \qquad (12.489)$$

or

$$r_n = \frac{1}{\sqrt{1 + \varepsilon}} \qquad (12.490)$$

The vibration amplitude of the system approaches infinity $\mu \to \infty$ when the excitation frequency approaches the natural frequency $\omega \to \omega_{n_i}$ or $r \to 1/(1 + \varepsilon)$.

Using Equations (12.487) and (12.488), we find

$$\frac{\left(r^2 - \alpha^2\right)^2}{\left(\varepsilon\alpha^2 r^2 - (r^2 - 1)(r^2 - \alpha^2)\right)^2} = \frac{1}{\left[r^2(1 + \varepsilon) - 1\right]^2} \qquad (12.491)$$

which can be simplified to

$$\varepsilon\alpha^2 r^2 - \left(r^2 - 1\right)\left(r^2 - \alpha^2\right) = \pm\left(r^2 - \alpha^2\right)\left[r^2(1 + \varepsilon) - 1\right] \qquad (12.492)$$

The negative sign is equivalent to

$$r^4\varepsilon = 0$$

which indicates that there is a common point at $r = 0$. The plus sign produces a quadratic equation for r^2

$$(2 + \varepsilon)r^4 - r^2\left(2 + 2\alpha^2(1 + \varepsilon)\right) + 2\alpha^2 = 0 \qquad (12.493)$$

with two positive solutions r_1 and r_2 corresponding to nodes P and Q.

$$r_{1,2}^2 = \frac{1}{\varepsilon + 2}\left(\alpha^2 \pm \sqrt{(\varepsilon^2 + 2\varepsilon + 1)\alpha^4 - 2\alpha^2 + 1} + \alpha^2\varepsilon + 1\right) \qquad (12.494)$$

$$r_1 < r_n < r_2 \qquad (12.495)$$

Because the frequency response curves always pass through P and Q, the optimal situation would be when the nodes P and Q have equal height.

$$\mu(P) = \mu(Q) \qquad (12.496)$$

Because the value of μ^2 at P and Q are independent of ξ, we may substitute r_1 and r_2 in Equation (12.488) for μ corresponding to $\xi = \infty$. However, μ from Equation (12.488)

$$\mu = \frac{1}{\left[r^2(1 + \varepsilon) - 1\right]} \qquad (12.497)$$

produces a positive number for $r < r_n$ and a negative number for $r > r_n$. Therefore,

$$\mu(r_1) = -\mu(r_2) \qquad (12.498)$$

generates the equality

$$\frac{1}{1 - r_1^2 (1 + \varepsilon)} = \frac{-1}{1 - r_2^2 (1 + \varepsilon)} \qquad (12.499)$$

which can be simplified to

$$r_1^2 + r_2^2 = \frac{2}{1 + \varepsilon} \qquad (12.500)$$

The sum of the roots from Equation (12.493) is

$$r_1^2 + r_2^2 = \frac{2 + 2\alpha^2 (1 + \varepsilon)}{1 + \varepsilon} \qquad (12.501)$$

and therefore,

$$\frac{2}{1 + \varepsilon} = \frac{\left[2 + 2\alpha^2 (1 + \varepsilon)\right]}{1 + \varepsilon} \qquad (12.502)$$

which provides

$$\alpha = \frac{1}{1 + \varepsilon} \qquad (12.503)$$

Equation (12.503) is the required condition to make the height of the nodes P and Q equal, and hence, provides the optimal value of α. Having the optimal α is equivalent to designing the optimal stiffness k_2 for the secondary suspension, because,

$$\alpha = \frac{\omega_2}{\omega_1} = \sqrt{\frac{m_1}{m_2}} \sqrt{\frac{k_2}{k_1}} \qquad (12.504)$$

and Equation (12.503) simplifies to

$$\alpha = \frac{m_1}{m_1 + m_2} \qquad (12.505)$$

to provide the following condition for optimal $k_2^\star$:

$$k_2^\star = k_1 \frac{m_1 m_2}{(m_1 + m_2)^2} \qquad (12.506)$$

To determine the optimal damping ratio ξ, we force μ to have its maximum at P or Q. Having μ_{Max} at P guarantees that $\mu(r_1)$ is the highest value in a frequency domain around r_1, and having μ_{Max} at Q, guarantees that $\mu(r_2)$ is the highest value in a frequency domain around r_2. The position of μ_{Max} is controlled by ξ, so we may determine two optimal ξ at which μ_{Max} is at $\mu(r_1)$ and $\mu(r_2)$. An example of this situation is shown in Figure 12.52.

Using the optimal α from (12.503), the nodal frequencies are

$$r_{1,2}^2 = \frac{1}{1 + \varepsilon} \left(1 \pm \sqrt{\frac{\varepsilon}{2 + \varepsilon}}\right) \qquad (12.507)$$

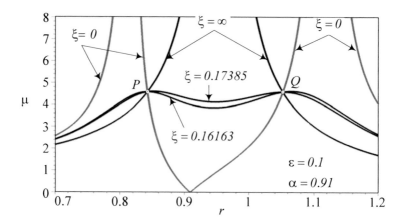

FIGURE 12.52. Optimal damping ratio ξ to have maximum of μ at P or Q.

To set the partial derivative $\partial\mu^2/\partial r^2$ equal to zero at the nodal frequencies

$$\left.\frac{\partial\mu^2}{\partial r^2}\right|_{r_1^2} = 0 \tag{12.508}$$

$$\left.\frac{\partial\mu^2}{\partial r^2}\right|_{r_2^2} = 0 \tag{12.509}$$

we write μ^2 by numerator $N(r)$ divided by denominator $D(r)$

$$\mu^2 = \frac{N(r)}{D(r)} \tag{12.510}$$

which helps to find the derivative easier.

$$\frac{\partial\mu^2}{\partial r^2} = \frac{1}{D^2}\left(D\frac{\partial N}{\partial r^2} - N\frac{\partial D}{\partial r^2}\right) = \frac{1}{D}\left(\frac{\partial N}{\partial r^2} - \frac{N}{D}\frac{\partial D}{\partial r^2}\right) \tag{12.511}$$

Differentiating gives

$$\frac{\partial N}{\partial r^2} = \frac{\partial N}{\partial r^2} = 4\xi^2 + Z_3 \tag{12.512}$$

$$\frac{\partial D}{\partial r^2} = 4\xi^2 Z_4 + 8\xi^2 Z_5 + Z_6 \tag{12.513}$$

Equations (12.512) and (12.513), along with (12.507), must be substituted in (12.511) to be solved for ξ. After substitution, the equation $\partial\mu^2/\partial r^2 = 0$ would be

$$\frac{\partial N}{\partial r^2} - \frac{N}{D}\frac{\partial D}{\partial r^2} = \left(4\xi^2 + Z_3\right)\left(4\xi^2 Z_8 + Z_9\right)$$
$$- \left(4\xi^2 r^2 + Z_7\right)\left(4\xi^2 Z_4 + 8\xi^2 Z_5 + Z_6\right) = 0 \tag{12.514}$$

because of
$$\frac{N}{D} = \frac{4\xi^2 r^2 + Z_7}{4\xi^2 Z_8 + Z_9} \tag{12.515}$$

Equation (12.514) is a quadratic for ξ^2

$$\left(16Z_8 - 4r^2\left(4Z_4 + 8Z_5\right)\right)\xi^4$$
$$+ \left(4Z_9 - 4Z_6 r^2 - Z_7\left(4Z_4 + 8Z_5\right) + 4Z_3 Z_8\right)\xi^2$$
$$+ (Z_3 Z_9 - Z_6 Z_7) = A\left(\xi^2\right)^2 + B\xi^2 + C = 0 \tag{12.516}$$

with the solution
$$\xi^2 = \frac{-B \pm \sqrt{B^2 - 4AC}}{2A} \tag{12.517}$$

The positive value of ξ from (12.517) for $r = r_1$ and $r = r_2$ provides the limiting values for $\xi_1^\star$ and $\xi_2^\star$. Figure 12.52 shows the behavior of μ for optimal α and $\xi = 0, \xi_1^\star, \xi_2^\star, \infty$. ■

Example 516 ★ *Optimal spring and damper for $\varepsilon = 0.1$.*
Consider a Frahm vibration absorber with

$$\varepsilon = \frac{m_2}{m_1} = 0.1 \tag{12.518}$$

We adjust the optimal frequency ratio α form Equation (12.503).

$$\alpha^\star = \frac{1}{1 + \varepsilon} \approx 0.9091 \tag{12.519}$$

and find the nodal frequencies $r_{1,2}^2$ from (12.507).

$$r_{1,2}^2 = \frac{1}{1 + \varepsilon}\left(1 \pm \sqrt{\frac{\varepsilon}{2 + \varepsilon}}\right) = 0.71071, \ 1.1075 \tag{12.520}$$

Now, we set $r = r_1 = \sqrt{0.71071} \approx 0.843$ and evaluate the parameters Z_3 to Z_9 from (12.455)-(12.461)

$$
\begin{aligned}
Z_3 &= -0.231470544 & Z_4 &= 0.0476190476 \\
Z_5 &= -0.170598842 & Z_6 &= 0.0246326501 \\
Z_7 &= 0.0133946532 & Z_8 &= 0.03384338136 \\
Z_9 &= 0.0006378406298
\end{aligned} \tag{12.521}
$$

and the coefficients A, B, and C from (12.452)-(12.454)

$$
\begin{aligned}
A &= 3.879887219 \\
B &= -0.08308086729 \\
C &= -0.0004775871233
\end{aligned} \tag{12.522}
$$

to find the first optimal damping ratio ξ_1.

$$\xi_1^\star = 0.1616320694 \qquad (12.523)$$

Using $r = r_2 = \sqrt{1.1075} \approx 1.05236$ we find the following numbers

$$
\begin{aligned}
Z_3 &= 0.562049056 & Z_4 &= 0.04761904752 \\
Z_5 &= 0.265836937 & Z_6 &= -0.375123324 \\
Z_7 &= 0.078974785 & Z_8 &= 0.05273670508 \\
Z_9 &= 0.003760704084 & &
\end{aligned}
\qquad (12.524)
$$

$$
\begin{aligned}
A &= -9.421012739 \\
B &= 0.1167823931 \\
C &= 0.005076228579
\end{aligned}
\qquad (12.525)
$$

to find the second optimal damping ratio ξ_1.

$$\xi_2^\star = 0.1738496023 \qquad (12.526)$$

Therefore, the optimal α is $\alpha^\star = 0.9091$, and the optimal ξ is between $0.1616320694 < \xi^\star < 0.1738496023$.

Example 517 ★ *The vibration absorber is most effective when $r = \alpha = 1$.*

When $\xi = 0$, then $\mu = 0$ at $r = 1$, which shows the amplitude of the primary mass reduces to zero if the natural frequency of the primary and secondary systems are equal to the excitation frequency $r = \alpha = 1$.

Example 518 ★ *The optimal nodal amplitude.*

Substituting the optimal α from (12.503) in Equation (12.493),

$$r^4 - \frac{2}{2+\varepsilon}r^2 + \frac{2}{(2+\varepsilon)(1+\varepsilon)^2} = 0 \qquad (12.527)$$

provides us with the following nodal frequencies:

$$r_{1,2}^2 = \frac{1}{1+\varepsilon}\left(1 \pm \sqrt{\frac{\varepsilon}{2+\varepsilon}}\right) \qquad (12.528)$$

Applying $r_{1,2}$ in Equation (12.497) shows that the common nodal amplitude $\mu(r_{1,2})$ is

$$\mu = \sqrt{\frac{2+\varepsilon}{\varepsilon}} \qquad (12.529)$$

Example 519 ★ *Optimal α and mass ratio ε.*

The optimal value of the natural frequency ratio, α, is only a function of mass ratio ε, as determined in Equation (12.503). Figure 12.53 depicts the behavior of α as a function of ε. The value of optimal α, and hence, the value of optimal k_2, decreases by increasing $\varepsilon = m_2/m_1$. Therefore, a smaller mass for the vibration absorber needs a stiffer spring.

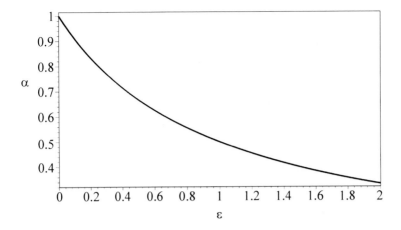

FIGURE 12.53. Optimal value of the natural frequency ratio, α, as a function of mass ratio ε.

Example 520 ★ *Nodal frequencies $r_{1,2}$ and mass ratio ε.*

As shown in Equation (12.507), the nodal frequencies $r_{1,2}$ for optimal α (12.503) are only a function of the mass ratio ε.

$$r_{1,2}^2 = \frac{1}{1+\varepsilon}\left(1 \pm \sqrt{\frac{\varepsilon}{2+\varepsilon}}\right) \tag{12.530}$$

Figure 12.54 illustrates the behavior of $r_{1,2}$ as a function of ε. When, $\varepsilon \to 0$, the vibration absorber m_2 vanishes, and hence, the system becomes a one-DOF primary oscillator. Such a system has only one natural frequency $r_n = 1$ as given by Equation (12.490). It is the frequency that $r_{1,2}$ will approach by vanishing m_2.

The nodal frequencies $r_{1,2}$ are always on both sides of the singe-DOF natural frequency r_n

$$r_1 < r_n < r_2 \tag{12.531}$$

while all of them are decreasing functions of the mass ratio ε.

Example 521 ★ *Natural frequencies for extreme values of damping.*

By setting $\xi = 0$ for $\varepsilon = 0.1$, we find

$$\mu = \left| \frac{r^2 - 1}{0.1r^2 - (r^2 - 1)^2} \right| \tag{12.532}$$

and by setting $\xi = \infty$, we find

$$\mu = \left| \frac{1}{1.1r^2 - 1} \right| \tag{12.533}$$

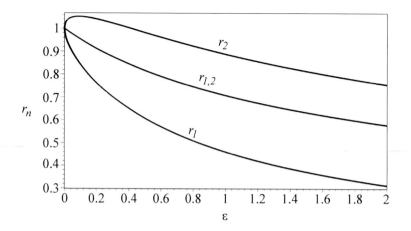

FIGURE 12.54. Behavior of nodal frequencies $r_{1,2}$ as a function of mass ratio ε.

Having $\xi = 0$ is equivalent to no damping. When there is no damping, μ approaches infinity at the real roots of its denominator, r_{n_1} and r_{n_2}, which are the natural frequencies of the system. As an example, the natural frequencies r_{n_1} and r_{n_2} for $\varepsilon = 0.1$, are

$$0.1r^2 - \left(r^2 - 1\right)^2 = 0 \qquad (12.534)$$
$$r_{n_1} = 0.854\,31 \qquad (12.535)$$
$$r_{n_2} = 1.170\,5 \qquad (12.536)$$

Having $\xi = \infty$ is equivalent to a rigid connection between m_1 and m_2. The system would have only one DOF and therefore, μ approaches infinity at the only roots of the denominator, r_n

$$1.1r^2 - 1 = 0 \qquad (12.537)$$
$$r_n = 0.953 \qquad (12.538)$$

where, r_n is always between r_{n_1} and r_{n_2}.

$$r_{n_1} < r_n < r_{n_2} \qquad (12.539)$$

12.7   Summary

Generally speaking, vibration is a harmful and unwanted phenomenon. Vibration is important when a non-vibrating system is connected to a vibrating system. To minimize the effects of vibration, we connect the systems by a damping elastic isolator. For simplicity, we model the isolator by a spring and damper parallel to each other. Such an isolator is called suspension.

Vibration can be physically expressed as a result of energy conversion. It can mathematically be expressed by solutions of a set of differential equations. If the system is linear, then its equations of motion can always be arranged in the following matrix form:

$$[M]\,\dot{\mathbf{x}} + [c]\,\dot{\mathbf{x}} + [k]\,\mathbf{x} = \mathbf{F}\,(\mathbf{x}, \dot{\mathbf{x}}, t) \tag{12.540}$$

Vibration can be separated into *free vibrations*, when $\mathbf{F} = 0$, and *forced vibrations*, when $\mathbf{F} \neq 0$. However, in applied vibrations, we usually separate the solution of the equations of motion into *transient* and *steady-state*. Transient response is the solution of the equations of motion when $\mathbf{F} = 0$ or $\mathbf{F}$ is active for a short period of time. Because most industrial machines are equipped with a rotating motor, periodic and harmonic excitation is very common. Frequency response is the steady-state solution of equations of motion when the system is harmonically excited. In frequency analysis we seek the steady-state response of the system, after the effect of initial conditions dies out.

Frequency response of mechanical systems, such as vehicles, is dominated by the natural frequencies of the system and by excitation frequencies. The amplitude of vibration increases when an excitation frequency approaches one of the natural frequencies of the system. Frequency domains around the natural frequencies are called the *resonance zone*. The amplitude of vibration in resonance zones can be reduced by introducing damping.

One-DOF, harmonically excited systems may be classified as *base excitation*, *eccentric excitation*, *eccentric base excitation*, and *forced excitation*. Every frequency response these systems can be expressed by one of the functions S_i, G_i, and Φ_i, each with a specific characteristic. We usually use a graphical illustration to see the frequency response of the system as a function of frequency ratio $r = \omega/\omega_n$ and damping ratio $\xi = c/\sqrt{4km}$.

12.8   Key Symbols

$a \equiv \ddot{x}$	acceleration
a_1	distance from mass center to front axle
a_2	distance from mass center to rear axle
$[a], [A]$	coefficient matrix
A, B, C	unknown coefficients for frequency responses
b_1	distance from mass center to left wheel
b_2	distance from mass center to right wheel
c	damping
$c^\star$	optimum damping
c_{eq}	equivalent damping
c_{ij}	element of row i and column j of a damping matrix
$[c]$	damping matrix
D	denominator
e	eccentricity arm
e	exponential function
E	mechanical energy
E	Young modulus of elasticity
$f = 1/T$	cyclic frequency [Hz]
f, $\mathbf{F}$	force
f_c	damper force
f_{eq}	equivalent force
f_k	spring force
f_m	required force to move a mass m
F	amplitude of a harmonic force $f = F \sin \omega t$
F_0	constant force
F_t	tension force
F_T	transmitted force
g	gravitational acceleration
G_0, G_1, G_2	amplitude frequency response
I	area moment of inertia for beams
I	mass moment of inertia for vehicles
$\mathbf{I}$	identity matrix
k	stiffness
$k^\star$	optimum stiffness
k_{eq}	equivalent stiffness
k_{ij}	element of row i and column j of a stiffness matrix
k_R	antiroll bar torsional stiffness
$[k]$	stiffness matrix
K	kinetic energy
l	length
m	mass
m_b	device mass

m_e	eccentric mass
m_{ij}	element of row i and column j of a mass matrix
m_s	sprung mass
m_s	mass of spring
m_u	unsprung mass
$[m]$	mass matrix
M	mass of platform
N	numerator
$\mathbf{p}$	momentum
Q	general amplitude
$r = \omega/\omega_n$	frequency ratio
$r,\ R$	radius
r_1, r_2	frequency ratio at nodes
$r_n = \omega_n/\omega_1$	dimensionless natural frequency
S	overshoot
S	quadrature
$S_0, S_1, \cdots, S_4$	amplitude frequency response
t	time
t_p	peak time
t_r	rise time
t_s	settling time
T	period
T_n	natural period
$v \equiv \dot{x},\ \mathbf{v}$	velocity
V	potential energy
w	track of a car
w_f	front track of a car
w_r	rear track of a car
$x, y, z,\ \mathbf{x}$	displacement
x_0	initial displacement
x_h	homogeneous solution
x_p	particular solution
x_P	peak displacement
$\dot{x}_0$	initial velocity
$X, Y, Z,$	amplitude
$Z_i, i = 1, 2, \cdots$	short notation parameters
$\alpha = \omega_2/\omega_1$	natural frequency ratio
δ	deflection
δ_s	static deflection
ε	mass ratio
θ	angular motion
Θ	amplitude of angular vibration
λ	eigenvalue
μ	amplitude frequency response

φ	phase angle
$\Phi_0, \Phi_1, \cdots, \Phi_3$	phase frequency response
$\omega = 2\pi f$	angular frequency $[\mathrm{rad/s}]$
ω_n	natural frequency
ξ	damping ratio
$\xi^\star$	optimum damping ratio

Subscriptions

d	driver
f	front
M	maximum
r	rear
s	sprung mass
u	unsprung mass

Exercises

1. Natural frequency and damping ratio.

   A *one* DOF mass-spring-damper has $m = 1\,\text{kg}$, $k = 1000\,\text{N/m}$ and $c = 100\,\text{N}\,\text{s/m}$. Determine the natural frequency, and damping ratio of the system.

2. Equivalent spring.

   Determine the equivalent spring for the vibrating system that is shown in Figure 12.55.

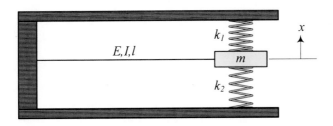

FIGURE 12.55. Spring connected cantilever beam.

3. ★ Equivalent mass for massive spring.

   Figure 12.56 illustrates an elastic cantilever beam with a tip mass m. The beam has characteristics: elasticity E, area moment of inertia I, mass m_s. Assume that when the tip mass m oscillates laterally, the beam gets a harmonic shape.

   $$y = Y \sin \frac{\pi x}{2l}$$

   Determine an equivalent mass m_e at the tip of a massless beam with the same E, I, l.

4. Ideal spring connected pendulum.

   Determine the kinetic and potential energies of the pendulum in Figure 12.57, at an arbitrary angle θ. The free length of the spring is $l = a - b$.

5. ★ General spring connected pendulum.

   Determine the potential energy of the pendulum in Figure 12.57 , at an angle θ, if:

   (a) The free length of the spring is $l = a - 1.2b$.

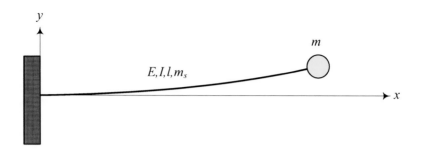

FIGURE 12.56. An elastic and massive cantilever beam with a tip mass m.

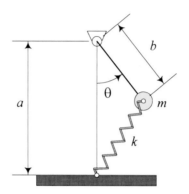

FIGURE 12.57. Spring connected pendulum.

(b) The free length of the spring is $l = a - 0.8b$.

6. ★ Spring connected rectilinear oscillator.

   Determine the kinetic and potential energies of the oscillator shown in Figure 12.58. The free length of the spring is a.

   (a) Express your answers in terms of the variable angle θ.

   (b) Express your answers in terms of the variable distance x.

   (c) Determine the equation of motion for large and small θ.

   (d) Determine the equation of motion for large and small x.

7. ★ Cushion mathematical model.

   Figure 12.59 illustrates a mathematical model for cushion suspension. Such a model can be used to analyze the driver's seat, or a rubbery pad suspension.

   (a) Derive the equations of motion for the variables x and z and using y as a known input function.

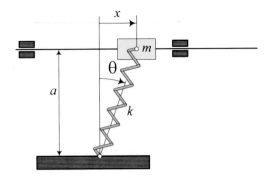

FIGURE 12.58. Spring connected rectilinear oscillator.

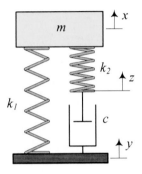

FIGURE 12.59. Mathematical model for cushion suspension.

(b) Eliminate z and derive a third-order equation for x.

8. Forced excitation and spring stiffness.

A forced excited mass-spring-damper system has

$$m = 200\,\text{kg} \qquad c = 1000\,\text{N}\,\text{s}/\,\text{m}$$

Determine the stiffness of the spring, k, such that the natural frequency of the system is one Hz. What would be the amplitude of displacement, velocity and acceleration of m if a force F is applied on the mass m.

$$F = 100\sin 10t$$

9. Forced excitation and system parameters.

A forced excited m-k-c system is under a force f.

$$F = 100\sin 10t$$

If the mass $m = 200\,\text{kg}$ should not have a dimensionless steady-state amplitude higher than two when it is excited at the natural frequency, determine c, k, φ_x, and F_T.

10. Base excited system and spring stiffness.

   A base excited m-k-c system has

   $$m = 200\,\text{kg} \qquad c = 1000\,\text{N s/m}$$

   Determine the stiffness of the spring, k, and the steady-state amplitude of m when the base is excited as

   $$y = 0.05\sin 2\pi t$$

   at the natural frequency of the system.

11. Base excited system and absolute acceleration.

   Assume a base excited m-k-c system is vibrating at the node of its absolute acceleration frequency response. If the base is excited by

   $$y = 0.05\sin 2\pi t$$

   determine ω_n, $\ddot{X}$, X.

12. Eccentric excitation and transmitted force.

   An engine with mass $m = 175\,\text{kg}$ and eccentricity $m_e e = 0.4{\times}0.1\,\text{kg m}$ is turning at $\omega_e = 4000\ rpm$.

   (a) Determine the steady-state amplitude of its vibration, if there are four engine mounts, each with $k = 10000\,\text{N/m}$ and $c = 100\,\text{N s/m}$.

   (b) Determine the transmitted force to the base.

13. ★ Eccentric base excitation and absolute displacement.

   An eccentric base excited system has $m = 3\,\text{kg}$, $m_b = 175\,\text{kg}$, $m_e e = 0.4 \times 0.1\,\text{kg m}$, and $\omega = 4000\ rpm$. If $Z/(e\varepsilon) = 2$ at $r = 1$, calculate X and Y.

14. Characteristic values and free vibrations.

   An m-k-c system has

   $$m = 250\,\text{kg} \qquad k = 8000\,\text{N/m} \qquad c = 1000\,\text{N s/m}$$

   Determine the characteristic values of the system and its free vibration response, for $x\,(0) = 1$, $\dot{x}\,(0) = 0$.

15. ★ Response to the step input.

    Consider an m-k-c system with

    $$m = 250\,\text{kg} \qquad k = 8000\,\text{N}/\,\text{m} \qquad c = 1000\,\text{N}\,\text{s}/\,\text{m}$$

    Determine the unit step input parameters, t_r, t_P, x_P, S, and t_s for 2% window.

16. Damping ratio determination.

    Consider a vibrating system that after $n = 100$ times oscillation, the peak amplitude drops by 2%. Determine the exact and approximate values of ξ.

17. The car lateral moment of inertia.

    Consider a car with the following characteristics:

b_1	746 mm
b_2	740 mm
mass	1245 kg
a_1	1100 mm
a_2	1323 mm
h	580 mm
I_x	335 kg m^2
I_y	1095 kg m^2

    Determine the period of oscillation when the car is on a solid steel platform with dimension 2000 mm × 3800 mm × 35 mm, and $h_1 = 3100\,\text{mm}$

    (a) laterally
    (b) longitudinally

18. ★ Optimal vibration absorber.

    Consider a primary system with $m_1 = 250\,\text{kg}$ and $k = 8000\,\text{N}/\,\text{m}$.

    Determine the best suspension for the secondary system with $m_2 = 1\,\text{kg}$ to act as a vibration absorber.

19. ★ Frequency response.

    Prove the following equations:

    $$G_2 = \frac{F_{T_B}}{kY}$$

    $$G_2 = \frac{F_{T_E}}{e\omega_n^2 m_e}$$

    $$G_2 = \frac{F_{T_R}}{e\omega_n^2 m_e}\left(1 + \frac{m_b}{m}\right)$$

13

Vehicle Vibrations

Vehicles are multiple-DOF systems as is shown in Figure 13.1. The vibration behavior of a vehicle, which is called *ride* or *ride comfort*, is highly dependent on the natural frequencies and mode shapes of the vehicle. In this chapter, we review and examine the applied methods of determining the equations of motion, natural frequencies, and mode shapes of different models of vehicles.

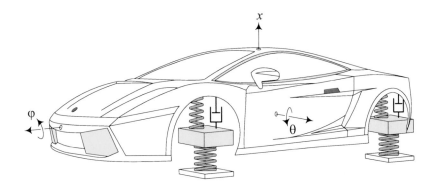

FIGURE 13.1. A full car vibrating model of a vehicle.

13.1 Lagrange Method and Dissipation Function

Lagrange equation,

$$\frac{d}{dt}\left(\frac{\partial K}{\partial \dot{q}_r}\right) - \frac{\partial K}{\partial q_r} = F_r \qquad r = 1, 2, \cdots n \qquad (13.1)$$

or,

$$\frac{d}{dt}\left(\frac{\partial \mathcal{L}}{\partial \dot{q}_r}\right) - \frac{\partial \mathcal{L}}{\partial q_r} = Q_r \qquad r = 1, 2, \cdots n \qquad (13.2)$$

as introduced in Equations (9.242) and (9.293), can both be applied to find the equations of motion for a vibrating system. However, for small and linear vibrations, we may use a simpler and more practical Lagrange equation

$$\frac{d}{dt}\left(\frac{\partial K}{\partial \dot{q}_r}\right) - \frac{\partial K}{\partial q_r} + \frac{\partial D}{\partial \dot{q}_r} + \frac{\partial V}{\partial q_r} = f_r \qquad r = 1, 2, \cdots n \qquad (13.3)$$

R.N. Jazar, *Vehicle Dynamics: Theory and Application*,
DOI 10.1007/978-1-4614-8544-5_13, © Springer Science+Business Media New York 2014

where K is the kinetic energy, V is the potential energy, and D is the *dissipation function* of the system

$$K = \frac{1}{2}\dot{\mathbf{x}}^T [m] \dot{\mathbf{x}} = \frac{1}{2}\sum_{i=1}^{n}\sum_{j=1}^{n} \dot{x}_i m_{ij} \dot{x}_j \qquad (13.4)$$

$$V = \frac{1}{2}\mathbf{x}^T [k] \mathbf{x} = \frac{1}{2}\sum_{i=1}^{n}\sum_{j=1}^{n} x_i k_{ij} x_j \qquad (13.5)$$

$$D = \frac{1}{2}\dot{\mathbf{x}}^T [c] \dot{\mathbf{x}} = \frac{1}{2}\sum_{i=1}^{n}\sum_{j=1}^{n} \dot{x}_i c_{ij} \dot{x}_j \qquad (13.6)$$

and f_r is the applied force on the mass m_r.

Proof. Consider a one-DOF mass-spring-damper vibrating system. When viscous damping is the only type of damping in the system, we may employ a function known as *Rayleigh dissipation function*

$$D = \frac{1}{2}c\dot{x}^2 \qquad (13.7)$$

to find the damping force f_c by differentiation.

$$f_c = -\frac{\partial D}{\partial \dot{x}} \qquad (13.8)$$

Remembering the elastic force f_k can be found from a potential energy V

$$f_k = -\frac{\partial V}{\partial x} \qquad (13.9)$$

then, the generalized force F can be separated to

$$F = f_c + f_k + f = -\frac{\partial D}{\partial \dot{x}} - \frac{\partial V}{\partial x} + f \qquad (13.10)$$

where f is the non-conservative applied force on mass m. Substituting (13.10) in (13.1)

$$\frac{d}{dt}\left(\frac{\partial K}{\partial \dot{x}}\right) - \frac{\partial K}{\partial x} = -\frac{\partial D}{\partial \dot{x}} - \frac{\partial V}{\partial x} + f \qquad (13.11)$$

gives us the Lagrange equation for a viscous damped vibrating system.

$$\frac{d}{dt}\left(\frac{\partial K}{\partial \dot{x}}\right) - \frac{\partial K}{\partial x} + \frac{\partial D}{\partial \dot{x}} + \frac{\partial V}{\partial x} = f \qquad (13.12)$$

When the vibrating system has n DOF, the kinetic energy K, potential energy V, and dissipating function D are as (13.4)-(13.6). Applying the Lagrange equation to the n-DOF system would result n second-order differential equations (13.3). ∎

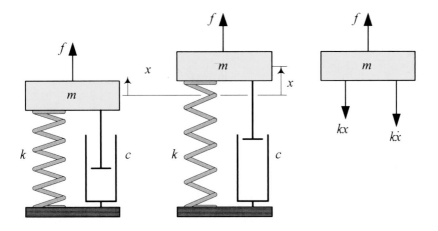

FIGURE 13.2. A one DOF forced mass-spring-damper system.

Example 522 *A one-DOF forced mass-spring-damper system.*

Figure 13.2 illustrates a single DOF mass-spring-damper system with an external force f applied on the mass m. The kinetic and potential energies of the system, when it is in motion, are

$$K = \frac{1}{2}m\dot{x}^2 \tag{13.13}$$

$$V = \frac{1}{2}kx^2 \tag{13.14}$$

and its dissipation function is

$$D = \frac{1}{2}c\dot{x}^2 \tag{13.15}$$

Substituting (13.13)-(13.15) in Lagrange equation (13.3), generates the following equation of motion:

$$\frac{d}{dt}(m\dot{x}) + c\dot{x} + kx = f \tag{13.16}$$

because

$$\frac{\partial K}{\partial \dot{x}} = m\dot{x} \qquad \frac{\partial K}{\partial x} = 0 \qquad \frac{\partial D}{\partial \dot{x}} = c\dot{x} \qquad \frac{\partial V}{\partial x} = kx \tag{13.17}$$

Example 523 *An undamped three-DOF system.*

Figure 13.3 illustrates an undamped three-DOF linear vibrating system. The kinetic and potential energies of the system are:

$$K = \frac{1}{2}m_1\dot{x}_1^2 + \frac{1}{2}m_2\dot{x}_2^2 + \frac{1}{2}m_3\dot{x}_3^2 \tag{13.18}$$

$$V = \frac{1}{2}k_1x_1^2 + \frac{1}{2}k_2(x_1 - x_2)^2 + \frac{1}{2}k_3(x_2 - x_3)^2 + \frac{1}{2}k_4x_3^2 \tag{13.19}$$

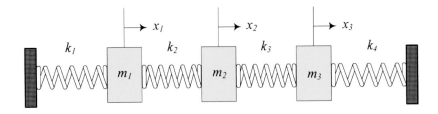

FIGURE 13.3. An undamped three-DOF system.

Because there is no damping in the system, we may find the Lagrangean $\mathcal{L}$

$$\mathcal{L} = K - V \tag{13.20}$$

and use Equation (13.2) with $Q_r = 0$

$$\frac{\partial \mathcal{L}}{\partial x_1} = -k_1 x_1 - k_2 (x_1 - x_2) \tag{13.21}$$

$$\frac{\partial \mathcal{L}}{\partial x_2} = k_2 (x_1 - x_2) - k_3 (x_2 - x_3) \tag{13.22}$$

$$\frac{\partial \mathcal{L}}{\partial x_3} = k_3 (x_2 - x_3) - k_4 x_3 \tag{13.23}$$

$$\frac{\partial \mathcal{L}}{\partial \dot{x}_1} = m_1 \dot{x}_1 \qquad \frac{\partial \mathcal{L}}{\partial \dot{x}_2} = m_2 \dot{x}_2 \qquad \frac{\partial \mathcal{L}}{\partial \dot{x}_3} = m_3 \dot{x}_3 \tag{13.24}$$

to find the equations of motion:

$$m_1 \ddot{x}_1 + k_1 x_1 + k_2 (x_1 - x_2) = 0 \tag{13.25}$$
$$m_2 \ddot{x}_2 - k_2 (x_1 - x_2) + k_3 (x_2 - x_3) = 0 \tag{13.26}$$
$$m_3 \ddot{x}_3 - k_3 (x_2 - x_3) + k_4 x_3 = 0 \tag{13.27}$$

These equations can be rewritten in matrix form for simpler calculation.

$$
\begin{bmatrix} m_1 & 0 & 0 \\ 0 & m_2 & 0 \\ 0 & 0 & m_3 \end{bmatrix}
\begin{bmatrix} \ddot{x}_1 \\ \ddot{x}_2 \\ \ddot{x}_3 \end{bmatrix}
$$
$$
+ \begin{bmatrix} k_1 + k_2 & -k_2 & 0 \\ -k_2 & k_2 + k_3 & -k_3 \\ 0 & -k_3 & k_3 + k_4 \end{bmatrix}
\begin{bmatrix} x_1 \\ x_2 \\ x_3 \end{bmatrix} = 0 \tag{13.28}
$$

Example 524 *An eccentric excited one-DOF system.*

   An eccentric excited one-DOF system is shown in Figure 12.31 with mass m supported by a suspension made up of a spring k and a damper c. There is also a mass m_e at a distance e that is rotating with an angular velocity ω. We may find the equation of motion by applying the Lagrange method.

The kinetic energy of the system is

$$K = \frac{1}{2}(m - m_e)\,\dot{x}^2 + \frac{1}{2}m_e\,(\dot{x} + e\omega\cos\omega t)^2 + \frac{1}{2}m_e\,(-e\omega\sin\omega t)^2 \tag{13.29}$$

because the velocity of the main vibrating mass $m - m_e$ is $\dot{x}$, and the velocity of the eccentric mass m_e has two components $\dot{x} + e\omega\cos\omega t$ and $-e\omega\sin\omega t$. The potential energy and dissipation function of the system are:

$$V = \frac{1}{2}kx^2 \qquad D = \frac{1}{2}c\dot{x}^2 \tag{13.30}$$

Applying the Lagrange equation (13.3),

$$\frac{\partial K}{\partial \dot{x}} = m\dot{x} + m_e e\omega\cos\omega t \tag{13.31}$$

$$\frac{d}{dt}\left(\frac{\partial K}{\partial \dot{x}}\right) = m\ddot{x} - m_e e\omega^2\sin\omega t \tag{13.32}$$

$$\frac{\partial D}{\partial \dot{x}} = c\dot{x} \tag{13.33}$$

$$\frac{\partial V}{\partial x} = kx \tag{13.34}$$

provides us with the equation of motion

$$m\ddot{x} + c\dot{x} + kx = m_e e\omega^2\sin\omega t \tag{13.35}$$

that is the same as Equation (12.200).

Example 525 *An eccentric base excited vibrating system.*

Figure 12.35 illustrates a one DOF eccentric base excited vibrating system. A mass m is mounted on an eccentric excited base by a spring k and a damper c. The base has a mass m_b with an attached unbalance mass m_e at a distance e. The mass m_e is rotating with an angular velocity ω.

We may derive the equation of motion of the system by applying Lagrange method. The required functions are:

$$\begin{aligned}
K &= \frac{1}{2}m\dot{x}^2 + \frac{1}{2}(m_b - m_e)\,\dot{y}^2 \\
&\quad + \frac{1}{2}m_e\,(\dot{y} - e\omega\cos\omega t)^2 + \frac{1}{2}m_e\,(e\omega\sin\omega t)^2 \tag{13.36}
\end{aligned}$$

$$V = \frac{1}{2}k(x - y)^2 \tag{13.37}$$

$$D = \frac{1}{2}c(\dot{x} - \dot{y})^2 \tag{13.38}$$

Applying the Lagrange method (13.3), provides us with

$$m\ddot{x} + c(\dot{x} - \dot{y}) + k(x - y) = 0 \tag{13.39}$$

$$m_b\ddot{y} + m_e e\omega^2\sin\omega t - c(\dot{x} - \dot{y}) - k(x - y) = 0 \tag{13.40}$$

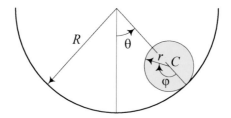

FIGURE 13.4. A uniform disc, rolling in a circular path.

because

$$\frac{\partial K}{\partial \dot{x}} = m\dot{x} \qquad (13.41)$$

$$\frac{d}{dt}\left(\frac{\partial K}{\partial \dot{x}}\right) = m\ddot{x} \qquad (13.42)$$

$$\frac{\partial D}{\partial \dot{x}} = c\left(\dot{x} - \dot{y}\right) \qquad (13.43)$$

$$\frac{\partial V}{\partial x} = k\left(x - y\right) \qquad (13.44)$$

$$\frac{\partial K}{\partial \dot{y}} = m_b \dot{y} - m_e e\omega \cos \omega t \qquad (13.45)$$

$$\frac{d}{dt}\left(\frac{\partial K}{\partial \dot{y}}\right) = m_b \ddot{y} + m_e e\omega^2 \sin \omega t \qquad (13.46)$$

$$\frac{\partial D}{\partial \dot{y}} = -c\left(\dot{x} - \dot{y}\right) \qquad (13.47)$$

$$\frac{\partial V}{\partial y} = -k\left(x - y\right) \qquad (13.48)$$

Using $z = x - y$, we may combine Equations (13.39) and (13.40) to find the equation of relative motion

$$\frac{mm_b}{m_b + m}\ddot{z} + c\dot{z} + kz = \frac{mm_e}{m_b + m}e\omega^2 \sin \omega t \qquad (13.49)$$

that is equal to

$$\ddot{z} + 2\xi\omega_n \dot{z} + \omega_n^2 z = \varepsilon e\omega^2 \sin \omega t \qquad (13.50)$$

$$\varepsilon = \frac{m_e}{m_b} \qquad (13.51)$$

Example 526 ★ *A rolling disc in a circular path.*

Figure 13.4 illustrates a uniform disc with mass m and radius r. The disc is rolling without slip in a circular path with radius R. The disc may have a free oscillation around $\theta = 0$.

When the oscillation is very small, we may substitute the oscillating disc with an equivalent mass-spring system. To find the equation of motion, we employ the Lagrange method. The energies of the system are

$$K = \frac{1}{2}mv_C^2 + \frac{1}{2}I_c\omega^2$$
$$= \frac{1}{2}m(R-r)^2\dot{\theta}^2 + \frac{1}{2}\left(\frac{1}{2}mr^2\right)\left(\dot{\varphi}-\dot{\theta}\right)^2 \tag{13.52}$$
$$V = -mg(R-r)\cos\theta \tag{13.53}$$

When there is no slip, there is a constraint between θ and φ

$$R\theta = r\varphi \tag{13.54}$$

which can be used to eliminate φ from K.

$$K = \frac{3}{4}m(R-r)^2\dot{\theta}^2 \tag{13.55}$$

Based on the following partial derivatives:

$$\frac{d}{dt}\left(\frac{\partial\mathcal{L}}{\partial\dot{\theta}}\right) = \frac{3}{2}m(R-r)^2\ddot{\theta} \tag{13.56}$$
$$\frac{\partial\mathcal{L}}{\partial\theta} = -mg(R-r)\sin\theta \tag{13.57}$$

we find the equation of motion for the oscillating disc.

$$\frac{3}{2}(R-r)\ddot{\theta} + g\sin\theta = 0 \tag{13.58}$$

When θ is very small, this equation is equivalent to a mass-spring system with $m_{eq} = 3(R-r)$ and $k_{eq} = 2g$.

Example 527 ★ *A double pendulum.*
Figure 13.5 illustrates a double pendulum made by a series of two pendulums. There are two massless rods with lengths l_1 and l_2, and two point masses m_1 and m_2. The variables θ_1 and θ_2 can be used as the generalized coordinates to express the system configuration. To calculate the Lagrangean of the system and find the equations of motion, we start by defining the global position of the masses.

$$x_1 = l_1\sin\theta_1 \tag{13.59}$$
$$y_1 = -l_1\cos\theta_1 \tag{13.60}$$
$$x_2 = l_1\sin\theta_1 + l_2\sin\theta_2 \tag{13.61}$$
$$y_2 = -l_1\cos\theta_1 - l_2\cos\theta_2 \tag{13.62}$$

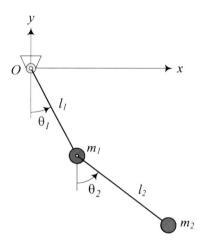

FIGURE 13.5. A double pendulum.

Time derivatives of the coordinates are

$$
\begin{aligned}
\dot{x}_1 &= l_1\dot{\theta}_1\cos\theta_1 & (13.63)\\
\dot{y}_1 &= l_1\dot{\theta}_1\sin\theta_1 & (13.64)\\
\dot{x}_2 &= l_1\dot{\theta}_1\cos\theta_1 + l_2\dot{\theta}_2\cos\theta_2 & (13.65)\\
\dot{y}_2 &= l_1\dot{\theta}_1\sin\theta_1 + l_2\dot{\theta}_2\sin\theta_2 & (13.66)
\end{aligned}
$$

and therefore, the squares of the masses' velocities are

$$
\begin{aligned}
v_1^2 &= \dot{x}_1^2 + \dot{y}_1^2 = l_1^2\dot{\theta}_1^2 & (13.67)\\
v_2^2 &= \dot{x}_2^2 + \dot{y}_2^2 = l_1^2\dot{\theta}_1^2 + l_2^2\dot{\theta}_2^2 + 2l_1l_2\dot{\theta}_1\dot{\theta}_2\cos(\theta_1 - \theta_2) & (13.68)
\end{aligned}
$$

The kinetic energy of the pendulum is then equal to

$$
\begin{aligned}
K &= \frac{1}{2}m_1v_1^2 + \frac{1}{2}m_2v_2^2\\
&= \frac{1}{2}m_1l_1^2\dot{\theta}_1^2 + \frac{1}{2}m_2\left(l_1^2\dot{\theta}_1^2 + l_2^2\dot{\theta}_2^2 + 2l_1l_2\dot{\theta}_1\dot{\theta}_2\cos(\theta_1 - \theta_2)\right) \quad (13.69)
\end{aligned}
$$

The potential energy of the pendulum is equal to sum of the potentials of each mass.

$$
\begin{aligned}
V &= m_1gy_1 + m_2gy_2\\
&= -m_1gl_1\cos\theta_1 - m_2g\left(l_1\cos\theta_1 + l_2\cos\theta_2\right) \quad (13.70)
\end{aligned}
$$

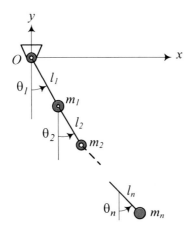

FIGURE 13.6. A chain pendulum.

The kinetic and potential energies make the following Lagrangean:

$$\begin{aligned}
\mathcal{L} &= K - V \\
&= \frac{1}{2}m_1 l_1^2 \dot{\theta}_1^2 + \frac{1}{2}m_2 \left(l_1^2 \dot{\theta}_1^2 + l_2^2 \dot{\theta}_2^2 + 2l_1 l_2 \dot{\theta}_1 \dot{\theta}_2 \cos(\theta_1 - \theta_2) \right) \\
&\quad + m_1 g l_1 \cos\theta_1 + m_2 g \left(l_1 \cos\theta_1 + l_2 \cos\theta_2 \right)
\end{aligned} \tag{13.71}$$

Employing Lagrange method (13.2) we find the following equations of motion:

$$\begin{aligned}
\frac{d}{dt}\left(\frac{\partial \mathcal{L}}{\partial \dot{\theta}_1} \right) - \frac{\partial \mathcal{L}}{\partial \theta_1} &= (m_1 + m_2) l_1^2 \ddot{\theta}_1 + m_2 l_1 l_2 \ddot{\theta}_2 \cos(\theta_1 - \theta_2) \\
&\quad + m_2 l_1 l_2 \dot{\theta}_2^2 \sin(\theta_1 - \theta_2) + (m_1 + m_2) l_1 g \sin\theta_1 \\
&= 0
\end{aligned} \tag{13.72}$$

$$\begin{aligned}
\frac{d}{dt}\left(\frac{\partial \mathcal{L}}{\partial \dot{\theta}_2} \right) - \frac{\partial \mathcal{L}}{\partial \theta_2} &= m_2 l_2^2 \ddot{\theta}_2 + m_2 l_1 l_2 \ddot{\theta}_1 \cos(\theta_1 - \theta_2) \\
&\quad - m_2 l_1 l_2 \dot{\theta}_1^2 \sin(\theta_1 - \theta_2) + m_2 l_2 g \sin\theta_2 \\
&= 0
\end{aligned} \tag{13.73}$$

Example 528 ★ *Chain pendulum.*

Consider an n-chain pendulum as shown in Figure 13.6. Each pendulum has a massless length l_i with a concentrated point mass m_i, and a generalized angular coordinate θ_i measured from the vertical direction.

The x_i and y_i components of the mass m_i are

$$x_i = \sum_{j=1}^{i} l_j \sin\theta_j \qquad y_i = -\sum_{j=1}^{i} l_j \cos\theta_j \tag{13.74}$$

We find their time derivatives

$$\dot{x}_i = \sum_{j=1}^{i} l_j \dot{\theta}_j \cos \theta_j \qquad \dot{y}_i = \sum_{j=1}^{i} l_j \dot{\theta}_j \sin \theta_j \qquad (13.75)$$

and the square of $\dot{x}_i$ and $\dot{y}_i$

$$\begin{aligned}
\dot{x}_i^2 &= \left(\sum_{j=1}^{i} l_j \dot{\theta}_j \cos \theta_j \right) \left(\sum_{k=1}^{i} l_k \dot{\theta}_k \cos \theta_k \right) \\
&= \sum_{j=1}^{n} \sum_{k=1}^{n} l_j l_k \dot{\theta}_j \dot{\theta}_k \cos \theta_j \cos \theta_k \qquad (13.76)
\end{aligned}$$

$$\begin{aligned}
\dot{y}_i^2 &= \left(\sum_{j=1}^{i} l_j \dot{\theta}_j \sin \theta_j \right) \left(\sum_{k=1}^{i} l_k \dot{\theta}_k \sin \theta_k \right) \\
&= \sum_{j=1}^{i} \sum_{k=1}^{i} l_j l_k \dot{\theta}_j \dot{\theta}_k \sin \theta_j \sin \theta_k \qquad (13.77)
\end{aligned}$$

to calculate the velocity v_i of the mass m_i.

$$\begin{aligned}
v_i^2 &= \dot{x}_i^2 + \dot{y}_i^2 \\
&= \sum_{j=1}^{i} \sum_{k=1}^{i} l_j l_k \dot{\theta}_j \dot{\theta}_k \left(\cos \theta_j \cos \theta_k + \sin \theta_j \sin \theta_k \right) \\
&= \sum_{j=1}^{i} \sum_{k=1}^{i} l_j l_k \dot{\theta}_j \dot{\theta}_k \cos \left(\theta_j - \theta_k \right) \\
&= \sum_{r=1}^{i} l_r^2 \dot{\theta}_r^2 + 2 \sum_{j=1}^{i} \sum_{k=j+1}^{i} l_j l_k \dot{\theta}_j \dot{\theta}_k \cos \left(\theta_j - \theta_k \right) \qquad (13.78)
\end{aligned}$$

Now, we may calculate the kinetic energy, K, of the chain.

$$\begin{aligned}
K &= \frac{1}{2} \sum_{i=1}^{n} m_i v_i^2 \\
&= \frac{1}{2} \sum_{i=1}^{n} m_i \left(\sum_{r=1}^{i} l_r^2 \dot{\theta}_r^2 + 2 \sum_{j=1}^{i} \sum_{k=j+1}^{i} l_j l_k \dot{\theta}_j \dot{\theta}_k \cos \left(\theta_j - \theta_k \right) \right) \\
&= \frac{1}{2} \sum_{i=1}^{n} \sum_{r=1}^{i} m_i l_r^2 \dot{\theta}_r^2 + \sum_{i=1}^{n} \sum_{j=1}^{i} \sum_{k=j+1}^{i} m_i l_j l_k \dot{\theta}_j \dot{\theta}_k \cos \left(\theta_j - \theta_k \right) \qquad (13.79)
\end{aligned}$$

The potential energy of the ith pendulum is related to m_i

$$V_i = m_i g y_i = -m_i g \sum_{j=1}^{i} l_j \cos \theta_j \qquad (13.80)$$

and therefore, the potential energy of the chain is

$$V = \sum_{i=1}^{n} m_i g y_i = -\sum_{i=1}^{n} \sum_{j=1}^{i} m_i g l_j \cos \theta_j \qquad (13.81)$$

To find the equations of motion for the chain, we may use the Lagrangean $\mathcal{L}$

$$\mathcal{L} = K - V \qquad (13.82)$$

and apply the Lagrange equation

$$\frac{d}{dt}\left(\frac{\partial \mathcal{L}}{\partial \dot{q}_s}\right) - \frac{\partial \mathcal{L}}{\partial q_s} = 0 \qquad s = 1, 2, \cdots n \qquad (13.83)$$

or

$$\frac{d}{dt}\left(\frac{\partial K}{\partial \dot{q}_s}\right) - \frac{\partial K}{\partial q_s} + \frac{\partial V}{\partial q_s} = 0 \qquad s = 1, 2, \cdots n. \qquad (13.84)$$

13.2 ★ Quadratures

If $[m]$ is an $n \times n$ square matrix and $\mathbf{x}$ is an $n \times 1$ vector, then S is a scalar function called *quadrature* and is defined by

$$S = \mathbf{x}^T [m] \mathbf{x} \qquad (13.85)$$

The derivative of the quadrature S with respect to the vector $\mathbf{x}$ is

$$\frac{\partial S}{\partial \mathbf{x}} = \left([m] + [m]^T\right) \mathbf{x} \qquad (13.86)$$

Kinetic energy K, potential energy V, and dissipation function D, are quadratures

$$K = \frac{1}{2}\dot{\mathbf{x}}^T [m] \dot{\mathbf{x}} \qquad (13.87)$$

$$V = \frac{1}{2}\mathbf{x}^T [k] \mathbf{x} \qquad (13.88)$$

$$D = \frac{1}{2}\dot{\mathbf{x}}^T [c] \dot{\mathbf{x}} \qquad (13.89)$$

and therefore,

$$\frac{\partial K}{\partial \dot{\mathbf{x}}} = \frac{1}{2} \left([m] + [m]^T \right) \dot{\mathbf{x}} \tag{13.90}$$

$$\frac{\partial V}{\partial \mathbf{x}} = \frac{1}{2} \left([k] + [k]^T \right) \mathbf{x} \tag{13.91}$$

$$\frac{\partial D}{\partial \dot{\mathbf{x}}} = \frac{1}{2} \left([c] + [c]^T \right) \dot{\mathbf{x}} \tag{13.92}$$

Employing quadrature derivatives and the Lagrange method,

$$\frac{d}{dt} \frac{\partial K}{\partial \dot{\mathbf{x}}} + \frac{\partial K}{\partial \mathbf{x}} + \frac{\partial D}{\partial \dot{\mathbf{x}}} + \frac{\partial V}{\partial \mathbf{x}} = \mathbf{F} \tag{13.93}$$

the equation of motion for a linear n degree-of-freedom vibrating system becomes

$$[m]\,\ddot{\mathbf{x}} + [c]\,\dot{\mathbf{x}} + [k]\,\mathbf{x} = \mathbf{F} \tag{13.94}$$

where, $[m]$, $[c]$, $[k]$ are symmetric matrices.

$$[m] = \frac{1}{2} \left([m] + [m]^T \right) \tag{13.95}$$

$$[c] = \frac{1}{2} \left([c] + [c]^T \right) \tag{13.96}$$

$$[k] = \frac{1}{2} \left([k] + [k]^T \right) \tag{13.97}$$

Quadratures are also called *Hermitian form*.

Proof. Let us define a general asymmetric quadrature as

$$S = \mathbf{x}^T [a]\, \mathbf{y} = \sum_i \sum_j x_i a_{ij} y_j \tag{13.98}$$

If the quadrature is symmetric, then $\mathbf{x} = \mathbf{y}$ and

$$S = \mathbf{x}^T [a]\, \mathbf{x} = \sum_i \sum_j x_i a_{ij} x_j \tag{13.99}$$

The vectors $\mathbf{x}$ and $\mathbf{y}$ may be functions of n generalized coordinates q_i and time t.

$$\mathbf{x} = \mathbf{x}\,(q_1, q_2, \cdots, q_n, t) \tag{13.100}$$

$$\mathbf{y} = \mathbf{y}\,(q_1, q_2, \cdots, q_n, t) \tag{13.101}$$

$$\mathbf{q} = \begin{bmatrix} q_1 & q_2 & \cdots & q_n \end{bmatrix}^T \tag{13.102}$$

The derivative of $\mathbf{x}$ with respect to $\mathbf{q}$ is a square matrix

$$\frac{\partial \mathbf{x}}{\partial \mathbf{q}} = \begin{bmatrix} \dfrac{\partial x_1}{\partial q_1} & \dfrac{\partial x_2}{\partial q_1} & \cdots & \dfrac{\partial x_n}{\partial q_1} \\ \dfrac{\partial x_1}{\partial q_2} & \dfrac{\partial x_2}{\partial q_2} & \cdots & \cdots \\ \cdots & \cdots & \cdots & \cdots \\ \dfrac{\partial x_1}{\partial q_n} & \cdots & \cdots & \dfrac{\partial x_n}{\partial q_n} \end{bmatrix} \tag{13.103}$$

that can also be expressed by

$$\frac{\partial \mathbf{x}}{\partial \mathbf{q}} = \begin{bmatrix} \dfrac{\partial \mathbf{x}}{\partial q_1} \\ \dfrac{\partial \mathbf{x}}{\partial q_2} \\ \cdots \\ \dfrac{\partial \mathbf{x}}{\partial q_n} \end{bmatrix} \tag{13.104}$$

or

$$\frac{\partial \mathbf{x}}{\partial \mathbf{q}} = \begin{bmatrix} \dfrac{\partial x_1}{\partial \mathbf{q}} & \dfrac{\partial x_2}{\partial \mathbf{q}} & \cdots & \dfrac{\partial x_n}{\partial \mathbf{q}} \end{bmatrix} \tag{13.105}$$

Now a derivative of S with respect to an element of q_k is

$$\begin{aligned} \frac{\partial S}{\partial q_k} &= \frac{\partial}{\partial q_k} \sum_i \sum_j x_i a_{ij} y_j \\ &= \sum_i \sum_j \frac{\partial x_i}{\partial q_k} a_{ij} y_j + \sum_i \sum_j x_i a_{ij} \frac{\partial y_j}{\partial q_k} \\ &= \sum_j \sum_i \frac{\partial x_i}{\partial q_k} a_{ij} y_j + \sum_i \sum_j \frac{\partial y_j}{\partial q_k} a_{ij} x_i \\ &= \sum_j \sum_i \frac{\partial x_i}{\partial q_k} a_{ij} y_j + \sum_j \sum_i \frac{\partial y_i}{\partial q_k} a_{ji} x_j \end{aligned} \tag{13.106}$$

and hence, the derivative of S with respect to $\mathbf{q}$ is

$$\frac{\partial S}{\partial \mathbf{q}} = \frac{\partial \mathbf{x}}{\partial \mathbf{q}} [a]\, \mathbf{y} + \frac{\partial \mathbf{y}}{\partial \mathbf{q}} [a]^T \mathbf{x}. \tag{13.107}$$

If S is a symmetric quadrature then,

$$\frac{\partial S}{\partial \mathbf{q}} = \frac{\partial}{\partial \mathbf{q}} \left(\mathbf{x}^T [a]\, \mathbf{x} \right) = \frac{\partial \mathbf{x}}{\partial \mathbf{q}} [a]\, \mathbf{x} + \frac{\partial \mathbf{x}}{\partial \mathbf{q}} [a]^T \mathbf{x} \tag{13.108}$$

and if $\mathbf{q} = \mathbf{x}$, then the derivative of a symmetric S with respect to $\mathbf{x}$ is

$$\begin{aligned} \frac{\partial S}{\partial \mathbf{x}} &= \frac{\partial}{\partial \mathbf{x}} \left(\mathbf{x}^T [a]\, \mathbf{x} \right) = \frac{\partial \mathbf{x}}{\partial \mathbf{x}} [a]\, \mathbf{x} + \frac{\partial \mathbf{x}}{\partial \mathbf{x}} [a]^T \mathbf{x} \\ &= [a]\, \mathbf{x} + [a]^T \mathbf{x} = \left([a] + [a]^T \right) \mathbf{x} \end{aligned} \tag{13.109}$$

If $[a]$ is a symmetric matrix, then

$$[a] + [a]^T = 2[a] \tag{13.110}$$

however, if $[a]$ is not a symmetric matrix, then $[\underline{a}] = [a] + [a]^T$ is a symmetric matrix because

$$\underline{a}_{ij} = a_{ij} + a_{ji} = a_{ji} + a_{ij} = \underline{a}_{ji} \tag{13.111}$$

and therefore,

$$[\underline{a}] = [\underline{a}]^T \tag{13.112}$$

Kinetic energy K, potential energy V, and dissipation function D can be expressed by quadratures.

$$K = \frac{1}{2}\dot{\mathbf{x}}^T [m] \dot{\mathbf{x}} \tag{13.113}$$

$$V = \frac{1}{2}\mathbf{x}^T [k] \mathbf{x} \tag{13.114}$$

$$D = \frac{1}{2}\dot{\mathbf{x}}^T [c] \dot{\mathbf{x}} \tag{13.115}$$

Substituting K, V, and D in the Lagrange equation provides the equations of motion:

$$
\begin{aligned}
\mathbf{F} &= \frac{d}{dt}\frac{\partial K}{\partial \dot{\mathbf{x}}} + \frac{\partial K}{\partial \mathbf{x}} + \frac{\partial D}{\partial \dot{\mathbf{x}}} + \frac{\partial V}{\partial \mathbf{x}} \\
&= \frac{1}{2}\frac{d}{dt}\frac{\partial}{\partial \dot{\mathbf{x}}} \left(\dot{\mathbf{x}}^T [m] \dot{\mathbf{x}}\right) + \frac{1}{2}\frac{\partial}{\partial \dot{\mathbf{x}}} \left(\dot{\mathbf{x}}^T [c] \dot{\mathbf{x}}\right) + \frac{1}{2}\frac{\partial}{\partial \mathbf{x}} \left(\mathbf{x}^T [k] \mathbf{x}\right) \\
&= \frac{1}{2}\left[\frac{d}{dt}\left(\left([m] + [m]^T\right)\dot{\mathbf{x}}\right) + \left([c] + [c]^T\right)\dot{\mathbf{x}} + \left([k] + [k]^T\right)\mathbf{x}\right] \\
&= \frac{1}{2}\left([m] + [m]^T\right)\ddot{\mathbf{x}} + \frac{1}{2}\left([c] + [c]^T\right)\dot{\mathbf{x}} + \frac{1}{2}\left([k] + [k]^T\right)\mathbf{x} \\
&= [\underline{m}]\ddot{\mathbf{x}} + [\underline{c}]\dot{\mathbf{x}} + [\underline{k}]\mathbf{x} \tag{13.116}
\end{aligned}
$$

where

$$[\underline{m}] = \frac{1}{2}\left([m] + [m]^T\right) \tag{13.117}$$

$$[\underline{c}] = \frac{1}{2}\left([k] + [k]^T\right) \tag{13.118}$$

$$[\underline{k}] = \frac{1}{2}\left([c] + [c]^T\right) \tag{13.119}$$

From now on, we assume that every equation of motion is found from the Lagrange method to have symmetric coefficient matrices. Hence, we show the equations of motion as,

$$[m]\ddot{\mathbf{x}} + [c]\dot{\mathbf{x}} + [k]\mathbf{x} = \mathbf{F} \tag{13.120}$$

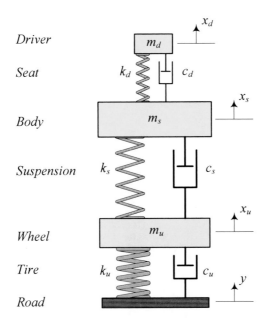

FIGURE 13.7. A quarter car model with driver.

and use $[m]$, $[c]$, $[k]$ as a substitute for $[\underline{m}]$, $[\underline{c}]$, $[\underline{k}]$

$$[m] \equiv [\underline{m}] \tag{13.121}$$

$$[c] \equiv [\underline{c}] \tag{13.122}$$

$$[k] \equiv [\underline{k}] \tag{13.123}$$

■

Example 529 ★ *A quarter car model with driver.*

Figure 13.7 illustrates a quarter car model plus a driver, which is modeled by a mass m_d over a linear cushion above the sprung mass m_s.

Assuming

$$y = 0 \tag{13.124}$$

we find the free vibration equations of motion by the Lagrange method and quadrature derivative.

The kinetic energy K of the system can be expressed by

$$
\begin{aligned}
K &= \frac{1}{2} m_u \dot{x}_u^2 + \frac{1}{2} m_s \dot{x}_s^2 + \frac{1}{2} m_d \dot{x}_d^2 \\
&= \frac{1}{2} \begin{bmatrix} \dot{x}_u & \dot{x}_s & \dot{x}_d \end{bmatrix} \begin{bmatrix} m_u & 0 & 0 \\ 0 & m_s & 0 \\ 0 & 0 & m_d \end{bmatrix} \begin{bmatrix} \dot{x}_u \\ \dot{x}_s \\ \dot{x}_d \end{bmatrix} \\
&= \frac{1}{2} \dot{\mathbf{x}}^T [m] \dot{\mathbf{x}}
\end{aligned} \tag{13.125}
$$

and the potential energy V can be expressed as

$$
\begin{aligned}
V &= \frac{1}{2}k_u\left(x_u\right)^2 + \frac{1}{2}k_s\left(x_s - x_u\right)^2 + \frac{1}{2}k_d\left(x_d - x_s\right)^2 \\
&= \frac{1}{2}\begin{bmatrix} x_u & x_s & x_d \end{bmatrix}
\begin{bmatrix}
k_u + k_s & -k_s & 0 \\
-k_s & k_s + k_d & -k_d \\
0 & -k_d & k_d
\end{bmatrix}
\begin{bmatrix} x_u \\ x_s \\ x_d \end{bmatrix} \\
&= \frac{1}{2}\mathbf{x}^T\left[k\right]\mathbf{x}
\end{aligned}
\tag{13.126}
$$

Similarly, the dissipation function D is

$$
\begin{aligned}
D &= \frac{1}{2}c_u\left(\dot{x}_u\right)^2 + \frac{1}{2}c_s\left(\dot{x}_s - \dot{x}_u\right)^2 + \frac{1}{2}c_d\left(\dot{x}_d - \dot{x}_s\right)^2 \\
&= \frac{1}{2}\begin{bmatrix} \dot{x}_u & \dot{x}_s & \dot{x}_d \end{bmatrix}
\begin{bmatrix}
c_u + c_s & -c_s & 0 \\
-c_s & c_s + c_d & -c_d \\
0 & -c_d & c_d
\end{bmatrix}
\begin{bmatrix} \dot{x}_u \\ \dot{x}_s \\ \dot{x}_d \end{bmatrix} \\
&= \frac{1}{2}\dot{\mathbf{x}}^T\left[c\right]\dot{\mathbf{x}}
\end{aligned}
\tag{13.127}
$$

Employing the quadrature derivative method, we find derivatives of K, V, and D with respect to their variable vectors as:

$$
\begin{aligned}
\frac{\partial K}{\partial \dot{\mathbf{x}}} &= \frac{1}{2}\left([m] + [m]^T\right)\dot{\mathbf{x}} = \frac{1}{2}\left([m] + [m]^T\right)\begin{bmatrix} \dot{x}_u \\ \dot{x}_s \\ \dot{x}_d \end{bmatrix} \\
&= \begin{bmatrix}
m_u & 0 & 0 \\
0 & m_s & 0 \\
0 & 0 & m_d
\end{bmatrix}
\begin{bmatrix} \dot{x}_u \\ \dot{x}_s \\ \dot{x}_d \end{bmatrix}
\end{aligned}
\tag{13.128}
$$

$$
\begin{aligned}
\frac{\partial V}{\partial \mathbf{x}} &= \frac{1}{2}\left([k] + [k]^T\right)\mathbf{x} = \frac{1}{2}\left([k] + [k]^T\right)\begin{bmatrix} x_u \\ x_s \\ x_d \end{bmatrix} \\
&= \begin{bmatrix}
k_u + k_s & -k_s & 0 \\
-k_s & k_s + k_d & -k_d \\
0 & -k_d & k_d
\end{bmatrix}
\begin{bmatrix} x_u \\ x_s \\ x_d \end{bmatrix}
\end{aligned}
\tag{13.129}
$$

$$
\begin{aligned}
\frac{\partial D}{\partial \dot{\mathbf{x}}} &= \frac{1}{2}\left([c] + [c]^T\right)\dot{\mathbf{x}} = \frac{1}{2}\left([c] + [c]^T\right)\begin{bmatrix} \dot{x}_u \\ \dot{x}_s \\ \dot{x}_d \end{bmatrix} \\
&= \begin{bmatrix}
c_u + c_s & -c_s & 0 \\
-c_s & c_s + c_d & -c_d \\
0 & -c_d & c_d
\end{bmatrix}
\begin{bmatrix} \dot{x}_u \\ \dot{x}_s \\ \dot{x}_d \end{bmatrix}
\end{aligned}
\tag{13.130}
$$

Therefore, we find the system's free vibration equations of motion.

$$[m]\,\ddot{\mathbf{x}} + [c]\,\dot{\mathbf{x}} + [k]\,\mathbf{x} = 0 \qquad (13.131)$$

$$
\begin{bmatrix} m_u & 0 & 0 \\ 0 & m_s & 0 \\ 0 & 0 & m_d \end{bmatrix}
\begin{bmatrix} \ddot{x}_u \\ \ddot{x}_s \\ \ddot{x}_d \end{bmatrix}
+
\begin{bmatrix} c_u + c_s & -c_s & 0 \\ -c_s & c_s + c_d & -c_d \\ 0 & -c_d & c_d \end{bmatrix}
\begin{bmatrix} \dot{x}_u \\ \dot{x}_s \\ \dot{x}_d \end{bmatrix}
$$

$$
+
\begin{bmatrix} k_u + k_s & -k_s & 0 \\ -k_s & k_s + k_d & -k_d \\ 0 & -k_d & k_d \end{bmatrix}
\begin{bmatrix} x_u \\ x_s \\ x_d \end{bmatrix} = 0 \qquad (13.132)
$$

Example 530 ★ *Different $[m]$, $[c]$, and $[k]$ arrangements.*

Mass, damping, and stiffness matrices $[m]$, $[c]$, $[k]$ for a vibrating system may be arranged in different forms with the same overall kinetic energy K, potential energy V, and dissipation function D. As an example, the potential energy V for the quarter car model that is shown in Figure 13.7 may be expressed by different $[k]$.

$$V = \frac{1}{2}k_u\left(x_u\right)^2 + \frac{1}{2}k_s\left(x_s - x_u\right)^2 + \frac{1}{2}k_d\left(x_d - x_s\right)^2 \qquad (13.133)$$

$$V = \frac{1}{2}\mathbf{x}^T \begin{bmatrix} k_u + k_s & -k_s & 0 \\ -k_s & k_s + k_d & -k_d \\ 0 & -k_d & k_d \end{bmatrix}\mathbf{x} \qquad (13.134)$$

$$V = \frac{1}{2}\mathbf{x}^T \begin{bmatrix} k_u + k_s & -2k_s & 0 \\ 0 & k_s + k_d & -2k_d \\ 0 & 0 & k_d \end{bmatrix}\mathbf{x} \qquad (13.135)$$

$$V = \frac{1}{2}\mathbf{x}^T \begin{bmatrix} k_u + k_s & 0 & 0 \\ -2k_s & k_s + k_d & 0 \\ 0 & -2k_d & k_d \end{bmatrix}\mathbf{x} \qquad (13.136)$$

The matrices $[m]$, $[c]$, $[k]$, in K, D, V, may not be symmetric however, the matrices $[\underline{m}]$, $[\underline{c}]$, and $[\underline{k}]$ in $\partial K/\partial\dot{\mathbf{x}}$, $\partial D/\partial\dot{\mathbf{x}}$, $\partial V/\partial\mathbf{x}$ are symmetric. When a matrix $[a]$ is diagonal, it is symmetric and

$$[a] = [\underline{a}] \qquad (13.137)$$

A diagonal matrix cannot be written in different forms. The matrix $[m]$ in Example 529 is diagonal and hence, K has only one form (13.125).

Example 531 ★ *Positive definite matrix.*

*A matrix $[a]$ is called **positive definite** if $\mathbf{x}^T[a]\mathbf{x} > 0$ for all $\mathbf{x} \neq 0$. A matrix $[a]$ is called **positive semidefinite** if $\mathbf{x}^T[a]\mathbf{x} \geq 0$ for all $\mathbf{x}$. Kinetic energy is positive definite and it means we cannot have $K = 0$ unless $\dot{\mathbf{x}} = 0$. Potential energy is positive semidefinite and it means we have $V \geq 0$ as long as $\mathbf{x} > 0$, however, it is possible to have a especial $\mathbf{x}_0 > 0$ at which $V = 0$.*

13.3 Natural Frequencies and Mode Shapes

Unforced and undamped vibrations of a system is a basic response of the system which expresses its natural behavior. We call a system with no damping and no external excitation, a *free system*. A free system is governed by the following set of differential equations.

$$[m]\,\ddot{\mathbf{x}} + [k]\,\mathbf{x} = \mathbf{0} \tag{13.138}$$

The response of the free system is harmonic

$$\mathbf{x} = \sum_{i=1}^{n} \mathbf{u}_i \left(A_i \sin \omega_i t + B_i \cos \omega_i t \right) \qquad i = 1, 2, 3, \cdots n$$

$$= \sum_{i=1}^{n} C_i \mathbf{u}_i \sin \left(\omega_i t - \varphi_i \right) \qquad i = 1, 2, 3, \cdots n \tag{13.139}$$

where, ω_i are the *natural frequencies* and $\mathbf{u}_i$ are the *mode shapes* of the system.

The natural frequencies ω_i are solutions of the characteristic equation of the system

$$\det\left[[k] - \omega^2\,[m]\right] = 0 \tag{13.140}$$

and the mode shapes $\mathbf{u}_i$, corresponding to ω_i, are solutions of the following equation.

$$\left[[k] - \omega_i^2\,[m]\right]\mathbf{u}_i = 0 \tag{13.141}$$

The unknown coefficients A_i and B_i, or C_i and φ_i, must be determined from the initial conditions.

Proof. By eliminating the force and damping terms from the general equations of motion

$$[m]\,\ddot{\mathbf{x}} + [c]\,\dot{\mathbf{x}} + [k]\,\mathbf{x} = \mathbf{F} \tag{13.142}$$

we find the equations for free systems.

$$[m]\,\ddot{\mathbf{x}} + [k]\,\mathbf{x} = \mathbf{0} \tag{13.143}$$

Let us search for a possible solution of the following form

$$\mathbf{x} = \mathbf{u}\,q(t) \tag{13.144}$$
$$x_i = u_i\,q(t) \qquad i = 1, 2, 3, \cdots n \tag{13.145}$$

This solution implies that the amplitude ratio of two coordinates during motion does not depend on time. Substituting (13.144) in Equation (13.143)

$$[m]\,\mathbf{u}\,\ddot{q}(t) + [k]\,\mathbf{u}\,q(t) = \mathbf{0} \tag{13.146}$$

and separating the time dependent terms, yields:

$$-\frac{\ddot{q}(t)}{q(t)} = [[m]\,\mathbf{u}]^{-1}\,[[k]\,\mathbf{u}] = \frac{\sum_{j=1}^{n} k_{ij}u_j}{\sum_{j=1}^{n} m_{ij}u_j} \qquad i = 1, 2, 3, \cdots n \qquad (13.147)$$

Because the right hand side of this equation is time independent and the left hand side is independent of the index i, both sides must be equal to a constant. Let us assume the constant be a positive number ω^2. Hence, Equation (13.147) can be separated into two equations

$$\ddot{q}(t) + \omega^2 q(t) = 0 \qquad (13.148)$$

and

$$[[k] - \omega^2\,[m]]\,\mathbf{u} = 0 \qquad (13.149)$$

or

$$\sum_{j=1}^{n} \left(k_{ij} - \omega^2 m_{ij}\right) u_j = 0 \qquad i = 1, 2, 3, \cdots n. \qquad (13.150)$$

The solution of (13.148) is

$$q(t) = \sin\omega t + \cos\omega t = \sin\left(\omega t - \varphi\right) \qquad (13.151)$$

which shows that all the coordinates of the system, x_i, have harmonic motion with identical frequency ω and identical phase angle φ. The frequency ω can be determined from Equation (13.149) which is a set of homogeneous equations for the unknown $\mathbf{u}$.

The set of equations (13.149) has a solution $\mathbf{u} = \mathbf{0}$, which is the *rest position* of the system and shows no motion. This solution is called *trivial solution* and is unimportant. To have a *nontrivial solution*, the determinant of the coefficient matrix must be zero.

$$\det\left[[k] - \omega^2\,[m]\right] = 0 \qquad (13.152)$$

Determining the constant ω, such that the set of equations (13.149) provide a nontrivial solution, is called *eigenvalue problem*. Expanding the determinant (13.152) provides an algebraic equation that is called the *characteristic equation*. The characteristic equation is an nth order equation in ω^2, and provides n natural frequencies ω_i. The natural frequencies ω_i can be set in the following order.

$$\omega_1 \leq \omega_2 \leq \omega_3 \leq \cdots \leq \omega_n \qquad (13.153)$$

Having n values for ω indicates that the solution (13.151) is possible with n different frequencies ω_i, $i = 1, 2, 3, \cdots n$.

We may multiply the Equation (13.143) by $[m]^{-1}$

$$\ddot{\mathbf{x}} + [m]^{-1}\,[k]\,\mathbf{x} = \mathbf{0} \qquad (13.154)$$

and find the the characteristic equation (13.152) as

$$\det\left[[A] - \lambda \mathbf{I}\right] = 0 \tag{13.155}$$

where

$$[A] = [m]^{-1}[k] \tag{13.156}$$

So, determination of the natural frequencies ω_i would be equivalent to determining the eigenvalues λ_i of the matrix $[A] = [m]^{-1}[k]$.

$$\lambda_i = \omega_i^2 \tag{13.157}$$

Determining the vectors $\mathbf{u}_i$ to satisfy Equation (13.149) is called the *eigenvector problem*. To determine $\mathbf{u}_i$, we may solve Equation (13.149) for every ω_i

$$\left[[k] - \omega_i^2 [m]\right]\mathbf{u}_i = 0 \tag{13.158}$$

and find n different $\mathbf{u}_i$. In vibrations and vehicle dynamics, the eigenvector $\mathbf{u}_i$ corresponding to the eigenvalue ω_i is called the *mode shape*.

Alternatively, we may find the eigenvectors of matrix $[A] = [m]^{-1}[k]$

$$\left[[A] - \lambda_i \mathbf{I}\right]\mathbf{u}_i = 0 \tag{13.159}$$

instead of finding the mode shapes from ((13.158)).

Equations (13.158) are homogeneous so, if $\mathbf{u}_i$ is a solution, then $a\mathbf{u}_i$ is also a solution. Hence, the eigenvectors are not unique and may be expressed with any length. However, the ratio of any two elements of an eigenvector is unique and therefore, $\mathbf{u}_i$ has a unique shape. If one of the elements of $\mathbf{u}_i$ is assigned, the remaining $n - 1$ elements are uniquely determined. The shape of an eigenvector indicates the relative amplitudes of the coordinates of the system in vibration.

Because the length of an eigenvector is not uniquely defined, there are many options to express $\mathbf{u}_i$. The most common expressions are:

1. normalization
2. normal form
3. high-unit
4. first-unit
5. last-unit

In the *normalization* expression, we may adjust the length of $\mathbf{u}_i$ such that

$$\mathbf{u}_i^T [m] \mathbf{u}_i = 1 \tag{13.160}$$

or

$$\mathbf{u}_i^T [k] \mathbf{u}_i = 1 \tag{13.161}$$

and call $\mathbf{u}_i$ a *normal mode* with respect to $[m]$ or $[k]$ respectively.

In the *normal form* expression, we adjust $\mathbf{u}_i$ such that its length has a unity value.

In the *high-unit* expression, we adjust the length of $\mathbf{u}_i$ such that the largest element has a unity value.

In the *first-unit* expression, we adjust the length of $\mathbf{u}_i$ such that the first element has a unity value.

In the *last-unit* expression, we adjust the length of $\mathbf{u}_i$ such that the last element has a unity value. ■

Example 532 *Eigenvalues and eigenvectors of a 2 × 2 matrix.*
Consider a 2 × 2 matrix is given as

$$[A] = \begin{bmatrix} 5 & 3 \\ 3 & 6 \end{bmatrix} \tag{13.162}$$

To find the eigenvalues λ_i of $[A]$, we find the characteristic equation of the matrix by subtracting an unknown λ from the main diagonal, and taking the determinant.

$$
\begin{aligned}
\det[[A] - \lambda \mathbf{I}] &= \det\left[\begin{bmatrix} 5 & 3 \\ 3 & 6 \end{bmatrix} - \lambda \begin{bmatrix} 1 & 0 \\ 0 & 1 \end{bmatrix}\right] \\
&= \det\begin{bmatrix} 5-\lambda & 3 \\ 3 & 6-\lambda \end{bmatrix} \\
&= \lambda^2 - 11\lambda + 21
\end{aligned}
\tag{13.163}
$$

The solution of the characteristic equation (13.163) are

$$\lambda_1 = 8.5414 \qquad \lambda_2 = 2.4586 \tag{13.164}$$

To find the corresponding eigenvectors $\mathbf{u}_1$ and $\mathbf{u}_2$ we solve the following equations.

$$[[A] - \lambda_1 \mathbf{I}] \mathbf{u}_1 = 0 \qquad [[A] - \lambda_2 \mathbf{I}] \mathbf{u}_2 = 0 \tag{13.165}$$

Let's denote the eigenvectors by

$$\mathbf{u}_1 = \begin{bmatrix} u_{11} \\ u_{12} \end{bmatrix} \qquad \mathbf{u}_2 = \begin{bmatrix} u_{21} \\ u_{22} \end{bmatrix} \tag{13.166}$$

therefore,

$$
\begin{aligned}
[[A] - \lambda_1 \mathbf{I}] \mathbf{u}_1 &= \left[\begin{bmatrix} 5 & 3 \\ 3 & 6 \end{bmatrix} - 8.5414 \begin{bmatrix} 1 & 0 \\ 0 & 1 \end{bmatrix}\right]\begin{bmatrix} u_{11} \\ u_{12} \end{bmatrix} \\
&= \begin{bmatrix} 3u_{12} - 3.5414u_{11} \\ 3u_{11} - 2.5414u_{12} \end{bmatrix} = 0
\end{aligned}
\tag{13.167}
$$

$$
\begin{aligned}
[[A] - \lambda_2 \mathbf{I}] \mathbf{u}_2 &= \left[\begin{bmatrix} 5 & 3 \\ 3 & 6 \end{bmatrix} - 2.4586 \begin{bmatrix} 1 & 0 \\ 0 & 1 \end{bmatrix}\right]\begin{bmatrix} u_{21} \\ u_{22} \end{bmatrix} \\
&= \begin{bmatrix} 2.5414u_{21} + 3u_{22} \\ 3u_{21} + 3.5414u_{22} \end{bmatrix} = 0
\end{aligned}
\tag{13.168}
$$

Assigning last-unit eigenvectors

$$u_{12} = 1 \qquad u_{22} = 1 \tag{13.169}$$

provides us with

$$\mathbf{u}_1 = \begin{bmatrix} -1.180\,5 \\ 1.0 \end{bmatrix} \qquad \mathbf{u}_2 = \begin{bmatrix} 0.84713 \\ 1.0 \end{bmatrix} \tag{13.170}$$

Example 533 ★ *Unique ratio of the eigenvectors' elements.*
   To show an example that the ratio of the elements of eigenvectors is unique, we examine the eigenvectors $\mathbf{u}_1$ and $\mathbf{u}_2$ in Example 532.

$$\mathbf{u}_1 = \begin{bmatrix} 3u_{12} - 3.5414u_{11} \\ 3u_{11} - 2.5414u_{12} \end{bmatrix} \tag{13.171}$$

$$\mathbf{u}_2 = \begin{bmatrix} 2.5414u_{21} + 3u_{22} \\ 3u_{21} + 3.5414u_{22} \end{bmatrix} \tag{13.172}$$

The ratio u_{11}/u_{12} may be found from the first row of $\mathbf{u}_1$ in (13.171)

$$\frac{u_{11}}{u_{12}} = \frac{3}{3.5414} = 0.84712 \tag{13.173}$$

or from the second row

$$\frac{u_{11}}{u_{12}} = \frac{2.5414}{3} = 0.84713 \tag{13.174}$$

to examine their equality.
   The ratio u_{21}/u_{22} may also be found from the first or second row of $\mathbf{u}_2$ in (13.172) to check their equality.

$$\frac{u_{21}}{u_{22}} = -\frac{3}{2.5414} = -\frac{3.5414}{3} = -1.1805 \tag{13.175}$$

Example 534 ★ *Characteristics of free systems.*
   Free systems have two characteristics:

1. *natural frequencies*

2. *mode shapes*

   An n DOF vibrating system will have n natural frequencies ω_i and n mode shapes $\mathbf{u}_i$. The natural frequencies ω_i are cores for the system's resonance zones, and the eigenvectors $\mathbf{u}_i$ show the relative vibration of different coordinates of the system at the resonance ω_i. The highest element of each mode shape $\mathbf{u}_i$, indicates the coordinate or the component of the system which is most willing to vibrate at ω_i.

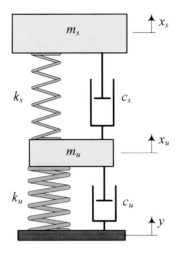

FIGURE 13.8. A quarter car model.

Example 535 *Importance of free systems.*

The response of free systems is the core for all other responses of the vibrating system. When there is some damping, then the response of the system is bounded by the free undamped solution. When there is a forcing function, the natural frequencies of the free response indicate the resonance zones at which the amplitude of the response may go to infinity if an excitation frequency of the force function matches.

Example 536 ★ *Sign of the separation constant ω^2.*

Both, left and right sides of Equation (13.147) must be equal to a constant. The sign of the constant is dictated by physical considerations. A free and undamped vibrating system is conservative and has a constant mechanical energy, so the amplitude of vibration must remain finite when $t \to \infty$. If the constant is positive, then the response is harmonic with a constant amplitude, however, if the constant is negative, the response is hyperbolic with an exponentially increasing amplitude.

Example 537 ★ *Quarter car natural frequencies and mode shapes.*

Figure 13.8 illustrates a quarter car model which is made of two solid masses m_s and m_u denoted as sprung and unsprung masses, respectively. The sprung mass m_s represents 1/4 of the body of the vehicle, and the unsprung mass m_u represents one wheel of the vehicle. A spring of stiffness k_s, and a shock absorber with viscous damping coefficient c_s support the sprung mass. The unsprung mass m_u is in direct contact with the ground through a spring k_u, and a damper c_u representing the tire stiffness and damping.

The governing differential equations of motion for the quarter car model *are*

$$m_s \ddot{x}_s = -k_s (x_s - x_u) - c_s (\dot{x}_s - \dot{x}_u) \tag{13.176}$$

$$m_u \ddot{x}_u = k_s (x_s - x_u) + c_s (\dot{x}_s - \dot{x}_u)$$
$$\qquad - k_u (x_u - y) - c_u (\dot{x}_u - \dot{y}) \tag{13.177}$$

which can be expressed in a matrix form

$$[M]\ddot{\mathbf{x}} + [c]\dot{\mathbf{x}} + [k]\mathbf{x} = \mathbf{F} \tag{13.178}$$

$$\begin{bmatrix} m_s & 0 \\ 0 & m_u \end{bmatrix} \begin{bmatrix} \ddot{x}_s \\ \ddot{x}_u \end{bmatrix} + \begin{bmatrix} c_s & -c_s \\ -c_s & c_s + c_u \end{bmatrix} \begin{bmatrix} \dot{x}_s \\ \dot{x}_u \end{bmatrix} +$$
$$\begin{bmatrix} k_s & -k_s \\ -k_s & k_s + k_u \end{bmatrix} \begin{bmatrix} x_s \\ x_u \end{bmatrix} = \begin{bmatrix} 0 \\ k_u y + c_u \dot{y} \end{bmatrix} \tag{13.179}$$

To find the natural frequencies and mode shapes of the quarter car model, we have to drop the damping and forcing terms and analyze the following set of equations.

$$\begin{bmatrix} m_s & 0 \\ 0 & m_u \end{bmatrix} \begin{bmatrix} \ddot{x}_s \\ \ddot{x}_u \end{bmatrix} + \begin{bmatrix} k_s & -k_s \\ -k_s & k_s + k_u \end{bmatrix} \begin{bmatrix} x_s \\ x_u \end{bmatrix} = 0 \tag{13.180}$$

Consider a vehicle with the following characteristics.

$$m_s = 375\,\text{kg} \qquad m_u = 75\,\text{kg}$$
$$k_u = 193000\,\text{N/m} \qquad k_s = 35000\,\text{N/m} \tag{13.181}$$

The equations of motion for this vehicle are

$$\begin{bmatrix} 375 & 0 \\ 0 & 75 \end{bmatrix} \begin{bmatrix} \ddot{x}_s \\ \ddot{x}_u \end{bmatrix} + \begin{bmatrix} 35000 & -35000 \\ -35000 & 2.28 \times 10^5 \end{bmatrix} \begin{bmatrix} x_s \\ x_u \end{bmatrix} = 0 \tag{13.182}$$

The natural frequencies of the vehicle can be found by solving its characteristic equation.

$$\begin{aligned} \det\left[[k] - \omega^2 [m]\right] &= \det\left[\begin{bmatrix} 35000 & -35000 \\ -35000 & 2.28 \times 10^5 \end{bmatrix} - \omega^2 \begin{bmatrix} 375 & 0 \\ 0 & 75 \end{bmatrix}\right] \\ &= \det\begin{bmatrix} 35\,000 - 375\omega^2 & -35\,000 \\ -35\,000 & 2.28 \times 10^5 - 75\omega^2 \end{bmatrix} \\ &= 28125\omega^4 - 8.8125 \times 10^7 \omega^2 + 6.755 \times 10^9 \end{aligned} \tag{13.183}$$

$$\omega_1 = 8.8671\,\text{rad/s} \approx 1.41\,\text{Hz} \tag{13.184}$$
$$\omega_2 = 55.269\,\text{rad/s} \approx 8.79\,\text{Hz} \tag{13.185}$$

To find the corresponding mode shapes, we use Equation (13.158).

$$\left[[k] - \omega_1^2 [m] \right] \mathbf{u}_1$$

$$= \left[\begin{bmatrix} 35000 & -35000 \\ -35000 & 2.28 \times 10^5 \end{bmatrix} - 3054.7 \begin{bmatrix} 375 & 0 \\ 0 & 75 \end{bmatrix} \right] \begin{bmatrix} u_{11} \\ u_{12} \end{bmatrix}$$

$$= \begin{bmatrix} -1.1105 \times 10^6 u_{11} - 35000 u_{12} \\ -35000 u_{11} - 1102.5 u_{12} \end{bmatrix} = 0 \qquad (13.186)$$

$$\left[[k] - \omega_2^2 [m] \right] \mathbf{u}_2$$

$$= \left[\begin{bmatrix} 35000 & -35000 \\ -35000 & 2.28 \times 10^5 \end{bmatrix} - 78.625 \begin{bmatrix} 375 & 0 \\ 0 & 75 \end{bmatrix} \right] \begin{bmatrix} u_{21} \\ u_{22} \end{bmatrix}$$

$$= \begin{bmatrix} 5515.6 u_{21} - 35000 u_{22} \\ 2.221 \times 10^5 u_{22} - 35000 u_{21} \end{bmatrix} = 0 \qquad (13.187)$$

Searching for the first-unit expression of $\mathbf{u}_1$ and $\mathbf{u}_2$ provides the following mode shapes.

$$\mathbf{u}_1 = \begin{bmatrix} 1 \\ 0.157\,58 \end{bmatrix} \qquad (13.188)$$

$$\mathbf{u}_2 = \begin{bmatrix} 1 \\ -3.1729 \times 10^{-3} \end{bmatrix} \qquad (13.189)$$

Therefore, the free vibrations of the quarter car is

$$\mathbf{x} = \sum_{i=1}^{n} \mathbf{u}_i \left(A_i \sin \omega_i t + B_i \cos \omega_i t \right) \qquad i = 1, 2 \qquad (13.190)$$

$$\begin{bmatrix} x_s \\ x_u \end{bmatrix} = \begin{bmatrix} 1 \\ -3.1729 \times 10^{-3} \end{bmatrix} \left(A_1 \sin 8.8671t + B_1 \cos 8.8671t \right)$$

$$+ \begin{bmatrix} 1 \\ 0.157\,58 \end{bmatrix} \left(A_2 \sin 55.269t + B_2 \cos 55.269t \right) \qquad (13.191)$$

13.4 Bicycle Car and Body Pitch Mode

Quarter car model is excellent to examine and optimize the body bounce mode of vibrations. However, we may expand the vibrating model of a vehicle to include pitch and other modes of vibrations as well. Figure 13.9 illustrates a bicycle vibrating model of a vehicle. This model includes the body bounce x, body pitch θ, wheels hop x_1 and x_2 and independent road excitations y_1 and y_2.

The equations of motion for the bicycle vibrating model of a vehicle are:

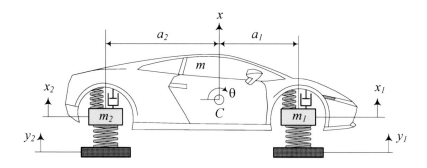

FIGURE 13.9. A bicycle vibrating model of a vehicle.

$$m\ddot{x} + c_1\left(\dot{x} - \dot{x}_1 - a_1\dot{\theta}\right) + c_2\left(\dot{x} - \dot{x}_2 + a_2\dot{\theta}\right)$$
$$+k_1\left(x - x_1 - a_1\theta\right) + k_2\left(x - x_2 + a_2\theta\right) = 0 \quad (13.192)$$
$$I_y\ddot{\theta} - a_1c_1\left(\dot{x} - \dot{x}_1 - a_1\dot{\theta}\right) + a_2c_2\left(\dot{x} - \dot{x}_2 + a_2\dot{\theta}\right)$$
$$-a_1k_1\left(x - x_1 - a_1\theta\right) + a_2k_2\left(x - x_2 + a_2\theta\right) = 0 \quad (13.193)$$

$$m_1\ddot{x}_1 - c_1\left(\dot{x} - \dot{x}_1 - a_1\dot{\theta}\right) + k_{t_1}\left(x_1 - y_1\right)$$
$$-k_1\left(x - x_1 - a_1\theta\right) = 0 \quad (13.194)$$
$$m_2\ddot{x}_2 - c_2\left(\dot{x} - \dot{x}_2 + a_2\dot{\theta}\right) + k_{t_2}\left(x_2 - y_2\right)$$
$$-k_2\left(x - x_2 + a_2\theta\right) = 0 \quad (13.195)$$

As a reminder, the definition of the employed parameters and variables are indicated in Table 13.1.

Table 13.1 - Parameters of a bicycle vibrating vehicle.

Parameter	Meaning
m	half of body mass
m_1	mass of a front wheel
m_2	mass of a rear wheel
x	body vertical motion coordinate
x_1	front wheel vertical motion coordinate
x_2	rear wheel vertical motion coordinate
θ	body pitch motion coordinate
y_1	road excitation at the front wheel
y_2	road excitation at the rear wheel
I_y	half of body lateral mass moment of inertia
a_1	distance of C from front axle
a_2	distance of C from rear axle

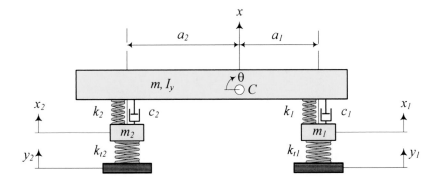

FIGURE 13.10. Bicycle model for a vehicle vibrations.

Proof. Figure 13.10 shows the vibrating model of the system. The body of the vehicle is assumed to be a rigid bar. This bar has a mass m, which is half of the total body mass, and a lateral mass moment I_y, which is half of the total body mass moment. The front and real wheels have a mass m_1 and m_2 respectively. The tires stiffness are indicated by parameters k_{t_1} and k_{t_2}. The difference is because the rear tires are usually stiffer than the fronts, although in a simpler model we may assume $k_{t_1} = k_{t_2}$. Damping of tires are much smaller than the damping of shock absorbers so, we may ignore the tire damping for simpler calculation.

To find the equations of motion for the bicycle vibrating model, we use the Lagrange method. The kinetic and potential energies of the system are

$$K = \frac{1}{2}m\dot{x}^2 + \frac{1}{2}m_1\dot{x}_1^2 + \frac{1}{2}m_2\dot{x}_2^2 + \frac{1}{2}I_y\dot{\theta}^2 \qquad (13.196)$$

$$V = \frac{1}{2}k_{t_1}(x_1 - y_1)^2 + \frac{1}{2}k_{t_2}(x_2 - y_2)^2$$
$$+ \frac{1}{2}k_1(x - x_1 - a_1\theta)^2 + \frac{1}{2}k_2(x - x_2 + a_2\theta) \qquad (13.197)$$

and the dissipation function is

$$D = \frac{1}{2}c_1\left(\dot{x} - \dot{x}_1 - a_1\dot{\theta}\right)^2 + \frac{1}{2}c_2\left(\dot{x} - \dot{x}_2 + a_2\dot{\theta}\right) \qquad (13.198)$$

Applying Lagrange method

$$\frac{d}{dt}\left(\frac{\partial K}{\partial \dot{q}_r}\right) - \frac{\partial K}{\partial q_r} + \frac{\partial D}{\partial \dot{q}_r} + \frac{\partial V}{\partial q_r} = f_r \qquad r = 1, 2, \cdots 4 \qquad (13.199)$$

provides the equations of motion (13.192)-(13.195). These set of equations may be rearranged in a matrix form

$$[m]\,\ddot{\mathbf{x}} + [c]\,\dot{\mathbf{x}} + [k]\,\mathbf{x} = \mathbf{F} \qquad (13.200)$$

where,

$$\mathbf{x} = \begin{bmatrix} x \\ \theta \\ x_1 \\ x_2 \end{bmatrix} \tag{13.201}$$

$$[m] = \begin{bmatrix} m & 0 & 0 & 0 \\ 0 & I_y & 0 & 0 \\ 0 & 0 & m_1 & 0 \\ 0 & 0 & 0 & m_2 \end{bmatrix} \tag{13.202}$$

$$[c] = \begin{bmatrix} c_1 + c_2 & a_2 c_2 - a_1 c_1 & -c_1 & -c_2 \\ a_2 c_2 - a_1 c_1 & c_1 a_1^2 + c_2 a_2^2 & a_1 c_1 & -a_2 c_2 \\ -c_1 & a_1 c_1 & c_1 & 0 \\ -c_2 & -a_2 c_2 & 0 & c_2 \end{bmatrix} \tag{13.203}$$

$$[k] = \begin{bmatrix} k_1 + k_2 & a_2 k_2 - a_1 k_1 & -k_1 & -k_2 \\ a_2 k_2 - a_1 k_1 & k_1 a_1^2 + k_2 a_2^2 & a_1 k_1 & -a_2 k_2 \\ -k_1 & a_1 k_1 & k_1 + k_{t_1} & 0 \\ -k_2 & -a_2 k_2 & 0 & k_2 + k_{t_2} \end{bmatrix} \tag{13.204}$$

$$\mathbf{F} = \begin{bmatrix} 0 \\ 0 \\ y_1 k_{t_1} \\ y_2 k_{t_2} \end{bmatrix} \tag{13.205}$$

■

Example 538 *Natural frequencies and mode shapes of a bicycle car model.*

Consider a vehicle with a heavy solid axle in the rear and independent suspensions in front. the vehicle has the following characteristics.

$$m = \frac{840}{2}\,\text{kg} \qquad I_y = \frac{1100}{2}\,\text{kg m}^2$$

$$m_1 = 53\,\text{kg} \qquad m_2 = \frac{152}{2}\,\text{kg} \tag{13.206}$$

$$a_1 = 1.4\,\text{m} \qquad a_2 = 1.47\,\text{m} \tag{13.207}$$

$$k_1 = 10000\,\text{N/m} \qquad k_2 = 13000\,\text{N/m}$$
$$k_{t_1} = k_{t_2} = 200000\,\text{N/m} \tag{13.208}$$

The natural frequencies should be found by using the undamped and free vibration equations of motion. The characteristic equation of the system is

$$\det\left[[k] - \omega^2[m]\right] =$$
$$8609 \times 10^9 \omega^8 - 1.2747 \times 10^{13} \omega^6$$
$$+2.1708 \times 10^{16} \omega^4 - 1.676 \times 10^{18} \omega^2 + 2.9848 \times 10^{19} \tag{13.209}$$

because

$$[m] = \begin{bmatrix} 420 & 0 & 0 & 0 \\ 0 & 550 & 0 & 0 \\ 0 & 0 & 53 & 0 \\ 0 & 0 & 0 & 76 \end{bmatrix} \qquad (13.210)$$

$$[k] = \begin{bmatrix} 23000 & 5110 & -10000 & -13000 \\ 5110 & 47692 & 14000 & -19110 \\ -10000 & 14000 & 210000 & 0 \\ -13000 & -19110 & 0 & 213000 \end{bmatrix} \qquad (13.211)$$

To find the natural frequencies we may solve the characteristic equation (13.209) or search for eigenvalues of $[A] = [m]^{-1}[k]$.

$$\begin{aligned} [A] &= [m]^{-1}[k] \\ &= \begin{bmatrix} 54.762 & 12.167 & -23.810 & -30.952 \\ 9.291 & 86.712 & 25.454 & -34.745 \\ -188.68 & 264.15 & 3962.3 & 0 \\ -171.05 & -251.45 & 0 & 2802.6 \end{bmatrix} \end{aligned} \qquad (13.212)$$

The eigenvalues of $[A]$ *are*

$$\lambda_1 = 48.91 \qquad \lambda_2 = 84.54 \qquad \lambda_3 = 2807.78 \qquad \lambda_4 = 3965.14 \qquad (13.213)$$

therefore, the natural frequencies of the bicycle car model are

$$\begin{aligned} \omega_1 &= \sqrt{\lambda_1} = 6.7\,\mathrm{rad/s} \approx 1.11\,\mathrm{Hz} \\ \omega_2 &= \sqrt{\lambda_2} = 9.19\,\mathrm{rad/s} \approx 1.46\,\mathrm{Hz} \\ \omega_3 &= \sqrt{\lambda_3} = 52.99\,\mathrm{rad/s} \approx 8.43\,\mathrm{Hz} \\ \omega_4 &= \sqrt{\lambda_4} = 62.96\,\mathrm{rad/s} \approx 10.02\,\mathrm{Hz} \end{aligned} \qquad (13.214)$$

The normal form of the mode shapes of the system are

$$\mathbf{u}_1 = \begin{bmatrix} 1.000 \\ -0.254 \\ 0.065 \\ 0.039 \end{bmatrix} \qquad \mathbf{u}_2 = \begin{bmatrix} 0.332 \\ 1.000 \\ -0.052 \\ 0.113 \end{bmatrix} \qquad (13.215)$$

$$\mathbf{u}_3 = \begin{bmatrix} -0.0113 \\ -0.0128 \\ 0.00108 \\ 1.000 \end{bmatrix} \qquad \mathbf{u}_4 = \begin{bmatrix} -0.00606 \\ 0.0065 \\ 1.000 \\ -0.00052 \end{bmatrix} \qquad (13.216)$$

The biggest element of the fourth mode shape $\mathbf{u}_4$ *belongs to* x_1. *It shows that in the fourth mode of vibrations at* $\omega_4 \approx 10.02\,\mathrm{Hz}$ *the front wheel will have the largest amplitude, while the amplitude of the other components are*

$$\Theta = \frac{u_{42}}{u_{43}} = 0.0065 X_1 \qquad (13.217)$$

$$X = \frac{u_{41}}{u_{43}} = -0.00606 X_1 \qquad (13.218)$$

$$X_2 = \frac{u_{44}}{u_{43}} = -0.00052 X_1 \qquad (13.219)$$

In this Example, the biggest element of the first mode shape $\mathbf{u}_1$ *belongs to* x, *the biggest element of the second mode shape* $\mathbf{u}_2$ *belongs to* θ, *and the biggest element of the third mode shape* $\mathbf{u}_3$ *belongs to* x_2. *Similar to the fourth mode shape* $\mathbf{u}_4$, *we can see the relative amplitude of different coordinates at each mode.*

Consider a car that starts to move on a bumpy road at a very small acceleration. By increasing the speed, the first resonance occurs at $\omega_1 \approx$ 1.11 Hz, *at which the bounce vibration is the most observable vibration. The second resonance occurs at* $\omega_2 \approx$ 1.46 Hz *when the pitch vibration of the body is the most observable vibration. The third and fourth resonances at* $\omega_3 \approx$ 8.43 Hz *and* $\omega_4 \approx$ 10.02 Hz *are related to rear and front wheels respectively.*

When the excitation frequency of a multiple DOF system increases, we will see that observable vibration moves from a coordinate to the others in the order of natural frequencies and associated mode shapes. When the excitation frequency is exactly at a natural frequency, the relative amplitudes of vibration are exactly similar to the associated mode shape. If the excitation frequency is not on a natural frequency, then vibration of the system is a combination of all modes of vibration. However, the weight factor of the closer modes are higher.

13.5   Half Car and Body Roll Mode

To examine and optimize the roll vibration of a vehicle we may use a half car vibrating model. Figure 13.11 illustrates a half car model of a vehicle. This model includes the body bounce x, body roll φ, wheels hop x_1 and x_2 and independent road excitations y_1 and y_2.

The equations of motion for the half car vibrating model of a vehicle are:

$$
\begin{aligned}
m\ddot{x} + c\left(\dot{x} - \dot{x}_1 + b_1\dot{\varphi}\right) + c\left(\dot{x} - \dot{x}_2 - b_2\dot{\varphi}\right) & \\
+ k\left(x - x_1 + b_1\varphi\right) + k\left(x - x_2 - b_2\varphi\right) &= 0 \qquad (13.220)
\end{aligned}
$$

$$
\begin{aligned}
I_x\ddot{\varphi} + b_1 c\left(\dot{x} - \dot{x}_1 + b_1\dot{\varphi}\right) - b_2 c\left(\dot{x} - \dot{x}_2 - b_2\dot{\varphi}\right) & \\
+ b_1 k\left(x - x_1 + b_1\varphi\right) - b_2 k\left(x - x_2 - b_2\varphi\right) + k_R\varphi &= 0 \qquad (13.221)
\end{aligned}
$$

$$
\begin{aligned}
m_1\ddot{x}_1 - c\left(\dot{x} - \dot{x}_1 + b_1\dot{\varphi}\right) + k_t\left(x_1 - y_1\right) & \\
- k\left(x - x_1 + b_1\varphi\right) &= 0 \qquad (13.222)
\end{aligned}
$$

$$
\begin{aligned}
m_2\ddot{x}_2 - c\left(\dot{x} - \dot{x}_2 - b_2\dot{\varphi}\right) + k_t\left(x_2 - y_2\right) & \\
- k\left(x - x_2 - b_2\varphi\right) &= 0 \qquad (13.223)
\end{aligned}
$$

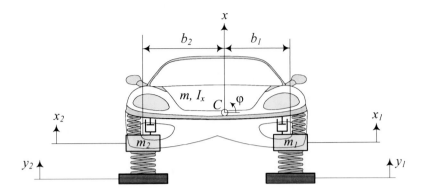

FIGURE 13.11. A half car vibrating model of a vehicle.

The half car model may be different for the front half and rear half due to different suspensions and mass distribution. Furthermore, different antiroll bars with different torsional stiffness may be used in the front and rear halves.

Proof. Figure 13.12 shows a better vibrating model of the system. The body of the vehicle is assumed to be a rigid bar. This bar has a mass m, which is the front or rear half of the total body mass, and a longitudinal mass moment I_x, which is half of the total body mass moment. The left and right wheels have a mass m_1 and m_2 respectively, although they are usually equal. The tires' stiffness are indicated by k_t. Damping of tires are much smaller than the damping of shock absorbers so, we may ignore the tire damping for simpler calculation. The suspension of the car has stiffness k and damping c for the left and right wheels. It is common to make the suspension of the left and right wheels mirror. So, their stiffness and damping are equal. However, the half car model has different k, c, and k_t for front or rear.

The vehicle may also have an antiroll bar with a torsional stiffness k_R in front and or rear. Using a simple model, the antiroll bar provides a torque M_R proportional to the roll angle φ.

$$M_R = -k_R \varphi \qquad (13.224)$$

However, a better model of the antiroll bar effect is

$$M_R = -k_R \left(\varphi - \frac{x_1 - x_2}{w} \right) \qquad (13.225)$$

To find the equations of motion for the half car vibrating model, we use the Lagrange method. The kinetic and potential energies of the system are:

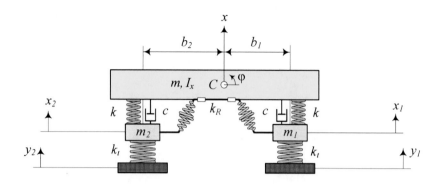

FIGURE 13.12. Half car model for a vehicle vibrations.

$$K = \frac{1}{2}m\dot{x}^2 + \frac{1}{2}m_1\dot{x}_1^2 + \frac{1}{2}m_2\dot{x}_2^2 + \frac{1}{2}I_x\dot{\varphi}^2 \qquad (13.226)$$

$$V = \frac{1}{2}k_t(x_1 - y_1)^2 + \frac{1}{2}k_t(x_2 - y_2)^2 + \frac{1}{2}k_R\varphi^2$$

$$+ \frac{1}{2}k(x - x_1 + b_1\varphi)^2 + \frac{1}{2}k(x - x_2 - b_2\varphi) \qquad (13.227)$$

and the dissipation function is

$$D = \frac{1}{2}c(\dot{x} - \dot{x}_1 + b_1\dot{\varphi})^2 + \frac{1}{2}c(\dot{x} - \dot{x}_2 - b_2\dot{\varphi}). \qquad (13.228)$$

Applying the Lagrange method

$$\frac{d}{dt}\left(\frac{\partial K}{\partial \dot{q}_r}\right) - \frac{\partial K}{\partial q_r} + \frac{\partial D}{\partial \dot{q}_r} + \frac{\partial V}{\partial q_r} = f_r \qquad r = 1, 2, \cdots 4 \qquad (13.229)$$

provides us with the following equations of motion (13.220)-(13.223). The set of equations of motion may be rearranged in a matrix form

$$[m]\ddot{\mathbf{x}} + [c]\dot{\mathbf{x}} + [k]\mathbf{x} = \mathbf{F} \qquad (13.230)$$

where,

$$\mathbf{x} = \begin{bmatrix} x \\ \varphi \\ x_1 \\ x_2 \end{bmatrix} \qquad (13.231)$$

$$[m] = \begin{bmatrix} m & 0 & 0 & 0 \\ 0 & I_x & 0 & 0 \\ 0 & 0 & m_1 & 0 \\ 0 & 0 & 0 & m_2 \end{bmatrix} \qquad (13.232)$$

$$[c] = \begin{bmatrix} 2c & cb_1 - cb_2 & -c & -c \\ cb_1 - cb_2 & cb_1^2 + cb_2^2 & -cb_1 & cb_2 \\ -c & -cb_1 & c & 0 \\ -c & cb_2 & 0 & c \end{bmatrix}$$ (13.233)

$$[k] = \begin{bmatrix} 2k & kb_1 - kb_2 & -k & -k \\ kb_1 - kb_2 & kb_1^2 + kb_2^2 + k_R & -kb_1 & kb_2 \\ -k & -kb_1 & k + k_t & 0 \\ -k & kb_2 & 0 & k + k_t \end{bmatrix}$$ (13.234)

$$\mathbf{F} = \begin{bmatrix} 0 \\ 0 \\ y_1 k_t \\ y_2 k_t \end{bmatrix}$$ (13.235)

■

Example 539 *Natural frequencies and mode shapes of a half car model.*
Consider a vehicle with the following characteristics.

$$m = \frac{840}{2}\,\text{kg} \qquad I_x = \frac{820}{2}\,\text{kg m}^2$$

$$m_1 = 53\,\text{kg} \qquad m_2 = 53\,\text{kg} \qquad (13.236)$$

$$b_1 = 0.7\,\text{m} \qquad b_2 = 0.75\,\text{m} \qquad (13.237)$$

$$k = 10000\,\text{N/ m} \qquad k_t = k_t = 200000\,\text{N/ m} \qquad k_R = 0\,\text{N m/ rad} \quad (13.238)$$

The natural frequency of the vehicle is found by using the undamped and free vibration equations of motion.

$$[m]\,\ddot{\mathbf{x}} + [k]\,\mathbf{x} = \mathbf{0} \qquad (13.239)$$

The $[m]$ *and* $[k]$ *matrices of the system are:*

$$[m] = \begin{bmatrix} 420 & 0 & 0 & 0 \\ 0 & 410 & 0 & 0 \\ 0 & 0 & 53 & 0 \\ 0 & 0 & 0 & 53 \end{bmatrix}$$ (13.240)

$$[k] = \begin{bmatrix} 20000 & -500 & -10000 & -10000 \\ -500 & 10525 & -7000 & 7500 \\ -10000 & -7000 & 210000 & 0 \\ -10000 & 7500 & 0 & 210000 \end{bmatrix}$$ (13.241)

To find the natural frequencies we may solve the characteristic equation (13.209) or search for eigenvalues of $[A] = [m]^{-1}\,[k]$.

$$[A] = [m]^{-1}\,[k]$$
$$= \begin{bmatrix} 47.619 & -1.1905 & -23.81 & -23.81 \\ -1.219 & 25.67 & -17.07 & 18.29 \\ -188.68 & -132.08 & 3962.3 & 0 \\ -188.68 & 141.51 & 0 & 3962.3 \end{bmatrix}$$ (13.242)

The eigenvalues of $[A]$ *are*

$$\lambda_1 = 24.38 \qquad \lambda_2 = 45.39$$
$$\lambda_3 = 3963.49 \qquad \lambda_4 = 3964.56 \qquad (13.243)$$

therefore, the natural frequencies of the half car model are

$$\omega_1 = \sqrt{\lambda_1} = 4.93 \, \text{rad/s} \approx 0.78 \, \text{Hz}$$
$$\omega_2 = \sqrt{\lambda_2} = 6.73 \, \text{rad/s} \approx 1.07 \, \text{Hz}$$
$$\omega_3 = \sqrt{\lambda_3} = 62.95 \, \text{rad/s} \approx 10.02 \, \text{Hz}$$
$$\omega_4 = \sqrt{\lambda_4} = 62.96 \, \text{rad/s} \approx 10.03 \, \text{Hz} \qquad (13.244)$$

The normal form of the mode shapes of the system are

$$\mathbf{u}_1 = \begin{bmatrix} 0.054 \\ 1.000 \\ 0.036 \\ -0.033 \end{bmatrix} \qquad \mathbf{u}_2 = \begin{bmatrix} 1.000 \\ -0.0554 \\ 0.0463 \\ 0.0502 \end{bmatrix} \qquad (13.245)$$

$$\mathbf{u}_3 = \begin{bmatrix} -0.46 \times 10^{-3} \\ -0.86 \times 10^{-2} \\ 1.000 \\ -0.923 \end{bmatrix} \qquad \mathbf{u}_4 = \begin{bmatrix} -0.0117 \\ 0.65 \times 10^{-3} \\ 0.923 \\ 1.000 \end{bmatrix} \qquad (13.246)$$

Example 540 *Comparison of the mode shapes of a half car model.*
   In example 539, the biggest element of the first mode shape $\mathbf{u}_1$ *belongs to* φ, *the biggest element of the second mode shape* $\mathbf{u}_2$ *belongs to* x, *the biggest element of the third mode shape* $\mathbf{u}_3$ *belongs to* x_1, *and the biggest element of the fourth mode shape* $\mathbf{u}_4$ *belongs to* x_2. *Consider a car that starts to move on a bumpy road at a very small acceleration. By increasing the speed, the first resonance occurs at* $\omega_1 \approx 0.78 \, \text{Hz}$, *at which the roll vibration is the most observable vibration. The second resonance occurs at* $\omega_2 \approx 1.07 \, \text{Hz}$ *when the bounce vibration of the body is the most observable vibration. The third and fourth resonances at* $\omega_3 \approx 10.02 \, \text{Hz}$ *and* $\omega_4 \approx 10.03 \, \text{Hz}$ *are related to left and right wheels respectively.*

Example 541 *Antiroll bar affects only the roll mode.*
   If in example 539, we set the antiroll bar to

$$k_R = 10000 \, \text{N m/rad} \qquad (13.247)$$

the natural frequencies and mode shapes of the half car model would be as follow.

$$\lambda_1 = 44.983 \qquad \lambda_2 = 49.165$$
$$\lambda_3 = 3963.498 \qquad \lambda_4 = 3964.561 \qquad (13.248)$$

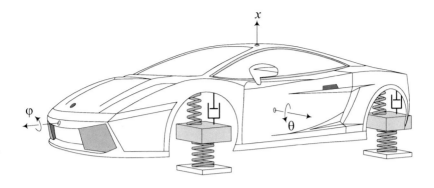

FIGURE 13.13. A full car vibrating model of a vehicle.

$$\omega_1 = \sqrt{\lambda_1} = 6.707\,\mathrm{rad/s} \approx 1.0674\,\mathrm{Hz}$$
$$\omega_2 = \sqrt{\lambda_2} = 7.012\,\mathrm{rad/s} \approx 1.1159\,\mathrm{Hz}$$
$$\omega_3 = \sqrt{\lambda_3} = 62.9563\,\mathrm{rad/s} \approx 10.0198\,\mathrm{Hz}$$
$$\omega_4 = \sqrt{\lambda_4} = 62.9647\,\mathrm{rad/s} \approx 10.0211\,\mathrm{Hz} \tag{13.249}$$

$$\mathbf{u}_1 = \begin{bmatrix} 1.000 \\ 0.9988 \\ 0.05835 \\ 0.03725 \end{bmatrix} \qquad \mathbf{u}_2 = \begin{bmatrix} -0.29488 \\ 1.000 \\ 0.01953 \\ -0.05038 \end{bmatrix} \tag{13.250}$$

$$\mathbf{u}_3 = \begin{bmatrix} -.4723 \times 10^{-3} \\ -.8671 \times 10^{-2} \\ 1.000 \\ -0.922 \end{bmatrix} \qquad \mathbf{u}_4 = \begin{bmatrix} -0.01168 \\ 0.655 \times 10^{-3} \\ 0.9219 \\ 1.000 \end{bmatrix} \tag{13.251}$$

Comparing these results with the results in example 539, shows the antiroll bar affects only the roll mode of vibration. A half car model needs a proper antiroll bar to increase the roll natural frequency.

It is recommended to have the roll mode as close as possible to the body bounce natural frequency to have a narrow resonance zone around the body bounce. Avoiding a narrow resonance zone would be simpler.

13.6   Full Car Vibrating Model

A general vibrating model of a vehicle is called the *full car* model. Such a model, that is shown in Figure 13.13, includes the body bounce x, body roll φ, body pitch θ, wheels hop x_1, x_2, x_3, and x_4 and independent road excitations y_1, y_2, y_3, and y_4.

A full car vibrating model has seven DOF with the following equations of motion.

$$m\ddot{x} + c_f \left(\dot{x} - \dot{x}_1 + b_1\dot{\varphi} - a_1\dot{\theta} \right) + c_f \left(\dot{x} - \dot{x}_2 - b_2\dot{\varphi} - a_1\dot{\theta} \right)$$
$$+c_r \left(\dot{x} - \dot{x}_3 - b_1\dot{\varphi} + a_2\dot{\theta} \right) + c_r \left(\dot{x} - \dot{x}_4 + b_2\dot{\varphi} + a_2\dot{\theta} \right)$$
$$+k_f \left(x - x_1 + b_1\varphi - a_1\theta \right) + k_f \left(x - x_2 - b_2\varphi - a_1\theta \right)$$
$$+k_r \left(x - x_3 - b_1\varphi + a_2\theta \right) + k_r \left(x - x_4 + b_2\varphi + a_2\theta \right) = 0 \quad (13.252)$$

$$I_x\ddot{\varphi} + b_1 c_f \left(\dot{x} - \dot{x}_1 + b_1\dot{\varphi} - a_1\dot{\theta} \right) - b_2 c_f \left(\dot{x} - \dot{x}_2 - b_2\dot{\varphi} - a_1\dot{\theta} \right)$$
$$-b_1 c_r \left(\dot{x} - \dot{x}_3 - b_1\dot{\varphi} + a_2\dot{\theta} \right) + b_2 c_r \left(\dot{x} - \dot{x}_4 + b_2\dot{\varphi} + a_2\dot{\theta} \right)$$
$$+b_1 k_f \left(x - x_1 + b_1\varphi - a_1\theta \right) - b_2 k_f \left(x - x_2 - b_2\varphi - a_1\theta \right)$$
$$-b_1 k_r \left(x - x_3 - b_1\varphi + a_2\theta \right) + b_2 k_r \left(x - x_4 + b_2\varphi + a_2\theta \right)$$
$$+k_R \left(\varphi - \frac{x_1 - x_2}{w} \right) = 0 \quad (13.253)$$

$$I_y\ddot{\theta} - a_1 c_f \left(\dot{x} - \dot{x}_1 + b_1\dot{\varphi} - a_1\dot{\theta} \right) - a_1 c_f \left(\dot{x} - \dot{x}_2 - b_2\dot{\varphi} - a_1\dot{\theta} \right)$$
$$+a_2 c_r \left(\dot{x} - \dot{x}_3 - b_1\dot{\varphi} + a_2\dot{\theta} \right) + a_2 c_r \left(\dot{x} - \dot{x}_4 + b_2\dot{\varphi} + a_2\dot{\theta} \right)$$
$$-a_1 k_f \left(x - x_1 + b_1\varphi - a_1\theta \right) - a_1 k_f \left(x - x_2 - b_2\varphi - a_1\theta \right)$$
$$+a_2 k_r \left(x - x_3 - b_1\varphi + a_2\theta \right) + a_2 k_r \left(x - x_4 + b_2\varphi + a_2\theta \right) = 0 \quad (13.254)$$

$$m_f\ddot{x}_1 - c_f \left(\dot{x} - \dot{x}_1 + b_1\dot{\varphi} - a_1\dot{\theta} \right) - k_f \left(x - x_1 + b_1\varphi - a_1\theta \right)$$
$$-k_R \frac{1}{w} \left(\varphi - \frac{x_1 - x_2}{w} \right) + k_{t_f} \left(x_1 - y_1 \right) = 0 \quad (13.255)$$

$$m_f\ddot{x}_2 - c_f \left(\dot{x} - \dot{x}_2 - b_2\dot{\varphi} - a_1\dot{\theta} \right) - k_f \left(x - x_2 - b_2\varphi - a_1\theta \right)$$
$$+k_R \frac{1}{w} \left(\varphi - \frac{x_1 - x_2}{w} \right) + k_{t_f} \left(x_2 - y_2 \right) = 0 \quad (13.256)$$

$$m_r\ddot{x}_3 - c_r \left(\dot{x} - \dot{x}_3 - b_1\dot{\varphi} + a_2\dot{\theta} \right)$$
$$-k_r \left(x - x_3 - b_1\varphi + a_2\theta \right) + k_{t_r} \left(x_3 - y_3 \right) = 0 \quad (13.257)$$

$$m_r\ddot{x}_4 - c_r \left(\dot{x} - \dot{x}_4 + b_2\dot{\varphi} + a_2\dot{\theta} \right)$$
$$-k_r \left(x - x_4 + b_2\varphi + a_2\theta \right) + k_{t_r} \left(x_4 - y_4 \right) = 0 \quad (13.258)$$

Proof. Figure 13.14 depicts the vibrating model of the system. The body of the vehicle is assumed to be a rigid slab. This slab has a mass m, which is the total body mass, a longitudinal mass moment I_x, and a lateral mass moment I_y. The mass moments are only the body mass moments and not the whole vehicle's mass moments. The wheels have a mass m_1, m_2, m_3, and m_4 respectively. However, it is common to have

$$m_1 = m_2 = m_f \tag{13.259}$$
$$m_3 = m_4 = m_r \tag{13.260}$$

The front and rear tires stiffness are indicated by k_{t_f} and k_{t_r} respectively. Because the damping of tires are much smaller than the damping of shock absorbers, we ignore the tires' damping for simpler calculation.

The suspension of the car has stiffness k_f and damping c_f in the front and stiffness k_r and damping c_r in the rear. It is common to make the suspension of the left and right wheels mirror. So, their stiffness and damping are equal. The vehicle may also have an antiroll bar in front and in the back, with a torsional stiffness k_{R_f} and k_{R_r}. Using a simple model, the antiroll bar provides a torque M_R proportional to the roll angle φ.

$$M_R = -\left(k_{R_f} + k_{R_r}\right)\varphi = -k_R\varphi \tag{13.261}$$

However, a better model of the antiroll bar reaction is

$$M_R = -k_{R_f}\left(\varphi - \frac{x_1 - x_2}{w_f}\right) - k_{R_f}\left(\varphi - \frac{x_4 - x_3}{w_r}\right) \tag{13.262}$$

Most cars only have an antiroll bar in front. For these cars, the moment of the antiroll bar simplifies to

$$M_R = -k_R\left(\varphi - \frac{x_1 - x_2}{w}\right) \tag{13.263}$$

if we use

$$w_f \equiv w = b_1 + b_2 \tag{13.264}$$
$$k_{R_f} \equiv k_R \tag{13.265}$$

To find the equations of motion for the full car vibrating model, we use the Lagrange method. The kinetic and potential energies of the system are

$$\begin{aligned} K &= \frac{1}{2}m\dot{x}^2 + \frac{1}{2}I_x\dot{\varphi}^2 + \frac{1}{2}I_y\dot{\theta}^2 \\ &+ \frac{1}{2}m_f\left(\dot{x}_1^2 + \dot{x}_2^2\right) + \frac{1}{2}m_r\left(\dot{x}_3^2 + \dot{x}_4^2\right) \end{aligned} \tag{13.266}$$

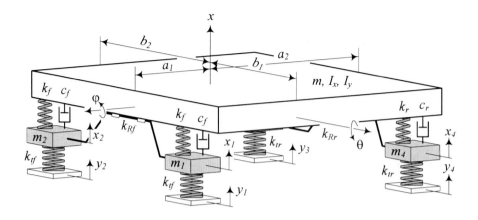

FIGURE 13.14. Full car model for a vehicle vibrations.

$$V = \frac{1}{2}k_f\left(x - x_1 + b_1\varphi - a_1\theta\right)^2 + \frac{1}{2}k_f\left(x - x_2 - b_2\varphi - a_1\theta\right)^2$$
$$+ \frac{1}{2}k_r\left(x - x_3 - b_1\varphi + a_2\theta\right)^2 + \frac{1}{2}k_r\left(x - x_4 + b_2\varphi + a_2\theta\right)^2$$
$$+ \frac{1}{2}k_R\left(\varphi - \frac{x_1 - x_2}{w}\right)^2 + \frac{1}{2}k_{t_f}\left(x_1 - y_1\right)^2 + \frac{1}{2}k_{t_f}\left(x_2 - y_2\right)^2$$
$$+ \frac{1}{2}k_{t_r}\left(x_3 - y_3\right)^2 + \frac{1}{2}k_{t_r}\left(x_4 - y_4\right)^2 \tag{13.267}$$

and the dissipation function is

$$D = \frac{1}{2}c_f\left(\dot{x} - \dot{x}_1 + b_1\dot{\varphi} - a_1\dot{\theta}\right)^2 + \frac{1}{2}c_f\left(\dot{x} - \dot{x}_2 - b_2\dot{\varphi} - a_1\dot{\theta}\right)^2$$
$$+ \frac{1}{2}c_r\left(\dot{x} - \dot{x}_3 - b_1\dot{\varphi} + a_2\dot{\theta}\right)^2$$
$$+ \frac{1}{2}c_r\left(\dot{x} - \dot{x}_4 + b_2\dot{\varphi} + a_2\dot{\theta}\right)^2 \tag{13.268}$$

Applying Lagrange method

$$\frac{d}{dt}\left(\frac{\partial K}{\partial \dot{q}_r}\right) - \frac{\partial K}{\partial q_r} + \frac{\partial D}{\partial \dot{q}_r} + \frac{\partial V}{\partial q_r} = f_r \qquad r = 1, 2, \cdots 7 \tag{13.269}$$

provides us with equations of motion (13.252)-(13.258).

The set of equations of motion may be rearranged in a matrix form

$$[m]\ddot{\mathbf{x}} + [c]\dot{\mathbf{x}} + [k]\mathbf{x} = \mathbf{F} \tag{13.270}$$

where,

$$\mathbf{x} = \begin{bmatrix} x & \varphi & \theta & x_1 & x_2 & x_3 & x_4 \end{bmatrix}^T \tag{13.271}$$

$$[m] = \begin{bmatrix} m & 0 & 0 & 0 & 0 & 0 & 0 \\ 0 & I_x & 0 & 0 & 0 & 0 & 0 \\ 0 & 0 & I_y & 0 & 0 & 0 & 0 \\ 0 & 0 & 0 & m_f & 0 & 0 & 0 \\ 0 & 0 & 0 & 0 & m_f & 0 & 0 \\ 0 & 0 & 0 & 0 & 0 & m_r & 0 \\ 0 & 0 & 0 & 0 & 0 & 0 & m_r \end{bmatrix} \quad (13.272)$$

$$[c] = \begin{bmatrix} c_{11} & c_{12} & c_{13} & -c_f & -c_f & -c_r & -c_r \\ c_{21} & c_{22} & c_{23} & -b_1 c_f & b_2 c_f & b_1 c_r & -b_2 c_r \\ c_{31} & c_{32} & c_{33} & a_1 c_f & a_1 c_f & -a_2 c_r & -a_2 c_r \\ -c_f & -b_1 c_f & a_1 c_f & c_f & 0 & 0 & 0 \\ -c_f & b_2 c_f & a_1 c_f & 0 & c_f & 0 & 0 \\ -c_r & b_1 c_r & -a_2 c_r & 0 & 0 & c_r & 0 \\ -c_r & -b_2 c_r & -a_2 c_r & 0 & 0 & 0 & c_r \end{bmatrix}$$
$$(13.273)$$

$$
\begin{aligned}
c_{11} &= 2c_f + 2c_r \\
c_{21} &= c_{12} = b_1 c_f - b_2 c_f - b_1 c_r + b_2 c_r \\
c_{31} &= c_{13} = 2a_2 c_r - 2a_1 c_f \\
c_{22} &= b_1^2 c_f + b_2^2 c_f + b_1^2 c_r + b_2^2 c_r \\
c_{32} &= c_{23} = a_1 b_2 c_f - a_1 b_1 c_f - a_2 b_1 c_r + a_2 b_2 c_r \\
c_{33} &= 2c_f a_1^2 + 2c_r a_2^2 \qquad (13.274)
\end{aligned}
$$

$$[k] = \begin{bmatrix} k_{11} & k_{12} & k_{13} & -k_f & -k_f & -k_r & -k_r \\ k_{21} & k_{22} & k_{23} & k_{24} & k_{25} & b_1 k_r & -b_2 k_r \\ k_{31} & k_{32} & k_{33} & a_1 k_f & a_1 k_f & -a_2 k_r & -a_2 k_r \\ -k_f & k_{42} & a_1 k_f & k_{44} & -\dfrac{k_R}{w^2} & 0 & 0 \\ -k_f & k_{52} & a_1 k_f & -\dfrac{k_R}{w^2} & k_{55} & 0 & 0 \\ -k_r & b_1 k_r & -a_2 k_r & 0 & 0 & k_r + k_{t_r} & 0 \\ -k_r & -b_2 k_r & -a_2 k_r & 0 & 0 & 0 & k_r + k_{t_r} \end{bmatrix}$$
$$(13.275)$$

$$
\begin{aligned}
k_{11} &= 2k_f + 2k_r \\
k_{21} &= k_{12} = b_1 k_f - b_2 k_f - b_1 k_r + b_2 k_r \\
k_{31} &= k_{13} = 2a_2 k_r - 2a_1 k_f \qquad (13.276)
\end{aligned}
$$

$$
\begin{aligned}
k_{22} &= k_R + b_1^2 k_f + b_2^2 k_f + b_1^2 k_r + b_2^2 k_r \\
k_{32} &= k_{23} = a_1 b_2 k_f - a_1 b_1 k_f - a_2 b_1 k_r + a_2 b_2 k_r \\
k_{42} &= k_{24} = -b_1 k_f - \frac{1}{w} k_R \\
k_{52} &= k_{25} = b_2 k_f + \frac{1}{w} k_R \qquad (13.277)
\end{aligned}
$$

$$k_{33} = 2k_f a_1^2 + 2k_r a_2^2$$

$$k_{44} = k_f + k_{t_f} + \frac{1}{w^2} k_R$$

$$k_{55} = k_f + k_{t_f} + \frac{1}{w^2} k_R \qquad (13.278)$$

$$\mathbf{F} = \begin{bmatrix} 0 & 0 & 0 & y_1 k_{t_f} & y_2 k_{t_f} & y_3 k_{t_r} & y_4 k_{t_r} \end{bmatrix}^T \qquad (13.279)$$

■

Example 542 *Natural frequencies and mode shapes of a full car model.*
Consider a vehicle with a the following characteristics.

$$m = 840 \,\mathrm{kg} \qquad m_f = 53 \,\mathrm{kg} \qquad m_r = 76 \,\mathrm{kg}$$
$$I_x = 820 \,\mathrm{kg\,m^2} \qquad I_y = 1100 \,\mathrm{kg\,m^2} \qquad (13.280)$$

$$a_1 = 1.4 \,\mathrm{m} \qquad a_2 = 1.47 \,\mathrm{m}$$
$$b_1 = 0.7 \,\mathrm{m} \qquad b_2 = 0.75 \,\mathrm{m} \qquad (13.281)$$

$$k_f = 10000 \,\mathrm{N/m} \qquad\qquad k_r = 13000 \,\mathrm{N/m}$$
$$k_{t_f} = k_{t_r} = 200000 \,\mathrm{N/m} \qquad k_R = 10000 \,\mathrm{N\,m/rad} \quad (13.282)$$

Using the matrix $[A] = [m]^{-1}[k]$ *and solving the associated eigenvalue and eigenvector problems, we find the following natural frequencies, and mode shapes for the full car model.*

$$\omega_1 = 0.989 \,\mathrm{Hz} \qquad \omega_2 = 1.113 \,\mathrm{Hz} \qquad \omega_3 = 1.464 \,\mathrm{Hz}$$
$$\omega_4 = 8.427 \,\mathrm{Hz} \qquad \omega_5 = 8.433 \,\mathrm{Hz} \qquad \omega_6 = 10.021 \,\mathrm{Hz}$$
$$\omega_7 = 10.245 \,\mathrm{Hz} \qquad\qquad\qquad (13.283)$$

$$\mathbf{u}_1 = \begin{bmatrix} 0.0177 \\ 1.000 \\ -0.0361 \\ 0.0671 \\ -0.06297 \\ -0.0455 \\ 0.0442 \end{bmatrix} \qquad \mathbf{u}_2 = \begin{bmatrix} 1.000 \\ -0.0303 \\ -0.253 \\ 0.0633 \\ 0.0673 \\ 0.0403 \\ 0.0376 \end{bmatrix} \qquad (13.284)$$

$$\mathbf{u}_3 = \begin{bmatrix} 0.332 \\ 0.0424 \\ 1.000 \\ -0.0492 \\ -0.0548 \\ 0.1115 \\ 0.1154 \end{bmatrix} \qquad \mathbf{u}_4 = \begin{bmatrix} -0.989 \times 10^{-4} \\ 0.822 \times 10^{-2} \\ -0.1086 \times 10^{-3} \\ 0.1616 \times 10^{-2} \\ -0.166 \times 10^{-2} \\ 1.000 \\ -0.982 \end{bmatrix} \qquad (13.285)$$

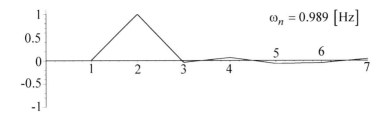

$$\omega_n = 0.989 \left[\text{Hz}\right]$$

FIGURE 13.15. 1st mode shape of a full car model.

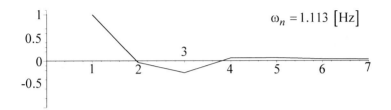

$$\omega_n = 1.113 \left[\text{Hz}\right]$$

FIGURE 13.16. 2nd mode shape of a full car model.

$$
\mathbf{u}_5 = \begin{bmatrix} -0.0112 \\ -0.366 \times 10^{-3} \\ -0.01268 \\ 0.998 \times 10^{-3} \\ 0.1145 \times 10^{-2} \\ 0.982 \\ 1.000 \end{bmatrix}
\qquad
\mathbf{u}_6 = \begin{bmatrix} -0.606 \times 10^{-2} \\ 0.156 \times 10^{-3} \\ 0.655 \times 10^{-2} \\ 1.000 \\ 0.999 \\ -0.509 \times 10^{-3} \\ -0.542 \times 10^{-3} \end{bmatrix}
\qquad (13.286)
$$

$$
\mathbf{u}_7 = \begin{bmatrix} -0.725 \times 10^{-6} \\ 0.841 \times 10^{-2} \\ 0.477 \times 10^{-5} \\ -0.999 \\ 1.000 \\ 0.75 \times 10^{-3} \\ -0.805 \times 10^{-3} \end{bmatrix}
\qquad (13.287)
$$

A visual illustration of the mode shapes are shown in Figures 13.15 to 13.21. The biggest element of the mode shapes $\mathbf{u}_1$ to $\mathbf{u}_7$ are φ, x, θ, x_3, x_4, x_1, x_2 respectively. These figures depict the relative amplitude of each coordinate of the full car model at a resonance frequency.

The natural frequencies of a full car can be separated in two classes. The first class is the natural frequencies of the body: body bounce, body roll, and body pitch. Body-related natural frequencies are always around 1 Hz. The second class is the natural frequencies of the wheels bounce. Wheel-related natural frequencies are always around 10 Hz.

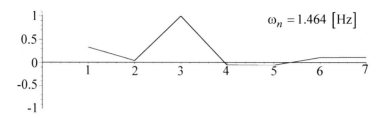

FIGURE 13.17. 3rd mode shape of a full car model.

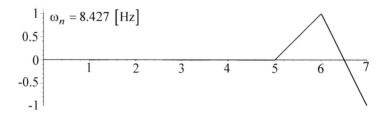

FIGURE 13.18. 4th mode shape of a full car model.

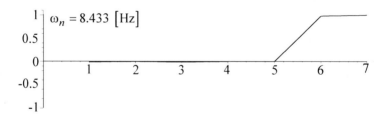

FIGURE 13.19. 5th mode shape of a full car model.

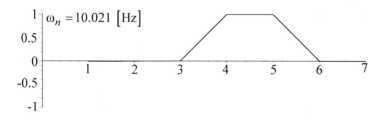

FIGURE 13.20. 6th mode shape of a full car model.

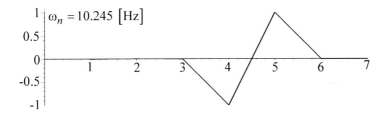

FIGURE 13.21. 7th mode shape of a full car model.

In this example, we assumed the car has independent suspension in front and rear. So, each wheel has only vertical displacement. In case of a solid axle, the left and right wheels make a rigid body with roll and bounce motions. The energies and hence, the equations of motion should be revised accordingly to show the bounce and roll of a solid axle.

13.7 Summary

Vehicles are connected multibody dynamic systems and hence, their vibrating model has multiple DOF system. The vibrating behavior of multiple DOF systems are very much dependent to their natural frequencies and mode shapes. These characteristics can be determined by solving an eigenvalue and an eigenvector problem.

The most practical vibrating model of vehicles, starting from the simplest to more complex, are the one-eight car, quarter car, bicycle car, half car, and full car models.

Having symmetric mass, stiffness, and damping matrices of multiple DOF system simplifies the calculation of the eigenvalue and eigenvector problems. To have symmetric coefficient matrices, we define the kinetic energy, potential energy, and dissipation function of the system by quadratures and derive the equations of motion by applying the Lagrange method.

13.8   Key Symbols

$a, \ddot{x}$	acceleration
a_1	distance from mass center to front axle
a_2	distance from mass center to rear axle
$[a], [A]$	coefficient matrix
$[A] = [m]^{-1}[k]$	coefficient matrix of characteristic equation
b_1	distance from mass center to left wheel
b_2	distance from mass center to right wheel
c	damping
c_{eq}	equivalent damping
c_{ij}	element of row i and column j of $[c]$
$[c]$	damping matrix
$[\underline{c}]$	symmetric damping matrix
C	mass center
D	dissipation function
e	eccentricity arm
E	mechanical energy
$f, \mathbf{F}$	harmonic force
$f = \frac{1}{T}$	cyclic frequency [Hz]
f_c	damper force
f_k	spring force
F	amplitude of a harmonic force f
F_r, Q_r	generalized force
g	gravitational acceleration
I	mass moment of inertia
$\mathbf{I}$	identity matrix
k	stiffness
k_{eq}	equivalent stiffness
k_{ij}	element of row i and column j of $[k]$
k_R	antiroll bar torsional stiffness
$[k]$	stiffness matrix
$[\underline{k}]$	symmetric stiffness matrix
K	kinetic energy
l	length
l	wheelbase
$\mathcal{L}$	Lagrangean
m	mass
m_e	eccentric mass
m_{ij}	element of row i and column j of $[m]$
m_s	sprung mass
m_u	unsprung mass
$[m]$	mass matrix
$[\underline{m}]$	symmetric mass matrix

n	number of DOF
$\mathbf{p}$	momentum
q_i, Q_i	generalized force
$r = \frac{\omega}{\omega_n}$	frequency ratio
r, R	radius
S	quadrature
t	time
T	period
u_{ij}	jth element of the ith mode shape
$\mathbf{u}$	mode shape, eigenvector
$\mathbf{u}_i$	ith eigenvector
$v, \mathbf{v}, \dot{x}, \dot{\mathbf{x}}$	velocity
V	potential energy
w	track
x	absolute displacement
X	steady-state amplitude of x
y	base excitation displacement
Y	steady-state amplitude of y
z	relative displacement
Z	steady-state amplitude of z
Z_i	short notation parameter

δ	deflection
$\xi = \frac{c}{2\sqrt{km}}$	damping ratio
λ	eigenvalue
λ_i	ith eigenvalue
$\omega = 2\pi f$	angular frequency $[\,\mathrm{rad/\,s}]$
ω_n	natural frequency
ω_i	ith natural frequency

Subscript

d	driver
f	front
r	rear
s	sprung
u	unsprung

Exercises

1. Equation of motion of a multiple DOF system.

    Figure 13.22 illustrates a two DOF vibrating system.

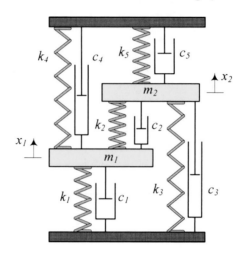

FIGURE 13.22. A two DOF vibrating system.

(a) Determine K, V, and D functions.

(b) Determine the equations of motion using the Lagrange method.

(c) ★ Rewrite K, V, and D in quadrature form.

(d) Determine the natural frequencies and mode shapes of the system.

2. ★ Absolute and relative coordinates.

    Figure 13.23 illustrates two similar double pendulums. We express the motion of the left one using absolute coordinates θ_1 and θ_2, and express the motion of the right one with absolute coordinate θ_1 and relative coordinate θ_2.

(a) Determine the equation of motion of the absolute coordinate double pendulum.

(b) Determine the equation of motion of the relative coordinate double pendulum.

(c) Compare their mass and stiffness matrices.

(d) Assume small values of θ_1 and θ_2 to linearize the equations of motion of both system.

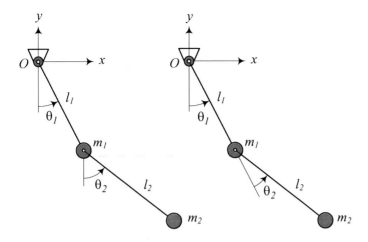

FIGURE 13.23. Two similar double pendulums, expressed by absolute and relative coordinates.

(e) Rewrite the linearized equations of motion in matrix form.

(f) Are the mass and stiffness of the systems equal? Explain the reason if they are different.

3. One-eight car model.

Consider a one-eight car model as a base excited one DOF system. Determine its natural w_n and damped natural frequencies w_d if

$$m = 1245\,\mathrm{kg} \qquad k = 60000\,\mathrm{N/m} \qquad c = 2400\,\mathrm{N\,s/m}$$

4. Quarter car model.

Consider a quarter car model. Determine its natural frequencies and mode shapes if

$$
\begin{aligned}
m_s &= 1085/4\,\mathrm{kg} & m_u &= 40\,\mathrm{kg} \\
k_s &= 10000\,\mathrm{N/m} & k_u &= 150000\,\mathrm{N/m} \\
c_s &= 800\,\mathrm{N\,s/m}
\end{aligned}
$$

5. Bicycle car model.

Consider a bicycle car model with the following characteristics:

$$
\begin{aligned}
m &= 1085/2\,\mathrm{kg} & I_y &= 1100\,\mathrm{kg\,m^2} \\
m_1 &= 40\,\mathrm{kg} & m_2 &= 40\,\mathrm{kg}
\end{aligned}
$$

$$a_1 = 1.4\,\mathrm{m} \qquad a_2 = 1.47\,\mathrm{m}$$

$$k_1 = 10000\,\mathrm{N/m} \qquad k_{t_1} = k_{t_2} = 150000\,\mathrm{N/m}$$

Determine its natural frequencies and mode shapes for

(a) $k_2 = 8000\,\text{N}/\text{m}$

(b) $k_2 = 10000\,\text{N}/\text{m}$

(c) $k_2 = 12000\,\text{N}/\text{m}$.

(d) Compare the natural frequencies for different k_1/k_2 and express the effect of increasing stiffness ratio on the pitch mode.

6. Half car model.

Consider a half car model with the following characteristics:

$$
\begin{aligned}
m &= 1085/2\,\text{kg} & I_x &= 820\,\text{kg m}^2 \\
m_1 &= 40\,\text{kg} & m_2 &= 40\,\text{kg}
\end{aligned}
$$

$$ b_1 = 0.7\,\text{m} \qquad b_2 = 0.75\,\text{m} $$

$$ k_1 = 10000\,\text{N}/\text{m} \qquad k_{t_1} = k_{t_2} = 150000\,\text{N}/\text{m} $$

Determine its natural frequencies and mode shapes for

(a) $k_R = 0$

(b) $k_R = 10000\,\text{N m}/\text{rad}$

(c) $k_R = 50000\,\text{N m}/\text{rad}$.

(d) Compare the natural frequencies for different k_R and express the effect of increasing roll stiffness on the roll mode.

(e) Determine k_R such that the roll natural frequency be equal to the bounce natural frequency and determine the mode shapes of the half car for that k_R.

7. Full car model.

Consider a full car model with the following characteristics:

$$
\begin{aligned}
m &= 1085\,\text{kg} & m_f &= 40\,\text{kg} & m_r &= 40\,\text{kg} \\
I_x &= 820\,\text{kg m}^2 & I_y &= 1100\,\text{kg m}^2
\end{aligned}
$$

$$
\begin{aligned}
a_1 &= 1.4\,\text{m} & a_2 &= 1.47\,\text{m} \\
b_1 &= 0.7\,\text{m} & b_2 &= 0.75\,\text{m}
\end{aligned}
$$

$$
\begin{aligned}
k_f &= 10000\,\text{N}/\text{m} & k_r &= 10000\,\text{N}/\text{m} \\
k_{t_f} &= k_{t_r} = 150000\,\text{N}/\text{m} & k_R &= 20000\,\text{N m}/\text{rad}
\end{aligned}
$$

(a) Determine its natural frequencies and mode shapes.

(b) Change k_R such that the roll mode and pitch modes have the closest possible frequency.

(c) Determine the mode shapes of the car for that k_R.

14

Suspension Optimization

In this chapter, we examine a linear one degree-of-freedom (DOF) base excited vibration isolator system as the simplest model for a vibration isolator and vehicle suspension. Based on a root mean square (RMS) optimization method, we develop a design chart to determine the optimal damper and spring for the best vibration isolation and ride comfort.

14.1 Mathematical Model

Figure 14.1 illustrates a single-DOF base excited linear vibrating system. It can represent a model for the vertical vibrations of a vehicle.

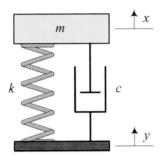

FIGURE 14.1. A base excited linear suspension.

A one-fourth (1/4) of the mass of the body is modeled as a solid mass m denoted as *sprung mass*. A spring of stiffness k, and a shock absorber with viscous damping c, support the sprung mass and represent the main suspension of the vehicle. The suspension parameters k and c are the equivalent stiffness and damping for one wheel, measured at the center of the wheel. Because we ignore the wheel mass and tire stiffness, this model is sometimes called one-eighth (1/8) car model.

The equation of motion for the system is

$$m\ddot{x} + c\dot{x} + kx = c\dot{y} + ky \qquad (14.1)$$

which can be transformed to

$$m\ddot{z} + c\dot{z} + kz = -m\ddot{y} \qquad (14.2)$$

R.N. Jazar, *Vehicle Dynamics: Theory and Application*,
DOI 10.1007/978-1-4614-8544-5_14, © Springer Science+Business Media New York 2014

using a relative displacement variable z.

$$z = x - y \tag{14.3}$$

The variable x is the absolute displacement of the body, and y is the absolute displacement of the ground.

The equation of motion (14.1) and (14.2), which are dependent on three parameters (m, c, k) can be transformed to the following equations:

$$\ddot{x} + 2\xi\omega_n\,\dot{x} + \omega_n^2\,x = 2\xi\omega_n\,\dot{y} + \omega_n^2\,y \tag{14.4}$$
$$\ddot{z} + 2\xi\omega_n\,\dot{z} + \omega_n^2\,z = -\ddot{y} \tag{14.5}$$

by introducing *natural frequency* ω_n and *damping ratio* ξ.

$$\xi = \frac{c}{2\sqrt{km}} \tag{14.6}$$

$$\omega_n = \sqrt{\frac{k}{m}} = 2\pi f_n \tag{14.7}$$

Proof. The kinetic energy, potential energy, and dissipation function of the system are:

$$K = \frac{1}{2}m\dot{x}^2 \tag{14.8}$$

$$V = \frac{1}{2}k\,(x-y)^2 \tag{14.9}$$

$$D = \frac{1}{2}c\,(\dot{x}-\dot{y})^2 \tag{14.10}$$

Employing the Lagrange method,

$$\frac{d}{dt}\left(\frac{\partial K}{\partial \dot{x}}\right) - \frac{\partial K}{\partial x} + \frac{\partial D}{\partial \dot{x}} + \frac{\partial V}{\partial x} = 0 \tag{14.11}$$

we find the equation of motion

$$\frac{d}{dt}\,(m\,\dot{x}) + c\,(\dot{x} - \dot{y}) + k\,(x - y) = 0 \tag{14.12}$$

which can be transformed to Equation (14.1). Introducing a relative position variable, $z = x - y$, we have

$$\dot{z} = \dot{x} - \dot{y} \tag{14.13}$$
$$\ddot{z} = \ddot{x} - \ddot{y} \tag{14.14}$$

to write Equation (14.12) as

$$m\frac{d}{dt}\,(\ddot{z} + \ddot{y}) + c\dot{z} + kz = 0 \tag{14.15}$$

which is equivalent to (14.2).

Dividing Equations (14.1) and (14.2) by m and using (14.6) and (14.7), generate their equivalent Equations (14.4) and (14.5), respectively. ∎

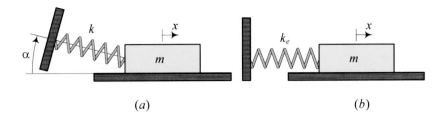

FIGURE 14.2. A tilted spring and its equivalent stiffness.

Example 543 *Different model for front and rear parts of a vehicle.*
Consider a car with the following information:

$$car\ mass\ =\ 1500\,\text{kg} \qquad wheel\ mass = 50\,\text{kg} \qquad (14.16)$$
$$F_{z_1}\ =\ 3941.78\,\text{N} \qquad F_{z_2} = 3415.6\,\text{N} \qquad (14.17)$$

where F_{z_1} and F_{z_2} are the front and rear tire loads, respectively. The mass m of a 1/8 vibrating model for the front of the car must be

$$m = \frac{F_{z_1}}{F_{z_1} + F_{z_2}} \times (1500 - 4 \times 50) = 696.49\,\text{kg} \qquad (14.18)$$

and for the rear of the car must be

$$m = \frac{F_{z_2}}{F_{z_1} + F_{z_2}} \times (1500 - 4 \times 50) = 603.51\,\text{kg} \qquad (14.19)$$

Example 544 *Tilted spring.*
Consider a mass-spring system such that the spring makes an angle α with the axis of mass translation, as shown in Figure 14.2(a). We may substitute such a tilted spring with an equivalent spring k_{eq} that is on the same axis of mass translation, as shown in Figure 14.2(b).

$$k_{eq} \approx k \cos^2 \alpha \qquad (14.20)$$

When the mass m is in motion, such as is shown in Figure 14.3(a), its free body diagram is as shown in Figure 14.3(b). If the motion of mass m is very small, $x \ll 1$, we may ignore any changes in α and then, as shown in Figure 14.3(c), the spring elongation is

$$\delta \approx x \cos \alpha \qquad (14.21)$$

Therefore, the spring force f_k is

$$f_k = k\delta \approx kx \cos \alpha \qquad (14.22)$$

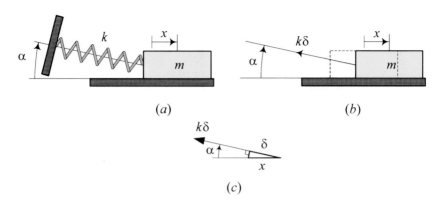

FIGURE 14.3. A mass-spring system such that the spring makes and angle α with directing of mass translation.

The spring force may be projected on the x-axis to find the x component, f_x, that moves the mass m.

$$f_x = f_k \cos \alpha \approx \left(k \cos^2 \alpha \right) x \qquad (14.23)$$

The tilted spring can be substituted with an equivalent spring k_{eq} on the x-axis that needs the same force f_x to elongate the same amount as the mass moves.

$$f_x = k_{eq} x \qquad k_{eq} \approx k \cos^2 \alpha \qquad (14.24)$$

Example 545 *Alternative proof for a tilted spring.*
Consider a spring that makes an angle α with the direction of motion as shown in Figure 14.3(a). When the mass translates x, the elongation of the spring is

$$\delta \approx x \cos \alpha \qquad (14.25)$$

The potential energy of such a spring would be

$$V = \frac{1}{2} k \delta^2 = \frac{1}{2} \left(k \cos^2 \alpha \right) x^2 \qquad (14.26)$$

An equivalent spring with stiffness k_{eq} must collect the same amount of potential energy for the same displacement x.

$$V = \frac{1}{2} k_{eq} x^2 \qquad (14.27)$$

Therefore, the equivalent stiffness k_{eq} is

$$k_{eq} = k \cos^2 \alpha \qquad (14.28)$$

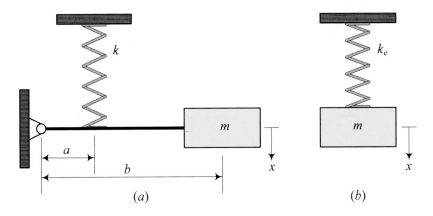

FIGURE 14.4. A mass m attached to the tip of a massless bar with length b.

Example 546 *Displaced spring.*

Figure 14.4(a) illustrates a mass m attached to the tip of a massless bar with length b. The bar is pivoted to the wall and a spring k is attached to the bar at a distance a from the pivot.

When the mass oscillates with displacement $x \ll 1$, the elongation δ of the spring is

$$\delta \approx \frac{a}{b}x \tag{14.29}$$

We may substitute the system with a translational mass-spring system such as shown in Figure 14.4(b). The new system has the same mass m and an equivalent spring k_{eq}.

$$k_{eq} = \left(\frac{a}{b}\right)^2 k \tag{14.30}$$

The equivalent spring provides the same potential energy as the original spring, when the mass moves.

$$V = \frac{1}{2}k_{eq}x^2 = \frac{1}{2}k\delta^2 = \frac{1}{2}k\left(\frac{a}{b}x\right)^2 = \frac{1}{2}k\left(\frac{a}{b}\right)^2 x^2 \tag{14.31}$$

Example 547 *Equivalent spring and damper for a McPherson suspension.*

Figure 14.5 illustrates a McPherson strut mechanism and its equivalent vibrating system.

We assume the tire is stiff and therefore, the wheel center gets the same motion y. Furthermore, we assume the wheel and body of the vehicle move only vertically.

To find the equivalent parameters for a 1/8 vibrating model, we set m to be equal to 1/4 of the body mass. The spring k and damper c make an angle α with the direction of wheel motion. They are also displaced $b - a$

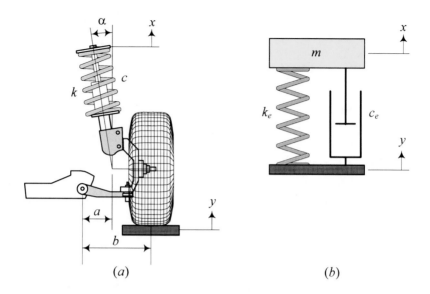

FIGURE 14.5. A MacPherson suspenssion and its equivalent vibrating system.

from the wheel center. So, the equivalent spring k_{eq} and damper c_{eq} are

$$k_{eq} = k \left(\frac{a}{b} \cos \alpha \right)^2 \qquad c_{eq} = c \left(\frac{a}{b} \cos \alpha \right)^2 \qquad (14.32)$$

For example assume that we have determined the following stiffness and damping as a result of optimization.

$$k_{eq} = 9869.6 \, \mathrm{N/m} \qquad c_{eq} = 87.965 \, \mathrm{N\,s/m} \qquad (14.33)$$

The actual k and c for a McPherson suspension with

$$a = 19 \, \mathrm{cm} \qquad b = 32 \, \mathrm{cm} \qquad \alpha = 27 \, \mathrm{deg} \qquad (14.34)$$

would be

$$k = 28489 \, \mathrm{N/m} \qquad c = 253.9 \, \mathrm{N\,s/m} \qquad (14.35)$$

Example 548 *Wavy road and excitation frequency.*

Figure 14.6 illustrates a 1/8 car model moving with speed v on a wavy road with length d_1 and peak-to-peak height d_2. Assuming a stiff tire with a small radius compared to the road waves, we may consider y as the fluctuation of the road.

The required time to pass one length d_1 is the period of the excitation

$$T = \frac{d_1}{v} \qquad (14.36)$$

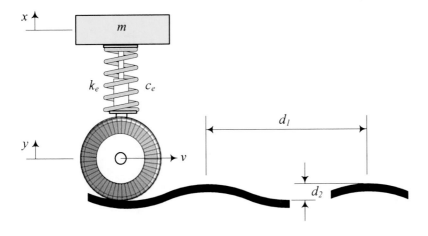

FIGURE 14.6. A 1/8 car model moving with speed v on a wavy road.

which can be used to find the frequency of excitation

$$\omega = \frac{2\pi}{T} = \frac{2\pi v}{d_1} \qquad (14.37)$$

Therefore, the excitation $y = Y \sin \omega t$ is

$$y = \frac{d_2}{2} \sin \frac{2\pi v}{d_1} t \qquad (14.38)$$

Example 549 *Function of an isolator*

The function of an isolator is to reduce the magnitude of motion transmitted from a vibrating foundation to the equipment, or to reduce the magnitude of transmitted force from the equipment to its foundation, both in time and frequency domain.

In the simplest approach to suspension analysis, the parameters m, k, and c are considered constant and independent of the excitation frequency or behavior of the foundation. This assumption is equivalent to considering an infinitely stiff and massive foundation. For rubber mounts, the damping coefficient usually decreases, and the stiffness coefficient increases with excitation frequency. Moreover, neither the engine nor body can be assumed an infinitely stiff rigid body at high frequencies.

14.2 Frequency Response

The most important frequency responses of a 1/8 car model, shown in Figure 14.1, are: absolute displacement G_0, relative displacement S_2, and absolute acceleration G_2

$$G_0 = \left| \frac{X}{Y} \right| = \frac{\sqrt{1 + (2\xi r)^2}}{\sqrt{(1 - r^2)^2 + (2\xi r)^2}} \tag{14.39}$$

$$S_2 = \left| \frac{Z}{Y} \right| = \frac{r^2}{\sqrt{(1 - r^2)^2 + (2\xi r)^2}} \tag{14.40}$$

$$G_2 = \left| \frac{\ddot{X}}{Y\omega_n^2} \right| = \frac{r^2\sqrt{1 + (2\xi r)^2}}{\sqrt{(1 - r^2)^2 + (2\xi r)^2}} \tag{14.41}$$

where

$$r = \frac{\omega}{\omega_n} \qquad \xi = \frac{c}{2\sqrt{km}} \qquad \omega_n = \sqrt{\frac{k}{m}} \tag{14.42}$$

Proof. Applying a harmonic excitation

$$y = Y \sin \omega t \tag{14.43}$$

the equation of motion (14.5) reduces to

$$\ddot{z} + 2\xi\omega_n \dot{z} + \omega_n^2 z = \omega^2 Y \sin \omega t \tag{14.44}$$

Now, we may consider a harmonic solution such as

$$z = A_3 \sin \omega t + B_3 \cos \omega t \tag{14.45}$$

to substitute in the equation of motion

$$-A_3\omega^2 \sin \omega t - B_3\omega^2 \cos \omega t + 2\xi\omega_n (A_3\omega \cos \omega t - B_3\omega \sin \omega t)$$
$$+\omega_n^2 (A_3 \sin \omega t + B_3 \cos \omega t) = \omega^2 Y \sin \omega t \tag{14.46}$$

and find a set of equations to calculate A_3 and B_3.

$$\begin{bmatrix} \omega_n^2 - \omega^2 & -2\xi\omega\omega_n \\ 2\xi\omega\omega_n & \omega_n^2 - \omega^2 \end{bmatrix} \begin{bmatrix} A_3 \\ B_3 \end{bmatrix} = \begin{bmatrix} Y\omega^2 \\ 0 \end{bmatrix} \tag{14.47}$$

The first row of the set (14.47) is a balance of the coefficients of $\sin \omega t$ in Equation (14.46), and the second row is a balance of the coefficients of $\cos \omega t$. Therefore, the coefficients A_3 and B_3 can be found as follow.

$$\begin{bmatrix} A_3 \\ B_3 \end{bmatrix} = \begin{bmatrix} \omega_n^2 - \omega^2 & -2\xi\omega\omega_n \\ 2\xi\omega\omega_n & \omega_n^2 - \omega^2 \end{bmatrix}^{-1} \begin{bmatrix} Y\omega^2 \\ 0 \end{bmatrix}$$
$$= \begin{bmatrix} -\dfrac{\omega^2 - \omega_n^2}{4\xi^2\omega^2\omega_n^2 + \omega^4 - 2\omega^2\omega_n^2 + \omega_n^4} Y\omega^2 \\ -\dfrac{2\xi\omega\omega_n}{4\xi^2\omega^2\omega_n^2 + \omega^4 - 2\omega^2\omega_n^2 + \omega_n^4} Y\omega^2 \end{bmatrix} \tag{14.48}$$

These equations may be transformed to this simpler form, by using r and ξ.

$$\begin{bmatrix} A_3 \\ B_3 \end{bmatrix} = \begin{bmatrix} \dfrac{1-r^2}{\left(1-r^2\right)^2 + \left(2\xi r\right)^2} r^2 Y \\[2ex] \dfrac{-2\xi r}{\left(1-r^2\right)^2 + \left(2\xi r\right)^2} r^2 Y \end{bmatrix} \tag{14.49}$$

The relative displacement amplitude Z is then equal to

$$Z = \sqrt{A_3^2 + B_3^2} = \frac{r^2}{\sqrt{\left(1-r^2\right)^2 + \left(2\xi r\right)^2}} Y \tag{14.50}$$

which provides $S_2 = |Z/Y|$ of Equation (14.40).

To find the absolute frequency response G_0, we may assume

$$x = A_2 \sin \omega t + B_2 \cos \omega t = X \sin\left(\omega t - \varphi_x\right) \tag{14.51}$$

and write

$$z = x - y \tag{14.52}$$
$$A_3 \sin \omega t + B_3 \cos \omega t = A_2 \sin \omega t + B_2 \cos \omega t - Y \sin \omega t \tag{14.53}$$

which shows

$$A_2 = A_3 + Y \tag{14.54}$$
$$B_2 = B_3 \tag{14.55}$$

The absolute displacement amplitude is then equal to

$$X = \sqrt{A_2^2 + B_2^2} = \sqrt{\left(A_3 + Y\right)^2 + B_3^2}$$
$$= \frac{\sqrt{1 + \left(2\xi r\right)^2}}{\sqrt{\left(1 - r^2\right)^2 + \left(2\xi r\right)^2}} Y \tag{14.56}$$

which provides $G_0 = |X/Y|$ in Equation (14.39).

The absolute acceleration frequency response

$$\ddot{x} = -X\omega^2 \sin\left(\omega t - \varphi_x\right) = -\ddot{X} \sin\left(\omega t - \varphi_x\right) \tag{14.57}$$

can be found by twice differentiating from the displacement frequency response (14.51). If we show the amplitude of the absolute acceleration by $\ddot{X}$, then we may define $\ddot{X}$ by

$$\left| \frac{\ddot{X}}{Y\omega_n^2} \right| = \frac{r^2 \sqrt{1 + \left(2\xi r\right)^2}}{\sqrt{\left(1 - r^2\right)^2 + \left(2\xi r\right)^2}} \tag{14.58}$$

which provides $G_2 = \left| \ddot{X} / \left(\omega_n^2 Y\right) \right|$ as in Equation (14.41). ∎

Example 550 *Principal method for absolute motion X.*
To find the absolute frequency response G_0, we substitute

$$y = Y \sin \omega t \tag{14.59}$$

and a harmonic solution for x

$$x = A_2 \sin \omega t + B_2 \cos \omega t \tag{14.60}$$

in Equation (14.4)

$$\ddot{x} + 2\xi\omega_n \dot{x} + \omega_n^2 x = 2\xi\omega_n \dot{y} + \omega_n^2 y \tag{14.61}$$

and solve for $X = \sqrt{A_2^2 + B_2^2}$.

$$-\omega^2 A_2 \sin \omega t - \omega^2 B_2 \cos \omega t + 2\xi\omega_n \omega \left(A_2 \cos \omega t - B_2 \sin \omega t\right)$$
$$+\omega_n^2 \left(A_2 \sin \omega t + B_2 \cos \omega t\right) = 2\xi\omega_n \omega Y \cos \omega t + \omega_n^2 Y \sin \omega t \tag{14.62}$$

The set of equations for A_2 and B_2 from the coefficients of sin *and* cos

$$\begin{bmatrix} \omega_n^2 - \omega^2 & -2\xi\omega\omega_n \\ 2\xi\omega\omega_n & \omega_n^2 - \omega^2 \end{bmatrix} \begin{bmatrix} A_2 \\ B_2 \end{bmatrix} = \begin{bmatrix} Y\omega_n^2 \\ 2Y\xi\omega\omega_n \end{bmatrix} \tag{14.63}$$

results in the following solution:

$$\begin{bmatrix} A_2 \\ B_2 \end{bmatrix} = \begin{bmatrix} \omega_n^2 - \omega^2 & -2\xi\omega\omega_n \\ 2\xi\omega\omega_n & \omega_n^2 - \omega^2 \end{bmatrix}^{-1} \begin{bmatrix} Y\omega_n^2 \\ 2Y\xi\omega\omega_n \end{bmatrix}$$

$$= \begin{bmatrix} \dfrac{-\left(\omega^2 - \omega_n^2\right)\omega_n^2 + 4\xi^2\omega^2\omega_n^2}{4\xi^2\omega^2\omega_n^2 + \omega^4 - 2\omega^2\omega_n^2 + \omega_n^4} Y \\[3mm] \dfrac{-2\xi\omega\omega_n^3}{4\xi^2\omega^2\omega_n^2 + \omega^4 - 2\omega^2\omega_n^2 + \omega_n^4} Y \end{bmatrix}$$

$$= \begin{bmatrix} \dfrac{(2\xi r)^2 - \left(1 - r^2\right)}{\left(1 - r^2\right)^2 + (2\xi r)^2} Y \\[3mm] \dfrac{-2\xi r^3}{\left(1 - r^2\right)^2 + (2\xi r)^2} Y \end{bmatrix} \tag{14.64}$$

Therefore, the amplitude of the absolute displacement X would be the same as (14.56).

Example 551 $G_0 \neq S_2 + 1$
We may try to find the absolute frequency response $G_0 = |X/Y|$, from the result for S_2. The frequency response S_2 is

$$S_2 = \frac{Z}{Y} \tag{14.65}$$

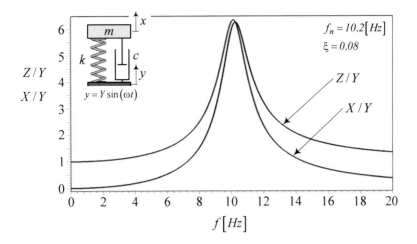

FIGURE 14.7. Absolute and relative displacement frequency responses for a vehicle.

however,

$$S_2 \neq \frac{X}{Y} - 1 \neq G_0 - 1 \tag{14.66}$$

because the amplitude of the relative displacement Z is not equal to the amplitude of absolute displacement X minus the amplitude of road excitation Y.

$$Z \neq X - Y \tag{14.67}$$

Example 552 *A sample of frequency responses.*
    Consider a vehicle with given natural frequency f_n and damping ratio ξ:

$$f_n = 10.2\,\text{Hz} \qquad \xi = 0.08 \tag{14.68}$$

The absolute and relative displacements frequency responses of the vehicle are shown in Figure 14.7. The relative displacement starts at zero and ends up at one, while the absolute displacement starts at one and ends up at zero.

14.3   RMS Optimization

Figure 14.8 is a design chart for optimal suspension parameters of base excited systems. The horizontal axis is the root mean square of relative displacement, $S_Z = RMS(S_2)$, and the vertical axis is the root mean square of absolute acceleration, $S_{\ddot{X}} = RMS(G_2)$. There are two sets of curves that make a mesh. The first set, which is almost parallel at the right end, is constant natural frequency f_n, and the second set, which spread from

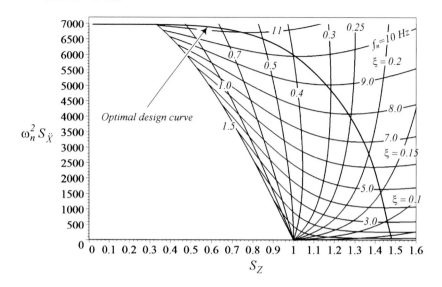

FIGURE 14.8. Design chart for optimal suspension parameters of equipments.

$S_Z = 1$, is a constant damping ratio ξ. There is a curve, called *optimal design curve*, which indicates the optimal suspension parameters.

Most equipment that are mounted on vehicles have natural frequencies around $f_n = 10\,\text{Hz}$, while the main natural frequencies of the vehicle are around $f_n = 1\,\text{Hz}$. So, we use Figure 14.8 to design the suspension of base excited equipment, and use the magnified chart shown in the Figure 14.9 to design vehicle suspensions.

The optimal design curve is the result of the following optimization strategy:

$$Minimize\ S_{\ddot{X}}\ with\ respect\ to\ S_Z \qquad (14.69)$$

which states that the minimum absolute acceleration with respect to the relative displacement, if there is any, makes a suspension optimal. Mathematically it is equivalent to the following minimization problem:

$$\frac{\partial S_{\ddot{X}}}{\partial S_Z} = 0 \qquad (14.70)$$

$$\frac{\partial^2 S_{\ddot{X}}}{\partial S_Z^2} > 0 \qquad (14.71)$$

To determine the optimal stiffness k and damping c, we start from an estimated value for S_X on the horizontal axis and draw a vertical line to hit the optimal curve. The intersection point indicates the optimal f_n and ξ for the S_X, to have the best vibration isolation. Figure 14.10 illustrates a sample application for $S_X = 1$, which indicates $\xi \approx 0.4$ and $f_n \approx 10\,\text{Hz}$ make

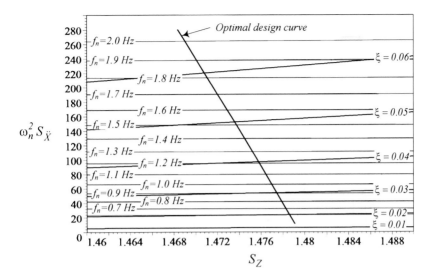

FIGURE 14.9. Design chart for optimal suspension parameters of vehicles.

the optimal suspension. f_n, ξ, and the mass of the equipment determine the optimal value of k and c.

Proof. Let us define a working frequency range $0 < f < 20\,\mathrm{Hz}$ to include almost all ground vehicles, especially road vehicles, and show the RMS of S_2 and G_2 by

$$S_Z = RMS(S_2) \tag{14.72}$$
$$S_{\ddot{X}} = RMS(G_2) \tag{14.73}$$

In applied vehicle dynamics, we usually measure frequencies in $[\mathrm{Hz}]$, instead of $[\mathrm{rad/s}]$, so we perform design calculations based on cyclic frequencies f and f_n in $[\mathrm{Hz}]$, and analytic calculations based on angular frequencies ω and ω_n in $[\mathrm{rad/s}]$.

To calculate S_Z and $S_{\ddot{X}}$ over the working frequency range

$$S_Z = \sqrt{\frac{1}{40\pi}\int_0^{40\pi} S_2^2\,d\omega} \tag{14.74}$$

$$S_{\ddot{X}} = \sqrt{\frac{1}{40\pi}\int_0^{40\pi} G_2\,d\omega} \tag{14.75}$$

we first find integrals of S_2^2 and G_2.

$$\int S_2^2\,d\omega = Z_1\omega - \frac{Z_2}{Z_3\sqrt{Z_4}}\tan^{-1}\frac{\omega}{\sqrt{Z_4}} + \frac{Z_5}{Z_6\sqrt{Z_7}}\tan^{-1}\frac{\omega}{\sqrt{Z_7}} \tag{14.76}$$

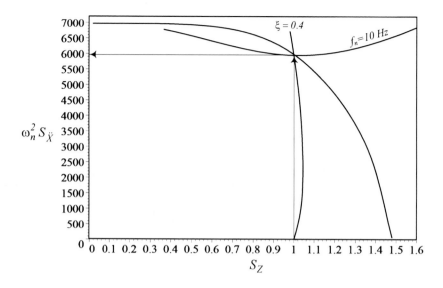

FIGURE 14.10. Application of the design chart for $S_X = 1$, that indicates the optimal values $\xi \approx 0.4$ and $f_n \approx 10\,\text{Hz}$.

$$\omega_n^4 \int G_2 d\omega = Z_8\omega + \frac{1}{3}Z_9\omega^3 + \frac{Z_{10}}{Z_{11}\sqrt{Z_{12}}}\tan^{-1}\frac{\omega}{\sqrt{Z_{12}}}$$
$$+ \frac{Z_{13}}{Z_{14}\sqrt{Z_{15}}}\tan^{-1}\frac{\omega}{\sqrt{Z_{15}}} \qquad (14.77)$$

The parameters Z_1 through Z_{15} are:

$$Z_1 = 1 \qquad (14.78)$$
$$Z_2 = \omega_n^2\left(8\xi^6 - 12\xi^4 + 4\xi^2 - \left(-8\xi^4 + 8\xi^2 - 1\right)\xi\sqrt{1-\xi^2}\right) \qquad (14.79)$$
$$Z_3 = -4\xi^2\left(1-\xi^2\right) \qquad (14.80)$$
$$Z_4 = \omega_n^2\left(-1+2\xi^2+2\xi\sqrt{1-\xi^2}\right) \qquad (14.81)$$

$$Z_5 = \omega_n^2\left(8\xi^6 - 12\xi^4 + 4\xi^2 - \left(8\xi^4 - 8\xi^2 + 1\right)\xi\sqrt{1-\xi^2}\right) \qquad (14.82)$$
$$Z_6 = -4\xi^2\left(1-\xi^2\right) \qquad (14.83)$$
$$Z_7 = \omega_n^2\left(-1+2\xi^2-2\xi\sqrt{1-\xi^2}\right) \qquad (14.84)$$

$$Z_8 = \omega_n^4 \left(-16\xi^4 + 8\xi^2 + 1\right) \tag{14.85}$$

$$Z_9 = 4\omega_n^2 \xi^2 \tag{14.86}$$

$$Z_{10} = \omega_n^6 \left(128\xi^{10} - 256\xi^8 + 144\xi^6 - 12\xi^4 - 4\xi^2\right) \tag{14.87}$$

$$-\omega_n^6 \left(-128\xi^8 + 192\xi^6 - 64\xi^4 - 4\xi^2 + 1\right)\xi\sqrt{1 - \xi^2} \tag{14.88}$$

$$Z_{11} = -4\xi^2 \left(1 - \xi^2\right) \tag{14.89}$$

$$Z_{12} = \omega_n^2 \left(-1 + 2\xi^2 + 2\xi\sqrt{1 - \xi^2}\right) \tag{14.90}$$

$$Z_{13} = \omega_n^6 \left(128\xi^{10} - 256\xi^8 + 144\xi^6 - 12\xi^4 - 4\xi^2\right) \tag{14.91}$$

$$-\omega_n^6 \left(128\xi^8 - 192\xi^6 + 64\xi^4 + 4\xi^2 - 1\right)\xi\sqrt{1 - \xi^2} \tag{14.92}$$

$$Z_{14} = -4\xi^2 \left(1 - \xi^2\right) \tag{14.93}$$

$$Z_{15} = \omega_n^2 \left(-1 + 2\xi^2 - 2\xi\sqrt{1 - \xi^2}\right) \tag{14.94}$$

Therefore, S_Z and $S_{\ddot{X}}$ over the frequency range $0 < f < 20\,\mathrm{Hz}$ can be calculated analytically from Equations (14.74) and (14.75).

Equations (14.76) and (14.77) show that both $S_{\ddot{X}}$ and S_Z are functions of only two variables ω_n and ξ.

$$S_{\ddot{X}} = S_{\ddot{X}}(\omega_n, \xi) \tag{14.95}$$

$$S_Z = S_Z(\omega_n, \xi) \tag{14.96}$$

Therefore, any pair of design parameters (ω_n, ξ) determines $S_{\ddot{X}}$ and S_Z uniquely. It is also possible theoretically to define ω_n and ξ as two functions of the variables $S_{\ddot{X}}$ and S_Z.

$$\omega_n = \omega_n(S_{\ddot{X}}, S_Z) \tag{14.97}$$

$$\xi = \xi(S_{\ddot{X}}, S_Z) \tag{14.98}$$

So, we would be able to determine the required ω_n and ξ for a specific value of $S_{\ddot{X}}$ and S_Z.

Using Equations (14.95) and (14.96), we may draw Figure 14.11 to illustrate how $S_{\ddot{X}}$ behaves with respect to S_Z when f_n and ξ vary. By keeping f_n constant and varying ξ, it is possible to minimize $S_{\ddot{X}}$ with respect to S_Z. The minimum points make the optimal curve and determine the best f_n and ξ. The key to use the optimal design curve is to adjust, determine, or estimate a value for S_Z or $S_{\ddot{X}}$ and find the associated point on the design curve.

To justify the optimization principle (14.69), we plot $\omega_n^2 S_{\ddot{X}}/S_Z$ versus f_n in Figure 14.12 for different values of ξ. It shows that increasing either one

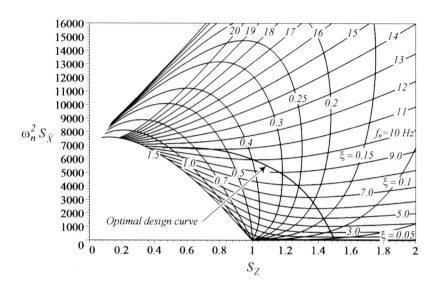

FIGURE 14.11. Behavior of $S_{\ddot{X}}$ with respect to S_Z when f_n and ξ are varied.

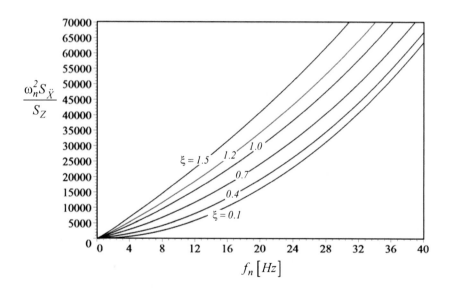

FIGURE 14.12. A plot of ratio $\omega_n^2 S_{\ddot{X}}/S_Z$ versus f_n for different values of ξ.

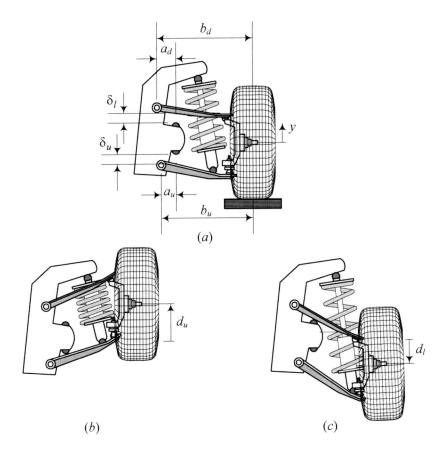

FIGURE 14.13. A double A-arm suspension mechanism at (a)- equilibrium, (b)- upper limit, and (c)- lower limit.

of ξ or f_n increases the value of $\omega_n^2 S_{\ddot{X}}/S_Z$. It is equivalent to making the suspension more rigid, which causes an increase in acceleration and decrease in relative displacement. On the contrary, decreasing ξ or f_n decreases the value of $\omega_n^2 S_{\ddot{X}}/S_Z$, which is equivalent to making the suspension softer.

Softening of a suspension decreases the body acceleration, however it requires a large room for relative displacement. Due to physical constraints the wheel travel is limited and hence, we must design the suspension to use the available suspension travel as much as possible, and decrease the body acceleration as low as possible. Mathematically it is equivalent to (14.70) and (14.71). ■

Example 553 *Wheel travel calculation.*
Figure 14.13(a) illustrates a double A-arm suspension mechanism at its equilibrium position. To limit the motion of the wheel with respect to the

body, two stoppers must be employed. There are many possible options for the type and position of stoppers. Most stoppers are made of stiff rubber balls and mounted somewhere on the body or suspension mechanism or both. It is also possible that the damper acts as a stopper. Figure 14.13(a) shows an example.

The gap sizes δ_u and δ_l indicate the upper and lower distances that a mechanism can move. However, the maximum motion of the wheel must be calculated at the center of the wheel. So, we transfer δ_u and δ_l to the center of the wheel and show them by d_u and d_l.

$$d_u \approx \frac{b_u}{a_u}\delta_u \qquad d_l \approx \frac{b_l}{a_l}\delta_l \qquad\qquad (14.99)$$

Figure 14.13(b) and (c) show the mechanism at the upper and lower limits respectively. The distance d_u is called the **upper wheel travel**, and d_l is called the **lower wheel travel**. The upper wheel travel is important in ride comfort and the lower wheel travel is important for safety. To have better ride comfort, the upper wheel travel should be as high as possible to make the suspension as soft as possible.

Although the upper and lower wheel travels may be different, for practical purposes, we may assume $d_l = d_u$ and design the suspension based on a unique wheel travel. Wheel travel is also called suspension travel, suspension room, and suspension clearance.

Example 554 *Soft and hard suspensions.*

Consider two pieces of equipment, A and B, under a base excitation with an average amplitude $Y = 1\,\text{cm} \approx 0.5\,\text{in}$. Suspension A has a suspension travel $d_A = 1.2\,\text{cm} \approx 0.6\,\text{in}$ and suspension B has $d_B = 0.8\,\text{cm} \approx 0.4\,\text{in}$. Let us assume $S_Z = d_u/Y$. Therefore,

$$S_{Z_A} = 1.2 \qquad S_{Z_B} = 0.8 \qquad\qquad (14.100)$$

Using the design chart in Figure 14.14, the optimal suspensions for A and B are

$$f_{n_A} \approx 8.53\,\text{Hz} \qquad \xi_A \approx 0.29 \qquad\qquad (14.101)$$
$$f_{n_B} \approx 10.8\,\text{Hz} \qquad \xi_B \approx 0.56 \qquad\qquad (14.102)$$

Assuming

$$m = 300\,\text{kg} \approx 660\,\text{lb} \qquad\qquad (14.103)$$

we calculate the optimal spring and dampers as:

$$k_A = (2\pi f_{n_A})^2 m = 8.6175 \times 10^5\,\text{N/m} \qquad\qquad (14.104)$$
$$k_B = (2\pi f_{n_B})^2 m = 13.814 \times 10^5\,\text{N/m} \qquad\qquad (14.105)$$

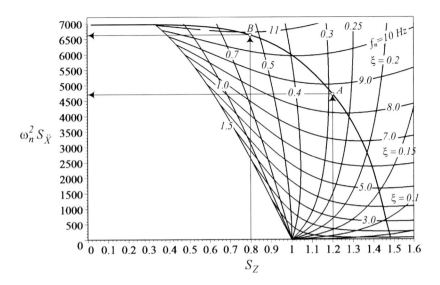

FIGURE 14.14. Comparing two suspensions A and B with $S_{Z_A} = 1.2$ and $S_{Z_B} = 0.8$.

$$c_A = 2\xi_A\sqrt{k_A m} = 9325.7\,\text{N s/ m} \qquad (14.106)$$

$$c_B = 2\xi_B\sqrt{k_B m} = 22800\,\text{N s/ m} \qquad (14.107)$$

Suspension B is harder compared to suspension A. This is because suspension B has less wheel travel, and hence, it has more acceleration level $\omega_n^2 S_{\ddot{X}}$. *Figure 14.14 shows that*

$$\omega_n^2 S_{\ddot{X}_A} \approx 4700\ \ 1/\text{s}^2 \qquad \omega_n^2 S_{\ddot{X}_B} \approx 6650\ \ 1/\text{s}^2 \qquad (14.108)$$

Example 555 *Soft and hard vehicle suspensions.*

Consider two vehicles A and B that are moving on a bumpy road with an average amplitude $Y = 10\,\text{cm} \approx 3.937\,\text{in}$. *Vehicle A has a suspension travel* $d_A = 14.772\,\text{cm} \approx 5.816\,\text{in}$ *and vehicle B has* $d_B = 14.714\,\text{cm} \approx 5.793\,\text{in}$. *Let us assume* $S_Z = d_u/Y$. *Therefore,*

$$S_{Z_A} = 1.4772 \qquad S_{Z_B} = 1.4714 \qquad (14.109)$$

Using design chart 14.15, the optimal suspensions for vehicles A and B are:

$$f_{n_A} \approx 0.7\,\text{Hz} \qquad \xi_A \approx 0.023 \qquad (14.110)$$

$$f_{n_B} \approx 1.85\,\text{Hz} \qquad \xi_B \approx 0.06 \qquad (14.111)$$

Assuming a mass m

$$m = 300\,\text{kg} \approx 660\,\text{lb} \qquad (14.112)$$

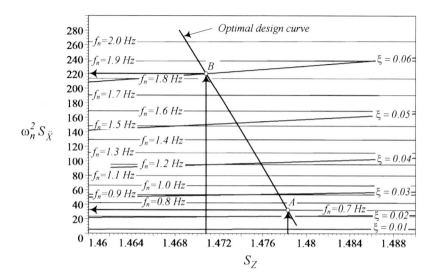

FIGURE 14.15. Comparing two suspensions, A and B, with $S_{Z_A} = 1.4772$ and $S_{Z_B} = 1.4714$.

we calculate the optimal spring and dampers as follows:

$$k_A = (2\pi f_{n_A})^2 m \approx 5803 \,\text{N}/\text{m} \tag{14.113}$$
$$k_B = (2\pi f_{n_B})^2 m \approx 40534 \,\text{N}/\text{m} \tag{14.114}$$

$$c_A = 2\xi_A \sqrt{k_A m} \approx 60.7 \,\text{N}\,\text{s}/\text{m} \tag{14.115}$$
$$c_B = 2\xi_B \sqrt{k_B m} \approx 418.5 \,\text{N}\,\text{s}/\text{m} \tag{14.116}$$

These are equivalent dampers and springs at the center of the wheel. The actual value of the suspension parameters depends on the geometry of the suspension mechanism and installment of the spring and damper. Because $k_B > k_A$ and $c_B > c_A$, suspension of vehicle B is harder than that of vehicle A. This is because vehicle B has less wheel travel, and hence, it has more acceleration level $\omega_n^2 S_{\ddot{X}}$. Figure 14.15 shows that

$$\omega_n^2 S_{\ddot{X}_B} \approx 220 \;\; 1/\text{s}^2 \qquad \omega_n^2 S_{\ddot{X}_A} \approx 28 \;\; 1/\text{s}^2 \tag{14.117}$$

Example 556 *Average vehicle suspension design.*

Most street cars with good ride comfort have a natural frequency equal or less than one Hertz. Optimal suspension characteristics of such a car are

$$f_n \approx 1\,\text{Hz} \qquad\qquad \xi \approx 0.028 \tag{14.118}$$
$$S_Z \approx 1.47644 \qquad \omega_n^2 S_{\ddot{X}_B} \approx 66 \;\; 1/\text{s}^2 \tag{14.119}$$

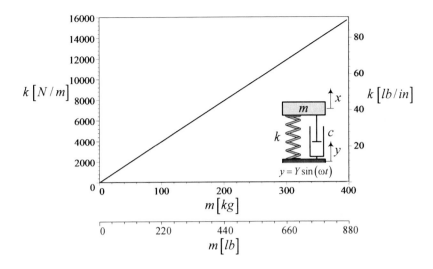

FIGURE 14.16. Optimal k as a function of m for a car with $f_n = 1$ Hz.

and therefore,

$$k = (2\pi f_n)^2 m \approx 4\pi^2 m \qquad (14.120)$$

$$c = 2\xi\sqrt{km} = 4\pi\xi m \approx 0.112\pi m \qquad (14.121)$$

Both k and c are proportional to the mass of the car, m. So, as a good estimate, we may use Figures 14.16 and 14.17 to design a car suspension.

For example, the optimal k and c for a car with $m = 250$ kg and $f_n = 1$ Hz are

$$k = 9869.6\,\mathrm{N/m} \qquad c = 87.96\,\mathrm{N\,s/m} \qquad (14.122)$$

Example 557 *Graphical representation of optimal characteristics.*

To visualize how the optimal parameters vary relatively, let us draw them in different coordinates. Figure 14.18 illustrates the optimal curve in plane $(S_{\ddot{x}}, S_Z)$. Figure 14.19 shows the optimal f_n and ξ versus S_Z, and Figure 14.20 shows the optimal f_n and ξ versus each other. The optimal ξ increases slightly with f_n for $f_n \lesssim 10$ Hz and it increases rapidly for $f_n \gtrsim 10$ Hz. So, as a general rule, when we change the spring of an optimal suspension with a harder spring, the damper should also be changed for a harder one.

Example 558 *Examination of the optimization of the design curve.*

To examine the optimal design curve and compare practical ways to make a suspension optimal, we assume that there is equipment with an off-optimal suspension, indicated by point P_1 in Figure 14.21.

$$f_n = 10\,\mathrm{Hz} \qquad \xi = 0.15 \qquad (14.123)$$

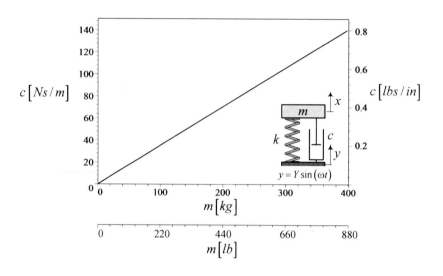

FIGURE 14.17. Optimal c as a function of m for a car with $f_n = 1\,\mathrm{Hz}$.

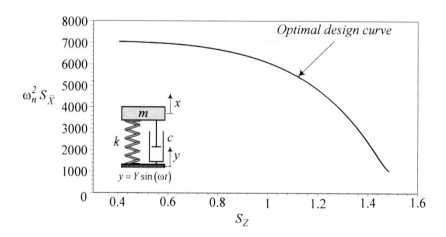

FIGURE 14.18. The optimal design curve in plane $(S_{\ddot{X}}, S_Z)$.

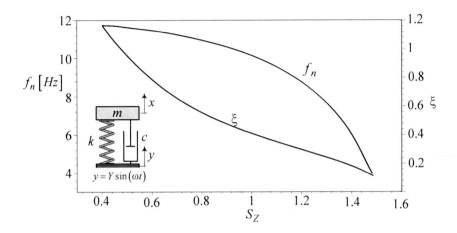

FIGURE 14.19. The optimal f_n and ξ versus S_Z.

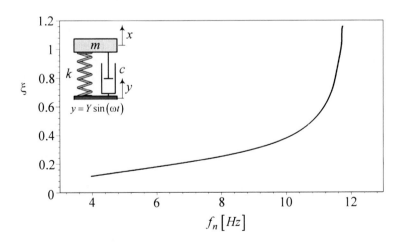

FIGURE 14.20. The optimal ξ versus optimal f_n.

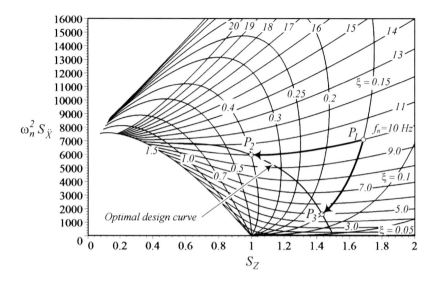

FIGURE 14.21. Two alternative optimal designs at points P_2 and P_3 for an off-optimal design at point P_1.

To optimize the suspension practically, we may keep the stiffness constant and change the damper to a corresponding optimal value, or keep the damping constant and change the stiffness to a corresponding optimal value. However, if it is possible, we may change both the stiffness and damping to a point on the optimal curve depending on the physical constraints and requirements.

Point P_2 in Figure 14.21 has the same f_n as point P_1 with an optimal damping ratio $\xi \approx 0.4$. Point P_3 in Figure 14.21 has the same ξ as point p_1 with an optimal natural frequency $f_n \approx 5\,Hz$. Hence, points P_2 and P_3 are two alternative optimal designs for the off-optimal point P_1.

Figure 14.22 compares the acceleration frequency response G_2 for the three points P_1, P_2, and P_3. Point P_3 has the minimum acceleration frequency response. Figure 14.23 depicts the absolute displacement frequency response G_0 and Figure 14.24 compares the relative displacement frequency response S_2 for points P_1, P_2, and P_3. These Figure show that both points P_2 and P_3 introduce better suspension than point P_1. Suspension P_2 has a higher level of acceleration but needs less relative suspension travel. Suspension P_3 has a lower acceleration, however it needs more room for suspension travel.

Example 559 *Sensitivity of $S_{\ddot{X}}$ with respect to S_Z on the optimal curve.*

Because $S_{\ddot{X}}$ is minimum on the optimal curve, the sensitivity of acceleration RMS with respect to relative displacement RMS is minimum at any point on the optimal curve. Therefore, an optimal suspension has the least

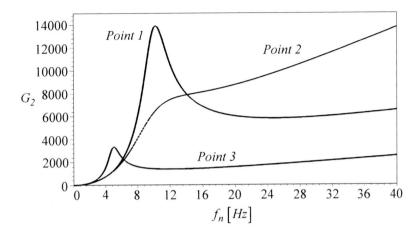

FIGURE 14.22. Acceleration frequency response G_2 for points P_1, P_2, and P_3 shown in Figure 14.21.

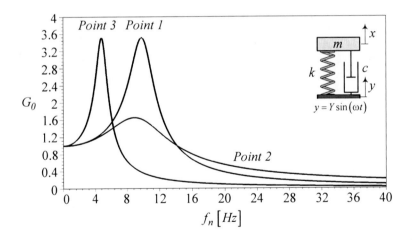

FIGURE 14.23. Absolute displacement frequency response G_0 for points P_1, P_2, and P_3 shown in Figure 14.21.

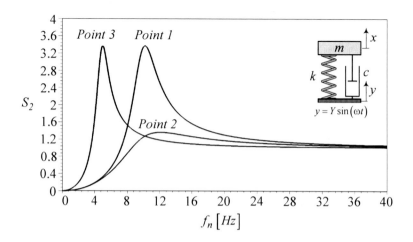

FIGURE 14.24. Relative displacement frequency response S_2 for points P_1, P_2, and P_3 shown in Figure 14.21.

sensitivity to mass variation which will change the available wheel travel. If a suspension is optimized for one passenger, it is still near optimal when the number of passengers changes.

Example 560 *Application of the optimal chart.*

Select a desired value for the relative displacement as a traveling space (or a desired value for the maximum absolute acceleration), and find the associated values for ω_n and ξ at the intersection of the associated vertical (or horizontal) line with the optimal curve.

Example 561 ★ *Three-dimensional view of the optimal curve.*

Figure 14.25 illustrates a 3D view of $S_{\ddot{X}}$ for different S_Z and f_n, to show the optimal curve in 3D.

Theoretically, we may show the surface by

$$S_{\ddot{X}} = S_{\ddot{X}}(S_Z, f_n) \tag{14.124}$$

and therefore, the optimal curve can be shown by the condition

$$\nabla S_{\ddot{X}} \cdot \hat{u}_{S_Z} = 0 \tag{14.125}$$

where $\hat{u}_{S_Z}$ is the unit vector along the S_Z-axis and $\nabla S_{\ddot{X}}$ is the gradient of the surface $S_{\ddot{X}}$.

Example 562 ★ *Suspension trade-off and trivial optimization.*

Reduction of the absolute acceleration is the main goal in the optimization of suspensions because it represents the transmitted force to the body. A vibration isolator reduces the absolute acceleration by increasing deflection

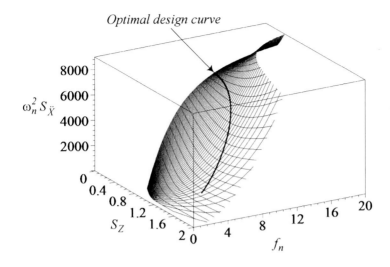

FIGURE 14.25. A three-dimensional view of $S_{\ddot{X}}$ for different S_Z and f_n, to show the optimal curve.

*of the isolator. The relative deflection is a measure of the clearance known as the **working space** of the isolator. The clearance should be minimized due to safety and the physical constraints in the mechanical design.*

There is a trade-off between the acceleration and relative motion. The ratio of $\omega_n^4 S_{\ddot{X}}$ to S_Z is a monotonically increasing function of ω_n and ξ. Keeping S_Z constant increases $\omega_n^4 S_{\ddot{X}}$ by increasing both ω_n and ξ. However, keeping $\omega_n^4 S_{\ddot{X}}$ constant, decrease S_Z by increasing ω_n and ξ. Hence, $\omega_n^4 S_{\ddot{X}}$ and S_Z have opposite behaviors. These behaviors show that $\omega_n = 0$ and $\xi = 0$ are the trivial and non practical solutions for the best isolation.

Example 563 ★ *Plot for RMS of absolute acceleration $RMS(G_2) = S_{\ddot{X}}$.*

Figures 14.26 and 14.27 illustrate the root mean square of absolute acceleration $RMS(G_2) = S_{\ddot{X}}$ graphically. In Figure 14.26, $S_{\ddot{X}}$ is plotted versus ξ with f_n as a parameter, and in Figure 14.27, $S_{\ddot{X}}$ is plotted versus f_n with ξ as a parameter.

Example 564 ★ *Plot for RMS of relative displacement $RMS(S_2) = S_Z$.*

Figures 14.28 and 14.29 illustrate the root mean square of relative displacement $RMS(S_2) = S_Z$. In Figure 14.28, S_Z is plotted versus ξ with f_n as a parameter and in Figure 14.29, S_Z is plotted versus f_n with ξ as a parameter.

Example 565 ★ *$RMS(G_0) \equiv RMS(X/Y)$.*

RMS of the absolute displacement, S_X, needs the integral of $G_0 \equiv$

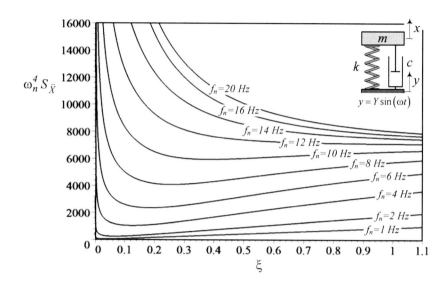

FIGURE 14.26. Plot of root mean square of absolute acceleration $RMS(S_3) = S_{\ddot{X}}$ versus ξ with f_n as a parameter.

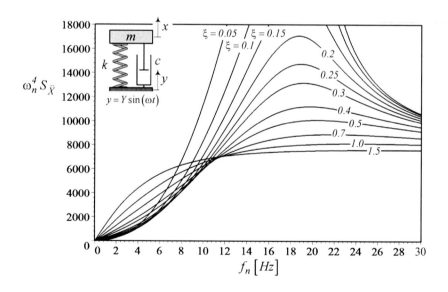

FIGURE 14.27. Plot of root mean square of absolute acceleration $RMS(S_3) = S_{\ddot{X}}$ versus f_n with ξ as a parameter.

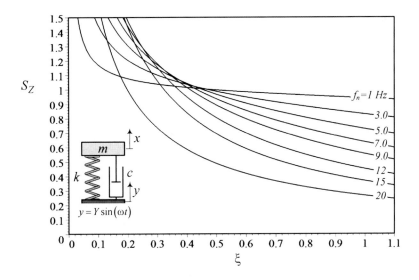

FIGURE 14.28. Plot of root mean square of relative displacement $RMS(S_2) = S_Z$ versus ξ with f_n as a parameter.

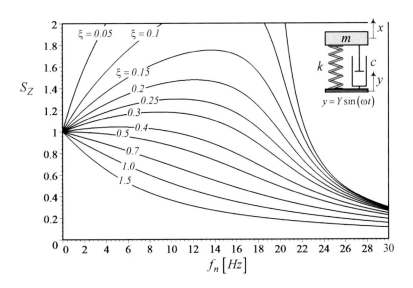

FIGURE 14.29. Plot of root mean square of relative displacement $RMS(S_2) = S_Z$ versus f_n with ξ as a parameter.

$(X/Y)^2$, which is determined as follows:

$$\int G_0 d\omega = \frac{Z_{16}}{Z_{17}\sqrt{Z_{18}}} \tan^{-1} \frac{\omega}{\sqrt{Z_{18}}} + \frac{Z_{19}}{Z_{20}\sqrt{Z_{21}}} \tan^{-1} \frac{\omega}{\sqrt{Z_{21}}} \qquad (14.126)$$

$$Z_{16} = \omega_n^2 \left(-8\xi^6 + 8\xi^4 - \left(8\xi^4 - 4\xi^2 - 1\right)\xi\sqrt{1-\xi^2} \right) \quad (14.127)$$

$$Z_{17} = -4\xi^2 \left(1 - \xi^2\right) \qquad (14.128)$$

$$Z_{18} = \omega_n^2 \left(1 - 2\xi^2 - 2\xi\sqrt{1-\xi^2}\right) \qquad (14.129)$$

$$Z_{19} = \omega_n^2 \left(8\xi^6 - 8\xi^4 - \left(8\xi^4 - 4\xi^2 - 1\right)\xi\sqrt{1-\xi^2} \right) \quad (14.130)$$

$$Z_{20} = Z_{17} = -4\xi^2 \left(1 - \xi^2\right) \qquad (14.131)$$

$$Z_{21} = -\omega_n^2 \left(1 - 2\xi^2 - 2\xi\sqrt{1-\xi^2}\right) \qquad (14.132)$$

Now the RMS of absolute displacement S_X can be determined analytically. Figures 14.30 and 14.31 illustrate the root mean square of relative displacement $RMS(G_0) = S_X$. In Figure 14.30, S_X is plotted versus ξ with f_n as a parameter and in Figure 14.31, S_X is plotted versus f_n with ξ as a parameter.

Example 566 ★ *Plot of $RMS(G_2) = S_{\ddot{X}}$ versus $RMS(S_2) = S_Z$.*

Figures 14.32 and 14.33 show $RMS(G_2) = S_{\ddot{X}}$ versus $RMS(S_2) = S_Z$ graphically. In Figure 14.32, $\omega_n^2 S_{\ddot{X}}$ is plotted for constant natural frequencies f_n, and in Figure 14.33 for constant ξ. Some of the curves in Figure 14.32 have a minimum, which shows that we may minimize $S_{\ddot{X}}$ versus S_Z for constant f_n. Such a minimum is the goal of optimization.

Figure 14.33 shows that there is a maximum on some of the constant ξ curves. These maximums indicate the worst suspension design.

Figure 14.34 illustrates the behavior of $S_{\ddot{X}}$, instead of $\omega_n^2 S_{\ddot{X}}$, versus S_Z. The minimum point on each curve occurs at the same S_Z as in Figure 14.32.

Example 567 ★ *Alternative optimization methods.*

There are various approaches and suggested methods for vibration isolator optimization, depending on the application. However, there is not a universally accepted method applicable to every application. Every optimization strategy can be transformed to a minimization of a function called the **cost function** or **objective function**. Considerable attention has been given to minimization of the absolute displacement, known as the **main transmissibility**. However, for a vibration isolator, the cost function may

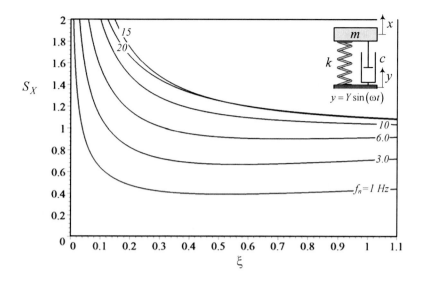

FIGURE 14.30. Plot of root mean square of absolute displacement $RMS(S_1) = S_X$ versus ξ with f_n.

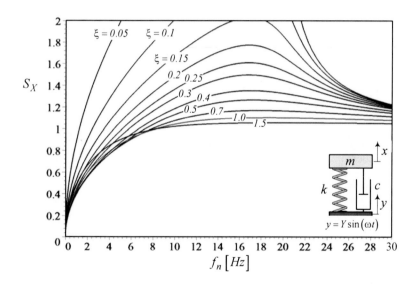

FIGURE 14.31. Plot of root mean square of absolute displacement $RMS(S_1) = S_X$ versus f_n with ξ as a parameter.

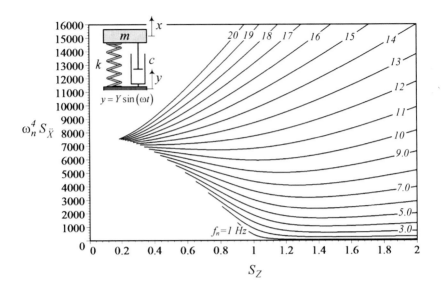

FIGURE 14.32. Plot of $\omega_n^2 RMS(G_2) = \omega_n^2 S_{\ddot{X}}$ versus $RMS(S_2) = S_Z$ for constant natural frequencies f_n.

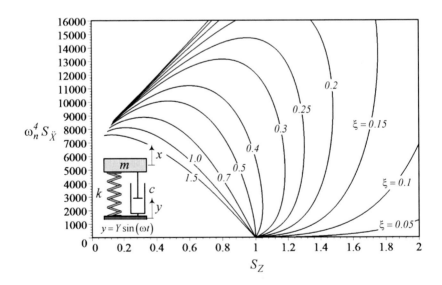

FIGURE 14.33. Plot of $\omega_n^2 RMS(G_2) = \omega_n^2 S_{\ddot{X}}$ versus $RMS(S_2) = S_Z$ for constant damping ratio ξ.

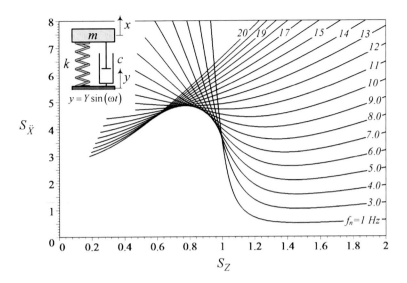

FIGURE 14.34. Plot of $RMS(G_2) = S_{\ddot{X}}$ versus $RMS(S_2) = S_Z$ for constant natural frequencies f_n.

include any state variables such as absolute and relative displacements, ve-locities, accelerations, and jerks.

Constraints may determine the domain of acceptable design parameters by dictating an upper and lower limit for ω_n and ξ. For vehicle suspension, it is generally desired to select ω_n and ξ such that the absolute acceleration of the system is minimized and the relative displacement does not exceed a prescribed level. The most common optimization strategies are:

Minimax absolute acceleration $S_{\ddot{X}}$ for specified relative dis-placement S_{Z_0}. Specify the allowable relative displacement, and then find the minimax of absolute acceleration

$$\frac{\partial S_{\ddot{X}}}{\partial \omega_n} = 0 \qquad \frac{\partial S_{\ddot{X}}}{\partial \xi} = 0 \qquad S_Z = S_{Z_0} \qquad (14.133)$$

Minimax relative displacement S_Z for specified absolute accel-eration $S_{\ddot{X}_0}$. Specify the allowable absolute acceleration, and then find the minimax relative displacement.

$$\frac{\partial S_Z}{\partial \omega_n} = 0 \qquad \frac{\partial S_Z}{\partial \xi} = 0 \qquad S_{\ddot{X}} = S_{\ddot{X}_0} \qquad (14.134)$$

Example 568 ★ *More application of the design chart.*

The optimization criterion

$$\frac{\partial S_{\ddot{X}}}{\partial S_Z} = 0 \qquad \frac{\partial^2 S_{\ddot{X}}}{\partial S_Z^2} > 0 \qquad (14.135)$$

is based on the root mean square of S_2 and G_2 over a working frequency range, in this case, zero to $20\,\mathrm{Hz}$.

$$S_Z = \sqrt{\frac{1}{40\pi} \int_0^{40\pi} S_2^2 d\omega} \tag{14.136}$$

$$S_{\ddot{X}} = \sqrt{\frac{1}{40\pi} \int_0^{40\pi} G_2 d\omega} \tag{14.137}$$

The optimal design curve is the optimal condition for suspension of a base excited system using the relative displacement and absolute acceleration:

$$S_2 = \frac{Z_B}{Y} \qquad G_2 = \frac{\ddot{X}_B}{\omega_n^2 Y}$$

However, because

$$S_2 = \frac{\ddot{X}_F}{F/m} = \frac{Z_B}{Y} = \frac{X_E}{e\varepsilon_E} = \frac{Z_R}{e\varepsilon_R} \tag{14.138}$$

$$G_2 = \frac{\ddot{X}_B}{\omega_n^2 Y} = \frac{F_{T_B}}{kY} = \frac{F_{T_E}}{e\omega^2 m_e} = \frac{F_{T_R}}{e\omega^2 m_e}\left(1 + \frac{m_a}{m}\right) \tag{14.139}$$

the optimal design curve can also be expressed as a minimization condition for any other G_2-function with respect to any other S_2-function, such as transmitted force to the base $\frac{F_{T_E}}{e\omega^2 m_e}$ for an eccentric excited system $\frac{X_E}{e\varepsilon_E}$. This minimization is equivalent to the optimization of an engine mount.

14.4 ★ Time Response Optimization

Transient response optimization depends on the type of transient excitation, as well as cost function definition. Figure 14.35 illustrates a 1/8 car model and a unit step displacement.

$$y = \begin{cases} 1 & t > 0 \\ 0 & t \le 0 \end{cases} \tag{14.140}$$

If the transient excitation is a step function, and the optimization criteria is minimization of the peak value of the acceleration versus peak value of the relative displacement, then there is an optimal $\xi^\star$ for any f_n that provides the best transient behavior of a 1/8 car model. This behavior is shown in Figure 14.36.

$$\xi^\star = 0.4 \tag{14.141}$$

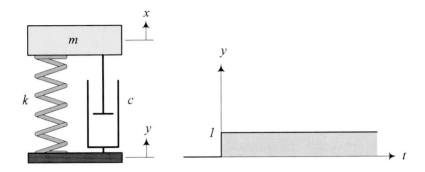

FIGURE 14.35. A 1/8 car model and a unit step displacement base excitation.

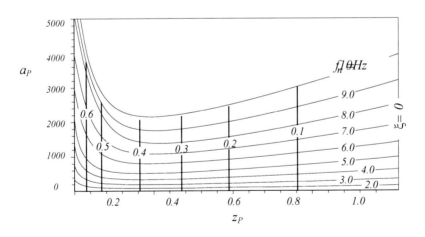

FIGURE 14.36. Peak value of acceleration versus peak value of relative displacement for different ξ and f_n.

Proof. The equation of motion for the base excited one-DOF system shown in Figure 14.35, is

$$\ddot{x} + 2\xi\omega_n\,\dot{x} + \omega_n^2\,x = 2\xi\omega_n\,\dot{y} + \omega_n^2\,y \tag{14.142}$$

Substituting $y = 1$ in Equation (14.142) provides the following initial value problem to determine the absolute outplacement of the mass m:

$$\ddot{x} + 2\xi\omega_n\,\dot{x} + \omega_n^2\,x \ = \ \omega_n^2 \tag{14.143}$$
$$y(0) \ = \ 0 \tag{14.144}$$
$$\dot{y}(0) \ = \ 0 \tag{14.145}$$

Solution of the differential equation with zero initial conditions is

$$x = 1 - \frac{1}{2}\frac{A}{ib}e^{-A\omega_n t} + \frac{1}{2}\frac{\overline{A}}{ib}e^{-\overline{A}\omega_n t} \tag{14.146}$$

where A and $\overline{A}$ are two complex conjugate numbers.

$$A \ = \ \xi + i\sqrt{1 - \xi^2} \tag{14.147}$$
$$\overline{A} \ = \ \xi - i\sqrt{1 - \xi^2} \tag{14.148}$$

Having x and $y = 1$ are enough to calculate the relative displacement $z = x - y$.

$$z = x - y = -\frac{1}{2}\frac{A}{ib}e^{-A\omega_n t} + \frac{1}{2}\frac{\overline{A}}{ib}e^{-\overline{A}\omega_n t} \tag{14.149}$$

The absolute velocity and acceleration of the mass m can be obtained from equation (14.146).

$$\dot{x} \ = \ \frac{1}{2}\frac{A^2\omega_n}{ib}e^{-A\omega_n t} - \frac{1}{2}\frac{\overline{A}^2\omega_n}{ib}e^{-\overline{A}\omega_n t} \tag{14.150}$$
$$\ddot{x} \ = \ -\frac{1}{2}\frac{A^3\omega_n^2}{ib}e^{-A\omega_n t} + \frac{1}{2}\frac{\overline{A}^3\omega_n^2}{ib}e^{-\overline{A}\omega_n t} \tag{14.151}$$

The peak value of the relative displacement is

$$z_P = \exp\left(\frac{\cos^{-1}\left(2\xi^2 - 1\right)}{\omega_n\sqrt{1 - \xi^2}}\right) \tag{14.152}$$

which occurs when $\dot{z} = 0$ at time t_1

$$t_1 = \frac{-\xi\cos^{-1}\left(2\xi^2 - 1\right)}{\sqrt{1 - \xi^2}} \tag{14.153}$$

The peak value of the absolute acceleration is

$$a_P = \omega_n^2 \exp\left(-\xi \frac{2\cos^{-1}\left(2\xi^2 - 1\right) - \pi}{\sqrt{1 - \xi^2}}\right) \tag{14.154}$$

which occurs at the beginning of the excitation, $t = 0$, or at the time instant when $\ddot{x} = 0$ at time t_2

$$t_2 = \frac{2\cos^{-1}\left(2\xi^2 - 1\right) - \pi}{\omega_n\sqrt{1 - \xi^2}} \tag{14.155}$$

Figure 14.36 is a plot for a_P versus z_P for different ξ and f_n. The minimum of the curves occur at $\xi \approx 0.4$ for every f_n. The optimal ξ can be found analytically by finding the minimum point of a_P versus z_P. The optimal ξ is the solution of the transcendental equation

$$2\xi \cos^{-1}\left(2\xi^2 - 1\right) - \pi - 4\xi\sqrt{1 - \xi^2} = 0 \tag{14.156}$$

which is $\xi \approx 0.4$. The minimum peak value of the absolute acceleration with respect to relative displacement is independent of the value of natural frequency f_n. ∎

Example 569 ★ *Optimal design curve and time response.*

To examine transient response of suspensions on the optimal design curve, we compare a base excited equipment having an off-optimal suspension, at point P_1, with optimal suspensions at points P_2 and P_3 in Figure 14.21. Point P_1 is at

$$f_n \approx 10\,\text{Hz} \qquad \xi \approx 0.15 \tag{14.157}$$

Points P_2 and P_3 are two alternative optimizations for point P_1. Point P_2 has $\xi \approx 0.4$ with the same natural frequency as P_1 and point P_2 has $f_n \approx 5\,\text{Hz}$ with the same damping as point P_1.

Figure 14.37 illustrates a base excited one-DOF system and a sine square bump input.

$$y = \begin{cases} d_2 \sin^2 \dfrac{2\pi v}{d_1} t & 0 < t < 0.1 \\ 0 & t \le 0, t \ge 0.1 \end{cases} \tag{14.158}$$

$$d_1 = 1\,\text{m} \qquad d_2 = 0.05\,\text{m} \tag{14.159}$$

$$v = 10\,\text{m/s} \tag{14.160}$$

The absolute and relative displacement time responses of the system at points 1, 2, and 3 are shown in Figures 14.38, and 14.39, respectively. The absolute acceleration of m is shown in Figure 14.40.

System 3 has a lower relative displacement peak value and a lower absolute acceleration peak value, but it takes more time to settle down.

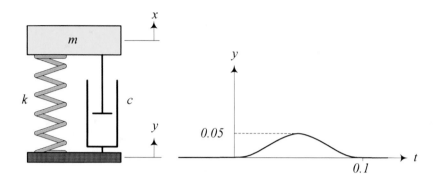

FIGURE 14.37. A base excited *one*-DOF system and a sine square bump input.

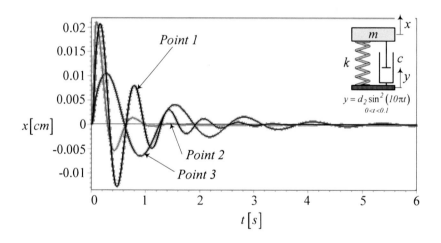

FIGURE 14.38. Absolute displacement time response of the system for three different suspensions.

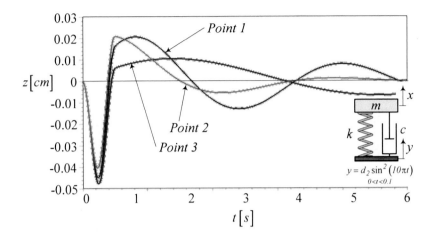

FIGURE 14.39. Relative displacement time response of the system for three different suspensions.

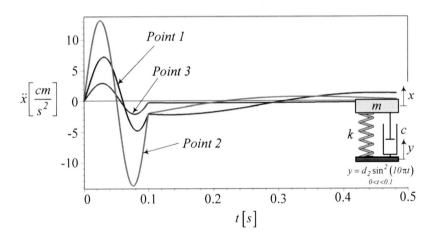

FIGURE 14.40. Absolute acceleration time response of the system for three different suspensions.

14.5 Summary

A one-DOF base excited system with the equation of motion

$$\ddot{x} + 2\xi\omega_n\,\dot{x} + \omega_n^2\,x = 2\xi\omega_n\,\dot{y} + \omega_n^2\,y \qquad (14.161)$$

is an applied model for equipment mounted on a vibrating base, as well as a model for vertical vibration of vehicles. Assuming a variable excitation frequency, we may determine the relative displacement $S_2 = |Z/Y|$ and absolute acceleration $G_2 = \left| \ddot{X} / \left(Y\omega_n^2 \right) \right|$ frequency responses to optimize the system. The optimization criterion is

$$\frac{\partial S_{\ddot{X}}}{\partial S_Z} = 0 \qquad (14.162)$$

$$\frac{\partial^2 S_{\ddot{X}}}{\partial S_Z^2} > 0 \qquad (14.163)$$

where S_Z and $S_{\ddot{X}}$ are the root mean square of S_2 and G_2 over a working frequency range, $0 \le \omega \le 20\,\mathrm{Hz}$.

$$S_Z = \sqrt{\frac{1}{40\pi} \int_0^{40\pi} S_2^2 d\omega} \qquad (14.164)$$

$$S_{\ddot{X}} = \sqrt{\frac{1}{40\pi} \int_0^{40\pi} G_2 d\omega} \qquad (14.165)$$

The optimization criterion states that the minimum absolute acceleration RMS with respect to the relative displacement RMS, makes a suspension optimal. The result of optimization may be cast in a design chart to visualize the relationship of optimal ξ and ω_n.

14.6   Key Symbols

$a, \ddot{x}$	acceleration
a, b	arm length of displaced spring
c	damping
$c^{\star}$	optimum damping
c_{eq}	equivalent damping
d_1	road wave length
d_2	road wave amplitude
D	dissipation function
$f, \mathbf{F}$	force
$f = \frac{1}{T}$	cyclic frequency [Hz]
f_c	damper force
f_k	spring force
f_n	cyclic natural frequency [Hz]
F	amplitude of a harmonic force f
g	gravitational acceleration
$G_0 = \lvert X/Y \rvert$	absolute displacement frequency response
$G_2 = \left\lvert \ddot{X}/Y\omega_n^2 \right\rvert$	absolute acceleration frequency response
k	stiffness
$k^{\star}$	optimum stiffness
k_{eq}	equivalent stiffness
K	kinetic energy
$\mathcal{L}$	Lagrangean
m	mass
$r = \frac{\omega}{\omega_n}$	frequency ratio
$S_2 = \lvert Z/Y \rvert$	relative displacement frequency response
S_Z	RMS of S_2
$S_{\ddot{X}}$	RMS of G_2
t	time
T	period
$v, \mathbf{v}, \dot{x}, \dot{\mathbf{x}}$	velocity
V	potential energy
x	absolute displacement
X	steady-state amplitude of x
y	base excitation displacement
Y	steady-state amplitude of y
z	relative displacement
Z	steady-state amplitude of z
Z_i	short notation parameter
α	tilted spring angle
δ	spring deflection
δ	displacement

$\xi = \frac{c}{2\sqrt{km}}$ damping ratio

$\omega = 2\pi f$ angular frequency $[\,\mathrm{rad/s}\,]$

ω_n natural frequency

Subscript

eq equivalent

f front

l low

r rear

s sprung

u unsprung

u up

Exercises

1. Equivalent McPherson suspension parameters.

   Figure 14.41(a) illustrates a McPherson suspension. Its equivalent vibrating system is shown in Figure 14.41(b).

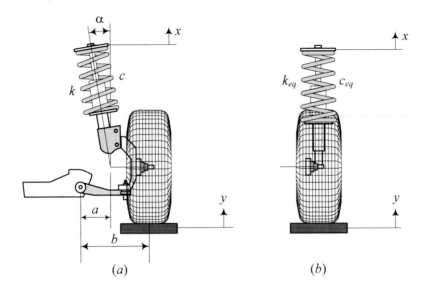

   FIGURE 14.41. A McPherson suspension and its equivalent vibrating system.

   (a) Determine k_{eq} and c_{eq} if

   $$a \; = \; 22\,\text{cm} \qquad b = 45\,\text{cm} \qquad c = 1000\,\text{N s}/\,\text{m} \qquad \alpha = 12\,\text{deg}$$
   $$k \; = \; 10000\,\text{N}/\,\text{m}$$

   (b) Determine the stiffness k such that the natural frequency of the vibrating system is $f_n = 1\,\text{Hz}$, if

   $$a \; = \; 22\,\text{cm} \qquad b = 45\,\text{cm} \qquad c = 1000\,\text{N s}/\,\text{m} \qquad \alpha = 12\,\text{deg}$$
   $$m \; = \; 1000/4\,\text{kg}$$

   (c) Determine the damping c such that the damping ratio of the vibrating system is $\xi = 0.4$, if

   $$a \; = \; 22\,\text{cm} \qquad b = 45\,\text{cm} \qquad c = 1000\,\text{N s}/\,\text{m} \qquad \alpha = 12\,\text{deg}$$
   $$m \; = \; 1000/4\,\text{kg} \qquad\qquad\qquad f_n = 1\,\text{Hz}$$

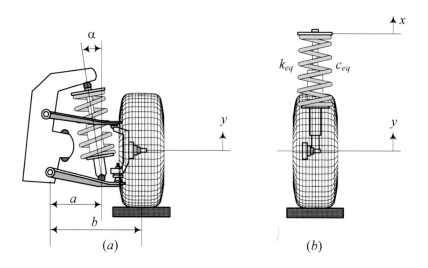

FIGURE 14.42. A double A-arm suspension and its equivalent vibrating system.

2. Equivalent double A-arm suspension parameters.

Figure 14.42(a) illustrates a double A-arm suspension. Its equivalent vibrating system is shown in Figure 14.42(b).

(a) Determine k_{eq} and c_{eq} if

$$a \;=\; 32\,\text{cm} \qquad b = 45\,\text{cm} \qquad c = 1000\,\text{N}\,\text{s}/\text{m} \qquad \alpha = 10\,\text{deg}$$
$$k \;=\; 8000\,\text{N}/\text{m}$$

(b) Determine the stiffness k such that the natural frequency of the vibrating system is $f_n = 1\,\text{Hz}$, if

$$a \;=\; 32\,\text{cm} \qquad b = 45\,\text{cm} \qquad \alpha = 10\,\text{deg}$$
$$m \;=\; 1000/4\,\text{kg}$$

(c) Determine the damping c such that the damping ratio of the vibrating system is $\xi = 0.4$, if

$$a \;=\; 32\,\text{cm} \qquad b = 45\,\text{cm} \qquad \alpha = 10\,\text{deg}$$
$$f_n \;=\; 1\,\text{Hz} \qquad m = 1000/4\,\text{kg}$$

3. Road excitation frequency.

A car is moving on a wavy road. What is the wave length d_1 if the excitation frequency is $f_n = 5\,\text{Hz}$ and

(a) $v = 30\,\text{km}/\text{h}$

(b) $v = 60\,\text{km/h}$

(c) $v = 100\,\text{km/h}$

4. ★ Road excitation frequency and wheelbase.

A car is moving on a wavy road.

(a) What is the wave length d_1 if the excitation frequency is $f_n = 8\,\text{Hz}$ at $v = 60\,\text{km/h}$?

(b) What is the phase difference between the front and rear wheel excitations if car's wheelbase is $l = 2.82\,\text{m}$?

(c) At what speed the front and rear wheel excitations have no phase difference?

5. ★ Road excitation amplitude.

A car is moving on a wavy road with a wave length $d_1 = 25\,\text{m}$. What is the damping ratio ξ if $S_2 = Z/Y = 1.02$ when the car is moving with $v = 120\,\text{km/h}$?

$$k = 10000\,\text{N/m} \qquad m = 1000/4\,\text{kg}$$

6. Optimized suspension comparison.

A car with $m = 1000/4\,\text{kg}$ is moving on a wavy road with a wave length $d_1 = 45\,\text{m}$ and wave amplitude $d_2 = 8\,\text{cm}$ at $v = 120\,\text{km/h}$. What is the best suspension parameters if the equivalent wheel travel of the car at the wheel center is

(a) $5\,\text{cm}$

(b) $8\,\text{cm}$

(c) $12\,\text{cm}$

(d) Calculate G_0, G_2, S_2, S_Z, $S_{\ddot{X}}$, X, Z, and $\ddot{X}$ in each case.

7. Suspension optimization and keeping k or c.

Consider a base excited system with

$$k = 10000\,\text{N/m} \qquad m = 1000/4\,\text{kg} \qquad c = 1000\,\text{N s/m}.$$

(a) Determine the best k for the same ξ.

(b) Determine the best c for the same k.

(c) Determine the value of S_Z and $S_{\ddot{X}}$ in each case.

8. Suspension optimization for minimum S_Z.

   A base excited system has

   $$k = 250000 \,\text{N}/\text{m} \qquad m = 2000 \,\text{kg} \qquad c = 2000 \,\text{N s}/\text{m}$$

   Determine the level of acceleration RMS $S_{\ddot{X}}$ that transfers to the system.

9. Peak values and step input.

   A base excited system has

   $$k = 10000 \,\text{N}/\text{m} \qquad m = 1000/4 \,\text{kg} \qquad c = 1000 \,\text{N s}/\text{m}$$

   What is the acceleration and relative displacement peak values for a unit step input?

10. ★ Acceleration peak value and spring stiffness.

   Explain why the ξ-constant curves are vertical lines in the plane (a_P, z_P).

15

★ Quarter Car Model

The most employed and useful model of a vehicle suspension system is a quarter car model, shown in Figure 15.1. We introduce, examine, and optimize the quarter car model in this chapter.

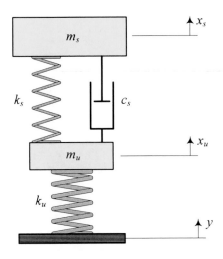

FIGURE 15.1. A quarter car model.

15.1 Mathematical Model

We may represent the vertical vibration of a vehicle using a quarter car model made of two solid masses m_s and m_u denoted as *sprung* and *unsprung* masses, respectively. The sprung mass m_s represents $1/4$ of the body of the vehicle, and the unsprung mass m_u represents one wheel of the vehicle. A spring of stiffness k_s, and a shock absorber with viscous damping coefficient c_s, support the sprung mass and are called the *main suspension*. The unsprung mass m_u is in direct contact with the ground through a spring k_u, representing the tire stiffness.

The governing differential equations of motion for the quarter car model shown in Figure 15.1, are:

$$m_s \ddot{x}_s + c_s (\dot{x}_s - \dot{x}_u) + k_s (x_s - x_u) = 0 \qquad (15.1)$$
$$m_u \ddot{x}_u + c_s (\dot{x}_u - \dot{x}_s) + (k_u + k_s) x_u - k_s x_s = k_u y \qquad (15.2)$$

R.N. Jazar, *Vehicle Dynamics: Theory and Application*,
DOI 10.1007/978-1-4614-8544-5_15 © Springer Science+Business Media New York 2014

Proof. The kinetic energy, potential energy, and dissipation function of the system are:

$$K = \frac{1}{2}m_s\dot{x}_s^2 + \frac{1}{2}m_u\dot{x}_u^2 \tag{15.3}$$

$$V = \frac{1}{2}k_s\left(x_s - x_u\right)^2 + \frac{1}{2}k_u\left(x_u - y\right)^2 \tag{15.4}$$

$$D = \frac{1}{2}c_s\left(\dot{x}_s - \dot{x}_u\right)^2 \tag{15.5}$$

Employing the Lagrange method,

$$\frac{d}{dt}\left(\frac{\partial K}{\partial \dot{x}_s}\right) - \frac{\partial K}{\partial x_s} + \frac{\partial D}{\partial \dot{x}_s} + \frac{\partial V}{\partial x_s} = 0 \tag{15.6}$$

$$\frac{d}{dt}\left(\frac{\partial K}{\partial \dot{x}_u}\right) - \frac{\partial K}{\partial x_u} + \frac{\partial D}{\partial \dot{x}_u} + \frac{\partial V}{\partial x_u} = 0 \tag{15.7}$$

we find the equations of motion

$$m_s\ddot{x}_s = -k_s\left(x_s - x_u\right) - c_s\left(\dot{x}_s - \dot{x}_u\right) \tag{15.8}$$

$$m_u\ddot{x}_u = k_s\left(x_s - x_u\right) + c_s\left(\dot{x}_s - \dot{x}_u\right) - k_u\left(x_u - y\right) \tag{15.9}$$

which can be expressed in a matrix form

$$[m]\ddot{\mathbf{x}} + [c]\dot{\mathbf{x}} + [k]\mathbf{x} = \mathbf{F} \tag{15.10}$$

$$\begin{bmatrix} m_s & 0 \\ 0 & m_u \end{bmatrix}\begin{bmatrix} \ddot{x}_s \\ \ddot{x}_u \end{bmatrix} + \begin{bmatrix} c_s & -c_s \\ -c_s & c_s \end{bmatrix}\begin{bmatrix} \dot{x}_s \\ \dot{x}_u \end{bmatrix} +$$
$$\begin{bmatrix} k_s & -k_s \\ -k_s & k_s + k_u \end{bmatrix}\begin{bmatrix} x_s \\ x_u \end{bmatrix} = \begin{bmatrix} 0 \\ k_u y \end{bmatrix} \tag{15.11}$$

■

Example 570 *Tire damping.*
 We may add a damper c_u in parallel to k_u, as shown in Figure 15.1, to model any damping in tires. However, the value of c_u for tires, compared to c_s, are very small and hence, we may ignore c_u to simplify the model. Having the damper c_u in parallel to k_u makes the equation of motion the same as Equations (12.44) and (12.45) with a matrix form as Equation (12.47).

Example 571 *Mathematical model's limitations.*
 The quarter car model contains no representation of the geometric effects of the full car and offers no possibility of studying longitudinal and lateral interconnections. However, it contains the most basic features of the real problem and includes a proper representation of the problem of controlling wheel and wheel-body load variations.

In the quarter car model, we assume that the tire is always in contact with the ground, which is true at low frequency but might not be true at high frequency. A better model must be able to include the possibility of separation between the tire and ground.

Optimal design of two-DOF vibration systems, including a quarter car model, is the subject of numerous investigations since the invention of the vibration absorber theory by Frahm in 1909. It seems that the first analytical investigation on the damping properties of two-DOF systems is due to Den Hartog (1901 − 1989).

15.2   Frequency Response

To find the frequency response, we consider a harmonic excitation,

$$y = Y \cos \omega t \tag{15.12}$$

and look for a harmonic solution in the form

$$x_s = A_1 \sin \omega t + B_1 \cos \omega t = X_s \sin (\omega t - \varphi_s) \tag{15.13}$$

$$x_u = A_2 \sin \omega t + B_2 \cos \omega t = X_u \sin (\omega t - \varphi_u) \tag{15.14}$$

$$z = x_s - x_u = A_3 \sin \omega t + B_3 \cos \omega t = Z \sin (\omega t - \varphi_z) \tag{15.15}$$

where X_s, X_u, and Z are complex amplitudes.

By introducing the following dimensionless characters:

$$\varepsilon = \frac{m_s}{m_u} \tag{15.16}$$

$$\omega_s = \sqrt{\frac{k_s}{m_s}} \tag{15.17}$$

$$\omega_u = \sqrt{\frac{k_u}{m_u}} \tag{15.18}$$

$$\alpha = \frac{\omega_s}{\omega_u} \tag{15.19}$$

$$r = \frac{\omega}{\omega_s} \tag{15.20}$$

$$\xi = \frac{c_s}{2m_s\omega_s} \tag{15.21}$$

we search for the absolute and relative frequency responses:

$$\mu = \left| \frac{X_s}{Y} \right| \tag{15.22}$$

$$\tau = \left| \frac{X_u}{Y} \right| \tag{15.23}$$

$$\eta = \left| \frac{Z}{Y} \right| \tag{15.24}$$

and obtain the following functions:

$$\mu^2 = \frac{4\xi^2 r^2 + 1}{Z_1^2 + Z_2^2} \tag{15.25}$$

$$\tau^2 = \frac{4\xi^2 r^2 + 1 + r^2 \left(r^2 - 2\right)}{Z_1^2 + Z_2^2} \tag{15.26}$$

$$\eta^2 = \frac{r^4}{Z_1^2 + Z_2^2} \tag{15.27}$$

$$Z_1 = \left[r^2 \left(r^2 \alpha^2 - 1\right) + \left(1 - (1 + \varepsilon) r^2 \alpha^2\right) \right] \tag{15.28}$$

$$Z_2 = 2\xi r \left(1 - (1 + \varepsilon) r^2 \alpha^2\right) \tag{15.29}$$

The absolute acceleration of sprung mass and unsprung mass may be defined by the following equations:

$$u = \left| \frac{\ddot{X}_s}{Y \omega_u^2} \right| = r^2 \alpha^2 \mu \tag{15.30}$$

$$v = \left| \frac{\ddot{X}_u}{Y \omega_u^2} \right| = r^2 \alpha^2 \tau \tag{15.31}$$

Proof. To find the frequency responses, let us apply a harmonic excitation

$$y = Y \cos \omega t \tag{15.32}$$

and assume that the solutions are harmonic functions with unknown coefficients.

$$x_s = A_1 \sin \omega t + B_1 \cos \omega t \tag{15.33}$$

$$x_u = A_2 \sin \omega t + B_2 \cos \omega t \tag{15.34}$$

Substituting the solutions in the equations of motion (15.1)-(15.2) and collecting the coefficients of $\sin \omega t$ and $\cos \omega t$ in both equations provides the following set of algebraic equations for A_1, B_1, A_2, B_2:

$$[A] \begin{bmatrix} A_1 \\ A_2 \\ B_1 \\ B_2 \end{bmatrix} = \begin{bmatrix} 0 \\ 0 \\ k_u Y \\ 0 \end{bmatrix} \tag{15.35}$$

where $[A]$ is the coefficient matrix.

$$[A] = \begin{bmatrix} k_s - m_s \omega^2 & -k_s & -c_s \omega & c_s \omega \\ c_s \omega & -c_s \omega & k_s - m_s \omega^2 & -k_s \\ -k_s & k_s + k_u - m_u \omega^2 & c_s \omega & -c_s \omega \\ -c_s \omega & c_s \omega & -k_s & k_s + k_u - m_u \omega^2 \end{bmatrix} \tag{15.36}$$

The unknowns may be found by matrix inversion

$$
\begin{bmatrix} A_1 \\ A_2 \\ B_1 \\ B_2 \end{bmatrix} = [A]^{-1} \begin{bmatrix} 0 \\ 0 \\ k_u Y \\ 0 \end{bmatrix}
\tag{15.37}
$$

and therefore, the amplitudes X_s and X_u can be found.

$$
X_s^2 = A_1^2 + B_1^2 = \frac{k_u}{k_s} \frac{\left(\omega^2 c_s^2 + k_s^2 \right)}{Z_3^2 + Z_4^2} Y^2
\tag{15.38}
$$

$$
X_u^2 = A_2^2 + B_2^2 = \frac{k_u}{k_s} \frac{\left(\omega^4 m_s^2 + \omega^2 c_s^2 - 2\omega^2 k_s m_s + k_s^2 \right)}{Z_3^2 + Z_4^2} Y^2
\tag{15.39}
$$

$$
Z_3 = -\left(\omega^2 \left(k_s m_s + k_s m_u + k_u m_s \right) - k_s k_u - \omega^4 m_s m_u \right)
\tag{15.40}
$$

$$
Z_4 = -\left(\omega^3 \left(c_s m_s + c_s m_u \right) - \omega c_s k_u \right)
\tag{15.41}
$$

Having X_s and X_u helps us to calculate z and its amplitude Z.

$$
\begin{aligned}
z &= x_s - x_u = (A_1 - A_2) \sin \omega t + (B_1 - B_2) \cos \omega t \\
&= A_3 \sin \omega t + B_3 \cos \omega t = Z \sin \left(\omega t - \varphi_z \right)
\end{aligned}
\tag{15.42}
$$

$$
Z^2 = A_3^2 + B_3^2 = \frac{k_u}{k_s} \frac{\omega^4 m_s^2}{Z_3^2 + Z_4^2} Y^2
\tag{15.43}
$$

Taking derivative from μ and τ provides the acceleration frequency responses u and v for the unsprung and sprung masses. Equations (15.30)-(15.31) express u and v.

Using the definitions (15.16)-(15.21), we may transform Equations (15.38), (15.39), (15.43) to (15.25), (15.26), (15.27). Figures 15.2, 15.3, 15.4, are samples of the frequency responses μ, τ, and η for $\varepsilon = 3$, and $\alpha = 0.2$. ■

Example 572 *Average value of parameters for street cars.*

Equations (15.25)-(15.27) indicate that the frequency responses μ, τ, and η are functions of four parameters: mass ratio ε, damping ratio ξ, natural frequency ratio α, and excitation frequency ratio r. The average, minimum, and maximum of practical values of the parameters are indicated in Table 14.1.

For a quarter car model, it is known that $m_s > m_u$, and therefore, $\varepsilon > 1$. Typical mass ratio, ε, for vehicles lies in the range 3 to 8, with small cars closer to 8 and large cars near 3. The excitation frequency ω is equal to ω_u, when $r = 1/\alpha$, and equal to ω_s, when $r = 1$. For a real model, the order of magnitude of the stiffness is $k_u > k_s$, so $\omega_u > \omega_s$, and $\alpha < 1$. Therefore, $r > 1$ at $\omega = \omega_u$. So, we expect to have two resonant frequencies greater

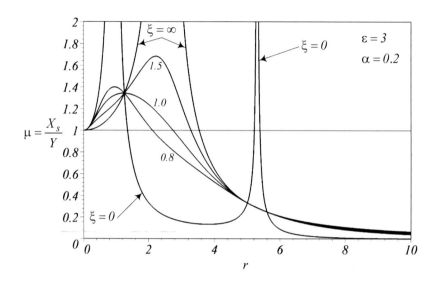

FIGURE 15.2. A sample for the sprung mass displacement frequency response, $\mu = \left| \frac{X_s}{Y} \right|$.

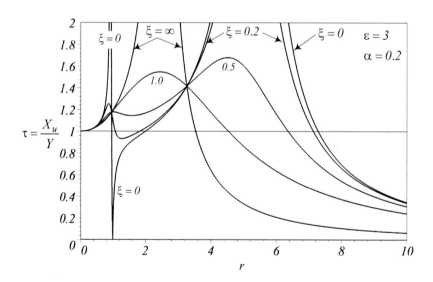

FIGURE 15.3. A sample of the unsprung mass displacement frequency response, $\tau = \left| \frac{X_u}{Y} \right|$.

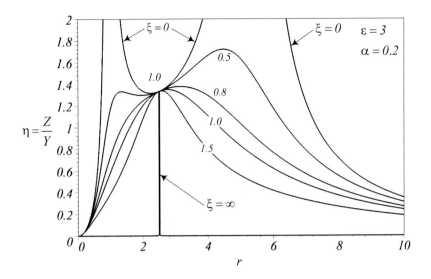

FIGURE 15.4. A sample of the relative displacement frequency response, $\eta = \left| \frac{Z}{Y} \right|$.

than $r = 1$.

Table 14.1 - *Average value of quarter car parameters.*

Parameter	Average	Minimum	Maximum
$\varepsilon = \dfrac{m_s}{m_u}$	$3 - 8$	2	20
$\omega_s = \sqrt{\dfrac{k_s}{m_s}}$	1	0.2	1
$\omega_u = \sqrt{\dfrac{k_u}{m_u}}$	10	2	20
$r = \dfrac{\omega}{\omega_s}$	$0 - 20\,\text{Hz}$	0	$200\,\text{Hz}$
$\alpha = \dfrac{\omega_s}{\omega_u}$	0.1	0.01	1
$\xi = \dfrac{c_s}{2m_s\omega_s}$	0.55	0	2

Example 573 ★ *Three-dimensional visualization for frequency responses.*

To get a sense about the behavior of different frequency responses of a quarter car model, Figures 15.5 to 15.8 are plotted for

$$m_s = 375\,\text{kg} \qquad k_s = 35000\,\text{N/m}$$
$$m_u = 75\,\text{kg} \qquad k_u = 193000\,\text{N/m} \qquad (15.44)$$

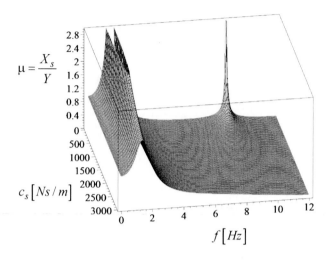

FIGURE 15.5. Three-dimensional view of the frequency response $\mu = \left| \frac{X_s}{Y} \right|$.

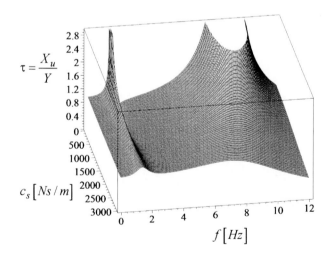

FIGURE 15.6. Three-dimensional view of the frequency response $\tau = \left| \frac{X_u}{Y} \right|$.

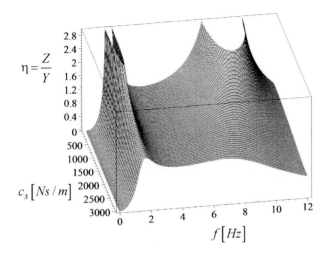

FIGURE 15.7. Three-dimensional view of the frequency response $\eta = \left| \frac{Z}{Y} \right|$.

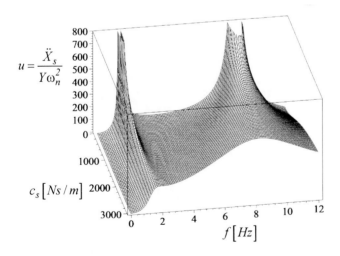

FIGURE 15.8. Three-dimensional view of the frequency response $u = \left| \frac{\ddot{X}_s}{Y\omega_u^2} \right|$.

15.3 ★ Natural and Invariant Frequencies

The quarter car system is a two-DOF system and therefore it has two natural frequencies r_{n_1}, r_{n_2}:

$$r_{n_1} = \sqrt{\frac{1}{2\alpha^2}\left(1 + (1+\varepsilon)\,\alpha^2 - \sqrt{(1+(1+\varepsilon)\,\alpha^2)^2 - 4\alpha^2}\right)} \quad (15.45)$$

$$r_{n_2} = \sqrt{\frac{1}{2\alpha^2}\left(1 + (1+\varepsilon)\,\alpha^2 + \sqrt{(1+(1+\varepsilon)\,\alpha^2)^2 - 4\alpha^2}\right)} \quad (15.46)$$

The family of response curves for the displacement frequency response of the sprung mass, μ, are obtained by keeping ε and ξ constant, and varying ξ. This family has several points in common, which are at frequencies r_1, r_2, r_3, r_4, and μ_1, μ_2, μ_3, μ_4,

$$\begin{cases} r_1 = 0 & \mu_1 = 1 \\ r_3 = \dfrac{1}{\alpha} & \mu_3 = \dfrac{1}{\varepsilon} \\ r_2 & \mu_2 = \dfrac{1}{1 - (1+\varepsilon)\,r_2^2\alpha^2} \\ r_4 & \mu_4 = \dfrac{-1}{1 - (1+\varepsilon)\,r_2^2\alpha^2} \end{cases} \quad (15.47)$$

$$r_2 = \sqrt{\frac{1}{2\alpha^2}\left(1 + 2\,(1+\varepsilon)\,\alpha^2 - \sqrt{(1+2\,(1+\varepsilon)\,\alpha^2)^2 - 8\alpha^2}\right)} \quad (15.48)$$

$$r_4 = \sqrt{\frac{1}{2\alpha^2}\left(1 + 2\,(1+\varepsilon)\,\alpha^2 + \sqrt{(1+2\,(1+\varepsilon)\,\alpha^2)^2 - 8\alpha^2}\right)} \quad (15.49)$$

where

$$r_1\,(=0) < r_2 < \frac{1}{\alpha\sqrt{1+\varepsilon}} < r_3\left(=\frac{1}{\alpha}\right) < r_4 \quad (15.50)$$

The corresponding transmissivities at r_2 and r_4 are

$$\mu_2 = \frac{1}{1 - (1+\varepsilon)\,r_2^2\alpha^2} \quad (15.51)$$

$$\mu_4 = \frac{-1}{1 - (1+\varepsilon)\,r_2^2\alpha^2} \quad (15.52)$$

The frequencies r_1, r_2, r_3, and r_4 are called *invariant frequencies*, and their corresponding amplitudes are called *invariant amplitudes* because they are not dependent on ξ. However, they are dependent on the values of ε and α. The order of magnitude of the natural and invariant frequencies are:

$$r_1\,(=0) < r_{n_1} < r_2 < \frac{1}{\alpha\sqrt{1+\varepsilon}} < r_3 < \left(=\frac{1}{\alpha^2}\right) < r_{n_2} < r_4 \quad (15.53)$$

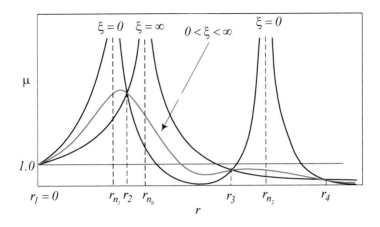

FIGURE 15.9. Schematic illustration of the amplitude μ versus excitation frequency ratio r.

The curves for μ have no other common points except r_1, r_2, r_3, r_4. The order of frequencies along with the order of corresponding amplitudes can be used to predict the shape of the frequency response curves of the sprung mass μ. Figure 15.9 shows schematically the shape of the amplitude μ versus excitation frequency ratio r.

Proof. The natural and resonant frequencies of a system are at positions where the amplitude goes to infinity when damping is zero. Hence, the natural frequencies would be the roots of the denominator of the μ function.

$$
\begin{aligned}
g\left(r^2\right) &= r^2\left(r^2\alpha^2 - 1\right) + \left(1 - (1+\varepsilon)\,r^2\alpha^2\right) \\
&= \alpha^2 r^2 - \left(1 + (1+\varepsilon)\,\alpha^2\right) + 1 = 0
\end{aligned}
\tag{15.54}
$$

The solution of this equation are the natural frequencies given in Equations (15.45) and (15.46).

The invariant frequencies are independent of ξ, so they can be found by intersecting the μ curves for $\xi = 0$ and $\xi = \infty$.

$$
\lim_{\xi \to 0} \mu^2 = \pm \frac{1}{\left(r^2\left(r^2\alpha^2 - 1\right) - r^2\alpha^2\left(\varepsilon + 1\right) + 1\right)^2}
\tag{15.55}
$$

$$
\lim_{\xi \to \infty} \mu^2 = \pm \frac{1}{\left(r^2\alpha^2\left(\varepsilon + 1\right) - 1\right)^2}
\tag{15.56}
$$

Therefore, the invariant frequencies, r_i, can be determined by solving the following equation:

$$
r^2\left(r^2\alpha^2 - 1\right) + \left(1 - (1+\varepsilon)\,r^2\alpha^2\right) = \pm\left(1 - (1+\varepsilon)\,r^2\alpha^2\right)
\tag{15.57}
$$

Using the (+) sign, we find r_1 and r_3 with their corresponding transmissivities μ_1, and μ_3,

$$r_1 \;=\; 0 \qquad \mu_1 = 1 \tag{15.58}$$

$$r_3 \;=\; \frac{1}{\alpha} \qquad \mu_3 = \frac{1}{\varepsilon} \tag{15.59}$$

and, with the (−) sign, we find the following equation for r_2, and r_4:

$$\alpha^2 r^4 - \left(1 + 2\left(1 + \varepsilon\right)\alpha^2\right) r^2 + 2 = 0 \tag{15.60}$$

Equation (15.60) has two real positive roots, r_2 and r_4,

$$r_2 \;=\; \sqrt{\frac{1}{2\alpha^2}\left(1 + 2\left(1 + \varepsilon\right)\alpha^2 - \sqrt{\left(1 + 2\left(1 + \varepsilon\right)\alpha^2\right)^2 - 8\alpha^2}\right)} \tag{15.61}$$

$$r_4 \;=\; \sqrt{\frac{1}{2\alpha^2}\left(1 + 2\left(1 + \varepsilon\right)\alpha^2 + \sqrt{\left(1 + 2\left(1 + \varepsilon\right)\alpha^2\right)^2 - 8\alpha^2}\right)} \tag{15.62}$$

with the following relative order of magnitude:

$$r_1 \left(= 0\right) < r_2 < \frac{1}{\alpha\sqrt{1 + \varepsilon}} < r_3 \left(= \frac{1}{\alpha}\right) < r_4 \tag{15.63}$$

The corresponding amplitudes at r_2, and r_4 can be found by substituting Equations (15.61) and (15.62) in (15.25).

$$\mu_2 \;=\; \frac{1}{1 - \left(1 + \varepsilon\right) r_2^2 \alpha^2} \tag{15.64}$$

$$\mu_4 \;=\; \frac{-1}{1 - \left(1 + \varepsilon\right) r_4^2 \alpha^2} \tag{15.65}$$

It can be checked that

$$\left(1 + \varepsilon\right)\alpha^2 r_4^2 - 1 > \varepsilon > 1 \tag{15.66}$$

and hence,

$$|r_4| < \frac{1}{\varepsilon} \left(= \mu_3\right) < 1 < |r_2| \tag{15.67}$$

and therefore,

$$\mu_2 \;>\; 1 \tag{15.68}$$

$$\mu_4 \;<\; 1 \tag{15.69}$$

Using Equation (15.54), we can evaluate $g\left(r_2^2\right)$, $g\left(r_4^2\right)$, and $g\left(r_3^2\right)$ as

$$g\left(r_2^2\right) \;=\; \left(1 + \varepsilon\right)\alpha^2 r_2^2 - 1 < 0 \tag{15.70}$$

$$g\left(r_4^2\right) \;=\; \left(1 + \varepsilon\right)\alpha^2 r_4^2 - 1 > 0 \tag{15.71}$$

$$g\left(r_3^2\right) \;=\; g\left(\frac{1}{\alpha^2}\right) \tag{15.72}$$

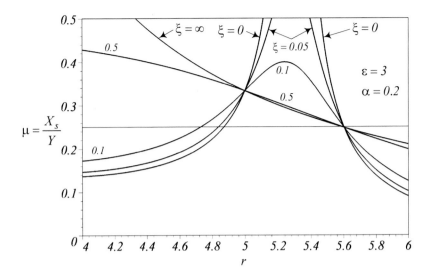

FIGURE 15.10. A magnification around the nodes for the sprung mass displacement frequency response, $\mu = \left| \frac{X_s}{Y} \right|$.

therefore, the two positive roots of Equation (15.54), r_{n_1} and r_{n_2} $\left(> \sqrt{2} > r_2 \right)$, have the order of magnitudes as:

$$r_1 \left(= 0 \right) < r_{n_1} < r_2 < \frac{1}{\alpha \sqrt{1 + \varepsilon}} < r_3 < \left(= \frac{1}{\alpha^2} \right) < r_{n_2} < r_4 \qquad (15.73)$$

■

Example 574 ★ *Nodes of the absolute frequency response μ.*

There are four nodes in the absolute displacement frequency response of a quarter car. The first node is at a trivial point $(r_1 = 0, \mu_1 = 1)$, which shows that $X_s = Y$ when the excitation frequency is zero. The fourth node is at $(r_4, \mu_4 < 1)$. There are also two middle nodes at $(r_2, \mu_2 > 1)$ and $\left(r_3 = \frac{1}{\alpha}, \mu_3 = \frac{1}{\varepsilon} \right)$.

Because $\mu_1 \leq 1$ and $\mu_4 \leq 1$, the middle nodes are important in optimization. To have a better view at the middle nodes, Figure 15.10 illustrates a magnification of the sprung mass displacement frequency response, $\mu = \left| \frac{X_s}{Y} \right|$ around the middle nodes.

Example 575 ★ *There is no Frahm optimal quarter car.*

Reduction in absolute amplitude is the first attempt for optimization. If the amplitude frequency response $\mu = \mu(r)$ contains fixed points with respect to some parameters, then using the Frahm method, the optimization process is carried out in two steps:

1. We select the parameters that control the position of the invariant

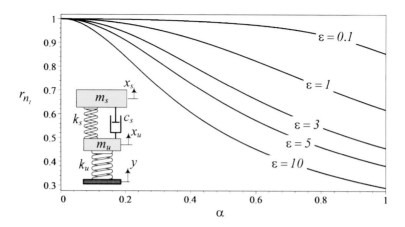

FIGURE 15.11. The natural frequency r_{n_1} as a function of ε and α.

points to equalize the corresponding height at the invariant frequencies, and minimize the height of the fixed points as much as possible.

2. We find the remaining parameters such that the maximum amplitude coincides precisely at the invariant points.

For a real problem, the values of mass ratio ε, and wheel frequency ω_u are fixed and we are trying to find the optimum values of α and ξ. The parameters α and ξ include the unknown stiffness of the main spring and the unknown damping of the main shock absorber, respectively.

The amplitude μ_i at invariant frequencies r_i, show that the first invariant point $(r_1 = 0, \mu_1 = 1)$ is always fixed, and the fourth one $(r_4, \mu_4 < 1)$ happens after the natural frequencies. Therefore, the second and third nodes are the suitable nodes for applying the above optimization steps. However,

$$\mu_2 \leq 1 \leq \mu_3 \qquad \forall \varepsilon > 1 \qquad (15.74)$$

and hence, we cannot apply the above optimization method. It is because μ_2 and μ_3 can never be equated by varying α. Ever so, we can still find the optimum value of ξ by evaluating α based on other constraints.

Example 576 *Natural frequency variation.*

The natural frequencies r_{n_1} and r_{n_2}, as given in Equations (15.45) and (15.46), are functions of ε and α. Figures 15.11 and 15.12 illustrate the effect of these two parameters on the variation of the natural frequencies.

The first natural frequency $r_{n_1} \leq 1$ decreases by increasing the mass ratio ε. r_{n_1} is close to the natural frequency of a $1/8$ car model and indicates the **principal natural frequency** of a car. Hence, it is called the **body bounce natural frequency**. The second natural frequency, r_{n_2}, approaches infinity when α decreases. However, $r_{n_2} \approx 10\,\mathrm{Hz}$ for street cars with acceptable ride

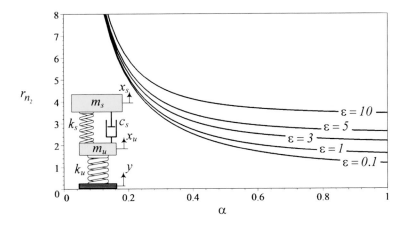

FIGURE 15.12. The natural frequency r_{n_2} as a function of ε and α.

comfort. r_{n_1} relates to the unsprung mass, and is called the **wheel hop natural frequency**.

Figure 15.13 that plots the natural frequency ratio r_{n_1}/r_{n_2} shows their relative behavior.

Example 577 *Invariant frequencies variation.*

The invariant frequencies r_2, r_3, and r_4, as given in Equation (15.47), are functions of ε and α. Figures 15.14 to 15.18 illustrate the effect of these two parameters on the invariant frequencies. The second invariant frequency r_2, as shown in Figure 15.14, is always less than $\sqrt{2}$ because

$$\lim_{\alpha \to 0} r_2 = \sqrt{2} \qquad (15.75)$$

So, whatever the value of the mass ratio is, r_2 cannot be greater than $\sqrt{2}$. Such a behavior does not let us control the position of second node freely. The third invariant frequency r_3 as shown in Figure 15.15 is not a function of the mass ratio and may have any value depending on α. The fourth invariant frequency r_4 is shown in Figure 15.15. r_4 increases when α decreases. However, r_4 settles when $\alpha \gtrsim 0.6$.

$$\lim_{\alpha \to 0} r_2 = \infty \qquad (15.76)$$

To have a better picture about the behavior of invariant frequencies, Figures 15.17 and 15.18 depict the relative frequency ratio r_4/r_3 and r_3/r_2.

Example 578 *Frequency response at invariant frequencies.*

The frequency response μ is a function of α, ε, and ξ. Damping always diminishes the amplitude of vibration, so at first we set $\xi = 0$ and plot the

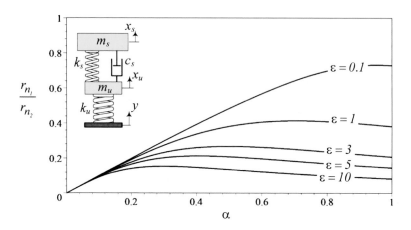

FIGURE 15.13. The natural frequency ratio r_{n_1}/r_{n_2} as a function of ε and α.

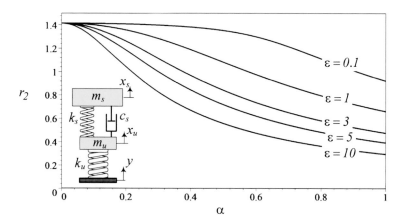

FIGURE 15.14. The second invariant frequency r_2 as a function of ε and α.

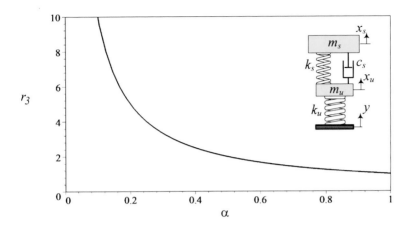

FIGURE 15.15. The third invariant frequency r_3 as a function of ε and α.

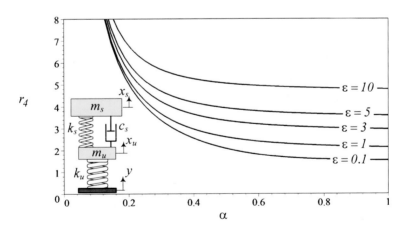

FIGURE 15.16. The fourth invariant frequency r_4 as a function of ε and α.

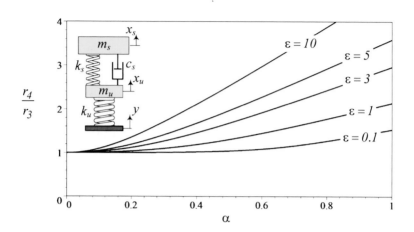

FIGURE 15.17. The ratio of r_4/r_3 as a function of ε and α.

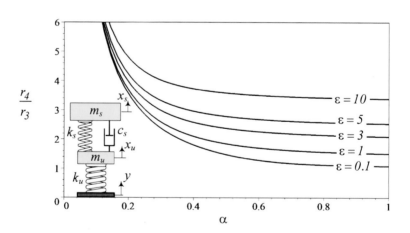

FIGURE 15.18. The ratio of r_3/r_2 as a function of ε and α.

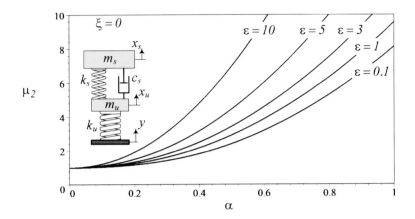

FIGURE 15.19. Behavior of μ_2 as a function of α, ε.

behavior of μ as a function of α, ε. Figure 15.19 illustrates the behavior of μ at the second invariant frequency r_2. Because

$$\lim_{\alpha \to 0} \mu_2 = 1 \qquad\qquad (15.77)$$

μ_2 starts at one, regardless of the value of ε. The value of μ_2 is always greater than one.

Figure 15.20 shows that μ_3 is not a function of α and is a decreasing function of ε. Figure 15.21 shows that $\mu_4 \leq 1$ regardless of the value of α and ε. The relative behavior of μ_2, μ_3, and μ_4 is shown in Figures 15.22 and 15.23.

Example 579 *Natural frequencies and vibration isolation of a quarter car.*

For a modern typical passenger car, the values of natural frequencies are around 1 Hz and 10 Hz respectively. The former is due to the bounce of sprung mass and the latter belongs to the unsprung mass. At average speed, bumps with wavelengths much greater than the wheelbase of the vehicle, will excite bounce motion of the body. At at higher speed, wavelength of the bumps become shorter than a wheelbase length and cause heavy vibrations of the unsprung. Therefore, when the wheels hit a single bump on the road, the impulse will set the wheels into oscillation at the natural frequency of the unsprung mass around 10 Hz. In turn, for the sprung mass, the excitation will be the frequency of vibration of the unsprung around 10 Hz. Because the natural frequency of the sprung is approximately 1 Hz, the excellent isolation for sprung mass occurs and the frequency range around 10 Hz has no essential influence on the sprung discomfort. When the wheel runs over a rough undulating surface, the excitation will consists of a wide range of frequencies. Again, high excitation frequency at 5 Hz to 20 Hz means high frequency input to the sprung mass, which can effectively be isolated. Low frequency excitation, however, will cause resonance in the sprung mass.

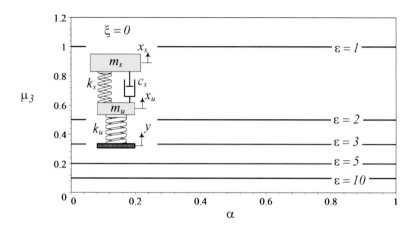

FIGURE 15.20. Behavior of μ_3 as a function of ε.

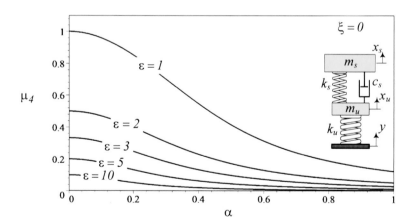

FIGURE 15.21. Behavior of μ_4 as a function of α, ε.

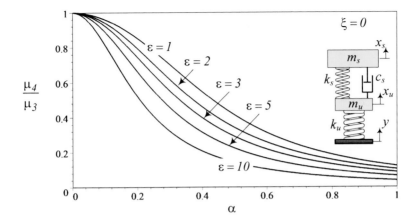

FIGURE 15.22. Behavior of $\dfrac{\mu_4}{\mu_3}$ as a function of α, ε.

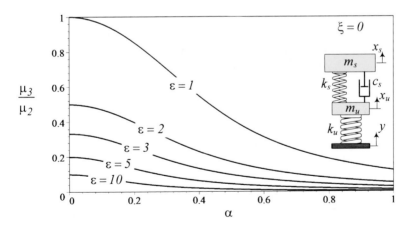

FIGURE 15.23. Behavior of $\dfrac{\mu_3}{\mu_2}$ as a function of α, ε.

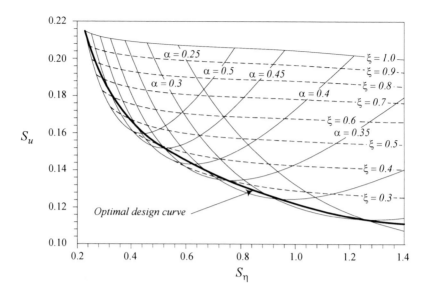

FIGURE 15.24. Root mean square of absolute acceleration, $S_u = RMS(u)$ vreus root mean square of relative displacement $S_\eta = RMS(\eta)$, for a quarter car model and the optimal curve.

15.4 ★ RMS Optimization

Figure 15.24 is a design chart for optimal suspension parameters of a base excited two-DOF system such as a quarter car model. The horizontal axis is the root mean square of relative displacement, $S_\eta = RMS(\eta)$, and the vertical axis is the root mean square of absolute acceleration, $S_u = RMS(u)$.

There are two sets of curves that make a mesh. The first set, which is almost parallel at the right end, are constant damping ratio ξ, and the second set is constant natural frequency ratio α. There is a curve, called the *optimal design curve*, which indicates the optimal main suspension parameters:

The optimal design curve is the result of the RMS optimization strategy

$$Minimize\ S_{\ddot{X}}\ with\ respect\ to\ S_Z \qquad (15.78)$$

which states that the minimum absolute acceleration with respect to the relative displacement, if there is any, makes the suspension of a quarter car optimal. Mathematically, it is equivalent to the following minimization

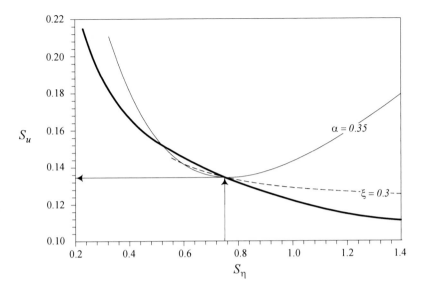

FIGURE 15.25. Application of the design chart for $S_\eta = 1$, which indicates the optimal values $\xi \approx 0.3$ and $\alpha \approx 0.35$.

problem:

$$\frac{\partial S_u}{\partial S_\eta} = 0 \qquad (15.79)$$

$$\frac{\partial^2 S_u}{\partial S_\eta^2} > 0 \qquad (15.80)$$

To use the design curve and determine optimal stiffness k_s and damping c_s for the main suspension of the system, we start from an estimate value for S_η on the horizontal axis and draw a vertical line to hit the optimal curve. The intersection point indicates the optimal α and ξ for the S_η.

Figure 15.25 illustrates a sample application for $S_\eta = 0.75$, which indicates $\xi \approx 0.3$ and $\alpha \approx 0.35$ for optimal suspension. Having α and ξ, determines the optimal value of k_s and c_s.

$$k_s = \alpha^2 \frac{m_s}{m_u} k_u \qquad (15.81)$$

$$c_s = 2\xi \sqrt{k_s m_s} \qquad (15.82)$$

Proof. The RMS of a continues function $g(\alpha, \xi, \varepsilon, \omega)$ is defined by

$$RMS(g) = \sqrt{\frac{1}{\omega_2 - \omega_1} \int_{\omega_1}^{\omega_2} g^2(\alpha, \xi, \varepsilon, \omega)\, d\omega} \qquad (15.83)$$

where $\omega_2 \leq \omega \leq \omega_1$ is called the *working frequency range*. Let us consider a working range for the excitation frequency $0 \leq f \left(= \frac{\omega}{2\pi}\right) \leq 20\,\mathrm{Hz}$ to include almost all ground vehicles, especially road vehicles, and show the RMS of η and u by

$$S_\eta = RMS(\eta) \tag{15.84}$$

$$S_u = RMS(u) \tag{15.85}$$

In applied vehicle dynamics, we usually measure frequencies in $[\mathrm{Hz}]$, instead of $[\mathrm{rad/s}]$, we perform design calculations based on cyclic frequencies f and f_n in $[\mathrm{Hz}]$, and we do analytic calculation based on angular frequencies ω and ω_n in $[\mathrm{rad/s}]$.

To calculate S_η and S_u over the working frequency range

$$S_\eta = \sqrt{\frac{1}{40\pi} \int_0^{40\pi} \eta^2 dr} \tag{15.86}$$

$$S_u = \sqrt{\frac{1}{40\pi} \int_0^{40\pi} u^2 dr} = \alpha^2 \sqrt{\frac{1}{40\pi} \int_0^{40\pi} r^2 \mu^2 dr} \tag{15.87}$$

we first find integrals of η^2 and u^2.

$$\begin{aligned}
\int u^2 dr &= \frac{1}{2Z_6}\left(\frac{1}{Z_1} + Z_1 Z_5\right)\ln\left(\frac{r - Z_1}{r + Z_1}\right) \\
&\quad + \frac{1}{2Z_7}\left(\frac{1}{Z_2} + Z_2 Z_5\right)\ln\left(\frac{r - Z_2}{r + Z_2}\right) \\
&\quad + \frac{1}{2Z_8}\left(\frac{1}{Z_3} + Z_3 Z_5\right)\ln\left(\frac{r - Z_3}{r + Z_3}\right) \\
&\quad + \frac{1}{2Z_9}\left(\frac{1}{Z_4} + Z_4 Z_5\right)\ln\left(\frac{r - Z_4}{r + Z_4}\right)
\end{aligned} \tag{15.88}$$

$$\begin{aligned}
\int \eta^2 dr &= \frac{Z_1^3}{2Z_6}\ln\left(\frac{r - Z_1}{r + Z_1}\right) + + \frac{Z_2^3}{2Z_7}\ln\left(\frac{r - Z_2}{r + Z_2}\right) \\
&\quad + \frac{Z_3^3}{2Z_8}\ln\left(\frac{r - Z_3}{r + Z_3}\right) + \frac{Z_4^3}{2Z_9}\ln\left(\frac{r - Z_4}{r + Z_4}\right)
\end{aligned} \tag{15.89}$$

The parameters Z_1 through Z_9 are:

$$Z_1 = \frac{1}{2}\frac{-Z_{19} + \sqrt{Z_{23}}}{Z_{19}}\frac{1}{4}\frac{Z_{15}}{Z_{14}} \tag{15.90}$$

$$Z_2 = \frac{1}{2}\frac{-Z_{19} - \sqrt{Z_{23}}}{Z_{19}}\frac{1}{4}\frac{Z_{15}}{Z_{14}} \tag{15.91}$$

$$Z_3 = \frac{1}{2} \frac{-Z_{19} + \sqrt{Z_{24}}}{Z_{19}} \frac{1}{4} \frac{Z_{15}}{Z_{14}} \tag{15.92}$$

$$Z_4 = \frac{1}{2} \frac{-Z_{19} - \sqrt{Z_{24}}}{Z_{19}} \frac{1}{4} \frac{Z_{15}}{Z_{14}} \tag{15.93}$$

$$Z_5 = 4\xi^2 \tag{15.94}$$

$$Z_6 = \left(Z_1^2 - Z_2^2\right)\left(Z_1^2 - Z_3^2\right)\left(Z_1^2 - Z_4^2\right) \tag{15.95}$$

$$Z_7 = \left(Z_2^2 - Z_3^2\right)\left(Z_2^2 - Z_3^2\right)\left(Z_2^2 - Z_1^2\right) \tag{15.96}$$

$$Z_8 = \left(Z_3^2 - Z_4^2\right)\left(Z_3^2 - Z_1^2\right)\left(Z_3^2 - Z_2^2\right) \tag{15.97}$$

$$Z_9 = \left(Z_4^2 - Z_1^2\right)\left(Z_4^2 - Z_2^2\right)\left(Z_4^2 - Z_3^2\right) \tag{15.98}$$

$$Z_{10} = \frac{1}{6}\sqrt[3]{Z_{20}} + \frac{8Z_{13} + \frac{2}{3}Z_{11}^2}{\sqrt[3]{Z_{20}}} + \frac{1}{3}Z_{11} \tag{15.99}$$

$$Z_{11} = \frac{8Z_{16}Z_{14} - 3Z_{15}^3}{8Z_{14}^3} \tag{15.100}$$

$$Z_{12} = -\frac{4Z_{16}Z_{14}Z_{15} - Z_{15}^3 - 8Z_{14}^2Z_{17}}{8Z_{14}^3} \tag{15.101}$$

$$Z_{13} = \frac{-64Z_{14}^2Z_{17}Z_{15} + 256Z_{14}^3Z_{18} + 16Z_{14}Z_{15}^2Z_{16} - 3Z_{15}^4}{256Z_{14}^4} \tag{15.102}$$

$$Z_{14} = \alpha^4 \tag{15.103}$$

$$Z_{15} = -2\alpha^4\left(1 + \varepsilon\right) - 2\alpha^2 + 4\left(1 + \varepsilon\right)^2\alpha^4\xi^2 \tag{15.104}$$

$$Z_{16} = -8\alpha^2\xi^2\left(1 + \varepsilon\right) + \left(1 + \varepsilon\right)^2\alpha^4 - 2\alpha^2\left(2 + \varepsilon\right) + 1 \tag{15.105}$$

$$Z_{17} = 4\xi^2 - 2\alpha^2\left(1 + \varepsilon\right) - 2 \tag{15.106}$$

$$Z_{18} = 1 \tag{15.107}$$

$$Z_{19} = Z_{10} - Z_{11} \tag{15.108}$$

$$Z_{20} = Z_{21} + 12\sqrt{Z_{22}} \tag{15.109}$$

$$Z_{21} = -288Z_{11}Z_{13} + 108Z_{12}^2 + 8Z_{11}^3 \tag{15.110}$$

$$Z_{22} = -768Z_{13}^3 + 384Z_{11}^2Z_{13}^2 - 48Z_{13}Z_{11}^4$$
$$-432Z_{11}Z_{12}^2Z_{13} + 81Z_{12}^4 + 12Z_{11}^3Z_{12}^2 \tag{15.111}$$

$$Z_{23} = Z_{19}\left(Z_{11} - Z_{10}\right) - 2Z_{12}Z_{19}^{3/2} \tag{15.112}$$

$$Z_{24} = Z_{19}\left(Z_{11} + Z_{10}\right) + 2Z_{12}Z_{19}^{3/2} \tag{15.113}$$

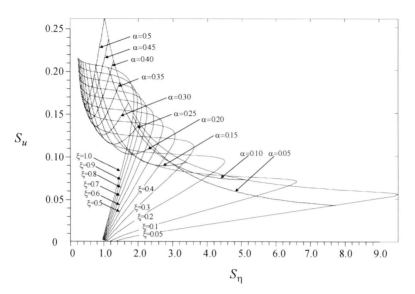

FIGURE 15.26. RMS of absolute acceleration, $S_u = RMS(u)$ vreus RMS of relative displacement $S_\eta = RMS(\eta)$, for a quarter car model.

Now the required RMS, S_η, and S_u, over the frequency range $0 < f < 20\,\mathrm{Hz}$, can be calculated analytically from Equations (15.86) and (15.87).

Equations (15.86) and (15.87) show that both S_η and S_u are functions of only three variables: ε, α, and ξ.

$$S_\eta = S_\eta (\varepsilon, \alpha, \xi) \qquad (15.114)$$
$$S_u = S_u (\varepsilon, \alpha, \xi) \qquad (15.115)$$

In applied vehicle dynamics, ε is usually a fixed parameter, so, any pair of design parameters (α, ξ) determines S_η and S_u uniquely. Let us set

$$\varepsilon = 3 \qquad (15.116)$$

Using Equations (15.86) and (15.87), we may draw Figure 15.26 to illustrate how S_u behaves with respect to S_η when α and ξ vary. Keeping α constant and varying ξ, it is possible to minimize S_u with respect to S_η. The minimum points make the optimal curve and determine the best α and ξ. The way to use the optimal design curve is to estimate a value for S_η or S_u and find the associated point on the design curve. A magnified picture is shown in Figure 15.24.

The horizontal axis is the root mean square of relative displacement, $S_\eta = RMS(\eta)$, and the vertical axis is the root mean square of absolute acceleration, $S_u = RMS(u)$. The optimal curve indicates that softening a suspension decreases the body acceleration, however, it requires a large

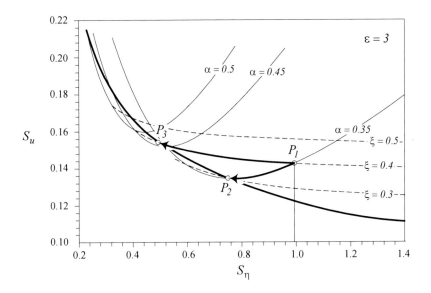

FIGURE 15.27. Two optimal designs at points P_2 and P_3 for an off-optimal design quarter car at point P_1.

room for relative displacement. Due to physical constraints, the wheel travel is limited, and hence, we must design the suspension such that to use the available suspension travel, and decrease the body acceleration as low as possible. Mathematically it is equivalent to (15.79) and (15.80). ∎

Example 580 *Examination of the optimal quarter car model.*

To examine the optimal design curve and compare practical ways to make a suspension optimal, we assume that there is a quarter car with an off-optimal suspension, indicated by point P_1 in Figure 15.27.

$$\varepsilon = 3 \qquad \alpha = 0.35 \qquad \xi = 0.4 \qquad (15.117)$$

To optimize the suspension practically, we may keep the stiffness constant and change the damper to a corresponding optimal value, or keep the damping constant and change the stiffness to a corresponding optimal value. However, if it is possible, we may change both, stiffness and damping to a point on the optimal curve depending on the physical constraints and requirements.

Point P_2 in Figure 15.27 has the same α as point P_1 with an optimal damping ratio $\xi \approx 0.3$. Point P_3 in Figure 15.27 has the same ξ as point P_1 with an optimal natural frequency ratio $\alpha \approx 0.452$. Hence, points P_2 and P_3 are two alternative optimal designs for the off-optimal point P_1.

Figure 15.28 compares the acceleration frequency response $\log u$ for the three points P_1, P_2, and P_3. Point P_3 has the minimum acceleration frequency response. Figure 15.29 depicts the absolute displacement frequency

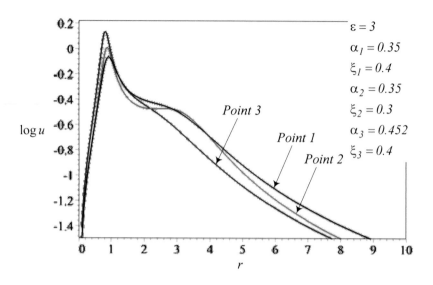

FIGURE 15.28. Absolute displacement frequency response μ for points P_1, P_2, and P_3 shown in Figure 15.27.

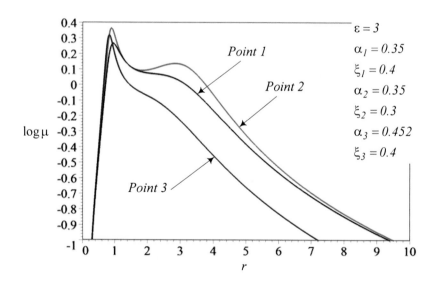

FIGURE 15.29. Relative displacement frequency response η for points P_1, P_2, and P_3 shown in Figure 15.27.

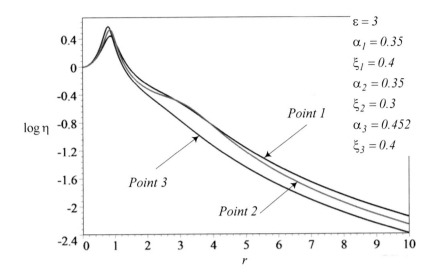

FIGURE 15.30. Absolute acceleration frequency response u for points P_1, P_2, and P_3 shown in Figure 15.27.

response $\log \mu$ and Figure 15.30 compares the relative displacement frequency response $\log \eta$ for the there points P_1, P_2, P_3. These figures show
that both points P_2 and P_3 introduce better suspension than point P_1. Suspension P_2 has a higher level of acceleration but needs less relative suspension travel than suspension P_3. Suspension P_3 has a lower level of acceleration, but it needs more room for suspension travel than suspension
P_2.

Example 581 *Comparison of an off-optimal quarter car with two opti-
mals.*

An alternative method to optimize an off-optimal suspension is to keep
the RMS of relative displacement S_η or absolute acceleration S_u constant
and find the associated point on the optimal design curve. Figure 15.31
illustrates two alternative optimal designs, points P_2 and P_3, for an off-
optimal design at point P_1.

The mass ratio is assumed to be

$$\varepsilon = 3 \qquad (15.118)$$

and the suspension characteristics at P_1 are

$$\xi = 0.0465 \qquad \alpha = 0.265 \qquad S_\eta = 2 \qquad S_u = 0.15 \qquad (15.119)$$

The optimal point corresponding to P_1 with the same S_u is at P_2 with the
characteristics

$$\xi = 0.23 \qquad \alpha = 0.45 \qquad S_\eta = 0.543 \qquad S_u = 0.15 \qquad (15.120)$$

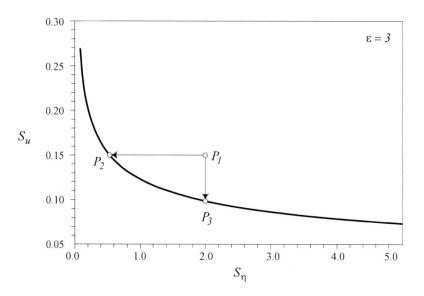

FIGURE 15.31. Two alternative optimal designs at points P_2 and P_3 for an off-optimal design quarter car at point P_1.

and the optimal point with the same S_η as point P_1 is a point at P_2 with the characteristics:

$$\xi = 0.0949 \qquad \alpha = 0.1858 \qquad S_\eta = 2 \qquad S_u = 0.0982 \qquad (15.121)$$

Figure 15.32 depicts the sprung mass vibration amplitude μ, which shows that both points P_2 and P_3 have lower overall amplitude specially at second resonance. Figure 15.33 shows the amplitude of relative displacement η between sprung and unsprung masses. The amplitude of absolute acceleration of the sprung mass u is shown in Figure 15.34.

Example 582 ★ *Natural frequencies and vibration isolation requirements.*
Road irregularities are the most common source of excitation for passenger cars. Therefore, the natural frequencies of vehicle system are the primary factors in determining design requirements for conventional isolators. The natural frequency of the vehicle body supported by the primary suspension is usually between 0.2 Hz and 2 Hz, and the natural frequency of the unsprung mass, called wheel hop frequency, usually is between 2 Hz and 20 Hz. The higher values generally apply to military vehicles.

The isolation of sprung mass from the uneven road can be improved by using a soft spring, which reduces the primary natural frequency. Lowering the natural frequency always improves the ride comfort, however it causes a design problem due to the large relative motion between the sprung and unsprung masses. One of the most important constraints that suspension

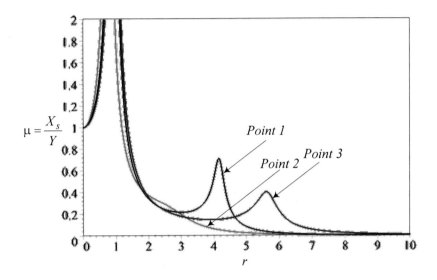

FIGURE 15.32. Absolute displacement frequency response μ for points P_1, P_2, and P_3 shown in Figure 15.31.

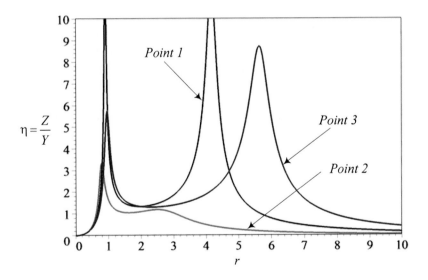

FIGURE 15.33. Relative displacement frequency response η for points P_1, P_2, and P_3 shown in Figure 15.31.

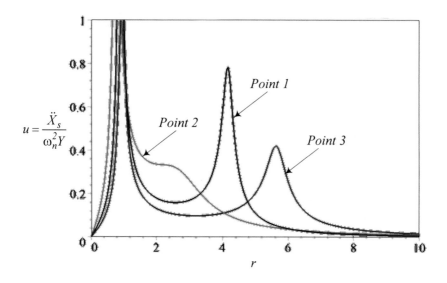

FIGURE 15.34. Absolute acceleration frequency response u for points P_1, P_2, and P_3 shown in Figure 15.31.

system designers have to consider is the rattle-space constraint, the maximum allowable relative displacement. Additional factors are imposed by the overall stability, reliability, and economic or cost factors.

Example 583 *Optimal characteristics variation.*
   We may collect the optimal α and ξ and plot them as shown in Figures 15.35 and 15.36. These figures illustrate the trend of their variation. The optimal value of both α and ξ are decreasing functions of relative displacement RMS S_η. So, when more room is available, we may reduce α and ξ and have a softer suspension for better ride comfort. Figure 15.37 shows how the optimal α and ξ change with each other.

15.5 ★ Optimization Based on Natural Frequency and Wheel Travel

Assume a fixed value for the mass ratio ε and natural frequency ratio α to fix the position of the nodes in the frequency response plot. Then, an optimal value for damping ratio ξ is

$$\xi^\star = \frac{\sqrt{Z_{35}}}{Z_{36}} \sqrt{\sqrt{Z_{37}^2 - 8\alpha^2} + Z_{37} - \frac{8\alpha^2}{Z_{35}}} \qquad (15.122)$$

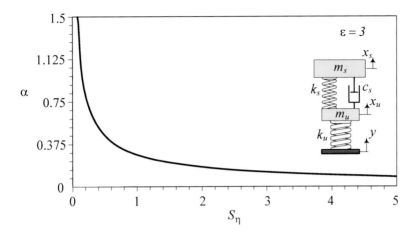

FIGURE 15.35. The optimal value of α as a function of relative displacement RMS S_η.

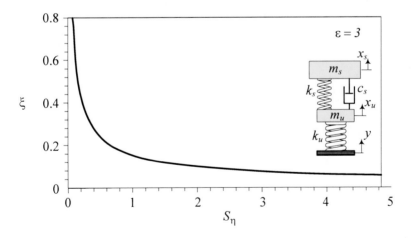

FIGURE 15.36. The optimal value of ξ as a function of relative displacement RMS S_η.

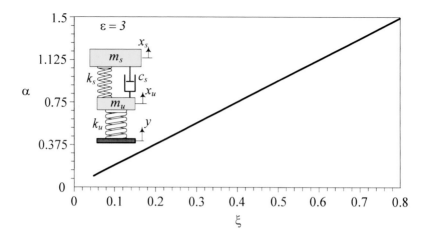

FIGURE 15.37. The optimal α versus optimal ξ for a quarter car with $\varepsilon = 3$.

where

$$Z_{35} = \alpha^2 (1 + \varepsilon) + 1 \tag{15.123}$$
$$Z_{36} = 4\alpha\sqrt{1 + \varepsilon} \tag{15.124}$$
$$Z_{37} = \left(2\alpha^2 (1 + \varepsilon) + 1\right) \tag{15.125}$$

The optimal damping ratio $\xi^{\star}$ causes the second resonant amplitude μ_2 to occur at the second invariant frequency r_2. The value of relative displacement η at $r = r_2$ for $\xi = \xi^{\star}$ is,

$$\eta_2 = \sqrt{\frac{\left(\sqrt{Z_{37}^2 - 8\alpha^2} - Z_{35}\right)\sqrt{1 + \varepsilon}}{2\alpha^2 \left(Z_{28}\sqrt{Z_{37}^2 - Z_{29}}\right)}} \tag{15.126}$$

where,

$$Z_{28} = 4\alpha^4 (1 + \varepsilon)^4 - 4\alpha^2 (1 + \varepsilon)^2 (1 - \varepsilon) + (1 + \varepsilon^2) \tag{15.127}$$
$$\begin{aligned} Z_{29} = {} & -8\alpha^6 (1 + \varepsilon)^5 + 12\alpha^4 (1 + \varepsilon)^3 (1 - \varepsilon) \\ & -2\alpha^4 (1 + \varepsilon) (1 + 3\varepsilon^2 - 2\varepsilon) + (1 + \varepsilon^2) \end{aligned} \tag{15.128}$$

Proof. Natural frequencies of the sprung and unsprung masses, as given in Equations (15.45) and (15.46), are related to ε and α. When ε is given, we can evaluate α by considering the maximum permissible static deflection, which in turn adjusts the value of natural frequencies. If the values of α and ε are determined and kept fixed, then the value of damping ratio ξ which cause the first resonant amplitude to occur at the second node, can be determined as optimum damping. For a damping ratio less or greater than the optimum, the resonant amplitude would be greater.

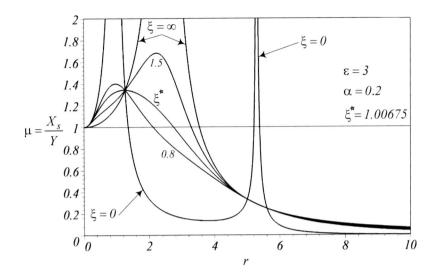

FIGURE 15.38. A sample of frequency response μ for different ξ including $\xi = \xi^{\star}$.

The frequencies related to the maximum of μ are obtained by differentiating μ with respect to r and setting the result equal to zero

$$\frac{\partial \mu}{\partial r} = \frac{1}{2\mu} \frac{\partial \mu^2}{\partial r} = \frac{1}{Z_{25}^2} \left(8\xi^2 r Z_{25} - Z_{26} - Z_{27} \right) = 0 \qquad (15.129)$$

where

$$
\begin{aligned}
Z_{25} &= \left[r^2 \left(r^2\alpha^2 - 1 \right) + \left(1 - (1+\varepsilon) r^2\alpha^2 \right) \right]^2 \\
&\quad + 4\xi^2 r^2 \left(1 - (1+\varepsilon) r^2\alpha^2 \right)^2 \qquad (15.130) \\
Z_{26} &= 8\xi^2 r \left(4\xi^2 r^2 + 1 \right) \left(3r^2\alpha^2 (1+\varepsilon) - 1 \right) \\
&\quad \times \left(r^2\alpha^2 (1+\varepsilon) - 1 \right) \qquad (15.131) \\
Z_{27} &= 4r \left(4\xi^2 r^2 + 1 \right) \left[r^2\alpha^2 (1+\varepsilon) + r^2 \left(1 - r^2\alpha^2 \right) - 1 \right] \\
&\quad \times \left[\alpha^2 (1+\varepsilon) - 2r^2\alpha^2 + 1 \right]. \qquad (15.132)
\end{aligned}
$$

Now, the optimal value $\xi^{\star}$ in Equation (15.122) is obtained if the frequency ratio r in Equation (15.129) is replaced with r_2 given by Equation (15.61). The optimal damping ratio $\xi^{\star}$ makes μ have a maximum at the second invariant frequency r_2. Figure 15.38 illustrates an example of frequency response μ for different ξ including $\xi = \xi^{\star}$.

Figure 15.39 shows the sensitivity of $\xi^{\star}$ to α and ε.

Substituting $\xi^{\star}$ in the general expression of μ, the absolute maximum value of μ would be equal to μ_2 given by equation (15.51). Substituting $r = r_2$ and $\xi = \xi^{\star}$ in Equation (15.25) gives us Equation (15.126) for η_2.

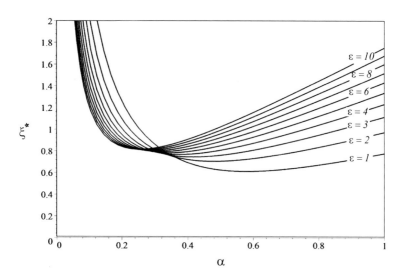

FIGURE 15.39. the optimal value $\xi^\star$ as a function of α and ε.

The lower the natural frequency of the suspension, the more effective the isolation from road irregularities. So, the stiffness of the main spring must be as low as possible. Figure 15.40 shows the behavior of η_2 for $\xi = \xi^\star$. ■

Example 584 *Nodes in η_2 for $\xi = \xi^\star$.*
The relative displacement at second node, η_2, is a monotonically increasing function of α and has two invariant points. The invariant points of η may be found from

$$\pm r^2 \left[\left(r^2 \alpha^2 - 1 \right) + \left(1 - (1 + \varepsilon)\,\alpha^2 \right) \right]$$
$$= \left[r^2 \left(r^2 \alpha^2 - 1 \right) + \left(1 - (1 + \varepsilon)\,r^2\alpha^2 \right) \right]^2$$
$$+ r^2 \left(1 - (1 + \varepsilon)\,r^2\alpha^2 \right) \tag{15.133}$$

that are,

$$r_1 \;=\; 0 \qquad \eta_1 = 0 \tag{15.134}$$

$$r_{n_0} \;=\; \frac{1}{\alpha\sqrt{1+\varepsilon}} \qquad \eta_{n_0} = 1 + \frac{1}{\varepsilon} \tag{15.135}$$

The value of μ at r_{n_0} is,

$$\mu_{n_0} = \frac{\alpha^2}{\varepsilon^2}\,(1+\varepsilon)^3 \left[4\xi^2 + \alpha^2\,(1+\varepsilon) \right] \tag{15.136}$$

Example 585 ★ *Maximum value of η.*
Figure 15.4 shows that η has a node at the intersection of the curves for $\xi = 0$ and $\xi = \infty$. There might be a specific damping ratio to make η have

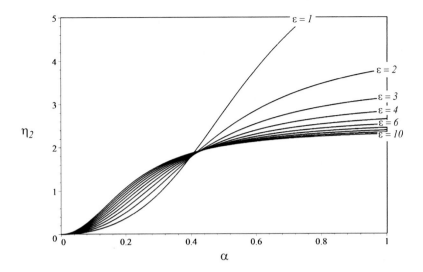

FIGURE 15.40. The behavior of η_2 for $\xi = \xi^\star$ as a function of α and ε.

a maximum at the node. To find the maximum value of η, we have to solve the following equation for r:

$$\frac{\partial \eta}{\partial r} = \frac{1}{2\eta} \frac{\partial \eta^2}{\partial r} = \frac{1}{Z_{23}^2} \left(4r^3 Z_{23} - Z_{30} - Z_{31} \right) \tag{15.137}$$

where

$$
\begin{aligned}
Z_{23} &= \left[r^2 \left(r^2 \alpha^2 - 1 \right) + \left(1 - (1+\varepsilon) r^2 \alpha^2 \right) \right]^2 \\
&\quad + 4\xi^2 r^2 \left(1 - (1+\varepsilon) r^2 \alpha^2 \right)^2 \tag{15.138} \\
Z_{30} &= 8\xi^2 r^5 \left[3r^2 \alpha^2 \left(1 + \varepsilon \right) - 1 \right] \left(r^2 \alpha^2 \left(1 + \varepsilon \right) - 1 \right) \tag{15.139} \\
Z_{31} &= 4r^5 \left[r^2 \alpha^2 \left(1 + \varepsilon \right) + r^2 \left(1 - r^2 \alpha^2 \right) - 1 \right] \\
&\quad \times \left[\alpha^2 \left(1 + \varepsilon \right) - 2r^2 \alpha^2 + 1 \right] \tag{15.140}
\end{aligned}
$$

Therefore, the maximum η occurs at the roots of the equation:

$$Z_{32} r^8 + Z_{33} r^6 + Z_{34} r^2 - 1 = 0 \tag{15.141}$$

where

$$
\begin{aligned}
Z_{32} &= \alpha^4 \tag{15.142} \\
Z_{33} &= 2\alpha^4 \xi^2 \left(1 + \varepsilon \right)^2 + \alpha^4 \left(1 + \varepsilon \right) - \alpha^2 \tag{15.143} \\
Z_{34} &= \alpha^2 \left(1 + \varepsilon \right) + 1 - 2\xi^2 \tag{15.144}
\end{aligned}
$$

Equation (15.141) has two positive roots when ξ is less than a specific value of damping ratio, ξ_η, and one positive root when ξ is greater than ξ_η, where,

$$\xi_\eta = \xi_\eta \left(\alpha, \xi, \varepsilon \right) \tag{15.145}$$

The positive roots of Equation (15.141) are r_5 and r_6, and the correspond-
ing relative displacements are denoted by η_5 and η_6, where $r_5 < r_6$. The
invariant frequencies r_5 and r_6 would be equal when $\xi \geq \xi_\eta$, and they ap-
proach r_{n_0} when ξ goes to infinity. The invariant frequency η_6 is greater
than η_5 as long as $\xi \leq \xi_\eta$, and they are equal when $\xi \geq \xi_\eta$. The relative
displacements η_5 and η_6 are monotonically decreasing functions of ξ and
they approach η_{n_0} when ξ goes to infinity.

It is seen from (15.135) that the invariant point at r_{n_0} depends on α and
ε but the value of η_{n_0} depends only on ε. If ε is given, then η_{n_0} is fixed.
Therefore, the maximum value of the relative displacement, η, cannot be less
than η_{n_0} and we cannot find any real value for ξ that causes the maximum
of η to occur at r_{n_0}. The optimum value of ξ could be found when we adjust
the maximum value of η_6, to be equal to the allowed wheel travel.

15.6   Summary

The vertical vibration of vehicles may be modeled by a two-DOF linear
system called *quarter car model*. One-fourth of the body mass, known as
sprung mass, is suspended by the main suspension of the vehicle k_s and
c_s. The main suspension k_s and c_s are mounted on a wheel of the vehicle,
known as *unsprung mass*. The wheel is sitting on the road by a tire with
stiffness k_u.

Assuming the vehicle is running on a harmonically bumped road we are
able to find the frequency responses of the sprung and unsprung masses,
and relative displacement can be found analytically by taking advantage
of the linearity of the system. The frequency response of the sprung mass
has four nodes. The first and fourth nodes are usually out of resonance
or out of working frequency range. The middle nodes sit at different sides
of $\mu = 1$, and therefore, they cannot be equated and Frahm optimization
cannot be applied.

The root mean square of the absolute acceleration and relative displacement can be found analytically by applying the RMS optimization method.
The RMS optimization method is based on minimizing the absolute acceleration RMS with respect to the relative displacement RMS. The result
of RMS optimization introduces an optimal design curve for a fixed mass
ratio.

15.7   Key Symbols

$a, \ddot{x}$	acceleration
c	damping
c_s	main suspension damper
$[c]$	damping matrix
d_1	road wave length
d_2	road wave amplitude
D	dissipation function
$f, \mathbf{F}$	force
$f = \frac{1}{T}$	cyclic frequency [Hz]
f_c	damper force
f_k	spring force
f_n	cyclic natural frequency [Hz]
$g\left(r^2\right)$	characteristic equation
k	stiffness
k_s	main suspension spring stiffness
k_u	tire stiffness
k_{eq}	equivalent stiffness
$[k]$	stiffness matrix
K	kinetic energy
$\mathcal{L}$	Lagrangean
m	mass
m_s	sprung mass
m_u	unsprung mass
$[m]$	mass matrix
$r = \omega/\omega_s$	excitation frequency ratio
$r_i,\ i \in N$	nodal frequency ratio
$r_n = \omega_n/\omega_s$	natural frequency ratio
$S_u = RMS(u)$	RMS of u
$S_\eta = RMS(\eta)$	RMS of η
t	time
T	period
$u = r^2\alpha^2\mu$	sprung mass acceleration frequency response
$v = r^2\alpha^2\tau$	unsprung mass acceleration frequency response
V	potential energy
x	absolute displacement
x_s	sprung mass displacement
x_u	unsprung mass displacement
X	steady-state amplitude of x
X_s	steady-state amplitude of x_s
X_u	steady-state amplitude of x_u
y	base excitation displacement
Y	steady-state amplitude of y

z	relative displacement
Z	steady-state amplitude of z
Z_i	short notation parameter

$\alpha = \omega_s/\omega_u$	sprung mass ratio
$\varepsilon = m_s/m_u$	sprung mass ratio
$\eta = |Z/Y|$	sprung mass relative frequency response
$\mu = |X_s/Y|$	sprung mass frequency response
$\xi = c_s/\left(2\sqrt{k_s m_s}\right)$	damping ratio
$\xi^\star$	optimal damping ratio
$\tau = |X_u/Y|$	unsprung mass frequency response
$\omega = 2\pi f$	angular frequency $[\,\mathrm{rad/\,s}]$
$\omega_s = \sqrt{k_s/m_s}$	sprung mass frequency
$\omega_u = \sqrt{k_u/m_u}$	unsprung mass frequency
ω_n	natural frequency

Subscript

$i \in N$	node number
n	natural
s	sprung
u	unsprung

Exercises

1. Quarter car natural frequencies.

   Determine the natural frequencies of a quarter car with the following characteristics:

   $$m_s = 275\,\text{kg} \qquad m_u = 45\,\text{kg} \qquad k_u = 200000\,\text{N}/\text{m} \qquad k_s = 10000\,\text{N}/\text{m}$$

2. Equations of motion.

   Derive the equations of motion for the quarter car model that is shown in Figure 15.1, using the relative coordinates:

   (a)
   $$z_s = x_s - y \qquad z_u = x_u - y$$

   (b)
   $$z = x_s - x_u \qquad z_u = x_u - y$$

   (c)
   $$z = x_s - x_u \qquad z_s = x_s - y$$

3. ★ Natural frequencies for different coordinates.

   Determine and compare the natural frequencies of the three cases in Exercise 2 and check their equality by employing the numerical data of Exercise 1.

4. Quarter car nodal frequencies.

   Determine the nodal frequencies of a quarter car with the following characteristics:

   $$m_s = 275\,\text{kg} \qquad m_u = 45\,\text{kg} \qquad k_u = 200000\,\text{N}/\text{m} \qquad k_s = 10000\,\text{N}/\text{m}$$

   Check the order of the nodal frequencies with the natural frequencies found in Exercise 1.

5. Frequency responses of a quarter car.

   A car is moving on a wavy road with a wave length $d_1 = 20\,\text{m}$ and wave amplitude $d_2 = 0.08\,\text{m}$.

   $$\begin{aligned} m_s &= 200\,\text{kg} & m_u &= 40\,\text{kg} \\ k_u &= 220000\,\text{N}/\text{m} & k_s &= 8000\,\text{N}/\text{m} & c_s &= 1000\,\text{N}\,\text{s}/\text{m} \end{aligned}$$

   Determine the steady-state amplitude X_s, X_u, and Z if the car is moving at:

(a) $v = 30 \, \text{km}/\text{h}$

(b) $v = 60 \, \text{km}/\text{h}$

(c) $v = 120 \, \text{km}/\text{h}.$

6. Quarter car suspension optimization.

   Consider a car with

   $$m_s = 200 \, \text{kg} \qquad m_u = 40 \, \text{kg} \qquad k_u = 220000 \, \text{N}/\text{m} \qquad S_\eta = 0.75$$

   and determine the optimal suspension parameters.

7. A quarter car has $\alpha = 0.45$ and $\xi = 0.4$. What is the required wheel travel if the road excitation has an amplitude $Y = 1 \, \text{cm}$?

8. ★ Quarter car and time response.

   Find the optimal suspension of a quarter car with the following characteristics:

   $$m_s = 220 \, \text{kg} \qquad m_u = 42 \, \text{kg} \qquad k_u = 150000 \, \text{N}/\text{m} \qquad S_\eta = 0.75$$

   and determine the response of the optimal quarter car to a unit step excitation.

9. ★ Quarter car mathematical model.

   In the mathematical model of the quarter car, we assumed the tire is always sticking to the road. Determine the condition at which the tire leaves the surface of the road.

10. Optimal damping.

    Consider a quarter car with $\alpha = 0.45$ and $\varepsilon = 0.4$. Determine the optimal damping ratio $\xi^\star$.

References

Abe, M., 2009, *Vehicle Handling Dynamics: Theory and Application*, Butterworth-Heinemann, Oxford, UK.

Alkhatib, R., Jazar, R. N., and Golnaraghi, M. F., Optimal Design of Passive Linear Mounts with Genetic Algorithm Method, *Journal of Sound and Vibration,* **275***(3-5), 665-691, 2004.*

American Association of State Highway Officials, AASHO, Highway Definitions, June 1968.

American National Standard, Manual on Classification of Motor Vehicle Traffic Accidents, Sixth Edition, National Safety Council, Itasca, Illinois, 1996.

Andrzejewski, R., and Awrejcewicz, J., 2005, *Nonlinear Dynamics of a Wheeled Vehicle*, Springer-Verlag, New York.

Asada, H., and Slotine, J. J. E., 1986, *Robot Analysis and Control*, John Wiley & Sons, New York.

Balachandran, B., Magrab, E. B., 2003, *Vibrations*, Brooks/Cole, Pacific Grove, CA.

Beatty, M. F., 1986, *Principles of Engineering Mechanics, Vol. 1, Kinematics-The Geometry of Motion*, Plenum Press, New York.

Benaroya, H., 2004, *Mechaniscal Vibration: Analysis, Uncertainities, and Control*, Marcel Dekker, New York.

Bourmistrova, A., Simic, M., Hoseinnezhad, R., and Jazar, Reza N., 2011, Autodriver Algorithm, *Journal of Systemics, Cybernetics and Informatics,* **9***(1), 56-66.*

Bottema, O., and Roth, B., 1979, *Theoretical Kinematics*, North-Holland Publication, Amsterdam, The Netherlands.

Cossalter, V., 2002, *Motorcycle Dynamics*, Race Dynamic Publishing, Greendale, WI.

Del Pedro, M., and Pahud, P., 1991, *Vibration Mechanics*, Kluwer Academic Publishers, The Netherland.

Den Hartog, J. P., 1934, *Mechanical Vibrations*, McGraw-Hill, New York.

Dixon, J. C., 1996, *Tire, Suspension and Handling*, SAE Inc.

Dukkipati, R. V., Pang, J. Qatu, M. S., Sheng, G., and Shuguang, Z., 2008, *Road Vehicle Dynamics*, SAE Inc.

Ellis, J. R., 1994, *Vehicle Handling Kinematics*, Mechanical Engineering Publications Limited, London.

Esmailzadeh, E., 1978, Design Synthesis of a Vehicle Suspension System Using Multi-Parameter Optimization, *Vehicle System Dynamics,* **7***, 83-96.*

Genta, G., 2007, *Motor Vehicle Dynamics, Modeling and Simulation*, World Scientific, Singapore.

Genta, G., and Morello, L., 2009, *The Automotive Chassis: Volume 1: Components Design*, Springer, New York.

Genta, G., and Morello, L., 2009, *The Automotive Chassis: Volume 2: System Design*, Springer, New York.

Goldstein, H., Poole, C., and Safko, J., 2002, *Classical Mechanics*, 3rd ed., Addison Wesley, New York.

Haney, P., 2003, *The Racing and High–Performance Tire*, SAE Inc.

Harris, C. M., and Piersol, A. G., 2002, *Harris' Shock and Vibration Handbook*, McGraw-Hill, New York.

Hartenberg, R. S., and Denavit, J., 1964, *Kinematic Synthesis of Linkages*, McGraw-Hill Book Co.

Hunt, K. H., 1978, *Kinematic Geometry of Mechanisms*, Oxford University Press, London.

Inman, D., 2007, *Engineering Vibrations*, Prentice Hall, New York.

Jazar, Reza. N., 2010, *Theory of Applied Robotics: Kinematics, Dynamics, and Control*, second ed., Springer, New York.

Jazar, Reza N., 2010, Mathematical Theory of Autodriver for Autonomous Vehicles, *Journal of Vibration and Control, 16(2), 253-279.*

Jazar, Reza. N., 2011, *Advanced Dynamics: Rigid Body, Multibody, and Aerospace Applications*, Wiley, New York.

Jazar, Reza. N., 2013, *Advanced Vibrations: A Modern Approach*, Springer, New York.

Jazar, Reza N., 2012, Derivative and Coordinate Frames, *Journal of Nonlinear Engineering, 1(1), p25-34*, DOI: 10.1515/nleng-2012-0001.

Jazar, Reza. N., and Golnaraghi, M. F., 2002, Engine Mounts for Automotive Applications: A Survey, *The Shock and Vibration Digest, 34(5), 363-379.*

Jazar, Reza. N., Alkhatib, R., and Golnaraghi, M. F., 2006, Root Mean Square Optimization Criterion for Vibration Behavior of Linear Quarter Car Using Analytical Methods, *Journal of Vehicle System Dynamics, 44(6), 477–512.*

Jazar, Reza. N., Kazemi, M., and Borhani, S., 1992, *Mechanical Vibrations*, Ettehad Publications, Tehran. (in Persian).

Jazar, Reza. N., Narimani, A., and Golnaraghi, M. F., and Swanson, D. A., 2003, Practical Frequency and Time Optimal Design of Passive Linear Vibration Isolation Mounts, *Journal of Vehicle System Dynamics, 39(6), 437-466.*

Jazar, Reza N., Subic A., Zhong N., 2012, Kinematics of a Smart Variable Caster Mechanism for a Vehicle Steerable Wheel, *Vehicle System Dynamics.*

Karnopp, D., 2013, *Vehicle Dynamics, Stability, and Control*, 2nd ed., CRC Press, London, UK.

Kane, T. R., Likins, P. W., and Levinson, D. A., 1983, *Spacecraft Dynamics*, McGraw-Hill, New York.

MacMillan, W. D., 1936, *Dynamics of Rigid Bodies*, McGraw-Hill, New York.

Marzbani H., and Jazar, Reza N., 2013, *Smart Flat Ride Tuning, Book Chapter, Nonlinear Approaches in Engineering Applications 2*, Liming Dai, Reza N. Jazar, Eds., Springer, New York.

Marzbani H., Jazar, Reza N., and Fard M., *2012*, Hydraulic Engine Mounts: A Survey, *Journal of Vibration and Control*, DOI: 10.1177/1077546 312456724.

Marzbani H., Jazar, Reza N., and Khazaei A., 2012, Smart Passive Vibration Isolation: Requirements and Unsolved Problems, *Journal of Applied Nonlinear Dynamics, 1(4), p341-386*, DOI:10.5890/JAND.2012.09.002.

Mason, M. T., 2001, *Mechanics of Robotic Manipulation*, MIT Press, Cambridge, Massachusetts.

Meirovitch, L., 2002, *Fundamentals of Vibrations*, McGraw-Hill, New York.

Meirovitch, L., 1967, *Analytical Methods in Vibrations*, Macmillan, New York.

Milliken, W. F., and Milliken, D. L., 2002, *Chassis Design*, SAE Inc.

Milliken, W. F., and Milliken, D. L., 1995, *Race Car Vehicle Dynamics*, SAE Inc.

Murray, R. M., Li, Z., and Sastry, S. S. S., 1994, *A Mathematical Introduction to Robotic Manipulation*, CRC Press, Boca Raton, Florida.

National Committee on Uniform Traffic Laws and Ordinances, Uniform Vehicle Code and Model Traffic Ordinance, 1992.

Nikravesh, P., 1988, *Computer-Aided Analysis of Mechanical Systems*, Prentice Hall, New Jersey.

Norbe, J. P., 1980, *The Car and its Weels, A Guide to Modern Suspension Systems*, TAB Books Inc.

Pacejka, H, 2012, *Tire and Vehicle Dynamics*, 3rd ed., Butterworth-Heinemann, Oxford, UK.

Paul, R. P., 1981, *Robot Manipulators: Mathematics, Programming, and Control*, MIT Press, Cambridge, Massachusetts.

Pawlowski, J., 1969, *Vehicle Body Engineering*, Business Books Limited, London.

Rajamani, R., 2006, *Vehicle Dynamics and Control*, Springer-Verlag, New York.

Rao, S. S., 2003, *Mechanical Vibrations*, Prentice Hall, New York.

Roseau, M., 1987, *Vibrations in Mechanical Systems*, Springer-Verlag, Berlin.

Rosenberg, R. M., 1977, *Analytical Dynamics of Discrete Systems*, Plenum Publishing Co., New York.

Schaub, H., and Junkins, J. L., 2003, *Analytical Mechanics of Space Systems*, AIAA Educational Series, American Institute of Aeronautics and Astronautics, Inc., Reston, Virginia.

Shabana, A. A., 1997, *Vibration of Discrete and Continuous Systems*, Springer-Verlag, New York.

Skalmierski, B., 1991, *Mechanics*, Elsevier, Poland.

Snowdon, J. C., 1968, *Vibration and shock in damped mechanical systems*, John Wiley, New York.

Spong, M. W., Hutchinson, S., and Vidyasagar, M., 2006, *Robot Modeling and Control*, John Wiley & Sons, New York.

Soni, A. H., 1974, *Mechanism Synthesis and Analysis*, McGraw-Hill Book Co.

Tsai, L. W., 1999, *Robot Analysis*, John Wiley & Sons, New York.

United States Code, Title 23. Highways. Washington: U.S. Government Printing Office.

Wittacker, E. T., 1947, *A Treatise on the Analytical Dynamics of Particles and Rigid Bodies*, 4th ed., Cambridge University Press, New York.

Wong, J. Y., 2008, *Theory of Ground Vehicles*, 4th ed., John Wiley & Sons, New York.

Appendix A

Frequency Response Curves

There are four types of *one*-DOF harmonically excited systems as shown in Figure 12.39:

1. base excitation,
2. eccentric excitation,
3. eccentric base excitation,
4. forced excitation.

The frequency responses of the four systemscan be summarized, labeled and shown as follows:

$$S_0 = \frac{X_F}{F/k} \tag{A.1}$$

$$= \frac{1}{\sqrt{(1-r^2)^2 + (2\xi r)^2}} \tag{A.2}$$

$$S_1 = \frac{\dot{X}_F}{F/\sqrt{km}} \tag{A.3}$$

$$= \frac{r}{\sqrt{(1-r^2)^2 + (2\xi r)^2}} \tag{A.4}$$

$$S_2 = \frac{\ddot{X}_F}{F/m} = \frac{Z_B}{Y} = \frac{X_E}{e\varepsilon_E} = \frac{Z_R}{e\varepsilon_R} \tag{A.5}$$

$$= \frac{r^2}{\sqrt{(1-r^2)^2 + (2\xi r)^2}} \tag{A.6}$$

$$S_3 = \frac{\dot{Z}_B}{\omega_n Y} = \frac{\dot{X}_E}{e\varepsilon_E \omega_n} = \frac{\dot{Z}_R}{e\varepsilon_R \omega_n} \tag{A.7}$$

$$= \frac{r^3}{\sqrt{(1-r^2)^2 + (2\xi r)^2}} \tag{A.8}$$

$$S_4 = \frac{\ddot{Z}_B}{\omega_n^2 Y} = \frac{\ddot{X}_E}{e\varepsilon_E \omega_n^2} = \frac{\ddot{Z}_R}{e\varepsilon_R \omega_n^2} \tag{A.9}$$

$$= \frac{r^4}{\sqrt{(1-r^2)^2 + (2\xi r)^2}} \tag{A.10}$$

R.N. Jazar, *Vehicle Dynamics: Theory and Application*,
DOI 10.1007/978-1-4614-8544-5, © Springer Science+Business Media New York 2014

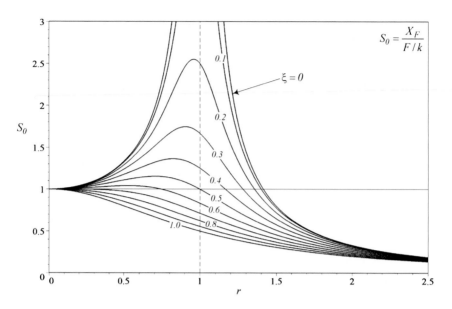

FIGURE A.1. Frequency response for S_0.

$$G_0 \;=\; \frac{F_{T_F}}{F} = \frac{X_B}{Y} \tag{A.11}$$

$$=\; \frac{\sqrt{1 + (2\xi r)^2}}{\sqrt{(1 - r^2)^2 + (2\xi r)^2}} \tag{A.12}$$

$$G_1 \;=\; \frac{\dot{X}_B}{\omega_n Y} \tag{A.13}$$

$$=\; \frac{r\sqrt{1 + (2\xi r)^2}}{\sqrt{(1 - r^2)^2 + (2\xi r)^2}} \tag{A.14}$$

$$G_2 \;=\; \frac{\ddot{X}_B}{\omega_n^2 Y} = \frac{F_{T_B}}{kY} = \frac{F_{T_E}}{e\omega_n^2 m_e} = \frac{F_{T_R}}{e\omega_n^2 m_e}\left(1 + \frac{m_b}{m}\right) \tag{A.15}$$

$$=\; \frac{r^2\sqrt{1 + (2\xi r)^2}}{\sqrt{(1 - r^2)^2 + (2\xi r)^2}} \tag{A.16}$$

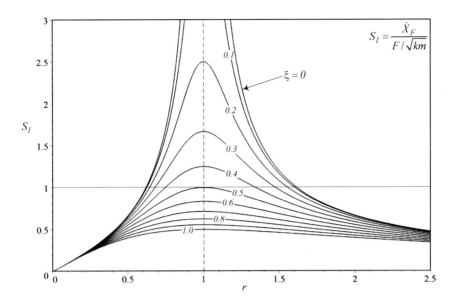

FIGURE A.2. Frequency response for S_1.

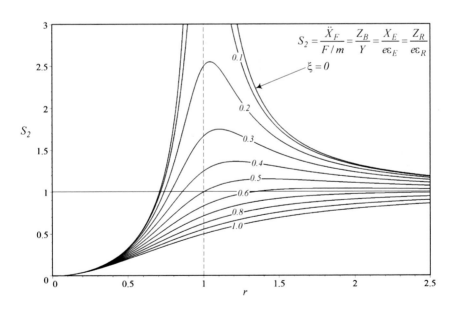

FIGURE A.3. Frequency response for S_2.

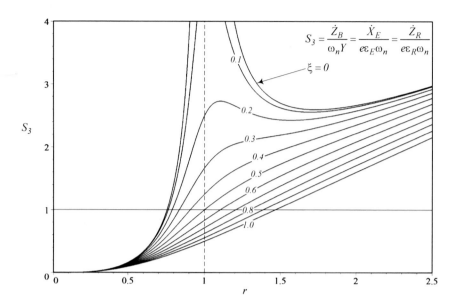

FIGURE A.4. Frequency response for S_3.

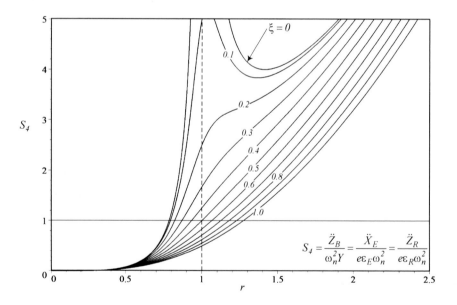

FIGURE A.5. Frequency response for S_4.

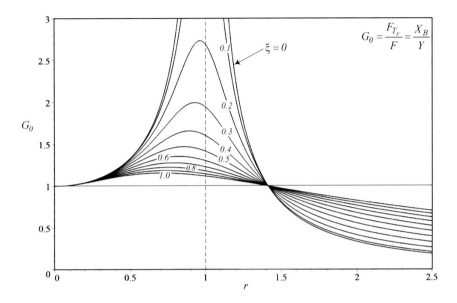

FIGURE A.6. Frequency response for G_0.

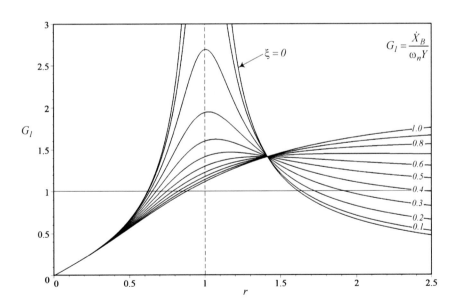

FIGURE A.7. Frequency response for G_1.

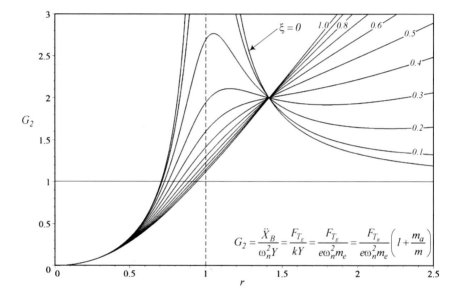

FIGURE A.8. Frequency response for G_2.

Appendix B

Trigonometric Formulas

Definitions in Terms of Exponentials

$$\cos z = \frac{e^{iz} + e^{-iz}}{2} \tag{B.1}$$

$$\sin z = \frac{e^{iz} - e^{-iz}}{2i} \tag{B.2}$$

$$\tan z = \frac{e^{iz} - e^{-iz}}{i\left(e^{iz} + e^{-iz}\right)} \tag{B.3}$$

$$e^{iz} = \cos z + i \sin z \tag{B.4}$$

$$e^{-iz} = \cos z - i \sin z \tag{B.5}$$

Angle Sum and Difference

$$\sin(\alpha \pm \beta) \;=\; \sin\alpha\cos\beta \pm \cos\alpha\sin\beta \tag{B.6}$$

$$\cos(\alpha \pm \beta) \;=\; \cos\alpha\cos\beta \mp \sin\alpha\sin\beta \tag{B.7}$$

$$\tan(\alpha \pm \beta) = \frac{\tan\alpha \pm \tan\beta}{1 \mp \tan\alpha\tan\beta} \tag{B.8}$$

$$\cot(\alpha \pm \beta) = \frac{\cot\alpha\cot\beta \mp 1}{\cot\beta \pm \cot\alpha} \tag{B.9}$$

Symmetry

$$\sin(-\alpha) \;=\; -\sin\alpha \tag{B.10}$$

$$\cos(-\alpha) \;=\; \cos\alpha \tag{B.11}$$

$$\tan(-\alpha) \;=\; -\tan\alpha \tag{B.12}$$

Multiple Angles

$$\sin(2\alpha) = 2\sin\alpha\cos\alpha = \frac{2\tan\alpha}{1 + \tan^2\alpha} \tag{B.13}$$

$$\cos(2\alpha) = 2\cos^2\alpha - 1 = 1 - 2\sin^2\alpha = \cos^2\alpha - \sin^2\alpha \tag{B.14}$$

$$\tan(2\alpha) = \frac{2\tan\alpha}{1 - \tan^2\alpha} \tag{B.15}$$

R.N. Jazar, *Vehicle Dynamics: Theory and Application*,
DOI 10.1007/978-1-4614-8544-5, © Springer Science+Business Media New York 2014

$$\cot(2\alpha) = \frac{\cot^2 \alpha - 1}{2\cot \alpha} \tag{B.16}$$

$$\sin(3\alpha) = -4\sin^3 \alpha + 3\sin \alpha \tag{B.17}$$

$$\cos(3\alpha) = 4\cos^3 \alpha - 3\cos \alpha \tag{B.18}$$

$$\tan(3\alpha) = \frac{-\tan^3 \alpha + 3\tan \alpha}{-3\tan^2 \alpha + 1} \tag{B.19}$$

$$\sin(4\alpha) = -8\sin^3 \alpha \cos \alpha + 4\sin \alpha \cos \alpha \tag{B.20}$$

$$\cos(4\alpha) = 8\cos^4 \alpha - 8\cos^2 \alpha + 1 \tag{B.21}$$

$$\tan(4\alpha) = \frac{-4\tan^3 \alpha + 4\tan \alpha}{\tan^4 \alpha - 6\tan^2 \alpha + 1} \tag{B.22}$$

$$\sin(5\alpha) = 16\sin^5 \alpha - 20\sin^3 \alpha + 5\sin \alpha \tag{B.23}$$

$$\cos(5\alpha) = 16\cos^5 \alpha - 20\cos^3 \alpha + 5\cos \alpha \tag{B.24}$$

$$\sin(n\alpha) = 2\sin((n-1)\alpha)\cos \alpha - \sin((n-2)\alpha) \tag{B.25}$$

$$\cos(n\alpha) = 2\cos((n-1)\alpha)\cos \alpha - \cos((n-2)\alpha) \tag{B.26}$$

$$\tan(n\alpha) = \frac{\tan((n-1)\alpha) + \tan \alpha}{1 - \tan((n-1)\alpha)\tan \alpha} \tag{B.27}$$

Half Angle

$$\cos\left(\frac{\alpha}{2}\right) = \pm\sqrt{\frac{1 + \cos \alpha}{2}} \tag{B.28}$$

$$\sin\left(\frac{\alpha}{2}\right) = \pm\sqrt{\frac{1 - \cos \alpha}{2}} \tag{B.29}$$

$$\tan\left(\frac{\alpha}{2}\right) = \frac{1 - \cos \alpha}{\sin \alpha} = \frac{\sin \alpha}{1 + \cos \alpha} = \pm\sqrt{\frac{1 - \cos \alpha}{1 + \cos \alpha}} \tag{B.30}$$

$$\sin \alpha = \frac{2\tan \frac{\alpha}{2}}{1 + \tan^2 \frac{\alpha}{2}} \tag{B.31}$$

$$\cos \alpha = \frac{1 - \tan^2 \frac{\alpha}{2}}{1 + \tan^2 \frac{\alpha}{2}} \tag{B.32}$$

Powers of Functions

$$\sin^2 \alpha = \frac{1}{2}\left(1 - \cos(2\alpha)\right) \tag{B.33}$$

$$\sin \alpha \cos \alpha = \frac{1}{2}\sin(2\alpha) \tag{B.34}$$

$$\cos^2 \alpha = \frac{1}{2}\left(1 + \cos(2\alpha)\right) \tag{B.35}$$

$$\sin^3 \alpha = \frac{1}{4} \left(3 \sin(\alpha) - \sin(3\alpha)\right) \tag{B.36}$$

$$\sin^2 \alpha \cos \alpha = \frac{1}{4} \left(\cos \alpha - 3 \cos(3\alpha)\right) \tag{B.37}$$

$$\sin \alpha \cos^2 \alpha = \frac{1}{4} \left(\sin \alpha + \sin(3\alpha)\right) \tag{B.38}$$

$$\cos^3 \alpha = \frac{1}{4} \left(\cos(3\alpha) + 3 \cos \alpha\right) \tag{B.39}$$

$$\sin^4 \alpha = \frac{1}{8} \left(3 - 4 \cos(2\alpha) + \cos(4\alpha)\right) \tag{B.40}$$

$$\sin^3 \alpha \cos \alpha = \frac{1}{8} \left(2 \sin(2\alpha) - \sin(4\alpha)\right) \tag{B.41}$$

$$\sin^2 \alpha \cos^2 \alpha = \frac{1}{8} \left(1 - \cos(4\alpha)\right) \tag{B.42}$$

$$\sin \alpha \cos^3 \alpha = \frac{1}{8} \left(2 \sin(2\alpha) + \sin(4\alpha)\right) \tag{B.43}$$

$$\cos^4 \alpha = \frac{1}{8} \left(3 + 4 \cos(2\alpha) + \cos(4\alpha)\right) \tag{B.44}$$

$$\sin^5 \alpha = \frac{1}{16} \left(10 \sin \alpha - 5 \sin(3\alpha) + \sin(5\alpha)\right) \tag{B.45}$$

$$\sin^4 \alpha \cos \alpha = \frac{1}{16} \left(2 \cos \alpha - 3 \cos(3\alpha) + \cos(5\alpha)\right) \tag{B.46}$$

$$\sin^3 \alpha \cos^2 \alpha = \frac{1}{16} \left(2 \sin \alpha + \sin(3\alpha) - \sin(5\alpha)\right) \tag{B.47}$$

$$\sin^2 \alpha \cos^3 \alpha = \frac{1}{16} \left(2 \cos \alpha - 3 \cos(3\alpha) - 5 \cos(5\alpha)\right) \tag{B.48}$$

$$\sin \alpha \cos^4 \alpha = \frac{1}{16} \left(2 \sin \alpha + 3 \sin(3\alpha) + \sin(5\alpha)\right) \tag{B.49}$$

$$\cos^5 \alpha = \frac{1}{16} \left(10 \cos \alpha + 5 \cos(3\alpha) + \cos(5\alpha)\right) \tag{B.50}$$

$$\tan^2 \alpha = \frac{1 - \cos(2\alpha)}{1 + \cos(2\alpha)} \tag{B.51}$$

Products of sin and cos

$$\cos \alpha \cos \beta = \frac{1}{2} \cos(\alpha - \beta) + \frac{1}{2} \cos(\alpha + \beta) \tag{B.52}$$

$$\sin \alpha \sin \beta = \frac{1}{2} \cos(\alpha - \beta) - \frac{1}{2} \cos(\alpha + \beta) \tag{B.53}$$

$$\sin \alpha \cos \beta = \frac{1}{2} \sin(\alpha - \beta) + \frac{1}{2} \sin(\alpha + \beta) \tag{B.54}$$

$$\cos \alpha \sin \beta = \frac{1}{2} \sin(\alpha + \beta) - \frac{1}{2} \sin(\alpha - \beta) \tag{B.55}$$

$$\sin(\alpha + \beta) \sin(\alpha - \beta) = \cos^2 \beta - \cos^2 \alpha = \sin^2 \alpha - \sin^2 \beta \tag{B.56}$$

$$\cos(\alpha + \beta) \cos(\alpha - \beta) = \cos^2 \beta + \sin^2 \alpha \tag{B.57}$$

Sum of Functions

$$\sin \alpha \pm \sin \beta = 2 \sin \frac{\alpha \pm \beta}{2} \cos \frac{\alpha \pm \beta}{2} \tag{B.58}$$

$$\cos \alpha + \cos \beta = 2 \cos \frac{\alpha + \beta}{2} \cos \frac{\alpha - \beta}{2} \tag{B.59}$$

$$\cos \alpha - \cos \beta = -2 \sin \frac{\alpha + \beta}{2} \sin \frac{\alpha - \beta}{2} \tag{B.60}$$

$$\tan \alpha \pm \tan \beta = \frac{\sin(\alpha \pm \beta)}{\cos \alpha \cos \beta} \tag{B.61}$$

$$\cot \alpha \pm \cot \beta = \frac{\sin(\beta \pm \alpha)}{\sin \alpha \sin \beta} \tag{B.62}$$

$$\frac{\sin \alpha + \sin \beta}{\sin \alpha - \sin \beta} = \frac{\tan \frac{\alpha + \beta}{2}}{\tan \frac{\alpha - + \beta}{2}} \tag{B.63}$$

$$\frac{\sin \alpha + \sin \beta}{\cos \alpha - \cos \beta} = \cot \frac{-\alpha + \beta}{2} \tag{B.64}$$

$$\frac{\sin \alpha + \sin \beta}{\cos \alpha + \cos \beta} = \tan \frac{\alpha + \beta}{2} \tag{B.65}$$

$$\frac{\sin \alpha - \sin \beta}{\cos \alpha + \cos \beta} = \tan \frac{\alpha - \beta}{2} \tag{B.66}$$

Trigonometric Relations

$$\sin^2 \alpha - \sin^2 \beta = \sin(\alpha + \beta) \sin(\alpha - \beta) \tag{B.67}$$

$$\cos^2 \alpha - \cos^2 \beta = -\sin(\alpha + \beta) \sin(\alpha - \beta) \tag{B.68}$$

Appendix C

Unit Conversions

General Conversion Formulas

$$\begin{aligned}
\mathrm{N}^a\,\mathrm{m}^b\,\mathrm{s}^c &\approx 4.448^a \times 0.3048^b \times \mathrm{lb}^a\,\mathrm{ft}^b\,\mathrm{s}^c \\
&\approx 4.448^a \times 0.0254^b \times \mathrm{lb}^a\,\mathrm{in}^b\,\mathrm{s}^c \\
\mathrm{lb}^a\,\mathrm{ft}^b\,\mathrm{s}^c &\approx 0.2248^a \times 3.2808^b \times \mathrm{N}^a\,\mathrm{m}^b\,\mathrm{s}^c \\
\mathrm{lb}^a\,\mathrm{in}^b\,\mathrm{s}^c &\approx 0.2248^a \times 39.37^b \times \mathrm{N}^a\,\mathrm{m}^b\,\mathrm{s}^c
\end{aligned}$$

Conversion Factors

Acceleration

$$1\,\mathrm{ft/s^2} \approx 0.3048\,\mathrm{m/s^2} \qquad 1\,\mathrm{m/s^2} \approx 3.2808\,\mathrm{ft/s^2}$$

Angle

$$1\,\mathrm{deg} \approx 0.01745\,\mathrm{rad} \qquad 1\,\mathrm{rad} \approx 57.307\,\mathrm{deg}$$

Area

$$1\,\mathrm{in^2} \approx 6.4516\,\mathrm{cm^2} \qquad 1\,\mathrm{cm^2} \approx 0.155\,\mathrm{in^2}$$
$$1\,\mathrm{ft^2} \approx 0.09290304\,\mathrm{m^2} \qquad 1\,\mathrm{m^2} \approx 10.764\,\mathrm{ft^2}$$
$$1\,\mathrm{acre} \approx 4046.86\,\mathrm{m^2} \qquad 1\,\mathrm{m^2} \approx 2.471 \times 10^{-4}\,\mathrm{acre}$$
$$1\,\mathrm{acre} \approx 0.4047\,\mathrm{hectare} \qquad 1\,\mathrm{hectare} \approx 2.471\,\mathrm{acre}$$

Damping

$$1\,\mathrm{N\,s/m} \approx 6.85218 \times 10^{-2}\,\mathrm{lb\,s/ft} \qquad 1\,\mathrm{lb\,s/ft} \approx 14.594\,\mathrm{N\,s/m}$$
$$1\,\mathrm{N\,s/m} \approx 5.71015 \times 10^{-3}\,\mathrm{lb\,s/in} \qquad 1\,\mathrm{lb\,s/in} \approx 175.13\,\mathrm{N\,s/m}$$

Energy and Heat

$$1\,\mathrm{Btu} \approx 1055.056\,\mathrm{J} \qquad 1\,\mathrm{J} \approx 9.4782 \times 10^{-4}\,\mathrm{Btu}$$
$$1\,\mathrm{cal} \approx 4.1868\,\mathrm{J} \qquad 1\,\mathrm{J} \approx 0.23885\,\mathrm{cal}$$
$$1\,\mathrm{kW\,h} \approx 3600\,\mathrm{kJ} \qquad 1\,\mathrm{MJ} \approx 0.27778\,\mathrm{kW\,h}$$
$$1\,\mathrm{ft\,lbf} \approx 1.355818\,\mathrm{J} \qquad 1\,\mathrm{J} \approx 0.737562\,\mathrm{ft\,lbf}$$

R.N. Jazar, *Vehicle Dynamics: Theory and Application*,
DOI 10.1007/978-1-4614-8544-5, © Springer Science+Business Media New York 2014

Force

$$1\,\text{lb} \approx 4.448222\,\text{N} \qquad 1\,\text{N} \approx 0.22481\,\text{lb}$$

Fuel Consumption

$1\ 1/100\,\text{km} \approx 235.214583\,\text{mi}/\text{gal}$ $1\,\text{mi}/\text{gal} \approx 235.214583\,\text{l}/100\,\text{km}$

$1\ 1/100\,\text{km} = 100\,\text{km}/\text{l}$ $1\,\text{km}/\text{l} = 100\ 1/100\,\text{km}$

$1\,\text{mi}/\text{gal} \approx 0.425144\,\text{km}/\text{l}$ $1\,\text{km}/\text{l} \approx 2.352146\,\text{mi}/\text{gal}$

Length

$$1\,\text{in} \approx 25.4\,\text{mm} \qquad 1\,\text{cm} \approx 0.3937\,\text{in}$$
$$1\,\text{ft} \approx 30.48\,\text{cm} \qquad 1\,\text{m} \approx 3.28084\,\text{ft}$$
$$1\,\text{mi} \approx 1.609347\,\text{km} \qquad 1\,\text{km} \approx 0.62137\,\text{mi}$$

Mass

$$1\,\text{lb} \approx 0.45359\,\text{kg} \qquad 1\,\text{kg} \approx 2.204623\,\text{lb}$$
$$1\,\text{slug} \approx 14.5939\,\text{kg} \qquad 1\,\text{kg} \approx 0.068522\,\text{slug}$$
$$1\,\text{slug} \approx 32.174\,\text{lb} \qquad 1\,\text{lb} \approx 0.03.1081\,\text{slug}$$

Moment and Torque

$$1\,\text{lb ft} \approx 1.35582\,\text{N m} \qquad 1\,\text{N m} \approx 0.73746\,\text{lb ft}$$
$$1\,\text{lb in} \approx 8.85075\,\text{N m} \qquad 1\,\text{N m} \approx 0.11298\,\text{lb in}$$

Mass Moment

$$1\,\text{lb ft}^2 \approx 0.04214\,\text{kg m}^2 \qquad 1\,\text{kg m}^2 \approx 23.73\,\text{lb ft}^2$$

Power

$1\,\text{Btu}/\text{h} \approx 0.2930711\,\text{W}$ $1\,\text{W} \approx 3.4121\,\text{Btu}/\text{h}$

$1\,\text{hp} \approx 745.6999\,\text{W}$ $1\,\text{kW} \approx 1.341\,\text{hp}$

$1\,\text{hp} \approx 550\,\text{lb ft}/\text{s}$ $1\,\text{lb ft}/\text{s} \approx 1.8182 \times 10^{-3}\,\text{hp}$

$1\,\text{lb ft}/\text{h} \approx 3.76616 \times 10^{-4}\,\text{W}$ $1\,\text{W} \approx 2655.2\,\text{lb ft}/\text{h}$

$1\,\text{lb ft}/\text{min} \approx 2.2597 \times 10^{-2}\,\text{W}$ $1\,\text{W} \approx 44.254\,\text{lb ft}/\text{min}$

Pressure and Stress

$1\,\text{lb}/\text{in}^2 \approx 6894.757\,\text{Pa}$ $1\,\text{MPa} \approx 145.04\,\text{lb}/\text{in}^2$

$1\,\text{lb}/\text{ft}^2 \approx 47.88\,\text{Pa}$ $1\,\text{Pa} \approx 2.0886 \times 10^{-2}\,\text{lb}/\text{ft}^2$

$1\,\text{Pa} \approx 0.00001\,\text{atm}$ $1\,\text{atm} \approx 101325\,\text{Pa}$

Stiffness

$$1\,\mathrm{N/m} \approx 6.85218 \times 10^{-2}\,\mathrm{lb/ft} \qquad 1\,\mathrm{lb/ft} \approx 14.594\,\mathrm{N/m}$$
$$1\,\mathrm{N/m} \approx 5.71015 \times 10^{-3}\,\mathrm{lb/in} \qquad 1\,\mathrm{lb/in} \approx 175.13\,\mathrm{N/m}$$

Temperature

$$^\circ\mathrm{C} = (\,^\circ\mathrm{F} - 32)/1.8$$
$$^\circ\mathrm{F} = 1.8\,^\circ\mathrm{C} + 32$$

Velocity

$$1\,\mathrm{mi/h} \approx 1.60934\,\mathrm{km/h} \qquad 1\,\mathrm{km/h} \approx 0.62137\,\mathrm{mi/h}$$
$$1\,\mathrm{mi/h} \approx 0.44704\,\mathrm{m/s} \qquad 1\,\mathrm{m/s} \approx 2.2369\,\mathrm{mi/h}$$
$$1\,\mathrm{ft/s} \approx 0.3048\,\mathrm{m/s} \qquad 1\,\mathrm{m/s} \approx 3.2808\,\mathrm{ft/s}$$
$$1\,\mathrm{ft/min} \approx 5.08 \times 10^{-3}\,\mathrm{m/s} \qquad 1\,\mathrm{m/s} \approx 196.85\,\mathrm{ft/min}$$

Volume

$$1\,\mathrm{in}^3 \approx 16.39\,\mathrm{cm}^3 \qquad 1\,\mathrm{cm}^3 \approx 0.0061013\,\mathrm{in}^3$$
$$1\,\mathrm{ft}^3 \approx 0.02831685\,\mathrm{m}^3 \qquad 1\,\mathrm{m}^3 \approx 35.315\,\mathrm{ft}^3$$
$$1\,\mathrm{gal} \approx 3.785\,\mathrm{l} \qquad 1\,\mathrm{l} \approx 0.2642\,\mathrm{gal}$$
$$1\,\mathrm{gal} \approx 3785.41\,\mathrm{cm}^3 \qquad 1\,\mathrm{l} \approx 1000\,\mathrm{cm}^3$$

Index

R.N. Jazar, *Vehicle Dynamics: Theory and Application*,
DOI 10.1007/978-1-4614-8544-5, © Springer Science+Business Media New York 2014